RNA Worlds

New Tools for Deep Exploration

ALSO FROM COLD SPRING HARBOR LABORATORY PRESS

ADVANCED TEXTBOOKS

The Cytoskeleton
Epigenetics, Second Edition
Mammalian Development: Networks, Switches, and Morphogenetic Processes
RNA Worlds: From Life's Origins to Diversity in Gene Regulation
Signal Transduction: Principles, Pathways, and Processes

SUBJECT COLLECTIONS FROM *COLD SPRING HARBOR PERSPECTIVES IN BIOLOGY*

Translation Mechanisms and Control
Cytokines
Circadian Rhythms
Immune Memory and Vaccines: Great Debates
Cell–Cell Junctions, Second Edition
Prion Biology
The Biology of the TGF-β Family
Synthetic Biology: Tools for Engineering Biological Systems
Cell Polarity
Cilia
Microbial Evolution
Learning and Memory
DNA Recombination

SUBJECT COLLECTIONS FROM *COLD SPRING HARBOR PERSPECTIVES IN MEDICINE*

RAS and Cancer in the 21st Century
Enteric Hepatitis Viruses
Bone: A Regulator of Physiology
Multiple Sclerosis

OTHER RELATED TITLES

CRISPR–Cas: A Laboratory Manual
RNA: A Laboratory Manual

RNA Worlds

New Tools for Deep Exploration

EDITED BY

Thomas R. Cech
Howard Hughes Medical Institute
University of Colorado Boulder

Joan A. Steitz
Howard Hughes Medical Institute
Yale University

John F. Atkins
University College Cork
The University of Utah

www.cshperspectives.org

COLD SPRING HARBOR LABORATORY PRESS
Cold Spring Harbor, New York • www.cshlpress.com

RNA Worlds: New Tools for Deep Exploration

Chapters online at www.cshperspectives.org

Printed in the United States of America

Publisher	John Inglis
Acquisition Editor	Richard Sever
Director of Editorial Development	Jan Argentine
Project Manager	Inez Sialiano
Permissions Coordinator	Carol Brown
Director of Publication Services	Linda Sussman
Production Editor	Kathleen Bubbeo
Production Manager	Denise Weiss
Cover Designers	Susan Roberts and Mike Albano

Front cover artwork: An ancient RNA World is represented by RNA self-replication at the center of the spiral with the diversity of modern RNA and tools to investigate it represented at the near end of the spiral. Some RNA structures are shown in the layers at the sides.

Library of Congress Cataloging-in-Publication Data

Names: Cech, Thomas, editor. | Steitz, Joan A., editor. | Atkins, John F. (John Fuller), editor.
Title: RNA worlds : new tools for deep exploration / edited by Thomas R. Cech, Joan A. Steitz, John F. Atkins.
Other titles: Cold Spring Harbor perspectives in biology.
Description: Second edition. | Cold Spring Harbor, New York : Cold Spring Harbor Laboratory Press, [2019] | "A Subject Collection from Cold Spring Harbor Perspectives in Biology." | Includes bibliographical references and index.
Identifiers: LCCN 2018038607 (print) | LCCN 2018041261 (ebook) | ISBN 9781621823537 (ePub3) | ISBN 9781621823544 (Kindle-Mobi) | ISBN 9781621822240 (printed hardcover)
Subjects: | MESH: RNA | Gene Expression Regulation
Classification: LCC QH450 (ebook) | LCC QH450 (print) | NLM QU 58.7 | DDC 572.8/8--dc23
LC record available at https://lccn.loc.gov/2018038607.

All World Wide Web addresses are accurate to the best of our knowledge at the time of printing.

For a complete catalog of all Cold Spring Harbor Laboratory Press publications, visit our website at www.cshlpress.org.

Contents

Preface

As this fifth edition of *RNA Worlds* goes to press, it is the eve of the 90th anniversary of the discovery of RNA. To gain perspective on the quickening pace of RNA research, consider the rate of subsequent seminal discoveries. After strong earlier indications, definitive proof that RNA can carry genetic information emerged 62 years ago. Insightful suggestions about a primordial role for RNA were made 50 years ago but languished until the discovery of catalytic RNA 36 years ago. RNA splicing, unknown before 1977, quickly exploded to dominate RNA research in the 1980s. The discovery of telomerase and its essential RNA subunit provided an important new example of RNP enzymes (1986). Forms of RNA interference, including siRNAs, microRNA, and piRNAs, were discovered 20 years ago and became household terms within a few years. The structural biology of RNA, initially limited to small motifs, exploded in the 1990s, leading to high-resolution views of the RNA basis of ribosome function. And CRISPR genome editing burst onto the scene around 2010.

This quickening pace of RNA research has been echoed in this series. The first editions, entitled *The RNA World*, were firmly centered on the RNA World hypothesis for origins of life, to the extent that all authors were instructed to relate their systems to the origins problem. By the 2011 edition, this vision had evolved with more chapters on contemporary RNA biology and technology, thus justifying the new title—*RNA Worlds*.

So, what's new in the 2018 edition? The RNA World hypothesis still fascinates, and it is well represented. Yet the explosion of new information about RNAs, especially noncoding RNAs, stimulated us to rethink the thrust of this edition. Understanding RNA function has been continually propelled by new technologies—next-generation sequencing and genomics, imaging down to the single-molecule level, CRISPR to precisely perturb function, and cryo-EM to visualize large RNP complexes that have not been amenable to X-ray crystallography. Thus, we chose authors who are experts in these and other transformative technologies. We asked them to explain their technology and how it is driving understanding of RNA biology, structure, and function. Although different authors have been differentially faithful to this mandate, it serves to tie the chapters together, as well as providing, in many cases, a glimpse into the future.

This fifth edition has also involved editor transitions. A founding editor, Ray Gesteland, has said farewell to the series. As we wish Ray well, we welcome Joan Steitz, whose knowledge and wisdom of all things ribonucleic has helped shape this volume. Together we have recruited a superb group of authors, the majority being first-timers to the series.

In the last edition we noted the passing of our friend and colleague Leslie Orgel, a giant of the RNA World. Since then, we have lost other major figures, including Aaron Shatkin (2012), Don Crothers (2014), Alex Rich (2015), Nacho Tinoco (2016), and Elisa Izaurralde (2018). The students and postdocs they trained and inspired continue to move the field forward.

Finally, we thank Cold Spring Harbor Laboratory Press, and especially John Inglis, for long and steady support for the *RNA Worlds* series. Special thanks to Inez Sialiano and Kathleen Bubbeo, who kept this fast-moving train on track, and to Susan K. Roberts and Denise Weiss for their roles in preparing the cover.

Thomas R. Cech
Joan A. Steitz
John F. Atkins

Foreword to the First Edition of *The RNA World*

THE TERM "RNA WORLD" ORIGINALLY REFERRED TO a hypothetical time in the evolution of earthly life when there was no elaborate mechanism for protein synthesis such as we have today. As described by Joyce and Orgel (Chapter 1), there were speculations in the 1960s that RNA catalysts existed at that stage in evolution, that RNA was the sole genetic material, and that the standard Watson-Crick pairing was the basis of genetic replication. None of these early authors was smart enough to suggest that relics of these hypothetical catalytic RNAs might still be around today. Indeed, it was speculated that the original ribosomes might have been made solely of RNA but not that their main catalytic activity (the formation of the peptide bond) might still be performed today by RNA alone, as recent evidence seems to suggest.

This hypothesis of an RNA world without protein was largely forgotten but has now become fashionable again because of the remarkable discoveries by Altman and by Cech (described by Cech, Chapter 11) of RNA molecules that do indeed have catalytic activity of one sort or another. These discoveries have removed the chief objection to the RNA world—that RNA by itself cannot act catalytically.

This volume on the RNA world covers a somewhat wider field. It was realized in the 1960s that not all present-day RNA was messenger RNA or viral RNA. The ribosome was known to contain several distinct types of "structural" RNA molecules; tRNAs were familiar, and there were isolated examples of other small RNA molecules. What was not realized was the richness of the present-day RNA world, including snRNPs (as described by Baserga and Steitz, Chapter 14) and now many others. We have come to realize that RNA molecules can occur with many different shapes and can thus have many different functions. Several of the chapters in this book deal with what these structures are, what their functions may be, what makes them stable, and how to predict their structure. Just how many of these RNAs are, by themselves, truly catalytic remains to be seen. Some of them may use divalent cations to produce both their folding and their catalytic activities, as discussed by Pan et al. (Chapter 12).

Some of the properties of present-day RNA molecules were quite unexpected. One might perhaps have guessed that some limited form of messenger-RNA editing could occur very rarely as a freak, but who would have predicted that such editing would take so many distinct forms and occur as widely as Bass describes in Chapter 15?

That a retrovirus contains a sequence rather like part of a tRNA molecule was a surprise when it was first discovered. In 1987, Maizels and Weiner proposed that such structures tagged RNA genomes for replication in the early RNA world. In Chapter 23, they explore the many possible ramifications of this genome tag hypothesis. In particular, they propose that both tRNA and tRNA-aminoacylation activity were subject to selection before protein synthesis arrived on the scene. Whether one can safely extrapolate from a present-day property of rapidly evolving RNA viruses to as far back in time as the early RNA world remains to be seen.

A good case can be made (Gold et al., Chapter 19) that small lengths of RNA can often form relatively rigid structures more easily than can polypeptides of similar length. New techniques are now being used (Chapter 19 and Chapter 20 [Szostak and Ellington]) to explore this vast "space" of small RNA sequences for interesting properties such as the binding of one specific substrate or another and, in some cases, a particular type of catalytic activity.

The recombinant-DNA revolution of the 1970s was possible because molecular biologists could use the sophisticated products of billions of years of natural evolution—the replicases, restriction enzymes, and so forth—as precise and delicate chemical tools. These molecules were all proteins that acted on nucleic acids in one way or another. Now experimentalists (Chapters 19 and 20) are using a combination of biochemical processes, related to nucleic acid replication, which together embody the mechanism of natural selection. These powerful tools can provide them with new and desirable RNA molecules in the laboratory while they wait. Because these methods do not need whole cells, but only cellular components, they can handle many more individuals at one time and so explore the RNA "space" more quickly. The message is clear: Where possible, let Nature do the work, and if she proves a little slow, take your whip to her.

The exact nature of the early RNA world is now a matter of active debate. What can we learn from the present "relics" and, in particular, can we deduce the composition and nature of each particular type of early RNA from the detailed study of its many descendants today? Can we learn anything from small RNA-related molecules (such as coenzymes) about the general nature of the early RNA world? Early speculations had assumed that the RNA world was rather simple and rather inaccurate. Benner and his colleagues (Chapter 2) now believe it was much more complex and more accurate. As more sequence data accumulate, one would expect these suggestions to become less speculative.

It may be possible to deduce something about some of the activities in the RNA world by a detailed study of the present mechanisms of protein synthesis. For example, Weiss and Cherry (Chapter 3) have proposed an ingenious model for the origins of the smaller ribosomal subunit. They suggest that it was originally an RNA replicase that used oligonucleotides as a substrate, cannibalizing triplets from them for its own replication.

We may, in time, arrive at a rather plausible picture of this early stage in evolution, even if true molecular fossils (that is, actual specimens of the molecules that then existed) are forever unavailable to us because of the ceaseless battering of thermal motion over billions of years.

The details of the transition from the RNA world to the present protein-dominated world are also debatable. For example, was there some sort of protein synthesis before the existence of a primitive ribosome? Did these first ribosomes have some protein, or did the ribosomal proteins arrive a little later?

It may turn out that we will eventually be able to see how this RNA world got started. At present, the gap from the primal "soup" to the first RNA system capable of natural selection looks forbiddingly wide. Was there perhaps a pre-RNA world, with replication based on some even more primitive system?

Such a pre-RNA world might have been of several types. It would presumably have been based on some form of genetic polymer or sheet that was formed more easily than RNA under prebiotic conditions. It might eventually have transcribed its detailed genetic information directly onto RNA to form the RNA world. Alternatively, it might have acted solely as a midwife, setting up the basis of RNA synthesis for some use of its own, with RNA replication then taking over and replacing it. Cairns-Smith has already suggested that this pre-RNA system might have been based on clays, or on some organic polymer (Cairns-Smith AG, *Genetic takeover and the mineral origins of life*, Cambridge University Press, Cambridge [1982]). This time we should be smart enough to ask if there are any relics of this pre-RNA world still with us today.

Finally, a word of heresy: As Moore has pointed out (Chapter 5), the real fossil record suggests that our present form of protein-based life was already in existence 3.6 billion years ago and evolved rather slowly for a billion or so years after that. This leaves an astonishingly short time to get life started. Moreover, the three main lines of descent (see Fig. 3 in Woese and Pace, Chapter 4) seem a very long distance from their hypothetical common ancestor.

The rather far-fetched hypothesis of Directed Panspermia would predict that life was sent here in the form of "bacteria" suitable for growth in anaerobic conditions, and that several somewhat different forms would probably have been sent at the same time, in the hope that at least one would survive. All are likely to have evolved originally from a common ancestor (on another planet) that existed some billions of years before the formation of our solar system. Therefore, the final question about the RNA world and the pre-RNA world (if it existed) is: Where did it occur? Are we totally confident that our form of life started here, or did it perhaps originate elsewhere in the universe? It might have been easier to start elsewhere because, for example, the atmosphere there was more reducing than the Earth's early atmosphere appears to have been.

The lively and authoritative chapters of this book deal comprehensively both with the hypothetical RNA world and with the complexities of RNA structure and function we find around us today. I recommend it to all molecular biologists and especially to anyone fascinated by the baroque complexity of the nucleic acids and of RNA in particular.

FRANCIS CRICK

Prologue to the First Edition of *The RNA World*

Early Speculations and Facts about RNA Templates

RNA FIRST CAME ALIVE TO ME DURING THE FALL OF 1947 at Indiana University when I took Salvador Luria's course on viruses. There I first learned that whereas the then-known phage, pox and papilloma viruses, contained DNA, this molecule was totally absent in several purified plant viruses as well as in the viruses that caused encephalitis and polio, which instead contained RNA. Apparently a given virus had either RNA or DNA, in contrast to cells which contained both. But whether it was the nucleic acid component that carried their genetic specificity was still unclear. At that time, most scientists wanted Avery, MacLeod, and McCarty's experiment on pneumococcal transformation by purified DNA to be extended to other life forms before jumping on the nucleic acid bandwagon. Then in the spring of 1952 came the report from Al Hershey that the DNA component of phage T2 carried genetic specificity. This immediately thrilled me, but I remember well the audience's indifference when in mid-April I read Hershey's letter at an Oxford meeting of The Society for General Microbiology.

When I had arrived at Cambridge in the fall of 1951, I started taking seriously the work of Brachet and his collaborators in Brussels, who emphasized the correlation between the RNA content and the protein-synthesizing capacity of cells. Those cells making large amounts of protein possessed large numbers of virus-sized ribonucleoprotein particles, known initially as microsomal particles but since 1958 as ribosomes. Most importantly, these particles had been pinpointed as the actual sites of protein synthesis by means of the then just-developed cell-free systems for protein synthesis. Here the key lab was that of Paul Zamecnik at Massachusetts General Hospital. Equally important was Brachet and Chantrenne's demonstration that the nucleus, and hence DNA, had no direct participation in protein synthesis. To show this, they cut the giant alga Acetabularia in half and observed that the half without a nucleus could maintain almost normal protein synthesis for more than a month. Yet from the one geneone enzyme (protein) results of Beadle and Tatum, the ultimate source of the genetic information that specifies the amino acid sequences of proteins had to be the genes found in the nucleus. I thus postulated a two-stage scheme for protein synthesis in which DNA first serves as a template for nucleus-located synthesis of RNA, and this RNA then in turn moves to the cytoplasm where it functions as the template for protein synthesis.

No one then had any compelling reason to take my hypothesis seriously, but by November 1952 I liked it well enough to print DNA→RNA→protein on a small piece of paper which I taped on the wall above my writing table in my rooms at Clare College. From the day of our first meeting, Francis Crick and I thought it highly likely that the genetic information of DNA is conveyed by the sequence of its four bases, but we knew it was premature to promote this idea before the structure of DNA was known. However, from the moment we first saw how to build a double helix out of the four base pairs, it was clear that the essential uniqueness of a gene must reside in its respective sequence of base pairs. Moreover, not only could base pairing provide the way for genes to be copied exactly during gene duplication, but it was also very likely to underlie the process by which the genetic information of a DNA molecule is transferred to its RNA product.

Still totally unclear, however, was how RNA might serve as the template for ordering the amino acids in their respective polypeptide products. Emboldened by our fantastic good luck in so simply finding the structural essence of the gene duplication process, I saw no reason not to take on the challenge of finding out what RNA molecules looked like in three dimensions. Such knowledge, I felt, would be indispensable to understanding how they functioned in protein synthesis. I took on this task when I moved to Pasadena in the fall of 1953. There I joined forces with Alex Rich, who had started working on DNA just before Francis and I found the double helix. There was no difficulty in getting him to move on to a better pasture, and soon I was collecting RNA samples and drawing them into fibers that Alex exposed to X-ray beams. But despite much travail, even those fibers displaying high birefringence never gave rise to ordered diffraction patterns like those of DNA. Although we thought we saw reflections that might have come from short sections of double helices, we could never be sure and saw no way to decide whether RNA was a one- or two-chain molecule. Here the base composition was a bad tease. Viral RNAs clearly did not show equivalence of A with U or G with C, but the RNA from cells had A/U and G/C ratios sometimes closely approaching 1/1. But we could see no difference in the general features of the X-ray diagrams from viral or cellular RNAs. To our annoyance, RNA, no matter from what source, showed identical X-ray diagrams characterized by strong reflections at 3.36 Å and 4.00 Å. Clearly, there was some ordered structure in RNA, but we saw no way to get to it. After some six months of such frustration, we gave up.

A very different approach to understanding protein synthesis came from the very clever Russian-born theoretical physicist George Gamow, who was struck by the fact that the 3.5-Å distance between adjacent amino acids in extended polypeptide chains was very similar to the 3.4-Å separation between adjacent base pairs in the B form of the double helix. Not then cognizant of RNA's primary role in protein synthesis, Gamow proposed that amino acids are directly ordered by contacts with the DNA base pairs, with the polypeptide products containing the same number of amino acids as there are base pairs in their DNA templates. To deal with the fact that DNA has only four letters in its alphabet, whereas 20 different letters (amino acids) are used to specify proteins, Gamow assumed that each amino acid must be coded by several adjacent base pairs. Because there are only 16 (4×4) combinations of the four bases, taken two at a time, Gamow made the assumption that each amino acid must be specified by groups of three adjacent base pairs (a codon) along DNA chains. To deal with the fact that there are 64 ($4 \times 4 \times 4$) such triplets, he assumed that many amino acids must be specified by more than one triplet (redundant triplets). In such an "overlapping" code, adjacent amino acid codons have two out of the three base pairs in common, thereby restricting which amino acids can lie adjacent to each other. Gamow's overlapping code was only the first of several that would later be devised, each leading to different combinations of forbidden amino acid neighbors.

When George first told me his scheme, I quickly dismissed it since DNA was not the template that ordered the amino acids. But he was having combinatorial fun, and besides I could not rule out the possibility that some RNA molecules were double helices. With time, moreover, I realized there was a real virtue to Gamow-like codes. In disproving them, the possibility of overlapping codes could be ruled out, pointing the way to codes in which adjacent groups of most likely three bases specified successive amino acids along a polypeptide chain. In fact, the first known amino acid sequence, that determined by Sanger for insulin, disproved the first code, although Gamow did not at first realize this because his initial list of the 20 amino acids had several embarrassing mistakes (e.g., he included both cystine and cysteine). Other such overlapping codes devised over the next year were also ruled out as more proteins became sequenced. It was during the first coding rush that Leslie Orgel and I, on a trip to Berkeley where Gamow was spending the spring of 1954, suggested that we form a club of 20 members whose purpose was to crack the RNA structure and in so doing reveal how the genetic code operated. Soon to be known as the "RNA Tie Club," with its members reflecting Gamow's eclectic taste, it never had a formal meeting, nor did all its members ever cough up the money to purchase their RNA ties and tiepins bearing their respective amino acid code letters. Gamow's tiepin sported ALA (alanine), and mine, PRO (proline). Much more important in the long run was the opportunity it provided to exchange ideas about the code through "Notes to the RNA Tie Club."

Several of these communications, the most important of which came from Francis Crick, Leslie Orgel, and Sydney Brenner, later became incorporated into published manuscripts. Among these was the definitive disproof by Brenner of any form of overlapping code, which he wrote in Johannesburg in the fall of 1956 just before he returned to England to join Francis Crick. Equally important was the communication by Crick, John Griffiths, and Orgel suggesting a comma-less code as a device to let non-overlapping triplets be read in the appropriate reading frame. But the most influential paper intellectually was the RNA Tie Club's first Note, written by Francis Crick and sent out early in 1955 under the title On Degenerate Templates and the Adaptor Hypothesis. After spending the previous August in Woods Hole batting about potential codes with Gamow and Brenner, Crick began questioning the basic assumption that a nucleic acid template provided specific cavities complementary in shape and charge to the amino acid side groups. Here he argued that the nucleic acid bases want to hydrogen-bond and are not at all suitable for forming cavities that could attract the hydrophobic side groups of amino acids such as valine, leucine, or isoleucine. Equally tricky to imagine was any structural basis for degenerative codes in which many amino acid side groups are specified by more than one set of triplets. Faced with what he considered insuperable obstacles, Crick made the radical proposal that prior to peptide bond formation, amino acids are enzymatically attached to small "adaptor" molecules that have surface specifically tailored to bond to nucleic acid triplets. Here Crick suggested that the "adaptor" molecules might be tiny polynucleotides that base pair to RNA templates.

My reaction to the adaptor hypothesis was initially very negative, even though I had spent the fall of 1954 in futile efforts trying to fold RNA chains into shapes bearing cavities appropriate for the amino acid side groups.

The adaptor idea seemed much too complicated to me to have ever gotten started at the origin of life several billion years ago. Francis, too, had his moments of doubt; he even concluded his now famous Tie Club Note with the phrase, "In the comparative isolation of Cambridge I must confess there are times when I have no stomach for decoding." In fact, by the time I returned to Cambridge for the year beginning in June 1955, Francis and Alex Rich were immersed in building new three-dimensional models for collagen to compete with one earlier proposed by Linus Pauling, this despite the fact that Francis and I had often referred to collagen as the most boring of macromolecules.

Equally frustrated, I reverted to thinking about plant viruses, in particular tobacco mosaic virus (TMV), whose helical construction I had worked out in the spring of 1952 after Francis and I had been told by Sir Lawrence Bragg to stop trying to work out the structure of DNA through model building. Always troublesome to me was the apparent necessity to postulate both genetic and protein synthesis roles for RNA. Knowing that only a tiny fraction of TMV particles are actually infectious, I speculated whether in fact these rare particles contained DNA, not RNA, chains. But this was ruled out when it became possible to reconstitute infectious TMV particles from their purified RNA and protein components. This experiment, first successfully accomplished in the spring of 1955 in Berkeley by Heinz Fraenkel-Conrat and Robley Williams, generated much newspaper publicity which gave uninformed readers the idea that life itself had been created. Francis, however, put the matter in its proper light, being quoted in the English press as saying that this was a finding he had anticipated.

Reconstitution by itself, however, did not answer the question of whether the protein component played any more than a protective-coat role for the genetic-information-bearing RNA component. Less than a year later, however, Alfred Gierer, working in Gerhard Schramm's lab in Tübingen, clearly showed that the RNA alone was infectious. This primacy of nucleic acids as bearers of genetic information then lay at the heart of the way Francis and I thought about cells and viruses. But this was far from an acceptable paradigm for many of the attendees at the key, late March 1956, CIBA Foundation meeting on the Structure of Viruses. They were not at home with the concept that information flows unidirectionally from nucleic acids to proteins and never backward. This was awkwardly shown when André Lwoff and I passed on to Robley Williams a telegram message supposedly from Wendell Stanley reading "TMV protein infectious—be cautious!" To our amazement, Robley did not question the result until we revealed the hoax.

At this gathering of some 30 scientists, Francis presented our ideas on why the protective coats of viruses are made up of protein subunits. In our view, it was a consequence of the fact that no viral nucleic acid had sufficient coding capacity to specify a single polypeptide chain large enough to surround a more centrally located core of nucleic acid. The 2-million-molecular-weight RNA of TMV, for example, contains only 6000 bases and, assuming a coding ratio of three bases per amino acid, is only capable of specifying a 2000-amino-acid polypeptide or about 230,000 molecular weight. To make up the 38-million-molecular-weight protein coat, at least 165 subunits would be needed. In fact, the TMV subunit contains only some 150 amino acids, suggesting that the TMV RNA codes for several proteins or that the coding ratio is very much larger than three. Initially, we thought that tomato bushy stunt virus (TBSV), with a much higher RNA content, might be more useful in giving a realistic value for the coding ratio. It contains only four bases for every amino acid in the crystallographic subunit. Later, direct analysis of its protein subunit size suggested that each crystallographic repeat contains some five protein subunits, very likely implying that TBSV RNA, like TMV RNA, also codes for several proteins.

After that 1956 meeting, I again had a go at the RNA structure, taking advantage of the newly discovered enzyme polynucleotide phosphorylase that Marianne Grunberg-Manago and Severo Ochoa found could be used to make synthetic RNA molecules. By then back at the National Institutes of Health, Alex Rich had shown the month before that random AU and AGCU co-polymers gave X-ray diffraction patterns similar to those we had obtained using purified cellular and viral RNAs. I focused instead on poly(A) (adenine) fibers drawn from material prepared in the Molteno Institute by Roy Markham and David Lipkin. To my delight, they generated clean helical X-ray diagrams that were best interpreted as base-paired double helices built up from two parallel poly(A) chains. Initially, I was disturbed by the fact that many of the key reflections overlapped with those generated by purified RNA, which by then we had every reason to believe was single-chained. Later, this apparent paradox was possibly resolved by the finding that sections of hydrogen-bonded hairpins form along most RNA chains. Conceivably, it is these short sections of imperfect double helices that generate the DNA-like feature of the RNA X-ray patterns.

By then discouraged that the study of purified RNA would lead toward understanding how protein synthesis occurs, I decided to concentrate on the structure of ribosomal particles, at that time believing that they must carry the genetic information for ordering amino acids in proteins. In the spring of 1956 I convinced Alfred Tissières, then a Fellow at Kings College working in the Molteno Institute on oxidative phosphorylation, to join me at Harvard, where I would be moving in the fall of 1956. Alfred had in fact already done some preliminary experiments on the then-called microsomal particles and was keen to follow them up.

I arrived at Harvard some six months before Alfred but was preoccupied most of this time trying to be an effective teacher for the seniors and beginning graduate students who were taking my course on viruses. As soon as my lectures were under control, however, I went across the Charles River to see Paul Zamecnik whom I had first met the year before while briefly stopping at Harvard on my way back to England. There, in the building where Fritz Lipmann also had his lab, I first appreciated the importance of the discovery made there a year earlier by Mahlon Hoagland of the activated high-energy acyl-amino acid intermediates for protein synthesis. Even more important, I first learned of the more recent observation by Mahlon, Mary Stephenson, and Paul of a soluble RNA (sRNA) fraction to which the activated amino acids are transferred prior to protein synthesis. Quickly, I realized that these sRNA molecules might be the polynucleotide adaptors postulated two years before by Francis in his first Note to the RNA Tie Club. Prior to my visit, Crick's ideas were unknown to the Massachusetts General group and they eagerly sought out Francis when they came together at the 1957 Gordon Conference on Nucleic Acids and Proteins.

Tissières and I commenced our molecular characterization of *Escherichia coli* ribosomes in the spring of 1957, suspecting that they might have structural plans like those of the small RNA viruses, whose isocahedral-shaped protein shells are formed by the regular aggregation of a single protein building block. We could not have been more wrong about how they are organized. To start with, we found that the *E. coli* ribosomes, like those from all other organisms, are formed by the aggregation of two RNA-containing subunits, the larger 50S subunit approximately twice the size of the smaller 30S subunit. Each subunit contains a single major RNA chain, with the 50S ribosomal subunits possessing 23S RNA chains and the 30S subunits possessing 16S RNA chains. Moreover, both subunits contain a large number of different small proteins that for the most part are subunit-specific. At low Mg^{++} levels, the 30S and 50S subunits do not associate with each other, but when the Mg^{++} concentration is raised, they come together to form the 70S complex that subsequent work has shown to be the ribosomal form that carries out protein synthesis.

Naively, at first, we assumed that either the 16S RNA or the 23S RNA, or both, were the actual templates for protein synthesis. Puzzling to us, however, was why the templates existed in only two size classes while there was great variation in the sizes of their putative polypeptide products. Equally disturbing was why the base composition of ribosomal RNAs barely varied between bacterial species with highly different AT/GC ratios. A priori we had expected to find that the base compositions of the RNA templates would reflect those of their DNA templates. Luckily, there was one powerful exception. In 1956 the phage T2-specific RNA made after T2 infection of *E. coli* was shown by Volkan and Astrachan to have a T2 DNA-like base composition. Moreover, in contrast to the metabolically very stable ribosomal RNA chains, T2 RNA had been found to have a half-life of only several minutes.

In retrospect, Tissières and I should have gravitated early to T2 RNA, but in fact not until the fall of 1959 were its molecular properties investigated anywhere. Then, Masayasu Nomura and Ben Hall, working with Sol Spiegelman at the University of Illinois, provided tentative evidence for the incorporation of T2 RNA into abnormally small ribosomes. In trying to follow up this observation early in 1960, my graduate student Bob Risebrough came to a radically different conclusion. After T2 RNA is synthesized, it does not become part of a ribosomal subunit by aggregating with newly made ribosomal proteins. What in fact happens is that in the presence of Mg^{++}, the T2 RNA becomes attached to the smaller 30S ribosomal subunit, which in turn binds the larger 50S subunit to form the >70S complex that actually carries out protein synthesis. This result instantly changed the way we visualized protein synthesis. Instead of serving template roles, ribosomes function as stable assemblage sites for protein synthesis. The true template had to be a new RNA class unknown until that moment both because it comprises such a small percentage of the total RNA and because it is heterogeneous in length. Later, these metabolically unstable templates, whose amounts respond to cellular needs, would be named messenger RNA (mRNA) by Jacques Monod and François Jacob.

The first lab outsider to whom I revealed this conceptual breakthrough was Leo Szilard, whom I had gone down to see in New York where he was successfully plotting the radiation therapy that would cure his bladder cancer. Leo's reaction was entirely negative, not being convinced that we had the right interpretation for Risebrough's experimental results. Predictably, he wanted us to show that mRNA existed in uninfected *E. coli* cells before he would change his mind-set. These were experiments that we had in fact already planned to start several weeks later, as soon as François Gros arrived from Paris to spend several months working with Alfred and me. Also soon to join this effort was Wally Gilbert, then still teaching theoretical physics to Harvard students, but who was increasingly tempted to move into molecular biology by our excitement about mRNA. By the time of the June Gordon Conference on Nucleic Acids we were virtually convinced that *E. coli* mRNA also existed, and by the summer's end we had data for a convincing publication. Already at the Gordon Conference we had heard rumors that Sydney Brenner and François Jacob, using very different arguments, had also postulated the existence of mRNA and that their idea was being tested by Sydney that week in Matt Meselson's lab at Caltech. Eventually, we were to publish our independent proofs for mRNA's existence early in 1961 in back-to-back articles in *Nature*.

With the basic scheme for how RNA participates in protein synthesis known, the path became open for definitive experiments on the exact nature of the genetic code. Using enzymatically synthesized RNA as messengers for in vitro protein synthesis, the correct assignments for all of the triplet codons were determined by early 1966. With this major goal achieved, the time had clearly come to ask how the DNA→RNA→protein flow of information had ever gotten started. Here, Francis was again far ahead of his time. In 1968 he argued that RNA must have been the first genetic molecule, further suggesting that RNA, besides acting as a template, might also act as an enzyme and, in so doing, catalyze its own self-replication. As the chapters in this book will show, how right he was!

James D. Watson

Imaging the Life and Death of mRNAs in Single Cells

Jeffrey A. Chao[1] and Timothée Lionnet[2]

[1]Friedrich Miescher Institute for Biomedical Research, CH-4058 Basel, Switzerland

[2]Institute for Systems Genetics, Department of Cell Biology, NYU Langone Health, New York, New York 10016

Correspondence: jeffrey.chao@fmi.ch; timothee.lionnet@nyumc.org

SUMMARY

RNA plays a central role in gene expression from its transcription in the nucleus through translation and degradation in the cytoplasm. Technological advances in fluorescent microscopy and labeling methodologies have made it possible to detect single molecules of RNA in both fixed and living cells. Here, we focus on the recent developments in RNA imaging that have allowed quantitatively measuring the lives of individual transcripts from birth to death and all the events in between in single cells and tissues. Direct observation of RNAs within their native cellular environment has revealed a complex layer of spatial and temporal regulation that has profoundly impacted our understanding of RNA biology.

Outline

Cite as *Cold Spring Harb Perspect Biol* doi: 10.1101/cshperspect.a032086

1 VISUALIZING RNA AT THE SINGLE-CELL AND SINGLE-MOLECULE SCALE

The ability to visualize messenger RNA (mRNA) provides a powerful means to probe its role in gene expression, and methods to directly image mRNAs in situ have long been sought (Harrison et al. 1973). The sensitivity and resolution of these techniques are continuously improving. Initial radioactivity-based methods have largely been replaced by fluorescence microscopy, and it is now possible to accurately measure the number and position of individual molecules of RNA in cells and tissue. Detection of RNAs in situ allows these molecules to be studied in their native context and bypasses common biochemical steps of extraction and purification. These procedures have variable efficiency and are prone to artifacts because RNAs are exquisitely sensitive to RNases (endogenous or environmental). Different RNA species vary in their sensitivity to degradation, which can engender serious biases in data collection and interpretation. Imaging is also applicable to limited samples (e.g., rare cell types) that are challenging to characterize using biochemical approaches that rely on ensemble measurements. The main advantage of mRNA imaging, however, resides in its inherent ability to capture RNAs within the environment of the cell or tissue, thereby allowing RNAs to be readily assigned to their relevant organelles or subcellular position. This information is extremely valuable because RNA location usually provides immediate clues to RNA function. The importance of obtaining gene expression data at the single-cell resolution is becoming more apparent, and imaging naturally provides cellular resolution, making it possible to identify specific cell types and capture cell-to-cell variability. Finally, genetically encoded fluorescent labels now enable live tracking of mRNA molecules, and visualizing sequences of events in living cells has provided fundamental insights into the life cycle of RNA. Here, we review the recent developments in single-molecule RNA imaging in fixed and living samples and outline the lessons learned on the way about the complex lives of RNAs.

2 smFISH REVEALS THE BASIC FEATURES OF GENE EXPRESSION REGULATION

Fluorescent in situ hybridization (FISH) relies on labeling a sample with fluorescent nucleic acid probes complementary to an RNA species of interest. After washing, only the probes bound to their RNA target remain, and the distribution of the fluorescent signal reflects that of the underlying RNAs. Although initial incarnations of the techniques used complementary DNA (cDNA) as a probe, progress in microscopy technology coupled with the use of DNA oligonucleotides coupled to fluorophores as probes has enabled tremendous increases in sensitivity. The first demonstration of single-molecule sensitivity FISH (smFISH) was published 20 years ago (Femino et al. 1998), using 50-nt-long DNA antisense probes (five to 10 different probes) containing four to five fluorophores. Although this technique worked robustly to detect single RNAs, a major barrier to its wide adoption was the requirement for each probe to be custom synthesized containing aminoallyl thymidine nucleotides and labeled with fluorescent dyes. The subsequent introduction of 20-nt probes coupled to a single fluorophore (25 to 40 different probes) that could readily be purchased from commercial sources (Raj et al. 2008) turned the technology into a popular tool that is now routinely used in a variety of laboratories, working on diverse sample types (Fig. 1A). Various image analysis tools can automatically detect the fluorescent spots emitted by individual RNA molecules and quantify their position and fluorescence intensity (Raj et al. 2008; Mueller et al. 2013; Trcek et al. 2017). The data typically collected in these experiments provides the absolute numbers of RNA molecules in each cell imaged, making the technique the gold standard in the field of quantitative gene expression. Sensitivity routinely ranges in the 80%–100% range, with low levels of false positives (Raj et al. 2008).

Early work took advantage of the high-sensitivity measurements of single-cell expression levels offered by smFISH to test quantitative models of gene expression (Raj et al. 2006; Zenklusen et al. 2008; Taniguchi et al. 2010). The results obtained in a variety of model systems showed that the transcription process, which was long thought to be finely regulated and predictable, is actually highly stochastic (Lionnet and Singer 2012): transcripts are generated in bursts of activity interspersed by periods of inactivity. Although the average bursting probabilities are biologically regulated and differ between genes, individual transcripts in a cell are randomly expressed over time, which means that genetically identical cells can express dramatically different levels of the same mRNA (Fig. 1B). A unified understanding of the molecular regulation of bursting kinetics is still lacking, although different factors have been suggested to play a role, from chromatin accessibility at the promoter (Bai et al. 2010; Lionnet and Singer 2012) to transcription-induced supercoiling accumulation (Chong et al. 2014), enhancer looping (Bartman et al. 2016), and binding of transcription factors or the transcription machinery at transcription hubs (Cho et al. 2016).

Early attempts at multiplex smFISH experiments labeled distinct mRNA species in the same sample with unique fluorophore labels (Fig. 1A). These experiments provided a way to explore the correlation between the mRNA numbers from different genes across a cell population. Interestingly, little correlation was observed between the transcription levels of different genes, even between functionally related ones and alleles of the same gene (Raj et al. 2006;

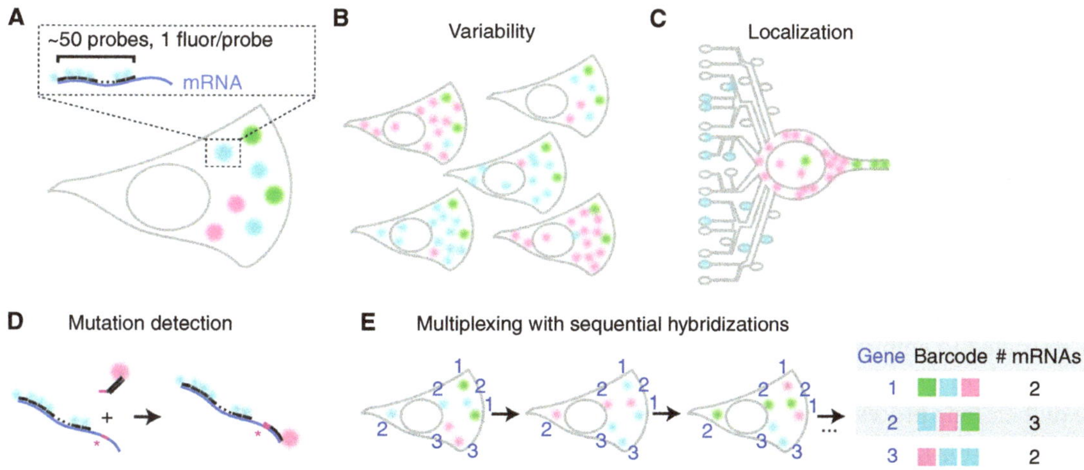

Figure 1. Single-molecule RNA fluorescent in situ hybridization (smFISH) and variants. (*A*) Principle of the technique: A probe library consisting of typically 50 20-mer oligonucleotides is hybridized to a sample. After washing excess probes, individual mRNAs appear as bright fluorescent spots. Multiplexing libraries in separate colors provides access to the expression of three separate genes in this example. (*B*) smFISH provides access to naturally occurring variability within a cell population, also called noise. Here, the products of two genes (pink and blue) display high variability and are anticorrelated. (*C*) smFISH provides access to the subcellular localization of mRNAs, separating for instance dendrite-targeted (blue) and axon-targeted mRNAs (green) from other species localized in this neuron's soma. (*D*) Combining a "guide" probe library (blue spots) with a specifically designed toehold probe labeled in a separate color (pink spot) provides a reading of the fraction of a given species carrying a mutation of interest. (*E*) Multiplexing via sequential hybridizations: RNAs in the same sample are hybridized sequentially to distinct library probes and imaged after each hybridization cycle. A given mRNA appears in a separate color at each cycle, and the sequence of colors detected at each mRNA position gives the molecule identity away.

Gandhi et al. 2011). The only genes displaying high correlation in their transcript levels were genes closely located on the genome, suggesting the importance of the chromatin environment in regulating expression. The immediate corollary of bursting is that expression levels between cells can vary substantially, setting obvious challenges to orderly cellular function. One simple strategy that seems to have evolved to generate precise mRNA levels from highly variable transcription sources is the temporal (or spatial) averaging of transcripts: If bursting occurs on a fast timescale (e.g., minutes) compared with the lifetime of its mRNA product (e.g., hours), the fluctuation of mRNA copy numbers are greatly attenuated compared with the variability of the transcription dynamics (Little et al. 2013). mRNA decay was recently shown to occur homogeneously in the cytoplasm, confirming its possible role as a dampener of expression noise (Horvathova et al. 2017). Besides its intrinsic importance as a fundamental constraint on cellular function, the control of expression variability in a cell population also has crucial consequences for disease. For instance, drug resistance in cancer cell lines has been shown to occur via transient expression of resistance genes (Shaffer et al. 2017). These transient bursts of expression eventually turn into stable expression profiles during drug treatment, indicating their capacity to function as a selection pressure.

3 LOCALIZATION AS A GENE EXPRESSION REGULATOR

Imaging provides immediate access to the spatial distribution of mRNAs in tissues and cell. Many studies have shown that the transcription, transport, and degradation of mRNA are used to generate spatial patterns of gene expression (Martin and Ephrussi 2009). The first step for most mRNAs involves export through the nuclear pore toward the cytoplasm where they can be translated. The regulation of nuclear export has been studied by various laboratories using smFISH. In principle, export from the nucleus could either amplify or buffer transcription-induced fluctuations of RNA levels depending on the relative kinetics of transcriptional bursting, export, and mRNA degradation. Indeed, a variety of scenarios have been observed that depend on the specific transcript (Bahar Halpern et al. 2015; Stoeger et al. 2016). As not all RNAs are destined to be translated, the localization of noncoding RNAs emerged as an important clue in their function. FISH experiments probing the localization of more than 60 long noncoding RNAs (lncRNAs) showed a wide variability of spatial patterns, from accumulation in nuclear foci to cytoplasmic localization, suggesting a wide range of function for these molecules (Cabili et al. 2015).

Developmental patterning provides striking examples of spatial restriction of mRNA expression at the tissue level, even before the advent of cellularization. In *Drosophila* embryos, ∼70% of mRNA species observed by FISH were found to have a regulated spatial pattern of expression (Lécuyer et al. 2007). Subcellular localization of mRNA transcripts is not restricted to embryogenesis, and perhaps the most striking example of spatial regulation is found in neurons. Because of their long processes, these cells face unique challenges in transporting gene expression products to their precise intended location. Part of this challenge is overcome by transporting select RNAs along dendrites and axons in which they can become translated near synapses on reception of the appropriate signal (Fig. 1C). Although most RNA species are found at or in the vicinity of the cell body, multiple examples of RNA localization into neuronal processes have been observed by FISH, from cultured neurons (Batish et al. 2012; Buxbaum et al. 2014) to intact brain tissue (Long et al. 2017). Together, these studies support a model of selective packaging of mRNA species into translationally repressive RNA–protein granules for their transport, and release from these structures then activates translation locally on demand. Interestingly, some of those structures can render RNAs invisible in standard smFISH experiments, presumably by blocking access of the probes to their target sequences (Buxbaum et al. 2014). A mild protease treatment can uncover those packaged transcripts and allow the study of the granule dynamics induced by neurostimulation. The recent discovery that one of the main activity-regulated neuronal genes, *Arc*, encodes a viral capsid-like protein that catalyzes its transfer to neighboring cells adds a new dimension to the role and regulation of RNA packaging and transport in neuronal function and communication (Ashley et al. 2018; Pastuzyn et al. 2018).

4 INCREASING SPATIAL RESOLUTION: TOWARD STRUCTURAL RESOLUTION OF mRNPs WITH LIGHT MICROSCOPY

As fluorescence microscopy technologies are improving at a fast pace, individual probes and molecules can now be localized with resolutions approaching the size of a macromolecular complexes (∼10 nm) (Sydor et al. 2015). Such progress opens the way to address structural questions left until now to the realm of electron microscopy. For instance, FISH combined with the stochastic optical reconstruction microscopy (STORM) super-resolution technique enabled the localization of individual probes in different colors tiling the same molecule of mRNA. The result was a spatial order of fluorescent spots consistent with the position of the probes along the sequence of a transcript (Lubeck and Cai 2012). A similar approach was recently taken to revisit the long-standing model positing that translating mRNAs adopt a circular conformation in which 5′ and 3′ ends of the transcripts are in close proximity, thereby enabling rapid reinitiation of translation (Wells et al. 1998). Surprisingly, those recent experiments challenge the notion of a circular conformation and suggest instead a somewhat extended mRNA (Adivarahan et al. 2017). In addition to these findings, recent work characterizing the RNA granules that form in the *Drosophila* oocyte using smFISH combined with structured illumination microscopy (SIM) showed a nonrandom spatial organization of the granules that contain the RNAs involved in instructing the fate of the future germ cells (Trcek et al. 2015). mRNAs of the same species tend to cluster together within a granule, and different RNAs occupy separate domains, pointing to a possible regulation of the granule organization. As the prevalence and importance of membrane-free granules and droplets is becoming more appreciated (Shin and Brangwynne 2017) and RNA binding proteins are disproportionately enriched in low complexity domains, smFISH will provide an important tool to probe the structure and assembly mechanism of these structures.

5 OPTIMIZED IN SITU TECHNIQUES TO VISUALIZE AND QUANTIFY THE DIVERSITY OF RNAs AND THEIR PROCESSING AT THE SINGLE-MOLECULE LEVEL

The smFISH technique relies on the binding of dozens (usually 20 to 100) of short oligonucleotide probes to the same transcript molecule. The large number of probes generate a signal amplification at the RNA location that allows detection of single RNA molecules from the background of single probes binding nonspecifically to other sites. Using multiple probes constitutes a limitation in many cases, for instance when imaging shorter RNA species, specific splicing isoforms or when single-nucleotide variants need to be resolved. Several adaptations to the smFISH technology now allow the detection of important classes of RNAs and variants. microRNAs (miRNAs) present a difficult challenge because of their short length, making it impossible to bind many probes. An optimized fixation protocol combined with specific probes made it possible to detect miRNAs in paraffin-embedded tissue, a notoriously difficult sample for RNA studies (Lu and Tsourkas 2009; Renwick et al. 2013). lncRNAs can be shorter than typical mRNAs and studies of localizations included sets of orthogonal probe library controls to validate the lower signals observed (Cabili et al. 2015). In the case of splicing isoforms, the design of probes targeting different exons in distinct colors allows the detection of splicing events based on the observations of the colocalization of the two colors (Vargas et al. 2011; Waks et al. 2011). Finally, perhaps the most challenging questions to be addressed with smFISH are the detection of single-nucleotide variants (SNVs) and RNA editing events, such as adenosine to inosine conversion (Nishikura 2010). Detection of such minimal changes in RNA sequence have been made possible using a two-color approach, in which probes in a first color label the conserved sequence while a unique toehold probe selective for the desired mutation is labeled with a second color (Fig. 1D) (Levesque et al. 2013; Mellis et al. 2017). Fluorescence in situ sequencing (FISSEQ, see below) is, in principle, capable of detecting individual mutations, but is technically challenging and typically only captures a small fraction (<1%) of the transcripts in a sample (Lee et al. 2014).

6 INCREASING RNA VISUALIZATION THROUGHPUT

One of the main limitations of classic smFISH-based approaches is its throughput. Traditionally, different species of RNAs can be distinguished by labeling them with spectrally distinguishable fluorophores. However, the small number of separable fluorescent labels (typically about five channels across the visible spectrum) severely constrains the number of RNA species that can be probed simultaneously in a given sample, setting a hard "multiplexing barrier" to studying large numbers of RNAs. Methods to overcome this barrier are now expanding at an accelerated pace. The first attempt at spectral multiplexing used a barcode approach. Instead of labeling each RNA species in a single color, it was assigned a combination of colors. In the case of five possible labels, in which the traditional labeling approach would provide access to five species only, using combinations of two colors and three colors per RNA gives access to 20 species. This was first shown on nascent transcripts because sites of active transcription give bright signals caused by the accumulation of nascent transcripts in one location (Levsky et al. 2002), and further refined for individual transcripts (Lubeck and Cai 2012; Levesque and Raj 2013). These findings offered insights into nuclear organization and showed that the transcription state of genes within a chromosomal translocation were different from those in the normal copy of the chromosome. Interestingly, correlations in expression between genes on the same chromosome changed sign in the translocation relative to the normal genomic context (Levesque et al. 2013). A conceptually similar barcoding approach is based on sequential hybridizations. mRNAs are first hybridized to a first probe library and imaged. The probes are then washed away on the microscope and the sample is hybridized to a second library with distinct color assignments and imaged again. Multiple cycles are performed, yielding a series of images of the same sample. A given mRNA appears in each image with a different color according to the design of the probe library barcodes. The sequence of colors at each spot in the series of images allow the assignment of specific transcript identities (Fig. 1E) (Lubeck et al. 2014; Chen et al. 2015; Shah et al. 2016b). Importantly, the number of molecular species accessible in this sequential approach increases exponentially with the number of hybridization cycles; hundreds to thousands of mRNA species have been probed simultaneously with these techniques. These experiments have helped probe expression correlations at a deeper level, for instance, finding that some functionally related genes displayed positively correlated fluctuations, and that some groups (e.g., cell motility– vs. extracellular matrix–related genes) tended to be anticorrelated with one another in cultured cells.

One of the practical limits to throughput lies in the requirement for single-molecule resolution. If each molecule needs to be spatially separable from its neighbor to be reliably detected, this sets a limitation on experimentally tractable spot densities. Optical methods (super-resolution, expansion microscopy [Chen et al. 2016]) might help overcome this limitation. Novel image analysis tools able to quantify expression even for dense samples also provide increased throughput (Coskun and Cai 2016). Other avenues to multiplexing adapt sequencing-based methods are using in situ rolling circle amplification and sequencing of the transcripts in a cell; FISSEQ provides, in principle, access to genome-wide coverage (Lee et al. 2014). However, the practical limitations on achievable densities of amplicons limit capture to a small fraction of the transcriptome (<1%). Another application of spatially resolved sequencing is spatial transcriptomics, in which a thin slice of sample is hybridized to a coverslip coupled to capture oligonucleotides. Each oligonucleotide carries a poly-T stretch to capture polyadenylated transcripts, as well as a sequence barcode that encodes its precise position along the coverslip (Ståhl et al. 2016). The resulting RNAs are reverse-transcribed and sequenced. This technique has genome-wide coverage and can deconvolve the initial position of each RNA molecule through the capture barcode of the initial capture strand. It remains limited in spatial resolution to ~100 µm pixel size and therefore provides a local average of RNA levels over a few cells. Although no technique is yet capable of providing genome-wide coverage at cellular and subcellular resolution with high efficiency, many of the technologies developed in recent years approach this goal, and combinations of image-based approach with sequencing provide exciting directions to investigate the spatiotemporal regulation of gene expression (Satija et al. 2015).

7 IMAGING RNA BEYOND CULTURED CELLS: RECENT DEVELOPMENTS FOR SINGLE-MOLECULE RNA IMAGING IN TISSUE

Acquiring images deep within biological samples is challenging, as absorption and scattering by biological tissue can hinder the precise excitation and collection of relevant signals. This limitation is especially acute in the case of smFISH, in which weak signals from individual molecules need to be accurately detected. As imaging RNAs in their tissue of origin is a highly desirable application, adaptations of the classic FISH techniques have emerged. One way to overcome optical challenges in fixed tissue is to clear the tissue by chemically dissolving the light scattering compounds from the tissue, while retaining its spatial integrity. These techniques include the use of denaturants (urea), solvents (xylene), or surfactants, alone or in combination with hydrogels, that help retain the tissue architecture and can be expanded to increase the effective spatial resolution (Richardson and Lichtman 2015; Chen et al. 2016). Adaptations of the FISH protocol to these samples have achieved quantification of RNAs at depths of hundreds of microns in various sample types (Shah et al. 2016a; Long et al. 2017). The coupling with light sheet microscopy provides effective z-sectioning without excessive photobleaching typically encountered when using confocal microscopy. Finally, biological tissues typically generate high amounts of autofluorescence, even when cleared. This interferes with the detection of the typically weak signals emitted by individual RNA molecules hybridized to fluorescent oligonucleotide probes. New signal amplification strategies have emerged to increase the fluorescence emitted by individual mRNA molecules. These rely on the hybridization of branched DNA probes (Wang et al. 2012) or the hybridization chain reaction (HCR) in which a set of primary probes hybridize to their target RNA (Choi et al. 2014; Shah et al. 2016a,b). These probes carry a single-stranded DNA (ssDNA) tail that triggers the hybridization of multiple fluorescent probes, thereby amplifying the signal over that of a single probe. These techniques provide increased sensitivity in large tissue and have been combined with multiplexing to offer high barcode accuracy (Shah et al. 2016b).

8 IMAGING RNAs IN LIVING CELLS

Biological processes can be misinterpreted or undetected when one is unable to directly follow events unfolding in real time. A variety of fluorescent labeling methodologies have been devised that enable mRNAs to be detected and quantified in living cells (Czaplinski 2017). The most widely used of these techniques that enables detection of single molecules is based on the highly specific interaction between the MS2 bacteriophage coat protein (MCP) and its cognate RNA stem loop (Bertrand et al. 1998). In this approach, multiple copies of the MS2 stem loop are inserted into a transcript (usually the 3′ untranslated region [UTR]) of interest that can then be bound by MCP fused to a fluorescent protein. The accumulation of many fluorescent proteins on an individual transcript provides sufficient signal to allow robust detection of single RNA molecules. Using this system, many aspects of the mRNA life cycle have been investigated. However, the reliance on a single labeling technique inherently limits the complexity of the questions that can be addressed (Tutucci et al. 2018).

Recently, a number of orthogonal labeling strategies based on the PP7 bacteriophage coat protein (Larson et al. 2011), λN bacteriophage (Daigle and Ellenberg 2007), U1A (Takizawa and Vale 2000), and BglG (Chen et al. 2009) have also been developed that can be used in combination for multicolor single-molecule RNA imaging when paired with spectrally distinct fluorescent proteins. The possibilities of multicolor RNA imaging experiments have just begun to be explored and have opened numerous new avenues of research. The following sections focus on the insights into the lives of RNAs that have been gained from two-color RNA imaging in living cells.

9 TRANSCRIPTION

Although fixed cell techniques such as smFISH make it possible to characterize cell-to-cell variability in mRNA and protein levels in genetically identical cells, multicolor live imaging provides a powerful tool to address the origin of expression fluctuations. The noise in gene expression can be separated into intrinsic (intracellular) and extrinsic (intercellular) factors that both contribute to the observed variability. Hocine and colleagues directly measured the contribution of intrinsic noise to transcription by labeling the two alleles of the essential gene MDN1, with either MS2 or PP7 stem loops in diploid yeast (Fig. 2A) (Hocine et al. 2013). By measuring the fluctuations in the number of MS2 and PP7-labeled transcripts over time in the same cell, the investigators found that the changes between alleles were not strongly correlated despite being transcribed in the same nucleus. These results highlight the stochastic nature of transcription and the contribution of intrinsic noise to this variability.

A similar approach was taken to measure the transcriptional activity of the pluripotency transcription factor *Nanog*, whose variability in expression regulates the balance between self-renewal and differentiation in mouse embryonic stem cells (Ochiai et al. 2014). Using transcription activator–like effector nuclease (TALEN)-mediated targeted integration, one *Nanog* allele was labeled by MS2 stem loops in its 3′ UTR and the other allele with PP7. Importantly, only a slight destabilization of the half-life of *Nanog* transcripts was observed on introduction of the stem loops, confirming that the introduction of these cassettes to endogenous genes can be minimally perturbing (Lionnet et al. 2011). Quantification of transcript numbers from the *Nanog*-MS2 and *Nanog*-PP7 alleles showed that ∼45% of the variability can be attributed to intrinsic noise, which again shows the highly variable nature of transcription.

Elucidating the molecular underpinning of bursting and its role in transcription variability has also been aided by two-color RNA imaging. Using *Drosophila* embryos as a model system, Fukaya and colleagues measured the transcription of two distinct MS2 and PP7-labeled genes controlled by the same *snail* shadow enhancer (Fukaya et al. 2016). When placed in a symmetric orientation with the enhancer equidistant from MS2 and PP7 reporter genes, these investigators observed similar bursting frequencies for both reporters. In the asymmetric orientation with the enhancer placed much closer to the PP7 reporter, the enhancer-proximal PP7 gene underwent sustained transcription, while transcriptional bursts of the enhancer-distal MS2 reporter occurred less often. These results showed that competition between genes for the same enhancer can modulate bursting frequencies. The investigators then used their two-color RNA imaging system to characterize the coordination of transcription by a shared enhancer. If an enhancer is only able to interact with a single promoter at any given moment, then one would anticipate that MS2 and PP7 reporter transcription should be anticorrelated in time. Interestingly, the investigators observed that approximately two-thirds of bursts were coordinated between MS2 and PP7 and this synchrony was detected with enhancers of varying strength. Furthermore, addition of the *gypsy* insulator resulted as expected in fewer bursts of the reporter insulated from the enhancer, but the remaining bursts still occurred in sync with the other target reporter of the enhancer. These results hint at the importance of higher order chromosome topology in regulating transcriptional activity and highlight the power of multicolor, temporally resolved measurements to understand gene expression regulation.

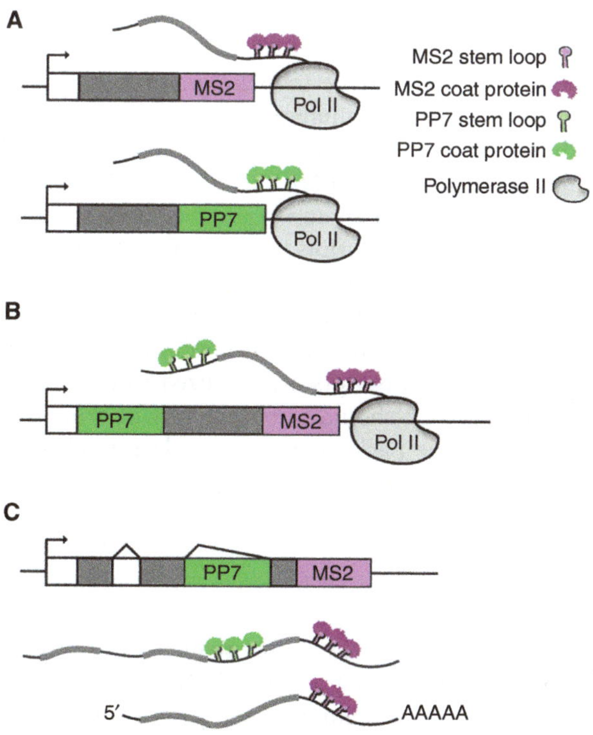

Figure 2. Single-molecule imaging of transcription and splicing in the nucleus of live cells. (*A*) Schematic representation of two-color labeling of alleles of the same gene with either MS2 or PP7 stem loops. (*B*) Schematic representation of two-color intramolecular labeling of a transcript at the 5′ and 3′ ends to directly measure RNA Pol II transcription rates. (*C*) Schematic representation of two-color intron and exon labeling for measurement of splicing.

Another mechanism by which transcription of a gene can be regulated is through the expression of antisense noncoding RNAs. Lenstra and colleagues explored the regulatory function of antisense transcription in yeast by tagging the *GAL10* sense transcript with MS2 stem loops and antisense *GAL10* ncRNAs with PP7 stem loops (Lenstra et al. 2015). Simultaneous observation of *GAL10-MS2* mRNAs and *GAL10-PP7* ncRNAs showed that during induction in galactose media, *GAL10-MS2* transcription occurred in stochastic bursts that were unaffected by antisense *GAL10-PP7* ncRNAs, even when detected simultaneously at the same locus. Inhibition of *GAL10-PP7* ncRNA transcription by strand-specific CRSPRi in cells grown in glucose, however, showed that its expression helps to repress leaky transcription of *GAL10* and *GAL1* in uninduced cells. These experiments showed that expression of an antisense ncRNA can function differently depending on the metabolic state of the cell, indicating that environmental conditions can alter the functionality of ncRNAs.

To directly measure the dynamics of transcription elongation in single living cells, the MS2 and PP7 stem loops can be positioned in the 5′ and 3′ ends of the same gene (Fig. 2B). If the intervening nucleotide sequence between the two labels is sufficiently long, then the time between the appearance of the 5′ and 3′ labels can be used to directly measure the rate of Pol II elongation. In yeast, Hocine and colleagues labeled a reporter driven by the *GAL* promoter that contained the MDN1 open reading frame (14.7 kb) flanked by PP7 stem loops in the 5′ end and MS2 stem loops in the 3′ end (Hocine et al. 2013). On induction of GAL transcription with galactose, these investigators measured an average Pol II elongation rate of 25 ± 2 bases sec^{-1};

however, there was considerable cell-to-cell variation with rates ranging from 14 to 61 bases sec^{-1}. A similar experiment in *Drosophila* embryos gave an average Pol II elongation rate of ~40 bases sec^{-1} (Fukaya et al. 2017). Although cell-to-cell variation was also observed as in yeast, this rate was not affected by the maternal Bicoid activator gradient as cells across the anterior–posterior axis had similar Pol II elongation rates despite differences in overall transcriptional output. Interestingly, placing the same reporter under the control of the *snail* enhancer or the inclusion of an intron within the body of a reporter gene resulted in only subtle changes (<25%) in Pol II elongation rates. Additionally, in human U2OS osteosarcoma cells, a Ponasterone A inducible promoter was used to induce expression of a reporter containing CFP-24xPP7 stem loops-IRES-Hygromycin-24xMS2 stem loops that places the stem loop cassettes 3 kb apart (Palangat and Larson 2016). In this experiment, the Pol II elongation rate was measured to be faster at 100 bases sec^{-1}. These studies provide a foundation to further explore the regulation of Pol II dynamics, either elongation rate or pausing, by site-specific integration of PP7 and MS2 stem loops into positions within endogenous genes. These single-molecule live cell measurements will enable gene-specific *cis* and *trans* effects that have been observed in ensemble measurements to be characterized in greater detail (Jonkers and Lis 2015).

10 SPLICING

In eukaryotes, intronic sequences must be removed from pre-mRNAs to generate mature transcripts. Using the well-studied β-globin gene that contains two introns as a model system, Martin and colleagues inserted either the λN boxB stem loops or MS2 stem loops into the first or second intron (Martin et al. 2013). This design allowed the splicing of the two introns to be followed independently in real time. These investigators measured different splicing times for the two introns with the first intron taking on average 20 sec, whereas the second intron took slightly longer at 30 sec. In these experiments, splicing of both introns was observed to finish before release of the transcript from the transcription site. Incorporation of the mouse IgM intron that contains a weak splice site resulted in a longer intron lifetime of 43 sec.

A similar approach using the β-globin gene was also used by Coulon and colleagues, except that PP7 stem loops were placed only into the second intron and MS2 stem loops were inserted into the 3′ UTR of a two-color splicing reporter gene (Fig. 2C) (Coulon et al. 2014). These investigators observed splicing events both before and after release of the pre-mRNA transcript from the transcription site. Their observations are consistent with a kinetic competition between stochastic splicing, elongation, and termination, suggesting that there is no deterministic quality control checkpoint that ensures pre-mRNAs are fully spliced before release. Surprisingly, these investigators found that post-transcriptional splicing of pre-mRNAs in the nucleoplasm is an order of magnitude faster than cotranscriptional splicing of the same intron. Combined, these studies provide a foundation to further explore the regulation of splicing and alternative splicing using multicolored RNA reporters.

Day and colleagues adapted the β-globin PP7 and MS2 splicing reporter to develop a fluorescence microscopy image-based screen to identify inhibitors of transcription, splicing, and termination (Day et al. 2016). These investigators used a library of small molecules that had been previously identified as inhibitors of chromatin readers and writers to show proof-of-concept for their approach. Several compounds that target bromodomains caused a reduction in the Pol II elongation rate. Although limitations in the sensitivity of high-throughput spinning-disk microscopes may have resulted in many false negatives in these initial experiments, this is not likely to constitute a permanent obstacle to cell-based phenotypic screens using single-molecule RNA imaging.

11 LOCALIZATION

The ability to monitor the trafficking and localization of multiple RNA species simultaneously in live cells will enable the mechanisms responsible for establishing and maintaining the polarization of RNAs to be further characterized. To this aim, Abbaszadeh and colleagues have shown the detection of *bicoid* transcripts labeled with MS2 stem loops and *oskar* transcripts labeled with PP7 stem loops in *Drosophila* oocytes (Abbaszadeh and Gavis 2016). In stage 10 oocytes, *bicoid-MS2* transcripts are localized to the anterior while *oskar-PP7* transcripts are localized to the posterior pole. This approach should be particularly informative for understanding the packaging of multiple mRNA species into granules and to dissect the specific *cis*-acting elements required for distinct RNA localization patterns.

12 TRANSLATION

Ribosomes move along the coding sequence of a transcript to decode the information contained within codons to synthesize proteins. This requires that any RNA–protein complexes (e.g., exon junction complex [EJC]) that are bound to the open reading frame must be displaced by the ribosome during the first round of translation. Consequently, the EJC can be used to specifically identify transcripts that have never been translated (Popp and Maquat 2013). Halstead and colleagues used this principle to design a two-color fluorescent microscopy methodology that enables the translation status of individual transcripts to be quantified in single cells. These investigators referred to this technique as translation RNA imaging by coat protein knockoff (TRICK) (Halstead et al. 2015).

TRICK relies on the detection of fluorescent labels that uniquely mark the coding sequence and the 3′ UTR in different colors. PP7 stem loops were engineered that could be efficiently translated by the ribosome and were placed within coding sequence of an mRNA, whereas MS2 stem loops were positioned in the 3′ UTR (Fig. 3A). Imaging of this reporter transcript in cells that also express the PP7 and MS2 coat proteins fused to spectrally distinct fluorescent proteins (e.g., NLS-PCP-GFP and NLS-MCP-RFP) results in untranslated mRNAs being detected as dual-colored spots (NLS-PCP-GFP and NLS-MCP-RFP). During the first round of translation, the NLS-PCP-GFP signal is removed from the coding sequence by the ribosome that is then detected as a single-labeled spot (NLS-MCP-RFP) (Fig, 3B). The incorporation of a nuclear localization sequence into NLS-PCP-GFP maintains a low concentration of unbound fluorescent protein in the cytoplasm, which prevents rebinding of NLS-PCP-GFP after translation. Consequently, TRICK cannot be used to monitor subsequent rounds of translation; however, the recently reported techniques for nascent polypeptide imaging (see Morisaki and Stasevich 2019) enable the dynamics of protein synthesis to be quantified in further detail.

The binary readout of the TRICK assay makes it an extremely sensitive assay for characterizing the spatial and temporal regulation of the initiation of protein synthesis. Halstead and colleagues found no evidence for translation of their TRICK reporter transcripts in the nucleus and that

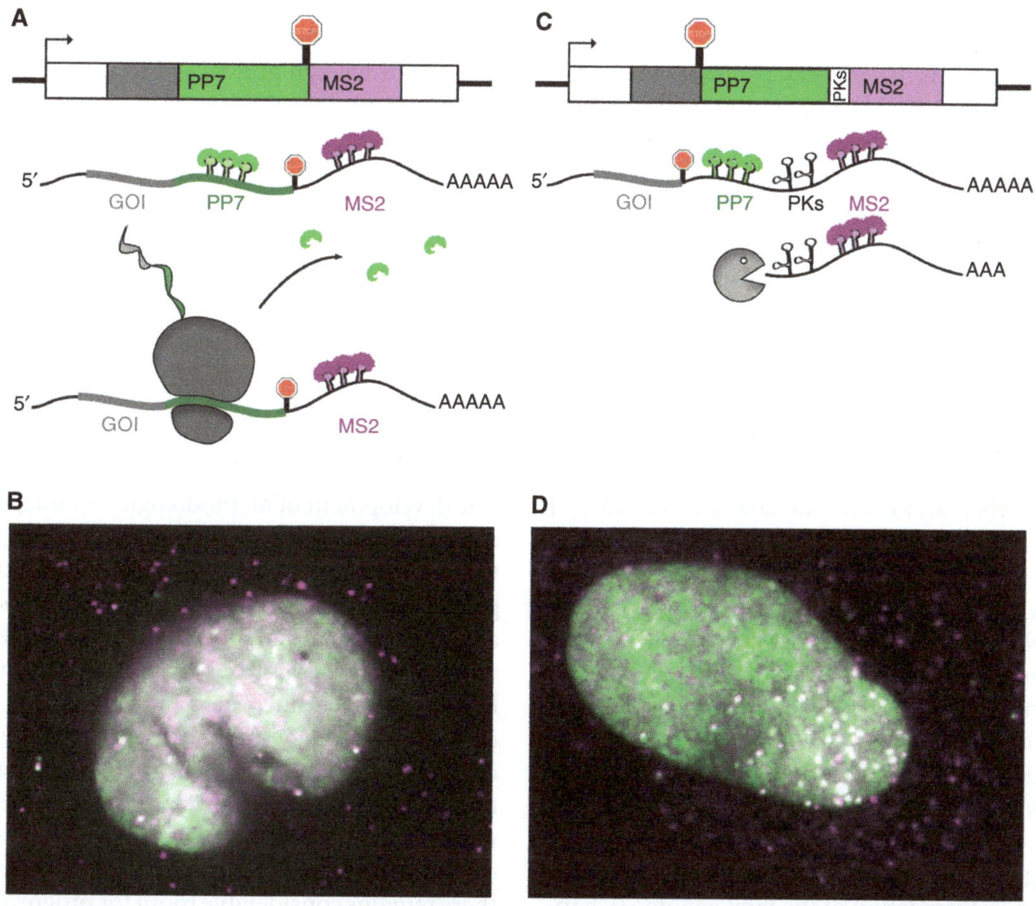

Figure 3. Single-molecule imaging of translation and mRNA decay in the cytoplasm of live cells. (*A*) Schematic representation of translation RNA imaging by coat protein knockoff (TRICK) for measuring the first round of translation by loss of PP7 label from the coding sequence during the first round of translation. (*B*) Representative image of TRICK experiment showing untranslated (PP7 and MS2; white) and translated (MS2-only; magenta) messenger RNAs (mRNAs). (*C*) Schematic representation of 3′-RNA end accumulation during turnover (TREAT) for measuring mRNA degradation using viral pseudoknots to stabilize a decay intermediate. (*D*) Representative image of TREAT experiment showing intact transcripts (PP7 and MS2; white) and stabilized 3′ end decay intermediates (MS2-only; magenta) arising from Xrn1-mediated degradation.

the first round of translation was distributed through the cytoplasm. The sequestration of mRNAs within messenger ribonucleoprotein (mRNP) granules either during stress maintained the granule-associated transcripts in a translationally repressed state. In *Drosophila* oocytes, *oskar* mRNA is localized to the posterior pole at late stages and TRICK confirmed that *oskar* transcripts are translationally repressed during transport. Voigt and colleagues used TRICK to show that mRNAs encoding secreted proteins are indeed localized to the endoplasmic reticulum (ER) during the first round of translation, which is consistent with the signal recognition particle (SRP)-mediated model for targeting (Voigt et al. 2017). Interestingly, it was also shown that mRNAs encoding cytosolic proteins could similarly be targeted as early as the first round of translation to the ER, albeit at much lower incidence.

13 mRNA DECAY

The nucleases involved in RNA decay pathways are highly processive enzymes that cause rapid disappearance of degradation intermediates within cells, which has prevented the spatial and temporal dynamics of mRNA turnover from being investigated in detail (Garneau et al. 2007). The direct measurement of mRNA decay by microscopy is additionally challenging because degradation must be unequivocally distinguished from photobleaching or movement of the transcript out of the focal volume that could also result in a loss of signal. Previously, it was shown that small interfering RNA (siRNA)-mediated knockdown of either Xrn1 or the exosome resulted in the accumulation of RISC-generated degradation intermediates that could be detected by northern blot (Orban and Izaurralde 2005). In the case of Xrn1 depletion, the half-life of the 3′-degradation intermediate increased from >15 min to >4 h. This result indicated that these RNA fragments were not immediately recognized as aberrant RNAs within the cell that could be efficiently cleared by alternative decay pathways and suggested that stabilizing degradation intermediates might be an entry point for imaging decay.

Although depletion of the decay machinery precludes studying physiological mRNA decay, a viral RNA pseudoknot structure was identified in flaviviruses that blocks Xrn1 by sequestering the 5′-phosphate from the enzyme, thereby preventing further degradation of the viral genome (Chapman et al. 2014). A single-molecule mRNA turnover biosensor was engineered that contains a 3′ UTR with the tandem pseudoknots from the Kunjin strain of the West Nile virus placed between the orthogonal PP7 and MS2 bacteriophage stem loops (Fig. 3C) (Horvathova et al. 2017). Because the MS2 stem loops are protected from Xrn1-mediated degradation by the pseudoknots, this biosensor allows discrimination between intact (PP7 and MS2) transcripts and stabilized 3′ end degradation intermediates (MS2 only) in living cells that express the PCP and MCP fused to spectrally distinct fluorescent proteins (Fig. 3D). Similarly, smFISH probes can be targeted to the two distinct regions of the transcript to image mRNA decay in fixed cells. This technique was named 3′-RNA end accumulation during turnover (TREAT).

Using TREAT, these investigators quantified the dynamics of mRNA degradation in single cells. Although transcription can occur in bursts, mRNA degradation was found to be consistent with a simple Poisson process indicating that individual decay events occur independently in time. Many of the eukaryotic proteins involved in mRNA decay have been found to accumulate in cytosolic structures called P-bodies suggesting that spatial compartmentalization may play a role in degradation (Sheth and Parker 2003). Using TREAT, these investigators did not detect either decay events or the accumulation of degradation intermediates within P-bodies during normal cell growth or stress, indicating that these compartments may not be general sites of mRNA decay.

The addition of a sequence targeted by an siRNA between the PP7 stem loops and the viral pseudoknots enabled monitoring of RISC-mediated slicing of mRNAs. Real-time observation of the severing of a transcript into distinct 5′ and 3′ ends that could be independently tracked revealed that cleavage was not coupled to degradation of the slicing intermediates and that slicing was not detected in the nucleus but throughout the cytosol.

14 OUTLOOK

The development of methodologies for imaging single molecules of RNA in fixed and living cells has profoundly altered our understanding of the regulation of this level of gene expression. For instance, biological phenomena such as transcriptional bursting, cellular heterogeneity in transcript abundances, mRNA localization, and localized translation have been uncovered in part because of progress in imaging technologies. These discoveries provide unique access to observe the molecular mechanisms underlying cell fate decisions, morphogenesis, drug resistance, and memory formation. Although the technological accomplishments of the field over the past two decades have been dramatic, there remains considerable room for progress. The throughput of imaging technologies lags far behind the capabilities of sequencing in terms of genomic coverage. Ongoing efforts in multiplexing RNA measurements in fixed cells is increasing at a rapid pace, but applications in live cell imaging have remained limited because of the low number of available orthogonal labeling techniques and spectrally separable fluorophores. Progress in genome editing and alter-

Cite as *Cold Spring Harb Perspect Biol* doi: 10.1101/cshperspect.a032086

native approaches for fluorescent labeling of RNAs should drive the field forward. Further developments in technologies able to visualize mRNA dynamics inside living organisms and tissues would certainly prove transformational in our understanding of the processes that regulate gene expression in complex cellular environments.

ACKNOWLEDGMENTS

This work was supported by the Novartis Research Foundation (J.A.C.), Swiss National Science Foundation (SNF) grant 31003A_156477 (J.A.C.), and the EMBO Young Investigator Program (J.A.C.). We thank F. Voigt for assistance with illustrations.

REFERENCES

**Reference is in this book.*

Abbaszadeh EK, Gavis ER. 2016. Fixed and live visualization of RNAs in *Drosophila* oocytes and embryos. *Methods* **98:** 34–41.

Adivarahan S, Rahman S, Zenklusen D. 2017. Spatial organization of single mRNPs at different stages of the gene expression pathway. *bioRxiv* doi: 10.1101/237008.

Ashley J, Cordy B, Lucia D, Fradkin LG, Budnik V, Thomson T. 2018. Retrovirus-like gag protein Arc1 binds RNA and traffics across synaptic boutons. *Cell* **172:** 262–274.e11.

Bahar Halpern K, Caspi I, Lemze D, Levy M, Landen S, Elinav E, Ulitsky I, Itzkovitz S. 2015. Nuclear retention of mRNA in mammalian tissues. *Cell Rep* **13:** 2653–2662.

Bai L, Charvin G, Siggia ED, Cross FR. 2010. Nucleosome-depleted regions in cell-cycle-regulated promoters ensure reliable gene expression in every cell cycle. *Dev Cell* **18:** 544–555.

Bartman CR, Hsu SC, Hsiung CC-S, Raj A, Blobel GA. 2016. Enhancer regulation of transcriptional bursting parameters revealed by forced chromatin looping. *Mol Cell* **62:** 237–247.

Batish M, van den Bogaard P, Kramer FR, Tyagi S. 2012. Neuronal mRNAs travel singly into dendrites. *Proc Natl Acad Sci* **109:** 4645–4650.

Bertrand E, Chartrand P, Schaefer M, Shenoy SM, Singer RH, Long RM. 1998. Localization of ASH1 mRNA particles in living yeast. *Mol Cell* **2:** 437–445.

Buxbaum AR, Wu B, Singer RH. 2014. Single β-actin mRNA detection in neurons reveals a mechanism for regulating its translatability. *Science* **343:** 419–422.

Cabili MN, Dunagin MC, McClanahan PD, Biaesch A, Padovan-Merhar O, Regev A, Rinn JL, Raj A. 2015. Localization and abundance analysis of human lncRNAs at single-cell and single-molecule resolution. *Genome Biol* **16:** 20.

Chapman EG, Costantino DA, Rabe JL, Moon SL, Wilusz J, Nix JC, Kieft JS. 2014. The structural basis of pathogenic subgenomic flavivirus RNA (sfRNA) production. *Science* **344:** 307–310.

Chen J, Nikolaitchik O, Singh J, Wright A, Bencsics CE, Coffin JM, Ni N, Lockett S, Pathak VK, Hu W-S. 2009. High efficiency of HIV-1 genomic RNA packaging and heterozygote formation revealed by single virion analysis. *Proc Natl Acad Sci* **106:** 13535–13540.

Chen KH, Boettiger AN, Moffitt JR, Wang S, Zhuang X. 2015. RNA imaging. Spatially resolved, highly multiplexed RNA profiling in single cells. *Science* **348:** aaa6090.

Chen F, Wassie AT, Cote AJ, Sinha A, Alon S, Asano S, Daugharthy ER, Chang J-B, Marblestone A, Church GM, et al. 2016. Nanoscale imaging of RNA with expansion microscopy. *Nat Methods* **13:** 679–684.

Cho W-K, Jayanth N, English BP, Inoue T, Owen Andrews J, Conway W, Grimm JB, Spille J-H, Lavis LD, Lionnet T, et al. 2016. RNA polymerase II cluster dynamics predict mRNA output in living cells. *eLife Sciences* **5:** e13617.

Choi HMT, Beck VA, Pierce NA. 2014. Next-generation in situ hybridization chain reaction: Higher gain, lower cost, greater durability. *ACS Nano* **8:** 4284–4294.

Chong S, Chen C, Ge H, Xie XS. 2014. Mechanism of transcriptional bursting in bacteria. *Cell* **158:** 314–326.

Coskun AF, Cai L. 2016. Dense transcript profiling in single cells by image correlation decoding. *Nat Methods* **13:** 657–660.

Coulon A, Ferguson ML, de Turris V, Palangat M, Chow CC, Larson DR. 2014. Kinetic competition during the transcription cycle results in stochastic RNA processing. *eLife* doi: 10.7554/eLife.03939.

Czaplinski K. 2017. Techniques for single-molecule mRNA imaging in living cells. *Adv Exp Med Biol* **978:** 425–441.

Daigle N, Ellenberg J. 2007. λ_N-GFP: An RNA reporter system for live-cell imaging. *Nat Methods* **4:** 633–636.

Day CR, Chen H, Coulon A, Meier JL, Larson DR. 2016. High-throughput single-molecule screen for small-molecule perturbation of splicing and transcription kinetics. *Methods* **96:** 59–68.

Femino AM, Fay FS, Fogarty K, Singer RH. 1998. Visualization of single RNA transcripts in situ. *Science* **280:** 585–590.

Fukaya T, Lim B, Levine M. 2016. Enhancer control of transcriptional bursting. *Cell* **166:** 358–368.

Fukaya T, Lim B, Levine M. 2017. Rapid rates of Pol II elongation in the *Drosophila* embryo. *Curr Biol* **27:** 1387–1391.

Gandhi SJ, Zenklusen D, Lionnet T, Singer RH. 2011. Transcription of functionally related constitutive genes is not coordinated. *Nat Struct Mol Biol* **18:** 27–34.

Garneau NL, Wilusz J, Wilusz CJ. 2007. The highways and byways of mRNA decay. *Nat Rev Mol Cell Biol* **8:** 113–126.

Halstead JM, Lionnet T, Wilbertz JH, Wippich F, Ephrussi A, Singer RH, Chao JA. 2015. Translation. An RNA biosensor for imaging the first round of translation from single cells to living animals. *Science* **347:** 1367–1671.

Harrison PR, Conkie D, Paul J, Jones K. 1973. Localisation of cellular globin messenger RNA by in situ hybridisation to complementary DNA. *FEBS Lett* **32:** 109–112.

Hocine S, Raymond P, Zenklusen D, Chao JA, Singer RH. 2013. Single-molecule analysis of gene expression using two-color RNA labeling in live yeast. *Nat Methods* **10:** 119–121.

Horvathova I, Voigt F, Kotrys AV, Zhan Y, Artus-Revel CG, Eglinger J, Stadler MB, Giorgetti L, Chao JA. 2017. The dynamics of mRNA turnover revealed by single-molecule imaging in single cells. *Mol Cell* **68:** 615–625.e9.

Jonkers I, Lis JT. 2015. Getting up to speed with transcription elongation by RNA polymerase II. *Nat Rev Mol Cell Biol* **16:** 167–177.

Larson DR, Zenklusen D, Wu B, Chao JA, Singer RH. 2011. Real-time observation of transcription initiation and elongation on an endogenous yeast gene. *Science* **332:** 475–478.

Lécuyer E, Yoshida H, Parthasarathy N, Alm C, Babak T, Cerovina T, Hughes TR, Tomancak P, Krause HM. 2007. Global analysis of mRNA localization reveals a prominent role in organizing cellular architecture and function. *Cell* **131:** 174–187.

Lee JH, Daugharthy ER, Scheiman J, Kalhor R, Yang JL, Ferrante TC, Terry R, Jeanty SS, Li C, Amamoto R, et al. 2014. Highly multiplexed subcellular RNA sequencing in situ. *Science* **343:** 1360–1363.

Lenstra TL, Coulon A, Chow CC, Larson DR. 2015. Single-molecule imaging reveals a switch between spurious and functional ncRNA transcription. *Mol Cell* **60:** 597–610.

Levesque MJ, Raj A. 2013. Single-chromosome transcriptional profiling reveals chromosomal gene expression regulation. *Nat Methods* **10:** 246–248.

Levesque MJ, Ginart P, Wei Y, Raj A. 2013. Visualizing SNVs to quantify allele-specific expression in single cells. *Nat Methods* **10:** 865–867.

Levsky JM, Shenoy SM, Pezo RC, Singer RH. 2002. Single-cell gene expression profiling. *Science* **297:** 836–840.

Lionnet T, Singer RH. 2012. Transcription goes digital. *EMBO Rep* **13:** 313–321.

Lionnet T, Czaplinski K, Darzacq X, Shav-Tal Y, Wells AL, Chao JA, Park HY, de Turris V, Lopez-Jones M, Singer RH. 2011. A transgenic mouse for in vivo detection of endogenous labeled mRNA. *Nat Methods* **8:** 165–170.

Little SC, Tikhonov M, Gregor T. 2013. Precise developmental gene expression arises from globally stochastic transcriptional activity. *Cell* **154:** 789–800.

Long X, Colonell J, Wong AM, Singer RH, Lionnet T. 2017. Quantitative mRNA imaging throughout the entire *Drosophila* brain. *Nat Methods* **14:** 703–706.

Lu J, Tsourkas A. 2009. Imaging individual microRNAs in single mammalian cells in situ. *Nucleic Acids Res* **37:** e100.

Lubeck E, Cai L. 2012. Single-cell systems biology by super-resolution imaging and combinatorial labeling. *Nat Methods* **9:** 743–748.

Lubeck E, Coskun AF, Zhiyentayev T, Ahmad M, Cai L. 2014. Single-cell in situ RNA profiling by sequential hybridization. *Nat Methods* **11:** 360–361.

Martin KC, Ephrussi A. 2009. mRNA localization: Gene expression in the spatial dimension. *Cell* **136:** 719–730.

Martin RM, Rino J, Carvalho C, Kirchhausen T, Carmo-Fonseca M. 2013. Live-cell visualization of pre-mRNA splicing with single-molecule sensitivity. *Cell Rep* **4:** 1144–1155.

Mellis IA, Gupte R, Raj A, Rouhanifard SH. 2017. Visualizing adenosine-to-inosine RNA editing in single mammalian cells. *Nat Methods* **14:** 801–804.

* Morisaki T, Stasevich TJ. 2019. Quantifying single mRNA translation kinetics in living cells. *Cold Spring Harb Persp Biol* **11:** a032078.

Mueller F, Senecal A, Tantale K, Marie-Nelly H, Ly N, Collin O, Basyuk E, Bertrand E, Darzacq X, Zimmer C. 2013. FISH-quant: Automatic counting of transcripts in 3D FISH images. *Nat Methods* **10:** 277–278.

Nishikura K. 2010. Functions and regulation of RNA editing by ADAR deaminases. *Annu Rev Biochem* **79:** 321–349.

Ochiai H, Sugawara T, Sakuma T, Yamamoto T. 2014. Stochastic promoter activation affects Nanog expression variability in mouse embryonic stem cells. *Sci Rep* **4:** 7125.

Orban TI, Izaurralde E. 2005. Decay of mRNAs targeted by RISC requires XRN1, the Ski complex, and the exosome. *RNA* **11:** 459–469.

Palangat M, Larson DR. 2016. Single-gene dual-color reporter cell line to analyze RNA synthesis in vivo. *Methods* **103:** 77–85.

Pastuzyn ED, Day CE, Kearns RB, Kyrke-Smith M, Taibi AV, McCormick J, Yoder N, Belnap DM, Erlendsson S, Morado DR, et al. 2018. The neuronal gene Arc encodes a repurposed retrotransposon Gag protein that mediates intercellular RNA transfer. *Cell* **172:** 275–288.e18.

Popp MW-L, Maquat LE. 2013. Organizing principles of mammalian nonsense-mediated mRNA decay. *Annu Rev Genet* **47:** 139–165.

Raj A, Peskin CS, Tranchina D, Vargas DY, Tyagi S. 2006. Stochastic mRNA synthesis in mammalian cells. *PLoS Biol* **4:** e309.

Raj A, van den Bogaard P, Rifkin SA, van Oudenaarden A, Tyagi S. 2008. Imaging individual mRNA molecules using multiple singly labeled probes. *Nat Methods* **5:** 877–879.

Renwick N, Cekan P, Masry PA, McGeary SE, Miller JB, Hafner M, Li Z, Mihailovic A, Morozov P, Brown M, et al. 2013. Multicolor microRNA FISH effectively differentiates tumor types. *J Clin Invest* **123:** 2694–2702.

Richardson DS, Lichtman JW. 2015. Clarifying tissue clearing. *Cell* **162:** 246–257.

Satija R, Farrell JA, Gennert D, Schier AF, Regev A. 2015. Spatial reconstruction of single-cell gene expression data. *Nat Biotechnol* **33:** 495–502.

Shaffer SM, Dunagin MC, Torborg SR, Torre EA, Emert B, Krepler C, Beqiri M, Sproesser K, Brafford PA, Xiao M, et al. 2017. Rare cell variability and drug-induced reprogramming as a mode of cancer drug resistance. *Nature* **546:** 431–435.

Shah S, Lubeck E, Schwarzkopf M, He T-F, Greenbaum A, Sohn CH, Lignell A, Choi HMT, Gradinaru V, Pierce NA, et al. 2016a. Single-molecule RNA detection at depth by hybridization chain reaction and tissue hydrogel embedding and clearing. *Development* **143:** 2862–2867.

Shah S, Lubeck E, Zhou W, Cai L. 2016b. In situ transcription profiling of single cells reveals spatial organization of cells in the mouse hippocampus. *Neuron* **92:** 342–357.

Sheth U, Parker R. 2003. Decapping and decay of messenger RNA occur in cytoplasmic processing bodies. *Science* **300:** 805–808.

Shin Y, Brangwynne CP. 2017. Liquid phase condensation in cell physiology and disease. *Science* **357:** aaf4382.

Ståhl PL, Salmén F, Vickovic S, Lundmark A, Navarro JF, Magnusson J, Giacomello S, Asp M, Westholm JO, Huss M, et al. 2016. Visualization and analysis of gene expression in tissue sections by spatial transcriptomics. *Science* **353:** 78–82.

Stoeger T, Battich N, Pelkmans L. 2016. Passive noise filtering by cellular compartmentalization. *Cell* **164:** 1151–1161.

Sydor AM, Czymmek KJ, Puchner EM, Mennella V. 2015. Super-resolution microscopy: From single molecules to supramolecular assemblies. *Trends Cell Biol* **25:** 730–748.

Takizawa PA, Vale RD. 2000. The myosin motor, Myo4p, binds Ash1 mRNA via the adapter protein, She3p. *Proc Natl Acad Sci* **97:** 5273–5278.

Taniguchi Y, Choi PJ, Li GW, Chen H, Babu M, Hearn J, Emili A, Xie XS. 2010. Quantifying *E. coli* proteome and transcriptome with single-molecule sensitivity in single cells. *Science* **329:** 533–538.

Trcek T, Grosch M, York A, Shroff H, Lionnet T, Lehmann R. 2015. *Drosophila* germ granules are structured and contain homotypic mRNA clusters. *Nat Commun* **6:** 7962.

Trcek T, Lionnet T, Shroff H, Lehmann R. 2017. mRNA quantification using single-molecule FISH in *Drosophila* embryos. *Nat Protoc* **12:** 1326–1348.

Tutucci E, Livingston NM, Singer RH, Wu B. 2018. Imaging mRNA in vivo, from birth to death. *Annu Rev Biophys* doi: 10.1146/annurev-biophys-070317-033037.

Vargas DY, Shah K, Batish M, Levandoski M, Sinha S, Marras SA, Schedl P, Tyagi S. 2011. Single-molecule imaging of transcriptionally coupled and uncoupled splicing. *Cell* **147:** 1054–1065.

Voigt F, Zhang H, Cui XA, Triebold D, Liu AX, Eglinger J, Lee ES, Chao JA, Palazzo AF. 2017. Single-molecule quantification of translation-dependent association of mRNAs with the endoplasmic reticulum. *Cell Rep* **21:** 3740–3753.

Waks Z, Klein AM, Silver PA. 2011. Cell-to-cell variability of alternative RNA splicing. *Mol Syst Biol* **7:** 506.

Wang F, Flanagan J, Su N, Wang L-C, Bui S, Nielson A, Wu X, Vo H-T, Ma X-J, Luo Y. 2012. RNAscope: A novel in situ RNA analysis platform for formalin-fixed, paraffin-embedded tissues. *J Mol Diagn* **14:** 22–29.

Wells SE, Hillner PE, Vale RD, Sachs AB. 1998. Circularization of mRNA by eukaryotic translation initiation factors. *Mol Cell* **2:** 135–140.

Zenklusen D, Larson DR, Singer RH. 2008. Single-RNA counting reveals alternative modes of gene expression in yeast. *Nat Struct Mol Biol* **15:** 1263–1271.

Quantifying Single mRNA Translation Kinetics in Living Cells

Tatsuya Morisaki[1] and Timothy J. Stasevich[1,2]

[1]Institute of Genome Architecture and Function and Department of Biochemistry & Molecular Biology, Colorado State University, Fort Collins, Colorado 80523

[2]Cell Biology Unit, Institute of Innovative Research, Tokyo Institute of Technology, Midori-ku, Yokohama, Kanagawa, 226-8503, Japan

Correspondence: tim.stasevich@colostate.edu; tatsuya.morisaki@colostate.edu

SUMMARY

One of the last hurdles to quantifying the full central dogma of molecular biology in living cells with single-molecule resolution has been the imaging of single messenger RNA (mRNA) translation. Here we describe how recent advances in protein tagging and imaging technologies are being used to precisely visualize and quantify the synthesis of nascent polypeptide chains from single mRNA in living cells. We focus on recent applications of repeat-epitope tags and describe how they enable quantification of single mRNA ribosomal densities, translation initiation and elongation rates, and translation site mobility and higher-order structure. Together with complementary live-cell assays to monitor translation using fast-maturing fluorophores and mRNA-binding protein knockoff, single-molecule studies are beginning to uncover striking and unexpected heterogeneity in gene expression at the level of translation.

Outline

Cite as *Cold Spring Harb Perspect Biol* doi: 10.1101/cshperspect.a032078

1 INTRODUCTION

The translation of messenger RNA (mRNA) into protein is a defining feature of life. All cells must tightly regulate translation to ensure proteins are correctly synthesized in the right place, at the right levels, and at the right time. This allows cells to quickly establish and maintain specific phenotypes in response to internal and external cues as well as environmental pressures (Sonenberg and Hinnebusch 2009; Buxbaum et al. 2015).

Given the dynamic nature of translation, the ability to quantify precisely when, where, and with what kinetics individual mRNA are translated in living cells and organisms provides the most direct means of dissecting complex translational regulatory mechanisms. Although bulk (Ingolia 2014) and single-cell analyses (Han et al. 2014) can provide exquisite snapshots of the average translational state of a specific type of transcript, their lack of spatiotemporal resolution makes it difficult to detect and monitor translational dynamics and heterogeneity at the single mRNA level. This leaves many basic questions about the nature of single mRNA expression unresolved (Chao et al. 2012). First, it remains unclear how heterogeneity in the translation of genetically identical transcripts is established. A variety of posttranscriptional mechanisms are implicated, including acquisition of higher-order structure (Babendure et al. 2006; Wen et al. 2008) and/or chemical modifications (Simms et al. 2014; Hoernes et al. 2016), binding of regulatory factors (Wu et al. 2015; Simsek and Barna 2017), and localization to subcellular structures (Jung et al. 2014; Buxbaum et al. 2015). However, without the ability to image the translation of individual mRNA, distinguishing these factors is extremely difficult. Second, it remains unclear how an individual mRNA is expressed over time (Chao et al. 2012). Is translation constitutive or bursty, localized or homogenous, motored or diffusive? How quickly does translation shut down or start up in response to environmental changes? To what extent does translation dynamics contribute to mRNA decay? Single-molecule imaging of mRNA translation in living cells provides the most direct route to answering these and many other related questions.

In what follows, we describe how technological advances in protein tagging and single-molecule imaging are beginning to revolutionize our understanding of single mRNA translation in living cells. We begin by reviewing breakthroughs in the field, followed by an in-depth discussion of recent applications of repeat-epitope tags. Our emphasis will be on the underlying tagging and imaging technologies and how they synergize to enable the quantification of ribosomal densities, translation initiation and elongation rates, and translation site mobility and higher-order structure (Fig. 1).

2 BREAKING THE SINGLE-MOLECULE BARRIER FOR LIVE-CELL IMAGING OF TRANSLATION

The quest to image protein expression in living cells has driven a tremendous amount of research and technology development, particularly in the field of fluorescence microscopy. A milestone was the advent of the green fluorescent protein (GFP), which made it possible to tag and monitor the expression and dynamics of any protein of interest in vivo (Giepmans et al. 2006; Chalfie 2009). Although GFP works tremendously well for most studies, it is not well suited for single-molecule imaging of mRNA translation. This is because GFP takes too long to mature and fluoresce to capture the translation process in real-time (Chao et al. 2012). Furthermore, GFP is not particularly bright or photostable, so it is difficult to see a single GFP fluorophore, even after the lights come on.

To get around this problem, researchers have turned to fluorophores that light up faster than GFP. The Venus fluorescent protein, for example, lights up just minutes after being translated in bacteria. This property, together with the low copy number of proteins in *Escherichia coli*, made it possible to break the live-cell single-molecule barrier for imaging mRNA translation in living cells (Yu et al. 2006). To achieve this remarkable feat, Venus was fused to a membrane-anchoring protein (serine chemoreceptor protein [Tsr]) and inserted into the native lacZ locus. Expression of Venus was then monitored intermittently between repeated intentional photobleaches of the entire cell. Single Tsr-Venus proteins anchored to the cell membrane could be detected above background by fluorescence immobilization. This revealed that translation occurred in random and uncorrelated bursts, with around one burst per cell cycle and around four Tsr-Venus proteins translated per burst.

Since this initial report, follow-up studies have used Venus fluorescence to quantify translation kinetics in living neurons with single-molecule resolution (Tatavarty et al. 2012; Ifrim et al. 2015; Ströhl et al. 2017). The first of these used cyanine 5-labeled uridine triphosphate (Cy5-UTP) to label mRNA encoding ARC (activity-regulated cytoskeletal-associated protein) and FMRP (fragile X mental retardation protein) tagged with Venus (Tatavarty et al. 2012). Labeled mRNA, introduced to neurons by injection, clustered together and defined the positions of neuronal granules. After whole-cell photobleaching was performed to remove preexisting mature Venus, local translation events were detected by the appearance and rapid single-step photobleaching of individual Venus molecules. A map was then generated showing the positions of all Venus appearance events, which revealed strong clustering near neuronal granules. Like in *E. coli*, translation was found to be bursty, with multiple translation events occurring over a relatively

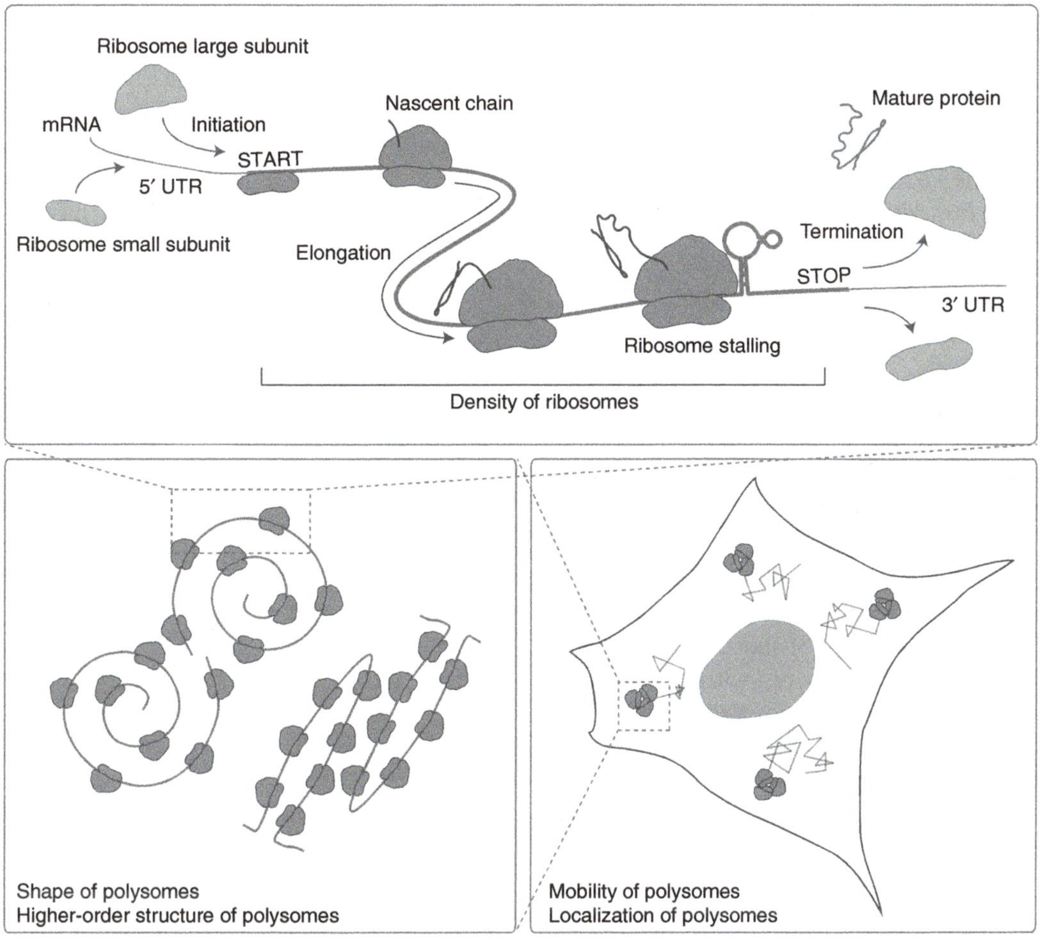

Figure. 1. Quantifying messenger RNA (mRNA) translation at the single-molecule level. Many mechanistic details of mRNA translation can now be measured in living cells. (*Top*) The rates of ribosome initiation, nascent chain elongation, ribosome stalling, and translation termination can be measured, as can the density of ribosomes within polysomes. (*Bottom*) Polysome shape, higher-order structure, mobility, and localization can also be measured.

short time. Here, bursts were defined as brief periods of high Venus translation, but strictly speaking, it is not possible to know if bursts originated from multiple mRNA in a single granule or from a single mRNA. As burst duration and fluorophore maturation times were convolved, the exact length of an individual burst was difficult to determine. Roughly, bursts were estimated to last around one to two minutes, shorter than Venus maturation times. In the two later studies using Venus in neurons, a similar strategy was used to show that the translation of immobilized Venus-PSD-95 (postsynaptic density protein 95) in neuronal dendrites and spines is dysregulated in fragile X syndrome and that synchronous translation of multiple β-actin transcripts is initiated by Netrin-1 (Ströhl et al. 2017).

A limitation of single-molecule studies of translation based on the synthesis of Venus (or any other fluorescent fusion protein) are the requirements that (1) nascent proteins be relatively immobile on the timescale of fluorophore maturation and (2) whole-cell photobleaching is required, which could have deleterious side effects (Dobrucki et al. 2007). Photoblinking of Venus also complicates the analysis. A powerful alternative strategy is to shift focus away from the main protein product and instead focus on the impact of translation. For example, as ribosomes elongate, they strip proteins off the coding sequence of mRNA. This effect is beautifully exploited by the TRICK (translation RNA imaging by coat protein knockoff) assay to detect the first round of mRNA translation (Halstead et al. 2015). Specifically, multiple fluorescent probes bound to the coding sequence of an mRNA are irreversibly stripped off by the first ribosome that translates through the mRNA. This loss of signal can then be used to distinguish translated from untranslated mRNA (Chao and Lionnet 2018). One disadvantage of the TRICK assay is that multiple rounds of translation cannot be monitored, because the signal is lost after the first round. However, a unique advantage is that TRICK only requires an RNA-binding fluorescent probe. With the advent of clustered regularly interspaced short palindromic repeats (CRISPR)-associated protein 13 (Cas13) (Abudayyeh et al. 2017), Pumilio homology domains (Adamala et al. 2016), and other programmable sequence-specific RNA binders, the TRICK assay could be adapted in the future to detect the translation of endogenous transcripts.

3 AMPLIFYING LIVE-CELL SINGLE mRNA TRANSLATION SIGNALS WITH REPEAT-EPITOPE TAGS

Recently, a number of laboratories have independently developed technology to image single mRNA translation dynamics over multiple rounds of translation in living cells (Morisaki et al. 2016; Pichon et al. 2016; Wang et al. 2016; Wu et al. 2016; Yan et al. 2016). As illustrated in Figure 2A, the technologies use fluorescent antibody-like probes to label tagged nascent peptides as they are being translated. Each tag contains a tandem array of short, linear epitopes that are individually bound by probes within seconds of their translation. By repeating epitopes in a single tag, multiple fluorescent probes can label a single nascent polypeptide chain. The signal is further amplified within polysomes, which contain multiple ribosomes translating multiple nascent chains from a single mRNA. The resulting bright signal makes it possible to track single translation sites within living cells for minutes or even hours at a time, not possible with single fluorophores like GFP or Venus.

So far, three epitopes have been used to image single mRNA translation in living cells: the SunTag epitope (EELLSKNYHLENEVARLKK) (Pichon et al. 2016; Wang et al. 2016; Wu et al. 2016; Yan et al. 2016) and the classic FLAG (DYKDDDDK) (Hopp et al. 1988) and HA (hemagglutinin; YPYDVPDYA) (Wilson et al. 1984) epitopes (Morisaki et al. 2016). Within the SunTag (Tanenbaum et al. 2014), individual epitopes are followed by linkers (GSGSG) and the resulting epitope-linker motif is repeated 24×, for a total length of 571 amino acids (aa). Smaller and larger repeats have also been tested, including 4× (Voigt et al. 2017), 5×, 10× (Yan et al. 2016), 32×, and 56× repeats (Pichon et al. 2016). In contrast to the uniform distribution of epitopes within SunTags, FLAG and HA epitopes are repeated nonuniformly within "spaghetti monster" tags (Viswanathan et al. 2015). Spaghetti monsters consist of a stable, nonfluorescent GFP-like core structure with strings of epitopes hanging off. In total, spaghetti monsters are ~325 aa in length, with 3× epitope repeats at both the amino and carboxyl termini of the core and a 4× repeat in the inner loop where GFP is commonly split.

Antibody-like probes are required to co-translationally light up the repeat epitopes within the SunTag and spaghetti monsters. In the case of the SunTag, a genetically encoded single-chain variable fragment (scFv) fused to a fluorescent protein like GFP is used (~50 kDa total size). The SunTag scFv was evolved via ribosome display from an anti-general control protein (GCN4) antibody and optimized for stability in the reducing cytoplasm of living cells (Hanes et al. 1998; Tanenbaum et al. 2014). In the case of spaghetti monsters, anti-FLAG or anti-HA antibody fragments (Fabs) conjugated to synthetic dyes like Alexa 488 are used (~50 kDa total size). Whereas fluorescent Fab needs to be generated using commercially available kits and parental antibodies, SunTag scFv can be stably expressed in living cells or organisms, a convenient and cost-effective advantage. On the other hand, spaghetti monsters come in many "flavors" for multicolor applications. Besides FLAG and HA, V5, Myc, and OLLAS spaghetti monsters are also available (Viswanathan et al. 2015). Thus, in principle, translation could be imaged in up to five colors with only spaghetti monsters.

4 LIMITATIONS OF REPEAT-EPITOPE TAGS FOR TRANSLATION IMAGING

The perfect repeat-epitope tag would light up brightly as soon as translation begins. Two fundamental biophysical constraints prevent this. First, temporal resolution is limited by how long it takes a fluorescent probe to label an epitope. Using microinjection, anti-FLAG Fab were shown to label FLAG spaghetti monster epitopes in <3 sec (Morisaki et al. 2016). SunTag probes also appear to label their epitopes quickly. When cells were treated with cycloheximide to stall ribosomes, polysome fluorescence did not increase much, suggesting steady state fluorescence had already been reached (Yan et al. 2016). A second factor limiting temporal resolution is epitope length. The longer the epitope, the longer it takes to be translated and detected. The FLAG and HA epitopes are respectively 8 aa and 9 aa in length, whereas the SunTag epitope is 19 aa in length. Provided that probes bind fast enough, the FLAG and HA epitopes can therefore be detected roughly two times faster than SunTag epitopes. However, it may be difficult to detect this difference in practice because of the high background signal from freely diffusing probes and the limited sensitivity of light microscopes.

A major concern of repeat-epitope tags is their large molecular mass and size. When fully labeled, the 24× SunTag repeat is 1.4 MDa, whereas the 10× spaghetti monster is 0.54 MDa. This extra mass should impact translation. Remarkably, however, measured elongation rates using SunTags and spaghetti monsters largely agree (Table 1). Furthermore, polysome mobility is also largely unaffected by tag size (Table 1). Placing either fewer (5× and 10×)

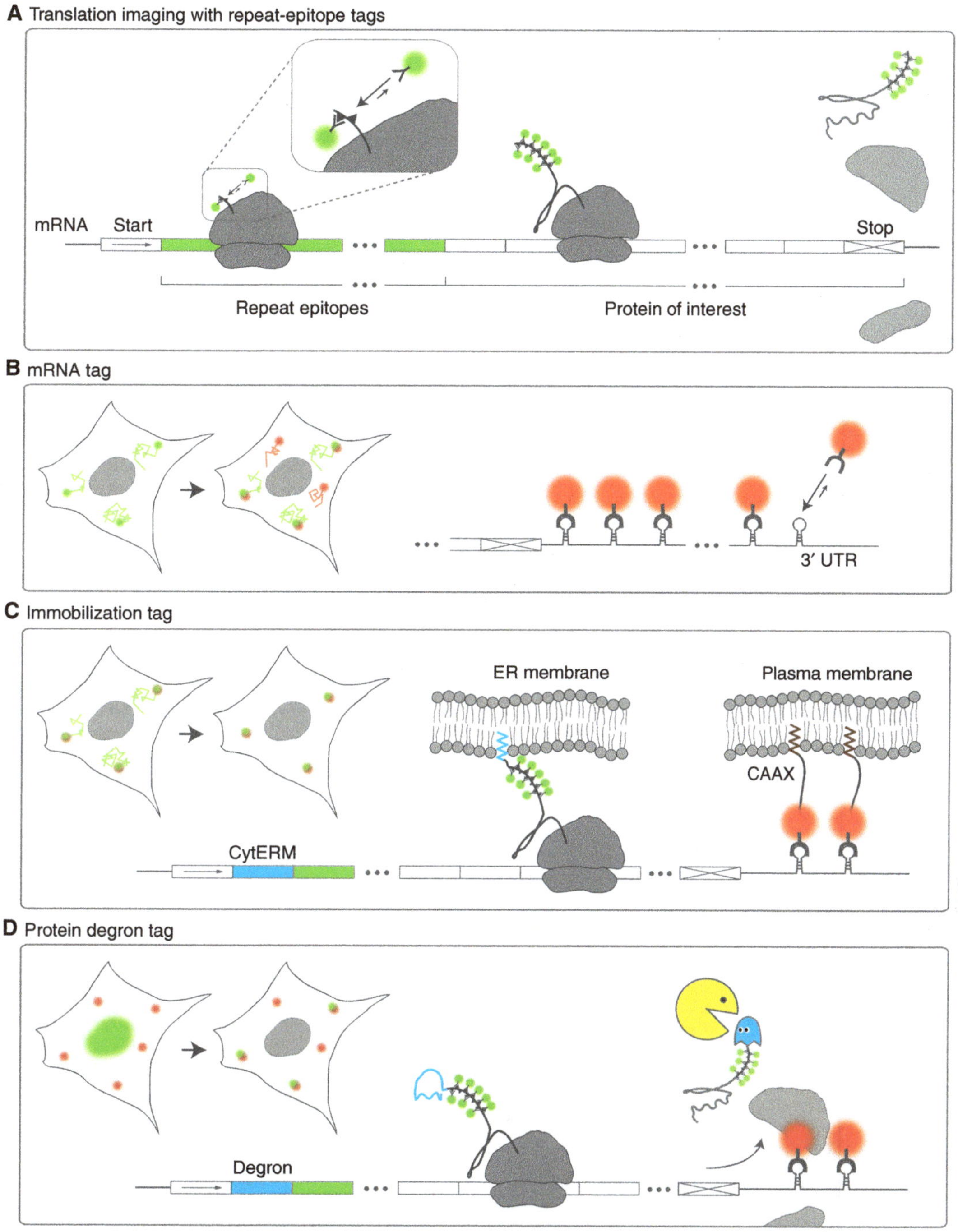

Figure 2. Repeat-epitope and accessory tags for single-molecule imaging of mRNA translation. A schematic showing how repeat-epitope tags are used to visualize single translation sites in living cells. (*A*) Nascent peptides (small triangles) encoded by repeat epitopes (green mRNA) placed in the amino terminus of a protein of interest are co-translationally labeled by fluorescent antibody fragments (green spheres). (*B*) An mRNA tag can facilitate single mRNA translation imaging. Stem loops within the 3′ untranslated region (UTR) of mRNA are co-transcriptionally labeled by coat proteins (red spheres). In combination with a repeat-epitope tag, translation sites are marked in green and red. (*C*) Immobilization tags can restrict the movement of mRNA so that translation sites can be tracked with less frequent imaging. The cytoplasmic end of an endoplasmic reticulum signal-anchor membrane protein (CytERM) tag (blue) embeds nascent chains into the endoplasmic reticulum (ER) co-translationally, whereas a CAAX domain embeds transcripts with an mRNA tag into the plasma membrane via bound coat proteins (red spheres). (*D*) A degron (blue mRNA) encoded upstream of the repeat epitopes enhances the degradation of proteins, so nascent chains can be preferentially imaged (rather than mature proteins).

Table 1. A comprehensive summary of single-molecule measurements of translation dynamics in living cells

System									Results					
Reference(s)	Cell	Protein of interest (POI)	aa	Tag	aa	Probe	RNA-tag	Probe	Initiation rate (sec)	Elongation rate (aa/sec)	Ribosome density (nt/ribo)	Expression	Mobility	Selected heterogeneity highlight
Repeat-epitope tags														
Voigt et al. 2017	U2OS	*Renilla*-FKBP	418	4×Sun	96	scFV-sfGFP	24×MS2	NLS-MCP-Halo-JF549	-	-	~380 (cytosol), ~260–380 (ER)	71% ± 2%	-	ER translation efficient; small fraction of cytosolic protein translated in ER
		Gaussia	~180	-	-	-			-	-	-	-	D (<0.06) ER[a]	
		Renilla	~330						-	-	-	-		
Pichon et al. 2016	HeLa	Ki67	3177	56×Sun	1344	scFV-sfGFP (± NLS)	24×MS2	NLS-MCP-TagRFP-T	~13	13.2 - 18	~770	47% B	D (0.042)	Translation efficiency is higher in translation factories than out
		POLR2A[b]	1970				-	-	~19[c]	13.8	~770	91% B	D (~0.04)	
		DYNC1H1[b]	4646	32×Sun	768				-	-	~480	~73% B	D (0.01) 19% M (1)	
Morisaki and McNally 2014; Morisaki et al. 2016	U2OS	KDM5B[c]	1544	10×HA/ FLAG	325	Fab-Cy3/A488[e]	24×MS2	tdMCP-Halo-JF646	~30	10 ± 1.3	~910[d]	86% ± 2% C[d]	D (~0.05)	Translation factories can be stable
		β-actin	374							12 ± 2.2	~360[d]	19% ± 4% C[d]	D (<0.05)	
		Histone H2B	125							7.7 ± 3.2	~180[d]	4% ± 1% C[d]	D (~0.13)	
Wu et al. 2016	U2OS	AID[c]	241	24×Sun	576	scFV-sfGFP	24×MS2	stdMCP-TagRFP-T/ stdMCP-stdGFP/stdMCP-Halo-JF549	~45	5.6 ± 1.4[e]	~610	~70% C	untethered: ~50% SD/ ~50% D CytERM: ~100% SD	β-actin 3′ UTR burst in neurons, not in U2OS; translation efficiency lower in distal dendrite
		oxBFP-AID	484							4.7 ± 0.6[e]	~640	-		
		FLuc-AID	496							5.0 ± 1.1[e]	~540	-		
		FLuc-oxBFP-AID	1037						-	-	~690	-		
	Neuron	oxBFP-AID	467				-	-	~29	5	-	80% B/20% C	C in distal M to dendrite	
Yan et al. 2016	U2OS	Kif18b	850	24×Sun	576	scFV-sfGFP	24×PP7	tdPCP-3xmCherry/ tdPCP-2xmCherry-CAAX	17–43	3.5 ± 1.1	200–400	70% B	untethered: D (0.047) CAAX: D (1.06 × 10^{-3})	5% of mRNA produce 90% of protein (Emi1 5′ UTR)
				10×Sun	240					-	200–400	-		
				5×Sun	120					3.1 ± 0.4	160–320	-		
		BFP	236	24×Sun	576			-	-	4.9[h]	-	-		
Wang et al. 2016	HeLa	ODC	473	24×Sun	576	scFV-sfGFP	24×PP7	tdPCP-2xTomato/-	~20[c]	4	~260	-	D (0.021) D (0.0053) perinuclear D (0.0013) ER[i] D (0.0017) ER[i]	Location-dependent mobility; transient burst of translation following stress (ATF4 uORFs)
		BFP-ODC	~700						-		-	-	D (0.021)	
		MBP-ODC	~850						-		-	-	-	
	Neuron	ODC	473	24xSun	576	scFV-sfGFP	24xPP7	-	-	-	-	-	D (0.006) 7% M (2.0)	

TRICK														
Halstead et al. 2015	U2OS	NLS-3xHA-dNP799	368	6xPP7	405 nt	NLS-PCP-GFP	24xMS2	NLS-MCP-RFP	-	-	-	-	-	Oskar transcript translated posterior pole of oocyte
	HeLa	*Renilla*	~330	12xPP7	810 nt			NLS-MCP-Halo-JF549	-	-	-	-	-	
	Oocyte	Osk	~600				6xMS2	NLS-MCP-RFP	-	-	-	-	-	
Venus														
Ströhl et al. 2017	RGC	B-actin	374	Venus	239	-	-	-	-	-	-	Synchronized C	-	Netrin-1-induced synchronized translation
Ifrim et al. 2015	Neuron	PSD95	~700	Venus	239	-	4×BoxB	λN22-CFP	-	-	-	-	-	PSD-95 translation dysregulated in dendrites and spines of FXS
Tatavarty et al. 2012	Neuron	ARC	~400	Venus	239	-	~10× UTP Cy5	-	~10–15	-	~200	12% B/88% C	-	Bursts in neuronal RNA granules
		FMRP	~630									52% B/48% C	-	
	Fmr1 KO neuron	ARC	~400									24% B/76% C	-	
Yu et al. 2006	*Escherichia coli*	Tsr	450	Venus	239	-	-	-	-	-	-	1.2 B per cell cycle, 4.2 molecule per B	-	Bursty translation

TRICK, translation RNA imaging by coat protein knockoff; U2OS, human osteosarcoma; HeLa, Henrietta Lacks; RGC, retinal ganglion cell; Fmr1 KO, fragile X mental retardation 1 knockout; FKBP, FK506-binding protein; POLR2A, RNA polymerase II subunit A; DYNC1H1, cytoplasmic dynein 1 heavy chain 1; KDM5B, lysine demethylase 5B; AID, activation-induced cytidine deaminase; oxBFP, oxidizing environment variant of blue fluorescent protein; FLuc, firefly luciferase; Kif18b, kinesin family member 18b; BFP, blue fluorescent protein; ODC, ornithine decarboxylase; MBP, maltose-binding protein; NLS-3xHA-dNP799, three copies of the hemagglutinin tag fused to a nuclear localization sequence and fluorogen-activating protein; Osk, oskar maternal effect protein; PSD95, postsynaptic density protein 95; ARC, activity-regulated cytoskeleton-associated protein; FMRP, fragile X mental retardation protein; Tsr, serine chemoreceptor protein; HA, hemagglutinin; PP7, *Pseudomonas* phage 7; scFV, single-chain variable fragment; sfGFP, superfolder green fluorescent protein; NLS, nuclear localization sequence; Fab, antibody fragment; Cy3/A488. cyanine 3/Alexa 488; NLS-PCP-GFP, PP7 coat protein fused to a nuclear localization sequence and green fluorescent protein; MS2, bacteriophage MS2; UTP Cy 5, uridine triphosphate labeled by cyanine 5; NLS-MCP-Halo-JF549, MS2 coat protein fused to a nuclear localization sequence and a Janelia Fluor 549 HaloTag ligand; NLS-MCP-TagRFP-T, MS2 coat protein fused to a nuclear localization sequence and the variant of red fluorescent protein; tdMCP-Halo-JF646, tandem dimer MS2 coat protein fused to a Janelia Fluor 646 HaloTag ligand; stdMCP-TagRFP-T, synonymous tandem dimer MS2 coat protein fused to the variant of red fluorescent protein; tdPCP-3xmCherry, tandem dimer PP7 coat protein fused to three copies of the monomeric mCherry protein; λN22-CFP, λN22 peptide fused to cyan fluorescent protein; ER, endoplasmic reticulum; C, constitutive; B, bursty; S, subdiffusive; D, diffusive (μm^2/sec); M, motored (μm/sec); UTR, untranslated region; mRNA, messenger RNA; Emi1, early mitotic inhibitor; ATF4 uORFs, activating transcription factor 4 upstream open reading frames; FXS, fragile X syndrome.

[a]Diffusion constant measured with RNA tag only.

[b]Endogenous locus.

[c]Not reported but calculated here from elongation rate and ribosomal density.

[d]Polysomes.

[e]Measured with CytERM.

[f]Estimated from the graph.

[g]Calculated from the ratio.

[h]Codon optimized.

[i]Targeted to ER by coding sBFP or SMTOM upstream via P2A.

(Yan et al. 2016) or more (32× and 56×) (Pichon et al. 2016) repeats within the SunTag leads to little difference in measured translation kinetics (Table 1). Importantly, reporter proteins tagged with the SunTag were produced at the same levels as untagged proteins (Yan et al. 2016). Together, these data suggest translation sites can cope with the extra mass.

5 ACCESSORY TAGS FOR IMAGING SINGLE mRNA TRANSLATION IN LIVING CELLS

In addition to the repeat-epitope tag and probes, accessory tags facilitate the imaging and quantification of single mRNA translation, as summarized in Figure 2. First, a tag for tracking single mRNA is highly recommended. An mRNA tag provides a reference position for translation. Without it, distinguishing translation sites from mature protein or aggregates is difficult. An mRNA tag also makes it possible to (1) count how many mRNA are within a translation site, (2) track bursts of translation on a single mRNA, and (3) determine the fraction of mRNA translating.

So far all groups have used MS2 bacteriophage (Bertrand et al. 1998; Fusco et al. 2003) or *Pseudomonas* phage 7 (PP7) mRNA tags (Chao et al. 2008) to monitor translation (Fig. 2B). These tags were reviewed recently in Buxbaum et al. 2015. Although extensively tested in living cells and organisms (Lionnet et al. 2011; Lionnet and Singer 2012), mRNA tags may interfere with underlying biology. In yeast, MS2 can artificially stabilize transcripts (Garcia and Parker 2015; Heinrich et al. 2017). Although this may not be a problem for long-lived transcripts in higher eukaryotes, a gentler MS2 tag has been developed to address this problem (Tutucci et al. 2018). An alternative to MS2 or PP7 tags is single molecule RNA FISH (fluorescence in situ hybridization) (Pichon et al. 2016), but this requires fixation.

Another useful accessory for translation imaging is an mRNA immobilization tag (Fig. 2C). The high mobility of mRNA makes it difficult to track their translation kinetics. To get around this issue, mRNA can be tethered to immobile substrates like the endoplasmic reticulum (ER) or the plasma membrane (PM). This makes it possible to image mRNA less frequently and acquire good images without photobleaching. To tether transcripts to the PM, a CAAX domain can be fused to MS2 coat protein (MCP) or PP7 coat protein (PCP) (Fig. 2C) (Yan et al. 2016). To tether to the ER, a cytoplasmic end of an endoplasmic reticulum signal-anchor membrane protein (CytERM) tag can be fused to the amino terminus of the protein of interest (Fig. 2C) (Costantini et al. 2012; Wu et al. 2016). The downside is tethering can alter translation kinetics and prevents the analysis of local translation.

A less essential accessory for translation imaging is a degron tag (Fig. 2D). Degrons are quickly degraded, so any attached proteins do not accumulate. Without a degron, mature proteins sequester probes and eventually prevent translation imaging. Three degrons have been tested so far: ornithine decarboxylase (ODC; 467 aa) (Wang et al. 2016), an auxin-inducible degron (AID; 68 aa) (Wu et al. 2016), and an unstable version of the FK506-binding protein (FKBP; 107 aa) (Voigt et al. 2017). Although degrons are convenient, their degradation may interfere with translation (Hurley et al. 2016). If this is a concern, alternatives to degrons include using an inducible promoter (Yan et al. 2016) or simply imaging soon after transfection (Morisaki et al. 2016), when mature protein levels are low.

6 TAGGING ENDOGENOUS PROTEINS WITH REPEAT-EPITOPE TAGS

With the advent of CRISPR-based genome editing, it has become easier to add tags directly to endogenous genes, including repeat epitopes for imaging single mRNA translation (Pichon et al. 2016). The major advantage of this approach is that mRNA and protein arise from their natural, endogenous contexts and are thus maintained at appropriate endogenous levels. The drawback for translation imaging is that endogenous mRNA and protein levels may be high, which could make it hard to detect single translation sites.

To insert 32× or 56× amino-terminal SunTag epitopes into endogenous loci, Pichon et al. (2016) used a homologous recombination donor containing a start codon, a selection marker (puromycin), a porcine teschovirus-1 2A (P2A) peptide cleavage sequence (Kim et al. 2011), and the repeat epitopes, all sandwiched between two homology arms. Properly inserted heterozygous cells were selected using puromycin, and translation beyond the P2A was detected via SunTag probe binding to nascent chains. Two genes were targeted with this strategy, the large subunit of RNA polymerase II (POLR2A) and the heavy chain of dynein 1 (DYNC1H1). Although this strategy is generalizable, both POLR2A (1970 aa) and DYNC1H1 (4646 aa) are relatively large proteins. In combination with the large number of repeat epitopes, tagged transcripts were very long, so translation sites were exceptionally bright. If shorter genes are targeted with fewer repeats, signal-to-noise may become limiting.

Although CRISPR editing is an appealing strategy to examine single mRNA translation kinetics in a nearly endogenous manner, there may be alternatives. Because probes are based on antibodies, endogenous repeat epitopes could in theory be targeted. It has been estimated that roughly one third of all human proteins performing fundamental functions contain tandem repeats (Pellegrini et al.

 Cite as *Cold Spring Harb Perspect Biol* doi: 10.1101/cshperspect.a032078

1999; Jorda et al. 2010). A well-known example is the heptad repeat in the carboxy-terminal domain of RNA polymerase II, which is repeated around 50 times in humans and for which antibodies and fluorescent Fab already exist (Stasevich et al. 2014). Even if repeats are not available, "programmable" antibodies may someday allow the imaging of any set of epitopes (Ernst and Plückthun 2017), similar to how dCas9 is currently used to light up specific DNA loci in living cells (Chen et al. 2013).

7 COUNTING RIBOSOMES IN POLYSOMES WITH REPEAT-EPITOPE TAGS

A strength of imaging single mRNA translation dynamics with repeat-epitope tags is the ability to precisely quantify a number of key biophysical parameters associated with single translation sites, including their ribosomal occupancy, their rates of elongation and initiation, as well as their mobility, localization, and shape (Fig. 1). To calculate how many ribosomes are in polysomes, polysome fluorescence must be normalized to single mature protein fluorescence F. A complication is that ribosomes translating repeat-epitope tags on average have translated only half the epitopes. This gives the open reading frame (ORF) an effective length $L' = (L_{POI} + 0.5\ L_{Tag})$, in which L_{Tag} is the length of the tag and L_{POI} is the length of the protein of interest. Then, polysome fluorescence is the product $F \times d \times L'$, in which d is the density of ribosomes along the transcript (Pichon et al. 2016; Wang et al. 2016; Yan et al. 2016). If nascent chains cannot be seen above background until a certain number of epitopes have been translated, an additional correction factor makes L' smaller (Yan et al. 2016). If tags cannot be observed until they are fully translated or if $L_{POI} > L_{Tag}$, then $L' \cong L_{POI}$ (Morisaki et al. 2016).

To accurately determine F, sites of mature protein can be identified by their resistance to puromycin and their lack of colocalization with mRNA. Tagging a control protein that does not aggregate or form higher-order structures, such as the maltose-binding protein MBP, may also help (Wang et al. 2016). Alternatively, a second control construct with a single epitope per nascent chain can be imaged (Morisaki et al. 2016). Each fluorophore in a polysome then represents a single nascent chain or ribosome. The number of ribosomes is minimally the total fluorescence divided by the fluorescence of a single fluorophore, the latter being determined unambiguously via single-step photobleaching.

8 QUANTIFYING TRANSLATION ELONGATION WITH REPEAT-EPITOPE TAGS

The translation elongation rate is another important parameter that can be quantified from repeat-epitope tag experiments. As shown in Figure 3, this can be performed via ribosomal runoff, fluorescence recovery after photobleaching (FRAP), or fluorescence correlation spectroscopy (FCS). Fundamentally, these all measure the same thing: the fluorescence dwell time at a translation site. To equate the dwell time to the total elongation time, some common assumptions are made. The main technical assumptions are that probes bind epitopes quickly and without interference, as discussed above. In addition, there are a few biological assumptions. First, it is assumed that ribosomes are in steady state so that, on average, when one finishes translation it is replaced by another that starts translation. Second, ribosomes are assumed to always complete translation, so no partial proteins are produced. Third, completed nascent chains are assumed to quickly leave the translation sites. When these assumptions are met, the elongation rate is approximately the effective length L' of the ORF divided by the dwell time.

A straightforward way to determine the elongation rate is via ribosomal runoff. The basic idea is that on inhibition of translation initiation, the remaining elongating ribosomes will finish translating one-by-one (i.e., "run off" the mRNA). Natural runoff can be seen if translation is bursty (Pichon et al. 2016; Wu et al. 2016; Yan et al. 2016). Runoff can also be induced by adding the translational inhibitor harringtonine to cell media (Pichon et al. 2016; Wang et al. 2016; Yan et al. 2016). As shown in Figure 3 (top, left), runoff leads to a loss of fluorescence at the translation site in four stages (Pichon et al. 2016; Wang et al. 2016; Yan et al. 2016). The fluorescence starts out in steady state. Harringtonine kicks in during phase II, when fluorescence loss due to runoff is countered by the translation and labeling of new epitopes. Eventually, ribosomes translate past all epitopes, at which point phase III begins and the fluorescence loss is maximal and linear. If many single-transcript runoff curves are averaged, a fourth artifactual phase appears because of heterogeneity (Yan et al. 2016). Fitting phase III of the runoff to a line is an easy way to estimate the elongation rate. The time at which this fitted line extrapolates to zero postharringtonine gives the total dwell or elongation time. Dividing the effective length L' of the ORF by the dwell time yields the elongation rate (Wang et al. 2016). Alternatively, more sophisticated models can be used that simultaneously account for the different phases (Pichon et al. 2016; Yan et al. 2016) or translation stochasticity (Wu et al. 2016). If translation is inefficient, single ribosomal runoffs can be observed. Although challenging to capture, the fluorescence signature from these single ribosome events also yields the elongation rate, as illustrated in Figure 3 (bottom, left) (Yan et al. 2016).

A second technique to estimate the elongation rate is FRAP (Fig. 3, top, right). The idea behind FRAP is that

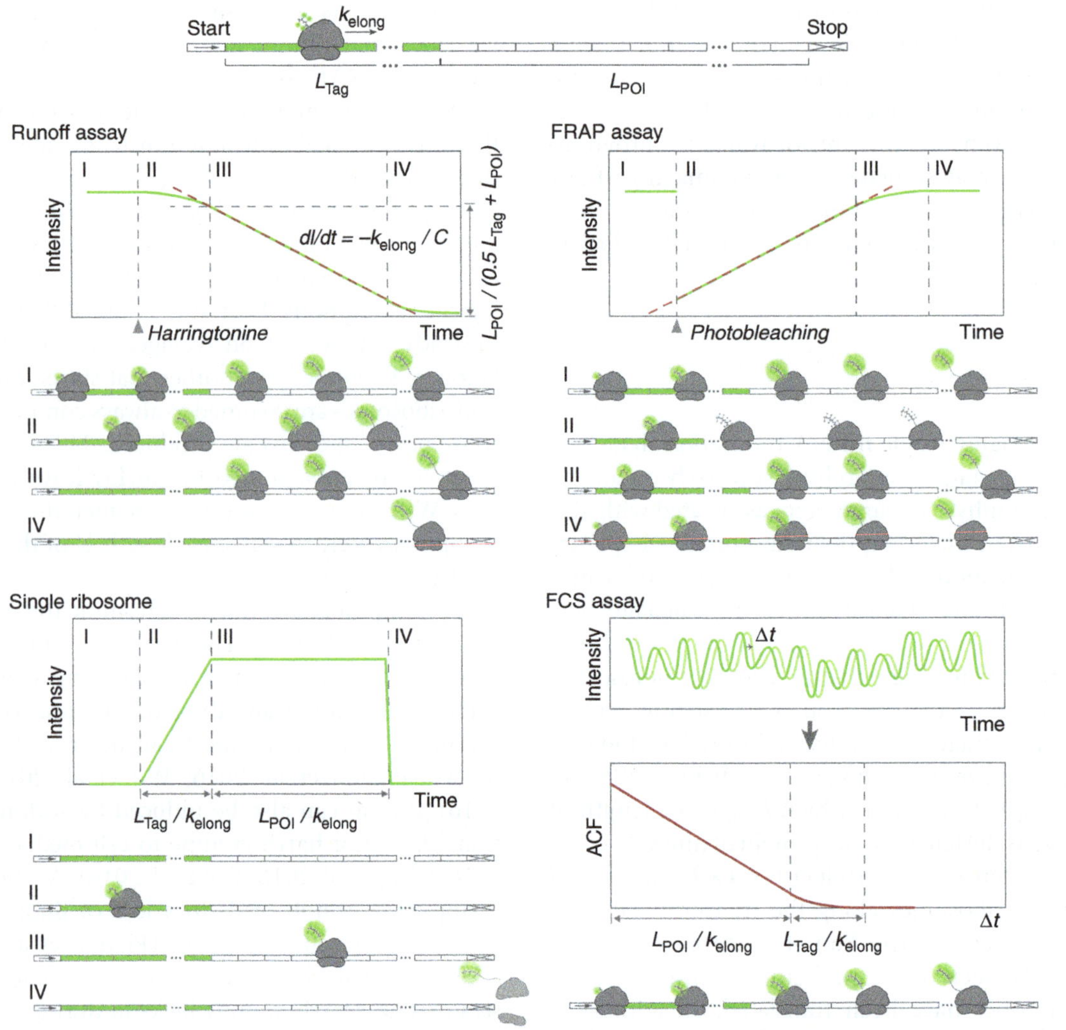

Figure 3. Quantifying translation elongation rates with repeat-epitope tags. The translation elongation rate k_{elong} can be measured using repeat-epitope tags in multiple ways. Common assumptions are (1) every ribosome that finishes translation is on average replaced by another that starts translation; (2) ribosomes are assumed to always complete translation, so no partial proteins are produced; and (3) completed nascent chains quickly leave the translation sites. L_{Tag} is the length of the repeat-epitope tag and L_{POI} is the length of the tagged protein of interest. (*Top, left*) Addition of the translation initiation inhibitor harringtonine leads to ribosome runoff. During runoff, polysome fluorescence on average drops in four phases (I–IV). The rate of fluorescence decay depends on the elongation rate. (*Bottom, left*) The fluorescence signal during monosome translation runoff also depends on the elongation rate. (*Top, right*) Fluorescence recovery after photobleaching (FRAP) of polysomes leads to an average fluorescence recovery that is approximately the inverse of the average polysome runoff fluorescence decay. (*Bottom, right*) Fluorescence correlation spectroscopy (FCS) at the translation sites produces a decaying autocorrelation function (ACF) that depends on the elongation rate.

nascent chains labeled by photobleached probes will eventually be replaced by newly translated nascent chains labeled by nonphotobleached probes. This requires an additional assumption compared with runoff: The probes should bind irreversibly on the timescale of the FRAP experiment. If not, the fluorescence recovery will be due to probe turnover on single nascent chains rather than nascent chain turnover from translation. Both SunTag probes and anti-FLAG Fab bind epitopes for tens of minutes at a time, making them suitable for estimating translation dwell times on the order of a few minutes (Morisaki et al. 2016; Yan et al. 2016). Provided ribosomes are in steady state, the fluorescence recovery curve is just the inverse of the runoff curve and can be fit accordingly (Pichon et al. 2016).

Like FRAP, FCS experiments use changes in fluorescence at translation sites to extract translation kinetics

(Fig. 3, bottom, right) (Stasevich et al. 2010). FCS therefore shares the same common assumptions as FRAP. Unlike FRAP, however, FCS does not require a strong laser pulse to induce an artificial fluctuation in the fluorescence, which could damage cells (Dobrucki et al. 2007). Instead, FCS relies on natural fluctuations that arise from the continual steady state creation and departure of nascent chains. To illustrate, if multiple ribosomes happen to initiate quickly on a single transcript in sequence, then the fluorescence there will get significantly brighter as multiple new nascent chains are translated and labeled by probes. Later, this fluorescence will depart the transcript as the ribosomes terminate in sequence. The time of departure is stochastic, but on average the total dwell time of the fluorescence should be a good approximation for the total elongation time (with all the common assumptions discussed above). If enough natural fluctuations are observed, then the dwell time can be accurately determined from the fluorescence autocorrelation (Morisaki et al. 2016; Wu et al. 2016). The fluorescence autocorrelation can be interpreted as the probability of seeing the same fluorescence at a later time. As translation leads to fluorescence turnover, the autocorrelation generally decays with time, analogous to autocorrelation analysis in the field of transcription (Coulon and Larson 2016). The time at which the autocorrelation drops to zero gives the dwell time of a nascent chain at the mRNA, from which the elongation rate can again be calculated as in FRAP or runoff. More sophisticated stochastic models can also be used to fit the full curve (Wu et al. 2016).

An advantage of FRAP and FCS relative to runoff is they can be performed on any transcript without the need for drugs or the requirement of bursty expression. On the other hand, both FRAP and FCS require care when performing and interpreting data. For example, the intentional FRAP photobleach has many potential pitfalls, including photodamage (Dobrucki et al. 2007), photoinduced cross-linking or unbinding (Heinze et al. 2009), and photoswitching (Mueller et al. 2012; Morisaki and McNally 2014). Likewise, FCS is technically demanding (requiring many hundreds or thousands of individual images), and sensitive to experimental noise from photobleaching, nonhomogeneous illumination, tracking errors, diffusion of free probe, and focusing issues (Coulon and Larson 2016).

9 QUANTIFYING TRANSLATION INITIATION WITH REPEAT-EPITOPE TAGS

Once the number of ribosomes in transcripts and the total elongation time are measured, it is relatively straightforward to calculate the translation initiation rate. The calculation again relies on the steady state assumption, in which ribosomes enter and exit translation sites at a constant, balanced rate. Then, the total elongation time divided by the number of ribosomes on the transcript gives the time between ribosomes. Assuming uniform movement along the transcript, this time is equal to the initiation time (i.e., the inverse initiation rate) (Morisaki et al. 2016; Pichon et al. 2016; Wang et al. 2016; Wu et al. 2016; Yan et al. 2016).

10 QUANTIFYING POLYSOME MOBILITY AND SHAPE WITH REPEAT-EPITOPE TAGS

Besides ribosome density, initiation, and elongation kinetics, the strong fluorescence of repeat-epitope-tagged nascent chains makes it possible to track them with great precision in living cells and characterize their mobility and shape. To quantify the mobility of mRNAs, mean-squared displacement is a standard metric (Saxton 1996). Four categories of mobility have been observed, in order of increasing dynamics: confined, corralled, diffusive, and motored (Park et al. 2010). Confined polysomes are bound to an immobile substrate. Some cutoff is generally required to define "bound." A threshold can be obtained by independently measuring the mobility of the confining structure. For example, the mobility of histone H2B has been used to define chromatin-bound transcription factors (Mazza et al. 2012; Morisaki et al. 2014). A similar approach could be used to define mRNA confinement to the PM or ER using CAAX domains (Hancock et al. 1991; Yan et al. 2016) or CytERM (Costantini et al. 2012; Wu et al. 2016) immobilizers, respectively. In contrast to total confinement, corralled polysomes are able to move about freely within a restricted subcellular volume. The mean-squared displacement for corralled polysomes starts off linear, but eventually plateaus to a peak value that marks the boundaries of the corralled zone. If the restricted volume is the entire cytoplasm, then polysome movement is purely diffusive and the mean-squared displacement is linear. Finally, motored or "super-diffusive" polysomes move about in an erratic fashion with brief periods of rapid movement in random directions at roughly constant velocity (Pichon et al. 2016; Wu et al. 2016). The mean squared displacement in this case is described by a curve with an exponent between one (pure diffusion) and two (constant velocity).

To quantify the shape of translation sites encoding repeat-epitope-tagged proteins, the intramolecular fluorescence distribution can be analyzed (Morisaki et al. 2016). As the fluorescence originates from nascent chains labeled by probes, the spatial extent of the chains constrains the spatial resolution of imaging, in addition to diffraction. Because three nucleotides totaling ~1 nm in length encode a single amino acid ~0.35 nm in length, the ORF of an mRNA is around three times longer than the nascent chain it encodes when both are stretched out. Including both untranslated regions (UTRs), a transcript can therefore be up to an order of magnitude longer than the protein it encodes. To illustrate, the β-actin transcript is ~370 nm when spread out (not diffraction-limited with 488 nm light), whereas the nascent chains would only be 374 aa × 0.35 nm ~ 130 nm (diffraction-limited). Nascent chain fluorescence, therefore, has the potential to provide subdiffractive structural information about a single translation site if ribosomes are sufficiently spread out within the site.

To accurately capture the intramolecular fluorescence of translation sites, imaging exposure times should be fast compared with the movement of the translation site. Faster exposures will create a better snapshot and prevent artifacts due to movement. Fixation could be used as a movement control as well. Spots may be diffraction-limited, in which case they will appear the same shape and size as images of immobilized 100-nm beads on a coverslip. If so, superresolution can be achieved by fitting the fluorescence distribution to a Gaussian function. The center of the Gaussian distribution function gives the average coordinates of the nascent chains (Morisaki et al. 2016). Deviations from the expected diffraction-limited spot may indicate the translation site is spread out in some manner. If an RNA tag is also included, the distribution of fluorescence within nascent chains can also be compared with the distribution of fluorescence within the RNA tag. This approach was used to show that translation sites composed of two mRNA had dimensions roughly twice the size of single mRNA translation sites, indicating the individual mRNAs retained their spatial independence despite co-movement (Morisaki et al. 2016)

11 EMERGING CONSENSUS IN THE FIELD OF LIVE-CELL IMAGING OF SINGLE RNA TRANSLATION

With the advent of repeat-epitope tags, fast-maturing fluorescent proteins like Venus, and RNA-based assays like TRICK (and, more recently, 3′-RNA end accumulation during turnover [TREAT]; Horvathova et al. 2017) to unambiguously spot hard-to-see molecular events in living cells, we expect the number of single-molecule studies of translation to grow significantly in the near future. Given the early state of the field, it is rather remarkable that there is already some consensus in terms of the quantification of translation kinetics. First, single mRNA studies of translation in eukaryotes have found good agreement in measured translation elongation rates. The average is close to 10 aa/sec, with an experimental range between 3 and 18 aa/sec (Table 1). Whereas this range is in part explained by codon usage (Yan et al. 2016), the overall order-of-magnitude agreement probably reflects underlying biophysical constraints on molecular motors. Consistent with this, 10 aa/sec corresponds to 1800 nt/min, which is close to what has been measured for transcription elongation by RNA polymerase II (Darzacq et al. 2009; Larson et al. 2011; Stasevich et al. 2014). In the future, it will be interesting to see how tightly coupled transcription and translation elongation rates are for specific genes in eukaryotic cells.

In addition to translation elongation rates, the measured density of ribosomes along active transcripts is also fairly consistent between studies, being one ribosome every 200 to 1000 nt. This order-of-magnitude agreement becomes even better when calculating ribosomal initiation rates from the measured ribosomal densities and elongation rates. In this case, all studies agree within a factor of just three, finding between two and five ribosomes initiating per minute on translationally active transcripts (Table 1). Again, this excellent agreement in the initiation rates may reflect some underlying biophysical constraint that limits the number of ribosomes that can initiate in a set amount of time. Alternatively, it could be an artifact of small sample size. More studies will be needed to get a better sampling of the full distribution of ribosome initiation and elongation rates.

12 HETEROGENEITY IN SINGLE mRNA TRANSLATION EFFICIENCY

In what follows, we highlight some of the interesting heterogeneity that has so far been uncovered by imaging the translation of single mRNA. As Table 1 summarizes, the heterogeneity takes a variety of forms, including heterogeneity in translation efficiency, ribosome dynamics, and polysome mobility and localization.

One of the most striking observations of single mRNA translation is that individual transcripts can switch between translationally active and inactive states (Table 1). This bursty behavior is likely a universal feature of gene expression, as it has now been observed in both living bacteria (Yu et al. 2006) and eukaryotes (Tatavarty et al. 2012; Pichon et al. 2016; Wang et al. 2016; Wu et al. 2016; Yan et al. 2016). In contrast to bursty translation, some transcripts appear to be constitutively expressed, i.e., perpetually in an active state in which initiation is always possible (Morisaki et al. 2016). Whereas the line between bursty and constitutive expression is somewhat gray, it seems clear that more abundant proteins are more likely to be translated in a constitutive fashion. For example, there are hundreds of thousands of RNA polymerase II proteins per cell (Kimura et al. 1999; Stasevich et al. 2014). Not surprisingly then, POL2R, the RNA polymerase II transcript, is constitutively translated at very high levels, with upward of 90% of mRNA being translated at any given time. In contrast, just 40%–70% of transcripts encoding components of the molecular motors dynein and kinesin are translated at any given time. These

Cite as *Cold Spring Harb Perspect Biol* doi: 10.1101/cshperspect.a032078

bursty transcripts stochastically switch between active and inactive states on the timescale of minutes to hours. These examples suggest that burstiness might be genetically encoded, but this is certainly not always the case. A counterexample is the β-actin transcript, which is constitutively expressed in human osteosarcoma (U2OS) cells (Morisaki et al. 2016; Wu et al. 2016), but bursty in neurons (Wu et al. 2016; Ströhl et al. 2017). Transcripts specifically in neurons seem especially prone to bursty behavior (Tatavarty et al. 2012; Ifrim et al. 2015; Wang et al. 2016; Wu et al. 2016; Ströhl et al. 2017). This could reflect a requirement to shut down or ramp up translation in a localized manner (Holt and Schuman 2013). Thus, both genetic and posttranscriptional factors dictate the expression patterns of individual transcripts in vivo.

An interesting consequence of heterogenous mRNA translation efficiency is that a relatively small subset of mRNA may be responsible for the vast majority of proteins produced in a cell. A good illustration of this point is the Emi1 5′ UTR, which has a long isoform with several upstream ORFs (Yan et al. 2016). The upstream ORFs sequester ribosomes. Because of this, only 20% of reporter mRNA harboring the long 5′-UTR isoform were observed to be actively translated, most with low efficiency. Remarkably, a small subset of these active mRNA was robustly translated, creating >50% of the final protein product. Thus, even among a presumably genetically identical pool of mRNA, translation may be heterogeneous on multiple levels, with a minor subset of a small fraction producing most proteins against the odds. How general this phenomenon is and what specific factors facilitate it remain open questions.

13 HETEROGENEITY IN RIBOSOME MOVEMENT ALONG SINGLE mRNA

Accurate measurements of translation elongation rates at the single mRNA level make it possible to directly address questions of codon usage and ribosome stalling. The most direct test of codon usage to date was an examination of the blue fluorescent protein BFP (Yan et al. 2016). A codon-optimized version of this construct elongated at an average rate of 4.9 aa/sec, whereas an unoptimized version (encoding Kif18b) had a rate of just 3.1 aa/sec. Although the optimized and unoptimized transcripts have different average rates, it is important to point out that there may be great variability within each pool. For example, FCS experiments on genetically identical individual transcripts showed some could be translated over twice as fast as others (Morisaki et al. 2016). This is an emerging theme: Variation in the translation kinetics within a pool of presumably identical transcripts can be as great or even greater than the average variation seen between two distinct pools of transcripts.

Other evidence for heterogeneity in ribosome movement along single RNA comes from ribosomal runoff assays (Pichon et al. 2016; Wang et al. 2016; Yan et al. 2016). Here, individual runoff curves rarely look like the mean, with some fast and others quite slow. Also, there may be a small subset of transcripts that shut down slowly, if at all (Yan et al. 2016). Presumably these transcripts have ribosomes that have either slowed down significantly or altogether stalled. The slowdown was attributed to damaged mRNA, as the percentage of slow runoff transcripts could be increased by treatment with 4NQO (4-nitroquilone-1-oxide) to oxidize RNA nucleotides (Yan et al. 2016). These data provide evidence that specific posttranscriptional modifications to nucleotides can significantly impact the elongation rate of ribosomes along transcripts.

Finally, genetically encoded pause sites can be directly inserted into transcripts to measure heterogeneity of ribosomal movement along single mRNA. The stress-related factor Xbp1 has a genetically encoded pause that helps target mRNA to the membrane (Ingolia et al. 2011; Yanagitani et al. 2011). When inserted into a reporter SunTag construct, polysome runoffs were observed to be nonlinear. This nonlinearity indicates heterogeneity in elongation among individual ribosomes within single polysomes. Also, a subset of polysomes showed a delayed runoff in response to harringtonine. This was interpreted as a traffic jam of ribosomes that was suddenly relieved.

14 HETEROGENEITY IN THE LOCALIZATION AND MOBILITY OF TRANSLATION SITES

One possible explanation for the translational heterogeneity seen among transcripts in single cells is their local subcellular environment. Local translation is thought to occur in special regions of the cell where translation efficiency is enhanced. For example, it was recently shown that individual transcripts translated in the ER have more ribosomes than those translated in the cytoplasm (Voigt et al. 2017), suggesting the efficiency of translation is higher in the ER. Furthermore, interaction with the ER is not limited exclusively to secreted proteins, as transcripts encoding cytoplasmic proteins also appear to occasionally associate with the ER when being translated. Considering the position of the ER around the cell nucleus, these interactions are likely responsible for the generally slower mobility of perinuclear translation sites observed (Wang et al. 2016).

In addition to localization within specific organelles, active transcripts also appear to occasionally colocalize (and comove) with one another to form higher-order structures. Such "translation factories" have been observed for both exogenous (Morisaki et al. 2016) and endogenous constructs (Pichon et al. 2016). These sites probably have higher local concentrations of translation factors and are therefore translated more efficiently. Consistent with this, endogenously tagged dynein transcripts within factories were translated with 90% efficiency, compared with just 70% outside factories. Importantly, the factories are translation-dependent, as treatment with a rapamycin inhibitor, which decreases translation initiation efficiency by two- to threefold (Thoreen et al. 2012), led to a significant loss of mRNA within the factories (Pichon et al. 2016). Similarly, treatment with puromycin results in the breakup of factories (TJ Stasevich, T Morisaki, K Lyon, pers. observation). It remains unclear what molecular mechanisms promote these higher-order structures. Future work is therefore needed to definitively prove the sites are bona fide translation factories rather than just transient associations by chance.

The motored transport of transcripts to specific sites is one mechanism by which local translation may be established. Motored transport of both actively translating and inactive single transcripts has been observed in living eukaryotic cells (Pichon et al. 2016; Wang et al. 2016; Wu et al. 2016). The motored motion is from 1 to 3 µm/sec and is typically sporadic, being interspersed between purely diffusive movements (Pichon et al. 2016; Wang et al. 2016). In neurons, motored transport is an efficient way to localize transcripts to dendrites and axons that may be physically separated from the nucleus by distances as large as a meter. Consistent with this, ARC transcripts harboring a 3′-UTR "zip code" that targets them to dendrites were motored with much higher probability in neurons than in HeLa cells (Wang et al. 2016). The motored movement was preferentially in an anterograde direction, resulting in a buildup of ARC polysomes in dendrites (Wang et al. 2016). Once in dendrites, ARC transcripts are translated in granules that undergo bursts of translation in an action potential-dependent manner (Tatavarty et al. 2012). This spatiotemporal regulation of translation is thought to be important for the integration of electrical signals into long-term phenotypic neuronal changes that are necessary for memory and learning. The observation that ARC and other transcripts can be motored when actively translated goes against the conventional view that transcripts are silenced until they reach their destination. For example, the tracking of single oskar transcripts during *Drosophila* development with the TRICK assay showed that translation does not occur until they reach the posterior poles of the oocyte (Halstead et al. 2015). As more transcripts are studied in different cellular contexts, it will be interesting to see how common active translation is during motored transport. The newer data suggest that transcripts may by default be in an active state amenable to translation, but the active state is switched off in specific contexts when bursty expression or complete silencing is required.

15 OUTLOOK

The last 20 years has seen tremendous progress in visualizing and quantifying single mRNA translation dynamics in living cells. Over the next 20 years, what can we expect from this technology? There is certainly plenty of room for improvement. In terms of repeat-epitope tag technology, more and smaller probes are needed for multiplexing and reduced invasiveness. Potential candidates that could replace scFv and Fab include chromobodies/nanobodies (Rothbauer et al. 2006), monobodies (Koide and Koide 2007), affibodies (Gao et al. 2011), DARPins (Plückthun 2015), thermobodies (Zhao et al. 2016), and split fluorescent reporters (Kamiyama et al. 2016). An ideal probe might be a single amino acid, which could bring live-cell imaging of translation on par with in vitro imaging (Perez and Gonzalez 2011), which already enjoys single-codon resolution. How this will be achieved is not yet clear, but promising starting points are noncanonical amino acids. Recently, Cy dyes were site-specifically incorporated into proteins in *Xenopus laevis* oocytes as noncanonical amino acids (Leisle et al. 2016), so it is not a stretch to imagine this being performed with fluorescently tagged mRNA to monitor translation dynamics. The ultimate goal would be in situ sequencing of protein synthesis in vivo. If achievable, this would place us directly behind the ribosome, so that we could see what it sees and fully comprehend its extraordinary behavior in translational regulation.

ACKNOWLEDGMENTS

We thank Amanda Koch and Charlotte Cialek for their helpful discussion and comments. T.J.S. and T.M. are supported by the W.M. Keck Foundation and the National Institutes of Health (NIH) grant number R35GM119728. T.J.S. is also supported by funds from the Boettcher Foundation's Webb–Waring Biomedical Research Program.

REFERENCES

** Reference is in this book.*

Abudayyeh OO, Gootenberg JS, Essletzbichler P, Han S, Joung J, Belanto JJ, Verdine V, Cox DBT, Kellner MJ, Regev A, et al. 2017. RNA targeting with CRISPR-Cas13. *Nature* **550:** 280–284.

Adamala KP, Martin-Alarcon DA, Boyden ES. 2016. Programmable RNA-binding protein composed of repeats of a single modular unit. *Proc Natl Acad Sci* **113:** E2579–E2588.

Babendure JR, Babendure JL, Ding J-H, Tsien RY. 2006. Control of mammalian translation by mRNA structure near caps. *RNA* **12:** 851–861.

Bertrand E, Chartrand P, Schaefer M, Shenoy SM, Singer RH, Long RM. 1998. Localization of ASH1 mRNA particles in living yeast. *Mol Cell* **2:** 437–445.

Buxbaum AR, Haimovich G, Singer RH. 2015. In the right place at the right time: Visualizing and understanding mRNA localization. *Nat Rev Mol Cell Biol* **16:** 95–109.

Chalfie M. 2009. GFP: Lighting up life. *Proc Natl Acad Sci* **106:** 10073–10080.

* Chao JA, Lionnet T. 2018. Imaging the life and death of mRNAs in single cells. *Cold Spring Harb Perspect Biol* **10:** a032086.

Chao JA, Patskovsky Y, Almo SC, Singer RH. 2008. Structural basis for the coevolution of a viral RNA–protein complex. *Nat Struct Mol Biol* **15:** 103–105.

Chao JA, Yoon YJ, Singer RH. 2012. Imaging translation in single cells using fluorescent microscopy. *Cold Spring Harb Perspect Biol* **4:** a012310.

Chen B, Gilbert LA, Cimini BA, Schnitzbauer J, Zhang W, Li G-W, Park J, Blackburn EH, Weissman JS, Qi LS, et al. 2013. Dynamic imaging of genomic loci in living human cells by an optimized CRISPR/Cas system. *Cell* **155:** 1479–1491.

Costantini LM, Fossati M, Francolini M, Snapp EL. 2012. Assessing the tendency of fluorescent proteins to oligomerize under physiologic conditions. *Traffic* **13:** 643–649.

Coulon A, Larson DR. 2016. Chapter Seven—Fluctuation analysis: Dissecting transcriptional kinetics with signal theory. In *Methods in enzymology* (eds. Filonov GS, Jaffrey SR), Vol. 572, *Visualizing RNA dynamics in the cell*, pp. 159–191. Academic, Cambridge, MA.

Darzacq X, Yao J, Larson DR, Causse SZ, Bosanac L, de Turris V, Ruda VM, Lionnet T, Zenklusen D, Guglielmi B, et al. 2009. Imaging transcription in living cells. *Annu Rev Biophys* **38:** 173–196.

Dobrucki JW, Feret D, Noatynska A. 2007. Scattering of exciting light by live cells in fluorescence confocal imaging: Phototoxic effects and relevance for FRAP studies. *Biophys J* **93:** 1778–1786.

Ernst P, Plückthun A. 2017. Advances in the design and engineering of peptide-binding repeat proteins. *Biol Chem* **398:** 23–29.

Fusco D, Accornero N, Lavoie B, Shenoy SM, Blanchard J-M, Singer RH, Bertrand E. 2003. Single mRNA molecules demonstrate probabilistic movement in living mammalian cells. *Curr Biol CB* **13:** 161–167.

Gao J, Chen K, Miao Z, Ren G, Chen X, Gambhir SS, Cheng Z. 2011. Affibody-based nanoprobes for HER2-expressing cell and tumor imaging. *Biomaterials* **32:** 2141–2148.

Garcia JF, Parker R. 2015. MS2 coat proteins bound to yeast mRNAs block 5′ to 3′ degradation and trap mRNA decay products: Implications for the localization of mRNAs by MS2-MCP system. *RNA NYN* **21:** 1393–1395.

Giepmans BNG, Adams SR, Ellisman MH, Tsien RY. 2006. The fluorescent toolbox for assessing protein location and function. *Science* **312:** 217–224.

Halstead JM, Lionnet T, Wilbertz JH, Wippich F, Ephrussi A, Singer RH, Chao JA. 2015. An RNA biosensor for imaging the first round of translation from single cells to living animals. *Science* **347:** 1367–1671.

Han K, Jaimovich A, Dey G, Ruggero D, Meyuhas O, Sonenberg N, Meyer T. 2014. Parallel measurement of dynamic changes in translation rates in single cells. *Nat Methods* **11:** 86–93.

Hancock JF, Cadwallader K, Paterson H, Marshall CJ. 1991. A CAAX or a CAAL motif and a second signal are sufficient for plasma membrane targeting of ras proteins. *EMBO J* **10:** 4033–4039.

Hanes J, Jermutus L, Weber-Bornhauser S, Bosshard HR, Plückthun A. 1998. Ribosome display efficiently selects and evolves high-affinity antibodies in vitro from immune libraries. *Proc Natl Acad Sci* **95:** 14130–14135.

Heinrich S, Sidler CL, Azzalin CM, Weis K. 2017. Stem–loop RNA labeling can affect nuclear and cytoplasmic mRNA processing. *RNA* **23:** 134–141.

Heinze KG, Costantino S, De Koninck P, Wiseman PW. 2009. Beyond photobleaching, laser illumination unbinds fluorescent proteins. *J Phys Chem B* **113:** 5225–5233.

Hoernes TP, Hüttenhofer A, Erlacher MD. 2016. mRNA modifications: Dynamic regulators of gene expression? *RNA Biol* **13:** 760–765.

Holt CE, Schuman EM. 2013. The central dogma decentralized: New perspectives on RNA function and local translation in neurons. *Neuron* **80:** 648–657.

Hopp TP, Prickett KS, Price VL, Libby RT, March CJ, Cerretti DP, Urdal DL, Conlon PJ. 1988. A short polypeptide marker sequence useful for recombinant protein identification and purification. *Nat Biotechnol* **6:** 1204–1210.

Horvathova I, Voigt F, Kotrys AV, Zhan Y, Artus-Revel CG, Eglinger J, Stadler MB, Giorgetti L, Chao JA. 2017. The dynamics of mRNA turnover revealed by single-molecule imaging in single cells. *Mol Cell* **68:** 615–625.e9.

Hurley JM, Loros JJ, Dunlap JC. 2016. Circadian oscillators: Around the transcription-translation feedback loop and on to output. *Trends Biochem Sci* **41:** 834–846.

Ifrim MF, Williams KR, Bassell GJ. 2015. Single-molecule imaging of PSD-95 mRNA translation in dendrites and its dysregulation in a mouse model of fragile X syndrome. *J Neurosci* **35:** 7116–7130.

Ingolia NT. 2014. Ribosome profiling: New views of translation, from single codons to genome scale. *Nat Rev Genet* **15:** 205–213.

Ingolia NT, Lareau LF, Weissman JS. 2011. Ribosome profiling of mouse embryonic stem cells reveals the complexity and dynamics of mammalian proteomes. *Cell* **147:** 789–802.

Jorda J, Xue B, Uversky VN, Kajava AV. 2010. Protein tandem repeats—The more perfect, the less structured. *FEBS J* **277:** 2673–2682.

Jung H, Gkogkas CG, Sonenberg N, Holt CE. 2014. Remote control of gene function by local translation. *Cell* **157:** 26–40.

Kamiyama D, Sekine S, Barsi-Rhyne B, Hu J, Chen B, Gilbert LA, Ishikawa H, Leonetti MD, Marshall WF, Weissman JS, et al. 2016. Versatile protein tagging in cells with split fluorescent protein. *Nat Commun* **7:** 11046.

Kim JH, Lee S-R, Li L-H, Park H-J, Park J-H, Lee KY, Kim M-K, Shin BA, Choi S-Y. 2011. High cleavage efficiency of a 2A peptide derived from porcine teschovirus-1 in human cell lines, zebrafish and mice. *PLoS One* **6:** e18556.

Kimura H, Tao Y, Roeder RG, Cook PR. 1999. Quantitation of RNA polymerase II and its transcription factors in an HeLa cell: Little soluble holoenzyme but significant amounts of polymerases attached to the nuclear substructure. *Mol Cell Biol* **19:** 5383–5392.

Koide A, Koide S. 2007. Monobodies: Antibody mimics based on the scaffold of the fibronectin type III domain. In *Methods in molecular biology* (ed. Walker JM), Vol. 352, *Protein engineering protocols* (eds. Arndt KM, Müller KM), pp. 95–109. Humana, Totowa, NJ.

Larson DR, Zenklusen D, Wu B, Chao JA, Singer RH. 2011. Real-time observation of transcription initiation and elongation on an endogenous yeast gene. *Science* **332:** 475–478.

Leisle L, Chadda R, Lueck JD, Infield DT, Galpin JD, Krishnamani V, Robertson JL, Ahern CA. 2016. Cellular encoding of Cy dyes for single-molecule imaging. *eLife* **5:** pii: e19088.

Lionnet T, Singer RH. 2012. Transcription goes digital. *EMBO Rep* **13:** 313–321.

Lionnet T, Czaplinski K, Darzacq X, Shav-Tal Y, Wells AL, Chao JA, Park HY, de Turris V, Lopez-Jones M, Singer RH. 2011. A transgenic mouse for in vivo detection of endogenous labeled mRNA. *Nat Methods* **8:** 165–170.

Mazza D, Abernathy A, Golob N, Morisaki T, McNally JG. 2012. A benchmark for chromatin binding measurements in live cells. *Nucleic Acids Res* **40:** e119.

Morisaki T, McNally JG. 2014. Photoswitching-free FRAP analysis with a genetically encoded fluorescent tag. *PLoS One* **9:** e107730.

Morisaki T, Müller WG, Golob N, Mazza D, McNally JG. 2014. Single-molecule analysis of transcription factor binding at transcription sites in live cells. *Nat Commun* **5:** 4456.

Morisaki T, Lyon K, DeLuca KF, DeLuca JG, English BP, Zhang Z, Lavis LD, Grimm JB, Viswanathan S, Looger LL, et al. 2016. Real-time quantification of single RNA translation dynamics in living cells. *Science* **352:** 1425–1429.

Mueller F, Morisaki T, Mazza D, McNally JG. 2012. Minimizing the impact of photoswitching of fluorescent proteins on FRAP analysis. *Biophys J* **102:** 1656–1665.

Park HY, Buxbaum AR, Singer RH. 2010. Chapter 18—Single mRNA tracking in live cells. In *Methods in enzymology* (ed. Walter NG), Vol. 472, *Single molecule tools: Fluorescence based approaches, Part A*, pp. 387–406, Academic, Cambridge, MA.

Pellegrini M, Marcotte EM, Yeates TO. 1999. A fast algorithm for genome-wide analysis of proteins with repeated sequences. *Proteins* **35:** 440–446.

Perez CE, Gonzalez RL. 2011. In vitro and in vivo single-molecule fluorescence imaging of ribosome-catalyzed protein synthesis. *Curr Opin Chem Biol* **15:** 853–863.

Pichon X, Bastide A, Safieddine A, Chouaib R, Samacoits A, Basyuk E, Peter M, Mueller F, Bertrand E. 2016. Visualization of single endogenous polysomes reveals the dynamics of translation in live human cells. *J Cell Biol* **214:** 769–781.

Plückthun A. 2015. Designed ankyrin repeat proteins (DARPins): Binding proteins for research, diagnostics, and therapy. *Annu Rev Pharmacol Toxicol* **55:** 489–511.

Rothbauer U, Zolghadr K, Tillib S, Nowak D, Schermelleh L, Gahl A, Backmann N, Conrath K, Muyldermans S, Cardoso MC, et al. 2006. Targeting and tracing antigens in live cells with fluorescent nanobodies. *Nat Methods* **3:** 887–889.

Saxton MJ. 1996. Anomalous diffusion due to binding: A Monte Carlo study. *Biophys J* **70:** 1250–1262.

Simms CL, Hudson BH, Mosior JW, Rangwala AS, Zaher HS. 2014. An active role for the ribosome in determining the fate of oxidized mRNA. *Cell Rep* **9:** 1256–1264.

Simsek D, Barna M. 2017. An emerging role for the ribosome as a nexus for post-translational modifications. *Curr Opin Cell Biol* **45:** 92–101.

Sonenberg N, Hinnebusch AG. 2009. Regulation of translation initiation in eukaryotes: Mechanisms and biological targets. *Cell* **136:** 731–745.

Stasevich TJ, Mueller F, Michelman-Ribeiro A, Rosales T, Knutson JR, McNally JG. 2010. Cross-validating FRAP and FCS to quantify the impact of photobleaching on in vivo binding estimates. *Biophys J* **99:** 3093–3101.

Stasevich TJ, Hayashi-Takanaka Y, Sato Y, Maehara K, Ohkawa Y, Sakata-Sogawa K, Tokunaga M, Nagase T, Nozaki N, McNally JG, et al. 2014. Regulation of RNA polymerase II activation by histone acetylation in single living cells. *Nature* **516:** 272–275.

Ströhl F, Lin JQ, Laine RF, Wong HH-W, Urbančič V, Cagnetta R, Holt CE, Kaminski CF. 2017. Single molecule translation imaging visualizes the dynamics of local β-actin synthesis in retinal axons. *Sci Rep* **7:** 709.

Tanenbaum ME, Gilbert LA, Qi LS, Weissman JS, Vale RD. 2014. A protein-tagging system for signal amplification in gene expression and fluorescence imaging. *Cell* **159:** 635–646.

Tatavarty V, Ifrim MF, Levin M, Korza G, Barbarese E, Yu J, Carson JH. 2012. Single-molecule imaging of translational output from individual RNA granules in neurons. *Mol Biol Cell* **23:** 918–929.

Thoreen CC, Chantranupong L, Keys HR, Wang T, Gray NS, Sabatini DM. 2012. A unifying model for mTORC1-mediated regulation of mRNA translation. *Nature* **485:** 109–113.

Tutucci E, Vera M, Biswas J, Garcia J, Parker R, Singer RH. 2018. An improved MS2 system for accurate reporting of the mRNA life cycle. *Nat Methods* **15:** 81–89.

Viswanathan S, Williams ME, Bloss EB, Stasevich TJ, Speer CM, Nern A, Pfeiffer BD, Hooks BM, Li W-P, English BP, et al. 2015. High-performance probes for light and electron microscopy. *Nat Methods* **12:** 568–576.

Voigt F, Zhang H, Cui XA, Triebold D, Liu AX, Eglinger J, Lee ES, Chao JA, Palazzo AF. 2017. Single-molecule quantification of translation-dependent association of mRNAs with the endoplasmic reticulum. *Cell Rep* **21:** 3740–3753.

Wang C, Han B, Zhou R, Zhuang X. 2016. Real-time imaging of translation on single mRNA transcripts in live cells. *Cell* **165:** 990–1001.

Wen J-D, Lancaster L, Hodges C, Zeri A-C, Yoshimura SH, Noller HF, Bustamante C, Tinoco I. 2008. Following translation by single ribosomes one codon at a time. *Nature* **452:** 598–603.

Wilson IA, Niman HL, Houghten RA, Cherenson AR, Connolly ML, Lerner RA. 1984. The structure of an antigenic determinant in a protein. *Cell* **37:** 767–778.

Wu B, Buxbaum AR, Katz ZB, Yoon YJ, Singer RH. 2015. Quantifying protein-mRNA interactions in single live cells. *Cell* **162:** 211–220.

Wu B, Eliscovich C, Yoon YJ, Singer RH. 2016. Translation dynamics of single mRNAs in live cells and neurons. *Science* **352:** 1430–1435.

Yan X, Hoek TA, Vale RD, Tanenbaum ME. 2016. Dynamics of translation of single mRNA molecules in vivo. *Cell* **165:** 976–989.

Yanagitani K, Kimata Y, Kadokura H, Kohno K. 2011. Translational pausing ensures membrane targeting and cytoplasmic splicing of XBP1u mRNA. *Science* **331:** 586–589.

Yu J, Xiao J, Ren X, Lao K, Xie XS. 2006. Probing gene expression in live cells, one protein molecule at a time. *Science* **311:** 1600–1603.

Zhao N, Schmitt MA, Fisk JD. 2016. Phage display selection of tight specific binding variants from a hyperthermostable Sso7d scaffold protein library. *FEBS J* **283:** 1351–1367.

Illuminating Genomic Dark Matter with RNA Imaging

Arjun Raj[1,3] and John L. Rinn[2,3]

[1]Department of Biology, University of Pennsylvania, Philadelphia, Pennsylvania 19104

[2]Department of Biochemistry, University of Colorado Boulder and BioFrontiers Institute, Boulder, Colorado 80303

Correspondence: arjunrajlab@gmail.com; john.rinn@colorado.edu

SUMMARY

In the postgenomic era, it is clear that the human genome encodes thousands of long noncoding RNAs (lncRNAs). Along the way, RNA imaging (e.g., RNA fluorescence in situ hybridization [RNA-FISH]) has been instrumental in identifying powerful roles for lncRNAs based on their subcellular localization patterns. Here, we explore how RNA imaging technologies have shed new light on how, when, and where lncRNAs may play functional roles. Specifically, we will synthesize the underlying principles of RNA imaging techniques by exploring several landmark lncRNA imaging studies that have illuminated key insights into lncRNA biology.

Outline

[3]Authorship was listed alphabetically.

Cite as *Cold Spring Harb Perspect Biol* doi: 10.1101/cshperspect.a032094

1 INTRODUCTION

It is becoming increasingly clear that RNA localization can influence numerous biological processes. Examples include specialized localization of mRNAs at opposing cellular poles resulting, on cellular division, in two different cell types, and the widespread rearrangement of RNA transcripts, on cellular stress, to subcellular compartments such as stress granules or P-bodies (Martin and Ephrussi 2009; Parton et al. 2014; Buxbaum et al. 2015; Van Treeck and Parker 2019). Yet, the information content present in subcellular RNA localization patterns is absent in most nonimaging-based expression measurements (e.g., RNA sequencing). Thus, the "when, where, and how" of RNA, which is critical to unraveling its biology, often remains a mystery.

The advent of RNA fluorescence in situ hybridization (RNA-FISH) provided an approach to address these questions. Moreover, such methods allow the measurement of how these patterns and properties vary between individual cells in a population. Together, this information provides powerful insights into the nature and mechanism of RNA-mediated cellular processes, including organization of nuclear subcompartments and the widespread interface between RNA and epigenetic regulation.

RNA-FISH is particularly relevant for noncoding RNA biology, in which the RNA itself is typically the final molecular effector. A classic example is the RNA-FISH visualization of the long noncoding RNA (lncRNA) Xist (Brown et al. 1992). Although it was known that Xist was sufficient to inactivate an entire female X chromosome during dosage compensation, its mechanism remained mysterious. Yet, with a single RNA-FISH experiment, it became clear that Xist remains in the nucleus and coats the inactive X chromosome in *cis* (Brown et al. 1992). This "seeing is believing" approach shows how visualization can provide insights critical for deciphering the mechanism of lncRNAs. Since Xist, many lncRNAs have been investigated using RNA imaging (Cabili et al. 2015; Dunagin et al. 2015; Orjalo and Johansson 2016). These studies have revealed many unexpected lncRNA localization patterns and in turn new biological insights (e.g., identification of new subcellular compartments).

Equally, as important as the spatiotemporal information provided by RNA imaging is the advantage of monitoring individual cells in a population. Specifically, RNA imaging can elucidate the heterogeneity of transcript abundance and localization within a population. The single-cell resolution of RNA-FISH can also provide key biological insights that would otherwise be missed in bulk population measurements (e.g., RNA sequencing). For example, a lncRNA could be highly correlated in expression with its neighboring gene according to RNA-seq, which could indicate a regulatory interaction. Alternatively, it could be that the lncRNA is expressed in half the cells and messenger RNA (mRNA) in the other half, in a mutually exclusive manner. Thus, on a population average, a positive interaction could be predicted but on a single-cell level, a repressive relationship would be revealed. Indeed, this is the case for at least two lncRNAs (linc-HOXA1 and lincRNA-Bxd) that have been implicated in promoter competition between the lncRNA and mRNA (Petruk et al. 2006; Maamar et al. 2013).

The single-cell nature of RNA imaging allows important insights into regulatory dynamics of RNA transcripts by looking at correlations at the single-cell level across subpopulations of cells. Yet, this is only the tip of the iceberg; one can—in parallel—glean information and correlate lncRNA expression patterns with many other cellular processes by simultaneously measuring, for example, at which stage in the cell cycle an RNA is expressed (e.g., cyclin mRNA), whether an RNA moves between the nucleus and cytoplasm (e.g., NRON is a lncRNA that sequesters the transcription factor NFAT in the cytoplasm and on release of RNA, NFAT localizes to the nucleus) or if an RNA is restricted to a subcellular compartment such as the nucleolus (pre-rRNA) (Willingham 2005; Cabili et al. 2015; Falahati et al. 2016). In other words, as the old aphorism goes: A picture is worth a thousand words.

Here, we will survey these examples and many more while highlighting different approaches and technologies for RNA-FISH and their application to lncRNA detection and visualization. We will explore lncRNA-FISH studies by introducing the basic experimental approaches, limitations, and controls. Then we will explore the application of lncRNA-FISH in the context of studies that have provided important insights into lncRNA biology. We will synthesize RNA-FISH, single-molecule RNA-FISH and live-cell RNA imaging by focusing on a few of these seminal examples.[4]

2 THE BASICS: COMPLEMENTARY PROBE PREPARATION FOR RNA-FISH

Before exploring the biological insights of lncRNA-FISH we will discuss the basic principles used to uncover lncRNA localization patterns. A common principle across most RNA-FISH approaches is the use of a DNA or RNA molecule(s) with sequence complementary to the endogenous lncRNA (Fig. 1). The first lncRNA-FISH studies used single

[4]We apologize in advance for not being able to cover all the existing and emerging RNA-FISH studies targeting lncRNA. Here we aim to focus only on a few examples to cover basic principles of RNA-FISH in the context of insights into lncRNA biology.

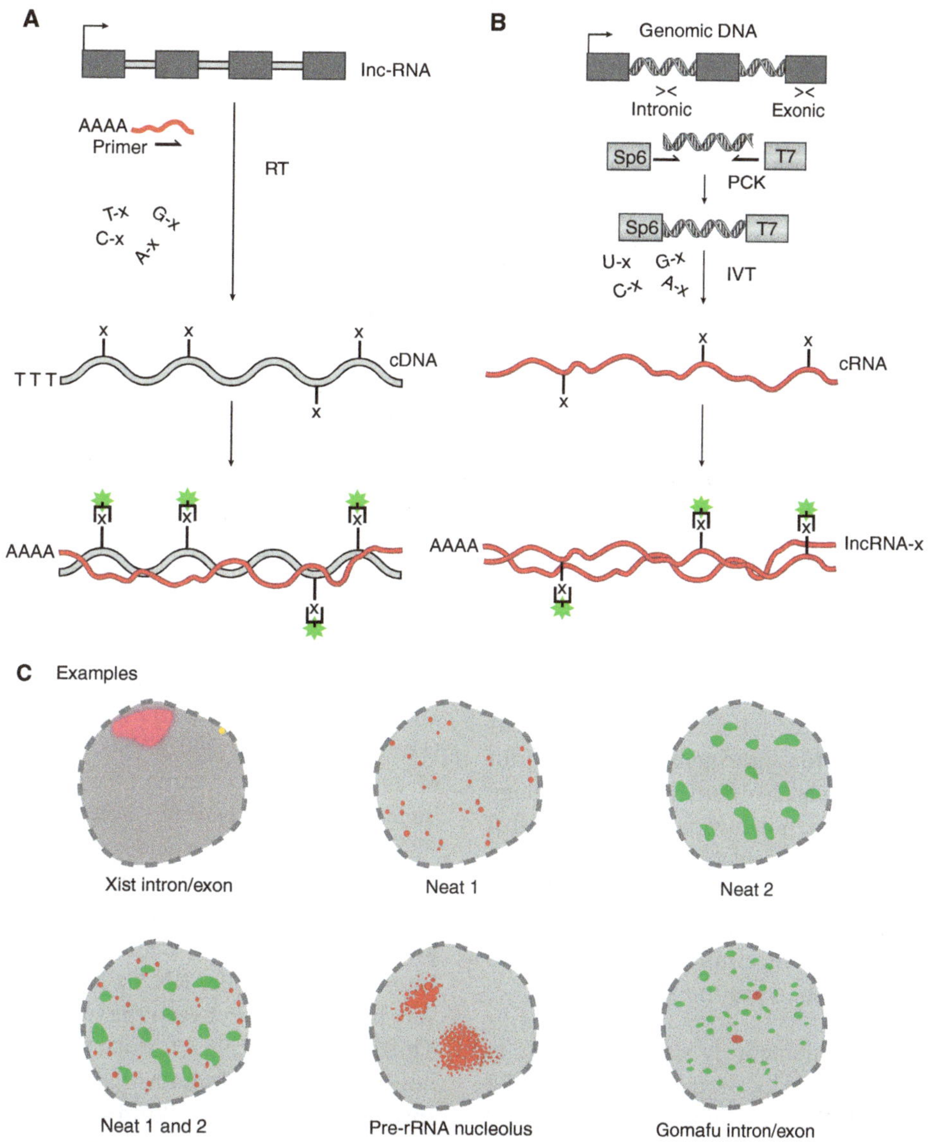

Figure 1. Classic RNA fluorescence in situ hybridization (RNA-FISH) probe design using complementary DNA (cDNA) and complementary RNA (cRNA) probes. (*A*) To generate cDNA probes for RNA-FISH, one performs reverse transcription (RT) with gene-specific primers to generate a cDNA to the targeted long noncoding RNA (lncRNA). During the RT reaction, nucleotides are randomly incorporated that have modifications (e.g., "x" = biotin or digoxigenin [DIG]). The labeled cDNA is hybridized to the endogenous lncRNA (red line), which can be imaged via addition of fluorescent secondary conjugates targeting the probe label. (*B*) Generation of cRNA RNA-FISH probes requires targeted polymerase chain reaction (PCR) of genomic DNA, targeting either the intronic or exonic regions of the lncRNA locus. PCR primers have 5′/3′ sequences appended encoding in vitro transcription (IVT) promoters that are incorporated into the final double-stranded DNA (dsDNA) product. The resulting dsDNA product is subjected to IVT either antisense (complementary) or antisense (noncomplementary) in the presence of modified nucleotides (same as in *A*) to produce a labeled cRNA probe (red line with "x"). The cRNA probe is then hybridized to the endogenous lncRNA target, which can be imaged via addition of fluorescent secondary conjugates targeting the probe label. (*C*) lncRNA-FISH examples of classic studies revealing new subnuclear structures (all images showing nucleus). (*Top left*) Xist showing intron and exon labeling of the Xist lncRNA showing a punctate site of transcription (yellow) and mature transcript as a "cloud" in red on the same chromosome as the site of transcription. (*Top middle* and *right*, *bottom left*) Cartoon rendition of the RNA-FISH nuclear localization pattern for Neat1, Neat2, and Neat1/2, respectively, with nonoverlapping "speckled" localization patterns. (*Bottom middle*) rRNA-FISH demarcates the nucleolus, the site of rRNA transcription. (*Bottom right*) Gomafu RNA-FISH with intronic (red) and exonic (green) RNA-FISH probes shows that the mature transcript localizes in large domains away from the site of transcription (intronic signal).

probe designs that roughly distill into two principal methods used today.

1. *cDNA probe*: A complementary DNA (cDNA) can be made through reverse transcription of a target lncRNA with incorporation of "labeled" (e.g., digoxigenin [DIG] or biotin) DNA nucleotides (Fig. 1A). This complementary sequence or "probe" can hybridize in situ to the target lncRNA with tagged nucleotides; that in turn can be detected with secondary fluorescent binding molecules (e.g., antibodies for digoxigenin or streptavidin for biotin). These cDNA probes often span exon junctions, and thus are more likely to hybridize to the mature lncRNA and not the unspliced immature transcripts (producing large gaps in probe hybridization).
2. *cRNA probe*: Hybridization probes can also be generated by double-stranded DNA (dsDNA) templates with flanking in vitro transcription (IVT) promoters (Fig. 1B). This promoter is used in an IVT reaction that incorporates labeled RNA nucleotides that produces a complementary RNA (cRNA) probe (Fig. 1B). This cRNA probe is then targeted by secondary fluorescent molecules (e.g., antibody for digoxigenin or streptavidin for biotin). This approach also confers the advantage of generating a "sense" or noncomplementary control probe if an IVT promoter is placed on the opposite strand of the complementary probe. Thus, the same dsDNA template can generate a complementary probe for specific signal and a noncomplementary probe to determine nonspecific or background signal.

Often, it is useful to simultaneously visualize the site of transcription relative to the final location of mature lncRNA transcript. For example, this was useful in determining that Xist localizes to the same chromosome from which it is transcribed, thus working in *cis*. Probes to determine the site of transcription are generated by IVT using dsDNA templates comprising intronic sequences from genomic DNA (Fig. 1B) (Fremeau et al. 1986; Brown et al. 1992; Soler et al. 2017). Collectively, these approaches provide flexible and cost-efficient methods to generate RNA-FISH probes to visualize lncRNAs.

After probes are generated, there are many protocols and procedures for permeabilizing (or opening) cells followed by hybridization of probes and microscopy-based analyses. We will not discuss these aspects here other than to mention this often requires experimental optimization to account for many factors such as cell type and abundance of the lncRNA. The experimental details of these optimization protocols are nicely covered in the primary literature, textbooks, and reviews (see Lawrence and Singer 1986; Lawrence et al. 1988, 1989; Dirks et al. 1993; Tripathi et al. 2015; Querido et al. 2017). We do note, however, that permeability issues often affect longer cRNA probes more than oligonucleotide ssDNA probes, in particular for detection in the nucleus (discussed below).

3 NEAT RNA-FISH: cDNA AND cRNA PATTERN RECOGNITION OF lncRNAs

The RNA-FISH approaches described above were seminal in understanding lncRNA biology, as they revealed fascinating and informative localization patterns. As early as 1992, it was clear from RNA-FISH studies that subnuclear compartments express lncRNAs (preribosomal RNA in the nucleolus and Xist in the Barr body; Fig. 1C), thus raising the question: could there be other subnuclear compartments demarcated by lncRNAs? It was not until the postgenomic era when two neat RNA-FISH experiments revealed that the answer was "yes."

A seminal example of RNA-FISH leading to a new understanding of nuclear biology came from the discovery of nuclear-enriched abundant transcripts 1 and 2 (Neat1 and Neat2, the latter more recently referred to as Malat1) (Hutchinson et al. 2007). Fractionation RNA-sequencing of several primary human cell lines showed that these lncRNAs are exclusively nuclear. However, it was RNA-FISH of where in the nucleus these lncRNAs are expressed that led to an important insight. The cRNA probes targeting either Neat1 and Malat1 both showed "speckled" (Fig. 1C) localization patterns across a large expanse of the nucleus. Interestingly, parallel RNA-FISH of Neat1 and 2 revealed very little overlap between the two sets of speckles (Fig. 1C). Thus, Neat1 and Neat2/Malat1 lncRNAs demarcated two novel subnuclear domains: "nuclear speckles" and "paraspeckles," respectively (Bond and Fox 2009; Mao et al. 2011; Spector and Lamond 2011). Thus, this single RNA-FISH experiment was able to reveal two different subnuclear compartments demarcated by a lncRNA.

RNA-FISH experiments led to our current understanding that many lncRNAs are fundamental constituents of subnuclear compartments (Quinodoz and Guttman 2014; Rinn and Guttman 2014; Vance and Ponting 2014; Cheng et al. 2016; Engreitz et al. 2016; Melé and Rinn 2016). Consistent with this notion, a similar and contemporary study of the lncRNA Gomafu (MIAT/Rncr2) showed a stunning nuclear localization pattern (Sone et al. 2007). Specifically, Gomafu shows widespread but solely nuclear localization that is associated with factors in the nuclear matrix. This study also used intronic probes that revealed two punctate foci representing the site of transcription for Gomafu (Fig. 1C). Together, these RNA-FISH images alone suggest a role for the mature Gomafu transcript in facilitating nuclear organization in *trans*, a property that would be very hard to identify by conventional population-based or even single-cell RNA sequencing.

Collectively, these and other seminal RNA-FISH studies suggested that lncRNAs are fundamental components of dynamic nuclear substructures that influence a plethora of biological processes.

4 SINGLE-MOLECULE RNA-FISH APPROACHES

Today, it is clear that the mammalian genome encodes thousands of lncRNAs. Along the way RNA-FISH has undergone many technical advances; including single-molecule detection sensitivity, accessible analysis software, and scalable probe generation techniques to target newly discovered lncRNAs. These single-molecule RNA-FISH (smRNA-FISH) studies provide quantification of individual RNA molecules, their subcellular position, and how those properties vary from cell to cell (Femino et al. 1998; Raj et al. 2008, 2010; Zenklusen et al. 2008; Larson et al. 2009; Raj and van Oudenaarden 2009; Batish et al. 2011; Trcek et al. 2011, 2012; Levesque et al. 2013; Shaffer et al. 2013; Dunagin et al. 2015; Querido et al. 2017; Tutucci et al. 2018a).

There are many variations of smRNA-FISH probe schemes, but most share a common principle of using DNA oligonucleotides (oligos) that are conjugated to fluorophores with defined stoichiometry. There are two basic approaches: (1) use of a small number of highly specific DNA oligos that provide a scaffold from which multiple sequences are hybridized in a "branched" manner termed a dendrimer (Wang et al. 2012; Sinnamon and Czaplinski 2014), in which the dendrimer is targeted by a complementary sequence conjugated to a fluorophore resulting in a signal that represents one molecule of the targeted lncRNA (Fig. 2A); and (2) use of pools of short (18–22-nt) DNA oligos—each with a 5′ or 3′-conjugated fluorophore—that tile individual lncRNA molecules (Fig. 2B) (Raj et al. 2008; Cabili et al. 2015).

Both dendrimer and oligo pool approaches can benefit greatly from computational design of optimal targeting oligos, as well as careful control experiments to determine the specificity of the smRNA-FISH probes. We will hereafter focus on smRNA-FISH oligo pools as a case study to discuss the basic principles in designing and validating smRNA-FISH probes.

4.1 Step 1: Oligo Design

Typically, each targeting oligo or probe is ~18–22-nt long, with a 5′ or 3′ nucleotide amenable to conjugation. Approximately 20–48 oligos are pooled to target the lncRNA of interest (Fig. 2). Each probe is designed to target a unique region of the RNA of interest. Oligos can be designed using several online software programs, but their specificity requires empirical validation, as small subsequences within an oligo can cross-hybridize to highly abundant off-target transcripts, leading to spurious signals (see below).

This oligo-based scheme can easily accommodate the design and modification of probe sets for a specific question. For example, it is possible to visualize both the site of transcription and subcellular localization of a mature RNA transcript using individually labeled probe sets targeting the introns and exons, respectively. Furthermore, individual oligos in the pool can be tested for specificity (see below), and "rogue" oligonucleotides (that are often responsible for background signal) can easily be eliminated from the pool. It is critical to validate oligo specificity in new cellular contexts, as the expression of high abundance off-target RNAs may differ between cell lines or tissues.

4.2 Step 2: Oligo Conjugation

Following oligo design, the next step is to conjugate specific fluorophores to the modified 5′ or 3′ nucleotide (Raj et al. 2008; Orjalo and Johansson 2016) of each probe (or enzymatically labeled) (Gaspar et al. 2017). The unconjugated oligonucleotides must then be removed by high-performance liquid chromatography (HPLC)-based methods (unless the conjugation efficiency is reliably very high). The choice of fluorophores depends on the specific filter sets available in the microscope and the degree of spectral multiplexing required for the experiment. Photostable fluorophores should be used where possible, as high-powered lasers (such as those on a confocal microscope) can easily bleach most fluorophores. These parameters could also vary from cell type to cell type.

4.3 Step 3: Probe Set Validation

A nice feature of the pooled approach is that each individual oligo can be tested for specificity. For practical reasons, this specificity is usually tested by taking "odd" and "even" probes from the pool of 20–48 oligos and conjugating the odds and even with different fluorophores. These odd and even pools are then hybridized simultaneously. If the probes are specific to the intended target, the odd and even probes will overlap (Fig. 2C), and so the merged image of the odd and even fluorophores will show strong signal overlap (Fig. 2C) (Raj et al. 2008; Raj and van Oudenaarden 2009; Cabili et al. 2015; Dunagin et al. 2015).

However, if any nonoverlapping signals are present, this indicates that one or more oligos within the pool are hybridizing to off-target transcripts. If this is the case, then the rogue probe(s) can be identified by removing subsets of the pool until only overlapping signal is detected (Fig. 2C).

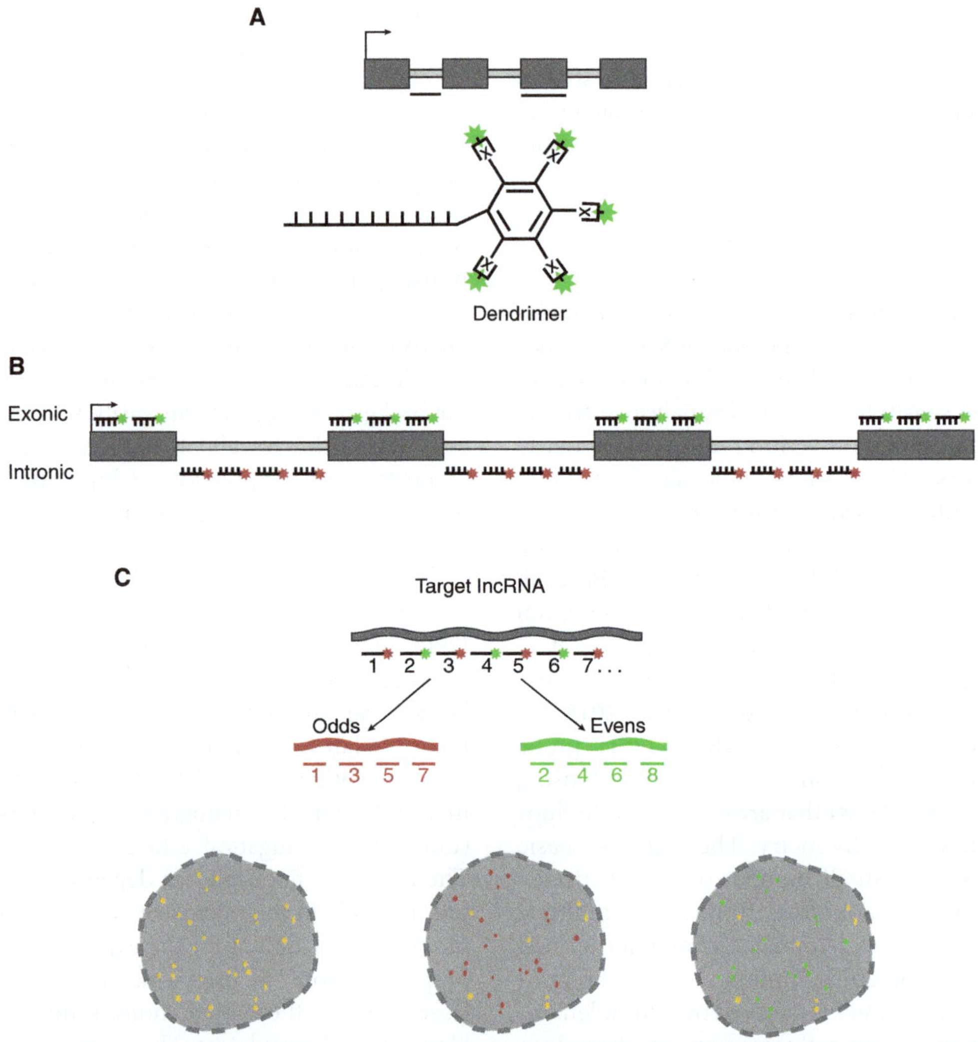

Figure 2. Probe design for single-molecule RNA fluorescence in situ hybridization (smRNA-FISH). (*A*) The dendrimer or branched oligo approach. Oligo(s) are designed to specify an intronic or exonic region of the target long noncoding RNA (lncRNA). The dendrimer approach appends either multiple conjugation sites (x) to the 5′ or 3′ end of the oligo. Alternatively, "branched-oligos" are similar to dendrimers using multiple "branched" sequences that form a tree-like structure of conjugatable epitopes. (*B*) smRNA-FISH probe design using "oligo pools." Each oligo ranges in length between 18 and 22 nucleotides (nt), and between 20 and 48 oligos are mixed into an oligo pool. Each oligo has modifications on the 5′ or 3′ end to allow conjugation of a fluorophore of choice. smRNA-FISH oligo pools are generated targeting either the intronic regions (site of lncRNA transcription) or exonic regions of the mature transcript. (*C*) smRNA-FISH probe validation. As a first test for oligo specificity, the pool can be split evenly and each half labeled with distinct fluorophores. If all oligos from the odd (red) and even (green) probe sets are specific to the targeted long noncoding RNA (lncRNA), the patterns will overlap producing yellow signal (*left*). If one or more of the odd (red) oligos hybridize to an off-target RNA, but most of the probes are specific, then yellow and red signal would be observed (*middle*). Similarly, if one of the even (green) probes had an off-target RNA, but some probes were specific, green and yellow signal would be observed (*right*).

Controlling for off target hybridization is critical when investigating lncRNAs, which often have much lower abundance than mRNAs and unique localization patterns that can be easily misidentified because of background. In particular, highly abundant off-target transcripts can produce strong, false signals that can sometimes look like nuclear blobs (Cabili et al. 2015). This can even occur within small subsections of an individual oligo. For example, it was previously found that the putative expression pattern of a lowly expressed lncRNA could be attributed almost entirely to a single rogue oligo with a 15-nt region complementary to an off-target RNA with several 1000-fold higher abundance.

(Cabili et al. 2015) Thus, even with the current best computational design tools, it is strongly recommended that one empirically validate the oligo pool as a whole using this "odds and evens" approach.

Often probe sets are designed using the above-described process to target intronic regions. Because introns degrade very rapidly after being spliced out of the nascent pre-mRNA, these probes often result in punctate foci surrounding the site of transcription. This provides a nice control, as signal at the site of transcription should be easily recognizable as a small "blob" of signal, whereas the mature transcripts may have more complex localization patterns with punctate foci. In summary, once a validated probe set is generated for a target lncRNA it allows for numerous single-cell and single-molecule analyses to quantitate the spatiotemporal properties of that target lncRNA.

5 LOCATION-TRACKING lncRNAs: ONE MOLECULE AT A TIME

There are many advantages to single-molecule quantitation of lncRNAs using smRNA-FISH at the level of single cells. This includes the ability of smRNA-FISH to quantify the number of molecules of a lncRNA that are present, where they are present, and how they change across a population of cells. These properties can provide key insights into lncRNA biology that may otherwise be missed by other methods (e.g., single-cell RNA sequencing).

Imagine a scenario in which a lncRNA shows context-dependent expression in a small subset of a cell population, a phenomenon called "jackpotting" (Cabili et al. 2015; Shaffer et al. 2017). RNA-seq of the bulk population would reveal a very low average expression value, thus masking robust expression in a small number of cells. Fortunately, this scenario is rare, but it highlights the usage of transcript visualization by smRNA-FISH. It is conceivable that single-cell RNA-sequencing (SCS) methods would also identify "jackpotting" cells. However, SCS methods often do not sample lncRNAs, because typically only the most abundant mRNAs are measured in SCS, especially in the context of rare-cell analysis (Torre et al. 2018).

Overall, smRNA-FISH provides a unique three-dimensional perspective on lncRNA abundance and localization within single cells. Below, we discuss recent applications of smRNA-FISH that have uncovered fundamental and unexpected insights into lncRNA biology.

5.1 Promoter Tug-of-War

A good example of where smRNA-FISH can reveal surprising subpopulation dynamics came from the investigation of linc-HOXA1 in mouse embryonic stem cells (mESCs). Population-level RNA sequencing showed linc-HOXA1 (HAUNT) strongly correlated in expression with its neighboring gene, *HOXA1*, suggesting that the lncRNA could positively regulate *HOXA1* (Maamar et al. 2013; Yin et al. 2015). However, smRNA-FISH revealed that the lncRNA and mRNA are expressed in a mutually exclusive manner: cells expressing HOXA1 did not express linc-HOXA1 and vice versa (Maamar et al. 2013).

Thus, on a single-cell level it was revealed that linc-HOXA1 may function as a repressor—rather than activator—of *HOXA1*. This repressive role was experimentally validated with antisense oligonucleotide (ASO) knockdown of linc-HOXA1, resulting in increased *HOXA1* expression. Thus, linc-HOXA1 promoter activation inhibits *HOXA1* transcription and vice versa (Maamar et al. 2013). A similar phenomenon was resolved in *Drosophila* between the Hox gene *Ubx* and an upstream lncRNA (lncRNA-bxd) that results in promoter competition (Petruk et al. 2006). Similarly, lncRNA-bxd is expressed in mutually exclusive subsets of cells from *Ubx*; again correlated in bulk, but repressive relationships are revealed at the single-cell level. Together, these examples show how smRNA-FISH can resolve complex regulatory relationships within subpopulations of cells.

5.2 lncRNAs Do Not Hit the Jackpot

Although some lncRNAs are highly abundant, the vast majority are estimated to be expressed at less than one molecule per cell (Cabili et al. 2011; Wang et al. 2011; Derrien et al. 2012; Hangauer et al. 2013). One possibility, described above, is "jackpotting" of lncRNA expression sub populations of cells. Thus, one hypothesis that lingered in the field was that lncRNAs are in fact highly abundant in a small subset of cells but are lowly expressed in most other cells. To address this question, one study examined the smRNA-FISH localization patterns of 61 lncRNAs and 49 mRNAs in three populations of different human cell lines (Cabili et al. 2015). It is important to note that more than 100 lncRNAs were targeted in this study, but only 61 had consistent signal in "odds and evens" probe validation. This further underscores the importance of probe optimization for lncRNAs (Fig. 3B). The 61 validated smRNA-FISH probe sets represented lncRNAs with a spectrum of cellular abundance (as measured by RNA-sequencing) and other properties (e.g., tissue specificity, which could be indicative of subpopulation expression dynamics). By quantifying lncRNA and mRNA abundance from cell to cell within a population, it was observed that overall, lncRNAs and mRNAs show similar cell-to-cell variability in expression and do not show any jackpot behavior. The scale and systematic approach of

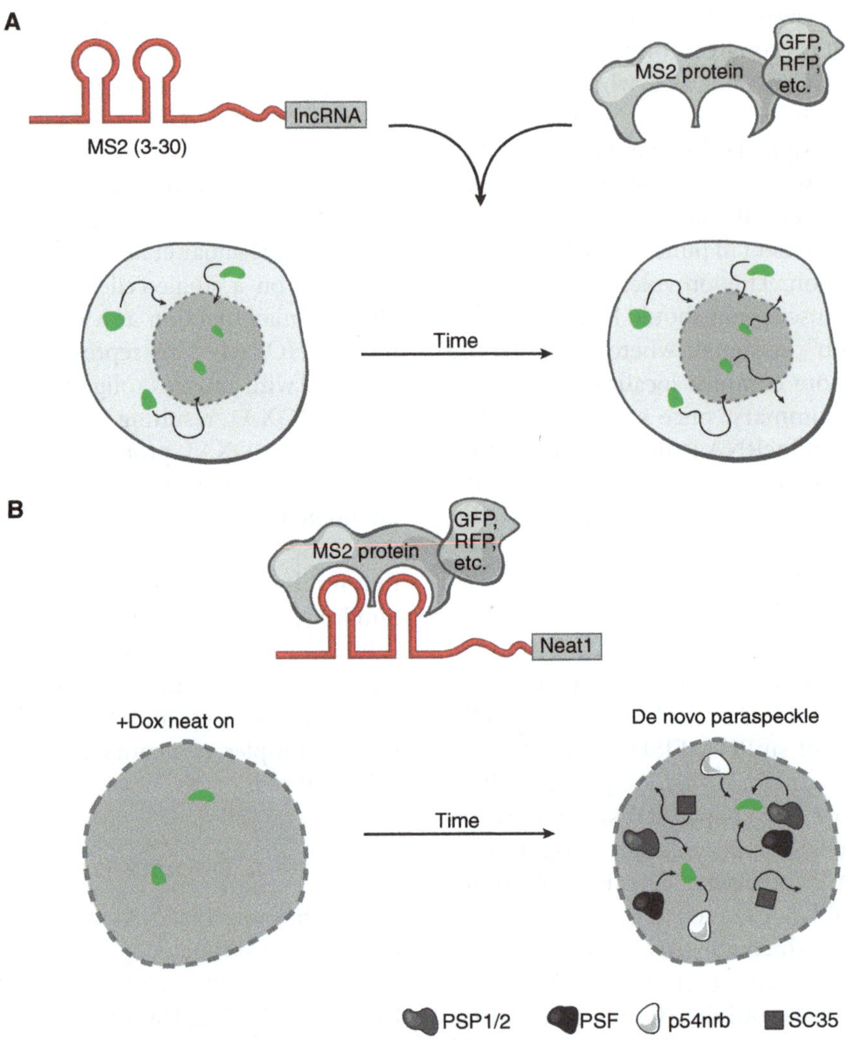

Figure 3. Live-cell RNA imaging. (*A*) The targeted long noncoding RNA (lncRNA) has a specific RNA sequence/structure (aptamer) appended (e.g., a series of MS2 hairpins). In parallel, a protein that specifically and tightly binds to the RNA aptamer is fused to a fluorescent marker. These constructs can be integrated at the endogenous locus or ectopically expressed in cells. Conjugation of the RNA aptamer with the cognate fusion protein provides the ability to track the fluorescent signal in live cells through time. This highlights the movements (arrows) of a long noncoding RNA (lncRNA), such as its rate of nuclear and cytoplasmic transport (bottom nuclei). (*B*) Neat1 live-cell RNA imaging. Live-cell RNA imaging of Neat1 revealed key insights into the role of its transcription in the establishment and maintenance of nuclear paraspeckles. The Neat1 RNA was tracked by appending 24× MS2 hairpins and cognate MS2 protein fused to a fluorescent protein. Neat1 expression was induced (green foci) by doxycycline (+Dox) and over time, paraspeckle proteins (PSP1/2, PSF, p54nrb) were recruited to the de novo expression sites of Neat1 RNA. In contrast, nuclear speckle protein (SC35) was not recruited to the de novo paraspeckle. When Neat1 expression was turned off in the absence of Dox, the paraspeckle disassembles. Together, these observations show that expression of Neat1 precedes paraspeckle formation and is required to maintain paraspeckles in the nucleus.

this smRNA-FISH study identified other fundamental lncRNA localization properties.

1. lncRNAs are more nuclear in their localization than mRNAs. For example, 55% of lncRNAs showed a significant bias for nuclear localization compared with only one mRNA that was found predominantly nuclear (Cabili et al. 2015). In contrast, some lncRNAs had exclusively cytoplasmic localization patterns, suggesting diverse subcellular functions for this class of RNA. Interestingly, these localization patterns were the same in three different human cell lines (85% of tested lncRNAs had the same nuclear to cytoplasmic ratios).
2. Nuclear chromatin localization of lncRNAs does not persist during mitosis. It remains a long-standing mys-

tery how two chromosomes reestablish their transcriptional identity after DNA replication and mitosis. One hypothesis is that chromatin-bound lncRNAs influence this process. smRNA-FISH is uniquely suited to address this question by simultaneously determining subcellular abundance of a lncRNA. For example, how much of a lncRNA resides on chromatin and in which phase of the cell cycle (by simultaneous multicolor smRNA-FISH visualization of the lncRNA and cell cycle marker genes). Surprisingly, each of the six-known chromatin-associating lncRNAs show altered localization during mitosis: in almost all cases, the lncRNAs are released from chromatin and reside in the cytoplasm (Cabili et al. 2015). This included the well-known nuclear and chromatin associated lncRNAs (e.g., Xist), suggesting that they do not play a role in organizing chromatin after mitosis.

3. RNAs that are highly correlated are not necessarily involved in positive regulatory relationships. As in the case of linc-HOXA1 and lncRNA-Ubx above, correlations between lncRNAs and putative regulatory targets via smRNA-FISH can implicate a regulatory relationship. Indeed, certain highly correlated lncRNAs and mRNAs (including several that share the same promoter) show very different localization patterns (Cabili et al. 2015). This suggests diverse independent functions for these lncRNAs despite coregulation with neighboring mRNAs. Collectively, these findings and many emerging examples have provided unique insights about lncRNA expression and localization properties from smRNA-FISH.

6 ENTERING THE FOURTH DIMENSION: LIVE-CELL RNA IMAGING

Although much has been learned from the above lncRNA-FISH studies, the temporal dynamics of lncRNA localization in a living cell has been less well understood. For several decades, live-cell RNA imaging has been possible in many model organisms (e.g., *Saccharomyces cerevisiae*), cell lines and even living mice (Tyagi 2009; Larson et al. 2011; Park et al. 2014; Lee et al. 2016a; Nelles et al. 2016; Wu et al. 2016; Tutucci et al. 2018a). Although most of these studies have focused on mRNA there are ever-emerging examples of live-cell imaging of lncRNAs. For example, the telomeric lncRNA, telomeric repeat-containing RNA (TERRA), is activated in space and time specifically at chromosomes with short telomeres (Cusanelli et al. 2013). Another elegant live-cell RNA imaging study identified a lncRNA that over time switches between two transcriptional states to repress the neighboring GAL1 mRNA (Lenstra et al. 2015). These live-cell studies and others have begun to uncover the temporal properties of lncRNA synthesis, degradation, and localization (Tutucci et al. 2018a).

There are many approaches to live-cell RNA imaging but here, we will focus on one: appending an RNA aptamer that is recognized by a cognate protein (Fig. 3). One of the most frequently appended aptamers is the MS2 RNA sequence, which is specifically and tightly recognized by the MS2 protein (Beach et al. 1999; Eliscovich and Singer 2017; Lyon and Stasevich 2017; Tutucci et al. 2018a,b). By fusing MS2 protein to a fluorescent protein (e.g., GFP or RFP) the cognate RNA–protein interaction can be imaged in real time (Fig. 3). There are several other similar RNA–protein aptamers (e.g., PP7 RNA/protein); these can be multiplexed to allow for multiple RNAs to be uniquely tagged and tracked in living cells.

These tagged lncRNAs are introduced as exogenous copies in living cells followed by imaging of the targeted RNA's spatiotemporal movements. One disadvantage of this approach is that endogenous (nontagged) copies of the lncRNA cannot be visualized. Alternatively, CRISPR–Cas9 can be used to engineer MS2 hairpins at the 3′ end of the endogenous lncRNA locus to overcome this limitation (see Engreitz et al. 2019).

As we discussed above, Neat1 and Neat2/Malat1, Xist, Gomafu, and pre-rRNA RNA-FISH each lead to the discovery of new nuclear sub compartments (Fig. 1). This raises several questions that can only be addressed using live-cell RNA imaging: (1) Do paraspeckles form first and then Neat1 localizes? (2) Does transcriptional activation of Neat1 result in paraspeckle formation? (3) Is the RNA a required component of maintaining the paraspeckle? Thus, only exploring these questions temporally in living cells will provided the needed insights into the functional role of Neat1 in paraspeckle formation.

To address these questions, a seminal study tagged Neat1 (both Menε/β isoforms) with 24 copies of the MS2 RNA aptamer on the 5′ end of Neat1 (Mao et al. 2010). This system was further engineered to induce Neat1 transcription via an inducible promoter (Tet-on: addition of Doxycycline [DOX] results in activation of Neat1). Importantly, the inducible-tagged Neat1 was placed in a dormant region of the genome (in mouse C2C12 cells). Together, these constructs allowed the visualization and tracking of Neat1 RNA at its site of transcription in parallel with protein constituents of the paraspeckle. This study found that indeed on induction of Neat1 transcription resulted in a concomitant formation of paraspeckles (Fig. 3B). In fact, the transcription of Neat1 recruited most paraspeckle proteins (e.g., PSP1, p54nrb, PSF, and PSP2), but not the protein components of the nuclear speckle associated with Malat1 (e.g., SC35). Moreover, it was shown that the newly formed paraspeckles were functional and stably formed over time. The reverse was true that shutdown of Neat1 transcription resulted, over time, in the disassembly of newly formed paraspeckles.

Together, these experiments brought our understanding of Neat1 into the fourth dimension and revealed the underlying dynamics of Neat1 induction and in turn the formation of paraspeckles. It was the aspect of time and living cells that provided these simple yet profound insights. Together, these results show that transcribing lncRNAs could facilitate changes in nuclear substructures and perhaps represents a more general role for nuclear lncRNAs.

However, many more questions remain about lncRNA biology that are suited to be tested by live-cell RNA imagining. For example, do lncRNAs that have both nuclear and cytoplasmic localization patterns shuttle between the nucleus and cytoplasm or is it the case that some molecules remain trapped in the nucleus whereas others shuttle to the cytoplasm (Fig. 3A)? If shuttling occurs, then one could dissect down the lncRNA to find the minimal RNA domain sufficient to shuttle RNA. These and many other questions are in the offing by applying live-cell imaging to lncRNA biology.

7 CRISPR-BASED VISUALIZATION OF lncRNA LOCI

With the advent of CRISPR-based technologies, it has become possible to visualize the four-dimensional dynamics of lncRNA genomic loci (Chen et al. 2013, 2016; Ma et al. 2015; Shechner et al. 2015; Shao et al. 2016; Tu et al. 2016; Dreissig et al. 2017; Qin et al. 2017; Zhou et al. 2017; Maass et al. 2018a,b). These studies have made it possible to track DNA in living cells using a combination of CRISPR–Cas9 guide sequences (see Engreitz et al. 2019) modified to contain RNA aptamers (such as MS2) and catalytically inactive, or dead Cas9 (dCas9) (Fig. 4). More recently, these approaches have been used to explore the dynamic movements and interactions of noncoding DNA including lncRNA loci. For example, how often, stable and frequent are lncRNA-mediated intra- and interchromosomal interactions in the nucleus (Maass et al. 2018a,b) and the underlying dynamics of DNA enhancers finding their targets (Gu et al. 2018).

If only one color and one position are being monitored, then simply fusing a fluorescent protein to the Cas9 protein with a guide to a specific DNA location is suitable. Yet, as soon as two different locations need to be monitored simultaneously, the guide sequence needs to be "coded" with an RNA aptamer (such as MS2 RNA described above) (Fig. 4). Otherwise, combining two different Cas9-fluorescent fusions with two different guides would result in a random mixing of guides and fluorescent Cas9. In other words, the "color" would not be specified by the guide.

By way of analogy, each Cas9 is like a drone with cargo. The location of the drone is determined by the guide sequence. The RNA sequences (e.g., MS2 RNA sequence) appended to the guide specify the cargo that is to be delivered to that location (e.g., MS2 protein fused to a fluorescent protein or other fusions) (Fig. 4). Thus, each Cas9 drone brings a specific cargo to specific genomic location. Finally, several drones can be deployed in a single-cell, each with a specific genomic location and cargo (Fig. 4). Interestingly, guide RNA sequences are amenable to adding very long RNA cargos, even entire lncRNAs.

Overall, the CRISPR–Cas systems are universally applicable to deliver varieties of RNA cargo to and/or demarcate specific genomic locations. In principle, this simplifies to CRISPR–Cas9-based live-cell DNA imaging, for coding and noncoding regions alike. Several recent studies have used the MS2 and PP7 RNA aptamers to monitor DNA dynamics in three and four dimensions. This has shed new light on diverse aspects of nuclear biology: from how telomeres are organized in the living nucleus, to how specific chromosomal regions move around the nucleus, to DNA packaging dynamics. More recently, these approaches have been applied to monitor the dynamics of intra- and interchromosomal interactions by lncRNA loci (Chen et al. 2013, 2016; Ma et al. 2015; Shechner et al. 2015; Shao et al. 2016; Tu et al. 2016; Dreissig et al. 2017; Qin et al. 2017; Zhou et al. 2017; Maass et al. 2018a,b). Collectively, live-cell RNA and DNA imaging are synergistic approaches to understanding the dynamic roles of RNA, facilitating nuclear subcompartments in parallel with the resulting reorganization DNA by lncRNAs.

8 CONCLUDING REMARKS

Herein, we have covered several RNA imaging approaches and in turn how they have illuminated fundamental properties of lncRNA biology. As we hope to have highlighted, RNA imaging has a unique advantage of "seeing" where, when, and how a lncRNA is expressed in each cell. Thus, RNA imaging is often the best place to start before moving forward with mechanistic studies. For example, if a lncRNA has an exclusively nuclear localization versus a predominantly cytoplasmic pattern, which would lead to very different mechanistic hypotheses. Indeed, a wide variety of localization patterns and lncRNA functional insights have been gleaned from RNA imaging: from NORAD, with its exquisite cytoplasmic localization led to an understanding of its role in translation regulation (Lee et al. 2016b; Tichon et al. 2016); to CONCR, which is specifically activated in the nucleus during S-phase and regulates sister chromatid pairing (Marchese et al. 2016); to SLERT, which localizes to the nucleolus and forms ring-shaped structures

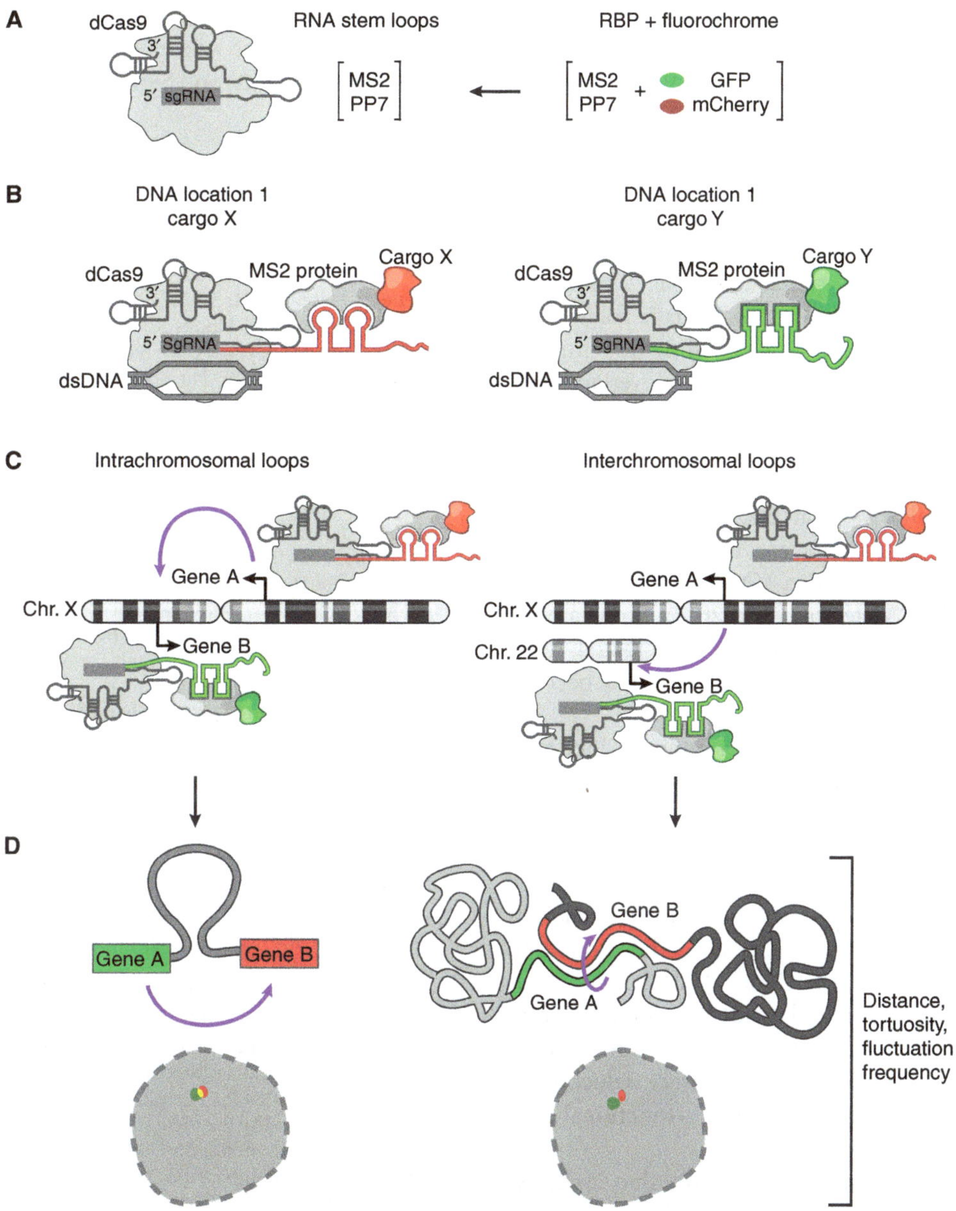

Figure 4. Live-cell imaging of long noncoding RNA (lncRNA) loci. (*A*) Catalytically inactive CRISPR–Cas9 constructs with extended single-guide RNA (sgRNA) sequences that contain RNA aptamers (e.g., MS2 or PP7). The RNA binding protein (RBP) that recognizes sgRNA aptamers is fused to a fluorescent protein. (*B*) A CRISPR–dCas9/cargo complex, in which the modified sgRNA specifies the genomic location and acts as a scaffold for delivery of chosen "cargo" fused to the MS2 coat protein. (*C*) Application of guide sequence aptamer for CRISPR live-cell imaging (CLING). Each guide has a unique RNA aptamer that forms a defined complex with RBP:fluorescent protein fusion. This allows for real-time monitoring of lncRNA loci (or any genomic location) forming intra- or interchromosomal interactions in space and time. (*D*) Examples of using CLING to monitor the biophysical properties (tortuosity, distance, fluctuation rates) of DNA folding within and across chromosomes.

that influence RNA polymerase I transcription (Xing et al. 2017); to 7SK that localizes to promoters to regulate transcription (Prasanth et al. 2010). With each emerging lncRNA-FISH study, fascinating new localization patterns, dynamics, and biological functions are being revealed for lncRNAs.

Collectively, the key findings from these and other RNA imaging studies raises a fundamental question about lncRNA biology. Could certain lncRNA localization patterns be indicative of specific functional lncRNA "classes" or "categories"? A general bifurcation is cytoplasmic versus nuclear localized lncRNAs that appear to a first approximation to indicate common functionalities within these subcellular regions. The screening of a small panel of lncRNAs revealed all possible lncRNA localization patterns, ranging from exclusively nuclear, nucleoplasmic, nuclear, and cytoplasmic or exclusively cytoplasmic (Cabili et al. 2015). Yet, in each category there were at least a few lncRNAs sharing similar expression patterns that were consistent across several cell lines. Thus, it is tantalizing to speculate that lncRNAs sharing specific spatial-temporal localization patterns could also have common mechanistic roles. Consistent with this notion, many lncRNAs show two nuclear foci in which intronic and exonic probes overlap. This localization pattern could suggest that such lncRNAs are acting in *cis* (see Engreitz et al. 2019). Indeed, this localization feature is shared by many lncRNAs that either function as enhancer RNAs (eRNAs) that "loop" enhancers to gene targets or facilitate global nuclear architecture through interactions with nuclear lamina (Ørom and Shiekhattar 2011; Lam et al. 2014; Melé and Rinn 2016; Nozawa et al. 2017). As more lncRNA mechanisms are elucidated, these association between localization and function may become more evident.

Teasing apart associations between different lncRNAs will further require RNA imaging advancements to simultaneously image numerous RNA species in situ. These limitations are starting to be overcome, with several multiplexing methods now available for the detection of hundreds or even thousands of RNAs in individual cells (Levsky et al. 2002; Lubeck and Cai 2012; Levesque and Raj 2013; Lubeck et al. 2014; Chen et al. 2015). These methods will greatly expand and enable a more discovery-oriented approach to lncRNA biology. Yet, perhaps more importantly, these methods will allow direct detection of common localization patterns and perhaps functional classes across hundreds of lncRNAs at once.

Moving forward, it is important to note that lncRNAs have a long-standing track record of nuclear regulatory roles. This may perhaps be caused by observational bias, but from the limited examples covered here we have seen that lncRNAs play roles in establishing nuclear domains such as the nucleolus (pre-rRNA), paraspeckles (Neat1), nuclear speckles (Malat1), nuclear matrix (Gomafu), and the Barr body (Xist). Moreover, many lncRNAs have been found to have focal localization patterns that could be indicative of nuclear organization on a smaller scale. Thus, understanding how these exquisite lncRNA localization patterns relate to function is a key goal of future studies.

ACKNOWLEDGMENTS

We thank all the students and postdocs that have provided numerous insights into lncRNA biology using RNA imaging techniques. We also thank Tom Cech for editorial advice and editing, Alexandra and Mark McCorkindale for manuscript edits, and Sigrid Knemeyer for illustrations.

REFERENCES

**Reference is in this book.*

Batish M, Raj A, Tyagi S. 2011. Single molecule imaging of RNA in situ. *Methods Mol Biol* **714:** 3–13.

Beach DL, Salmon ED, Bloom K. 1999. Localization and anchoring of mRNA in budding yeast. *Curr Biol* **9:** 569–578.

Bond CS, Fox AH. 2009. Paraspeckles: Nuclear bodies built on long noncoding RNA. *J Cell Biol* **186:** 637–644.

Brown CJ, Hendrich BD, Rupert JL, Lafrenière RG, Xing Y, Lawrence J, Willard HF. 1992. The human *XIST* gene: Analysis of a 17 kb inactive X-specific RNA that contains conserved repeats and is highly localized within the nucleus. *Cell* **71:** 527–542.

Buxbaum AR, Haimovich G, Singer RH. 2015. In the right place at the right time: Visualizing and understanding mRNA localization. *Nat Rev Mol Cell Biol* **16:** 95–109.

Cabili MN, Trapnell C, Goff L, Koziol M, Tazon-Vega B, Regev A, Rinn JL. 2011. Integrative annotation of human large intergenic noncoding RNAs reveals global properties and specific subclasses. *Genes Dev* **25:** 1915–1927.

Cabili MN, Dunagin MC, McClanahan PD, Biaesch A, Padovan-Merhar O, Regev A, Rinn JL, Raj A. 2015. Localization and abundance analysis of human lncRNAs at single-cell and single-molecule resolution. *Genome Biol* **16:** 20.

Chen B, Gilbert LA, Cimini BA, Schnitzbauer J, Zhang W, Li G-W, Park J, Blackburn EH, Weissman JS, Qi LS, et al. 2013. Dynamic imaging of genomic loci in living human cells by an optimized CRISPR/Cas system. *Cell* **155:** 1479–1491.

Chen KH, Boettiger AN, Moffitt JR, Wang S, Zhuang X. 2015. RNA imaging. Spatially resolved, highly multiplexed RNA profiling in single cells. *Science* **348:** aaa6090.

Chen B, Guan J, Huang B. 2016. Imaging specific genomic DNA in living cells. *Annu Rev Biophys* **45:** 1–23.

Cheng L, Ming H, Zhu M, Wen B. 2016. Long noncoding RNAs as organizers of nuclear architecture. *Sci China Life Sci* **59:** 236–244.

Cusanelli E, Romero CAP, Chartrand P. 2013. Telomeric noncoding RNA TERRA is induced by telomere shortening to nucleate telomerase molecules at short telomeres. *Mol Cell* **51:** 780–791.

Derrien T, Johnson R, Bussotti G, Tanzer A, Djebali S, Tilgner H, Guernec G, Martin D, Merkel A, Knowles DG, et al. 2012. The GENCODE v7 catalog of human long noncoding RNAs: Analysis of their gene structure, evolution, and expression. *Genome Res* **22:** 1775–1789.

Dirks RW, van de Rijke FM, Fujishita S, van der Ploeg M, Raap AK. 1993. Methodologies for specific intron and exon RNA localization in cultured cells by haptenized and fluorochromized probes. *J Cell Sci* **104:** 1187–1197.

Dreissig S, Schiml S, Schindele P, Weiss O, Rutten T, Schubert V, Gladilin E, Mette MF, Puchta H, Houben A. 2017. Live-cell CRISPR imaging in plants reveals dynamic telomere movements. *Plant J* **91:** 565–573.

Dunagin M, Cabili MN, Rinn J, Raj A. 2015. Visualization of lncRNA by single-molecule fluorescence in situ hybridization. *Methods Mol Biol* **1262:** 3–19.

Eliscovich C, Singer RH. 2017. RNP transport in cell biology: The long and winding road. *Curr Opin Cell Biol* **45:** 38–46.

Engreitz JM, Ollikainen N, Guttman M. 2016. Long non-coding RNAs: Spatial amplifiers that control nuclear structure and gene expression. *Nat Rev Mol Cell Biol* **17:** 756–770.

* Engreitz J, Abudayyeh O, Gootenberg J, Zhang F. 2019. CRISPR tools for systematic studies of RNA regulation. *Cold Spring Harb Perspect Biol* **11:** a035386.

Falahati H, Pelham-Webb B, Blythe S, Wieschaus E. 2016. Nucleation by rRNA dictates the precision of nucleolus assembly. *Curr Biol* **26:** 277–285.

Femino AM, Fay FS, Fogarty K, Singer RH. 1998. Visualization of single RNA transcripts in situ. *Science* **280:** 585–590.

Fremeau RT, Lundblad JR, Pritchett DB, Wilcox JN, Roberts JL. 1986. Regulation of pro-opiomelanocortin gene transcription in individual cell nuclei. *Science* **234:** 1265–1269.

Gaspar I, Wippich F, Ephrussi A. 2017. Enzymatic production of single-molecule FISH and RNA capture probes. *RNA* **23:** 1582–1591.

Gu B, Swigut T, Spencley A, Bauer MR, Chung M, Meyer T, Wysocka J. 2018. Transcription-coupled changes in nuclear mobility of mammalian *cis*-regulatory elements. *Science* **359:** 1050–1055.

Hangauer M J, Vaughn IW, McManus MT. 2013. Pervasive transcription of the human genome produces thousands of previously unidentified long intergenic noncoding RNAs. *PLoS Genet* **9:** e1003569.

Hutchinson JN, Ensminger AW, Clemson CM, Lynch CR, Lawrence JB, Chess A. 2007. A screen for nuclear transcripts identifies two noncoding RNAs associated with SC35 splicing domains. *BMC Genomics* **8:** 39.

Lam MTY, Li W, Rosenfeld MG, Glass CK. 2014. Enhancer RNAs and regulated transcriptional programs. *Trends Biochem Sci* **39:** 170–182.

Larson DR, Singer RH, Zenklusen D. 2009. A single molecule view of gene expression. *Trends Cell Biol* **19:** 630–637.

Larson DR, Zenklusen D, Wu B, Chao JA, Singer RH. 2011. Real-time observation of transcription initiation and elongation on an endogenous yeast gene. *Science* **332:** 475–478.

Lawrence JB, Singer RH. 1986. Intracellular localization of messenger RNAs for cytoskeletal proteins. *Cell* **45:** 407–415.

Lawrence JB, Singer RH, Villnave CA, Stein JL, Stein GS. 1988. Intracellular distribution of histone mRNAs in human fibroblasts studied by in situ hybridization. *Proc Natl Acad Sci* **85:** 463–467.

Lawrence JB, Singer RH, Marselle LM. 1989. Highly localized tracks of specific transcripts within interphase nuclei visualized by in situ hybridization. *Cell* **57:** 493–502.

Lee BH, Bae S-W, Shim JJ, Park SY, Park HY. 2016a. Imaging single-mRNA localization and translation in live neurons. *Mol Cells* **39:** 841–846.

Lee S, Kopp F, Chang T-C, Sataluri A, Chen B, Sivakumar S, Yu H, Xie Y, Mendell JT. 2016b. Noncoding RNA NORAD regulates genomic stability by sequestering PUMILIO proteins. *Cell* **164:** 69–80.

Lenstra TL, Coulon A, Chow CC, Larson DR. 2015. Single-molecule imaging reveals a switch between spurious and functional ncRNA transcription. *Mol Cell* **60:** 597–610.

Levesque MJ, Raj A. 2013. Single-chromosome transcriptional profiling reveals chromosomal gene expression regulation. *Nat Methods* **10:** 246–248.

Levesque MJ, Ginart P, Wei Y, Raj A. 2013. Visualizing SNVs to quantify allele-specific expression in single cells. *Nat Meth* **10:** 865–867.

Levsky JM, Shenoy SM, Pezo RC, Singer RH. 2002. Single-cell gene expression profiling. *Science* **297:** 836–840.

Lubeck E, Cai L. 2012. Single-cell systems biology by super-resolution imaging and combinatorial labeling. *Nat Methods* **9:** 743–748.

Lubeck E, Coskun AF, Zhiyentayev T, Ahmad M, Cai L. 2014. Single-cell in situ RNA profiling by sequential hybridization. *Nat Methods* **11:** 360–361.

Lyon K, Stasevich TJ. 2017. Imaging translational and post-translational gene regulatory dynamics in living cells with antibody-based probes. *Trends Genet* **33:** 322–335.

Ma H, Naseri A, Reyes-Gutierrez P, Wolfe SA, Zhang S, Pederson T. 2015. Multicolor CRISPR labeling of chromosomal loci in human cells. *Proc Natl Acad Sci* **112:** 3002–3007.

Maamar H, Cabili MN, Rinn J, Raj A. 2013. *linc-HOXA1* is a noncoding RNA that represses *Hoxa1* transcription in *cis*. *Genes Dev* **27:** 1260–1271.

Maass PG, Barutcu AR, Shechner DM, Weiner CL, Melé M, Rinn JL. 2018a. Spatiotemporal allele organization by allele-specific CRISPR live-cell imaging (SNP-CLING). *Nat Struct Mol Biol* **25:** 176–184.

Maass PG, Barutcu AR, Weiner CL, Rinn JL. 2018b. Inter-chromosomal contact properties in live-cell imaging and in Hi-C. *Mol Cell* **70:** 188–189.

Mao YS, Sunwoo H, Bin Zhang, Spector DL. 2010. Direct visualization of the co-transcriptional assembly of a nuclear body by noncoding RNAs. *Nat Cell Biol* **13:** 95–101.

Mao YS, Zhang B, Spector DL. 2011. Biogenesis and function of nuclear bodies. *Trends Genet* **27:** 295–306.

Marchese FP, Grossi E, Marín-Béjar O, Bharti SK, Raimondi I, González J, Martínez-Herrera DJ, Athie A, Amadoz A, Brosh RM, et al. 2016. A long noncoding RNA regulates sister chromatid cohesion. *Mol Cell* **63:** 397–407.

Martin KC, Ephrussi A. 2009. mRNA localization: Gene expression in the spatial dimension. *Cell* **136:** 719–730.

Melé M, Rinn JL. 2016. "Cat's cradling" the 3D genome by the act of lncRNA transcription. *Mol Cell* **62:** 657–664.

Nelles DA, Fang MY, O'Connell MR, Xu JL, Markmiller SJ, Doudna JA, Yeo GW. 2016. Programmable RNA tracking in live cells with CRISPR/Cas9. *Cell* **165:** 488–496.

Nozawa R-S, Boteva L, Soares DC, Naughton C, Dun AR, Buckle A, Ramsahoye B, Bruton PC, Saleeb RS, Arnedo M, et al. 2017. SAF-A regulates interphase chromosome structure through oligomerization with chromatin-associated RNAs. *Cell* **169:** 1214–1227.e18.

Orjalo AV, Johansson HE. 2016. Stellaris® RNA fluorescence in situ hybridization for the simultaneous detection of immature and mature long noncoding RNAs in adherent cells. *Methods Mol Biol* **1402:** 119–134.

Ørom UA, Shiekhattar R. 2011. Noncoding RNAs and enhancers: Complications of a long-distance relationship. *Trends Genet* **27:** 433–439.

Park HY, Lim H, Yoon YJ, Follenzi A, Nwokafor C, Lopez-Jones M, Meng X, Singer RH. 2014. Visualization of dynamics of single endogenous mRNA labeled in live mouse. *Science* **343:** 422–424.

Parton RM, Davidson A, Davis I, Weil TT. 2014. Subcellular mRNA localisation at a glance. *J Cell Sci* **127:** 2127–2133.

Petruk S, Sedkov Y, Riley KM, Hodgson J, Schweisguth F, Hirose S, Jaynes JB, Brock HW, Mazo A. 2006. Transcription of *bxd* noncoding RNAs promoted by trithorax represses *Ubx* in *cis* by transcriptional interference. *Cell* **127:** 1209–1221.

Prasanth KV, Camiolo M, Chan G, Tripathi V, Denis L, Nakamura T, Hübner MR, Spector DL. 2010. Nuclear organization and dynamics of 7SK RNA in regulating gene expression. *Mol Biol Cell* **21:** 4184–4196.

Qin P, Parlak M, Kuscu C, Bandaria J, Mir M, Szlachta K, Singh R, Darzacq X, Yildiz A, Adli M. 2017. Live cell imaging of low- and non-repetitive chromosome loci using CRISPR-Cas9. *Nat Commun* **8:** 14725.

Querido E, Dekakra-Bellili L, Chartrand P. 2017. RNA fluorescence in situ hybridization for high-content screening. *Methods* **126:** 149–155.

Quinodoz S, Guttman M. 2014. Long noncoding RNAs: An emerging link between gene regulation and nuclear organization. *Trends Cell Biol* **24:** 651–663.

Raj A, van Oudenaarden A. 2009. Single-molecule approaches to stochastic gene expression. *Annu Rev Biophys* **38:** 255–270.

Raj A, van den Bogaard P, Rifkin SA, van Oudenaarden A, Tyagi S. 2008. Imaging individual mRNA molecules using multiple singly labeled probes. *Nat Methods* **5:** 877–879.

Raj A, Rifkin SA, Andersen E, van Oudenaarden A. 2010. Variability in gene expression underlies incomplete penetrance. *Nature* **463:** 913–918.

Rinn J, Guttman M. 2014. RNA Function. RNA and dynamic nuclear organization. *Science* **345:** 1240–1241.

Shaffer SM, Wu M-T, Levesque MJ, Raj A. 2013. Turbo FISH: A method for rapid single molecule RNA FISH. *PLoS ONE* **8:** e75120.

Shaffer SM, Dunagin MC, Torborg SR, Torre EA, Emert B, Krepler C, Beqiri M, Sproesser K, Brafford PA, Xiao M, et al. 2017. Rare cell variability and drug-induced reprogramming as a mode of cancer drug resistance. *Nature* **546:** 431–435.

Shao S, Zhang W, Hu H, Xue B, Qin J, Sun C, Sun Y, Wei W, Sun Y. 2016. Long-term dual-color tracking of genomic loci by modified sgRNAs of the CRISPR/Cas9 system. *Nucleic Acids Res* **19:** e86.

Shechner DS, Hacisuleyman E, Younger ST, Rinn JL. 2015. Multiplexable, locus-specific targeting of long RNAs with CRISPR-display. *Nat Methods* **12:** 664–670.

Sinnamon JR, Czaplinski K. 2014. RNA detection in situ with FISH-STICs. *RNA* **20:** 260–266.

Soler M, Boque-Sastre R, Guil S. 2017. RNA-FISH to study regulatory RNA at the site of transcription. *Methods Mol Biol* **1543:** 221–229.

Sone M, Hayashi T, Tarui H, Agata K, Takeichi M, Nakagawa S. 2007. The mRNA-like noncoding RNA Gomafu constitutes a novel nuclear domain in a subset of neurons. *J Cell Sci* **120:** 2498–2506.

Spector DL, Lamond AI. 2011. Nuclear speckles. *Cold Spring Harb Perspect Biol* **3:** a000646.

Tichon A, Gil N, Lubelsky Y, Havkin Solomon T, Lemze D, Itzkovitz S, Stern-Ginossar N, Ulitsky I. 2016. A conserved abundant cytoplasmic long noncoding RNA modulates repression by Pumilio proteins in human cells. *Nat Commun* **7:** 12209.

Torre E, Dueck H, Shaffer S, Gospocic J, Gupte R, Bonasio R, Kim J, Murray J, Raj A. 2018. Rare cell detection by single-cell RNA sequencing as guided by single-molecule RNA FISH. *Cell Syst* **6:** 171–179.e5.

Trcek T, Larson DR, Moldón A, Query CC, Singer RH. 2011. Single-molecule mRNA decay measurements reveal promoter-regulated mRNA stability in yeast. *Cell* **147:** 1484–1497.

Trcek T, Chao JA, Larson DR, Park HY, Zenklusen D, Shenoy SM, Singer RH. 2012. Single-mRNA counting using fluorescent in situ hybridization in budding yeast. *Nature Protoc* **7:** 408–419.

Tripathi V, Fei J, Ha T, Prasanth KV. 2015. RNA fluorescence in situ hybridization in cultured mammalian cells. *Methods Mol Biol* **1206:** 123–136.

Tu L-C, Naseri A, Huisman M, Zhang S, Grunwald D, Pederson T, Ma H. 2016. Multiplexed labeling of genomic loci with dCas9 and engineered sgRNAs using CRISPRainbow. *Nat Biotechnol* **34:** 1–4.

Tutucci E, Livingston N M, Singer RH, Wu B. 2018a. Imaging mRNA in vivo, from birth to death. *Annu Rev Biophys* **47:** 85–106.

Tutucci E, Vera M, Biswas J, Garcia J, Parker R, Singer RH. 2018b. An improved MS2 system for accurate reporting of the mRNA life cycle. *Nat Methods* **15:** 81–89.

Tyagi S. 2009. Imaging intracellular RNA distribution and dynamics in living cells. *Nat Methods* **6:** 331–338.

Vance KW, Ponting CP. 2014. Transcriptional regulatory functions of nuclear long noncoding RNAs. *Trends Genet* **30:** 348–355.

* Van Treeck B, Parker R. 2019. Principles of stress granules revealed by imaging approaches. *Cold Spring Harb Perspect Biol* **11:** a033068.

Wang KC, Yang YW, Liu B, Sanyal A, Corces-Zimmerman R, Chen Y, Lajoie BR, Protacio A, Flynn RA, Gupta RA, et al. 2011. A long noncoding RNA maintains active chromatin to coordinate homeotic gene expression. *Nature* **472:** 120–124.

Wang DO, Matsuno H, Ikeda S, Nakamura A, Yanagisawa H, Hayashi Y, Okamoto A. 2012. A quick and simple FISH protocol with hybridization-sensitive fluorescent linear oligodeoxynucleotide probes. *RNA* **18:** 166–175.

Willingham AT. 2005. A strategy for probing the function of noncoding RNAs finds a repressor of NFAT. *Science* **309:** 1570–1573.

Wu B, Eliscovich C, Yoon YJ, Singer RH. 2016. Translation dynamics of single mRNAs in live cells and neurons. *Science* **352:** 1430–1435.

Xing Y-H, Yao R-W, Zhang Y, Guo C-J, Jiang S, Xu G, Dong R, Yang L, Chen L-L. 2017. *SLERT* regulates *DDX21* rings associated with Pol I transcription. *Cell* **169:** 664–678.e16.

Yin Y, Yan P, Lu J, Song G, Zhu Y, Li Z, Zhao Y, Shen B, Huang X, Zhu H, et al. 2015. Opposing roles for the lncRNA *Haunt* and its genomic locus in regulating *HOXA* gene activation during embryonic stem cell differentiation. *Cell Stem Cell* **16:** 504–516.

Zenklusen D, Larson DR, Singer RH. 2008. Single-RNA counting reveals alternative modes of gene expression in yeast. *Nat Struct Mol Biol* **15:** 1263–1271.

Zhou Y, Wang P, Tian F, Gao G, Huang L, Wei W, Xie XS. 2017. Painting a specific chromosome with CRISPR/Cas9 for live-cell imaging. *Cell Res* **27:** 1–4.

Cite as *Cold Spring Harb Perspect Biol* doi: 10.1101/cshperspect.a032094

Principles of Stress Granules Revealed by Imaging Approaches

Briana Van Treeck[1] and Roy Parker[1,2]

[1]Department of Chemistry and Biochemistry, University of Colorado, Boulder, Colorado 80303

[2]Howard Hughes Medical Institute, Boulder, Colorado 80303

Correspondence: roy.parker@colorado.edu

SUMMARY

Eukaryotic cells contain a large number of RNA–protein assemblies, generically referred to as ribonucleoprotein (RNP) granules. Such RNP granules include stress granules and P-bodies in the cytosol and the nucleolus, Cajal bodies, and paraspeckles in the nucleus. A variety of imaging approaches have been used to reveal different components, structural features, and dynamics of RNP granules. In this review, we discuss imaging approaches that have been used to study stress granules and the insights gained from these experiments. A general theme is that these approaches can be transferred to other RNP granules to examine similar aspects of their composition, ultrastructure, dynamics and control.

Outline

Cite as *Cold Spring Harb Perspect Biol* doi: 10.1101/cshperspect.a033068

1 INTRODUCTION

Ribonucleoprotein (RNP) granules are higher-order assemblies composed of various RNAs and proteins, and they are ubiquitous in eukaryotes. RNP granules are diverse and vary in their components, regulation, and dynamics. Cytoplasmic RNP granules include stress granules, P-bodies, germ granules, and neuronal granules, whereas nuclear RNP granules include paraspeckles, the nucleolus, Cajal bodies, and RNA foci formed from repeat expansion RNAs in disease contexts (Gall 2000; Fox et al. 2002; Kiebler and Bassell 2006; Wojciechowska and Krzyzosiak 2011; Voronina et al. 2012; Buchan 2014). Different environmental or gene expression cues modulate each type of RNP granule. For example, although P-bodies are ubiquitously present at a basal level (Yang et al. 2004; Jain and Parker 2013), stress granules only form under conditions that disrupt translation initiation and lead to ribosomes running off the messenger RNA (mRNA) (Kedersha et al. 1999). Specialized RNP granules can also be found in specific cell types, such as neuronal granules and germ granules present in neurons and oocytes, respectively (Kiebler and Bassell 2006; Voronina et al. 2012).

Despite differences in the cellular context of assembly, RNP granules share a number of common characteristics. First, all RNP granules examined to date are nonmembrane-bound organelles. Second, by definition all RNP granules include a population of specific RNAs, which can vary between different RNP granules. For example, the nucleolus is dependent on pre-rRNA transcripts and Alu-containing RNAs for assembly, paraspeckles contain and are dependent on the NEAT1 long noncoding RNA (lncRNA), and cytoplasmic P-bodies and stress granules contain and are dependent on nontranslating mRNAs (Bond and Fox 2009; Protter and Parker 2016; Németh and Grummt 2018; Fox et al. 2018). Third, although RNP granules are often characterized and visualized by their specific components, many protein components can be found in more than one type of RNP granule, blurring the distinct lines that separate them. For example, several proteins, including the exoribonuclease Xrn1 and the RNA helicase and translational repressor Rck, are present in both P-bodies and stress granules and are also found in neuronal transport or germinal granules (Barbee et al. 2006; Buchan and Parker 2009; Buchan 2014). Fourth, RNP granules assemble through a combination of protein–protein, RNA–protein, and RNA–RNA interactions, leading to a networked assembly of RNPs, although the relative importance of these interactions can vary depending on the specific RNP granule (Kaiser et al. 2008; Jain and Vale 2017; Van Treeck et al. 2018).

RNP granules have been implicated in a number of biological processes. In some cases, RNP granules appear to be sites of localized activity, such as the nucleolus that functions as the site of ribosome biogenesis (Scheer et al. 1993) or Cajal bodies that play a role in small nuclear ribonucleoprotein (snRNP) assembly (Staněk and Neugebauer 2004; Strzelecka et al. 2010). Similarly, by concentrating components of the mRNA decay machinery, P-bodies have been proposed to increase the rates of mRNA decapping and/or 5′ to 3′ exonuclease decay (Sheth and Parker 2003). In contrast, germinal and neuronal RNP granules have been proposed to modulate translation rates through the storage of mRNAs until proper temporal or spatial release (Barbee et al. 2006; Kiebler and Bassell 2006; Strome and Updike 2015; Hubstenberger et al. 2017). In addition, the formation of stress granules, and possibly other RNP granules, can modulate signaling pathways (Kedersha et al. 2013).

Stress granules are particularly interesting RNP granules for a number of reasons. First, stress granules typically form during a variety of stress responses, including responses to heat shock, nutrient deprivation, hypo-osmotic conditions, and oxidative stress, all of which inhibit bulk translation initiation, leading to the release of most mRNAs from ribosomes (reviewed in Protter and Parker 2016). Subsets of these nontranslating mRNAs then coalesce into stress granules with their associated proteins (see below). Second, in mammalian cells in culture, a number of different genetic perturbations that limit stress granule formation lead to increased cell death during stress (Baguet et al. 2007; Eisinger-Mathason et al. 2008; Hofmann et al. 2012). Thus, stress granule formation seems to improve cell viability during the stress response, although the molecular mechanism for this effect is not clear. Finally, stress granules are very similar to germinal granules and RNP granules in neurons, the latter of which are important for at least some forms of synaptic plasticity (Barbee et al. 2006; McCann et al. 2011; Bakthavachalu et al. 2018).

Stress granules also have a number of connections to human diseases. For example, the ability to form stress granules can contribute to tumor progression, perhaps by helping tumor cells survive the insults of chemotherapeutic agents or transitions in the process of forming metastases (Anderson et al. 2015; Somasekharan et al. 2015; Grabocka and Bar-Sagi 2016). Stress granules also affect a number of viral infections; for example, the ability to form stress granules can enhance the innate immune response to dsRNA (reviewed in McCormick and Khaperskyy 2017). Moreover, mutations that affect stress granule disassembly or formation can be causative in degenerative diseases such as amyotrophic lateral sclerosis, frontotemporal lobar degeneration, and some muscle myopathies (Li et al. 2013; Ramaswami et al. 2013). One class of such mutations are dominant mutations in abundant RNA-binding proteins found in

Cite as *Cold Spring Harb Perspect Biol* doi: 10.1101/cshperspect.a033068

stress granules; the mutations increase the propensity of the proteins to form amyloid-like structures (Guo et al. 2011; Kim et al. 2013, Schwartz et al. 2015). This has led to the model in which stress granules create a high local concentration of these proteins, which then enhances their transition to a toxic amyloid-like state (Li et al. 2013; Ramaswami et al. 2013).

Given their biological importance and connection to human disease, stress granules have become one of the most studied RNP granules. In this review, we take advantage of the diverse work on stress granules to illustrate how RNP granules can be studied using classical and modern microscopy-based approaches. These studies provide a wealth of information about the principles of stress granule formation and dynamics as well as highlight approaches that could be applied to other RNP granules.

2 PROTEIN COMPONENTS OF STRESS GRANULES

A starting point of analysis for any RNP granule is to identify its components. A variety of different approaches and methods have been used to compile the stress granule proteome (Table 1A). For many RNP granules, the discovery of the granule and an initial component occurred concurrently, by observing intracellular foci marked with a specific protein of interest. For example, stress granules were first defined by noticing that heat shock proteins and the predominantly nuclear mRNA-binding protein TIA1 formed cytoplasmic foci during heat shock and oxidative stress, respectively (Collier and Schlesinger 1986; Kedersha et al. 1999). Similarly, P-bodies were discovered by the observation that components of the mRNA decay machinery concentrated in cytoplasmic foci (Sheth and Parker 2003). Some RNP granules were first identified through contrast imaging, such as the discovery of Cajal bodies by silver staining (Cajal 1903), and although this provided information on their overall size and structure, identification of key protein components could only occur after the advent of molecular markers. Even now, punctate nuclear or cytoplasmic localization of a protein often leads to colocalization studies to determine if this protein is localized to a previously identified granule or is a member of a novel assembly.

The discovery of protein components of stress granules has gone through two phases. In the initial experiments, multiple new components of stress granules were identified either by immunofluorescence (IF) using specific antibodies or by observing the accumulation of a given fluorescently tagged protein in stress granules (e.g., Kedersha et al. 2002; Tourrière et al. 2003; Antar et al. 2005; Nonhoff et al. 2007). This piecemeal approach led to the identification of approximately 75 different proteins or protein complexes that accumulate in stress granules (reviewed in Buchan and Parker 2009). The main types of stress granule components found in these targeted experiments were RNA-binding proteins and components of the translation machinery.

The identification of an initial set of stress granule localized proteins allowed further interrogation of stress granule assembly, disassembly, and modulation. Deletion or knockdown of some protein components has resulted in an inability of stress granules to form in certain cellular contexts. For example, cells lacking G3BP1 and G3BP2 no longer form stress granules in response to eIF2α phosphorylation or eIF4A inhibition but still assemble stress granules under heat shock or osmotic stress (Kedersha et al. 2016). Conversely, overexpression of certain RNA-binding proteins highly enriched in stress granules can lead to the accumulation of constitutive stress granules, which argues that these proteins play an assembly role in stress granule formation (Kedersha et al. 1999, 2005; Mazroui 2002; Wilczynska 2005; Baguet et al. 2007). The identification of robust protein markers for stress granules has also allowed genetic screens to identify modulators of stress granule assembly and disassembly globally (Ohn et al. 2008; Buchan et al. 2013; Yang et al. 2014).

In addition, analysis of stress granule components has permitted the elucidation of compositional heterogeneity in certain contexts. Stress granules are generally conserved in eukaryotes, as orthologs of multiple mammalian stress granule proteins are present in yeast and *Drosophila* stress granules (Buchan et al. 2008; Aguilera-Gomez and Rabouille 2017). However, stress granule composition can vary under different stresses (Buchan et al. 2008; 2010; Aulas et al. 2017). Specifically, stress granules induced in yeast cells from NaN_3 treatment contain a number of initiation factors that are absent in stress granules induced from glucose deprivation (Buchan et al. 2010). Similarly, in mammalian cells, treatment with RocA, an eIF4A inhibitor, induces the formation of cytoplasmic foci that are positive for stress granule markers but lack accumulation of poly (A)-containing mRNAs, although exposure to ultraviolet (UV) forms noncanonical stress granules lacking eIF4G and eIF3 (Aulas et al. 2017).

The second phase of stress granule component discovery has been achieved through several systematic approaches. In one set of experiments, a stable substructure within stress granules (see below), referred to as a stress granule "core," was purified and subjected to mass spectroscopy analysis (Jain et al. 2016). This analysis identified 228 and 317 proteins enriched in yeast and mammalian stress granules, respectively. Validation of a number of hits via microscopy extended the list of stress granule protein components (Jain et al. 2016). This unbiased approach identified expected stress granule components, such as RNA-binding pro-

Table 1. Approaches to identify stress granule components

A. Methods to identify protein components

Method	Pros	Cons	Contribution	Reference
Immunofluorescence/tagging with fluorescent protein	Technically straightforward; can measure dynamics; can measure partition coefficients	Inefficient for global analysis; examined protein must be enriched over background; need to avoid overexpression	Identification of multiple stress granule components; measured exchange rates of many components	Reviewed in Buchan and Parker 2009
Purify stress granule cores followed by mass spec	Allows global analysis; able to identify proteins that are not enriched over the cytoplasm	Need to verify hits; may lose weakly associated factors during purification	Identified 228/317 components of yeast/ mammalian stress granule proteomes	Jain et al. 2016
APEX proximity labeling	Allows global analysis in vivo; able to identify proteins that are not enriched over the cytoplasm; able to identify transient interactors	Partitioning and dynamics of protein with stress granules increases noise and background	Identified ~150 possible new stress granule components; showed differences in stress granule composition in different stresses, different cell types	Markmiller et al. 2018
Bio-ID	Allows global analysis in vivo; able to identify proteins that are not enriched over the cytoplasm; able to identify transient interactors	Partitioning and dynamics of protein with stress granules increases noise and background	Identified new components of stress granules; provided evidence for subassemblies within stress granules	Youn et al. 2018

B. Methods to identify RNA components

Method	Pros	Cons	Contribution	Reference(s)
FISH	Techincally straightforward; can be performed for bulk mRNA (oligodT) or target-specific RNAs	Slow and expensive for high-throughput	mRNA enriched in stress granules; identified specific mRNAs enriched or depleted from stress granules	Kedersha et al. 1999, 2002; Damgaard and Lykke-Andersen 2011; Zurla et al. 2011
Purify stress granule cores and perform RNA-seq	Allows global analysis; coupled with smFISH can yield quantitative comparisons	Need to verify hits by smFISH; may lose weakly associated RNAs during purification	Length and translatability are key determinants for RNA targeting to stress granules; 10%–15% of total cytoplasmic mRNA is in stress granules	Khong et al. 2017
Pellet followed by RNA-seq	Allows global analysis	Not only stress granules pellet; need to validate	AU-rich elements enriched in stress granules; identified differences in stress granules during different stresses	Namkoong et al. 2018
smFISH	Validation is easy; allows quantification of numbers of mRNAs in stress granules	Expensive, low-throughput	Validation of RNAs in sequencing data sets	Khong et al. 2017

FISH, fluoresence in situ hybridization; RNA-seq, RNA sequencing; smFISH, single-molecule FISH.

teins and translation components, and also unexpected components involved in metabolism and signaling (Jain et al. 2016). Similar results were seen in a systematic approach to identify yeast stress granule components based on screening the genome-wide library of green fluorescent protein (GFP)-tagged proteins to identify proteins that concentrate into cytoplasmic stress granules during extreme heat shock (Cherkasov et al. 2015).

Another systematic approach to identify components of stress granules has used proximity-based proteomics, such as APEX and BioID, which were originally applied to identifying protein components within membrane-bound compartments (Roux et al. 2012; Rhee et al. 2013). In these experiments, a stress granule component is expressed as a fusion protein with either ascorbate peroxidase or biotin ligase, which labels neighboring proteins within the cell. Subsequent purification and mass spectroscopy can identify the labeled proteins. Utilizing proximity ligation for proteomics of nonmembrane bound organelles has a unique set of challenges. First, even for a highly enriched protein in stress granules, the percentage of that protein in a granule can be relatively small, which reduces the specificity for biotin modification of proteins that are in the RNP granule. For example, only 18% of G3BP1 is estimated to accumulate within stress granules in U-2 OS cells during oxidative stress (Wheeler et al. 2017). A second issue for BioID in particular is that the reactive species generated has a half-life of ~30 sec, which is similar to the exchange rate of many RNP granule proteins (Buchan and Parker 2009). Again, this will reduce the specificity of identifying strictly RNP granule components.

Despite these limitations, two recent studies have used BioID or APEX labeling to provide several new insights into the stress granule proteome (Markmiller et al. 2018; Youn et al. 2018). First, these experiments identified multiple new components interacting with stress granule components, some of which were verified to be localized within stress granules (Markmiller et al. 2018; Youn et al. 2018). Second, by examining stress granule formation in different cell types and under different stresses, it was observed that, although the majority of stress granule components are conserved, some protein components are stress-specific and cell type–specific (Markmiller et al. 2018). Third, by repeated BioID analysis of multiple RNP granule components, subgroups of proteins within stress granules that preferentially interacted with each other were identified (Youn et al. 2018). It will be of interest in future work to determine if these subgroups represent different regions within stress granules, different intermediates on the pathways of stress granule assembly or disassembly, or simply represent specific types of RNPs that can accumulate within stress granules.

Taken together, this compilation of work has led to a rich census of proteins enriched in stress granules. The components encompass a spectrum of protein functions including RNA-binding proteins, translation initiation factors, stress granule remodeling complexes such as protein chaperones and helicases (Jain et al. 2016), as well as unexpected proteins involved in metabolism and signaling (Kedersha et al. 2013). This provides a rich resource for the understanding of stress granule formation and disassembly, function, and regulation.

3 RNA COMPONENTS OF STRESS GRANULES

A second set of experiments has revealed the RNA content of stress granules (Table 1B). The first experiments on mRNAs within stress granules used oligo-dT probes to show that mRNAs were highly enriched in stress granules by in situ hybridization (Kedersha et al. 1999). Based on poly(A)-staining signals, anywhere from 5% to 50% of the total mRNA population was estimated to accumulate in stress granules (Kedersha et al. 1999; Anderson and Kedersha 2008; Mollet et al. 2008; Zurla et al. 2011). Similar to the work with proteins, initial experiments examining specific mRNAs by fluorescence in situ hybridization (FISH) showed that some mRNAs could be enriched in stress granules including 5′ terminal oligopyrimidine motif (5′ TOP) mRNAs (Damgaard and Lykke-Andersen 2011). In contrast, only a minor fraction of some mRNAs was observed to be present in stress granules (e.g., only 3% of β-actin mRNA was observed to be in stress granules during arsenite stress [Zurla et al. 2011]). Similarly, mRNAs encoding heat shock proteins were largely excluded from stress granules induced by heat shock (Kedersha and Anderson 2002). These observations suggested that the mRNA recruited to stress granules was a specific subset of the total mRNA, although the nature of that population remained to be determined.

A more comprehensive analysis of the RNAs in both yeast and mammalian stress granules has come from purifying and sequencing stress granule cores (Khong et al. 2017). Importantly, by standardizing the RNA-seq analysis of stress granule cores to the fraction of a given mRNA accumulating in stress granules as determined by single-molecule FISH (smFISH) (Fig. 1), this work allowed a quantitative description of the stress granule transcriptome. This quantitative analysis revealed that ~10–15% of cytoplasmic mRNA molecules accumulate within stress granules, and that this is a highly diverse population of RNAs with no single RNA accounting for >0.5% of the total stress granule RNA population (Khong et al. 2017). The stress granule transcriptome also revealed that ~85% is mRNAs, with some lncRNAs also accumulating in stress granules (Khong et al. 2017). Similar results have been observed in

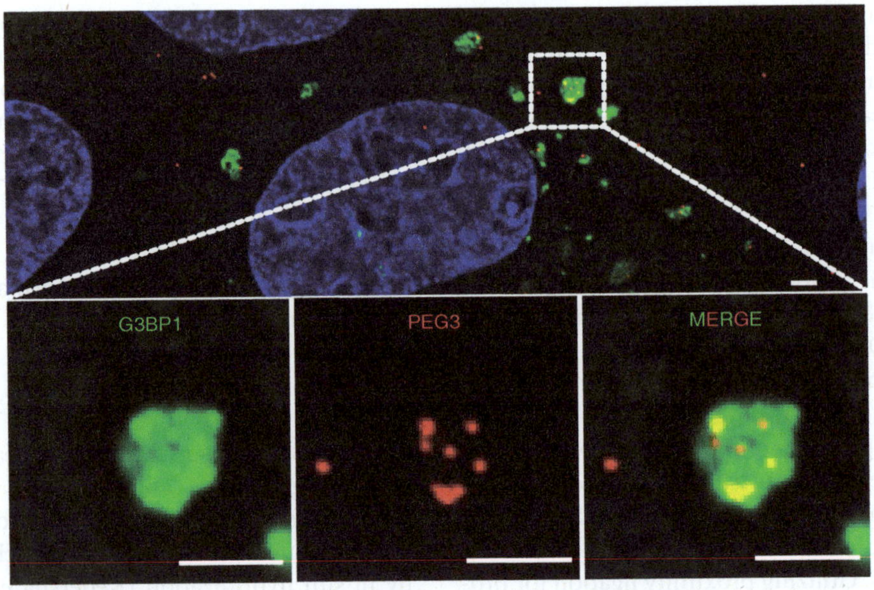

Figure 1. Immunofluorescence (IF) and single-molecule fluorescence in situ hybridization (smFISH) reveal stress granule proteins and RNAs, respectively. IF of G3BP (green) delineates arsenite-induced stress granules. smFISH using probes to PEG3 (red) show robust localization of PEG3 RNAs to stress granules. Scale bar, 2 µm.

the sequencing of mRNAs that pellet in response to stress (Namkoong et al. 2018), which is based on the premise that stress granules are large and pellet at moderate centrifugation. Thus, stress granules represent a diverse mixture of cellular RNAs, with essentially every mRNA being present in stress granules to some degree.

A striking observation was that individual mRNAs could vary from having 1% to 95% of their molecules in stress granules (Khong et al. 2017). Analysis of mRNA parameters revealed that long transcript length and poor translatability were the major determinants of recruitment to stress granules (Khong et al. 2017). In addition, mRNAs associated with the endoplasmic reticulum (ER) or mitochondrial membranes were generally excluded from stress granules, perhaps because they remain tethered to these membranes even after exiting translation. Additional sequence elements that correlate with increased mRNA accumulation in pellets during stress have been identified, such as AU-rich elements, but whether these directly target mRNAs into stress granules will require future experiments (Namkoong et al. 2018).

An emerging possibility is that long RNAs accumulate in stress granules simply because they can form more nonspecific RNA–RNA interactions in *trans* (Van Treeck et al. 2018). This possibility is first suggested by the limited correlation of the binding of stress granule proteins and RNA enrichment in stress granules (Khong et al. 2017). For example, on average, long mRNAs found in stress granules are not enriched in binding sites for stress granule proteins based on eCLIP analysis (Khong et al. 2017). Moreover, the NORAD lncRNA, which contains 19 binding sites for the stress granule component pumilio, accumulates in stress granules independent of pumilio (Namkoong et al. 2018). Evidence that RNA–RNA interactions could provide the basis for RNA enrichment in stress granules is that purified total yeast RNA will self-assemble in vitro under physiological conditions into large assemblies with the same transcript composition seen in the yeast stress granule transcriptome (Van Treeck et al. 2018). Thus, RNA–RNA interactions in *trans* are likely to play a contributing role to stress granule assembly in cells.

4 MICROSCOPIC ANALYSES OF STRESS GRANULES REVEAL INTERNAL ULTRASTRUCTURE

A variety of imaging approaches with higher resolution have been used to reveal that stress granules are irregular assemblies composed of dense "cores" surrounded by regions in which stress granule components are at lower concentrations. This organization was first suggested by examination of stress granules by electron microscopy (EM), which showed stress granules to be an electron-dense material with substructures of higher density (Souquere et al. 2009). Subsequent super-resolution microscopy using both STORM and SIM approaches showed that G3BP1, the poly(A)-binding protein PAB1, and mRNAs as assessed by oligo-dT FISH were all distributed nonuniformly in stress granules with clusters of "cores" surrounded by regions of lower density, referred to collectively as the "shell" (Jain et al. 2016). Refinement of stress granule core images by double-helix point spread function showed that cores have diameters of 175–190 nm on average, with G3BP and poly (A) mRNA concentrated ~30× more in the core than the shell, and proteins concentrated ~10× more in the shell than the bulk cytosol (Jain et al. 2016). Moreover, stress granule cores are stable in lysates and can be purified to identify their protein and RNA components (Jain et al. 2016; Khong et al. 2017). Additional evidence for cores within stress granules has come from superresolution imaging revealing that G3BP1 and IMP1 show restrained dynamics within 150–200-nm regions of the stress granule (Niewidok et al. 2018).

Similar microscopic approaches have revealed that other RNP granules also have an irregular shape with distinct substructures. For example, using lattice light sheet microscopy, the Meg-3 protein in *C. elegans* has been shown to form a dynamic domain that surrounds the bulk of the P-granule (Wang et al. 2014). Similarly, P-bodies in mammalian cells have been shown to have an irregular shape by EM tomography (Cougot et al. 2012) and to have subdomains with some fibrillar nature by EM (Yang 2004; Souquere et al. 2009). An interesting goal of future work will be to determine the nature of these subdomains and the assembly mechanisms that allow their formation and interchange of components.

5 IMAGING OF STRESS GRANULE ASSEMBLY AND DISASSEMBLY IN REAL TIME

The use of fluorescently tagged stress granule proteins and imaging in real time has allowed the determination of the assembly and disassembly pathways of stress granules with multiple distinct steps (Fig. 2). When cells are first exposed to stress, there is a lag period of ~10 min before the appearance of small stress granules (Kedersha et al. 2000; Ohshima

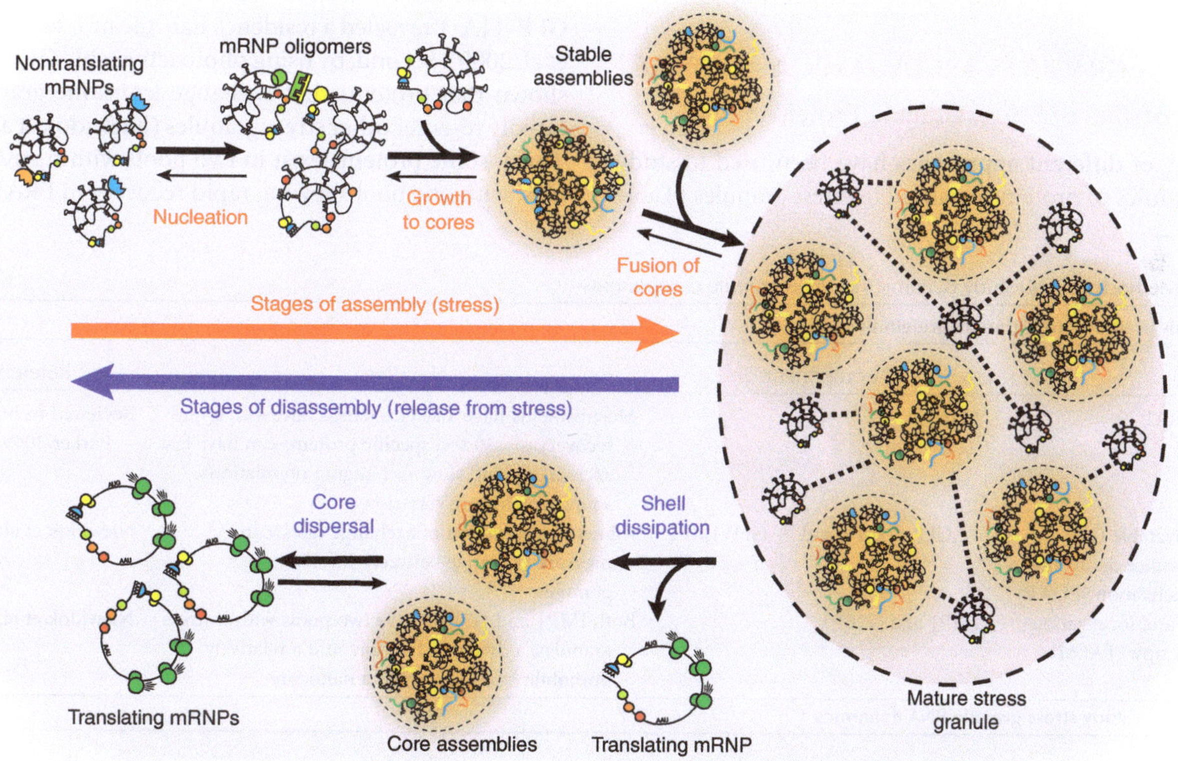

Figure 2. Distinct steps of stress granule assembly and disassembly. Microscopy has revealed several distinct stages of stress granule assembly (orange, *top*) and disassembly (purple, *bottom*). In brief, during stress, translation is inhibited, and ribosomes run off their corresponding messenger ribonucleoproteins (mRNPs). These nontranslating mRNPs then oligomerize through either protein–protein, protein–RNA, or RNA–RNA interactions. As more RNAs enter the nontranslating pool, oligomers grow into stable core assemblies. Through fusion and RNP recruitment, a mature stress granule is produced with clear substructure. On stress release, mRNPs re-enter translation and stress granules begin to disassemble by a loss of the shell followed by core dispersal.

et al. 2015; Wheeler et al. 2016). This lag may represent the time it takes for mRNAs to exit translation by ribosome runoff, as well as additional time-dependent events in the coalescing of nontranslating messenger ribonucleoproteins (mRNPs). The small stress granules that form early in the time course are biochemically stable in cell lysates, which is consistent with the first step in stress granule assembly being the coalescence of a group of nontranslating mRNPs into a stress granule core structure (Wheeler et al. 2016). Subsequently, the small stress granules increase in size both by accretion of new material and by fusion, resulting in a decrease of the total number of stress granules over time (Kedersha et al. 2000; Ohshima et al. 2015; Wheeler et al. 2016). During recovery from stress, stress granules disassemble in a reverse of this process, wherein large stress granules show a decrease in the concentration of components and fracture into smaller stress granules, which then disappear in a final step (Kedersha et al. 2000; Ohshima et al. 2015; Wheeler et al. 2016). An interesting area of current research is examining how perturbations to key stress granule components or remodeling machines affect the different stages of this assembly and disassembly pathway.

6 STRESS GRANULES ARE DYNAMIC ASSEMBLIES

6.1 Dynamics of Stress Granule Proteins

A variety of different approaches have been used to study the dynamics of proteins localized to stress granules (Table 2A). The most common has been fluorescence recovery after photobleaching (FRAP), in which a given component of stress granules is tagged with a fluorescent protein, bleached within a specific granule using a focused laser, and then the recovery of fluorescence is measured as a marker for exchange of bleached components for protein molecules that were originally outside the granule. Inverse FRAP (iFRAP) has also been used wherein proteins outside the granule are bleached and the rate of fluorescence loss from the granule gives a measure of protein exchange. A third approach has been to tag a stress granule protein with photoactivatable GFP, and then activate the fluorescence within an individual stress granule, which allows the experimenter to measure both the exchange of the protein from the activated granule and to determine if that protein transfers to other specific regions or granules within the cell.

Examination of numerous stress granule components by these approaches has revealed several principles of their dynamics. First, many protein components are highly dynamic and exchange with the surrounding cytosol with $t_{1/2}$ exchange rates on the order of 1–30 sec (reviewed in depth in Buchan and Parker 2009). For example, FRAP analysis of GFP-TIA-1 revealed a residence half-life of 2 sec (Kedersha et al. 2000). Second, by using photoactivatable GFP it can be shown that proteins that exchange from one granule can rapidly re-enter other stress granules (Niewidok et al. 2018). Third, some proteins exist in two pools within stress granules, with one pool showing rapid recovery in FRAP exper-

Table 2. Methods used to study dynamics of stress granule components

A. Methods to study stress granule protein dynamics

Method	Component	Finding	Reference
FRAP/iFRAP	Many	Most proteins have a fast exchange rate, with $t_{1/2}$ of recovery of <30 sec; specific proteins can have fast exchanging and slow exchanging populations; some proteins are very slow	Reviewed in Buchan and Parker 2009
Photoactivatable GFP, fluorescence decay after photoactivation (FDAP)	PAGFP-G3BP1 and PAGFP-IMP1	Allows measurement of exchange rates; can measure exchange between different stress granules	Niewidok et al. 2018
Tracking and localization microscopy (TALM)	IMP1 and G3BP1	Both IMP1 and G3BP1 have two pools within stress granules: a dynamic fraction, and a relatively immobile fraction termed a nanocore	Niewidok et al. 2018

B. Methods to study stress granule RNA dynamics

Method	Component	Finding	Reference(s)
FRAP/iFRAP	MS2-GFP tagged β-Gal mRNA; fluorescent poly(U)$_{22}$ for mRNA detection	Mostly mobile, short residence time in stress granules; ~30% of mRNA in stress granules is immobile, mobile fraction has two exchange rates (40 sec and 275 sec)	Mollet et al. 2008; Zhang et al. 2011
Single-molecule RNA tracking	Reporter RNA with MS2 stem loops to visualize	Two interaction types: brief and long-lived; larger stress granules hold RNAs longer; reporter RNAs relatively immobile within granules	Moon et al. 2018

Cite as *Cold Spring Harb Perspect Biol* doi: 10.1101/cshperspect.a033068

iments, and the other failing to recover, indicative of a stably associated fraction of a given protein within stress granules (Buchan and Parker 2009). The difference between the stable and dynamic pool of proteins within stress granules could be the presence of different posttranslational modifications, different exchange rates based on being in the core or shell region of the stress granule, or differences in engaged individual protein–protein interactions.

Although many protein components exchange quickly, some show dramatically slower exchange rates. For example, the stress granule proteins FAST and Pab1 show slower exchange rates from stress granules than that of TIA1 or TTP, with FAST being extremely slow at exchanging (Kedersha et al. 2005). In principle, a slow exchange rate can occur through multiple molecular mechanisms. First, often in large assemblies, an increased number of interactions (e.g., higher valency) can lead to a slower off-rate simply because the off-rate requiring the loss of all interactions (Banani et al. 2016). Alternatively, a slow off-rate can occur if there are multiple binding sites within the stress granule, such that the probability of local rebinding is very high before exchange. Finally, a slow off-rate can also occur by formation of a single very stable interaction. An interesting area of future work will be in examining how different exchange rates of stress granule components are altered by specific protein–protein and/or protein–RNA interactions.

Although poorly understood at this time, exchange rates of proteins in stress granules do not just represent the biochemical on–off rates of proteins with their binding partners within stress granules; we should expect that these dynamics will be influenced by protein modification enzymes and energy-dependent stress granule remodeling complexes. For example, stress granule assembly and disassembly are known to be affected by a number of protein modification enzymes (e.g., Hilliker and Parker 2008), as well as protein chaperones such as the VCP or Hsp70/40 complexes, and RNA helicases, which can enhance or promote stress granule stability (Walters et al. 2015; Jain et al. 2016). Consistent with ATP-driven machines affecting stress granule dynamics, depletion of ATP from cells reduces the dynamics of G3BP1 protein exchange (Jain et al. 2016). Similarly, although the RHAU helicase exchanges from stress granules within 10 sec, a point mutation disrupting ATP hydrolysis by RHAU leads to a much slower exchange rate (Challupníková et al. 2008). This is consistent with the properties of DEAD box helicases, in which ATP hydrolysis generally leads to decreased RNA binding. An important area of future research will be to understand how different stress granule remodeling complexes affect the exchange rates of specific components.

One powerful approach for studying the dynamics of stress granules will be to use single-particle tracking and localization microscopy (TALM) (Appelhans et al. 2012). In this approach, proteins are tagged with protein modules such as the Halo or SNAP tags (Juillerat et al. 2003; Kogure et al. 2006) that can be labeled with exogenously provided chemical dyes. When the chemical dyes are provided at substoichiometric levels the limited number of molecules that are fluorescently labeled allows the experimenter to track the behavior of individual molecules. An initial application of TALM to stress granules illustrates its ability to yield new insights. Specifically, this approach shows that individual molecules of both G3BP1 and IMP1 show prolonged residence times within subregions of stress granules, referred to as nanocores (Niewidok et al. 2018). Such nanocores were measured to be ~150–200 nm in diameter, similar to the size of cores defined by superresolution microscopy (Jain et al. 2016), and provide evidence from living cells for an internal core structure of stress granules. The power of tracking individual proteins also provided evidence for different diffusion rates of the mobile fraction of G3BP1 within stress granules and the cytosol, providing evidence for a different viscosity within stress granules (Niewidok et al. 2018).

The real power of TALM for the analysis of stress granules and other RNP assemblies will come from the analysis of how perturbations of the system affect the behavior of individual molecules. For example, deletion of an amino-terminal dimerization and caprin-interacting domain of G3BP1 led to a higher fraction of G3BP1 molecules in the bound nanocore region, although the molecular basis for this difference remains to be determined (Niewidok et al. 2018). Future experiments applying TALM to individual proteins and mRNAs and determining how their dynamics are altered by various perturbations to stress granule components, modifying enzymes, and stress granule remodeling complexes promises to be a rich area of investigation.

6.2 Dynamics of RNAs in Stress Granules

To date, few experiments have addressed the dynamics of mRNAs within stress granules (Table 2B). Given the large size of mRNAs, we should expect an mRNP within stress granules to have multiple interaction sites with other stress granule mRNPs, either through multivalent mRNA-binding proteins (Protter and Parker 2016), or through direct RNA–RNA interactions between mRNAs (Van Treeck et al. 2018). A high valency of interaction for a specific mRNA within stress granules would be expected to give a slow rate of exchange (Banani et al. 2016). Some observations consistent with this expectation have come from examining the exchange rates of bulk poly(A)+ mRNAs within stress granules by the injection of fluorescent oligo(U) probes into COS7 cells. Specifically, these experiments revealed that

~30% of the mRNAs within stress granules are immobile and exchange only very slowly, if at all (Zhang et al. 2011). Additionally, two exchanging populations of mRNAs were observed. Approximately 35% of the mRNAs exchanged with half-times of ~30 sec, whereas a second population (~30% of mRNAs) exchanged with a half-time of ~275 sec (Zhang et al. 2011). These results show that mRNAs can interact with stress granules in different modes and that overall, the exchange rates of mRNAs from stress granules are significantly slower than those of most stress granule proteins. Multiple modes of mRNA interaction with stress granules were also observed in the FRAP analysis of a β-Gal reporter mRNA tagged with multiple MS2-GFP fusion proteins, although the half-times for exchange, 2 and 58 sec, were much faster (Mollet et al. 2008). Whether these faster rates are caused by the specific mRNA being examined or differences in the experimental setup remain to be determined.

7 PERSPECTIVES AND FUTURE DIRECTIONS

Much progress has been made since the discovery of stress granules approximately 30 years ago. The ultrastructure of stress granules, as well as the protein and RNAs that accumulate in stress granules, are now known. Specific protein–protein and RNA–RNA interactions that contribute to their assembly have been defined and provide an intellectual framework for understanding perturbations that affect stress granule assembly. In addition, a variety of modification enzymes and stress granule remodeling complexes are known to alter assembly and disassembly. Finally, a number of mutations in stress granule components that lead to human disease have been identified.

Many of the outstanding key issues in understanding stress granules can be addressed by imaging approaches. For example, understanding how specific stress granule components are affected by alterations in stress granule remodeling complexes can be achieved using TALM and chemical or genetic inhibition of stress granule remodeling complexes. Similarly, understanding the dynamics of mRNA interaction with stress granules, and how that interaction is modulated, will be achieved with live single-molecule imaging of mRNAs and stress granules over time. Furthermore, understanding how pathogenic mutations affect the behavior of key components of stress granules may yield insight into how perturbations of stress granules can influence human disease.

Finally, it should be noted that all of the approaches discussed in this review are applicable to other RNP granules, and can be used to understand their composition, ultrastructure, dynamics, and regulation. All of these areas promise rich investigation in the near future.

REFERENCES

Aguilera-Gomez A, Rabouille C. 2017. Membrane-bound organelles versus membrane-less compartments and their control of anabolic pathways in *Drosophila*. *Dev Biol* **428:** 310–317.

Anderson P, Kedersha N. 2008. Stress granules: The Tao of RNA triage. *Trends Biochem Sci* **33:** 141–150.

Anderson P, Kedersha N, Ivanov P. 2015. Stress granules, P-bodies and cancer. *Biochimica et Biophysica Acta* **1849:** 861–870.

Antar LN, Dictenberg JB, Plociniak M, Afroz R, Bassell GJ. 2005. Localization of FMRP-associated mRNA granules and requirement of microtubules for activity-dependent trafficking in hippocampal neurons. *Genes Brain Behav* **4:** 350–359.

Appelhans T, Richter CP, Wilkens V, Hess ST, Piehler J, Busch KB. 2012. Nanoscale organization of mitochondrial microcompartments revealed by combined tracking and localization microscopy. *Nano Lett* **12:** 610–616.

Aulas A, Fay MM, Lyons SM, Achorn CA, Kedersha N, Anderson P, Ivanov P. 2017. Stress-specific differences in assembly and composition of stress granules and related foci. *J Cell Sci* **130:** 927–937.

Baguet A, Degot S, Cougot N, Bertrand E, Chenard MP, Wendling C, Kessler P, Le Hir H, Rio MC, Tomasetto C. 2007. The exon-junction-complex-component metastatic lymph node 51 functions in stress-granule assembly. *J Cell Sci* **120:** 2774–2784.

Banani SF, Rice AM, Peeples WB, Lin Y, Jain S, Parker R, Rosen MK. 2016. Compositional control of phase-separated cellular bodies. *Cell* **166:** 651–663.

Barbee SA, Estes PS, Cziko A-M, Hillebrand J, Luedeman RA, Coller JM, Johnson N, Howlett IC, Geng C, Ueda R, et al. 2006. Staufen- and FMRP-containing neuronal RNPs are structurally and functionally related to somatic P bodies. *Neuron* **52:** 997–1009.

Bakthavachalu B, Huelsmeier J, Sudhakaran IP, Hillebrand J, Singh A, Petrauskas A, Thiagarajan D, Sankaranarayanan M, Mizoue L, Anderson EN, et al. 2018. RNP-granule assembly via Ataxin-2 disordered domains is required for long-term memory and neurodegeneration. *Neuron* **98:** 754–766.

Bond CS, Fox AH. 2009. Paraspeckles: Nuclear bodies built on long noncoding RNA. *J Cell Biol* **186:** 637–644.

Buchan JR. 2014. mRNP granules: Assembly, function, and connections with disease. *RNA Biol* **11:** 1–12.

Buchan JR, Parker R. 2009. Eukaryotic stress granules: The ins and outs of translation. *Mol Cell* **36:** 932–941.

Buchan JR, Muhlrad D, Parker R. 2008. P bodies promote stress granule assembly in Saccharomyces cerevisiae. *J Cell Biol* **183:** 441–455.

Buchan JR, Yoon JH, Parker R. 2010. Stress-specific composition, assembly and kinetics of stress granules in *Saccharomyces cerevisiae*. *J Cell Sci* **124:** 228–239.

Buchan JR, Kolaitis R-M, Taylor JP, Parker R. 2013. Eukaryotic stress granules are cleared by autophagy and Cdc48/VCP function. *Cell* **153:** 1461–1474.

Cajal SR. 1903. Un sencillo metodo de coloracion selectiva del reticulo protoplasmico y sus efectos en lo diversos organos nerviosos de vertebrados e invertebrados. *Trab Lab Investig Biol Univ Madr* **2:** 129–221.

Challupníková K, Lattmann S, Selak N, Iwamoto F, Fujiki Y, Nagamine Y. 2008. Recruitment of the RNA helicase RHAU to stress granules via a unique RNA-binding domain. *J Biol Chem* **283:** 35186–35198.

Cherkasov V, Grousl T, Theer P, Vainshtein Y, Gläßer C, Mongis C, Kramer G, Stoecklin G, Knop M, Mogk A, et al. 2015. Systemic control of protein synthesis through sequestration of translation and ribosome biogenesis factors during severe heat stress. *FEBS Lett* **589:** 3654–3664.

Collier NC, Schlesinger MJ. 1986. The dynamic state of heat shock proteins in chicken embryo fibroblasts. *J Cell Biol* **103:** 1495–1507.

Cougot N, Cavalier A, Thomas D, Gillet R. 2012. The dual organization of P-bodies revealed by immunoelectron microscopy and electron tomography. *J Mol Biol* **420:** 17–28.

Damgaard CK, Lykke-Andersen J. 2011. Translational coregulation of 5′TOP mRNAs by TIA-1 and TIAR. *Genes Dev* **25:** 2057–2068.

Eisinger-Mathason TSK, Andrade J, Groehler AL, Clark DE, Muratore-Schroeder TL, Pasic L, Smith JA, Shavanowitz J, Hunt DF, Macara IG, et al. 2008. Codependent functions of RSK2 and the apoptosis-promoting factor TIA-1 in stress granule assembly and cell survival. *Mol Cell* **31:** 722–736.

Fox AH, Lam YW, Leung AKL, Lyon CE, Andersen J, Mann M, Lamond AI. 2002. Paraspeckles: A novel nuclear domain. *Curr Biol* **12:** 13–25.

Fox AH, Nakagawa S, Hirose T, Bond CS. 2018. Paraspeckles: Where long noncoding RNA meets phase separation. *Trends Biochem Sci* **43:** 124–135.

Gall JG. 2000. Cajal bodies: The first 100 years. *Annu Rev Cell Dev Biol* **16:** 273–300.

Grabocka E, Bar-Sagi D. 2016. Mutant KRAS enhances tumor cell fitness by upregulating stress granules. *Cell* **167:** 1803–1806.

Guo W, Chen Y, Zhou X, Kar A, Ray P, Chen X, Rao EJ, Yang M, Ye H, Zhu L, et al. 2011. An ALS-associated mutation affecting TDP-43 enhances protein aggregation, fibril formation and neurotoxicity. *Nat Struct Mol Biol* **18:** 822–830.

Hilliker A, Parker R. 2008. Stressed out? Make some modifications! *Nat Cell Biol* **10:** 1129–1130.

Hofmann S, Cherkasova V, Bankhead P, Bukau B, Stoecklin G. 2012. Translation suppression promotes stress granule formation and cell survival in response to cold shock. *Mol Biol Cell* **23:** 3775–3924.

Hubstenberger A, Courel M, Bénard M, Souquere S, Ernoult-Lange M, Chouaib R, Yi Z, Morlot J-B, Munier A, Fradet M, et al. 2017. P-body purification reveals the condensation of repressed mRNA regulons. *Mol Cell* **68:** 1–14.

Jain S, Parker R. 2013. The discovery and analysis of P bodies. *Adv Exp Med Biol* **768:** 23–44.

Jain A, Vale RD. 2017. RNA phase transitions in repeat expansion disorders. *Nature* **546:** 243–247.

Jain S, Wheeler JR, Walters RW, Agrawal A, Barsic A, Parker R. 2016. ATPase-modulated stress granules contain a diverse proteome and substructure. *Cell* **164:** 1–12.

Juillerat A, Gronemeyer T, Keppler A, Gendreizig S, Pick H, Vogel H, Johnsson K. 2003. Directed evolution of O6-alkylguanine-DNA alkyltransferase for efficient labeling of fusion proteins with small molecules in vivo. *Chem Biol* **10:** 313–317.

Kaiser TE, Intine RV, Dundr M. 2008. De novo formation of a subnuclear body. *Science* **322:** 1713–1717.

Kedersha N, Anderson P. 2002. Stress granules: Sites of mRNA triage that regulate mRNA stability and translatability. *Biochem Soc Trans* **30:** 963–969.

Kedersha NL, Gupta M, Li W, Miller I, Anderson P. 1999. RNA-binding proteins TIA-1 and TIAR link the phosphorylation of eIF-2α to the assembly of mammalian stress granules. *J Cell Biol* **147:** 1431–1441.

Kedersha N, Cho MR, Li W, Yacono PW, Chen S, Gilks N, Golan DE, Anderson P. 2000. Dynamic shuttling of Tia-1 accompanies the recruitment of mRNA to mammalian stress granules. *J Cell Biol* **151:** 1257–1268.

Kedersha N, Chen S, Gilks N, Li W, Miller IJ, Stahl J, Anderson P. 2002. Evidence that ternary complex (eIF2–GTP–tRNA$_i^{Met}$)-deficient preinitiation complexes are core constituents of mammalian stress granules. *Mol Biol Cell* **13:** 195–210.

Kedersha N, Stoecklin G, Ayodele M, Yacono P, Lykke-Andersen J, Fritzler MJ, Scheuner D, Kaufman RJ, Golan DE, Anderson P. 2005. Stress granules and processing bodies are dynamically linked sites of mRNP remodeling. *J Cell Biol* **169:** 871–884.

Kedersha N, Ivanov P, Anderson P. 2013. Stress granules and cell signaling: More than just a passing phase? *Trends Biochem Sci* **38:** 494–506.

Kedersha N, Panas MD, Achorn CA, Lyons S, Tisdale S, Hickman T, Thomas M, Lieberman J, McInerney GM, Ivanov P, et al. 2016. G3BP–Caprin1–USP10 complexes mediate stress granule condensation and associate with 40S subunits. *J Cell Biol* **212:** 845–860.

Khong A, Matheny T, Jain S, Mitchell SF, Wheeler JR, Parker R. 2017. The stress granule transcriptome reveals principles of mRNA accumulation in stress granules. *Mol Cell* **68:** 808–820.

Kiebler MA, Bassell GJ. 2006. Neuronal RNA granules: Movers and makers. *Neuron* **51:** 685–690.

Kim HJ, Kim NC, Wang Y-D, Scarborough EA, Moore J, Diaz Z, MacLea KS, Freibaum B, Li S, Molliex A, et al. 2013. Mutations in prion-like domains in hnRNPA2B1 and hnRNPA1 cause multisystem proteinopathy and ALS. *Nature* **495:** 467–473.

Kogure T, Karasawa S, Araki T, Saito K, Kinjo M, Miyawaki A. 2006. A fluorescent variant of a protein from the stony coral *Montipora* facilitates dual-color single-laser fluorescence cross-correlation spectroscopy. *Nat Biotechnol* **24:** 577–581.

Li YR, King OD, Shorter J, Gitler AD. 2013. Stress granules as crucibles of ALS pathogenesis. *J Cell Biol* **201:** 361–372.

Markmiller S, Soltanieh S, Server KL, Mak R, Jin W, Fang MY, Luo E-C, Krach F, Yang D, Sen A, et al. 2018. Context-dependent and disease-specific diversity in protein interactions within stress granules. *Cell* **172:** 590–598.e13.

Mazroui R. 2002. Trapping of messenger RNA by fragile X mental retardation protein into cytoplasmic granules induces translation repression. *Hum Mol Gen* **11:** 3007–3017.

McCann C, Holohan EE, Das S, Dervan A, Larkin A, Lee JA, Rodrigues V, Parker R, Ramaswami M. 2011. The ataxin-2 protein is required for microRNA function and synapse-specific long-term olfactory habituation. *Proc Natl Acad Sci* **108:** E655–E662.

McCormick C, Khaperskyy DA. 2017. Translation inhibition and stress granules in the antiviral immune response. *Nat Rev Immunol* **17:** 647–660.

Mollet S, Cougot N, Wilczynska A, Dautry F, Kress M, Bertrand E, Weil D. 2008. Translationally repressed mRNA transiently cycles through stress granules during stress. *Mol Biol Cell* **19:** 4469–4479.

Moon SL, Morisaki T, Khong A, Lyon K, Parker R, Stasevich TJ. 2018. Imaging of single mRNA translation repression reveals diverse interactions with mRNP granules. *bioRxiv* doi: 10.1101/332692.

Namkoong S, Ho A, Woo YM, Kwak H, Lee JH. 2018. Systematic characterization of stress-induced RNA granulation. *Mol Cell* **70:** 1–13.

Németh A, Grummt I. 2018. Dynamic regulation of nucleolar architecture. *Curr Opin Cell Biol* **52:** 105–111.

Niewidok B, Igaev M, Pereira da Graca A, Strassner A, Lenzen C, Richter CP, Piehler J, Kurre R, Brandt R. 2018. Single-molecule imaging reveals dynamic biphasic partition of RNA-binding proteins in stress granules. *J Cell Biol* **217:** 1303–1318.

Nonhoff U, Ralser M, Welzel F, Piccini I, Balzereit D, Yaspo M-L, Lehrach H, Krobitsch S. 2007. Ataxin-2 interacts with the DEAD/H-Box RNA helicase DDX6 and interferes with P-bodies and stress granules. *Mol Biol Cell* **18:** 1385–1396.

Ohn T, Kedersha N, Hickman T, Tisdale S, Anderson P. 2008. A functional RNAi screen links O-GlcNAc modification of ribosomal proteins to stress granule and processing body assembly. *Nat Cell Biol* **10:** 1224–1231.

Ohshima D, Arimoto-Matsuzaki K, Tomida T, Takekawa M, Ichikawa K. 2015. Spatio-temporal dynamics and mechanisms of stress granule assembly. *PLoS Comput Biol* **11:** 1–22.

Protter DSW, Parker R. 2016. Principles and properties of stress granules. *Trends Cell Biol* **26:** 668–679.

Ramaswami M, Taylor JP, Parker R. 2013. Altered ribostasis: RNA–protein granules in degenerative disorders. *Cell* **154:** 727–736.

Rhee HW, Zou P, Udeshi ND, Martell JD, Mootha VK, Carr SA, Ting AY. 2013. Proteomic mapping of mitochondria in living cells via spatially restricted enzymatic tagging. *Science* **339:** 1328–1331.

Roux KJ, Kim DI, Raida M, Burke B. 2012. A promiscuous biotin ligase fusion protein identifies proximal and interacting proteins in mammalian cells. *J Cell Biol* **196:** 801–810.

Scheer U, Thiry M, Goessens G. 1993. Structure, function and assembly of the nucleolus. *Trends Cell Biol* **3:** 236–241.

Schwartz JC, Cech TR, Parker RR. 2015. Biochemical properties and biological functions of FET proteins. *Annu Rev Biochem* **84:** 355–379.

Sheth U, Parker R. 2003. Decapping and decay of messenger RNA occur in cytoplasmic processing bodies. *Science* **300:** 805–808.

Somasekharan SP, El-Naggar A, Leprivier G, Cheng H, Hajee S, Grunewald TGP, Zhang F, Ng T, Delattre O, Evdokimova V, et al. 2015. YB-1 regulates stress granule formation and tumor progression by translationally activating G3BP1. *J Cell Biol* **208:** 913–929.

Souquere S, Mollet S, Kress M, Dautry F, Pierron G, Weil D. 2009. Unravelling the ultrastructure of stress granules and associated P-bodies in human cells. *J Cell Sci* **122:** 3619–3626.

Staněk D, Neugebauer KM. 2004. Detection of snRNP assembly intermediates in Cajal bodies by fluorescence resonance energy transfer. *J Cell Biol* **166:** 1015–1025.

Strome S, Updike D. 2015. Specifying and protecting germ cell fate. *Nat Rev Mol Cell Biol* **16:** 406–416.

Strzelecka M, Trowitzsch S, Weber G, Lührmann R, Oates AC, Neugebauer KM. 2010. Coilin-dependent snRNP assembly is essential for zebrafish embryogenesis. *Nat Struct Mol Biol* **17:** 403–409.

Tourrière H, Chebli K, Zekri L, Courselaud B, Blanchard JM, Bertrand E, Tazi J. 2003. The RasGAP-associated endoribonuclease G3BP assembles stress granules. *J Cell Biol* **160:** 823–831.

Van Treeck B, Protter DSW, Matheny T, Khong A, Link CD, Parker R. 2018. RNA self-assembly contributes to stress granule formation and defining the stress granule transcriptome. *Proc Natl Acad Sci* **115:** 2734–2739.

Voronina E, Paix A, Seydoux G. 2012. The P granule component PGL-1 promotes the localization and silencing activity of the PUF protein FBF-2 in germline stem cells. *Development* **139:** 3732–3740.

Walters RW, Muhlrad D, Garcia J, Parker R. 2015. Differential effects of Ydj1 and Sis1 on Hsp70-mediated clearance of stress granules in *Saccharomyces cerevisiae*. *RNA* **21:** 1660–1671.

Wang JT, Smith J, Chen B-C, Schmidt H, Rasoloson D, Paix A, Lambrus BG, Calidas D, Betzig E, Seydoux G. 2014. Regulation of RNA granule dynamics by phosphorylation of serine-rich, intrinsically disordered proteins in *C. elegans*. *eLife* **3:** e04591.

Wheeler JR, Matheny T, Jain S, Abrisch R, Parker R. 2016. Distinct stages in stress granule assembly and disassembly. *eLife* **5:** e18413.

Wheeler JR, Jain S, Khong A, Parker R. 2017. Isolation of yeast and mammalian stress granule cores. *Methods* **126:** 12–17.

Wilczynska A. 2005. The translational regulator CPEB1 provides a link between dcp1 bodies and stress granules. *J Cell Sci* **118:** 981–992.

Wojciechowska M, Krzyzosiak WJ. 2011. Cellular toxicity of expanded RNA repeats: Focus on RNA foci. *Hum Mol Genet* **20:** 3811–3821.

Yang Z, Jakymiw A, Wood MR, Eystathioy T, Rubin RL, Fritzler MJ, Chan EKL. 2004. GW182 is critical for the stability of GW bodies expressed during the cell cycle and cell proliferation. *J Cell Sci* **117:** 5567–5578.

Yang X, Shen Y, Garre E, Hao X, Krumlinde D, Cvijović M, Arens C, Nyström T, Liu B, Sunnerhagen P. 2014. Stress granule-defective mutants deregulate stress responsive transcripts. *PLoS Genet* **10:** e1004763.

Youn J-Y, Dunham WH, Hong SJ, Knight JDR, Bashkurov M, Chen GI, Bagci H, Rathod B, MacLeod G, Eng SWM, et al. 2018. High-density proximity mapping reveals the subcellular organization of mRNA-associated granules and bodies. *Mol Cell* **69:** 517–532.e11.

Zhang J, Okabe K, Tani T, Funatsu T. 2011. Dynamic association-dissociation and harboring of endogenous mRNAs in stress granules. *J Cell Sci* **124:** 4087–4095.

Zurla C, Lifland AW, Santangelo PJ. 2011. Characterizing mRNA interactions with RNA granules during translation initiation inhibition. *PLoS ONE* **6:** e19727.

Nascent RNA and the Coordination of Splicing with Transcription

Karla M. Neugebauer

Molecular Biophysics and Biochemistry, Yale University, New Haven, Connecticut 06520

Correspondence: karla.neugebauer@yale.edu

SUMMARY

At each active protein-encoding gene, nascent RNA is tethered to the DNA axis by elongating RNA polymerase II (Pol II) and is continuously altered by splicing and other processing events during its synthesis. This review discusses the development of three major methods that enable us to track the conversion of precursor messenger RNA (pre-mRNA) to messenger RNA (mRNA) products in vivo: live-cell imaging, metabolic labeling of RNA, and RNA-seq of purified nascent RNA. These approaches are complementary, addressing distinct issues of transcription rates and intron lifetimes alongside spatial information regarding the gene position of Pol II at which spliceosomes act. The findings will be placed in the context of active transcription units, each of which—because of the presence of nascent RNA, Pol II, and features of the chromatin environment—will recruit a potentially gene-specific constellation of RNA binding proteins and processing machineries.

Outline

Cite as *Cold Spring Harb Perspect Biol* doi: 10.1101/cshperspect.a032227

1 INTRODUCTION

Gene expression begins with the synthesis of RNA from the DNA template by RNA polymerase. The steady state level of any RNA in the cell reflects the balance between its synthesis and degradation by specific enzymes; yet, many reactions take place between these two end points. One implicit goal of next-generation RNA sequencing (RNA-seq) is to characterize all possible RNAs expressed in cells from their 5′ ends—indicative of transcription start sites (TSSs)—to their 3′ ends—indicative of 3′-end formation. RNA-seq also reveals massive variation throughout transcript bodies. In the case of eukaryotic messenger RNA (mRNA), this variation is the result of splicing, editing, nucleotide modification, and 3′-end cleavage and polyadenylation, each of which is subject to regulation. A second goal of RNA-seq is to quantify RNA levels in cells. Standard analysis of mRNA reports on steady state levels of the products of all of these reactions and can generate hypotheses about what may have occurred during an mRNA's lifetime. Because each step of mRNA processing at least begins co-transcriptionally (Fig. 1), sequencing of nascent RNA provides evidence of the precursors, intermediates, and products. In the past decade, these next-generation sequencing applications have complemented live-cell imaging approaches to provide key insights into the coordination of precursor messenger RNA (pre-mRNA) processing with transcription.

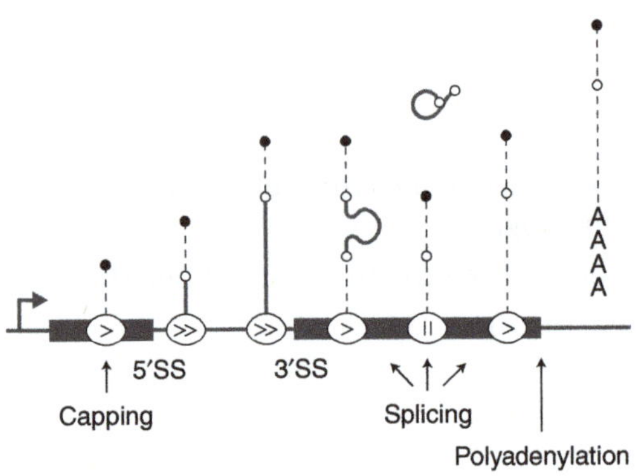

Figure 1. Co-transcriptional RNA processing of eukaryotic protein-coding genes. Schematic diagram depicting a eukaryotic gene with its transcription start site (forward arrow), two annotated exons (thick boxes), and one intron (line). The 5′ end of the annotated intron, or 5′ splice site, and the 3′ end, the 3′ splice site, are marked 5′SS and 3′SS, respectively. RNA polymerase II (Pol II) (oval) moves from *left* to *right* with varying speeds indicated by the forward arrowheads, including pausing marked with a pause symbol. The nascent RNA (exons depicted with dashed lines and intron with solid line) emerges from Pol II as elongation proceeds. The nascent RNA 5′ end immediately receives a 7-methylguanosine cap (black ball), and cleavage at the polyadenylation site releases the nascent RNA from Pol II. Many splicing events occur after capping and before poly(A) cleavage. The 5′SS and 3′SS are depicted in the nascent RNA as open circles, and spliceosome assembly is indicated by looping out of the intron. The intron lariat and spliced nascent RNA products are shown.

What proportion of introns are removed concurrently with transcription? Global studies have approached this question by purifying nascent RNA from a chromatin fraction and preparing RNA-seq libraries or probes for high-density tiling arrays. Co-transcriptional splicing frequencies were similarly high in budding yeast (75%), fly (83%), and human (74%–85%) cell lines and tissues, although a lower proportion (45%) was detected by analysis of mouse liver nascent RNA (Carrillo Oesterreich et al. 2010; Ameur et al. 2011; Khodor et al. 2011; Khodor et al. 2012; Tilgner et al. 2012). Validation of RNA-seq and array data by quantitative reverse transcription polymerase chain reaction (RT-qPCR) strengthened and extended these findings (Carrillo Oesterreich et al. 2010; Ameur et al. 2011). General agreement with these numbers came from metabolic labeling (Windhager et al. 2012) and protein biochemistry and immunofluorescence, showing that most active spliceosomes are associated with chromatin (Girard et al. 2012). That said, not all intron removal is co-transcriptional. In particular, terminal introns are least well removed co-transcriptionally, in agreement with RT-PCR studies of individual endogenous genes (Schmidt et al. 2011; Tilgner et al. 2012). Furthermore, 20% of activated spliceosomes in the cell are not chromatin-associated (Girard et al. 2012), and constitutive splicing appears to be more co-transcriptional than alternative splicing (Ameur et al. 2011; Khodor et al. 2011; Tilgner et al. 2012). Moreover, particular cell types use intron retention or detention as a regulatory step (Pandya-Jones et al. 2013; Wong et al. 2013; Braunschweig et al. 2014; Boutz et al. 2015; Pimentel et al. 2016). Taken together, global studies indicate that most introns are removed co-transcriptionally despite differences in species, cell types, and analysis methods (Brugiolo et al. 2013).

Co-transcriptional pre-mRNA processing events can be coupled physically or temporally to one another, to chromatin, and/or to RNA polymerase II (Pol II). Early evidence that transcription-splicing coupling is an important determinant of gene expression came from the analysis of mRNA products in response to changes in coupling. For example, the 5′-end capping enzymes physically associate with Pol II carboxy-terminal domain (CTD) heptad repeats that are phosphorylated on serine 5 and are allosterically stimulated by this interaction (Cho et al. 1997; McCracken et al. 1997a; Cho et al. 1998). When the CTD is truncated, capping fails to occur and mRNA becomes degraded (McCracken et al. 1997b). It is currently unclear how the exosome gains access to unspliced RNA and whether some proportion of RNA escapes capping. An example of "kinetic" coupling between splicing and transcription is the observation that global patterns of alternative splicing are dependent on Pol II elongation rates, which are slowed or accelerated on ultraviolet (UV) damage, chromatin modification, and direct mutation of Pol II (Munoz et al. 2009; Luco et al. 2011; Hnilicova et al. 2013; Schor et al. 2013; Dujardin et al. 2014; Fong et al. 2014). In considering how splicing and transcription are coordinated with one another, an obvious goal would be to determine in vivo transcription rates and, independently, in vivo splicing rates. Quantifying the rate of co-transcriptional splicing and the relative position of Pol II when splicing occurs is discussed in the following sections. These splicing rates will inevitably be confounded by the transcription rates.

Pol II transcription elongation rates have been calculated from metabolic labeling data, timed chromatin immunoprecipitation (ChIP) experiments, or fluorescence imaging and are available for specific genes or gene regions. Globally, mammalian Pol II transcribes 1–4.5 kb/min, with faster elongation rates within introns (Fig. 1) (Singh and Padgett 2009; Jonkers et al. 2014; Veloso et al. 2014). Complementary to these approaches, the distribution of Pol II along genes can be determined at relatively low resolution by ChIP or with strand specificity and higher resolution through UV cross-linking of Pol II to nascent RNA or sequencing the 3′ ends of nascent RNA (Churchman and Weissman 2011; Mayer et al. 2015; Nojima et al. 2015; Carrillo Oesterreich et al. 2016; Mayer and Churchman 2016; Milligan et al. 2016; Nojima et al. 2016). Changes in Pol II density inform on relative local rates of elongation. Accumulation of Pol II in gene regions, such as TSSs, transcription termination sites, splice sites, and/or over exons indicates that Pol II can pause locally when capping, splicing, and poly(A) cleavage occur (Harlen and Churchman 2017; Herzel et al. 2017; Mayer et al. 2017). Thus, one cannot assume that transcription rates are constant when measuring splicing. The following sections detail how the field has explored splicing kinetics in vivo using methods that can be divided into two categories: distance-based, which measure the gene location of Pol II when splicing has occurred, and time-based, which measure the time taken for splicing to be completed (Fig. 2).

2 QUANTIFICATION OF SPLICING RELATIVE TO TRANSCRIPTION USING CHROMATIN-BASED ASSAYS

Co-transcriptional splicing was initially discovered by analyzing electron micrographs of chromatin spreads from *Drosophila melanogaster*, in which shortening of the nascent RNA and concomitant loss of the attached particles provided evidence for co-transcriptional splicing (Osheim et al. 1985). The DNA axis, variable lengths of attached nascent RNA, and electron-dense ribonucleoprotein (RNP) particles attached to the nascent RNA were visualized in these striking images. RNP particles 25 and 40 nm in diameter were separated by an RNA loop and appeared at predictable sites along nascent RNA, suggesting they are components of the assembling spliceosome. These studies measured DNA distance in µm to deduce that splicing occurred when Pol II had traveled 4.5 kb past the 3′SS (Fig. 2) (Beyer and Osheim 1988). Many of the observations made using chromosome spreading methods are consistent with findings in the Balbiani ring of *Chironomus tentans,* where it has been possible to dissect out chromosomal regions and infer co-transcriptional splicing by RT-PCR and other methods (Bauren and Wieslander 1994; Bjork and Wieslander 2015). Although the harsh conditions of chromatin spreading unfold nascent RNA for analysis, electron tomography images of nascent RNPs in the Balbiani ring system showed the U2 small-nuclear ribonucleoprotein (snRNP) and Pol II in the same electron-dense particle adjacent to the DNA axis (Wetterberg et al. 2001). These early observations of co-transcriptional splicing in model systems raised the possibility that the physical proximity of Pol II, chromatin, the splicing machinery, and the nascent RNP provide opportunities for regulation. There is currently no existing in vitro method that recapitulates splicing in the context of transcription of chromatin, making in vivo approaches obligatory.

2.1 ChIP-Based Detection of Spliceosome Assembly and Splicing within Gene Bodies

Given the proximity of nascent RNA to the DNA axis and chromatin, it was natural to extend ChIP protocols to investigate whether and how spliceosomes assemble co-transcriptionally on individual genes. The key concept is that if RNA binding proteins and higher-order complexes, such as assembling spliceosomes, are present on nascent RNA, it should be possible to preserve interactions with formaldehyde cross-linking and reveal the regions of the DNA template where nascent RNPs contain factors of interest (Bieberstein et al. 2014). The cap binding complex (CBC) and mRNA export factors were among the first to be analyzed. Capping factors and CBC accumulated in promoter-proximal regions (Zenklusen et al. 2002; Listerman et al. 2006; Glover-Cutter et al. 2008), as predicted by the demonstration that nascent RNA is only ~20 nt long when the 7-methylguanosine cap is added (Rasmussen and Lis 1993; Martinez-Rucobo et al. 2015). In contrast, mRNA export factors were detected in downstream gene regions where nascent messenger ribonucleoproteins (mRNPs) might be expected to mature for export (Lei et al. 2001; Zenklusen et al. 2002), and 3′-end processing factors similarly were

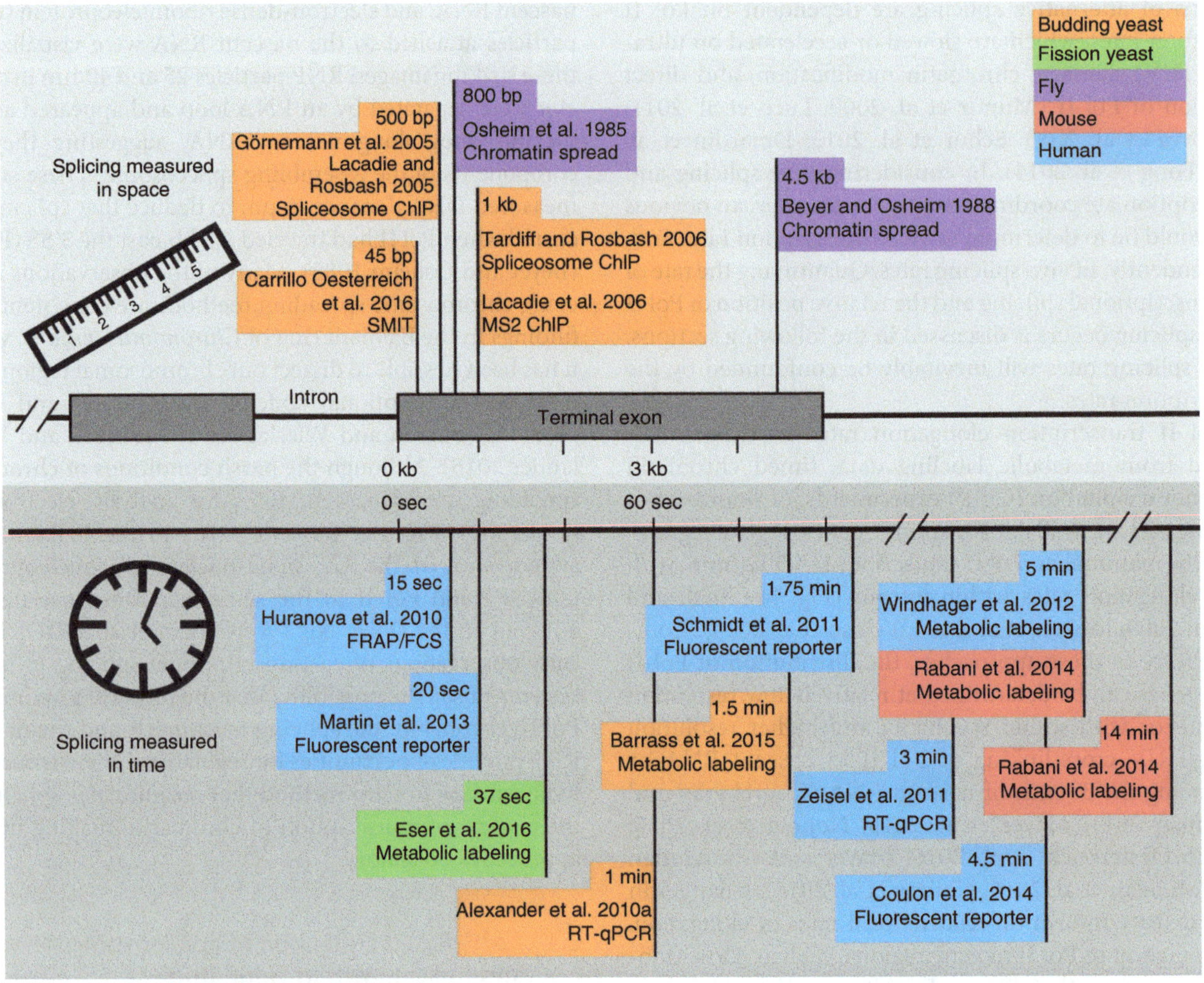

Figure 2. Precursor messenger RNA (pre-mRNA) splicing kinetics in living cells can be measured in space or time. *Upper* panel: Using chromatin-based assays—chromatin spreads, chromatin immunoprecipitation (ChIP), or nascent RNA-seq—the removal of introns can be associated with a gene position, representing the distance RNA polymerase II (Pol II) has traveled by the time the spliceosome has assembled. *Lower* panel: Live-cell fluorescence imaging of model genes and metabolic labeling followed by RNA-seq report on splicing rates in units of time. Note that co-transcriptional splicing kinetics vary among species, cell types, genes examined, and methods used for analysis. (Modified from Alpert et al. 2017.)

detected by ChIP associated with downstream gene regions extending past the poly A cleavage site (Glover-Cutter et al. 2008). Thus, ChIP is capable of identifying molecular components of RNA processing machineries previously inferred from chromatin spreads.

Application of ChIP to spliceosome assembly in yeast revealed that the spliceosomal snRNPs accumulate in distinct patterns along gene regions, consistent with stepwise assembly (Kotovic et al. 2003; Görnemann et al. 2005; Lacadie and Rosbash 2005; Tardiff et al. 2006; Tardiff and Rosbash 2006; Harlen et al. 2016): The U1 snRNP, which recognizes the 5′ splice site (5′SS), was detected over introns, consistent with its recruitment to 5′SSs as soon as they are transcribed. Accumulation of U2 and U5 snRNPs as well as active spliceosomal components downstream from 3′ splice sites (3′SSs) was accompanied by the loss of U1 snRNP, as expected from in vitro biochemical analyses (Wahl 2009). The findings also indicate that U1 snRNP and CBC independently promote later steps of spliceosome assembly, as observed in vitro. In yeast, a ChIP assay was also used to detect the occurrence of splicing relative to transcription of a reporter gene, HZ18, in which the formation of MS2 stem loops either in the intron or after exon–exon ligation would be bound by MS2-binding protein (Lacadie et al. 2006). The MS2 ChIP signal 1.5 kb downstream from the intron suggested that splicing takes place long after intron synthesis in HZ18 (Fig. 2). It appears that splicing delays are present in HZ18, given later findings that indicate more rapid splicing in yeast (see below).

The ChIP approach also revealed co-transcriptional recruitment of spliceosomal snRNPs to mammalian genes, although mapping spliceosome assembly along gene lengths has been less clear-cut in metazoans at least in part because of more complex gene architectures (Listerman et al. 2006; Pabis et al. 2013). Intriguingly, histone posttranslational modifications and the variant histone H2Az modify co-transcriptional spliceosome assembly patterns in yeast, indicating that regulation can be mediated by chromatin as well as by more traditional *trans*-acting splicing factors (Kress et al. 2008; Gunderson and Johnson 2009; Gunderson et al. 2011; Moehle et al. 2012; Herissant et al. 2014; Patrick et al. 2015; Neves et al. 2017; Nissen et al. 2017). Roles for specific histone posttranslational modifications (e.g., H3K4me3 and H3K36me3) in promoting splicing and responding to splicing activity have also been proposed (Sims et al. 2007; de Almeida et al. 2011; Hnilicova et al. 2011; Bieberstein et al. 2012; Kfir et al. 2015). ChIP and UV cross-linking followed by immunopurification of RNA (CLIP) of members of the SR protein family of splicing factors revealed recruitment to actively transcribed intron-containing genes, consistent with their roles in co-transcriptional splicing (Sapra et al. 2009; Brugiolo et al. 2017). In addition, SR proteins were detectable by ChIP at gene promoters because of the association of SR proteins with the regulator of transcription initiation, 7SK (Ji et al. 2013). Thus, ChIP can reveal splicing-related events in the in vivo context of chromatin; yet, it also carries the caveat that other, splicing-unrelated processes are being measured at the same time.

2.2 Single-Molecule Nascent RNA-seq Relates Splicing to Pol II State and Position

There are two methods for obtaining nascent RNA for the analysis of spliceosome assembly and splicing (Fig. 3A): (1) purification of Pol II and the nascent RNPs bound to it and (2) purification of nascent RNA from a chromatin preparation. Changes in posttranslational modifications (PTMs) of the Pol II CTD mirror and influence the different phases of transcription and nascent RNA processing, owing to the interaction of the CTD with factors that regulate transcription, mRNA processing, and downstream steps (Custodio and Carmo-Fonseca 2016; Saldi et al. 2016; Zaborowska et al. 2016; Harlen and Churchman 2017). The CTD consists of repeats of nearly the same seven amino acids, $Tyr_1Ser_2Pro_3Thr_4Ser_5Pro_6Ser_7$, mainly modified by phosphorylation of Ser2, Ser5, Ser7, Thr4, and Tyr1. In budding yeast, Ser5 and Ser7 phosphorylation (yielding Ser5P and Ser7P) occurs in promoter-proximal gene regions, whereas Ser2P, Thr4P and Tyr1P appear in downstream regions. PTM transitions have been mapped to pause sites along yeast gene bodies and in particular at 3′SSs, consistent with changes in Pol II elongation rate around intron–exon boundaries (Harlen et al. 2016; Herzel et al. 2017). Although the Ser5P levels are highest at the beginning of transcription units, a link between this PTM and splicing has been found in both yeast and human cells (Fig. 3B), particularly during alternative splicing (Mayer et al. 2015; Nojima et al. 2015; Mayer and Churchman 2016; Nojima et al. 2016). In yeast, nascent RNP protein interactomes obtained by pulldown of the differently phosphorylated forms of Pol II are enriched for splicing factors with the exception of Thr4P (Harlen et al. 2016), consistent with co-transcriptional spliceosome assembly, splicing, and the dissociation of the spliceosome from nascent RNA in downstream gene regions after splicing and before termination.

Chromatin fractionation provides an alternative source of nascent RNA (Fig. 3A), which has been analyzed using single-molecule RNA-seq methods that determine the position of Pol II when splicing is completed (Carrillo Oesterreich et al. 2016). Direct, long-read sequencing of nascent transcripts with Pol II positions marked by linker ligation to their 3′ ends provides images reminiscent of chromatin spreads (Fig. 3C). A second method, single-molecule intron tracking (SMIT), uses paired-end sequencing to associate splicing status with Pol II position; when compiled, the data indicate the percentage of molecules spliced at a given Pol II position with approximately 300 reads per position (Fig. 3D). SMIT analysis of 87 endogenous genes and long-read sequencing of nascent RNA from *Saccharomyces cerevisiae* and *Schizosaccharomyces pombe* reveal detectable exon–exon ligation when Pol II has transcribed ~50 bp downstream from 3′SSs (Fig. 2). These findings indicate that the active spliceosome is physically very close to Pol II, harkening back to tomography images of nascent RNPs in the Balbiani ring system, in which the U2 snRNP and Pol II were shown to be in the same electron-dense particle adjacent to the DNA axis (Wetterberg et al. 2001). Because 15 nt of nascent RNA is embedded within the elongating polymerase (Martinez-Rucobo et al. 2015), it seems likely that the active spliceosome is positioned at the exit channel of Pol II similar to the 5′-end capping enzymes (see above). Thus, direct interactions between Pol II and spliceosome components appear feasible (Saldi et al. 2016). In mammalian cells, U2AF and FUS interact directly with Pol II and components of the splicing machinery, potentially bridging the two machineries during splicing (Ujvari and Luse 2004; Yu et al. 2015). The observed association of particular Pol II CTD phosphorylation states with spliceosome assembly and splicing suggests that regulation of such interactions occurs during transcription elongation.

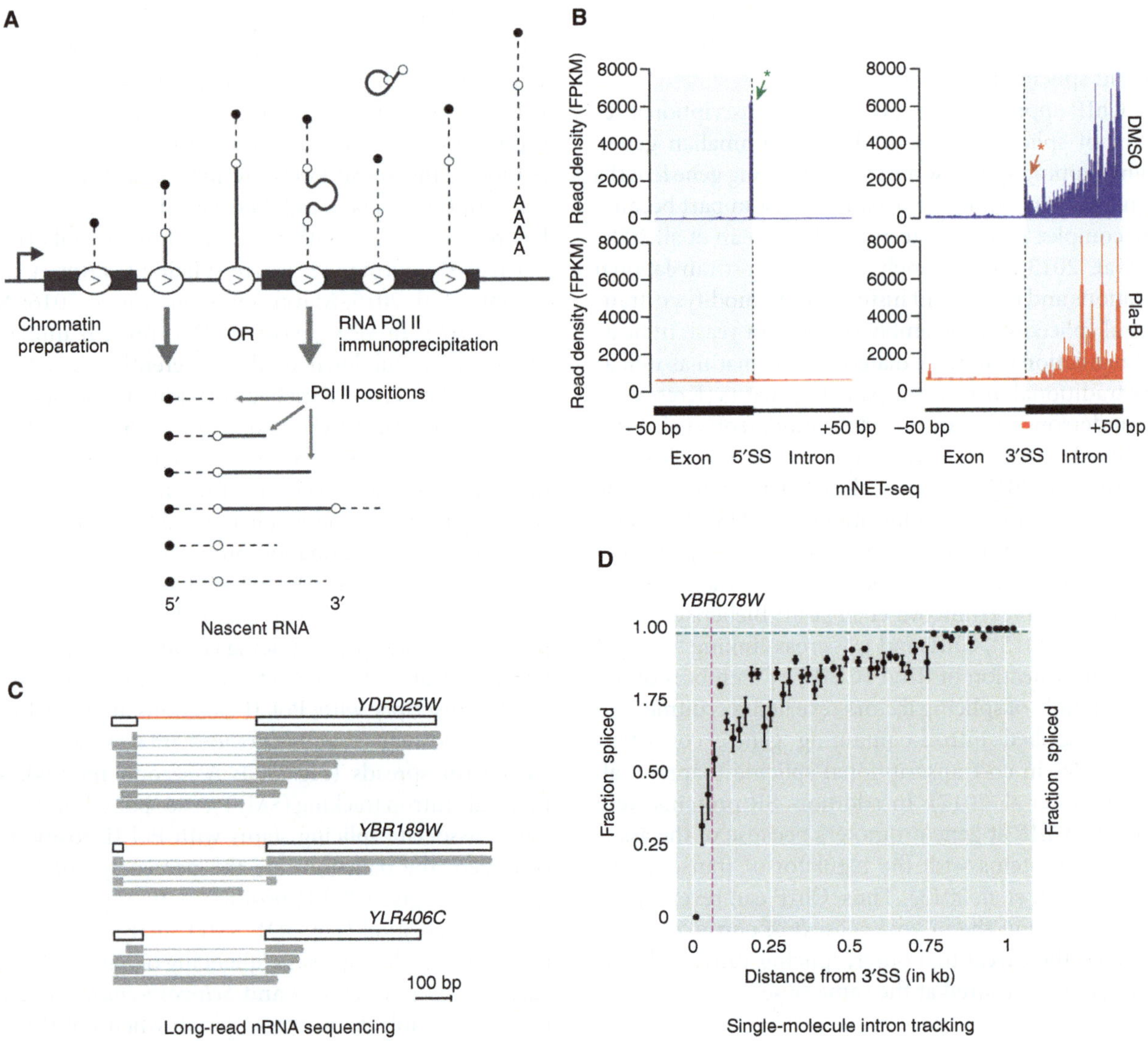

Figure 3. Analysis of nascent RNA captures splicing precursors, intermediates, and products. (*A*) Schematic illustration showing that nascent RNA can be prepared through chromatin purification and/or RNA polymerase II (Pol II) immunoprecipitation. These nascent RNAs vary in length and complexity because of lengthening by Pol II and shortening through splicing. The 3′ end of nascent RNA marks the position of Pol II when the cells were lysed. (*B*) Alignment of 3′ end reads from mNET-seq shows that immunoprecipitation of Pol II with antibodies specific for phosphorylated Ser5 of the carboxy-terminal domain (CTD) enriches splicing first step intermediates (the 3′ end of exon 1, green asterisk). When splicing is blocked chemically by pladienolide B (Pla-B), this splicing intermediate is lost. (Data from Nojima et al. 2015.) (*C*) Genome-wide, full-length, long-read nascent RNA sequencing using the Pacific Biosciences or Oxford Nanopore platform reveals the splicing status of individual transcripts relative to the Pol II position (3′ end). (*D*) Single-molecule intron tracking (SMIT) is a targeted sequencing strategy that detects splicing status relative to Pol II position at base-pair resolution, in which zero indicates Pol II at the 3′SS. A representative budding yeast gene is shown, including the half maximum of splicing (pink dotted line) and the saturation level (green). (Data for panels *C* and *D* are from Carrillo Oesterreich et al. 2016.)

3 QUANTIFICATION OF SPLICING RATES THROUGH RNA METABOLIC LABELING

To measure splicing in terms of time rather than relative to the progress of transcription, some studies have implemented inducible transcription to allow tracking of (pre-)mRNA intermediates over a time course. To date, the highest time resolution achieved with RT-qPCR analysis is 30 sec, during which an integrated, tetracycline-inducible reporter gene in budding yeast yielded detectable spliced transcripts 60 sec

 Cite as *Cold Spring Harb Perspect Biol* doi: 10.1101/cshperspect.a032227

after pre-mRNA transcription was induced (Alexander et al. 2010a). In human mammary epithelial MCF10A cells stimulated with EGF, pre-mRNA half-lives of 2–3 min were measured by RT-qPCR (Zeisel et al. 2011). Reversible treatment with the elongation inhibitor 5,6-dichloro-1-β-D-ribofuranosylbenzimidazole (DRB) enables the analysis of RNA undergoing transcription without induction, using RT-qPCR. In a study designed to analyze co-transcriptional splicing in very long genes, all >100-kb introns were spliced within 5–10 min in human Tet21 cells (Singh and Padgett 2009). This study confirmed that splicing is co-transcriptional, even in endogenous human genes with extremely long introns.

In traditional metabolic labeling, nucleotide analogs, such as 4-thiouracil (4tU) or 4-thiouridine (4sU), can be fed to cells and incorporated into newly synthesized RNA (Fig. 4). Thiol-specific biotinylation and purification of labeled RNA on streptavidin-coated magnetic beads can be followed by sample preparation for RNA-seq, microarray analysis, or RT-qPCR. This approach has been used to quantify transcription and degradation dynamics by labeling newly transcribed RNA for minutes to hours. Recent technical advances allow for shorter labeling times from 1.5 to 5 min and more efficient biotinylation (Rabani et al. 2011, 2014; Windhager et al. 2012; Barrass et al. 2015; Duffy et al. 2015; Eser et al. 2016). The amount of metabolically labeled RNA obtained depends on cellular uptake of nucleotide analogs, labeling time, incorporation rate, transcription rate, processing rate, and degradation rate. The first three are predefined or can be determined by measuring

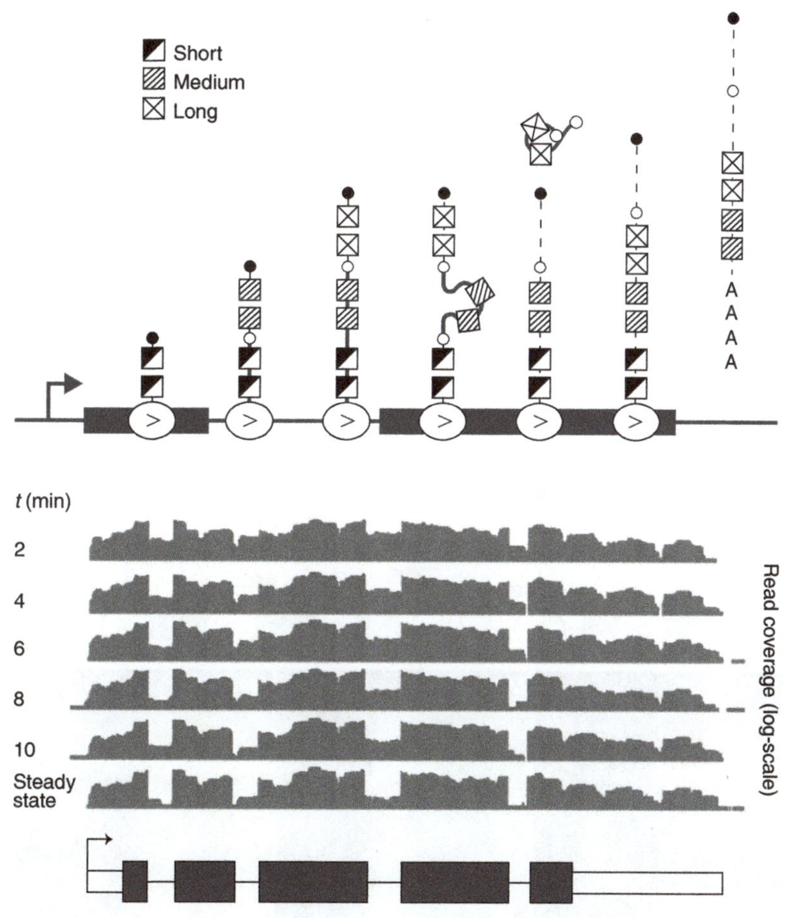

Figure 4. Metabolic labeling measures intron half-lives. *Upper* panel: Schematic representation of nascent RNA after incorporation of a nucleotide analog, such as 4-thio-UTP (4tU), administered to cells for short, medium and long periods of time as indicated. A short pulse will label nucleotides close to RNA polymerase II (Pol II), whereas longer analog feeds lead to incorporation farther along the nascent transcript. *Lower* panel: Example of a 4tU-seq experiment in *Schizosaccharomyces pombe*, showing that at short times (e.g., 2 min) read density across the gene is relatively flat and may approximate Pol II transcription, whereas introns are removed and degraded as time progresses (Eser et al. 2016).

the proportion of labeled versus unlabeled RNA. The latter can be inferred by mathematical modeling and/or analyzing the labeled RNA fraction over multiple time points.

All five of the above-mentioned studies detect substantial amounts of pre-mRNA splicing at the very first labeling time point: 1.5 min in budding yeast, 2 min in *S. pombe* (with a calculated median intron splicing time of 37 sec), 5 min in human B cells, and 10 min in lipopolysaccharide (LPS)-stimulated mouse dendritic cells (with a median intron splicing time of 14 min) (Windhager et al. 2012; Rabani et al. 2014; Barrass et al. 2015; Eser et al. 2016). Such genome-scale data sets identify broad ranges in splicing duration from 4 seconds to many minutes (Fig. 2). Direct comparison of these values is difficult, given differences in experimental systems and analysis methods. It is also important to note that these time windows reflect the convolution of transcription, processing, and degradation, as is the case for live-cell imaging approaches addressed below. Nevertheless, the global nature of metabolic labeling has enabled the discovery of gene architecture and sequence motif correlations linked to synthesis, degradation, and splicing kinetics with which to analyze the coordination between transcription and splicing (Rabani et al. 2011; Windhager et al. 2012; Barrass et al. 2015; Eser et al. 2016).

4 QUANTIFICATION OF SPLICING RATES THROUGH LIVE-CELL FLUORESCENCE IMAGING

Early confirmation that splicing begins co-transcriptionally came from high-resolution fluorescence imaging experiments, which showed the presence of endogenous spliced transcripts at their sites of transcription in mammalian tissue culture cells (Zhang et al. 1994). Consistent with these findings and those of the ChIP experiments discussed above, splicing factors were also localized to sites of transcription by imaging of fixed and living cells (Jimenez-Garcia and Spector 1993; Huang and Spector 1996; Misteli et al. 1997; Neugebauer and Roth 1997). An important development toward determining the in vivo kinetics of splicing through imaging came from the implementation of fluorescent tags and reporter genes to allow the visualization of transcription and splicing within the three-dimensional space of the living cell nucleus (Fig. 5). One study took a

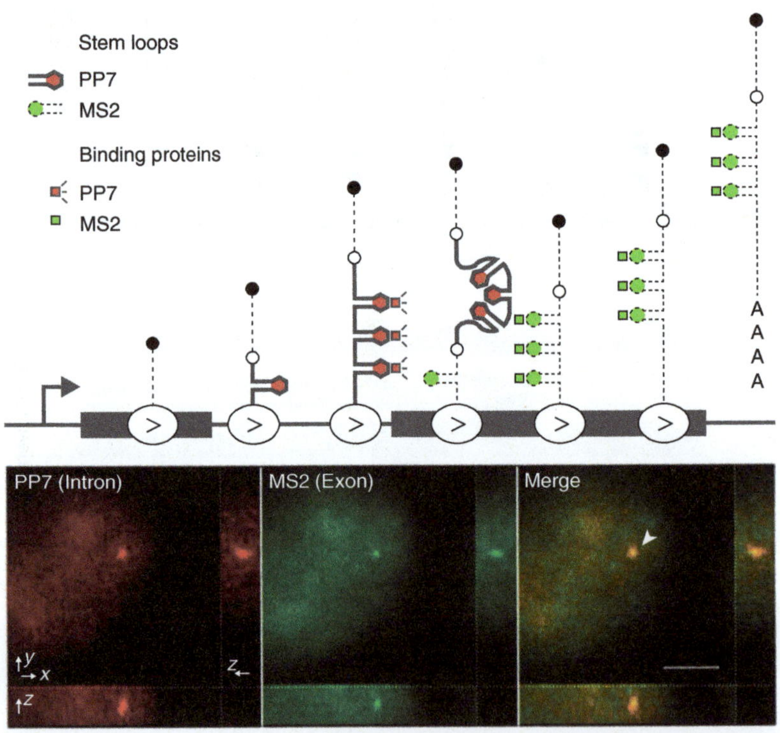

Figure 5. Measurement of transcription and splicing times with fluorescent live-cell imaging. *Upper* panel: Diagram of a representative experimental setup, utilizing a reporter gene in which intronic PP7 stem loops bind to fluorescent PP7 binding protein (red) when the intron is transcribed. MS2 stem loops in the last exon bind to MS2 coat protein (green) and allow quantification of transcripts present at the transcription site. Fluorescence detection of both fluorophores reports on transcription elongation time, splicing, and release after poly(A) cleavage. *Lower* panel: Example showing overlapping PP7 and MS2 signals in human U2OS cells in three dimensions (Coulon et al. 2014).

Cite as *Cold Spring Harb Perspect Biol* doi: 10.1101/cshperspect.a032227

global approach by measuring the residency times of fluorescently labeled spliceosomal snRNPs on pre-mRNA transcripts using fluorescence recovery after photobleaching (FRAP) and fluorescence correlation spectroscopy (FCS) in HeLa cells (Huranova et al. 2010). U1 and U4 snRNPs, which transiently associate with the assembling spliceosome, display shorter residency times; subunits of the active spliceosome—U2 and U5 snRNPs—reside on pre-mRNA for 15–30 sec. This bulk value indicates that the average duration of the splicing process at steady state in HeLa cells lies within a 30-sec window. In contrast, most groups measure specific intron half-lives with fluorescent probes. Results from these experiments vary widely (Fig. 2), which is difficult to explain. Chromosomally integrated reporter genes do not report the diversity of kinetics observed for endogenous genomes, and reporters may not be spliced with the same kinetics as the endogenous, parental gene (Martin et al. 2013). However, the attraction of live-cell imaging is that real-time measurements provide higher time resolution than time points taken in metabolic labeling or gene induction analyses. Single-cell information can also provide insight into cell-to-cell variation within the population.

Comparison among experiments points to several sources of variability. Two studies used stably integrated β-globin reporter genes with MS2 or PP7 stem loops inserted into intronic or exonic sequences to track pre-mRNA transcription and splicing (Fig. 5). One reported splicing within 20 and 30 sec for the two introns in HEK293 cells; their analysis identified short bursts of fluorescence caused by transcription, and the number and lifetime of introns in each burst yielded median intron lifetimes that were considered to be the time window in which splicing occurred (Martin et al. 2013). The second study detected splicing 267 sec after transcription of the last intron in U2OS cells, applying an autocorrelation function to the fluorescent fluctuations to determine how the intronic and exonic signals correlated after a time delay (Coulon et al. 2014). An advantage of this system was the second label placed in the downstream exon, which reports on the amount of RNA at the transcription site before poly(A) cleavage and release. Other possible explanations for the differences between these two values include differences in the cell lines, model genes, and/or data analysis methods. A third study in U2OS cells measured the fluorescent half-life of the MINX reporter intron labeled with MS2-GFP as 105 sec, an intermediate value (Schmidt et al. 2011). It is important to note that measurements of intron half-lives by fluorescence microscopy or metabolic labeling encompass the transcription of intron and exon elements, spliceosome assembly, splicing, spliceosome disassembly, and/or intron release, intron debranching and degradation. Thus, the overall time range observed—0.5 to 3 min—could include variation in the rates of any one or more of these processes.

5 THE COORDINATION OF SPLICING WITH TRANSCRIPTION

It would seem that co-transcriptional RNA processing should aid the efficiency of eukaryotic gene expression by allowing processing to take place within the same time frame and protecting against degradation of the product. In higher metazoans, splicing is a burden. 5′-end capping and poly(A) cleavage happen only once per transcript in all species; yet, splicing occurs typically seven times per human transcript. Moreover, the spliceosome assembles de novo each time an intron is spliced out. Thus, introns hiding within the body of the gene have the time provided by transcription elongation to work out which splice sites will be used and how efficiently splicing will occur. One possibility is that splicing might be facilitated by high local concentrations of splicing, transcription, and chromatin factors, which tend to be intrinsically disordered and may even phase separate at active genes (Harlen and Churchman 2017; Herzel et al. 2017). From the stand point of timing, typical human genes (~30 kb) will require ~15 min for transcription elongation given the known range of transcription elongation rates. As we have seen, in vivo distance-based or time-based measurements of splicing support the possibility that introns "can" be completely removed before poly(A) cleavage (Fig. 2). Perhaps the challenge of co-transcriptional splicing is greatest for terminal introns, which will have the least amount of time for splicing before poly(A) cleavage. Interestingly, last exons are on average 10 times longer than internal exons (940 bp and 120 bp, respectively) in humans, providing an extra ~30 sec for splicing to finish. Indeed, gene architecture may have evolved to take advantage of co-transcriptional splicing and vice versa (Davis-Turak et al. 2015; Hollander et al. 2016). Although splicing can continue posttranscriptionally, the nuclear exosome competes with splicing co-transcriptionally and additional degradation mechanisms await intron-containing mRNAs after transcription (Bousquet-Antonelli et al. 2000; Vargas et al. 2011; Braunschweig et al. 2014). Taken together, attaining high mRNA levels may rely on the ability of the splicing to keep up with transcription.

Transcription and splicing rates seem remarkably well-matched. Although numbers vary (Fig. 2), the spliceosome is capable of acting on the 3′SS shortly after it emerges from the exit channel of Pol II. When transcription is made faster through mutagenesis of Pol II, splicing lags behind in budding yeast (Braberg et al. 2013; Carrillo Oesterreich et al. 2016; Aslanzadeh et al. 2018), suggesting that splicing and transcription rates are matched in wild-type conditions.

Naïvely, we would expect slower transcription to favor both fidelity and orderly splice site choices predicated by the order of intron synthesis. However, analyses of a series of slow and fast Pol II variants have revealed that when the rate of transcription elongation is increased or decreased, alternative splicing patterns are disrupted in human cells and splicing fidelity is compromised in yeast (Fong et al. 2014; Aslanzadeh et al. 2018). These recent findings are consistent with an earlier study that altered transcription rates without mutating Pol II (Howe et al. 2003) and with the reported global correlation between intron retention and Pol II pausing (Braunschweig et al. 2014), implicating transcription rate as a general determinant of splice site choice and fidelity. There are many other factors that can impinge on co-transcriptional splicing efficiency, and how these are integrated in a gene-specific manner is less well understood. Included are RNA modifications, editing, and folding. For example, slowing transcription elongation rate inhibits proper nascent RNA folding, thereby preventing 3′-end processing of histone transcripts (Saldi et al. 2018). Considering the diversity of gene architecture and sequences involved, these observations hint that the kinetics of transcription, spliceosome assembly, and splicing catalysis likely evolved together and with the chromatin landscape.

As described in the Introduction, accumulating evidence that elongation rates can vary in vivo suggests that local regulation of transcription might control co-transcriptional splicing, leading to different alternative splicing outcomes. The higher density of Pol II over internal and terminal exons in mammalian cells and correspondingly slower elongation rates may be due to the preferential exonic position of nucleosomes, which can be obstacles to transcription elongation (Hodges et al. 2009; Schwartz et al. 2009; Tilgner et al. 2009). In addition, nascent RNA folding and the formation of RNA–DNA hybrids (R-loops) within the transcription bubble can promote or impede transcription elongation (Huertas and Aguilera 2003; Klopper et al. 2010). These fluctuations in Pol II elongation rate relative to intron–exon architecture within mammalian genes could be directly related to splicing activity or indirectly related to sequence biases in introns and exons. On the other hand, evidence that splicing directly influences Pol II elongation behavior comes from budding yeast, in which transcriptional pausing within terminal exons is specifically detected in short genes that undergo efficient co-transcriptional splicing (Carrillo Oesterreich et al. 2010). In addition, interference with splicing stimulates pausing at 5′ and 3′SSs (Alexander et al. 2010b; Chathoth et al. 2014). A recent study has shown that the large subunit of Pol II (Rbp1) in yeast is modified by ubiquitination on lysines 452, 695, and 1246 and that Bre5, an RNA binding protein, recruits de-ubiquitinase activity to remove this modification on K1246 (Milligan et al. 2017). A viable deletion of Bre5 or mutation of K1246 to arginine alters Pol II density in intron-containing gene bodies, raising the possibility that de-ubiquitination releases Pol II from splicing-associated pausing and affects the efficiency of splicing. These findings suggest that regulation of Pol II by Bre5 may provide an important link between splicing and transcription.

6 CONCLUDING REMARKS

To bring the concepts discussed here into even better focus and to facilitate biomedical understanding, more fundamental knowledge is needed. First, we need specific, comprehensive knowledge of potential direct contacts between spliceosomal complexes and chromatin components and between spliceosomal complexes and the transcriptional machinery. Second, we need to determine comprehensively the rates of intron removal in higher metazoans in which alternative splicing is common. There are many reasons to suspect that these splicing rates will vary enormously, likely depending on the many factors discussed above and possibly factors yet to be discovered. Third, we need more information about splicing intermediates to obtain a fuller picture of in vivo kinetics. In general, current assays exploit either detection of exon–exon junctions or determination of intron half-lives. Because each has biases, new methods for detecting splicing intermediates would enhance accuracy (Wallace and Beggs 2017). Related to this, it will be important to identify steps in spliceosome assembly that are rate limiting in vivo, in which the multiple, potentially competing interactions between nascent RNA and RNA binding proteins and spliceosomal components may be modulated by the chromatin environment, transcription rates, RNA editing and modification, and the rates of co-transcriptional nascent RNA folding.

These steps forward will bring us closer to the unexplored frontier of how the coordination of splicing and transcription impacts disease and can be exploited for therapeutics. *Cis* mutations that affect gene splicing, *trans* mutations that affect splicing factors, and the overexpression of splicing factors can be drivers of cancer, diseases like myelodysplastic syndrome, and neurodegenerative conditions like spinal muscular atrophy. Today, standard RNA-seq approaches can reveal faulty products of splicing but fail to tell us how those products arose and what additional downstream consequences may accompany these mutations in living cells. Does a given mutation cause splicing rates to slow down or accelerate? Will kinetic changes in splicing lead to changes in transcription elongation and/or output? Will chromatin signatures respond to such changes and drive broader effects? Clearly, the spectacular advances in methodologies and the insights they have already provided point us in numerous, exciting future directions.

ACKNOWLEDGMENTS

I thank past and present members of my laboratory and especially Manuel Ares, Jean Beggs, David Bentley, Tracy Johnson, and Alberto Kornblihtt for many illuminating discussions. I thank Tucker Carrocci and Tara Alpert for discussions and comments on the manuscript. I am grateful to Olivia Howard for artwork and preparing the figures. Apologies to colleagues whose work may not have been cited because of limitation in the number of references. This work was supported by the National Institutes of Health (NIH R01 GM112766). Its contents are solely the responsibility of the author and do not necessarily represent the official views of the NIH.

REFERENCES

Alexander RD, Barrass JD, Dichtl B, Kos M, Obtulowicz T, Robert MC, Koper M, Karkusiewicz I, Mariconti L, Tollervey D, et al. 2010a. RiboSys, a high-resolution, quantitative approach to measure the in vivo kinetics of pre-mRNA splicing and 3′-end processing in *Saccharomyces cerevisiae*. *RNA* **16:** 2570–2580.

Alexander RD, Innocente SA, Barrass JD, Beggs JD. 2010b. Splicing-dependent RNA polymerase pausing in yeast. *Mol Cell* **40:** 582–593.

Alpert T, Herzel L, Neugebauer KM. 2017. Perfect timing: Splicing and transcription rates in living cells. *Wiley Interdiscip Rev RNA* **8:** e1401.

Ameur A, Zaghlool A, Halvardson J, Wetterbom A, Gyllensten U, Cavelier L, Feuk L. 2011. Total RNA sequencing reveals nascent transcription and widespread co-transcriptional splicing in the human brain. *Nat Struct Mol Biol* **18:** 1435–1440.

Aslanzadeh V, Huang Y, Sanguinetti G, Beggs JD. 2018. Transcription rate strongly affects splicing fidelity and cotranscriptionality in budding yeast. *Genome Res* **28:** 203–213.

Barrass JD, Reid JE, Huang Y, Hector RD, Sanguinetti G, Beggs JD, Granneman S. 2015. Transcriptome-wide RNA processing kinetics revealed using extremely short 4tU labeling. *Genome Biol* **16:** 282.

Bauren G, Wieslander L. 1994. Splicing of Balbiani ring 1 gene pre-mRNA occurs simultaneously with transcription. *Cell* **76:** 183–192.

Beyer AL, Osheim YN. 1988. Splice site selection, rate of splicing, and alternative splicing on nascent transcripts. *Genes Dev* **2:** 754–765.

Bieberstein NI, Carrillo Oesterreich F, Straube K, Neugebauer KM. 2012. First exon length controls active chromatin signatures and transcription. *Cell Rep* **2:** 62–68.

Bieberstein NI, Straube K, Neugebauer KM. 2014. Chromatin immunoprecipitation approaches to determine co-transcriptional nature of splicing. *Methods Mol Biol* **1126:** 315–323.

Bjork P, Wieslander L. 2015. The Balbiani ring story: Synthesis, assembly, processing, and transport of specific messenger RNA–protein complexes. *Annu Rev Biochem* **84:** 65–92.

Bousquet-Antonelli C, Presutti C, Tollervey D. 2000. Identification of a regulated pathway for nuclear pre-mRNA turnover. *Cell* **102:** 765–775.

Boutz PL, Bhutkar A, Sharp PA. 2015. Detained introns are a novel, widespread class of post-transcriptionally spliced introns. *Genes Dev* **29:** 63–80.

Braberg H, Jin H, Moehle EA, Chan YA, Wang S, Shales M, Benschop JJ, Morris JH, Qiu C, Hu F, et al. 2013. From structure to systems: High-resolution, quantitative genetic analysis of RNA polymerase II. *Cell* **154:** 775–788.

Braunschweig U, Barbosa-Morais NL, Pan Q, Nachman EN, Alipanahi B, Gonatopoulos-Pournatzis T, Frey B, Irimia M, Blencowe BJ. 2014. Widespread intron retention in mammals functionally tunes transcriptomes. *Genome Res* **24:** 1774–1786.

Brugiolo M, Herzel L, Neugebauer KM. 2013. Counting on co-transcriptional splicing. *F1000Prime Rep* **5:** 9.

Brugiolo M, Botti V, Liu N, Muller-McNicoll M, Neugebauer KM. 2017. Fractionation iCLIP detects persistent SR protein binding to conserved, retained introns in chromatin, nucleoplasm and cytoplasm. *Nucleic Acids Res* **45:** 10452–10465.

Carrillo Oesterreich F, Preibisch S, Neugebauer KM. 2010. Global analysis of nascent RNA reveals transcriptional pausing in terminal exons. *Mol Cell* **40:** 571–581.

Carrillo Oesterreich F, Herzel L, Straube K, Hujer K, Howard J, Neugebauer KM. 2016. Splicing of nascent RNA coincides with intron exit from RNA polymerase II. *Cell* **165:** 372–381.

Chathoth KT, Barrass JD, Webb S, Beggs JD. 2014. A splicing-dependent transcriptional checkpoint associated with prespliceosome formation. *Mol Cell* **53:** 779–790.

Cho EJ, Takagi T, Moore CR, Buratowski S. 1997. mRNA capping enzyme is recruited to the transcription complex by phosphorylation of the RNA polymerase II carboxy-terminal domain. *Genes Dev* **11:** 3319–3326.

Cho EJ, Rodriguez CR, Takagi T, Buratowski S. 1998. Allosteric interactions between capping enzyme subunits and the RNA polymerase II carboxy-terminal domain. *Genes Dev* **12:** 3482–3487.

Churchman LS, Weissman JS. 2011. Nascent transcript sequencing visualizes transcription at nucleotide resolution. *Nature* **469:** 368–373.

Coulon A, Ferguson ML, de Turris V, Palangat M, Chow CC, Larson DR. 2014. Kinetic competition during the transcription cycle results in stochastic RNA processing. *eLife* **3:** e03939.

Custodio N, Carmo-Fonseca M. 2016. Co-transcriptional splicing and the CTD code. *Crit Rev Biochem Mol Biol* **51:** 395–411.

Davis-Turak JC, Allison K, Shokhirev MN, Ponomarenko P, Tsimring LS, Glass CK, Johnson TL, Hoffmann A. 2015. Considering the kinetics of mRNA synthesis in the analysis of the genome and epigenome reveals determinants of co-transcriptional splicing. *Nucleic Acids Res* **43:** 699–707.

de Almeida SF, Grosso AR, Koch F, Fenouil R, Carvalho S, Andrade J, Levezinho H, Gut M, Eick D, Gut I, et al. 2011. Splicing enhances recruitment of methyltransferase HYPB/Setd2 and methylation of histone H3 Lys36. *Nat Struct Mol Biol* **18:** 977–983.

Duffy EE, Rutenberg-Schoenberg M, Stark CD, Kitchen RR, Gerstein MB, Simon MD. 2015. Tracking distinct RNA populations using efficient and reversible covalent chemistry. *Mol Cell* **59:** 858–866.

Dujardin G, Lafaille C, de la Mata M, Marasco LE, Munoz MJ, Le Jossic-Corcos C, Corcos L, Kornblihtt AR. 2014. How slow RNA polymerase II elongation favors alternative exon skipping. *Mol Cell* **54:** 683–690.

Eser P, Wachutka L, Maier KC, Demel C, Boroni M, Iyer S, Cramer P, Gagneur J. 2016. Determinants of RNA metabolism in the *Schizosaccharomyces pombe* genome. *Mol Syst Biol* **12:** 857.

Fong N, Kim H, Zhou Y, Ji X, Qiu J, Saldi T, Diener K, Jones K, Fu XD, Bentley DL. 2014. Pre-mRNA splicing is facilitated by an optimal RNA polymerase II elongation rate. *Genes Dev* **28:** 2663–2676.

Girard C, Will CL, Peng J, Makarov EM, Kastner B, Lemm I, Urlaub H, Hartmuth K, Luhrmann R. 2012. Post-transcriptional spliceosomes are retained in nuclear speckles until splicing completion. *Nat Commun* **3:** 994.

Glover-Cutter K, Kim S, Espinosa J, Bentley DL. 2008. RNA polymerase II pauses and associates with pre-mRNA processing factors at both ends of genes. *Nat Struct Mol Biol* **15:** 71–78.

Görnemann J, Kotovic KM, Hujer K, Neugebauer KM. 2005. Cotranscriptional spliceosome assembly occurs in a stepwise fashion and requires the cap binding complex. *Mol Cell* **19:** 53–63.

Gunderson FQ, Johnson TL. 2009. Acetylation by the transcriptional coactivator Gcn5 plays a novel role in co-transcriptional spliceosome assembly. *PLoS Genet* **5:** e1000682.

Gunderson FQ, Merkhofer EC, Johnson TL. 2011. Dynamic histone acetylation is critical for cotranscriptional spliceosome assembly and spliceosomal rearrangements. *Proc Natl Acad Sci* **108:** 2004–2009.

Harlen KM, Churchman LS. 2017. The code and beyond: Transcription regulation by the RNA polymerase II carboxy-terminal domain. *Nat Rev Mol Cell Biol* **18:** 263–273.

Harlen KM, Trotta KL, Smith EE, Mosaheb MM, Fuchs SM, Churchman LS. 2016. Comprehensive RNA polymerase II interactomes reveal distinct and varied roles for each phospho-CTD residue. *Cell Rep* **15:** 2147–2158.

Herissant L, Moehle EA, Bertaccini D, Van Dorsselaer A, Schaeffer-Reiss C, Guthrie C, Dargemont C. 2014. H2B ubiquitylation modulates spliceosome assembly and function in budding yeast. *Biol Cell* **106:** 126–138.

Herzel L, Ottoz DSM, Alpert T, Neugebauer KM. 2017. Splicing and transcription touch base: Co-transcriptional spliceosome assembly and function. *Nat Rev Mol Cell Biol* **18:** 637–650.

Hnilicova J, Hozeifi S, Duskova E, Icha J, Tomankova T, Stanek D. 2011. Histone deacetylase activity modulates alternative splicing. *PLoS ONE* **6:** e16727.

Hnilicova J, Hozeifi S, Stejskalova E, Duskova E, Poser I, Humpolickova J, Hof M, Stanek D. 2013. The C-terminal domain of Brd2 is important for chromatin interaction and regulation of transcription and alternative splicing. *Mol Biol Cell* **24:** 3557–3568.

Hodges C, Bintu L, Lubkowska L, Kashlev M, Bustamante C. 2009. Nucleosomal fluctuations govern the transcription dynamics of RNA polymerase II. *Science* **325:** 626–628.

Hollander D, Naftelberg S, Lev-Maor G, Kornblihtt AR, Ast G. 2016. How are short exons flanked by long introns defined and committed to splicing? *Trends Genet* **32:** 596–606.

Howe KJ, Kane CM, Ares M Jr. 2003. Perturbation of transcription elongation influences the fidelity of internal exon inclusion in *Saccharomyces cerevisiae*. *RNA* **9:** 993–1006.

Huang S, Spector DL. 1996. Intron-dependent recruitment of pre-mRNA splicing factors to sites of transcription. *J Cell Biol* **133:** 719–732.

Huertas P, Aguilera A. 2003. Cotranscriptionally formed DNA:RNA hybrids mediate transcription elongation impairment and transcription-associated recombination. *Mol Cell* **12:** 711–721.

Huranova M, Ivani I, Benda A, Poser I, Brody Y, Hof M, Shav-Tal Y, Neugebauer KM, Stanek D. 2010. The differential interaction of snRNPs with pre-mRNA reveals splicing kinetics in living cells. *J Cell Biol* **191:** 75–86.

Ji X, Zhou Y, Pandit S, Huang J, Li H, Lin CY, Xiao R, Burge CB, Fu XD. 2013. SR proteins collaborate with 7SK and promoter-associated nascent RNA to release paused polymerase. *Cell* **153:** 855–868.

Jimenez-Garcia LF, Spector DL. 1993. In vivo evidence that transcription and splicing are coordinated by a recruiting mechanism. *Cell* **73:** 47–59.

Jonkers I, Kwak H, Lis JT. 2014. Genome-wide dynamics of Pol II elongation and its interplay with promoter proximal pausing, chromatin, and exons. *eLife* **3:** e02407.

Kfir N, Lev-Maor G, Glaich O, Alajem A, Datta A, Sze SK, Meshorer E, Ast G. 2015. SF3B1 association with chromatin determines splicing outcomes. *Cell Rep* **11:** 618–629.

Khodor YL, Rodriguez J, Abruzzi KC, Tang CH, Marr MT 2nd, Rosbash M. 2011. Nascent-seq indicates widespread cotranscriptional pre-mRNA splicing in *Drosophila*. *Genes Dev* **25:** 2502–2512.

Khodor YL, Menet JS, Tolan M, Rosbash M. 2012. Cotranscriptional splicing efficiency differs dramatically between *Drosophila* and mouse. *RNA* **18:** 2174–2186.

Klopper AV, Bois JS, Grill SW. 2010. Influence of secondary structure on recovery from pauses during early stages of RNA transcription. *Phys Rev E Stat Nonlin Soft Matter Phys* **81:** 030904.

Kotovic KM, Lockshon D, Boric L, Neugebauer KM. 2003. Cotranscriptional recruitment of the U1 snRNP to intron-containing genes in yeast. *Mol Cell Biol* **23:** 5768–5779.

Kress TL, Krogan NJ, Guthrie C. 2008. A single SR-like protein, Npl3, promotes pre-mRNA splicing in budding yeast. *Mol Cell* **32:** 727–734.

Lacadie SA, Rosbash M. 2005. Cotranscriptional spliceosome assembly dynamics and the role of U1 snRNA:5′SS base pairing in yeast. *Mol Cell* **19:** 65–75.

Lacadie SA, Tardiff DF, Kadener S, Rosbash M. 2006. In vivo commitment to yeast cotranscriptional splicing is sensitive to transcription elongation mutants. *Genes Dev* **20:** 2055–2066.

Lei EP, Krebber H, Silver PA. 2001. Messenger RNAs are recruited for nuclear export during transcription. *Genes Dev* **15:** 1771–1782.

Listerman I, Sapra AK, Neugebauer KM. 2006. Cotranscriptional coupling of splicing factor recruitment and precursor messenger RNA splicing in mammalian cells. *Nat Struct Mol Biol* **13:** 815–822.

Luco RF, Allo M, Schor IE, Kornblihtt AR, Misteli T. 2011. Epigenetics in alternative pre-mRNA splicing. *Cell* **144:** 16–26.

Martin RM, Rino J, Carvalho C, Kirchhausen T, Carmo-Fonseca M. 2013. Live-cell visualization of pre-mRNA splicing with single-molecule sensitivity. *Cell Rep* **4:** 1144–1155.

Martinez-Rucobo FW, Kohler R, van de Waterbeemd M, Heck AJ, Hemann M, Herzog F, Stark H, Cramer P. 2015. Molecular basis of transcription-coupled pre-mRNA capping. *Mol Cell* **58:** 1079–1089.

Mayer A, Churchman LS. 2016. Genome-wide profiling of RNA polymerase transcription at nucleotide resolution in human cells with native elongating transcript sequencing. *Nat Protoc* **11:** 813–833.

Mayer A, di Iulio J, Maleri S, Eser U, Vierstra J, Reynolds A, Sandstrom R, Stamatoyannopoulos JA, Churchman LS. 2015. Native elongating transcript sequencing reveals human transcriptional activity at nucleotide resolution. *Cell* **161:** 541–554.

Mayer A, Landry HM, Churchman LS. 2017. Pause & go: From the discovery of RNA polymerase pausing to its functional implications. *Curr Opin Cell Biol* **46:** 72–80.

McCracken S, Fong N, Rosonina E, Yankulov K, Brothers G, Siderovski D, Hessel A, Foster S, Shuman S, Bentley DL. 1997a. 5′-Capping enzymes are targeted to pre-mRNA by binding to the phosphorylated carboxy-terminal domain of RNA polymerase II. *Genes Dev* **11:** 3306–3318.

McCracken S, Fong N, Yankulov K, Ballantyne S, Pan G, Greenblatt J, Patterson SD, Wickens M, Bentley DL. 1997b. The C-terminal domain of RNA polymerase II couples mRNA processing to transcription. *Nature* **385:** 357–361.

Milligan L, Huynh-Thu VA, Delan-Forino C, Tuck A, Petfalski E, Lombrana R, Sanguinetti G, Kudla G, Tollervey D. 2016. Strand-specific, high-resolution mapping of modified RNA polymerase II. *Mol Syst Biol* **12:** 874.

Milligan L, Sayou C, Tuck A, Auchynnikava T, Reid JE, Alexander R, Alves FL, Allshire R, Spanos C, Rappsilber J, et al. 2017. RNA polymerase II stalling at pre-mRNA splice sites is enforced by ubiquitination of the catalytic subunit. *eLife* **6:** e27082.

Misteli T, Caceres JF, Spector DL. 1997. The dynamics of a pre-mRNA splicing factor in living cells. *Nature* **387:** 523–527.

Moehle EA, Ryan CJ, Krogan NJ, Kress TL, Guthrie C. 2012. The yeast SR-like protein Npl3 links chromatin modification to mRNA processing. *PLoS Genet* **8:** e1003101.

Munoz MJ, Perez Santangelo MS, Paronetto MP, de la Mata M, Pelisch F, Boireau S, Glover-Cutter K, Ben-Dov C, Blaustein M, Lozano JJ, et al. 2009. DNA damage regulates alternative splicing through inhibition of RNA polymerase II elongation. *Cell* **137:** 708–720.

Neugebauer KM, Roth MB. 1997. Distribution of pre-mRNA splicing factors at sites of RNA polymerase II transcription. *Genes Dev* **11:** 1148–1159.

Neves LT, Douglass S, Spreafico R, Venkataramanan S, Kress TL, Johnson TL. 2017. The histone variant H2A.Z promotes efficient cotranscriptional splicing in *S. cerevisiae*. *Genes Dev* **31:** 702–717.

Nissen KE, Homer CM, Ryan CJ, Shales M, Krogan NJ, Patrick KL, Guthrie C. 2017. The histone variant H2A.Z promotes splicing of weak introns. *Genes Dev* **31:** 688–701.

Nojima T, Gomes T, Grosso ARF, Kimura H, Dye MJ, Dhir S, Carmo-Fonseca M, Proudfoot NJ. 2015. Mammalian NET-Seq reveals genome-wide nascent transcription coupled to RNA processing. *Cell* **161:** 526–540.

Nojima T, Gomes T, Carmo-Fonseca M, Proudfoot NJ. 2016. Mammalian NET-seq analysis defines nascent RNA profiles and associated RNA processing genome-wide. *Nat Protoc* **11:** 413–428.

Osheim YN, Miller OL Jr., Beyer AL. 1985. RNP particles at splice junction sequences on *Drosophila* chorion transcripts. *Cell* **43:** 143–151.

Pabis M, Neufeld N, Steiner MC, Bojic T, Shav-Tal Y, Neugebauer KM. 2013. The nuclear cap-binding complex interacts with the U4/U6.U5 tri-snRNP and promotes spliceosome assembly in mammalian cells. *RNA* **19:** 1054–1063.

Pandya-Jones A, Bhatt DM, Lin CH, Tong AJ, Smale ST, Black DL. 2013. Splicing kinetics and transcript release from the chromatin compartment limit the rate of Lipid A-induced gene expression. *RNA* **19:** 811–827.

Patrick KL, Ryan CJ, Xu J, Lipp JJ, Nissen KE, Roguev A, Shales M, Krogan NJ, Guthrie C. 2015. Genetic interaction mapping reveals a role for the SWI/SNF nucleosome remodeler in spliceosome activation in fission yeast. *PLoS Genet* **11:** e1005074.

Pimentel H, Parra M, Gee SL, Mohandas N, Pachter L, Conboy JG. 2016. A dynamic intron retention program enriched in RNA processing genes regulates gene expression during terminal erythropoiesis. *Nucleic Acids Res* **44:** 838–851.

Rabani M, Levin JZ, Fan L, Adiconis X, Raychowdhury R, Garber M, Gnirke A, Nusbaum C, Hacohen N, Friedman N, et al. 2011. Metabolic labeling of RNA uncovers principles of RNA production and degradation dynamics in mammalian cells. *Nat Biotechnol* **29:** 436–442.

Rabani M, Raychowdhury R, Jovanovic M, Rooney M, Stumpo DJ, Pauli A, Hacohen N, Schier AF, Blackshear PJ, Friedman N, et al. 2014. High-resolution sequencing and modeling identifies distinct dynamic RNA regulatory strategies. *Cell* **159:** 1698–1710.

Rasmussen EB, Lis JT. 1993. In vivo transcriptional pausing and cap formation on three *Drosophila* heat shock genes. *Proc Natl Acad Sci* **90:** 7923–7927.

Saldi T, Cortazar MA, Sheridan RM, Bentley DL. 2016. Coupling of RNA polymerase II transcription elongation with pre-mRNA splicing. *J Mol Biol* **428:** 2623–2635.

Saldi T, Fong N, Bentley DL. 2018. Transcription elongation rate affects nascent histone pre-mRNA folding and 3′ end processing. *Genes Dev* **32:** 297–308.

Sapra AK, Anko ML, Grishina I, Lorenz M, Pabis M, Poser I, Rollins J, Weiland EM, Neugebauer KM. 2009. SR protein family members display diverse activities in the formation of nascent and mature mRNPs in vivo. *Mol Cell* **34:** 179–190.

Schmidt U, Basyuk E, Robert MC, Yoshida M, Villemin JP, Auboeuf D, Aitken S, Bertrand E. 2011. Real-time imaging of cotranscriptional splicing reveals a kinetic model that reduces noise: Implications for alternative splicing regulation. *J Cell Biol* **193:** 819–829.

Schor IE, Fiszbein A, Petrillo E, Kornblihtt AR. 2013. Intragenic epigenetic changes modulate NCAM alternative splicing in neuronal differentiation. *EMBO J* **32:** 2264–2274.

Schwartz S, Meshorer E, Ast G. 2009. Chromatin organization marks exon–intron structure. *Nat Struct Mol Biol* **16:** 990–995.

Sims RJ 3rd, Millhouse S, Chen CF, Lewis BA, Erdjument-Bromage H, Tempst P, Manley JL, Reinberg D. 2007. Recognition of trimethylated histone H3 lysine 4 facilitates the recruitment of transcription postinitiation factors and pre-mRNA splicing. *Mol Cell* **28:** 665–676.

Singh J, Padgett RA. 2009. Rates of in situ transcription and splicing in large human genes. *Nat Struct Mol Biol* **16:** 1128–1133.

Tardiff DF, Rosbash M. 2006. Arrested yeast splicing complexes indicate stepwise snRNP recruitment during in vivo spliceosome assembly. *RNA* **12:** 968–979.

Tardiff DF, Lacadie SA, Rosbash M. 2006. A genome-wide analysis indicates that yeast pre-mRNA splicing is predominantly posttranscriptional. *Mol Cell* **24:** 917–929.

Tilgner H, Nikolaou C, Althammer S, Sammeth M, Beato M, Valcarcel J, Guigo R. 2009. Nucleosome positioning as a determinant of exon recognition. *Nat Struct Mol Biol* **16:** 996–1001.

Tilgner H, Knowles DG, Johnson R, Davis CA, Chakrabortty S, Djebali S, Curado J, Snyder M, Gingeras TR, Guigo R. 2012. Deep sequencing of subcellular RNA fractions shows splicing to be predominantly co-transcriptional in the human genome but inefficient for lncRNAs. *Genome Res* **22:** 1616–1625.

Ujvari A, Luse DS. 2004. Newly Initiated RNA encounters a factor involved in splicing immediately upon emerging from within RNA polymerase II. *J Biol Chem* **279:** 49773–49779.

Vargas DY, Shah K, Batish M, Levandoski M, Sinha S, Marras SA, Schedl P, Tyagi S. 2011. Single-molecule imaging of transcriptionally coupled and uncoupled splicing. *Cell* **147:** 1054–1065.

Veloso A, Kirkconnell KS, Magnuson B, Biewen B, Paulsen MT, Wilson TE, Ljungman M. 2014. Rate of elongation by RNA polymerase II is associated with specific gene features and epigenetic modifications. *Genome Res* **24:** 896–905.

Wahl MC, Will CL, Lührmann R. 2009. The spliceosome: Design principles of a dynamic RNP machine. *Cell* **136:** 701–718.

Wallace EWJ, Beggs JD. 2017. Extremely fast and incredibly close: Cotranscriptional splicing in budding yeast. *RNA* **23:** 601–610.

Wetterberg I, Zhao J, Masich S, Wieslander L, Skoglund U. 2001. In situ transcription and splicing in the Balbiani ring 3 gene. *EMBO J* **20:** 2564–2574.

Windhager L, Bonfert T, Burger K, Ruzsics Z, Krebs S, Kaufmann S, Malterer G, L'Hernault A, Schilhabel M, Schreiber S, et al. 2012. Ultrashort and progressive 4sU-tagging reveals key characteristics of RNA processing at nucleotide resolution. *Genome Res* **22:** 2031–2042.

Wong JJ, Ritchie W, Ebner OA, Selbach M, Wong JW, Huang Y, Gao D, Pinello N, Gonzalez M, Baidya K, et al. 2013. Orchestrated intron retention regulates normal granulocyte differentiation. *Cell* **154:** 583–595.

Yu Y, Chi B, Xia W, Gangopadhyay J, Yamazaki T, Winkelbauer-Hurt ME, Yin S, Eliasse Y, Adams E, Shaw CE, et al. 2015. U1 snRNP is mislocalized in ALS patient fibroblasts bearing NLS mutations in FUS and is required for motor neuron outgrowth in zebrafish. *Nucleic Acids Res* **43:** 3208–3218.

Zaborowska J, Egloff S, Murphy S. 2016. The pol II CTD: New twists in the tail. *Nat Struct Mol Biol* **23:** 771–777.

Zeisel A, Kostler WJ, Molotski N, Tsai JM, Krauthgamer R, Jacob-Hirsch J, Rechavi G, Soen Y, Jung S, Yarden Y, et al. 2011. Coupled pre-mRNA and mRNA dynamics unveil operational strategies underlying transcriptional responses to stimuli. *Mol Syst Biol* **7:** 529.

Zenklusen D, Vinciguerra P, Wyss JC, Stutz F. 2002. Stable mRNP formation and export require cotranscriptional recruitment of the mRNA export factors Yra1p and Sub2p by Hpr1p. *Mol Cell Biol* **22:** 8241–8253.

Zhang G, Taneja KL, Singer RH, Green MR. 1994. Localization of pre-mRNA splicing in mammalian nuclei. *Nature* **372:** 809–812.

U1 snRNP Telescripting: Suppression of Premature Transcription Termination in Introns as a New Layer of Gene Regulation

Christopher C. Venters,[1] Jung-Min Oh,[1] Chao Di, Byung Ran So, and Gideon Dreyfuss

Howard Hughes Medical Institute, Department of Biochemistry and Biophysics, University of Pennsylvania School of Medicine, Philadelphia, Pennsylvania 19104

Correspondence: gdreyfuss@hhmi.upenn.edu

SUMMARY

Recent observations showed that nascent RNA polymerase II transcripts, pre-mRNAs, and noncoding RNAs are highly susceptible to premature 3′-end cleavage and polyadenylation (PCPA) from numerous intronic cryptic polyadenylation signals (PASs). The importance of this in gene regulation was not previously appreciated as PASs, despite their prevalence, were thought to be active in terminal exons at gene ends. Unexpectedly, antisense oligonucleotide interference with U1 snRNA base-pairing to 5′ splice sites, which is necessary for U1 snRNP's (U1) function in splicing, caused widespread PCPA in metazoans. This uncovered U1's PCPA suppression activity, termed telescripting, as crucial for full-length transcription in thousands of vertebrate genes, providing a general role in transcription elongation control. Progressive intron-size expansion in metazoan evolution greatly increased PCPA vulnerability and dependence on U1 telescripting. We describe how these observations unfolded and discuss U1 telescripting's role in shaping the transcriptome.

Outline

[1]These authors contributed equally to this work.

Cite as *Cold Spring Harb Perspect Biol* doi: 10.1101/cshperspect.a032235

1 INTRODUCTION

Studies on the survival of motor neuron (SMN) complex have shown that it is an assembly chaperone for RNA–protein complexes (RNPs) best characterized for outfitting uridine-rich, noncoding small nuclear RNAs (snRNAs) with seven-membered Sm protein rings (Sm cores) (Fischer et al. 1997; Liu et al. 1997; Yong et al. 2004b; Cauchi 2010; Fischer et al. 2011; So et al. 2017). The assembly of Sm cores is a rate-limiting step in the biogenesis of small nuclear ribonucleo proteins (snRNPs), designated U1, U2, U4, U5, U11, U12, U4atac, which are well-characterized for their functions in splicing (Guthrie and Patterson 1988; Patel and Steitz 2003; Yong et al. 2004a; Battle et al. 2006; Wahl et al. 2009; Cauchi 2010; Yong et al. 2010; Fischer et al. 2011; Matera and Wang 2014). However, SMN deficiency, which causes spinal muscular atrophy (SMA), changes snRNP levels nonuniformly; rather than reducing snRNPs proportionally, SMA cells and cells engineered to have low SMN have an altered snRNP repertoire (Gabanella et al. 2007; Zhang et al. 2008; Workman et al. 2009). It had been noted earlier that snRNPs' abundances vary despite their 1:1 stoichiometry in spliceosomes (Baserga and Steitz 1993). For example, at around 1,000,000 copies per human (HeLa) cell, U1 snRNP (U1) is several-fold more abundant than U4 and U6. However, the potential role of snRNP abundance in pre-messenger RNA (mRNA) processing was unknown. As we had also observed numerous splicing abnormalities in SMA (Zhang et al. 2008), we set out to determine what effects altering snRNP repertoire might have on the transcriptome. Both the snRNP repertoire and splicing changes were complex and varied in SMA mouse tissues (Zhang et al. 2008, 2013), suggesting a role for cell-specific factors and making it difficult to recapitulate in an experimental cell system. To circumvent this, we chose to systematically inhibit individual snRNPs one at a time.

Decreasing individual snRNPs levels to varying degrees by targeted RNA interference (RNAi) with shRNA or short interfering RNAs (siRNAs) did not seem suitable, as it takes more than 48 hours and is difficult to control. Instead, we chose to systematically inhibit individual snRNPs' splicing activity using antisense morpholino oligonucleotides (AMOs) to mask the sequences in their snRNAs that are necessary for base-pairing with pre-mRNA (Kaida et al. 2010). Transfection of 25-mer AMOs to U2 and U12, tested on select introns, had been previously shown by Harald Konig and colleagues to inhibit splicing (Matter and Konig 2005). Based on this, we designed a series of snRNA targeting AMOs, including U1, U2, U5, U6, U11, and U6atac. RT-PCR on select introns validated their efficacy and guided optimization of transfection conditions and selection of doses and times.

Several features of AMOs make them an advantageous tool for probing function and interactions of specific RNA sequences. Their small size facilitates cell penetration, ensuring transfection of all cells in the experimental population. They are fast-acting, binding target RNAs nearly instantaneously, which minimizes the likelihood of indirect effects, and they are nondestructive to the RNP target (unlike antisense deoxy oligonucleotides that elicit RNase H-mediated RNA degradation), thereby preventing changes in the bound versus free RNP proteins pools. The morpholino chemistry also increases base-pairing avidity for RNA targets, and they are more RNase resistant compared with RNA antisense oligonucleotides (Summerton 1999; Heasman 2002), allowing the effects of a single transfection to be measured 24–72 hours later. Importantly, AMO dose can be readily varied over a large range and nontargeting AMOs can be used as controls.

2 U1 snRNP SUPPRESSES PASs IN NASCENT TRANSCRIPTS

U1 AMO, which emerged as the key probe in this study, illustrates the principles that guided the design of splicing-interfering AMOs. U1 is a ubiquitous RNP in eukaryotes, comprising U1 snRNA (164 nucleotides in human) and 11 proteins, including seven Sm proteins and three U1-specific proteins: U1-70K, U1A, and U1C. U1-70K and U1A bind stem-loops 1 and 2, designated in 5′ to 3′ order, out of U1 snRNA's four stem-loops. In addition to its potential to advance understanding of SMA pathogenesis, U1 was a target of interest in the context of snRNPs' stoichiometry because it is the most abundant small noncoding RNA in vertebrates, which was unexplained. Although U1, along with U2 snRNA, had been detected in the 1960s (Hodnett and Busch 1968; Weinberg and Penman 1968), they received little attention and there was no biological framework in which to consider them. That changed with a series of landmark discoveries by Joan A. Steitz's group, beginning in 1979 (Lerner and Steitz 1979), of a new class of RNPs—the snRNPs, which contained U1 and U2 snRNAs as well as additional snRNPs they discovered, U4, U5, and U6. The same group then put forward the transformative insight (Lerner et al. 1980), confirmed shortly thereafter, by direct evidence (Mount et al. 1983; Padgett et al. 1983), that U1 snRNP recognizes the 5′ splice site (SS) by RNA:RNA base-pairing between U1 snRNA's 5′-end and the pre-mRNA, thus identifying the first factors and initiating step in the removal of introns. Based on this, to inhibit splicing, we have designed U1 AMO, a 25-nt AMO complementary to U1 snRNA's 5′-end, thereby providing a functional knockdown of U1's known function.

Cite as *Cold Spring Harb Perspect Biol* doi: 10.1101/cshperspect.a032235

Several assays were performed to verify that the transfected U1 AMO bound U1's 5′-end, determine the dose–response profile and identify the dose required to mask all, or nearly all, of U1 in these cells (Kaida et al. 2010). An RNase H protection assay was performed on extracts from cells transfected with various U1 AMO doses. This assay, in which RNase H and an antisense DNA oligonucleotide probe complementary to U1 snRNA's 5′-sequence was added to the extract, cleaves U1 snRNA that has this sequence accessible. The results showed a U1 AMO dose-dependent decrease in U1 snRNA cleavage, indicating that U1 AMO prevented the antisense DNA oligonucleotide probe from binding. This determined the U1 AMO dose at which complete or near complete interference with U1 snRNA 5′-base-pairing in cells was achieved (hereafter, high U1 AMO). The same dose–response was observed in a fluorescent in situ hybridization assay using a fluorescent labeled locked nucleic acid (LNA) probe complementary to U1 snRNA's 5′ sequence on cells transfected with various U1 AMO doses. In addition, U1 AMO inhibited splicing of several test introns in vitro. Thus, U1 AMO functionally inactivated U1 snRNP both in vivo and in vitro (Kaida et al. 2010).

As these initial experiments (ca. 2008/2009) predated high-throughput RNA sequencing (RNA-seq), we used high density genomic tiling arrays, consisting of glass slide-immobilized oligonucleotides covering three human chromosomes harboring more than 3600 annotated genes, the method of choice at the time for surveying transcriptome differences between samples. For this, cDNA was prepared from total RNA from HeLa cells 8-hours post-transfection with U1 AMO at a dose that masks all, or nearly all, of U1's 5′-end, and from cells transfected with nontargeting AMO control in the same experiment. Hybridization of the cDNAs to the tiling arrays identified differential RNA expression at each chromosomal location.

Genome browser visualization of the first series of experiments produced a surprising result, which was challenging to explain (Fig. 1). Expecting U1 AMO to inhibit splicing, we performed in parallel the same analysis on cells treated with spliceostatin A (SSA), a small molecule splicing inhibitor that targets SF3B1 (Kaida et al. 2007; Kotake et al.

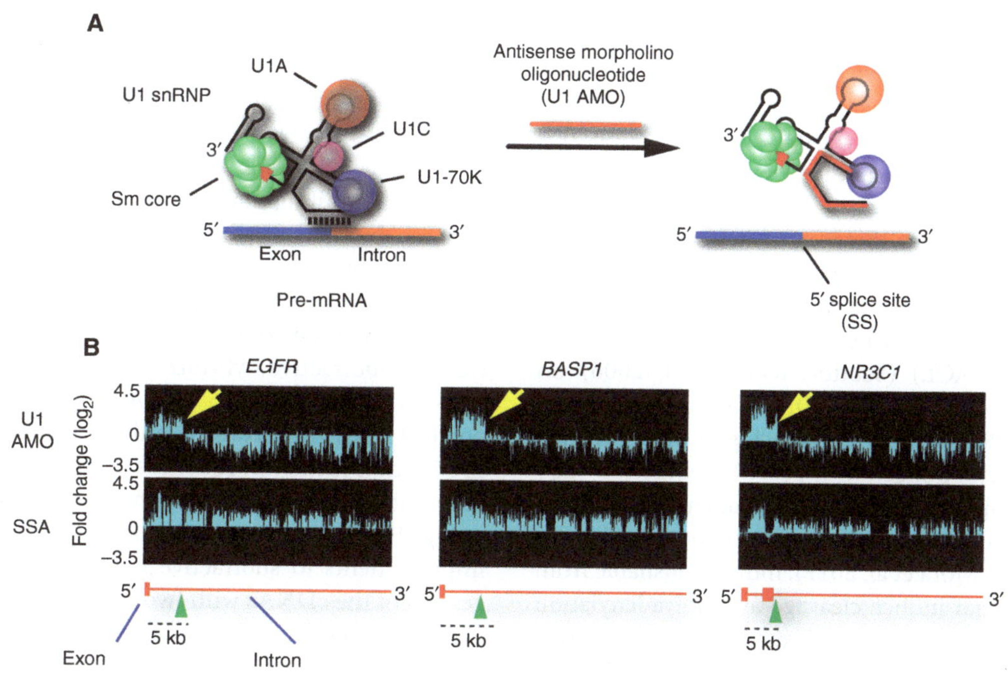

Figure 1. Genomic tiling arrays detect, in U1 antisense morpholino oligonucleotide (AMO) transfected cells, transcripts extending from the transcription start site (TSS) of the genes EGFR, BASP1, and NR3C1 into the first part of the intron in which they abruptly end. (*A*) AMO (25-mer) to U1 snRNA's 5′-end (U1 AMO) interferes with U1 snRNP (U1) base-pairing with 5′ splice sites, necessary for splicing of introns, and with other sequences on nascent transcripts. (*B*) Transcriptome profiling with genomic tiling arrays show the fold change in RNA signals compared with control. The *top* panel shows that, in addition to inhibiting splicing, U1 AMO also induces premature 3′-end cleavage and polyadenylation (PCPA) as RNA reads downstream from these end points are strongly decreased. In contrast (*bottom* panels), splicing inhibition with spliceostatin A (SSA) shows RNA reads increasing over the full length of introns, as expected for unspliced pre-mRNAs. Schematic gene structures (based on RefSeq, hg19) are depicted in red, with horizontal lines indicating introns and boxes indicating exons.

2007), and with U2 AMO. Like AMOs, SSA is fast-acting, and it produced patterns consistent with splicing inhibition, including accumulation of signals covering entire introns. However, in the majority of genes, U1 AMO produced a remarkably different pattern, consisting of reads extending several kilobases from the transcription start site (TSS) into an intron were they abruptly ended, followed by a precipitous drop over the rest of the gene. Confirming U1's expected activity, widespread splicing inhibition was evident by the accumulation of unspliced introns upstream of the point where the transcripts precipitously ended. U2 AMO produced a similar pattern to SSA, indicating that the phenomenon reflected U1-specific activity.

However, we could not discern from the pattern what the mechanism might be. Possibilities included that U1 is a transcription factor, particularly for elongation; that RNA polymerases (Pol II) stalled at road blocks that required U1 base-pairing to overcome (namely, termination did not occur; polymerases remained engaged but stopped moving); that transcription termination occurred early; or that transcription was in fact full-length, but nascent transcripts were being rapidly degraded leaving only 5′-side RNAs. It was puzzling why such a dramatic effect had not been previously detected or reported considering that U1 has been one of the most intensively investigated RNAs over decades and in many organisms. The likelihood of a technical artifact seemed remote, as the same result was observed in three separate biological experiments, but it could not be ruled out.

Significant progress in this respect came from application of another tool, which we used to define the 3′-end points of the U1 AMO-induced transcripts (Kaida et al. 2010). For this, we performed 3′ rapid amplification of cDNA ends (3′-RACE) (Scotto-Lavino et al. 2006) on RNAs using primers to produce cDNAs for 3′-end location of several genes. Surprisingly, sequencing of these revealed that they had 3′-end nongenomic poly(A) sequences ~20–60 nt downstream from a consensus polyadenylation signal, typically AAUAAA and its variants (Proudfoot and Brownlee 1976; Magana-Mora et al. 2017), indistinguishable from canonical PASs that induce cleavage and polyadenylation (CPA) (Shi and Manley 2015; Tian and Manley 2017) at genes' ends. This suggested that they were produced by premature cleavage and polyadenylation (PCPA) from cryptic polyadenylation signals (PASs) in introns, counter to the prevailing view that PASs marked the end of genes. Mutational inactivation of intronic PASs from which PCPA is elicited with U1 AMO (actionable PASs, used hereafter to indicate locations that not only have consensus PAS hexamer, but also show CPA activity) showed the PAS-dependent nature of this phenomenon, confirming that these intronic signals function like PASs at the canonical 3′-end of genes' full-length transcripts. Interestingly, PAS inactivation did not necessarily prevent PCPA; instead, PCPA occurred from downstream PASs in the same intron or pre-mRNA. Having the same 5′ to 3′ polarity as transcription suggested that PCPA and its suppression by U1 is a directional, cotranscriptional mechanism (Berg et al. 2012).

These experiments revealed that, in addition to its splicing role, U1 is also a PCPA suppressor. PCPA is not a secondary effect of splicing inhibition as it frequently occurs in introns at great distances before transcription reaches the 3′SS. Early, promoter proximal PCPA is the default state for the majority of human genes, and full-length transcription requires U1. For brevity, we refer to this activity as telescripting because it is necessary for long-distance, full-length synthesis.

As there was nothing known about this new process, much more information was needed to understand its generality, potential role in biology and regulation. However, genomic tiling arrays were a significant bottleneck in terms of throughput, cost, labor, resolution, and availability of arrays for probing diverse organisms. RNA-seq became more widely available, but sequencing capacity, access to instruments and informatics expertise were major constraints. The expense of transcriptome sequencing to a depth required to identify differences between experimental samples also limited its application more broadly. To overcome these limitations, we devised a strategy for rapid transcriptome profiling that greatly accelerated the pace of our research by applying high-throughput RNA-seq only to differentially expressed transcripts (HIDE-seq; Berg et al. 2012). Reasoning that the most relevant information we sought was about what changed on U1 snRNP inhibition, we used subtractive hybridization (Diatchenko et al. 1996; Gurskaya et al. 1996) on cDNA fragments prepared from poly(A) RNA from control and U1 AMO-treated samples to eliminate identical sequences and thus only PCR amplify differentially expressed fragments for sequencing. In developing HIDE-seq we have introduced several technical improvements to subtractive hybridization, including digestion of the cDNAs with two or more 4-cutter restriction enzymes, which refined resolution, and ligation of bar-coded adaptors. These adaptors allowed samples to be combined and sequenced together, drastically reducing cost and controlling for sample-to-sample variation. By physically eliminating sequences that did not change, HIDE-seq increased sequencing depth of differential effects and greatly simplified data analysis as the number of reads directly reflected transcriptome change at any location. Algorithms developed to process the HIDE-seq datasets helped classify transcriptome changes, detect PCPA locations and visualize them on genome browsers. It confirmed the finding from

 Cite as *Cold Spring Harb Perspect Biol* doi: 10.1101/cshperspect.a032235

genomic tiling arrays and provided a wealth of new information.

Applying HIDE-seq to human, mouse, and fruit fly (using U1 AMO matching the U1 snRNA 5′-sequence of each organism), we showed that U1 telescripting is evolutionarily conserved in metazoans (Berg et al. 2012). Genome-wide maps of PCPA locations at high U1 AMO doses further showed that it generally occurs in one of the first introns, frequently within 1 kb from the intron's 5′SS. Assuming that U1 base-paired at that 5′SS for the purpose of splicing also provided telescripting, then effective U1 telescripting could be estimated to have a range of around 1 kb. However, in many genes, U1 AMO-induced PCPA occurred at much greater distances from the intron's 5′SS, raising the possibility that U1 bound to 5′SS alone may be insufficient to ensure telescripting of introns longer than 1 kb. Mutating a 5′SS to a sequence that is incompatible with splicing in cases in which PCPA occurred from a PAS <1 kb downstream, caused constitutive PCPA from that PAS. However, U1 AMO increased that PCPA several-fold, suggesting that the PAS received additional suppression from U1 base-paired elsewhere. Furthermore, transfection of synthetic U1 snRNA with 5′-end complementary to the mutated 5′SS restored PAS suppression. This separated U1 splicing and telescripting requirements and indicated that U1 base-paired to any sequence could potentially function in telescripting, even if it is nonfunctional in splicing. U1 base-pairing to 5′SS is highly degenerate and depends on various pre-mRNA binding proteins, which likely facilitates its binding to cryptic 5′SS and many other RNA sequences. An important conclusion from these experiments was that U1 snRNP bound at 5′SS alone would likely be insufficient to protect long introns (>1 kb), which are common in complex organisms, requiring additional U1 base-paired in the intron to fully suppress PASs within them.

The genomic HIDE-seq maps revealed that telescripting is a physiological process, as many PCPA sites coincide with 3′-poly(A) sequence tags from diverse biological specimens, including normal human tissues. Therefore, while telescripting occurs naturally the PCPA elicited artificially in our experiments with U1 AMO enhance PCPA detection of actionable PASs that are widespread nature. Thus, serendipitously, U1 AMO uncovered telescripting and its ability to recapitulate it in a controlled experimental setting makes it a useful research tool for this process.

Testing the effect of various U1 AMO doses showed an interesting and dose-dependent effect on PCPA locations. While high U1 AMO triggered drastic TSS-proximal PCPA, generally in the first quarter of the gene, at low U1 AMO doses (masking <15% of U1) PCPA occurred at greater distances downstream, resulting in widespread 3′ untranslated regions (3′ UTRs) shortening. These PCPAs corresponded to an alternative 3′-end processing and polyadenylation (APA) shift to usage of a more proximal PAS in 3′ UTRs that contain tandem PASs. Such 3′ UTR shortening, which maintains the full-length coding sequence (CDS) but could de-regulate mRNA translation, stability, and localization, is associated with stimulated immune cells and neurons, cell proliferation, and cancer (Niibori et al. 2007; Flavell et al. 2008; Sandberg et al. 2008; Mayr and Bartel 2009; Lianoglou et al. 2013). Thus, U1 AMO recapitulated 3′ UTR APA. Knockdown of several 3′-end CPA factors (CPAFs), CFIm25/CPSF5, CFIm68/CFSF6, or PABPN1, also shifts APA to proximal PASs in the terminal exon, but apparently not in introns (Gruber et al. 2012, 2014; Jenal et al. 2012; Elkon et al. 2013; Masamha et al. 2014; Li et al. 2015; Zhu et al. 2018).

In addition, in many genes low U1 AMO induced a shift to usage of a PAS in an intron, resulting in shorter mRNA isoforms that lack the carboxy-terminal portion of the full-length CDS. These events are frequently attributed to alternative splicing that gives rise to an alternative last exon that has a PAS. However, as it emerged that potentially PAS usage anywhere in the transcript is under U1 control we considered a scenario whereby the primary event is PCPA in an intron. Mutational inactivation of the intronic PAS supported the PCPA-first scenario as U1 AMO no longer induced the splicing of the alternative last exon. An example of such a case, *homer-1*, illustrates the biological importance of this type of U1-controlled PCPA. Homer-1 encodes a scaffold protein critical for synaptogenesis and synapse strengthening in neurons. Neuronal stimulation, which can be experimentally mimicked in mouse neuronal-type PC12 cells, rapidly shifts synthesis from full-length (homer-ll) to a shorter isoform lacking the protein's carboxy-terminal domain (CTD) (homer-1s). homer-1s antagonizes homer-ll, which is necessary to prevent overstimulation causing epilepsy (Niibori et al. 2007). Interestingly, low U1 AMO recapitulated homer-1l to homer-1s shift by PCPA in the intron downstream from homer-1s' alternative last exon (Berg et al. 2012). Using several low U1 AMO concentrations showed a dose-dependent increase in the fraction of homer-1s and a reciprocal decrease in homer-1l.

3 U1 AMO PREMATURELY TERMINATES ELONGATING Pol II IN GENE BODIES

We used chromatin immunoprecipitation-sequencing (ChIP-seq) (Gilmour and Lis 1984) with antibodies to RNA Pol II to investigate the relationship between PCPA and Pol II transcription (Oh et al. 2017). ChIP-seq maps of Pol II chromosomal locations in controls extended from the TSS to the canonical 3′-end of genes, from which point it progressively declined to background level within a few kb

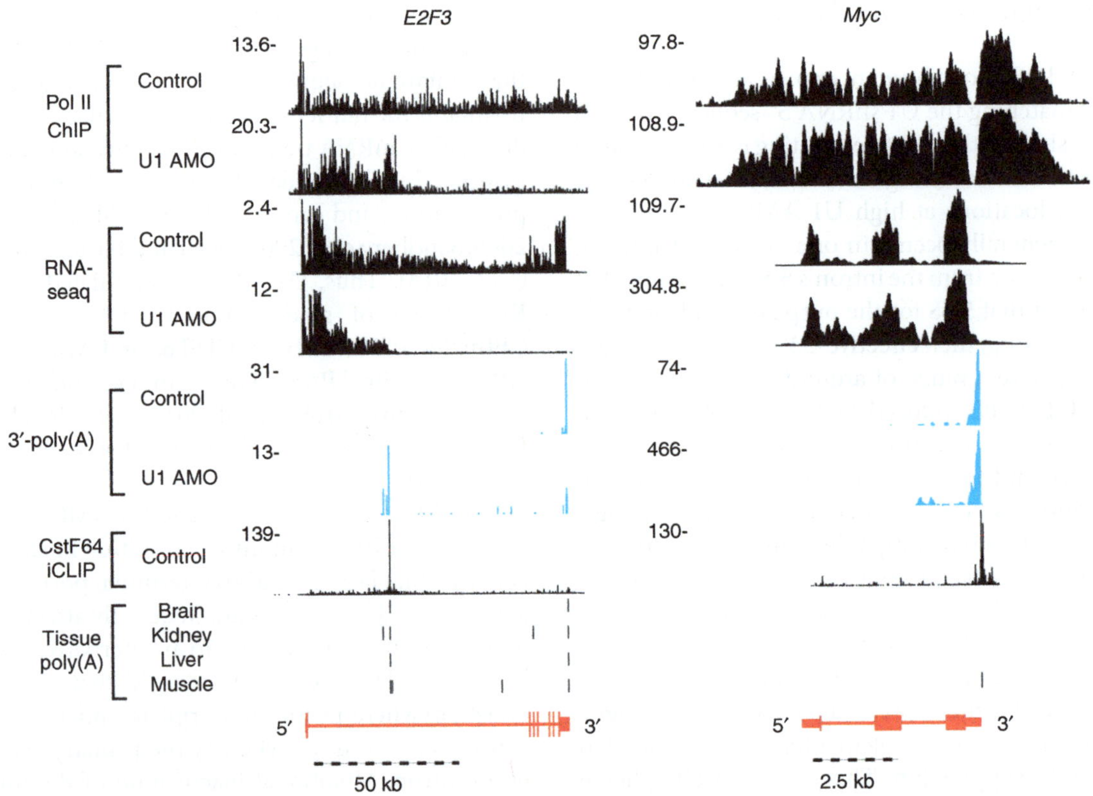

Figure 2. Genome browser images of a medium/large, PCPAed gene, *E2F3*, and a small, up-regulated gene, *Myc*, in control and U1 antisense morpholino oligonucleotide (AMO)-transfected cells. The distinct RNA-seq read changes in these two genes show the gene-size-dependent transcriptome changes after U1 AMO treatment. Further evidence for these peaks being premature 3′-end cleavage and polyadenylation (PCPA) is shown with the overlap between nongenomically encoded 3′-poly(A) reads from our own poly(A) selected RNA-seq samples shown in blue, known iCLIP binding sites for the cleavage and polyadenylation (CPA) factor CstF64 (Yao et al. 2012) shown in black, and poly(A) sites found in various human tissues (brain, kidney, liver, and muscle; Derti et al. 2012) shown as vertical black bars. Schematic gene structures (based on RefSeq, hg19) are depicted in red, with horizontal lines indicating introns and boxes indicating exons.

downstream from the PAS (Fig. 2). The gradual decline in the post-3′-end section, described as the termination zone (TZ) (Fong et al. 2015), is consistent with the torpedo termination model in which XRN2 exonuclease degrades, in the 5′ to 3′ direction, the unprotected 5′-end of the transcript (because of a lack of a 5′-methylated G cap) trailing from Pol II before it catches up and causes the polymerase to release (Connelly and Manley 1988; Fong et al. 2015; Proudfoot 2016). In PCPAed genes, Pol II patterns were the same, including a TZ, except that the PCPA point marked the 3′-end. Pol II occupancy downstream, through the rest of the gene, was sharply reduced or eliminated. These observations indicated that PCPA causes premature transcription termination, supporting a cotranscriptional U1 telescripting model as opposed to an alternative mechanism whereby PCPA products are processed from full-length transcripts, posttranscriptionally. The Pol II ChIP-seq patterns, which tracked with nascent RNAs also indicated that transcription initiation was not inhibited despite PCPA; rather, transcription continued to flow into genes.

4 SELECTIVE TELESCRIPTING DEPENDENCE OF LONG GENES

Improvements in RNA-seq instruments and greater availability made next-generation RNA-seq a tool of choice for whole-transcriptome sequencing, which we used to obtain more comprehensive definition of U1 AMO's effect at nucleotide resolution (Fig. 2). Knowing that PCPA and telescripting are cotranscriptional, we used a brief (5–30 min at 3.5 to 7.5 h post-AMO transfection) metabolic labeling with 4-thiouridine and selected thiol-labeled RNAs for RNA-seq, which enhanced detection of nascent transcripts, short-lived PCPA products and newly spliced mRNAs (Dolken et al. 2008; Younis et al. 2013; Oh et al. 2017). Of note, in terms of technologies, HIDE-seq remains a powerful tool detecting segmental differences, especially rare ones, among RNAs in multiple samples. Although dedicated 3′-poly(A)-seq is more suitable for pinpointing locations of this specific processing step, it does not inform about any other transcriptome changes.

PCPA calling algorithms developed to capture characteristic RNA-seq features of PCPAed transcripts, consisting of increased reads in the 5′-portions of introns and decreased reads in downstream exons, detected PCPA in transcripts of thousands of genes in high U1 AMO-treated human (HeLa) cells. Multiple 3′-poly(A) reads were frequently detected in the same gene along with clusters of closely spaced 3′-poly(A) reads in the same intron, suggesting that full-length transcription in some genes requires multiple actionable PASs to be suppressed. Actionable PASs in introns and throughout pre-mRNAs are therefore transcription elongation checkpoints that depend on U1 and other factors to clear. The higher number of PASs in large genes, that arose stochastically, makes them more dependent on telescripting, which makes PCPA a gene regulation mechanism based on gene size, which to our knowledge is unprecedented.

Many U1 AMO-induced PCPAs coincided with 3′-poly(A) reads in normal (control) cells and tissues (Derti et al. 2012), indicating that, despite U1's abundance, telescripting is insufficient to permanently suppress all PASs (Fig. 2). The resulting loss of polymerases in midgenes is a process we call transcription attrition. The extent of this unanticipated phenomenon is probably much greater than the 3′-poly(A) reads suggest because PCPAed transcripts are generally unstable and rapidly cleared away by the exosome (Lubas et al. 2015; Ogami et al. 2018). Although it seems wasteful, transcription attrition likely has a biological purpose.

However, around 1000 genes showed no evidence of PCPA under the same U1 AMO conditions (e.g., Myc; Fig. 2). A clear difference between this group and the PCPAed genes was gene size, which in higher eukaryotes is derived almost entirely from intron size; as PCPAed genes had a median size of 39 kb versus 14.2 kb for non-PCPAed genes (median size of all expressed genes was 22.8 kb). This showed that U1 telescripting is selectively required for full-length transcription of large genes. Remarkably, many of the PCPA-resistant genes were up-regulated (median 6.8 kb), producing full-length, spliced mRNAs, in the same environment where widespread splicing inhibition was evident and large genes were PCPAed (Oh et al. 2017).

The ability of acute stimuli to elicit PCPA, including in 3′ UTRs, could be the result of a transient U1 shortage caused by transcription up-regulation that draws U1 to an increasing number of nascent transcripts. Increasing U1 levels to keep up with greater demand would inevitably be slower because it entails an elaborate assembly of U1 snRNA with Sm proteins by the SMN complex in the cytoplasm and additional processing and transport steps. Alternatively, cell stimulation might change the balance between PCPA and telescripting in some other way. Regardless of its mechanism, the shift to shorter transcripts suggests that transient U1 telescripting insufficiency is a built-in aspect of immediate/early response.

The U1 AMO transfections have also tested the effect of U1 base-pairing interference on splicing in human cells. As expected, high U1 AMO strongly reduced splicing, but not uniformly. Intron retention, a reflection of splicing inhibition, was apparent in small and intermediate size introns. In large genes, the number of splicing events was sharply lower, however much of it was secondary to PCPA, which eliminated the opportunity of splicing from downstream introns. However, splicing in small genes that were up-regulated with U1 AMO was surprisingly robust in the same environment. How might this be explained? One potential explanation is a residual amount of uninhibited U1 remained and is sufficient for splicing in these genes, typically in all their introns. Another possibility is that these introns can use U1 that has U1 AMO bound, without U1 snRNA base-pairing. U1's ability to interact with nascent transcripts independent of base-pairing has been described (Spiluttini et al. 2010). A third potential explanation is that these introns could splice even without U1 at all. Previous studies have shown that splicing in U1-depleted extracts can be restored by the addition of excess of SR proteins, a group of splicing-activating hnRNP proteins (Crispino et al. 1994; Tarn and Steitz 1994; Fukumura et al. 2009). It is plausible that by decreasing the number of introns that compete for the same factors, PCPA creates similar conditions of excess SR proteins in the nuclei of cells.

For this model to make sense, sustaining or increasing mRNA productivity from small genes that are up-regulated with U1 AMO must be critically important, or, more generally, up-regulation of these genes and simultaneous PCPA in large genes must be beneficial. Several arguments support this proposition. The small genes that are up-regulated with U1 AMO are ubiquitously expressed and enriched in primary response genes that are induced during acute cell stimulation. They include genes such as Myc, Cyr61, and GADD45B whose functions are necessary to adapt to adverse environmental changes and stressors. In contrast, these functions are underrepresented in large genes, which encode more diverse functions generally expressed in differentiated tissues. For example, neuronal and developmental genes are among the largest (Bertagnolli et al. 2013; Gabel et al. 2015), which makes them highly susceptible to PCPA (Fig. 3). Under circumstances that threaten cell

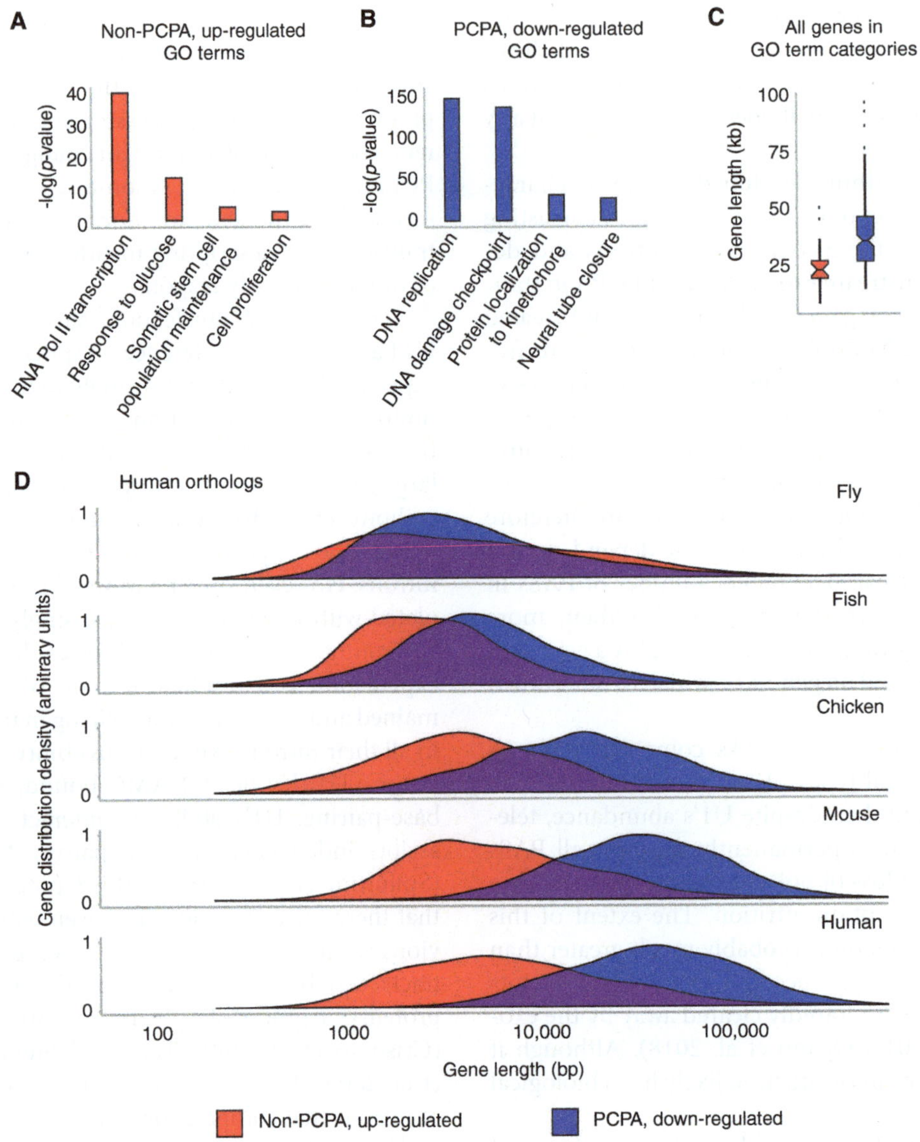

Figure 3. Gene ontology (GO) term enrichment and gene size expansion in premature 3′-end cleavage and polyadenylation (PCPA) sensitive genes compared with non-PCPAed, up-regulated genes. The enrichment of the GO terms (Supek et al. 2011; Fang et al. 2016) for the top 50% of genes, ranked by fold change, in either non-PCPAed, up-regulated (*A*) or PCPAed, down-regulated (*B*) genes are shown. The gene size for all human genes in these functional categories, not just expressed in our samples, are shown as box plots (*C*) that were found enriched in either PCPAed genes (blue, median size 38 kb) or non-PCPAed, up-regulated genes (red, median size 23 kb). Density plots showing the gene size (*D*) of human gene orthologs among non-PCPAed, up-regulated genes and PCPAed genes in five metazoans. The sizes of non-PCPAed and up-regulated genes did not change across *Drosophila melanogaster* (median 3.5 kb), *Takifugu rubripes* (median 2.5 kb), *Gallus gallus* (median 4.3 kb), *Mus musculus* (median 5.9 kb), and *Homo sapiens* (median 6.8 kb); but PCPAed and down-regulated genes expanded significantly in these same species (medians 4.4 kb, 5.2 kb, 18.4 kb, 38 kb, and 45.0 kb, respectively).

survival PCPA of large genes transiently might be a strategic sacrifice to enhance expression of cell survival genes.

The architecture of large and small genes seems well-suited for such a model. Small genes take less time to transcribe and their small size reduces susceptibility to PCPA. Interestingly, comparison of orthologous genes in several organisms showed that the size of PCPAed genes increased in vertebrates, especially in mammals, whereas small genes that are up-regulated with U1 AMO maintained a relatively small size (Fig. 3). Striking intron size expansion occurred in the course of evolution of vertebrates, especially in mammals (Catania and Lynch 2008; Gelfman et al. 2012; Rogo-

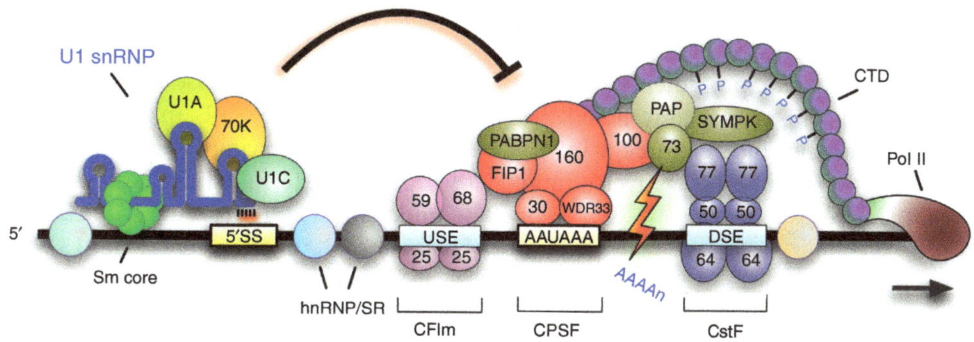

Figure 4. A schematic representing the role of U1 telescripting, suppression of premature transcription termination, by 3′ cleavage and polyadenylation (CPA) from polyadenylation signals (PASs) scattered throughout genes' nascent transcripts. Shown are the common components known to carry out 3′ cleavage and polyadenylation processing at the canonical terminal PAS in the last exon; it is unknown which of these play a role in premature 3′-end cleavage and polyadenylation (PCPA) in introns. The precise location of U1 small nuclear ribonucleo protein (snRNP) binding relative to PASs is unknown but is likely to be generally within 1 kb. Transcription length depends on the balance between available U1 and CPA factors, and on the density of potential U1 base-pairing sites, which include 5′SS, cryptic 5′SS, and other sequences that U1 can bind, as well as PASs.

zin et al. 2012). The significance of this expansion and its impact on gene expression are unknown. However, our studies suggest that intron size expansion was a strategic process, rather than random. It was selected against in essential acute response and survival genes but permissible in other genes. The selective intron size expansion stratified genes according to size and function, made progressively more polarized in vertebrate evolution. Its adoption suggests that it was beneficial.

We suggest that an important benefit of large introns is that they increase the opportunity to use PCPA to rapidly decrease competition from large genes for both transcription and pre-mRNA processing factors to boost expression from small, PCPA-insensitive genes. Many studies suggest that competition among pre-mRNAs for various binding factors can have strong effects on mRNA synthesis of many other genes (Timchenko et al. 1996; Miller et al. 2000; Cooper et al. 2009; Berg et al. 2012; Elkon et al. 2013; Munding et al. 2013). Another interesting and unexplained aspect of intron expansion is its positional bias for introns in the 5′ portion of genes. In humans, the first or second introns are frequently the largest (Bradnam and Korf 2008), and it is there that PCPA typically occurs. As large genes tend to have more exons, PCPA in the first or second introns eliminates multiple downstream exons and introns. We speculate that this architecture was selected for because it leverages PCPA to maximize the loss of competition for splicing resources and thus the potential gain for non-PCPAed genes.

A potential advantage of a polarized genome with genes stratified by size and function is that provides a layer of regulation and an ability to shift gene expression priorities nearly instantly, without a need for de novo synthesis of transcription and pre-mRNA processing factors. It is difficult to imagine that metazoan genome evolution could have taken such a course without U1 telescripting. Figure 4 presents a schematic overview of the role of U1 telescripting in gene expression regulation.

5 BEYOND PROTEIN-CODING GENES: TELESCRIPTING'S ROLE IN SHAPING THE TRANSCRIPTOME

As the shift to shorter mRNA and protein isoforms, represented by homer-1, illustrated, PCPA is not necessarily destructive and is also potentially useful for increasing proteome diversity (Berg et al. 2012). An additional way in which PCPA in introns contributes to protein isoform diversity is by the translation of an open reading frame that extends into the intron. Cartegni's group described such cases for the EGF receptor and related receptor protein kinases (Vorlova et al. 2011). Like in homer-1, the short protein isoform lacks or has a different CTD and is antagonistic to the full-length protein.

Adding an important dimension, several studies from Phillip Sharp's and Torben Jensen's groups discovered that Pol II upstream antisense RNAs (uaRNAs) (Flynn et al. 2011, also known as PROMPTs; Preker et al. 2008) generally receive insufficient U1 telescripting relative to their prevalence of PASs (referred to in that study as U1-PAS axis), resulting in their rapid elimination by the nuclear exosome targeting complex, thereby reinforcing transcription in the sense direction (Almada et al. 2013; Andersen et al. 2013; Ntini et al. 2013; Lubas et al. 2015). This highlighted U1's important role in transcription regulation, placing it as a central and bidirectional regulator of transcription elongation from divergent promoters (Almada et al. 2013; Ntini et al. 2013). This same study noted the evolutionary pressure to maintain and reinforce a U1 to PAS ratio in the sense direction and a low or inverse ratio in the antisense direction.

6 TELESCRIPTING MECHANISM

The molecular mechanism by which U1 telescripting suppresses actionable PASs particularly in introns remains unknown. CPAFs bind directly to nascent transcripts' PASs cotranscriptionally with the transcription elongation complex and are dependent on Pol II CTD phosphorylation state (Buratowski 2009; Perales and Bentley 2009; Eick and Geyer 2013; Bentley 2014). We have noted that ultraviolet (UV) cross-linking studies from other groups (Martin et al. 2012; Yao et al. 2012; Chan et al. 2014; Schonemann et al. 2014) show several CPAFs binding in the vicinity of prominent PCPA locations in introns in control cells (e.g., Fig. 2) (Oh et al. 2017), indicating that intronic actionable PASs are recognized by at least some of the same or very similar factors that cleave and polyadenylate canonical 3′-end PASs. This makes it unlikely that U1 shields PASs. However, there is lack of critical information on where U1 binds such that it suppresses a PAS. Eric Lander's group (Engreitz et al. 2014) has identified U1 base-pairing at 5′SS and in introns in mouse embryonic stem cells, but the dataset is not sufficiently deep and is from a different organism and cell type (our studies have focused on human cells) to address this question. It is therefore not possible to distinguish among various general models, including whether U1 telescripting is mediated by direct U1 interaction with CPAFs.

Previous studies showing CPA-inhibitory activity of two U1 proteins, U1A and U1-70K albeit in specific contexts, may be relevant. U1A as a U1 snRNP-free protein dimer inhibits CPA from an adjacent PAS in the terminal exon of both its own pre-mRNA and a few others that contain tandem stem-loops highly similar to U1 snRNA SL2 (Boelens et al. 1993; Gunderson et al. 1994, 1997; Klein Gunnewiek et al. 2000; Phillips and Gunderson 2003; Workman et al. 2014). In vitro studies showed that this inhibition is mediated by U1A binding to poly(A) polymerase (PAP). U1-70K can also inhibit PAP bound to U1 using the similar PAP-inhibitory motifs in U1A (Gunderson et al. 1998). As a U1-free protein, U1-70K can also associate with the PAS' upstream element (USE)-binding CFIm subunit (Ruegsegger et al. 1996; Awasthi and Alwine 2003). It remains to be determined whether U1A and U1-70K also have such CPA-inhibitory activities in the context of U1 snRNP and, if so, whether such activities applies to PASs in introns. In specific cases, U1 base-paired to a 5′SS-like sequence adjacent to the PAS in a late-phase transcript in bovine papilloma virus (BPV) and 5′LTRs in HIV-1 and foamy viruses or tethered by engineering extensive base-pairing in proximity of a PAS in the terminal exon can inhibit usage of that PAS (Gunderson et al. 1998; Ashe et al. 2000; Vagner et al. 2000; Fortes et al. 2003; Schrom et al. 2013).

U1 base-paired upstream and in the vicinity of the PAS (<1 kb) could either prevent formation of an active CPA complex or inhibit its activity. CPA occurs at the first actionable PAS that is not suppressed by U1. U1 shortage thus increases the likelihood of PCPA. Like U1 AMO, knockdowns of CFIm25/CPSF5, CFIm68/CFSF6 and PABPN1 can elicit proximal PAS usage in tandem PASs in 3′ UTRs (Gruber et al. 2012; Jenal et al. 2012; Elkon et al. 2013; Masamha et al. 2014; Li et al. 2015). It will be interesting to determine whether these knockdowns and U1 AMO induce the same APA changes, which will help determine whether these CPAFs act in concert with U1 in the last exon.

Few PCPAs were detected in internal exons in our studies. It is possible that exons are bound by factors that shield PASs and that their small size and relative GC-richness (Amit et al. 2012) decreased the likelihood of having AU-rich PAS hexanucleotides. Small genes and introns are less susceptible both because their small introns decrease the opportunity for PASs to arise and because kinetic competition from splicing removes introns more quickly.

As PCPA is cotranscriptional and PASs induce Pol II pausing (Rigo et al. 2005; Davidson et al. 2014; Kaida 2016), we investigated the effect of small molecule reagents, 5,6-dichloro-1-β-D-ribofuranosylbenzimidazole (DRB) (Chodosh et al. 1989) and camptothecin (CP) (Pommier 2006), which slow down Pol II transcription on PCPA. DRB inhibits CDK9 and related kinases that phosphorylate the Pol II CTD, which is required to release Pol II pausing at promoter proximal pause sites (PPP; within 100 nt from the TSS) and at PASs (Guo and Price 2013; Kwak and Lis 2013; Laitem et al. 2015). Immediately following DRB washout, as PPP is released, polymerases rapidly flow into genes and are likely slowed down as the leading polymerases encounter a barrier or checkpoint, forcing lagging polymerases to also slow. CP slows transcription through inhibition of topoisomerase I, which relieves transcription-obstructing DNA supercoiling. Despite different chemistries and mechanisms of action, both DRB and CP caused PCPA at similar locations as U1 AMO (Fig. 5), suggesting that slower transcription and/or pausing increases the propensity for PCPA. Interestingly, CP and DNA damage, which create transcription roadblocks, have been shown previously to selectively decrease expression from large genes (McKay

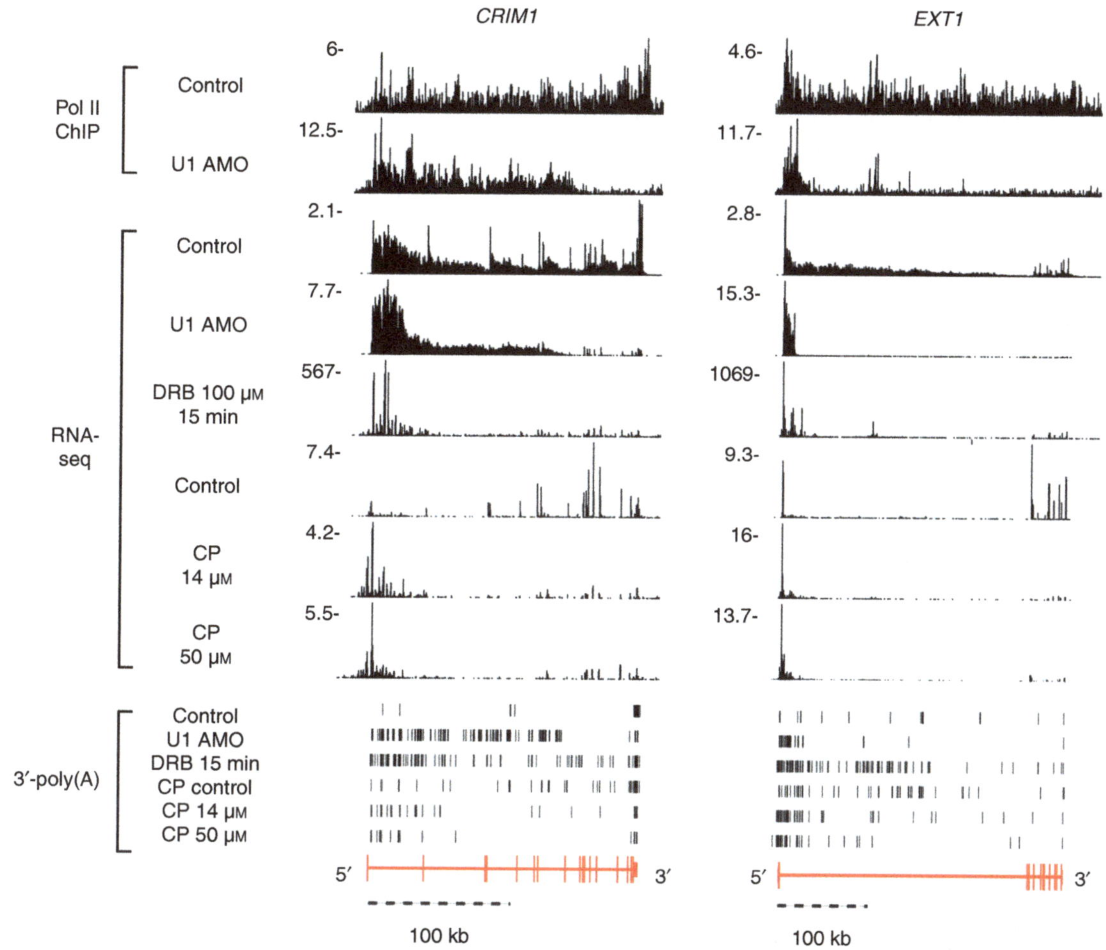

Figure 5. Genome browser views showing that, in the example genes *CRIM1* and *EXT1*, DRB and CP induce premature 3′-end cleavage and polyadenylation (PCPA), suggesting that slower Pol II elongation speed or pausing increases the propensity for PCPA. PCPA locations induced by DRB and CP overlap with U1 antisense morpholino oligonucleotide (AMO)-induced PCPAs and their nature is confirmed by nongenomic 3′-poly(A) reads from samples, shown as vertical black bars, indicating that these termination sites are true PCPAs, and not just random nuclease cleavage sites. RNAs produced during the 15-min DRB washout phase and 30-min CP treatment were metabolically labeled, selected, and sequenced (RNA-seq). The signal beyond the PCPA point seen in the Pol II ChIP-seq sample represents a transcription termination zone. Pol II occupancy beyond that site was strongly decreased.

et al. 2004; Andrade-Lima et al. 2015). PCPA could provide checkpoints for orderly removal of polymerases to prevent overcrowding and transcription errors.

A recent study from Sharp's group showed that PCPA frequently occurs at the first stable nucleosomes relative to the TSS (Chiu et al. 2018). This study showed that U1 AMO-induced PCPA sites tend to cluster at the edge of CpG islands at the first stable nucleosomes downstream from the TSS. Global run-on sequencing and use of flavopiridol (a CDK9 inhibitor, similar to DRB) linked PCPAs to Pol II pausing in these locations, thus showing the influence of transcription elongation speed on telescripting. These first stable nucleosomes act as speed bumps, slowing Pol II and allowing more time for the CPAF machinery to act on a PAS. Chromatin structure, through its effect on transcription elongation speed and resulting changes in Pol II CTD phosphorylation, could therefore also play a role in the balance between telescripting and PCPA.

7 HOW CELLS MAKE A GREAT ABUNDANCE OF U1

The demand for high U1 abundance to use for telescripting required cells of higher organisms to produce more U1 than other spliceosomal snRNPs (with the possible exception of U2, which is also very abundant). To some extent, this is helped by the greater number (potentially >100) of U1 genes, although it is unknown how many of those are transcriptionally active (O'Reilly et al. 2013). However, pre-snRNA transcription may not be a limiting factor for U1 synthesis. The rate-limiting step in snRNP biogenesis is likely the assembly of the Sm core. Interestingly, recent studies revealed that the U1-specific RBP, U1-70K can hijack the SMN–Gemin2–Sm pentamer complex, a key intermediate of the Sm core assembly, and promote access of U1 snRNA to the SMN complex over other snRNAs (So et al. 2016). This reveals an additional and a U1-exclusive Sm core assembly pathway at the expense of other snRNPs, which explains at least in part how U1 abundance is achieved in cells.

8 CONCLUDING REMARKS

The discovery that U1 snRNA 5′-end base-pairing (to pre-mRNAs' 5′SS) provides crucial recognition for the removal of introns was a conceptual breakthrough later generalized to other *trans*-acting small RNPs. It is a marvel of nature's ingenuity that U1, through the same U1 snRNA 9-nucleotide sequence, is also crucial for ensuring that transcription continues far enough to produce introns. By suspending termination, U1 increases its own chances to find a 3′SS, and thereby remove itself from the pre-mRNA. The much earlier discovery of U1's role in splicing influences a more splicing-centric perspective, but an evolutionary perspective invites the question of which came first: splicing or telescripting? Although the answer is unknown and U1's origin is uncertain, this abundant noncoding small RNP has emerged as a central regulator of the Pol II transcriptome.

Technological advances, application of new tools to study RNA and RNPs have made possible the discoveries on telescripting. Understanding the mechanism by which U1 telescripting suppresses PASs to avert premature transcription termination now presents exciting new challenges. Additional tools will be necessary to dissect the biochemical and structural basis of U1 telescripting, including U1 interactions with components of the transcription and CPA machineries, and to precisely determine its binding locations on nascent transcripts and visualize PCPA in cells in real time, and identify modulators by high-throughput screening.

Although physiological circumstances in which telescripting is drastically inhibited, as occurs with high U1 AMO doses, have not been described, it can be reasonably predicted that some viruses have evolved mechanisms to inhibit telescripting, including by decoying U1, to benefit from resources that are tied up by long-distance transcription.

ACKNOWLEDGMENTS

We thank our laboratory colleagues, present and past, whose shared curiosity to understand unexpected observations and team work contributed to this research. This work was supported by a grant from the National Institutes of Health (R01GM112923 to G.D.). G.D. is an Investigator of the Howard Hughes Medical Institute.

REFERENCES

Almada AE, Wu X, Kriz AJ, Burge CB, Sharp PA. 2013. Promoter directionality is controlled by U1 snRNP and polyadenylation signals. *Nature* **499:** 360–363.

Amit M, Donyo M, Hollander D, Goren A, Kim E, Gelfman S, Lev-Maor G, Burstein D, Schwartz S, Postolsky B, et al. 2012. Differential GC content between exons and introns establishes distinct strategies of splice-site recognition. *Cell Rep* **1:** 543–556.

Andersen PR, Domanski M, Kristiansen MS, Storvall H, Ntini E, Verheggen C, Schein A, Bunkenborg J, Poser I, Hallais M, et al. 2013. The human cap-binding complex is functionally connected to the nuclear RNA exosome. *Nat Struct Mol Biol* **20:** 1367–1376.

Andrade-Lima LC, Veloso A, Paulsen MT, Menck CF, Ljungman M. 2015. DNA repair and recovery of RNA synthesis following exposure to ultraviolet light are delayed in long genes. *Nucleic Acids Res* **43:** 2744–2756.

Ashe MP, Furger A, Proudfoot NJ. 2000. Stem-loop 1 of the U1 snRNP plays a critical role in the suppression of HIV-1 polyadenylation. *RNA* **6:** 170–177.

Awasthi S, Alwine JC. 2003. Association of polyadenylation cleavage factor I with U1 snRNP. *RNA* **9:** 1400–1409.

Baserga SJ, Steitz JA. 1993. The diverse world of small ribonucleoproteins. In *The RNA world* (ed. Gesteland RF, Atkins JF), pp. 359–382. Cold Spring Harbor Laboratory Press, Cold Spring Harbor, New York.

Battle DJ, Kasim M, Yong J, Lotti F, Lau CK, Mouaikel J, Zhang Z, Han K, Wan L, Dreyfuss G. 2006. The SMN complex: An assembly machine for RNPs. *Cold Spring Harb Symp Quant Biol* **71:** 313–320.

Bentley DL. 2014. Coupling mRNA processing with transcription in time and space. *Nat Rev Genet* **15:** 163–175.

Berg MG, Singh LN, Younis I, Liu Q, Pinto AM, Kaida D, Zhang Z, Cho S, Sherrill-Mix S, Wan L, et al. 2012. U1 snRNP determines mRNA length and regulates isoform expression. *Cell* **150:** 53–64.

Bertagnolli NM, Drake JA, Tennessen JM, Alter O. 2013. SVD identifies transcript length distribution functions from DNA microarray data and reveals evolutionary forces globally affecting GBM metabolism. *PLoS ONE* **8:** e78913.

Boelens WC, Jansen EJ, van Venrooij WJ, Stripecke R, Mattaj IW, Gunderson SI. 1993. The human U1 snRNP-specific U1A protein inhibits polyadenylation of its own pre-mRNA. *Cell* **72:** 881–892.

Bradnam KR, Korf I. 2008. Longer first introns are a general property of eukaryotic gene structure. *PLoS ONE* **3:** e3093.

Buratowski S. 2009. Progression through the RNA polymerase II CTD cycle. *Mol Cell* **36:** 541–546.

Catania F, Lynch M. 2008. Where do introns come from? *PLoS Biol* **6:** e283.

Cauchi RJ. 2010. SMN and Gemins: "We are family" … or are we? Insights into the partnership between Gemins and the spinal muscular atrophy disease protein SMN. *Bioessays* **32:** 1077–1089.

Chan SL, Huppertz I, Yao C, Weng L, Moresco JJ, Yates JR III, Ule J, Manley JL, Shi Y. 2014. CPSF30 and Wdr33 directly bind to AAUAAA in mammalian mRNA 3′ processing. *Genes Dev* **28:** 2370–2380.

Chiu AC, Suzuki HI, Wu X, Mahat DB, Kriz AJ, Sharp PA. 2018. Transcriptional pause sites delineate stable nucleosome-associated premature polyadenylation suppressed by U1 snRNP. *Mol Cell* **69:** 648–663. e647.

Chodosh LA, Fire A, Samuels M, Sharp PA. 1989. 5,6-Dichloro-1-β-D-ribofuranosylbenzimidazole inhibits transcription elongation by RNA polymerase II in vitro. *J Biol Chem* **264:** 2250–2257.

Connelly S, Manley JL. 1988. A functional mRNA polyadenylation signal is required for transcription termination by RNA polymerase II. *Genes Dev* **2:** 440–452.

Cooper TA, Wan L, Dreyfuss G. 2009. RNA and disease. *Cell* **136:** 777–793.

Crispino JD, Blencowe BJ, Sharp PA. 1994. Complementation by SR proteins of pre-mRNA splicing reactions depleted of U1 snRNP. *Science* **265:** 1866–1869.

Davidson L, Muniz L, West S. 2014. 3′ end formation of pre-mRNA and phosphorylation of Ser2 on the RNA polymerase II CTD are reciprocally coupled in human cells. *Genes Dev* **28:** 342–356.

Derti A, Garrett-Engele P, Macisaac KD, Stevens RC, Sriram S, Chen R, Rohl CA, Johnson JM, Babak T. 2012. A quantitative atlas of polyadenylation in five mammals. *Genome Res* **22:** 1173–1183.

Diatchenko L, Lau YF, Campbell AP, Chenchik A, Moqadam F, Huang B, Lukyanov S, Lukyanov K, Gurskaya N, Sverdlov ED, et al. 1996. Suppression subtractive hybridization: A method for generating differentially regulated or tissue-specific cDNA probes and libraries. *Proc Natl Acad Sci* **93:** 6025–6030.

Dolken L, Ruzsics Z, Radle B, Friedel CC, Zimmer R, Mages J, Hoffmann R, Dickinson P, Forster T, Ghazal P, et al. 2008. High-resolution gene expression profiling for simultaneous kinetic parameter analysis of RNA synthesis and decay. *RNA* **14:** 1959–1972.

Eick D, Geyer M. 2013. The RNA polymerase II carboxy-terminal domain (CTD) code. *Chem Rev* **113:** 8456–8490.

Elkon R, Ugalde AP, Agami R. 2013. Alternative cleavage and polyadenylation: Extent, regulation and function. *Nat Rev Genet* **14:** 496–506.

Engreitz JM, Sirokman K, McDonel P, Shishkin AA, Surka C, Russell P, Grossman SR, Chow AY, Guttman M, Lander ES. 2014. RNA–RNA interactions enable specific targeting of noncoding RNAs to nascent pre-mRNAs and chromatin sites. *Cell* **159:** 188–199.

Fang H, Knezevic B, Burnham KL, Knight JC. 2016. XGR software for enhanced interpretation of genomic summary data, illustrated by application to immunological traits. *Genome Med* **8:** 129.

Fischer U, Liu Q, Dreyfuss G. 1997. The SMN–SIP1 complex has an essential role in spliceosomal snRNP biogenesis. *Cell* **90:** 1023–1029.

Fischer U, Englbrecht C, Chari A. 2011. Biogenesis of spliceosomal small nuclear ribonucleoproteins. *Wiley Interdiscip Rev RNA* **2:** 718–731.

Flavell SW, Kim TK, Gray JM, Harmin DA, Hemberg M, Hong EJ, Markenscoff-Papadimitriou E, Bear DM, Greenberg ME. 2008. Genome-wide analysis of MEF2 transcriptional program reveals synaptic target genes and neuronal activity-dependent polyadenylation site selection. *Neuron* **60:** 1022–1038.

Flynn RA, Almada AE, Zamudio JR, Sharp PA. 2011. Antisense RNA polymerase II divergent transcripts are P-TEFb dependent and substrates for the RNA exosome. *Proc Natl Acad Sci* **108:** 10460–10465.

Fong N, Brannan K, Erickson B, Kim H, Cortazar MA, Sheridan RM, Nguyen T, Karp S, Bentley DL. 2015. Effects of transcription elongation rate and Xrn2 exonuclease activity on RNA polymerase II termination suggest widespread kinetic competition. *Mol Cell* **60:** 256–267.

Fortes P, Cuevas Y, Guan F, Liu P, Pentlicky S, Jung SP, Martinez-Chantar ML, Prieto J, Rowe D, Gunderson SI. 2003. Inhibiting expression of specific genes in mammalian cells with 5′ end-mutated U1 small nuclear RNAs targeted to terminal exons of pre-mRNA. *Proc Natl Acad Sci* **100:** 8264–8269.

Fukumura K, Taniguchi I, Sakamoto H, Ohno M, Inoue K. 2009. U1-independent pre-mRNA splicing contributes to the regulation of alternative splicing. *Nucleic Acids Res* **37:** 1907–1914.

Gabanella F, Butchbach ME, Saieva L, Carissimi C, Burghes AH, Pellizzoni L. 2007. Ribonucleoprotein assembly defects correlate with spinal muscular atrophy severity and preferentially affect a subset of spliceosomal snRNPs. *PLoS ONE* **2:** e921.

Gabel HW, Kinde B, Stroud H, Gilbert CS, Harmin DA, Kastan NR, Hemberg M, Ebert DH, Greenberg ME. 2015. Disruption of DNA-methylation-dependent long gene repression in Rett syndrome. *Nature* **522:** 89–93.

Gelfman S, Burstein D, Penn O, Savchenko A, Amit M, Schwartz S, Pupko T, Ast G. 2012. Changes in exon–intron structure during vertebrate evolution affect the splicing pattern of exons. *Genome Res* **22:** 35–50.

Gilmour DS, Lis JT. 1984. Detecting protein–DNA interactions in vivo: Distribution of RNA polymerase on specific bacterial genes. *Proc Natl Acad Sci* **81:** 4275–4279.

Gruber AR, Martin G, Keller W, Zavolan M. 2012. Cleavage factor I_m is a key regulator of 3′ UTR length. *RNA Biol* **9:** 1405–1412.

Gruber AR, Martin G, Keller W, Zavolan M. 2014. Means to an end: Mechanisms of alternative polyadenylation of messenger RNA precursors. *Wiley Interdiscip Rev RNA* **5:** 183–196.

Gunderson SI, Beyer K, Martin G, Keller W, Boelens WC, Mattaj LW. 1994. The human U1A snRNP protein regulates polyadenylation via a direct interaction with poly(A) polymerase. *Cell* **76:** 531–541.

Gunderson SI, Vagner S, Polycarpou-Schwarz M, Mattaj IW. 1997. Involvement of the carboxyl terminus of vertebrate poly(A) polymerase in U1A autoregulation and in the coupling of splicing and polyadenylation. *Genes Dev* **11:** 761–773.

Gunderson SI, Polycarpou-Schwarz M, Mattaj IW. 1998. U1 snRNP inhibits pre-mRNA polyadenylation through a direct interaction between U1 70K and poly(A) polymerase. *Mol Cell* **1:** 255–264.

Guo J, Price DH. 2013. RNA polymerase II transcription elongation control. *Chem Rev* **113:** 8583–8603.

Gurskaya NG, Diatchenko L, Chenchik A, Siebert PD, Khaspekov GL, Lukyanov KA, Vagner LL, Ermolaeva OD, Lukyanov SA, Sverdlov ED. 1996. Equalizing cDNA subtraction based on selective suppression of polymerase chain reaction: Cloning of Jurkat cell transcripts induced by phytohemaglutinin and phorbol 12-myristate 13-acetate. *Anal Biochem* **240:** 90–97.

Guthrie C, Patterson B. 1988. Spliceosomal snRNAs. *Annu Rev Genet* **22:** 387–419.

Heasman J. 2002. Morpholino oligos: Making sense of antisense? *Dev Biol* **243:** 209–214.

Hodnett JL, Busch H. 1968. Isolation and characterization of uridylic acid-rich 7 S ribonucleic acid of rat liver nuclei. *J Biol Chem* **243:** 6334–6342.

Jenal M, Elkon R, Loayza-Puch F, van Haaften G, Kuhn U, Menzies FM, Vrielink JA, Bos AJ, Drost J, Rooijers K, et al. 2012. The poly(A)-binding protein nuclear 1 suppresses alternative cleavage and polyadenylation sites. *Cell* **149:** 538–553.

Kaida D. 2016. The reciprocal regulation between splicing and 3′-end processing. *Wiley Interdiscip Rev RNA* **7:** 499–511.

Kaida D, Motoyoshi H, Tashiro E, Nojima T, Hagiwara M, Ishigami K, Watanabe H, Kitahara T, Yoshida T, Nakajima H, et al. 2007. Spliceostatin A targets SF3b and inhibits both splicing and nuclear retention of pre-mRNA. *Nat Chem Biol* **3:** 576–583.

Kaida D, Berg MG, Younis I, Kasim M, Singh LN, Wan L, Dreyfuss G. 2010. U1 snRNP protects pre-mRNAs from premature cleavage and polyadenylation. *Nature* **468:** 664–668.

Klein Gunnewiek JM, Hussein RI, van Aarssen Y, Palacios D, de Jong R, van Venrooij WJ, Gunderson SI. 2000. Fourteen residues of the U1 snRNP-specific U1A protein are required for homodimerization, cooperative RNA binding, and inhibition of polyadenylation. *Mol Cell Biol* **20:** 2209–2217.

Kotake Y, Sagane K, Owa T, Mimori-Kiyosue Y, Shimizu H, Uesugi M, Ishihama Y, Iwata M, Mizui Y. 2007. Splicing factor SF3b as a target of the antitumor natural product pladienolide. *Nat Chem Biol* **3:** 570–575.

Kwak H, Lis JT. 2013. Control of transcriptional elongation. *Annu Rev Genet* **47:** 483–508.

Laitem C, Zaborowska J, Isa NF, Kufs J, Dienstbier M, Murphy S. 2015. CDK9 inhibitors define elongation checkpoints at both ends of RNA polymerase II-transcribed genes. *Nat Struct Mol Biol* **22:** 396–403.

Lerner MR, Steitz JA. 1979. Antibodies to small nuclear RNAs complexed with proteins are produced by patients with systemic lupus erythematosus. *Proc Natl Acad Sci* **76:** 5495–5499.

Lerner MR, Boyle JA, Mount SM, Wolin SL, Steitz JA. 1980. Are snRNPs involved in splicing? *Nature* **283:** 220–224.

Li W, You B, Hoque M, Zheng D, Luo W, Ji Z, Park JY, Gunderson SI, Kalsotra A, Manley JL, et al. 2015. Systematic profiling of poly(A)$^+$ transcripts modulated by core 3′ end processing and splicing factors reveals regulatory rules of alternative cleavage and polyadenylation. *PLoS Genet* **11:** e1005166.

Lianoglou S, Garg V, Yang JL, Leslie CS, Mayr C. 2013. Ubiquitously transcribed genes use alternative polyadenylation to achieve tissue-specific expression. *Genes Dev* **27:** 2380–2396.

Liu Q, Fischer U, Wang F, Dreyfuss G. 1997. The spinal muscular atrophy disease gene product, SMN, and its associated protein SIP1 are in a complex with spliceosomal snRNP proteins. *Cell* **90:** 1013–1021.

Lubas M, Andersen PR, Schein A, Dziembowski A, Kudla G, Jensen TH. 2015. The human nuclear exosome targeting complex is loaded onto newly synthesized RNA to direct early ribonucleolysis. *Cell Rep* **10:** 178–192.

Magana-Mora A, Kalkatawi M, Bajic VB. 2017. Omni-PolyA: A method and tool for accurate recognition of poly(A) signals in human genomic DNA. *BMC Genomics* **18:** 620.

Martin G, Gruber AR, Keller W, Zavolan M. 2012. Genome-wide analysis of pre-mRNA 3′ end processing reveals a decisive role of human cleavage factor I in the regulation of 3′ UTR length. *Cell Rep* **1:** 753–763.

Masamha CP, Xia Z, Yang J, Albrecht TR, Li M, Shyu AB, Li W, Wagner EJ. 2014. CFIm25 links alternative polyadenylation to glioblastoma tumour suppression. *Nature* **510:** 412–416.

Matera AG, Wang Z. 2014. A day in the life of the spliceosome. *Nat Rev Mol Cell Biol* **15:** 108–121.

Matter N, Konig H. 2005. Targeted "knockdown" of spliceosome function in mammalian cells. *Nucleic Acids Res* **33:** e41.

Mayr C, Bartel DP. 2009. Widespread shortening of 3′UTRs by alternative cleavage and polyadenylation activates oncogenes in cancer cells. *Cell* **138:** 673–684.

McKay BC, Stubbert LJ, Fowler CC, Smith JM, Cardamore RA, Spronck JC. 2004. Regulation of ultraviolet light–induced gene expression by gene size. *Proc Natl Acad* **101:** 6582–6586.

Miller JW, Urbinati CR, Teng-Umnuay P, Stenberg MG, Byrne BJ, Thornton CA, Swanson MS. 2000. Recruitment of human muscleblind proteins to $(CUG)_n$ expansions associated with myotonic dystrophy. *EMBO J* **19:** 4439–4448.

Mount SM, Pettersson I, Hinterberger M, Karmas A, Steitz JA. 1983. The U1 small nuclear RNA–protein complex selectively binds a 5′ splice site in vitro. *Cell* **33:** 509–518.

Munding EM, Shiue L, Katzman S, Donohue JP, Ares M Jr. 2013. Competition between pre-mRNAs for the splicing machinery drives global regulation of splicing. *Mol Cell* **51:** 338–348.

Niibori Y, Hayashi F, Hirai K, Matsui M, Inokuchi K. 2007. Alternative poly(A) site-selection regulates the production of alternatively spliced vesl-1/homer1 isoforms that encode postsynaptic scaffolding proteins. *Neurosci Res* **57:** 399–410.

Ntini E, Jarvelin AI, Bornholdt J, Chen Y, Boyd M, Jorgensen M, Andersson R, Hoof I, Schein A, Andersen PR, et al. 2013. Polyadenylation site-induced decay of upstream transcripts enforces promoter directionality. *Nat Struct Mol Biol* **20:** 923–928.

Ogami K, Chen Y, Manley JL. 2018. RNA surveillance by the nuclear RNA exosome: Mechanisms and significance. *Noncoding RNA* **4:** 8.

Oh JM, Di C, Venters CC, Guo J, Arai C, So BR, Pinto AM, Zhang Z, Wan L, Younis I, et al. 2017. U1 snRNP telescripting regulates a size-function-stratified human genome. *Nat Struct Mol Biol* **24:** 993–999.

O'Reilly D, Dienstbier M, Cowley SA, Vazquez P, Drozdz M, Taylor S, James WS, Murphy S. 2013. Differentially expressed, variant U1 snRNAs regulate gene expression in human cells. *Genome Res* **23:** 281–291.

Padgett RA, Mount SM, Steitz JA, Sharp PA. 1983. Splicing of messenger RNA precursors is inhibited by antisera to small nuclear ribonucleoprotein. *Cell* **35:** 101–107.

Patel AA, Steitz JA. 2003. Splicing double: Insights from the second spliceosome. *Nat Rev Mol Cell Biol* **4:** 960–970.

Perales R, Bentley D. 2009. "Cotranscriptionality": The transcription elongation complex as a nexus for nuclear transactions. *Mol Cell* **36:** 178–191.

Phillips C, Gunderson S. 2003. Sequences adjacent to the 5′ splice site control U1A binding upstream of the IgM heavy chain secretory poly(A) site. *J Biol Chem* **278:** 22102–22111.

Pommier Y. 2006. Topoisomerase I inhibitors: Camptothecins and beyond. *Nat Rev Cancer* **6:** 789–802.

Preker P, Nielsen J, Kammler S, Lykke-Andersen S, Christensen MS, Mapendano CK, Schierup MH, Jensen TH. 2008. RNA exosome depletion reveals transcription upstream of active human promoters. *Science* **322:** 1851–1854.

Proudfoot NJ. 2016. Transcriptional termination in mammals: Stopping the RNA polymerase II juggernaut. *Science* **352:** aad9926.

Proudfoot NJ, Brownlee GG. 1976. 3′ Non-coding region sequences in eukaryotic messenger RNA. *Nature* **263:** 211–214.

Rigo F, Kazerouninia A, Nag A, Martinson HG. 2005. The RNA tether from the poly(A) signal to the polymerase mediates coupling of transcription to cleavage and polyadenylation. *Mol Cell* **20:** 733–745.

Rogozin IB, Carmel L, Csuros M, Koonin EV. 2012. Origin and evolution of spliceosomal introns. *Biol Direct* **7:** 11.

Ruegsegger U, Beyer K, Keller W. 1996. Purification and characterization of human cleavage factor I_m involved in the 3′ end processing of messenger RNA precursors. *J Biol Chem* **271:** 6107–6113.

Sandberg R, Neilson JR, Sarma A, Sharp PA, Burge CB. 2008. Proliferating cells express mRNAs with shortened 3′ untranslated regions and fewer microRNA target sites. *Science* **320:** 1643–1647.

Schonemann L, Kuhn U, Martin G, Schafer P, Gruber AR, Keller W, Zavolan M, Wahle E. 2014. Reconstitution of CPSF active in polyadenylation: Recognition of the polyadenylation signal by WDR33. *Genes Dev* **28:** 2381–2393.

Schrom EM, Moschall R, Hartl MJ, Weitner H, Fecher D, Langemeier J, Bohne J, Wohrl BM, Bodem J. 2013. U1snRNP-mediated suppression of polyadenylation in conjunction with the RNA structure controls poly(A) site selection in foamy viruses. *Retrovirology* **10:** 55.

Scotto-Lavino E, Du G, Frohman MA. 2006. 3′ end cDNA amplification using classic RACE. *Nat Protoc* **1:** 2742–2745.

Shi Y, Manley JL. 2015. The end of the message: Multiple protein–RNA interactions define the mRNA polyadenylation site. *Genes Dev* **29:** 889–897.

So BR, Wan L, Zhang Z, Li P, Babiash E, Duan J, Younis I, Dreyfuss G. 2016. A U1 snRNP-specific assembly pathway reveals the SMN complex as a versatile hub for RNP exchange. *Nat Struct Mol Biol* **23:** 225–230.

So BR, Zhang Z, Dreyfuss G. 2017. The function of survival motor neuron complex and its role in spinal muscular atrophy pathogenesis. In *Spinal muscular atrophy: Disease mechanisms and therapy* (ed. Sumner CJ, Paushkin S, Ko C-P), pp. 99–111. Academic, San Diego.

Spiluttini B, Gu B, Belagal P, Smirnova AS, Nguyen VT, Hebert C, Schmidt U, Bertrand E, Darzacq X, Bensaude O. 2010. Splicing-independent recruitment of U1 snRNP to a transcription unit in living cells. *J Cell Sci* **123:** 2085–2093.

Summerton J. 1999. Morpholino antisense oligomers: The case for an RNase H-independent structural type. *Biochim Biophys Acta* **1489:** 141–158.

Supek F, Bosnjak M, Skunca N, Smuc T. 2011. REVIGO summarizes and visualizes long lists of gene ontology terms. *PLoS ONE* **6:** e21800.

Tarn WY, Steitz JA. 1994. SR proteins can compensate for the loss of U1 snRNP functions in vitro. *Genes Dev* **8:** 2704–2717.

Tian B, Manley JL. 2017. Alternative polyadenylation of mRNA precursors. *Nat Rev Mol Cell Biol* **18:** 18–30.

Timchenko LT, Miller JW, Timchenko NA, DeVore DR, Datar KV, Lin L, Roberts R, Caskey CT, Swanson MS. 1996. Identification of a $(CUG)_n$ triplet repeat RNA-binding protein and its expression in myotonic dystrophy. *Nucleic Acids Res* **24:** 4407–4414.

Vagner S, Ruegsegger U, Gunderson SI, Keller W, Mattaj IW. 2000. Position-dependent inhibition of the cleavage step of pre-mRNA 3′-end processing by U1 snRNP. *RNA* **6:** 178–188.

Vorlova S, Rocco G, Lefave CV, Jodelka FM, Hess K, Hastings ML, Henke E, Cartegni L. 2011. Induction of antagonistic soluble decoy receptor tyrosine kinases by intronic polyA activation. *Mol Cell* **43:** 927–939.

Wahl MC, Will CL, Luhrmann R. 2009. The spliceosome: Design principles of a dynamic RNP machine. *Cell* **136:** 701–718.

Weinberg RA, Penman S. 1968. Small molecular weight monodisperse nuclear RNA. *J Mol Biol* **38:** 289–304.

Workman E, Saieva L, Carrel TL, Crawford TO, Liu D, Lutz C, Beattie CE, Pellizzoni L, Burghes AH. 2009. A SMN missense mutation complements SMN2 restoring snRNPs and rescuing SMA mice. *Hum Mol Genet* **18:** 2215–2229.

Workman E, Veith A, Battle DJ. 2014. U1A regulates 3′ processing of the survival motor neuron mRNA. *J Biol Chem* **289:** 3703–3712.

Yao C, Biesinger J, Wan J, Weng L, Xing Y, Xie X, Shi Y. 2012. Transcriptome-wide analyses of CstF64–RNA interactions in global regulation of mRNA alternative polyadenylation. *Proc Natl Acad Sci* **109:** 18773–18778.

Yong J, Golembe TJ, Battle DJ, Pellizzoni L, Dreyfuss G. 2004a. snRNAs contain specific SMN-binding domains that are essential for snRNP assembly. *Mol Cell Biol* **24:** 2747–2756.

Yong J, Wan L, Dreyfuss G. 2004b. Why do cells need an assembly machine for RNA–protein complexes? *Trends Cell Biol* **14:** 226–232.

Yong J, Kasim M, Bachorik JL, Wan L, Dreyfuss G. 2010. Gemin5 delivers snRNA precursors to the SMN complex for snRNP biogenesis. *Mol Cell* **38:** 551–562.

Younis I, Dittmar K, Wang W, Foley SW, Berg MG, Hu KY, Wei Z, Wan L, Dreyfuss G. 2013. Minor introns are embedded molecular switches regulated by highly unstable U6atac snRNA. *eLife* **2:** e00780.

Zhang Z, Lotti F, Dittmar K, Younis I, Wan L, Kasim M, Dreyfuss G. 2008. SMN deficiency causes tissue-specific perturbations in the repertoire of snRNAs and widespread defects in splicing. *Cell* **133:** 585–600.

Zhang Z, Pinto AM, Wan L, Wang W, Berg MG, Oliva I, Singh LN, Dengler C, Wei Z, Dreyfuss G. 2013. Dysregulation of synaptogenesis genes antecedes motor neuron pathology in spinal muscular atrophy. *Proc Natl Acad Sci* **110:** 19348–19353.

Zhu Y, Wang X, Forouzmand E, Jeong J, Qiao F, Sowd GA, Engelman AN, Xie X, Hertel KJ, Shi Y. 2018. Molecular mechanisms for CFIm-mediated regulation of mRNA alternative polyadenylation. *Mol Cell* **69:** 62–74.e4.

Approaches for Understanding the Mechanisms of Long Noncoding RNA Regulation of Gene Expression

Patrick McDonel[1] and Mitchell Guttman[2]

[1]Program in Bioinformatics and Integrative Biology, University of Massachusetts Medical School, Worcester, Massachusetts 01655

[2]Division of Biology and Biological Engineering, California Institute of Technology, Pasadena, California 91125

Correspondence: mguttman@caltech.edu

SUMMARY

Mammalian genomes encode tens of thousands of long noncoding RNAs (lncRNAs) that have been implicated in a diverse array of biological processes and human diseases. In recent years, the development of new tools for studying lncRNAs has enabled important progress in defining the mechanisms by which Xist and other lncRNAs function. This collective work provides a framework for how to define the mechanisms by which lncRNAs act. This includes defining lncRNA function, identifying and characterizing lncRNA–protein interactions, and lncRNA localization in the cell. In this review, we discuss various experimental approaches for deciphering lncRNA mechanisms and discuss issues and limitations in interpreting these results. We explore what these data can reveal about lncRNA function and mechanism as well as emerging insights into lncRNA biology that have been derived from these studies.

Outline

Cite as *Cold Spring Harb Perspect Biol* doi: 10.1101/cshperspect.a032151

1 INTRODUCTION

Over the last decade, tens of thousands of long noncoding RNAs (lncRNAs) have been discovered in mammalian genomes, with their numbers likely to eclipse the number of proteins (Guttman et al. 2009; Derrien et al. 2012). lncRNAs generally have similar properties to messenger RNAs in that they are often spliced, 5′-capped, and polyadenylated (Cabili et al. 2011; Derrien et al. 2012), yet are not translated into a protein product and instead function as an RNA molecule (Guttman et al. 2013). lncRNAs range in size from hundreds to tens of thousands of nucleotides in length, are expressed across a wide range of levels (Cabili et al. 2011; Derrien et al. 2012), and can localize to target sites within the nucleus or cytoplasm of the cell (Khalil et al. 2009; Cabili et al. 2015; Werner and Ruthenburg 2015). lncRNAs have begun to take a center stage in many aspects of biology, having been shown to play essential roles in diverse biological processes from development to immune responses (Rinn and Chang 2012; Sauvageau et al. 2013), and hundreds of lncRNAs are amplified, deleted, or otherwise disrupted in various human cancers and diseases (Wapinski and Chang 2011). Yet, despite the increasing prominence of lncRNAs in biology and disease, we still know very little about the mechanisms by which most lncRNAs act (Rinn and Chang 2012; Engreitz et al. 2016b), largely because they remain difficult to study.

The best-studied lncRNA is Xist, which orchestrates X-chromosome inactivation (XCI), an essential developmental process that ensures proper expression balance of genes on the X chromosome in female mammals (da Rocha and Heard 2017). Xist localizes across the entire X chromosome (Clemson et al. 1996; Engreitz et al. 2013), recruits repressive chromatin proteins to these sites (Plath et al. 2003; McHugh et al. 2015), triggers chromosome-wide transcriptional silencing (Penny et al. 1996; Wutz et al. 2002; McHugh et al. 2015), and mediates a cascade of chromosome structure changes and chromatin modifications to maintain this silenced state (da Rocha and Heard 2017).

Deciphering the mechanism by which Xist orchestrates XCI by modulating chromatin structure and silencing gene expression required the work of many laboratories and nearly 25 years of effort (da Rocha and Heard 2017). In addition to chromatin regulation, we now know that lncRNAs can play other roles in posttranscriptional gene regulation. For example, Malat1 localizes with nascent precursor messenger RNA (pre-mRNA) (Engreitz et al. 2014; West et al. 2014) around nuclear speckles (Spector and Lamond 2011), interacts with various splicing and mRNA processing proteins, and may act to control alternative splicing (Tripathi et al. 2010). Neat1 can act to nucleate the establishment of a nuclear territory called the paraspeckle (Clemson et al. 2009), which contains various mRNA species of unknown identity (Fox et al. 2018). In the cytoplasm, NORAD can interact with mRNA translation proteins (e.g., Pumilio) and may act to titrate these proteins away from specific mRNA targets to regulate mRNA translation (Lee et al. 2016). These examples highlight the diverse roles that lncRNAs can play by interacting with different molecules in different cellular locations (Kopp and Mendell 2018).

In recent years, the development of new tools for studying lncRNAs has enabled important progress and has begun to uncover key aspects of the mechanisms by which Xist and other lncRNAs work (Engreitz et al. 2016b; Kopp and Mendell 2018). This collective work provides a framework for how to define lncRNA function and mechanism. Specifically, deciphering these mechanisms required (1) determining that the lncRNA acts as a functional RNA molecule, (2) comprehensively defining the proteins that interact with the lncRNA, (3) defining where the lncRNA binds (to genomic DNA or RNA) and in what precise location within the cell, and (4) determining how these RNA–protein components are integrated to control cellular function.

In this review, we will explore how information about lncRNA protein binding, genomic and cellular localization, and function can be integrated to understand the mechanisms of lncRNAs in gene regulation, and we will discuss the methods used for exploring these questions.

2 METHODS FOR DEFINING lncRNA FUNCTION

Understanding how any lncRNA controls cellular function requires direct perturbation followed by observations of the functional consequences in vivo (e.g., cells, tissues, or animal models). Yet, this remains challenging in practice—more so than disrupting protein-coding genes. Whereas protein-coding genes can be disrupted through small frameshift mutations in the coding sequence, we do not know a priori which parts of a lncRNA sequence are necessary for its function and therefore whether short insertions/deletions should disrupt its function. This necessitates larger deletions, such as deletion of the entire genomic locus, to confidently disrupt the lncRNA function. Indeed, this was precisely how Xist was initially characterized. Shortly after Xist was identified as a transcript expressed from the inactive X chromosome (Xi) but not the active X (Brockdorff et al. 1991; Brown et al. 1991), it was shown that cells containing a deletion of the Xist locus failed to initiate XCI (Penny et al. 1996).

However, large deletions frequently have unintended consequences, such as by removing DNA regulatory elements contained within the lncRNA locus that can affect nearby genes (Groff et al. 2016; Engreitz et al. 2016a). These considerations must be taken into account when designing deletion experiments and interpreting their results. In fact, a recent study genetically deleted 12 lncRNA loci and found that five regulate the expression of a neighboring gene through mechanisms that do not involve the lncRNA transcript itself. Instead, several of these lncRNA promoters acted at the DNA level as *cis*-regulatory elements (enhancers) that controlled another neighboring gene, whereas others controlled expression of neighboring genes in a manner that was dependent on the active transcription—but not the sequence—of the lncRNA gene (Engreitz et al. 2016a).

Accordingly, orthogonal approaches are required to eliminate lncRNA expression to show whether a genetic locus functions through an encoded lncRNA (Engreitz et al. 2016a). In the case of Xist, additional observations were required to show that it acts as a functional RNA molecule. Specifically, (1) the Xist RNA was shown to form a unique nuclear compartment that "coats" the Xi (Clemson et al. 1996); (2) forced expression of Xist on X chromosomes or autosomes, in females or males, was shown to be sufficient to trigger gene silencing in *cis* (Wutz et al. 2002); and (3) deletion of a single region of the Xist RNA, called the A-repeat, ablates its silencing role (Wutz et al. 2002) but not its localization to Xi. Together, these sets of experiments established that the Xist RNA itself is required for initiating the process of XCI.

In this section, we will describe the various perturbation strategies for studying lncRNAs (Fig. 1) and discuss the considerations associated with interpreting each of these approaches to establish the functional role of a lncRNA in vivo.

2.1 Inhibition of Transcriptional Initiation

To account for the issues that arise from deleting a lncRNA locus, which can often entail deletion of thousands of nucleotides, one approach is to make a more precise deletion that simply removes the promoter region of a lncRNA transcript. This entails ablating transcription of the locus using CRISPR to delete the promoter. A similar approach is to use CRISPR interference (CRISPRi), in which a repressive protein is recruited to the promoter region of a lncRNA to block transcriptional initiation of the locus (Liu et al. 2016). Although these approaches can define whether a lncRNA-encoding locus has an effect on gene expression, it cannot distinguish between an RNA-independent (e.g., *cis*-regulatory elements or RNA transcription) and RNA-directed mechanism because CRISPRi is known to also inhibit the activity of *cis*-regulatory elements and preclude transcription (Fulco et al. 2016).

2.2 Premature Transcriptional Termination

Another approach is to retain the promoter and transcriptional initiation of the locus while eliminating the mature RNA product. One way to achieve this is through the insertion of a polyadenylation site downstream from the transcription start site, which will lead to premature termination of transcription, reduce transcription in the locus, and eliminate the mature RNA while leaving the promoter sequence intact. If a phenotype is observed in this way, one can conclude that the phenotype is not dependent on a *cis*-regulatory element but instead requires transcription activity at the locus. Yet, this approach does not fully discriminate between a sequence-dependent role for the lncRNA and the active process of transcription across the lncRNA locus.

2.3 Precise Sequence-Specific Mutagenesis and Functional Rescue

One approach to directly determine whether a lncRNA has a sequence-specific function is to delete precise sequences within the lncRNA. For example, deleting or mutating a specific protein binding site contained within a lncRNA should distinguish between sequence-independent effects of lncRNA transcription and sequence-specific effects of the lncRNA itself. This approach also has the benefit of allowing specific functions to be assigned to the mutated domains. Because this approach is often associated with identification of the proteins that interact with a specific region of a lncRNA, such perturbations can be directly complemented by functional rescue experiments that establish a direct relationship between the function of a sequence specific domain and RNA–protein function. This can be achieved by expressing a mutant RNA containing an RNA aptamer sequence (e.g., MS2 or BoxB) that can interact with the protein of interest through an engineered protein tag (e.g., MCP or λN) (Keryer-Bibens et al. 2008). Although this approach was critical for defining the functions of several Xist domains (Chen et al. 2016; Pintacuda et al. 2017), it generally requires knowledge of the protein binding sites on a lncRNA.

2.4 Posttranscriptional Degradation

A more direct way to determine whether a lncRNA acts directly as an RNA molecule is to degrade the mature RNA after it is fully transcribed. There are several approaches for doing this, including antisense DNA oligonucleotides (ASOs) that anneal to the lncRNA and trigger RNase H–mediated degradation (Liang et al. 2017) and RNA interference (RNAi) in which an antisense RNA anneals to

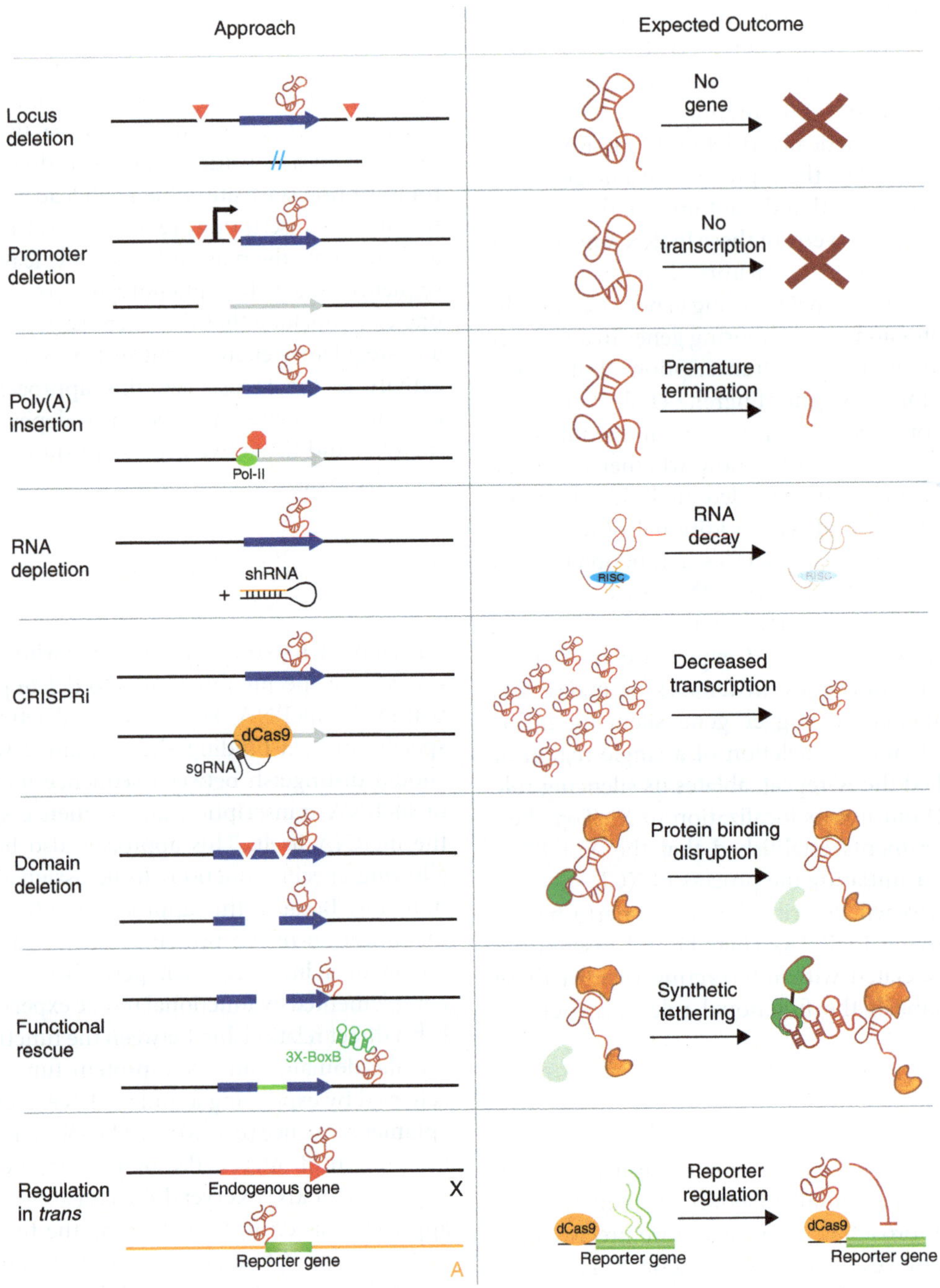

Figure 1. Genetic strategies to study long noncoding RNA (lncRNA) function. Locus and promoter deletions (e.g., by CRISPR [red arrows]) can generate null alleles such that the gene that encodes the lncRNA is removed from the genome or the promoter is removed preventing lncRNA transcription. However, both of these approaches could also delete functional regulatory DNA sequences contained within these DNA sequences. Inserting a poly(A) transcription termination signal does not disrupt transcription initiation at the lncRNA locus but terminates transcription prematurely. Antisense oligonucleotides or RNA interference (RNAi) can degrade the lncRNA posttranscriptionally, but there may be off-target effects that are often hard to predict. Recruitment of dCas9 (usually fused to a repressor domain such as KRAB) by short guide RNAs using CRISPR leads to decreased transcription of the lncRNA without mutating the DNA sequence, but because it also precludes recruitment of PolII to enhancer elements, it can also impact functions mediated through DNA elements. Deleting a specific lncRNA–protein interaction domain can impact protein binding and can be complemented with synthetic rescue experiments that tether the protein to the mutant RNA. Expressing a *cis*-acting lncRNA from a different locus and chromatin context can reveal mechanisms of action and gene regulatory function in *trans*.

Cite as *Cold Spring Harb Perspect Biol* doi: 10.1101/cshperspect.a032151

the lncRNA and triggers argonaute-mediated degradation (Ghildiyal and Zamore 2009). Although these represent important methods, in practice they are often less precise than genetic engineering because hybridization of these short oligonucleotide sequences can have off-target effects that are difficult to identify. Indeed, several lncRNAs that were initially characterized using RNAi, ASOs, or similar methods, have recently been reevaluated by genetic studies and shown to have distinct functions from those previously reported. These include several examples that have since been shown to act as *cis*-regulatory elements rather than functional RNA molecules (Engreitz et al. 2016a; Groff et al. 2016). For example, lincRNA-P21 was initially described as acting in *trans* based on functions identified by RNAi-mediated depletion of the lncRNA (Huarte et al. 2010). However, subsequent perturbation of this lncRNA using genetic dissection suggested that the RNA product may be dispensable and the observed phenotype may simply be explained by the *cis*-regulatory function of the DNA element (Groff et al. 2016). Indeed, several lncRNAs previously defined to contain functional roles based on RNAi in embryonic stem cells (ESCs) (Guttman et al. 2011) have been shown to have functions that occur in *cis*, independent of their RNA products (Engreitz et al. 2016a). Although these results do not exclude the possibility that these RNA products may still have other functional roles, they do highlight the importance of orthogonal approaches and validation to directly confirm the functional role of a lncRNA.

Although the individual approaches above have caveats regarding their interpretation, when performed in combinations they can provide high confidence about the function of a lncRNA. One orthogonal approach to gain additional confidence is to complement genetic perturbation with a rescue construct in *trans*. This may often fail, however, because many lncRNA functions are associated with 3D proximity and therefore simply expressing the lncRNA from an exogenous site may fail to rescue the defect. Other approaches, such as targeted recruitment of a lncRNA to a genomic DNA locus, or rescue of individual RNA–protein interactions, can overcome this issue, but requires more precise knowledge of the mechanisms by which a lncRNA may work. Recently, several new technologies have been developed that enable the direct assessment of the functional role of a lncRNA through targeted recruitment of the lncRNA in *trans* to a reporter gene. These approaches have been used for studying HOTAIR by recruitment through the GAL4-UAS system (Portoso et al. 2017). More recently, a method called CRISPR-Display has generalized this approach in a manner that enables programmable recruitment of lncRNAs to different DNA regions on larger scales (Shechner et al. 2015). Because these approaches recruit a lncRNA independently of its transcriptional and local environment, presence of a function in these assays indicates a role for the RNA transcript itself. However, the absence of a phenotype in such an assay does not preclude a critical role for the lncRNA in its endogenous chromatin environment, which may be distinct from the chromatin context that is present at the reporter gene.

3 DEFINING lncRNA–PROTEIN INTERACTIONS AND THEIR FUNCTIONAL CONTRIBUTIONS

Determining how Xist mediates transcriptional silencing required advances in biochemical purification approaches and mass spectrometry to comprehensively define the proteins that directly interact with Xist (Chu et al. 2015; McHugh et al. 2015). These studies identified numerous proteins that begin to explain key aspects of how Xist works to orchestrate the process of XCI. Mapping in which these proteins bind across the Xist lncRNA showed that they bind at discrete locations (Chen et al. 2016; Chu et al. 2015; Cirillo et al. 2016; Pintacuda et al. 2017) and enabled genetic studies to delete individual domains to study the functional contribution of each lncRNA–protein complex in orchestrating XCI. Together, these studies uncovered the Xist–protein interactions that are required for transcriptional silencing, recruitment of various chromatin regulators (e.g., PRC1 and PRC2), localization to genomic DNA, and chromosome-wide spreading.

In this section, we will discuss methods for defining (1) which proteins interact with a lncRNA, (2) where on the RNA each of these proteins binds, and (3) the functional role of each lncRNA–protein interaction. We will highlight important issues and consideration that are needed for designing and interpreting lncRNA–protein interactions.

3.1 Methods for Comprehensively Defining Proteins That Bind to a Specific lncRNA

There are several methods for purifying a specific lncRNA and defining associated proteins. These methods differ in how the RNA is purified after incubation with proteins in solution or after cross-linking of RNA–protein complexes in vivo.

One general approach for purifying RNA is to attach an RNA sequence that forms a tight interaction with a defined protein complex. For example, a hairpin structure from the MS2 RNA forms a tight interaction with the bacteriophage MS2 viral coat protein (Keryer-Bibens et al. 2008). In this approach, repeats of the MS2-binding RNA are attached to an RNA of interest and the RNA is captured by immobilizing the associated RNA onto a protein-bound resin. There are many variants of this approach using different combinations of RNA–protein substrates that have various affinities and therefore enable different stringencies (McHugh et al. 2014). In addition, there are variants such as the S1 RNA aptamer that mimics biotin in that it can interact directly with streptavidin. Because it has a lower affinity than biotin, it can be selectively eluted through the addition of free biotin (Srisawat and Engelke 2002). This specific elution strategy provides important benefits for reducing the background, and other similar approaches have been developed for this reason. For example, the Csy4 protein, derived from the bacterial CRISPR system, binds tightly to its cognate RNA, more tightly than MS2 or PP7 coat protein interactions, while enabling specific elution through the addition of imidazole (Lee et al. 2013).

The other general approach for RNA purification is based on hybridization of antisense oligonucleotides (Chu et al. 2011; Engreitz et al. 2013; Simon et al. 2013; Rogell et al. 2017). These approaches generally use biotin labeled oligonucleotides to hybridize to a specific RNA sequence and capture it, and its associated proteins, from a complex mixture. These strategies vary in several ways: (1) some use individual oligonucleotides (Simon et al. 2013; Rogell et al. 2017) whereas others use pools of oligonucleotides that tile across the RNA of interest (Chu et al. 2011; Engreitz et al. 2013); (2) some methods use short oligonucleotides (Chu et al. 2011; Simon et al. 2013; Rogell et al. 2017) and others use long oligonucleotides (Engreitz et al. 2013); and (3) some use unmodified nucleotides (Chu et al. 2011; Engreitz et al. 2013; Simon et al. 2013), whereas others use locked nucleic acids (Rogell et al. 2017). The key consideration in choosing between these parameters comes down to specificity of the hybrid and the stringency of the purification conditions used in the purification procedure.

Until recently, these RNA purification approaches were generally performed after incubating a labeled RNA with protein obtained from total cell extracts. Although such an approach had proven successful for defining proteins that interact with classical noncoding RNAs (ncRNAs) (such as snRNPs and 7SK), they proved more problematic for studying lncRNAs. For example, early attempts to purify Xist and its associated proteins used an affinity tag coupled to the RNA (Brown and Baldry 1996) to identify an interaction between Xist and heterogeneous nuclear ribonucleoprotein C (hnRNPC); yet, this interaction was insufficient to explain the silencing role mediated by Xist. Similar approaches led to the detection of specific proteins that interact with several lncRNAs (e.g., NORAD, linc-P21) (Huarte et al. 2010; Lee et al. 2016), but they are usually limited to identifying abundant RNA binding proteins (e.g., Pumilio, heterogeneous nuclear ribonucleoprotein K [hnRNPK]) and do not provide a comprehensive picture of the proteins that interact with a lncRNA.

These results highlight one of the key challenges with RNA purification methods that are performed in solution: Many biologically relevant RNA-binding proteins (RBPs) that interact with a lncRNA are not identified. The issue is that when incubated in solution, abundant RBPs can outcompete lower abundance proteins. Thus, it is easier to observe interactions with abundant RBPs, but not necessarily other bona fide RBPs that interact with the lncRNA in cells, which may be present at much lower concentrations. This is true even if specific RBPs have significantly higher affinity because the differences in concentration are often large enough to overcome differences in affinity. Moreover, because these approaches break open the cell, such that the concentrations of proteins are uniform throughout solution, the results may not accurately reflect the actual local concentration of protein complexes near an lncRNA in vivo (Darnell 2010; Mili and Steitz 2004; McHugh et al. 2014). Consequently, promiscuous RNA–protein interactions might form in vitro that do not occur in vivo, because of compartmentalization, sequestration, and phase separation. Accordingly, despite numerous attempts to define the mechanism of how Xist silences transcription, until recently it remained unclear what RBPs interact with Xist and what components are required for Xist-mediated transcriptional silencing.

To address this challenge, several groups developed RNA purification methods (e.g., ChIRP-MS, CHART-MS, and RAP-MS) (Fig. 2) that can both comprehensively define the proteins that interact with a specific lncRNA and at the same time achieve the specificity required to only identify interactions that occur in vivo (West et al. 2014; Chu et al. 2015; McHugh et al. 2015). These methods achieve this goal by using in vivo cross-linking to fix RNA–protein interactions that occur inside cells and then purifying the RNA of interest using extremely stringent, even denaturing, conditions to remove background associating proteins. These methods differ in the type of cross-linking that is used to fix RNA–protein interactions, which impacts the types of proteins that are identified by each method. For example, RAP uses ultraviolet (UV) cross-linking to fix only those proteins that directly interact with an RNA, whereas ChIRP uses formaldehyde cross-linking which also cross-links interacting proteins. After purification, proteins that are covalently cross-linked to a given RNA are identified using mass spectrometry.

These approaches led to the discovery of dozens of proteins that interact with Xist. Although there are differences between the proteins identified by these approaches, the proteins identified using UV cross-linking would be expected to represent proteins that directly interact with

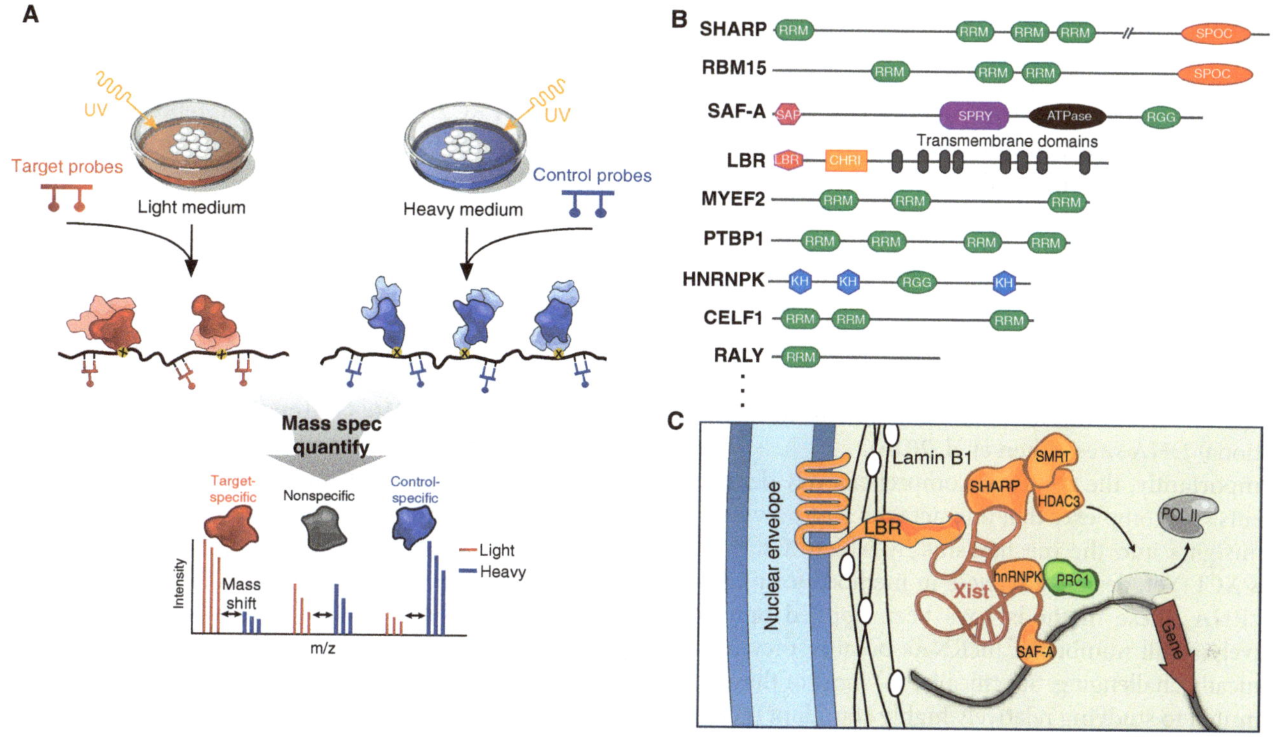

Figure 2. Comprehensive identification of proteins that interact with long noncoding RNAs (lncRNAs). (*A*) Antisense capture probes are hybridized to the target lncRNA in cross-linked lysate that has been SILAC-labeled with heavy or light amino acids. Following capture and stringent washes, the purified proteins are identified using mass spectroscopy. Nonspecific background peptides found in the control sample are subtracted out, identifying specific interacting proteins. (*B*) Example list of Xist-interacting proteins found by multiple Xist purification and mass spectrometry (MS) methods, and their associated protein domains. (*C*) Current model for how Xist interacts directly and indirectly with proteins in the cell to silence transcription on Xi, revealed by recent MS studies. Xist RNA associates with genomic DNA through a direct interaction with SAF-A/hnRNPU, and binds SHARP/SPEN to recruit SMRT and HDAC3 to deacetylate histone H3 and exclude RNA polymerase II from nearby loci. Xist also binds hnRNP-K which recruits the PRC1 complex, leading to histone ubiquitination and subsequent recruitment of the PRC2 complex and its H3K27me3 histone methyltransferase activity. Lamin B receptor (LBR) binding of Xist recruits Xi to the nuclear lamina, creating a repressive nuclear territory and enabling further spreading by Xist. Proteins depicted as directly touching Xist RNA in this figure were identified by MS studies as directly bound to Xist; the other proteins depicted were subsequently identified by orthogonal studies. (Adapted, with permission, from McHugh et al. 2015.)

an RNA, whereas those identified by formaldehyde would also include proteins that associate with an RNA via protein–protein interactions. Importantly, the core protein components identified are similar between both methods (Fig. 2) (Chu et al. 2015; McHugh et al. 2015).

Cross-linking studies identified the proteins that explain eviction of RNA polymerase II and transcriptional silencing of the X chromosome. Specifically, numerous analyses identified SHARP (also known as SPEN) as a direct Xist-interacting protein that is required for chromosome-wide transcriptional silencing on the X chromosome (Chu et al. 2015; McHugh et al. 2015; Moindrot et al. 2015; Monfort et al. 2015). Determination of the extended protein complex interacting with SHARP revealed an essential role for SMRT and HDAC3 in the exclusion of RNA polymerase II from the X-chromosome territory and subsequent transcriptional silencing (McHugh et al. 2015). These results indicated that Xist initiates transcriptional silencing by directly binding to SHARP, recruiting SMRT and HDAC3 to trigger deacetylation on the X chromosome. Additionally, these studies identified hnRNPK as a direct Xist-interacting protein (Chu et al. 2015; Pintacuda et al. 2017). Analysis of the extended interaction network showed that hnRNPK acts to recruit the PRC1 complex to the X chromosome through the noncanonical PCGF3/5 complex (Pintacuda et al. 2017). This leads to ubiquitylation of chromatin and subsequent recruitment of PRC2 and its associated H3K27me3 modifications across the X chromosome (Almeida et al. 2017), thus reconciling prior observations that PRC2 is recruited to Xi but is not recruited through a direct interaction between PRC2 and the Xist RNA. These studies also identified SAF-A (also known as heterogeneous nuclear ribonucleoprotein U [hnRNPU]) and showed that this interaction is required for tethering the Xist RNA to genomic DNA sites on the X chromosome (Hasegawa et al. 2010; Chu et al. 2015; McHugh et al. 2015). Finally, these studies identified the lamin B receptor (LBR) and showed that this interaction is required for enabling Xist to spread across the X chromosome (McHugh et al. 2015; Chen et al. 2016). Specifically, the Xist–LBR interaction is required for recruiting the X chromosome to the nuclear lamina and by doing so leads to three-dimensional (3D) structural changes on the X chromosome that allow Xist to spread to additional DNA sites (Chen et al. 2016).

Importantly, the ability to comprehensively define the proteins and complexes that interact with Xist provided the first insights into the mechanisms by which Xist orchestrates XCI. Yet, despite advances in methods for purifying a lncRNA, these methods have been applied only to a relatively small number of lncRNAs because they remain technically challenging. Specifically, (1) most of these tools are limited to studying relatively highly abundant lncRNAs (i.e., Xist, Malat1) and accordingly do not work well for the vast majority of lncRNAs; (2) these methods often require large numbers of cells, limiting their usage to cell types that can be easily grown in culture; and accordingly will not work for primary cells used for studying human disease, and (3) these methods often require extensive optimization for each lncRNA and thus are not amenable to systematic characterization of many lncRNAs. Future work will be needed to adapt these approaches to enable greater sensitivity and scalability for studying a larger number of lncRNAs across broader contexts.

3.2 Mapping lncRNAs That Interact with Specific Proteins

To overcome the limitation in detection sensitivity imposed by RNA-centric purifications, attempts were made to use information about XCI to guide identification of the proteins that might be involved. One important discovery was that the process of XCI is associated with recruitment of the PRC2 complex and its associated H3K27me3 marks across the X chromosome (Plath et al. 2003). Based on this observation, the interaction between Xist and PRC2 was tested using RNA immunoprecipitation (RIP) experiments. RIP uses native purification conditions in cross-linked or non-cross-linked cells to immunoprecipitate proteins and measure their associated RNAs (Zhao et al. 2008). Indeed, an interaction between Xist and PRC2 was shown to occur using these methods. Moreover, PRC2 was shown to interact with a truncated Xist RNA that just contained the A-repeat region of Xist (Zhao et al. 2008), the RNA region required for transcriptional silencing (Wutz et al. 2002). These observations led to a model whereby Xist directly binds to PRC2 and recruits this repressive chromatin complex to the X chromosome to mediate transcriptional silencing (Zhao et al. 2008).

More generally, observation of an interaction between Xist and PRC2 led to intense study of the PRC2 complex as an RNA binding protein, with recent studies arguing that PRC2 binds to hundreds of different lncRNAs. This insight led to the proposal that lncRNAs act to recruit PRC2 to genomic DNA targets, among other roles (Khalil et al. 2009; Zhao et al. 2010; Guttman et al. 2011).

Although this result seemed to indicate a potential mechanism for how Xist-mediated transcriptional silencing might occur, several subsequent observations are incompatible with this model. Specifically, deletion of PRC2 components that prevent its recruitment to the X chromosome have no impact on Xist-mediated transcriptional silencing (Kalantry and Magnuson 2006; Schoeftner et al. 2006) and deletion of the A-repeat from Xist does not preclude PRC2 recruitment to the X chromosome (Plath et al. 2003; Kohlmaier et al. 2004; da Rocha et al. 2014; McHugh et al. 2015). Moreover, the distance between Xist and PRC2 on the inactive X chromosome as measured by super-resolution microcopy appears to be incompatible with direct binding (Cerase et al. 2014). In addition, recent studies purifying Xist using denaturing conditions failed to identify an interaction between Xist and any of the PRC2 components in vivo (Chu et al. 2015; McHugh et al. 2015; Minajigi et al. 2015). Instead, these studies identified other RBPs that directly bind to Xist, and are essential for PRC2 recruitment across the X chromosome. Furthermore, it is now clear that PRC2 is also dispensable for the functions of other lncRNAs, including HOTAIR, which it had been previously reported to bind (Portoso et al. 2017).

This discrepancy between the biochemical evidence supporting lncRNA–PRC2 interactions, and functional evidence showing that PRC2 is dispensable for Xist function, highlight important considerations for studying lncRNA–protein interactions. Specifically, interactions that are observed using immunoprecipitation methods do not necessarily reflect interactions that occur in vivo, but could instead reflect those that form in solution after cell lysis. Indeed, in a classic experiment, Mili and Steitz showed that immunoprecipitation methods can indeed identify RNA–protein interactions that do not occur in vivo, but rather form in solution after lysis (Mili and Steitz 2004). Consistent with this explanation, several recent in vitro experiments have shown that PRC2 components bind with high affinity to all RNAs, including bacterial RNAs

(Davidovich et al. 2013), suggesting that detecting their interactions with Xist—and possibly other lncRNAs—may represent high-affinity associations that may not be functionally required.

To date, most studies of lncRNA–protein interactions have used RIP, which has led to reports of numerous additional chromatin complexes that interact with various lncRNAs (Guttman et al. 2011). Accordingly, these same considerations must be taken into account when exploring the functional relevance of these interactions.

To account for the limited stringency afforded by the RIP method, the cross-linking and immunoprecipitation (CLIP) method was developed (Darnell 2010). CLIP uses UV cross-linking to form covalent interactions inside cells between directly interacting RNA and protein, followed by purification using more stringent wash conditions (i.e., 1 M salt) and subsequent separation through a denaturing gel and transfer to a nitrocellulose membrane to enrich for in vivo cross-linked RNA–protein interactions (Fig. 3A) (Darnell 2010). Importantly, the key difference between RIP and CLIP is the stringency of the wash conditions and denaturing gel separation used rather than the cross-linking itself. In fact, RIP can be performed in UV cross-linked cells, but does not achieve the same stringency as CLIP without incorporating the purification of RNA–protein complexes from a denaturing gel. Because of this increased stringency, CLIP methods have emerged as the gold-standard for defining in vivo RNA–protein interactions.

Despite the critical advantages of the CLIP method, simply observing sequencing reads in a CLIP experiment does not necessarily indicate an in vivo RNA–protein interaction, and careful controls need to be designed, especially when studying proteins that may not bind RNA or that may bind to only a few RNAs in vivo.

3.3 Methods for Mapping the Precise Protein Binding Sites on RNA

CLIP also enables the identification of the direct RNA binding site of a protein—a necessary prerequisite for functional characterization of lncRNA–protein interactions. CLIP entails controlled RNase digestion of RNA into small fragments before immunoprecipitation of the protein. Accordingly, the only RNA fragment that will be purified is the region that is directly cross-linked to the protein of interest. Although this approach hones in on the region of the RNA

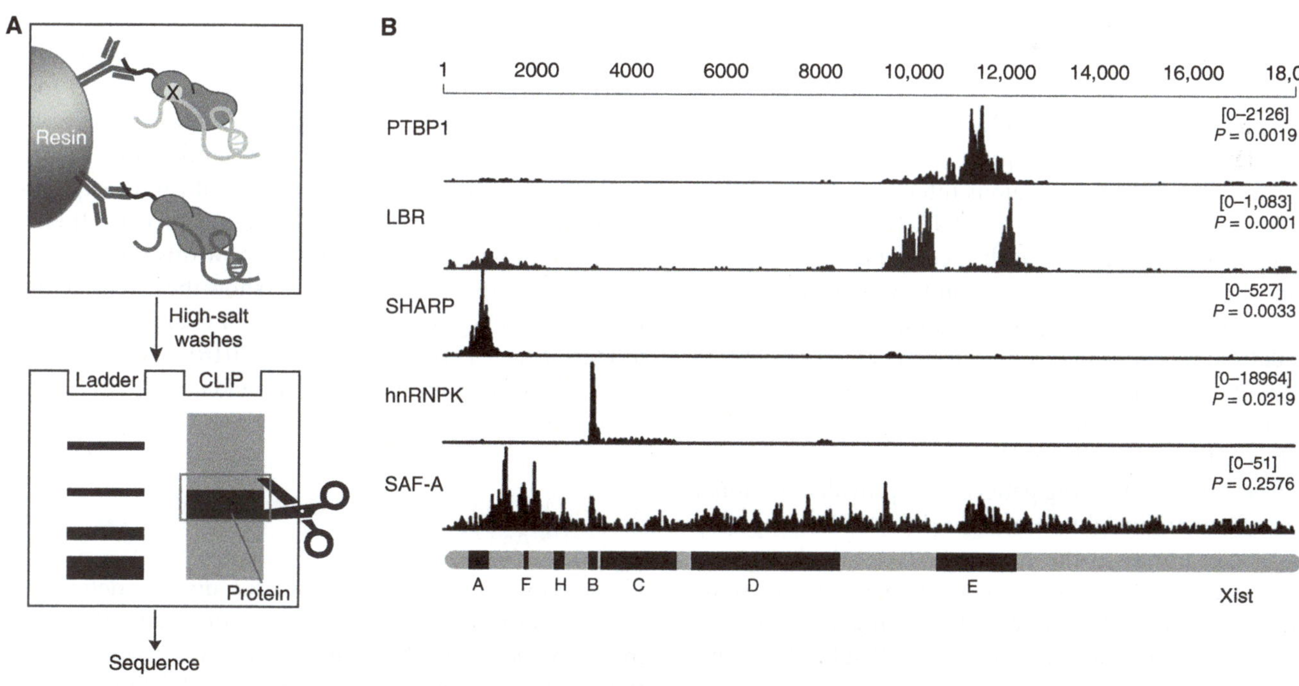

Figure 3. Cross-linking and immunoprecipitation (CLIP) methods identify protein binding sites on long noncoding RNAs (lncRNAs). (*A*) An overview of the CLIP method. RNA–protein interactions are cross-linked using ultraviolet (UV) light (depicted by X). Purification and stringent washes are followed by gel extraction of cross-linked RNA–protein complexes, and the purified RNA fragments are sequenced. (*B*) Example of protein interaction domain identification on Xist by CLIP (Cirillo et al. 2016). Although SAF-A/hnRNP-U has a broad binding distribution on Xist, other interacting proteins have clearly defined binding domains. (*B*, From Cirillo et al. 2016, adapted, with permission.)

that is bound by a protein, it does not resolve the precise nucleotides that are bound by a protein. To surmount this problem, several groups have developed adaptations of the CLIP method that more accurately resolve the sites of interaction. Although several such methods have been developed, one of the most popular strategies takes advantage of the fact that UV cross-linking creates an irreversible RNA–protein linkage, producing a peptide adduct present on the purified RNA after immunoprecipitation. Because reverse transcriptase often pauses at these adducts, the locations of the resulting truncated cDNA products can be mapped using sequencing to identify the likely cross-link sites (Konig et al. 2010).

CLIP has been successfully used to define the precise RNA binding sites of numerous RNA binding proteins (Darnell 2010), including those of several on Xist (Fig. 3B) (Chen et al. 2016; Cirillo et al. 2016). For example, mapping the SHARP protein by CLIP showed that it binds to the highly-conserved A-repeat domain of Xist, the domain that was previously shown to be required for transcriptional silencing (Chu et al. 2015; Chen et al. 2016). Interestingly, CLIP studies of hnRNPK showed that it interacts specifically with the B-repeat region of Xist (Cirillo et al. 2016; Pintacuda et al. 2017), a domain that was previously showed to be required for PRC2 recruitment to the X chromosome (da Rocha et al. 2014). CLIP studies of LBR identified three distinct sites on the Xist RNA, including a region overlapping the A-repeat deletion (Chen et al. 2016; Cirillo et al. 2016). Interestingly, the interaction between Xist and LBR appears to require each of the individual sites, suggesting that these regions interact to form a higher-order structured region of the RNA (Chen et al. 2016). Finally, CLIP studies of SAF-A showed that it binds to many regions across the Xist lncRNA and appears to recognize the RNA in a promiscuous manner, lacking precise sequences or structural recognition motifs (Cirillo et al. 2016). This is consistent with recent in vitro studies showing that the RGG domains of SAF-A bind broadly to many different RNA sequences (Ozdilek et al. 2017).

These different binding patterns can provide functional and mechanistic insights into how these RBP complexes work. For example, in the case of SAF-A, which appears to bind broadly across the Xist RNA, many interactions between Xist and SAF-A may lead to high-avidity interactions that create a tight interaction of the Xist lncRNA with chromatin.

Importantly, all of these biochemical measurements represent an average profile of interactions occurring across millions of cells and do not provide information about whether these interactions occur simultaneously with a single RNA molecule. Novel methods will be needed to determine the structures and stoichiometries of these complexes, such information will provide critical new insights into lncRNA mechanisms.

3.4 Methods for Studying the Function of Individual lncRNA–Protein Interactions

There are several approaches to defining the function of a lncRNA–protein interaction. For example, targeted deletion of the RNA binding site that is recognized by a protein enables studies of the specific RNA–protein interaction without impacting other roles that an RNA binding protein might play. Indeed, this approach was used to study the Xist interactions with SHARP, LBR, and hnRNPK (Chen et al. 2016; Almeida et al. 2017; Pintacuda et al. 2017). However, this method only works when there is a clearly defined binding site for a protein on the lncRNA. If a protein binds broadly across an RNA, such as SAF-A on Xist, targeted mutagenesis of the RNA will not work. In such cases, deletion of the RNA binding domain of the protein can also be used to disrupt this interaction. However, in this case, the mutation will also impact the interaction between the RBP and other RNAs with which it might associate.

Importantly, in this mutagenesis approach, any observed phenotypes can be confirmed by performing rescue experiments in which the RBP and RNA are tethered together independently to determine if their interaction can explain the observed phenotype. For example, a mutant RNA (lacking a protein binding site) can be fused to the BoxB RNA apatamer and coexpressed with the protein containing a λN peptide that will tightly bind to BoxB. In this case, if the deleted domain is critical for explaining a phenotype because of the interacting protein, then the synthetic rescue should no longer show this observed phenotype. This approach was used to show the essential role of LBR in enabling Xist spreading (Chen et al. 2016), RBM15 in recruiting YTHDC1 (Patil et al. 2016), and in hnRNPK binding and recruiting PRC1 (Almeida et al. 2017; Pintacuda et al. 2017).

4 METHODS FOR DETERMINING WHERE lncRNAs LOCALIZE

Understanding how lncRNAs work requires a detailed understanding of what the direct targets of each lncRNA are—including genomic DNA and RNA targets—and where in the cell these interactions occur. The subcellular localizations of lncRNAs provide some of the greatest clues about their biological roles and molecular functions. As functional molecules, lncRNAs must localize to their sites of action within the cell. For instance, a lncRNA that localizes to subnuclear compartments beyond its site of transcription could regulate gene expression or RNA biogenesis in *trans*,

Cite as *Cold Spring Harb Perspect Biol* doi: 10.1101/cshperspect.a032151

but is unlikely to be translated by ribosomes, whereas an exclusively cytoplasmic lncRNA would not be expected to have a direct role in altering transcription or chromatin structure, but instead might be involved in regulating proteins or other RNAs after they are produced. Moreover, the subcellular localization of lncRNAs can be a consequence of interacting with proteins, or they can recruit proteins to specific sites, structures, and territories to integrate functions or sequester molecules away from their targets.

Indeed, one of the earliest and most significant insights into Xist function was the visual observation that the Xist RNA localizes to a single chromosome in female cell nuclei, apparently coating the inactive X (Brockdorff et al. 1991; Brown et al. 1991; Clemson et al. 1996). It was this observation that led to the initial proposal that Xist acts through a functional lncRNA—not its protein product—by spreading in *cis* across the chromosome from which it is expressed, to broadly silence transcription. Subsequent experiments in which Xist was integrated into autosomal regions (Lee et al. 1996; Herzing et al. 1997) showed that Xist similarly spreads across these autosomal regions, showing that the affinity of Xist RNA is not for particular DNA sequences but instead is related to the direct physical connection between the gene expressing Xist RNA and the DNA it binds. These observations established the core model for how the Xist RNA works. More recently, high-resolution mapping of Xist localization by molecular methods revealed in detail the underlying relationships of Xist spreading with 3D spatial proximity, chromatin structure, and the kinetics of gene silencing across X (Engreitz et al. 2013).

In this section, we will discuss the various methods for studying lncRNA localization within the cell (Fig. 4).

4.1 Direct Visualization of lncRNAs by Microscopy

Fluorescence microscopy has proven to be one of the most powerful tools for studying lncRNA localization, principally through fluorescence in situ hybridization (FISH) and immunofluorescence (IF). In FISH, fixed cells are hybridized with fluorescently labeled probes antisense to the RNA target and visualized by light microscopy. In fact, Xist was initially shown to localize to the Barr body at the nuclear periphery and to broadly coat the inactive X using FISH (Clemson et al. 1996). These observations directly linked the RNA molecule itself with the subnuclear territory of X inactivation. Importantly, these studies used probes that distinguished nascent Xist RNA—produced at the XIC—from the spliced, polyadenylated, mature Xist RNA that spreads over the chromosome. Subsequent FISH studies performed throughout development and in differentiating female ESCs observed the individual stages of Xist-induced silencing of X: initiation of Xist expression from the XIC on the presumptive Xi, widespread localization of spliced Xist concomitant with propagation of inactivation in *cis*, comigration with the Xi to the nuclear periphery, and continued association with Xi throughout most of the cell cycle, even in the absence of continued Xist transcription. Immunofluorescence, wherein specific proteins in fixed cells are labeled using fluorescent antibodies, combined with Xist FISH, showed that Xist localization and spreading precedes hallmarks of transcriptional repression, such as histone deacetylation and methylation, DNA methylation, and exclusion of Pol II. These observations, when combined with genetic experiments described earlier in this review, strongly implicated the RNA itself as required for establishing X inactivation, even though the mechanistic details remained mysterious.

Improvements in microscopy and staining revealed subchromosomal enrichment of Xist RNA at gene-rich "G-light" bands on Xi, and an apparently low affinity for gene-poor heterochromatic regions (Duthie et al. 1999), with spreading into active gene regions requiring the A-repeat domain (Wutz et al. 2002), leading to the model that Xist initially localizes to high-affinity sites before spreading to the rest of the chromosome. Recently, super-resolution 3D structured illumination microscopy (3D-SIM) of the Xi confirmed that Xist RNA initially localizes to discrete foci before chromatin compaction, followed by spreading to occupy discrete chromatin domains in close association with the nuclear matrix protein SAF-A (previously shown to be required for Xist association with the Xi [Hasegawa et al. 2010]) and preceding Pol II exclusion but not involving repressive chromatin modifications, such as H3K27me3 (Cerase et al. 2014).

4.2 Methods for Genome-Wide Mapping of lncRNA Localization

Although imaging provided key insights into how Xist works, these approaches could not define where an RNA binds to genomic DNA at high resolution. Accordingly, it was still unclear how Xist spread across the X chromosome on initiation of XCI.

Recently, several methods to map lncRNA localization at high resolution using antisense purification coupled with sequencing of genomic DNA have been developed. These methods take advantage of labeled antisense probes to purify target RNAs from cross-linked cell lysates and identify DNA or RNA interactions by sequencing (Chu et al. 2011; Engreitz et al. 2013; Simon et al. 2013), an approach that is conceptually similar to the ChIP-seq method used for mapping protein–DNA interactions. Indeed, the use of different cross-linking agents, such as formaldehyde or pso-

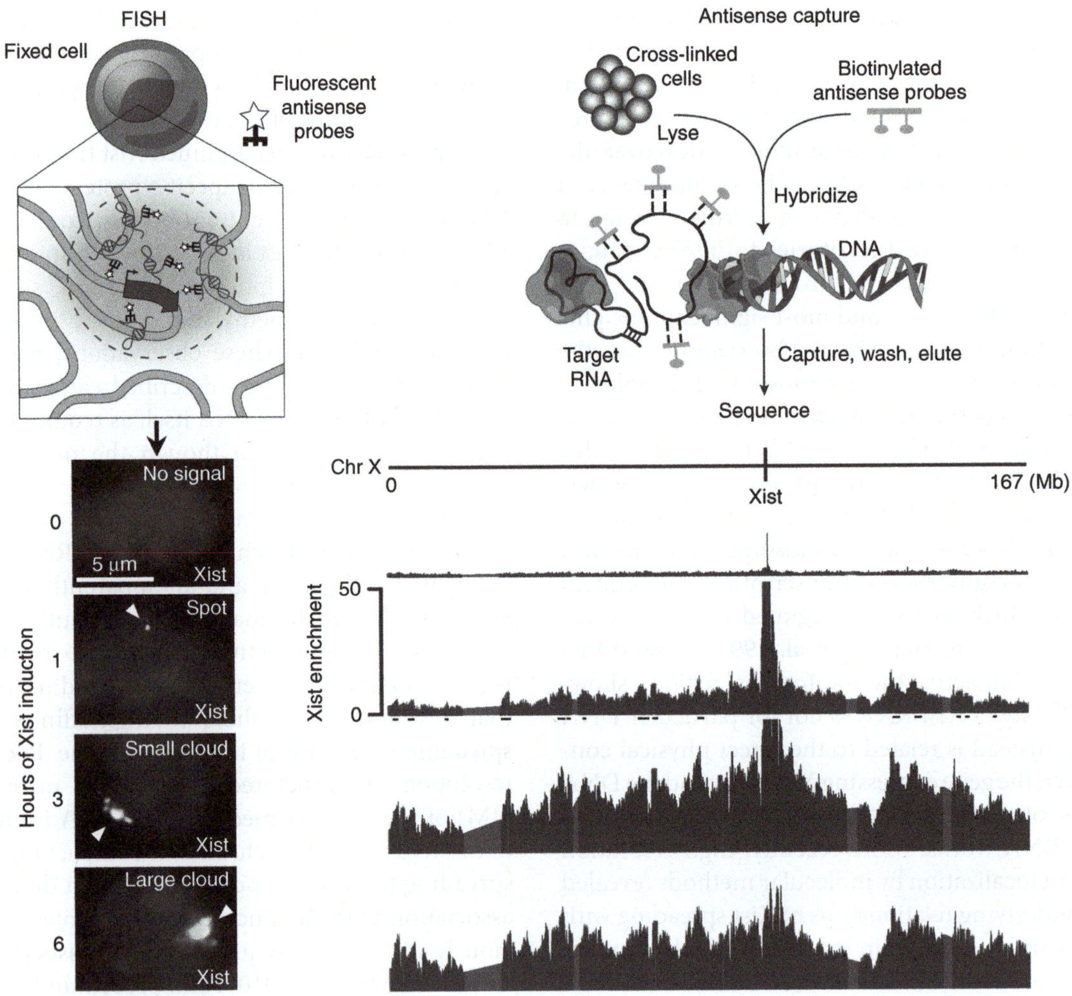

Figure 4. Methods to study long noncoding RNA (lncRNA) localization. Fluorescence in situ hybridization (FISH) has proven to be a powerful and general method to understand lncRNA localization. In this method, fluorescent antisense probes are hybridized to RNA targets in fixed cells and visualized using fluorescence microscopy. Shown on the *left* by FISH is the Xist RNA cloud coating the Xi after Xist induction. However, the resolution provided by this method is limited to relatively coarse observations. Depicted on the *right* is the same Xist induction time course assayed by RNA antisense purification (RAP) of Xist RNA using biotinylated antisense probes and sequencing of the bound X chromosome DNA. Such methods revealed that Xist exploits the 3D structure of the X chromosome to initially bind sites spatially proximal to the Xist locus and then spread to nearby active regions. (From Engreitz et al. 2013, adapted, with permission.)

ralen, enables the mapping of different interaction types, including protein-mediated RNA localization and direct nucleic acid hybridization (Engreitz et al. 2014).

Application of these methods to map Xist interactions with chromatin shed new light on how Xist spreads across the X chromosome during XCI in inducible Xist cell lines and female ESCs (Engreitz et al. 2013; Simon et al. 2013). These studies identified discrete and linearly distal sites on the Xi where the Xist RNA first localizes before spreading broadly across the X chromosome. Interestingly, the initial Xist localization sites lack enrichment for specific sequence motifs or chromatin modifications. Instead, comparison to high-throughput chromatin conformation capture (Hi-C) interaction data revealed a striking relationship between the initial Xist binding sites and DNA regions that are in close 3D physical proximity to the XIC in the nucleus. This observation led to the model that Xist exploits 3D chromosome conformation, rather than specific high-affinity sequences, to identify its initial target sites on the Xi. In fact, determining the localization of Xist when it is expressed from a transgene located on a distinct region of the X chromosome showed that the RNA then localized at sites that were closest in 3D proximity to the transgene integration site—displaying an entirely distinct pattern of localization from that observed when Xist was expressed from its endogenous locus (Engreitz et al. 2013).

These genomic mapping tools were also essential for uncovering how Xist spreads from its site of transcription to other sites on the Xi. Specifically, Xist spreads from its initial target sites to more distal sites on the X chromosome. This spreading to actively transcribed regions on the X chromosome requires the A repeat region of Xist (Chaumeil et al. 2006; Engreitz et al. 2013; Chen et al. 2016), a region that is necessary for transcriptional silencing (Wutz et al. 2002). These studies revealed that Xist initially localizes to DNA regions that most frequently contact the *Xist* genomic locus in 3D proximity (Engreitz et al. 2013). At each of these sites, Xist interacts with the nuclear matrix protein SAF-A, which provides the physical link that tethers the Xist lncRNA to genomic DNA (McHugh et al. 2015). One of the most striking architectural changes associated with the induction of Xist is the repositioning of the entire Xi adjacent to the nuclear lamina. Recent proteomic studies have revealed that Xist directly interacts with lamin B receptor (LBR), and disruption of the Xist–LBR interaction leads to defective recruitment of the X to the lamina, as well as preventing Xist from spreading (McHugh et al. 2015; Minajigi et al. 2015). These results showed that Xist interacting with LBR leads to a change in chromosome architecture and enables Xist spreading from its site of transcription to other sites on the Xi (Engreitz et al. 2013; Chen et al. 2016).

Although these methods have provided critical insights for studying lncRNAs, there are still many important issues that need to be taken into account when interpreting the data. Specifically, capture of any RNA will enrich its transcribed DNA locus and accordingly, it is extremely challenging to discriminate between lncRNA interactions that occur in *cis* from those that simply reflect nascent transcription. In the case of Xist, this was addressed by using a doxycycline inducible promoter to perform a pulse-chase experiment and prevent nascent transcription of Xist before mapping its localization, ensuring that the DNA sites identified accurately represented the lncRNA targets (Engreitz et al. 2013). Given that many lncRNAs are thought to act in *cis*, future work is required to develop more generally applicable solutions that address this challenge directly.

4.3 How Localization Provides Insights into lncRNA Mechanisms

These same microscopy and sequencing approaches have been used to study the localization of other lncRNAs and derive key insights into their mechanisms.

For example, FISH experiments revealed that the imprinting lncRNAs Air (Nagano et al. 2008) and Kcnq1ot1 (Terranova et al. 2008) accumulate along the chromosome near their sites of transcription only on the same allele, colocalizing with repressive chromatin modifiers and marks, suggesting that these RNAs may mediate repression of neighboring genes in *cis*. Conversely, FISH against the abundant highly conserved lncRNAs MALAT1 and NEAT1 revealed that they localize in many different nuclear sites corresponding to nuclear speckles and paraspeckles, respectively (Clemson et al. 2009; Mao et al. 2011). Sequencing studies showed that both Malat1 and Neat1 localize broadly to hundreds of transcribed DNA loci (Engreitz et al. 2014; West et al. 2014). In the case of Malat1, the lncRNA specifically associates with the nascent pre-mRNA at these sites and the interaction is rapidly abrogated on transcriptional inhibition (Engreitz et al. 2014; West et al. 2014). Finally, FIRRE was shown to accumulate over an ~5 Mbp region at its site of transcription on the X chromosome, and it also localizes to a number of seemingly distal autosomal sites. FISH on FIRRE showed that it concentrates in only one nuclear focus and that the *trans* targets are spatially organized within this nuclear compartment (Hacisuleyman et al. 2014). Despite these important insights derived from localization studies, it is still unclear what roles most of these lncRNAs play in organizing these precise locations within the nucleus (Engreitz et al. 2016b).

These examples highlight two strategies by which a lncRNA can localize in the nucleus—spatial proximity and affinity to targets. The balance between lncRNA localization through spatial proximity or DNA affinity may depend on the abundance of a lncRNA. For example, low-abundance lncRNAs may preferentially interact with regions that are in close proximity because they can attain high local concentration at these regions, whereas abundant lncRNAs may be able to diffuse through the nucleus to achieve high concentration at multiple target regions in *trans*. As an example, the HOTTIP lncRNA is expressed at low levels (about one copy per cell) and localizes in direct proximity to its transcription locus (Wang et al. 2011). In contrast, Malat1 is highly expressed (thousands of copies per cell) and can localize broadly to many different actively transcribed DNA regions throughout the nucleus (Engreitz et al. 2014; West et al. 2014). Although these lncRNAs represent examples of these respective localization strategies, other lncRNAs can integrate features of both strategies. For example, the roX2 lncRNA binds to specific DNA regions that are bound by the CLAMP protein, but does so preferentially at binding sites that are in close spatial proximity to its genomic locus (Soruco et al. 2013).

Although most well-studied lncRNAs accumulate in the nucleus, a number have been found in the cytoplasm, suggesting functions in posttranscriptional cellular processes. The most comprehensive lncRNA localization study to date used single-molecule FISH to examine the subcellular localization and abundance of 61 lncRNAs in three cell types, and showed that nearly half exist in the cytoplasm (Cabili et al. 2015). Although cytoplasmic functions of lncRNAs are not well understood, a number of studies have used the methods described here to begin to describe the diverse roles lncRNAs can play in regulating cellular processes outside the nucleus. These include blocking posttranslational modifications on signaling proteins, acting as decoys for proteins and miRNAs, and regulating mRNA stability and translation (Rinn and Chang 2012; Engreitz et al. 2016b; Kopp and Mendell 2018).

5 FUTURE DIRECTIONS: lncRNAs AND PHASE SEPARATION IN THE NUCLEUS

An emerging theme from studies of Xist, Malat1, Neat1, Firre, and other lncRNAs is their role in organizing nuclear architecture. lncRNA expression creates a high local concentration near its site of transcription in 3D space, scaffolding the assembly of chromatin and RNPs into a structural domain (Fig. 5). Such domains could serve to locally coordinate shared regulatory processes (e.g., nucleoli, Cajal bodies, nuclear speckles) (Shevtsov and Dundr

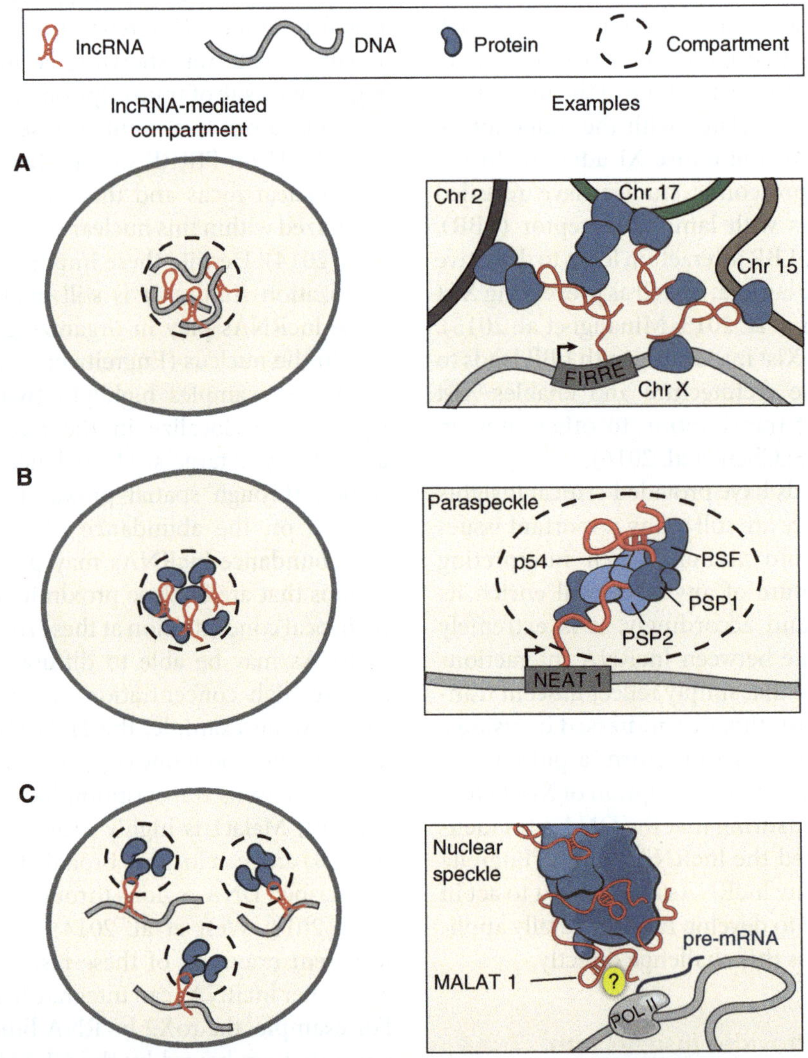

Figure 5. Examples of long noncoding RNAs (lncRNAs) forming dynamic spatial compartments in the nucleus by interacting with DNA, RNA, and protein. (*A*) Firre interacts with multiple functionally related loci on different chromosomes, brought together into a single nuclear territory for co-regulation. A high local concentration of Firre is required to form this structure. (*B*) Neat1 expression nucleates paraspeckle formation in the nucleus, creating phase-separated domains that may also contain additional RNA species. It is unclear if the role of paraspeckles is to coordinate RNA biogenesis and processing, or to trap and sequester molecules away from the rest of the nucleus. (*C*) Malat1 binds nascent pre-RNAs and splicing factors in nuclear speckles and also interacts with hundreds of actively transcribed DNA loci, potentially integrating transcription and splicing within the same nuclear structure.

2011), bring together multiple chromosomal loci for common chromatin modifications within nuclear territories (e.g., Polycomb bodies and the inactive X) (Delest et al. 2012), or sequester RNA binding proteins and regulatory RNAs to titrate them away from other regions of the nucleus (e.g., paraspeckles) (Mao et al. 2011; Fox et al. 2018). In addition to nucleating functional domains within the nucleoplasm, lncRNAs can also tether these structures to chromatin through association with DNA-binding proteins such as SAF-A/hnRNPU, and localize them to particular nuclear territories, as is the case with Xist interacting with LBR to associate the Xi with the nuclear lamina (Chen et al. 2016), coordinating higher levels of organization.

A model of liquid–liquid phase separation in the nucleoplasm has recently been proposed to explain the dynamic formation of membraneless nuclear compartments to coordinately regulate different aspects of gene expression (Hyman et al. 2014; Engreitz et al. 2016b; Hnisz et al. 2017). A central tenant of this model is that cooperative interactions between high local concentrations of nucleic acids and proteins—particularly those with intrinsically disordered domains, present in many RBPs—can lead to the formation of phase-separated bodies that compartmentalize biochemical functions in cells. High local concentrations of lncRNAs recruiting and concentrating proteins into lncRNA–protein structures could nucleate such events, thus dynamically organizing nuclear structures and functions into discrete domains (Fig. 5). In addition, lncRNAs may also act to create phase separated compartments within the nucleoplasm or cytoplasm. Such phase separated structures, however, are large and complex, containing dozens to thousands of distinct chromatin, RNA, and protein components occupying relatively large volumes in cells. Current methods such as microscopy, RNA purification, and Hi-C can be combined to study the individual components of nuclear bodies and territories, but are incapable of comprehensively characterizing how molecules simultaneously associate to function within an individual nucleus. A new method recently developed uses split-and-pool barcoding of cross-linked nucleic acids and proteins to define their associations in higher-order structures (Quinodoz et al. 2018). In the future, such methods will allow simultaneous mapping of multiple RNAs and DNA targets that co-occur within 3D structures in the nucleus. In addition, new developments in CRISPR-based live-cell imaging will make it possible to map large numbers of RNA and DNA structural dynamics as they occur in real time (Maass et al. 2018), a critical requirement for studying the role of RNA in phase separation, shaping genome organization, and gene regulation across time.

REFERENCES

Almeida M, Pintacuda G, Masui O, Koseki Y, Gdula M, Cerase A, Brown D, Mould A, Innocent C, Nakayama M, et al. 2017. PCGF3/5-PRC1 initiates Polycomb recruitment in X chromosome inactivation. *Science* **356:** 1081–1084.

Brockdorff N, Ashworth A, Kay GF, Cooper P, Smith S, McCabe VM, Norris DP, Penny GD, Patel D, Rastan S. 1991. Conservation of position and exclusive expression of mouse Xist from the inactive X chromosome. *Nature* **351:** 329–331.

Brown CJ, Baldry SE. 1996. Evidence that heteronuclear proteins interact with XIST RNA in vitro. *Somat Cell Mol Genet* **22:** 403–417.

Brown CJ, Ballabio A, Rupert JL, Lafreniere RG, Grompe M, Tonlorenzi R, Willard HF. 1991. A gene from the region of the human X inactivation centre is expressed exclusively from the inactive X chromosome. *Nature* **349:** 38–44.

Cabili MN, Trapnell C, Goff L, Koziol M, Tazon-Vega B, Regev A, Rinn JL. 2011. Integrative annotation of human large intergenic noncoding RNAs reveals global properties and specific subclasses. *Genes Dev* **25:** 1915–1927.

Cabili MN, Dunagin MC, McClanahan PD, Biaesch A, Padovan-Merhar O, Regev A, Rinn JL, Raj A. 2015. Localization and abundance analysis of human lncRNAs at single-cell and single-molecule resolution. *Genome Biol* **16:** 20.

Cerase A, Smeets D, Tang YA, Gdula M, Kraus F, Spivakov M, Moindrot B, Leleu M, Tattermusch A, Demmerle J, et al. 2014. Spatial separation of Xist RNA and Polycomb proteins revealed by superresolution microscopy. *Proc Natl Acad Sci* **111:** 2235–2240.

Chaumeil J, Le Baccon P, Wutz A, Heard E. 2006. A novel role for Xist RNA in the formation of a repressive nuclear compartment into which genes are recruited when silenced. *Genes Dev* **20:** 2223–2237.

Chen C-KC-K, Blanco M, Jackson C, Aznauryan E, Ollikainen N, Surka C, Chow A, Cerase A, McDonel P, Guttman M. 2016. Xist recruits the X chromosome to the nuclear lamina to enable chromosome-wide silencing. *Science* **354:** 468–472.

Chu C, Qu K, Zhong FL, Artandi SE, Chang HY. 2011. Genomic maps of long noncoding RNA occupancy reveal principles of RNA–chromatin interactions. *Mol Cell* **44:** 667–678.

Chu C, Zhang QCC, da Rocha ST, Flynn RAA, Bharadwaj M, Calabrese JM, Magnuson T, Heard E, Chang HYY, da Rocha ST, et al. 2015. Systematic discovery of Xist RNA binding proteins. *Cell* **161:** 404–416.

Cirillo D, Blanco M, Armaos A, Buness A, Avner P, Guttman M, Cerase A, Tartaglia GG. 2016. Quantitative predictions of protein interactions with long noncoding RNAs. *Nat Methods* **14:** 5–6.

Clemson CM, McNeil JA, Willard HF, Lawrence JB. 1996. XIST RNA paints the inactive X chromosome at interphase: Evidence for a novel RNA involved in nuclear/chromosome structure. *J Cell Biol* **132:** 259–275.

Clemson CM, Hutchinson JN, Sara SA, Ensminger AW, Fox AH, Chess A, Lawrence JB. 2009. An architectural role for a nuclear noncoding RNA: NEAT1 RNA is essential for the structure of paraspeckles. *Mol Cell* **33:** 717–726.

da Rocha ST, Heard E. 2017. Novel players in X inactivation: Insights into Xist-mediated gene silencing and chromosome conformation. *Nat Struct Mol Biol* **24:** 197–204.

da Rocha ST, Boeva V, Escamilla-Del-Arenal M, Ancelin K, Granier C, Matias NR, Sanulli S, Chow J, Schulz E, Picard C, et al. 2014. Jarid2 is implicated in the initial Xist-induced targeting of PRC2 to the inactive X chromosome. *Mol Cell* **53:** 301–316.

Darnell RB. 2010. HITS-CLIP: Panoramic views of protein–RNA regulation in living cells. *Wiley Interdiscip Rev RNA* **1:** 266–286.

Davidovich C, Zheng L, Goodrich KJ, Cech TR. 2013. Promiscuous RNA binding by Polycomb repressive complex 2. *Nat Struct Mol Biol* **20:** 1250–1257.

Delest A, Sexton T, Cavalli G. 2012. Polycomb: A paradigm for genome organization from one to three dimensions. *Curr Opin Cell Biol* **24:** 405–414.

Derrien T, Johnson R, Bussotti G, Tanzer A, Djebali S, Tilgner H, Guernec G, Martin D, Merkel A, Knowles DG, et al. 2012. The GENCODE v7 catalog of human long noncoding RNAs: Analysis of their gene structure, evolution, and expression. *Genome Res* **22:** 1775–1789.

Duthie SM, Nesterova TB, Formstone EJ, Keohane AM, Turner BM, Zakian SM, Brockdorff N. 1999. Xist RNA exhibits a banded localization on the inactive X chromosome and is excluded from autosomal material in *cis*. *Hum Mol Genet* **8:** 195–204.

Engreitz JMM, Pandya-Jones A, McDonel P, Shishkin A, Sirokman K, Surka C, Kadri S, Xing J, Goren A, Lander ESS, et al. 2013. The Xist lncRNA exploits three-dimensional genome architecture to spread across the X chromosome. *Science* **341:** 1237973.

Engreitz JM, Sirokman K, McDonel P, Shishkin AA, Surka C, Russell P, Grossman SR, Chow AY, Guttman M, Lander ES. 2014. RNA–RNA interactions enable specific targeting of noncoding RNAs to nascent pre-mRNAs and chromatin sites. *Cell* **159:** 188–199.

Engreitz JM, Haines JE, Perez EM, Munson G, Chen J, Kane M, McDonel PE, Guttman M, Lander ES. 2016a. Local regulation of gene expression by lncRNA promoters, transcription and splicing. *Nature* **539:** 452–455.

Engreitz JM, Ollikainen N, Guttman M. 2016b. Long non-coding RNAs: Spatial amplifiers that control nuclear structure and gene expression. *Nat Rev Mol Cell Biol* **17:** 756–770.

Fox AH, Nakagawa S, Hirose T, Bond CS. 2018. Paraspeckles: Where long noncoding RNA meets phase separation. *Trends Biochem Sci* **43:** 124–135.

Fulco CP, Munschauer M, Anyoha R, Munson G, Grossman SR, Perez EM, Kane M, Cleary B, Lander ES, Engreitz JM. 2016. Systematic mapping of functional enhancer–promoter connections with CRISPR interference. *Science* **354:** 769–773.

Ghildiyal M, Zamore PD. 2009. Small silencing RNAs: An expanding universe. *Nat Rev Genet* **10:** 94–108.

Groff AF, Sanchez-Gomez DB, Soruco MML, Gerhardinger C, Barutcu AR, Li E, Elcavage L, Plana O, Sanchez LV, Lee JC, et al. 2016. In vivo characterization of *Linc-p21* reveals functional *cis*-regulatory DNA elements. *Cell Rep* **16:** 2178–2186.

Guttman M, Amit I, Garber M, French C, Lin MF, Feldser D, Huarte M, Zuk O, Carey BW, Cassady JP, et al. 2009. Chromatin signature reveals over a thousand highly conserved large non-coding RNAs in mammals. *Nature* **458:** 223–227.

Guttman M, Donaghey J, Carey BW, Garber M, Grenier JKK, Munson G, Young G, Lucas AB, Ach R, Bruhn L, et al. 2011. lincRNAs act in the circuitry controlling pluripotency and differentiation. *Nature* **477:** 295–300.

Guttman M, Russell P, Ingolia NT, Weissman JS, Lander ES. 2013. Ribosome profiling provides evidence that large noncoding RNAs do not encode proteins. *Cell* **154:** 240–251.

Hacisuleyman E, Goff LA, Trapnell C, Williams A, Henao-Mejia J, Sun L, McClanahan P, Hendrickson DG, Sauvageau M, Kelley DR, et al. 2014. Topological organization of multichromosomal regions by the long intergenic noncoding RNA Firre. *Nat Struct Mol Biol* **21:** 198–206.

Hasegawa Y, Brockdorff N, Kawano S, Tsutui K, Tsutui K, Nakagawa S. 2010. The matrix protein hnRNP U is required for chromosomal localization of Xist RNA. *Dev Cell* **19:** 469–476.

Herzing LB, Romer JT, Horn JM, Ashworth A. 1997. Xist has properties of the X-chromosome inactivation centre. *Nature* **386:** 272–275.

Hnisz D, Shrinivas K, Young RA, Chakraborty AK, Sharp PA. 2017. A phase separation model for transcriptional control. *Cell* **169:** 13–23.

Huarte M, Guttman M, Feldser D, Garber M, Koziol MJ, Kenzelmann-Broz D, Khalil AM, Zuk O, Amit I, Rabani M, et al. 2010. A large intergenic noncoding RNA induced by p53 mediates global gene repression in the p53 response. *Cell* **142:** 409–419.

Hyman A, Weber CA, Jülicher F. 2014. Liquid-liquid phase separation in biology. *Annu Rev Cell Dev Biol* **30:** 39–58.

Kalantry S, Magnuson T. 2006. The Polycomb group protein EED is dispensable for the initiation of random X-chromosome inactivation. *PLoS Genet* **2:** e66.

Keryer-Bibens C, Barreau C, Beverley Osborne H, Osborne HB. 2008. Tethering of proteins to RNAs by bacteriophage proteins. *Biol Cell* **100:** 125–138.

Khalil AM, Guttman M, Huarte M, Garber M, Raj A, Rivea Morales D, Thomas K, Presser A, Bernstein BE, Van Oudenaarden A, et al. 2009. Many human large intergenic noncoding RNAs associate with chromatin-modifying complexes and affect gene expression. *Proc Natl Acad Sci* **106:** 11667–11672.

Kohlmaier A, Savarese F, Lachner M, Martens J, Jenuwein T, Wutz A. 2004. A chromosomal memory triggered by Xist regulates histone methylation in X inactivation. *PLoS Biol* **2:** E171.

Konig J, Zarnack K, Rot G, Curk T, Kayikci M, Zupan B, Turner DJ, Luscombe NM, Ule J. 2010. iCLIP reveals the function of hnRNP particles in splicing at individual nucleotide resolution. *Nat Struct Mol Biol* **17:** 909–915.

Kopp F, Mendell JT. 2018. Functional classification and experimental dissection of long noncoding RNAs. *Cell* **172:** 393–407.

Lee JT, Strauss WM, Dausman JA, Jaenisch R. 1996. A 450 kb transgene displays properties of the mammalian X-inactivation center. *Cell* **86:** 83–94.

Lee HY, Haurwitz RE, Apffel A, Zhou K, Smart B, Wenger CD, Laderman S, Bruhn L, Doudna JA. 2013. RNA–protein analysis using a conditional CRISPR nuclease. *Proc Natl Acad Sci* **110:** 5416–5421.

Lee S, Kopp F, Chang T-C, Sataluri A, Chen B, Sivakumar S, Yu H, Xie Y, Mendell JT. 2016. Noncoding RNA NORAD regulates genomic stability by sequestering PUMILIO proteins. *Cell* **164:** 69–80.

Liang X-H, Sun H, Nichols JG, Crooke ST. 2017. RNase H1-dependent antisense oligonucleotides are robustly active in directing RNA cleavage in both the cytoplasm and the nucleus. *Mol Ther* **25:** 2075–2092.

Liu SJ, Liu SJ, Horlbeck MA, Cho SW, Birk HS, Malatesta M, Attenello FJ, Villalta JE, Cho MY, Chen Y, et al. 2016. CRISPRi-based genome-scale identification of functional long noncoding RNA loci in human cells. *Science* **355:** aah7111.

Maass PG, Barutcu AR, Shechner DM, Weiner CL, Melé M, Rinn JL. 2018. Spatiotemporal allele organization by allele-specific CRISPR live-cell imaging (SNP-CLING). *Nat Struct Mol Biol* **25:** 176–184.

Mao YS, Sunwoo H, Zhang B, Spector DL. 2011. Direct visualization of the co-transcriptional assembly of a nuclear body by noncoding RNAs. *Nat Cell Biol* **13:** 95–101.

McHugh CA, Russell P, Guttman M. 2014. Methods for comprehensive experimental identification of RNA–protein interactions. *Genome Biol* **15:** 203.

McHugh CA, Chen C-K, Chow A, Surka CF, Tran C, McDonel P, Pandya-Jones A, Blanco M, Burghard C, Moradian A, et al. 2015. The Xist lncRNA interacts directly with SHARP to silence transcription through HDAC3. *Nature* **521:** 232–236.

Mili S, Steitz JA. 2004. Evidence for reassociation of RNA-binding proteins after cell lysis: Implications for the interpretation of immunoprecipitation analyses. *RNA* **10:** 1692–1694.

Minajigi A, Froberg JE, Wei C, Sunwoo H, Kesner B, Colognori D, Lessing D, Payer B, Boukhali M, Haas W, et al. 2015. Chromosomes. A comprehensive Xist interactome reveals cohesin repulsion and an RNA-directed chromosome conformation. *Science* **349:** aab2276.

Moindrot B, Cerase A, Coker H, Masui O, Grijzenhout A, Pintacuda G, Schermelleh L, Nesterova TB, Brockdorff N. 2015. A pooled shRNA screen identifies Rbm15, Spen, and Wtap as factors required for Xist RNA-mediated silencing. *Cell Rep* **12:** 562–572.

Monfort A, Minin G Di, Postlmayr A, Arieti F, Wutz A. 2015. Identification of Spen as a crucial factor for Xist function through forward genetic screening in haploid embryonic stem cells. *Cell Rep* **12:** 554–561.

Nagano T, Mitchell JA, Sanz LA, Pauler FM, Ferguson-Smith AC, Feil R, Fraser P. 2008. The *Air* noncoding RNA epigenetically silences transcription by targeting G9a to chromatin. *Science* **322:** 1717–1720.

Ozdilek BA, Thompson VF, Ahmed NS, White CI, Batey RT, Schwartz JC. 2017. Intrinsically disordered RGG/RG domains mediate degenerate specificity in RNA binding. *Nucleic Acids Res* **45:** 7984–7996.

Patil DP, Chen CK, Pickering BF, Chow A, Jackson C, Guttman M, Jaffrey SR. 2016. M6A RNA methylation promotes XIST-mediated transcriptional repression. *Nature* **537:** 369–373.

Penny GD, Kay GF, Sheardown SA, Rastan S, Brockdorff N. 1996. Requirement for Xist in X chromosome inactivation. *Nature* **379:** 131–137.

Pintacuda G, Wei G, Roustan C, Kirmizitas BA, Solcan N, Cerase A, Castello A, Mohammed S, Moindrot B, Nesterova TB, et al. 2017. hnRNPK recruits PCGF3/5-PRC1 to the Xist RNA B-repeat to establish Polycomb-mediated chromosomal silencing. *Mol Cell* **68:** 955–969.e10.

Plath K, Fang J, Mlynarczyk-Evans SK, Cao R, Worringer KA, Wang H, de la Cruz CC, Otte AP, Panning B, Zhang Y. 2003. Role of histone H3 lysine 27 methylation in X inactivation. *Science* **300:** 131–135.

Portoso M, Ragazzini R, Brenčič Ž, Moiani A, Michaud A, Vassilev I, Wassef M, Servant N, Sargueil B, Margueron R. 2017. PRC2 is dispensable for *HOTAIR*-mediated transcriptional repression. *EMBO J* **36:** 981–994.

Quinodoz SA, Ollikainen N, Tabak B, Palla A, Schmidt JM, Detmar E, Lai MM, Shishkin AA, Bhat P, Takei Y, et al. 2018. Higher-order interchromosomal Hubs shape 3D genome organization in the nucleus. *Cell* **174:** 744–757.e24.

Rinn JL, Chang HY. 2012. Genome regulation by long noncoding RNAs. *Annu Rev Biochem* **81:** 145–166.

Rogell B, Fischer B, Rettel M, Krijgsveld J, Castello A, Hentze MW. 2017. Specific RNP capture with antisense LNA/DNA mixmers. *RNA* **23:** 1290–1302.

Sauvageau M, Goff LA, Lodato S, Bonev B, Groff AF, Gerhardinger C, Sanchez-Gomez DB, Hacisuleyman E, Li E, Spence M, et al. 2013. Multiple knockout mouse models reveal lincRNAs are required for life and brain development. *Elife* **2:** e01749.

Schoeftner S, Sengupta AK, Kubicek S, Mechtler K, Spahn L, Koseki H, Jenuwein T, Wutz A. 2006. Recruitment of PRC1 function at the initiation of X inactivation independent of PRC2 and silencing. *EMBO J* **25:** 3110–3122.

Shechner DM, Hacisuleyman E, Younger ST, Rinn JL. 2015. Multiplexable, locus-specific targeting of long RNAs with CRISPR-Display. *Nat Methods* **12:** 664–670.

Shevtsov SP, Dundr M. 2011. Nucleation of nuclear bodies by RNA. *Nat Cell Biol* **13:** 167–173.

Simon MD, Pinter SF, Fang R, Sarma K, Rutenberg-Schoenberg M, Bowman SK, Kesner BA, Maier VK, Kingston RE, Lee JT. 2013. High-resolution Xist binding maps reveal two-step spreading during X-chromosome inactivation. *Nature* **504:** 465–469.

Soruco MML, Chery J, Bishop EP, Siggers T, Tolstorukov MY, Leydon AR, Sugden AU, Goebel K, Feng J, Xia P, et al. 2013. The CLAMP protein links the MSL complex to the X chromosome during *Drosophila* dosage compensation. *Genes Dev* **27:** 1551–1556.

Spector DL, Lamond AI. 2011. Nuclear speckles. *Cold Spring Harb Perspect Biol* **3:** a000646.

Srisawat C, Engelke DR. 2002. RNA affinity tags for purification of RNAs and ribonucleoprotein complexes. *Methods* **26:** 156–161.

Terranova R, Yokobayashi S, Stadler MB, Otte AP, van Lohuizen M, Orkin SH, Peters AHFM. 2008. Polycomb group proteins Ezh2 and Rnf2 direct genomic contraction and imprinted repression in early mouse embryos. *Dev Cell* **15:** 668–679.

Tripathi V, Ellis JD, Shen Z, Song DY, Pan Q, Watt AT, Freier SM, Bennett CF, Sharma A, Bubulya PA, et al. 2010. The nuclear-retained noncoding RNA MALAT1 regulates alternative splicing by modulating SR splicing factor phosphorylation. *Mol Cell* **39:** 925–938.

Wang KC, Yang YW, Liu B, Sanyal A, Corces-Zimmerman R, Chen Y, Lajoie BR, Protacio A, Flynn RA, Gupta RA, et al. 2011. A long noncoding RNA maintains active chromatin to coordinate homeotic gene expression. *Nature* **472:** 120–124.

Wapinski O, Chang HY. 2011. Long noncoding RNAs and human disease. *Trends Cell Biol* **21:** 354–361.

Werner MS, Ruthenburg AJ. 2015. Nuclear fractionation reveals thousands of chromatin-tethered noncoding RNAs adjacent to active genes. *Cell Rep* **12:** 1089–1098.

West JA, Davis CP, Sunwoo H, Simon MD, Sadreyev RI, Wang PI, Tolstorukov MY, Kingston RE. 2014. The long noncoding RNAs NEAT1 and MALAT1 bind active chromatin sites. *Mol Cell* **55:** 791–802.

Wutz A, Rasmussen TP, Jaenisch R. 2002. Chromosomal silencing and localization are mediated by different domains of Xist RNA. *Nat Genet* **30:** 167–174.

Zhao J, Sun BK, Erwin JA, Song JJ, Lee JT. 2008. Polycomb proteins targeted by a short repeat RNA to the mouse X chromosome. *Science* **322:** 750–756.

Zhao J, Ohsumi TK, Kung JT, Ogawa Y, Grau DJ, Sarma K, Song JJ, Kingston RE, Borowsky M, Lee JT. 2010. Genome-wide identification of Polycomb-associated RNAs by RIP-seq. *Mol Cell* **40:** 939–953.

What Are 3′ UTRs Doing?

Christine Mayr

Department of Cancer Biology and Genetics, Memorial Sloan Kettering Cancer Center, New York, New York 10065

Correspondence: mayrc@mskcc.org

SUMMARY

3′ untranslated regions (3′ UTRs) of messenger RNAs (mRNAs) are best known to regulate mRNA-based processes, such as mRNA localization, mRNA stability, and translation. In addition, 3′ UTRs can establish 3′ UTR-mediated protein–protein interactions (PPIs), and thus can transmit genetic information encoded in 3′ UTRs to proteins. This function has been shown to regulate diverse protein features, including protein complex formation or posttranslational modifications, but is also expected to alter protein conformations. Therefore, 3′ UTR-mediated information transfer can regulate protein features that are not encoded in the amino acid sequence. This review summarizes both 3′ UTR functions—the regulation of mRNA and protein-based processes—and highlights how each 3′ UTR function was discovered with a focus on experimental approaches used and the concepts that were learned. This review also discusses novel approaches to study 3′ UTR functions in the future by taking advantage of recent advances in technology.

Outline

Cite as *Cold Spring Harb Perspect Biol* doi: 10.1101/cshperspect.a034728

1 INTRODUCTION

Genetic information is stored in DNA and transmitted via messenger RNA (mRNA) to proteins (Crick 1958). For many years, it was thought that the information transfer from DNA to proteins happens exclusively through translation of the coding region of mRNAs into amino acids of proteins. Although it was well known that mRNAs also contain untranslated regions (UTRs) at their 5′ and 3′ ends, the prevailing view was that these regions largely regulate mRNA localization or protein abundance, either through the regulation of translation or stability of mRNAs.

With such a protein-centric view, it was surprising to find that the number of protein-coding genes was similar in the human genome and the genomes of more simple eukaryotic organisms (Lander et al. 2001; Mayr 2017). Also, protein size is largely similar across organisms (Milo and Phillips 2016). This points to a remarkable conservation of protein sequence and reveals the tremendous constraints that act on proteins. However, it also begs the question of what enables biological complexity of higher organisms. In this context, it is interesting to note that the sequence space occupied by 3′ UTRs has substantially expanded during the evolution of higher organisms and correlates with cellular complexity of organisms (Chen et al. 2012; Mayr 2016). Importantly, there are 3′ UTR regions ranging from a few to a few hundred nucleotides (nt) that are often highly conserved (Siepel et al. 2005; Xie et al. 2005). The expansion of 3′ UTR sequence during evolution, together with the recent finding that genetic information stored in 3′ UTRs can also be transmitted to proteins via the formation of 3′ UTR-mediated protein–protein interactions (PPIs) (Berkovits and Mayr 2015), suggests that 3′ UTRs may play important roles in the regulation of biological complexity. If this is so, why are 3′ UTRs not at the center of everyone's attention?

One contributing factor is the scarcity of suitable and robust methods for the study of 3′ UTR functions as the availability of experimental approaches is often crucial for progress and discoveries. Compared with research on transcription or translation regulation, or on microRNA (miRNA)-mediated posttranscriptional regulation, research on miRNA-independent functions of 3′ UTRs lags far behind. Most functions of 3′ UTRs are mediated by RNA-binding proteins (RBPs). But, in contrast to miRNA target sites, the binding motifs for many RBPs are still not known (Baltz et al. 2012; Ray et al. 2013; Dominguez et al. 2018). And even if they are known, clear functional effects often require the motifs to be present several times. As the motifs are often spread out over large distances, functional motifs can often not be identified using deletion mutants (Besse et al. 2009; Kristjansdottir et al. 2015; Ma and Mayr 2018). Moreover, many RBPs bind to the same motif (Ray et al. 2013; Dominguez et al. 2018), and it is usually unknown whether they compete or cooperate (Hennig and Sattler 2015). Finally, more than half of human genes use alternative cleavage and polyadenylation to generate mRNA isoforms that differ only in their 3′ UTRs (Lianoglou et al. 2013). This makes research on 3′ UTR isoform–specific functions challenging as the amino acid sequences of the proteins generated from the alternative 3′ UTR isoforms are identical.

This review summarizes how research on 3′ UTR functions has been approached in the last 25 years, what has been learned, and how it could be addressed in the future. Several topics relevant for 3′ UTR biology are being reviewed elsewhere in this collection, including regulation by miRNAs, RNA editing, RNA structure, the identification of RBPs, and RNA–protein interactions. Also, several "noncanonical" functions of 3′ UTRs were reviewed recently and are not included here (Mayr 2017). Thus, this review is not comprehensive but tries to focus on experimental approaches and concepts to show how regulation by 3′ UTRs is accomplished.

2 3′ UTRs THAT CONTAIN AU-RICH ELEMENTS REGULATE mRNA STABILITY

Nucleotide sequencing was developed in the 1970s and revealed the existence of untranslated sequence between the coding region and the polyadenylation signal. Having the DNA sequences from genes of different organisms in hand allowed the first comparative genomic analyses (Needleman and Wunsch 1970). As the rate of base substitution reflects the degree of functional constraint, the evaluation of sequence divergence across distant organisms then allowed the evaluation of functionally import specific elements. Although these early comparative genomic analyses found a higher degree of base substitutions in untranslated regions than in coding regions, they also revealed a high degree of sequence homology in 3′ UTRs (Miyata et al. 1980). Early on, one of the most intriguing findings was that 3′ UTR sequences of homologous genes coding for actin proteins are highly conserved across organisms, but the 3′ UTR sequences are highly divergent when similar actin isoforms with different functions or tissue distributions were compared (Yaffe et al. 1985). The high conservation indicated that 3′ UTRs have important regulatory roles, and the divergence of the 3′ UTR sequence of isoforms, on the other hand, suggested that 3′ UTRs contain additional genetic information to distinguish the functions of highly similar proteins.

The first indication that 3′ UTRs contain specific functional elements was found through investigation of the transforming ability of the *c-fos* gene (Miller et al. 1984).

Whereas the viral *fos* (*v-fos*) gene was able to transform fibroblasts in culture, the cellular *fos* (*c-fos*) lacked this feature. Interestingly, the functional difference was not the result of a slightly different amino acid sequence of the two fos proteins. Instead, it was shown that *c-fos* could also transform fibroblasts when the 3′ UTR was omitted (Miller et al. 1984). This finding motivated the search for the responsible 3′ UTR *cis*-element and led to the discovery of AU-rich elements which were found to destabilize mRNAs (Meijlink et al. 1985). The proof for this function was obtained through transfer of the *cis*-element to an otherwise stable, heterologous mRNA, the *β-actin* mRNA (Shaw and Kamen 1986). At around the same time, it was noticed that AU-rich elements were predominantly found in 3′ UTRs of a certain class of genes that encodes short-lived factors, including cytokines, lymphokines, growth factors, and oncogenes (Caput et al. 1986). Intriguingly, the AU-rich elements were sometimes more conserved than the coding regions of these early response genes (Shaw and Kamen 1986). The identification of the *trans*-acting factors that bind to AU-rich elements, however, took many years and is still ongoing (Ray et al. 2013; Dominguez et al. 2018). Also, the mechanism of AU-rich element–induced mRNA decay was elucidated much later, as will be described below.

3 3′ UTRs REGULATE mRNA LOCALIZATION

At around the time of discovery of AU-rich elements, 3′ UTRs were also identified as important regulators of subcellular mRNA localization. Early studies on the regulation of mRNA localization were mostly performed in fly or frog oocytes. But mRNA localization is not only essential in germ cells and early development as it also occurs in other polarized cells, including fibroblasts, myoblasts, and neurons (Lawrence and Singer 1986; Melton 1987). However, germ cells and the early stages of development are especially advantageous for RNA research as gene regulation in early animal development depends on maternally deposited mRNAs from oocytes before the onset of embryonic transcription. For example, the anterior body pattern of flies is specified predominantly by the protein product of the *bicoid* gene as mutants containing a disrupted gene lack the head and thorax (Macdonald and Struhl 1988). Bicoid protein localizes mainly to the anterior pole of the fly embryo and generates a protein gradient. As bicoid is a transcription factor, it drives expression of a large number of downstream genes. Importantly, the gradient in the embryo depends on prior localization of *bicoid* mRNA to the anterior pole of oocytes. Subcellular localization of *bicoid* mRNA to the anterior poles of oocytes and early embryos was revealed by RNA in situ hybridization. To identify the responsible *cis*-element, deletion constructs were tested using transgenic flies. RNase protection assays from the anterior and posterior poles of early fly embryos revealed that a 630 nt fragment of the *bicoid* 3′ UTR was necessary and sufficient for the subcellular mRNA localization (Macdonald and Struhl 1988).

4 3′ UTRs REGULATE TRANSLATION OF mRNAs

mRNA localization allows spatial as well as temporal regulation of protein production. In neurons, 3′ UTRs are well-known to regulate local protein synthesis in dendrites and synapses (Miller et al. 2002; An et al. 2008; Martin and Ephrussi 2009). But, in addition to the spatial organization of protein production, 3′ UTRs are also important regulators of temporal protein production. The initial observation in the discovery that 3′ UTRs regulate translation was a discrepancy between *oskar* mRNA and protein expression in fly oocytes. *oskar* mRNA determines the posterior fly body pattern (Ephrussi et al. 1991; Kim-Ha et al. 1993). Although *oskar* mRNA was present throughout oogenesis, oskar protein was only detectable after localization of *oskar* mRNA to the posterior pole (Kim-Ha et al. 1995). An ultraviolet (UV) cross-linking assay identified the RBP Bruno as the responsible factor for translation regulation. Binding of Bruno to *oskar* mRNA prevents premature translation of oskar protein and is necessary for oskar expression at the correct developmental stage as *oskar* mRNA is translated only after reaching the posterior pole of the oocytes (Kim-Ha et al. 1995). The exact mechanism of translation repression of *oskar* mRNA by Bruno was elucidated much later (Chekulaeva et al. 2006).

5 RBPs BIND TO 3′ UTR *cis*-ELEMENTS AND MEDIATE MULTIPLE AND OFTEN OPPOSING 3′ UTR FUNCTIONS

Initially, AU-rich elements were considered to be mRNA decay elements. Then, they were shown to also repress translation (Shaw and Kamen 1986; Kruys et al. 1989). After that, it was found that AU-rich elements can also increase protein production (Lindstein et al. 1989; Kontoyiannis et al. 1999). For example, in resting immune cells, AU-rich elements repress translation, but after lipopolysaccharide (LPS) stimulation of T cells, they were required for fast induction of protein expression (Lindstein et al. 1989; Kontoyiannis et al. 1999). This led to the notion that not the *cis*-element itself, but rather the binding of specific *trans*-acting factors, determines the function of a particular 3′ UTR *cis*-element.

This regulatory principle was nicely shown by deletion of the 62-nt-long, conserved AU-rich element of tumor necrosis factor (TNF)-α in the mouse (Kontoyiannis et al.

1999). Importantly, the remaining gene regulation of TNF-α expression was preserved because the endogenous promoter and polyadenylation signal were retained. Sole absence of the *cis*-regulatory AU-rich element resulted in inflammatory arthritis and inflammatory bowel disease with death of the mice at 5 to 12 weeks. In unstimulated conditions, AU-rich elements destabilize mRNAs. The AU-rich element of TNF-α represses translation permanently in nonhematopoietic cells and transiently in hematopoietic cells. Importantly, their function is not only repressive, as after activation by LPS they initially increase mRNA stability and translation in hematopoietic cells, whereas in later stages, they become repressive. In summary, AU-rich elements were found to be important regulatory elements that enable fast and highly dynamic regulation of protein production with a sharp on and off switch (Kontoyiannis et al. 1999).

There are currently more than 10 known RBPs that bind to AU-rich elements (Brennan and Steitz 2001; Chen et al. 2001; Barreau et al. 2005; Lebedeva et al. 2011). The binding of tristetraprolin (TTP) or KHSRP leads to destabilization through recruitment of the exosome complex that degrades mRNAs (Chen et al. 2001; Lykke-Andersen and Wagner 2005). In contrast, binding of HuR stabilizes AU-rich containing mRNAs, likely through its inability to recruit the exosome (Chen et al. 2001). These findings revealed an important regulatory concept for 3′ UTR functions: RBPs bind to 3′ UTR *cis*-elements and serve as adapters for the recruitment of effector proteins. The recruited effector proteins are responsible for the observed effects (Fig. 1A). The adapter function of RBPs is critical as effector proteins are unable to directly bind to AU-rich-containing mRNAs (Chen et al. 2001).

3′ UTR elements can accomplish opposing effects in environments where different signaling pathways are active. This can be caused by altering the relative abundance or activity of RBPs that bind to the same *cis*-element. Thus, the function of a particular 3′ UTR *cis*-element is "context"-dependent, as competition and cooperativity between RBPs will ultimately determine the functional outcome of a specific 3′ UTR (Mayr 2017).

Similarly, a single RBP can accomplish diverse functions at different stages of development. In early development, translation is predominantly regulated through poly(A) tail length, which is determined by the RBP CPEB (cytoplasmic polyadenylation element [CPE]-binding protein) (Richter 2007). CPEB bound to the CPE can regulate mRNA polyadenylation, deadenylation, and translation. CPEB achieves these functions through binding to several different factors. It simultaneously binds to the deadenylase poly(A)-specific ribonuclease (PARN) and to the poly(A)polymerase Gld2 (PAPD4). In oocytes, mRNAs

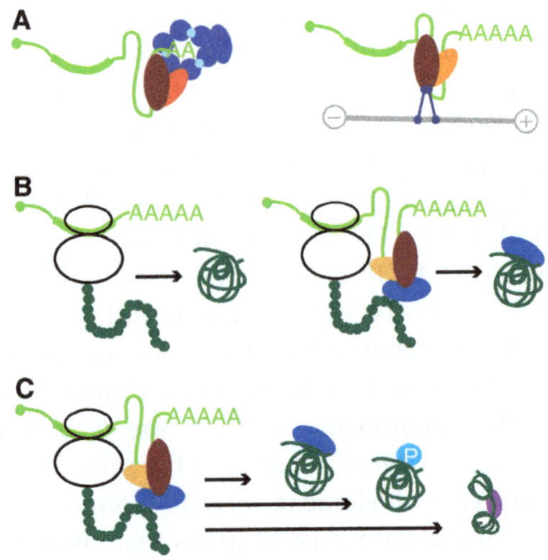

Figure 1. Functions of 3′ untranslated regions (UTRs). RNA-binding proteins (RBPs) bind to the 3′ UTR and recruit diverse effector proteins that determine 3′ UTR functions. (*A*) 3′ UTRs regulate processes at the messenger RNA (mRNA) level. RBPs (red, orange) bind to 3′ UTRs of mRNAs (light green) and recruit diverse effector proteins. The recruitment of the exosome (blue) results in mRNA destabilization (*left* panel), whereas the recruitment of a motor protein (blue) results in the regulation of mRNA localization using movement on a microtubule (gray line; *right* panel). (*B*) 3′ UTRs regulate protein features by mediating 3′ UTR-dependent protein–protein interactions (PPIs). Alternative 3′ UTRs can determine alternative protein functions despite encoding proteins with identical amino acid sequences. This results from 3′ UTR-dependent PPIs that are mediated only by the long 3′ UTR isoform (*right* panel) and not by the short 3′ UTR isoform (*left* panel). The RBPs that bind to the 3′ UTR, as well as the recruited effector protein, are color-coded as in *A*. (*C*) 3′ UTRs regulate diverse protein features by mediating 3′ UTR-dependent PPIs. This can result in 3′ UTR-dependent protein complex formation, 3′ UTR-dependent posttranslational modifications (P), and 3′ UTR-dependent protein folding.

that contain the CPE have short poly(A) tails resulting in translation inhibition as in these cells the activity of PARN is higher than the activity of PAPD4 (Kim and Richter 2006). After oocyte fertilization, CPEB is phosphorylated, which leads to exclusion of PARN from the complex. This is followed by poly(A) tail lengthening, leading to translation initiation (Mendez et al. 2000). In summary, during early development, temporal regulation of protein synthesis occurs via the 3′ UTR and is mediated by the RBP CPEB and its associated effector proteins. Therefore, to elucidate different mechanisms of action, the associated effector proteins of RBPs need to be identified and studied.

6 MORE THAN HALF OF HUMAN GENES GENERATE ALTERNATIVE 3′ UTRs

Transcription of DNA by RNA polymerase II produces primary transcripts that need to be processed to generate mature mRNAs. The steps of mRNA processing include addition of a 5′ cap, splicing of introns and cleavage/polyadenylation at the 3′ end (reviewed in Tian and Manley 2017). Transcription does not stop at the 3′ ends of mRNAs as primary transcripts are extended beyond the polyadenylation signal by several thousand nucleotides (Core et al. 2008). Some factors of the cleavage and polyadenylation machinery travel with RNA polymerase II and the encounter of a functional polyadenylation signal results in endonucleolytic cleavage, followed by the addition of a poly(A) tail (Chan et al. 2014; Schonemann et al. 2014). Thus, 3′-end cleavage and polyadenylation determine 3′ UTR length as well as the *cis*-elements present within a 3′ UTR.

Approximately 10 years after the discovery that 3′ UTRs regulate several aspects of gene expression, a literature survey revealed that many genes produce alternative mRNA isoforms that differ in their 3′ ends. At the time, 95 genes were known to generate alternative 3′ UTRs, whereas 31 genes were known that used intronic polyadenylation sites to change the carboxyl terminus of the protein as well as the 3′ UTR (Edwalds-Gilbert et al. 1997). At around the same time, new technologies that allowed for shotgun sequencing of complementary DNA (cDNA) libraries enabled the creation of expressed sequence tag (EST) databases. The mining of these databases revealed that a sizable fraction of the human and mouse transcriptomes use alternative polyadenylation signals to generate mRNAs with alternative 3′ UTRs (Gautheret et al. 1998). The development of deep sequencing technology enabled the establishment of methods that were able to map all the 3′ ends of mRNAs transcriptome-wide (Derti et al. 2012; Hoque et al. 2013; Lianoglou et al. 2013; Gruber et al. 2016). These 3′-end sequencing methods revealed that more than half of human and mouse genes generate alternative mRNA isoforms that differ in their 3′ UTRs but encode proteins with identical amino acid sequences.

7 ALTERNATIVE 3′ UTR ISOFORM RATIOS ARE CELL TYPE–SPECIFIC AND CHANGE ON ACTIVATION OF SIGNALING PATHWAYS

3′-end sequencing methods provide information on 3′ UTR length and measure alternative 3′ UTR isoform levels quantitatively. 3′-end sequencing of samples from diverse tissues revealed that the location of functional polyadenylation sites does not vary across cell types, but the expression ratios of alternative 3′ UTR isoforms are highly tissue- and cell type–specific (Lianoglou et al. 2013). 3′-end sequencing data are available in databases and can be accessed and browsed (3′-seq, see cbio.mskcc.org/leslielab/ApA/atlas; polyA_DB, see exon.njms.rutgers.edu/polya_db/v3/; polyAsite, see polyasite.unibas.ch) (Lianoglou et al. 2013; Gruber et al. 2016; Wang et al. 2017; Singh et al. 2018).

Compared with RNA-seq data that exist for a large variety of samples, the availability of 3′-end sequencing data is still limited. Therefore, it would be very helpful if the expression of alternative 3′ UTRs could be extracted from RNA-seq data. Although this has been tried by several research groups (Masamha et al. 2014; Chang et al. 2015; Shenker et al. 2015; Grassi et al. 2016), the data obtained need to be used with caution as RNA-seq read coverage over 3′ UTRs is highly variable. If strict criteria are used to identify differences in 3′ UTR isoform expression, only a small number of changes (10%–15%) that are identified by 3′-end sequencing methods will be found by RNA-seq. On the other hand, less stringent analyses contain many artifacts, shown by the fact that there is little overlap in the locations of 3′-ends inferred from the different methods. Also, the direction of change (shortening or lengthening of 3′ UTRs) agreed only in 15% of the events considered to be significant by each method. Therefore, currently, 3′-end sequencing is the preferred method to determine alternative 3′ UTR expression.

In addition to being cell type–specific, alternative 3′ UTR isoform expression responds to receptor stimulation and subsequent activation of signaling pathways as was shown during immune cell stimulation or synaptic activation. A change in 3′ UTR isoform ratio can be achieved through exposure of the cells to extracellular stimuli and can be mimicked experimentally by the addition of growth factors, cytokines, or neurotransmitters (Flavell et al. 2008; Sandberg et al. 2008; Rhinn et al. 2012; Gruber et al. 2014; Jia et al. 2017). Expression ratios of alternative 3′ UTRs also change during normal development and differentiation (Lackford et al. 2014; Brumbaugh et al. 2018; Freimer et al. 2018) as well as in disease (Mayr and Bartel 2009; Rhinn et al. 2012; Batra et al. 2014; Masamha et al. 2014). A dramatic change in 3′ UTR length was recently found during mouse oocyte maturation from the germinal vesicle stage to metaphase II, which takes 9 hours and results in substantial lengthening of 3′ UTRs (Freimer et al. 2018). For ~20% of genes, the functional consequence of changes in alternative 3′ UTR isoform expression is a difference in protein levels, as the availability of binding sites of miRNAs and RBPs is altered (Gupta et al. 2014; Nam et al. 2014; Brumbaugh et al. 2018).

8 3′ UTRs REGULATE PROTEIN–PROTEIN INTERACTIONS

Alternative 3′ UTRs are widespread and the 3′ UTR sequences have expanded during the evolution of higher organisms (Mayr 2017). Therefore, we wondered whether

3′ UTRs might be able to regulate cellular processes beyond protein abundance. In this context, we discovered that some PPIs are established only when one of the interaction partners is recruited by the 3′ UTR (Fig. 1B) (Berkovits and Mayr 2015). This was first shown for the PPI between SET and CD47 and was then expanded to additional plasma membrane proteins in human cells and yeast (Berkovits and Mayr 2015; Chartron et al. 2016; Ma and Mayr 2018).

CD47 membrane protein is generated from alternative mRNA isoforms either containing a short or long 3′ UTR (SU or LU) (Lianoglou et al. 2013). The encoded proteins have identical amino acid sequences and are called CD47-SU and CD47-LU, respectively (Berkovits and Mayr 2015). It was found that only CD47-LU was able to interact with the protein SET in a 3′ UTR-dependent manner. As a result, CD47-SU and CD47-LU reside in different protein complexes, and thus, can have different functions. SET binding to CD47-LU protein resulted in more efficient plasma membrane localization, protected cells better from phagocytosis by macrophages, activated RAC1, enabled formation of lamellipodia, and improved cell survival after exposure to γ-irradiation (Berkovits and Mayr 2015). This list of functions is likely not exhaustive as RAC1 activation has multiple downstream consequences, including the activation of signaling pathways and increased cell migration.

The fact that 3′ UTRs are able to regulate protein complex formation and determine protein functions indicated that genetic information encoded in 3′ UTRs can be transmitted to proteins. This information transfer can regulate protein features not encoded in the amino acid sequence. As the sequence space contained within 3′ UTRs has expanded during the evolution of higher organisms, 3′ UTRs may contribute to the higher organismal complexity by providing additional information regarding proteins, including information about PPIs, posttranslational modifications, protein conformations, and protein multifunctionality (Fig. 1B,C). Thus, information about protein features can be genetically encoded in the untranslated mRNA sequence. It has not been shown yet that 3′ UTRs can regulate protein conformations but this is likely to occur as binding of the 3′ UTR-dependent interaction partner to the amino terminus of the nascent peptide chain could change the folding of the nascent protein.

Following the discovery of this new 3′ UTR function, it remained unclear how widespread this function is. It seems that 3′ UTR-mediated PPIs are common as they also occur on cytosolic and nuclear proteins (Halbach et al. 2009; Duncan and Mata 2011; Lee and Mayr 2017). This was shown in detail for the 3′ UTR-mediated interaction between IQGAP1 and the E3 ubiquitin ligase BIRC3, also called cIAP2 (Lee and Mayr 2017). BIRC3 protein is generated by alternative 3′ UTRs. Again, BIRC3 protein that was generated from the long 3′ UTR isoform was called BIRC3-LU, whereas BIRC3 protein encoded by the short 3′ UTR isoform was called BIRC3-SU. The long *BIRC3* 3′ UTR facilitated the assembly of a protein complex consisting of BIRC3-LU, the kinase scaffold IQGAP1, and the Ras-GTPase RALA. Importantly, this signaling complex could not be formed by BIRC3-SU. The 3′ UTR-dependent signaling complex associates with the G-protein-coupled receptor CXCR4 and influences plasma membrane expression of CXCR4 through the regulation of receptor recycling after ligand binding (Lee and Mayr 2017). The 3′ UTR-dependent recruitment of the signaling complex to CXCR4 also had biological consequences, as it was shown to be necessary for CXCR4-mediated B-cell migration. Importantly, despite similar overall *BIRC3* mRNA levels observed in normal and malignant B cells, BIRC3-LU is up-regulated relative to BIRC3-SU in malignant B cells. This up-regulation is predicted to promote leukemia cell survival in vivo, as CXCR4-dependent B-cell migration was shown to allow the leukemia cells to home to bone marrow niches, which provide survival and proliferation signals (Lee and Mayr 2017).

Last, the 3′ UTR-mediated protein complex formation involving the E3 ubiquitin ligase BIRC3-LU has additional molecular consequences as it controls substrate specificity of the enzyme. Posttranslational modification of RALA was accomplished only by BIRC3-LU and not by BIRC3-SU, as substrate modification required prior protein interaction (Lee and Mayr 2017).

9 MECHANISM OF TRANSFER OF EFFECTOR PROTEINS FROM 3′ UTRs TO PROTEINS

How genetic information concerning protein features that go beyond the amino acid sequence is transmitted from mRNAs to proteins is still largely unclear. But the requirements for establishment of the 3′ UTR-mediated interaction between CD47-LU and SET proteins were recently identified (Ma and Mayr 2018). It was found that two RBPs that bind to AU-rich elements, HuR and TIS11B, are required for this process. HuR's function is to recruit SET, whereas TIS11B creates a permissive environment that enables the 3′ UTR-mediated binding of SET to CD47 and other membrane proteins (Berkovits and Mayr 2015; Ma and Mayr 2018). TIS11B forms RNA granules that enrich membrane protein-encoding mRNAs containing AU-rich elements. TIS11B granules form a reticular meshwork that is intertwined with the endoplasmic reticulum. The functional interplay between TIS11B granules and the endoplasmic reticulum creates a subcellular compartment with a biophysically and biochemically environment distinct from that of the cytoplasm. Only within this special com-

Cite as *Cold Spring Harb Perspect Biol* doi: 10.1101/cshperspect.a034728

partment are specific PPIs formed that cannot be established outside. This compartment enables 3′ UTR-mediated interaction of SET with CD47-LU, and also other membrane proteins (Ma and Mayr 2018). Moreover, these findings show that AU-rich elements have roles other than the control of protein abundance, as they regulate the formation of functionally relevant 3′ UTR-mediated PPIs.

10 3′ UTR *cis*-ELEMENTS ARE USUALLY REPEATED AND OFTEN ACT SYNERGISTICALLY

There are several common features characteristic of 3′ UTR *cis*-elements. Generally, a motif needs only 3–8 nt to bind an RBP (Lambert et al. 2014). Such short motifs often occur multiple times in a given mRNA and can act in a synergistic or even cooperative manner (Kruys et al. 1989; Chao et al. 2010; Hennig et al. 2014; Hennig and Sattler 2015). The characteristics of individual 3′ UTR *cis*-elements are detailed in previous reviews (see Johnstone and Lasko 2001; Jambhekar and Derisi 2007).

However, mRNA localization signals tend to be much longer (50–500 nt), often contain stem-loops, and have no recognizable or recurrent motifs (Johnstone and Lasko 2001; Jambhekar and Derisi 2007). An exception to this rule is the zip-code element in the *β-actin* 3′ UTR, which initially was thought to be 54 nt long and is recognized by the RBP IGF2BP1, also called ZBP1 (Kislauskis et al. 1994). However, the motif was narrowed down to a bipartite element that required a spacer of defined length with each element recognized by the KH3 or KH4 domains of IGF2BP1 (Chao et al. 2010). Narrowing down the sequence to a motif was important as it enabled genome-wide screening for additional motif-containing 3′ UTRs (Patel et al. 2012).

The identification of *cis*-elements is important as it is required for the identification of RBPs that bind and mediate specific 3′ UTR functions. Traditionally, *cis*-elements were determined using deletion constructs (Macdonald and Struhl 1988; Kim-Ha et al. 1993; Kislauskis et al. 1994; Kim-Ha et al. 1995). However, often, the phenotypic effect of the entire 3′ UTR cannot be recapitulated using isolated parts of the 3′ UTR. This was shown for the translation regulation element of *oskar* mRNA and for the repressive element in the *Hmga2* 3′ UTR (Mayr et al. 2007; Besse et al. 2009; Kristjansdottir et al. 2015). To identify the element in the 2800-nt-long *Hmga2* 3′ UTR, which is able to fully recapitulate the repressive effect on protein production, a large set of 3′ UTR fragments was tested, but none conferred full activity. In the end, it was found that the cooperative action of three 3′ UTR subregions was necessary for the full repressive function (Kristjansdottir et al. 2015).

Another example is the 4200-nt long *CD47* 3′ UTR in which the *cis*-elements responsible for SET binding are distributed across the whole 3′ UTR and none of the tested fragments contained the full activity (Ma and Mayr 2018). To find the *cis*-element that fully recapitulates the desired phenotype, instead of using deletion mutants, 50–200-nt-long, diverse and unrelated 3′ UTR elements were screened. This new strategy identified the conserved AU-rich element from TNF-α as being fully able to recapitulate the effects of the full-length *CD47* 3′ UTR with respect to SET binding. Interestingly, comparable SET binding was accomplished whether 19 AUUUA motifs were spread out over the entire *CD47* 3′ UTR, or when six AU-rich elements were concatenated as occurs in the AU-rich element of TNF-α. This is similar to building arrays of transcription factor binding sites to study transcription factor functions. The identification of the short *cis*-element then enabled the discovery of the interacting RBPs using RNA-oligonucleotide pull-down followed by mass spectrometry (Ma and Mayr 2018).

In addition to AU-rich elements, many other *cis*-regulatory motifs are known (Ray et al. 2013; Dominguez et al. 2018), including those bound by IGF2BP1, Pumilio, PTB, and MBNL. However, AU-rich elements were the first *cis*-elements discovered and remain the most extensively studied.

11 PRIMARY CELLS SHOW MORE EXTENSIVE REGULATION OF 3′ UTR ISOFORM EXPRESSION THAN CELL LINES

3′ UTR isoform ratios are substantially influenced by signaling pathways. Therefore, it was not surprising that primary cells show significantly greater changes in 3′ UTR isoform expression, when primary cells or tissues were compared with cell lines. In addition to the number of changes, the magnitude of change is higher in primary cells (Fig. 2) (Lianoglou et al. 2013). Alternative cleavage and polyadenylation can also occur in introns (Edwalds-Gilbert et al. 1997), and the recognition of intronic polyadenylation signals occurs substantially more frequently in primary cells than in cell lines (Lee et al., in press; Singh et al. 2018). Therefore, primary cells are beneficial for profiling alternative cleavage and polyadenylation.

Because of the cell type–specific nature of 3′ UTR isoform expression, cell sorting into pure populations before profiling provides better quantification of cell type–specific 3′ UTR expression patterns than profiling complex tissues (Lianoglou et al. 2013). If sorting is not an option, cTag-PAPERCLIP was recently developed to measure 3′ UTR isoform expression of specific cell types in intact tissues. Based on co-IP of mRNA 3′ ends associated with green fluorescent protein (GFP)-tagged poly(A)-binding protein

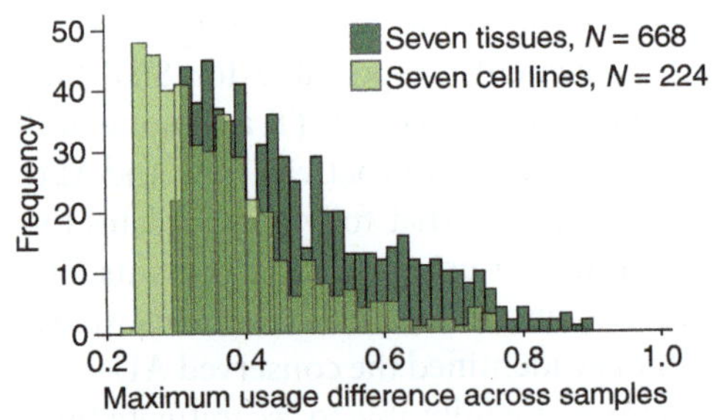

Figure 2. Primary cells show a higher number and a greater magnitude of cell type–specific changes in 3′ untranslated region (3′ UTR) isoform expression. 3′-seq data are from Lianoglou et al. (2013) and were used to calculate the maximum difference in usage of alternative polyadenylation sites across seven tissues and seven cell lines. Usage is the fraction of reads mapping to a single polyadenylation site out of all the reads mapping to the 3′ UTR. Shown are the 50% of genes with most variable 3′ UTR isoform expression. Only genes that generated two 3′ UTR isoforms were included in the analysis.

(PABP), the method requires mice that express tagged PABP in a tissue-specific manner (Hwang et al. 2017).

Metabolic differences between cell lines and primary cells not only cause differences in 3′ UTR isoform profiles, but also alter the regulatory effects of miRNAs and AU-rich elements (Kontoyiannis et al. 1999; La Rocca et al. 2015). For miRNAs to repress target mRNA expression, they need to be loaded into a megadalton-sized complex, called RISC (RNA-induced silencing complex). In cell lines and proliferating cells, most miRNAs reside in these large complexes. However, in primary cells, the majority of miRNAs are bound to Argonaute, a component of RISC, but are not part of the larger RISC complex (La Rocca et al. 2015). Assembly of the larger RISC complexes is mediated by the activation of signal transduction pathways, including the PI3K pathway. Thus, in primary cells, miRNAs are able to repress their targets only after pathway activation (La Rocca et al. 2015). This principle is likely to hold true for the function of RBPs, as signaling pathways may influence protein complex formation.

12 3′ UTR-MEDIATED REGULATION IN VIVO

The powerful influence of 3′ UTRs on gene expression in whole organisms has been best shown using mouse models. One of the first examples was the *c-fos* transgenic mouse (Ruther et al. 1987). No mRNA expression was detectable when c-fos cDNA that contained its endogenous 3′ UTR and polyadenylation signal was expressed from a strong promoter. However, after replacement of the 3′ UTR with a retroviral long terminal repeat high mRNA expression was observed throughout tissues (Ruther et al. 1987). Thus, the *c-fos* 3′ UTR is necessary for repression of c-fos protein production in vivo.

In many additional mouse models, this principle was adopted as genes were expressed from their endogenous promoters, but their 3′ UTRs and polyadenylation signals were replaced with strong polyadenylation signals. Although the influence of 3′ UTRs on phenotypes can be assessed in these mice, researchers need to be aware that the resulting phenotype results from deletion of 3′ UTR *cis*-elements and overexpression of the protein through more efficient mRNA processing. The gene encoding the Camk2a protein kinase generates a short and long 3′ UTR isoform (Lianoglou et al. 2013). Replacement of the two isoforms by a single strong polyadenylation signal prevented Camk2a mRNA localization to the dendrites of hippocampal neurons. This impairs long-term memory formation as dendritic mRNA localization and local protein production are necessary for memory consolidation (Miller et al. 2002). The neurotrophic factor Bdnf is also encoded by an mRNA with a short or long 3′ UTR isoform. Bdnf is expressed in hippocampal neurons and exclusive disruption of the long 3′ UTR isoform results in impaired memory formation (An et al. 2008). Bdnf is also expressed in hypothalamic neurons and disruption of the long 3′ UTR isoform in these cells results in severe obesity and phenocopies a complete Bdnf knockout in neurons (Liao et al. 2012).

The 3′ UTR functions of additional growth factors, neurotrophic factors, and enzymes were studied in mice and included *TGFβ1*, *Gdnf*, and aldosterone synthase (Kakoki et al. 2004; Andressoo et al. 2006; Kakoki et al. 2013). The phenotypes observed range from organ abnormalities, increased blood pressure, to embryonic lethality. These experiments revealed that usage of different 3′ UTRs can change specific protein levels by 100-fold, which enables the generation of mice with graded mRNA expression levels ranging from 10% to 900% of normal. This is particularly useful if the complete gene knockout is embryonic-lethal (Kakoki et al. 2013).

13 NEW METHODS TO STUDY 3′ UTR FUNCTIONS

13.1 CRISPR Technology Facilitates New Approaches for Studying 3′ UTR Isoform–Specific Functions in Cell Lines and Organisms

From the 3′ UTR studies in mice, we learned that preserving regulation by the endogenous promoter and polyadenylation signal should be an experimental priority. Furthermore, whereas 3′ UTR isoform–specific functions have so far been studied mostly using tagged, 3′ UTR isoform–specific overexpression constructs (Sandberg et al. 2008; Mayr and Bartel 2009; Rhinn et al. 2012; Berkovits and Mayr 2015), recent advances in CRISPR (clustered regularly interspaced short palindromic repeats) technology allow more sophisticated manipulations of 3′ UTR isoforms in cell lines and animal models at endogenous loci (Jinek et al. 2012; Diao et al. 2017).

In the following paragraphs and in Figure 3, experimental approaches to study 3′ UTR functions using CRISPR technology are shown. If a gene generates alternative 3′ UTRs and the function of only the long 3′ UTR isoform is of interest, the strategy shown in Figure 3A is useful. In the experimental set-up, the following cell lines or mice are compared: (1) wild-type, (2) CRISPR-mediated disruption of the protein by adding a frame-shift in the coding region, (3) stable expression of a short hairpin RNA (shRNA) that targets the long 3′ UTR isoform, and (4) stable expression of a control shRNA. This approach was used to study the functions of CD47-LU and BIRC3-LU and is most suitable if the long 3′ UTR isoform is expressed at low levels, as its knockdown will not change overall protein levels much (Berkovits and Mayr 2015; Lee and Mayr 2017).

However, if the long 3′ UTR isoform is expressed at substantial levels and overall protein expression levels

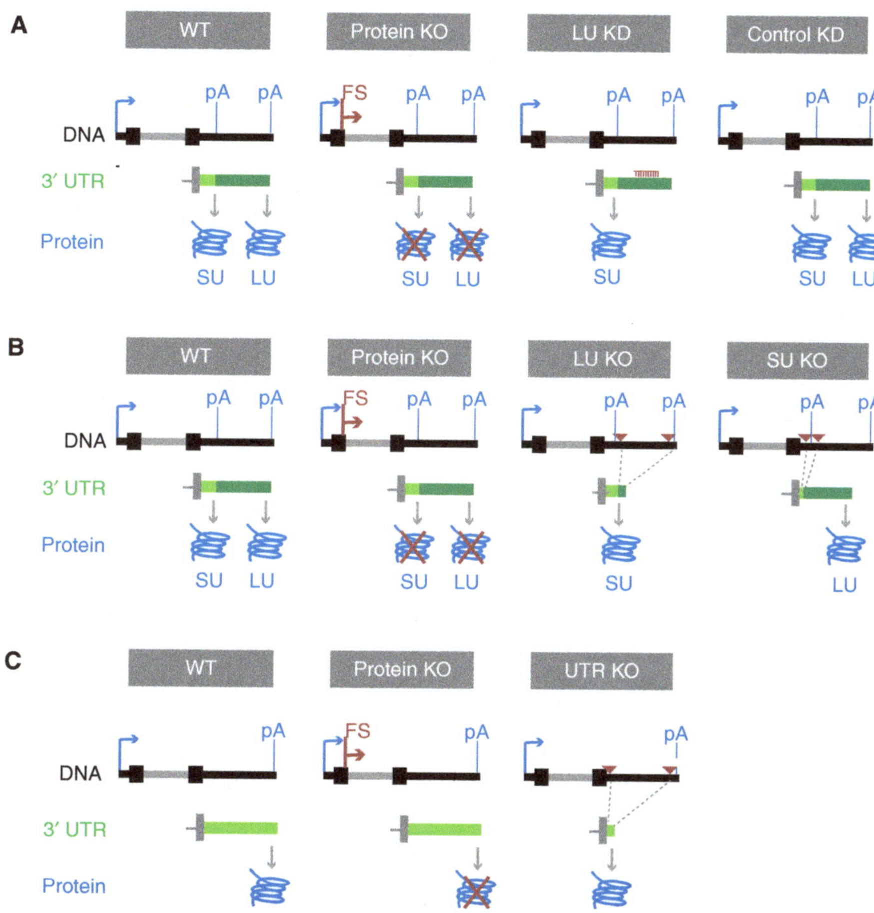

Figure 3. Experimental approaches using CRISPR (clustered regularly interspaced short palindromic repeats) technology at the endogenous locus to study 3′ untranslated region (3′ UTR) functions in vivo. (*A*) Approach that allows phenotypes obtained for exclusive loss of the LU-generated protein compared with the total protein knockout (KO). Loss of the LU isoform is accomplished through short hairpin RNA (shRNA)-mediated knockdown (KD). At the DNA level, the arrow depicts the transcription start site. pA, Polyadenylation site; FS, frame-shift mutation. In the 3′ UTR panel, the last exon of the messenger RNA (mRNA) isoform is shown with the short 3′ UTR depicted in light green and the long 3′ UTR shown in dark green. The generated proteins are shown in blue. SU, Protein generated from the short 3′ UTR isoform; LU, protein generated from the long 3′ UTR isoform. (*B*) Approach that allows phenotypes obtained from the total protein KO to be compared to with exclusive expression of the protein generated from either the short (SU) or long 3′ UTR (LU) isoform. Shown as in *A*. The red triangles represent paired guide RNAs to delete 3′ UTR fragments. (*C*) Approach that allows the comparison of phenotypes obtained from the total protein KO to those generated by exclusive loss of regulatory elements in the 3′ UTR of genes that generate mRNAs with constitutive 3′ UTRs. Shown as in *B*. The regulation of mRNA abundance through the endogenous polyadenylation signal is preserved.

need to be preserved, a modified strategy, shown in Figure 3B, can be used. The phenotypes produced from the (1) wild-type and (2) protein knockout are compared with cell lines or animals that express only the (3) short or (4) long 3′ UTR isoforms. To obtain cells that express only the short 3′ UTR isoform, a pair of guide RNAs is used to delete the sequence between the proximal and distal polyadenylation signals. This moves the distal polyadenylation signal close to the proximal polyadenylation signal and should preserve overall protein expression as mRNA processing remains intact. To obtain cells that express only the long 3′ UTR isoform, the proximal polyadenylation site can either be deleted (using a pair of guide RNAs) or disrupted by point mutations (Li et al. 2017). As 3′ UTR deletions can also affect DNA-regulatory elements located at the 3′ end of genes (Sanjana et al. 2016), it is useful to include in the experiment tagged cDNA constructs that contain the coding region and either the short or long 3′ UTR isoform. Expression of these constructs will enable the rescue of the observed phenotypes and ensure that the phenotypes indeed are caused by the 3′ UTRs and not by deletion of DNA-regulatory elements (Lee and Mayr 2017).

To study the regulatory role of the 3′ UTR of a gene with a single 3′ UTR isoform, the wild-type and protein knockout can be compared with a cell line or animal with CRISPR-mediated deletion of the 3′ UTR (Fig. 3C). In this case, a pair of guide RNAs is used to delete the 3′ UTR sequence between the stop codon and the polyadenylation signal. However, ~100 nucleotides of sequence located upstream of the polyadenylation signal should be preserved to enable proper mRNA processing from the endogenous polyadenylation signal (Zhao et al. 2017).

Instead of DNA manipulation, changes in mRNA processing through transfection of antisense morpholinos can also switch 3′ UTR isoform expression. This strategy has been used so far to alter inclusion or exclusion of alternative exons, including the splicing of terminal exons (Vorlova et al. 2011). However, similarly, alternative polyadenylation sites can be masked to shift alternative 3′ UTR isoform ratios (Marsollier et al. 2016).

13.2 3′ UTR Isoform-Specific Translation

Several novel approaches were recently developed to study translation, but none of them allows assessment of translation in a 3′ UTR isoform-specific manner (reviewed in Chekulaeva and Landthaler 2016; King and Gerber 2016). Although it is now possible to monitor the translation of single endogenous mRNAs through simultaneous imaging of the mRNA and newly made protein, to do so requires disruption of the 3′ UTR by the addition of binding sites for either MS2 or PP7 coat protein (Chekulaeva and Landthaler 2016).

The only currently available method that can inspect translation in a 3′ UTR isoform-specific manner uses ribo-tag or TRAP (translating ribosome affinity purification) (Doyle et al. 2008; Heiman et al. 2014; Kocabas et al. 2015). Ribo-tag/TRAP adds a tag to a ribosomal protein of the large ribosomal subunit in a cell type-specific manner by using either bac transgenics or the Cre-lox system in mice (Doyle et al. 2008; Heiman et al. 2014; Hupe et al. 2014), but can be applied to worms, flies, plants, and yeast (King and Gerber 2016). In addition to quantifying the transcriptome in a cell type-specific manner, the pulldown of ribosome-associated mRNAs also provides an approximation of the translatome without the necessity of purifying the cells. Ribo-tag/TRAP enables the identification of ribosome-associated transcripts in a 3′ UTR-specific manner if, in addition to RNA-seq, 3′-end sequencing is performed.

13.3 Temporal Regulation of Translation of Alternative 3′ UTR Isoforms

Application of the ribo-tag technique during meiosis of mouse oocytes revealed that translation of alternative 3′ UTR isoforms is regulated temporally (Yang et al. 2017). This was shown in detail for *Cyclin B1*, which expresses three 3′ UTR isoforms. Combining RNA-seq of ribosome-associated mRNAs with luciferase reporter assays revealed that the short 3′ UTR isoform is constitutively translated in all phases of meiosis. In contrast, the longer 3′ UTR isoforms—which contain binding sites for CPEB1—are translationally repressed in a CPEB1-dependent manner in dormant oocytes. However, during progression through meiosis and cell-cycle entry, these longer isoforms become polyadenylated, thus increasing loading into ribosomes, and leading to increased Cyclin B1 protein expression during oocyte maturation (Yang et al. 2017).

Another beautiful example of temporal regulation of translation of alternative 3′ UTR isoforms in nongerm cells was revealed by studying the differentiation of R cells into presynaptic terminals in *Drosophila melanogaster* (Zhang et al. 2016). This is a particularly good model system to study gene expression changes during differentiation as R cells differentiate in a synchronous fashion within 72 hours from growth cones to presynaptic terminals. Performing transcriptome and TRAP analysis during this time course revealed only modest changes in mRNA abundance, but a large number of alternative 3′ UTR isoforms changed their ribosome association in a temporal manner (Zhang et al. 2016). Moreover, mRNAs encoding presynaptic proteins changed their 3′ UTR isoform expression

Cite as *Cold Spring Harb Perspect Biol* doi: 10.1101/cshperspect.a034728

twice as often as randomly selected genes. Thus, changes in alternative 3′ UTR isoform expression as well as dynamic ribosome association of alternative transcripts are major contributors to the regulation of synaptic differentiation (Zhang et al. 2016).

A similar temporal regulation of translation of alternative 3′ UTR isoforms was observed in differentiated neurons during synaptic activation (Lau et al. 2010; Rhinn et al. 2012). In resting neurons, the short 3′ UTR isoforms of *Bdnf* and α-synuclein (*Snca*) are translated to maintain basal levels of protein expression, whereas the long 3′ UTR isoforms are translationally repressed. In contrast, upon activation, either through neuronal activity or exposure to dopamine, only the long 3′ UTR isoforms become associated with polyribosomes and are actively translated. Thus, the alternative 3′ UTRs of *Bdnf* or *Snca* are translated differentially in resting and activated neurons (Lau et al. 2010; Rhinn et al. 2012).

13.4 Identification of RNA-Binding Proteins That Cooperate with Each Other

In the end, the functions of 3′ UTRs are determined by the bound RBPs and their associated effector proteins. Whereas cross-linking immunoprecipitation (CLIP) identifies all mRNAs bound by a single RBP, complementary approaches detect all RBPs that bind single 3′ UTRs (Chu et al. 2015; Matia-Gonzalez et al. 2017).

However, the identification of functionally relevant RBP-binding sites is complicated as the motifs are often degenerate. Therefore, in vivo, strong and specific binding of an RBP is often achieved through oligomerization of the RBP, which enables binding to several sites simultaneously (Besse et al. 2009). Stronger binding and increased specificity can also be accomplished through cooperativity with another RBP (Hennig and Sattler 2015; Mayr 2017). However, the identification of cooperating pairs of RBPs has been accomplished for only a few select cases (Leeper et al. 2010; Campbell et al. 2012; Hennig et al. 2014; Hennig and Sattler 2015; Lee and Mayr 2017; Ma and Mayr 2018).

To approach this problem systematically, an experimental procedure based on CLIP was recently developed (Gregersen et al. 2014). First, the binding sites for the RBP MOV10 were identified by PAR-CLIP. Next, cross-linked cells were mildly treated with RNase to produce RNA fragments of ~400 nt. To identify proteins on the MOV10-bound RNA fragments, the MOV10 IP was repeated and the enriched proteins were identified by mass spectrometry (Gregersen et al. 2014). This method found UPF1 to be the RBP mostly highly enriched on the MOV10-bound fragments and suggested synergistic action by the two RBPs.

13.5 Identification of Effector Proteins of RNA-Binding Proteins

It has been shown in many instances that RBPs can interact with diverse effector proteins. This suggests that RBPs have considerable multifunctionality that goes beyond the regulation of RNA metabolism. In some cases, the interaction partners of RBPs bind simultaneously, as shown for CPEB (Kim and Richter 2006). In other cases, the RBP was found to be part of different protein subcomplexes, and each subcomplex accomplished a different function. Different subcomplexes were identified using co-IP of 3′ UTR-mediated protein complexes or through affinity purification of native complexes (Hildebrandt et al. 2017; Lee and Mayr 2017).

One of the best-studied examples is HuR, which is known to regulate mRNA stability, alternative splicing, and alternative polyadenylation (Fan and Steitz 1998; Oktaba et al. 2015). In addition to regulating mRNA-based processes, HuR is necessary for 3′ UTR-mediated PPIs, and thus is involved in regulating surface expression of several membrane proteins (Berkovits and Mayr 2015; Lee and Mayr 2017; Ma and Mayr 2018). Last, UPF1, which is a well-known regulator of nonsense mediated decay, was discovered to also regulate protein decay of MYOD1, a master regulator of muscle differentiation (Feng et al. 2017).

These examples show that investigating the diverse functions of 3′ UTRs requires determination of the effector proteins of RBPs. One study identified RNA-dependent as well as RNA-independent protein-interaction partners of 12 RBPs, including hnRNPs, RBFOX2, and UPF1 (Brannan et al. 2016). Through intersection with known RBPs, the majority of interaction partners (72%) were found to be proteins other than RBPs. However, as this was not the focus of the study, it is currently unclear how many of the identified factors were true interaction partners.

RBPs have rich interactomes (Baltz et al. 2012). This may be owing to the enrichment of disordered regions, which are known to promote PPIs. To lower the rate of false-positive protein-interaction partners of RBPs, the following strategy was used to identify the protein-interaction partners of E3 ubiquitin ligases that contain domains for RNA binding (Hildebrandt et al. 2017). Stable isotope labeling with amino acids in cell culture (SILAC)-based mass spectrometry compared the interactomes of GFP and GFP-tagged RBPs after co-IP. Importantly, the two experimental conditions were already pooled before cell lysis and co-IP, rather than after cell lysis, which is usually the case. As a result, transient interactors reach an equilibrium consisting of light and heavy labeled forms, which should result in SILAC ratios close to background ratios (Hildebrandt et al. 2017). Moreover, performing replicate experiments allows identification of robust interaction partners (Brannan et al.

2016; Hildebrandt et al. 2017). For a specific RNA-binding ubiquitin ligase, between 9 and 42 stable protein-interaction partners were identified. The majority (69%) bound in an RNA-independent manner and belong to diverse functional categories (Hildebrandt et al. 2017), suggesting additional regulatory roles of RBPs beyond mRNA-based regulation.

14 FUTURE DIRECTIONS

1. As 3′ UTRs can regulate protein features, readouts for testing 3′ UTR functions need to go beyond the regulation of protein abundance.
2. As 3′ UTRs can accomplish a wide range of functions through the regulation of protein complex formation, the protein–interaction partners of RBPs need to be identified. Approximately half of identified PPIs do not appear to play a role in RNA biology (Hildebrandt et al. 2017). Therefore, the opportunity to study 3′ UTR functions goes beyond the regulation of protein abundance.
3. One of the greatest bottlenecks in the evaluation of 3′ UTR functions is the difficulty of reliably identifying responsible *cis*-elements, especially if the regulatory elements are distributed across the entire 3′ UTR. In cases in which 3′ UTR structure brings the scattered *cis*-elements together, a CRISPR deletion screen with high-resolution tiling across the entire 3′ UTR may help to identify such elements, as disruption of structure may suffice to detect a phenotype. Such a screen was initially developed to assess functional protein domains, but was recently also applied to discover enhancers or functional miRNA-binding sites (Shi et al. 2015; Sanjana et al. 2016; Wu et al. 2017).
4. As the majority of RBPs bind with low affinity to mostly degenerate motifs located in 3′ UTRs, methods need to be developed that allow identification of binding sites for pairs of RBPs. Identification of cobound sites will be important for network analyses and may help to disentangle opposing functions known to be accomplished by RBPs.

ACKNOWLEDGMENTS

The author apologizes to all colleagues whose work could not be cited because of space constraints. She is grateful to all the members of the Mayr laboratory for helpful discussions and for critical reading of the manuscript, with special thanks to Sarah Tisdale. Funding for work on 3′ UTRs was obtained from the Innovator Award of the Damon Runyon-Rachleff Cancer Foundation and the Island Outreach Foundation, the Pershing Square Sohn Cancer Research Alliance, the MSC core grant (P30 CA008748), the National Cancer Institute (NCI) Grant U01-CA164190, and the National Institutes of Health (NIH) Director's Pioneer Award (DP1-GM123454).

REFERENCES

An JJ, Gharami K, Liao GY, Woo NH, Lau AG, Vanevski F, Torre ER, Jones KR, Feng Y, Lu B, et al. 2008. Distinct role of long 3′ UTR BDNF mRNA in spine morphology and synaptic plasticity in hippocampal neurons. *Cell* **134:** 175–187.

Andressoo JO, Mitchell JR, de Wit J, Hoogstraten D, Volker M, Toussaint W, Speksnijder E, Beems RB, van Steeg H, Jans J, et al. 2006. An Xpd mouse model for the combined xeroderma pigmentosum/Cockayne syndrome exhibiting both cancer predisposition and segmental progeria. *Cancer Cell* **10:** 121–132.

Baltz AG, Munschauer M, Schwanhausser B, Vasile A, Murakawa Y, Schueler M, Youngs N, Penfold-Brown D, Drew K, Milek M, et al. 2012. The mRNA-bound proteome and its global occupancy profile on protein-coding transcripts. *Mol Cell* **46:** 674–690.

Barreau C, Paillard L, Osborne HB. 2005. AU-rich elements and associated factors: Are there unifying principles? *Nucleic Acids Res* **33:** 7138–7150.

Batra R, Charizanis K, Manchanda M, Mohan A, Li M, Finn DJ, Goodwin M, Zhang C, Sobczak K, Thornton CA, et al. 2014. Loss of MBNL leads to disruption of developmentally regulated alternative polyadenylation in RNA-mediated disease. *Mol Cell* **56:** 311–322.

Berkovits BD, Mayr C. 2015. Alternative 3′ UTRs act as scaffolds to regulate membrane protein localization. *Nature* **522:** 363–367.

Besse F, Lopez de Quinto S, Marchand V, Trucco A, Ephrussi A. 2009. *Drosophila* PTB promotes formation of high-order RNP particles and represses *oskar* translation. *Genes Dev* **23:** 195–207.

Brannan KW, Jin W, Huelga SC, Banks CA, Gilmore JM, Florens L, Washburn MP, Van Nostrand EL, Pratt GA, Schwinn MK, et al. 2016. SONAR discovers RNA-binding proteins from analysis of large-scale protein–protein interactomes. *Mol Cell* **64:** 282–293.

Brennan CM, Steitz JA. 2001. HuR and mRNA stability. *Cell Mol Life Sci* **58:** 266–277.

Brumbaugh J, Di Stefano B, Wang X, Borkent M, Forouzmand E, Clowers KJ, Ji F, Schwarz BA, Kalocsay M, Elledge SJ, et al. 2018. Nudt21 controls cell fate by connecting alternative polyadenylation to chromatin signaling. *Cell* **172:** 106–120.e21.

Campbell ZT, Bhimsaria D, Valley CT, Rodriguez-Martinez JA, Menichelli E, Williamson JR, Ansari AZ, Wickens M. 2012. Cooperativity in RNA–protein interactions: Global analysis of RNA binding specificity. *Cell Rep* **1:** 570–581.

Caput D, Beutler B, Hartog K, Thayer R, Brown-Shimer S, Cerami A. 1986. Identification of a common nucleotide sequence in the 3′-untranslated region of mRNA molecules specifying inflammatory mediators. *Proc Natl Acad Sci* **83:** 1670–1674.

Chan SL, Huppertz I, Yao C, Weng L, Moresco JJ, Yates JR 3rd, Ule J, Manley JL, Shi Y. 2014. CPSF30 and Wdr33 directly bind to AAUAAA in mammalian mRNA 3′ processing. *Genes Dev* **28:** 2370–2380.

Chang JW, Zhang W, Yeh HS, de Jong EP, Jun S, Kim KH, Bae SS, Beckman K, Hwang TH, Kim KS, et al. 2015. mRNA 3′-UTR shortening is a molecular signature of mTORC1 activation. *Nat Commun* **6:** 7218.

Chao JA, Patskovsky Y, Patel V, Levy M, Almo SC, Singer RH. 2010. ZBP1 recognition of β-actin zipcode induces RNA looping. *Genes Dev* **24:** 148–158.

Chartron JW, Hunt KC, Frydman J. 2016. Cotranslational signal-independent SRP preloading during membrane targeting. *Nature* **536:** 224–228.

Chekulaeva M, Landthaler M. 2016. Eyes on translation. *Mol Cell* **63:** 918–925.

Chekulaeva M, Hentze MW, Ephrussi A. 2006. Bruno acts as a dual repressor of *oskar* translation, promoting mRNA oligomerization and formation of silencing particles. *Cell* **124:** 521–533.

Chen CY, Gherzi R, Ong SE, Chan EL, Raijmakers R, Pruijn GJ, Stoecklin G, Moroni C, Mann M, Karin M. 2001. AU binding proteins recruit the exosome to degrade ARE-containing mRNAs. *Cell* **107:** 451–464.

Chen CY, Chen ST, Juan HF, Huang HC. 2012. Lengthening of 3′UTR increases with morphological complexity in animal evolution. *Bioinformatics* **28:** 3178–3181.

Chu C, Zhang QC, da Rocha ST, Flynn RA, Bharadwaj M, Calabrese JM, Magnuson T, Heard E, Chang HY. 2015. Systematic discovery of Xist RNA binding proteins. *Cell* **161:** 404–416.

Core LJ, Waterfall JJ, Lis JT. 2008. Nascent RNA sequencing reveals widespread pausing and divergent initiation at human promoters. *Science* **322:** 1845–1848.

Crick FH. 1958. On protein synthesis. *Symp Soc Exp Biol* **12:** 138–163.

Derti A, Garrett-Engele P, Macisaac KD, Stevens RC, Sriram S, Chen R, Rohl CA, Johnson JM, Babak T. 2012. A quantitative atlas of polyadenylation in five mammals. *Genome Res* **22:** 1173–1183.

Diao Y, Fang R, Li B, Meng Z, Yu J, Qiu Y, Lin KC, Huang H, Liu T, Marina RJ, et al. 2017. A tiling-deletion-based genetic screen for *cis*-regulatory element identification in mammalian cells. *Nat Methods* **14:** 629–635.

Dominguez D, Freese P, Alexis MS, Su A, Hochman M, Palden T, Bazile C, Lambert NJ, Van Nostrand EL, Pratt GA, et al. 2018. Sequence, structure, and context preferences of human RNA binding proteins. *Mol Cell* **70:** 854–867.

Doyle JP, Dougherty JD, Heiman M, Schmidt EF, Stevens TR, Ma G, Bupp S, Shrestha P, Shah RD, Doughty ML, et al. 2008. Application of a translational profiling approach for the comparative analysis of CNS cell types. *Cell* **135:** 749–762.

Duncan CD, Mata J. 2011. Widespread cotranslational formation of protein complexes. *PLoS Genet* **7:** e1002398.

Edwalds-Gilbert G, Veraldi KL, Milcarek C. 1997. Alternative poly(A) site selection in complex transcription units: Means to an end? *Nucleic Acids Res* **25:** 2547–2561.

Ephrussi A, Dickinson LK, Lehmann R. 1991. *oskar* organizes the germ plasm and directs localization of the posterior determinant *nanos*. *Cell* **66:** 37–50.

Fan XC, Steitz JA. 1998. Overexpression of HuR, a nuclear-cytoplasmic shuttling protein, increases the in vivo stability of ARE-containing mRNAs. *EMBO J* **17:** 3448–3460.

Feng Q, Jagannathan S, Bradley RK. 2017. The RNA surveillance factor UPF1 represses myogenesis via its E3 ubiquitin ligase activity. *Mol Cell* **67:** 239–251.e236.

Flavell SW, Kim TK, Gray JM, Harmin DA, Hemberg M, Hong EJ, Markenscoff-Papadimitriou E, Bear DM, Greenberg ME. 2008. Genome-wide analysis of MEF2 transcriptional program reveals synaptic target genes and neuronal activity-dependent polyadenylation site selection. *Neuron* **60:** 1022–1038.

Freimer JW, Krishnakumar R, Cook MS, Blelloch R. 2018. Expression of alternative Ago2 isoform associated with loss of microRNA-driven translational repression in mouse oocytes. *Curr Biol* **28:** 296–302.e3

Gautheret D, Poirot O, Lopez F, Audic S, Claverie JM. 1998. Alternate polyadenylation in human mRNAs: A large-scale analysis by EST clustering. *Genome Res* **8:** 524–530.

Grassi E, Mariella E, Lembo A, Molineris I, Provero P. 2016. Roar: Detecting alternative polyadenylation with standard mRNA sequencing libraries. *BMC Bioinformatics* **17:** 423.

Gregersen LH, Schueler M, Munschauer M, Mastrobuoni G, Chen W, Kempa S, Dieterich C, Landthaler M. 2014. MOV10 is a 5′ to 3′ RNA helicase contributing to UPF1 mRNA target degradation by translocation along 3′ UTRs. *Mol Cell* **54:** 573–585.

Gruber AR, Martin G, Muller P, Schmidt A, Gruber AJ, Gumienny R, Mittal N, Jayachandran R, Pieters J, Keller W, et al. 2014. Global 3′ UTR shortening has a limited effect on protein abundance in proliferating T cells. *Nat Commun* **5:** 5465.

Gruber AJ, Schmidt R, Gruber AR, Martin G, Ghosh S, Belmadani M, Keller W, Zavolan M. 2016. A comprehensive analysis of 3′ end sequencing data sets reveals novel polyadenylation signals and the repressive role of heterogeneous ribonucleoprotein C on cleavage and polyadenylation. *Genome Res* **26:** 1145–1159.

Gupta I, Clauder-Munster S, Klaus B, Jarvelin AI, Aiyar RS, Benes V, Wilkening S, Huber W, Pelechano V, Steinmetz LM. 2014. Alternative polyadenylation diversifies post-transcriptional regulation by selective RNA–protein interactions. *Mol Syst Biol* **10:** 719.

Halbach A, Zhang H, Wengi A, Jablonska Z, Gruber IM, Halbeisen RE, Dehe PM, Kemmeren P, Holstege F, Geli V, et al. 2009. Cotranslational assembly of the yeast SET1C histone methyltransferase complex. *EMBO J* **28:** 2959–2970.

Heiman M, Kulicke R, Fenster RJ, Greengard P, Heintz N. 2014. Cell type-specific mRNA purification by translating ribosome affinity purification (TRAP). *Nat Protoc* **9:** 1282–1291.

Hennig J, Sattler M. 2015. Deciphering the protein–RNA recognition code: Combining large-scale quantitative methods with structural biology. *Bioessays* **37:** 899–908.

Hennig J, Gebauer F, Sattler M. 2014. Breaking the protein–RNA recognition code. *Cell Cycle* **13:** 3619–3620.

Hildebrandt A, Alanis-Lobato G, Voigt A, Zarnack K, Andrade-Navarro MA, Beli P, Konig J. 2017. Interaction profiling of RNA-binding ubiquitin ligases reveals a link between posttranscriptional regulation and the ubiquitin system. *Sci Rep* **7:** 16582.

Hoque M, Ji Z, Zheng D, Luo W, Li W, You B, Park JY, Yehia G, Tian B. 2013. Analysis of alternative cleavage and polyadenylation by 3′ region extraction and deep sequencing. *Nat Methods* **10:** 133–139.

Hupe M, Li MX, Gertow Gillner K, Adams RH, Stenman JM. 2014. Evaluation of TRAP-sequencing technology with a versatile conditional mouse model. *Nucleic Acids Res* **42:** e14.

Hwang HW, Saito Y, Park CY, Blachere NE, Tajima Y, Fak JJ, Zucker-Scharff I, Darnell RB. 2017. cTag-PAPERCLIP reveals alternative polyadenylation promotes cell-type specific protein diversity and shifts *Araf* isoforms with microglia activation. *Neuron* **95:** 1334–1349.e1335.

Jambhekar A, Derisi JL. 2007. *cis*-acting determinants of asymmetric, cytoplasmic RNA transport. *RNA* **13:** 625–642.

Jia X, Yuan S, Wang Y, Fu Y, Ge Y, Ge Y, Lan X, Feng Y, Qiu F, Li P, et al. 2017. The role of alternative polyadenylation in the antiviral innate immune response. *Nat Commun* **8:** 14605.

Jinek M, Chylinski K, Fonfara I, Hauer M, Doudna JA, Charpentier E. 2012. A programmable dual-RNA-guided DNA endonuclease in adaptive bacterial immunity. *Science* **337:** 816–821.

Johnstone O, Lasko P. 2001. Translational regulation and RNA localization in *Drosophila* oocytes and embryos. *Annu Rev Genet* **35:** 365–406.

Kakoki M, Tsai YS, Kim HS, Hatada S, Ciavatta DJ, Takahashi N, Arnold LW, Maeda N, Smithies O. 2004. Altering the expression in mice of genes by modifying their 3′ regions. *Dev Cell* **6:** 597–606.

Kakoki M, Pochynyuk OM, Hathaway CM, Tomita H, Hagaman JR, Kim HS, Zaika OL, Mamenko M, Kayashima Y, Matsuki K, et al. 2013. Primary aldosteronism and impaired natriuresis in mice underexpressing TGFβ1. *Proc Natl Acad Sci* **110:** 5600–5605.

Kim JH, Richter JD. 2006. Opposing polymerase-deadenylase activities regulate cytoplasmic polyadenylation. *Mol Cell* **24:** 173–183.

Kim-Ha J, Webster PJ, Smith JL, Macdonald PM. 1993. Multiple RNA regulatory elements mediate distinct steps in localization of *oskar* mRNA. *Development* **119:** 169–178.

Kim-Ha J, Kerr K, Macdonald PM. 1995. Translational regulation of *oskar* mRNA by bruno, an ovarian RNA-binding protein, is essential. *Cell* **81:** 403–412.

King HA, Gerber AP. 2016. Translatome profiling: Methods for genome-scale analysis of mRNA translation. *Brief Funct Genomics* **15:** 22–31.

Kislauskis EH, Zhu X, Singer RH. 1994. Sequences responsible for intracellular localization of β-actin messenger RNA also affect cell phenotype. *J Cell Biol* **127:** 441–451.

Kocabas A, Duarte T, Kumar S, Hynes MA. 2015. Widespread differential expression of coding region and 3′ UTR sequences in neurons and other tissues. *Neuron* **88:** 1149–1156.

Kontoyiannis D, Pasparakis M, Pizarro TT, Cominelli F, Kollias G. 1999. Impaired on/off regulation of TNF biosynthesis in mice lacking TNF AU-rich elements: Implications for joint and gut-associated immunopathologies. *Immunity* **10:** 387–398.

Kristjansdottir K, Fogarty EA, Grimson A. 2015. Systematic analysis of the *Hmga2* 3′ UTR identifies many independent regulatory sequences and a novel interaction between distal sites. *RNA* **21:** 1346–1360.

Kruys V, Marinx O, Shaw G, Deschamps J, Huez G. 1989. Translational blockade imposed by cytokine-derived UA-rich sequences. *Science* **245:** 852–855.

Lackford B, Yao C, Charles GM, Weng L, Zheng X, Choi EA, Xie X, Wan J, Xing Y, Freudenberg JM, et al. 2014. Fip1 regulates mRNA alternative polyadenylation to promote stem cell self-renewal. *EMBO J* **33:** 878–889.

Lambert N, Robertson A, Jangi M, McGeary S, Sharp PA, Burge CB. 2014. RNA Bind-n-Seq: Quantitative assessment of the sequence and structural binding specificity of RNA binding proteins. *Mol Cell* **54:** 887–900.

Lander ES, Linton LM, Birren B, Nusbaum C, Zody MC, Baldwin J, Devon K, Dewar K, Doyle M, FitzHugh W, et al. 2001. Initial sequencing and analysis of the human genome. *Nature* **409:** 860–921.

La Rocca G, Olejniczak SH, Gonzalez AJ, Briskin D, Vidigal JA, Spraggon L, DeMatteo RG, Radler MR, Lindsten T, Ventura A, et al. 2015. In vivo, Argonaute-bound microRNAs exist predominantly in a reservoir of low molecular weight complexes not associated with mRNA. *Proc Natl Acad Sci* **112:** 767–772.

Lau AG, Irier HA, Gu J, Tian D, Ku L, Liu G, Xia M, Fritsch B, Zheng JQ, Dingledine R, et al. 2010. Distinct 3′UTRs differentially regulate activity-dependent translation of brain-derived neurotrophic factor (BDNF). *Proc Natl Acad Sci* **107:** 15945–15950.

Lawrence JB, Singer RH. 1986. Intracellular localization of messenger RNAs for cytoskeletal proteins. *Cell* **45:** 407–415.

Lebedeva S, Jens M, Theil K, Schwanhausser B, Selbach M, Landthaler M, Rajewsky N. 2011. Transcriptome-wide analysis of regulatory interactions of the RNA-binding protein HuR. *Mol Cell* **43:** 340–352.

Lee SH, Mayr C. 2017. 3′UTR-dependent signalosome assembly determines substrate specificity of BIRC3 E3 ligase during B cell migration. In EMBO conference, RNA localisation and local translation. Barga, Italy.

Lee SH, Singh I, Tisdale S, Abdel-Wahab O, Leslie CS, Mayr C. Widespread intronic polyadenylation inactivates tumor suppressor genes in leukemia. *Nature* doi: 10.1038/s41586-018-0465-8.

Leeper TC, Qu X, Lu C, Moore C, Varani G. 2010. Novel protein–protein contacts facilitate mRNA 3′-processing signal recognition by Rna15 and Hrp1. *J Mol Biol* **401:** 334–349.

Li R, Harvey AR, Hodgetts SI, Fox AH. 2017. Functional dissection of NEAT1 using genome editing reveals substantial localization of the NEAT1_1 isoform outside paraspeckles. *RNA* **23:** 872–881.

Lianoglou S, Garg V, Yang JL, Leslie CS, Mayr C. 2013. Ubiquitously transcribed genes use alternative polyadenylation to achieve tissue-specific expression. *Genes Dev* **27:** 2380–2396.

Liao GY, An JJ, Gharami K, Waterhouse EG, Vanevski F, Jones KR, Xu B. 2012. Dendritically targeted *Bdnf* mRNA is essential for energy balance and response to leptin. *Nat Med* **18:** 564–571.

Lindstein T, June CH, Ledbetter JA, Stella G, Thompson CB. 1989. Regulation of lymphokine messenger RNA stability by a surface-mediated T cell activation pathway. *Science* **244:** 339–343.

Lykke-Andersen J, Wagner E. 2005. Recruitment and activation of mRNA decay enzymes by two ARE-mediated decay activation domains in the proteins TTP and BRF-1. *Genes Dev* **19:** 351–361.

Ma W, Mayr C. 2018. A subcellular compartment defined by TIS granules and the ER enables 3′UTR-mediated protein–protein interactions. Society for Developmental Biology, 77th Annual Meeting, Portland, OR (July 20–24, 2018).

Macdonald PM, Struhl G. 1988. *cis*-acting sequences responsible for anterior localization of *bicoid* mRNA in *Drosophila* embryos. *Nature* **336:** 595–598.

Marsollier AC, Ciszewski L, Mariot V, Popplewell L, Voit T, Dickson G, Dumonceaux J. 2016. Antisense targeting of 3′ end elements involved in DUX_4 mRNA processing is an efficient therapeutic strategy for facioscapulohumeral dystrophy: A new gene-silencing approach. *Hum Mol Genet* **25:** 1468–1478.

Martin KC, Ephrussi A. 2009. mRNA localization: Gene expression in the spatial dimension. *Cell* **136:** 719–730.

Masamha CP, Xia Z, Yang J, Albrecht TR, Li M, Shyu AB, Li W, Wagner EJ. 2014. CFIm25 links alternative polyadenylation to glioblastoma tumour suppression. *Nature* **510:** 412–416.

Matia-Gonzalez AM, Iadevaia V, Gerber AP. 2017. A versatile tandem RNA isolation procedure to capture in vivo formed mRNA–protein complexes. *Methods* **118-119:** 93–100.

Mayr C. 2016. Evolution and biological roles of alternative 3′UTRs. *Trends Cell Biol* **26:** 227–237.

Mayr C. 2017. Regulation by 3′-untranslated regions. *Annu Rev Genet* **51:** 171–194.

Mayr C, Bartel DP. 2009. Widespread shortening of 3′UTRs by alternative cleavage and polyadenylation activates oncogenes in cancer cells. *Cell* **138:** 673–684.

Mayr C, Hemann MT, Bartel DP. 2007. Disrupting the pairing between *let-7* and *Hmga2* enhances oncogenic transformation. *Science* **315:** 1576–1579.

Meijlink F, Curran T, Miller AD, Verma IM. 1985. Removal of a 67-base-pair sequence in the noncoding region of protooncogene fos converts it to a transforming gene. *Proc Natl Acad Sci* **82:** 4987–4991.

Melton DA. 1987. Translocation of a localized maternal mRNA to the vegetal pole of *Xenopus* oocytes. *Nature* **328:** 80–82.

Mendez R, Hake LE, Andresson T, Littlepage LE, Ruderman JV, Richter JD. 2000. Phosphorylation of CPE binding factor by Eg2 regulates translation of *c-mos* mRNA. *Nature* **404:** 302–307.

Miller AD, Curran T, Verma IM. 1984. *c-fos* protein can induce cellular transformation: A novel mechanism of activation of a cellular oncogene. *Cell* **36:** 51–60.

Miller S, Yasuda M, Coats JK, Jones Y, Martone ME, Mayford M. 2002. Disruption of dendritic translation of CaMKIIα impairs stabilization of synaptic plasticity and memory consolidation. *Neuron* **36:** 507–519.

Milo R, Phillips R. 2016. *Cell biology by the numbers*. Taylor & Francis, New York.

Miyata T, Yasunaga T, Nishida T. 1980. Nucleotide sequence divergence and functional constraint in mRNA evolution. *Proc Natl Acad Sci* **77:** 7328–7332.

Nam JW, Rissland OS, Koppstein D, Abreu-Goodger C, Jan CH, Agarwal V, Yildirim MA, Rodriguez A, Bartel DP. 2014. Global analyses of the effect of different cellular contexts on microRNA targeting. *Mol Cell* **53:** 1031–1043.

Needleman SB, Wunsch CD. 1970. A general method applicable to the search for similarities in the amino acid sequence of two proteins. *J Mol Biol* **48:** 443–453.

Oktaba K, Zhang W, Lotz TS, Jun DJ, Lemke SB, Ng SP, Esposito E, Levine M, Hilgers V. 2015. ELAV links paused Pol II to alternative polyadenylation in the *Drosophila* nervous system. *Mol Cell* **57:** 341–348.

Patel VL, Mitra S, Harris R, Buxbaum AR, Lionnet T, Brenowitz M, Girvin M, Levy M, Almo SC, Singer RH, et al. 2012. Spatial arrangement of an RNA zipcode identifies mRNAs under post-transcriptional control. *Genes Dev* **26:** 43–53.

Ray D, Kazan H, Cook KB, Weirauch MT, Najafabadi HS, Li X, Gueroussov S, Albu M, Zheng H, Yang A, et al. 2013. A compendium of RNA-binding motifs for decoding gene regulation. *Nature* **499:** 172–177.

Rhinn H, Qiang L, Yamashita T, Rhee D, Zolin A, Vanti W, Abeliovich A. 2012. Alternative α-synuclein transcript usage as a convergent mechanism in Parkinson's disease pathology. *Nat Commun* **3:** 1084.

Richter JD. 2007. CPEB: A life in translation. *Trends Biochem Sci* **32:** 279–285.

Ruther U, Garber C, Komitowski D, Muller R, Wagner EF. 1987. Deregulated *c-fos* expression interferes with normal bone development in transgenic mice. *Nature* **325:** 412–416.

Sandberg R, Neilson JR, Sarma A, Sharp PA, Burge CB. 2008. Proliferating cells express mRNAs with shortened 3′ untranslated regions and fewer microRNA target sites. *Science* **320:** 1643–1647.

Sanjana NE, Wright J, Zheng K, Shalem O, Fontanillas P, Joung J, Cheng C, Regev A, Zhang F. 2016. High-resolution interrogation of functional elements in the noncoding genome. *Science* **353:** 1545–1549.

Schonemann L, Kuhn U, Martin G, Schafer P, Gruber AR, Keller W, Zavolan M, Wahle E. 2014. Reconstitution of CPSF active in polyadenylation: Recognition of the polyadenylation signal by WDR33. *Genes Dev* **28:** 2381–2393.

Shaw G, Kamen R. 1986. A conserved AU sequence from the 3′ untranslated region of GM-CSF mRNA mediates selective mRNA degradation. *Cell* **46:** 659–667.

Shenker S, Miura P, Sanfilippo P, Lai EC. 2015. IsoSCM: Improved and alternative 3′ UTR annotation using multiple change-point inference. *RNA* **21:** 14–27.

Shi J, Wang E, Milazzo JP, Wang Z, Kinney JB, Vakoc CR. 2015. Discovery of cancer drug targets by CRISPR-Cas9 screening of protein domains. *Nat Biotechnol* **33:** 661–667.

Siepel A, Bejerano G, Pedersen JS, Hinrichs AS, Hou M, Rosenbloom K, Clawson H, Spieth J, Hillier LW, Richards S, et al. 2005. Evolutionarily conserved elements in vertebrate, insect, worm, and yeast genomes. *Genome Res* **15:** 1034–1050.

Singh I, Lee SH, Sperling AS, Samur MK, Tai YT, Fulciniti M, Munshi NC, Mayr C, Leslie CL. 2018. Widespread intronic polyadenylation diversifies immune cell transcriptomes. *Nat Commun* **9:** 1716.

Tian B, Manley JL. 2017. Alternative polyadenylation of mRNA precursors. *Nat Rev Mol Cell Biol* **18:** 18–30.

Vorlova S, Rocco G, Lefave CV, Jodelka FM, Hess K, Hastings ML, Henke E, Cartegni L. 2011. Induction of antagonistic soluble decoy receptor tyrosine kinases by intronic polyA activation. *Mol Cell* **43:** 927–939.

Wang R, Nambiar R, Zheng D, Tian B. 2017. PolyA_DB 3 catalogs cleavage and polyadenylation sites identified by deep sequencing in multiple genomes. *Nucleic Acids Res* **46:** D315–D319.

Wu Q, Ferry QRV, Baeumler TA, Michaels YS, Vitsios DM, Habib O, Arnold R, Jiang X, Maio S, Steinkraus BR, et al. 2017. In situ functional dissection of RNA *cis*-regulatory elements by multiplex CRISPR-Cas9 genome engineering. *Nat Commun* **8:** 2109.

Xie X, Lu J, Kulbokas EJ, Golub TR, Mootha V, Lindblad-Toh K, Lander ES, Kellis M. 2005. Systematic discovery of regulatory motifs in human promoters and 3′ UTRs by comparison of several mammals. *Nature* **434:** 338–345.

Yaffe D, Nudel U, Mayer Y, Neuman S. 1985. Highly conserved sequences in the 3′ untranslated region of mRNAs coding for homologous proteins in distantly related species. *Nucleic Acids Res* **13:** 3723–3737.

Yang Y, Yang CR, Han SJ, Daldello EM, Cho A, Martins JPS, Xia G, Conti M. 2017. Maternal mRNAs with distinct 3′ UTRs define the temporal pattern of Ccnb1 synthesis during mouse oocyte meiotic maturation. *Genes Dev* **31:** 1302–1307.

Zhang KX, Tan L, Pellegrini M, Zipursky SL, McEwen JM. 2016. Rapid changes in the translatome during the conversion of growth cones to synaptic terminals. *Cell Rep* **14:** 1258–1271.

Zhao W, Siegel D, Biton A, Tonqueze OL, Zaitlen N, Ahituv N, Erle DJ. 2017. CRISPR-Cas9-mediated functional dissection of 3′-UTRs. *Nucleic Acids Res* **45:** 10800–10810.

Discovering and Mapping the Modified Nucleotides That Comprise the Epitranscriptome of mRNA

Bastian Linder[1,2] and Samie R. Jaffrey[2]

[1]Genome Biology Unit, European Molecular Biology Laboratory, Heidelberg 69117, Germany

[2]Department of Pharmacology, Weill Cornell Medicine, New York, New York 10065

Correspondence: srj2003@med.cornell.edu

SUMMARY

An important mechanism of gene expression regulation is the regulated modification of nucleotides in messenger RNA (mRNA). These modified nucleotides affect mRNA translation, stability, splicing, and other processes. A cluster of nucleotide modifications is found adjacent to the mRNA cap structure and another set can be found internally within transcripts. The most prominent modifications are methylations of adenosine to form either N^6-methyladenosine (m^6A), an internal modified nucleotide, or N^6,2′-*O*-dimethyladenosine (m^6Am), which is found exclusively at the first templated nucleotide of certain mRNAs. In addition, other rare modified nucleotides have been identified and together these form the epitranscriptomic code of mRNA. In the case of some modified nucleotides, the presence, location, or abundance is a subject of debate. Here, we review the methods that enable the discovery of modified nucleotides and how these approaches can be used to map epitranscriptomic modifications in mRNA.

Outline

Cite as *Cold Spring Harb Perspect Biol* doi: 10.1101/cshperspect.a032201

1 INTRODUCTION

Chemically modified nucleotides were identified in messenger RNAs (mRNAs) more than 40 years ago. These modifications were found within the mRNA cap structure at the 5′ end of all mRNAs and as modifications of internal nucleotides. However, until recently these modifications were generally overlooked or thought to be constitutive modifications that occur during mRNA biogenesis. Only recently has it become clear that nucleotide modifications can be selectively distributed on specific mRNAs, can vary depending on disease state or cellular context, and are endowed with regulatory potential for mRNA stability, translation, splicing, localization, and other functions. These effects account for the ability of nucleotide modifications to affect diverse physiological and pathological processes such as differentiation, viral infection, and cancer progression (for recent reviews, see Gokhale and Horner 2017; Jaffrey and Kharas 2017; Klungland et al. 2017).

The reemergence of the concept of epitranscriptomic regulation (i.e., the idea that the number, location, and type of nucleotide modifications contain information that influences mRNA fate in cells) has been a consequence of new technologies that allow nucleotide modifications to be mapped throughout the transcriptome. The initial epitranscriptomic mapping studies developed by us and others focused on N^6-methyladenosine (m^6A) (Dominissini et al. 2012; Meyer et al. 2012) and showed the power of mapping technologies for revealing the dynamics of epitranscriptomic variation in diverse cellular conditions and disease states. Since that time, newer m^6A-mapping methods have been introduced that provide single-nucleotide resolution m^6A maps of the transcriptome (Linder et al. 2015), enabling researchers to use mutagenesis to ascertain the function of specific m^6A residues in mRNA.

The studies on m^6A spurred researchers to discover additional modifications that comprise the epitranscriptomic code. Currently, the most abundant and well-documented modifications are m^6A, N^6,2′-*O*-dimethyladenosine (m^6Am), and pseudouridine. In the case of m^6A, thousands of m^6A sites have been mapped in mRNA, with the number of sites depending on the study, cell type, and, most importantly, sensitivity of the assay for detecting low-stoichiometry modification sites. Other modifications have also been mapped in mRNA, including 5-methylcytidine (m^5C), N^1-methyladenosine (m^1A), and 2′-*O*-methylated nucleotides (N_m). It should be noted that there has been some disagreement in the field regarding the abundance of these modifications, largely focusing on high false-positive rates in some of the mapping approaches for m^5C, m^1A, and N_m (Zaringhalam and Papavasiliou 2016; Legrand et al. 2017; Grozhik and Jaffrey 2017). Additional work will be required to understand the prevalence of these modifications and the basis for the disagreements in the field.

Among the abundant modifications in mRNA, the only ones mapped to date are m^6A and m^6Am. Other modifications, especially at mRNA cap structures, are also abundant and may have regulatory potential. Currently, a major goal is to discover all the modifications that comprise the epitranscriptomic code and to develop techniques to map these modifications. Additionally, in many cases, mapping the known modifications can provide key insights into whether a signaling pathway or disease process involves epitranscriptomic regulation and if a specific mRNA of interest is potentially controlled by nucleotide modifications. Below, we describe how mRNA modifications are detected and localized throughout the transcriptome.

2 THE HISTORY OF THE DETECTION OF METHYLATED NUCLEOTIDES IN mRNA

So far, nine methylated nucleotides have been identified in mammalian mRNA (Fig. 1). Most of these methylated nucleotides were initially discovered in the 1970s when researchers were investigating the composition of the mRNA 5′ cap, which is now known to be an untemplated guanosine that is methylated to form 7-methylguanosine (m^7G). m^7G is linked to the first transcribed nucleotide by an unconventional 5′-5′ triphosphate bond.

These experiments also led to the discovery of m^6A. In contrast to m^7G, m^6A is an internal methylated nucleotide. In the experiments that led to the discovery of m^6A, cells were metabolically labeled with [^{3}H-methyl]-methionine, which allows methyl modifications in mRNA to be labeled. Hydrolysis of cellular poly(A) RNA revealed evidence of a methylated nucleotide, which was eventually determined to be m^6A. Early biochemical studies found that as many as 0.5% of all adenosines in mRNAs are modified to m^6A, with an average frequency of 1 to 5 m^6A nucleotides per capped mRNA molecule (Sommer et al. 1978).

Studies of the mRNA cap also led to the discovery of the 2′-*O*-methylated versions of A, G, C, and U, which are designated A_m, G_m, C_m, and U_m (jointly referred to as N_m in this text). N_m was identified as part of the "extended mRNA cap," which comprises the m^7G and the first or the first two transcribed nucleotides. Labeling studies using [^{3}H-methyl]-methionine as the radioactive methyl donor revealed that either the first or the first and second transcribed nucleotide can be methylated at the 2′-hydroxyl group of the ribose (Wei et al. 1975a). The corresponding extended cap structures are referred to as cap1 and cap2, respectively. Recent mapping studies have suggested that

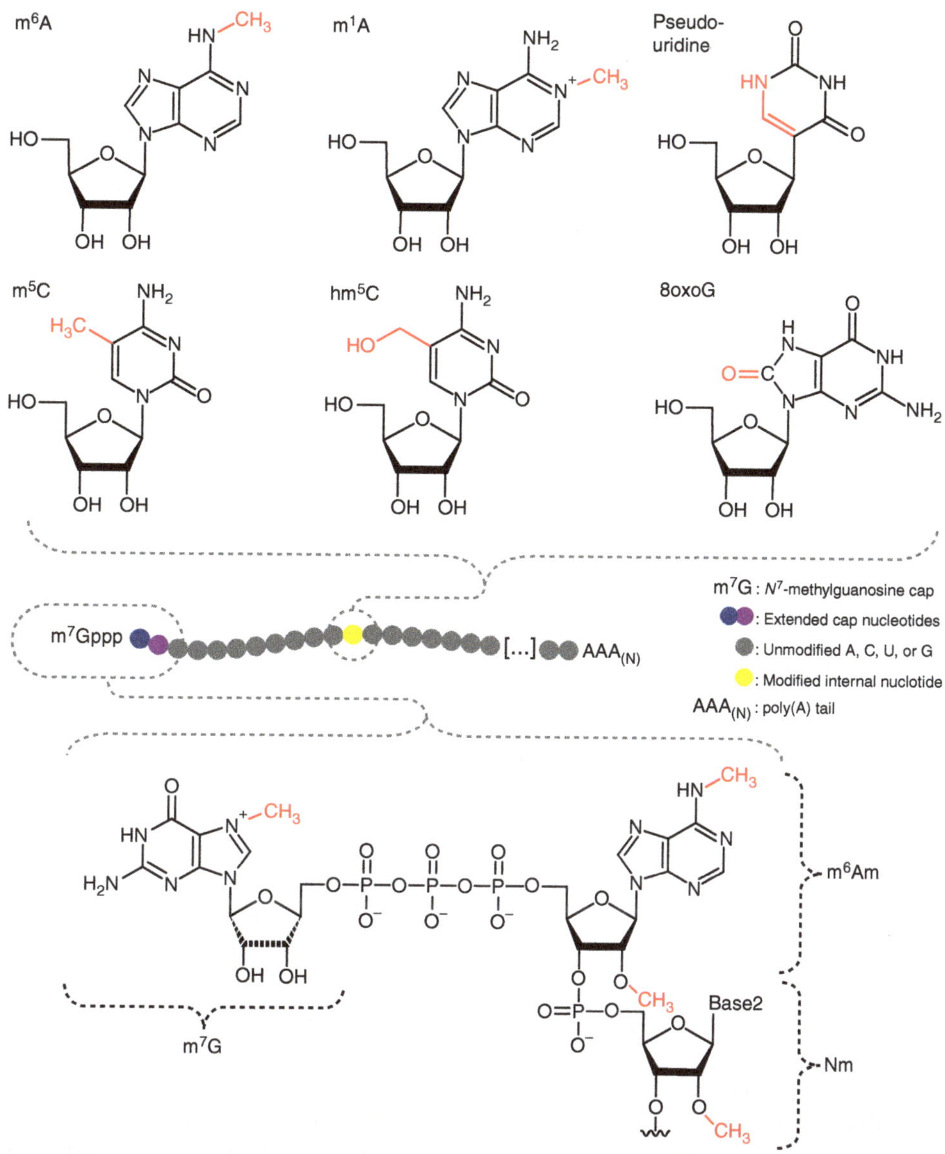

Figure 1. Modified nucleotides of eukaryotic messenger RNAs (mRNAs). (*Upper* panel) Structures of the six base-modified nucleotides identified in eukaryotic mRNA. m^{6}A, N^6-methyladenosine; m^{1}A, N^1-methyladenosine; m^{5}C, 5-methylcytidine; hm^{5}C, 5-hydroxymethylcytidine; 8-oxoG, 8-oxoguanosine. Chemical groups that have been added to or isomerized in canonical nucleotides are indicated in red. (*Middle* panel) Schematic of a modified mRNA molecule. The m^{7}G cap with its unusual 5′-5′-triphosphate bond and the poly(A) tail are indicated. Unmodified nucleotides are represented as gray beads. Modified first and second nucleotides are indicated in blue and purple, respectively. Internal modifications are indicated in yellow. (*Lower* panel) The extended cap of eukaryotic mRNA in its highest methylation state contains four methyl groups (red). The N^7-methylguanosine (m^{7}G), N^6, 2′-*O*-dimethyl (m^{6}Am), and 2′-*O*-methyl (N$_m$) moieties are indicated. Note that the m^{6}Am at the first position can also be a 2-*O*-methylated nucleotide (A$_m$, G$_m$, C$_m$, U$_m$), whereas unmodified nucleotides are rare at this position. At the second position, although N$_m$ is depicted here, unmodified nucleotides are also frequently present.

N$_m$ may also be present internally in mammalian and yeast mRNA (Dai et al. 2017; Bartoli et al. 2018).

The next modified nucleotide to be identified was a dimethylated nucleotide, m^{6}Am. When analyzing the 2′-*O*-methylated cap nucleotides, Wei and colleagues noticed a base-methylated version of Am and shortly after were able to confirm that the additional methylation occurred at the N^6 position of this nucleotide (Wei et al. 1975a, 1975b).

The eighth and ninth methylated nucleotides discovered in mRNA are m^{5}C and m^{1}A, respectively. One study that used [^{3}H-methyl]-methionine labeling showed evidence for m^{5}C in mRNA in the 1970s (Dubin and Taylor

1975), with the first transcriptome-wide m^5C mapping study in 2012 (Squires et al. 2012). In contrast, evidence for m^1A in mRNA came only recently (Dominissini et al. 2016; Li et al. 2016). Both m^5C and m^1A were initially described as highly abundant (Squires et al. 2012; Dominissini et al. 2016), but other researchers have argued that mass spectrometry-based detection of these modifications in poly(A) RNA reflects contamination from ribosomal RNA (rRNA) and transfer RNA (tRNA) (Legrand et al. 2017), and it has been argued that problems in the mapping approaches have overestimated the prevalence by ~1000-fold (Safra et al. 2017). Nevertheless, it appears likely that these modifications are present in rare and select mRNAs (Safra et al. 2017) but may not represent a general epitranscriptomic mechanism as originally proposed.

It should be noted that not all modified nucleotides in mRNA are methylated. Inosine is the product of RNA editing, and pseudouridine, an abundant modification in rRNA and tRNA, was recently shown to be present in mammalian and yeast mRNA (Carlile et al. 2014; Lovejoy et al. 2014; Schwartz et al. 2014). 5-hydoxymethyl cytosine is an oxidized product of m^5C and was originally mapped in *Drosophila* mRNA (Delatte et al. 2016). 5-hydoxymethyl cytosine has recently been proposed to be present also in mammalian mRNA (Shen et al. 2018). 8-oxo-G has also been detected in mRNA, and is largely thought to represent a product of oxidative damage (Shan et al. 2003).

3 ADVANTAGES AND CHALLENGES OF BIOCHEMICAL METHODS

Traditionally, modified mRNA nucleotides were identified by purifying poly(A) RNA, followed by the detection of modified nucleotides by classical biochemical methods like chromatography. This approach does not definitively establish that the modified nucleotide is in mRNA, largely because poly(A) RNA can have low levels of rRNA and tRNA contamination and can contain poly(A)-containing transcripts that are not mRNA. As a result, transcriptome-wide mapping has become the major method for detecting nucleotide modifications in mRNA. This approach can reveal the precise mRNAs that contain the modified nucleotide and the location of the modified nucleotide within the transcript.

Nevertheless, biochemical methods are useful and have revealed key principles of m^6A biology. The advantage of chromatography is that it is inherently quantitative because the levels of the modified and nonmodified nucleotides can usually be measured in parallel. This allows the fractional abundance of the modified nucleotide to be readily determined.

In addition, the quantitative readout of the purification/detection strategy allows relatively straightforward kinetic studies of the modification dynamics. For example, it was shown that m^6A methylation is rapidly lost from polyadenylated mRNA in the cytosol, suggesting a fast turnover of m^6A-containing mRNA (Friderici et al. 1976; Sommer et al. 1978).

The power of these chromatographic methods is also underscored by the use of clever purification and digestion strategies that revealed which parts of the mRNA are modified. When polyadenylated mRNA was partially hydrolyzed and repurified using poly(A) selection, it was found that m^6A is enriched in the 3′ portion of transcripts (Perry et al. 1975; Beemon and Keith 1977), a finding that was validated by more recent transcriptome-wide m^6A maps (Dominissini et al. 2012; Meyer et al. 2012; Linder et al. 2015). Also, systematic digestion of the modified mRNAs with nucleases with different nucleotide specificities revealed that m^6A occurs in a consensus motif which best described as N_1-R-A-C-N_2, with R being a purine, N_1 being mostly a purine and N_2 being rarely a G (Wei et al. 1976; Schibler et al. 1977).

The high sensitivity of the radioactive readout also allowed the detection of modifications that are less abundant than m^6A. For example, m^5C was detected in mRNA that underwent two rounds of poly(A)-enrichment at levels of ~0.18 m^5C residues/1000 nt (Dubin and Taylor 1975). However, these investigators noted that others did not detect m^5C in mRNA and it was possible that it originated from a contaminating non-mRNA species (Dubin and Taylor 1975).

Indeed, contaminating tRNA or rRNA can lead to spurious detection of modified nucleotides or an overestimation of the modified nucleotide in mRNA. This reflects the high abundance of modified nucleotides in tRNA and rRNA. Even a small amount of these contaminating species can lead to the erroneous conclusion that a modified nucleotide is present in mRNA. In line with this, sequencing of poly(A) RNA subjected to two rounds of purification using oligo-dT beads showed that ~40% of the RNA can be rRNA (Legrand et al. 2017), consistent with other studies showing that poly(A) RNA preparations contain diverse noncoding RNAs including rRNA (Cui et al. 2010; Sultan et al. 2014; Zhao et al. 2014). Thus, any measurement of modified nucleotides in poly(A) RNA may reflect the presence of the modified nucleotide in these ubiquitous contaminants.

It should be noted, however, that contaminating RNA can also give an impression of reduced m^6A levels. Because 18S and 28S rRNA contain approximately 420 and approximately 850 unmodified adenosines, respectively, their average m^6A/A ratio is 2 per 1270 or 0.16%, which is below the 0.5% m^6A/A ratio of mRNA (Sommer et al. 1978).

Accordingly, if the ratio of m^6A/A is measured, contamination of a human mRNA sample with rRNA leads to an underestimation of its m^6A/A ratio.

The problem of contaminating tRNA or rRNA is similarly a problem for more recent techniques such as mass spectrometry and ELISA-based detection methods.

To overcome the problem of contaminants, Fray and colleagues developed an approach that allows m^6A in the mRNA to be selectively detected, even in the presence of contaminating rRNA and small nuclear RNA (snRNA) (Zhong et al. 2008). In this approach, RNA is treated with ribonuclease T1, resulting in cleavage of RNA after G residues. The resulting RNA fragments are radiolabeled at their 5′ ends. After RNA hydrolysis the labeled nucleotides are detected by thin-layer chromatography. Because m^6A in snRNA and rRNA exist in a C-m^6A-G context (Shimba et al. 1995) and A-m^6A-C context (Boccaletto et al. 2018), respectively, these m^6A residues are not at a 5′ end after ribonuclease T1 treatment and therefore these m^6A residues are not radiolabeled. In contrast, m^6A in mRNA is predominantly found in a G-m^6A-C context (Schibler et al. 1977), and therefore is selectively measured.

Nevertheless, the concern about contamination makes it essential to determine if the source of a modified nucleotide is mRNA or a contaminant species. For this reason, transcriptome-wide mapping has become the method of choice for studying RNA modifications in mRNA.

4 NEXT-GENERATION SEQUENCING–BASED TECHNOLOGIES FOR MAPPING INTERNAL NUCLEOTIDE MODIFICATIONS

4.1 m^6A

4.1.1 *Mapping m^6A Using Antibody-Based Enrichment*

The original maps of m^6A were generated by using m^6A-specific antibodies to immunoprecipitate m^6A-containing RNA fragments from poly(A) RNA fractions (Fig. 2A) (Dominissini et al. 2012; Meyer et al. 2012). The antibody-bound RNA fragments were then subjected to RNA sequencing (RNA-seq). This technique, called MeRIP-seq (methylated RNA immunoprecipitation and sequencing) (Meyer et al. 2012) or m^6A-seq (Dominissini et al. 2012), has been extended to other modified nucleotide-specific antibodies to map hm^5C in *Drosophila* (Delatte et al. 2016), m^1A in mammals (Dominissini et al. 2016; Li et al. 2016), and m^5C in yeast and bacteria (Edelheit et al. 2013).

In the m^6A-mapping studies, m^6A sites throughout the transcriptome were predicted based on the presence of multiple overlapping aligned sequencing reads, or "peaks" (Fig. 3A). To call peaks, computational strategies are used to identify transcriptomic locations where the read accumulations are enriched in the immunoprecipitated sample, but not the input sample. Together, these strategies have identified 13,471 m^6A peaks in HEK 293T cells (Meyer et al. 2012) and 12,769 m^6A peaks in HepG2 cells (Dominissini et al. 2012).

4.1.2 *Detecting m^6A at Single-Nucleotide Resolution by Covalent Cross-Linking to Antibodies*

One of the key challenges with distinguishing m^6A from A is that there is no chemical reagent that selectively reacts with m^6A or A. This contrasts with m^5C and pseudouridine (see below), which each benefit from chemoselective reactions that enabled the development of chemical-based mapping approaches that map the modified nucleotides at single-nucleotide resolution.

To achieve chemoselective detection of m^6A, we developed miCLIP (m^6A individual nucleotide cross-linking and immunoprecipitation). miCLIP exploits the ability of m^6A antibodies to selectively bind m^6A (Linder et al. 2015). To generate a chemical modification at m^6A sites, ultraviolet (UV) light is used to create a covalent bond between RNA and a bound m^6A-specific antibody (Fig. 2A). After proteolysis, reverse transcription and high-throughput sequencing, cross-link-induced single-nucleotide substitutions, and cDNA truncations are characterized and used to call potential m^6A sites throughout the transcriptome (Fig. 3B,C). This provided the first single-nucleotide maps of m^6A (Linder et al. 2015).

Importantly, not all m^6A-binding antibodies are suitable for miCLIP. Characterization of multiple m^6A antibodies showed that only one antibody resulted in efficient termination of reverse transcription at m^6A, whereas another caused nucleotide misincorporations at the +1 position relative to the m^6A (Linder et al. 2015). Other m^6A-binding antibodies did not produce m^6A mutations that were efficient or located at a consistent position relative to diverse m^6A sites.

It is difficult to distinguish cross-link-induced misincorporations from sequencing errors, so statistical modeling is often applied (Zhang and Darnell 2011; Weyn-Vanhentenryck et al. 2014). However, prior knowledge about m^6A helped to validate the specificity of miCLIP. First, m^6A is known to occur in a specific consensus motif, so most sites called by miCLIP outside of this motif are likely to be false positives. Second, a number of methylatable sites had been tested for their methylation status (and stoichiometry) by thin layer chromatography (Liu et al. 2013). These features allowed us to establish criteria for calling of m^6A sites from cross-link-induced mutations that identified 11,832 sites with an estimate of ~12% false positives (Linder et al. 2015).

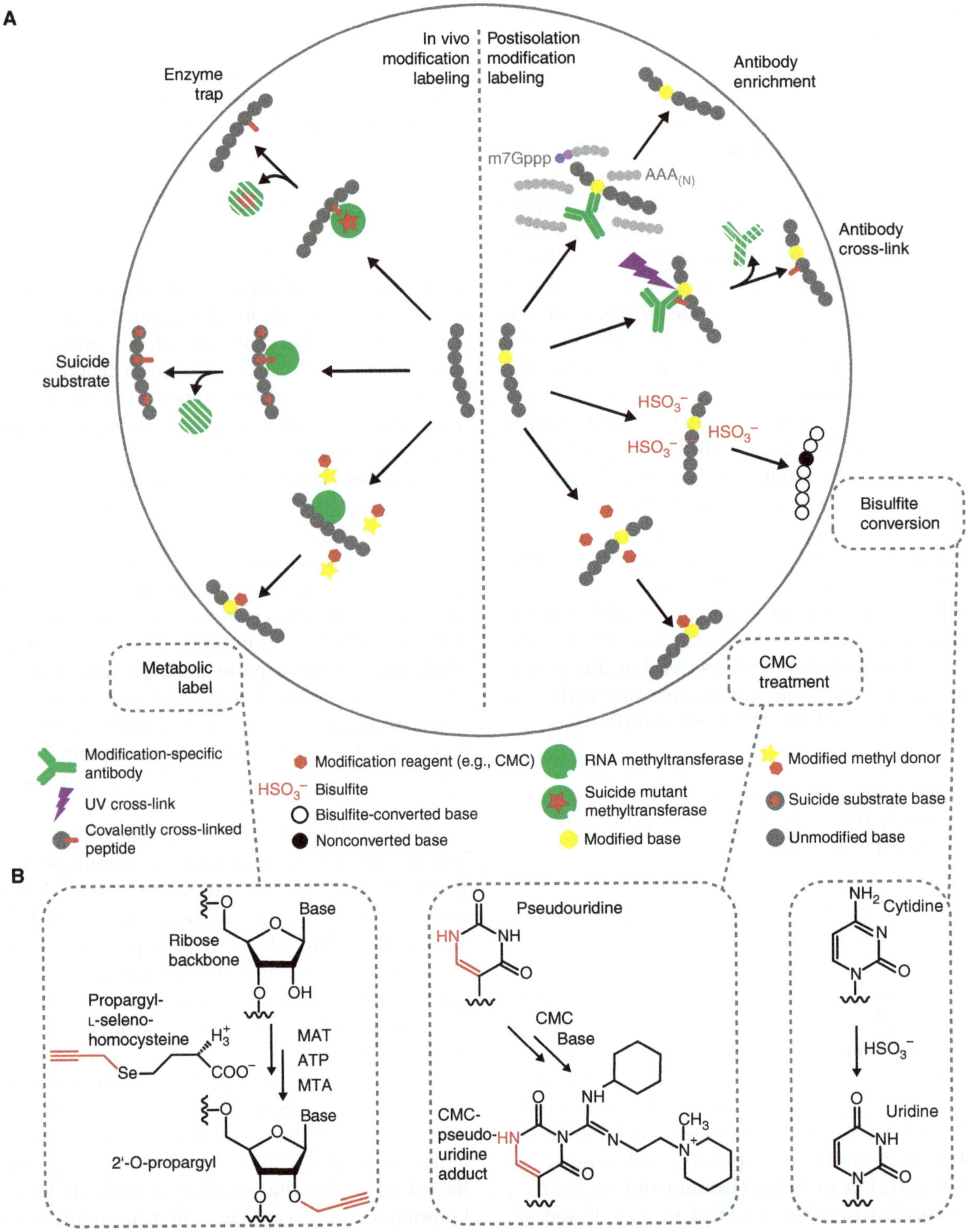

Figure 2. Approaches for the labeling and detection of modified nucleotides. (*A*) As many modified nucleotides are refractory to detection by classical molecular biology techniques, a variety of approaches have been developed that facilitate their discovery (see main text for details and references). Strategies to label modified nucleotides in vivo (*left*). A mutant methyltransferase can be introduced into cells that traps a covalent reaction intermediate. After proteolysis, a covalently attached peptide remnant marks the substrate nucleotide (*upper left*). A similar remnant peptide can be obtained if a nucleotide analog is incorporated into RNA that serves as a suicide substrate for the methyltransferase (*middle left*). (*Legend continues on following page.*)

Cite as *Cold Spring Harb Perspect Biol* doi: 10.1101/cshperspect.a032201

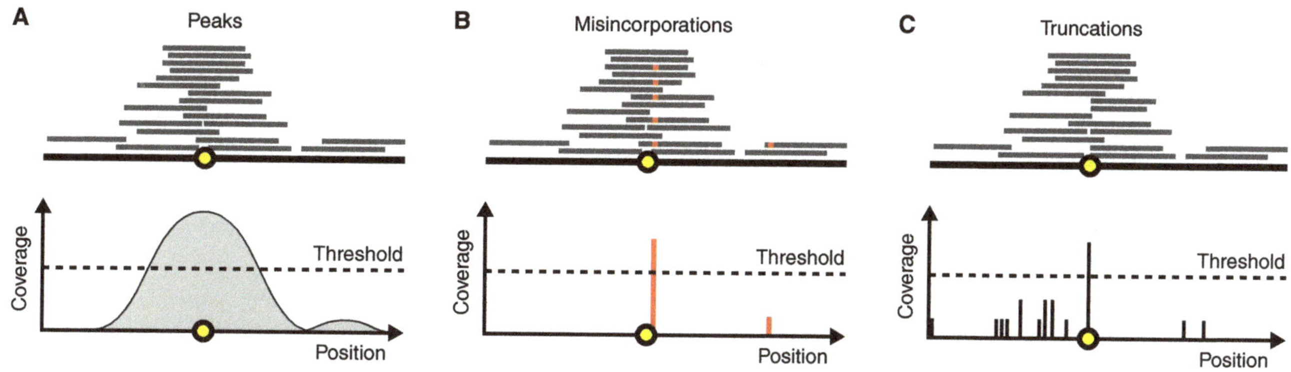

Figure 3. Strategies to identify modifications in high-throughput sequencing data. After enrichment and/or labeling, modified RNAs are subjected to reverse transcription, library preparation, and sequencing. Sequenced reads (horizontal gray lines) are then mapped to a reference genome or transcriptome (horizontal black line). (*A*) A peak calling approach is often used in antibody-based enrichment experiments. It tries to identify regions in the transcriptome where more reads are captured than expected from background noise. Although the shape of the peak and the observed midpoint can sometimes be used to infer the position of the methylated nucleotide (yellow circle), the resolution of this approach is limited. (*B*) Modifications or chemical treatments that alter the base-pairing properties of a nucleotide often lead to nucleotide misincorporations during reverse transcription. After mapping, these mutations are identified as mismatches to the reference sequence (orange squares). By counting these mismatches and comparing them to the mismatch distribution expected by background noise (e.g., from sequencing errors), modified nucleotides can be identified with single-nucleotide resolution. (*C*) Some modifications cause the reverse transcriptase to abort cDNA synthesis. The resulting accumulation of read 5′ ends can be compared with the background distribution to identify sites with significant enrichment of such truncations. Although these truncations in principle also mark modification sites with base resolution, this method is more prone to false positives because the background noise is typically higher than what is observed for misincorporations.

4.2 Pseudouridine

Pseudouridine is a highly abundant nucleotide modification in ribosomal RNA and is also present in tRNA (Modomics database; Boccaletto et al. 2018). A series of studies in 2014 showed that pseudouridines are also found at specific locations in mammalian and yeast mRNA (Carlile et al. 2014; Lovejoy et al. 2014; Schwartz et al. 2014).

Pseudouridine is an enzymatically isomerized version of uridine in which the uracil is detached from the ribose, undergoes a rotation, and then is reattached to the ribose. Despite this isomerization, pseudouridine behaves like uri-

Figure 2. (*Continued*) Reactive analogs of methyl-group donors can be used to metabolically label target nucleotides (*lower left*). This approach is promising because methyl-donor analogs have been developed that make use of versatile propargyl groups that can be further reacted using click chemistry. Also, metabolic labeling in principle could be used to study methylation kinetics on a transcriptome-wide scale. Many techniques use chemical strategies to "hyper-modify" RNAs after they are isolated from cells (*right*). Modification-specific antibodies are used to enrich modified RNA fragments (*top right*). By using ultraviolet (UV) light to cross-link them to the bound RNA, modification-specific antibodies can also be used as chemical agents to mark modified nucleotides (or their close vicinity; *upper right*). Treatment of RNA with bisulfite leads to the conversion of cytidine to uridine (*lower right*). This deamination is far less efficient for m^5C, and such nonconversion events can be used as a signal to detect m^5C. Pseudouridine can be converted to a bulky nucleotide species using *N*-cyclohexyl-*N′*-(2-morpholinoethyl)carbodiimide (CMC; *bottom right*). Uridine also reacts with CMC but is eliminated from this unmodified nucleotide by treatment with base. (*B*) Chemical reactions used to label modified nucleotides. (*Left* panel) The methionine analog propargyl-L-seleno-homocysteine can be used to label positions of methyl groups. Propargyl-L-seleno-homocysteine is readily taken up by cells and converted into a *S*-adenosyl-methionine analog by cellular methionine adenosyltransferases (MAT). The bioorthogonal propargyl group is then transferred to the 2′-hydroxyl position of nucleotides in ribosomal RNA (rRNA) in cells. In in vitro reactions, the N6 position of adenosine can also be modified by an N6-adenosine methyltransferase (MTA). In both cases, the propargyl-modified nucleotide can be biotinylated by click chemistry for selective recovery of RNA that contains the modified nucleotide. (*Middle* panel) Labeling of pseudouridine with CMC (*N*-cyclohexyl-*N′*-(β-[*N*-methylmorpholino]ethyl)carbodiimide). Note that CMC reacts with other nucleotides as well but is eliminated from the other nucleotides with treatment by base. (*Right* panel) Deamination of cytidine by bisulfite (HSO_3^-) to yield uridine. Because of the methylation at the N5 position, this reaction is inefficient for m^5C.

dine during reverse transcription. As a result, it does not produce a reverse transcription signature that lends itself to straightforward detection by sequencing (Ryvkin et al. 2013).

To make pseudouridine detectable, the most common method is to use *N*-cyclohexyl-*N'*-(β-[*N*-methylmorpholino]ethyl)carbodiimide (CMC) (Fig. 2B, middle panel). CMC reacts to form an adduct with U, pseudouridine, and to a lesser extent G (Ho and Gilham 1971). However, the CMC adduct is readily reversible by alkali treatment for U and G. In contrast, the CMC adduct on pseudouridine is unaffected, leaving a bulky CMC moiety that induces reverse transcription stops, which can be detected by next-generation sequencing.

Ψ-seq, Pseudo-seq (pseudouridine sequencing), and PSI-seq (pseudouridine site identification sequencing) used this approach to map pseudouridines throughout the transcriptome in yeast and mammals (Carlile et al. 2014; Lovejoy et al. 2014; Schwartz et al. 2014). Pseudouridine sites were called based on recurrent reverse transcription terminations, resulting in approximately 50 to 300 mapped sites in yeast and 100 to 400 mapped sites in human cell lines (Carlile et al. 2014; Lovejoy et al. 2014; Schwartz et al. 2014). Together, these studies led to single-nucleotide resolution maps of pseudouridine in yeast and mammals.

Pseudouridine in mRNA was typically found to exist in sequence motifs that resembled the motifs in tRNA that are modified with pseudouridine. In some cases, pseudouridine was found in motifs consistent with snoRNA-mediated pseudouridine formation by the dyskerin enzyme; however, these were <25% of all sites (Schwartz et al. 2014). Importantly, the investigators validated sites by mapping pseudouridine in yeast lacking specific pseudouridine-forming enzymes, thereby providing independent proof that a significant fraction of mapped sites indeed represent pseudouridine (Carlile et al. 2014; Schwartz et al. 2014).

A more sensitive version of this method was described in which CMC was replaced with a clickable analog of CMC (Li et al. 2015). In this method, an alkyne-modified form of CMC is reacted with RNA. After the CMC-moiety is removed from U and G, the remaining adducts are biotinylated using click chemistry, and the RNA is recovered for sequencing. This approach, which allows greater sequencing depth because only pseudouridine-containing RNA is recovered, is particularly valuable because it allows pseudouridine residues in low abundance mRNAs to be detected (Li et al. 2015).

The function of pseudouridine remains elusive. Pseudouridine does not impair protein synthesis by the ribosome, as shown by the translation of transfected pseudouridine-containing reporter mRNAs into functional enzymes (Karikó et al. 2008). Pseudouridine is known to promote the stability of helices by increasing the stability of base-stacking interactions (Davis 1995). Therefore, it is conceivable that the function of pseudouridine is to stabilize specific double-stranded RNA structures.

4.3 m^{5}C

4.3.1 *Bisulfite Mapping of m^{5}C in mRNA*

Methylated cytosine is most widely known as a modified nucleobase in DNA, forming the m^{5}dC nucleotide (5-methyl deoxycytidine). This modification is usually referred to by the name of the modified base (i.e., 5-methylcytosine [5mC]) rather than by the name of the nucleotide (m^{5}dC). However, 5mC is also a constituent of m^{5}C, an RNA nucleotide that is found in rRNA and tRNA (Modomics database; Boccaletto et al. 2018).

5mC in DNA can be detected based on its selective resistance to modification by bisulfite (Fig. 2B, right panel) (Frommer et al. 1992). Bisulfite treatment leads to deamination of unmodified C, but does not efficiently deaminate 5mC. Thus, after bisulfite treatment and the subsequent deamination step, cytosines are sequenced as thymidines. Because 5mC residues are not deaminated, they are read as cytosines during DNA sequencing.

The bisulfite-based 5mC detection approach for DNA was applied to RNA to map m^{5}C in diverse types of RNA species (Schaefer et al. 2009). Most notably, this method was recently applied to poly(A) RNA to map m^{5}C in mRNA. Based on global mapping of bisulfite-resistant cytidine residues, 10,275 m^{5}C sites were mapped at single-nucleotide resolution in HeLa cell mRNA and long-noncoding RNA (lncRNA) (Squires et al. 2012).

The high abundance of m^{5}C in mRNA has been questioned (Legrand et al. 2017). The primary concern is that the bisulfite mapping approach can be associated with some cytidine residues showing lack of bisulfite-mediated deamination, but not actually being methylated (Legrand et al. 2017). Why might this occur? The most likely reason is that some cytidines may be inaccessible to bisulfite because they are located within a structured region of RNA. Another reason is that the bisulfite modification step and deamination steps may not be performed to completion. Unlike DNA, which is fairly stable, RNA is susceptible to degradation by hydrolysis. Therefore, milder conditions are used with RNA, which can lead to incomplete cytidine deamination and thus the appearance of unconverted cytidines at non-m^{5}C positions (Schaefer et al. 2009). Lastly, read mapping can be difficult because bisulfite-converted RNA reads are typically short (to facilitate narrow peaks) and of low complexity (because most cytidines are converted to thymidine). Therefore, unambiguous alignment of these reads to the genome is difficult to achieve, which could alter the measurements of the bisulfite-mediated conversion rates (Legrand et al. 2017).

Lyko and colleagues have argued that the appearance of unmodified cytidines does not reflect m^5C, but instead reflects the statistical variation in bisulfite modification that would normally occur in RNA (Legrand et al. 2017). In earlier studies, a threshold cutoff was used to call m^5C sites based on whether the residue showed a reduction in the conversion rate. Lyko and colleagues argued that the use of a cutoff can be misleading because the conversion rate should be modeled as a statistical distribution. As with any distribution, there will be tails reflecting cytosines with increased or decreased conversion rates. Lyko and colleagues calculated that the number of cytidines that showed reduced conversions in m^5C mapping experiments was not significantly larger than what would be expected based on a statistical distribution of random nonconversions. Because these various questions that have been raised concerning m^5C mapping, it will be important to biochemically validate previously predicted m^5C sites (Squires et al. 2012; Amort et al. 2013). Biochemical validation can be used to assess the false-positive rates in m^5C mapping studies and would therefore enable the true abundance of m^5C in mRNA to be extrapolated.

4.3.2 Enzyme-Trapping Methods to Map m^5C

In addition to bisulfite mapping, another method to map m^5C takes advantage of the relatively unique enzymatic mechanism of cytosine methyltransferases. Cytosine methyltransferases form a covalent adduct with cytosine as an initial step before methylation (Liu and Santi 2000). Two approaches have been described to trap this intermediate and thereby identify the RNAs and their cytosines that are targets of cytosine methyltransferases (Fig. 2A) (Hussain et al. 2013; Khoddami and Cairns 2013).

In Aza-IP (5-azacytidine-mediated RNA immunoprecipitation), a modified cytosine is used to trap cytosine methyltransferases (Khoddami and Cairns 2013). In this method, cells are cultured with 5-azacytidine (5-azaC) in the culture media. This nucleotide is incorporated as if it were cytidine during RNA transcription. When cytosine methyltransferases attempt to methylate 5-azaC, they remain covalently bound to 5-azaC, resulting in a covalent enzyme-RNA intermediate that cannot be resolved. In the Aza-IP study, the cytosine methyltransferases DNMT2 or NSUN2 were immunoprecipitated to detect covalently bound RNA. Hydrolysis by divalent metal ions was used to release the bound RNA, resulting in opening of the 5-azaC ring that could be detected by the presence of nucleotide misincorporations seen during sequencing of the cDNA (Khoddami and Cairns 2013). The misincorporations also allow high-resolution mapping of the cytosines targeted by these enzymes.

In a related approach, no modified nucleotide was used. Instead, an NSun2 mutant that cannot resolve the covalent enzyme-RNA intermediate was expressed in cells (Hussain et al. 2013). This approach, termed miCLIP (methylation iCLIP; different from m^6A miCLIP above), involves immunoprecipitating the NSun2–RNA complexes and sequencing bound RNAs. Peptide remnants of the covalently bound enzyme cause termination of reverse transcription. These terminations are then used for the high-resolution mapping of m^5C in RNA (Hussain et al. 2013).

In both approaches, m^5C was detected in tRNA and other noncoding RNAs. However, m^5C was not readily detected in mRNA, in contrast to m^5C mapping with bisulfite. m^5C miCLIP detected 312 m^5C sites in mRNA (Hussain et al. 2013), while just one high-confidence m^5C in mRNA was detected with Aza-IP (Khoddami and Cairns 2013). The ability of these methods to detect m^5C in mRNA may however have been limited, because the enzyme-trapping method did not enrich for mRNA, which is much lower in abundance than tRNA or rRNA. Thus, these studies do not exclude the possibility that m^5C is prevalent in mRNA.

4.4 m^1A

4.4.1 Antibody-Enrichment-Based Mapping of m^1A

Two antibody-based m^1A mapping approaches, m^1A-seq (m^1A methylated RNA immunoprecipitation and sequencing) and m^1A-ID-seq (m^1A-specific RNA immunoprecipitation with demethylase-assisted RNA sequencing), provided evidence that m^1A, a modified nucleotide previously known to occur in tRNA and rRNA (Modomics database; Boccaletto et al. 2018) is also present in mRNA (Dominissini et al. 2016; Li et al. 2016).

These two initial studies were similar to the original m^6A mapping approach in that they relied on mRNA fragment pulldown using a modification-specific antibody (Dominissini et al. 2016; Li et al. 2016). In addition, they took advantage of unique features of m^1A. As m^1A is unable to form a Watson–Crick base pair with thymidine, this nucleotide blocks reverse transcription. Therefore, both of the initial transcriptome-wide maps of m^1A used the tendency of m^1A to introduce reverse transcription stops to help distinguish between nonspecific peaks and peaks that were indeed caused by m^1A. m^1A peaks were enriched in "trough"-like peaks (m^1A-seq) or show a drop in sequencing coverage upstream of the m^1A (m^1A-ID-seq). Furthermore, the RNA was either treated with the m^1A

demethylase AlkB (m^1A-ID-seq) or subjected to Dimroth rearrangement (m^1A-seq) to convert m^1A into m^6A, which does not induce reverse transcription stops. These studies identified more than 7000 m^1A peaks in 4000 mRNAs and estimated an average stoichiometry of 20%, sometimes exceeding 50% (Dominissini et al. 2016). Based on the localization of m^1A near start codons and bioinformatic analysis of published ribosome profiling data, m^1A was proposed to enhance mRNA translation efficiency (Dominissini et al. 2016). These studies suggested that m^1A is a previously unrecognized, yet highly prevalent internal base modification in mRNA that regulates mRNA translation.

Two subsequent mapping studies reported that m^1A is present at fewer sites and in different transcripts than initially reported. These studies sought to map the exact sites of m^1A in the antibody-immunoprecipitated RNA (Li et al. 2017; Safra et al. 2017), and took advantage of the ability of m^1A to induce nucleotide misincorporations during reverse transcription (Hauenschild et al. 2015). This effect can be amplified by using specific polymerases such as the thermostable group II intron reverse transcriptase (TGIRT), which has an increased frequency of misincorporation when encountering m^1A (Safra et al. 2017). Thus, TGIRT was used to analyze RNA immunoprecipitated with the m^1A antibody to determine if and where m^1A-induced misincorporations were present in the recovered cDNA (Li et al. 2017; Safra et al. 2017) Therefore, rather than relying on m^1A peaks, this approach added specificity by searching for m^1A-induced misincorporations.

Intriguingly, this high-resolution approach showed that most mRNA fragments immunoprecipitated with the m^1A antibody lacked m^1A-induced misincorporations. In fact, only seven cytosolic mRNAs contained misincorporations consistent with the presence of m^1A (Safra et al. 2017). Surprisingly, the investigators identified five m^1A-containing mitochondrial mRNAs (Safra et al. 2017). For most of these 13 mRNAs, the stoichiometry of m^1A was very low. Indeed, the stoichiometry was so low that many cells are unlikely to contain even a single copy of a m^1A-modified mRNA. None of the mapped sites were localized near start codons. Thus, the use of TGIRT in m^1A-seq revealed that m^1A is approximately 1000-fold less prevalent and in most cases much lower in stoichiometry than originally described (Safra et al. 2017).

Only one mRNA showed a high stoichiometry m^1A, the mitochondrial mRNA that encodes the enzyme NADH dehydrogenase-5 (*ND5*). The investigators asked whether the stoichiometry of this modification changes in different cellular states, which might suggest that this modification has a regulatory function, at least in *ND5*. The authors found that m^1A modification was high in stoichiometry in egg cells and up to the four-cell stage but dropped in the late blastocyst stage (Safra et al. 2017). Thus, m^1A may have a regulatory effect on *ND5*. The data were most consistent with an inhibitory effect on translation, consistent with the general idea that m^1A would interfere with translation, rather than enhance translation (Grozhik and Jaffrey 2018).

The finding that m^1A is a rare modified nucleotide in mRNA is consistent with the recent mass spectrometry study that found that m^1A was not detectable in highly purified poly(A) RNA (Legrand et al. 2017). These investigators were able to detect m^1A in poly(A) RNA fractions that were prepared using standard oligo-dT purification. However, they showed that these poly(A) RNA preparations contain rRNA. Notably, this poly(A) RNA purification protocol is commonly used to search for the presence of modified nucleotides in mRNA. When rigorous purification protocols were used and rRNA depletion was validated by next-generation sequencing, the investigators found that m^1A and other modified nucleotides such as m^5C and 5hmC were depleted from the poly(A) RNA.

This study pointed out that mass spectrometry-based analysis of poly(A) RNA preparations require that the poly(A) RNA be analyzed by next-generation sequencing to rigorously quantify the presence of any rRNA or tRNA; otherwise it would not be clear if the modified nucleotide reflects low-level contamination from tRNA or if it indeed reflects the presence of an internal modified nucleotide in mRNA.

Because m^1A is especially abundant in tRNAs, it is important to determine the levels of contaminating tRNA when quantifying m^1A and other modified nucleotides in a poly(A)-enriched sample by mass spectrometry. Sequencing methods that use the TGIRT polymerase now allow the sequencing-based quantification of tRNA levels despite the high level of modified nucleotides in tRNA (Zheng et al. 2015).

Taken together, these results show that m^1A is indeed present in mRNA, but it is likely to be rare rather than highly prevalent under normal cellular conditions. Importantly, m^1A and other modifications might be of low abundance if they are transiently introduced and trigger mRNA degradation by no-go decay. Such ribosome-mediated removal of modified RNA has been found for 8-oxo-G (Simms et al. 2014). It will thus be important to determine if translational inhibition increases the abundance of m^1A.

One of the important findings of the m^1A mapping studies is that the few transcripts that contain m^1A tend to contain m^1A within a structured sequence that resembles the "T-loop" of tRNA. The T-loop of tRNA contains a single m^1A in the majority of tRNA molecules in the cell. This is a form of "tRNA mimicry" that allows an mRNA to become modified by taking advantage of the structural and

Cite as *Cold Spring Harb Perspect Biol* doi: 10.1101/cshperspect.a032201

sequence specificity of tRNA-modifying enzymes (Grozhik and Jaffrey 2017). It is possible that other tRNA-modifying enzymes similarly modify mRNAs that contain structures that resemble conserved sequence or structural motifs normally found in tRNA.

Additionally, it is interesting that the more recent m^1A mapping studies focus on mitochondrial transcripts as the major target of m^1A. However, unlike *ND5* mRNA most of these mitochondrial transcripts are modified with exceptionally low stoichiometry. It is possible that these transcripts become methylated because of the high concentration of m^1A-forming tRNA-modifying enzymes in the mitochondrial cytoplasm, which may induce spurious modification of mRNA sequences that show weak similarity to T-loops. Therefore, the functional significance of these low stoichiometry m^1A modifications may be minimal.

4.5 2′-*O*-Methylated Nucleotides (N_m)

2′-*O*-methylated nucleotides, referred to as N_m, are highly prevalent in rRNA and tRNA (Modomics database; Boccaletto et al. 2018) and are also found immediately adjacent to the m^7G mRNA cap (Furuichi et al. 1975). Recent studies have suggested that they are present at internal sites in mRNA in yeast (Bartoli et al. 2018) and potentially in mammalian cells (Dai et al. 2017).

The initial experiments that pointed to the presence of abundant N_m at internal sites were performed in mammalian cells. Using a high-throughput sequencing-based approach termed N_m-seq, the investigators mapped 3515 N_m sites internally in mRNA and ncRNA (Dai et al. 2017). N_m-seq exploits a unique feature of N_m (i.e., the lack of a 2′-hydroxyl) to map N_m. In this approach, periodate is used to oxidize 3′-terminal nucleotides of RNAs that contain hydroxyl groups at both the 2′ and 3′ positions on the ribose portion of the nucleotide. Coupled with removal of the 3′-terminal phosphate, this process removes the 3′-terminal nucleotide. However, if the terminal nucleotide contains a 2′-*O*-methyl, the 3′ end is protected from this reaction. To map N_m, the investigators therefore fragmented poly(A) RNA and subjected it to multiple rounds of elimination. This led to an enrichment of RNA fragment 3′ ends at N_m sites that could be detected by high-throughput sequencing (Dai et al. 2017).

The validity of the original N_m-seq data has been questioned, as many mapped N_m sites were nearly identical to the sequence of the 3′ adapter used for preparing the RNA library (Grozhik and Jaffrey 2018). This type of sequencing result is usually the result of a "mispriming" artifact in which the primers used for reverse transcription inadvertently amplify sequences within unligated RNA, as previously observed in CLIP libraries (Gillen et al. 2016). Consequently, a corrigendum with a revised protocol was subsequently published, reporting a new set of putative N_m sites (Dai et al. 2018). Notably, the new sites were not enriched in any consensus motif. Also, the predicted sites did not cluster in any specific region of mRNAs, and no function was proposed for these modifications. None of the N_m sites were individually biochemically validated, making the false-positive rate of this mapping technology currently unknown.

In contrast to the N_m mapping results in human cells, a recent study showed that N_m sites in yeast mRNA are in sequence contexts that match the consensus sequences associated with N_m in rRNA and tRNA. Notably, the N_m sites in yeast were mapped using a different approach (Bartoli et al. 2018). This approach exploits the property of reverse transcriptases to pause or stall at N_m when nucleotides or magnesium are limiting. This method, termed MeTH-seq (sequencing of methylations at two prime hydroxyls), allowed the identification of thousands of RT pause sites on yeast mRNA (Bartoli et al. 2018). Importantly, to call an RT pause site identified by MeTH-seq as a potential site of 2′-*O*-methyl modification, it had to disappear in a tRNA or rRNA 2′-*O*-methyltransferase knockout strain. This strategy identified 690 N_m sites in yeast mRNAs and assigned them to a methyltransferase enzyme that installs it. In many cases, the N_m sites were found in mRNA within sequence motifs that match consensus sites seen in tRNA or ribosomal RNA, further supporting the idea that N_m sites are generated by the methyltransferases normally thought to target tRNA or rRNA. This markedly contrasts with the N_m sites mapped in mammalian cells, which lack evidence for tRNA or rRNA-like N_m sequence motifs (Dai et al. 2018).

Notably, a subset of the yeast N_m sites was dependent on the rRNA methyltransferase Spb1. These sites were mapped to mRNAs encoding ribosomal proteins. In addition, Spb1 was necessary to maintain high expression levels of these transcripts, prompting the investigators to hypothesize that the 2′-*O*-methylation of rRNA and ribosomal protein mRNAs might promote the formation of a regulon that spans different classes of RNAs. Another important finding was that levels of N_m in mRNA appear to be regulated. Using MeTH-seq, several thousand sites were identified that were differentially methylated in cells undergoing exponential growth versus postdiauxic growth. This variation raises the possibility that different cellular states might mediate their effects by altering N_m levels in cellular RNA. It will be important to establish how the signals influence N_m levels, and if N_m also shows dynamic variations in mammalian cells.

4.5.1 *Metabolic Labeling of N_m*

Another recently developed approach to map methylated nucleotides is to identify the targets of methyltransferase enzymes by using a modified precursor of *S*-adenosyl-methionine, the cosubstrate that is the methyl donor in nucleotide methylation reactions (Hartstock et al. 2018). In this approach, a *S*-adenosyl-methionine analog that contains propargyl-L-selenohomocysteine in place of methionine is used. This causes a propargyl group, rather than a methyl group, to be added to nucleotides (Fig. 2B, left panel). The propargyl group can then be conjugated to biotin by "click"-chemistry, enabling the purification of the targets of any RNA methyltransferase that can use this *S*-adenosyl-methionine analog. This method was used to label N_m in HeLa cell rRNA. In addition, the investigators have shown that this approach can be used to identify m^6A methyltransferase targets in vitro, indicating that it could be developed into a method to label various ribose and base modifications in cells.

4.6 Other Internal Modified Nucleotides in mRNA

A variety of other modified nucleotides might be present in mRNA. Although most of the early chromatography-based studies from the 1970s did not detect additional modified nucleotides beyond m^6A, these may have been present at levels that were too low to detect. It is important to note that although these modifications may be rare, they might still have important regulatory functions. Additionally, any modified nucleotide that is highly transient, associated with rapid mRNA decay, or present only in certain disease or stress states, would not have been identified in these earlier studies.

One such modification is m^7G. m^7G is typically thought of as a nucleotide found at the mRNA cap structure. However, recent liquid chromatography-tandem mass spectrometry analyses of purified poly(A) RNA prepared from human and plant cells and rat tissues show that m^7G is present internally in mRNA (Chu et al. 2018). The levels of m^7G were found to be altered in rice exposed to heavy metals, raising the possibility that this modification could be regulated.

Another RNA modification is 8-oxo-G. Unlike the modifications described above, which are introduced via specific nucleotide-modifying enzymes, 8-oxo-G is formed as a result of oxidative stress. The DNA version of this modified nucleotide, 8-oxo-dG, is commonly used as a marker of oxidative stress. However, RNA is also susceptible to oxidation, and 8-oxo-G is detected in purified mRNA fraction using 8-oxo-G antibodies (Li et al. 2006; Shan et al. 2003). As for m^7G, single-nucleotide resolution maps of 8-oxo-G are not yet available.

The function of 8-oxo-G and RNA is not known. Some studies suggest that 8-oxo-G induces translational impairment, which would subsequently lead to no-go decay (Simms et al. 2014). Whether 8-oxo-G preferentially forms in specific mRNAs, and whether 8-oxo-G has any specific effect on mRNA besides impairing translation and subsequently inducing degradation is not yet known.

5-hydroxymethylcytosine (hm^5C) can be formed in RNA by the oxidization of m^5C by deoxygenases of the Ten-eleven translocation (Tet) family (Fu et al. 2014). The first transcriptome-wide approach to map hm^5C was performed in *Drosophila melanogaster* (Delatte et al. 2016). In an approach termed hMeRIP-seq (hydroxymethylated RNA immunoprecipitation followed by sequencing), the investigators use an hm^5C-specific antibody to enrich modified RNA fragments and then use a peak calling approach to identify regions containing hm^5C. These hm^5C-containing regions were found to be slightly enriched in the CDS and were proposed to restore translation of mRNAs whose translation was otherwise inhibited by the 5mC.

A recent study investigating the function of the Tet2 dioxygenase in mice used bisulfite sequencing to identify m^5C sites that were not converted to 5hmC in *Tet2* knockout animals (Shen et al. 2018). This revealed that m^5C and hm^5C might be used to control mRNA secondary structure and thereby regulate recognition by RNA binding proteins during the immune response (Shen et al. 2018).

The abundance of 5hmC in *Drosophila* mRNA has been challenged based on the inability to detect this modification using mass spectrometry analysis of highly purified *Drosophila* mRNA (Legrand et al. 2017). However, as with other mapping approaches, direct biochemical measurements of 5hmC at specific sites can resolve these controversies and establish the false-positive rate of these mapping methods to assess the overall prevalence and stoichiometry of this modification in mRNA.

5 MAPPING THE CAP EPITRANSCRIPTOME

5.1 m^7G and N_m in the Extended mRNA Cap

Although much of the recent interest in epitranscriptomics has focused on modified nucleotides located at internal sites (i.e., between the mRNA cap and the poly(A) tail), the level of RNA modification is considerably higher at the mRNA cap compared with internal sites. Besides m^6A, there are four other abundant methyl modifications in mRNA. These are the N-7 methyl in the m^7G cap, the two 2′-*O*-methyl modifications that are found at the first and second templated nucleotides, and the N^6-methyl that is found along with a 2′-*O*-methyl in m^6A_m that can be found at the first templated nucleotide. Together, these methyl marks comprise the five major methyl modifications in mRNA.

As of yet, no transcriptome-wide analysis has been reported for the methyl modification in the m^7G cap or the 2′-*O*-methyl modifications at the first and second templated nucleotides. Nevertheless, there is evidence that these modifications might be regulated. In the case of m^7G, the enzyme that methylates the guanosine to *N*7-methylguanosine, RNMT, is regulated in cancer, and overexpression of RNMT promotes cancer development (Cowling 2010a, 2010b). This suggests that under normal conditions, some mRNAs are not efficiently methylated to form the m^7G cap, and expression of RNMT allows these mRNAs to mature to acquire a fully functional m^7G cap. Because the m^7G cap is needed for mRNA nuclear export (Inesta-Vaquera and Cowling 2017), the unmethylated guanosine-capped mRNAs may be retained in the nucleus or degraded.

The cap-associated 2′-*O*-methyl modifications may also be regulated. It should be noted that some mRNAs may lack the 2′-*O*-methyl modifications altogether. These mRNAs are referred to as cap 0. Other mRNAs may only have the 2′-*O*-methyl modification on the first encoded nucleotide, but not the second (referred to as "cap 1"). This 2′-*O*-methyl modification is installed by CMTR1, which requires a m^7G-capped RNA as a substrate (Bélanger et al. 2010; Inesta-Vaquera and Cowling 2017). CMTR1 is physically associated with RNA Pol II (Haline-Vaz et al. 2008), which suggests that this modification may be cotranscriptionally introduced into mRNA.

mRNAs may also have both the first and second transcribed nucleotide methylated at the 2′-hydroxyl, which is referred to as cap 2 (Furuichi et al. 1975). mRNAs are not likely to have only the second templated nucleotide 2′-*O*-methylated because CMTR2, the enzyme that introduces the 2′-*O*-methyl on the second nucleotide shows increased methyltransferase activity toward capped mRNA with a 2′-*O*-methyl on the first nucleotide (Werner et al. 2011).

One major function of 2′-*O*-methylation is that this modification may contribute to an antiviral defense pathway. Viral infection is associated with the induction of CMTR1 which converts cap 0 RNAs to cap 1 and potentially cap 2 mRNAs (Daffis et al. 2010; Züst et al. 2011). The presence of 2′-*O*-methyl modifications prevents these mRNAs from binding IFIT proteins, which are induced by interferon in response to viral infection (Fensterl and Sen 2015). IFITs bind their target mRNAs at the extended cap structure and suppress translation by competing for binding by cap-binding proteins (Habjan et al. 2013; Kimura et al. 2013). Thus, 2′-*O*-methylation protects host mRNAs from the antiviral response, which is intended to be directed toward viral RNAs.

However, 2′-*O*-methylation is detected in cells even in the absence of viral infection, and the levels of cap 1 and cap 2 appear to differ between cells (Furuichi et al. 1975). Thus, 2′-*O*-methylation at the cap may have functions even in the absence of viral infection. As of yet, the distribution of 2′-*O*-methyl modifications associated with mRNA caps has not been mapped in the transcriptome.

5.2 m^6Am

The only cap-associated methylation that has been profiled on a transcriptome-wide level is m^6Am. m^6Am is mapped simultaneously with m^6A in the m^6A miCLIP protocol (described above), which uses an m^6A antibody. m^6A antibodies also bind m^6Am (Munns et al. 1979) and therefore are more accurately described as 6-methyladenine antibodies to reflect their binding to the two 6-methyladenine-containing nucleotides, m^6A and m^6Am.

Before the development of miCLIP, m^6A and m^6Am could not be distinguished. Therefore MeRIP-seq peaks found in the 5′ UTR could not be easily assigned as an m^6A or an m^6Am. Conceivably the location of the MeRIP-seq peak within the 5′ UTR could hint that it reflects an m^6A, rather than an m^6Am, which is exclusively found at the transcription-start nucleotide. A peak in the middle of the 5′ UTR is clearly not at the transcription-start nucleotide. Unfortunately, this reasoning is not valid. Many transcripts have alternative transcription start sites, some of which may be downstream from annotated transcription start sites (Shiraki et al. 2003). Therefore, an m^6Am from an isoform with a shorter 5′ UTR than the annotated 5′ UTR isoform will produce a peak that appears to be in the middle of the annotated 5′ UTR despite the fact that it is located at the transcription-start site of a different mRNA isoform.

In MeRIP-seq, RNA fragments are immunoprecipitated, and the RNA is reverse transcribed to make the first strand of cDNA. Next, the second strand of cDNA is made by nicking the RNA and using the free end as a primer for DNA synthesis. In this protocol, the 5′ end of the RNA is not preserved, because the second strand of cDNA starts somewhat randomly based on the position of the nick.

In the case of m^6A, the lack of preservation of the 5′ end is not important because m^6A can be anywhere in the immunoprecipitated RNA. Therefore, the m^6A will still be in the middle of the resulting peak. However, for m^6Am, the RNA fragments will be different—in each case, the m^6Am will always be exactly at the 5′ end of the RNA. Therefore, the loss of the 5′ end means that the MeRIP-seq will be positioned downstream from the actual site of the m^6Am. The peak will look like an m^6A peak downstream from the actual site of the m^6Am.

miCLIP overcomes this problem by preserving the exact 5′ end of immunoprecipitated RNAs. Because all reads that contain m^6Am will have the same 5′ end (i.e., the transcription start nucleotide), this will result in the appearance of a "cliff" or edge that clearly demarcates the position of m^6Am.

Library methods that preserve the 5′ end of RNAs reduce the ambiguity of assigning a peak to m^6A or m^6Am.

Based on the mapping of m^6Am using miCLIP, a function for m^6Am could be identified. Our analysis showed that m^6Am-containing mRNAs show longer mRNA half-lives than mRNAs that begin with other nucleotides. The stability of m^6Am appears to be mediated, at least in part, by reduced susceptibility of m^6Am-containing mRNAs to decapping. Biochemical analysis of RNAs containing m^6Am showed that they were decapped by the Dcp2 decapping enzyme with markedly reduced efficiency compared with similar RNAs that contained Am. Thus, a single N-6 methyl modification causes the RNA to show reduced decapping. Notably, m^6Am mRNAs also show reduced susceptibility to microRNAs, which induce a decapping step as part of their mechanism to induce RNA degradation. Thus, m^6Am may have a role in stabilizing mRNAs.

At present, the enzyme that forms m^6Am from Am has been biochemically purified (Keith et al. 1978) but not yet cloned. Identification of this enzyme will enable a more comprehensive analysis of the function of this modification in mRNA. Additionally, the "readers" of m^6Am are not yet known, but they may contribute to the mechanism of m^6Am-mediated translational enhancement or other effects of m^6Am.

6 CONCLUDING REMARKS AND FURTHER DIRECTIONS

As the field moves forward a major emphasis is validating modification sites that are predicted by mapping studies. Although it is likely that certain transcripts contain m^5C, m^1A, and N_m in mammalian transcriptomes, it is not clear how many of the sites predicted by mapping methods reflect actual modification sites or reflect noise inherent in these assays. Methods such as SCARLET (Liu et al. 2013), which can detect the amounts of modified and nonmodified nucleotide in a site-specific manner in mRNAs, will be useful to detect modified nucleotides at specific sites.

In the case of m^6A, m^6A_m, and pseudouridine, a major goal will be to move from simple mapping methods to new approaches that reveal the stoichiometry of these modifications. Newer methods for transcriptome-wide quantification of the stoichiometry of these modifications are needed to determine how these modifications are regulated at each site in the transcriptome in disease states and in response to signaling pathways.

Lastly, a major underexplored area of investigation is the three unmapped methyl modifications in the mRNA cap structure—that is, the methyl group in the m^7G and the 2′-*O*-methyl modifications at the first and second templated nucleotide. Mapping studies for these modifications will be important for determining if these modifications are dynamic and encode epitranscriptomic information that determines the fate of mRNAs in the cell.

ACKNOWLEDGMENTS

This work was supported by National Institutes of Health (NIH) Grants R01DA037755 and R01CA186702 (S.R.J.) and German Research Foundation (DFG) research fellowship LI 2385/1-1 (B.L.).

REFERENCES

Amort T, Soulière MF, Wille A, Jia X-Y, Fiegl H, Wörle H, Micura R, Lusser A. 2013. Long non-coding RNAs as targets for cytosine methylation. *RNA Biol* **10:** 1002–1008.

Bartoli KM, Schaening C, Carlile T, Gilbert WV. 2018. Conserved methyltransferase Spb1 targets mRNAs for regulated modification with 2′-O-methyl ribose. *bioRxiv* doi: 10.1101/271916.

Beemon K, Keith J. 1977. Localization of N^6-methyladenosine in the Rous sarcoma virus genome. *J Mol Biol* **113:** 165–179.

Bélanger F, Stepinski J, Darzynkiewicz E, Pelletier J. 2010. Characterization of hMTr1, a human Cap1 2′-*O*-ribose methyltransferase. *J Biol Chem* **285:** 33037–33044.

Boccaletto P, Machnicka MA, Purta E, Piątkowski P, Bagiński B, Wirecki TK, de Crécy-Lagard V, Ross R, Limbach PA, Kotter A, et al. 2018. MODOMICS: A database of RNA modification pathways. 2017 update. *Nucleic Acids Res* **46:** D303–D307.

Carlile TM, Rojas-Duran MF, Zinshteyn B, Shin H, Bartoli KM, Gilbert WV. 2014. Pseudouridine profiling reveals regulated mRNA pseudouridylation in yeast and human cells. *Nature* **515:** 143–146.

Chu J-M, Ye T-T, Ma C-J, Lan M-D, Liu T, Yuan B-F, Feng Y-Q. 2018. Existence of internal N^7-methylguanosine modification in mRNA determined by differential enzyme treatment coupled with mass spectrometry analysis. *ACS Chem Biol* doi: 10.1021/acschembio.7b00906.

Cowling VH. 2010a. Enhanced mRNA cap methylation increases cyclin D1 expression and promotes cell transformation. *Oncogene* **29:** 930–936.

Cowling VH. 2010b. Myc up-regulates formation of the mRNA methyl cap. *Biochem Soc Trans* **38:** 1598–1601.

Cui P, Lin Q, Ding F, Xin C, Gong W, Zhang L, Geng J, Zhang B, Yu X, Yang J, et al. 2010. A comparison between ribo-minus RNA-sequencing and polyA-selected RNA-sequencing. *Genomics* **96:** 259–265.

Daffis S, Szretter KJ, Schriewer J, Li J, Youn S, Errett J, Lin T-Y, Schneller S, Zust R, Dong H, et al. 2010. 2′-*O* methylation of the viral mRNA cap evades host restriction by IFIT family members. *Nature* **468:** 452–456.

Dai Q, Moshitch-Moshkovitz S, Han D, Kol N, Amariglio N, Rechavi G, Dominissini D, He C. 2017. Nm-seq maps 2′-*O*-methylation sites in human mRNA with base precision. *Nat Methods* **14:** 695–698.

Dai Q, Moshitch-Moshkovitz S, Han D, Kol N, Amariglio N, Rechavi G, Dominissini D, He C. 2018. Corrigendum: Nm-seq maps 2′-*O*-methylation sites in human mRNA with base precision. *Nat Methods* **15:** 226–227.

Davis DR. 1995. Stabilization of RNA stacking by pseudouridine. *Nucleic Acids Res* **23:** 5020–5026.

Delatte B, Wang F, Ngoc LV, Collignon E, Bonvin E, Deplus R, Calonne E, Hassabi B, Putmans P, Awe S, et al. 2016. Transcriptome-wide distribution and function of RNA hydroxymethylcytosine. *Science* **351:** 282–285.

Dominissini D, Moshitch-Moshkovitz S, Schwartz S, Salmon-Divon M, Ungar L, Osenberg S, Cesarkas K, Jacob-Hirsch J, Amariglio N, Kupiec M, et al. 2012. Topology of the human and mouse m6A RNA methylomes revealed by m6A-seq. *Nature* **485:** 201–206.

Dominissini D, Nachtergaele S, Moshitch-Moshkovitz S, Peer E, Kol N, Ben-Haim MS, Dai Q, Di Segni A, Salmon-Divon M, Clark WC, et al. 2016. The dynamic N^1-methyladenosine methylome in eukaryotic messenger RNA. *Nature* **530:** 441–446.

Dubin DT, Taylor RH. 1975. The methylation state of poly A-containing messenger RNA from cultured hamster cells. *Nucleic Acids Res* **2:** 1653–1668.

Edelheit S, Schwartz S, Mumbach MR, Wurtzel O, Sorek R. 2013. Transcriptome-wide mapping of 5-methylcytidine RNA modifications in bacteria, archaea, and yeast reveals m5C within archaeal mRNAs. *PLOS Genet* **9:** e1003602.

Fensterl V, Sen GC. 2015. Interferon-induced Ifit proteins: Their role in viral pathogenesis. *J Virol* **89:** 2462–2468.

Friderici K, Kaehler M, Rottman F. 1976. Kinetics of Novikoff cytoplasmic messenger RNA methylation. *Biochemistry* **15:** 5234–5241.

Frommer M, McDonald LE, Millar DS, Collis CM, Watt F, Grigg GW, Molloy PL, Paul CL. 1992. A genomic sequencing protocol that yields a positive display of 5-methylcytosine residues in individual DNA strands. *Proc Natl Acad Sci* **89:** 1827–1831.

Fu L, Guerrero CR, Zhong N, Amato NJ, Liu Y, Liu S, Cai Q, Ji D, Jin S-G, Niedernhofer LJ, et al. 2014. Tet-mediated formation of 5-hydroxymethylcytosine in RNA. *J Am Chem Soc* **136:** 11582–11585.

Furuichi Y, Morgan M, Shatkin AJ, Jelinek W, Salditt-Georgieff M, Darnell JE. 1975. Methylated, blocked 5 termini in HeLa cell mRNA. *Proc Natl Acad Sci* **72:** 1904–1908.

Gillen AE, Yamamoto TM, Kline E, Hesselberth JR, Kabos P. 2016. Improvements to the HITS-CLIP protocol eliminate widespread mispriming artifacts. *BMC Genomics* **17: 338.**

Gokhale NS, Horner SM. 2017. RNA modifications go viral. *PLoS Pathog* **13:** e1006188.

Grozhik AV, Jaffrey SR. 2017. Epitranscriptomics: Shrinking maps of RNA modifications. *Nature* **551:** 174–176.

Grozhik AV, Jaffrey SR. 2018. Distinguishing RNA modifications from noise in epitranscriptome maps. *Nat Chem Biol* **14:** 215–225.

Habjan M, Hubel P, Lacerda L, Benda C, Holze C, Eberl CH, Mann A, Kindler E, Gil-Cruz C, Ziebuhr J, et al. 2013. Sequestration by IFIT1 impairs translation of 2′*O*-unmethylated capped RNA. *PLoS Pathog* **9:** e1003663.

Haline-Vaz T, Lima Silva TC, Zanchin NIT. 2008. The human interferon-regulated ISG95 protein interacts with RNA polymerase II and shows methyltransferase activity. *Biochem Biophys Res Commun* **372:** 719–724.

Hartstock K, Nilges Benedikt S, Ovcharenko A, Cornelissen Nicolas V, Püllen N, Lawrence-Dörner A-M, Leidel Sebastian A, Rentmeister A. 2018. Enzymatic or in vivo installation of propargyl groups in combination with click chemistry for the enrichment and detection of methyltransferase target sites in RNA. *Angew Chem Int Ed* 21: 6342–6346.

Hauenschild R, Tserovski L, Schmid K, Thüring K, Winz M-L, Sharma S, Entian K-D, Wacheul L, Lafontaine DLJ, Anderson J, et al. 2015. The reverse transcription signature of *N*-1-methyladenosine in RNA-seq is sequence dependent. *Nucleic Acids Res* **43:** 9950–9964.

Ho NWY, Gilham PT. 1971. Reaction of pseudouridine and inosine with *N*-cyclohexyl-*N*′-β-(4-methylmorpholinium) ethylcarbodiimide. *Biochemistry* **10:** 3651–3657.

Hussain S, Sajini AA, Blanco S, Dietmann S, Lombard P, Sugimoto Y, Paramor M, Gleeson JG, Odom DT, Ule J, et al. 2013. NSun2-mediated cytosine-5 methylation of vault noncoding RNA determines its processing into regulatory small RNAs. *Cell Rep* **4:** 255–261.

Inesta-Vaquera F, Cowling VH. 2017. Regulation and function of CMTR1-dependent mRNA cap methylation. *Wiley Interdiscip Rev RNA* **8:** e1450.

Jaffrey SR, Kharas MG. 2017. Emerging links between m^6A and misregulated mRNA methylation in cancer. *Genome Med* **9:** 2.

Karikó K, Muramatsu H, Welsh FA, Ludwig J, Kato H, Akira S, Weissman D. 2008. Incorporation of pseudouridine into mRNA yields superior nonimmunogenic vector with increased translational capacity and biological stability. *Mol Ther* **16:** 1833–1840.

Keith JM, Ensinger MJ, Mose B. 1978. HeLa cell RNA (2′-*O*-methyladenosine-N^6-)-methyltransferase specific for the capped 5′-end of messenger RNA. *J Biol Chem* **253:** 5033–5039.

Khoddami V, Cairns BR. 2013. Identification of direct targets and modified bases of RNA cytosine methyltransferases. *Nat Biotechnol* **31:** 458–464.

Kimura T, Katoh H, Kayama H, Saiga H, Okuyama M, Okamoto T, Umemoto E, Matsuura Y, Yamamoto M, Takeda K. 2013. Ifit1 inhibits Japanese encephalitis virus replication through binding to 5′ capped 2′-*O* unmethylated RNA. *J Virol* **87:** 9997–10003.

Klungland A, Dahl JA, Greggains G, Fedorcsak P, Filipczyk A. 2017. Reversible RNA modifications in meiosis and pluripotency. *Nat Methods* **14:** 18–22.

Legrand C, Tuorto F, Hartmann M, Liebers R, Jacob D, Helm M, Lyko F. 2017. Statistically robust methylation calling for whole-transcriptome bisulfite sequencing reveals distinct methylation patterns for mouse RNAs. *Genome Res* **27:** 1589–1596.

Li Z, Wu J, Deleo CJ. 2006. RNA damage and surveillance under oxidative stress. *IUBMB Life* **58:** 581–588.

Li X, Zhu P, Ma S, Song J, Bai J, Sun F, Yi C. 2015. Chemical pulldown reveals dynamic pseudouridylation of the mammalian transcriptome. *Nat Chem Biol* **11:** 592–597.

Li X, Xiong X, Wang K, Wang L, Shu X, Ma S, Yi C. 2016. Transcriptome-wide mapping reveals reversible and dynamic N^1-methyladenosine methylome. *Nat Chem Biol* **12:** 311–316.

Li X, Xiong X, Zhang M, Wang K, Chen Y, Zhou J, Mao Y, Lv J, Yi D, Chen X-W, et al. 2017. Base-resolution mapping reveals distinct m^1A methylome in nuclear- and mitochondrial-encoded transcripts. *Mol Cell* **68:** 993–1005.e9.

Linder B, Grozhik AV, Olarerin-George AO, Meydan C, Mason CE, Jaffrey SR. 2015. Single-nucleotide-resolution mapping of m6A and m6Am throughout the transcriptome. *Nat Methods* **12:** 767–772.

Liu Y, Santi DV. 2000. m5C RNA and m5C DNA methyl transferases use different cysteine residues as catalysts. *Proc Natl Acad Sci* **97:** 8263–8265.

Liu N, Parisien M, Dai Q, Zheng G, He C, Pan T. 2013. Probing N^6-methyladenosine RNA modification status at single nucleotide resolution in mRNA and long noncoding. *RNA* **19:** 1848–1856.

Lovejoy AF, Riordan DP, Brown PO. 2014. Transcriptome-wide mapping of pseudouridines: Pseudouridine synthases modify specific mRNAs in *S. cerevisiae. PLoS ONE* **9:** e110799.

Meyer KD, Saletore Y, Zumbo P, Elemento O, Mason CE, Jaffrey SR. 2012. Comprehensive analysis of mRNA methylation reveals enrichment in 3′ UTRs and near stop codons. *Cell* **149:** 1635–1646.

Munns TW, Oberst RJ, Sims HF, Liszewski MK. 1979. Antibody-nucleic acid complexes. Immunospecific recognition of 7-methylguanine- and N^6-methyladenine-containing 5′-terminal oligonucleotides of mRNA. *J Biol Chem* **254:** 4327–4330.

Perry RP, Kelley DE, Friderici K, Rottman F. 1975. The methylated constituents of L cell messenger RNA: Evidence for an unusual cluster at the 5′ terminus. *Cell* **4:** 387–394.

Ryvkin P, Leung YY, Silverman IM, Childress M, Valladares O, Dragomir I, Gregory BD, Wang L-S. 2013. HAMR: High-throughput annotation of modified ribonucleotides. *RNA* **19:** 1684–1692.

Safra M, Sas-Chen A, Nir R, Winkler R, Nachshon A, Bar-Yaacov D, Erlacher M, Rossmanith W, Stern-Ginossar N, Schwartz S. 2017. The m^1A landscape on cytosolic and mitochondrial mRNA at single-base resolution. *Nature* **551:** 251–255.

Schaefer M, Pollex T, Hanna K, Lyko F. 2009. RNA cytosine methylation analysis by bisulfite sequencing. *Nucleic Acids Res* **37:** e12.

Schibler U, Kelley DE, Perry RP. 1977. Comparison of methylated sequences in messenger RNA and heterogeneous nuclear RNA from mouse L cells. *J Mol Biol* **115:** 695–714.

Schwartz S, Bernstein DA, Mumbach MR, Jovanovic M, Herbst RH, León-Ricardo BX, Engreitz JM, Guttman M, Satija R, Lander ES, et al. 2014. Transcriptome-wide mapping reveals widespread dynamic-regulated pseudouridylation of ncRNA and mRNA. *Cell* **159:** 148–162.

Shan X, Tashiro H, Lin CG. 2003. The identification and characterization of oxidized RNAs in Alzheimer's disease. *J Neurosci* **23:** 4913–4921.

Shen Q, Zhang Q, Shi Y, Shi Q, Jiang Y, Gu Y, Li Z, Li X, Zhao K, Wang C, et al. 2018. Tet2 promotes pathogen infection-induced myelopoiesis through mRNA oxidation. *Nature* **554:** 123–127.

Shimba S, Bokar JA, Rottman F, Reddy R. 1995. Accurate and efficient *N*-6-adenosine methylation in spliceosomal U6 small nucelar RNA by HeLa cell extract in vitro. *Nucleic Acids Res* **23:** 2421–2426.

Shiraki T, Kondo S, Katayama S, Waki K, Kasukawa T, Kawaji H, Kodzius R, Watahiki A, Nakamura M, Arakawa T, et al. 2003. Cap analysis gene expression for high-throughput analysis of transcriptional starting point and identification of promoter usage. *Proc Natl Acad Sci* **100:** 15776–15781.

Simms CL, Hudson BH, Mosior JW, Rangwala AS, Zaher HS. 2014. An active role for the ribosome in determining the fate of oxidized mRNA. *Cell Rep* **9:** 1256–1264.

Sommer S, Lavi U, Darnell JE. 1978. The absolute frequency of labeled *N*-6-methyladenosine in HeLa cell messenger RNA decreases with label time. *J Mol Biol* **124:** 487–499.

Squires JE, Patel HR, Nousch M, Sibbritt T, Humphreys DT, Parker BJ, Suter CM, Preiss T. 2012. Widespread occurrence of 5-methylcytosine in human coding and non-coding RNA. *Nucleic Acids Res* **40:** 5023–5033.

Sultan M, Amstislavskiy V, Risch T, Schuette M, Dökel S, Ralser M, Balzereit D, Lehrach H, Yaspo M-L. 2014. Influence of RNA extraction methods and library selection schemes on RNA-seq data. *BMC Genomics* **15:** 675.

Wei C-M, Gershowitz A, Moss B. 1975a. Methylated nucleotides block 5′ terminus of HeLa cell messenger RNA. *Cell* **4:** 379–386.

Wei C-M, Gershowitz A, Moss B. 1975b. N6, O2′-dimethyladenosine a novel methylated ribonucleoside next to the 5′ terminal of animal cell and virus mRNAs. *Nature* **257:** 251–253.

Wei CM, Gershowitz A, Moss B. 1976. 5′-Terminal and internal methylated nucleotide sequences in HeLa cell mRNA. *Biochemistry* **15:** 397–401.

Werner M, Purta E, Kaminska KH, Cymerman IA, Campbell DA, Mittra B, Zamudio JR, Sturm NR, Jaworski J, Bujnicki JM. 2011. 2′-*O*-ribose methylation of cap2 in human: Function and evolution in a horizontally mobile family. *Nucleic Acids Res* **39:** 4756–4768.

Weyn-Vanhentenryck SM, Mele A, Yan Q, Sun S, Farny N, Zhang Z, Xue C, Herre M, Silver PA, Zhang MQ, et al. 2014. HITS-CLIP and integrative modeling define the Rbfox splicing-regulatory network linked to brain development and autism. *Cell Rep* **6:** 1139–1152.

Zaringhalam M, Papavasiliou FN. 2016. Pseudouridylation meets next-generation sequencing. *Methods* **107:** 63–72.

Zhang C, Darnell RB. 2011. Mapping in vivo protein-RNA interactions at single-nucleotide resolution from HITS-CLIP data. *Nat Biotech* **29:** 607–614.

Zhao W, He X, Hoadley KA, Parker JS, Hayes DN, Perou CM. 2014. Comparison of RNA-seq by poly(A) capture, ribosomal RNA depletion, and DNA microarray for expression profiling. *BMC Genomics* **15:** 419.

Zheng G, Qin Y, Clark WC, Dai Q, Yi C, He C, Lambowitz AM, Pan T. 2015. Efficient and quantitative high-throughput tRNA sequencing. *Nat Methods* **12:** 835.

Zhong S, Li H, Bodi Z, Button J, Vespa L, Herzog M, Fray RG. 2008. MTA is an arabidopsis messenger RNA adenosine methylase and interacts with a homolog of a sex-specific splicing factor. *Plant Cell Online* **20:** 1278–1288.

Züst R, Cervantes-Barragan L, Habjan M, Maier R, Neuman BW, Ziebuhr J, Szretter KJ, Baker SC, Barchet W, Diamond MS, et al. 2011. Ribose 2′-*O*-methylation provides a molecular signature for the distinction of self and non-self mRNA dependent on the RNA sensor Mda5. *Nat Immunol* **12:** 137–143.

Cite as *Cold Spring Harb Perspect Biol* doi: 10.1101/cshperspect.a032201

Mapping the dsRNA World

Daniel P. Reich and Brenda L. Bass

Department of Biochemistry, University of Utah, Salt Lake City, Utah 84112

Correspondence: bbass@biochem.utah.edu

SUMMARY

Long double-stranded RNAs (dsRNAs) are abundantly expressed in animals, in which they frequently occur in introns and 3′ untranslated regions of mRNAs. Functions of long, cellular dsRNAs are poorly understood, although deficiencies in adenosine deaminases that act on RNA, or ADARs, promote their recognition as viral dsRNA and an aberrant immune response. Diverse dsRNA-binding proteins bind cellular dsRNAs, hinting at additional roles. Understanding these roles is facilitated by mapping the genomic locations that express dsRNA in various tissues and organisms. ADAR editing provides a signature of dsRNA structure in cellular transcripts. In this review, we detail approaches to map ADAR editing sites and dsRNAs genome-wide, with particular focus on high-throughput sequencing methods and considerations for their successful application to the detection of editing sites and dsRNAs.

Outline

Cite as *Cold Spring Harb Perspect Biol* doi: 10.1101/cshperspect.a035352

1 INTRODUCTION

The ability of RNA to copy itself, or replicate, is a key feature of an RNA world, allowing Darwinian evolution and what we call life (Rich 1962; Gilbert 1986; Joyce 2002; Robertson and Joyce 2012). The RNA World hypothesis assumes that Watson–Crick base-pairing was key to replication in a primordial cell. Thus, by definition, replication of an RNA genome involved double-stranded RNA (dsRNA) intermediates, and evolution of modern-day cells likely occurred in the constant presence of dsRNA. Our goal in this review is to describe the unique properties dsRNA confers on a biological system and the techniques used for the genome-wide mapping of dsRNA (i.e., determining a dsRNAome).

In modern-day cells, nucleic acids play key and obvious roles, but they are also fundamental to self versus nonself discrimination (Crowl et al. 2017). The diverse RNA editing and modifications found in all kingdoms of life, in recent years referred to as the Epitranscriptome (Saletore et al. 2012), have myriad functions in modern-day cells, but it seems very plausible, and others have proposed (Eigenbrod et al. 2015; O'Connell et al. 2015), that they originated as a means to discriminate self from nonself. Selfish elements and viruses of the early RNA world also may have had an RNA genome that replicated through a dsRNA intermediate. Perhaps reflecting an ancient and ongoing conflict between cells and selfish elements, in modern cells, dsRNA-mediated pathways play critical roles in immune defense in response to viral infection.

Viruses have long been known to give rise to dsRNA (Ehrenfeld and Hunt 1971), which acts in the cytoplasm as a pathogen-associated molecular pattern (PAMP), to trigger an immune response. The source of viral dsRNA is attributed to genomes of dsRNA viruses, replication intermediates and copy-back structures of single-stranded RNA viruses, and bidirectional transcription of certain DNA viruses (Schlee 2013). When assayed, this dsRNA is typically ≥100 base pairs (Pfaller et al. 2015) and thus distinct from smaller silencing RNAs, such as microRNAs (miRNAs) and small-interfering RNAs (siRNAs). It is now clear that our own cells encode and express long dsRNA, in both vertebrates and invertebrates (Whipple et al. 2015; Blango and Bass 2016; Reich et al. 2018), and that the adenosine deaminases that act on RNA (ADARs), by converting adenosine to inosine (A-to-I), mark a dsRNA as self (Hartner et al. 2009; Mannion et al. 2014; Liddicoat et al. 2015; Pestal et al. 2015; George et al. 2016; Reich et al. 2018). To understand how cells discriminate between self dsRNA and nonself viral dsRNA, it is important to understand the dsRNAome.

2 THE dsRNA WORLD IS A WORLD GOVERNED BY DIFFERENT PROPERTIES THAN THE RNA WORLD

There are several properties of the dsRNA world that are distinct from those of the RNA world. First, the RNAs that comprise the dsRNA world are rod-shaped molecules that can stretch for hundreds of base pairs with few branches (Fig. 1). Short dsRNA helices are the secondary structural elements that assemble to create the three-dimensional shapes of RNA molecules. However, with few exceptions (Rieder et al. 2013), dsRNA-binding proteins (dsRBPs) do not bind dsRNA helices that are buried in tertiary structure.

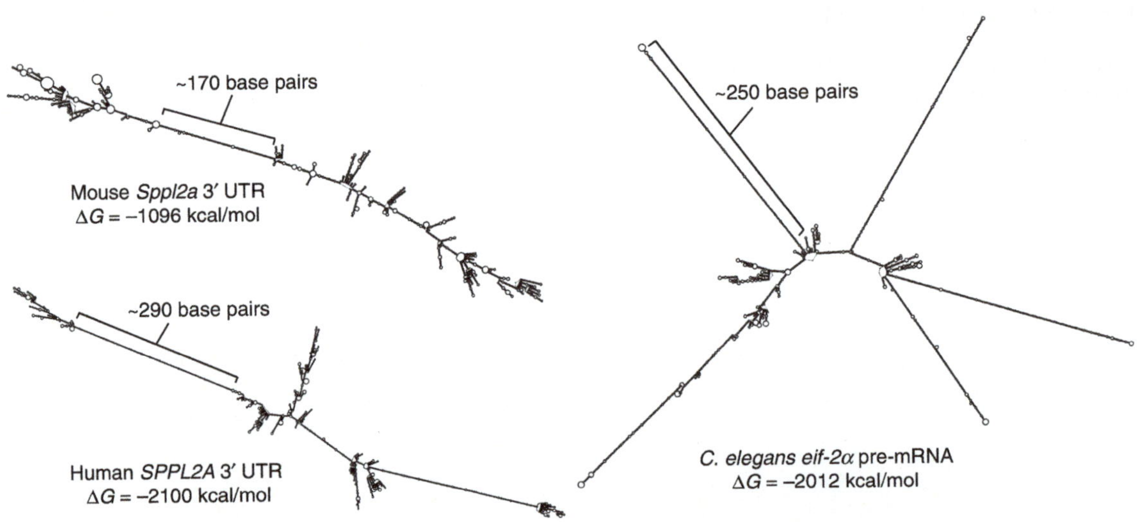

Figure 1. Representative double-stranded RNAs (dsRNAs) from three organisms. UNAfold-predicted RNA structures are shown for mouse *Sppl2a* 3′ untranslated region (3′ UTR), human *SPPL2A* 3′ UTR, and *Caenorhabditis elegans eif-2α* pre-mRNA. Approximate lengths of highly base-paired regions are shown for scale, and minimum (most stable) predicted folding free energies (ΔG) are reported beneath structures.

ADAR and PKR do not bind transfer RNA (tRNA) (Bass and Weintraub 1987; Schmedt et al. 1995), and similarly, the A-to-I modifications created by ADARs are not observed in ribosomal RNA (rRNA) (Paul and Bass 1998). The preference for binding to dsRNA that is not buried in tertiary structure is likely important for self:nonself discrimination by dsRBPs. Although wild-type RIG-I, a dsRBP essential for the mammalian innate immune response, shows limited association with rRNA, a RIG-I mutation observed in Singleton–Merten syndrome allows the protein to associate with dsRNA expansion sequences in the large rRNA subunit (Lässig et al. 2015).

Because of the A-form structure of dsRNA, which includes a deep major groove, it is difficult for dsRBPs to recognize dsRNA sequence (Tian et al. 2004). In fact, although dsRBPs bind tightly to dsRNA, typically showing dissociation constants in the nanomolar range (Kim et al. 1994; Schmedt et al. 1995; Ohman et al. 2000; Ma et al. 2008; Parker et al. 2008; Sinha et al. 2015), they are not sequence-specific. This second property of the dsRNA world is key to functions of dsRBPs, which interact with dsRNAs of diverse sequences. For example, Dicer must bind pre-miRNAs of many different sequences (Daugaard and Hansen 2017), ADARs must edit numerous distinct dsRNAs (Bajad et al. 2017), and antiviral dsRBPs must bind diverse viral RNAs (Yoo et al. 2014). Consistent with the idea that dsRBPs have structural specificity for the A-form helix, disruptions to the helix, such as mismatches and loops, can establish a register for dsRBP binding (Lehmann and Bass 1999). Similarly, although dsRBPs cannot access functional groups of bases within the major groove of completely base-paired dsRNA, sequence-specific contacts in the minor groove can also establish register (Stefl et al. 2010).

The final property of the dsRNA world is derivative of the first two. Because dsRBPs all bind dsRNA, and they are not sequence-specific, the dsRNA world is a world of constant competition. During its lifetime, a given dsRNA may interact with multiple dsRBPs.

3 FEATURES OF THE dsRNAome

This review focuses on methods used to determine dsRNAomes, and with few exceptions, these involve the determination of ADAR RNA editing sites. ADARs are a family of RNA editing enzymes, present in all animals (Deffit and Hundley 2016; Nishikura 2016; Walkley and Li 2017). ADARs target only dsRNA, and the longer a dsRNA is, the more editing sites it will acquire. The observation of ADAR editing sites in an endogenous RNA is proof that the RNA is double-stranded in vivo, a gold standard used not only in determining dsRNAomes, but also as proof for specific dsRNA structures (Sijen and Plasterk 2003).

Figure 2 illustrates three dsRNAomes, with genomic locations that express dsRNA, sometimes called editing enriched regions (EERs), indicated on chromosomes of mouse, human, and *Caenorhabditis elegans*. Based on dsRNAomes determined to date (Table 1), dsRNA is predominantly encoded within protein-coding genes, rather than intergenic regions, with 5.9% of protein-coding genes encoding structures in *C. elegans*, 13% in human monocytes, and 1.3% in mouse bone marrow–derived macrophages. (Repetitive elements in mouse are more divergent, resulting in fewer pairing partners and lower free energies [Neeman et al. 2006; Blango and Bass 2016].) For protein-coding genes, structures mostly inhabit noncoding regions of mRNAs, introns and 3′ untranslated regions (UTRs). Base-pairing usually occurs intramolecularly (Fig. 1), with complementary sequences within a single transcript folding back on themselves. The dsRNAs are predicted to fold into remarkably stable structures (Table 1) and, depending on the organism, encompass an average of 546–845 nucleotides (nt).

The chromosome maps of Figure 2 highlight interesting trends—some that are easily explained and others that are enigmatic, possibly hinting at EER functions. For example, human chromosome 19 has a dense concentration of EERs, likely because it is gene-rich, with a high number of repetitive elements (Grimwood et al. 2004). Consistent with their ability to base-pair, repeats, especially those that are inverted, inhabit many EERs (Fig. 3). Abundant repeat classes are represented in EERs, like *Alu* elements, which occur in more than a million copies in the human genome (Bazak et al. 2014) and DNA transposons, comprising 12.6% of the *C. elegans* genome (Ahringer and Gasser 2018). Comparisons of transcripts with the greatest number of editing sites, or "hyperedited" dsRNAs, in diverse metazoa, indicate that repeats, particularly transposon-derived repeats, give rise to most cellular dsRNAs (Porath et al. 2017).

In the *C. elegans* dsRNAome, EERs are enriched on distal arms, or chromosome ends, of autosomes (Fig. 2). This trend is not observed on the X chromosome, which overall has fewer EERs. Distal arms of *C. elegans* chromosomes show properties of heterochromatin and, compared with central regions, have fewer essential genes, more repetitive elements, lower levels of gene expression, and longer than average introns (Prachumwat et al. 2004; Liu et al. 2011; Ho et al. 2014; Ahringer and Gasser 2018). Intriguingly, although EERs also are enriched for repetitive elements and inhabit longer than average introns and 3′ UTRs, they typically inhabit genes with higher than average expression (Fig. 4). Except for the correlation with repetitive elements, as observed on human chromosome 19, an enrichment of EERs on specific regions of mammalian chromosomes has not been observed.

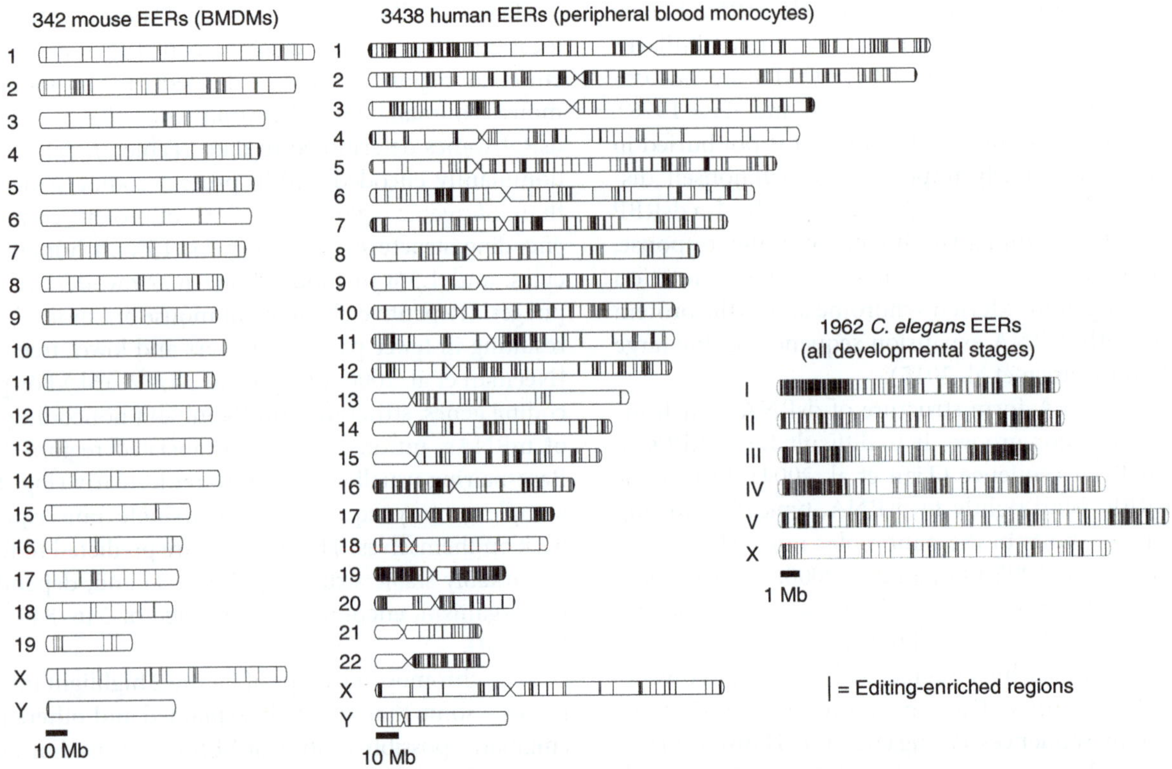

Figure 2. Mouse, human, and *Caenorhabditis elegans* dsRNAomes. Vertical black lines denote positions of editing enriched regions (EERs) on chromosomes of mouse, human, and *C. elegans*. Chromosomes are not drawn to scale, so the horizontal black bars at the *bottom* display relative chromosome length. Maps of mouse and human dsRNAomes were generated with Idiographica and that for *C. elegans* with PhenoGram (Kin and Ono 2007; Wolfe et al. 2013). BMDMs, bone marrow–derived macrophages.

Although EERs are not conserved at a sequence level, of the 285 mouse genes that contain EERs, 74 (26%) have an orthologous human gene with an EER ($p < 0.0001$, χ^2 test) (Blango and Bass 2016). Similarly, whereas a *Caenorhabditis briggsae* dsRNAome has not been determined, of 1092 *C. elegans* genes with a structured intron (ΔG/nt < –0.5 kcal/mole*nt at 20°C), 147 have an orthologous *C. briggsae* gene with a structured intron, a significant enrichment above the expected number ($p < 0.0001$, χ^2 test). Structured introns cluster on autosome distal arms in *C. briggsae*, as in *C. elegans* (Fig. 5). EERs and structured introns do not occur in the same locations in orthologous genes, and an EER within an intron can be in a UTR in the ortholog. Regardless, the presence of EERs and structured introns in orthologous genes and chromosomal domains raises the possibility of a conserved function in gene regulation.

4 HISTORICAL PERSPECTIVE ON METHODS TO FIND ADAR EDITING SITES

ADARs were discovered when perfectly base-paired dsRNA was injected into *Xenopus laevis* embryos. Researchers no-

Table 1. Properties of long dsRNAs (EERs)

	EERs			EER-associated genes			
Organism (cell type)	#	Length (nt)	ΔG/nt[a]	#	% of all genes[b]	% with intronic EERs	% with 3′-UTR EERs
Caenorhabditis elegans	1962	563	−0.349	1196	5.9	65.0	41.6
Mouse (BMDMs)	342	546	−0.307	285	1.3	20.4	69.1
Human (monocytes)	3438	845	−0.346	2792	13.0	57.4	23.9

EERs, editing enriched regions; UTR, untranslated region; ΔG, predicted folding free energy; nt, nucleotide; BMDMs, bone marrow–derived macrophages.
[a]Predicted at 37°C; kcal/mol*nt.
[b]Protein-coding genes only.

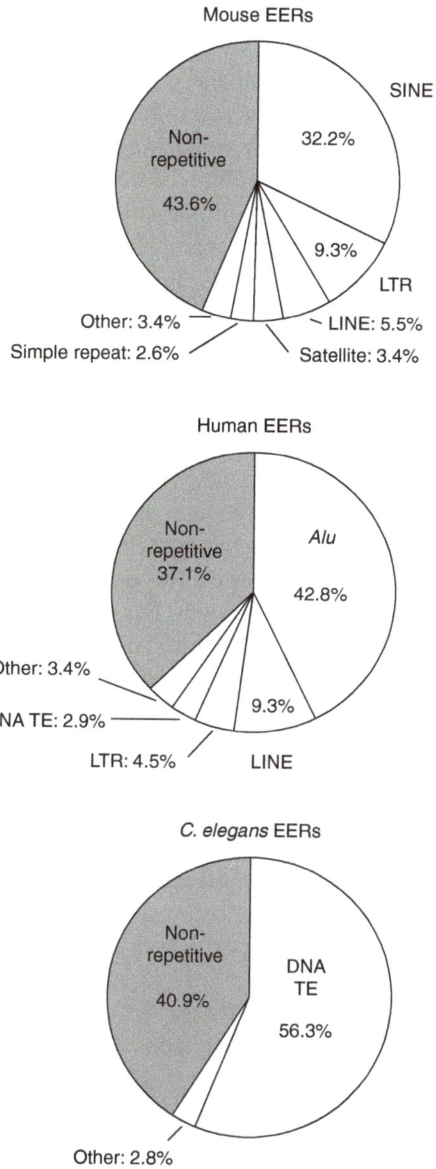

Figure 3. Repeat content of mouse, human, and *Caenorhabditis elegans* editing enriched regions (EERs). Pie charts depict percentage of total EER sequences that overlap RepeatMasker-annotated repeats (mouse: mm10; human: hg19; *C. elegans*: ce10). Major classes of repetitive elements (>2% of total EER sequence) are labeled, and classes comprising <2% are grouped as "Other." Nonrepetitive sequences did not overlap any sequences annotated as repetitive by RepeatMasker. SINE, short interspersed nuclear element; LINE, long interspersed nuclear element; LTR, long terminal repeat; DNA TE, DNA transposable element.

ticed that after incubation in *Xenopus* cells, the dsRNA migrated aberrantly on native gels (Bass and Weintraub 1987) and showed altered sensitivity to single-strand-specific ribonucleases (Rebagliati and Melton 1987). Thus, ADARs were initially called an unwinding activity, and this activity was also observed in extracts of mammalian cultured cells (Wagner and Nishikura 1988). Subsequent studies, using thin-layer chromatography or high-performance liquid chromatography (HPLC) to analyze nucleotides in treated dsRNA, revealed the activity involved covalent modification of adenosine to inosine (Bass and Weintraub 1988; Wagner et al. 1989). Only then was it realized that dsRNA strands were not being unwound. Rather, the RNA was becoming more single-stranded in character as AU base pairs were changed to IU mismatches. Although ADARs indeed change the stability of dsRNA, to date there is no example in which editing causes dsRNA to completely separate into two single strands. Subsequent studies revealed that A-to-I conversion occurred by hydrolytic deamination (Polson et al. 1991), and the activity was briefly called dsRAD or DRADA. In 1997, researchers agreed to rename the enzyme ADAR, based on recommendations by the HUGO committee (Bass et al. 1997).

Although ADAR activity in the above studies was detected using dsRNA prepared in vitro, it was not long before in vivo editing within naturally occurring transcripts was reported. The first example was identified serendipitously by Sanger sequencing of cloned cDNAs made from measles virus transcripts (Cattaneo et al. 1988; Bass et al. 1989). Inosine base-pairs like guanosine and prefers to pair with cytidine. Thus, reverse transcription of RNA containing

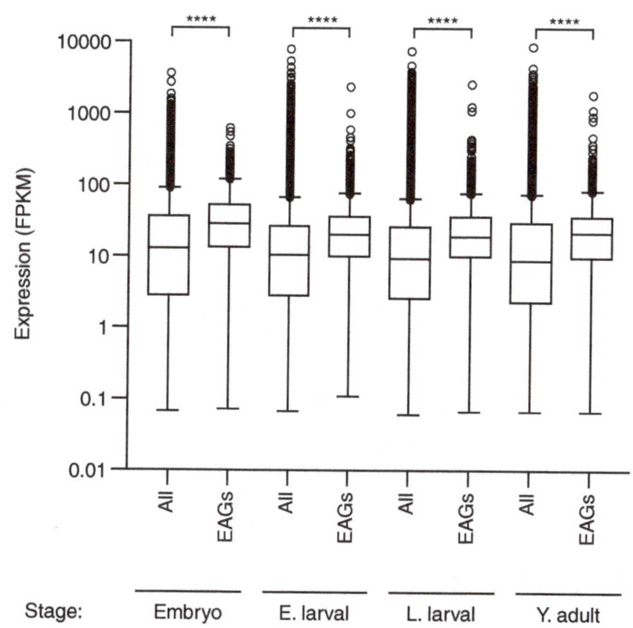

Figure 4. Expression of *Caenorhabditis elegans* editing enriched region (EER)-associated genes (EAGs). Tukey box plot shows gene expression, in fragments per kilobase*million reads (FPKM), for all expressed genes or EAGs in RNA-seq of four *C. elegans* developmental stages: embryo, early larval (E. larval; L1–L2), late larval (L. larval; L3–L4), and young adult (Y. adult) stages. ****, $P < 0.0001$, Mann–Whitney *U*-test.

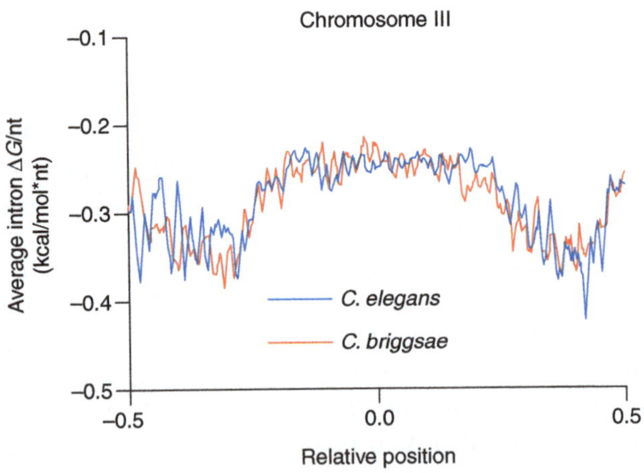

Figure 5. Predicted intron structure along chromosome III for two nematode species. Length-normalized UNAFold-predicted folding free energies are plotted by relative position on chromosome III of *Caenorhabditis elegans* (blue) and *Caenorhabditis briggsae* (red). Trends observed on chromosome III are representative of all *C. elegans* autosomes. Average intronic ΔG/nt values were calculated by splitting chromosomes into 1000 equal-length segments and averaging ΔG/nt values of introns in each segment. Lower ΔG/nt values indicate presence of more stable intronic structures. ΔG, predicted folding free energy; nt, nucleotide.

A-to-I editing shows T-to-C changes in the first-strand cDNA, revealed as A-to-G changes in the second-strand cDNA. Initial examples of in vivo ADAR editing showed numerous editing events in individual, cloned cDNAs (Bass 1997), reiterating the nonselective hyperediting observed in the perfectly paired dsRNAs used in the first in vitro studies. In these early days of ADAR research, it was hard to imagine an important biological activity for ADARs.

This all changed with the observation of editing events in specific codons. The first reported example involved a specific codon in an α-amino-3-hydroxy-5-methylisoxazole-4-propionic acid (AMPA) glutamate receptor, GRIA2 (previously called GluR-B) (Sommer et al. 1991). cDNA analyses indicated a specific amino acid was encoded as a glutamine or arginine, and the arginine was known to be important for ion transport properties of channels assembled with glutamate receptors. Researchers were puzzled when only a single gene could be identified for GRIA2, and its genomic sequence specified a glutamine codon. The investigators hypothesized that ADAR editing changed a glutamine (Q) codon to an arginine (R) codon (Q/R editing site). Indeed, subsequent studies revealed complementary sequence in the intron adjacent to the Q/R site that could fold back to encompass the edited codon (Higuchi et al. 1993; Egebjerg et al. 1994). The predicted structure was disrupted by mismatches and loops, and the idea that such disruptions could promote more selective editing began to emerge (Hurst et al. 1995; Bass 1997; Lehmann and Bass 1999), as well as the realization that ADARs constituted a family of enzymes (Melcher et al. 1996), with some organisms having a single ADAR and others several (Bass 2002).

An astounding finding was that lethality of *ADAR2*$^{-/-}$ mice could be rescued by replacing the unedited *GRIA2* allele with the edited one (Higuchi et al. 2000). These studies foreshadowed the many important ADAR editing events that occur in the nervous system of vertebrates and invertebrates (Morse and Bass 1999; Rosenthal 2015; Behm and Ohman 2016; Deffit and Hundley 2016; Nishikura 2016) and focused the field on finding and characterizing ADAR editing in codons. The nonselective editing observed in early studies was put aside by most.

Sanger sequencing of cloned cDNAs was instrumental in the identification of the first ADAR editing sites. However, this method required sequencing cDNAs on a case-by-case basis and was not amenable to the systematic discovery of ADAR editing sites. Toward this goal, a method for inosine-specific cleavage (Morse and Bass 1997) was developed and, when coupled with differential display, allowed the first unbiased identification of A-to-I edited RNAs in *C. elegans* (Morse and Bass 1999) and human brain (Morse et al. 2002). These efforts were facilitated by large-scale genome sequencing projects (Berks et al. 1995; *C. elegans* Sequencing Consortium 1998; Lander et al. 2001; Venter et al. 2001) that allowed identified transcripts to be cross-referenced with genomic sequences and specific genes. At the time, the field assumed that the primary role of ADARs was to edit codons, and it was perplexing that this "systematic" method identified inosine only in noncoding sequences, largely introns and 3′ UTRs, in which base-pairing often occurred between repetitive elements. This hinted that ADAR editing in codons, or mRNA recoding, is a rare event, something now known to be true for studied organisms, excepting some coleoid cephalopods (Liscovitch-Brauer et al. 2017). Although inosine-specific cleavage provided a systematic way to find editing sites, the method required isolation of differentially cleaved bands from a gel and was limited compared with the high-throughput protocols used today. Similar limitations applied to other methods developed to identify ADAR editing sites without Sanger sequencing. These included approaches to identify recoding sites by immunoprecipitating ADAR2 and detecting enriched transcripts by microarray analysis (Ohlson et al. 2005) and methods using high-resolution melt analysis, denaturing HPLC, or allele-specific polymerase chain reaction (PCR) to detect and measure editing in cDNA amplicons (Gallo et al. 2002; Chateigner-Boutin and Small 2007; Chen et al. 2008).

The first attempt to identify inosine-containing transcripts focused on *C. elegans*, because at the time it had the most complete genome sequence (*C. elegans* Sequencing Consortium 1998). As sequencing of other genomes progressed (Adams et al. 2000; Lander et al. 2001; Venter et al. 2001; Mouse Genome Sequencing Consortium et al. 2002), comparative analyses became feasible. Phylogenetic comparisons of glutamate receptor pre-mRNAs from six vertebrates revealed surprising conservation of exonic and intronic sequences around the R/G recoding site, which results in the conversion of an arginine (R) to a glycine (G) (Aruscavage and Bass 2000). Similar regions of strong conservation between different *Drosophila* species allowed identification of 16 genes edited in coding sequences (Hoopengardner et al. 2003), and comparative analyses in mammals identified a handful of previously unrecognized editing sites (Clutterbuck et al. 2005; Levanon et al. 2005).

Around this time, large numbers of cDNAs and cDNA fragments, many as expressed sequence tags (ESTs), were sequenced and mapped to genomic DNA sequences (Boguski et al. 1993; Kikuno et al. 2002; Ota et al. 2004). Several research groups recognized that cDNA and EST libraries, when compared with genomic sequences, could be used to identify editing-dependent RNA–DNA differences (Athanasiadis et al. 2004; Blow et al. 2004; Kim et al. 2004; Levanon et al. 2004) and found tens of thousands of unique editing sites in human transcripts. These studies indicated that A-to-G conversions, the most abundant class of RNA–DNA mismatches, were abundant in repetitive sequences, especially *Alu* retroelements, and displayed nearest neighbor preferences similar to those observed for ADAR in vitro (Polson and Bass 1994; Lehmann and Bass 2000). Approaches to align and compare libraries of cDNA sequences proved powerful, but ultimately were limited by the number of cDNAs and ESTs available. As described below, using modern sequencing technology, one can sequence millions of cDNAs per sample and detect editing using rapid and robust pipelines.

5 FINDING EDITED RNAs IN HIGH-THROUGHPUT DATA SETS

Although early cDNA and EST sequencing projects compiled tens to hundreds of thousands of cellular sequences, high-throughput technologies like RNA-seq rapidly produce tens to hundreds of millions of sequences per sample (Wang et al. 2009). RNA-seq facilitates genome-wide profiling of RNA expression and editing patterns and can easily be applied to different organisms, cell types, and conditions, for comparative analyses. Several previous reviews provide useful information for detecting edited RNAs with high-throughput protocols (Eisenberg et al. 2010; Ramaswami and Li 2016).

Like Sanger sequencing and cDNA/EST approaches, RNA-seq is used to identify editing sites by searching for A-to-G differences between RNA-derived cDNA libraries and matched genomic sequences. First and foremost, finding RNA editing sites requires that edited RNAs are well represented among the transcripts sequenced in an experiment. Samples dominated by abundant transcripts like ribosomal RNA (rRNA) will not have sufficient sequencing coverage of edited transcripts to reliably distinguish editing sites from sequencing errors and genomic variants. Equally important is the approach used to align sequencing reads to the genome. By definition, edited reads will not perfectly align to the genome, and mismatch parameters for alignment and read filtering must be carefully chosen to avoid discarding edited transcripts (Lee et al. 2013).

Editing sites that occur at low frequency are particularly problematic. A site naturally edited in only 10% of cellular transcripts on average will appear as G in only one of ten reads. Increasing the number of sequencing reads will increase the average number of times a particular nucleotide is represented, or the depth of coverage (Sims et al. 2014). Given a site edited at frequency f, one can calculate the number of reads, n, that provide probability P of observing editing at that site using the formula

$$n = \log_{1-f}(1 - P).$$

For instance, one needs approximately 18 reads to have a 90% chance of observing editing at a site edited in 12% of endogenous transcripts, whereas coverage of approximately five reads only provides an ~50% chance. In practice, costs of high-throughput sequencing are balanced with coverage demands. Regardless, coverage of edited transcripts will increase by removing abundant unedited transcripts or by biochemically enriching for edited RNAs, without significant increase in costs. Thus, this is standard protocol in analyses of edited transcripts.

5.1 Enriching for Edited Transcripts

High-abundance transcripts, especially rRNAs, present the major obstacle to obtaining deep coverage of edited RNAs. Because rRNA makes up 80%–90% of most total RNA samples (O'Neil et al. 2013), its inclusion in sequencing libraries substantially reduces the fraction of RNA-seq reads that contain editing information. Typically, rRNA is removed from an RNA sample before library preparation, either by selecting polyadenylated RNAs with oligo(dT) or depleting rRNAs with bead-conjugated antisense oligos (ribo-minus) (O'Neil et al. 2013; Zhao et al. 2014). Although oligo(dT) capture removes rRNA, it also depletes transcripts that lack a poly(A) tail, including nascent pre-mRNAs that have not yet acquired a poly(A) tail, spliced intron lariats, and abundant polymerase III transcripts like tRNAs. Ribo-minus protocols, although they remove rRNA less efficiently than oligo(dT) capture, do not deplete intronic and intergenic sequences and other abundant noncoding RNAs (Cui et al. 2010). If the goal is to look for editing sites in mature mRNAs, poly(A) selection is an excellent method. However, if the goal is a comprehensive determination of editing sites, including those that frequently occur in introns (Table 1) (Whipple et al. 2015; Zhao et al. 2015; Blango and Bass 2016), ribo-minus treatment to remove rRNA is preferable.

In addition to increasing the complexity of RNA samples by removing abundant, unedited transcripts, biochemical approaches can be used to enrich for ADAR substrates in RNA-seq libraries. An immunoprecipitation approach to enrich for dsRNA with the J2 anti-dsRNA antibody, raised against L-dsRNA, from a "killer" virus of *Saccharomyces cerevisiae* (Schönborn et al. 1991), has been used in studies of human Dicer (Kaneko et al. 2011), TDP-1, the *C. elegans* ortholog of TDP-43 (Saldi et al. 2014), and in RNA-seq approaches to map dsRNAs using ADAR editing sites (Whipple et al. 2015; Blango and Bass 2016; Reich et al. 2018). Although this immunoprecipitation approach can improve coverage of dsRNAs, its reported efficiency is modest. In *C. elegans*, edited sequences were enriched about twofold (Reich et al. 2018), and transposon sequences, which often form inverted repeat structures, were enriched four- or fivefold (Saldi et al. 2014). When applied to mouse RNAs, J2 immunoprecipitation improved the number of EERs detected by at most 1.8-fold, suggesting it may have limited benefit in certain systems (Blango and Bass 2016). Other research groups have immunoprecipitated ADARs to isolate bound RNAs and compare substrate specificity of different ADAR isoforms and paralogs (Wang et al. 2013). Although not all ADAR-bound RNAs were edited, these groups observed edited RNAs enriched among ADAR-bound transcripts and in one case identified thousands of novel editing sites (Wang et al. 2013). Still, immunoprecipitation methods run risks. For example, it is unknown whether hyperedited transcripts with numerous IU mismatches are efficiently immunoprecipitated with the J2 antibody. Nonetheless, although not essential to define edited sequences, enrichment protocols can improve coverage of edited transcripts.

Sequencing RNA from genetic mutants that either accumulate dsRNA or lack proteins that compete with ADARs for binding dsRNA can also increase coverage of edited sequences and/or elevate editing levels. Several *C. elegans* studies used mutant strains, including *dcr-1(mg375)*, which contains a mutation in the helicase domain of *C. elegans* Dicer that results in deficient endo-siRNA processing (Whipple et al. 2015), and *tdp-1(ok803)*, which accumulates excess cellular dsRNA (Saldi et al. 2014). Indeed, the *tdp-1* mutant showed increased editing in ~40% of well-represented edited RNAs. In mammalian cells, reduction in N^6-methyladenosine RNA modification by *METTL3* or *METTL14* knockdown likewise results in elevated A-to-I editing at many sites (Xiang et al. 2018). Performing RNA-seq on mutants with elevated editing will enable determination of a more comprehensive dsRNAome. However, if the goal is to determine biologically relevant editing sites, it is important to remember that such genetic perturbations will identify editing sites that may not exist in the wild-type context.

A-to-I editing only occurs in dsRNA structures, which may impede efficient reverse transcription. The use of thermostable reverse transcriptases at high temperatures, or addition of organic solvents like dimethylsulfoxide (DMSO), can relax secondary structure to promote cDNA synthesis from long RNA duplexes (Yasukawa et al. 2010; Mohr et al. 2013; Whipple et al. 2015; Nottingham et al. 2016). However, these adaptations are usually unnecessary, because RNA fragmentation during library preparation reduces the length, and thus the stability, of base-paired regions, and further, dsRNAs containing many IU mismatches are easier to reverse transcribe with conventional enzymes (Whipple et al. 2015). Importantly, library preparation protocols should incorporate high-fidelity enzymes to maintain low error rates during cDNA synthesis (Lee et al. 2013).

5.2 Sequencing and Alignment Protocols

Most RNA editing studies use Illumina sequencing platforms because of their ability to sequence relatively long reads at deep coverage, providing around 200 million reads per lane (Diroma et al. 2017). The high sequence information content of long reads (>100 nt) enables accurate mapping to a reference genome, especially across splice junctions and in repetitive regions. Paired-end protocols effectively double the information per read by sequencing from both 5′ and 3′ ends (Chhangawala et al. 2015). Stranded library construction protocols further refine mapping information and distinguish between A-to-G and T-to-C conversions (Mills et al. 2013). We typically perform stranded, paired-end RNA-seq with >100-nt reads for editing detection studies, because paired-end reads with ~200-nt total sequence rarely align to more than one genomic location, even if they include repetitive sequences.

Early studies used cloned cDNAs to determine the number of editing sites per transcript and defined transcripts with A-to-G changes at >10% of adenosines as nonselectively edited or hyperedited (Bass 1997). Although a number of viral transcripts were found to be hyperedited, with a few exceptions (Morse and Bass 1999; Morse et al.

2002), endogenous transcripts showed editing at <10% of the adenosines in a single base-paired region. Information about the number of editing sites in a single transcript is often limited when using Illumina RNA-seq protocols, because individual reads are comparatively short. Compilation of all reads for a given transcript indicates the number of potential editing sites, but not the number of sites that can occur in a single transcript. High-throughput protocols suited for sequencing longer reads, such as the 300-nt paired-end reads provided by Illumina MiSeq platforms, have been used to determine how many editing events occur in a single transcript (Wheeler et al. 2015). Such studies provide information about how editing at one position influences editing at other sites.

After library construction and sequencing, quality control software such as FastQC (see bioinformatics.babraham.ac.uk/projects/fastqc/) assess base quality and sequence complexity in the resulting RNA-seq reads. Preprocessing programs like Cutadapt (Martin 2011) trim reads to remove adapter sequences, poly(A) tails, and low-quality bases, after which reads are aligned to a reference genome. Alignment algorithms differ in run time and accuracy (Ruffalo et al. 2011; Borozan et al. 2013). Many aligners detect exon–intron junctions to map spliced RNA-seq reads, although aligners without this feature map spliced reads effectively if provided a table of splice junctions (Borozan et al. 2013; Diroma et al. 2017).

Two ways are commonly used to align reads so as to avoid discarding those with ADAR editing sites. The first uses standard alignment algorithms, but increases the number of mismatches allowed in a single read, whereas the second uses "editing aware" alignment algorithms to align cDNA sequences with either an A or G to a genomic A. Using a standard aligner, one must allow for mismatched bases to map edited reads, because the edited sequence varies from the genomic sequence. The optimal number of mismatches allowed depends on read length, because a 100-nt read with four possible mismatches will generally align to fewer locations than a 50-nt read with the same number of mismatches. Because most *C. elegans* edited RNAs have four or fewer editing events per molecule (Wheeler et al. 2015), we permit four mismatches when using a standard aligner to map 100-nt paired-end reads from *C. elegans*. For other organisms, different mismatch parameters should be tested to find those that accurately map edited reads while minimizing mapping to multiple locations.

The second approach to mapping ADAR-edited sequences relies on alignment programs like GSNAP and GNUMAP that include "editing aware" modes. These aligners are particularly adept at mapping highly edited sequences (Clement et al. 2010; Wu and Nacu 2010; Hong et al. 2013). In one comparison, GNUMAP identified more editing sites and editing clusters than the "editing unaware" aligner, Novoalign, which aligned 100-nt reads with up to four mismatches (Whipple et al. 2015). Applying a similar "editing aware" principle, several research groups used a "three-base" approach, so-called because it takes highly edited reads unmapped by conventional alignment and realigns them after converting all As to Gs in read and reference genome sequences, thus restricting the genome to three bases: G, C, and T (Wu et al. 2011; St Laurent et al. 2013; Porath et al. 2014; Zhao et al. 2015). Three-base methods involve additional alignment and computational steps and risk mapping reads that contain G-to-A mismatches (Porath et al. 2014). Thus, editing-aware algorithms (e.g., GNUMAP) that accurately map highly edited sequences in a single alignment step are usually preferable.

5.3 Tools and Approaches to Identify Editing Sites

A growing number of bioinformatics tools and pipelines are publicly available for detecting ADAR editing sites (Porath et al. 2014; Whipple et al. 2015; Blango and Bass 2016; Deffit and Hundley 2016; Diroma et al. 2017). Editing detection pipelines use variant calling to identify A-to-G conversions in RNA-seq reads, and then assess if A-to-G changes represent true editing events. Assessment approaches aim to distinguish editing sites from genetic polymorphisms and sequencing errors, typically through statistical analyses or filtering based on location and frequency of editing (Porath et al. 2014; Deffit and Hundley 2016; Wang et al. 2016; Diroma et al. 2017; John et al. 2017). Our laboratory uses applications in the USeq (see github.com/HuntsmanCancerInstitute/USeq) and SAMtools (see github.com/samtools/) sequencing analysis suites to identify edited sites and define editing clusters (Whipple et al. 2015; Blango and Bass 2016; Reich et al. 2018). In addition to variant-calling (SAMtools mpileup) and editing-detection functions (USeq RNAEditingPileupParser), these analysis suites provide useful, intuitive programs to work with sequencing data sets and compare editing sites to other genomic features. Other pipelines are specifically for detecting RNA editing, and a recent review evaluated five of these for accuracy and sensitivity (Diroma et al. 2017). Results varied based on alignment algorithm used, but overall the RNAEditor, JACUSA, and REDItools pipelines predicted the most editing sites, whereas GIREMI predicted fewer sites but with a lower false discovery rate. All-inclusive programs like RNAEditor and RES-Scanner are useful for those with little experience to quickly and easily identify editing sites, as they incorporate alignment, detection, and filtering steps to define editing sites from raw sequencing files (Wang et al. 2016; John et al. 2017). However, for determination of dsRNAomes, we prefer the EER pipeline (described below), which defines editing clusters to identify long dsRNAs (Whipple et al. 2015; Blango and Bass 2016; Reich et al. 2018).

In addition to RNA-seq-based approaches that identify editing through A-to-G changes, high-throughput sequencing can be applied to inosine-specific detection methods. Inosine-specific cleavage with RNase T1, an early tool for editing detection (Morse and Bass 1999), was paired with RNA-seq to identify 665 editing sites in mouse brain, many of which were novel (Cattenoz et al. 2013). Inosine reactivity with acrylonitrile underlies another chemical detection method, termed inosine chemical erasing (ICE) (Sakurai et al. 2010). High-throughput sequencing improves the scale and sensitivity of these methods (Cattenoz et al. 2013; Sakurai et al. 2014; Suzuki et al. 2015), which distinguish inosine from A-to-G changes caused by genetic polymorphisms or sequencing errors. However, these methods rely on chemical conversion methods that are not 100% efficient; the recommended conditions for ICE convert only 80%–90% of inosines at the highly edited *GRIA2* Q/R site (Suzuki et al. 2015). Further, because inosine-specific detection methods require complex analysis pipelines and suffer from the same read coverage issues as conventional editing detection pipelines (Cattenoz et al. 2013; Sakurai et al. 2014; Suzuki et al. 2015), they do not provide a strong advantage over other approaches.

5.4 Excluding False Positives

Effective editing pipelines must distinguish true RNA editing events from editing-independent mismatches that arise because of single-nucleotide polymorphisms (SNPs) and errors in sequencing and alignment. Controlling for SNPs requires accurately identifying genetic variants and removing them from analysis. Sequencing genomic DNA from the same samples used for RNA-seq provides a definitive solution (Picardi and Pesole 2013; Wang et al. 2016) by providing a more accurate genomic sequence than a published reference. Alternatively, RNA-seq analyses of ADAR mutant strains can validate that identified A-to-G conversions require ADAR (Bahn et al. 2012; Zhao et al. 2015). Although these approaches effectively control for genomic variation, they are expensive and resource-intensive, and SNPs typically make up a tiny fraction of A-to-G changes identified (<0.1% in Zhao et al. 2015). A cheap alternative is to simply remove SNPs recorded in public variant databases from potential editing sites (Ramaswami et al. 2013; Bazak et al. 2014; Whipple et al. 2015; Blango and Bass 2016; Diroma et al. 2017; Tan et al. 2017). Several research groups have shown that true editing sites can be accurately identified using only RNA-seq data, without the need for sample-matched genomic sequence (Ramaswami et al. 2013; Zhu et al. 2013; Zhang and Xiao 2015). However, without an available reference sequence to align reads, DNA-seq is important to determine RNA–DNA differences (Alon et al. 2015).

Filtering out sequencing errors is critical for accurate editing detection. Higher Illumina error rates at read ends, and mismatches introduced by random hexamer priming, lead to clustering of sequencing errors at the ends of reads (Minoche et al. 2011; van Gurp et al. 2013). Thus, 5′ and 3′ read ends are typically trimmed before variant calling either by a fixed amount or according to mismatch density (Bazak et al. 2014; Whipple et al. 2015; Zhao et al. 2015; Wang et al. 2016). Reads with multiple non-A-to-G mismatches are also typically filtered out, because additional mismatches are an indication of poor sequence quality or inaccurate mapping (Blango and Bass 2016; Deffit et al. 2017). Repetitive genomic regions, in particular, risk issues of mismapping (Eisenberg 2012). Although reads that map to multiple repetitive regions can be removed, editing is most abundant in repetitive sequences (Bazak et al. 2014; Whipple et al. 2015; Blango and Bass 2016; Porath et al. 2017). Filtering out repetitively mapped reads removes sequences carrying true editing sites. Sequencing protocols that provide longer reads facilitate more accurate mapping, including within repetitive sequences, so these protocols reduce the number of reads needed to be filtered due to ambiguous alignment. Even with long reads, we recommend including repetitively mapped reads, as they are rich sources of editing information.

Several approaches can be used to validate that alignment and detection parameters are appropriate for editing discovery. The most straightforward is to choose candidate edited regions for editing validation by another method, typically cDNA amplification and Sanger sequencing (Li et al. 2009). This approach is accurate and sensitive, but it can be time-consuming to test many candidates. We recommend initially interrogating data sets for the presence of known ADAR targets. By comparing known editing sites to experimentally determined patterns, one can identify and address potential issues in bioinformatic pipelines. For instance, the absence of editing within well-expressed ADAR substrates might indicate that edited reads are being discarded during alignment or filtering. Publicly available databases, including DARNED, RADAR, and REDIportal, provide curated information on RNA editing sites, incorporating annotation, editing level, and tissue-specific editing information (Kiran et al. 2013; Ramaswami and Li 2014; Picardi et al. 2017). However, these databases do not always include up-to-date information; whereas DARNED and RADAR only provide data for human, mouse, and *Drosophila*, REDIportal currently includes only human infor-

Cite as *Cold Spring Harb Perspect Biol* doi: 10.1101/cshperspect.a035352

mation (Kiran et al. 2013; Ramaswami and Li 2014; Picardi et al. 2017). Without curated data, we have used published editing targets (Morse et al. 2002; Hellwig and Bass 2008) to validate parameters for detecting editing in *C. elegans* data sets (Whipple et al. 2015). Once the pipeline is considered optimal, it is standard practice to verify editing sites in a subset of newly identified targets using cDNA amplification and Sanger sequencing, especially those chosen for more in-depth study.

5.5 Defining Editing Clusters and Determining a dsRNAome

In highly base-paired dsRNAs, ADARs nonselectively deaminate many adenosines to inosine (Nishikura 2010; Samuel 2011; Deffit and Hundley 2016), resulting in clusters of A-to-G changes in sequencing reads. Defining editing clusters provides increased sensitivity to identify true editing events, because sequencing errors or genomic variants are unlikely to result in clusters of a single variant (A-to-G) (Bazak et al. 2014; Zhao et al. 2015). Importantly, editing clusters also indicate the presence of long, highly base-paired dsRNA (Fig. 1) and, when mapped on a genome-wide scale, allow the determination of a dsRNAome (Fig. 2). We have determined dsRNAomes for specific cell types of human and mouse, as well as *C. elegans* (Whipple et al. 2015; Blango and Bass 2016).

Our EER-detection pipeline uses a window-scanning approach to find genomic regions covered by RNA-seq reads carrying clustered A-to-G changes (Whipple et al. 2015; Blango and Bass 2016; Reich et al. 2018). Genomic regions with sufficient read coverage (≥5 reads) are scanned in small windows, typically 50 nt, to identify windows containing ≥3 sites with A-to-G changes in >1% of reads. We chose read coverage and editing thresholds to optimize the sensitivity of EER detection without markedly increasing the false discovery rate. Alternative parameters could be used to predict greater numbers of clusters, or to identify clusters with extremely low false-positive rates.

Once edited windows are determined, overlapping windows are merged, and then combined with other merged windows separated by a predetermined "gap" distance to define the EER. The gap parameter connects complementary regions of an intramolecular dsRNA that may be separated by unedited, intervening sequences. The optimal gap distance varies between organisms. For *C. elegans*, a gap of 1 kb merges complementary sequences of a single dsRNA without also combining dsRNAs from separate, independent transcripts (Whipple et al. 2015), whereas mouse and human dsRNAomes require a longer 2.5-kb gap (Blango and Bass 2016). We typically determine optimal gap distance by testing several lengths and manually surveying several dozen EERs in a genome browser to determine if single structures are merged without also merging separate transcripts. Including a gap parameter occasionally causes closely juxtaposed independent structures (as in *eIF-2α* pre-mRNA in Fig. 1) to be merged into one EER; however, it is rare that structures from different genes are merged (Whipple et al. 2015; Blango and Bass 2016).

Once EERs are defined, additional methods are used to validate that dsRNAs have been accurately determined (Whipple et al. 2015; Blango and Bass 2016). RNA secondary structure prediction algorithms like UNAFold (Markham and Zuker 2008) provide a measure of the thermodynamic stability of predicted EER structures. Length-matched random regions provide a control to confirm that predicted dsRNAs are more structured than expected by chance. Length-matched control sets should sample properties of the transcriptome that meet the same criteria used to define dsRNAs. For instance, if the pipeline dictates that EERs are defined only in regions covered by more than five reads, control regions should be restricted to the same read coverage threshold. We use the BEDtools2 (see github.com/arq5x/bedtools2) application shuffleBed to make length-matched control regions, because it can randomly permute regions across the genome or within specified areas. For edited sequences that do not have predicted duplex structure, we use the BLAT sequence alignment tool to find complementary genomic sequences that may facilitate intermolecular dsRNA formation (Whipple et al. 2015). Similar approaches to detect proximal (within 10 kb) complementary sequences with the *bl2seq* algorithm have been used in other editing cluster analyses (Porath et al. 2014; 2017). Examining nearest neighbor preferences of EER editing sites validates that sites show characteristics of ADAR editing (Whipple et al. 2015; Blango and Bass 2016).

6 LOOKING TOWARD THE FUTURE

6.1 Determining Other dsRNAomes

To date, ADAR editing sites have been determined for approximately 23 metazoan species (Liscovitch-Brauer et al. 2017; Porath et al. 2017) and 53 human tissues (Tan et al. 2017). In all cases, editing shows similar properties (Table 1). It predominates in noncoding sequences, particularly introns and 3′ UTRs (Whipple et al. 2015; Blango and Bass 2016; Liscovitch-Brauer et al. 2017; Porath et al. 2017), and frequently occurs in mobile element–derived repetitive sequences predicted to form stable intramolecular structures (Fig. 3) (Bazak et al. 2014; Whipple et al. 2015; Blango and Bass 2016; Porath et al. 2017).

In most cases, editing information from these studies has not been used to map structures and determine a dsRNAome. Such analyses would be straightforward because existing data can simply be mined and analyzed with available pipelines. EERs and editing sites can be viewed in a genome browser, allowing researchers to evaluate their favorite genes. BED files of existing dsRNAomes (Whipple et al. 2015; Blango and Bass 2016; Reich et al. 2018) are freely available in the Gene Expression Omnibus repository.

An unanswered question is whether all long dsRNAs are edited. Most dsRNAomes were determined using ADAR editing sites as an in vivo signature of dsRNA, so dsRNAs without editing sites were not identified. Possibly, certain dsRNAs are protected from ADAR editing by other dsRBPs, and indeed, there are clear examples of dsRBPs competing for the same substrates (Warf et al. 2012; Elbarbary et al. 2013; Sakurai et al. 2017). In a few cases, cellular dsRNAs were identified by other criteria, including accessibility to nuclease (Li et al. 2012), chemical probes (Lucks et al. 2011), or in vivo cross-linking to dsRBPs (Rybak-Wolf et al. 2014). About 18% of *C. elegans* EERs (Whipple et al. 2015) overlap 9972 ssRNA nuclease-resistant sites (Li et al. 2012), validating the double-stranded character of EERs. Nonoverlapping sites could include dsRNA that is protected from ADAR editing, but the latter study was not focused on long dsRNAs, and many nonoverlapping sites are short dsRNA regions buried in tertiary structure that are inaccessible to ADARs. About 26% of human EER-associated genes overlap DICER-bound human genes (Rybak-Wolf et al. 2014; Blango and Bass 2016), consistent with the idea that dsRBPs compete for dsRNA structures. These two studies were performed in different cell types, so again, further analyses are required to evaluate whether nonoverlapping sites are dsRNAs that are protected from editing. In future studies, it will be interesting to obtain definitive information about competition between dsRBPs and determine if competition is affected by dsRBP tissue-specificity, abundance, or intracellular localization.

6.2 What Is the Function of the dsRNAome?

The dsRNAomes determined so far indicate that EERs do not show a predilection for the type of gene they inhabit; they are typically enriched in genes expected to be expressed in the particular cell type or condition being studied (Blango and Bass 2016). Is the dsRNAome just a vestige of an ongoing battle with mobile elements? Silencing mobile element expression is essential for viability (Friedli and Trono 2015), and studies in *C. elegans* (Reich et al. 2018) and *D. melanogaster* (Savva et al. 2013) are consistent with the idea that ADARs exist to allow expression of mRNAs that contain repetitive elements and otherwise would be silenced.

Selfish elements that integrate into a genome over time can become recognized as "self," so the above idea is closely tied with studies indicating that ADARs function to mark cellular dsRNA as self (Hartner et al. 2009; Mannion et al. 2014; Liddicoat et al. 2015; Pestal et al. 2015; George et al. 2016; Reich et al. 2018). The simple idea is that ADARs, which are typically in the nucleus, target cellular dsRNA, whereas viruses, which often replicate in the cytoplasm, are protected from deamination. However, the actual situation is clearly more complicated. Some viruses replicate in the nucleus, such as HDV, which uses editing for regulating its life cycle (Polson et al. 1996). The ADAR1 isoform most closely associated with an immune response, ADAR1p150, is expressed in response to interferon and is found in both the nucleus and cytoplasm (Patterson and Samuel 1995). Mounting evidence indicates additional regulation during viral infection. ADAR1p110 is degraded in response to interferon (Li et al. 2016), whereas the timing of interferon induction of ADAR1p150 is balanced with induction of MDA5 and other proteins that mediate the mammalian immune response (Ahmad et al. 2018). As discussed earlier, some viral transcripts are edited by ADARs (Pfaller et al. 2015), and ADARs can have both proviral and antiviral effects (Samuel 2011). Repurposing ADARs as antiviral factors emphasizes their roles in innate immunity. Indeed, *ADAR1* shows evidence of positive selection, suggesting it has adapted through genetic conflict with viruses (Forni et al. 2015).

ADARs likely arose, and continue to function, for these global roles, but have also been co-opted for additional purposes, such as mRNA recoding (Higuchi et al. 2000; Jepson and Reenan 2009; Garrett and Rosenthal 2012; Liscovitch-Brauer et al. 2017), creating or destroying splice sites or altering splicing regulatory sequences (Rueter et al. 1999; Solomon et al. 2013). A key issue in future studies will be to definitively connect observed phenotypes with specific editing events. In *C. elegans* lacking ADARs, most unedited dsRNAs, rather than a small subset, are processed into siRNAs by antiviral RNAi machinery (Reich et al. 2018). Similarly, in ADAR1-deficient human cells, MDA5 oligomerizes on hundreds of inverted-repeat *Alu* elements and triggers downstream immune signaling (Ahmad et al. 2018). However, in *C. elegans* and mammals, it is unclear if all, or only some, dsRNAs are relevant to mutant phenotypes. The early lethality of $ADAR2^{-/-}$ mutant mice and chemotaxis defects of *adr-1;adr-2* mutant *C. elegans* are largely ascribed to editing of single targets (Higuchi et al. 2000; Deffit et al. 2017). Understanding if the same applies to immune-relevant ADAR-dependent pathologies could allow potential treatments to specifically target relevant substrates without disrupting global ADAR functions.

Cite as *Cold Spring Harb Perspect Biol* doi: 10.1101/cshperspect.a035352

REFERENCES

Adams MD, Celniker SE, Holt RA, Evans CA, Gocayne JD, Amanatides PG, Scherer SE, Li PW, Hoskins RA, Galle RF, et al. 2000. The genome sequence of *Drosophila melanogaster*. *Science* **287:** 2185–2195.

Ahmad S, Mu X, Yang F, Greenwald E, Park JW, Jacob E, Zhang C-Z, Hur S. 2018. Breaching self-tolerance to Alu duplex RNA underlies MDA5-mediated inflammation. *Cell* **172:** 797–810.e13.

Ahringer J, Gasser SM. 2018. Repressive chromatin in *Caenorhabditis elegans*: Establishment, composition, and function. *Genetics* **208:** 491–511.

Alon S, Garrett SC, Levanon EY, Olson S, Graveley BR, Rosenthal JJC, Eisenberg E. 2015. The majority of transcripts in the squid nervous system are extensively recoded by A-to-I RNA editing. *eLife* **4:** e05198.

Aruscavage PJ, Bass BL. 2000. A phylogenetic analysis reveals an unusual sequence conservation within introns involved in RNA editing. *RNA* **6:** 257–269.

Athanasiadis A, Rich A, Maas S. 2004. Widespread A-to-I RNA editing of Alu-containing mRNAs in the human transcriptome. *PLoS Biol* **2:** e391.

Bahn JH, Lee J-H, Li G, Greer C, Peng G, Xiao X. 2012. Accurate identification of A-to-I RNA editing in human by transcriptome sequencing. *Genome Res* **22:** 142–150.

Bajad P, Jantsch MF, Keegan L, O'Connell M. 2017. A to I editing in disease is not fake news. *RNA Biol* **14:** 1223–1231.

Bass BL. 1997. RNA editing and hypermutation by adenosine deamination. *Trends Biochem Sci* **22:** 157–162.

Bass BL. 2002. RNA editing by adenosine deaminases that act on RNA. *Annu Rev Biochem* **71:** 817–846.

Bass BL, Weintraub H. 1987. A developmentally regulated activity that unwinds RNA duplexes. *Cell* **48:** 607–613.

Bass BL, Weintraub H. 1988. An unwinding activity that covalently modifies its double-stranded RNA substrate. *Cell* **55:** 1089–1098.

Bass BL, Weintraub H, Cattaneo R, Billeter MA. 1989. Biased hypermutation of viral RNA genomes could be due to unwinding/modification of double-stranded RNA. *Cell* **56:** 331.

Bass BL, Nishikura K, Keller W, Seeburg PH, Emeson RB, O'Connell MA, Samuel CE, Herbert A. 1997. A standardized nomenclature for adenosine deaminases that act on RNA. *RNA* **3:** 947–949.

Bazak L, Haviv A, Barak M, Jacob-Hirsch J, Deng P, Zhang R, Isaacs FJ, Rechavi G, Li JB, Eisenberg E, et al. 2014. A-to-I RNA editing occurs at over a hundred million genomic sites, located in a majority of human genes. *Genome Res* **24:** 365–376.

Behm M, Ohman M. 2016. RNA editing: A contributor to neuronal dynamics in the mammalian brain. *Trends Genet* **32:** 165–175.

Berks M, Mapping CEG, Consortium S. 1995. The *C. elegans* genome sequencing project. *Genome Res* **5:** 99–104.

Blango MG, Bass BL. 2016. Identification of the long, edited dsRNAome of LPS-stimulated immune cells. *Genome Res* **26:** 852–862.

Blow M, Futreal PA, Wooster R, Stratton MR. 2004. A survey of RNA editing in human brain. *Genome Res* **14:** 2379–2387.

Boguski MS, Lowe TM, Tolstoshev CM. 1993. dbEST—Database for "expressed sequence tags". *Nat Genet* **4:** 332–333.

Borozan I, Watt SN, Ferretti V. 2013. Evaluation of alignment algorithms for discovery and identification of pathogens using RNA-Seq. *PLoS ONE* **8:** e76935.

C. elegans Sequencing Consortium. 1998. Genome sequence of the nematode *C. elegans*: A platform for investigating biology. *Science* **282:** 2012–2018.

Cattaneo R, Schmid A, Eschle D, Baczko K, Meulen ter V, Billeter MA. 1988. Biased hypermutation and other genetic changes in defective measles viruses in human brain infections. *Cell* **55:** 255–265.

Cattenoz PB, Taft RJ, Westhof E, Mattick JS. 2013. Transcriptome-wide identification of A > I RNA editing sites by inosine specific cleavage. *RNA* **19:** 257–270.

Chateigner-Boutin A-L, Small I. 2007. A rapid high-throughput method for the detection and quantification of RNA editing based on high-resolution melting of amplicons. *Nucleic Acids Res* **35:** e114.

Chen Y-C, Kao S-C, Chou H-C, Lin W-H, Wong F-H, Chow W-Y. 2008. A real-time PCR method for the quantitative analysis of RNA editing at specific sites. *Anal Biochem* **375:** 46–52.

Chhangawala S, Rudy G, Mason CE, Rosenfeld JA. 2015. The impact of read length on quantification of differentially expressed genes and splice junction detection. *Genome Biol* **16:** 131.

Clement NL, Snell Q, Clement MJ, Hollenhorst PC, Purwar J, Graves BJ, Cairns BR, Johnson WE. 2010. The GNUMAP algorithm: Unbiased probabilistic mapping of oligonucleotides from next-generation sequencing. *Bioinformatics* **26:** 38–45.

Clutterbuck DR, Leroy A, O'Connell MA, Semple CAM. 2005. A bioinformatic screen for novel A-I RNA editing sites reveals recoding editing in BC10. *Bioinformatics* **21:** 2590–2595.

Crowl JT, Gray EE, Pestal K, Volkman HE, Stetson DB. 2017. Intracellular nucleic acid detection in autoimmunity. *Annu Rev Immunol* **35:** 313–336.

Cui P, Lin Q, Ding F, Xin C, Gong W, Zhang L, Geng J, Zhang B, Yu X, Yang J, et al. 2010. A comparison between ribo-minus RNA-sequencing and polyA-selected RNA-sequencing. *Genomics* **96:** 259–265.

Daugaard I, Hansen TB. 2017. Biogenesis and function of Ago-associated RNAs. *Trends Genet* **33:** 208–219.

Deffit SN, Hundley HA. 2016. To edit or not to edit: Regulation of ADAR editing specificity and efficiency. *Wiley Interdiscip Rev RNA* **7:** 113–127.

Deffit SN, Yee BA, Manning AC, Rajendren S, Vadlamani P, Wheeler EC, Domissy A, Washburn MC, Yeo GW, Hundley HA. 2017. The *C. elegans* neural editome reveals an ADAR target mRNA required for proper chemotaxis. *eLife* **6:** e28625.

Diroma MA, Ciaccia L, Pesole G, Picardi E. 2017. Elucidating the editome: Bioinformatics approaches for RNA editing detection. *Brief Bioinform* doi: 10.1093/bib/bbx129.

Egebjerg J, Kukekov V, Heinemann SF. 1994. Intron sequence directs RNA editing of the glutamate receptor subunit GluR2 coding sequence. *Proc Natl Acad Sci* **91:** 10270–10274.

Ehrenfeld E, Hunt T. 1971. Double-stranded poliovirus RNA inhibits initiation of protein synthesis by reticulocyte lysates. *Proc Natl Acad Sci* **68:** 1075–1078.

Eigenbrod T, Keller P, Kaiser S, Rimbach K, Dalpke AH, Helm M. 2015. Recognition of specified RNA modifications by the innate immune system. *Meth Enzymol* **560:** 73–89.

Eisenberg E. 2012. Bioinformatic approaches for identification of A-to-I editing sites. *Curr Top Microbiol Immunol* **353:** 145–162.

Eisenberg E, Li JB, Levanon EY. 2010. Sequence based identification of RNA editing sites. *RNA Biol* **7:** 248–252.

Elbarbary RA, Li W, Tian B, Maquat LE. 2013. STAU1 binding 3′ UTR IR*Alus* complements nuclear retention to protect cells from PKR-mediated translational shutdown. *Genes Dev* **27:** 1495–1510.

Forni D, Mozzi A, Pontremoli C, Vertemara J, Pozzoli U, Biasin M, Bresolin N, Clerici M, Cagliani R, Sironi M. 2015. Diverse selective regimes shape genetic diversity at ADAR genes and at their coding targets. *RNA Biol* **12:** 149–161.

Friedli M, Trono D. 2015. The developmental control of transposable elements and the evolution of higher species. *Annu Rev Cell Dev Biol* **31:** 429–451.

Gallo A, Thomson E, Brindle J, O'Connell MA, Keegan LP. 2002. Micro-processing events in mRNAs identified by DHPLC analysis. *Nucleic Acids Res* **30:** 3945–3953.

Garrett S, Rosenthal JJC. 2012. RNA editing underlies temperature adaptation in K^+ channels from polar octopuses. *Science* **335:** 848–851.

George CX, Ramaswami G, Li JB, Samuel CE. 2016. Editing of cellular self-RNAs by adenosine deaminase ADAR1 suppresses innate immune stress responses. *J Biol Chem* **291:** 6158–6168.

Gilbert W. 1986. The RNA world. *Nature* **319:** 618.

Grimwood J, Gordon LA, Olsen A, Terry A, Schmutz J, Lamerdin J, Hellsten U, Goodstein D, Couronne O, Tran-Gyamfi M, et al. 2004. The DNA sequence and biology of human chromosome 19. *Nature* **428:** 529–535.

Hartner JC, Walkley CR, Lu J, Orkin SH. 2009. ADAR1 is essential for the maintenance of hematopoiesis and suppression of interferon signaling. *Nat Immunol* **10:** 109–115.

Hellwig S, Bass BL. 2008. A starvation-induced noncoding RNA modulates expression of Dicer-regulated genes. *Proc Natl Acad Sci* **105:** 12897–12902.

Higuchi M, Single FN, Köhler M, Sommer B, Sprengel R, Seeburg PH. 1993. RNA editing of AMPA receptor subunit GluR-B: A base-paired intron–exon structure determines position and efficiency. *Cell* **75:** 1361–1370.

Higuchi M, Maas S, Single FN, Hartner J, Rozov A, Burnashev N, Feldmeyer D, Sprengel R, Seeburg PH. 2000. Point mutation in an AMPA receptor gene rescues lethality in mice deficient in the RNA-editing enzyme ADAR2. *Nature* **406:** 78–81.

Ho JWK, Jung YL, Liu T, Alver BH, Lee S, Ikegami K, Sohn K-A, Minoda A, Tolstorukov MY, Appert A, et al. 2014. Comparative analysis of metazoan chromatin organization. *Nature* **512:** 449–452.

Hong C, Clement NL, Clement S, Hammoud SS, Carrell DT, Cairns BR, Snell Q, Clement MJ, Johnson WE. 2013. Probabilistic alignment leads to improved accuracy and read coverage for bisulfite sequencing data. *BMC Bioinformatics* **14:** 337.

Hoopengardner B, Bhalla T, Staber C, Reenan R. 2003. Nervous system targets of RNA editing identified by comparative genomics. *Science* **301:** 832–836.

Hurst SR, Hough RF, Aruscavage PJ, Bass BL. 1995. Deamination of mammalian glutamate receptor RNA by *Xenopus* dsRNA adenosine deaminase: Similarities to in vivo RNA editing. *RNA* **1:** 1051–1060.

Jepson JEC, Reenan RA. 2009. Adenosine-to-inosine genetic recoding is required in the adult stage nervous system for coordinated behavior in *Drosophila*. *J Biol Chem* **284:** 31391–31400.

John D, Weirick T, Dimmeler S, Uchida S. 2017. RNAEditor: Easy detection of RNA editing events and the introduction of editing islands. *Brief Bioinform* **18:** 993–1001.

Joyce GF. 2002. The antiquity of RNA-based evolution. *Nature* **418:** 214–221.

Kaneko H, Dridi S, Tarallo V, Gelfand BD, Fowler BJ, Cho WG, Kleinman ME, Ponicsan SL, Hauswirth WW, Chiodo VA, et al. 2011. DICER1 deficit induces *Alu* RNA toxicity in age-related macular degeneration. *Nature* **471:** 325–330.

Kikuno R, Nagase T, Waki M, Ohara O. 2002. HUGE: A database for human large proteins identified in the Kazusa cDNA sequencing project. *Nucleic Acids Res* **30:** 166–168.

Kim U, Garner TL, Sanford T, Speicher D, Murray JM, Nishikura K. 1994. Purification and characterization of double-stranded RNA adenosine deaminase from bovine nuclear extracts. *J Biol Chem* **269:** 13480–13489.

Kim DDY, Kim TTY, Walsh T, Kobayashi Y, Matise TC, Buyske S, Gabriel A. 2004. Widespread RNA editing of embedded alu elements in the human transcriptome. *Genome Res* **14:** 1719–1725.

Kin T, Ono Y. 2007. Idiographica: A general-purpose web application to build idiograms on-demand for human, mouse and rat. *Bioinformatics* **23:** 2945–2946.

Kiran AM, O'Mahony JJ, Sanjeev K, Baranov PV. 2013. Darned in 2013: Inclusion of model organisms and linking with Wikipedia. *Nucleic Acids Res* **41:** D258–D261.

Lander ES, Linton LM, Birren B, Nusbaum C, Zody MC, Baldwin J, Devon K, Dewar K, Doyle M, FitzHugh W, et al. 2001. Initial sequencing and analysis of the human genome. *Nature* **409:** 860–921.

Lässig C, Matheisl S, Sparrer KMJ, de Oliveira Mann CC, Moldt M, Patel JR, Goldeck M, Hartmann G, García-Sastre A, Hornung V, et al. 2015. ATP hydrolysis by the viral RNA sensor RIG-I prevents unintentional recognition of self-RNA. *eLife* **4:** e10859.

Lee J-H, Ang JK, Xiao X. 2013. Analysis and design of RNA sequencing experiments for identifying RNA editing and other single-nucleotide variants. *RNA* **19:** 725–732.

Lehmann KA, Bass BL. 1999. The importance of internal loops within RNA substrates of ADAR1. *J Mol Biol* **291:** 1–13.

Lehmann KA, Bass BL. 2000. Double-stranded RNA adenosine deaminases ADAR1 and ADAR2 have overlapping specificities. *Biochemistry* **39:** 12875–12884.

Levanon EY, Eisenberg E, Yelin R, Nemzer S, Hallegger M, Shemesh R, Fligelman ZY, Shoshan A, Pollock SR, Sztybel D, et al. 2004. Systematic identification of abundant A-to-I editing sites in the human transcriptome. *Nat Biotechnol* **22:** 1001–1005.

Levanon EY, Hallegger M, Kinar Y, Shemesh R, Djinovic-Carugo K, Rechavi G, Jantsch MF, Eisenberg E. 2005. Evolutionarily conserved human targets of adenosine to inosine RNA editing. *Nucleic Acids Res* **33:** 1162–1168.

Li JB, Levanon EY, Yoon J-K, Aach J, Xie B, Leproust E, Zhang K, Gao Y, Church GM. 2009. Genome-wide identification of human RNA editing sites by parallel DNA capturing and sequencing. *Science* **324:** 1210–1213.

Li F, Zheng Q, Ryvkin P, Dragomir I, Desai Y, Aiyer S, Valladares O, Yang J, Bambina S, Sabin LR, et al. 2012. Global analysis of RNA secondary structure in two metazoans. *Cell Rep* **1:** 69–82.

Li L, Qian G, Zuo Y, Yuan Y, Cheng Q, Guo T, Liu J, Liu C, Zhang L, Zheng H. 2016. Ubiquitin-dependent turnover of adenosine deaminase acting on RNA 1 (ADAR1) is required for efficient antiviral activity of Type I interferon. *J Biol Chem* **291:** 24974–24985.

Liddicoat BJ, Piskol R, Chalk AM, Ramaswami G, Higuchi M, Hartner JC, Li JB, Seeburg PH, Walkley CR. 2015. RNA editing by ADAR1 prevents MDA5 sensing of endogenous dsRNA as nonself. *Science* **349:** 1115–1120.

Liscovitch-Brauer N, Alon S, Porath HT, Elstein B, Unger R, Ziv T, Admon A, Levanon EY, Rosenthal JJC, Eisenberg E. 2017. Trade-off between transcriptome plasticity and genome evolution in cephalopods. *Cell* **169:** 191–202.

Liu T, Rechtsteiner A, Egelhofer TA, Vielle A, Latorre I, Cheung M-S, Ercan S, Ikegami K, Jensen M, Kolasinska-Zwierz P, et al. 2011. Broad chromosomal domains of histone modification patterns in *C. elegans*. *Genome Res* **21:** 227–236.

Lucks JB, Mortimer SA, Trapnell C, Luo S, Aviran S, Schroth GP, Pachter L, Doudna JA, Arkin AP. 2011. Multiplexed RNA structure characterization with selective 2′-hydroxyl acylation analyzed by primer extension sequencing (SHAPE-Seq). *Proc Natl Acad Sci* **108:** 11063–11068.

Ma E, MacRae IJ, Kirsch JF, Doudna JA. 2008. Autoinhibition of human dicer by its internal helicase domain. *J Mol Biol* **380:** 237–243.

Mannion NM, Greenwood SM, Young R, Cox S, Brindle J, Read D, Nellåker C, Vesely C, Ponting CP, McLaughlin PJ, et al. 2014. The RNA-editing enzyme ADAR1 controls innate immune responses to RNA. *Cell Rep* **9:** 1482–1494.

Markham NR, Zuker M. 2008. UNAFold: Software for nucleic acid folding and hybridization. *Methods Mol Biol* **453:** 3–31.

Martin M. 2011. Cutadapt removes adapter sequences from high-throughput sequencing reads. *EMBnetjournal* **17:** 10–12.

Melcher T, Maas S, Herb A, Sprengel R, Seeburg PH, Higuchi M. 1996. A mammalian RNA editing enzyme. *Nature* **379:** 460–464.

Mills JD, Kawahara Y, Janitz M. 2013. Strand-specific RNA-seq provides greater resolution of transcriptome profiling. *Curr Genomics* **14:** 173–181.

Minoche AE, Dohm JC, Himmelbauer H. 2011. Evaluation of genomic high-throughput sequencing data generated on Illumina HiSeq and genome analyzer systems. *Genome Biol* **12:** R112.

Mohr S, Ghanem E, Smith W, Sheeter D, Qin Y, King O, Polioudakis D, Iyer VR, Hunicke-Smith S, Swamy S, et al. 2013. Thermostable group II intron reverse transcriptase fusion proteins and their use in cDNA synthesis and next-generation RNA sequencing. *RNA* **19:** 958–970.

Morse DP, Bass BL. 1997. Detection of inosine in messenger RNA by inosine-specific cleavage. *Biochemistry* **36:** 8429–8434.

Morse DP, Bass BL. 1999. Long RNA hairpins that contain inosine are present in *Caenorhabditis elegans* poly(A)$^+$ RNA. *Proc Natl Acad Sci* **96:** 6048–6053.

Morse DP, Aruscavage PJ, Bass BL. 2002. RNA hairpins in noncoding regions of human brain and *Caenorhabditis elegans* mRNA are edited by adenosine deaminases that act on RNA. *Proc Natl Acad Sci* **99:** 7906–7911.

Mouse Genome Sequencing Consortium, Waterston RH, Lindblad-Toh K, Birney E, Rogers J, Abril JF, Agarwal P, Agarwala R, Ainscough R, Alexandersson M, et al. 2002. Initial sequencing and comparative analysis of the mouse genome. *Nature* **420:** 520–562.

Neeman Y, Levanon EY, Jantsch MF, Eisenberg E. 2006. RNA editing level in the mouse is determined by the genomic repeat repertoire. *RNA* **12:** 1802–1809.

Nishikura K. 2010. Functions and regulation of RNA editing by ADAR deaminases. *Annu Rev Biochem* **79:** 321–349.

Nishikura K. 2016. A-to-I editing of coding and non-coding RNAs by ADARs. *Nat Rev Mol Cell Biol* **17:** 83–96.

Nottingham RM, Wu DC, Qin Y, Yao J, Hunicke-Smith S, Lambowitz AM. 2016. RNA-seq of human reference RNA samples using a thermostable group II intron reverse transcriptase. *RNA* **22:** 597–613.

O'Connell MA, Mannion NM, Keegan LP. 2015. The epitranscriptome and innate immunity. *PLoS Genet* **11:** e1005687.

Ohlson J, Enstërö M, Sjöberg B-M, Ohman M. 2005. A method to find tissue-specific novel sites of selective adenosine deamination. *Nucleic Acids Res* **33:** e167.

Ohman M, Källman AM, Bass BL. 2000. In vitro analysis of the binding of ADAR2 to the pre-mRNA encoding the GluR-B R/G site. *RNA* **6:** 687–697.

O'Neil D, Glowatz H, Schlumpberger M. 2013. Ribosomal RNA depletion for efficient use of RNA-seq capacity. *Curr Protoc Mol Biol* doi: 10.1002/0471142727.mb0419s103.

Ota T, Suzuki Y, Nishikawa T, Otsuki T, Sugiyama T, Irie R, Wakamatsu A, Hayashi K, Sato H, Nagai K, et al. 2004. Complete sequencing and characterization of 21,243 full-length human cDNAs. *Nat Genet* **36:** 40–45.

Parker GS, Maity TS, Bass BL. 2008. dsRNA binding properties of RDE-4 and TRBP reflect their distinct roles in RNAi. *J Mol Biol* **384:** 967–979.

Patterson JB, Samuel CE. 1995. Expression and regulation by interferon of a double-stranded-RNA-specific adenosine deaminase from human cells: Evidence for two forms of the deaminase. *Mol Cell Biol* **15:** 5376–5388.

Paul MS, Bass BL. 1998. Inosine exists in mRNA at tissue-specific levels and is most abundant in brain mRNA. *EMBO J* **17:** 1120–1127.

Pestal K, Funk CC, Snyder JM, Price ND, Treuting PM, Stetson DB. 2015. Isoforms of RNA-editing enzyme ADAR1 independently control nucleic acid sensor MDA5-driven autoimmunity and multi-organ development. *Immunity* **43:** 933–944.

Pfaller CK, Mastorakos GM, Matchett WE, Ma X, Samuel CE, Cattaneo R. 2015. Measles virus defective interfering RNAs are generated frequently and early in the absence of C protein and can be destabilized by adenosine deaminase acting on RNA-1-like hypermutations. *J Virol* **89:** 7735–7747.

Picardi E, Pesole G. 2013. REDItools: High-throughput RNA editing detection made easy. *Bioinformatics* **29:** 1813–1814.

Picardi E, D'Erchia AM, Giudice Lo C, Pesole G. 2017. REDIportal: A comprehensive database of A-to-I RNA editing events in humans. *Nucleic Acids Res* **45:** D750–D757.

Polson AG, Bass BL. 1994. Preferential selection of adenosines for modification by double-stranded RNA adenosine deaminase. *EMBO J* **13:** 5701–5711.

Polson AG, Crain PF, Pomerantz SC, McCloskey JA, Bass BL. 1991. The mechanism of adenosine to inosine conversion by the double-stranded RNA unwinding/modifying activity: A high-performance liquid chromatography-mass spectrometry analysis. *Biochemistry* **30:** 11507–11514.

Polson AG, Bass BL, Casey JL. 1996. RNA editing of hepatitis delta virus antigenome by dsRNA-adenosine deaminase. *Nature* **380:** 454–456.

Porath HT, Carmi S, Levanon EY. 2014. A genome-wide map of hyper-edited RNA reveals numerous new sites. *Nat Commun* **5:** 4726.

Porath HT, Knisbacher BA, Eisenberg E, Levanon EY. 2017. Massive A-to-I RNA editing is common across the metazoa and correlates with dsRNA abundance. *Genome Biol* **18:** 185.

Prachumwat A, DeVincentis L, Palopoli MF. 2004. Intron size correlates positively with recombination rate in *Caenorhabditis elegans*. *Genetics* **166:** 1585–1590.

Ramaswami G, Li JB. 2014. RADAR: A rigorously annotated database of A-to-I RNA editing. *Nucleic Acids Res* **42:** D109–D113.

Ramaswami G, Li JB. 2016. Identification of human RNA editing sites: A historical perspective. *Methods* **107:** 42–47.

Ramaswami G, Zhang R, Piskol R, Keegan LP, Deng P, O'Connell MA, Li JB. 2013. Identifying RNA editing sites using RNA sequencing data alone. *Nat Meth* **10:** 128–132.

Rebagliati MR, Melton DA. 1987. Antisense RNA injections in fertilized frog eggs reveal an RNA duplex unwinding activity. *Cell* **48:** 599–605.

Reich DP, Tyc KM, Bass BL. 2018. *C. elegans* ADARs antagonize silencing of cellular dsRNAs by the antiviral RNAi pathway. *Genes Dev* **32:** 271–282.

Rich A. 1962. On the problems of evolution and biochemical information transfer. In *Horizons in biochemistry* (ed. KashaM, Pullman B), pp. 103–126, Academic, New York.

Rieder LE, Staber CJ, Hoopengardner B, Reenan RA. 2013. Tertiary structural elements determine the extent and specificity of messenger RNA editing. *Nat Commun* **4:** 2232.

Robertson MP, Joyce GF. 2012. The origins of the RNA world. *Cold Spring Harb Perspect Biol* **4:** a003608.

Rosenthal JJC. 2015. The emerging role of RNA editing in plasticity. *J Exp Biol* **218:** 1812–1821.

Rueter SM, Dawson TR, Emeson RB. 1999. Regulation of alternative splicing by RNA editing. *Nature* **399:** 75–80.

Ruffalo M, LaFramboise T, Koyutuerk M. 2011. Comparative analysis of algorithms for next-generation sequencing read alignment. *Bioinformatics* **27:** 2790–2796.

Rybak-Wolf A, Jens M, Murakawa Y, Herzog M, Landthaler M, Rajewsky N. 2014. A variety of dicer substrates in human and *C. elegans*. *Cell* **159:** 1153–1167.

Sakurai M, Yano T, Kawabata H, Ueda H, Suzuki T. 2010. Inosine cyanoethylation identifies A-to-I RNA editing sites in the human transcriptome. *Nat Chem Biol* **6:** 733–740.

Sakurai M, Ueda H, Yano T, Okada S, Terajima H, Mitsuyama T, Toyoda A, Fujiyama A, Kawabata H, Suzuki T. 2014. A biochemical landscape of A-to-I RNA editing in the human brain transcriptome. *Genome Res* **24:** 522–534.

Sakurai M, Shiromoto Y, Ota H, Song C, Kossenkov AV, Wickramasinghe J, Showe LC, Skordalakes E, Tang H-Y, Speicher DW, et al. 2017. ADAR1 controls apoptosis of stressed cells by inhibiting Staufen1-mediated mRNA decay. *Nat Struct Mol Biol* **24:** 534–543.

Saldi TK, Ash PE, Wilson G, Gonzales P, Garrido-Lecca A, Roberts CM, Dostal V, Gendron TF, Stein LD, Blumenthal T, et al. 2014. TDP-1, the *Caenorhabditis elegans* ortholog of TDP-43, limits the accumulation of double-stranded RNA. *EMBO J* **33:** 2947–2966.

Saletore Y, Meyer K, Korlach J, Vilfan ID, Jaffrey S, Mason CE. 2012. The birth of the Epitranscriptome: Deciphering the function of RNA modifications. *Genome Biol* **13:** 175.

Samuel CE. 2011. Adenosine deaminases acting on RNA (ADARs) are both antiviral and proviral. *Virology* **411:** 180–193.

Savva YA, Jepson JEC, Chang Y-J, Whitaker R, Jones BC, St Laurent G, Tackett MR, Kapranov P, Jiang N, Du G, et al. 2013. RNA editing regulates transposon-mediated heterochromatic gene silencing. *Nat Commun* **4:** 2745.

Schlee M. 2013. Master sensors of pathogenic RNA—RIG-I like receptors. *Immunobiol* **218:** 1322–1335.

Schmedt C, Green SR, Manche L, Taylor DR, Ma Y, Mathews MB. 1995. Functional characterization of the RNA-binding domain and motif of the double-stranded RNA-dependent protein kinase DAI (PKR). *J Mol Biol* **249:** 29–44.

Schönborn J, Oberstaß J, Breyel E, Tittgen J, Schumacher J, Lukacs N. 1991. Monoclonal antibodies to double-stranded RNA as probes of RNA structure in crude nucleic acid extracts. *Nucleic Acids Res* **19:** 2993–3000.

Sijen T, Plasterk RHA. 2003. Transposon silencing in the *Caenorhabditis elegans* germ line by natural RNAi. *Nature* **426:** 310–314.

Sims D, Sudbery I, Ilott NE, Heger A, Ponting CP. 2014. Sequencing depth and coverage: Key considerations in genomic analyses. *Nat Rev Genet* **15:** 121–132.

Sinha NK, Trettin KD, Aruscavage PJ, Bass BL. 2015. *Drosophila* Dicer-2 cleavage is mediated by helicase- and dsRNA termini-dependent states that are modulated by Loquacious-PD. *Mol Cell* **58:** 406–417.

Solomon O, Oren S, Safran M, Deshet-Unger N, Akiva P, Jacob-Hirsch J, Cesarkas K, Kabesa R, Amariglio N, Unger R, et al. 2013. Global regulation of alternative splicing by adenosine deaminase acting on RNA (ADAR). *RNA* **19:** 591–604.

Sommer B, Köhler M, Sprengel R, Seeburg PH. 1991. RNA editing in brain controls a determinant of ion flow in glutamate-gated channels. *Cell* **67:** 11–19.

St Laurent G, Tackett MR, Nechkin S, Shtokalo D, Antonets D, Savva YA, Maloney R, Kapranov P, Lawrence CE, Reenan RA. 2013. Genome-wide analysis of A-to-I RNA editing by single-molecule sequencing in *Drosophila*. *Nat Struct Mol Biol* **20:** 1333–1339.

Stefl R, oberstrass FC, Hood JL, Jourdan M, Zimmermann M, Skrisovska L, Maris C, Peng L, Hofr C, Emeson RB, et al. 2010. The solution structure of the ADAR2 dsRBM-RNA complex reveals a sequence-specific readout of the minor groove. *Cell* **143:** 225–237.

Suzuki T, Ueda H, Okada S, Sakurai M. 2015. Transcriptome-wide identification of adenosine-to-inosine editing using the ICE-seq method. *Nat Protoc* **10:** 715–732.

Tan MH, Li Q, Shanmugam R, Piskol R, Kohler J, Young AN, Liu KI, Zhang R, Ramaswami G, Ariyoshi K, et al. 2017. Dynamic landscape and regulation of RNA editing in mammals. *Nature* **550:** 249–254.

Tian B, Bevilacqua PC, Diegelman-Parente A, Mathews MB. 2004. The double-stranded-RNA-binding motif: Interference and much more. *Nat Rev Mol Cell Biol* **5:** 1013–1023.

van Gurp TP, McIntyre LM, Verhoeven KJF. 2013. Consistent errors in first strand cDNA due to random hexamer mispriming. *PLoS ONE* **8:** e85583.

Venter JC, Adams MD, Myers EW, Li PW, Mural RJ, Sutton GG, Smith HO, Yandell M, Evans CA, Holt RA, et al. 2001. The sequence of the human genome. *Science* **291:** 1304–1351.

Wagner RW, Nishikura K. 1988. Cell cycle expression of RNA duplex unwindase activity in mammalian cells. *Mol Cell Biol* **8:** 770–777.

Wagner RW, Smith JE, Cooperman BS, Nishikura K. 1989. A double-stranded RNA unwinding activity introduces structural alterations by means of adenosine to inosine conversions in mammalian cells and *Xenopus* eggs. *Proc Natl Acad Sci* **86:** 2647–2651.

Walkley CR, Li JB. 2017. Rewriting the transcriptome: Adenosine-to-inosine RNA editing by ADARs. *Genome Biol* **18:** 205.

Wang Z, Gerstein M, Snyder M. 2009. RNA-Seq: A revolutionary tool for transcriptomics. *Nat Rev Genet* **10:** 57–63.

Wang IX, So E, Devlin JL, Zhao Y, Wu M, Cheung VG. 2013. ADAR regulates RNA editing, transcript stability, and gene expression. *Cell Rep* **5:** 849–860.

Wang Z, Lian J, Li Q, Zhang P, Zhou Y, Zhan X, Zhang G. 2016. RES-Scanner: A software package for genome-wide identification of RNA-editing sites. *Gigascience* **5:** 37.

Warf MB, Shepherd BA, Johnson WE, Bass BL. 2012. Effects of ADARs on small RNA processing pathways in *C. elegans*. *Genome Res* **22:** 1488–1498.

Wheeler EC, Washburn MC, Major F, Rusch DB, Hundley HA. 2015. Noncoding regions of *C. elegans* mRNA undergo selective adenosine to inosine deamination and contain a small number of editing sites per transcript. *RNA Biol* **12:** 162–174.

Whipple JM, Youssef OA, Aruscavage PJ, Nix DA, Hong C, Johnson WE, Bass BL. 2015. Genome-wide profiling of the *C. elegans* dsRNAome. *RNA* **21:** 786–800.

Wolfe D, Dudek S, Ritchie MD, Pendergrass SA. 2013. Visualizing genomic information across chromosomes with PhenoGram. *BioData Min* **6:** 18.

Wu TD, Nacu S. 2010. Fast and SNP-tolerant detection of complex variants and splicing in short reads. *Bioinformatics* **26:** 873–881.

Wu D, Lamm AT, Fire AZ. 2011. Competition between ADAR and RNAi pathways for an extensive class of RNA targets. *Nat Struct Mol Biol* **18:** 1094–1101.

Xiang J-F, Yang Q, Liu C-X, Wu M, Chen L-L, Yang L. 2018. N^6-Methyladenosines modulate A-to-I RNA editing. *Mol Cell* **69:** 126–135.e6.

Yasukawa K, Konishi A, Inouye K. 2010. Effects of organic solvents on the reverse transcription reaction catalyzed by reverse transcriptases from avian myeloblastosis virus and Moloney murine leukemia virus. *Biosci Biotechnol Biochem* **74:** 1925–1930.

Yoo J-S, Kato H, Fujita T. 2014. Sensing viral invasion by RIG-I like receptors. *Curr Opin Microbiol* **20:** 131–138.

Zhao W, He X, Hoadley KA, Parker JS, Hayes DN, Perou CM. 2014. Comparison of RNA-Seq by poly (A) capture, ribosomal RNA depletion, and DNA microarray for expression profiling. *BMC Genomics* **15:** 419.

Zhao H-Q, Zhang P, Gao H, He X, Dou Y, Huang AY, Liu X-M, Ye AY, Dong M-Q, Wei L. 2015. Profiling the RNA editomes of wild-type *C. elegans* and ADAR mutants. *Genome Res* **25:** 66–75.

Zhang Q, Xiao X. 2015. Genome sequence–independent identification of RNA editing sites. *Nat Meth* **12:** 347–350.

Zhu S, Xiang J-F, Chen T, Chen L-L, Yang L. 2013. Prediction of constitutive A-to-I editing sites from human transcriptomes in the absence of genomic sequences. *BMC Genomics* **14:** 206.

Technique Development for Probing RNA Structure In Vivo and Genome-Wide

Philip C. Bevilacqua[1] **and Sarah M. Assmann**[2]

[1]Departments of Chemistry and Biochemistry & Molecular Biology, Pennsylvania State University, University Park, Pennsylvania 16802

[2]Department of Biology, Pennsylvania State University, University Park, Pennsylvania 16802

Correspondence: pcb5@psu.edu; sma3@psu.edu

SUMMARY

How organisms perceive and respond to their surroundings is one of the great questions in biology. It is clear that RNA plays key roles in sensing. Cellular and environmental cues that RNA responds to include temperature, ions, metabolites, and biopolymers. Recent advances in next-generation sequencing and in vivo chemical probing have provided unprecedented insights into RNA folding in vivo and genome-wide. Patterns of chemical reactivity have implicated control of gene expression by RNA and aided prediction of RNA structure. Central to these advances has been development of molecular biological and chemical techniques. Key advances are improvements in the quality, cost, and throughput of library preparation; availability of a wider array of chemicals for probing RNA structure in vivo; and robustness and user friendliness of data analysis. Insights from probing transcriptomes and future directions are provided.

Outline

Cite as *Cold Spring Harb Perspect Biol* doi: 10.1101/cshperspect.a032250

1 INTRODUCTION

This review focuses on the development of biological and chemical techniques for studying RNA folding in vivo. However, technique development in and of itself can ring hollow. Regarding the mere development of technique, a leader of the American studio glass movement, Harvey Littleton famously said, "Technique is cheap" (Byrd 2011). To give value to technique development in probing RNA folding genome-wide and in vivo, we root our motivation in the desire to understand RNA biology.

In many organisms, the majority of the genome appears to be transcribed, a concept known as "pervasive transcription" (Lybecker et al. 2014), leading to an astonishing but daunting array of RNA species, with hundreds of thousands of different RNAs in a living organism. For instance, in humans, 75%–85% of the genome is transcribed with just 1.5%–2% being protein coding (Pennisi 2012; Hangauer et al. 2013), and in *Escherichia coli* the majority of the genome is transcribed (Wade 2015) with pervasive transcription mostly occurring antisense to protein-coding genes (Lybecker et al. 2014). Adding to the challenge, the many varied functions of RNA typically manifest themselves in a meaningful way only in vivo. The RNA biologist is thus faced with the difficulty of probing vast numbers of RNA molecules in vivo and genome-wide to make sense of the natural world. This has been the primary driver of technique development in the RNA structural genomics field. To motivate probing of RNA structure in vivo and genome-wide, we describe some of the myriad roles RNA plays in nature, many of which remain to be explored, and then briefly discuss its range of structures and the intermolecular interactions.

Two broad functions of RNA are coding genetic information and regulating gene expression. Messenger RNAs (mRNAs) are unique in that they contain both coding and noncoding regions and so can partake in both roles. Standing as the intermediary between DNA and proteins, mRNAs comprise an ideal point in the flow of genetic information to control gene expression. That control often manifests itself in the 5′ and 3′ untranslated regions (UTRs) of mRNAs, which regulate translation initiation and mRNA turnover (Babitzke et al. 2009; Sonenberg and Hinnebusch 2009; Schoenberg and Maquat 2012). In addition, the coding sequence (CDS) itself can be a point of regulation; indeed, protein levels are often not directly proportional to mRNA levels (Ghaemmaghami et al. 2003; Taniguchi et al. 2010; Walley et al. 2016), suggesting additional functions for mRNAs, and splicing isoforms offer another type of regulatory control. Recent studies suggest that mRNAs can chaperone protein folding through RNA structure–based control of translation rate (Watts et al. 2009; Tang et al. 2016) or by interacting with proteins themselves to help them fold or retain their functional conformations (Docter et al. 2016; Horowitz and Bardwell 2016).

RNAs that do not encode proteins, referred to as noncoding RNAs (ncRNAs), make up all the RNAs in an organism other than mRNA. This broad class of RNAs is the majority both in terms of mass and diversity. The ncRNAs include familiar highly structured and abundant RNAs such as ribosomal RNAs (rRNAs) and transfer RNAs (tRNAs), as well as introns (Lambowitz and Belfort 2015), long noncoding RNAs (lncRNAs) (Kung et al. 2013), and small RNAs such as the ∼20–30-nt microRNAs (miRNAs), small interfering RNAs (siRNAs), and Piwi-interacting RNAs (piRNAs) in eukaryotes (Grosshans and Filipowicz 2008; Axtell 2013), and the ∼50–500-nt small RNAs in bacteria (Storz et al. 2011). Multiple lines of support, including sequence conservation and crystal structures, have shown that introns have complex structures that include catalytic domains and engender complex RNA–protein interactions (Marcia and Pyle 2012). Assessing structure–function relationships of lncRNAs is an active area of research. Some studies have concluded that lncRNAs have conserved and biologically important structures (Maenner et al. 2010; Novikova et al. 2012a; Fang et al. 2015; Somarowthu et al. 2015; Hawkes et al. 2016), although others have not found such support using tests for statistical significance (Rivas et al. 2017). This is an important topic in which more data are needed before universal paradigms can be considered (Novikova et al. 2012b). Other RNAs such as miRNAs and siRNAs form extensive base-pairing with mRNAs, leading to double-stranded regions that control RNA stability and translation.

Because it is single stranded, RNA can form an astounding array of structures that are more protein-like than DNA-like in terms of their complexity and hierarchical nature. These structures are often modular and can comprise motifs as simple as base-paired double-stranded regions that form in *trans* or in *cis* (Bloomfield et al. 2000; Leontis and Westhof 2001), and as complex as intricate tertiary structures such as pseudoknots, ribose zippers, and kink turns (Westhof and Auffinger 2006). These motifs can also lead to independent folding domains in sufficiently large RNAs like the group I intron and the ribosome (Noller and Woese 1981; Murphy and Cech 1993). Each of these structural classes has important biological functions. For example, in humans and other higher organisms, double-stranded RNA (dsRNA) is recognized as a pathogen-associated molecular pattern (PAMP) in innate immunity by proteins such as protein kinase R (PKR) and retinoic acid-inducible gene I (RIG-I) (Heinicke et al. 2011; Jensen and Thomsen 2012). Indeed, dsRNA structures and certain RNA modifications allow discrimination between self and nonself

Cite as *Cold Spring Harb Perspect Biol* doi: 10.1101/cshperspect.a032250

(Nallagatla et al. 2011). Other biological roles of dsRNA in eukaryotes include base-pairing by miRNA, siRNA, and piRNA, in which structures form in *trans* primarily by base-pairing with mRNA, leading to RNA silencing, regulation of gene expression, and epigenetic events (Grosshans and Filipowicz 2008; Axtell 2013). In prokaryotes, helical regions, typically in the 5′ UTR or early in the CDS, are critical to controlling transcription and translation (Merino and Yanofsky 2005; Babitzke et al. 2009). These regions comprise the expression platform of riboswitches (Fig. 1) and can switch their base-pairing between mutually exclusive folds to sequester or expose critical elements that interact with ligands to dictate transcription, termination, or initiation of translation (Nudler and Mironov 2004; Tucker and Breaker 2005; Smith et al. 2010). Additionally, simple helices often form upstream of heat shock proteins in microorganisms and serve as RNA thermometers, melting out at higher temperatures to expose the Shine–Dalgarno sequence and drive translation (Narberhaus et al. 2006).

The first unveiling of the complexity of RNA tertiary structures came with the structure of tRNA (Kim et al. 1973). This structure revealed that RNA can form globular shapes with diverse interactions including non-Watson–Crick base pairs and interhelical interactions. Since that time, hundreds of RNA structures have appeared, including aptamers, riboswitches, ribozymes, the ribosome, and the spliceosome. The details of these structures are beyond the scope of this review, but collectively they show that RNA can achieve positioning of atoms that allows it to be an enzyme and to recognize small molecules specifically and tightly to regulate gene expression.

The versatility of RNA in forming both simple and complex structures allows it to engage in diverse interactions with other molecules, ranging from base-pairing with small RNAs to forming binding pockets for diminutive cations and anions (e.g., Mg^{2+}, Mn^{2+}, and F^{-}) and complex metabolites (e.g., vitamins, coenzymes, nucleotide derivatives, and amino acids), as well as binding structural, processing and signaling proteins (e.g., ribosomal, capping/decapping, spliceosomal, and P-body proteins) (Grosshans and Filipowicz 2008; Re et al. 2014; McCown et al. 2017). Of course, mRNA and rRNA also form intermolecular interactions with each other and with tRNA to decode the genome. It is this dizzying array of structures and intermolecular interactions whose biology only makes sense in vivo that makes RNA an enticing yet challenging molecule for in-depth study. Uncovering the folds of RNAs in vivo and genome-wide has inspired our laboratories and those of others to develop new techniques that span biology and chemistry.

2 TECHNIQUE: THE CHEMISTRY OF PROBING RNA STRUCTURE

The structure of RNA can be determined by a variety of experimental methods. X-ray crystallography and nuclear magnetic resonance (NMR) spectroscopy offer atomic res-

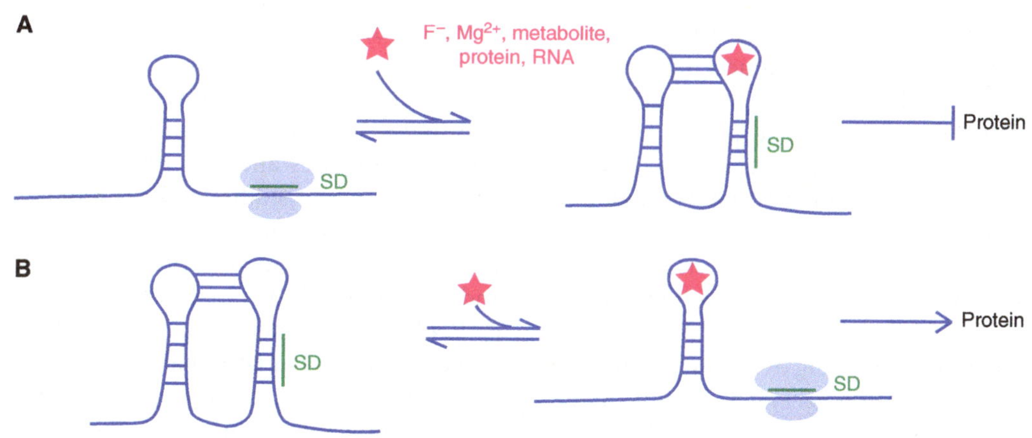

Figure 1. RNA can adopt diverse and multiple folds. Shown is an RNA transcript with a Shine–Dalgarno sequence (SD). The RNA can bind a ligand, which could be a cation, anion, metabolite, protein, or other RNA, depicted as a star. (*A*) RNA binds a ligand, shown here in a kissing loop/pseudoknot tertiary structure, that sequesters the SD and prevents ribosome binding resulting in the absence of protein production. This is the more common type of riboswitch found for negative feedback. (*B*) A different RNA binds a ligand in a loop structure that releases the SD and allows ribosome binding resulting in protein production. This less common type of riboswitch might be for a toxic ligand and the protein produced could pump the ligand out of the cell. Not shown is the effect of temperature, which might melt the right hairpin in *B* (*left*) and expose the SD to translate a heat shock protein. The examples here are for prokaryotes but RNAs adopt multiple base-paired and tertiary structures in eukaryotes as well. The diversity of RNA fold and function motivates development of in vivo genome-wide probing technologies.

olution views, whereas electron cryomicroscopy (cryoEM) is playing an increasingly important role in determining the structures of large and important RNAs at moderate resolution (Nguyen et al. 2016). Although deep insights have been generated from these approaches, they are low-throughput and cannot keep pace with the generation of genomic data, which can now be obtained in a matter of days via next-generation sequencing (Shendure et al. 2017). Moreover, crystallography cannot generally resolve dynamic molecules or all the states of structure-switching RNAs. Finally, most crystallography, NMR, and cryoEM structures obtained are not in the cell and so do not reflect cellular crowding, metabolites, and other RNA-interacting molecules (Leamy et al. 2016) or the potential kinetic effects of cotranscriptional folding (Watters et al. 2016a).

Methods for probing RNA structure with chemicals and enzymes have been known for nearly 40 years (Peattie and Gilbert 1980; Ehresmann et al. 1987). We begin by tracing the development of RNA structure probing, starting with the chemistry used; we then describe how studies moved from in vitro to in vivo; and finally we present more recent advances that have moved the field from gene-specific to genome-wide. Because enzymes have not generally been used to query RNA structure in vivo, our focus here is on chemical probes.

Before we describe the various chemical reactions and their specificities, we consider how the chemical modifications can be detected. Modifications can be read out several different ways. For chemicals that cleave the backbone—often accomplished via workup with a second reagent—RNA can be fractionated by polyacrylamide gel electrophoresis (PAGE) and analyzed alongside sequencing lanes. For chemicals that modify the RNA on the Watson–Crick face, as well as those that cleave the backbone, primer extension by reverse transcription (RT) can be used to query the RNA according to resultant RT termination or runoff (Inoue and Cech 1985; Ehresmann et al. 1987; Bloomfield et al. 2000). When combined with next-generation sequencing, primer extension is typically achieved genome-wide with the aid of random hexamers and can be used not only with RT stops but also with mutational profiling wherein polymerase readthrough of the modification generally leaves behind a mutation (Siegfried et al. 2014; Zubradt et al. 2017) (see Section 3). Reverse transcription has proven to be the most important of these readout methods for in vivo and genome-wide probing of RNA structures because the resulting complementary DNA (cDNA) libraries can be interrogated by next-generation sequencing methods.

Chemistry has been established that queries nearly every aspect of RNA structure. Figure 2 shows this chemistry and also denotes whether the reactions directly result in a stop during reverse transcription or whether additional reactions are needed to read out the base-pairing status of the nucleotide. Probes have been identified that react with the Watson–Crick, Hoogsteen, and sugar faces of the base, the ribose sugar, and the phosphate, as reviewed in Ehresmann et al. (1987) with recent advances presented in Merino et al. (2005), Spitale et al. (2013), Umeyama and Ito (2017), Feng et al. (2018), and Mitchell et al. (2018). Dimethyl sulfate (DMS) reacts with A(N1) and C(N3) on the Watson–Crick face, as well as G(N7) and A(N3) in the major and minor grooves, respectively, where it has also been used to probe dsDNA (Lawley and Brookes 1963; Umeyama and Ito 2017). Kethoxal is used to probe the Watson–Crick face of G, whereas selective 2′-hydroxyl (2′-OH) acylation analyzed by primer extension (SHAPE) reacts with the 2′-OH (Merino et al. 2005) and light-activated structural examination of RNA (LASER) modifies the C8 position of purines (G>A) (Feng et al. 2018). Importantly, the above reactions can all be used in vivo as well (Fig. 2B), except that glyoxal has been used more than kethoxal in vivo (Mitchell et al. 2018). In addition, the following reactions occur exclusively in vitro: diethylpyrocarbonate (DEPC) with A(N7), 1-cyclohexyl-3-(2-morpholinoethyl) carbodiimide metho-*p*-toluene sulfonate (CMCT) with U(N3), bisulfite with C (C4–C6), and ethylnitrosourea (ENU) with phosphates.

Early work probed the structures of individual RNAs in vitro and allowed secondary structure models to be developed (Ehresmann et al. 1987; Bloomfield et al. 2000). The importance of measuring RNA folding in vivo was realized relatively early on and applied to single, abundant transcripts. Seminal studies probed DNA–protein and RNA–protein structures in vivo, taking advantage of the ability of DMS to modify the Hoogsteen face of G in fully dsDNA (Nick and Gilbert 1985), as well as the Watson–Crick faces of A and C in single-stranded RNA (ssRNA) (Climie and Friesen 1988; Ares and Igel 1990; Senecoff and Meagher 1992). Subsequent studies by Zaug and Cech on three different RNA species in living *Tetrahymena thermophila* revealed that RNA structure is often altered by proteins (Zaug and Cech 1995): In the case of the U2 small nuclear RNA (snRNA) and telomerase RNA, very different folds were found in vivo and in vitro that could be explained in terms of protein binding in vivo, whereas the self-splicing rRNA intron had the same fold in vivo and in vitro supporting it acting alone as an enzyme.

The above in vitro chemistry has been applied in vivo, albeit in a more limited way, with four chemicals reported to date (Fig. 2B). Small, neutral, and somewhat hydrophobic chemicals seem to work best, as they can permeate the cell membrane/cell wall (Fig. 2C). At present, DMS, SHAPE (2-methylnicotinic acid imidazolide, NAI), glyoxal, and LASER (nicotinoyl azide, NAz) have been used in vivo in organisms that include plants, human cells, and bacteria.

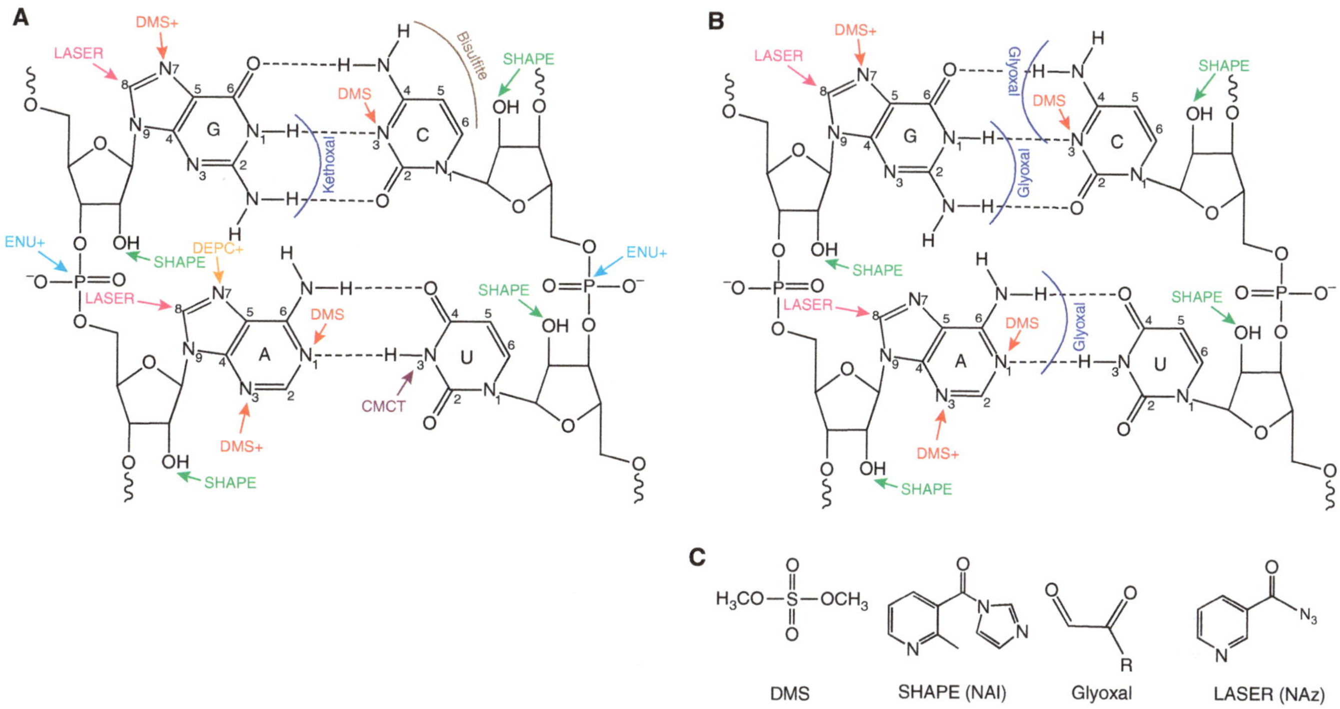

Figure 2. Current state of chemical probing of RNA structure in vitro and in vivo. Shown are the standard GC and AU Watson–Crick base pairs. Reaction of the Watson–Crick face will only occur if the base is not protected by base-pairing or interacting with a protein or small molecule. Sites of reaction of RNA structures (*A*) in vitro and (*B*) in vivo are provided. The following chemical reagents are common to in vitro and in vivo approaches: Dimethyl sulfate (DMS) reacts with A(N1) and C(N3) on the Watson–Crick face, as well as the G(N7) and A(N3); the latter two require additional workup (denoted with a plus sign) of the reaction in the presence of an amine to induce depurination (Ehresmann et al. 1987; Bloomfield et al. 2000; Gates 2009). Glyoxal (Kethoxal in vitro, see below) reacts with the amidine functionality on the Watson–Crick face of G, C, and A, with a preference for G, and can be directly detected by reverse transcription. The SHAPE reagents acylate the 2′-hydroxyl group (2′-OH) of the sugar of all four nucleosides, whereas the LASER reagent reacts with purines at the C8 position, with a preference for G>A. Both SHAPE and LASER reactions can be directly detected by reverse transcription. The following reagents are exclusive to in vitro reactions at this point: diethylpyrocarbonate (DEPC) reacts with A(N7); 1-cyclohexyl-3-(2-morpholinoethyl) carbodiimide metho-*p*-toluene sulfonate (CMCT) reacts with U(N3); bisulfite reacts with C(C4–C6), and ethylnitrosourea (ENU) reacts with the phosphates. Generally, kethoxal has been used in vitro and is reported to be G-specific. Detection of DEPC and ENU require additional chemical treatment (denoted with a plus sign), whereas CMCT and bisulfite can be directly detected by reverse transcription. (These panels are adapted from Ehresmann et al. 1987.) (*C*) Four chemicals used to date for in vivo chemical probing of RNA. Note that these chemicals are small, neutral, and relatively hydrophobic (R = H, CH_3, phenyl).

Sometimes, small alterations to the reagents improve their performance. For instance, probing with SHAPE has been reported in living cells (Spitale et al. 2013; Smola et al. 2015a), with optimal probing and permeability reported with the SHAPE reagent NAI (Lee et al. 2016), whereas methyl- and phenyl-glyoxal provide stronger signals than glyoxal in vivo (Mitchell et al. 2018). For LASER, the small molecule NAz leads to a highly reactive nitrene species on exposure to ultraviolet (UV) light, which is detectable by reverse transcription (Feng et al. 2018), perhaps because it drives the *syn* conformation of the base.

Some of the above four in vivo chemical reagents are sensitive to base-pairing, although others are not. For instance, reaction of A(N1) and C(N3) with DMS, as well as reaction of G, C, and A with glyoxal are blocked by Watson–Crick base-pairing. Additionally, reaction of SHAPE is greatly reduced on base-pairing (McGinnis et al. 2012). On the other hand, reaction of G(N7) and A(N3) with DMS is not sensitive to Watson–Crick base-pairing but only to unusual base pairs and triples. Additionally, because LASER modification occurs at a non-hydrogen bonding position (C8 of G or A), it is not generally sensitive to any base-pairing (Feng et al. 2018). The latter two sets of reactions are potentially powerful because their reactivity should be impaired at nucleotides engaged in protein binding or tertiary structure, and thus absence of chemical modification should provide a readout of the presence of those phenomena.

Deeper insights into biology require a more holistic view than that provided by just one or a few transcripts. It was the confluence of the above seminal experiments, which worked out nucleic acid–probing chemistry and reverse transcription in the test tube and then living cells, and next-generation sequencing that set the stage for the development of techniques to query RNA folding in vivo and genome-wide. The next section provides the rationale behind the approaches taken at the level of library preparation. Each of these approaches relies on at least one of the above chemical modifications of the RNA, with readout by reverse transcription stops or mutational profiling.

3 TECHNIQUE: THE MOLECULAR BIOLOGY OF PROBING RNA STRUCTURE IN VIVO AND GENOME-WIDE

The previous section considered the various chemistries that allow RNA structure to be probed in vivo. We presented a growing list of RNA-reactive membrane-permeant chemicals. Probing RNA structure in vivo first began with the application of DMS 30 years ago (Climie and Friesen 1988), and more recently SHAPE, glyoxal, and LASER approaches were added (Spitale et al. 2013; Smola et al. 2015a; Feng et al. 2018; Mitchell et al. 2018). These four reagents have opened the door to current efforts at probing RNA structure genome-wide and in vivo. During a given treatment of a living organism, *all* of the RNAs in vivo to which the reactive base is exposed are amenable to modification by these chemicals. The challenge then becomes to decode the sites of reactivity and to score them quantitatively.

In this section we shift our focus from the chemistry of RNA to the reading out of the reactivities, which involves application of molecular biology and nucleic acid sequencing approaches. These techniques are multistep and generally quite complex, perhaps not unlike a many-step natural product synthesis for an organic chemist. The goal of this section of the review is not to provide a step-by-step accounting of these methods or to present protocols on how to perform these experiments; indeed, our laboratory and others have published such reviews and protocols previously and we direct the interested reader to these (Low and Weeks 2010; Weeks 2010; Wan et al. 2011; Spitale et al. 2014; Kwok et al. 2015; Bevilacqua et al. 2016; Ritchey et al. 2018; Tack et al. 2018). Rather, we wish to trace the rationale behind the development of the molecular biology aspects of the techniques, giving consideration to how the various approaches differ. We hope that this point of view will facilitate the future growth of this field, which is still ripe for innovation. However, before we describe in vivo probing of RNA genome-wide, it is important to consider early foundational studies that allowed RNA structure to be probed genome-wide in vitro.

Genome-wide in vitro studies took advantage of the favorable properties of ribonucleases (RNases) as reagents because the membrane impermeability inherent to enzymes is not an issue in vitro. Fragmentation sequencing, or FragSeq, probed RNA structure with the ss-specific RNase P1 and analyzed sites of reaction with next-generation sequencing (Underwood et al. 2010). In the same year, parallel analysis of RNA structure (PARS) was developed, which uses a combination of ss- and ds-specific RNases (Kertesz et al. 2010). This technique enabled development of parallel analysis of RNA structure with temperature elevation (PARTE), a temperature-dependent version of PARS (Wan et al. 2012). The unique feature of PARS and PARTE was the use of RNase V1, which allows direct detection of double-stranded regions of the RNA (Lockard and Kumar 1981; Lowman and Draper 1986). Protein footprinting was assessed genome-wide in vivo with the invention of protein interaction profile sequencing (PIP-seq) in which RNAs and proteins are first cross-linked in vivo followed by extraction and treatment with ss- and ds-specific RNases with and without treatment by proteinase K, which degrades proteins (Silverman et al. 2014). The issue of protein binding in vivo is an important one and this approach is notable for taking it on. The in vitro RNA structure-determining aspects of PIP-seq have several concerns, however, including buffer mismatch between RNase treatments and the potential for RNA refolding on protein degradation. Finally, at least one genome-wide in vitro study has taken advantage of membrane impermeant chemicals to probe RNA structure. In chemical interference of RNA structure sequencing (CIRS-Seq), CMCT, which does not react in vivo (D Mitchell, SM Assmann, and PC Bevilacqua, unpubl.), likely because its size and charge prevent cell membrane permeation, was used along with DMS to offer chemical information on all four bases in vitro and genome-wide because CMCT can react with G/U and DMS reacts on the Watson–Crick face with A/C (Incarnato et al. 2014). This provided an information-rich view of RNA folding in vitro. In sum, the above methods have taken advantage of a warehouse of known RNA-reactive enzymes and chemicals (Bloomfield et al. 2000), which allows extensive analysis of RNA structure genome-wide but in vitro. These studies have inspired, and continue to inspire, development of in vivo reagents that have the desirable properties of reacting with diverse functionalities on the RNA as well as with unique RNA structural motifs.

The first two reports of probing RNA structure genome-wide and in vivo appeared in a single journal issue in 2014. These studies included the Structure-seq method from our laboratories and the DMS-seq approach from the Weissman laboratory (Ding et al. 2014; Rouskin et al. 2014). Shortly thereafter, modification sequencing (Mod-seq) was reported from the McManus laboratory (Talkish et al. 2014). A diagram comparing these methods is provided in Figure 3. After in vivo treatment with the structure-probing chemical, the process of analyzing RNA structure genome-wide and in vivo begins with extraction of the treated RNA, typically using a commercial kit. Unless it is the focus of the study, rRNA needs to be removed from the extracted RNA because rRNA accounts for ~80% of all of the RNA in a cell and will otherwise waste an enormous number of next-generation sequencing reads. Removal of rRNA can be accomplished by a subtractive method such as Ribo-Zero, in which the rRNA binds to complementary oligonucleotides and is pulled down. The Ribo-Zero approach is critical in cases in which there is no common sequence, like a poly(A) tail, to capture the RNAs of interest. Alternatively, the RNA of interest can be pulled down leaving behind rRNA. The most common method uses an oligonucleotide deoxythymine (oligo-dT) column to collect polyadenylated mRNA (Ding et al. 2014; Rouskin et al. 2014), although one could alternatively extract only a single RNA or a specific collection of RNAs using appropriate capture oligonucleotides. It is also possible to use gene-specific primers to focus in on a single RNA without first removing rRNA or enriching for the RNA of interest. For instance, our laboratory used gene-specific primers to target single RNAs (U12 snRNA, 25S rRNA, 5.8S rRNA, and *PSBA*) in plants (*Arabidopsis*) and to probe their in vivo structures with both DMS and the SHAPE reagent NAI. Using a ligation-mediated polymerase chain reaction (LMPCR) approach, we increased the sensitivity of the method to attomole level detection (Kwok et al. 2013). Simons and colleagues probed the structure of lncRNA Xist in mouse cells by tiling 87 different gene-specific primers spaced ~200 nt apart along the ~17,000 nt RNA (Fang et al. 2015). They dubbed this method "targeted Structure-seq" because it is based on Structure-seq (Ding et al. 2014) and targets a single RNA. In addition, Lucks and colleagues have reported a pipeline for applying SHAPE in vitro or in vivo to a collection of RNAs and generating restraints for RNA folding via SHAPE-Seq (Watters et al. 2016b).

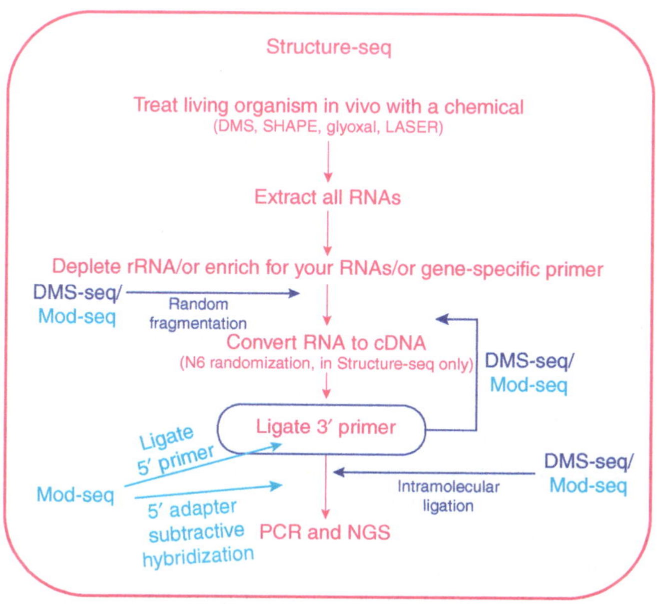

Figure 3. Comparison of the Structure-seq, DMS-seq, and modification sequencing (Mod-seq) in vivo genome-wide RNA structure probing techniques. Structure-seq is shown in magenta font. Additional steps present in DMS-seq and Mod-seq are depicted in dark blue and light blue font, respectively. Note the extra steps before conversion of RNA to cDNA in DMS-seq and Mod-seq as well as the different points of enhancing coverage in the various techniques. Mod-seq has the most steps of all techniques. PCR, Polymerase chain reaction; NGS, next-generation sequencing.

The Structure-seq, DMS-seq, and Mod-seq methodologies all proceed to form libraries but use quite different approaches. In an effort to minimize the universally negative impacts of RNA degradation, we designed Structure-seq to have its RT process of converting RNA into cDNA relatively early on in the pipeline (Ding et al. 2014). We convert RNA to cDNA immediately after poly(A) selection, whereas DMS-seq and Mod-seq have two or four additional RNA-handling steps, respectively, before RT (Fig. 3) (Kwok et al. 2015). To increase coverage in preparing the library, we conduct RT using a primer whose 3′ end is composed of a library of random hexanucleotide sequences (N6) to facilitate random priming, but whose 5′ end has an adapter sequence to bind to TruSeq reverse primers that barcode for a given condition and allow Illumina sequencing (Ding et al. 2014). It is then necessary to append a fixed sequence to the other end of the cDNA to enable PCR, which we do via intermolecular ligation of a ssDNA adapter. This sample then undergoes PCR and next-generation sequencing. Recently, we introduced "Structure-seq2"—a significantly improved version of the original Structure-seq (Ritchey et al. 2017).

DMS-seq and Mod-seq enhance coverage not via a randomized section of a primer, but with fragmentation of the RNA, conducted via Zn^{2+}-mediated hydrolysis (Rouskin et al. 2014; Talkish et al. 2014). Both approaches then ligate on a 3′-RNA adapter, followed by RT and a subsequent intramolecular ligation to form cDNA circles that are subject to PCR amplification and next-generation sequencing. The Mod-seq technique has an additional step of 5′-RNA adapter ligation that allows a later 5′-adapter subtractive hybridization to remove samples that have no chemical modification wherein the RT reads through to the 5′-adapter.

There are several key differences in the libraries produced by these three methods. In Structure-seq2 and Mod-seq, reads from a minus chemical lane are used to correct for background stops, which can arise from RNA secondary structure, covalent modification, or strand scission attributable to degradation or processing. In addition, both Structure-seq and Mod-seq use single-hit kinetics in which only about one in 200 RNA nucleotides is modified; this is generally desired because modification can refold the RNA and reduce reads near the 5′ end of the RNA. The DMS-seq method does not use a minus DMS correction nor single-hit kinetics. In addition, although Mod-seq libraries have fewer unmodified reads owing to its adapter subtractive hybridization, as a result it also does not have an accurate tally on the total number of reads in a given RNA and so may not wholly account for transcript abundance.

Several other improvements in the technique of library preparation have recently appeared. Chang, Spitale, and coworkers introduced an improvement to their NAI SHAPE reagent in the method of in vivo click SHAPE, or icSHAPE (Spitale et al. 2015). In this technique, their NAI reagent is fused to an azido group, which can be attached to a biotin functionality via copper-free click chemistry to enrich for SHAPE-modified RNAs via a streptavidin pull-down. This technique achieves a similar result as Mod-seq —removal of unmodified RNAs from the library—although it is achieved for Mod-seq by capturing the undesired RNAs, whereas for icSHAPE it is accomplished by capturing the desired RNAs.

Two other improvements in preparation of genome-wide in vivo RNA libraries involve the reading out of multiple modifications from a single sequencing read. Weeks and coworkers introduced mutational profiling in SHAPE-MaP, in which SHAPE-modified nucleotides induce a mutation during RT rather than a stop (Siegfried et al. 2014; Smola et al. 2015b; Busan and Weeks 2017). The mutation is achieved by altering the RT conditions, specifically using Mn^{2+} with the reverse transcriptase Superscript II (SSII). The primary advantage of SHAPE-MaP is that correlated modifications in a single RNA can be identified, essentially turning the method into a single-molecule experiment. For instance, two alleles can be monitored simultaneously, potentially overcoming sequence and structural polymorphism. Weeks and coworkers also used SHAPE-MaP correlations to predict RNA pseudoknots (Siegfried et al. 2014). Rouskin and coworkers adapted mutational profiling to DMS modifications in DMS-MaPseq, wherein a thermostable group II intron reverse transcriptase (TGIRT) is used to obtain mutational data (Zubradt et al. 2017). As compared with the SSII-Mn^{2+} method, these investigators achieved significantly higher detection of adenine methylation (48% vs. 1.4% in one instance) and significantly lower rate of insertions/deletions (indels) (6% vs. ∼33%), which is helpful for accurate mapping. Another technical advantage of mutational approaches is that library preparation is simplified and more closely follows standard RNA-seq protocols.

It is worth briefly considering some of the biological insights that have been gained via these techniques. These methods have identified RNA reactivity features in plants associated with the CDS and start and stop codons, alternative polyadenylation and alternative splice sites, and stress-related genes (Ding et al. 2014), as well as relationships of protein and mRNA structure (Watts et al. 2009; Tang et al. 2016). Studies in mammalian and yeast cells revealed that RNAs are more unfolded in vivo than in vitro, and that adenosine triphosphate (ATP)-dependent processes, such as helicase unwinding, are at least partially responsible for this (Rouskin et al. 2014). Additional studies in mouse embryonic stem cells revealed that certain RNA features are the same in vitro and in vivo, such as structural features prevailing at translational start and pause sites, whereas RNA-binding proteins and RNA modifications rearrange RNA structure in vivo and can provide reactivity patterns whose molecular features agree with crystal structures (Spitale et al. 2015). In other cases, identification of known and new RNA motifs in human immunodeficiency virus 1 (HIV-1) has been achieved, including several pseudoknots, and polymorphic RNAs have been studied in a single experiment (Siegfried et al. 2014). Translation efficiency in *E. coli* has been related to open reading frame structures (Burkhardt et al. 2017). Finally, functional structures involved in noncanonical translation initiation have been characterized, along with comparison of spliced and unspliced isoforms (Zubradt et al. 2017). This brief summary illustrates that these structure-probing methods are organism-agnostic and suggests that there is nothing inherent to any of these methods that would prevent their application, singly or multiplexed, to any organism.

All of the above in vivo and genome-wide techniques generate big datasets and so require careful computational analysis. A full review of computational techniques is beyond the scope of this review, but we point out some important advances. Each of the experimental methods described above had a computational approach published as part of the initial study. Some studies later published additional stand-alone companion computational studies and advances (Talkish et al. 2014; Tang et al. 2015; Busan and Weeks 2017; Tack et al. 2018). Key features of computational studies are subtraction of background reaction and normalization for RNA abundance. In addition, all of the methods require benchmarking, which is most commonly performed with structurally characterized RNAs such as rRNA (Spitale et al. 2013; Ding et al. 2014; Rouskin et al.

2014; Talkish et al. 2014). One important feature of computational analysis is consideration of the tendency of an RNA sequence to adopt more than one fold. Recently, several advances have been made in this arena, both experimentally with mutational profiling (Siegfried et al. 2014; Zubradt et al. 2017) and computationally by modeling folding ensembles (Li and Aviran 2018; Spasic et al. 2018). In addition, efforts are afoot to make the software accessible to scientists without extensive expertise in bioinformatics and to incorporate more downstream analysis (Tack et al. 2018). As the field moves forward, it will become increasingly important for the computational pipelines to have the capacity to handle datasets involving more than one chemical and multiple conditions.

4 THE FUTURE OF PROBING RNA STRUCTURE IN VIVO AND GENOME-WIDE

There are tremendous opportunities to advance RNA structure probing in vivo and genome-wide. First, there is the need to develop a more expansive chemical toolbox for probing RNA structure in vivo. As shown in Figure 2, there are several regions of the RNA that are presently not probed in vivo. These include the Watson–Crick face of uracil, as well as the phosphate backbone. Another area for advancement is increasing the sensitivity of probing. Although advances have been made (Kwok et al. 2013), simpler and more sensitive probing techniques might facilitate probing of smaller populations of RNA, such as might be found in biologically interesting areas of specific tissues and even single cells. The latter is an especially exciting area of modern biology and may simplify RNA structurome results, as the folding ensemble may be less diverse.

Another area ripe for development is prediction of RNA structure from sequence. Efforts have been made in this area using in vitro SHAPE chemical probing restraints, which generally manifest themselves as pseudo-free energy parameters (Deigan et al. 2009); however, prediction using in vivo chemical probing restraints is overall still inaccurate for many RNAs (Leamy et al. 2016). The increasing diversity of chemicals available for probing RNA structure in vivo will provide richer experimental data to help constrain RNA structure prediction. The recently developed psoralen cross-linking method (PARIS) from Chang and colleagues provides information on RNA duplexes and RNA–RNA interactions in vivo and genome-wide (Lu et al. 2016). Advances in predicting RNA 3D structure are being made (Miao et al. 2017), and with a plethora of experimental and proximity constraints it should become possible to model 3D structure of RNA genome-wide as it occurs in vivo. Finally, the ability of RNA to switch between conformations requires measuring and modeling folding ensembles (Siegfried et al. 2014; Zubradt et al. 2017; Li and Aviran 2018; Spasic et al. 2018), and further genome-wide advances in this area can be anticipated.

Yet another area for further advancement is probing protein–RNA interactions. Although high reactivity of a given site fully supports its single-stranded conformation, low reactivity observed in vivo does not necessarily support base-pairing, as protection can also be conferred by binding of proteins, small molecules, and ions; moreover, low reactivity does not distinguish intramolecular versus intermolecular base-pairing. Advances in protein protection have been reported in correlating SHAPE reactivities with RNA–protein 3D structures (Spitale et al. 2015) and in using uncertainties in SHAPE reactivity between states (Smola et al. 2015a). In addition, LASER probes the non-hydrogen bonding C8 position of purines and so is inherently sensitive to protein protection at this position (Feng et al. 2018), as is reactivity at other non-Watson–Crick positions (Umeyama and Ito 2017). Ultimately, a combination of these approaches, along with description of RNA-binding motifs (Ray et al. 2013) and selective deployment of cross-linking (Lu et al. 2016) may help with the assignment of protections.

Finally, there are a large number of organisms and conditions to probe. To date, in vivo genome-wide experiments have been conducted in living human and mouse cells, as well as in living whole organisms such as *Arabidopsis*, rice, yeast, and *E. coli* (Ding et al. 2014; Incarnato et al. 2014; Rouskin et al. 2014; Spitale et al. 2015; Ritchey et al. 2017; Deng et al. 2018). There is the potential to now perform comparative analysis of RNA structure between different states, for instance, between organisms that differ in genotype and phenotype, between closely or distantly related species, and indeed among all three domains of life. Moreover, there is the opportunity to perform comparative analysis of RNA structure in a single organism between two states, such as different environmental or stress conditions. This offers the chance to discover new functional RNAs and pathways. Finally, the issue of time comes into play, allowing one to ask how the structure of the transcriptome changes over long or even very short times. For example, the LASER reagent can react in less than a second and is self-quenching, allowing querying of rapid processes genome-wide (Feng et al. 2018). There have been tremendous advances in our understanding of RNA biology in vivo over just the last 5 years. All of these can be traced to advances in chemical and biological techniques. These advances have led to surprising new insights into the relationship of RNA structure and gene regulation. New biology awaits discovery. With additional technical advances, these discoveries will be accelerated, allowing greater insight into the natural world.

ACKNOWLEDGMENTS

We gratefully acknowledge the National Science Foundation (NSF)-Plant Genome Research Program (NSF-IOS-1339282) for support, and Dr. Elizabeth Jolley and Laura Ritchey for careful reading of the manuscript.

REFERENCES

Ares M Jr, Igel AH. 1990. Lethal and temperature-sensitive mutations and their suppressors identify an essential structural element in U2 small nuclear RNA. *Genes Dev* **4:** 2132–2145.

Axtell MJ. 2013. Classification and comparison of small RNAs from plants. *Annu Rev Plant Biol* **64:** 137–159.

Babitzke P, Baker CS, Romeo T. 2009. Regulation of translation initiation by RNA binding proteins. *Annu Rev Microbiol* **63:** 27–44.

Bevilacqua PC, Ritchey LE, Su Z, Assmann SM. 2016. Genome-wide analysis of RNA secondary structure. *Annu Rev Genet* **50:** 235–266.

Bloomfield VA, Crothers DM, Tinoco I Jr. 2000. *Nucleic acids: Structures, properties, and functions.* University Science Books, Sausalito, CA.

Burkhardt DH, Rouskin S, Zhang Y, Li GW, Weissman JS, Gross CA. 2017. Operon mRNAs are organized into ORF-centric structures that predict translation efficiency. *eLife* **6:** e22037.

Busan S, Weeks KM. 2017. Accurate detection of chemical modifications in RNA by mutational profiling (MaP) with ShapeMapper 2. *RNA* **24:** 143–148.

Byrd JF. 2011. *Harvey K. Littleton: A life in glass.* Rizzoli, New York.

Climie SC, Friesen JD. 1988. In vivo and in vitro structural analysis of the rplJ mRNA leader of *Escherichia coli*. Protection by bound L10-L7/L12. *J Biol Chem* **263:** 15166–15175.

Deigan KE, Li TW, Mathews DH, Weeks KM. 2009. Accurate SHAPE-directed RNA structure determination. *Proc Natl Acad Sci* **106:** 97–102.

Deng H, Cheema J, Zhang H, Woolfenden H, Norris M, Liu Z, Liu Q, Yang X, Yang M, Deng X, et al. 2018. Rice in vivo RNA structurome reveals RNA secondary structure conservation and divergence in plants. *Mol Plant* **11:** 607–622.

Ding Y, Tang Y, Kwok CK, Zhang Y, Bevilacqua PC, Assmann SM. 2014. In vivo genome-wide profiling of RNA secondary structure reveals novel regulatory features. *Nature* **505:** 696–700.

Docter BE, Horowitz S, Gray MJ, Jakob U, Bardwell JC. 2016. Do nucleic acids moonlight as molecular chaperones? *Nucleic Acids Res* **44:** 4835–4845.

Ehresmann C, Baudin F, Mougel M, Romby P, Ebel J-P, Ehresmann B. 1987. Probing the structure of RNAs in solution. *Nucleic Acids Res* **15:** 9109–9128.

Fang R, Moss WN, Rutenberg-Schoenberg M, Simon MD. 2015. Probing Xist RNA structure in cells using Targeted Structure-Seq. *PLoS Genet* **11:** e1005668.

Feng C, Chan D, Joseph J, Muuronen M, Coldren WH, Dai N, Correa IR Jr., Furche F, Hadad CM, Spitale RC. 2018. Light-activated chemical probing of nucleobase solvent accessibility inside cells. *Nat Chem Biol* **14:** 276–283.

Gates KS. 2009. An overview of chemical processes that damage cellular DNA: Spontaneous hydrolysis, alkylation, and reactions with radicals. *Chem Res Toxicol* **22:** 1747–1760.

Ghaemmaghami S, Huh WK, Bower K, Howson RW, Belle A, Dephoure N, O'Shea EK, Weissman JS. 2003. Global analysis of protein expression in yeast. *Nature* **425:** 737–741.

Grosshans H, Filipowicz W. 2008. Molecular biology: The expanding world of small RNAs. *Nature* **451:** 414–416.

Hangauer MJ, Vaughn IW, McManus MT. 2013. Pervasive transcription of the human genome produces thousands of previously unidentified long intergenic noncoding RNAs. *PLoS Genet* **9:** e1003569.

Hawkes EJ, Hennelly SP, Novikova IV, Irwin JA, Dean C, Sanbonmatsu KY. 2016. COOLAIR antisense RNAs form evolutionarily conserved elaborate secondary structures. *Cell Rep* **16:** 3087–3096.

Heinicke LA, Nallagatla SR, Hull CM, Bevilacqua PC. 2011. RNA helical imperfections regulate activation of the protein kinase PKR: Effects of bulge position, size, and geometry. *RNA* **17:** 957–966.

Horowitz S, Bardwell JC. 2016. RNAs as chaperones. *RNA Biol* **13:** 1228–1231.

Incarnato D, Neri F, Anselmi F, Oliviero S. 2014. Genome-wide profiling of mouse RNA secondary structures reveals key features of the mammalian transcriptome. *Genome Biol* **15:** 491.

Inoue T, Cech TR. 1985. Secondary structure of the circular form of the *Tetrahymena* rRNA intervening sequence: A technique for RNA structure analysis using chemical probes and reverse transcriptase. *Proc Natl Acad Sci* **82:** 648–652.

Jensen S, Thomsen AR. 2012. Sensing of RNA viruses: A review of innate immune receptors involved in recognizing RNA virus invasion. *J Virol* **86:** 2900–2910.

Kertesz M, Wan Y, Mazor E, Rinn JL, Nutter RC, Chang HY, Segal E. 2010. Genome-wide measurement of RNA secondary structure in yeast. *Nature* **467:** 103–107.

Kim SH, Quigley GJ, Suddath FL, McPherson A, Sneden D, Kim JJ, Weinzierl J, Rich A. 1973. Three-dimensional structure of yeast phenylalanine transfer RNA: Folding of the polynucleotide chain. *Science* **179:** 285–288.

Kung JT, Colognori D, Lee JT. 2013. Long noncoding RNAs: Past, present, and future. *Genetics* **193:** 651–669.

Kwok CK, Ding Y, Tang Y, Assmann SM, Bevilacqua PC. 2013. Determination of in vivo RNA structure in low-abundance transcripts. *Nat Commun* **4:** 2971.

Kwok CK, Tang Y, Assmann SM, Bevilacqua PC. 2015. The RNA structurome: Transcriptome-wide structure probing with next-generation sequencing. *Trends Biochem Sci* **40:** 221–232.

Lambowitz AM, Belfort M. 2015. Mobile bacterial group II introns at the crux of eukaryotic evolution. *Microbiol Spectr* **40:** MDNA3-0050-2014.

Lawley PD, Brookes P. 1963. Further studies on the alkylation of nucleic acids and their constituent nucleotides. *Biochem J* **89:** 127–138.

Leamy KA, Assmann SM, Mathews DH, Bevilacqua PC. 2016. Bridging the gap between in vitro and in vivo RNA folding. *Q Rev Biophys* **49:** e10.

Lee B, Flynn RA, Kadina A, Guo JK, Kool ET, Chang HY. 2016. Comparison of SHAPE reagents for mapping RNA structures inside living cells. *RNA* **23:** 169–174.

Leontis NB, Westhof E. 2001. Geometric nomenclature and classification of RNA base pairs. *RNA* **7:** 499–512.

Li H, Aviran S. 2018. Statistical modeling of RNA structure profiling experiments enables parsimonious reconstruction of structure landscapes. *Nat Commun* **9:** 606.

Lockard RE, Kumar A. 1981. Mapping tRNA structure in solution using double-strand-specific ribonuclease V_1 from cobra venom. *Nucleic Acids Res* **9:** 5125–5140.

Low JT, Weeks KM. 2010. SHAPE-directed RNA secondary structure prediction. *Methods* **52:** 150–158.

Lowman HB, Draper DE. 1986. On the recognition of helical RNA by cobra venom V1 nuclease. *J Biol Chem* **261:** 5396–5403.

Lu Z, Zhang QC, Lee B, Flynn RA, Smith MA, Robinson JT, Davidovich C, Gooding AR, Goodrich KJ, Mattick JS, et al. 2016. RNA duplex map in living cells reveals higher-order transcriptome structure. *Cell* **165:** 1267–1279.

Lybecker M, Bilusic I, Raghavan R. 2014. Pervasive transcription: Detecting functional RNAs in bacteria. *Transcription* **5:** e944039.

Maenner S, Blaud M, Fouillen L, Savoye A, Marchand V, Dubois A, Sanglier-Cianferani S, Van Dorsselaer A, Clerc P, Avner P, et al. 2010. 2-D structure of the A region of Xist RNA and its implication for PRC2 association. *PLoS Biol* **8:** e1000276.

Marcia M, Pyle AM. 2012. Visualizing group II intron catalysis through the stages of splicing. *Cell* **151:** 497–507.

McCown PJ, Corbino KA, Stav S, Sherlock ME, Breaker RR. 2017. Riboswitch diversity and distribution. *RNA* **23:** 995–1011.

McGinnis JL, Dunkle JA, Cate JH, Weeks KM. 2012. The mechanisms of RNA SHAPE chemistry. *J Am Chem Soc* **134:** 6617–6624.

Merino E, Yanofsky C. 2005. Transcription attenuation: A highly conserved regulatory strategy used by bacteria. *Trends Genet* **21:** 260–264.

Merino EJ, Wilkinson KA, Coughlan JL, Weeks KM. 2005. RNA structure analysis at single nucleotide resolution by selective 2′-hydroxyl acylation and primer extension (SHAPE). *J Am Chem Soc* **127:** 4223–4231.

Miao Z, Adamiak RW, Antczak M, Batey RT, Becka AJ, Biesiada M, Boniecki MJ, Bujnicki J, Chen SJ, Cheng CY, et al. 2017. RNA-Puzzles Round III: 3D RNA structure prediction of five riboswitches and one ribozyme. *RNA* **23:** 655–672.

Mitchell D, 3rd, Ritchey LE, Park H, Babitzke P, Assmann SM, Bevilacqua PC. 2018. Glyoxals as in vivo RNA structural probes of guanine base-pairing. *RNA* **24:** 114–124.

Murphy FL, Cech TR. 1993. An independently folding domain of RNA tertiary structure within the *Tetrahymena* ribozyme. *Biochemistry* **32:** 5291–5300.

Nallagatla SR, Toroney R, Bevilacqua PC. 2011. Regulation of innate immunity through RNA structure and the protein kinase PKR. *Curr Opin Struct Biol* **21:** 119–127.

Narberhaus F, Waldminghaus T, Chowdhury S. 2006. RNA thermometers. *FEMS Microbiol Rev* **30:** 3–16.

Nguyen TH, Galej WP, Fica SM, Lin PC, Newman AJ, Nagai K. 2016. CryoEM structures of two spliceosomal complexes: Starter and dessert at the spliceosome feast. *Curr Opin Struct Biol* **36:** 48–57.

Nick H, Gilbert W. 1985. Detection in vivo of protein-DNA interactions within the lac operon of *Escherichia coli*. *Nature* **313:** 795–798.

Noller HF, Woese CR. 1981. Secondary structure of 16S ribosomal RNA. *Science* **212:** 403–411.

Novikova IV, Hennelly SP, Sanbonmatsu KY. 2012a. Structural architecture of the human long non-coding RNA, steroid receptor RNA activator. *Nucleic Acids Res* **40:** 5034–5051.

Novikova IV, Hennelly SP, Sanbonmatsu KY. 2012b. Sizing up long non-coding RNAs: Do lncRNAs have secondary and tertiary structure? *Bioarchitecture* **2:** 189–199.

Nudler E, Mironov AS. 2004. The riboswitch control of bacterial metabolism. *Trends Biochem Sci* **29:** 11–17.

Peattie DA, Gilbert W. 1980. Chemical probes for higher-order structure in RNA. *Proc Natl Acad Sci* **77:** 4679–4682.

Pennisi E. 2012. ENCODE project writes eulogy for junk DNA. *Science* **337:** 1159–1161.

Ray D, Kazan H, Cook KB, Weirauch MT, Najafabadi HS, Li X, Gueroussov S, Albu M, Zheng H, Yang A, et al. 2013. A compendium of RNA-binding motifs for decoding gene regulation. *Nature* **499:** 172–177.

Re A, Joshi T, Kulberkyte E, Morris Q, Workman CT. 2014. RNA–protein interactions: An overview. In: *RNA sequence, structure, and function: Computational and bioinformatic methods* (ed. Gorodkin J, Ruzzo WL), pp. 491–521. Humana, Totowa, NJ.

Ritchey LE, Su Z, Tang Y, Tack DC, Assmann SM, Bevilacqua PC. 2017. Structure-seq2: Sensitive and accurate genome-wide profiling of RNA structure in vivo. *Nucleic Acids Res* **45:** e135.

Ritchey LE, Su Z, Assmann SM, Bevilacqua PC. 2018. In vivo genome-wide RNA structure probing with Structure-seq2. *Methods Mol Biol* (in press).

Rivas E, Clements J, Eddy SR. 2017. A statistical test for conserved RNA structure shows lack of evidence for structure in lncRNAs. *Nat Methods* **14:** 45–48.

Rouskin S, Zubradt M, Washietl S, Kellis M, Weissman JS. 2014. Genome-wide probing of RNA structure reveals active unfolding of mRNA structures in vivo. *Nature* **505:** 701–705.

Schoenberg DR, Maquat LE. 2012. Regulation of cytoplasmic mRNA decay. *Nat Rev Genet* **13:** 246–259.

Senecoff JF, Meagher RB. 1992. In vivo analysis of plant RNA structure: Soybean 18S ribosomal and ribulose-1,5-bisphosphate carboxylase small subunit RNAs. *Plant Mol Biol* **18:** 219–234.

Shendure J, Balasubramanian S, Church GM, Gilbert W, Rogers J, Schloss JA, Waterston RH. 2017. DNA sequencing at 40: Past, present and future. *Nature* **550:** 345–353.

Siegfried NA, Busan S, Rice GM, Nelson JA, Weeks KM. 2014. RNA motif discovery by SHAPE and mutational profiling (SHAPE-MaP). *Nat Methods* **11:** 959–965.

Silverman IM, Li F, Alexander A, Goff L, Trapnell C, Rinn JL, Gregory BD. 2014. RNase-mediated protein footprint sequencing reveals protein-binding sites throughout the human transcriptome. *Genome Biol* **15:** R3.

Smith AM, Fuchs RT, Grundy FJ, Henkin TM. 2010. Riboswitch RNAs: Regulation of gene expression by direct monitoring of a physiological signal. *RNA Biol* **7:** 104–110.

Smola MJ, Calabrese JM, Weeks KM. 2015a. Detection of RNA-protein interactions in living cells with SHAPE. *Biochemistry* **54:** 6867–6875.

Smola MJ, Rice GM, Busan S, Siegfried NA, Weeks KM. 2015b. Selective 2′-hydroxyl acylation analyzed by primer extension and mutational profiling (SHAPE-MaP) for direct, versatile and accurate RNA structure analysis. *Nat Protoc* **10:** 1643–1669.

Somarowthu S, Legiewicz M, Chillon I, Marcia M, Liu F, Pyle AM. 2015. HOTAIR forms an intricate and modular secondary structure. *Mol Cell* **58:** 353–361.

Sonenberg N, Hinnebusch AG. 2009. Regulation of translation initiation in eukaryotes: Mechanisms and biological targets. *Cell* **136:** 731–745.

Spasic A, Assmann SM, Bevilacqua PC, Mathews DH. 2018. Modeling RNA secondary structure folding ensembles using SHAPE mapping data. *Nucleic Acids Res* **46:** 314–323.

Spitale RC, Crisalli P, Flynn RA, Torre EA, Kool ET, Chang HY. 2013. RNA SHAPE analysis in living cells. *Nat Chem Biol* **9:** 18–20.

Spitale RC, Flynn RA, Torre EA, Kool ET, Chang HY. 2014. RNA structural analysis by evolving SHAPE chemistry. *Wiley Interdiscip Rev RNA* **5:** 867–881.

Spitale RC, Flynn RA, Zhang QC, Crisalli P, Lee B, Jung JW, Kuchelmeister HY, Batista PJ, Torre EA, Kool ET, et al. 2015. Structural imprints in vivo decode RNA regulatory mechanisms. *Nature* **519:** 486–490.

Storz G, Vogel J, Wassarman KM. 2011. Regulation by small RNAs in bacteria: Expanding frontiers. *Mol Cell* **43:** 880–891.

Tack DC, Tang Y, Ritchey LE, Assmann SM, Bevilacqua PC. 2018. StructureFold2: Bringing chemical probing data into the computational fold of RNA structural analysis. *Methods* doi: 10.1016/j.ymeth.2018.01.018.

Talkish J, May G, Lin Y, Woolford JL Jr., McManus CJ. 2014. Mod-seq: High-throughput sequencing for chemical probing of RNA structure. *RNA* **20:** 713–720.

Tang Y, Bouvier E, Kwok CK, Ding Y, Nekrutenko A, Bevilacqua PC, Assmann SM. 2015. StructureFold: Genome-wide RNA secondary structure mapping and reconstruction in vivo. *Bioinformatics* **31:** 2668–2675.

Tang Y, Assmann SM, Bevilacqua PC. 2016. Protein structure is related to RNA structural reactivity in vivo. *J Mol Biol* **428:** 758–766.

Taniguchi Y, Choi PJ, Li GW, Chen H, Babu M, Hearn J, Emili A, Xie XS. 2010. Quantifying *E. coli* proteome and transcriptome with single-molecule sensitivity in single cells. *Science* **329:** 533–538.

Tucker BJ, Breaker RR. 2005. Riboswitches as versatile gene control elements. *Curr Opin Struct Biol* **15:** 342–348.

Umeyama T, Ito T. 2017. DMS-seq for in vivo genome-wide mapping of protein–DNA interactions and nucleosome centers. *Cell Rep* **21:** 289–300.

Underwood JG, Uzilov AV, Katzman S, Onodera CS, Mainzer JE, Mathews DH, Lowe TM, Salama SR, Haussler D. 2010. FragSeq: Transcriptome-wide RNA structure probing using high-throughput sequencing. *Nat Methods* **7:** 995–1001.

Wade JT. 2015. Where to begin? Mapping transcription start sites genome-wide in *Escherichia coli*. *J Bacteriol* **197:** 4–6.

Walley JW, Sartor RC, Shen Z, Schmitz RJ, Wu KJ, Urich MA, Nery JR, Smith LG, Schnable JC, Ecker JR, et al. 2016. Integration of omic networks in a developmental atlas of maize. *Science* **353:** 814–818.

Wan Y, Kertesz M, Spitale RC, Segal E, Chang HY. 2011. Understanding the transcriptome through RNA structure. *Nat Rev Genet* **12:** 641–655.

Wan Y, Qu K, Ouyang Z, Kertesz M, Li J, Tibshirani R, Makino DL, Nutter RC, Segal E, Chang HY. 2012. Genome-wide measurement of RNA folding energies. *Mol Cell* **48:** 169–181.

Watters KE, Strobel EJ, Yu AM, Lis JT, Lucks JB. 2016a. Cotranscriptional folding of a riboswitch at nucleotide resolution. *Nat Struct Mol Biol* **23:** 1124–1131.

Watters KE, Yu AM, Strobel EJ, Settle AH, Lucks JB. 2016b. Characterizing RNA structures in vitro and in vivo with selective 2′-hydroxyl acylation analyzed by primer extension sequencing (SHAPE-Seq). *Methods* **103:** 34–48.

Watts JM, Dang KK, Gorelick RJ, Leonard CW, Bess JW Jr, Swanstrom R, Burch CL, Weeks KM. 2009. Architecture and secondary structure of an entire HIV-1 RNA genome. *Nature* **460:** 711–716.

Weeks KM. 2010. Advances in RNA structure analysis by chemical probing. *Curr Opin Struct Biol* **20:** 295–304.

Westhof E, Auffinger P. 2006. RNA tertiary structure. In *Encyclopedia of analytical chemistry* (ed. Meyers RA), pp. 5222–5232. Wiley, New York.

Zaug AJ, Cech TR. 1995. Analysis of the structure of Tetrahymena nuclear RNAs in vivo: Telomerase RNA, the self-splicing rRNA intron, and U2 snRNA. *RNA* **1:** 363–374.

Zubradt M, Gupta P, Persad S, Lambowitz AM, Weissman JS, Rouskin S. 2017. DMS-MaPseq for genome-wide or targeted RNA structure probing in vivo. *Nat Methods* **14:** 75–82.

The RNA Base-Pairing Problem and Base-Pairing Solutions

Zhipeng Lu and Howard Y. Chang

Center for Personal Dynamic Regulomes, Stanford University, Stanford, California 94305

Correspondence: howchang@stanford.edu

SUMMARY

RNA molecules are folded into structures and complexes to perform a wide variety of functions. Determination of RNA structures and their interactions is a fundamental problem in RNA biology. Most RNA molecules in living cells are large and dynamic, posing unique challenges to structure analysis. Here we review progress in RNA structure analysis, focusing on methods that use the "cross-link, proximally ligate, and sequence" principle for high-throughput detection of base-pairing interactions in living cells. Beginning with a comparison of commonly used methods in structure determination and a brief historical account of psoralen cross-linking studies, we highlight the important features of cross-linking methods and new biological insights into RNA structures and interactions from recent studies. Further improvement of these cross-linking methods and application to previously intractable problems will shed new light on the mechanisms of the "modern RNA world."

Outline

Cite as *Cold Spring Harb Perspect Biol* doi: 10.1101/cshperspect.a034926

1 THE EVERLASTING RNA STRUCTURE PROBLEM AND EVER-EVOLVING SOLUTIONS

Ever since the discovery of the DNA structure, specific base-pairing has been seen as a perfect mechanism of heredity, as Watson and Crick postulated "a possible copying mechanism for the genetic material" (Watson and Crick 1953). Although DNA structures are relatively limited in variety, RNA structures are more diverse, and their essential roles have been repeatedly discovered and shown in many fundamental RNA-centric processes that make up life on Earth. As the driving force in the formation of helices, RNA base-pairing underlies both intramolecular structures and intermolecular interactions. In the early 1950s and 1960s, it was shown that both transfer of information from DNA to RNA (transcription) and from RNA to protein (translation) use the same sequence complementarity mechanism. In the past three decades, structural analysis showed that it is the complex and dynamic structures in the spliceosome and the ribosome, rather than the protein components, that catalyze chemical reactions (Cech and Steitz 2014). RNA–RNA interactions also form the molecular basis of small RNA (sRNA)-mediated gene regulation in eukaryotes (small interfering RNAs [siRNAs] and microRNAs [miRNAs]) and bacteria (sRNAs). These classical studies have largely dispelled the long-held notion of RNA as passive carriers of genetic information and put RNA-based regulation at center stage of molecular biology.

A large variety of methods that use fundamentally different principles have been developed to study RNA structures (Table 1). X-ray crystallography provided the first atomic resolution picture of an RNA molecule, when the first transfer RNA (tRNA) crystal structure was solved in the 1970s (Kim et al. 1973; Robertus et al. 1974). Since then, X-ray crystallography has been the dominating tool in studying both protein and RNA structures (Shi 2014). Nuclear magnetic resonance (NMR) probes the conformation of molecules in solution, providing an alternative means to analyze RNA, especially flexible ones (Bothe et al. 2011). Cryogenic electron microscopy (cryo-EM), on the other hand, examines single frozen-hydrated particles in their native state, and therefore can be applied to larger and more dynamic molecules and assemblies. With new technological innovations in electron detection and image processing, the resolution of cryo-EM has dramatically improved in the past few years and in many cases approaches that of X-ray crystallography (Bai et al. 2015). Some of the most spectacular advances using cryo-EM include the

Table 1. Major categories of methods for the analysis of RNA structures and interactions

Categories	Variety of methods	Features	General reference(s)
X-ray crystallography		In vitro, atomic resolution, smaller, and nearly static molecules	Shi 2014
NMR		In vitro, atomic resolution, smaller molecules, can be either static or flexible	Bothe et al. 2011
Cryo-EM		In vitro, near atomic resolution, large, and heterogeneous complexes	Bai et al. 2015
Thermodynamic methods	Nussinov algorithm, Zuker algorithm, etc. RNA structure, Sfold, Mfold, UNAfold, ViennaRNA Package, etc.	Guarantees theoretically optimal structures, slow when permitting pseudoknotted structures	Eddy 2004; Mathews et al. 2010
Comparative sequence analysis	Fold and Align, Fold then Align, Align then Fold, etc.	Implies functional relevance, limited by alignments and efficiency (local structures)	Mathews et al. 2010
Chemical probing	DMS-seq, Structure-seq, SHAPE, icSHAPE, Mutate and Map, etc.	In vitro and in vivo, one-dimension averaged reactivity/flexibility/accessibility profile, can be used to probe secondary, tertiary structures and RNA–protein interactions	Kladwang et al. 2011; Ding et al. 2014; Rouskin et al. 2014; Spitale et al. 2015
Enzymatic probing	PARS, Frag-seq, etc.	In vitro, one-dimension averaged reactivity/flexibility/accessibility profile	Kertesz et al. 2010; Underwood et al. 2010
Cross-linking (based on physical proximity)	PARIS, LIGR-seq, SPLASH, CLASH, hiCLIP, MARIO	In vitro and in vivo, direct physical base-pairing contacts (except MARIO), captures alternative conformations	Helwak et al. 2013; Sugimoto et al. 2015; Aw et al. 2016; Lu and Chang 2016; Nguyen et al. 2016; Sharma et al. 2016

NMR, Nuclear magnetic resonance, Cryo-EM, cryogenic electron microscopy; DMS, dimethyl sulfate; SHAPE, selective 2′-hydroxyl acylation by primer extension; icSHAPE, in vivo click SHAPE; PARS, parallel analysis of RNA structures; Frag-seq, fragmentation sequencing; PARIS, psoralen analysis of RNA interactions and structures; LIGR-seq, ligation of interacting RNA and high-throughput sequencing; SPLASH, sequencing of psoralen cross-linked, ligated, and selected hybrids; CLASH, cross-linking, ligation, and sequencing of hybrids; hiCLIP, RNA hybrid and individual nucleotide resolution cross-linking and immunoprecipitation; MARIO, mapping RNA interactome in vivo.

Cite as *Cold Spring Harb Perspect Biol* doi: 10.1101/cshperspect.a034926

structures of spliceosomes corresponding to several intermediary steps in splicing, each containing multiple small nuclear RNAs (snRNAs) and proteins (Kastner et al. 2019; Plaschka et al. 2019; Yan et al. 2019). Messenger RNAs (mRNAs) and long noncoding (lnc)RNAs, which make up the majority of distinct RNA species, are generally much more flexible, with well-folded regions interspersed with amorphous linker regions, and folded together with a large collection of nonstoichiometric RNA partners, RNA-binding proteins (RBPs), as well as indirect binders. Although it is likely that further technological innovations will greatly enhance the power of X-ray crystallography, NMR, and cryo-EM, currently the vast majority of RNA molecules are beyond the reach of these in vitro methods.

In parallel with the in vitro experimental methods, two classes of computational algorithms were developed to address the RNA-folding problem, using thermodynamics and comparative sequence analysis, respectively (Table 1) (Fallmann et al. 2017). As RNA molecules are assumed to most likely fold into lowest free energy states, stable structures can be predicted based on free energy terms for each unit of stacked base pairs and loops, also called the Turner energy rules (Mathews et al. 2010). Given the time and space complexity of these algorithms, restrictions are commonly applied, such as prohibiting the formation of pseudoknots and long-range structures. In addition, our understanding of the basic energy rules in RNA structure formation, as well as the contributions from the cellular environment, is limited. Together, these issues have resulted in structure models that are "elegant and too often wrong" (Eddy 2004; Mathews et al. 2010).

RNA structures that have biological functions are under evolutionary pressure to preserve specific secondary or tertiary interactions, often in the form of invariant or covariant base pairs (Rivas and Eddy 2001). Therefore, another approach to determine RNA structures is by identifying these signatures of selection in multiple sequence alignments (Table 1). Comparative sequence analysis is not restricted by our limited understanding of the thermodynamic rules and, therefore, can reveal both standard base pairs and tertiary interactions. By the very nature of this approach, the structures derived from sequence comparison are probably functional. However, comparative sequence analysis is only applicable to high-quality alignments with certain sequence variation levels, and often restricted to small sequence windows in the interest of efficiency.

RNA folding brings nucleotides into specific chemical environments that change their reactivity properties (or solvent accessibility, flexibility). Therefore, reading out the reactivity of individual nucleotides reveals information about the specific folding. Based on this principle, a large number of methods have been developed in the past several decades, using various chemicals and enzymes to probe nucleotide reactivity (Table 1) (Ehresmann et al. 1987; Lu and Chang 2016). The chemicals and enzymes differentially react with the base, sugar, or the phosphodiester backbone, either in vitro or in vivo, resulting in modifications that can be read out as reverse transcription stops in gel electrophoresis or high-throughput sequencing. Reactivity profiles obtained from such experiments correlate with base-pairing status and can be used as constraints in secondary structure modeling, often as additional pseudoenergy terms. Probing data have been shown to improve the accuracy of structure modeling to some extent yet are still limited to simple structures. These experiments provide averaged signals for each RNA species, which does not reflect the dynamic nature of RNA in living cells (Lu and Chang 2016). Many of the inherent problems in RNA structure prediction persist even with the addition of chemical probing data. Statistically correlated modification patterns can be obtained from reverse transcriptase-induced mutations but are very low in efficiency and limited in length (Siegfried et al. 2014). A mutate-and-map strategy was recently developed to derive information about base pairs but is only applicable to one RNA at a time in vitro (Kladwang et al. 2011).

Contrary to the conventional chemical and enzymatic probing experiments, cross-linking-based methods, which are the topic of the current review, aim to directly identify the base-pairing regions and limit the "guesswork" of structure modeling to very short sequences that typically fold into a single possible duplex (Table 1). Originally developed in the 1970s, photoactivated psoralen cross-linking has provided mechanistic insights into many important and challenging problems in RNA biology. Recently, new technical improvements have led to a revival of this class of methods, bringing superior sensitivity, resolution, and throughput. Starting from a historical perspective, in this review we will discuss the experimental and analytical aspects of several psoralen cross-linking techniques that use the same principle of "cross-link, proximally ligate, and sequence." Current applications of psoralen cross-linking have resulted in the surprising discoveries of long range, alternative, and dynamic structures, new RNA–RNA interactions, and principles in the high-level organization and networks of the transcriptome. As no tool is perfect for any task, it is essential to integrate cross-linking methods with other types of structure determination techniques. We will discuss the limitations of current implementations and how to push the technologies to the next level.

2 HISTORICAL USE OF PSORALEN TO ANALYZE RNA STRUCTURES AND INTERACTIONS

Psoralen is a group of planar tricyclic compounds that can intercalate in double-stranded nucleic acids (Fig. 1A)

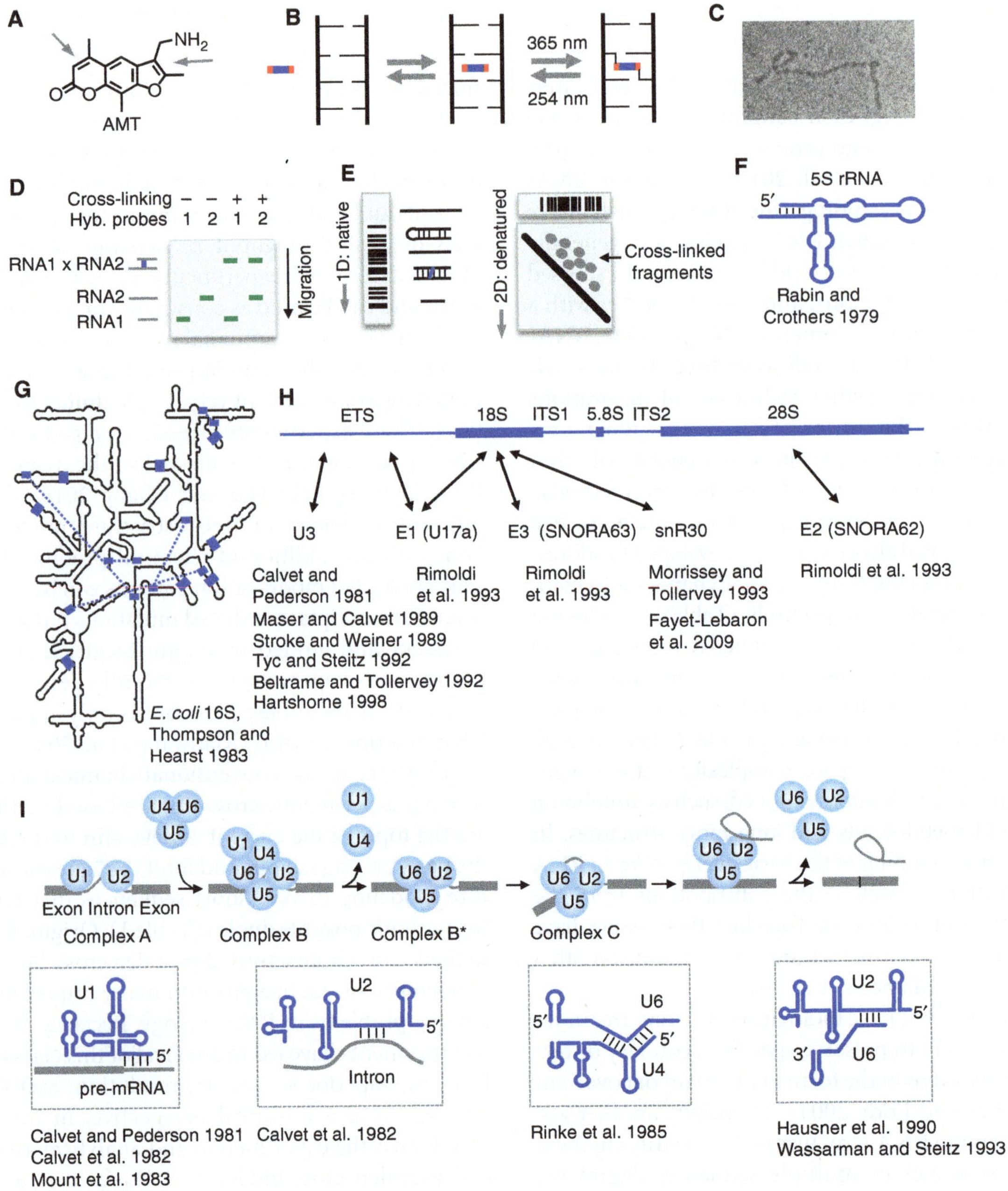

Figure 1. Psoralen chemistry and historical applications. (*A*) Structure of the commonly used psoralen AMT (4′-aminomethyltrioxsalen). Arrows point to the double bonds that undergo cycloaddition to pyrimidines. (*B*) Model of psoralen intercalation into nucleic acid helices and reversible cross-linking. (*C*) Detection of psoralen cross-linked RNA structures by electron microscopy. (Image courtesy of Paul L. Wollenzien et al.; reprinted, with permission, from Wollenzien et al. 1978, © National Academy of Sciences.) (*D*) Detection of RNA–RNA interactions using gel electrophoresis and northern blotting. Cross-linked RNA pairs can be identified by the simultaneous hybridization of probes to both RNAs. (*E*) The "native-denatured" 2D gel method for selecting psoralen cross-linked RNA. The first-dimension gel is native. In the second dimension, the first-dimension gel slices are embedded in a urea-denatured gel and run at high temperature, in which the cross-linked fragments assume an "X" shape and migrate much more slowly, above the diagonal, forming scattered dots for simple RNA species or a smear for complex RNA mixtures, such as total RNA. (Adapted from Lu et al. 2016, with permission, from Elsevier.) (*F*) Secondary structure of the *Escherichia coli* 5S ribosomal RNA (rRNA), based on psoralen cross-linking and enzymatic sequencing (Rabin and Crothers 1979). (*G*) Psoralen cross-linked duplexes (blue blocks and the blue dashed lines) superimposed on the *Escherichia coli* 16S rRNA model derived from crystal structure (Petrov et al. 2014). (Redrawn, with permission, from Thompson and Hearst 1983.) (*H*) Summary of small nucleolar RNA (snoRNA)–rRNA interactions based on psoralen cross-linking. ETS/ITS: external/internal transcribed sequence. Arrows point to cross-linking sites. (*I*) Summary of several RNA–RNA interactions in precursor messenger RNA (pre-mRNA) splicing based on psoralen cross-linking. (The splicing model was redrawn, with permission, from Will and Luhrmann 2011.)

(Hearst 1981). The extensively conjugated system undergoes cycloaddition with pyrimidines upon long-wavelength ultraviolet (UV) irradiation, forming monoadducts and then diadducts, or interstrand cross-links (Fig. 1B). Short-wavelength UV reverses the cross-link, permitting analysis of cross-linked RNA using various methods. Cross-linking presents definitive and specific evidence for base-pairing and has provided the first mechanistic insights into a variety of noncoding (nc)RNAs including the ribosomal RNAs (rRNAs), spliceosomal snRNAs, and small nucleolar and small Cajal body RNAs (sno/scaRNAs). Hearst and colleagues have extensively discussed the chemical properties and early applications of psoralen on nucleic acid structures (Hearst 1981; Cimino et al. 1985). Here we briefly review some of the classical studies that paved the way for recent breakthroughs in throughput, sensitivity, and resolution of cross-linking-based methods.

Initially, electron microscopy (EM) was used to detect psoralen cross-linking sites in denatured and psoralen cross-linked nucleic acids (Fig. 1C) (Cech and Pardue 1976; Wollenzien et al. 1978). EM is limited in resolution to dozens of base pairs and only shows the overall contour of the secondary structure. To analyze RNA–RNA interactions, denatured gel electrophoresis was used to resolve psoralen cross-linked samples, in which cross-linked RNAs were identified as slower bands by northern blotting (Fig. 1D) (Calvet and Pederson 1981; Calvet et al. 1982). Reversal of cross-linking further confirms the interaction. This setup works well for interactions involving sRNAs in which the partners can be exhaustively tested. To analyze complex interactions and structures, Brimacombe and colleagues developed the 2D gel method (Fig. 1E) (Zwieb and Brimacombe 1980). The 2D gel uses the differences in migration of the cross-linked RNA fragments in denatured gel versus native gel. After the first-dimension native gel, cross-linked fragments assume an "X" shape in the denatured gel and migrate slower than non-cross-linked fragments. The retarded fragments can be isolated from above the diagonal for sequencing. Together, these three methods were extensively used in the early studies of RNA structures and interactions.

The ribosome is a large RNA–protein complex that makes proteins from genetic information stored in mRNA. Several groups used psoralen cross-linking to study the folding of the rRNAs before crystallization was possible. Rabin and Crothers applied psoralen cross-linking to the 5S rRNA folded in solution and showed the existence of a terminal stem (Fig. 1F) (Rabin and Crothers 1979). Thompson and Hearst combined the in-solution psoralen cross-linking, 2D gel purification, photoreversal, and enzymatic sequencing to discover 13 base-paired regions in the *Escherichia coli* 16S rRNA (Fig. 1G) (Thompson and Hearst 1983). Although most of the duplexes from cross-linking are consistent with the evolutionary model built by Noller and Woese and the crystal structure model (Noller and Woese 1981; Petrov et al. 2014), a few of them could be artifacts of in vitro folding (for review, see Sergiev et al. 2001 and Whirl-Carrillo et al. 2002). Nevertheless, these heroic efforts showed the possibility of using psoralen to study complex RNAs.

The rRNAs are transcribed as a polycistronic precursor, then extensively processed and modified into the mature subunits. Psoralen cross-linking elucidated several critical steps in rRNA biogenesis and revealed functions of an abundant class of sRNAs, snoRNAs (Fig. 1H). Calvet and Pederson showed that the U3 snoRNA could be cross-linked to the large rRNAs in vivo, providing the first direct evidence of snoRNAs involvement in rRNA processing (Calvet and Pederson 1981). This interaction is required for the cleavage of the rRNAs to release the mature 18S rRNA (Kass et al. 1990). Later studies have continued to refine the cross-linking and mapping methods to identify the precise binding sites on rRNAs for U3, E1 (U17a), E2 (SNORA62), E3 (SNORA63), snR30, etc. (Calvet and Pederson 1981; Maser and Calvet 1989; Stroke and Weiner 1989; Beltrame and Tollervey 1992; Tyc and Steitz 1992; Morrissey and Tollervey 1993; Rimoldi et al. 1993; Hartshorne 1998; Fayet-Lebaron et al. 2009). Together with genetic experiments that specifically edit the identified binding sites, these studies showed the essential roles of snoRNAs in rRNA processing.

The psoralen cross-linking methods made critical contributions in elucidating the mechanisms of eukaryotic mRNA splicing (Fig. 1I). Steitz and colleagues hypothesized that the U-rich snRNAs are involved in the removal of introns from eukaryotic heterogeneous nuclear RNAs (hnRNAs; precursor messenger RNAs [pre-mRNAs]) based on sequence complementarity of U1 with the 5′ splice site (Lerner et al. 1980). Shortly after that, Calvet and Pederson showed that U1 and U2 small nuclear ribonucleoprotein complexes (snRNPs) can be reversibly cross-linked to pre-mRNAs, providing the first physical evidence for the direct interactions (Calvet and Pederson 1981; Calvet et al. 1982). Since then, a series of studies from several groups pinpointed the interaction sites between pre-mRNAs and snRNAs, revealing a complex and dynamic ribonucleoprotein complex (RNP) machine (Mount et al. 1983; Rinke et al. 1985; Hausner et al. 1990; Wassarman and Steitz 1993). The sequential formation and reorganization of distinct RNA–RNA interactions ensure efficient and precise removal of introns (Fig. 1I). The splicing mechanisms revealed by psoralen cross-linking have since been validated and refined by genetic analysis, X-ray crystallography, and cryo-EM (Will and Luhrmann 2011).

3 THE NEXT GENERATION: CROSS-LINKING WITH PROXIMITY LIGATION AND HIGH-THROUGHPUT SEQUENCING

Traditional methods for the analysis of cross-linked RNA, such as EM, 1D and 2D gels, and enzymatic sequencing, are laborious and inefficient (Fig. 1C,D). As a result, applications were limited to highly abundant RNAs one at a time, like the rRNAs, snRNAs, and snoRNAs (Fig. 1). Several recent technological advancements have transformed classical psoralen cross-linking into a powerful transcriptome-scale discovery tool (Fig. 2A, left side). In the new generation of methods, including psoralen analysis of RNA interactions and structures (PARIS), sequencing of psoralen cross-linked, ligated, and selected hybrids (SPLASH), and ligation of interacting RNA and high-throughput sequencing (LIGR-seq), cells are treated with psoralens and UV irradiation; then cross-linked RNA fragments are enriched and proximally ligated so that each RNA duplex can be uniquely identified via high-throughput sequencing (Aw et al. 2016; Lu et al. 2016; Sharma et al. 2016; Lu et al. 2018). This strategy determines RNA structure and interactions with near base-pair resolution and single molecule accuracy at the transcriptome level. Three related methods that use a similar strategy, but without psoralen cross-linking, were also reported recently (Fig. 2A, right side) (Kudla et al. 2011; Helwak et al. 2013; Sugimoto et al. 2015; Nguyen et al. 2016). Here these methods are compared side by side to illustrate their similarities and differences.

Several major differences exist among the psoralen cross-linking methods (Fig. 2B). First, biotinylated psoralen was used in SPLASH instead of the more commonly used 4′-aminomethyltrioxsalen (AMT), thus enabling convenient isolation of RNA fragments covalently bound to psoralen, either with monoadducts (cycloaddition on one side) or cross-links. The bigger size of the bio-psoralen molecule also reduces efficiency in the intercalation of RNA duplexes. The concentrations of psoralens used vary considerably. Conventional chemical probing usually selects chemical concentrations to aim for "single-hit kinetics" to avoid the induction of structural changes during probing (Spitale et al. 2015). However, high concentrations of psoralen may not be an issue because cross-linking fixes structures to prevent further changes.

Given the low efficiency of cross-linking, several different strategies were used to enrich cross-linked fragments from total RNA. The 2D gel technique used in PARIS, although laborious, enriches cross-linked duplexes up to 100% in purity. In LIGR-seq, an exoribonuclease, RNase R, was used to remove nonlinked single-stranded fragments. This approach is easier to perform but does not efficiently remove non-cross-linked fragments, especially highly structured ones. In SPLASH, purification of bio-psoralen-reacted fragments is simple with the streptavidin beads, but both monoadduct and cross-linked RNA fragments are extracted. The three methods select RNA fragments at different sizes, affecting both proximity ligation efficiency and structure model resolution. Longer fragments promote ligation but also result in ambiguous base-pairing models.

Short-wavelength UV and bifunctional chemical cross-linkers were used in cross-linking, ligation, and sequencing of hybrids (CLASH), RNA hybrid and individual nucleotide resolution cross-linking and immunoprecipitation (hiCLIP), and mapping RNA interactome in vivo (MARIO; Table 1, Fig. 2). In CLASH and hiCLIP, UV–cross-linked protein–RNA complexes were selected using antibodies; therefore, only the targets of a single RBP are identified. These methods achieve deeper coverage of specific interactions given the limited scope. MARIO fixes both specific (from UV cross-linking) and nonspecific (from chemical cross-linking) interactions in physical proximity, and therefore the identified RNA fragment pairs are not necessarily base-paired. The detection of all protein-associated interactions is possible with MARIO because all RNA–protein complexes are recovered from cell lysate. However, the large size of the RNA fragments in MARIO precludes accurate structure modeling (Fig. 2B).

A commonly used strategy in the analysis of chromatin conformations, proximity ligation joins two restriction-digested DNA fragments using DNA ligases to enable the identification of their physical proximity in 3D space in the nucleus (Dekker et al. 2002). Proximity ligation in RNA was first noticed in the analysis of RNA–protein cross-linking studies and then intentionally used to identify potential RNA–RNA interactions (Kudla et al. 2011). In contrast to DNA, in which longer fragments and the sticky ends ensure highly efficient ligation, the ends of shorter RNA duplexes are under steric constraint, resulting in much lower ligation efficiency (Fig. 2B). The ligation efficiency has not improved much despite extensive optimizations. The typical T4 RNA ligase 1 (Rnl1) is used in most applications, whereas in LIGR-seq, a thermostable T4 Rnl1 homolog CircLigase was used (Blondal et al. 2005). In hiCLIP and MARIO, a linker oligonucleotide was used to bridge the two ends, but with no obvious improvement, except in MARIO in which much longer fragments are ligated.

The various technical approaches used in these methods provide flexible choices that can be used in a "mix-and-match" manner depending on the nature of the specific problem to be solved. Users can choose psoralen derivatives, RNA fragmentation methods, enrichment techniques, ligases with or without linkers, and antibodies or antisense oligos that select specific subsets of RNAs. Together these methods

Cite as *Cold Spring Harb Perspect Biol* doi: 10.1101/cshperspect.a034926

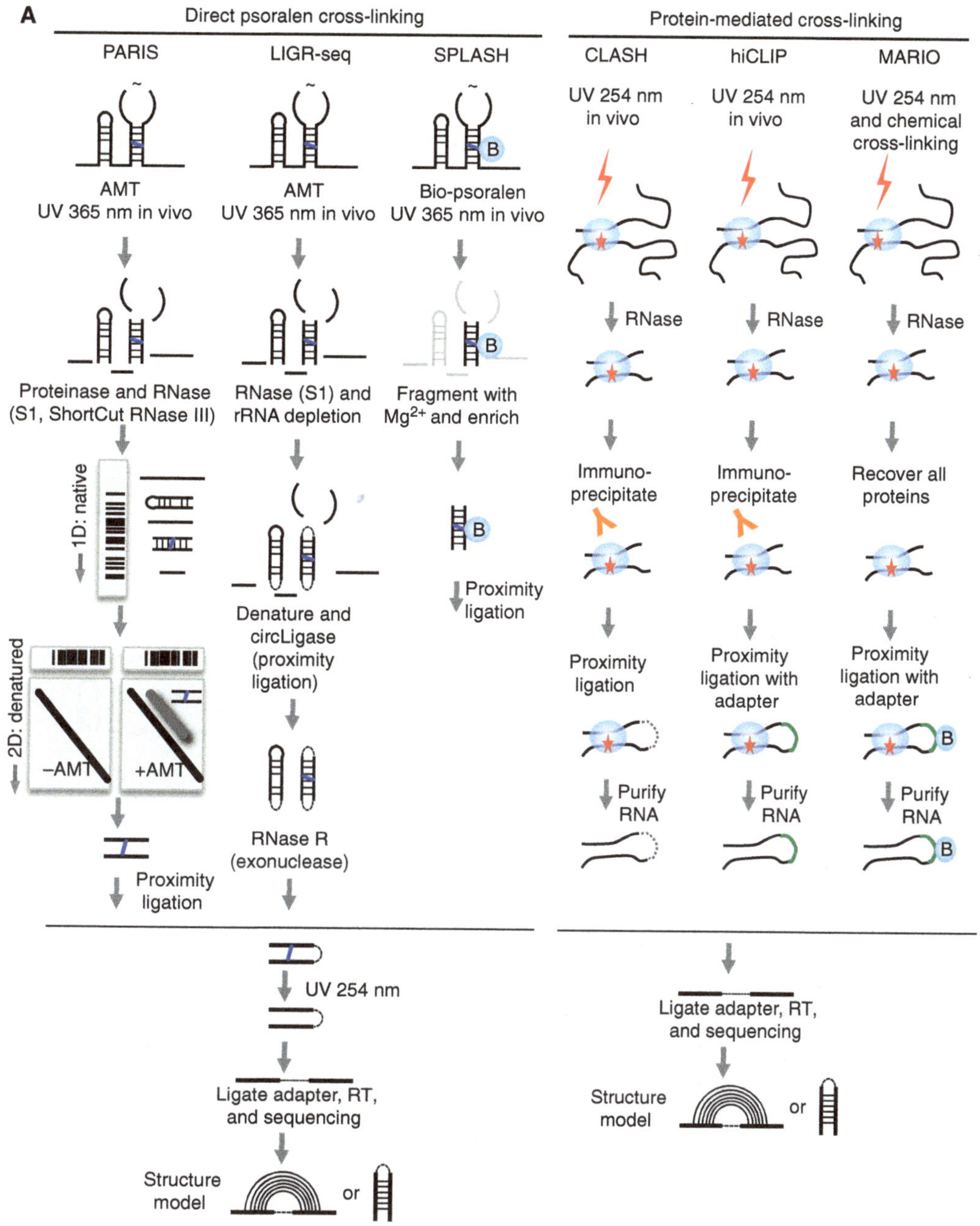

B Summary of comparison

Methods	PARIS	LIGR-seq	SPLASH	CLASH	hiCLIP	MARIO
Scale	Global	Global	Global	Targeted	Targeted	Global
Cross-linker	Psoralen AMT	Psoralen AMT	Bio-psoralen	UV 254 nm	UV 254 nm	Aldehyde, glyoxal
Fragment	S1, RNase III	S1	Mg^{2+} (90–110 nt)	RNase A/T1	RNase I	RNase I (>500 nt)
Enrichment	2D gel	Exonuclease	Biotin-streptavidin	Antibody + gel	Antibody + gel	Biotin on proteins
Ligation	T4 Rnl1	CircLigase	T4 Rnl1 + linker	T4 Rnl1	T4 Rnl1 + linker	T4 Rnl1 + linker
rRNA removal	No	Yes	No	No	No	No
Resolution	Near bp	Near bp	Near bp	Near bp	Near bp	Low resolution
Chimeric %	2.5%–6%	< 0.4%	1.7%–5.9%	< 1%	~ 2%	8%–30%

Figure 2. Recently developed cross-linking-based methods for the direct detection of RNA interactions and structures. Only the methods that use the "cross-link, proximally ligate, and sequence" principle are listed here. (*A*) Workflow of the methods. (*B*) Major differences among the methods. (Part of panel *A* is adapted from Lu et al. 2016, with permission, from Elsevier.)

have been applied to multiple species and cell lines, providing a rich resource for future studies (Gong et al. 2018).

4 BASIC ANALYSIS OF RNA STRUCTURE CROSS-LINKING DATA

The cross-linking (PARIS-like) experiments produce a small percentage of noncontinuous reads as a result of proximity ligation of base-paired duplexes (Fig. 3A,B). The noncontinuous reads can be regularly gapped or chiastic, depending on which side proximity ligation occurs (Fig. 3A,B). Short-read mappers, like the spliced transcripts alignment to a reference (STAR) software, are capable of directly mapping both types of sequences with appropriate parameters (Dobin et al. 2013). Although the noncontinuous reads can be directly used, assembly of duplex groups (DGs) is a critical step that standardizes and simplifies subsequent analysis of both structures and interactions

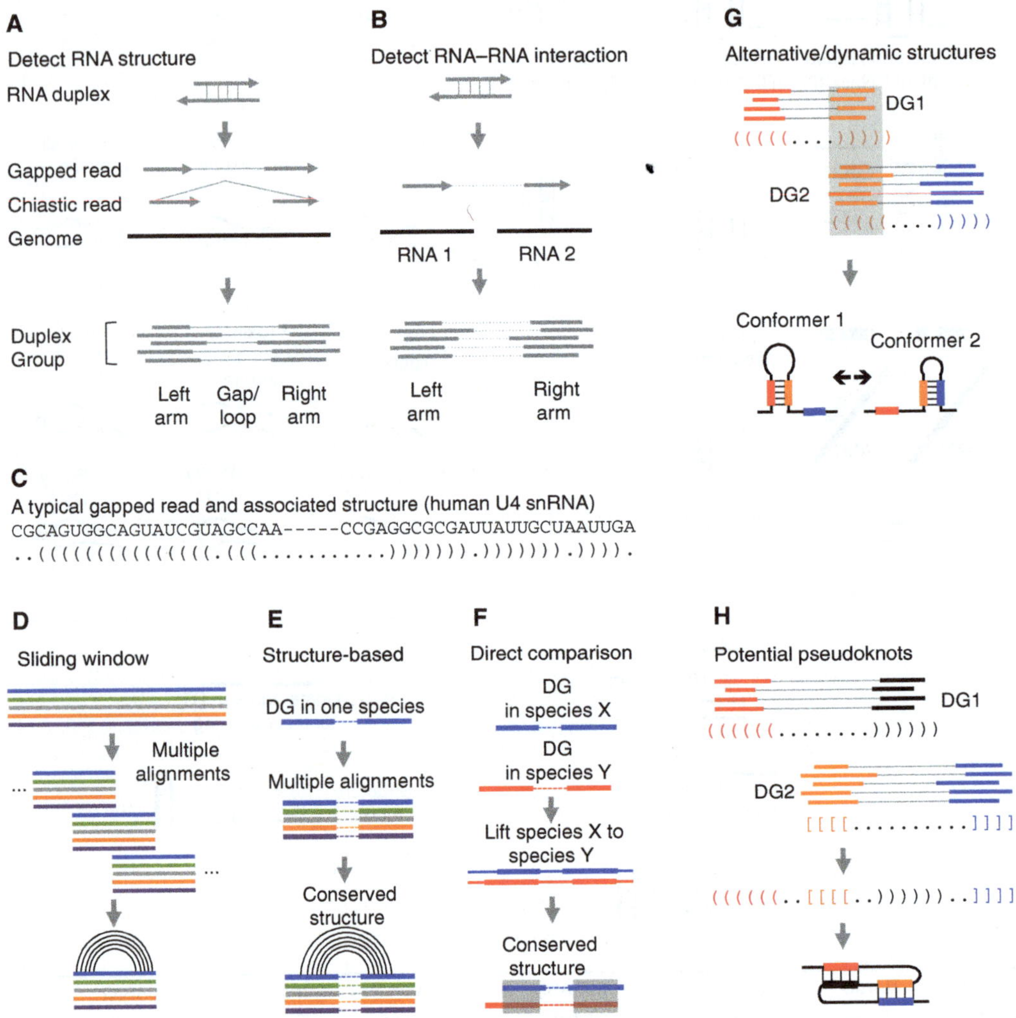

Figure 3. RNA structure and interaction analysis based on cross-linking data. (*A*) Detecting RNA structures. Discontinuous sequencing reads, including "regularly" gapped and chiastic, are assembled into duplex groups (DGs). (*B*) Detecting RNA–RNA interaction. The two arms of discontinuous reads mapped to two RNA molecules are assembled into DGs. (*C*) An example of a discontinuous read from the U4 small nuclear RNA (snRNA). Pairs of parentheses indicate base pairs. (*D*) Conventional sliding window screen for conserved structures from whole genome alignments. Multiple genome alignments are split into windows with fixed width and step. Conservation and covariation are then scored in each window. (*E*,*F*) DGs guide comparative sequence analysis. (*E*) DGs from one species are used to extract two blocks of alignments separated by arbitrary distance for structure conservation analysis. (*F*) Structures can be directly compared with DGs from multiple species, using sequence alignments. (*G*) Alternative/dynamic RNA structures are identified based on the conflicts of the base-pairing models between two DGs. (*H*) Interlocked DGs suggest either alternative conformations or pseudoknot structures. (Panels *E*–*G* are adapted from Lu et al. 2016, with permission, from Elsevier.)

Cite as *Cold Spring Harb Perspect Biol* doi: 10.1101/cshperspect.a034926

(Lu et al. 2016, 2018). A DG is a group of reads from RNA molecules in the same conformation, corresponding to one specific RNA duplex. Many structure prediction programs can be used to build the base-pairing model from the DGs (Table 1). Given the small arm size and the specific psoralen cross-linking of staggered uridines, a unique and unambiguous structure model can be established (Fig. 3C). The structure models can be validated using various orthogonal methods, including selective 2′-hydroxyl acylation by primer extension (SHAPE)-like chemical probing and conservation analysis (Lu et al. 2016). In addition, DGs enable facile analysis of significance, assembly of complex structures, including dynamic conformations, pseudoknots and high-level architectures, and analysis of RNA–protein interactions (Fig. 3C–H).

Whereas intramolecular structures are straightforward, intermolecular interactions are more difficult to determine, in part because of the complexity of the transcriptome, which contains homologous and repetitive sequences and frequent errors in the gene annotation. Genome-mapped reads need to be extensively filtered to remove artifacts. Alternatively, carefully curated transcriptome annotations can be used as reference to avoid spurious mapping but may miss ones not in the annotation. The significance of detected structures has been determined using different approaches. After DG assembly, a threshold was applied to the ratio of the number of reads in the DG divided by sequencing depth at the two arms (Lu et al. 2016). Alternatively, the Blencowe group used a probabilistic model to assess the significance of the detected interactions based on the observed and expected connections between the two arms (Sharma et al. 2016).

5 PREVALENT LONG-RANGE STRUCTURES IN THE TRANSCRIPTOME

One surprising observation in the psoralen cross-linking and hiCLIP studies is the prevalence of long-range structures (Fig. 4A). (Here we use "structure" to denote all intramolecular base-pairing interactions, and "interaction" to refer to only intermolecular interactions [Sugimoto et al. 2015; Aw et al. 2016; Lu et al. 2016; Sharma et al. 2016].) Many structures span hundreds to thousands of nucleotides, and in mRNAs, often span multiple exons. Computational predictions, including ones that incorporate chemical probing data, have focused on local structures in the interest of efficiency. The resulting models should now be viewed with caution in light of evidence showing prevalent long-range structures.

In both mRNAs and lncRNAs, structures that connect the 5′ and 3′ ends have been detected. Wan and colleagues showed that there is a positive correlation between the circularization score (tendency to form longer range structures) with translation efficiency for mRNAs (Fig. 4I) (Aw et al. 2016). In the case of mRNAs, the end-to-end connections are reminiscent of the mRNA circularization mechanism essential for translation, in which translation initiation factors on the 5′ end interact with the poly(A)-binding protein on the 3′ end (Wells et al. 1998). mRNA circularization can also be mediated by base-pairing, as shown in the p53 mRNA, and this long-range base-pairing between the 5′ and 3′ untranslated regions (UTRs) promotes translation (Chen and Kastan 2010). End-to-end base-paired structures are frequently observed in viruses and play critical roles in virus replication and translation (Nicholson and White 2014). As RNA structures form during 5′ to 3′ transcription, and are constantly reorganized by splicing, translation, and other processes, it is important to determine how long-range structures emerge given that local contacts are more likely to form by chance. More interestingly, how do long-range structures contribute to the various RNA metabolism events? Comprehensive dissection of each structure by mutagenesis is needed to study their functions.

6 CONSERVATION ANALYSIS OF THE RNA STRUCTURES

RNA structures form spontaneously as a result of thermodynamics; presence of RNA structures does not imply function. Base-pair covariation, however, is an important indicator of evolutionary pressure on functional structures. Sifting through sequence alignments for conserved and covaried base pairs is a classical strategy for identifying functional structures, and this can be applied to either individual RNAs or whole genomes (Washietl et al. 2005; Pedersen et al. 2006; Smith et al. 2013). In practice, the search is restricted to small sliding windows across the genome because of extraordinary computational requirements (Fig. 3D). Significance can be calculated by shuffling in each window (Gesell and von Haeseler 2006). However, these methods only examine local structures and lack experimental validation. Given the stringency of these methods, conserved structures cannot be identified from sequences lacking sufficient variation.

The direct determination of helices in the cross-linking methods enables targeted analysis of structure conservation without distance limit. Using the PARIS-determined duplexes as guides, we developed two approaches to directly search for structure conservation (Lu et al. 2016). First, the DGs from one species were used to extract alignment blocks from multiple genomes, in which the two arms can come from distant regions (Fig. 3E). Second, when structure data are available from more than one species, direct comparison

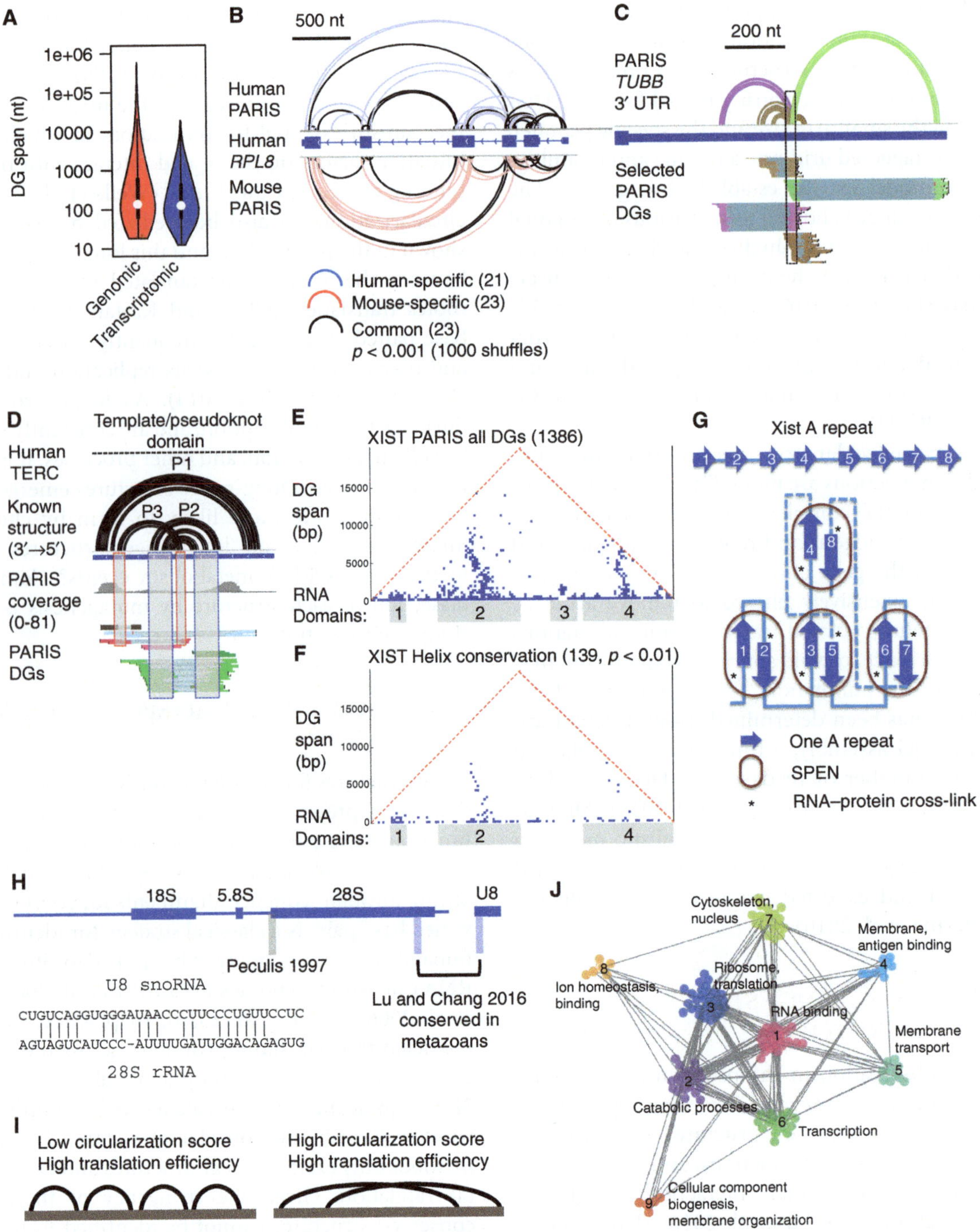

Figure 4. New insights into the RNA structurome and interactome. (*A*) Long-range structures are prevalent in the transcriptome. Duplex group (DG) spans were calculated from Henrietta Lacks cell line psoralen analysis of RNA interactions and structures (HeLa PARIS) data on genomic and transcriptomic coordinates. (*B*) Example of conserved structures in *RPL8* messenger RNA (mRNA) based on direct comparison of human (human embryonic kidney cells 293 [HEK293]) and mouse (J1 embryonic stem cells) PARIS data. Significance of the overlap was tested using random shuffling of DGs in the exons. (*C*) Example of alternative/dynamic structures in the 3′ untranslated region (UTR) of tubulin β class 1 (TUBB) mRNA from HeLa PARIS data. The corresponding structure models (first track) and DGs are color coded. (*D*) PARIS detects pseudoknot structure (interlocked DGs) in the telomerase RNA component (TERC). (*E*) PARIS determines the architecture of X-inactive specific transcript (XIST). Each point in the heat map shows the connection between the two regions indicated by the feet of the triangle. Four major domains are obvious from the clustered DGs. (*Legend continued on following page.*)

Cite as *Cold Spring Harb Perspect Biol* doi: 10.1101/cshperspect.a034926

of the DGs is performed on sequence alignments (Fig. 3F). The second approach is especially useful when variation is too low or too high for statistical analysis of significance. We found that a substantial fraction of structures that span more than 200 nt are conserved. These analyses also showed that overall architectures are conserved for both coding and noncoding long RNAs (Fig. 4B). On the other hand, structure data may also be helpful in correcting sequence-based alignments that often do not consider structure information. This approach will be especially useful for distantly related sequences that cannot be aligned properly by sequence alone.

Given the ubiquitous presence of RNA structures, they can be quickly adapted for specific functions without showing obvious signs of conservation. As a result, such functional structures would have been missed by conventional methods that rely on covariation. Bartel and colleagues showed that random structures in the 3′ UTR are required for mRNA stability (Wu and Bartel 2017). This study highlights the limits of conventional comparative sequence analysis and the need for direct determination of structures.

7 DYNAMIC AND COMPLEX STRUCTURES: CATCHING RNA IN ACTION

Original work in the 1950s by Anfinsen showed that protein structures are mostly determined by primary sequence (Anfinsen 1973). Dynamics in protein structures occurs locally, usually without breaking secondary structures. However, RNA folding is more promiscuous given the simpler alphabet and base-pairing rules. In addition to local structure dynamics, global rearrangements of structures are frequent in RNA and essential for their activities (Dethoff et al. 2012). As described earlier in this review, dramatic rearrangement of structures and interactions occurs multiple times during pre-mRNA splicing (Fig. 1H). Conventional chemical probing, although applicable to structure determination in vivo, results in averaged reactivity profiles that are hard to deconvolve into the individual conformations. In contrast, psoralen cross-linking and proximity ligation directly capture each conformation in a single sequencing read, thus providing a direct readout of all the "cross-linkable" conformations, like taking snapshots of the RNA in action.

Dynamic and alternative structures take the form of conflicting DGs, in which two DGs overlap on one arm (Figs. 3G and 4C). Significantly overlapped base pairs cannot form at the same time, with the exception of triplexes, and usually are the result of two mutually exclusive conformations. Using this simple criterion, a large number of dynamic structures were uncovered, some of which are conserved between human and mouse (Lu et al. 2016). Whereas some of the conformations could be simply the kinetically or thermodynamically trapped folding intermediates, others may represent important functional states in a dynamic regulatory process. Wan and colleagues applied the SPLASH method to the human embryonic stem cell differentiation process and found that the transition in cellular states is accompanied by alterations of both RNA structures and RNA–RNA interactions, suggesting regulatory mechanisms by the dynamic structures and interactions (Aw et al. 2016). The constant rearrangement of intra- and intermolecular base-pairing interactions and the dynamic protein interactions create a rugged free energy landscape that integrates genetic and environmental signals. Further investigation of RNA dynamics will likely uncover new "switches" in gene regulation pathways.

Pseudoknots are composite structures that involve at least two duplexes that are interlocked. Pseudoknots stabilize structures and participate in several critical cellular processes, including telomere maintenance and ribosomal frameshifting (Gilley and Blackburn 1999; Giedroc and Cornish 2009). Computational methods can be used to identify pseudoknots but are very slow (Rivas and Eddy 1999). Psoralen cross-linking directly identifies interlocked duplexes that could be either alternative conformations or pseudoknots (Fig. 3H). For example, PARIS identified the pseudoknots in telomerase RNA component (TERC; Fig. 4D), RNA component of mitochondrial RNA processing endoribonuclease (RMRP), and ribonuclease P RNA com-

Figure 4. (*Continued.*) (*F*) About 10% of PARIS-determined duplexes are conserved in eutherian mammals (p value < 0.01). (*G*) Model of SPEN-A-repeat complex in the XIST ribonucleoprotein (RNP). The base-pairing interactions among the 8.5 repeats are stochastic and only one specific family of conformations from PARIS data is shown here. SPEN binding involves both single- and double-stranded regions but is only cross-linked to the single-stranded region 3–5 nt upstream of the interrepeat duplex unit. (*H*) PARIS in human and mouse cells revealed the precise U8:28S interaction, which is conserved in metazoans. PARIS-determined interaction sites in blue, and the previously reported site in gray (Peculis 1997). The base-pairing model is shown in the inset. (*I*) Long-span structures as detected by sequencing of psoralen cross-linked, ligated, and selected hybrids (SPLASH) correlates with higher translation efficiency. (*J*) RNAs of related functions tend to interact more frequently than nonrelated ones. Global RNA–RNA interaction networks can be organized into modules and the network topology changes in response to stem cell differentiation. (Panels *A–H* are adapted from Lu et al. 2016; panels *I–J* are adapted from Aw et al. 2016, both, with permission, from Elsevier.)

ponent H1 (RPPH1) RNAs in both human and mouse (Lu et al. 2016). Careful analysis and validation of the interlocked duplexes may reveal additional pseudoknots in the transcriptome.

8 ARCHITECTURE OF THE XIST RNP: MODULAR STRUCTURES FOR MODULAR FUNCTIONS

The psoralen cross-linking methods identify physical contacts without a distance limit and therefore, reveal the overall architecture of RNA molecules. With few exceptions, whole transcript architectures have remained elusive for the vast majority of the transcriptome (Zappulla and Cech 2004). X-inactive specific transcript (XIST) is a lncRNA essential for X-chromosome inactivation (XCI) in placental mammals (eutheria). XIST serves as a scaffold to recruit a variety of protein complexes to orchestrate the complex XCI process. Using PARIS, we showed that the XIST RNA folds into compact and discrete domains, each spanning hundreds to thousands of nucleotides (Fig. 4E) (Lu et al. 2016, 2017). Despite the low sequence conservation, roughly 10% of the structures are conserved (Fig. 4F). The conserved duplexes support the demarcation of domains and suggest functional relevance of the high-level architecture. The XIST RNA coordinates multiple steps in XCI, including XIST localization to the inactive X, membrane attachment of the inactive X, histone and DNA modifications, and chromatin compaction. Targeted genetic analysis of each of the XIST domains will uncover the structural basis of the specific functions.

Stable and discrete domains may not be a common feature for all RNAs; for example, highly stable structures would not persist in mRNA coding regions because of the constant action of ribosomes. Even for lncRNAs, architectures differ greatly; for example, metastasis-associated lung adenocarcinoma transcript 1 (MALAT1) and XIST have more long-range structures that are organized into compact domains, whereas nuclear-enriched abundant transcript 1 (NEAT1) has more local structures, without clearly separated modular domains (Lu et al. 2016). Comparison of the duplex spans in general showed that mRNAs have more local structures than lncRNAs (Aw et al. 2016). These studies have revealed a complex picture for the high-level organization of RNA structures. However, with the exception of a few RNAs (such as XIST and MALAT1), the extent of domain formation in other RNAs is still unknown because of the limited sequencing coverage and the inherent bias of psoralen cross-linking.

The integration of PARIS with in vivo click SHAPE (icSHAPE), conservation analysis, and protein-binding assays (in vitro electrophoretic mobile shift assay [EMSA] and cross-linking and immunoprecipitation [CLIP]) led to the first structure-interaction model of the essential A-repeat domain complex in XIST RNP (Fig. 4G) (Lu et al. 2016). In this model, repeat units in the A-repeat region randomly form duplexes with each other, resulting in near identical structure units. These structure units associate with the essential XCI factor SPEN in a cooperative manner. The structure model is consistent with nucleotide resolution reactivity of the repeats determined by icSHAPE. Stringent tests for covariation showed no significance for any pair of repeats, consistent with the stochastic nature of the duplex formation, which dilutes and distributes the evolutionary pressure (Rivas et al. 2017). This study establishes a "top-down" approach for studying long RNAs, both coding and noncoding, in which modular RNA structure domains are identified and isolated for further studies.

9 NOVEL RNA–RNA INTERACTIONS: THE MOLECULAR SOCIAL NETWORK

Traditionally, the analysis of RNA–RNA interactions has required knowledge of at least one partner in the RNA–RNA complex, or the proteins involved. For example, in the CLASH and hiCLIP methods, the identification of new interactions was based on the proteins that bind the RNA duplexes (Fig. 2). Application of psoralen cross-linking methods has resulted in important de novo discoveries of RNA–RNA interactions, such as snoRNA:rRNA, scaRNA:snRNA, mRNA:mRNA, snoRNA:mRNA, and snRNA:lncRNA. These discoveries have both solved some long-standing questions in the field and revealed potentially new principles that govern coordinated gene expression programs. Similar to the analysis of intramolecular RNA structures, psoralen cross-linking achieves near base-pair level resolution and single-molecule accuracy, which greatly facilitates further mechanistic studies.

As discussed in the beginning, early studies of snoRNAs and scaRNAs led to the discovery of snoRNA-guided rRNA processing and modifications. Yet many sno/scaRNAs are still "orphans," with no identified targets. In certain cases, these RNAs may have more than one partner, but only one was known owing to the limited scope of conventional methods. An example of these poorly annotated snoRNAs is U8, which is highly expressed and conserved in metazoans (Fig. 4H). Previous studies suggested that the 5′ end of U8 base pairs with the 5′ end of the 28S rRNA to guide the cleavage of precursor ribosomal RNA (pre-rRNA; Fig. 4H, site labeled "Peculis 1997"). However, PARIS experiments in both human and mouse cells clearly identified an interaction site near the 3′ end of the 28S rRNA (Lu et al. 2016). The newly identified interaction is conserved in all metazoans and the duplex is more extensive than the pre-

Cite as *Cold Spring Harb Perspect Biol* doi: 10.1101/cshperspect.a034926

vious site (Fig. 4H, inset). This result highlights the critical contribution of psoralen cross-linking in providing direct physical evidence for RNA–RNA interactions.

In addition to the intermolecular interactions that involve sRNAs (snRNAs, snoRNAs, scaRNAs, etc.), psoralen cross-linking also detected mRNA:mRNA interactions, although with much fewer reads given their low abundance. Interestingly, the mRNA:mRNA interactions form extended networks, in which functionally related mRNAs are clustered together, indicating co-regulation on the RNA level (Fig. 4J) (Aw et al. 2016). Even more surprisingly, differentiation of human embryonic stem (ES) cells alters the topology of the interaction networks. It remains unclear what factors drive the formation and the alteration of the interaction networks and what functional consequences such physical contacts bring. Given the limited coverage of current cross-linking data, it is likely that future work will uncover other new interactions and mechanisms. Quantitative information about the interactions is also needed to further understand their relevance. For example, are these interactions kiss-and-run encounters or long-term committed relationships?

10 LIMITATIONS OF THE PSORALEN CROSS-LINKING METHODS AND FUTURE DIRECTIONS

Psoralen cross-linking has provided novel insights into the RNA structurome and interactome in living cells. Despite more than 40 years of work, current cross-linking-based methods still suffer from several limitations. Here we discuss the major issues, which will hopefully facilitate data interpretation and point to directions for further improvement.

Psoralen cross-linking of nucleic acid duplexes is slow and inefficient, often taking 30 min or more to achieve effective cross-linking (Lu et al. 2016). Prolonged cross-linking can change the cellular physiology to alter RNA structures. For example, long exposure to low temperature (needed to avoid UV-induced heating) may induce stress response. Light-activated psoralens also modify other cellular components such as lipids and proteins (Cimino et al. 1985). These side effects should be considered when designing experiments that examine the effects of certain conditions on RNA structures. In addition, the slow reaction makes it difficult to obtain dynamic information about RNA structures in response to environmental signals.

Psoralen cross-linking has significant bias on both the sequence and structure levels. Psoralens are preferentially (>90%) cross-linked to staggered uridines (Cimino et al. 1985; Boyer et al. 1988). Compact RNP structures may block cross-linking, and this effect is even more difficult to estimate. Therefore, we can only make semiquantitative conclusions about the structures that we observe in these experiments. For example, it is not possible to directly calculate the relative abundance of alternative conformations. In addition, conclusions regarding the overall topology of intra- and inter-RNA interaction networks are likely to be inaccurate because of the bias.

Psoralens are the only class of photoreversible cross-linkers available now for the general analysis of nucleic acid structure and interactions. New reversible cross-linkers with improved reaction speed and reduced bias will be of great value for the RNA field. For example, reversible RNA-modifying groups can be added to other classes of compounds that intercalate the DNA/RNA duplex. In fact, many cancer chemotherapy drugs act by modifying DNA, and can be potentially adapted to reversibly cross-link RNA (Deans and West 2011). Chemicals that cross-link tertiary contacts will provide important new information currently missed by psoralen cross-linking in living cells. For example, nitrogen mustard has been used to successfully identify tertiary contacts in the ribosome and snRNAs (Datta and Weiner 1992; Sergiev et al. 2001).

Another major bottleneck in the cross-linking methods is the low efficiency of proximity ligation. In general, the highest efficiency achieved is only ~5% for short fragments that permit near base-pair resolution structure modeling (Fig. 2B). Improving proximity ligation efficiency would greatly reduce sequencing cost and increase the amount of useful information obtained from high-throughput experiments. Potential solutions include new ligases that are less sensitive to the structural context, and ligation conditions that promote the "ligatable" conformations, such as higher temperature, chemical denaturation, and linkers that effectively increase the flexibility of the RNA ends.

Although sequencing cost will continue to drop, targeted analysis is a more cost-effective solution to analyzing RNAs of interest, especially low abundance RNAs from essential genes. In particular, the psoralen cross-linking-based methods could be combined with antibody enrichment of RBPs or antisense oligo enrichment of single or groups of RNAs.

11 INTEGRATING METHODS TO SOLVE COMPLEX RNA STRUCTURES AND INTERACTIONS

The cross-linking-based methods are particularly powerful in sorting through the complexities of RNA molecules in living cells, which are commonly depicted as a plate of "cooked spaghetti." X-ray crystallography, NMR, cryo-EM, computational modeling, chemical/enzymatic probing, and cross-linking-based methods, as summarized in Table 1, each have their own merits and limitations; there-

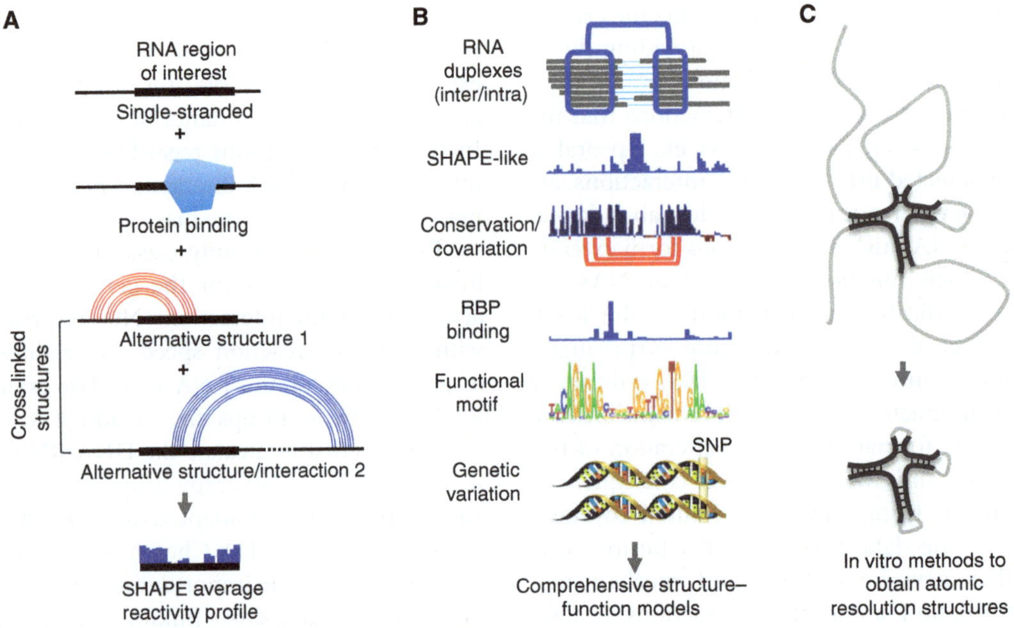

Figure 5. Integrated RNA structure analysis using multiple methods. (*A*) Comparison of cross-linking-based methods and accessibility/flexibility measurement (here using selective 2′-hydroxyl acylation with primer extension [SHAPE] as an example). The measured average accessibility profile is the sum of all possible RNA structure and interaction conformations plus the effect of protein binding. In contrast, cross-linking methods directly identify the structure conformations. (*B*) Cross-linking data provide an essential guide to place various types of RNA data in a structural context, thereby facilitating integrated functional analysis of RNA molecules. (*C*) From the large and flexible RNA molecules (gray lines), cross-linking-based methods identify small, compact, and stable structure domains (black lines), which can be further studied using higher resolution methods like X-ray crystallography and nuclear magnetic resonance (NMR).

fore, integration of diverse approaches will lead to the most comprehensive understanding of the basic properties of RNA (Fig. 5). In this context, the direct base-pairing information from cross-linking experiments provides an essential guide for the integrated analysis.

Many approaches can be envisioned to integrate structure data and other relevant information from the various methods described above. For example, incorporation of SHAPE-like data into cross-linking-based structure models achieves both high confidence and high resolution that is not possible by either method alone (Lu et al. 2016). As SHAPE-like methods provide a largely unbiased average reactivity profile of the structure ensemble and cross-linking methods nominate all cross-linkable conformations (see diagram in Fig. 5A), the integration of these two methods enables quantitative analysis of the alternative conformations. Analysis of RNA–protein interactions can now be performed in a structural context, which may explain the lack of apparent specificity on the sequence level (Fig. 5B). Also, as already shown (Fig. 3D,E), cross-linking-based structure models can guide comparative sequence analysis, and on the other hand, the conservation assigns functional significance to structures, prioritizing some of them for further studies (Fig. 5B). Over the years, extensive knowledge of RNA sequence motifs has been accumulated, such as ones that regulate splicing, translation, stability, and localization. Transcriptome-wide structures and interactions place such motifs in a larger context, which will help reveal previously unknown regulatory mechanisms and serve as examples for identifying novel functional motifs. A relevant example is the identification of long-range structures that bring RNA-binding Fox-1 Homolog 2 (Rbfox2) proteins to the vicinity of splicing motifs (Lovci et al. 2013). Genome-wide association studies have identified large numbers of genetic variations in the noncoding parts of the genome, and these are very likely to exert their effects through RNA structures (Halvorsen et al. 2010). Although conventional chemical probing has already been used in characterizing and interpreting genetic variations, cross-linking methods are more direct in revealing their structural contexts, especially for long-range, dynamic, and complex structures (Fig. 5B).

With the ability to define overall architectures of RNA by cross-linking, large RNA molecules refractory to atomic resolution structure analysis in vitro can be studied based on the "divide-and-conquer" principle (Fig. 5C). Stable and compact structure domains can be trimmed to remove flexible regions based on cross-linking data, purified and then subject to X-ray crystallography, NMR, and cryo-EM analysis. The identification of stable duplex units in the XIST A-repeat domain is one such example, in which the physiologically relevant structure units can be studied in great detail in vitro (see an example of the XIST A-repeat domain in Fig. 4G). De novo and data-assisted 3D structure modeling represent an alternative solution to directly solving the structure by in vitro atomic resolution methods. The modeling efforts will also benefit from the cross-linking methods that provide physical contact information.

The biogenesis and function of RNA molecules in cells are on a long journey, in which constant remodeling of structures impacts every aspect of their life. The dissection of these dynamic structures will certainly reveal new principles of gene regulation. Identification of RNA structure motifs that control gene expression may lead to new therapeutic opportunities by revealing potential drug targets. For example, RNA viruses represent a major health concern, and the mechanistic studies of viral RNA structures would provide important information for designing antiviral drugs. With this new generation of methods for direct identification of base-pairing interactions in nucleic acids, opportunities abound for many topics in the RNA field for both basic and translational research.

ACKNOWLEDGMENTS

This work was supported by National Institutes of Health (NIH) R01-HG004361, P50-HG007735 (H.Y.C.) and Stanford Jump Start Award (Z.L.). Z.L. is a Layton Family Fellow of the Damon Runyon-Sohn Foundation Pediatric Cancer Fellowship Award (DRSG-14–15). We thank Fan Liu for comments on the manuscript.

REFERENCES

*Reference is in this book.

Anfinsen CB. 1973. Principles that govern the folding of protein chains. *Science* **181:** 223–230.

Aw JG, Shen Y, Wilm A, Sun M, Lim XN, Boon KL, Tapsin S, Chan YS, Tan CP, Sim AY, et al. 2016. In vivo mapping of eukaryotic RNA interactomes reveals principles of higher-order organization and regulation. *Mol Cell* **62:** 603–617.

Bai XC, McMullan G, Scheres SH. 2015. How cryo-EM is revolutionizing structural biology. *Trends Biochem Sci* **40:** 49–57.

Beltrame M, Tollervey D. 1992. Identification and functional analysis of two U3 binding sites on yeast pre-ribosomal RNA. *EMBO J* **11:** 1531–1542.

Blondal T, Thorisdottir A, Unnsteinsdottir U, Hjorleifsdottir S, Aevarsson A, Ernstsson S, Fridjonsson OH, Skirnisdottir S, Wheat JO, Hermannsdottir AG, et al. 2005. Isolation and characterization of a thermostable RNA ligase 1 from a *Thermus scotoductus* bacteriophage TS2126 with good single-stranded DNA ligation properties. *Nucleic Acids Res* **33:** 135–142.

Bothe JR, Nikolova EN, Eichhorn CD, Chugh J, Hansen AL, Al-Hashimi HM. 2011. Characterizing RNA dynamics at atomic resolution using solution-state NMR spectroscopy. *Nat Methods* **8:** 919–931.

Boyer V, Moustacchi E, Sage E. 1988. Sequence specificity in photoreaction of various psoralen derivatives with DNA: Role in biological activity. *Biochemistry* **27:** 3011–3018.

Calvet JP, Pederson T. 1981. Base-pairing interactions between small nuclear RNAs and nuclear RNA precursors as revealed by psoralen cross-linking in vivo. *Cell* **26:** 363–370.

Calvet JP, Meyer LM, Pederson T. 1982. Small nuclear RNA U2 is base-paired to heterogeneous nuclear RNA. *Science* **217:** 456–458.

Cech TR, Pardue ML. 1976. Electron microscopy of DNA crosslinked with trimethylpsoralen: Test of the secondary structure of eukaryotic inverted repeat sequences. *Proc Nal Acad Sci* **73:** 2644–2648.

Cech TR, Steitz JA. 2014. The noncoding RNA revolution—Trashing old rules to forge new ones. *Cell* **157:** 77–94.

Chen J, Kastan MB. 2010. 5′-3′-UTR interactions regulate p53 mRNA translation and provide a target for modulating p53 induction after DNA damage. *Genes Dev* **24:** 2146–2156.

Cimino GD, Gamper HB, Isaacs ST, Hearst JE. 1985. Psoralens as photoactive probes of nucleic acid structure and function: Organic chemistry, photochemistry, and biochemistry. *Ann Rev Biochem* **54:** 1151–1193.

Datta B, Weiner AM. 1992. Cross-linking of U1 snRNA using nitrogen mustard. Evidence for higher order structure. *J Biol Chem* **267:** 4503–4507.

Deans AJ, West SC. 2011. DNA interstrand crosslink repair and cancer. *Nat Rev Cancer* **11:** 467–480.

Dekker J, Rippe K, Dekker M, Kleckner N. 2002. Capturing chromosome conformation. *Science* **295:** 1306–1311.

Dethoff EA, Chugh J, Mustoe AM, Al-Hashimi HM. 2012. Functional complexity and regulation through RNA dynamics. *Nature* **482:** 322–330.

Ding Y, Tang Y, Kwok CK, Zhang Y, Bevilacqua PC, Assmann SM. 2014. In vivo genome-wide profiling of RNA secondary structure reveals novel regulatory features. *Nature* **505:** 696–700.

Dobin A, Davis CA, Schlesinger F, Drenkow J, Zaleski C, Jha S, Batut P, Chaisson M, Gingeras TR. 2013. STAR: Ultrafast universal RNA-seq aligner. *Bioinformatics* **29:** 15–21.

Eddy SR. 2004. How do RNA folding algorithms work? *Nat Biotechnol* **22:** 1457–1458.

Ehresmann C, Baudin F, Mougel M, Romby P, Ebel JP, Ehresmann B. 1987. Probing the structure of RNAs in solution. *Nucleic Acids Res* **15:** 9109–9128.

Fallmann J, Will S, Engelhardt J, Gruning B, Backofen R, Stadler PF. 2017. Recent advances in RNA folding. *J Biotechnol* **261:** 97–104.

Fayet-Lebaron E, Atzorn V, Henry Y, Kiss T. 2009. 18S rRNA processing requires base pairings of snR30 H/ACA snoRNA to eukaryote-specific 18S sequences. *EMBO J* **28:** 1260–1270.

Gesell T, von Haeseler A. 2006. In silico sequence evolution with site-specific interactions along phylogenetic trees. *Bioinformatics* **22:** 716–722.

Giedroc DP, Cornish PV. 2009. Frameshifting RNA pseudoknots: Structure and mechanism. *Virus Res* **139:** 193–208.

Gilley D, Blackburn EH. 1999. The telomerase RNA pseudoknot is critical for the stable assembly of a catalytically active ribonucleoprotein. *Proc Natl Acad Sci* **96:** 6621–6625.

Gong J, Shao D, Xu K, Lu Z, Lu ZJ, Yang YT, Zhang QC. 2018. RISE: A database of RNA interactome from sequencing experiments. *Nucleic Acids Res* **46:** D194–D201.

Halvorsen M, Martin JS, Broadaway S, Laederach A. 2010. Disease-associated mutations that alter the RNA structural ensemble. *PLoS Genet* **6:** e1001074.

Hartshorne T. 1998. Distinct regions of U3 snoRNA interact at two sites within the 5′ external transcribed spacer of pre-rRNAs in *Trypanosoma brucei* cells. *Nucleic Acids Res* **26:** 2541–2553.

Hausner TP, Giglio LM, Weiner AM. 1990. Evidence for base-pairing between mammalian U2 and U6 small nuclear ribonucleoprotein particles. *Genes Dev* **4:** 2146–2156.

Hearst JE. 1981. Psoralen photochemistry. *Ann Rev Biophys Bioeng* **10:** 69–86.

Helwak A, Kudla G, Dudnakova T, Tollervey D. 2013. Mapping the human miRNA interactome by CLASH reveals frequent noncanonical binding. *Cell* **153:** 654–665.

Kass S, Tyc K, Steitz JA, Sollner-Webb B. 1990. The U3 small nucleolar ribonucleoprotein functions in the first step of preribosomal RNA processing. *Cell* **60:** 897–908.

* Kastner B, Will CL, Stark H, Lührmann R. 2019. Structural insights into nuclear pre-mRNA splicing in higher eukaryotes. *Cold Spring Harb Perspect Biol* doi: 10.1101/cshperspect.a032417.

Kertesz M, Wan Y, Mazor E, Rinn JL, Nutter RC, Chang HY, Segal E. 2010. Genome-wide measurement of RNA secondary structure in yeast. *Nature* **467:** 103–107.

Kim SH, Quigley GJ, Suddath FL, McPherson A, Sneden D, Kim JJ, Weinzierl J, Rich A. 1973. Three-dimensional structure of yeast phenylalanine transfer RNA: Folding of the polynucleotide chain. *Science* **179:** 285–288.

Kladwang W, VanLang CC, Cordero P, Das R. 2011. A two-dimensional mutate-and-map strategy for non-coding RNA structure. *Nat Chem* **3:** 954–962.

Kudla G, Granneman S, Hahn D, Beggs JD, Tollervey D. 2011. Cross-linking, ligation, and sequencing of hybrids reveals RNA–RNA interactions in yeast. *Proc Natl Acad Sci* **108:** 10010–10015.

Lerner MR, Boyle JA, Mount SM, Wolin SL, Steitz JA. 1980. Are snRNPs involved in splicing? *Nature* **283:** 220–224.

Lovci MT, Ghanem D, Marr H, Arnold J, Gee S, Parra M, Liang TY, Stark TJ, Gehman LT, Hoon S, et al. 2013. Rbfox proteins regulate alternative mRNA splicing through evolutionarily conserved RNA bridges. *Nat Struct Mol Biol* **20:** 1434–1442.

Lu Z, Chang HY. 2016. Decoding the RNA structurome. *Curr Opin Struct Biol* **36:** 142–148.

Lu Z, Zhang QC, Lee B, Flynn RA, Smith MA, Robinson JT, Davidovich C, Gooding AR, Goodrich KJ, Mattick JS, et al. 2016. RNA duplex map in living cells reveals higher-order transcriptome structure. *Cell* **165:** 1267–1279.

Lu Z, Carter AC, Chang HY. 2017. Mechanistic insights in X chromosome inactivation. *Philos Trans R Soc B* **372:** 20160356.

Lu Z, Gong J, Zhang QC. 2018. PARIS: Psoralen analysis of RNA interactions and structures with high throughput and resolution. *Methods Mol Biol* **1649:** 59–84.

Maser RL, Calvet JP. 1989. U3 small nuclear RNA can be psoralen-cross-linked in vivo to the 5′ external transcribed spacer of pre-ribosomal-RNA. *Proc Natl Acad Sci* **86:** 6523–6527.

Mathews DH, Moss WN, Turner DH. 2010. Folding and finding RNA secondary structure. *Cold Spring Harb Perspect Biol* **2:** a003665.

Morrissey JP, Tollervey D. 1993. Yeast snR30 is a small nucleolar RNA required for 18S rRNA synthesis. *Mol Cell Biol* **13:** 2469–2477.

Mount SM, Pettersson I, Hinterberger M, Karmas A, Steitz JA. 1983. The U1 small nuclear RNA-protein complex selectively binds a 5′ splice site in vitro. *Cell* **33:** 509–518.

Nguyen TC, Cao X, Yu P, Xiao S, Lu J, Biase FH, Sridhar B, Huang N, Zhang K, Zhong S. 2016. Mapping RNA–RNA interactome and RNA structure in vivo by MARIO. *Nat Commun* **7:** 12023.

Nicholson BL, White KA. 2014. Functional long-range RNA–RNA interactions in positive-strand RNA viruses. *Nat Rev Microbiol* **12:** 493–504.

Noller HF, Woese CR. 1981. Secondary structure of 16S ribosomal RNA. *Science* **212:** 403–411.

Peculis BA. 1997. The sequence of the 5′ end of the U8 small nucleolar RNA is critical for 5.8S and 28S rRNA maturation. *Mol Cell Biol* **17:** 3702–3713.

Pedersen JS, Bejerano G, Siepel A, Rosenbloom K, Lindblad-Toh K, Lander ES, Kent J, Miller W, Haussler D. 2006. Identification and classification of conserved RNA secondary structures in the human genome. *PLoS Comput Biol* **2:** e33.

Petrov AS, Bernier CR, Gulen B, Waterbury CC, Hershkovits E, Hsiao C, Harvey SC, Hud NV, Fox GE, Wartell RM, et al. 2014. Secondary structures of rRNAs from all three domains of life. *PloS One* **9:** e88222.

* Plaschka C, Newman AJ, Nagai K. 2019. Structural basis of nuclear pre-mRNA splicing: Lessons from yeast. *Cold Spring Harb Perspect Biol* doi: 10.1101/cshperspect.a032391.

Rabin D, Crothers DM. 1979. Analysis of RNA secondary structure by photochemical reversal of psoralen crosslinks. *Nucleic Acids Res* **7:** 689–703.

Rimoldi OJ, Raghu B, Nag MK, Eliceiri GL. 1993. Three new small nucleolar RNAs that are psoralen cross-linked in vivo to unique regions of pre-rRNA. *Mol Cell Biol* **13:** 4382–4390.

Rinke J, Appel B, Digweed M, Luhrmann R. 1985. Localization of a base-paired interaction between small nuclear RNAs U4 and U6 in intact U4/U6 ribonucleoprotein particles by psoralen cross-linking. *J Mol Biol* **185:** 721–731.

Rivas E, Eddy SR. 1999. A dynamic programming algorithm for RNA structure prediction including pseudoknots. *J Mol Biol* **285:** 2053–2068.

Rivas E, Eddy SR. 2001. Noncoding RNA gene detection using comparative sequence analysis. *BMC Bioinformatics* **2:** 8.

Rivas E, Clements J, Eddy SR. 2017. Lack of evidence for conserved secondary structure in long noncoding RNAs. *Nat Methods* **14:** 45–48.

Robertus JD, Ladner JE, Finch JT, Rhodes D, Brown RS, Clark BF, Klug A. 1974. Structure of yeast phenylalanine tRNA at 3 Å resolution. *Nature* **250:** 546–551.

Rouskin S, Zubradt M, Washietl S, Kellis M, Weissman JS. 2014. Genome-wide probing of RNA structure reveals active unfolding of mRNA structures in vivo. *Nature* **505:** 701–705.

Sergiev PV, Dontsova OA, Bogdanov AA. 2001. Chemical methods for the structural study of the ribosome: Judgment day. *Mol Biol* **35:** 472–496.

Sharma E, Sterne-Weiler T, O'Hanlon D, Blencowe BJ. 2016. Global mapping of human RNA–RNA interactions. *Mol Cell* **62:** 618–626.

Shi Y. 2014. A glimpse of structural biology through X-ray crystallography. *Cell* **159:** 995–1014.

Siegfried NA, Busan S, Rice GM, Nelson JA, Weeks KM. 2014. RNA motif discovery by SHAPE and mutational profiling (SHAPE-MaP). *Nat Methods* **11:** 959–965.

Smith MA, Gesell T, Stadler PF, Mattick JS. 2013. Widespread purifying selection on RNA structure in mammals. *Nucleic Acids Res* **41:** 8220–8236.

Spitale RC, Flynn RA, Zhang QC, Crisalli P, Lee B, Jung JW, Kuchelmeister HY, Batista PJ, Torre EA, Kool ET, et al. 2015. Structural imprints in vivo decode RNA regulatory mechanisms. *Nature* **519:** 486–490.

Stroke IL, Weiner AM. 1989. The 5′ end of U3 snRNA can be crosslinked in vivo to the external transcribed spacer of rat ribosomal RNA precursors. *J Mol Biol* **210:** 497–512.

Sugimoto Y, Vigilante A, Darbo E, Zirra A, Militti C, D'Ambrogio A, Luscombe NM, Ule J. 2015. hiCLIP reveals the in vivo atlas of mRNA secondary structures recognized by Staufen 1. *Nature* **519:** 491–494.

Thompson JF, Hearst JE. 1983. Structure of *E. coli* 16S RNA elucidated by psoralen crosslinking. *Cell* **32:** 1355–1365.

Tyc K, Steitz JA. 1992. A new interaction between the mouse 5′ external transcribed spacer of pre-rRNA and U3 snRNA detected by psoralen crosslinking. *Nucleic Acids Res* **20:** 5375–5382.

Underwood JG, Uzilov AV, Katzman S, Onodera CS, Mainzer JE, Mathews DH, Lowe TM, Salama SR, Haussler D. 2010. FragSeq: Transcriptome-wide RNA structure probing using high-throughput sequencing. *Nat Methods* **7:** 995–1001.

Washietl S, Hofacker IL, Lukasser M, Huttenhofer A, Stadler PF. 2005. Mapping of conserved RNA secondary structures predicts thousands of functional noncoding RNAs in the human genome. *Nat Biotechnol* **23:** 1383–1390.

Wassarman DA, Steitz JA. 1993. A base-pairing interaction between U2 and U6 small nuclear RNAs occurs in >150S complexes in HeLa cell extracts: Implications for the spliceosome assembly pathway. *Proc Natl Acad Sci* **90:** 7139–7143.

Watson JD, Crick FH. 1953. Molecular structure of nucleic acids: A structure for deoxyribose nucleic acid. *Nature* **171:** 737–738.

Wells SE, Hillner PE, Vale RD, Sachs AB. 1998. Circularization of mRNA by eukaryotic translation initiation factors. *Mol Cell* **2:** 135–140.

Whirl-Carrillo M, Gabashvili IS, Bada M, Banatao DR, Altman RB. 2002. Mining biochemical information: Lessons taught by the ribosome. *RNA* **8:** 279–289.

Will CL, Luhrmann R. 2011. Spliceosome structure and function. *Cold Spring Harb Perspect Biol* **3:** a003707.

Wollenzien PL, Youvan DC, Hearst JE. 1978. Structure of psoralen-cross-linked ribosomal RNA from *Drosophila melanogaster*. *Proc Natl Acad Sci* **75:** 1642–1646.

Wu X, Bartel DP. 2017. Widespread influence of 3′-end structures on mammalian mRNA processing and stability. *Cell* **169:** 905–917.e11.

* Yan C, Wan R, Shi Y. 2019. Molecular mechanisms of pre-mRNA splicing through structural biology of the spliceosome. *Cold Spring Harb Perspect* doi: 10.1101/cshperspect.a032409.

Zappulla DC, Cech TR. 2004. Yeast telomerase RNA: A flexible scaffold for protein subunits. *Proc Natl Acad Sci* **101:** 10024–10029.

Zwieb C, Brimacombe R. 1980. Localisation of a series of intra-RNA cross-links in 16S RNA, induced by ultraviolet irradiation of *Escherichia coli* 30S ribosomal subunits. *Nucleic Acids Res* **8:** 2397–2411.

Detecting RNA G-Quadruplexes (rG4s) in the Transcriptome

Chun Kit Kwok,[1,4] Giovanni Marsico,[2,3,4] and Shankar Balasubramanian[2,3]

[1]Department of Chemistry, City University of Hong Kong, Hong Kong SAR, China

[2]Department of Chemistry, University of Cambridge, Cambridge CB2 1EW, United Kingdom

[3]Cancer Research UK Cambridge Institute, University of Cambridge, Cambridge CB2 0RE, United Kingdom

Correspondence: sb10031@cam.ac.uk; ckkwok42@cityu.edu.hk

SUMMARY

RNA G-quadruplex (rG4) secondary structures are proposed to play key roles in fundamental biological processes that include the modulation of transcriptional, co-transcriptional, and posttranscriptional events. Recent methodological developments that include predictive algorithms and structure-based sequencing have enabled the detection and mapping of rG4 structures on a transcriptome-wide scale at high sensitivity and resolution. The data generated by these studies provide valuable insights into the potentially diverse roles of rG4s in biology and open up a number of mechanistic hypotheses. Herein we highlight these methodologies and discuss the associated findings in relation to rG4-related biological mechanisms.

Outline

[4]These authors contributed equally to this work.

Cite as *Cold Spring Harb Perspect Biol* doi: 10.1101/cshperspect.a032284

1 INTRODUCTION

In 1910 Bang reported that guanylic acid (GMP) could form a gel, suggestive of a self-assembly phenomenon (Bang 1910). Some 50 years later, Gellert and coworkers revealed that the fibers obtained from the dried GMP gel comprised what became known as the G-quartet motif (Gellert et al. 1962). A G-quartet involves four guanine bases interacting with each other via H-bonding, further stabilized by a central monovalent cation, such as K^+ or Na^+ (Fig. 1). The biological relevance of such non-Watson–Crick nucleic acid structures was largely ignored until in the late 1980s, when it was shown that G-rich sequences based on either the telomeric region of DNA or immunoglobulin switch region of DNA could form four-stranded structural motifs (Sen and Gilbert 1988), which became referred to as G-quadruplexes (G4s) (Fig. 1).

Cellular visualization of DNA G4s was reported in 2001 for the ciliate *Stylonychia lemnae*, by immunostaining the telomeric G4 (Schaffitzel et al. 2001). In 2013, a G4-specific single-chain antibody, blood group antigen H1 (BG4), was generated by phage display and used to visualize DNA G4s in the nuclei of human cells by immunofluorescence (Biffi et al. 2013). Shortly afterward, RNA G4s (rG4s) were visualized in the cytoplasm of human cells using BG4 and were also shown to be selectively stabilized by an rG4-specific ligand carboxypyridostatin (cPDS) (Biffi et al. 2014).

Sequence-based prediction approaches have enabled exploration of sequence and structural features characteristic of G4s in vitro at the level of the whole genome or transcriptome for a range of organisms. Early predictors (Huppert and Balasubramanian 2005; Todd et al. 2005) were informed by biophysical measurements conducted on G4 DNA oligonucleotides, and it has been largely assumed that rG4s conform to the same "rules." However, biophysical measurements have highlighted important differences between DNA and RNA G4s. The presence of the 2′-hydroxyl (2′-OH) group in the ribose can enable additional intramolecular interactions within the loops of rG4s and with water molecules, providing the rG4 with greater stability than the corresponding DNA G4 (Zhang et al. 2010). Furthermore, the 2′-OH causes a C3′-endo rather than a C2′-endo sugar pucker, generally favoring a parallel rG4 folding topology, whereas DNA G4s can fold into parallel, antiparallel, and mixed conformations (Zhang et al. 2010). These issues are described in detail elsewhere (Fay et al. 2017; Kwok and Merrick 2017). The higher propensity of RNA to fold into other single-strand secondary structures such as stem-loops and pseudoknots is a key feature for considering G4 formation in the context of transcripts.

Numerous studies suggest that rG4s may play significant roles in a myriad of biological processes (Bugaut and Balasubramanian 2012; Fay et al. 2017) and are linked to human diseases (Simone et al. 2015; Cammas and Millevoi 2017). rG4s in the 5′ untranslated region (UTR) of messenger RNA (mRNA) have been shown to impede translation (Kumari et al. 2007), and rG4s at 3′ UTR can suppress translation (Arora and Suess 2011; Crenshaw et al. 2015) and affect microRNA targeting (Stefanovic et al. 2015; Rouleau et al. 2017), alternative polyadenylation (Beaudoin and Perreault 2013), and RNA localization (Subramanian et al. 2011). rG4s in the coding DNA sequence (CDS) can suppress translation (Endoh et al. 2013) and stimulate ribosomal frameshifting (Endoh and Sugimoto 2013; Yu et al. 2014), whereas rG4 near splice junctions can influence alternative splicing (Marcel et al. 2011; Weldon et al. 2018), suggesting a co-transcriptional role. In addition, rG4 has also been shown to act like the canonical hairpin loop in the ρ-independent pathway and stimulate mitochondrial transcription termination (Wanrooij et al. 2010). It has also been shown that rG4 has stronger binding affinity to Polycomb repressive complex 2 (PRC2) than unstructured G-rich motif or duplex RNA (Wang et al. 2017). Recently, it has been reported that rG4s have potential to form in primary and precursor microRNAs (Mirihana Arachchilage et al. 2015; Kwok et al. 2016b; Rouleau et al. 2018), as well

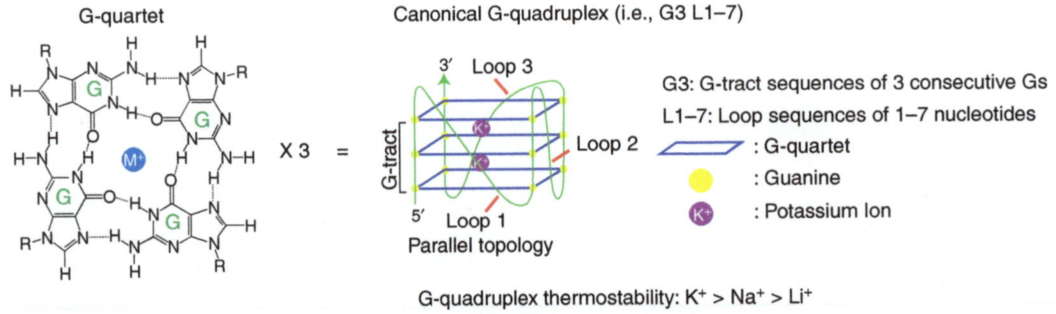

Figure 1. RNA G-quadruplex (rG4) structure. A G-quartet showing the hydrogen bonding and stabilizing cation. Three stacked G-quartet planes with connecting loop sequences form a canonical G-quadruplex (G4) (i.e., G3 L1–7).

Cite as *Cold Spring Harb Perspect Biol* doi: 10.1101/cshperspect.a032284

as in long-noncoding RNA (Matsumura et al. 2017), suggesting roles of rG4 in the noncoding RNA transcriptome.

Herein, we describe recent methodological advances to detect rG4s in the transcriptome.

2 COMPUTATIONAL APPROACHES FOR SEARCHING RNA G-QUADRUPLEXES

2.1 Early Predictors

Biophysical data led to a motif for unimolecular G4s comprising stretches of guanines (G runs) separated by linker sequences (loops) of limited length. The algorithm Quadparser (Fig. 2A) (Huppert and Balasubramanian 2005) was used to search the motif ($G_{3+}N_{1-7}G_{3+}N_{1-7}G_{3+}N_{1-7}G_{3+}$) predicted to fold into stable G4s under near-physiological conditions (Hazel et al. 2004). A parallel study using the same consensus motif (Todd et al. 2005) analyzed the specific sequence and loop lengths within G4s to illuminate preferred structural and sequence features in the human genome. Whereas sequences that form G4s and escape this rule were known (Patel and Hosur 1999), a conservative choice of parameters initially provided more than 360,000 predicted G4s in the human genome. Lower stringency—for example, allowing two quartets (G_{2+} instead of G_{3+})—gave more than 8.5 million predicted G4s. A limitation of Quadparser was the fixed motif definition and the binary nature of the G4 prediction, whereas biophysics would suggest a variation in thermal stability between different G4s (Hazel et al. 2004). Other G4 predictors were described at about the same time. The quadruplex-forming G-rich sequence (QGRS) mapper (Fig. 2) (D'Antonio and Bagga

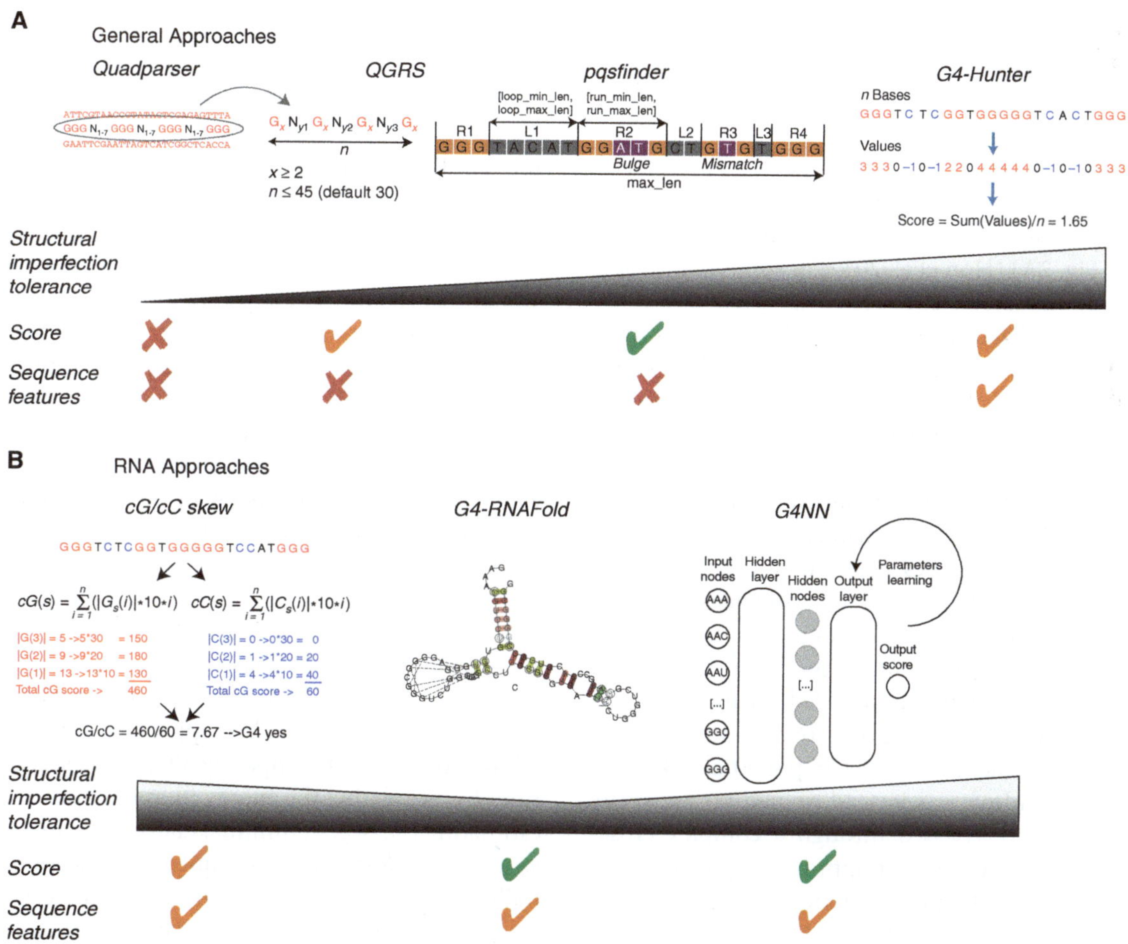

Figure 2. Computational prediction of G4s. (*A*) General appoaches: Approaches originally developed for DNA G4s. Key features are indicated below each method: Tolerance to structural "'imperfections'" (e.g., bulges, incomplete guanine run), with higher gray bar indicating more tolerance; if the method returns a score indicating the propensity to form a G4; if the method takes into account sequence features. Red cross, feature absent; orange tick, feature partially present; green tick, feature present. (*B*) RNA approaches: Approaches specifically developed for rG4s. Features and legend as for *A*.

2004) used an algorithm comprising a more flexible motif definition (G_xN_{y1} G_xN_{y2} G_xN_{y3} G_x, in which $x \geq 2$) and a ranking of predicted G4s according to the G score, defined as the likelihood to form a stable G4 based on favoring short loops, equal loop lengths, and a higher number of quartets. An alternative approach using a sliding window strategy was implemented in G-quadruplex potential (G4P) calculator (Eddy and Maizels 2006): The program computed the G4 DNA potential based on the density of G runs in a given sequence window and calculating the percentage that meets the searched criteria, returning a score independent of sequence length.

All such computational approaches have limitations. First, they did not use a large-scale experimental data set to validate the algorithm or optimize the critical parameters for scoring G4s. Rather, these decisions were based on extrapolating from a small number of biophysical studies with limited scope. Second, scoring functions are used interchangeably for DNA and RNA, ignoring critical differences between the respective G4s.

2.2 Recent Predictors

Most recent approaches allow greater motif flexibility, such as longer loops, bulges, or mismatches in the G4 motif (see Pqsfinder in Fig. 2A) (Hon et al. 2017) as supported by few experimentally validated structures (Mukundan and Phan 2013). Others penalized the presence of Cs proximal to or within the G4 motif (see G4-Hunter in Fig. 2A) (Bedrat et al. 2016), because Cs had been experimentally shown to base-pair with Gs to compete with G4 formation (Beaudoin et al. 2014). Quadron (Sahakyan et al. 2017) takes into account more than 200 sequence-based and structural features to classify via machine learning putative G4s into forming and nonforming ones. pqsfinder was trained using 392 in vitro experimentally validated sequences and validated using the larger G4-sequencing (G4-seq) data set generated by high-throughput sequencing (Chambers et al. 2015). G4-Hunter was also tested on these 392 sequences, plus experimental validation on the entire human mitochondrial genome, in which 165 potential G4s were tested in vitro. Quadron was instead trained on the entire G4-seq data set and assessed through a complex cross-validation scheme. Those recent approaches provided improved accuracy in predicting nonstandard G4 motifs for a more comprehensive identification of G4s in the genome.

2.3 RNA G-Quadruplex-Specific Predictors

The approaches described in the previous section are not RNA-specific and were not based on experimentally validated rG4-forming sequences. RNA-specific features include the 2′-OH, which confers higher stability, a preference for parallel conformation (Zhang et al. 2010), and the propensity for competing, alternative RNA secondary structures, and also long-range interactions. These key differences from DNA should be considered.

2.4 *cG/cC* Skew Approach

Cytosines proximal to guanines within the G4 motif can base-pair (C:G) and compete with the Hoogsteen base-pairing required for G-quartet formation (Beaudoin et al. 2014). An approach proposed to address this uses the *cG/cC* scoring scheme (Beaudoin et al. 2014), which penalizes the presence of Cs to account for their negative effect on G4 stability. This method calculates the ratio between two different factors, the *cG* and the *cC* score, each proportional to the number of G (or C) stretches, progressively weighted more for longer stretches, according to the formula

$$cG(s) = \sum_{i=1}^{n} (|G_s(i)| * 10 * i),$$

and similarly for the *cC* score (Fig. 2B). The experimental validation used two sets of more than 10 G4 sequences and led to an empirical threshold of 2–3 as *cG/cC* score for the formation of stable G4s. This scoring scheme overcomes the limitation of using rigid sequence motifs (i.e., two or three Gs and a loop of defined length). However, the parameterization is arbitrary: Both the scoring threshold and the multiplicative factors in the formulas are chosen based on heuristics that have not been rigorously justified. Another limitation is that only Gs and Cs are taken into account explicitly, whereas other nucleotides (A or T) or more complex sequence motifs (dinucleotides, k-mers) are not considered.

2.5 G-Quadruplex RNAfold

A thermodynamically based approach was introduced within the RNAfold tool of the Vienna package (Fig. 2B) (Lorenz et al. 2013). This approach aims to explicitly address the issue of competing alternative secondary structures by calculating the energy function due to the rG4 and estimating the overall minimal energy (i.e., higher stability) structure for a given RNA sequence. This approach essentially models the ΔG as a logarithmic function of the total length of linkers between G runs and incorporates this term into a simplified energy function. The investigators found that the majority of putative quadruplex-forming sequences in the human genome are likely to fold into non-G4 secondary structures instead. This is not in agreement with recent experimental reports showing rG4 formation in transcripts in vitro (Guo and Bartel 2016; Kwok et al.

Cite as *Cold Spring Harb Perspect Biol* doi: 10.1101/cshperspect.a032284

2016a). This is perhaps attributable to the folding algorithm having limited knowledge of the energy function for longer or asymmetric loops and to the limited training set. Moreover, kinetically trapped structures may be missed by a thermodynamic approach, especially if the Cs are downstream from the Gs.

2.6 G4RNA Screener

A recent data mining approach, G4RNA screener (see G4NN in Fig. 2B) (Garant et al. 2017), is based on training an artificial neural network on a compendium of RNA sequences (149 G4 and 179 non-G4), investigated by the literature for G4 folding, plus 200 sequences randomly taken from the transcriptome. The approach was tested on nearly 4000 in vitro detected rG4s (Kwok et al. 2016a), considered as a positive set, and compared for classification performance with the *cG/cC* and the G4-Hunter algorithms, yielding comparable or better outcomes. This approach paves the way to more comprehensive approaches in which complex sequence features other than just C stretches can be factored in the prediction. However, the neural network predictor is a black box, which does not readily provide insights into the predictive features determining G4 formation.

3 DETECTING RNA G-QUADRUPLEXES IN THE TRANSCRIPTOME IN VITRO

Broadly, three categories of transcript-specific methods have been developed for the detection of rG4s (Kwok and Merrick 2017): (i) biophysics on short synthetic oligos, including circular dichroism (CD) spectroscopy (Paramasivan et al. 2007), ultraviolet (UV) or fluorescence resonance energy transfer (FRET) thermal melting analysis (Mergny et al. 1998; Mergny et al. 2001), structural analysis by nuclear magnetic resonance (NMR) spectroscopy (Webba da Silva 2007), and fluorescence assay (Kwok et al. 2013c; Renaud de la Faverie et al. 2014); (ii) chemical approaches on longer in vitro transcribed transcripts, including in-line probing (Beaudoin et al. 2013), selective 2′-OH acylation with lithium ion-based primer extension (SHALiPE) (Kwok et al. 2016b), dimethyl sulfate with lithium ion-based primer extension (DMSLiPE) (Kwok et al. 2016b), footprinting of long 7-deazaguanine-substituted RNAs (FOLDeR) (Weldon et al. 2017); and (iii) approaches on transcripts extracted from cells, including reverse transcriptase stalling (RTS) and RTS-ligation mediated polymerase chain reaction (PCR) (Kwok and Balasubramanian 2015).

These approaches collectively provide insights into the structure and folding of rG4s for candidate sequences; however, they are low-throughput and preclude a global, transcriptome-wide analysis of rG4s.

The pausing by a DNA/RNA polymerase has been exploited to detect nucleic acid secondary structures, including G4s (Woodford et al. 1994). To clarify whether the structure is a G4, one can compare polymerase pausing under conditions that differentially stabilize folded G4 structures, for example physiological K^+ conditions (that help stabilize folded G4s) versus Li^+ or Na^+ conditions of the same ionic strength (which lead to less stable G4s). This principle has been adapted for high-throughput sequencing (G4-seq) to generate a map of DNA G4s in the human genome (Chambers et al. 2015). Sequencing rG4s was enabled by detecting RTS at rG4 structures under K^+ conditions, which is not observable under Li^+ conditions. There is also the option of extending the principle to strong RTS at rG4s on the inclusion of an rG4 stabilizing small molecule (Kwok and Balasubramanian 2015). This general approach was developed into an in vitro transcriptome-wide rG4 profiling method called rG4-sequencing (rG4-seq) and applied to purified poly(A)-enriched HeLa RNAs (Kwok et al. 2016a). Here, complementary DNA (cDNA) fragments are prepared into libraries for next-generation sequencing (Fig. 3A). At a similar time an in vitro method called reverse transcriptase (RT) stop profiling was reported (Fig. 3B), which used similar concepts to obtain an in vitro rG4 map in purified poly(A)-enriched RNAs from mouse embryonic stem cells, as well as human HEK293T and HeLa cells (Guo and Bartel 2016). This map was then used as reference in the rest of their study to assess rG4 formation in vivo via chemical mapping (Guo and Bartel 2016), which will be discussed later.

Here, we discuss the experimental workflows and bioinformatics pipelines for the rG4-seq and RT stop profiling (see also Fig. 3), and then we summarize biological implications of the in vitro studies. Although the two methods are conceptually similar and share some steps, they differ in aspects of the experimental design, as discussed in the next section.

3.1 Comparison of Experimental Procedures in rG4-seq and RT Stop Profiling

A key difference is that for RT stop profiling, 60–80-nt RNA fragments were selected, whereas for rG4-seq, an average size of 250-nt RNA was obtained. Given canonical rG4s (G3L1–7) are ∼25-nt-long, the length of the flanking sequences can affect the rG4 folding propensity (Beaudoin et al. 2014). As evidence suggests that RNA can have long-range intramolecular base-pairing (Sugimoto et al. 2015), as well as rG4 motifs that cover a long range (Jodoin et al. 2014), the use of a longer flanking sequence may help reveal

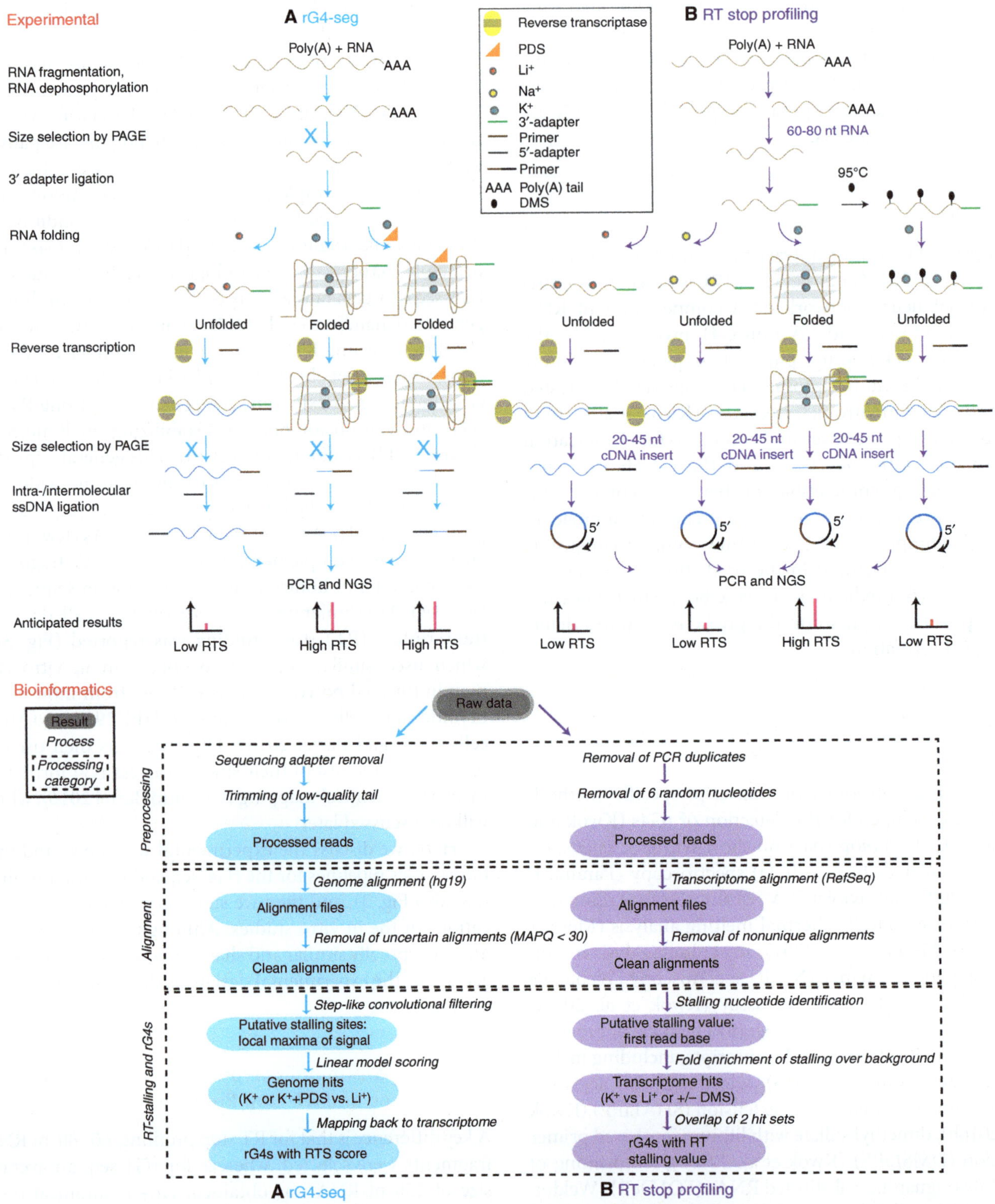

Figure 3. Reverse transcriptase stalling (RTS) sequencing approach to map rG4s. (*Top*) The experimental flowchart with relevant processes and results is illustrated for (*A*) rG4-sequencing (rG4-seq) (*left*, cyan) and for (*B*) reverse transcriptase (RT) stop profiling (*right*, purple). Key steps are shown. (*Bottom*) The step-by-step bioinformatics pipeline with relevant processes and results is illustrated for (*A*) rG4-seq (*left*, cyan) and for (*B*) RT stop profiling (*right*, purple). Starting from the raw data, the three major processing categories are shown (dashed boxes): Reads preprocessing; reads alignment; RTS score calculation, and identification of rG4s as hit regions.

Cite as *Cold Spring Harb Perspect Biol* doi: 10.1101/cshperspect.a032284

such interactions and better reflect the natural sequence context.

The second difference is that for rG4-seq, an rG4 stabilizing ligand, pyridostatin (PDS), was used, which can increase the RTS (Kwok and Balasubramanian 2015). For RT stop profiling, dimethyl sulfate (DMS) was used under denaturing conditions (95°C, 0 mM K^+). Under such conditions the N7 position of Gs, which would otherwise be hydrogen-bonded in an intact G4, is methylated to m7G, which is typically not inhibitory to the RT (Wells et al. 2000). After methylation of Gs, it is assumed that rG4s cannot be refolded.

The third major difference is the single-stranded DNA (ssDNA) ligation strategy used. For RT stop profiling, CircLigase-mediated intramolecular ssDNA ligation was performed. For rG4-seq, a T4 DNA ligase-mediated intermolecular ssDNA ligation was performed. The latter approach showed less bias in nucleotide preference and is more cost efficient than the CircLigase approach (Kwok et al. 2013a; Ritchey et al. 2017).

3.2 Comparison of Bioinformatics Procedures in rG4-seq and RT Stop Profiling

In rG4-seq, reads are aligned to the genome and significant stalling sites are later assigned to the most abundant isoform, an assignment that can introduce uncertainty. In RT stop profiling, transcriptome alignment is performed and ambiguous reads mapping to multiple isoforms are removed, potentially introducing coverage biases. For these reasons, both methods could suffer from false positives/negatives.

rG4-seq (Fig. 3A) computes RT stops with a two-step procedure: Coverage signal is processed by convolutional filters to identify candidate stalling positions, which are then statistically assessed using a linear model that contrasts the positive condition (i.e., K^+ or K^+ + PDS) against the Li^+ negative control. This approach is able to reduce false positives and identify bona fide G4 motifs but has the drawback that a small number of structures displaying high stability also under Li^+ conditions (estimated 72 putative rG4s) could be false negatives.

RT stop profiling (Fig. 3B) examines the nucleotide immediately adjacent to the stalling by calculating a fold enrichment value between the number of reads stalled and the background read density. RT-stalling sites specific to K^+ (the rG4-stabilizing condition) were identified by comparing to conditions using less rG4-stabilizing cations (Na^+ and Li^+). Independently, extracted RNA was treated with DMS under in vitro denaturing conditions (i.e., 95°C, 0 mM K^+), followed by RT with K^+, providing another diagnostic feature for rG4 presence, because Gs will not easily methylate within a rG4. The combination of these two approaches identified bona fide rG4 structures in vitro; however, the lack of replicates and the choice of an arbitrary threshold could potentially hamper confident detection.

3.3 Biological Validation and Findings

To support the approaches in the two methods, both laboratories have reported enrichment in Gs in the sequence immediately upstream of the RTS site, which would be expected for rG4-based stalling. In addition, for rG4-seq, in vitro SHALiPE was performed to verify several individual candidates, and for RT stop profiling, CD spectroscopic experiments were used to confirm selected rG4 candidates. Although the two methods use different experimental and bioinformatics steps, both appear to be robust.

Together, both methods identified thousands of rG4 structures in the human transcriptome, in vitro, for the first time. The rG4s in human mRNAs were enriched in UTRs, suggestive of role(s) in translation consistent with a number of publications on specific transcripts (Kumari et al. 2007; Arora and Suess 2011; Crenshaw et al. 2015). Also, rG4s were enriched near microRNA (miRNA) target sites and polyadenylation sites, suggesting potential roles in miRNA-mediated regulation and alternative polyadenylation, which is in agreement with recent findings on individual transcripts (Beaudoin and Perreault 2013; Stefanovic et al. 2015; Rouleau et al. 2017). Notably, rG4-seq revealed a cluster of G4 sequences that are conserved among eukaryotes and are overrepresented in genes that are annotated with RNA processing and RNA stability, which warrant further investigation. For RT stop profiling, 7852 nonoverlapping rG4 regions were detected in at least one of the two human cell lines (HEK293T and HeLa), and 4935 were identified in both. Factors such as sequencing coverage of transcripts, differences in natural RNA abundance, sequence difference in HEK293T and HeLa transcriptomes, in addition to technical variability could explain differences in rG4 detected across cell lines. The overlap between the HeLa sites and those identified by rG4-seq in the same cell line was ~45% when compared with K^+ in rG4-seq and >65% when compared with the larger rG4 data set generated on further stabilization in the K^+ + PDS conditions. This constitutes a reasonable overlap of rG4s measured in vitro considering the studies were performed in different laboratories using different experimental approaches and computational analyses.

The in vitro experimental rG4 maps provide a resource to further investigate rG4 structure and function in vivo. The main limitations in these data sets arise from rG4s that cannot yet be detected, such as those that are insufficiently stable or extremely stable, those that form from long-range rG4 interactions, and those in low abundance RNAs. It will also be interesting to go beyond the transcriptomes of mouse and human and map rG4s in other organisms to expand the repertoire of rG4s identified in vitro, which may enhance our ability to understand and predict the relationship between sequence and rG4 folding.

4 DETECTING RNA G-QUADRUPLEXES IN VIVO

The folding propensity of rG4s in cells has been recently explored by two methods (Guo and Bartel 2016). The first uses methylation by DMS, followed by RT stop profiling under K^+ conditions, and then next-generation sequencing. The second uses selective 2′-OH acylation analyzed by primer extension (SHAPE) using 2-methylnicotinic acid imidazolide (NAI), followed by RT stop profiling under Na^+ conditions, and then next-generation sequencing. Here, we discuss the experimental details of both methods (see also Fig. 4A,B), the bioinformatics pipelines applied to the data, and finally the biological findings.

4.1 Experimental Pipeline and Assessment of DMS-seq + RT Stop Profiling

To probe rG4 formation, mouse embryonic stem cell (mESC) cells were treated with ~800 mM (8%) DMS for 5 min at 37°C to methylate the N7 position of Gs (see Table 1 for comparison with related studies). For any rG4s folded in cells, Gs involved in rG4 G-quartet formation are protected from methylation, thus when RT profiling is performed on the extracted RNA under K^+ conditions bona fide rG4 structures will form, causing RTS. Conversely, if the rG4 motif is unfolded in cells, the associated Gs are methylated, and subsequently the extracted RNA does not form rG4 or cause RTS (Guo and Bartel 2016).

DMS can also methylate N1 of A and N3 of C causing rG4-independent RTS at those sites (Wells et al. 2000) and ~70% of global RT stops were at A or C for in vitro and in vivo DMS-treated samples. An individual rG4 exemplar that was ectopically expressed (G3A2) also showed similar A and C modification induced by DMS between in vitro and in vivo conditions along the gel. Given the RT stop signal at an rG4 site in vitro, at 0 mM K^+, was diminished (although not quite baseline) compared with in vitro, at 150 mM K^+, the in vitro DMS assay was deemed to be within the experimental dynamic range (i.e., not overmethylated). As RT stop at A and C was comparable in vitro and in vivo the in vivo DMS probing was also considered to be within the dynamic range of the assay (i.e., not overmethylated by DMS). It should be noted that the RT stop at G, primarily caused by rG4s stalling, constitutes 20% of all RT stops for in vitro DMS reaction (at both 0 mM K^+ and 150 mM K^+), whereas in vivo the RT stop at G is <10% of total, suggesting that the Gs are methylated to a greater extent in vivo, leading to less rG4-mediated RT stalling. As DMS methylation of N7 of G has much faster kinetics than methylation of N1 of A or N3 of C (Lawley and Brookes 1963). RNA structure probing should ideally be performed under single-hit kinetics conditions (see Table 1). Thus, the dynamic range of the DMS assay and the condition of single-hit kinetics should ideally be carefully balanced for each cell line and species under study. On considering reaction conditions used for in vivo RNA structurome experiments, the DMS concentration used by Guo and Bartel was relatively high leaving open the possibility of some overmethylation in that study (Table 1).

4.2 Experimental Pipeline and Assessment NAI-seq + RT Stop Profiling

Another in-cell structure-probing method involves the use of SHAPE reagents that acylate the RNA 2′-OH group with reaction kinetics influenced by its local flexibility (Wilkinson et al. 2006). Typically, unpaired and unconstrained nucleotides are kinetically more susceptible to 2′ acylation (Weeks 2010). As the SHAPE reagent reacts with all four nucleotides, it provides structural information on RNA at single-nucleotide resolution. One of the SHAPE reagents, NAI, was developed for in vivo probing of RNA structure (Kwok et al. 2013b; Spitale et al. 2013). Recently, we developed SHALiPE that used NAI, followed by lithium ion–mediated reverse transcription to probe rG4 in vitro (Kwok et al. 2016b). Interestingly, unlike Watson–Crick base pairs that typically lower the SHAPE reactivity, the in vitro formation of rG4 increased the SHAPE reactivity at the 3′ G position of the first three G-tracts (Kwok et al. 2016b).

In the study of Guo and Bartel (Guo and Bartel 2016), mESC cells were treated with 80 mM NAI for 15 min at 37°C (single-hit conditions, see Table 1), followed by RT stop profiling under Na^+ conditions, and then compared with in vitro NAI conditions (at both 0 mM K^+ and 150 mM K^+). Given Na^+ can also stabilize rG4, reverse transcription under Li^+ conditions would have been more optimal. Based on observations under rG4-forming conditions in vitro (150 mM K^+; Guo and Bartel 2016; Kwok et al. 2016b), if rG4 folding in cells is similar to in vitro, under the NAI reaction conditions (150 mM K^+), it would be expected that the 3′ G of each G-tract in rG4 and the loop residues of rG4s would be susceptible to modification, whereas the other Gs involved in rG4 formation would be modified to a lower degree. Guo and Bartel provided an explanation for the ectopically expressed rG4 examples by analysis of the crystal structure of the rG4 formed by telomeric repeat-containing

Cite as *Cold Spring Harb Perspect Biol* doi: 10.1101/cshperspect.a032284

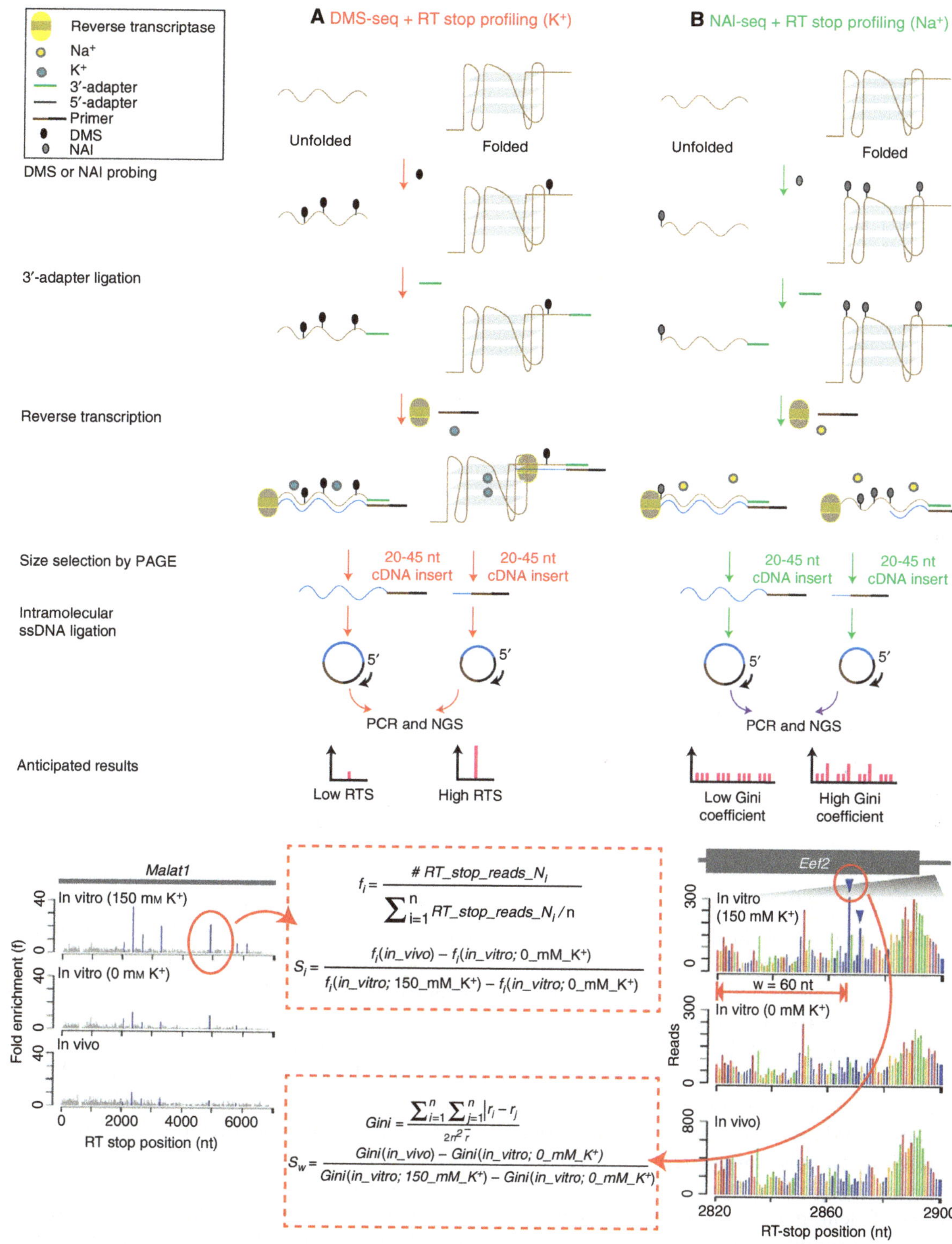

$$f_i = \frac{\#\,RT\_stop\_reads\_N_i}{\sum_{i=1}^{n} RT\_stop\_reads\_N_i/n}$$

$$S_i = \frac{f_i(in\_vivo) - f_i(in\_vitro;\ 0\_mM\_K^+)}{f_i(in\_vitro;\ 150\_mM\_K^+) - f_i(in\_vitro;\ 0\_mM\_K^+)}$$

$$Gini = \frac{\sum_{i=1}^{n}\sum_{j=1}^{n}|r_i - r_j|}{2n^2\bar{r}}$$

$$S_w = \frac{Gini(in\_vivo) - Gini(in\_vitro;\ 0\_mM\_K^+)}{Gini(in\_vitro;\ 150\_mM\_K^+) - Gini(in\_vitro;\ 0\_mM\_K^+)}$$

Figure 4. Chemical probing sequencing approach to map rG4s. (*Top*) Experimental flowchart of processes and results for (*A*) dimethyl sulfate sequencing (DMS-seq) + RT stop profiling (K^+) (*left*, red) and for (*B*) 2-methylnicotinic acid imidazolide sequencing (NAI-seq) + RT stop profiling (Na^+) (*right*, green). (*Bottom left*) Graphical representation of the scoring procedure for DMS-seq followed by RT stop profiling, with the formulas to calculate the fold enrichment score (f) at each nucleotide (i) and the in vivo folding scores (S), as indicated by the red arrow and red dashed box. (*Bottom right*) Graphical representation of the scoring procedure for NAI-seq followed by RT stop profiling, with the formulas to calculate the Gini index for a window (w) of size 60 upstream of stalling sites over putative rG4s previously detected by RT stop in vitro, as indicated by the red arrow and red dashed box below the DMS-seq one.

Table 1. Chemical probes and reaction conditions used in in vivo transcriptome-wide RNA structure probing studies

In vivo probe	Reaction conditions (Conc., time, temp.)	System	Single-hit kinetics?[a]	Reference
DMS	75 mM, 15 min, 22°C	*Arabidopsis thaliana*	Yes	Ding et al. 2014
DMS	200–400 mM, 2–4 min, 30°C; 260 mM, 40 min, 10°C	*Saccharomyces cerevisiae*	No	Rouskin et al. 2014
DMS	200–260 mM, 4 min, 37°C	Human K562, fibroblast	No	
DMS	100 mM, 2 min, 30°C	*S. cerevisiae*	Yes	Talkish et al. 2014
NAI-N3	100 mM, 15 min, 37°C	mESC	Yes	Spitale et al. 2015
DMS	500 mM, 4 min, 30°C	*S. cerevisiae*	No	Zubradt et al. 2017
DMS	200 mM, 4–5 min, 37°C	HEK293T	No	
DMS	5 M, 5 min, 26°C	*Drosophila melanogaster*	No	
DMS	800 mM, 5 min, 37°C	mESC, HEK293T, *Escherichia coli*	No	Guo and Bartel 2016
DMS	800 mM, 5 min, 30°C	*S. cerevisiae*	No	
NAI	80 mM, 15 min, 37°C	mESC	Yes	
NAI	80 mM, 15 min, 30°C	*S. cerevisiae*	Yes	
DMS	75 mM, 10 min, 22°C	*Oryza sativa*	Yes	Ritchey et al. 2017
DMS	340 mM, 5 min, 37°C	HEK293T	No	Wu and Bartel 2017
NAI	100 mM, 10 min, 22°C	*S. cerevisiae*	Yes	Selega et al. 2017
DMS	25 mM, 15 min, 28°C	*O. sativa*	Yes	Deng et al. 2018

DMS, dimethyl sulfate; NAI, 2-methylnicotinic acid imidazolide; Conc., concentration; temp., temperature; mESC, mouse embryonic stem cell.

[a]One modification every ~300 nt RNA region (McGinnis et al. 2009) or 75%–90% unmodified RNA (Wan et al. 2013; Ding et al. 2015). With chemical-based RNA structure probing experiments, single-hit kinetics is preferred as the modification of the first nucleotide/site that can induce conformational changes that can cause modification of non-native nucleotides/sites and, therefore, lead to experimental artifacts and inaccurate data interpretation. 100% DMS is ~10 M.

RNA (TERRA) (Collie et al. 2010), and found that the 2′-OH group of the 3′ G of the G-tracts was exposed (Guo and Bartel 2016). Conversely, if the rG4 is fully unfolded in cells, the Gs associated with the sequence motif will be modified more evenly by NAI.

4.3 Bioinformatic Pipeline of DMS-seq + RT Stop Profiling and NAI-seq + RT Stop Profiling

The computational analysis of the DMS-seq profiling is very similar to RT stop profiling (Fig. 4A). For the NAI-seq analysis, Gini coefficients were calculated for rG4-containing regions (Fig. 4B). Essentially, the Gini index measures inequality within a distribution: A value of 0 expresses perfect equality (e.g., nucleotides have the same stalling frequency), whereas a value of 1 expresses maximal inequality (i.e., nucleotide-specific stalling). However, outliers due to experimental noise could affect Gini estimates, and transcripts with vastly different reactivity profiles might have the same coefficient, making it difficult to use it as a comparative feature (Choudhary et al. 2017).

General-purpose methods and metrics for analyzing RT stop profiling data exist (Aviran and Pachter 2014; Choudhary et al. 2017; Li et al. 2017). These approaches use libraries from unmodified transcripts to distinguish RT drop-off noise from chemical incorporation and estimate of reactivity at single-nucleotide level. It may be worth revisiting rG4 mapping by chemical mapping using these methods and analyses.

4.4 Biological Validation and Findings

The most noteworthy outcome from the DMS-seq + RT stop profiling method in mouse, human, and yeast was that the RT stop profile for most rG4 motif regions was similar to what was observed at 0 mM K^+ in vitro (assumed, rG4-unfolded state), as opposed to what was observed at 150 mM K^+ in vitro (assumed, rG4-folded state), suggesting either that most rG4 regions are unfolded in these eukaryotes (Guo and Bartel 2016), or that the method samples the unfolded state during the time course of the reaction. In experiments intended to perturb rG4 folding, the rG4-stabilizing ligand PDS caused a small, but detectable increase in global rG4 folding, and the deletion of a known rG4 helicase, DEAH box protein 36 (DHX36), caused little detectable change in global rG4 folding. The absence of a "positive control" in the context of these experiments, despite several rational attempts to perturb the system or introduce rG4s exogenously, does leave open the possibility that rG4 structures in the transcript population may exist but were undetectable under the experimental conditions used, DMS and SHAPE structural probing assays (Guo and Bartel 2016) measure whole-cell, ensemble RNA structural conformations within the reaction time frame. Information on the structural conformation of individual RNAs and the dynamic structural interconversion, subpopulation, and heterogeneity within and across cells may be lost.

Previous computational analyses predicted a role of DNA G4 motifs in *Escherichia coli* gene regulation (Rawal et al. 2006) and a conserved DNA G4-hairpin-duplex-switch with a potential regulatory role, in the same species (Kaplan et al. 2016). In the study by Guo and Bartel (Guo and Bartel 2016) a bioinformatics search over several bacteria transcriptomes such as *E. coli*, *Pseudomonas putida*, and *Synechococcus* indicated that rG4 regions are generally depleted in bacteria, although this may reflect that bacteria do not comprise much noncoding RNA (UTRs, introns, lncRNA), whereas rG4s are typically enriched in mammals. As endogenous rG4 is rare, G4-forming RNAs comprising G3A2 (GGGAAGGGAAGGGAAGGG) or G3U (UUUGGGU GGGUGGGUGGG) were appended to the 3′ UTR of mCherry CDS and the PCR product was inserted to the expression vector plasmid (pCR2.1 backbone), and ectopically introduced to *E. coli*. The investigators showed that, in contrast to eukaryotes, ectopically introduced rG4s can be folded in *E. coli*.

4.5 Concluding Remarks

These recent approaches to detect and map rG4 structures have provided maps of rG4s that can form from in vitro refolded cellular transcripts. The global picture emerging from the initial in-cell mapping experiments suggest that rG4s are generally unfolded within a cellular context in mammalian cells, with a G4-stabilizing ligand and also knockdown of the G4-specific helicase not causing a detectable shift in global rG4 formation. This initial picture actually closely mirrors what has been observed for DNA G4s in the human genome for which of the totality of G4 structures mapped in vitro by sequencing (Chambers et al. 2015) only a very small minority (~1%) were detected in chromatin by chromatin immunoprecipitation (ChIP) using a G4-antibody (Hansel-Hertsch et al. 2016). Careful control of the global formation of rG4 structures, by protein binding or catalytic function, would be consistent with the various regulatory functions proposed for rG4s, that would logically require dynamic control between folded and unfolded states. There exists the possibility of heterogeneity between cells and also within transcript copies in a given cell, with regard to rG4 folding. While chemical mapping provides a useful means to assess the global picture, an approach with improved temporal resolution may be required to observe dynamic structures, and enrichment approaches might be necessary to select/detect structures that exist as subpopulations. Recent disclosures (Murat et al. 2017; Sauer et al. 2017) have reported measureable cellular effects of knocking down expression of rG4-targeting helicase DHX36 on translation and usage of alternative upstream open reading frames. A greater acknowledgment of rG4s as dynamic structures coupled with an improved understanding of functionally important protein–rG4 interactions is now needed for advancement of this area.

ACKNOWLEDGMENTS

We thank Professor Phil Bevilacqua and Professor Sharon Aviran for critically reading this manuscript and for their helpful comments. The Balasubramanian laboratory is supported by European Research Council Advanced Grant No. 339778, a Wellcome Trust Senior Investigator Award, and core funding from Cancer Research UK. The Kwok laboratory is supported by City University of Hong Kong Project No. 9610363, 7200520, Croucher Foundation Project No. 9500030, and Hong Kong RGC Project No. CityU 21302317, N_CityU110/17.

COMPETING FINANCIAL INTERESTS

The authors declare no competing financial interests.

REFERENCES

Arora A, Suess B. 2011. An RNA G-quadruplex in the 3′ UTR of the proto-oncogene PIM1 represses translation. *RNA Biol* **8:** 802–805.

Aviran S, Pachter L. 2014. Rational experiment design for sequencing-based RNA structure mapping. *RNA* **20:** 1864–1877.

Bang I. 1910. Untersuchungen über die Guanylsäure. *Biochem Zeit* **26:** 293–311.

Beaudoin JD, Perreault JP. 2013. Exploring mRNA 3′-UTR G-quadruplexes: Evidence of roles in both alternative polyadenylation and mRNA shortening. *Nucleic Acids Res* **41:** 5898–5911.

Beaudoin JD, Jodoin R, Perreault JP. 2013. In-line probing of RNA G-quadruplexes. *Methods* **64:** 79–87.

Beaudoin JD, Jodoin R, Perreault JP. 2014. New scoring system to identify RNA G-quadruplex folding. *Nucleic Acids Res* **42:** 1209–1223.

Bedrat A, Lacroix L, Mergny JL. 2016. Re-evaluation of G-quadruplex propensity with G4Hunter. *Nucleic Acids Res* **44:** 1746–1759.

Biffi G, Tannahill D, McCafferty J, Balasubramanian S. 2013. Quantitative visualization of DNA G-quadruplex structures in human cells. *Nat Chem* **5:** 182–186.

Biffi G, Di Antonio M, Tannahill D, Balasubramanian S. 2014. Visualization and selective chemical targeting of RNA G-quadruplex structures in the cytoplasm of human cells. *Nat Chem* **6:** 75–80.

Bugaut A, Balasubramanian S. 2012. 5′-UTR RNA G-quadruplexes: Translation regulation and targeting. *Nucleic Acids Res* **40:** 4727–4741.

Cammas A, Millevoi S. 2017. RNA G-quadruplexes: Emerging mechanisms in disease. *Nucleic Acids Res* **45:** 1584–1595.

Chambers VS, Marsico G, Boutell JM, Di Antonio M, Smith GP, Balasubramanian S. 2015. High-throughput sequencing of DNA G-quadruplex structures in the human genome. *Nat Biotechnol* **33:** 877–881.

Choudhary K, Deng F, Aviran S. 2017. Comparative and integrative analysis of RNA structural profiling data: Current practices and emerging questions. *Quant Biol* **5:** 3–24.

Collie GW, Haider SM, Neidle S, Parkinson GN. 2010. A crystallographic and modelling study of a human telomeric RNA (TERRA) quadruplex. *Nucleic Acids Res* **38:** 5569–5580.

Crenshaw E, Leung BP, Kwok CK, Sharoni M, Olson K, Sebastian NP, Ansaloni S, Schweitzer-Stenner R, Akins MR, Bevilacqua PC, et al. 2015. Amyloid precursor protein translation is regulated by a 3′UTR guanine quadruplex. *PLoS One* **10:** e0143160.

D'Antonio L, Bagga P. 2004. Computational methods for predicting intramolecular G-quadruplexes in nucleotide sequences. In *Proceedings of the IEEE Computational Systems Bioinformatics Conference*, pp. 590–591. IEEE, Stanford, CA.

Deng H, Cheema J, Zhang H, Woolfenden H, Norris M, Liu Z, Liu Q, Yang X, Yang M, Deng X, et al. 2018. Rice in vivo RNA structurome reveals RNA secondary structure conservation and divergence in plants. *Mol Plant* **11:** 602–622.

Ding Y, Tang Y, Kwok CK, Zhang Y, Bevilacqua PC, Assmann SM. 2014. In vivo genome-wide profiling of RNA secondary structure reveals novel regulatory features. *Nature* **505:** 696–700.

Ding Y, Kwok CK, Tang Y, Bevilacqua PC, Assmann SM. 2015. Genome-wide profiling of in vivo RNA structure at single-nucleotide resolution using structure-seq. *Nat Protoc* **10:** 1050–1066.

Eddy J, Maizels N. 2006. Gene function correlates with potential for G4 DNA formation in the human genome. *Nucleic Acids Res* **34:** 3887–3896.

Endoh T, Sugimoto N. 2013. Unusual -1 ribosomal frameshift caused by stable RNA G-quadruplex in open reading frame. *Anal Chem* **85:** 11435–11439.

Endoh T, Kawasaki Y, Sugimoto N. 2013. Suppression of gene expression by G-quadruplexes in open reading frames depends on G-quadruplex stability. *Angew Chem Int Ed* **52:** 5522–5526.

Fay MM, Lyons SM, Ivanov P. 2017. RNA G-quadruplexes in biology: Principles and molecular mechanisms. *J Mol Biol* **429:** 2127–2147.

Garant JM, Perreault JP, Scott MS. 2017. Motif independent identification of potential RNA G-quadruplexes by G4RNA screener. *Bioinformatics* **33:** 3532–3537.

Gellert M, Lipsett MN, Davies DR. 1962. Helix formation by guanylic acid. *Proc Natl Acad Sci* **48:** 2013–2018.

Guo JU, Bartel DP. 2016. RNA G-quadruplexes are globally unfolded in eukaryotic cells and depleted in bacteria. *Science* **353:** aaf5371.

Hansel-Hertsch R, Beraldi D, Lensing SV, Marsico G, Zyner K, Parry A, Di Antonio M, Pike J, Kimura H, Narita M, et al. 2016. G-quadruplex structures mark human regulatory chromatin. *Nat Genet* **48:** 1267–1272.

Hazel P, Huppert J, Balasubramanian S, Neidle S. 2004. Loop-length-dependent folding of G-quadruplexes. *J Am Chem Soc* **126:** 16405–16415.

Hon J, Martinek T, Zendulka J, Lexa M. 2017. Pqsfinder: An exhaustive and imperfection-tolerant search tool for potential quadruplex-forming sequences in R. *Bioinformatics* **33:** 3373–3379.

Huppert JL, Balasubramanian S. 2005. Prevalence of quadruplexes in the human genome. *Nucleic Acids Res* **33:** 2908–2916.

Jodoin R, Bauer L, Garant JM, Mahdi Laaref A, Phaneuf F, Perreault JP. 2014. The folding of 5′-UTR human G-quadruplexes possessing a long central loop. *RNA* **20:** 1129–1141.

Kaplan OI, Berber B, Hekim N, Doluca O. 2016. G-quadruplex prediction in *E. coli* genome reveals a conserved putative G-quadruplex-Hairpin-Duplex switch. *Nucleic Acids Res* **44:** 9083–9095.

Kumari S, Bugaut A, Huppert JL, Balasubramanian S. 2007. An RNA G-quadruplex in the 5′ UTR of the NRAS proto-oncogene modulates translation. *Nat Chem Biol* **3:** 218–221.

Kwok CK, Balasubramanian S. 2015. Targeted detection of G-quadruplexes in cellular RNAs. *Angew Chem Int Ed* **54:** 6751–6754.

Kwok CK, Merrick CJ. 2017. G-quadruplexes: Prediction, characterization, and biological application. *Trends Biotechnol* **35:** 997–1013.

Kwok CK, Ding Y, Sherlock ME, Assmann SM, Bevilacqua PC. 2013a. A hybridization-based approach for quantitative and low-bias single-stranded DNA ligation. *Anal Biochem* **435:** 181–186.

Kwok CK, Ding Y, Tang Y, Assmann SM, Bevilacqua PC. 2013b. Determination of in vivo RNA structure in low-abundance transcripts. *Nat Commun* **4:** 2971.

Kwok CK, Sherlock ME, Bevilacqua PC. 2013c. Effect of loop sequence and loop length on the intrinsic fluorescence of G-quadruplexes. *Biochemistry* **52:** 3019–3021.

Kwok CK, Marsico G, Sahakyan AB, Chambers VS, Balasubramanian S. 2016a. rG4-seq reveals widespread formation of G-quadruplex structures in the human transcriptome. *Nat Methods* **13:** 841–844.

Kwok CK, Sahakyan AB, Balasubramanian S. 2016b. Structural analysis using SHALiPE to reveal RNA G-quadruplex formation in human precursor microRNA. *Angew Chem Int Ed* **55:** 8958–8961.

Lawley P, Brookes P. 1963. Further studies on the alkylation of nucleic acids and their constituent nucleotides. *Biochem J* **89:** 127–138.

Li B, Tambe A, Aviran S, Pachter L. 2017. PROBer provides a general toolkit for analyzing sequencing-based toeprinting assays. *Cell Syst* **4:** 568–574.e7.

Lorenz R, Bernhart SH, Qin J, Honer zu Siederdissen C, Tanzer A, Amman F, Hofacker IL, Stadler PF. 2013. 2D meets 4G: G-quadruplexes in RNA secondary structure prediction. *IEEE/ACM Trans Comput Biol Bioinform* **10:** 832–844.

Marcel V, Tran PL, Sagne C, Martel-Planche G, Vaslin L, Teulade-Fichou MP, Hall J, Mergny JL, Hainaut P, Van Dyck E. 2011. G-quadruplex structures in TP53 intron 3: Role in alternative splicing and in production of p53 mRNA isoforms. *Carcinogenesis* **32:** 271–278.

Matsumura K, Kawasaki Y, Miyamoto M, Kamoshida Y, Nakamura J, Negishi L, Suda S, Akiyama T. 2017. The novel G-quadruplex-containing long non-coding RNA GSEC antagonizes DHX36 and modulates colon cancer cell migration. *Oncogene* **36:** 1191–1199.

McGinnis JL, Duncan CDS, Weeks KM, Daniel H. 2009. High-throughput SHAPE and hydroxyl radical analysis of RNA structure and ribonucleoprotein assembly. in *Methods in Enzymology*, pp. 67–89. Academic, New York.

Mergny JL, Phan AT, Lacroix L. 1998. Following G-quartet formation by UV-spectroscopy. *FEBS Lett* **435:** 74–78.

Mergny JL, Lacroix L, Teulade-Fichou MP, Hounsou C, Guittat L, Hoarau M, Arimondo PB, Vigneron JP, Lehn JM, Riou JF, et al. 2001. Telomerase inhibitors based on quadruplex ligands selected by a fluorescence assay. *Proc Natl Acad Sci* **98:** 3062–3067.

Mirihana Arachchilage G, Dassanayake AC, Basu S. 2015. A potassium ion-dependent RNA structural switch regulates human pre-miRNA 92b maturation. *Chem Biol* **22:** 262–272.

Mukundan VT, Phan AT. 2013. Bulges in G-quadruplexes: Broadening the definition of G-quadruplex-forming sequences. *J Am Chem Soc* **135:** 5017–5028.

Murat P, Marsico G, Herdy B, Ghanbarian A, Portella G, Balasubramanian S. 2017. RNA G-quadruplexes mark repressive upstream open reading frames in human mRNAs. *bioRxiv* doi: 10.1101/223073.

Paramasivan S, Rujan I, Bolton PH. 2007. Circular dichroism of quadruplex DNAs: Applications to structure, cation effects and ligand binding. *Methods* **43:** 324–331.

Patel PK, Hosur RV. 1999. NMR observation of T-tetrads in a parallel stranded DNA quadruplex formed by *Saccharomyces cerevisiae* telomere repeats. *Nucleic Acids Res* **27:** 2457–2464.

Rawal P, Kummarasetti VBR, Ravindran J, Kumar N, Halder K, Sharma R, Mukerji M, Das SK, Chowdhury S. 2006. Genome-wide prediction of G4 DNA as regulatory motifs: Role in *Escherichia coli* global regulation. *Genome Research* **16:** 644–655.

Renaud de la Faverie A, Guedin A, Bedrat A, Yatsunyk LA, Mergny JL. 2014. Thioflavin T as a fluorescence light-up probe for G4 formation. *Nucleic Acids Res* **42:** e65.

Ritchey LE, Su Z, Tang Y, Tack DC, Assmann SM, Bevilacqua PC. 2017. Structure-seq2: Sensitive and accurate genome-wide profiling of RNA structure in vivo. *Nucleic Acids Res* **45:** e135.

Rouleau S, Glouzon JS, Brumwell A, Bisaillon M, Perreault JP. 2017. 3′ UTR G-quadruplexes regulate miRNA binding. *RNA* **23:** 1172–1179.

Rouleau SG, Garant JM, Bolduc F, Bisaillon M, Perreault JP. 2018. G-quadruplexes influence pri-microRNA processing. *RNA Biol* **15:** 198–206.

Rouskin S, Zubradt M, Washietl S, Kellis M, Weissman JS. 2014. Genome-wide probing of RNA structure reveals active unfolding of mRNA structures in vivo. *Nature* **505:** 701–705.

Sahakyan AB, Chambers VS, Marsico G, Santner T, Di Antonio M, Balasubramanian S. 2017. Machine learning model for sequence-driven DNA G-quadruplex formation. *Sci Rep* **7:** 14535.

Sauer M, Juranek SA, Kazemier HG, Benhalevy D, Wang X, Hafner M, Paeschke K. 2017. DHX36 binding at G-rich sites in mRNA untranslated regions promotes translation. *bioRxiv* doi: 10.1101/215541.

Schaffitzel C, Berger I, Postberg J, Hanes J, Lipps HJ, Pluckthun A. 2001. In vitro generated antibodies specific for telomeric guanine-quadruplex DNA react with *Stylonychia lemnae* macronuclei. *Proc Natl Acad Sci* **98:** 8572–8577.

Selega A, Sirocchi C, Iosub I, Granneman S, Sanguinetti G. 2017. Robust statistical modeling improves sensitivity of high-throughput RNA structure probing experiments. *Nat Methods* **14:** 83–89.

Sen D, Gilbert W. 1988. Formation of parallel four-stranded complexes by guanine-rich motifs in DNA and its implications for meiosis. *Nature* **334:** 364–366.

Simone R, Fratta P, Neidle S, Parkinson GN, Isaacs AM. 2015. G-quadruplexes: Emerging roles in neurodegenerative diseases and the non-coding transcriptome. *FEBS Lett* **589:** 1653–1668.

Spitale RC, Crisalli P, Flynn RA, Torre EA, Kool ET, Chang HY. 2013. RNA SHAPE analysis in living cells. *Nat Chem Biol* **9:** 18–20.

Spitale RC, Flynn RA, Zhang QC, Crisalli P, Lee B, Jung JW, Kuchelmeister HY, Batista PJ, Torre EA, Kool ET, et al. 2015. Structural imprints in vivo decode RNA regulatory mechanisms. *Nature* **519:** 486–490.

Stefanovic S, Bassell GJ, Mihailescu MR. 2015. G quadruplex RNA structures in PSD-95 mRNA: Potential regulators of miR-125a seed binding site accessibility. *RNA* **21:** 48–60.

Subramanian M, Rage F, Tabet R, Flatter E, Mandel JL, Moine H. 2011. G-quadruplex RNA structure as a signal for neurite mRNA targeting. *EMBO Rep* **12:** 697–704.

Sugimoto Y, Vigilante A, Darbo E, Zirra A, Militti C, D'Ambrogio A, Luscombe NM, Ule J. 2015. hiCLIP reveals the in vivo atlas of mRNA secondary structures recognized by Staufen 1. *Nature* **519:** 491–494.

Talkish J, May G, Lin Y, Woolford JL Jr., McManus CJ. 2014. Mod-seq: High-throughput sequencing for chemical probing of RNA structure. *RNA* **20:** 713–720.

Todd AK, Johnston M, Neidle S. 2005. Highly prevalent putative quadruplex sequence motifs in human DNA. *Nucleic Acids Res* **33:** 2901–2907.

Wan Y, Qu K, Ouyang Z, Chang HY. 2013. Genome-wide mapping of RNA structure using nuclease digestion and high-throughput sequencing. *Nat Protoc* **8:** 849–869.

Wang X, Goodrich KJ, Gooding AR, Naeem H, Archer S, Paucek RD, Youmans DT, Cech TR, Davidovich C. 2017. Targeting of Polycomb repressive complex 2 to RNA by short repeats of consecutive guanines. *Mol Cell* **65:** 1056–1067.

Wanrooij PH, Uhler JP, Simonsson T, Falkenberg M, Gustafsson CM. 2010. G-quadruplex structures in RNA stimulate mitochondrial transcription termination and primer formation. *Proc Natl Acad Sci* **107:** 16072–16077.

Webba da Silva M. 2007. NMR methods for studying quadruplex nucleic acids. *Methods* **43:** 264–277.

Weeks KM. 2010. Advances in RNA structure analysis by chemical probing. *Curr Opin Struct Biol* **20:** 295–304.

Weldon C, Behm-Ansmant I, Hurley LH, Burley GA, Branlant C, Eperon IC, Dominguez C. 2017. Identification of G-quadruplexes in long functional RNAs using 7-deazaguanine RNA. *Nat Chem Biol* **13:** 18–20.

Weldon C, Dacanay JG, Gokhale V, Boddupally PVL, Behm-Ansmant I, Burley GA, Branlant C, Hurley LH, Dominguez C, Eperon IC. 2018. Specific G-quadruplex ligands modulate the alternative splicing of Bcl-X. *Nucleic Acids Res* **46:** 886–896.

Wells SE, Hughes JMX, Igel AH, Ares M. 2000. Use of dimethyl sulfate to probe RNA structure in vivo. *Method Enzymol* **318:** 479–493.

Wilkinson KA, Merino EJ, Weeks KM. 2006. Selective 2′-hydroxyl acylation analyzed by primer extension (SHAPE): Quantitative RNA structure analysis at single nucleotide resolution. *Nat Protoc* **1:** 1610–1616.

Woodford KJ, Howell RM, Usdin K. 1994. A novel K^+-dependent DNA synthesis arrest site in a commonly occurring sequence motif in eukaryotes. *J Biol Chem* **269:** 27029–27035.

Wu X, Bartel DP. 2017. Widespread influence of 3′-end structures on mammalian mRNA processing and stability. *Cell* **169:** 905–917.

Yu CH, Teulade-Fichou MP, Olsthoorn RC. 2014. Stimulation of ribosomal frameshifting by RNA G-quadruplex structures. *Nucleic Acids Res* **42:** 1887–1892.

Zhang D-H, Fujimoto T, Saxena S, Yu H-Q, Miyoshi D, Sugimoto N. 2010. Monomorphic RNA G-quadruplex and polymorphic DNA G-quadruplex structures responding to cellular environmental factors. *Biochemistry* **49:** 4554–4563.

Zubradt M, Gupta P, Persad S, Lambowitz AM, Weissman JS, Rouskin S. 2017. DMS-MaPseq for genome-wide or targeted RNA structure probing in vivo. *Nat Methods* **14:** 75–82.

Linking RNA Sequence, Structure, and Function on Massively Parallel High-Throughput Sequencers

Sarah K. Denny[1] **and William J. Greenleaf**[1,2,3]

[1]Stanford University Department of Genetics, Stanford, California 94305

[2]Stanford University Department of Applied Physics, Stanford, California 94025

[3]Chan Zuckerberg Biohub, San Francisco, California 94158

Correspondence: wjg@stanford.edu

SUMMARY

High-throughput sequencing methods have revolutionized our ability to catalog the diversity of RNAs and RNA–protein interactions that can exist in our cells. However, the relationship between RNA sequence, structure, and function is enormously complex, demonstrating the need for methods that can provide quantitative thermodynamic and kinetic measurements of macromolecular interaction with RNA, at a scale commensurate with the sequence diversity of RNA. Here, we discuss a class of methods that extend the core functionality of DNA sequencers to enable high-throughput measurements of RNA folding and RNA–protein interactions. Topics discussed include a description of the method and multiple applications to RNA-binding proteins, riboswitch design and engineering, and RNA tertiary structure energetics.

Outline

Cite as *Cold Spring Harb Perspect Biol* doi: 10.1101/cshperspect.a032300

1 BACKGROUND

Our ability to catalog the immense diversity of RNAs that exist in the cell has accelerated at a spectacular rate with the advent of high-throughput sequencing methods, allowing the mapping of a universe of previously uncharacterized transcripts. Likewise, sequencing methods have enabled the mapping of the diverse proteins that interact with RNA in vivo, providing a window into a network of protein–RNA interactions critical to gene expression regulation (Butter et al. 2009; Castello et al. 2012). However, our understanding of the detailed relationship between the sequence of RNA and its structure and function is limited, both in terms of how complex RNAs fold into 3D structures and also how RNA sequence and structures determine protein–RNA interactions. A predictive, quantitative understanding for these functional determinants (both 3D structure and protein interaction) will be needed before we can truly understand, and engineer, RNA functions.

However, the sequence–structure–function relationship in RNA is complex—perhaps not as fundamentally combinatorial and interactive as protein, but much more challenging than double-stranded, or even single-stranded, DNA. RNA makes diverse intramolecular interactions—Watson-Crick base pairs, noncanonical base pairs, tertiary interactions, etc.—that define the 3D structures RNA takes on (Turner et al. 1988; Tinoco and Bustamante 1999), which in turn sets the landscape for its function. For example, critical RNAs in translation, splicing, and chromatin organization and compaction must form precise tertiary structures to achieve their biological function (Staley and Guthrie 1998; Kieft et al. 2001; Noller 2005). In higher eukaryotes, these biological functions are rarely achieved in the absence of RNA-binding proteins, which help to coordinate and regulate gene expression at multiple stages in the life cycle of an RNA (Moore 2005; Keene 2007), and further entangle the RNA folding process with the physical determinants of RNA–protein interactions. Furthermore, RNA sequences form conformationally dynamic structures, with biologically functional conformational transitions and multiple nonnative folded states (Winkler and Breaker 2005; Solomatin et al. 2010). Protein binding can affect the occupancy and rate of exchange between these structural states, while conversely the structure of the RNA helps to define protein binding affinity and specificity. Therefore, there exists a need for methods that can begin to provide generalizable, broad insights into the complex relationship between RNA sequence, dynamic conformations, and protein binding, creating a principled framework for understanding RNA's diverse functions within the cell.

However, the challenge is vast. A productive approach to these problems requires measurements that can account for these multiple structural states at an immense scale—commensurate with the combinatorial sequence complexity of nucleic acids. And although a variety of high-throughput methods have been developed to measure RNA structure and RNA–protein binding, few allow for reporting of these interactions in terms of the thermodynamic and kinetic constants that must govern physical behaviors in the cell. Transcriptome-wide RNA structure measurements have relied on proximity cross-linking (Ramani et al. 2015) or base accessibility (Ding et al. 2014), each of which provides constraints on the behavior of the ensemble of RNA structures, but these methods require averaging over many different structures and thus cannot quantify occupancy across different structural states and cannot be easily read out in terms of binding energies and kinetic constants. In vivo approaches to quantify RNA–protein interactions through cross-linking and immunoprecipitation (CLIP, eCLIP, iCLIP, etc.) (Hafner et al. 2010; Van Nostrand et al. 2016) have mapped enrichment of RNA-binding protein (RBP) binding at high throughput, but often suffer from poor sequence resolution, biased cross-linking efficiency, large variability in the abundance of different potential binding sites, and indirect targeting of binding interactions with an antibody. In vitro approaches can solve a number of these uncertainties, and medium-throughput approaches for quantitative measurement have been enabled by microfluidic methods (Martin et al. 2012). Higher-throughput measurements of relative RBP affinity have been achieved with sequencing-enrichment approaches like RNAcompete, RNA-bind-and-seq, RNA SELEX, HiTS-Kin, and HiTS-Eq (Ellington and Szostak 1990; Ray et al. 2009; Lambert et al. 2014; Jankowsky and Harris 2017; Lou et al. 2017). These methods have enabled measurements of relative RBP affinity across randomized libraries; however, they often require immense sequencing resources for readout and the enrichment readout is not as generalizable as direct measurements of biophysical parameters.

A recently developed class of techniques has enabled direct, fluorescence-based, quantitative measurement of binding affinity and kinetics to RNA, with the same or larger experimental capacity as sequencing-based approaches. Two highly related approaches in this class are RNA-MaP (quantitative analysis of RNA on a massively parallel array) (Buenrostro et al. 2014) and HiTS-RAP (high-throughput sequencing and RNA affinity profiling) (Tome et al. 2014); In this review, we will refer to this class of methods collectively as RNA-HiTS, for an RNA array on a high-throughput sequencer.

These methods make use of the core functionality of the high-throughput DNA sequencers (i.e., clonal amplification and massively parallel fluorescence quantification) to enable high-throughput, fluorescence-based detection of

Cite as *Cold Spring Harb Perspect Biol* doi: 10.1101/cshperspect.a032300

nucleic acid substrates for diverse applications. RNA-HiTS was born from work on DNA–protein interactions that used a high-throughput sequencer to characterize protein binding to a vast swath of DNA sequences (Nutiu et al. 2011)—the first demonstration of the power of the sequencing instrument for quantitative determination of affinity landscapes of mutant nucleic acid sequences. Other applications of sequencing infrastructure include selection of specific oligonucleotide sequences within a pool of targets (Matzas et al. 2010), detection of mRNA sequence using fluorescently labeled transfer RNAs (tRNAs) (Uemura et al. 2010), and determination of the length of poly(A) tails among different classes of mRNAs (Subtelny et al. 2014). RNA-HiTS technology substituted diverse RNA targets for the original dsDNA targets in Nutiu et al. (2011), ultimately enabling equally comprehensive and quantitative characterization of binding to RNA (Buenrostro et al. 2014; Tome et al. 2014).

2 TECHNIQUES

2.1 Illumina Platforms Generate Arrays of Sequence-Identified DNA Clusters on the Surface of a Chip

During the process of high-throughput sequencing on an Illumina GAII(x) DNA Sequencer, millions of individual DNA strands are first immobilized on a flow cell surface, then clonally amplified to generate a "cluster" of DNA containing approximately 1000 identical molecules (Fig. 1A). During the sequencing process, the nucleic acid sequence of each clonal cluster is determined by step-wise sequencing-by-synthesis of one strand of each DNA molecule on the surface of the flow cell. Fluorescently labeled nucleotides are incorporated such that each incorporation cycle allows determination of the identity of one base of the DNA sequence (Fig. 1A). Typical sequencing reactions proceed for at least 30 cycles per read (up to 300 cycles), and multiple reads can be performed starting from different priming sequences to

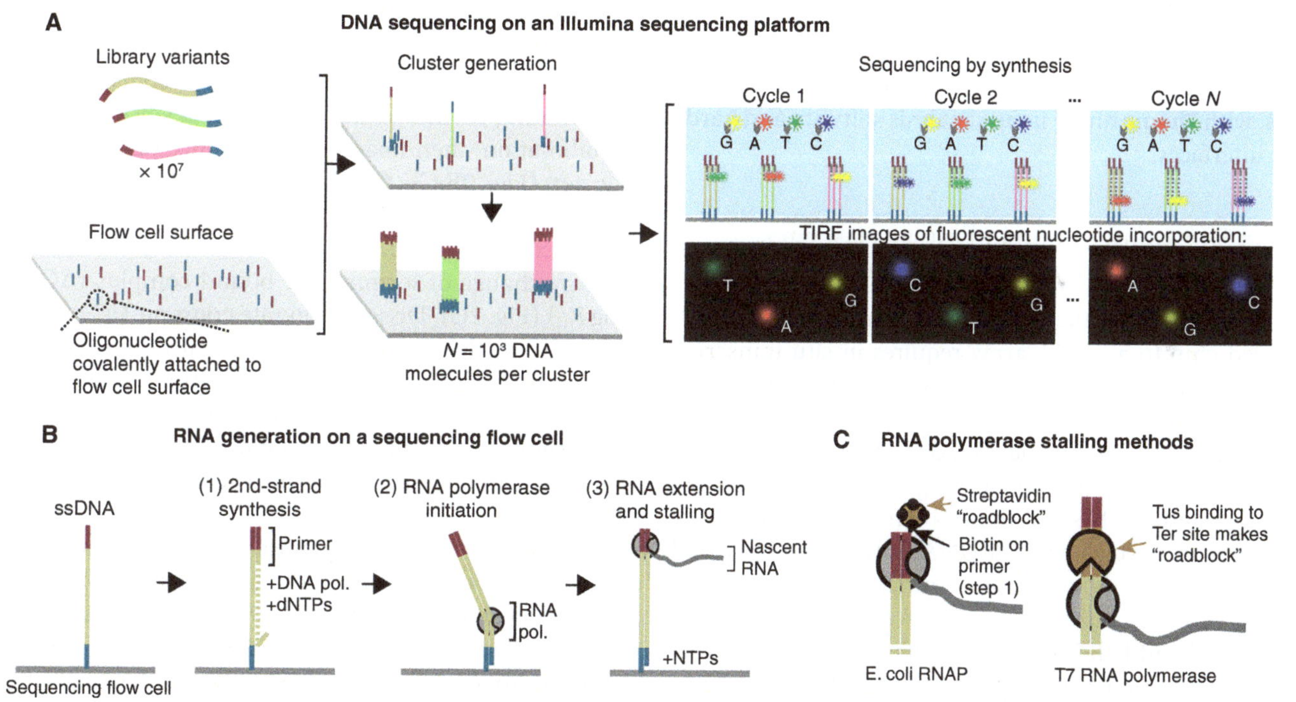

Figure 1. Harnessing the Illumina sequencing platform for systems-level biochemistry of RNA. (*A*) Schematic for DNA sequencing on an Illumina platform. DNA fragments with sequencing adapters on 5′ and 3′ ends anneal to complementary oligonucleotides covalently attached to the sequencing flow cell surface. Molecules are clonally amplified by polymerase chain reaction (PCR) to form "clusters" on the chip surface, which are then sequenced by sequential incorporation of reversibly terminated, fluorescent nucleotides. (*B*) RNA-HiTS extends the Illumina sequencing platform by transcribing clusters of DNA into RNA directly on the surface of the sequencing chip. (1) The second strand of DNA is regenerated with a DNA polymerase. (2) An RNA polymerase is initiated by allowing binding to an initiation sequence designed into the DNA fragment. (3) The RNA polymerase is allowed to extend until it stalls at a "roadblock," resulting in stable display of the nascent RNA transcript (gray). (*C*) Two methods for stalling the RNA polymerase. (*Left*) A biotinylated primer is used for second-strand synthesis (see *B*); streptavidin is then bound to this biotin before transcription, making a "roadblock" that stalls the *Escherichia coli* RNA polymerase during the extension step. (*Right*) A Tus protein binds a Ter sequence element designed into each DNA fragment, which stalls T7 RNA polymerase.

detect barcodes for demultiplexing or to read from both ends of the molecule for paired-end sequencing.

The GAII hardware platform integrates fluidics handling and thermal control that enables molecular biology reactions to take place on the surface of the chip, as well as total internal reflection fluorescence (TIRF) imaging for detection of nucleotide incorporation across clusters in massive throughput. Initial implementations of RNA-HiTS methods hijacked these instruments directly (Buenrostro et al. 2014; Tome et al. 2014). However, the GAII platform is antiquated and unsupported, and newer, more economical Illumina instruments like the MiSeq allow much more rapid sequencing turnaround. However, these newer instruments are highly integrated and do not support the facile implementation of nonstandard molecular biology and imaging workflows. Recently, our laboratory has developed a modified, custom-built imaging station based on the original Illumina GAII that decouples the platform used for sequencing from the instrument used for custom biochemistry applications (She et al. 2017). This imaging station is built using components from the original Illumina GAII(x) but is designed to interface with Illumina MiSeq chips, with "home brew" image analysis software and hardware interfaces.

2.2 In Situ Transcription to Display Nascent Transcripts of RNA at High Throughput

The conversion of the native DNA array of a postsequenced chip to an RNA array requires in situ transcription of the DNA clusters on the surface of the flow cell. Broadly, this in situ transcription is performed in three steps: beginning with single-stranded DNA molecules covalently attached to the sequencing flow cell surface, (1) second-strand synthesis of the DNA molecules by primer annealing and extension by a DNA polymerase is carried out, (2) an RNA polymerase is initiated in a sequence-specific manner, and then (3) a nascent RNA transcript is extended until the RNA polymerase reaches a "roadblock" that stably halts the polymerase on the DNA template (Fig. 1B).

This "roadblock" has been implemented in two distinct ways. In the first, a biotinylated primer can be used to prime second-strand synthesis, and subsequently this biotin may be bound by streptavidin before RNA polymerase initiation and extension (Fig. 1C) (Buenrostro et al. 2014). The RNA polymerase (in this case *Escherichia coli* RNAP) stalls when it encounters this terminal biotin–streptavidin roadblock (Greenleaf et al. 2008), allowing stable display of the nascent RNA transcript. Furthermore, this workflow aims to allow only a single RNAP molecule to initiate at the engineered promoter, producing a single piece of RNA stably tethered to its DNA template. In an alternate stalling scheme, the replication terminator protein Tus is bound to a 32-bp *Ter* sequence element engineered into the DNA library (Mohanty et al. 1996; Mulugu et al. 2001). T7 RNA polymerase then stalls when it encounters the bound Tus protein (Fig. 1C) (Tome et al. 2014).

After transcription, DNA oligos may be annealed to common sequences on all transcripts to help stabilize the desired secondary structure of the RNA, to provide a readout of transcription efficiency (Fig. 2A), or to otherwise label the transcribed RNA.

2.3 Performing Kinetic and Thermodynamic Measurements and Quantifying Images

The TIRF imaging platform enables direct measurement of association of any fluorescently labeled binding partner to every RNA cluster on the surface of the chip. This direct observation of binding allows relatively straightforward measurement of kinetic and thermodynamic parameters. For example, the equilibrium dissociation constants (K_d) can be measured by applying different concentrations of the binding partner to the chip, waiting sufficient time for equilibration, then imaging fluorescence observed at each cluster (Fig. 2B,C). Quantifying the bound fluorescence across concentrations and fitting these data to a binding curve enables inference of the K_d and the standard free-energy change of forming the bound complex ($\Delta G^o = RT \log K_d$) (Fig. 2C). Dissociation rate constant (k_{off}) can likewise be directly measured by subsequently diluting the fluorescently bound binding partner from solution, then sequentially imaging of the loss of bound fluorescence over time. In principle, association rate constants can be measured in a similar manner, or can be calculated from $k_{on} = k_{off}/K_d$.

A necessary step to quantifying fluorescence of each cluster is to align the positions of clusters obtained from the sequencing data with each fluorescent image (Fig. 2D). An initial cross-correlation provides a rough alignment that accounts for overall offsets between the sequencing data and the images, and then subsequent iterations of cross-correlation of progressively finer grids of subtiles then account for any optical aberrations introduced in different imaging stations. Together, these values are fit to a continuous offset map that can be used to obtain precise alignment of sequence data and experimental images (She et al. 2017). Subsequently, subtiles are fit to a sum of 2D Gaussian functions to determine the integrated fluorescence of each cluster. This strategy also allows for a subtile-specific background fluorescence value to be determined to account for any differences in overall fluorescence in different regions of the chip.

Cite as *Cold Spring Harb Perspect Biol* doi: 10.1101/cshperspect.a032300

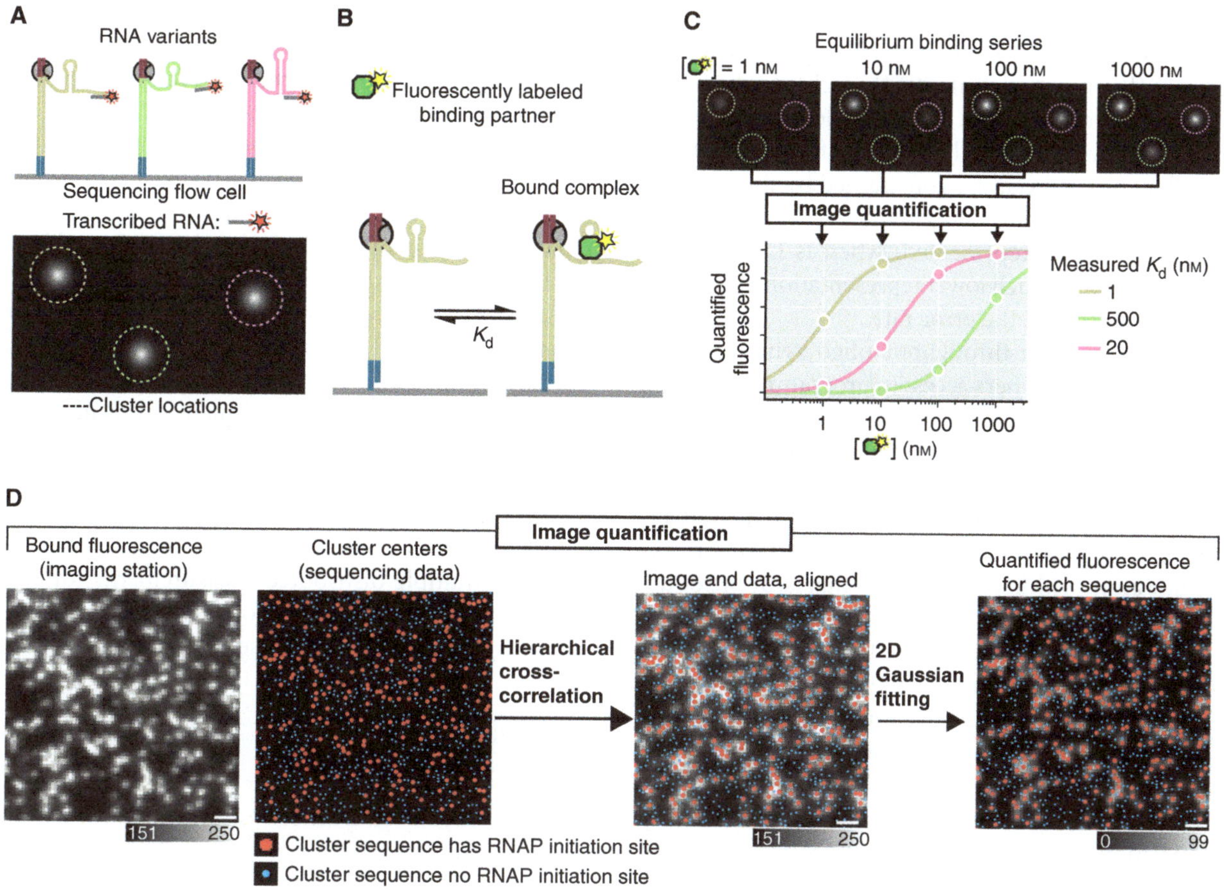

Figure 2. Performing kinetic and thermodynamic measurements and quantifying images. (*A*) After in situ transcription, transcribed RNAs may be labeled with fluorescent oligonucleotides complementary to a common sequence to all library variants (simulated data below). (*B*) Application of a fluorescently labeled binding partner to the flow cell enables detection of the thermodynamics and kinetics of binding to each transcribed library variant. (*C*) A typical equilibrium binding experiment proceeds by applying a specific concentration of the binding partner to the flow cell, waiting for equilibrium, detecting bound fluorescence at each cluster of RNA, and then repeating at multiple concentrations of binding partner. After image quantification, fluorescent measurements are fit to a binding isotherm to obtain the observed equilibrium dissociation constant, K_d. (*D*) Image quantification workflow. First, images are aligned with the sequencing data through hierarchical cross-correlation. Images are then fit to a sum of 2D Gaussians, with each Gaussian centered at the position of each cluster. Scale bar, 2.5 µm.

2.4 Library Considerations and Requirements

The first step in any RNA-HiTS experiment is to generate a diverse library of sequenceable DNA fragments that will serve as the basis of the RNA array. The generation of this DNA library has two distinct steps. First, one must design and synthesize the variable region that will be transcribed in situ. Next, this variant library must be converted into a sequenceable library. We will deal with each of these steps in turn.

Variable DNA sequences that serve as the core of a RNA variant library can be created in three ways: (1) a full or partial randomization of nucleic acid synthesis, (2) an array-based programmed library synthesis, and (3) the selection of a diverse set of natural sequences. A completely randomized nucleic acid sequence library is often not feasible—experiments can generally probe on the order of 10^6 different sequences at most (assuming an average of 10 replicate clusters per sequence variant and 10^7 total clusters per chip), which corresponds to a completely random sequence of only 10 bases. Structured RNA motifs like hairpin stems will often exceed this limit and have considerable sequence constraints to maintain base-pairing, making targeted sequence variations as opposed to random *N*-mers often more informative and interpretable. One way to focus investigational throughput around interesting functional variations is to produce a library of sequence variants "centered" on a known consensus sequence motif with error-prone polymerase chain reaction (PCR) (Cirino et al. 2003; Tome et al. 2014), or with doped-in synthesis (Buenrostro et al. 2014), each of which can introduce single, double, and some fraction of triple and higher-order mutations to the known motif. However, these classes of "dirty" synthesis are constrained by the statistics of random incorporation: In general, the consensus sequence will be highly represented, whereas higher-order mutations will have a much lower representation depending on sequence length and the error rate.

More recently, high-throughput oligo synthesis platforms have given researchers access to programmed parallel synthesis of up to 10^6 unique oligonucleotide sequences. These commercially available technologies allow researchers to design a library of sequence variants that will each be synthesized independently with approximately equal representation. This synthesis method allows the designer to ask multiple questions within each library—that is, by introducing variation in multiple known consensus sequences to assess evolutionary paths between the sequences or by changing the sequences flanking the consensus sequence to assess the context dependence of mutational effects. Currently, commercially available oligonucleotide pools are generally limited to sequences <200 nt in length, and whereas synthesis error rates can introduce additional unwanted variation, efforts for improving fidelity are ongoing.

A final method takes advantage of the diversity of sequence variants inherent to biological genomes. For example, the *Saccharomyces cerevisiae* genome can be enzymatically digested to form the variable transcribed region in an RNA-HiTS experiment (She et al. 2017). This approach enabled assessment of binding to targets embedded in their physiologically relevant, local sequence context that could affect local structure formation.

The second step in library generation requires the construction of a DNA fragment amenable to sequencing and transcription, which is achieved by adding common sequences to the entire library of variants at the 5′ and 3′ end. Each library variant must have an RNAP initiation site at the 5′ end of the variable transcribed region, and if using Ter/Tus stalling, the 32-bp Ter element at the 3′ end of the transcribed region. Both stalling methods require a sufficient number of base pairs between the stall site of the RNAP polymerase and the 3′ end of the RNA target, as the RNAP polymerase transcripts have a footprint that leaves approximately 25 bases of RNA within the exit channel (Greenleaf et al. 2008), preventing access by the fluorescently labeled binding partner.

The DNA sequencing error rate can present an issue for an RNA-HiTS experiment, especially for a library of highly related sequences, as sequencing errors can lead to clusters being assigned to the wrong molecular variant, producing spurious results. A per-base error rate of ~0.1% or more can be expected for Illumina sequencing. For a variable region of 100 nt, such a rate might produce misassignment of a substantial fraction (1/10th) of library variants. One strategy to remove the impact of sequencing error is to introduce a unique molecular identifier (UMI) to each molecule within the library, comprising a random sequence of 16 nt. On incorporation of this UMI to each member of the library, the library is diluted to $\sim 8 \times 10^5$ molecules and then reamplified, creating a "bottlenecked" population of PCR amplicons, with each UMI represented many times in the library. Sequencing this bottlenecked library allows for confident assessment of the specific sequence variant associated with each UMI, and downstream analysis requires only the sequence of the UMI to identify the molecular variant.

In general, the number of sequenced clusters should exceed the number of variants by at least 10-fold, in order to make multiple distributed measurements for each molecular variant in the library to enable a principled understanding of measurement noise.

3 APPLICATIONS

3.1 Sequence and Structural Determinants of Protein–RNA Interactions

3.1.1 MS2 Coat Protein to Mutated RNA Targets

An overarching goal of RNA-HiTS technology has been to make a generalizable, predictive model of protein–RNA interaction affinity across any possible RNA sequence variant. The first protein targeted for this analysis was the MS2 coat protein (Buenrostro et al. 2014), a protein that was initially discovered to play a bifunctional role in viral coat assembly and translational repression in the bacteriophage MS2. The high affinity and specificity of this interaction has made it useful in biotechnology applications including affinity purification, live-cell RNA imaging, and synthetic biology (Bardwell and Wickens 1990; Bertrand et al. 1998). The MS2 protein binds an RNA motif consisting of a stem loop structure with a single bulged residue (Fig. 3A). Starting with this consensus sequence, RNA-HiTS enabled the quantitative measurement of MS2 interaction with all single, double, and a subset of triple and higher-order mutants to comprehensively assess the contribution of each residue to the free-energy landscape of the binding process. The effects of each single mutant revealed high mutational sensitivity to a subset of single stranded residues important for docking into the single-stranded protein. Double-mutant effects revealed epistasis between specific base-paired residues, showing the importance of maintaining base-pairing throughout the stem for MS2 protein recognition and binding.

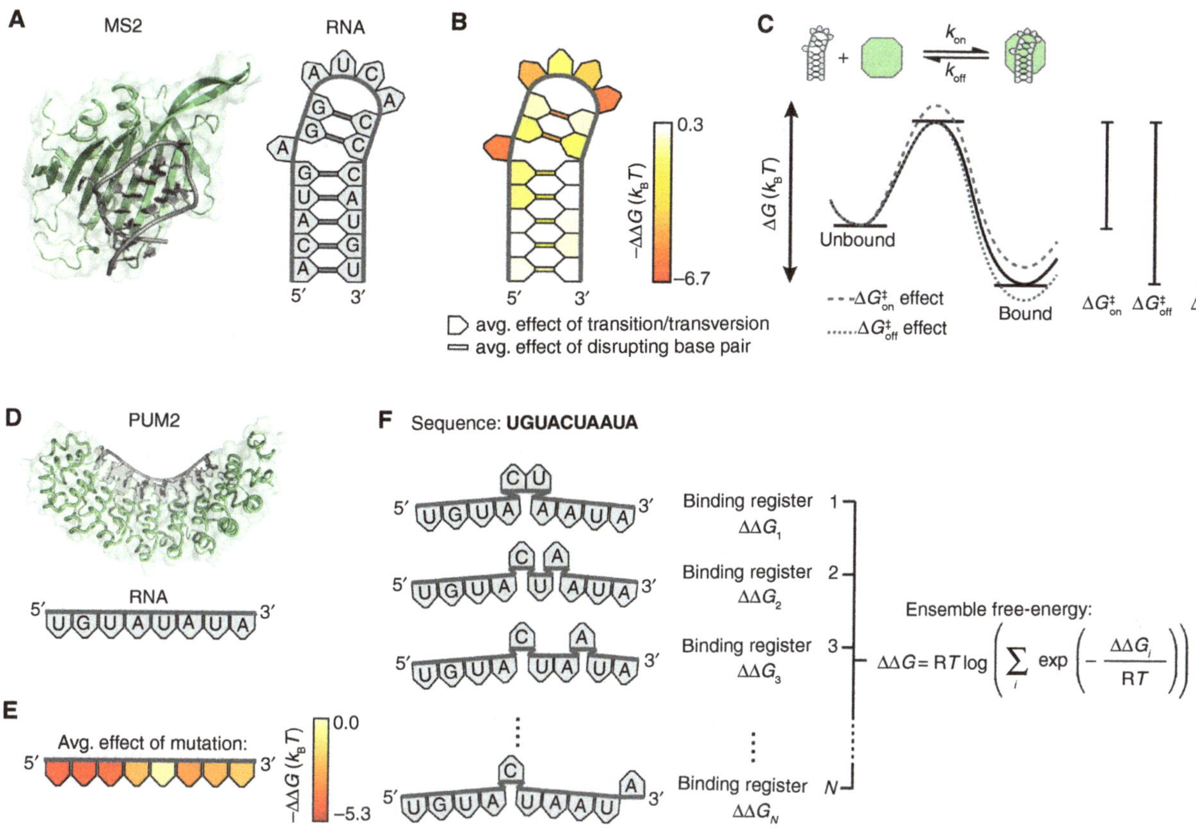

Figure 3. Sequence and structural determinants of protein–RNA interactions. (*A*) (*Left*) Crystal structure of the MS2 coat protein–RNA interaction (PDB: 2BU1) (Grahn et al. 2001). (*Right*) Primary sequence and secondary structure of the consensus MS2 RNA binding site. (*B*) Linear regression analysis of single and double mutants of the consensus MS2 RNA binding site attributed effects to either primary sequence (base colors) or to secondary structure (base pair colors) (Buenrostro et al. 2014). (*C*) Free-energy diagram showing the unbound, transition, and bound states. Each $\Delta G^‡$ term is linked to the kinetics of forming the complex, with $\Delta G = \Delta G^‡_{off} - \Delta G^‡_{on}$. Mutations can affect the ΔG by affecting $\Delta G^‡_{off}$ (dotted line), $\Delta G^‡_{on}$ (dashed line), or both. (*D*) Crystal structure of the RNA-binding domain of human PUM2 protein in complex with RNA (PDB: 3Q0Q) (Lu and Hall 2011), with the primary sequence of the consensus site indicated. (*E*) The average effect of each single mutation to the consensus site (data from I Jarmoskaite, SK Denny, P Vaidyanathan, et al., in prep.). (*F*) Multiple binding registers can give rise to a stable interaction between PUM2 and RNA for a given RNA sequence. Residues pointing down contribute to protein binding, whereas residues pointing up are "flipped out." The $\Delta\Delta G$ of each binding register is calculated with an energy term for each residue contacting the protein, with a penalty for having flipped out residues, as well as other coupling terms. The final effect for a given sequence is the ensemble free-energy of each possible binding register.

These observations allow direct quantification of primary sequence and secondary structure determinants of protein recognition and binding using a simple regression model that estimates the effect of a transition or a transversion of each residue, and the effect of converting a base pair to either a noncanonical base pair or to an otherwise disrupted base pair (Fig. 3B). This full model was predictive of the higher-order mutation effects.

The free-energy change in forming a bound complex can be decomposed into two contributions: the free-energy change between the starting state and the transition state ($\Delta G^‡_{on}$), and the free-energy change between the transition state and the final state ($\Delta G^‡_{off}$), such that $\Delta G = \Delta G^‡_{off} - \Delta G^‡_{on}$ (Fig. 3C). Each of these $\Delta G^‡$ terms is linked to the kinetics of forming the complex, with the first step affecting the association rate constant, and the second step affecting the dissociation rate constant: $\Delta G^‡_{on} = RT \log k_{on}$ and $\Delta G^‡_{off} = RT \log k_{off}$. Thus, measuring the kinetic rate constants in addition to equilibrium binding allows quantification of mutational effects on each of these steps of the binding process (Fig. 3C). In the MS2 coat protein experiment, dissociation rate constants (k_{off}) were measured for each mutant, revealing that mutation effects predominantly resulted from changes in the association rate constant, k_{on}, suggesting that MS2 must wait for the formation of a binding competent RNA secondary structure before association.

Mutational effects can be used to constrain the functional landscape in which evolution must operate to generate or maintain high affinity binding. Many double mutants within the MS2 hairpin had very small effects and thus represent equivalent "solutions" to the evolutionary "problem" of creating a stable RNA–protein interaction. However, the mutational paths that connect each pair of double mutants were observed to have very different "roughness" in this functional binding space, depending on the order in which the mutations accumulated. These types of data provide quantitative insight into complexities and path dependences of natural and engineered methods for evolving high-affinity binding partners.

3.1.2 RNA Aptamers for GFP and NELF

RNA-HiTS allows characterization of synthetic RNA–protein interactions. For example, Tome and colleagues characterized the specificity of two RNA aptamers that were each originally identified through in vitro selection: aptamers to the green fluorescent protein (GFP) and to the *Drosophila* negative elongation factor (NELF-E) (Shui et al. 2012; Pagano et al. 2014; Tome et al. 2014). Variants of each aptamer were generated with error-prone PCR, resulting in single, double, and some higher order mutants of each sequence. Examining the effects on binding affinity of each aptamer to their respective proteins identified regions with substantial sensitivity to single point mutants. The summed effects of these single point mutations were predictive of the effects of higher-order mutations in the NELF aptamer, supporting an additive model whereby each residue contributes additively to the affinity of this interaction. However, this linearity was not observed in the GFP aptamer, which had sensitivity to changes in base-paired regions, leading to significant epistasis between residues. These observations show the ability of RNA-HiTS to quantify and define the potentially diverse binding mechanisms that may come out of an in vitro selection. Mutational analysis of the GFP aptamer also revealed two point mutants with several-fold higher affinity than the original aptamer, showing RNA-HiTS's capacity to optimize affinity over the original selected aptamer. We envision that, in the future, the RNA-HiTS technology may be used as part of the in vitro selection process (i.e., to characterize a broader diversity of putative RNA targets at earlier rounds of selection).

3.1.3 Puf Binding to Designed Library of RNA

Puf family proteins are versatile posttranscriptional regulators found throughout eukaryotes (Quenault et al. 2011). The specificity of Puf–RNA interactions is thought to be driven by an ~8-nt consensus sequence found within the 3′ UTR of genes (Wickens et al. 2002; Van Nostrand et al. 2016). Binding preference for this consensus motif is conferred by a highly modular protein structure composed of eight amino acid repeats that each interact with a single RNA nucleotide (Fig. 3D) (Wang et al. 2001). Thus, this system provides an ideal testbed for developing an additive model aimed at predicting binding free-energy from primary nucleic acid sequence.

However, the high-precision measurements afforded by RNA-HiTS analysis revealed a substantially more complex picture of the binding process, even for this simple interaction, than could be captured by a simple linear model (I Jarmoskaite, SK Denny, P Vaidyanathan P, et al., in prep.). These complexities stemmed from three primary sources: (1) RNA secondary structure formation can affect the accessibility of the binding site; (2) alternate binding registers with "flipped-out" residues can form stable complexes; and finally (3) subtle, yet pervasive, energetic coupling can occur between residues within the binding site. Systematic identification and quantification of these distinct contributions resulted in a two-step model that first enumerates the possible binding configurations, then predicts the binding free-energy of each configuration using a linearly additive model with coupling terms (Fig. 3E,F). The final observed affinity for any sequence is thus derived from the ensemble of free energies for all binding configurations (Fig. 3F), modulated by secondary structure formation as predicted by algorithms such as Vienna fold (Gruber et al. 2008). This model predicted Puf occupancy that was linearly related to the observed occupancy of these sites as determined in an in vivo cross-linking experiment, highlighting both the usage of this fully predictive model and supporting the importance of thermodynamics in in vivo settings (Van Nostrand et al. 2016).

3.2 Protein–RNA Interactions across Transcriptome Targets

The investigation of the relationship between protein binding in the context of in vivo–derived RNAs is another exciting application of RNA-HiTS. The immense capacity of these instruments opens the possibility of placing an entire eukaryotic genome on the chip, and probing binding to every theoretically possible transcript. This approach, termed the transcribed genome array (TGA), was used to examine binding preferences of the RNA binding domain of Vts1 from *S. cerevisiae* (She et al. 2017), a member of the Smaug family of proteins with conserved RNA binding domains implicated in mediating RNA decay throughout eukaryotes (Aviv et al. 2003; Oberstrass et al. 2006; Rendl et al. 2008, 2012; Riordan et al. 2011). The entire *S. cerevisiae* genome was fragmented into ~100-nt fragments resulting in coverage of >30× per nucleotide. This library of overlapping fragments formed the substrate for in vitro transcription in the RNA-HiTS experiment.

Using this unbiased, transcriptome-wide approach, approximately 300 sites within the genome were expected to be significantly occupied within the cell, based on the apparent K_ds observed and the estimated physiological protein concentration (130 nM); two-thirds of these sites fell into the transcribed strand of an open reading frame. These gene targets were significantly overlapping with those from in vivo cross-linking and immunoprecipitation (RIP-seq) experiments, suggesting that in vitro identified targets recapitulate in vivo binding events in aggregate (Aviv et al. 2006; Hogan et al. 2008). The functional relevance of these binding events was assessed by comparing the expression between *vts1Δ* and wild-type cells by RNA-seq. These data showed that genes associated with TGA-identified binding sites had significantly increased average expression, although RIP-identified targets displayed no significant average expression change. Targets common to both analyses had the greatest expression increase, suggesting that TGA may help uncover the true positives within RIP-identified targets. In addition, TGA identified many new targets, virtually all of which were lowly expressed in vivo, highlighting a crucial difference between TGA and in vivo immunoprecipitation experiments: Cross-linking and pulldown experiments are often limited to probing only targets within genes with high expression in the conditions that the cells had been grown. Instead, TGA allows investigation of affinity to all possible transcripts simultaneously.

3.3 Design and Engineering of RNA Devices Using Massively Parallel Rational Design

The flexibility of RNA-HiTS enables investigation of RNA beyond its capacity to bind proteins. One exciting area of application is in the field of biological engineering wherein RNA is used as a programmable tool for control of biological systems. The motivation for this molecular engineering stemmed from the discovery of natural RNA riboswitches, which are conceptually composed of two pieces: the structured aptamer that binds a small-molecule "effector," and a second region that undergoes a conformational change in response to the effector, generally resulting in a change of gene expression through various mechanisms (Tucker and Breaker 2005). The separation between these two functions (i.e., the aptamer domain and the resulting functional consequence) supports that RNAs might enable modular programming of specific computations within the cell (Kim and Smolke 2017). However, the design of these devices is currently limited by imperfect predictive understanding of these aptamers and their coupling to conformational changes.

To make progress on this question, the RNA-HiTS platform was used to measure the response of tens of thousands of rationally designed RNA riboswitches. These designs came from scientific "enthusiasts" playing the online game EteRNA. This crowd-based rational approach enables massively parallel, yet hypothesis-driven science, as originally shown for RNA secondary structure folding (Lee et al. 2014). EteRNA players were tasked with designing RNA sequences that would satisfy the "challenge" of forming the prespecified aptamer domain and changing another domain's conformation in response to the effector molecule (see eternagame.org/web/). The player-generated designs were tested using RNA-HiTS to detect the response of each riboswitch, resulting in a score for each riboswitch; players could then use these scores in subsequent rounds to optimize their designs. In combination with RNA-HiTS, this crowd-sourced approach produced thermodynamically optimal sensors, with binding of the aptamer domain fully coupled to changes in the conformation of the readout domain (JOL Andresen et al., in prep.).

3.4 Understanding Sequence Contributions to RNA Tertiary Structure

A last compelling application of RNA-HiTS is in developing a predictive model for the RNA folding process. Although the first step of hierarchical RNA folding (i.e., secondary structure formation) may be predicted fairly accurately for the lowest free-energy state with readily available software (Turner et al. 1988; Gruber et al. 2008), models to predict RNA tertiary structure from this secondary structure are substantially more limited in scope. Current approaches have focused on obtaining structural characterization of the "modules" of RNA tertiary structure—that is, recurrent RNA elements like helices and junctions that have similar crystal structures in different RNAs—and then using these structures like Legos to infer the conformation of larger RNAs (Jaeger et al. 2001; Petrov et al. 2013; Miao and Westhof 2017). However, the inherent flexibility of RNA limits the applicability of these approaches to achieve thermodynamic understanding of the stability of tertiary structures (Herschlag et al. 2015). Furthermore, this approach is limited by the small number of elements for which high-quality crystallographic information exists.

To approach these challenges, RNA-HiTS was used to measure the sequence dependence of tertiary structure formation at high throughput (Denny et al. 2018; Yesselman et al. 2018). RNA elements like helices and two-way junctions were introduced as "guest" motifs into a "host" system—together, they formed a minimal RNA tertiary as-

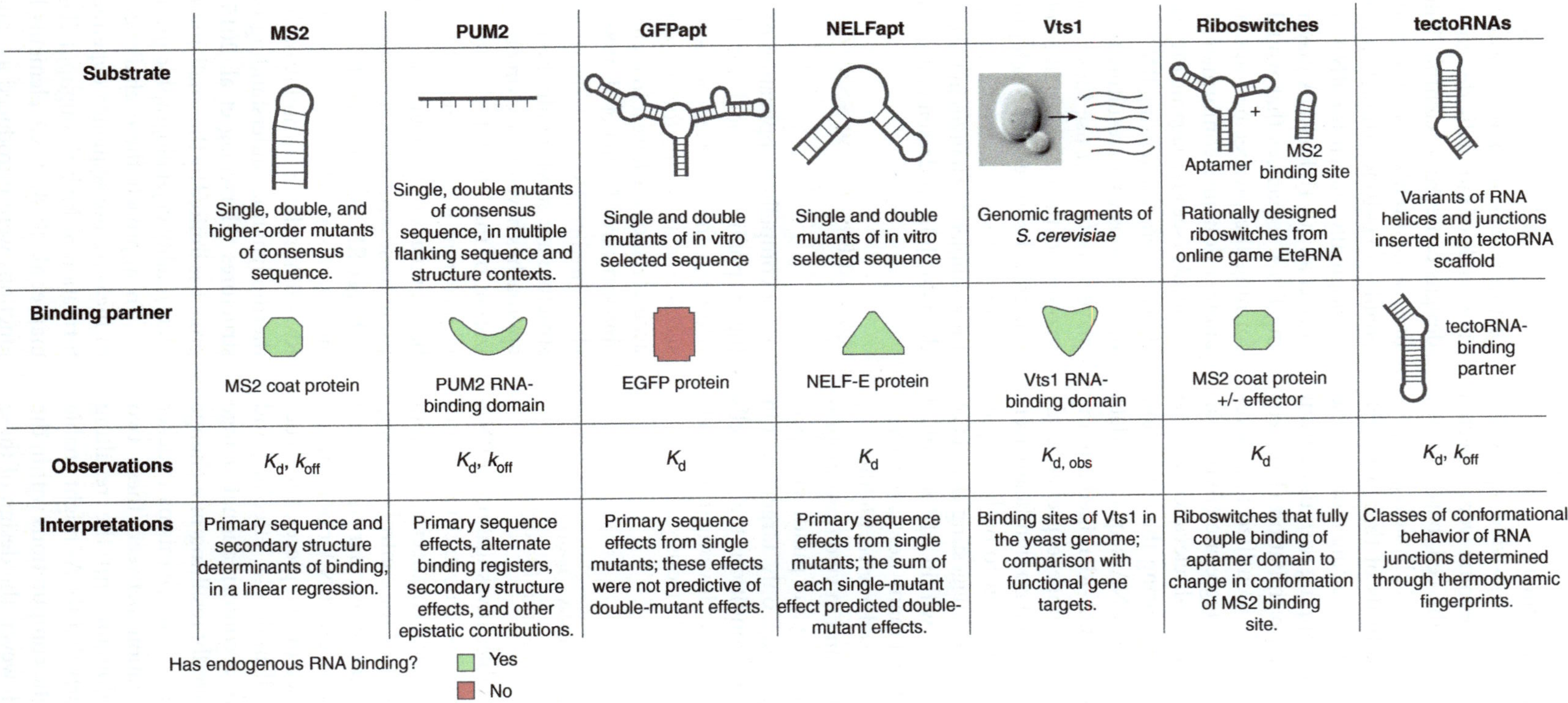

Figure 4. Summary of RNA-HiTS applications.

sembly (tectoRNA), whose formation quantitatively depends on the conformational preferences of the integrated element (Jaeger and Leontis 2000; Geary et al. 2008). For two-way junction elements, "thermodynamic fingerprints" were constructed that represented a multidimensional readout of the junction's conformational behavior. Unbiased clustering of junction thermodynamic fingerprints revealed classes of junctions with similar conformational preferences. Differences in conformational behavior were largely driven by differences in secondary structure (as expected from Bailor et al. 2010), but also by previously unrecognized primary sequence attributes like mismatch identity. Conformational classes were also useful for inferring structural behavior. Existing crystallographic conformations were associated with specific thermodynamic fingerprints, ultimately enabling the identification of structural ensembles describing the flexible and dynamic behavior of junctions with similar thermodynamic fingerprints. These "bootstrapped" ensembles, when combined with ensembles for RNA helices and a statistical mechanics model for tectoRNA assembly formation, were predictive of measured thermodynamic fingerprints and outperformed the use of static structures for predicting these behaviors.

4 CONCLUSION

Initial applications of the RNA-HiTS platform focused on dissecting the sequence and structural determinants of RNA–protein interactions. Collectively, these studies allowed accurate quantification of the subtle contributions to binding affinity of primary sequence effects, alternate binding registers, secondary structure effects, and other epistatic relationships between residues not directly involved in base-pairing (Fig. 4). The quantitative contributions of each of these elements were identified in the case of the GFP aptamer and Vts1; and these contributions were further modeled for any arbitrary sequence in the cases of MS2, NELF-E, and PUM2. In the future, diverse other protein–RNA interactions may be interrogated in such a system, allowing a more complete mapping of the diverse physical mechanisms that determine RNA–protein interactions.

RNA-HiTS data can complement in vivo cross-linking and immunoprecipitation experiments, as seen in the cases of Vts1 and Pum2, by identifying likely false-positive targets, as well as binding sites in lowly expressed transcripts that are not able to be probed with cross-linking methods. The manner by which the in vitro occupancy predicted by RNA-HiTS measurements is affected by additional effects in vivo—such as RNA localization and sequestration, cooperativity with other RNA binding proteins, as well as helicases that might periodically "reset" the binding state—remain to be systematically addressed. However, a systematic understanding of these other effects is fundamentally impossible without a "baseline" of expected occupancy of sites derived from thermodynamic parameters measured in the absence of these confounders.

More broadly, we anticipate other areas of applications of "systems biochemistry" approaches like RNA-HiTS. Multiple groups have used sequencing arrays to investigate DNA–protein interactions (Nutiu et al. 2011; Boyle et al. 2017), and we anticipate that extending this approach with multicolor imaging capabilities will begin to unravel the nature of transcription factor binding cooperativity, and nucleic acid binding cooperative more generally. Beyond transcription factors, proteins that can be engineered to bind diverse nucleic acid targets (e.g., Cas9, Cpf1, TALEN, Cas13, and AGO proteins) are of special interest. The potential binding targets of each of these programmable proteins span an immense combinatorial space, making high-throughput and quantitative investigations a necessary step in characterizing their behavior for applied and therapeutic purposes.

With methods such as RNA-HiTS, the biochemical and biophysical realms have been brought into the high-throughput era. Now, we as RNA biologists must direct the immense bandwidth afforded by these approaches to "carve nature at its joints." Thus, a significant challenge remains in designing and engineering RNA libraries to provide maximal insight into the physical underpinnings of complex RNA behaviors.

REFERENCES

Aviv T, Lin Z, Lau S, Rendl LM, Sicheri F, Smibert CA. 2003. The RNA-binding SAM domain of Smaug defines a new family of post-transcriptional regulators. *Nat Struct Biol* **10:** 614–621.

Aviv T, Lin Z, Ben-Ari G, Smibert CA, Sicheri F. 2006. Sequence-specific recognition of RNA hairpins by the SAM domain of Vts1p. *Nat Struct Mol Biol* **13:** 168–176.

Bailor MH, Sun X, Al-Hashimi HM. 2010. Topology links RNA secondary structure with global conformation, dynamics, and adaptation. *Science* **327:** 202–206.

Bardwell VJ, Wickens M. 1990. Purification of RNA and RNA–protein complexes by an R17 coat protein affinity method. *Nucleic Acids Res* **18:** 6587–6594.

Bertrand E, Chartrand P, Schaefer M, Shenoy SM, Singer RH, Long RM. 1998. Localization of ASH1 mRNA particles in living yeast. *Mol Cell* **2:** 437–445.

Boyle EA, Andreasson JOL, Chircus LM, Sternberg SH, Wu MJ, Guegler CK, Doudna JA, Greenleaf WJ. 2017. High-throughput biochemical profiling reveals sequence determinants of dCas9 off-target binding and unbinding. *Proc Natl Acad Sci* **114:** 5461–5466.

Buenrostro JD, Araya CL, Chircus LM, Layton CJ, Chang HY, Snyder MP, Greenleaf WJ. 2014. Quantitative analysis of RNA–protein interactions on a massively parallel array reveals biophysical and evolutionary landscapes. *Nat Biotechnol* **32:** 562–568.

Butter F, Scheibe M, Mörl M, Mann M. 2009. Unbiased RNA–protein interaction screen by quantitative proteomics. *Proc Natl Acad Sci* **106:** 10626–10631.

Castello A, Fischer B, Eichelbaum K, Horos R, Beckmann BM, Strein C, Davey NE, Humphreys DT, Preiss T, Steinmetz LM, et al. 2012. Insights into RNA biology from an atlas of mammalian mRNA-binding proteins. *Cell* **149:** 1393–1406.

Cirino PC, Mayer KM, Umeno D. 2003. Generating mutant libraries using error-prone PCR. *Methods Mol Biol* **231:** 3–9.

Denny SK, Bisaria N, Yesselman JD, Das R, Herschlag D, Greenleaf WJ. 2018. High-throughput investigation of diverse junction elements in RNA tertiary folding. *Cell* **174:** 377–390.e20.

Ding Y, Tang Y, Kwok CK, Zhang Y, Bevilacqua PC, Assmann SM. 2014. In vivo genome-wide profiling of RNA secondary structure reveals novel regulatory features. *Nature* **505:** 696–700.

Ellington AD, Szostak JW. 1990. In vitro selection of RNA molecules that bind specific ligands. *Nature* **346:** 818–822.

Geary C, Baudrey S, Jaeger L. 2008. Comprehensive features of natural and in vitro selected GNRA tetraloop-binding receptors. *Nucleic Acids Res* **36:** 1138–1152.

Grahn E, Moss T, Helgstrand C, Fridborg K, Sundaram M, Tars K, Lago H, Stonehouse NJ, Davis DR, Stockley PG, et al. 2001. Structural basis of pyrimidine specificity in the MS2 RNA hairpin-coat-protein complex. *RNA* **7:** 1616–1627.

Greenleaf WJ, Frieda KL, Foster DAN, Woodside MT, Block SM. 2008. Direct observation of hierarchical folding in single riboswitch aptamers. *Science* **319:** 630–633.

Gruber AR, Lorenz R, Bernhart SH, Neuböck R, Hofacker IL. 2008. The Vienna RNA websuite. *Nucleic Acids Res* **36:** W70–W74.

Hafner M, Landthaler M, Burger L, Khorshid M, Hausser J, Berninger P, Rothballer A, Ascano M, Jungkamp A-C, Munschauer M, et al. 2010. Transcriptome-wide identification of RNA-binding protein and microRNA target sites by PAR-CLIP. *Cell* **141:** 129–141.

Herschlag D, Allred BE, Gowrishankar S. 2015. From static to dynamic: The need for structural ensembles and a predictive model of RNA folding and function. *Curr Opin Struct Biol* **30:** 125–133.

Hogan DJ, Riordan DP, Gerber AP, Herschlag D, Brown PO. 2008. Diverse RNA-binding proteins interact with functionally related sets of RNAs, suggesting an extensive regulatory system. *PLoS Biol* **6:** e255.

Jaeger L, Leontis NB. 2000. Tecto-RNA: One-dimensional self-assembly through tertiary interactions. *Angew Chem Int Ed* **39:** 2521–2524.

Jaeger L, Westhof E, Leontis NB. 2001. TectoRNA: Modular assembly units for the construction of RNA nano-objects. *Nucleic Acids Res* **29:** 455–463.

Jankowsky E, Harris ME. 2017. Mapping specificity landscapes of RNA–protein interactions by high throughput sequencing. *Methods* **118–119:** 111–118.

Keene JD. 2007. RNA regulons: Coordination of post-transcriptional events. *Nat Rev Genet* **8:** 533–543.

Kieft JS, Zhou K, Jubin R, Doudna JA. 2001. Mechanism of ribosome recruitment by hepatitis C IRES RNA. *RNA* **7:** 194–206.

Kim CM, Smolke CD. 2017. Biomedical applications of RNA-based devices. *Curr Opin Biomed Eng* **4:** 106–115.

Lambert N, Robertson A, Jangi M, McGeary S, Sharp PA, Burge CB. 2014. RNA Bind-n-Seq: Quantitative assessment of the sequence and structural binding specificity of RNA binding proteins. *Mol Cell* **54:** 887–900.

Lee J, Kladwang W, Lee M, Cantu D, Azizyan M, Kim H, Limpaecher A, Yoon S, Treuille A, Das R, et al. 2014. RNA design rules from a massive open laboratory. *Proc Natl Acad Sci* **111:** 2122–2127.

Lou T-F, Weidmann CA, Killingsworth J, Tanaka Hall TM, Goldstrohm AC, Campbell ZT. 2017. Integrated analysis of RNA-binding protein complexes using in vitro selection and high-throughput sequencing and sequence specificity landscapes (SEQRS). *Methods* **118–119:** 171–181.

Lu G, Hall TMT. 2011. Alternate modes of cognate RNA recognition by human PUMILIO proteins. *Structure* **19:** 361–367.

Martin L, Meier M, Lyons SM, Sit RV, Marzluff WF, Quake SR, Chang HY. 2012. Systematic reconstruction of RNA functional motifs with high-throughput microfluidics. *Nat Methods* **9:** 1192–1194.

Matzas M, Stähler PF, Kefer N, Siebelt N, Boisguérin V, Leonard JT, Keller A, Stähler CF, Häberle P, Gharizadeh B, et al. 2010. High-fidelity gene synthesis by retrieval of sequence-verified DNA identified using high-throughput pyrosequencing. *Nat Biotechnol* **28:** 1291–1294.

Miao Z, Westhof E. 2017. RNA structure: Advances and assessment of 3D structure prediction. *Annu Rev Biophys* **46:** 483–503.

Mohanty BK, Sahoo T, Bastia D. 1996. The relationship between sequence-specific termination of DNA replication and transcription. *EMBO J* **15:** 2530–2539.

Moore MJ. 2005. From birth to death: The complex lives of eukaryotic mRNAs. *Science* **309:** 1514–1518.

Mulugu S, Potnis A, Shamsuzzaman, Taylor J, Alexander K, Bastia D. 2001. Mechanism of termination of DNA replication of *Escherichia coli* involves helicase–contrahelicase interaction. *Proc Natl Acad Sci* **98:** 9569–9574.

Noller HF. 2005. RNA structure: Reading the ribosome. *Science* **309:** 1508–1514.

Nutiu R, Friedman RC, Luo S, Khrebtukova I, Silva D, Li R, Zhang L, Schroth GP, Burge CB. 2011. Direct measurement of DNA affinity landscapes on a high-throughput sequencing instrument. *Nat Biotechnol* **29:** 659–664.

Oberstrass FC, Lee A, Stefl R, Janis M, Chanfreau G, Allain FH-T. 2006. Shape-specific recognition in the structure of the Vts1p SAM domain with RNA. *Nat Struct Mol Biol* **13:** 160–167.

Pagano JM, Kwak H, Waters CT, Sprouse RO, White BS, Ozer A, Szeto K, Shalloway D, Craighead HG, Lis JT. 2014. Defining NELF-E RNA binding in HIV-1 and promoter-proximal pause regions. *PLoS Genet* **10:** e1004090.

Petrov AI, Zirbel CL, Leontis NB. 2013. Automated classification of RNA 3D motifs and the RNA 3D Motif Atlas. *RNA* **19:** 1327–1340.

Quenault T, Lithgow T, Traven A. 2011. PUF proteins: Repression, activation and mRNA localization. *Trends Cell Biol* **21:** 104–112.

Ramani V, Qiu R, Shendure J. 2015. High-throughput determination of RNA structure by proximity ligation. *Nat Biotechnol* **33:** 980–984.

Ray D, Kazan H, Chan ET, Peña Castillo L, Chaudhry S, Talukder S, Blencowe BJ, Morris Q, Hughes TR. 2009. Rapid and systematic analysis of the RNA recognition specificities of RNA-binding proteins. *Nat Biotechnol* **27:** 667–670.

Rendl LM, Bieman MA, Smibert CA. 2008. *S. cerevisiae* Vts1p induces deadenylation-dependent transcript degradation and interacts with the Ccr4p-Pop2p-Not deadenylase complex. *RNA* **14:** 1328–1336.

Rendl LM, Bieman MA, Vari HK, Smibert CA. 2012. The eIF4E-binding protein Eap1p functions in Vts1p-mediated transcript decay. *PLoS ONE* **7:** e47121.

Riordan DP, Herschlag D, Brown PO. 2011. Identification of RNA recognition elements in the *Saccharomyces cerevisiae* transcriptome. *Nucleic Acids Res* **39:** 1501–1509.

She R, Chakravarty AK, Layton CJ, Chircus LM, Andreasson JOL, Damaraju N, McMahon PL, Buenrostro JD, Jarosz DF, Greenleaf WJ. 2017. Comprehensive and quantitative mapping of RNA–protein interactions across a transcribed eukaryotic genome. *Proc Natl Acad Sci* **114:** 3619–3624.

Shui B, Ozer A, Zipfel W, Sahu N, Singh A, Lis JT, Shi H, Kotlikoff MI. 2012. RNA aptamers that functionally interact with green fluorescent protein and its derivatives. *Nucleic Acids Res* **40:** e39.

Solomatin SV, Greenfeld M, Chu S, Herschlag D. 2010. Multiple native states reveal persistent ruggedness of an RNA folding landscape. *Nature* **463:** 681–684.

Staley JP, Guthrie C. 1998. Mechanical devices of the spliceosome: Motors, clocks, springs, and things. *Cell* **92:** 315–326.

Subtelny AO, Eichhorn SW, Chen GR, Sive H, Bartel DP. 2014. Poly(A)-tail profiling reveals an embryonic switch in translational control. *Nature* **508:** 66–71.

Tinoco I, Bustamante C. 1999. How RNA folds. *J Mol Biol* **293:** 271–281.

Cite as *Cold Spring Harb Perspect Biol* doi: 10.1101/cshperspect.a032300

Tome JM, Ozer A, Pagano JM, Gheba D, Schroth GP, Lis JT. 2014. Comprehensive analysis of RNA–protein interactions by high-throughput sequencing-RNA affinity profiling. *Nat Methods* **11:** 683–688.

Tucker BJ, Breaker RR. 2005. Riboswitches as versatile gene control elements. *Curr Opin Struct Biol* **15:** 342–348.

Turner DH, Sugimoto N, Freier SM. 1988. RNA structure prediction. *Ann Rev Biophys Biophys Chem* **17:** 167–192.

Uemura S, Aitken CE, Korlach J, Flusberg BA, Turner SW, Puglisi JD. 2010. Real-time tRNA transit on single translating ribosomes at codon resolution. *Nature* **464:** 1012–1017.

Van Nostrand EL, Pratt GA, Shishkin AA, Gelboin-Burkhart C, Fang MY, Sundararaman B, Blue SM, Nguyen TB, Surka C, Elkins K, et al. 2016. Robust transcriptome-wide discovery of RNA-binding protein binding sites with enhanced CLIP (eCLIP). *Nat Methods* **13:** 508–514.

Wang X, Zamore PD, Hall TM. 2001. Crystal structure of a Pumilio homology domain. *Mol Cell* **7:** 855–865.

Wickens M, Bernstein DS, Kimble J, Parker R. 2002. A PUF family portrait: 3′UTR regulation as a way of life. *Trends Genet* **18:** 150–157.

Winkler WC, Breaker RR. 2005. Regulation of bacterial gene expression by riboswitches. *Annu Rev Microbiol* **59:** 487–517.

Yesselman JD, Denny SK, Bisaria N, Herschlag D, Greenleaf WJ, Das R. 2018. RNA tertiary structure energetics predicted by an ensemble model of the RNA double helix. *bioRxiv* doi: 10.1101/341107.

The Future of Cross-Linking and Immunoprecipitation (CLIP)

Jernej Ule,[1,2] Hun-Way Hwang,[3] and Robert B. Darnell[4,5]

[1]The Francis Crick Institute, London NW1 1AT, United Kingdom

[2]Department of Molecular Neuroscience, Institute of Neurology, University College London, London WC1N 3BG, United Kingdom

[3]Department of Pathology, School of Medicine, University of Pittsburgh, Pittsburgh, Pennsylvania 15261

[4]Laboratory of Molecular Neuro-Oncology, The Rockefeller University, New York, New York 10065

[5]Howard Hughes Medical Institute, The Rockefeller University, New York, New York 10065

Correspondence: darnelr@rockefeller.edu

SUMMARY

To understand the assembly and functional outcomes of protein–RNA regulation, it is crucial to precisely identify the positions of such interactions. Cross-linking and immunoprecipitation (CLIP) serves this purpose by exploiting covalent protein–RNA cross-linking and RNA fragmentation, along with a series of stringent purification and quality control steps to prepare complementary DNA (cDNA) libraries for sequencing. Here we describe the core steps of CLIP, its primary variations, and the approaches to data analysis. We present the application of CLIP to studies of specific cell types in genetically engineered mice and discuss the mechanistic and physiologic insights that have already been gained from studies using CLIP. We conclude by discussing the future opportunities for CLIP, including studies of human postmortem tissues from disease patients and controls, RNA epigenetic modifications, and RNA structure. These and other applications of CLIP will continue to unravel fundamental gene regulatory mechanisms while providing important biologic and clinically relevant insights.

Outline

Cite as *Cold Spring Harb Perspect Biol* doi: 10.1101/cshperspect.a032243

1 INTRODUCTION

A crucial step in understanding RNA-related gene regulation and its relationship to disease is identifying how RNAs are bound and hence regulated by RNA-binding proteins (RBPs) in specific cell types and subcellular compartments. Driving our interest in this topic has been the growing list of human neurologic diseases that have RNA dysregulation at their core (Conlon and Manley 2017). The first method developed for this purpose used antibodies against the spliceosomal Sm proteins (lupus autoimmune sera) to identify the small nuclear RNAs (snRNAs), which interact with Sm proteins within the abundant small nuclear ribonucleoprotein complexes (snRNPs) (Lerner and Steitz 1979). This method, later referred to as RIP (RNP/RNA immunoprecipitation), relies on immunoprecipitation (IP) of an RBP under conditions that preserve RNPs (Niranjanakumari et al. 2002). However, RIP can suffer from low specificity, partly because it preserves protein–protein interactions and can therefore purify multiple RBPs in complex with their bound RNAs, and partly because RNA–protein complexes can reassociate in vitro (Mili and Steitz 2004).

To understand the assembly of RNPs, a method to identify direct protein–RNA interactions with high specificity was required. Moreover, it was important to also identify the positions on RNAs that are recognized by a specific RBP to identify the RNA sequences and structures where regulation takes place. We have developed cross-linking and immunoprecipitation (CLIP) for this purpose by exploiting zero-length covalent protein–RNA cross-linking and RNA fragmentation, along with a series of stringent purification and quality control steps, followed by the preparation of a complementary DNA (cDNA) library for sequencing (Ule et al. 2003; Darnell 2010). With CLIP, only RNA fragments that are cross-linked to the specific RBP are isolated under well-optimized conditions that avoid copurification of additional RBPs. CLIP data can be validated with the use of appropriate negative controls, such as the use of cells or tissue that lack the RBP of interest or have not been cross-linked (Licatalosi and Darnell 2010). CLIP is now generally accepted as a highly reliable method for identifying the positions of endogenous RNA–protein interactions (Conlon and Manley 2017).

Initially, CLIP relied on Sanger sequencing to identify 340 sequences corresponding to RNA interactions of splicing factors Nova1 and Nova2 in the mouse brain (Ule et al. 2003). Despite the limited number of these initial sequences, many were located next to alternative exons that turned out to be regulated by Nova proteins, thus demonstrating the capacity of CLIP to identify functionally important binding sites. Subsequent development of HITS-CLIP (high-throughput sequencing of RNA isolated by CLIP) was then applied to further studies of the functional interactions of Nova and several other RBPs with neuronal transcripts in the mouse brain (Licatalosi et al. 2008; Darnell 2013).

This work established a general strategy to link functional studies with a transcriptome-wide map of protein–RNA interactions. Such maps validated regulatory rules that had been bioinformatically predicted, in which the position of binding of RBPs within primary transcripts largely determines the outcome of RNA regulation (Ule et al. 2006; Licatalosi et al. 2008; Licatalosi and Darnell 2010). These rules are now recognized as a general feature of pre-mRNA processing, applicable to dozens of different RBPs (Witten and Ule 2011). Following these approaches, CLIP has been applied to a large number of different RBPs in a variety of cell types, organs, and species.

2 THE CLIP METHOD

2.1 The Fundamentals

Many variant protocols of CLIP have been derived during the last decade. Their detailed variations have been reviewed elsewhere (Darnell 2010; Lee and Ule 2018), and therefore we will focus here only on the primary steps that are common to most protocols (Fig. 1). A feature common to all protocols is that RBPs are cross-linked by ultraviolet (UV) light to their bound RNAs within the intact endogenous environment. Acutely harvested cells, triturated or pulverized tissues, or even whole organisms such as *Caenorhabditis elegans* are immediately exposed to UV light, which cross-links RBPs to their bound RNAs in their native (intracellular) state. We have adopted use of the term "CLIP" as a general way to refer to all protocols that use UV cross-linking and protein purification of directly bound native RNA–protein complexes.

Most CLIP variants exploit the capacity of UV light to promote formation of covalent bonds between RBPs and their direct RNA-binding sites. Unlike formaldehyde cross-linking that is used in chromatin immunoprecipitation (ChIP) and some variants of RIP, UV does not cross-link proteins to each other. UV cross-linking requires direct (Angstrom) contact between an amino acid and a nucleobase, and therefore ensures that only direct protein–RNA interactions are preserved. The high strength of the covalent bond allows a stringent buffer containing ionic detergents to be used for cell lysis, which disrupts most non-cross-linked protein–protein and protein–RNA interactions. The stringent buffer also increases the accessibility of RNA for RNase fragmentation and minimizes the chance of copurifying multiple associated RBPs during later IP, thus increasing data specificity. RNA fragmentation is cru-

Cite as *Cold Spring Harb Perspect Biol* doi: 10.1101/cshperspect.a032243

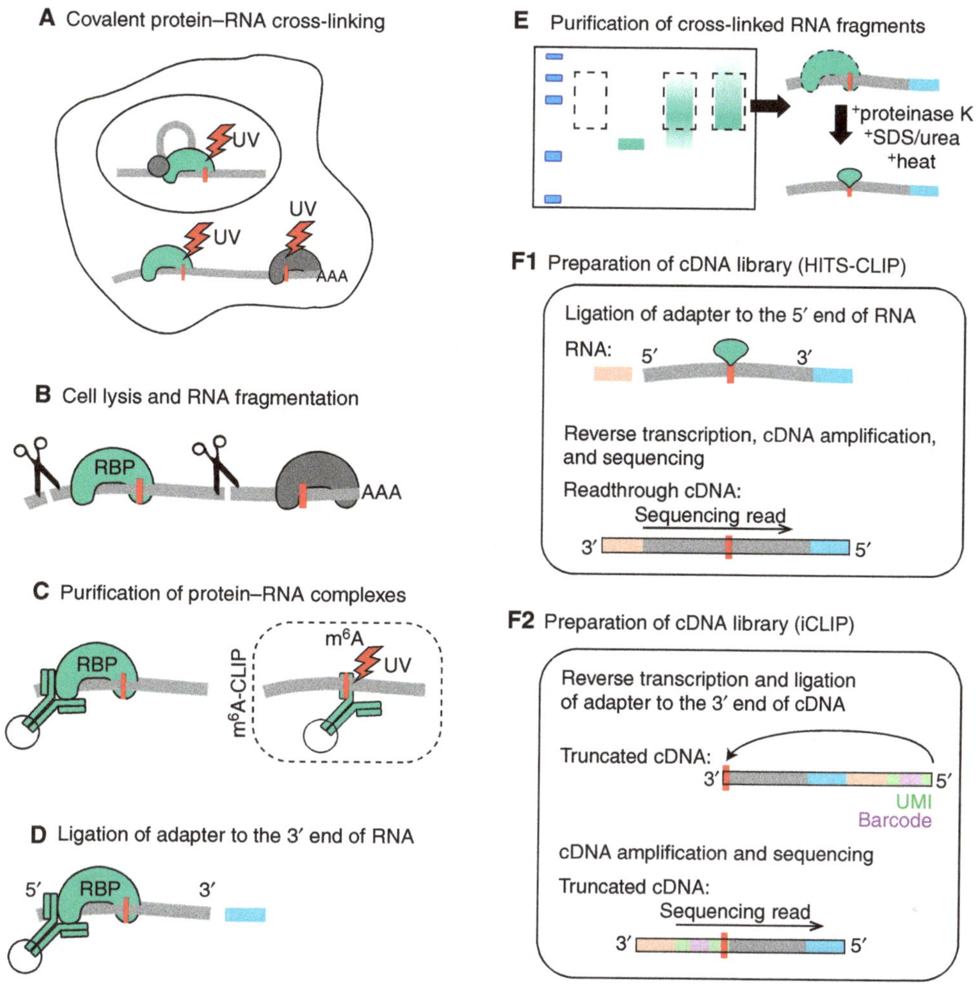

Figure 1. The core steps of cross-linking and immunoprecipitation (CLIP). The steps *A–E* are common across most variants of CLIP. Methods that identify the sites of RNA methylation cross-link purified RNA to an antibody specific to a methylated base, as in m^{6}A-CLIP, and then proceed from step *C* onward. The first option to prepare the complementary DNA (cDNA) library is as in high-throughput sequencing (HITS)-CLIP (*F1*), where the cDNA library is prepared by ligating an adapter to the 5′ end of RNA to allow amplification and sequencing of readthrough cDNAs, but not truncated cDNAs. The second option is as in individual nucleotide resolution CLIP (iCLIP) (*F2*), in which an adapter is ligated to the 3′ end of cDNAs, which allows amplification of both readthrough and truncated cDNAs (only a truncated cDNA is shown here, because this type most often represents >90% of the iCLIP cDNA library) (Sugimoto et al. 2012; Haberman et al. 2017). The unique molecular identifiers (UMIs) and experimental barcodes, which are introduced in iCLIP during reverse transcription (RT), are color coded in green and violet. The RNA fragments and their corresponding cDNA inserts are in gray. (Adapted based on previous publication by Lee and Ule 2018.)

cial to avoid copurifying multiple RBPs that cross-link to the same RNAs, and to provide insight into the position of RNA-binding sites, because the RNA fragment contains the cross-link site. A well-optimized RNase concentration is important to minimize the constraints at the sites of RNase cleavage, which can lead to biased binding site assignment (Haberman et al. 2017).

Purification of endogenous (i.e., untagged) RBPs normally requires that antibodies are available for efficient IP. The need for stringent purification and quality control varies to some extent on the type and expression of the RBP being studied, but in general this improves signal to noise of detection of directly bound RNAs. Some RBPs participate in stable RNPs that may not efficiently dissociate under standard CLIP IP conditions, thus increasing their risk of copurifying multiple RBPs, and careful protocols have articulated how to optimize the stringency of CLIP IPs (see Darnell 2012). Even a minor copurification of an abundant RBP that cross-links with high efficiency can lead to a major loss of specificity if the RBP of interest is of low abundance or cross-links poorly. In such cases, RBPs can be epitope-tagged to allow purification under denaturing conditions (Granneman et al. 2009; Huppertz et al. 2014). With epitope tagging, it is important to avoid overexpression, which can give artificial binding to unnatural RNA partners, and to test whether the function, stability, and localization of the tagged RBP remain unperturbed.

To prepare cDNA libraries from the CLIP RNA fragments, an adapter is ligated to the 3′ end of the RNA. The sequence of the adapter is complementary to the primers that are later used in reverse transcription (RT) and polymerase chain reaction (PCR). This is followed by purification of the protein–RNA complexes with the use of sodium dodecyl sulfate polyacrylamide gel electrophoresis (SDS–PAGE) and membrane transfer. Both steps reduce the contamination of non-cross-linked RNAs and excess adapters, which normally run at a lower molecular weight (MW) in the gel than fragments cross-linked to the RBP; moreover, free (contaminating) RNA does not bind well to the membrane. The labeled protein–RNA complex can be visualized on the membrane, serving to optimize RNA fragmentation and to control for the specificity of purified complexes. Labeling can be achieved radioactively with ^{32}P on the 5′ end of the RNA (Ule et al. 2003) or the adapter (Chi et al. 2009), or by covalently coupling an infrared dye to the adapter, which allows visualization with infrared imaging (irCLIP) (Zarnegar et al. 2016). Under well-optimized RNase conditions, the protein–RNA complexes migrate on the gel diffusely above the apparent MW of the immunoprecipitated RBP, because the diverse sizes of RNA fragments variably affect the migration of the RBP–RNA complex. These complexes are then excised from the appropriate region of the membrane according to recommended guidelines (Ule et al. 2005; Huppertz et al. 2014), and the RBP is digested with proteinase K, which leaves only a short peptide at the cross-link site and releases the RNA fragments into solution. To prepare a cDNA library from these RNA fragments, the original CLIP protocol ligated a second adapter to the 5′ end of RNA fragments. This approach has been used when CLIP was first combined with HITS-CLIP (Fig. 1). With the HITS-CLIP protocol, it is important that the cDNAs read through the cross-link site during the RT to ensure that they contain both adapters, which is required for the amplification and sequencing of cDNAs.

2.2 Primary Variations in CLIP Protocols

The first aspect that has been varied in the CLIP methods is the cross-linking of protein–RNA complexes. Most variants of CLIP expose cells or tissues to the UVC wavelength (254 nm), a procedure that usually takes less than a minute with cells placed on ice (Ule et al. 2003) or frozen in liquid nitrogen. Photoactivatable ribonucleoside-enhanced (PAR)-CLIP introduces a variation in the cross-linking strategy (Hafner et al. 2010), in which live cells are preincubated for hours with PARs4-thiouridine (4SU) or 6-thioguanosine (6SG), which label the RNA in vivo and enable protein–RNA cross-linking to be performed with UVA wavelength (365 nm). A mass spectrometry study has compared this protocol with UVC cross-linking to conclude that the efficiency of the two approaches is quite similar across all RBPs (Castello et al. 2012). The advantage of UVA cross-linking is its decreased capacity to induce DNA damage as compared with UVC, although in most CLIP experiments samples are chilled or frozen during ~30-sec-long UV cross-linking and then immediately solubilized, so that biologic consequences of DNA damage are minimal. The disadvantages of PAR-CLIP are its limitation to biological systems in which the photoactivatable nucleosides can be efficiently incorporated, and its recommended use of >100 million cells (Danan et al. 2016), whereas one million or less are used by standard variants of CLIP. Moreover, prolonged preincubation with 6SG or 4SU can cause cellular toxicity, including stress responses and inhibition of ribosomal RNA (rRNA) synthesis (Burger et al. 2013; Huppertz et al. 2014), and therefore care needs to be taken to monitor the cellular response to these ribonucleosides (Burger et al. 2013; Huppertz et al. 2014).

For methods aimed at studying RNA methylation, it is also possible to use a mutant RNA methylase enzyme, which forms a covalent bond with RNA spontaneously without the need for UV-induced cross-linking, as has been performed by 5-methylcytosine m^6A individual-nucleotide-resolution CLIP (m5C-miCLIP) for transcriptome-wide mapping of m5C modification sites (Sugimoto et al. 2012; Hussain et al. 2013). Alternatively, proteins can be cross-linked to purified RNA with UV light in vitro. This has been exploited by protocols that cross-link an antibody that binds to a methylated base to study RNA methylation, such as m^6A-CLIP (Fig. 1) (Chen et al. 2015; Ke et al. 2015; Linder et al. 2015). The antibody–RNA complex is then captured on protein A/G magnetic beads, and the sample continues to the rest of the CLIP protocol.

The second aspect of CLIP methods in which major variations have been explored involves ligating adapters to the fragmented RNA. By modifying the ligation conditions, it is possible to enable proximity ligation of the two strands of RNA duplexes that are bound by double-strand RNA-binding proteins (dsRBPs). When the two strands get ligated, the resulting sequencing reads are referred to as "hybrid reads," because they need to be split into two sequences that map to separate loci in the genome. This approach has been exploited to study intermolecular duplexes, such as small nucleolar RNA (snoRNA)–rRNA, messenger RNA–

microRNA (mRNA–miRNA) and long noncoding RNA (lncRNA)–mRNA, as well as intramolecular RNA–RNA duplexes bound by dsRBPs (Chi et al. 2009; Kudla et al. 2011; Imig et al. 2014; Moore et al. 2015; Sugimoto et al. 2015). Further technical details of these protocols, their data analysis, and relationship to other methods to study RNA secondary structure are reviewed in more depth elsewhere (Sugimoto et al. 2017).

A major advance in analysis of RNA structure was made with the development of ligation conditions to allow hybrid CLIP (hiCLIP), which ligates the two RNA strands with an additional adapter to gain control over the intermolecular ligation. This is achieved with the use of two types of RNA adapters, which are ligated to RNA duplexes: RNA adapter A with an irreversible 3′ block (dideoxynucleotide), and RNA adapter B with an enzymatically removable 3′ block (phosphate). Whereas the RNA duplexes remain bound to dsRBPs during IP, the two adapters are first ligated in equimolar concentration; thus, ~50% of the duplexes are expected to contain a different adapter at each strand. The 3′ block of adapter B is then removed, and the next ligation reaction is performed to ligate the two arms of RNA duplexes via the adapter B. This results in hybrid reads that contain the two strands of RNA duplexes separated by the adapter B, which allows unambiguous identification of the two arms of the hybrid reads. The method has been used to identify RNA duplexes bound by Staufen 1 in human embryonic kidney 293 (HEK293) cells, which identified intermolecular mRNA–mRNA and mRNA–lncRNA duplexes in addition to RNA duplexes that connect complementary parts of the same RNA. In the coding sequence, these complementary parts were generally proximally located, although in untranslated regions (UTRs) they were more often located >100 nucleotides (nt) apart. Initial studies with a reporter assay indicated a role for such "long-range" duplexes in the regulation of mRNA stability (Sugimoto et al. 2015), and their broader functional importance remains to be examined.

The final major variations in CLIP involve the preparation of the cDNA library. Individual nucleotide resolution CLIP (iCLIP) ligates the second adapter to the 3′ end of the cDNA via circularization (König et al. 2010). This enables amplification of cDNAs that prematurely truncate at the cross-linked nucleotide, in addition to the readthrough cDNAs, thus increasing the sensitivity of the method. Seventeen other published protocols similarly amplify truncated cDNAs, including bromodeoxyuridine (BrdU)-CLIP, enhanced CLIP (eCLIP), and irCLIP (Weyn-Vanhentenryck et al. 2014; Van Nostrand et al. 2016; Zarnegar et al. 2016). In iCLIP and most other CLIP methods, additional sequences are part of the adapter oligonucleotide, including an experimental barcode that enables multiplexing of different experiments, and a unique molecular identifier (UMI) that enables quantification of unique cDNAs to remove the artifacts of variable PCR amplification (Fig. 1) (König et al. 2010). Following adapter ligation, cDNAs are then purified and amplified by PCR, which is achieved in multiple ways by variant protocols, as described in detail elsewhere (Lee and Ule 2018).

2.3 CLIP Data Analysis

The primary advantage of CLIP is that it allows investigators to narrow down the very large number of computationally possible binding sites for a given RBP to a smaller subset of sites with evidence of direct RBP contact in the relevant cell type. This was clearly illustrated in the ability of Argonaute (Ago)-miRNA-mRNA CLIP to identify true miRNA-binding sites from among a larger set of computationally predicted sites, improving accuracy (Chi et al. 2009) and leading to unexpected discoveries (Chi et al. 2012; Luna et al. 2015; Moore et al. 2015).

The first steps in analyzing CLIP sequencing data involve using the experimental barcodes to demultiplex the cDNA libraries, mapping the data to the genome, and identifying unique cDNAs with the use of UMIs, if available. Moreover, multiple methods can be used to normalize the binding patterns relative to RNA abundance (Lee and Ule 2018). Depending on whether the protocol amplified truncated cDNAs, the peaks of RNA-binding sites can be assigned in two primary ways. For HITS-CLIP and other protocols that only amplify readthrough cDNAs, the full position of mapped reads is examined. The most likely position of the cross-link site can be identified within a proportion of the reads by identifying cross-linking-induced mutation sites, such as C to T transitions when 4SU is used in PAR-CLIP (Hafner et al. 2010), or deletions and other types of mutations in HITS-CLIP (Zhang and Darnell 2011). In iCLIP and 17 other protocols that amplify truncated cDNAs, cross-link sites are located at the start of these cDNAs. The truncated cDNAs strongly dominate the iCLIP libraries, and thus analysis of cDNA starts enables the nucleotide-resolution mapping of the cross-link sites, especially when combined with high-resolution peak-calling algorithms (Sugimoto et al. 2012; Haberman et al. 2017). The simplest peak-calling approach is to identify clusters of cross-linking events with cDNA counts that are significantly higher compared with randomized data in the same genomic regions (König et al. 2010).

Although peak analysis of RBP–RNA-binding sites has been of great value in understanding binding specificity and demarcating discrete points of regulation (Chakrabarti et al. 2017), not all regulatory factors have been shown to bind to individual discrete peaks. For example, heterogeneous nuclear ribonucleoprotein C (hnRNPC) binds widely across the transcriptome, with a periodicity suggesting the possibility of nucleosomal-like organization on RNA transcripts, which also reflects its binding to the dual U-rich sequences in antisense *Arthrobacter luteus* restriction endonuclease (Alu) elements (König et al. 2010; Zarnack et al. 2013). Fragile X mental retardation protein (FMRP), the protein lost in the commonest inherited form of intellectual disability, the fragile X syndrome, was found by CLIP to bind to a subset of transcripts across their entire coding length (Darnell et al. 2011). This was then shown to result from dual interactions of FMRP with the ribosome and the mRNA to cause ribosomal elongation arrest and translational pausing on its target transcripts. In such instances, more complex computational modeling may be needed in addition to peak analysis to understand how the RBPs recognize specific target RNAs.

Paralleling the genomic mapping and peak assignment, an essential concomitant is to examine the quality of data. Analysis of negative controls is particularly important for interpreting the data specificity. For example, purification can be performed from cells in which the RBP is absent, such as knockout (KO) cells, or when using tag-based purification, cells that do not express a tagged protein. As an alternative, nonspecific serum or immunoglobulin G (IgG) can be used for IP, or the RBP can be purified from non-cross-linked cells. Ideally, these controls should not produce any clearly detectable signal during SDS–PAGE analysis, and the number of unique cDNAs in their resulting libraries should be at least 100-fold fewer than in the specific experiments (König et al. 2010).

The sensitivity and specificity of data can be further examined by assessing the total number of unique cDNAs in the sequencing library and the peaks, respectively. To monitor the specificity of CLIP, a suitable peak-calling program needs to be chosen according to the CLIP protocol used to produce the data (Chakrabarti et al. 2017). A low number of peaks indicates either that the cDNAs are dispersed along the transcripts or that they are concentrated in a small number of abundant RNAs (such as rRNA)—this could either reflect low preference of the RBPs for specific sites on longer RNAs or low quality of data. Thus, some understanding of the expected binding properties of the RBP can help when interpreting CLIP data.

Finally, an extremely informative way to examine CLIP data sensitivity and specificity is to assess binding peaks with orthogonal data in the format of RNA maps (Licatalosi et al. 2008; Licatalosi and Darnell 2010), which allow the investigator to examine whether binding is enriched at exons or mRNAs that are differentially regulated by the RBP. Analysis of cross-link peaks in the format of RNA maps is particularly valuable when comparing multiple datasets for the same RBP. For example, data produced by iCLIP of polypyrimidine tract-binding protein 1 (PTBP1) led to a larger number of peaks around regulated exons than data produced by other protocols, although the number of unique cDNAs in iCLIP is equal or smaller (Chakrabarti et al. 2017).

3 CONDITIONALLY TAGGED (cTAG)-CLIP

A major forefront in biological research is to understand the functional roles of individual cell types in a multicellular organism. Cell type–specific regulation of RNA expression and processing emerges as a common phenomenon from recent RNA profiling studies in selected cell populations (Bhate et al. 2015; Yap et al. 2016; Zhang et al. 2016). Importantly, it is clear from these studies that many cell type–specific RNA regulatory events are not identifiable in whole tissue analysis. Therefore, developing new technologies with cell type resolutions is critical to advance our understanding of cell type–specific RNA regulation.

To meet the challenge of applying CLIP to study cell type–specific biology, we developed the conditionally tagged (cTag) CLIP technology, which uses a genetically engineered mouse that carries a latent green fluorescent protein (GFP)-tagged allele of the RBP of interest. The latent GFP-tagged allele is "off" by default, leaving the native allele unaffected, and only turns "on" in the presence of Cre recombinase (Fig. 2A) (Hwang et al. 2017). Importantly, the GFP-tagged allele is expressed by homologous recombination into the endogenous locus, such that the quantitative, qualitative (alternate isoforms), spatial, and temporal expression of GFP-tagged RBP is minimally disruptive to the biology of the system and can be precisely controlled through Cre expression to achieve cell type specificity. Cell type–specific CLIP analysis for the RBP of interest can then be performed using a mix of high-affinity GFP antibodies to immunoprecipitate the GFP-tagged RBPs under stringent conditions (Fig. 2B) only from cells expressing Cre recombinase.

The cTag-CLIP technology has several important features to ensure a proper representation of endogenous RNA–protein interactions but at the same time to maintain a broad applicability. First, because cTag mice are created using a "knock-in" approach, the proper cellular expression of RBPs and the RNA-RBP stoichiometry is likely to be maintained, both of which are difficult to achieve with either traditional or even bacterial artificial chromosome (BAC) transgenes (Fig. 2A) (He et al. 2012; Gerfen et al. 2013). Second, the utilization of a mix of high-affinity GFP antibodies (Heiman et al. 2014) allows cTag-CLIP analysis to be performed with a standardized and optimized high-stringency protocol for different RBPs (Hwang et al. 2017). The standardization eliminates the need for RBP-specific

Cite as *Cold Spring Harb Perspect Biol* doi: 10.1101/cshperspect.a032243

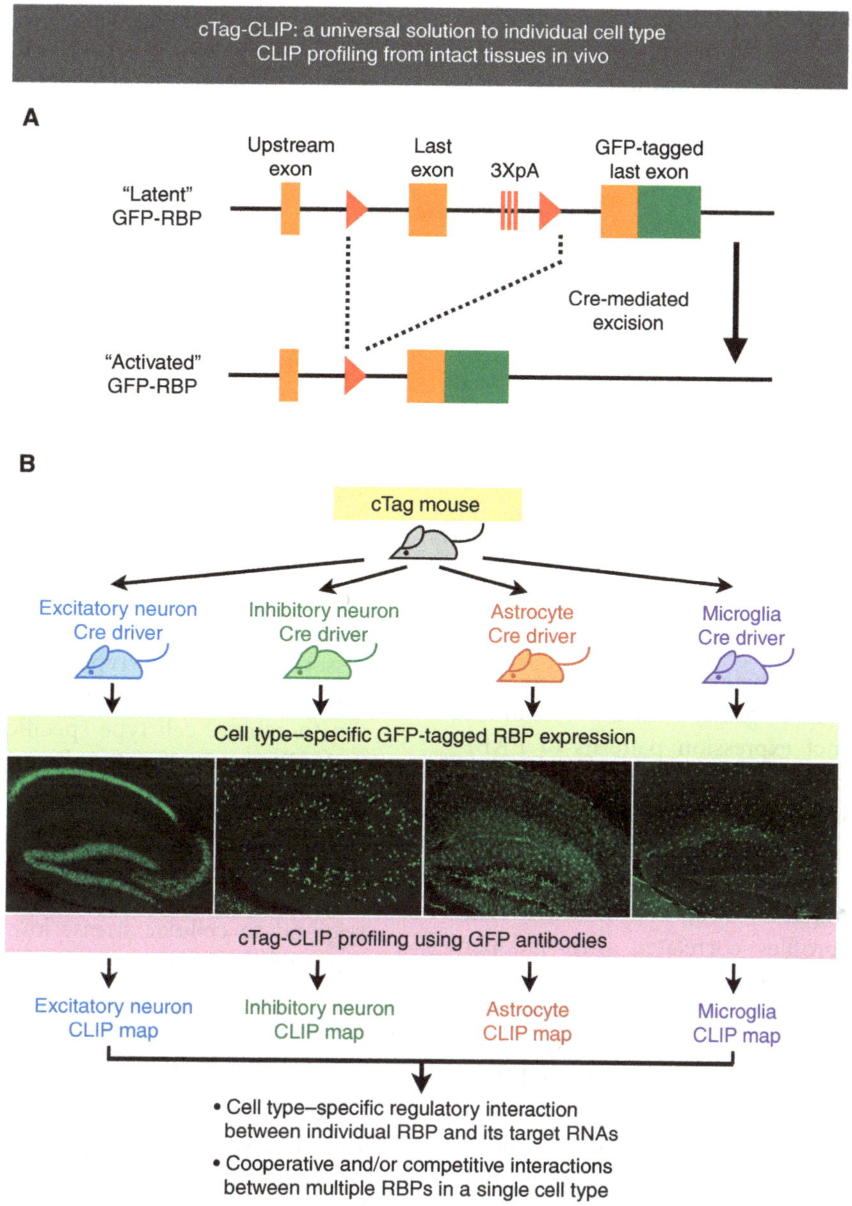

Figure 2. Conditionally tagged (cTag)-cross-linking and immunoprecipitation (CLIP). (*A*) The basic design of a cTag mouse. An extra copy of the last coding exon fused with green fluorescent protein (GFP) sequence (green) is inserted downstream from sequences encoding the RNA-binding protein (RBP) of interest. To prevent readthrough, triple poly(A) sites (3XpA) are inserted before the GFP-tagged exon. Cre-mediated excision of the endogenous last exon results in expression of GFP-tagged RBP from its native locus. Arrowheads indicate the locus of X-over P1 (loxP) sites. (*B*) Illustrative examples of cTag-CLIP with results from the cTag-poly(A)-binding protein (PABP) mouse (Hwang et al. 2017).

antibodies and the subsequent experimental optimization to achieve high-stringency IP for individual RBPs. Furthermore, it also provides an equal footing for cross-comparison of cTag-CLIP data from different RBPs to facilitate data integration. Third, its compatibility with the Cre/Lox system allows direct application of existing or newly developed Cre mouse lines, which makes it a versatile tool for wide-ranging biological investigations.

We have since successfully established a series of cTag mice to perform cTag-CLIP analysis for different RBPs in multiple cell types. We validated this approach using cTag poly(A)-binding protein (PABP; gene symbol: *Pabpc1*) to generate the cTag-PABP mouse and to establish the cTag-PAPERCLIP technique (Hwang et al. 2017). cTag-PAPERCLIP uses PABP as a biological filter to specifically clone distal mRNA 3′ ends to generate comprehensive mRNA alternative polyadenylation (APA) maps. With cTag-PAPERCLIP, we performed an in-depth APA analysis of individual cell types in vivo in the adult mouse brain to show the dynamics of APA regulation and discovered a previously unknown role of APA in diversifying protein expression in the brain.

Several observations from this cTag-PAPERCLIP study exemplify the technical advances and, more importantly, biological implications of the cTag-CLIP technology, which are detailed below.

1. *cTag-CLIP achieves desired cell type specificity to provide a powerful means to dissect cell type–specific biology:* The ubiquitous expression of PABPC1 presents an ideal test case for the capabilities of cTag-CLIP technology in achieving high cell type specificity. To evaluate the cell type specificity at the anatomical level, we examined the expression patterns of PABP-GFP in brains from four mouse models generated by crossing the cTag-PABP mouse to four Cre drivers with different cell type specificity (excitatory neurons, inhibitory neurons, astrocytes and microglia [Fig. 2B]). The analyses indeed confirmed the distinct expression patterns of PABP-GFP from the four mouse models and their consistency with the corresponding cell type markers. Next, we evaluated the cell type specificity at the molecular level by analyzing the four resulting RNA profiles from each mouse model generated by cTag-PAPERCLIP. These cTag-PAPERCLIP profiles correlated with independently generated cell type–specific RNA sequencing (RNA-seq) profiles and showed expected segregation of known cell type–specific marker genes by the Cre activities. Overall, these data show that cTag-CLIP indeed accomplishes its goals: (a) specific expression of the tagged RBP in the desired cell populations, and (b) generation of cell type–specific CLIP maps. Importantly, cTag-CLIP data obtained from cTag mouse models are very consistent with data generated in wild-type mouse models through a variety of methods in multiple occasions (Hwang et al. 2017; Jereb et al. 2018; Saito et al. 2018), showing that the cTag-CLIP approach in general does not interfere with normal RBP biology and hence provides a powerful means for dissecting cell type–specific biology.

2. *cTag-CLIP has excellent power to "zoom in" on less abundant cell types to discover cell type–specific regulatory events:* Our in-depth cTag-PAPERCLIP analysis of APA patterns across four major brain cell types and between different cellular states in microglia (quiescent vs. activated) revealed novel cell type–specific regulatory events. For example, we showed that cell type–specific RBPs directly shape the distinct APA patterns of *Itsn1*, *Map4*, and *Cdc42* between neurons and glia/nonneural tissues. Moreover, in activated microglia, we identified that *Araf* switches from a truncated APA isoform (which generates a kinase-deficient protein) to a full-length APA isoform (which generates a kinase-intact protein). Importantly, these findings nicely illustrate how gaining resolution at the cell type level allows cTag-CLIP to generate new biological insights, as these four genes are expressed in multiple cell types in the brain, and it would be difficult for whole tissue analysis to pick up these cell type–specific regulatory events.

3. *cTag-CLIP well preserves native transcriptome states and has an advantage in profiling sensitive cell types:* One distinctive feature of cTag-CLIP is its "built-in" cell type specificity imparted through genetic expression of cell type–restricted Cre in cTag mice (Fig. 2B). As a result, cTag-CLIP can use the whole tissue as the input material without going through the process of tissue dissociation and cell type purification, which is necessary for common RNA profiling methods such as RNA-seq to achieve cell type specificity. Tissue dissociation and cell type purification disrupts the tissue microenvironment, causes cellular stress, and may alter gene expression (Cardona et al. 2006; Okaty 2011). Most importantly, cTag-CLIP kept the expression of microglia activation markers and immediate early genes (which respond to cellular stress) low compared with RNA-seq profiling from ex vivo purified microglia. These results show that cTag-CLIP better preserves native transcriptome states of individual cell types and therefore is suitable for accurate in vivo RNA profiling.

The success of our initial cTag-CLIP studies starts to reveal a diversity of RNA regulatory events in individual cell types. Generation and integration of cTag-CLIP maps for different RBPs in multiple cell types will provide essential information to refine the existing rules of RNA–protein interaction for individual cell types. Importantly, with the cTag-CLIP maps, it is now feasible to deduce the cooperative and/or competitive interactions among different RBPs for the same RNA-binding sites in a single cell type to construct high-order RNA–protein interaction maps. Recent improvement in generating knock-in mouse models from advances in synthetic biology and gene-targeting technology will speed up the process of expanding the cTag mice repertoire (Shalem et al. 2015; Wright et al. 2016). Moreover, continued identification of cell type–specific markers and technical advancement enhancing the sensitivity of CLIP technique will broaden the application of cTag-CLIP to new and rare cell types (Tasic et al. 2016; Lee and Ule 2018).

4 INSIGHTS FROM CLIP AND FUTURE DIRECTIONS

The development of CLIP has been driven largely by the need to understand the regulation of pre-mRNA processing, although it can be applied to all forms of posttranscriptional RNA regulation, including translation as shown with CLIP studies of FMRP. For studies of pre-mRNA, it is crucial to map the binding of nuclear RBPs to introns, which are generally of low abundance and of transient nature. To allow studies of RBP interactions on introns, specificity of the method had to be maximized, which led to the inclusion of covalent cross-linking followed by several stringent purification and quality control steps. Although the success of this approach was shown by the very first studies of Nova, CLIP also led to an unexpected finding that nuclear RBPs have many thousands of high-affinity binding sites, most of which are far from exons (Ule et al. 2003; Licatalosi et al. 2008). In recent years, it has become clear that many abundant nuclear RBPs, such as hnRNPC, often bind to specific transposable elements or other repetitive sequences that are present in introns, where they repress processing of cryptic exons and polyadenylation sites (Zarnack et al. 2013; Sibley et al. 2016). While this repressive binding safeguards the transcriptome from newly emerging intronic sites of RNA processing, it may also create an opportunity for a gradual evolutionary emergence of new alternative exons on mutations that weaken the repressive binding (Attig et al. 2016). Thus, although the large number of intronic-binding sites of nuclear RBPs initially came as a surprise, it is becoming clear that this plays an important role in the control of rarely included cryptic exons, including those mediating nonsense-mediated decay (Eom et al. 2013), and in the evolution of gene expression (Sibley et al. 2016).

4.1 Molecular and Physiologic Insights from CLIP

Following identification of in vivo binding sites, further integration with orthogonal functional data can then allow interpretation of regulatory mechanisms that operate in cells and organs. For example, the integrative approach to examine binding profiles around regulated exons and polyadenylation sites has been visualized in the form of RNA maps (Ule et al. 2006; Licatalosi et al. 2008; Witten and Ule 2011). This union of biochemical CLIP data with functional evidence via the bioinformatic generation of global "RNA maps" led to the recognition that the position of protein-binding sites in transcripts provided information relevant to their mechanisms and regulatory outcomes (Ule et al. 2003; Licatalosi et al. 2008). For example, it revealed that nuclear RBPs often regulate splicing and 3′ processing via similar positional principles, with repressive binding located close to the sites of RNA processing, and enhancing binding located further downstream (Licatalosi et al. 2008; Rot et al. 2017). The insights from such analysis can complement predictive programs such as TargetScan, which predicts possible binding sites for miRNAs by incorporating information on evolutionary conservation and general binding rules; for example, CLIP could identify important miRNA-binding sites that are unique to human (Chi et al. 2009) or are located at noncanonical sequences that do not conform to general rules (Chi et al. 2012; Moore et al. 2015).

Many studies have shown that RNP assembly on the nascent and mature mRNA, and the resulting regulatory mechanisms are coordinated at many levels (Singh et al. 2015), and CLIP has solidified these observations. Many RBPs shuttle between nucleus and cytoplasm, which allows messenger ribonucleoprotein (mRNP) assembly to start in the nucleus on pre-mRNAs and then continue to influence cytoplasmic regulation. For example, although CLIP identified roles for Nova proteins in nuclear pre-mRNA processing (Ule et al. 2003; Licatalosi et al. 2008), it also revealed 3′ UTR-binding sites in the vicinity of factors involved in alternative polyadenylation; again using genetics, such binding was found to be functional in determining the use of APA sites (Ule et al. 2003; Licatalosi et al. 2008). Moreover, CLIP, together with immunoelectron microscopy (EM) localization of Nova in the dendrite, identified a new link between nuclear regulation and synaptic RNA localization within neurons (Racca et al. 2010). This showed that an RNA regulatory factor was governed by its binding affinity to motifs that can be located at various positions in the transcriptome, and that a mammalian splicing factor could also bind and regulate mature mRNA biology.

Biologically, some of the more powerful conclusions from CLIP biochemistry have come from pairing it with genetic manipulations. CLIP studies have been most definitive in vivo in mice in which individual (Darnell 2013) and, more recently, multiple (Saito et al. 2018) RBPs have been genetically deleted. Similarly, CLIP studies have been successful in tissue culture cells in which RBPs have been knocked down. Such genetic studies have been complemented by genetic rescue experiments. This was illustrated in early studies demonstrating the requirement of Nova for splicing of the Z+ isoform of the agrin gene in motor neurons (Ule et al. 2005), and as a result the Nova1 KO mice have abnormal neuromuscular junctions and are paralyzed (Ruggiu et al. 2009). Moreover, Nova proteins regulate an alternative exon in Dab1 (Ule et al. 2006), an important regulator of reelin signaling in newly born migrating neurons (Yano et al. 2010). Importantly, specific phenotypes of Nova KO mice can be rescued by expressing the appropriately spliced isoform; interestingly, when this was performed by mating Nova1 KO mice to transgenic mice constitutively expressing Z+ agrin, the normal neuromuscular junction was reestablished, but the mice remained paralyzed, illustrating the multilayered aspect of RBP-RNA regulation in vivo (Ruggiu et al. 2009). Beyond studies of Nova, CLIP has been successfully applied to characterize the function and assembly of many additional disease-relevant RBPs in the brain and other tissues of model organisms (Darnell 2013; Modic et al. 2013).

4.2 CLIP: Future Directions

One of the most recent applications of CLIP has been to undertake studies of epigenetic modifications of RNA. These were pioneered with studies of m^6A-RNA modifications, using a variant of CLIP in which the modified nucleotide itself was directly cross-linked, in vitro, to m^6A-specific antibodies (Ke et al. 2015; Linder et al. 2015; Patil et al. 2018). This provides a potentially general strategy for analyzing what is an expanding number of posttranscriptional RNA modifications important in cell function and human disease (Jonkhout et al. 2017).

The same general principles of CLIP experiments apply in analyses of RNA modifications as in traditional CLIP: unbiased, genome-wide genetic studies allow demarcation of specific sites, here using direct modification-specific antibodies to cross-link the modified RNA nucleotide. For example, m^6A-CLIP was used to identify terminal exons as the most common positions for m^6A modification on RNA transcripts, rather than exons harboring stop codons (Ke et al. 2015); to show primary functions of m^6A in RNA turnover rather than regulation of alternative splicing (Ke et al. 2017); and to clarify our current understanding of the biologic role of m^6A-RNA modifications (Ke et al. 2017; Darnell et al. 2018).

Importantly, CLIP has been shown to be of value also for studies of human postmortem tissues (Tollervey et al. 2011). This is important as such tissues are often frozen in liquid nitrogen. RNA–protein complexes appear to survive such conditions well, and allow CLIP to be applied to tissues relevant for a range of human disease (e.g., cancer, immunologic, or neurologic disease) using a range of human tissue (tumor [Vanharanta et al. 2014], blood [Loeb et al. 2012], and brain [Tollervey et al. 2011; Scheckel et al. 2016], respectively). The value of CLIP has also been shown to unravel the dynamics of RNPs under disease conditions by comparing tissue from disease patients and controls (Tollervey et al. 2011), or monitoring the impact of viral infection on host transcripts in biology and human disease (Flynn et al. 2015; Luna et al. 2015, 2017; Moore et al. 2018).

Demarcation of RNA structure using CLIP was originally pioneered with hiCLIP to analyze Staufen–RNA interactions (Sugimoto et al. 2015) and subsequently developed with a variety of complementary approaches such as PARIS (psoralen analysis of RNA interactions and structures; Lu et al. 2016, 2018) as discussed above. Such analyses are important in their own right and in their potential to allow more general understanding of structures accessible to RBP regulation, as determined from CLIP analyses. The combined understanding of protein-binding sites, RNA structure, and evolutionary constraints promises to expand our ability to predict genetic variants using Bayesian (Zhang et al. 2010) and machine-learning tools (Barash et al. 2010) and ultimately apply such understanding to interpretation of those variants that have an impact on disease phenotypes.

Since the very initial development of CLIP, a common theme of many CLIP studies has been that they combine discovery of fundamental mechanisms of gene regulation along with biologic and clinically relevant insights. CLIP holds great promise for studies aimed at defining the functionally important RNA sites across the transcriptome, unraveling the mechanisms of RNP assembly and function, and uncovering the role of these mechanisms in diseases.

ACKNOWLEDGMENTS

The authors thank their many colleagues for fruitful discussions and contributions over the years—too many to cite here despite our best efforts. In particular, J.U. thanks Flora Lee for her help and feedback on the cross-linking and immunoprecipitation (CLIP) methods section. This work was supported by the National Institutes of Health (NIH) grant 5RC2NS069473-02 (R.B.D.), the European Research Council (617837-Translate), and the Wellcome Trust (103760/Z/14/Z) (J.U.). R.B.D. is a Howard Hughes Medical Institute Investigator.

REFERENCES

Attig J, Ruiz de Los Mozos I, Haberman N, Wang Z, Emmett W, Zarnack K, König J, Ule J. 2016. Splicing repression allows the gradual emergence of new Alu-exons in primate evolution. *eLife* **5:** e19545.

Barash Y, Calarco JA, Gao W, Pan Q, Wang X, Shai O, Blencowe BJ, Frey BJ. 2010. Deciphering the splicing code. *Nature* **465:** 53–59.

Bhate A, Parker DJ, Bebee TW, Ahn J, Arif W, Rashan EH, Chorghade S, et al. 2015. ESRP2 controls an adult splicing programme in hepatocytes to support postnatal liver maturation. *Nat Commun* **6:** 8768.

Burger K, Mühl B, Kellner M, Rohrmoser M, Gruber-Eber A, Windhager L, Friedel CC, Dölken L, Eick D. 2013. 4-thiouridine inhibits rRNA synthesis and causes a nucleolar stress response. *RNA Biol* **10:** 1623–1630.

Cardona AE, Huang D, Sasse ME, Ransohoff RM. 2006. Isolation of murine microglial cells for RNA analysis or flow cytometry. *Nat Protoc* **1:** 1947–1951.

Castello A, Fischer B, Eichelbaum K, Horos R, Beckmann BM, Strein C, Davey NE, et al. 2012. Insights into RNA biology from an atlas of mammalian mRNA-binding proteins. *Cell* **149:** 1393–1406.

Chakrabarti AM, Haberman N, Praznik A, Luscombe NM, Ule J. 2017. Data science issues in understanding protein-RNA interactions. *bioRxiv* doi.org/10.1101/208124.

Chen K, Lu Z, Wang X, Fu Y, Luo GZ, Liu N, Han D, et al. 2015. High-resolution N^6-methyladenosine (m^6A) map using photo-crosslinking-assisted m^6A sequencing. *Angew Chem* **54:** 1587–1590.

Chi SW, Zang JB, Mele A, Darnell RB. 2009. Argonaute HITS-CLIP decodes microRNA-mRNA interaction maps. *Nature* **460:** 479–486.

Chi SW, Hannon GJ, Darnell RB. 2012. An alternative mode of micro-RNA target recognition. *Nat Struct Mol Biol* **19:** 321–327.

Conlon EG, Manley JL. 2017. RNA-binding proteins in neurodegeneration: Mechanisms in aggregate. *Genes Dev* **31:** 1509–1528.

Danan C, Manickavel S, Hafner M. 2016. PAR-CLIP: A method for transcriptome-wide identification of RNA binding protein interaction sites. *Methods Mol Biol* **1358:** 153–173.

Darnell RB. 2010. HITS-CLIP: Panoramic views of protein-RNA regulation in living cells. *Wiley Interdiscip Rev RNA* **1:** 266–286.

Darnell R. 2012. CLIP (cross-linking and immunoprecipitation) identification of RNAs bound by a specific protein. *Cold Spring Harb Protoc* **2012:** 1146–1160.

Darnell RB. 2013. RNA protein interaction in neurons. *Ann Rev Neurosci* **36:** 243–270.

Darnell JC, Van Driesche SJ, Zhang C, Hung KYS, Mele A, Fraser CE, Stone EF, et al. 2011. FMRP stalls ribosomal translocation on mRNAs linked to synaptic function and autism. *Cell* **146:** 247–261.

Darnell RB, Ke S, Darnell JE Jr. 2018. Pre-mRNA processing includes N^6 methylation of adenosine residues that are retained in mRNA exons and the fallacy of "RNA epigenetics." *RNA* **24:** 262–267.

Eom T, Zhang C, Wang H, Lay K, Fak J, Noebels JL, Darnell RB. 2013. NOVA-dependent regulation of cryptic NMD exons controls synaptic protein levels after seizure. *eLife* **2:** e00178.

Flynn RA, Martin L, Spitale RC, Do BT, Sagan SM, Zarnegar B, Qu K, et al. 2015. Dissecting noncoding and pathogen RNA–protein interactomes. *RNA* **21:** 135–143.

Gerfen CR, Paletzki R, Heintz N. 2013. GENSAT BAC Cre-recombinase driver lines to study the functional organization of cerebral cortical and basal ganglia circuits. *Neuron* **80:** 1368–1383.

Granneman S, Kudla G, Petfalski E, Tollervey D. 2009. Identification of protein binding sites on U3 snoRNA and pre-rRNA by UV cross-linking and high-throughput analysis of cDNAs. *Proc Natl Acad Sci* **106:** 9613–9618.

Haberman N, Huppertz I, Attig J, König J, Wang Z, Hauer C, Hentze MW, et al. 2017. Insights into the design and interpretation of iCLIP experiments. *Gen Biol* **18:** 7.

Hafner M, Landthaler M, Burger L, Khorshid M, Hausser J, Berninger P, Rothballer A, et al. 2010. Transcriptome-wide identification of RNA-binding protein and microRNA target sites by PAR-CLIP. *Cell* **141:** 129–141.

He M, Liu Y, Wang X, Zhang MQ, Hannon GJ, Huang ZJ. 2012. Cell-type-based analysis of microRNA profiles in the mouse brain. *Neuron* **73:** 35–48.

Heiman M, Kulicke R, Fenster RJ, Greengard P, Heintz N. 2014. Cell type–specific mRNA purification by translating ribosome affinity purification (TRAP). *Nat Protoc* **9:** 1282–1291.

Huppertz I, Attig J, D'Ambrogio A, Easton LE, Sibley CR, Sugimoto Y, Tajnik M, König J, Ule J. 2014. iCLIP: Protein–RNA interactions at nucleotide resolution. *Methods* **65:** 274–287.

Hussain S, Sajini AA, Blanco S, Dietmann S, Lombard P, Sugimoto Y, Paramor M, et al. 2013. NSun2-mediated cytosine-5 methylation of vault noncoding RNA determines its processing into regulatory small RNAs. *Cell Rep* **4:** 255–261.

Hwang HW, Saito Y, Park CY, Blachère NE, Tajima Y, Fak JJ, Zucker-Scharff I, Darnell RB. 2017. cTag-PAPERCLIP reveals alternative polyadenylation promotes cell-type specific protein diversity and shifts *Araf* isoforms with microglia activation. *Neuron* **95:** 1334–1349.e5.

Imig J, Brunschweiger A, Brümmer A, Guennewig B, Mittal N, Kishore S, Tsikrika P, Gerber AP, Zavolan M, Hall J. 2014. miR-CLIP capture of a miRNA targetome uncovers a lincRNA H19–miR-106a interaction. *Nat Chem Biol* **11:** 107–114.

Jereb S, Hwang HW, Van Otterloo E, Govek EE, Fak JJ, Yuan Y, Hatten ME, Darnell RB. 2018. Differential 3′ processing of specific transcripts expands regulatory and protein diversity across neuronal cell types. *bioRxiv* doi.org/10.1101/245886.

Jonkhout N, Tran J, Smith MA, Schonrock N, Mattick JS, Novoa EM. 2017. The RNA modification landscape in human disease. *RNA* **23:** 1754–1769.

Ke S, Alemu EA, Mertens C, Gantman EC, Fak JJ, Mele A, Haripal B, et al. 2015. A majority of m^6A residues are in the last exons, allowing the potential for 3′ UTR regulation. *Genes Dev* **29:** 2037–2053.

Ke S, Pandya-Jones A, Saito Y, Fak JJ, Vågbø CB, Geula S, Hanna JH, Black DL, Darnell JE Jr., Darnell RB. 2017. m^6A mRNA modifications are deposited in nascent pre-mRNA and are not required for splicing but do specify cytoplasmic turnover. *Genes Dev* **31:** 990–1006.

König J, Zarnack K, Rot G, Curk T, Kayikci M, Zupan B, Turner DJ, Luscombe NM, Ule J. 2010. iCLIP reveals the function of hnRNP particles in splicing at individual nucleotide resolution. *Nat Struct Mol Biol* **17:** 909–915.

Kudla G, Granneman S, Hahn D, Beggs JD, Tollervey D. 2011. Cross-linking, ligation, and sequencing of hybrids reveals RNA–RNA interactions in yeast. *Proc Natl Acad Sci* **108:** 10010–10015.

Lee FCY, Ule J. 2018. Advances in CLIP technologies for studies of protein-RNA interactions. *Mol Cell* **69:** 354–369.

Lerner MR, Steitz JA. 1979. Antibodies to small nuclear RNAs complexed with proteins are produced by patients with systemic lupus erythematosus. *Proc Natl Acad Sci* **76:** 5495–5499.

Licatalosi DD, Darnell RB. 2010. RNA processing and its regulation: Global insights into biological networks. *Nat Rev Gen* **11:** 75–87.

Licatalosi DD, Mele A, Fak JJ, Ule J, Kayikci M, Chi SW. 2008. HITS-CLIP yields genome-wide insights into brain alternative RNA processing. *Nature* **456:** 464–469.

Linder B, Grozhik AV, Olarerin-George AO, Meydan C, Mason CE, Jaffrey SR. 2015. Single-nucleotide-resolution mapping of m^6A and m^6Am throughout the transcriptome. *Nat Methods* **12:** 767–772.

Loeb GB, Khan AA, Canner D, Hiatt JB, Shendure J, Darnell RB, Leslie CS, Rudensky AY. 2012. Transcriptome-wide miR-155 binding map reveals widespread noncanonical microRNA targeting. *Mol Cell* **48:** 760–770.

Lu Z, Zhang QC, Lee B, Flynn RA, Smith MA, Robinson JT, Davidovich C, et al. 2016. RNA duplex map in living cells reveals higher-order transcriptome structure. *Cell* **165:** 1267–1279.

Lu Z, Gong J, Zhang QC. 2018. PARIS: Psoralen analysis of RNA interactions and structures with high throughput and resolution. *Methods Mol Biol* **1649:** 59–84.

Luna JM, Scheel TKH, Danino T, Shaw KS, Mele A, Fak JJ, Nishiuchi E, et al. 2015. Hepatitis C virus RNA functionally sequesters miR-122. *Cell* **160:** 1099–1110.

Luna JM, Barajas JM, Teng KY, Sun HL, Moore MJ, Rice CM, Darnell RB, Ghoshal K. 2017. Argonaute CLIP defines a deregulated miR-122-bound transcriptome that correlates with patient survival in human liver cancer. *Mol Cell* **67:** 400–410.e7.

Mili S, Steitz JA. 2004. Evidence for reassociation of RNA-binding proteins after cell lysis: Implications for the interpretation of immunoprecipitation analyses. *RNA* **10:** 1692–1694.

Modic M, Ule J, Sibley CR. 2013. CLIPing the brain: Studies of protein–RNA interactions important for neurodegenerative disorders. *Mol Cell Neurosci* **56:** 429–435.

Moore MJ, Scheel TKH, Luna JM, Park CY, Fak JJ, Nishiuchi E, Rice CM, Darnell RB. 2015. miRNA-target chimeras reveal miRNA 3′-end pairing as a major determinant of Argonaute target specificity. *Nat Comm* **6:** 8864.

Moore MJ, Blachère NE, Fak JJ, Park CY, Sawicka K, Parveen S, Zucker-Scharff I, Moltedo B, Rudensky A, Darnell R. 2018. ZFP36 RNA-binding proteins restrain T-cell activation and anti-viral immunity. *bioRxiv* doi.org/10.1101/247668.

Niranjanakumari S, Lasda E, Brazas R, Garcia-Blanco MA. 2002. Reversible cross-linking combined with immunoprecipitation to study RNA–protein interactions in vivo. *Methods* **26:** 182–190.

Okaty BW, Sugino K, Nelson SB. 2011. A quantitative comparison of cell-type-specific microarray gene expression profiling methods in the mouse brain. *PLoS One* **6:** e16493.

Patil DP, Pickering BF, Jaffrey SR. 2018. Reading m^6A in the transcriptome: m^6A-binding proteins. *Trends Cell Biol* **28:** 113–127.

Racca C, Gardiol A, Eom T, Ule J, Triller A, Darnell RB. 2010. The neuronal splicing factor Nova co-localizes with target RNAs in the dendrite. *Front Neural Circuits* **4:** 5.

Rot G, Wang Z, Huppertz I, Modic M, Lenče T, Hallegger M, Haberman N, Curk T, von Mering C, Ule J. 2017. High-resolution RNA maps suggest common principles of splicing and polyadenylation regulation by TDP-43. *Cell Rep* **19:** 1056–1067.

Ruggiu M, Herbst R, Kim N, Jevsek M, Fak JJ, Mann MA, Fischbach G, Burden SJ, Darnell RB. 2009. Rescuing Z^+ agrin splicing in *Nova* null mice restores synapse formation and unmasks a physiologic defect in motor neuron firing. *Proc Natl Acad Sci* **106:** 3513–3518.

Saito Y, Yuan Y, Zucker-Scharff I, Fak JJ, Tajima Y, Licatalosi DD, Darnell RB. 2018. Neuron-specific cTag-CLIP reveals cell-specific diversity of functional RNA regulation in the brain. *bioRxiv* doi.org/10.1101/244905.

Scheckel C, Drapeau E, Frias MA, Park CY, Fak J, Zucker-Scharff I, Kou Y, et al. 2016. Regulatory consequences of neuronal ELAV-like protein binding to coding and non-coding RNAs in human brain. *eLife* **5:** e10421.

Shalem O, Sanjana NE, Zhang F. 2015. High-throughput functional genomics using CRISPR-Cas9. *Nat Rev Gen* **16:** 299–311.

Sibley CR, Blazquez L, Ule J. 2016. Lessons from non-canonical splicing. *Nat Rev Gen* **17:** 407–421.

Singh G, Pratt G, Yeo GW, Moore MJ. 2015. The clothes make the mRNA: Past and present trends in mRNP fashion. *Ann Rev Biochem* **84:** 325–354.

Sugimoto Y, König J, Hussain S, Zupan B, Curk T, Frye M, Ule J. 2012. Analysis of CLIP and iCLIP methods for nucleotide-resolution studies of protein-RNA interactions. *Gen Biol* **13:** R67.

Sugimoto Y, Vigilante A, Darbo E, Zirra A, Militti C, D'Ambrogio A, Luscombe NM, Ule J. 2015. hiCLIP reveals the in vivo atlas of mRNA secondary structures recognized by Staufen 1. *Nature* **519:** 491–494.

Sugimoto Y, Chakrabarti AM, Luscombe NM, Ule J. 2017. Using hiCLIP to identify RNA duplexes that interact with a specific RNA-binding protein. *Nat Protoc* **12:** 611–637.

Tasic B, Menon V, Nguyen TN, Kim TK, Jarsky T, Yao Z, Levi B, et al. 2016. Adult mouse cortical cell taxonomy revealed by single cell transcriptomics. *Nat Neurosci* **19:** 335–346.

Tollervey JR, Curk T, Rogelj B, Briese M, Cereda M, Kayikci M, König J, et al. 2011. Characterizing the RNA targets and position-dependent splicing regulation by TDP-43. *Nat Neurosci* **14:** 452–458.

Ule J, Jensen KB, Ruggiu M, Mele A, Ule A, Darnell RB. 2003. CLIP identifies Nova-regulated RNA networks in the brain. *Science* **302:** 1212–1215.

Ule J, Ule A, Spencer J, Williams A, Hu JS, Cline M, Wang H, et al. 2005. Nova regulates brain-specific splicing to shape the synapse. *Nat Gen* **37:** 844–852.

Ule J, Stefani G, Mele A, Ruggiu M, Wang X, Taneri B, Gaasterland T, Blencowe BJ, Darnell RB. 2006. An RNA map predicting Nova-dependent splicing regulation. *Nature* **444:** 580–586.

Vanharanta S, Marney CB, Shu W, Valiente M, Zou Y, Mele A, Darnell RB, Massagué J. 2014. Loss of the multifunctional RNA-binding protein RBM47 as a source of selectable metastatic traits in breast cancer. *eLife* **3:** e02734.

Van Nostrand EL, Pratt GA, Shishkin AA, Gelboin-Burkhart C, Fang MY, Sundararaman B, Blue SM, et al. 2016. Robust transcriptome-wide discovery of RNA-binding protein binding sites with enhanced CLIP (eCLIP). *Nat Methods* **13:** 508–514.

Weyn-Vanhentenryck SM, Mele A, Yan Q, Sun S, Farny N, Zhang Z, Xue C, et al. 2014. HITS-CLIP and integrative modeling define the Rbfox splicing-regulatory network linked to brain development and autism. *Cell Rep* **6:** 1139–1152.

Witten JT, Ule J. 2011. Understanding splicing regulation through RNA splicing maps. *Trends Genet* **27:** 89–97.

Wright AV, Nuñez JK, Doudna JA. 2016. Biology and applications of CRISPR systems: Harnessing nature's toolbox for genome engineering. *Cell* **164:** 29–44.

Yano M, Hayakawa-Yano Y, Mele A, Darnell RB. 2010. Nova2 regulates neuronal migration through an RNA switch in disabled-1 signaling. *Neuron* **66:** 848–858.

Yap K, Xiao Y, Friedman BA, Je HS, Makeyev EV. 2016. Polarizing the neuron through sustained co-expression of alternatively spliced isoforms. *Cell Rep* **15:** 1316–1328.

Zarnack K, König J, Tajnik M, Martincorena I, Eustermann S, Stévant I, Reyes A, Anders S, Luscombe NM, Ule J. 2013. Direct competition between hnRNP C and U2AF65 protects the transcriptome from the exonization of Alu elements. *Cell* **152:** 453–466.

Zarnegar BJ, Flynn RA, Shen Y, Do BT, Chang HY, Khavari PA. 2016. irCLIP platform for efficient characterization of protein–RNA interactions. *Nat Methods* **13:** 489–492.

Zhang C, Darnell RB. 2011. Mapping in vivo protein-RNA interactions at single-nucleotide resolution from HITS-CLIP data. *Nat Biotech* **29:** 607–614.

Zhang C, Frias MA, Mele A, Ruggiu M, Eom T, Marney CB, Wang H, Licatalosi DD, Fak JJ, Darnell RB. 2010. Integrative modeling defines the Nova splicing-regulatory network and its combinatorial controls. *Science* **329:** 439–443.

Zhang X, Chen MH, Wu X, Kodani A, Fan J, Doan R, Ozawa M, et al. 2016. Cell-type-specific alternative splicing governs cell fate in the developing cerebral cortex. *Cell* **166:** 1147–1162.e15.

Principles and Practices of Hybridization Capture Experiments to Study Long Noncoding RNAs That Act on Chromatin

Matthew D. Simon[1,2] and Martin Machyna[1,2]

[1]Department of Molecular Biophysics and Biochemistry, Yale University, New Haven, Connecticut 06511

[2]Chemical Biology Institute, Yale University, West Haven, Connecticut 06516

Correspondence: matthew.simon@yale.edu

SUMMARY

The diverse roles of cellular RNAs can be studied by purifying RNAs of interest together with the biomolecules they bind. Biotinylated antisense oligonucleotides that hybridize specifically to the RNA of interest provide a general approach to develop affinity reagents for these experiments. Such oligonucleotides can be used to enrich endogenous RNAs from cross-linked chromatin extracts to study the genomic binding sites of RNAs. These hybridization capture protocols are evolving modular experiments that are compatible with a range of cross-linkers and conditions. This review discusses the principles of these hybridization capture experiments as well as considerations and controls necessary to interpret the resulting data without being misled by artifactual signals.

Outline

Cite as *Cold Spring Harb Perspect Biol* doi: 10.1101/cshperspect.a032276

1 INTRODUCTION

Long noncoding RNAs (lncRNAs) are important regulators of chromatin structure and function. For example, lncRNAs act as master regulators of dosage compensation in both flies and mammals (including mice and humans), binding across an entire chromosome to regulate transcription (Creamer and Lawrence 2017; Samata and Akhtar 2018). Although approaches to study the proteins involved in these processes such as chromatin immunoprecipitation (ChIP) have existed for decades (Gilmour and Lis 1985; Solomon and Varshavsky 1985), only recently have similar approaches been developed to study RNA–chromatin interactions (Machyna and Simon 2018). Expanding ChIP-style experiments to the biochemical enrichment of RNAs may seem like a trivial problem—after all, the predictability of Watson–Crick base-pairing means that it should be possible to use first principles to design biotinylated complementary oligonucleotides as affinity reagents that are specific for any RNA. As will be discussed in this review, hybridization capture methods have indeed proven to be powerful tools to understand RNAs (Fig. 1) and especially lncRNAs that bind chromatin, but designing these experiments is still far from trivial. These hybridization techniques, including CHART (Simon et al. 2011), ChIRP (Chu et al. 2011), RAP (Engreitz et al. 2013), and other protocols, are best thought of as a family of interrelated approaches that contain different modules. These modules can then be optimal-

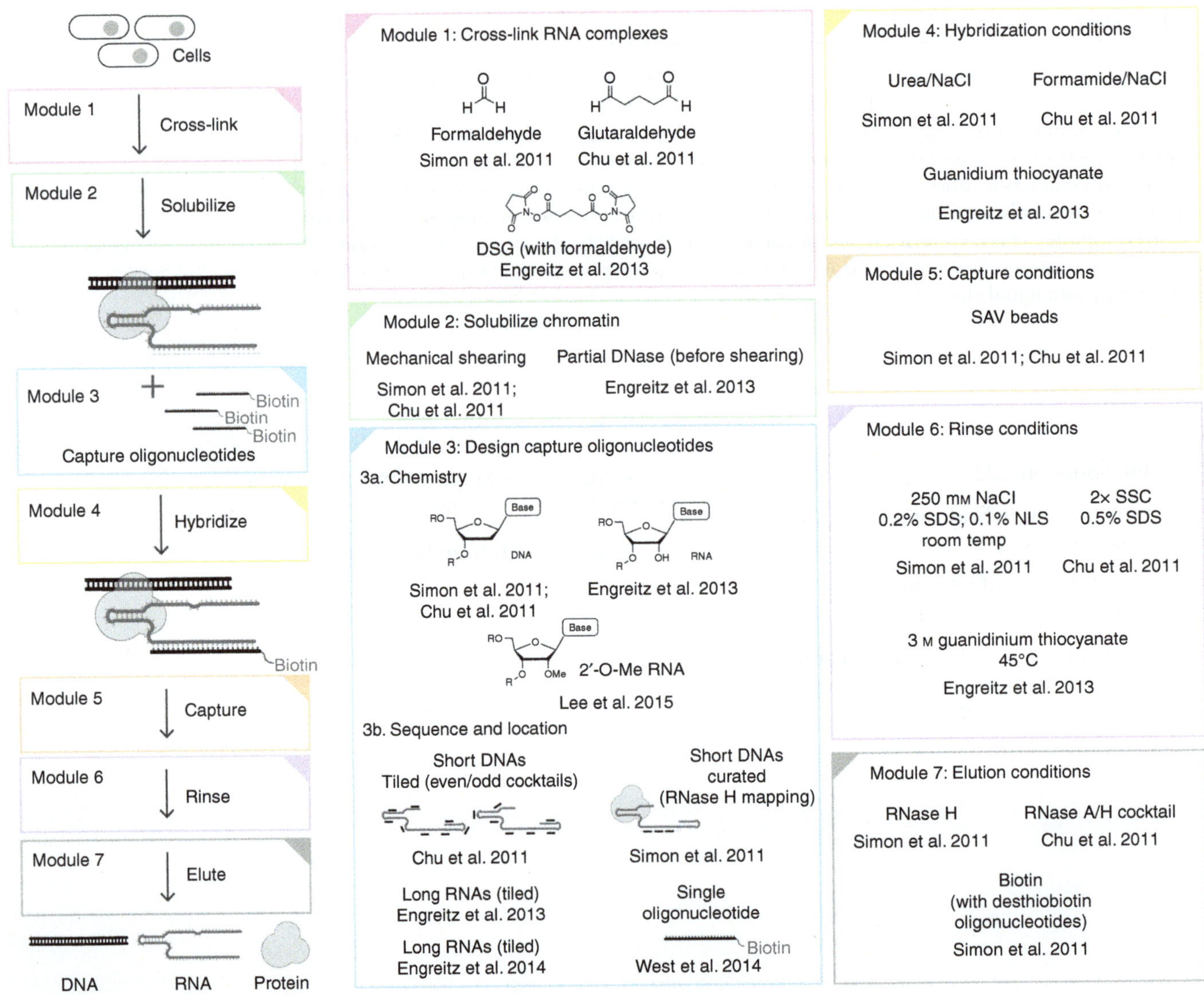

Figure 1. Overview of hybridization capture experiments. Different options for each module are indicated along with primary references.

Cite as *Cold Spring Harb Perspect Biol* doi: 10.1101/cshperspect.a032276

ly combined depending on the application. As the modular protocols for hybridization capture continue to mature, using them to gain biological insight requires a critical understanding of the challenges and opportunities they provide. This review will (1) review the biochemical and biophysical principles of hybridization capture experiments; (2) discuss the development of several modular protocols that have proven compatible with deep sequencing and modern proteomic workflows; (3) explore applications of these approaches in the context of RNAs involved in dosage compensation; (4) critically discuss the challenges inherent in interpreting data from capture hybridization experiments; and (5) discuss how these approaches are likely to improve and be integrated into the results from complementary technologies.

2 PRINCIPLES OF HYBRIDIZATION CAPTURE EXPERIMENTS

Oligonucleotides that hybridize to RNAs have been used as affinity reagents to enrich a target RNA together with other biomolecules from complex extracts. As with any enrichment technique, there are two primary considerations: what is the yield of the target RNA complex (sensitivity), and how much undesired contamination copurifies with the target RNA (specificity)? In practice, modern next-generation sequencing (NGS) and proteomic detection platforms have become exquisitely sensitive. Specificity is therefore generally a greater concern than sensitivity.

2.1 Tuning Hybridization Conditions

Although there are similarities between using antibodies to immunoprecipitate a cellular protein (as in ChIP) and using biotinylated oligonucleotides to capture a cellular RNA, there are important differences that impact the experimental design of hybridization capture experiments. Specifically, the biophysics of nucleic acid hybridization differ substantially from antigen–antibody interactions (Demidov and Frank-Kamenetskii 2004). The lock-and-key interactions of an antibody with an antigen (Pauling 1974) lead to the trend that tighter antibody-target binding correlates with higher specificity for its target peptide. In contrast, conditions that lead an oligonucleotide to bind more tightly to its target sequence tend to correlate with tighter binding to nontarget sequences as well, and therefore often lead to lower specificity (Fig. 2). Tight binding can be driven by the chemical composition of the oligonucleotide (e.g., increased length or inclusion of stabilizing nucleotide analogs) as shown in Figure 2, or through altering the conditions of the reaction (e.g., hybridization is favored by higher salt concentrations and lower temperatures). As will be discussed for specific hybridization capture protocols below, this relationship can lead to a complicated balance between the affinity and specificity of capture oligonucleotides for their targets.

Nucleic acid hybridization is more stringent at low ionic strength (low salt) because of electrostatic repulsion between the phosphates in a hybridized duplex (Schildkraut and Lifson 1965). In contrast, the binding of basic proteins to RNA has the opposite trend: At low ionic strength, positively charged proteins can bind nonspecifically to the negative charges on the RNA backbone, and using high ionic strength (high salt) can reduce these interactions. To study chromatin-bound RNA complexes, capture hybridization experiments tend to be performed at high ionic strength, and so the hybridization specificity needs to be tuned to retain specificity.

Although most molecular biologists think of nucleic acid hybridization in terms of the melting temperature (T_m, the temperature at which the duplex is half denatured), it is worth considering a traditional thermodynamic framework (SantaLucia and Hicks 2004). When the $\Delta G°$ of the hybridization reaction is large and negative, the fraction of RNA that is hybridized to the capture oligonucleotide will increase. Although high yields of the target RNA are desirable, it is worth considering the binding of undesirable off-target RNAs as well. At room temperature in standard biological buffers, even very short nucleic acid duplexes can have highly negative $\Delta G°$ values, meaning that off-target oligonucleotides with only partial complementarity to the capture oligonucleotide can be bound with high yield (Fig. 2, bottom left), decreasing the specificity of the reaction. By chance alone, unintended matches to short stretches of nucleotides in off-target RNAs are common. For shorter oligonucleotides, the number of RNAs that differ from target RNA by one or two nucleotides can be large, further exacerbating the challenge of achieving specificity. Even if an oligonucleotide displays a large difference in the binding energies between the target RNA and an off-target RNA (e.g., $\Delta G°_{target} < \Delta G°_{off\text{-}target}$), if both are substantially negative, then both RNAs will be retrieved with high yields. Higher specificity is achieved when the oligonucleotide composition is tuned to make the reaction less favorable, but less favorable reactions also decreased yield (Fig. 2, bottom right). Therefore, a balance between affinity and specificity is required to achieve specific hybridization (Fig. 2, yellow shaded area).

Specificity can be achieved by altering the composition of the oligonucleotide or by adjusting the reaction conditions. As hybridization involves a trade-off between enthalpy and entropy, increasing temperature is the traditional means to achieve higher specificity (e.g., polymerase chain reaction [PCR] annealing reactions). However, because it is

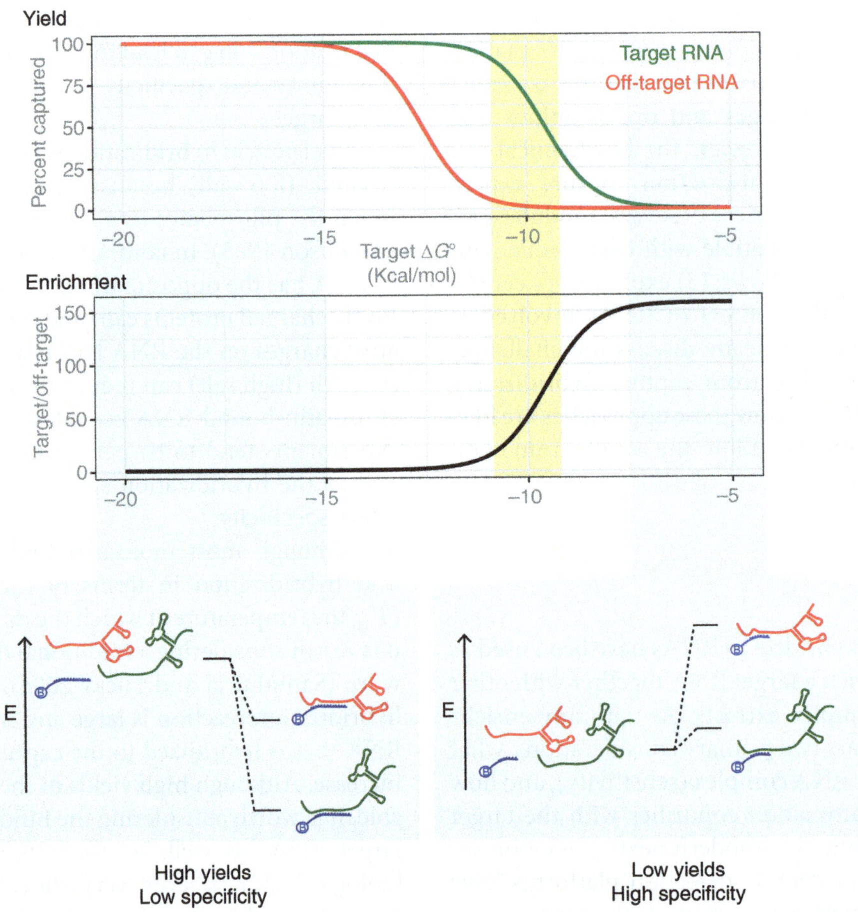

Figure 2. Illustration of the affinity-specificity trade-off in hybridization reactions. Under conditions in which the oligonucleotide binds very tightly (e.g., longer oligonucleotide) to the target RNA (*bottom left*), off-target RNAs will also be bound. Higher selectivity can be achieved under conditions in which on-target binding is only marginally favorable (e.g., shorter oligonucleotide, *bottom right*), but at the cost of yield. Optimal conditions (yellow shaded region) balance affinity and specificity. In this example, reaction conditions were simulated under fixed conditions ($[RNA_{target}] = [RNA_{off\text{-}target}] = 10$ pM, [capture oligonucleotide] = 100 nM). In this simulation, the capture oligonucleotide prefers the target RNA (green) over the off-target RNA (red) by a constant energetic difference (3 kcal/mol). This example depicts the trade-off when designing the chemical composition of capture oligonucleotides, but similar considerations apply when choosing conditions for hybridization.

not always desirable to handle complex cross-linked RNA-containing extracts at high temperatures, other parameters may also be used to tune binding and increase stringency, including using shorter oligonucleotides with lower GC content. Capture oligonucleotides with RNA and other stabilizing sugar modifications (e.g., O2′-methylated oligos or locked nucleic acids) favor binding, but under conditions that drive $\Delta G°_{off\text{-}target} < 0$, these modifications can degrade specificity. In some cases, sugar modifications can increase the energic costs of mismatched nucleotides (Appella 2009). Buffer composition can play a large role: Low ionic strength creates more stringent conditions, as does using solvents and additives that disrupt stacking and lower the dielectric of the solution (e.g., formamide or urea). Although there are many parameters that can be used to tune hybridization yields and selectivity, these parameters need to be carefully tuned.

2.2 Rinsing Immobilized Complexes to Remove Off-Target Contamination

The hybridization reaction is one of three principal opportunities to increase specificity in a hybridization capture experiment. The second opportunity arises after the hybridized complexes have been immobilized (e.g., biotin captured on streptavidin resin), in which rinses are used to remove off-target molecules. The molecules that bind the target RNA (e.g., protein and DNA) are covalently attached to the RNA and will not dissociate unless the cross-links are reversed. Therefore, the rinses can be as stringent as desired so long as the capture oligonucleotide remains bound to the target RNA. Most protocols involve high-salt rinses, which help remove basic proteins from binding nucleic acids on the beads. Low-salt rinses or high-temperature rinses can be added to remove weakly hybridized contaminating sequences. Although rinsing is a critical step for all hybridization capture methods, the local concentration of nucleic acids (and basic proteins) on the solid support can be high, leading to opportunities for high avidity (i.e., off-target sequences can bind in multiple places simultaneously decreasing the k_{off} and making them difficult to dislodge), and also biomolecules can precipitate on the surface of the resin. Rinses are an important opportunity to increase specificity of enrichment, but are best combined with conditions that minimize capturing background signal in the hybridization reaction.

2.3 Specific Elution of the Desired Complexes

The elution of the complexes from the beads provides a final opportunity for biochemical specificity. As will be discussed below, strategies have been developed to retrieve the desired complexes while leaving off-target complexes bound to the resin. Although selective elution can be powerful, conditions that maximize the elution specificity frequently compromise yields of the desired complexes (e.g., when using RNase H; see Sec. 3.3.6).

3 ESTABLISHMENT OF NGS HYBRIDIZATION CAPTURE TECHNIQUES

3.1 Precedent for Hybridization Capture Techniques

Hybridization capture protocols have been used in the context of purified nucleic acids for half a century (Riggsby 1969). Experiments capturing RNA from extracts to analyze copurifying factors have been central to our understanding of RNA biology for decades. Pioneering work on U2 (Barabino et al. 1989; Lamond et al. 1990) and 7SK (Wassarman and Steitz 1991) used biotinylated O2′-methylated oligonucleotides to capture native ribonucleoproteins (RNPs) from extracts. Important precedent for hybridization capture experiments also comes from approaches that use oligo(dT) to globally capture mRNAs (via their poly(A) tails) with their associated proteins (Lindberg and Sundquist 1974). Another impressive example of native hybridization capture was the purification of active preparations of the telomerase enzyme, which involved competitive hybridization to elute the desired complex (Lingner and Cech 1996). Adapting this family of experiments to chromatin poses several challenges, but luckily many of these challenges had been previously addressed during the development of ChIP protocols (Orlando 2000; Park 2009) and other similar protocols such as proteomics of isolated chromatin segments (PICh) (Dejardin and Kingston 2009). Chromatin itself is largely insoluble under most buffer conditions and must be fragmented to make a soluble extract. Because many complexes can be disrupted during fragmentation procedures, the complexes are generally cross-linked while the cell is intact. Although adding complexity to the experiment, cross-linking has a substantial advantage in that the covalent cross-links provide a record of the interactions that were present at the time of cross-linking, distinguishing them from noncovalent interactions that take place after chromatin solubilization when the components of complexes can sometimes shuffle.

3.2 Extending Hybridization Capture to Cross-Linked Extracts and Global Analyses

The first protocols to successfully adapt hybridization capture experiments to cross-linked extracts for the purposes of genome-wide sequencing both included analysis of the roX2 lncRNA from *Drosophila melanogaster* cells (Chu et al. 2011; Simon et al. 2011). These contemporaneous protocols are conceptually similar to each other and analogous to ChIP-seq (Fig. 1). In both, biotinylated DNA oligonucleotides were used to enrich target RNAs (e.g., roX2) from cross-linked chromatin extracts. After rinsing, the DNA that was bound to the RNA (via covalent cross-links) was eluted and sequenced, revealing the genome-wide distribution of chromatin binding sites (Fig. 3A). Although the details of many aspects of the protocols differed, both converged on hybridization conditions with high ionic strength counterbalanced by high levels of a denaturant. Despite the broad similarities between protocols, the differences can have important biochemical consequences that are worth examining.

As with other NGS methods, the proliferation of names for conceptually similar hybridization capture techniques can be confusing (Hadfield and Retief 2018). For example, in the first two reports establishing genome-wide mapping of RNAs, ChIRP established the use of glutaraldehyde cross-linking (Chu et al. 2011), whereas CHART established formaldehyde cross-linking (Simon et al. 2011). However, subsequent iterations of the ChIRP protocol including dChIRP (Quinn et al. 2014) and ChIRP-MS (Chu et al. 2015) are fusion techniques that use the formaldehyde cross-linking conditions established for the CHART protocol rather than glutaraldehyde but use the hybridization buffers from ChIRP. Although CHART generally is performed with curated capture oligonucleotides, using a ChIRP-like tiled oligonucleotide set was shown to achieve similar results in at least one case (Simon et al. 2013). Sim-

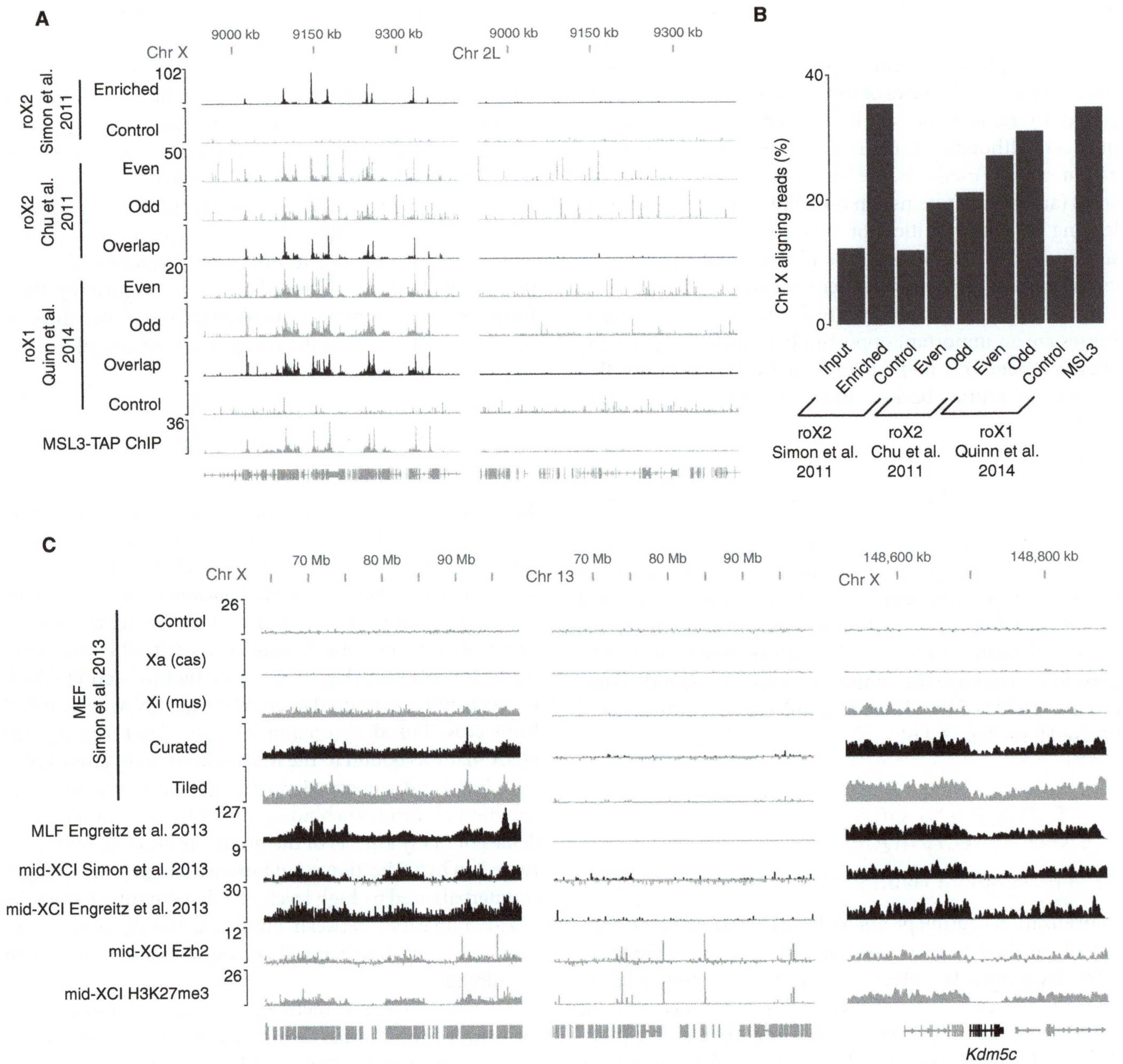

Figure 3. Comparison of different hybridization capture experiments. (*A*) roX RNAs enrichment on Chr X is highly reproducible among several published studies and agrees with ChIP experiments targeting subunits of the male specific lethal (MSL) complex (Alekseyenko et al. 2008). The roX1 profiles are from a fusion experiment that uses formaldehyde cross-linking and formamide hybridization buffers (Quinn et al. 2014). Autosomal data (Chr 2L) show how controls and postprocessing (overlap) can help to bioinformatically reduce off-target hybridization signals from odd and even oligonucleotide pools. (*B*) Bar graph illustrating the fraction of purified DNA from different enrichment strategies that originates from Chr X. (*C*) Xist long noncoding RNA (lncRNA) coats broad regions of inactive Chr X (Xi) except for several escapee genes including Kdm5c (*right* panel). Signals from tiled oligonucleotides and curated oligonucleotides agree well (Simon et al. 2013). Xist enrichment on chromosome X was found to highly correlate with PRC2 (Ezh2 ChIP; Pinter et al. 2012) and H3K27me3 ChIP (Pinter et al. 2012) during X-chromosome inactivation (XCI). Hybrid cell lines in which active (Xa) and inactive (Xi) X chromosomes can be distinguished based on their allele-specific differences validate that signals arise from biological cross-linking and not artifact (Simon et al. 2013).

Cite as *Cold Spring Harb Perspect Biol* doi: 10.1101/cshperspect.a032276

ilarly, a third protocol, RAP, originally established the use of RNA capture oligos and additional disuccinimidyl glutarate (DSG) cross-linking (Engreitz et al. 2013), yet an updated RAP protocol (Engreitz et al. 2014) retains the hybridization buffer from RAP (high guanidinium salts), but suggests formaldehyde cross-linking (similar to CHART) and DNA oligo sets (similar but longer than CHART and ChIRP), even/odd controls (from ChIRP) and RNase H elution (developed in CHART), or new cross-linkers altogether to capture RNA–RNA interactions. Although the cross-fertilization of strategies between different protocols is good for the field, names such as CHART, ChIRP, and RAP do not specify the relevant details of what experiment was performed. For this reason, this review will minimize the use of these names and instead treat these protocols as modular experiments containing the components described above (e.g., cross-linking, hybridization, and elution) that continue to be optimized and will be referred to generically as hybridization capture experiments. The modules are summarized in Figure 1 and discussed below.

3.3 Modules for Experimental Designs

3.3.1 *Cross-Linking*

There are several chemistries that have been used to stabilize RNA complexes through covalent cross-linking. As is standard in ChIP protocols, formaldehyde alone can be used as a cross-linker for capture hybridization experiments, but requires higher levels of cross-linking. This was the basis for one of the first hybridization capture protocols (Simon et al. 2011), and formaldehyde cross-linking has now been adopted as a standard module for most updated hybridization capture protocols (Engreitz et al. 2015; Chu and Chang 2016; Sexton et al. 2016; Sridhar et al. 2017). Not only does formaldehyde have a long history in chromatin biochemistry (Solomon and Varshavsky 1985), but also formaldehyde cross-links are reversible, thereby facilitating analysis of copurifying biomolecules (Hoffman et al. 2015).

Other cross-linkers that have been successfully added for monitoring RNA localization genome-wide include glutaraldehyde (Chu et al. 2011) and DSG (Engreitz et al. 2013). Unlike formaldehyde, which has a single electrophilic carbon, these cross-linkers have two electrophilic sites connected by three rotable carbons that lead them to cross-link residues that are a greater distance from each other. Unlike formaldehyde, these compounds cause cross-links that are not considered reversible.

3.3.2 *Extract Preparation*

Like ChIP, capture hybridization experiments generally involve shearing the chromatin to make soluble extracts. This can be accomplished with instruments such as sonicators (Chu et al. 2011; Simon et al. 2011). The higher degree of cross-linking required for hybridization capture experiments compared with ChIP leads to the need for more extensive shearing. Even with extensive shearing, DNA insert sizes are frequently at least two times longer than would be used in ChIP, which modestly impacts the resolution of these experiments. Extensive shearing of chromatin also leads to shearing of the target RNA. Although generally considered a liability, this latter property was used as a feature rather than a limitation to study RNA–RNA interactions between different parts of the same RNA (Quinn et al. 2014). Alternatively, enzymes such as DNase can be used to facilitate chromatin fragmentation reducing the levels of mechanical shearing (Engreitz et al. 2013). This has the advantage of minimizing RNA shearing, but it is harder to control and leads to signal in input samples that is uneven.

3.3.3 *Capture Oligonucleotide Design*

The first two capture hybridization experiments both used short (~20–25 nt) biotinylated DNA oligonucleotides (Chu et al. 2011; Simon et al. 2011), but were designed differently. Only certain regions of an RNA are available for hybridization in cross-linked extracts. Other regions are likely to be occluded by proteins or engaged in secondary structure. In the Simon study, the sites available for hybridization were determined in advance by performing RNase H mapping (Simon et al. 2011). Candidate DNAs were tested for their ability to hybridize to the target RNA in the context of cross-linked chromatin extracts. This led to a cocktail of three capture oligonucleotides that could successfully capture the target RNA in cross-linked extracts. In a more extreme example of this approach, West and colleagues showed it was possible to capture a lncRNA using a single capture oligonucleotide developed using RNase H mapping (West et al. 2014).

Alternatively, Chu and colleagues used software developed for RNA fluorescent in situ hybridization (FISH) experiments to design oligonucleotides that were specific to the target RNA and used these to tile the entire length of the RNA (Chu et al. 2011). As will be discussed below (see Sec. 5.3.3), Chu and colleagues also developed the use of orthogonal cocktails to help distinguish real signal from artifacts.

Unlike the first two studies that used short DNA oligonucleotides, Engreitz and colleagues developed capture oligonucleotides based on long (120-nt) biotinylated RNA nucleotides using in vitro transcription (Engreitz et al. 2013). This approach more closely resembles capture hybridization conditions that have been developed for isolated nucleic acids (Fisher et al. 2011). Further optimization of this module led Engreitz and colleagues to use long DNAs instead of RNA capture oligonucleotides (Engreitz et al. 2014). Although most experiments are now performed with DNA capture oligonucleotides, tighter-binding oligonucleotides with 2′-*O*-methylation (Lee et al. 2015) or locked nucleic acids (Rogell et al. 2017) have also been developed.

3.3.4 Hybridization Conditions

As mentioned above, these protocols require buffer conditions that maintain chromatin solubility while simultaneously providing the desired hybridization stringency. Both original reports used a combination of high ionic strength with high levels of duplex destabilizing additives such as formamide (Chu et al. 2011) or urea (Simon et al. 2011). The formamide conditions are more denaturing, leading to higher yields but at the cost of higher levels of artifactual signals (Fig. 3A). These artifacts can arise from denaturation of double-stranded DNA (dsDNA) that copurifies with RNA target. By using less denaturing conditions (Simon et al. 2011), melting of dsDNA can be minimized, thus reducing the extent to which the capture oligonucleotides have access to bind directly to the DNA. This leads to less signal from direct DNA binding of the capture oligonucleotides, and also requires less bioinformatic filtering to reveal the true RNA binding sites.

The conditions listed above were optimized for hybridization of short DNA oligonucleotides. An experimental design with longer RNA capture oligonucleotides required more denaturing conditions to achieve specificity but had the advantage that oligonucleotide-bound complexes were more stable to harsh conditions during rinses (Engreitz et al. 2013). In this case, a protocol was developed that uses high concentrations of guanidinium, which both increases the ionic strength and acts as a denaturant.

3.3.5 Rinses

Rinsing is an important step for all hybridization capture protocols, and the conditions to date have been similar across approaches. Streptavidin–biotin interactions are sufficiently tight (and streptavidin a sufficiently stable protein) that the immobilized capture oligonucleotides are retained through harsh rinses. Rinses generally include denaturing detergents and salts (to remove nonspecifically bound contaminants such as basic proteins). Higher temperatures have also been used to remove partially hybridized contaminants. Longer oligonucleotides withstand more stringent rinses.

It is tempting to expect that optimizing the temperature and conditions of the rinses could compensate for lower specificity in the hybridization step. However, as discussed above (see Sec. 2.2), once complexes are immobilized it can be difficult to remove contaminating species because a high concentration of biomolecules on the resin can lead to avidity effects (e.g., a contaminating RNA can hybridize to several sites simultaneously). Also, if denaturants are removed, the complexes can precipitate on the surface of the resin, trapping contaminants and preventing efficient elution.

3.3.6 Elution

When analyzing the genome-wide binding sites of an RNA, the DNA that is enriched by capture hybridization analysis can be eluted by digesting proteins (including streptavidin), but this elutes all material that is bound to the beads. Alternatively, a cocktail of RNases can be used (Chu et al. 2011), which releases all RNA-bound species. Although it produces a lower yield, a more specific elution can be achieved using RNase H to specifically degrade RNA–DNA hybrids (Simon et al. 2011). This strategy works when the capture oligonucleotides are DNA, and can dramatically boost specificity. Using RNase H to selectively elute the desired molecules appears to be general as it has been found to increase the specificity of other modular designs (Engreitz et al. 2014).

Additional elution strategies have also been successfully used. For example, capture oligonucleotides can be synthesized that contain desthiobiotin—a biotin analog that binds streptavidin more weakly than biotin. Chromatin complexes captured with desthiobiotin-labeled nucleotides can then be eluted using soluble biotin, an approach originally designed for a DNA hybridization capture protocol that produced good yields (Dejardin and Kingston 2009). This approach has been applied to RNA but led to higher levels of background than using RNase H (Simon et al. 2011).

3.4 Extension to Other Applications Including Global Proteomic Analyses

Although the discussion above has focused on using hybridization capture to study genomic binding sites, these protocols have been extended to other applications including proteomic analysis of protein targets. Some of these studies have been straightforward extensions of the protocols developed for mapping DNA targets (West et al. 2014) or fusion protocols that use combinations of the modules described above (Chu et al. 2015). Analysis of protein interactors also allows for alternative modules that are not compatible with genome-wide mapping. This is shown in specialized hybridization capture experiments designed to capture proteins bound to all mRNAs by using capture oligonucleotides complementary to poly(A) tails of mRNAs (Baltz et al. 2012; Castello et al. 2012). These studies used ultraviolet (UV) cross-linking to covalently cross-link proteins to the target RNAs. One advantage of UV cross-linking is that it captures direct interactions as opposed to the indirect interactions that can be captured with cross-linkers such as formaldehyde. Coupling a proteomic workflow to high levels of UV cross-linking combined with capture oligonucleotides specific for individual RNAs has been shown to be a powerful enrichment strategy for individual RNAs (McHugh et al. 2015). In general, both formaldehyde cross-linking and UV cross-linking modules have become widely used in hybridization capture experiments for proteomic analysis of copurifying proteins (for examples, see Baltz et al. 2012; Castello et al. 2012; Chu et al. 2012; West et al. 2014; McHugh et al. 2015; Minajigi et al. 2015).

The modules described above continue to be developed and optimized for various applications. For example, long-range RNA–RNA interactions were examined using a protocol with formaldehyde cross-linking, few DNA oligos, but formamide rinse buffers (Quinn et al. 2014). Alternatively, 4′-aminomethyltrioxalen—a psoralen derivative that is a UV-reactive reversible cross-linker that targets AU-rich regions of dsRNA—was developed as a cross-linker for similar purposes (Engreitz et al. 2014). These results show how the modules can be adapted to a wide range of experimental workflows beyond analysis of genomic binding sites.

4 APPLICATIONS OF HYBRIDIZATION CAPTURE TO STUDY lncRNAs INVOLVED IN DOSAGE COMPENSATION

The hybridization capture experiments generate data that reveal the binding patterns of lncRNAs that can be visualized in a genome browser just as ChIP-seq tracks are visualized (Fig. 3A,C). Xist and the roX2 RNAs have both been studied using more than one hybridization capture approach (Chu et al. 2011; Simon et al. 2011; Engreitz et al. 2013; Simon et al. 2013), providing an opportunity to examine similarities and differences between the technical approaches and the resulting biological observations.

The results from examination of roX1 and roX2 binding sites both revealed hundreds of peaks along the fly X chromosome (Fig. 3A) (Chu et al. 2011; Simon et al. 2011; Quinn et al. 2014). These roX1 and roX2 profiles agreed very well with the profile of proteins in the male specific lethal (MSL) complex (Alekseyenko et al. 2008), which is known to interact with the roX RNAs and affect dosage compensation by guiding the H4K16-acetyltransferase activity of the MOF subunit of this complex (Samata and Akhtar 2018). Motif analysis showed that roX2 enrichment occurred over a GA-rich motif that had been previously implicated in MSL targeting (Alekseyenko et al. 2008; Straub et al. 2008). These results are all consistent with roX2 binding to chromatin as a stoichiometric subunit of the MSL complex, which corresponds well with current models in the field (Samata and Akhtar 2018).

Although the dominant signals from roX2 hybridization capture experiments were relatively localized, this pattern is not general for all RNAs, as became clear on analysis of the spreading pattern of Xist lncRNA. Two studies examined the spreading patterns of Xist during X-chromosome inactivation in both differentiated cells and female mouse ESCs using distinct capture hybridization approaches (Engreitz et al. 2013; Simon et al. 2013). Despite differences in the protocols, the data from these two studies largely agree (Fig. 3C). The results revealed very broad spreading patterns of Xist over megabases of chromatin, yet Xist avoids the few genes that escape X-chromosome inactivation (Berletch et al. 2010) (e.g., Kdm5c in Fig. 3C). The two reports focused on different aspects of Xist biology. Engreitz and colleagues emphasized the similarities between the signals from Xist capture hybridization experiments with Hi-C signals along the X-chromosome, leading to the proposal that 3D contacts between the Xist endogenous locus and distal regions on the chromosome drive spreading (Engreitz et al. 2013). Simon and colleagues proposed a two-step spreading mechanism for Xist during X-chromosome inactivation based on the initial spreading patterns of Xist, in which Xist initially targets gene-rich regions on the future inactive X before spreading more broadly (Simon et al. 2013). This latter observation agrees well with recent results during mouse embryogenesis showing that genes in the early domains in which Xist initially targets are silenced first, following by spreading of repression to nearby genes (Borensztein et al. 2017). From a technical perspective, it is reassuring that the shared central observations of the Xist spreading patterns (correlation of Xist spreading with Hi-C signals and initial spreading of Xist into gene rich regions) were found in both studies using two different hybridization capture protocols.

5 SOURCES OF BACKGROUND AND CHALLENGES INTERPRETING DATA

Although hybridization capture experiments can capture real signal and reveal new biology, as with any recently developed family of techniques, it is important to understand how to distinguish real signals from artifacts in the data. This section focuses on the sources of background that can complicate the interpretation of data resulting from these techniques, and controls that can be used to help avoid being misled by the data.

5.1 Comparisons with ChIP-seq Data

Considerations such as sequencing depth and data processing have been well documented for ChIP-seq (Park 2009), and these considerations are also applicable to hybridization capture experiments. Although the majority of this discussion will be devoted to considerations specific to hybridization capture experiments, it is worth remarking on some of the limitations shared between ChIP and capture hybridization experiments. Quantitative interpretation of either method relies on the assumption that cross-linking efficiency and hybridization are consistent across the genome. Just as concerns about epitope masking and biased cross-linking can complicate analysis of ChIP data, these possibilities must also be considered when analyzing genome-wide hybridization capture data. Also, like ChIP, these experiments represent biochemical averages of localization data across the sample and thus cannot provide single-cell localization information. This type of information would require alternative techniques such as FISH.

5.2 Sources of Background

In most ChIP experiments, the majority of DNA arises from contaminating background signal rather than real signal. Sequencing tracks from hybridization capture experiments have background levels comparable to ChIP experiments (Fig. 3B). To illustrate this, we can examine roX2 signal. Assuming that this RNA acts exclusively on the X chromosome, we can use the enrichment of roX2 signal on the X chromosome as a proxy for true positive signal and enrichment of signal on autosomes as false positives. By this metric, roX2 hybridization capture experiments provide similar biochemical enrichment across different protocols and are comparable to high quality MSL3 ChIP signals (Fig. 3B).

Even though this similarity with ChIP is reassuring, unfortunately the profile of the background differs substantially between hybridization capture experiments and ChIP experiments. With ChIP, background signal is generally reasonably smooth across the genome. This makes it easy to distinguish even low levels of enriched signal relative to background. In contrast, experiments from all hybridization protocols lead to background that is uneven and peaky (see the autosomal signals from roX2 enrichment in Fig. 3A). There are many frustrations that result from this uneven background. One consequence is that it is very difficult to get a clear sense of enrichment by quantitative polymerase chain reaction (qPCR) experiments, as the background levels can be highly variable when using primers that happen to overlap with a background peak or trough. This uneven background is one of the reasons that this review focuses almost exclusively on global analyses such as sequencing—it is very hard to judge the levels and quality of enrichment obtained from pioneering hybridization capture experiments coupled only to qPCR rather than sequencing (Mariner et al. 2008). Now that sequencing experiments have been performed under a number of conditions with a range of RNAs, it is possible to explore and identify sources of this background (Fig. 4).

5.2.1 Nonspecific Binding to Beads

Nonspecific binding is inherent to all forms of biochemical enrichment. This background signal is generally even across the genome and thus the easiest to identify with appropriate

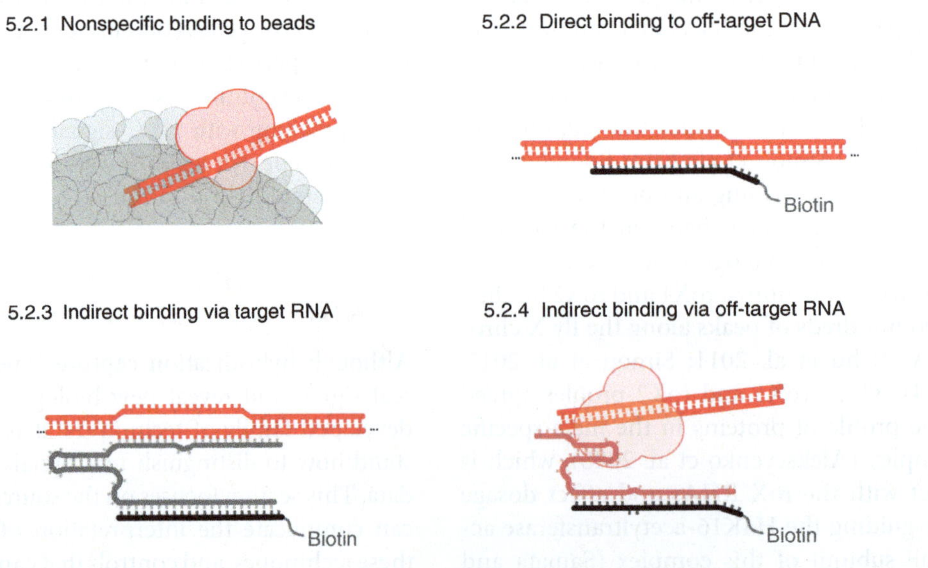

Figure 4. Examples of four sources of background that can arise in hybridization capture experiments.

experimental controls (see below). In practice, this signal is minimized by the denaturing rinses in all cross-linked hybridization capture experiments, and it appears to be lowest when using a mixture of rinses, with some that include high concentrations of guanidinium thiocyanate at increased temperatures (Engreitz et al. 2013).

5.2.2 *Direct Hybridization of Capture Oligonucleotides to DNA*

If the genomic DNA melts, capture oligonucleotides can directly hybridize to the genomic DNA and enrich sequences that are partially complementary to the capture oligonucleotides. Strategies to remove this background include (1) optimizing hybridization conditions in which these interactions are not thermodynamically favorable; (2) developing conditions that minimize dsDNA unwinding during hybridization, thereby slowing the kinetics of formation of these unwanted duplexes; (3) using sufficiently denaturing rinse conditions to remove DNA-capture oligonucleotide hybrids during the rinse steps; or (4) through selective elution by using DNA capture oligonucleotides and RNase H elution (because the background interactions are comprised of DNA–DNA duplexes, RNase H should not elute these interactions and only elute the RNA targets). These signals can also be diagnosed using extracts pretreated with RNase, as they are RNA-independent (Simon et al. 2011).

High degrees of cross-linking can covalently capture long range DNA–DNA interactions such as those observed in a Hi-C experiment. When hybridization capture experiments artifactually enrich DNA from a specific locus, all the DNA–DNA cross-links will also be enriched. Under such conditions, the signal is essentially a targeted Hi-C experiment that says little about the RNA of interest. This type of artifact is most likely when using particularly high levels of cross-linking and under highly denaturing hybridization conditions in which the genomic locus can be extensively denatured.

5.2.3 *Hybridization of DNA Indirectly via the Captured RNA*

In some situations, the target RNA is captured as desired, but the RNA also artifactually hybridizes to off-target DNA sequences during hybridization. Such signals are particularly problematic because they track with real signals from almost every control experiment (except allele-specific reagents, described in Sec. 5.3.5). For example, these hybridization-induced artifacts will disappear along with real signals in a cellular context lacking the target RNA. Like real signals, these signals will be consistently present even when using different capture oligonucleotide cocktails that target the same RNA. One key biochemical difference between the interactions that lead to real signals and these artifacts that can be leveraged to disentangle these results is that artifactual interactions are not stabilized by covalent cross-links. Therefore, with sufficiently denaturing conditions, it should be possible to remove them. As discussed above (in Sec. 2.1), this requires very tight-binding capture oligonucleotides, leading to a trade-off between specificity and affinity. To mitigate these problems, it is worth noting that these signals have some consistent features that can be identified in data: Specifically, they tend to be discrete signals over genomic sites that have some complementarity to the target RNA. The profile of bona fide peaks from roX2 or Xist enrichment, for example, tend to be broader (Fig. 3A, C). Extra care should be taken if motif analyses of peaks from hybridization capture data closely resemble sequences in the target RNA, as this increases the chances that these peaks are artifactual.

5.2.4 *Hybridization to Off-Target RNAs*

If capture oligonucleotides bind to other RNAs besides the target RNA, this can also lead to background. The transcriptome has a very wide dynamic range of concentrations, which means that even very low levels of binding to a highly abundant RNA (such as rRNA) can lead to background (which is even more problematic in cases of proteomic analyses). This type of background can be reduced by using well-curated capture oligonucleotides and can be controlled for by using different capture oligonucleotide sequences that target the same RNA.

5.3 Controls to Guide Analysis of Hybridization Capture Experiments

The cleaner and more dependable the biochemical purification associated with hybridization capture experiments, the easier the interpretation of the results. There is substantial reason to be attentive to potential artifactual signals. The following controls have been used to increase the reliability of interpretation.

5.3.1 *Using Capture Oligonucleotides That Do Not Target the RNA of Interest*

Many sources of background described above can be identified using control hybridization capture experiments using capture oligonucleotides that do not capture the RNA of interest (e.g., targeting LacZ) (Chu et al. 2011). Although any nontargeting oligonucleotide can control for nonspecific binding to the RNA, the sense oligonucleotide can also control for the GC content of the specific RNA and thus allow measurement of the extent of genomic DNA enrichment (at the locus of the target RNA in Simon et al. 2011). Importantly, this control does not address artifacts caused when the target RNA hybridizes to off-target sequences (see Sec. 5.2.3).

5.3.2 Using a Cell System That Does Not Contain the Target RNA

Using cells that lack the RNA of interest is a useful and intuitive control for off-target hybridization (see Secs. 5.2.1, 5.2.2, and 5.2.4). However, this does not control for any of the indirect signals retrieved via the target RNA (see Sec. 5.2.3).

5.3.3 Using Nonoverlapping Cocktails of Capture Oligonucleotides

One particularly powerful control established by Chu and colleagues is to perform the hybridization capture reactions using two independent sets of capture oligonucleotides (Chu et al. 2011). This control is similar to a technique used in FISH experiments (Dunagin et al. 2015), in which two different probe sets targeting the same RNA are used with different fluorophores. Off-target hybridization events are likely to appear in only a single channel, whereas real signal should appear in both. An analogous approach using different cocktails in a hybridization capture experiment can also help identify spurious signals that only occur with one set of capture oligonucleotides. Although this approach does not increase the biochemical enrichment, it does allow for a bioinformatic pipeline that can remove artifactual signals that are not present with both cocktails. The power of this approach is shown in Figure 3A in which the raw signals from roX2 ChIRP show many spurious signals on autosomes that do not agree between cocktails. By only considering signal present in both of the two tracks, this background can be removed bioinformatically. Note that the nonoverlapping signals tend to have the characteristics described above for those signals caused by direct or indirect hybridization to DNA. Also note that this approach addresses artifactual signals that are caused by off-target hybridization (see Sec. 5.2.4), but not those artifactual signals mediated by the target RNA (see Sec. 5.2.3, which is captured by both cocktails). Furthermore, it is not necessarily safe to assume that the background from different oligonucleotide cocktails is randomly distributed, meaning that chromatin structure (e.g., open chromatin at a transcription start site) or DNA sequence (e.g., high GC content) may be more likely to be artifactually captured by a range of oligonucleotide sequences. Nonetheless, as shown in Figure 3, this use of dual cocktails can substantially help separate real signal from contaminants that are retrieved during biochemical enrichment. This approach has been incorporated into other hybridization capture protocols (West et al. 2014; Engreitz et al. 2015).

5.3.4 Using RNase-Pretreated Extracts

One option to ensure that signals in hybridization capture experiments originate from RNA is to pretreat extracts with RNase before enrichment, with the expectation that all real signals from RNA capture will be removed (Simon et al. 2011). Although this control has provided some insight in past reports, complete removal of RNA can be challenging in cross-linked extracts, and so it can be challenging to interpret whether or not low levels of residual signals are artifactual or real signals resulting from incomplete removal of the RNA. Nonetheless, any real signal should at least be substantially reduced after RNase pretreatment of the extract. Unfortunately, aside from DNA binding, most other artifactual signals will also be removed.

5.3.5 Use of Allele-Specific Reagents

One of the central challenges discussed above is how to distinguish signals that arise from artifactual hybridization of the target RNA with DNA species. Most of the controls discussed above cannot separate real signal from these potential artifacts. In the case of hybridization capture experiment with Xist (Simon et al. 2013), allele-specific reagents provided a means to separate these different signals. Xist spreads in *cis* over the future inactive X chromosome. By using a hybrid cell line in which sequencing reads from the two different X chromosomes can be distinguished using SNPs and INDELs (Yildirim et al. 2011; Pinter et al. 2012), and including a mutation that genetically predisposes the same X chromosomes to always be inactivated, it was possible to identify allele-specific signals from the hybridization capture of Xist (Fig. 3C). This is a powerful system for biological interpretation of the different compositions of the two X chromosomes. It is also a powerful control because all artifactual hybridization that occurs after chromatin shearing should be agnostic as to the allele from which the nucleotide originated. Therefore, the active X acts as an internal control in the same cell for almost all artifacts, and it is possible to confidently interpret all allele-specific signals on the X chromosome as arising from bona fide Xist localization. Although this experiment shows the dependability of these specific hybridization capture conditions, unfortunately allele-specific biology and reagents are not frequently available.

5.3.6 *Controls for Proteomics*

Many of the controls above extend to analysis of proteins that bind RNAs when hybridization capture experiments are applied to proteomic pipelines. One particularly stringent experiment that was established for a proteomic analysis is the use of stable isotope labeling (SILAC) (Ong et al. 2002) to examine proteins that bind Xist after extensive UV cross-linking (McHugh et al. 2015). This control allows analysis of mixed samples, providing a way to control for artifactually enriched sequences. From these experiments, only 10 proteins were identified as enriched, which is much lower than the numbers from two other proteomic analyses of Xist-interacting proteins (Chu et al. 2015; Minajigi et al. 2015). It will be interesting to see whether these numbers differ because of different statistical criteria or whether this approach is better able to remove false positives caused by many of the same sources of background shown in Figure 4.

5.4 Agreement between Experiments

Although some of the most intriguing findings from hybridization capture experiments have yet to be validated or replicated by other methods, there are many corroborating findings that suggest that the hybridization capture experiments have the potential to mature into highly dependable techniques. First, the general principles of using capture oligonucleotides to enrich target RNAs from cross-linked chromatin extracts by using high ionic strength and high denaturants (Chu et al. 2011; Simon et al. 2011) agree well with what has been found in subsequent work. Second, in the few cases in which multiple techniques have been applied to study the same RNAs (e.g., roX2 and Xist discussed here), there is good agreement between the signals obtained across these methods. Importantly, in these cases the results also agree well with earlier FISH experiments (e.g., roX2) (Kelley et al. 1999). Finally, if an RNA acts together with proteins when bound to chromatin, the ChIP signal from those proteins should agree with the hybridization capture signal from the RNAs. This is exactly what was found for roX2 and the MSL complex (Chu et al. 2011; Simon et al. 2011).

One particularly interesting case of correspondence between ChIP signals and hybridization capture signals is the agreement between PRC2 and Xist localization during X-chromosome inactivation (Simon et al. 2013). It has been known for over a decade that H3K27me3, the product of PRC2, is enriched across the X chromosome and that Xist is genetically upstream of PRC2 activity (Plath et al. 2003). The simplest biochemical hypothesis, that PRC2 binds directly to Xist to spread across the X chromosome and promote H3K27me3, has been the subject of controversy (Davidovich et al. 2015). Examination of ChIP data on the autosomes shows that Ezh2 (the catalytic core of PRC2) binds across the genome mostly occupying discrete regions at transcription start sites, and occasionally spreading many kilobases along the chromatin. When examining PRC2 localization on the X chromosome during X-chromosome inactivation, however, PRC2 spreads broadly over megabases of chromatin in a pattern that corresponds very highly with the spreading of Xist (Fig. 3C) and in a manner that appears fundamentally different from spreading of PRC2 elsewhere in the genome (Simon et al. 2013). Although these results do not address whether the binding is direct, they clearly show that PRC2 spreads on the X chromosome in a pattern that matches Xist during X-chromosome inactivation. Whatever the more general roles for RNAs in regulating PRC2 localization (Davidovich and Cech 2015), Xist itself or another mechanism highly correlated with Xist localization leads to a distinctive spreading pattern of PRC2 that agrees almost perfectly between Xist and PRC2 on the inactivated X chromosome.

6 OUTLOOK AND FUTURE DIRECTIONS

Hybridization capture experiments continue to mature. This family of modular protocols creates opportunities to customize experiments, but the range of potential artifacts necessitates care in experimental design and appropriate controls. The dependability of these methods can be enhanced with more controls and sophisticated analyses to identify artifactual signals. Alternatively, higher levels of biochemical enrichment would also simplify data analysis. Because the integrity of the desired biological signal in these experiments is preserved through covalent cross-links, in principle it would be possible to minimize all the artifacts listed here through the use of capture reagents with higher affinity and specificity, allowing the removal of any noncovalently bound contaminants.

One exciting development has been the establishment of orthogonal genome-wide approaches to examine RNA localization, including experiments in which tagged RNAs are subjected to the equivalent of Dam-ID (Cheetham and Brand 2018) or through ligation-based methods that attach RNAs to genomic DNA using the equivalent of Hi-C (Li et al. 2017; Sridhar et al. 2017; Bell et al. 2018). These experiments promise to help discover RNA–chromatin interactions and to provide an independent means to validate surprising findings from hybridization capture experiments. Ligation-based approaches provide global analyses that are complementary to hybridization capture experiments, but are extremely unlikely to achieve the detailed binding profiles provided through the biochemical enrichment of hybridization capture methods. These new approaches are likely to help clarify if an RNA binds to a region of the genome, but observing spreading patterns and isolating copurifying factors will still require biochemical enrichment. Uncovering new RNA biology and mechanisms will always benefit from increasingly specific methods to biochemically enrich endogenous RNAs together with their targets.

REFERENCES

Alekseyenko AA, Peng S, Larschan E, Gorchakov AA, Lee OK, Kharchenko P, McGrath SD, Wang CI, Mardis ER, Park PJ, et al. 2008. A sequence motif within chromatin entry sites directs MSL establishment on the *Drosophila* X chromosome. *Cell* **134:** 599–609.

Appella DH. 2009. Non-natural nucleic acids for synthetic biology. *Curr Opin Chem Biol* **13:** 687–696.

Baltz AG, Munschauer M, Schwanhausser B, Vasile A, Murakawa Y, Schueler M, Youngs N, Penfold-Brown D, Drew K, Milek M, et al. 2012. The mRNA-bound proteome and its global occupancy profile on protein-coding transcripts. *Mol Cell* **46:** 674–690.

Barabino SM, Sproat BS, Ryder U, Blencowe BJ, Lamond AI. 1989. Mapping U2 snRNP—pre-mRNA interactions using biotinylated oligonucleotides made of 2′-OMe RNA. *EMBO J* **8:** 4171–4178.

Bell JC, Jukam D, Teran NA, Risca VI, Smith OK, Johnson WL, Skotheim JM, Greenleaf WJ, Straight AF. 2018. Chromatin-associated RNA sequencing (ChAR-seq) maps genome-wide RNA-to-DNA contacts. *eLife* **7:** e27024.

Berletch JB, Yang F, Disteche CM. 2010. Escape from X inactivation in mice and humans. *Genome Biol* **11:** 213.

Borensztein M, Syx L, Ancelin K, Diabangouaya P, Picard C, Liu T, Liang JB, Vassilev I, Galupa R, Servant N, et al. 2017. Xist-dependent imprinted X inactivation and the early developmental consequences of its failure. *Nat Struct Mol Biol* **24:** 226–233.

Castello A, Fischer B, Eichelbaum K, Horos R, Beckmann BM, Strein C, Davey NE, Humphreys DT, Preiss T, Steinmetz LM, et al. 2012. Insights into RNA biology from an atlas of mammalian mRNA-binding proteins. *Cell* **149:** 1393–1406.

Cheetham SW, Brand AH. 2018. RNA-DamID reveals cell-type-specific binding of roX RNAs at chromatin-entry sites. *Nat Struct Mol Biol* **25:** 109–114.

Chu C, Chang HY. 2016. Understanding RNA–chromatin interactions using chromatin isolation by RNA purification (ChIRP). *Methods Mol Biol* **1480:** 115–123.

Chu C, Qu K, Zhong FL, Artandi SE, Chang HY. 2011. Genomic maps of long noncoding RNA occupancy reveal principles of RNA–chromatin interactions. *Mol Cell* **44:** 667–678.

Chu C, Quinn J, Chang HY. 2012. Chromatin isolation by RNA purification (ChIRP). *J Vis Exp* **61:** 3912.

Chu C, Zhang QC, da Rocha ST, Flynn RA, Bharadwaj M, Calabrese JM, Magnuson T, Heard E, Chang HY. 2015. Systematic discovery of Xist RNA binding proteins. *Cell* **161:** 404–416.

Creamer KM, Lawrence JB. 2017. XIST RNA: A window into the broader role of RNA in nuclear chromosome architecture. *Philos Trans R Soc Lond B Biol Sci* **372:** 20160360.

Davidovich C, Cech TR. 2015. The recruitment of chromatin modifiers by long noncoding RNAs: Lessons from PRC2. *RNA* **21:** 2007–2022.

Davidovich C, Wang X, Cifuentes-Rojas C, Goodrich KJ, Gooding AR, Lee JT, Cech TR. 2015. Toward a consensus on the binding specificity and promiscuity of PRC2 for RNA. *Mol Cell* **57:** 552–558.

Dejardin J, Kingston RE. 2009. Purification of proteins associated with specific genomic Loci. *Cell* **136:** 175–186.

Demidov VV, Frank-Kamenetskii MD. 2004. Two sides of the coin: Affinity and specificity of nucleic acid interactions. *Trends Biochem Sci* **29:** 62–71.

Dunagin M, Cabili MN, Rinn J, Raj A. 2015. Visualization of lncRNA by single-molecule fluorescence in situ hybridization. *Methods Mol Biol* **1262:** 3–19.

Engreitz JM, Pandya-Jones A, McDonel P, Shishkin A, Sirokman K, Surka C, Kadri S, Xing J, Goren A, Lander ES, et al. 2013. The Xist lncRNA exploits three-dimensional genome architecture to spread across the X chromosome. *Science* **341:** 1237973.

Engreitz JM, Sirokman K, McDonel P, Shishkin AA, Surka C, Russell P, Grossman SR, Chow AY, Guttman M, Lander ES. 2014. RNA–RNA interactions enable specific targeting of noncoding RNAs to nascent pre-mRNAs and chromatin sites. *Cell* **159:** 188–199.

Engreitz J, Lander ES, Guttman M. 2015. RNA antisense purification (RAP) for mapping RNA interactions with chromatin. *Methods Mol Biol* **1262:** 183–197.

Fisher S, Barry A, Abreu J, Minie B, Nolan J, Delorey TM, Young G, Fennell TJ, Allen A, Ambrogio L, et al. 2011. A scalable, fully automated process for construction of sequence-ready human exome targeted capture libraries. *Genome Biol* **12:** R1.

Gilmour DS, Lis JT. 1985. In vivo interactions of RNA polymerase II with genes of *Drosophila melanogaster*. *Mol Cell Biol* **5:** 2009–2018.

Hadfield J, Retief J. 2018. A profusion of confusion in NGS methods naming. *Nat Methods* **15:** 7–8.

Hoffman EA, Frey BL, Smith LM, Auble DT. 2015. Formaldehyde crosslinking: A tool for the study of chromatin complexes. *J Biol Chem* **290:** 26404–26411.

Kelley RL, Meller VH, Gordadze PR, Roman G, Davis RL, Kuroda MI. 1999. Epigenetic spreading of the *Drosophila* dosage compensation complex from roX RNA genes into flanking chromatin. *Cell* **98:** 513–522.

Lamond AI, Barabino S, Blencowe BJ, Sproat B, Ryder U. 1990. Studying pre-mRNA splicing using antisense 2-OMe RNA oligonucleotides. *Mol Biol Rep* **14:** 201.

Lee N, Moss WN, Yario TA, Steitz JA. 2015. EBV noncoding RNA binds nascent RNA to drive host PAX5 to viral DNA. *Cell* **160:** 607–618.

Li X, Zhou B, Chen L, Gou LT, Li H, Fu XD. 2017. GRID-seq reveals the global RNA–chromatin interactome. *Nat Biotechnol* **35:** 940–950.

Lindberg U, Sundquist B. 1974. Isolation of messenger ribonucleoproteins from mammalian cells. *J Mol Biol* **86:** 451–468.

Lingner J, Cech TR. 1996. Purification of telomerase from *Euplotes aediculatus*: Requirement of a primer 3′ overhang. *Proc Natl Acad Sci* **93:** 10712–10717.

Machyna M, Simon MD. 2018. Catching RNAs on chromatin using hybridization capture methods. *Brief Funct Genomics* **17:** 96–103.

Mariner PD, Walters RD, Espinoza CA, Drullinger LF, Wagner SD, Kugel JF, Goodrich JA. 2008. Human Alu RNA is a modular transacting repressor of mRNA transcription during heat shock. *Mol Cell* **29:** 499–509.

McHugh CA, Chen CK, Chow A, Surka CF, Tran C, McDonel P, Pandya-Jones A, Blanco M, Burghard C, Moradian A, et al. 2015. The Xist lncRNA interacts directly with SHARP to silence transcription through HDAC3. *Nature* **521:** 232–236.

Minajigi A, Froberg J, Wei C, Sunwoo H, Kesner B, Colognori D, Lessing D, Payer B, Boukhali M, Haas W, et al. 2015. Chromosomes. A comprehensive Xist interactome reveals cohesin repulsion and an RNA-directed chromosome conformation. *Science* **349:** aab2276.

Ong SE, Blagoev B, Kratchmarova I, Kristensen DB, Steen H, Pandey A, Mann M. 2002. Stable isotope labeling by amino acids in cell culture, SILAC, as a simple and accurate approach to expression proteomics. *Mol Cell Proteomics* **1:** 376–386.

Orlando V. 2000. Mapping chromosomal proteins in vivo by formaldehyde-crosslinked-chromatin immunoprecipitation. *Trends Biochem Sci* **25:** 99–104.

Park PJ. 2009. ChIP-seq: Advantages and challenges of a maturing technology. *Nat Rev Genet* **10:** 669–680.

Cite as *Cold Spring Harb Perspect Biol* doi: 10.1101/cshperspect.a032276

Pauling L. 1974. Molecular basis of biological specificity. *Nature* **248:** 769–771.

Pinter SF, Sadreyev RI, Yildirim E, Jeon Y, Ohsumi TK, Borowsky M, Lee JT. 2012. Spreading of X chromosome inactivation via a hierarchy of defined Polycomb stations. *Genome Res* **22:** 1864–1876.

Plath K, Fang J, Mlynarczyk-Evans SK, Cao R, Worringer KA, Wang H, de la Cruz CC, Otte AP, Panning B, Zhang Y. 2003. Role of histone H3 lysine 27 methylation in X inactivation. *Science* **300:** 131–135.

Quinn JJ, Ilik IA, Qu K, Georgiev P, Chu C, Akhtar A, Chang HY. 2014. Revealing long noncoding RNA architecture and functions using domain-specific chromatin isolation by RNA purification. *Nat Biotechnol* **32:** 933–940.

Riggsby WS. 1969. A general method for fractionation of nucleic acids on the basis of sequence homology. *Biochemistry* **8:** 222–230.

Rogell B, Fischer B, Rettel M, Krijgsveld J, Castello A, Hentze MW. 2017. Specific RNP capture with antisense LNA/DNA mixmers. *RNA* **23:** 1290–1302.

Samata M, Akhtar A. 2018. Dosage compensation of the X chromosome: A complex epigenetic assignment involving chromatin regulators and long noncoding RNAs. *Annu Rev Biochem* **87:** 323–350.

SantaLucia J Jr, Hicks D. 2004. The thermodynamics of DNA structural motifs. *Annu Rev Biophys Biomol Struct* **33:** 415–440.

Schildkraut C, Lifson F. 1965. Dependence of the melting temperature of DNA on salt concentration. *Biopolymers* **3:** 195–208.

Sexton AN, Machyna M, Simon MD. 2016. Capture hybridization analysis of DNA targets. *Methods Mol Biol* **1480:** 87–97.

Simon MD, Wang CI, Kharchenko PV, West JA, Chapman BA, Alekseyenko AA, Borowsky ML, Kuroda MI, Kingston RE. 2011. The genomic binding sites of a noncoding RNA. *Proc Natl Acad Sci* **108:** 20497–20502.

Simon MD, Pinter SF, Fang R, Sarma K, Rutenberg-Schoenberg M, Bowman SK, Kesner BA, Maier VK, Kingston RE, Lee JT. 2013. High-resolution Xist binding maps reveal two-step spreading during X-chromosome inactivation. *Nature* **504:** 465–469.

Solomon MJ, Varshavsky A. 1985. Formaldehyde-mediated DNA–protein crosslinking: A probe for in vivo chromatin structures. *Proc Natl Acad Sci* **82:** 6470–6474.

Sridhar B, Rivas-Astroza M, Nguyen TC, Chen W, Yan Z, Cao X, Hebert L, Zhong S. 2017. Systematic mapping of RNA–chromatin interactions in vivo. *Curr Biol* **27:** 602–609.

Straub T, Grimaud C, Gilfillan GD, Mitterweger A, Becker PB. 2008. The chromosomal high-affinity binding sites for the *Drosophila* dosage compensation complex. *PLoS Genet* **4:** e1000302.

Wassarman DA, Steitz JA. 1991. Structural analyses of the 7SK ribonucleoprotein (RNP), the most abundant human small RNP of unknown function. *Mol Cell Biol* **11:** 3432–3445.

West JA, Davis CP, Sunwoo H, Simon MD, Sadreyev RI, Wang PI, Tolstorukov MY, Kingston RE. 2014. The long noncoding RNAs NEAT1 and MALAT1 bind active chromatin sites. *Mol Cell* **55:** 791–802.

Yildirim E, Sadreyev RI, Pinter SF, Lee JT. 2011. X-chromosome hyperactivation in mammals via nonlinear relationships between chromatin states and transcription. *Nat Struct Mol Biol* **19:** 56–61.

CRISPR Tools for Systematic Studies of RNA Regulation

Jesse Engreitz,[1,2,8] Omar Abudayyeh,[1,3,8] Jonathan Gootenberg,[1,4,8] and Feng Zhang[1,4,5,6,7]

[1]Broad Institute of MIT and Harvard, Cambridge, Massachusetts 02142
[2]Harvard Society of Fellows, Harvard University, Cambridge, Massachusetts 02139
[3]Department of Health Sciences and Technology, Massachusetts Institute of Technology, Cambridge, Massachusetts 02139
[4]Department of Systems Biology, Harvard Medical School, Boston, Massachusetts 02115
[5]McGovern Institute for Brain Research, Massachusetts Institute of Technology, Cambridge, Massachusetts 02139
[6]Department of Brain and Cognitive Science, Massachusetts Institute of Technology, Cambridge, Massachusetts 02139
[7]Department of Biological Engineering, Massachusetts Institute of Technology, Cambridge, Massachusetts 02139
Correspondence: engreitz@broadinstitute.org; zhang@broadinstitute.org

SUMMARY

RNA molecules perform diverse functions in mammalian cells, including transferring genetic information from DNA to protein and playing diverse regulatory roles through interactions with other cellular components. Here, we discuss how clustered regularly interspaced short palindromic repeat (CRISPR)-based technologies for directed perturbations of DNA and RNA are revealing new insights into RNA regulation. First, we review the fundamentals of CRISPR–Cas enzymes and functional genomics tools that leverage these systems. Second, we explore how these new perturbation technologies are transforming the study of regulation of and by RNA, focusing on the functions of DNA regulatory elements and long noncoding RNAs (lncRNAs). Third, we highlight an emerging class of RNA-targeting CRISPR–Cas enzymes that have the potential to catalyze studies of RNA biology by providing tools to directly perturb or measure RNA modifications and functions. Together, these tools enable systematic studies of RNA function and regulation in mammalian cells.

Outline

[8]These authors contributed equally to this work.

Cite as *Cold Spring Harb Perspect Biol* doi: 10.1101/cshperspect.a035386

1 INTRODUCTION

The transfer of genetic information from DNA to protein by RNA molecules is controlled by a number of regulatory processes (Fig. 1A). These include DNA sequence elements and DNA-binding proteins that control RNA production; RNA sequence elements and RNA-binding proteins that control RNA processing, degradation, localization, and translation; and small and long noncoding RNAs (lncRNAs) that guide and integrate the functions of protein effectors.

In the past decade, many advances in our understanding of RNA biology were facilitated by genome-scale observational studies, including the application of sequencing-based technologies to map the locations of RNA transcription, modifications, and RNA-binding proteins (Li et al. 2016; Hentze et al. 2018). The pace at which these observations were made far outstripped the pace at which they could be functionally characterized because there were no comparably powerful technologies for RNA perturbation. As a result, the precise functions and mechanisms of most of these observed sequences, interactions, and modifications remain unknown.

The recent development of DNA and RNA perturbation tools based on clustered regularly interspaced short palindromic repeat (CRISPR)-associated (Cas) enzymes has opened new opportunities for efficiently manipulating specific molecular features to test their functions (Fig. 1B)

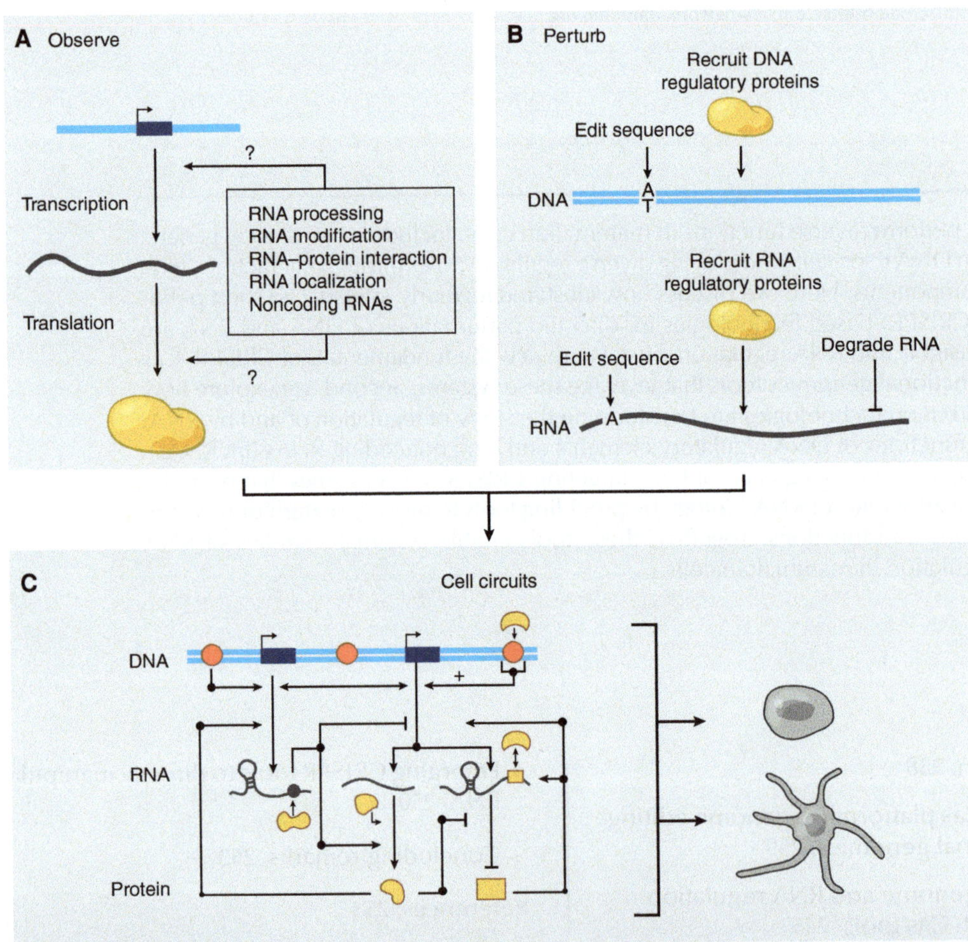

Figure 1. Systematic perturbations to understand RNA regulation. (*A*) The transfer of genetic information from DNA to RNA to protein is controlled by many molecular processes that regulate RNA. New sequencing technologies have allowed us to map some of these processes at genome-wide scale, but functional characterization has lagged. (*B*) Clustered regularly interspaced short palindromic repeat (CRISPR) perturbations at the level of DNA or RNA will help to test the functions of RNA regulatory mechanisms. (*C*) Combining technologies for observation and perturbation will allow us to build models of cell circuits and understand how different components (DNA and RNA regulatory elements, DNA- and RNA-binding proteins, and posttranscriptional modifications) are integrated to control cellular functions.

(Doudna and Charpentier 2014; Hsu et al. 2014). Just as the recombinant DNA revolution in the 1970s unleashed a wave of new biotechnologies and insights into disease biology by giving us the ability to clone and manipulate human genes in plasmids, CRISPR-based tools have ushered in a new era of studies to understand human genes and gene regulation via editing of DNA and RNA directly in mammalian cells. Here, we explore how this emerging suite of CRISPR tools will enable us to dissect the regulation of and by RNA and determine how these regulatory circuits influence cellular functions (Fig. 1C).

2 A CRISPR–Cas PLATFORM FOR GENOME EDITING AND FUNCTIONAL GENOMICS

Components from CRISPR–Cas systems can be engineered and optimized to enable programmable targeting of specific DNA or RNA sequences. Typically, a Cas protein effector (e.g., Cas9, Cas12a, or Cas13) is combined with a customized guide RNA to establish a ribonucleoprotein complex capable of targeting DNA or RNA in a cell type of interest (Gasiunas et al. 2012; Jinek et al. 2012; Cong et al. 2013; Mali et al. 2013b; Zetsche et al. 2015; Abudayyeh et al. 2016; Smargon et al. 2017). In contrast to earlier DNA-editing enzymes (Kim and Kim 2014)—such as meganucleases, zinc finger nucleases, and transcription activator–like effector nucleases—CRISPR–Cas systems are targeted by Watson–Crick base-pairing between the guide RNA and target DNA or RNA, allowing rapid and flexible reprogramming by adjusting the sequence of the guide RNA. By modifying the Cas protein or RNA guide, the system can be used to recruit other molecules to act on the target sequence in different ways. In combination with appropriate cellular or molecular assays, this flexibility enables systematic forward and reverse genetic studies in mammalian cells.

In this section, we review key aspects of CRISPR biology and ongoing technology development to expand this platform, including (1) exploring the natural diversity of CRISPR systems to identify a flexible suite of targeting enzymes, (2) using CRISPR–Cas proteins to edit DNA sequence in precise ways or perturb specific regulatory processes, and (3) combining these tools with pooled assays to connect genetic perturbations with specific cellular or molecular phenotypes.

2.1 Properties of CRISPR–Cas Systems

CRISPR and CRISPR-associated (CRISPR–Cas) proteins in bacteria and archaea constitute adaptive immune systems that capture fragments of genetic material from invading phages or mobile genetic elements and use these fragments to generate CRISPR RNAs (crRNAs) that guide the cleavage of matching viral sequences in future infections (Marraffini 2015). CRISPR–Cas systems display diverse sequence characteristics and architectural organization, providing a range of features that have utility in genome-editing applications (Fig. 2A). Here, we highlight the properties of a few of the most studied systems (Fig. 2B).

2.1.1 Cas9

The most characterized single-protein Cas effector (Class 2) is Cas9. Cas9 is a dual-guide RNA-dependent endonuclease that contains two nuclease domains, RuvC and HNH, that coordinate to cleave both strands of a double-stranded DNA (dsDNA) target complementary to the crRNA spacer sequence (Doudna and Charpentier 2014; Hsu et al. 2014). Cas9 can be engineered to facilitate genome editing in eukaryotic cells (Cong et al. 2013; Mali et al. 2013b). The dual-guide RNAs, crRNA and its *trans*-activating crRNA (tracrRNA) (Deltcheva et al. 2011), can be covalently linked to form a single-guide RNA (sgRNA) to simplify the RNA components (Jinek et al. 2012). Targeting is dependent on a protospacer adjacent motif (PAM) sequence (e.g., Cas9 from *Streptococcus pyogenes* requires a 5′-NGG-3′ PAM at the 3′ end of the target sequence), restricting editing to genomic sites that contain the appropriate PAM sequence.

2.1.2 Cas12

CRISPR–Cas12a systems (originally called Cpf1) also target DNA using a single-protein endonuclease and can be engineered to facilitate genome editing in eukaryotic cells (Zetsche et al. 2015). Cas12a cleaves dsDNA using two domains, the RuvC and nuclease domains, although the RuvC domain only contains the catalytic residues necessary to cleave both strands and generate a dsDNA break (Zetsche et al. 2015; Yamano et al. 2016; Swarts et al. 2017). In contrast to Cas9, Cas12a does not require a tracrRNA (Zetsche et al. 2015) and possesses its own RNase domain capable of cleaving a pre-crRNA array into individual mature crRNAs for convenient delivery of multiple guide RNAs (Fonfara et al. 2016; Zetsche et al. 2017). Cas12a-mediated dsDNA cleavage results in 5′ staggered ends (Zetsche et al. 2015), and the PAM sequence and location differ from those used by Cas9, with most Cas12a orthologs requiring a 5′-TTTV-3′ PAM at the 5′ end of the target site and cleaving distal to the PAM (Zetsche et al. 2015).

2.1.3 Cas13

The recent discovery of single-effector CRISPR systems that target RNA (Shmakov et al. 2015, 2017) has further expanded our conception of the diversity of these bacterial

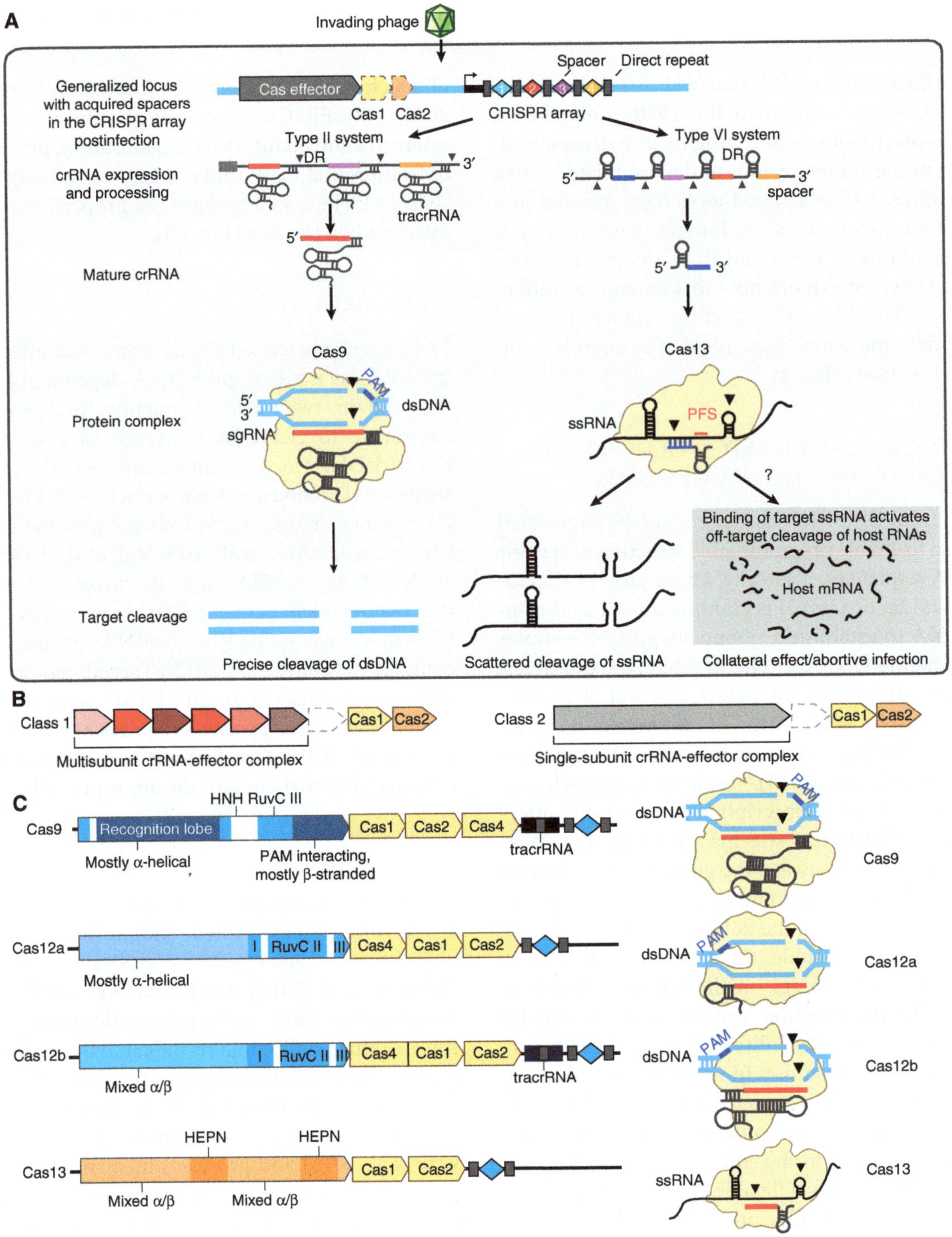

Figure 2. Structure and function of CRISPR systems. (*A*) Depiction of the three steps of CRISPR immunity (adaptation, expression, and interference) by DNA-targeting (e.g., Cas9-containing type II) and RNA-targeting (e.g., Cas13-containing type VI) systems. Cas9 interference is mediated by RNA-guided cleavage of double-stranded DNA (dsDNA) targets. Cas13 interference results in two modes of activation in which RNA-guided targeting of single-stranded RNAs (ssRNAs) results in targeted cleavage of the transcripts as well as collateral cleavage of noncomplementary RNAs that are in proximity to the activated Cas13 complex. The collateral cleavage of Cas13 systems is thought to be part of a programmed cell death pathway to further augment immunity against pathogen infection. (*B*) CRISPR systems can be divided into two classes: Class 1 systems consist of multiprotein interference complexes (types I, III, and IV systems), whereas Class 2 systems contain a single-protein interference effector (types II, V, and VI systems). (*C*) Genetic loci for four Class 2 CRISPR systems (*left*), including sequences coding for effector proteins annotated with key protein domains, accessory Cas proteins involved in adaptation colored in beige, *trans*-activating CRISPR RNAs (tracrRNAs) (if required), and CRISPR array. The protein complexes (*right*) show the nucleic acid target (dsDNA or ssRNA), locations of the protospacer adjacent motif (PAM) sequence and cleavage sites, and approximate structure of the guide RNA.

enzymes. Members of the Cas13 family contain two domains with homology to higher eukaryotes and prokaryotes nucleotide-binding (HEPN) RNase domains (Anantharaman et al. 2013). Cas13a, like DNA-targeting CRISPR systems, is guided to target RNA sequences by a complementary crRNA, and hybridization between the target RNA and its complementary crRNA leads to cleavage of the target RNA at multiple sites within single-stranded regions (Abudayyeh et al. 2016). Continued protein discovery has since uncovered many Cas13 subfamilies (Cas13a, Cas13b, Cas13c, and Cas13d) (Shmakov et al. 2015, 2017; Abudayyeh et al. 2016; East-Seletsky et al. 2016, 2017; Smargon et al. 2017; Konermann et al. 2018; Yan et al. 2018). In bacteria, both Cas13a and Cas13b have a sequence constraint known as the protospacer flanking site (PFS) (Abudayyeh et al. 2016; Smargon et al. 2017). In bacteria, Cas13a from *Leptotrichia shahii* has a 3′H (not G) requirement (Abudayyeh et al. 2016); in mammalian cells it appears that the PFS does not affect targeting (Abudayyeh et al. 2017; Cox et al. 2017). An additional unique feature of these enzymes observed in vitro is that, upon target RNA binding, the activated Cas13 complex is capable of cleaving both the targeted transcript and other nearby noncomplementary RNAs via a "collateral" activity (Abudayyeh et al. 2016), which has been harnessed for nucleic acid detection applications (East-Seletsky et al. 2016, 2017; Gootenberg et al. 2017, 2018).

These examples illustrate the value of continued exploration of the diversity of CRISPR systems for the development of molecular tools. Many CRISPR systems remain unexplored and may harbor novel combinations of sequence flexibility, endonuclease capabilities, or other unexpected functions to facilitate genome-engineering applications.

2.2 CRISPR-Mediated Recruitment of Proteins and RNAs to Specific Genomic Sites

CRISPR–Cas tools have transformed studies of molecular biology and genetics by enabling direct, easily reprogrammable genome editing in mammalian cells via endogenous repair of targeted double strand breaks (Doudna and Charpentier 2014; Hsu et al. 2014; Komor et al. 2017). Edits can include creation of small indels (using one guide RNA), precise gene editing (with one guide RNA and a repair template), or larger deletions (using two guide RNAs) (Fig. 3A). Beyond cleaving genomic sequences, CRISPR effectors can be used to direct protein or RNA cargo to specific locations in the genome. In such applications, functional proteins or RNAs are adjoined either to catalytically dead Cas9 (dCas9) (Gilbert et al. 2013; Mali et al. 2013a; Qi et al. 2013), in which the active sites of the nuclease domains have been mutated, or to the guide RNA via a number of different recruitment strategies (Fig. 3B). Emerging tools in this category include a variety of base editors, which precisely convert one nucleotide to another, and "epigenome editors," which recruit one or more proteins to alter gene regulatory processes.

2.2.1 *Base Editing to Modify Specific Nucleotides*

CRISPR-mediated base editing is performed by fusing certain enzymes—such as APOBEC, AID, or ADAT engineered to target DNA—to a Cas9 protein in which one or both catalytic domains have been genetically inactivated (Fig. 3A) (Komor et al. 2016; Nishida et al. 2016; Gaudelli et al. 2017). These fusions currently enable specific and programmable conversions of C:G to T:A base pairs or vice versa, enabling relatively efficient modification of just over half of all known pathogenic single-nucleotide variants annotated in ClinVar (Gaudelli et al. 2017), although the requirement for a PAM sequence nearby reduces this number. An advantage of these approaches is that they can edit DNA without introducing double-stranded DNA breaks, thereby avoiding certain error-prone repair mechanisms. For example, the APOBEC C-to-T base editor first enzymatically converts a target cytosine to uracil, and then creates a single-strand DNA break to engage the mismatch repair (MMR) machinery to repair the opposing guanine to adenine (Komor et al. 2016).

2.2.2 *CRISPR Interference (CRISPRi) for Gene Repression*

By itself, dCas9 delivery to gene promoters can physically block RNA polymerase or transcription factors from accessing DNA, leading to modest (20%–40%) repression of endogenous gene expression (Gilbert et al. 2013). Repressive activity can be dramatically improved by fusing dCas9 to repressive chromatin regulators (Gilbert et al. 2013). For example, fusion to the KRAB domain leads to deposition of repressive histone modifications (including H3K9me3) and loss of activating modifications (including H3K27ac) at gene promoters, leading to 50%–90% repression of gene expression when delivered in a critical window 0–200 bp downstream from a transcription start site (Fig. 3C) (Gilbert et al. 2013, 2014). Other repressive domains—including SID, LSD1, and DNMT3A proteins or protein domains (Konermann et al. 2013; Kearns et al. 2015; Bintu et al. 2016; Liu et al. 2016; Vojta et al. 2016)—have been explored and may have different kinetic properties for fine-tuning gene silencing.

2.2.3 *CRISPR Activation (CRISPRa)*

CRISPR–Cas9 tools for activation use Cas9 to recruit multiple transcriptional activators to achieve robust induction

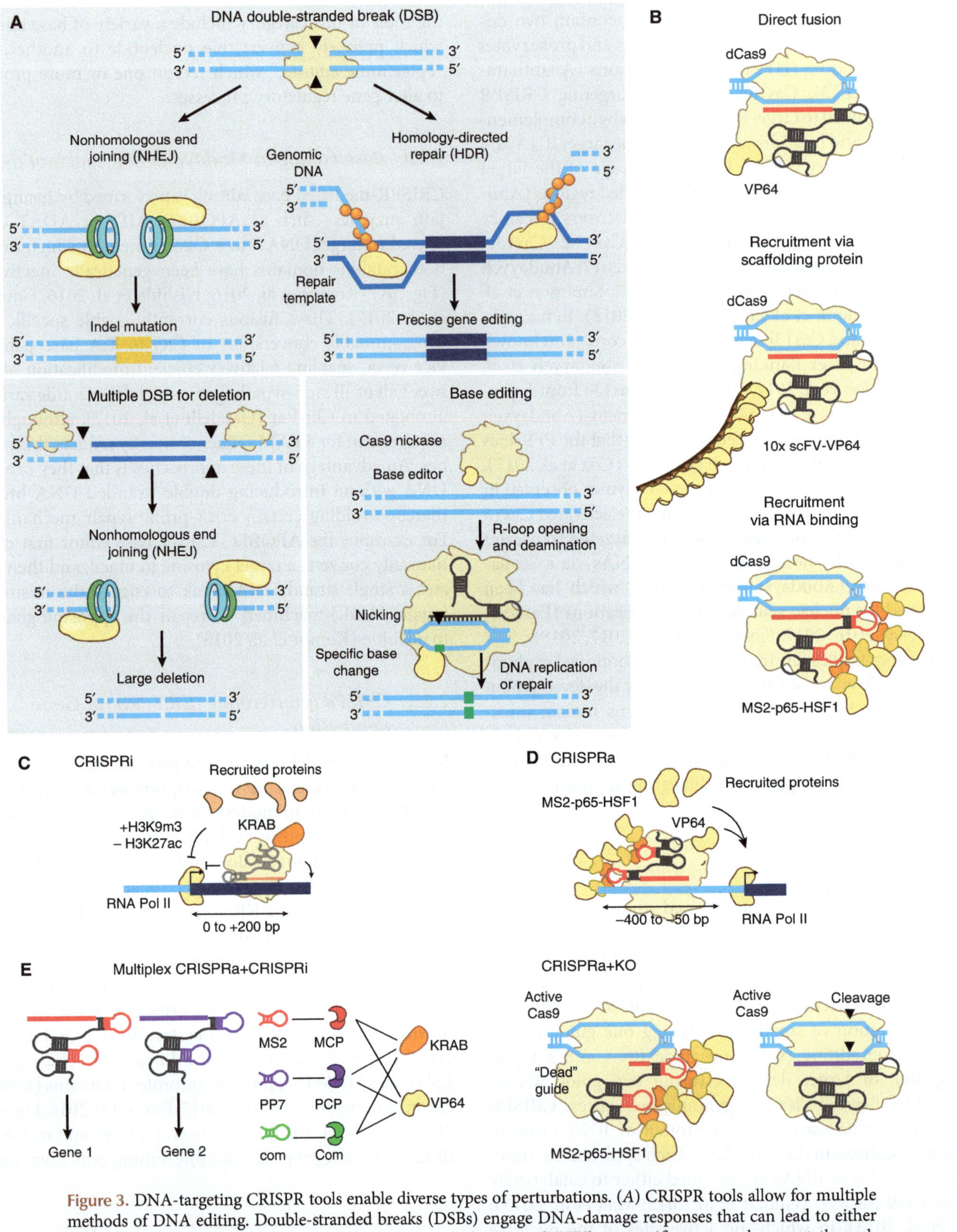

Figure 3. DNA-targeting CRISPR tools enable diverse types of perturbations. (*A*) CRISPR tools allow for multiple methods of DNA editing. Double-stranded breaks (DSBs) engage DNA-damage responses that can lead to either indels and gene knockout through nonhomologous end joining (NHEJ) or specific gene replacement through homology-directed repair (HDR). (Adapted from Hsu et al. 2014.) Two targeted cuts can also lead to removal of large stretches of DNA. (*Legend continues on following page.*)

of gene expression at promoters (Bikard et al. 2013; Cheng et al. 2013; Farzadfard et al. 2013; Gilbert et al. 2013, 2014; Maeder et al. 2013; Mali et al. 2013a; Perez-Pinera et al. 2013; Chavez et al. 2015; Konermann et al. 2015). Konermann and colleagues developed an approach called "synergistic activation mediator" (SAM), in which MS2 stem loops are fused to modified Cas9 guide RNAs to recruit multiple activating proteins to the same site (Fig. 3D) (Konermann et al. 2015). Recruiting multiple activating proteins (VP64, HSF1, and p65) led to synergistic effects on gene expression compared with recruiting each protein individually (Konermann et al. 2015). In an alternative approach ("SunTag"), Tanenbaum and colleagues fused dCas9 to a repetitive GCN4 peptide-domain scaffold that is recognized by a GCN4-targeting nanobody fused to VP64, enabling simultaneous recruitment of up to 10 copies of VP64 (Tanenbaum et al. 2014). Each of these approaches appears to vary in efficiency across different genes in a way that is not yet entirely predictable, and the precise location of targeting (ideally, 50–400 bp upstream of the transcription start site) has an important effect on the efficiency of activation (Gilbert et al. 2014; Konermann et al. 2015).

Building on these basic systems for CRISPR–Cas9 recruitment, new tools have now been developed to manipulate the localization of lncRNA molecules (CRISP-Disp) (Shechner et al. 2015), modify chromosome conformation (by linking multiple Cas9 molecules to dimerization domains) (Morgan et al. 2017), and visualize specific DNA loci in living cells (via Cas9 fusions to green fluorescent protein [GFP]) (Chen et al. 2013; Ma et al. 2016; Gu et al. 2018; Maass et al. 2018). Simultaneous activation and inhibition can be achieved at different targets in the same cell by, for example, using guide RNAs containing different RNA aptamers to recruit different protein effectors (Zalatan et al. 2015) or using variable guide spacer lengths to either enable or suppress catalytic Cas9 activity at different targets (Fig. 3E) (Dahlman et al. 2015; Kiani et al. 2015). Although the repertoire of CRISPR perturbations is expanding, much work remains to understand the specific regulatory processes perturbed by each approach and to determine how these tools operate in different genomic contexts.

2.3 CRISPR-Based Forward Genetics with Pooled Cellular Assays

Paired with various CRISPR effectors, pooled cellular assays are a key part of the CRISPR functional genomics toolkit, enabling experiments that assay the effects of up to millions of perturbations simultaneously. Although pooled experiments were previously applied in combination with RNA interference (RNAi) (Brummelkamp et al. 2006; Ngo et al. 2006; Iorns et al. 2007), the diversity and flexibility of CRISPR systems have led to a new period of rapid innovation in the development of pooled assays for forward genetics (Shalem et al. 2014; Wang et al. 2014). In such assays, libraries of CRISPR sgRNAs are typically delivered by viral transduction such that individual cells in a population receive only one sgRNA (Fig. 4A). Sorting, selecting, or phenotyping single cells then allows associating specific genotypes (specified by a sgRNA perturbation) with phenotypes at high throughput (Fig. 4B) (Joung et al. 2017b).

2.3.1 *Cell Sorting and Guide Counting*

Simple pooled assays involve selecting cells based on a single phenotype and counting the distribution of sgRNAs before and after the selection process using high-throughput DNA sequencing. For example, cell proliferation can serve as a selection parameter to identify genes that are essential for cell growth in culture, in vivo, in the presence of drugs, or in response to a viral infection (Shalem et al. 2014; Wang et al. 2014), and can be combined with genome-wide gene knockout or activation screens in both positive and negative selection screens (Doench et al. 2014; Gilbert et al. 2014; Koike-Yusa et al. 2014; Zhou et al. 2014; Wang et al. 2015). Cells can also be selected based on fluorescent sorting read-

Figure 3. (*Continued*) Catalytically inactive or nicking mutants of Cas9 can be used to recruit nucleoside deaminases or other base-editing enzymes for precise, single-nucleotide edits. As an example, this figure shows a C:G to T:A editor comprised of Cas9 nickase fused to cytidine deaminase. (*B*) Cas proteins can recruit other domains or effector proteins through multiple designs, including direct fusion to the Cas protein, indirect recruitment via a scaffolding or conditional dimerization domain fused to the Cas protein, and/or recruitment using a guide RNA modified with a sequence motif recognized by an exogenous RNA-binding protein. (*C*) In CRISPR interference (CRISPRi), recruitment of repressive chromatin regulators, such as Krüppel-associated box (KRAB), result in assembly of repressive complexes and silencing of active regulatory elements including promoters and enhancers. (*D*) In CRISPR activation (CRISPRa), recruitment of activating chromatin regulators to promoters or other *cis*-acting elements can promote gene expression. Recruitment of multiple different transactivators, such as combinations of VP64, p65, and HSF1, can lead to increased gene expression. (*E*) These tools can be multiplexed for simultaneous and differential perturbation of genes. Activation and inhibition can be targeted to different genes by guide RNAs containing orthogonal RNA binding domains and associated regulatory proteins, or Cas9 cleavage activity can be tuned on a per-guide basis by adjusting the length of the spacer on the guide RNA.

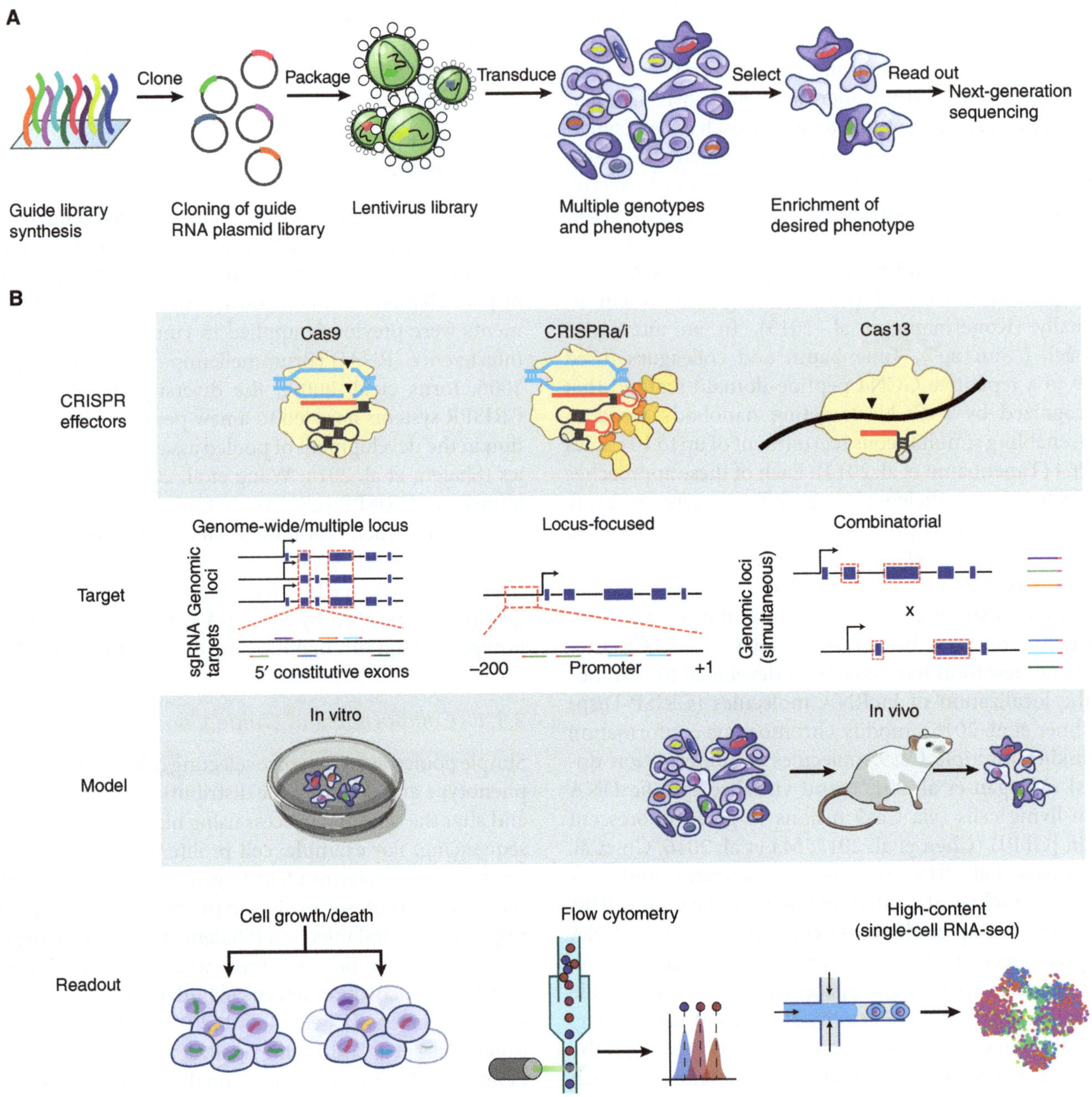

Figure 4. Forward genetics with pooled screening and CRISPR tools. (*A*) General workflow for pooled screening with CRISPR-based perturbations. A library of single-guide RNAs (sgRNAs) is synthesized, packaged into a viral vector, and delivered at a ratio of ~1 sgRNA per cell. This pool of cells, each with a different genotype, is selected or sorted according to a phenotype of interest. Sequencing of the sgRNA in each cell is used to map the genotypes to the resulting phenotypes. (*B*) A pooled screen can be constructed from multiple different perturbations, distributions of targeting, model contexts, and phenotyping methods. Screens are amenable to many of the CRISPR-based tools, such as knockout, CRISPRa/i, base editing, or RNA-targeting tools. The targeting of guides in the screen can be distributed across the genome or transcriptome to test many different genes or focused at a specific locus to dissect regulatory elements. Multiple perturbations can also be delivered simultaneously to test epistatic mechanisms. Screens can be performed in vitro or in animal models for in vivo phenotypes. Phenotyping approaches can include enrichment or depletion caused by cell death or changes in proliferation, fluorescence-based sorting to measure gene expression or protein localization, or high-content methods such as single-cell RNA sequencing.

Cite as *Cold Spring Harb Perspect Biol* doi: 10.1101/cshperspect.a035386

outs or other properties, enabling forward genetic screens that examine the expression, localization or stability of specific genes (Canver et al. 2015; Parnas et al. 2015; Fulco et al. 2016; Sanjana et al. 2016; Wu et al. 2016; Stewart et al. 2017; Doench 2018).

2.3.2 High-Content Phenotyping

Recent approaches leverage new single-cell technologies to simultaneously associate CRISPR sgRNAs with multiple phenotypes. In Perturb-seq, single-cell RNA sequencing is used to assay the transcriptional state of the cell as well as determine which sgRNA is expressed in that cell, allowing studies to perturb multiple genes and characterize their effects on cells in an unbiased way that does not require designing specific assays to read out specific phenotypes (Adamson et al. 2016; Dixit et al. 2016; Jaitin et al. 2016; Datlinger et al. 2017). Similar approaches may be developed to combine CRISPR screens with other high-content readouts such as cellular morphology profiling (Caicedo et al. 2016) or single-cell measurements of protein abundance (Peterson et al. 2017; Stoeckius et al. 2017).

The development of appropriate pooled assays can therefore leverage the power of CRISPR-based forward genetics to systematically connect genes, noncoding elements, and RNAs with specific phenotypes.

Together, these technologies—including targetable CRISPR systems, methods to modify DNA sequence or localize regulatory proteins, and pooled screening assays for forward genetics—form a flexible platform for perturbation-based studies in mammalian cells. In the next section, we explore how these approaches can be applied to systematically explore regulation of and by RNA.

3 DISSECTING GENOME AND RNA REGULATION WITH CRISPR–Cas TOOLS

To illustrate how CRISPR-based approaches can illuminate mechanisms of gene regulation, we examine recent studies of the functions of two classes of regulatory elements in mammalian genomes: (1) noncoding DNA regulatory elements and (2) lncRNAs. These examples show how systematic CRISPR perturbations can establish causal relationships between regulatory elements and effects on gene expression and identify the molecular features that specify their context-specific functions in different cell types and at different genomic loci.

3.1 Functions of DNA Elements and Proteins That Regulate Transcription

Mammalian genomes are estimated to encode millions of distal noncoding DNA elements that quantitatively control the expression of specific genes in specific cell types (Roadmap Epigenomics Consortium 2015). Major consortia efforts have generated genome-wide maps of histone modifications, transcription factor binding, and DNA accessibility to identify and annotate noncoding elements that appear to have the characteristics of enhancers, promoters, and topological boundary elements (Fig. 5A) (ENCODE Project Consortium 2012; Roadmap Epigenomics Consortium 2015). However, the ability of these annotations to predict the functional role of any given element in regulating gene expression is unclear. For many years, perturbational studies of these elements were limited to knockouts of individual regulatory elements, providing only a glimpse of the complex networks responsible for regulating gene expression in different cell types. The programmable nature of CRISPR tools has opened new avenues to study DNA regulatory elements in an unbiased manner, including methods to perturb three facets of these elements: DNA sequence, recruitment of DNA-binding proteins, and 3D contacts with target genes.

CRISPR-based technologies are now enabling forward genetics screening approaches to define which noncoding sequences in the genome functionally regulate gene expression (Fig. 5A). An important technical consideration is that noncoding elements appear to consist of short functional sequences, such as transcription factor binding sites, interspersed with nonfunctional ones (Bulger and Groudine 2011; Spitz and Furlong 2012), making them relatively robust to indels in some locations. As such, two categories of recent approaches explore the effect of perturbations at two levels of resolution: sequence motifs (1–10 bp) and full regulatory elements (100–1000 bp).

In the first category (1–10 bp resolution), researchers have designed pooled experiments in which each cell in the pool expresses catalytically active Cas9 as well as an sgRNA, with different sgRNAs in the population tiling densely across noncoding regions around a gene of interest. Cas9 cuts at the target sequences, typically resulting in short deletions or insertions (indels). When coupled to phenotypic assays for the expression of that gene (for example, fusing the gene to GFP and sorting cells based on fluorescence), these experiments can identify short sequences (1–10 bp) important for transcriptional control. In one of the first such studies, Canver and colleagues designed 1338 sgRNAs tiled across a 12-kb enhancer known to regulate *BCL11A* and identified multiple sgRNAs that disrupted conserved transcription factor motifs in one of these elements (Canver et al. 2015). These approaches have now been scaled up to include thousands of sgRNAs in a single experiment (Diao et al. 2016; Korkmaz et al. 2016; Rajagopal et al. 2016; Sanjana et al. 2016).

In the second category (100–1000 bp resolution), larger-scale perturbations can be achieved by (1) using epigenome

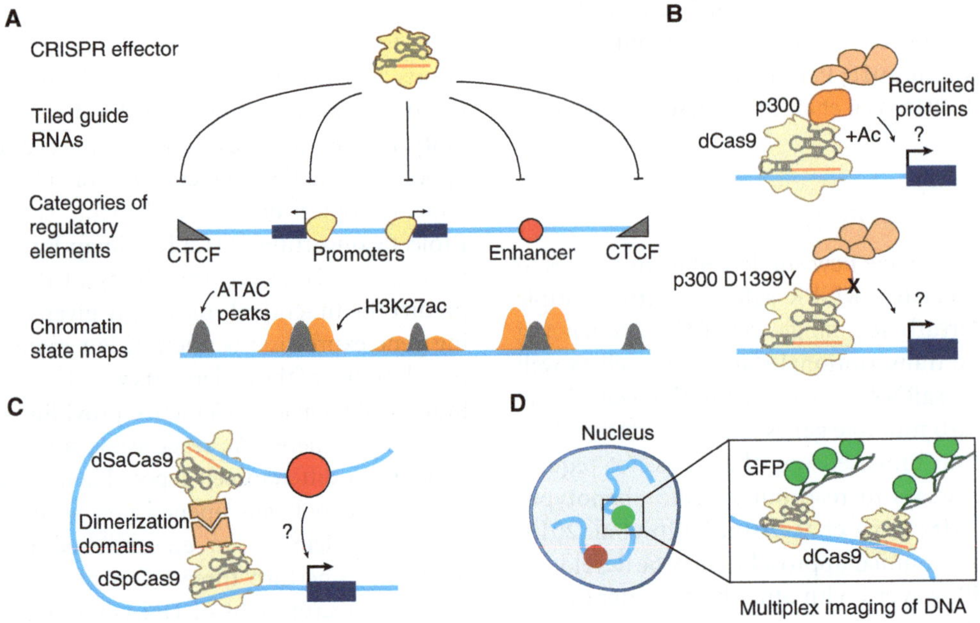

Figure 5. CRISPR tools to study transcriptional regulation by noncoding DNA elements. (*A*) High-throughput CRISPR screens of noncoding regulatory elements enable testing the functions of various categories of regulatory elements. In such screens, a CRISPR perturbation method (including deletions caused by one or two single-guide RNAs [sgRNAs], CRISPRi, or CRISPRa) are combined with forward genetic screening approaches to enable high-throughput screens targeting hundreds of thousands of sequences in parallel. Different categories of elements are marked by different chromatin features: CTCF sites are marked by chromatin accessibility (as measured in ATAC-seq [gray peaks] or DNase-seq experiments) and CTCF binding. Gene promoters and enhancers are marked by chromatin accessibility and H3K27ac occupancy (orange peaks). (*B*) Recruitment strategies can be used to study the functions of chromatin regulatory proteins that read or write chromatin modifications. For example, studies of a protein domain of the transcriptional activator p300 (an acetyltransferase that also interacts with many other regulatory proteins) have shown that its catalytic activity is required for gene activation at several loci. (*C*) CRISPR can be used to engineer 3D chromatin contacts, for example, by fusing dimerization domains to dead SpCas9 and SaCas9 and recruiting each to nearby sites. (*D*) dCas9 can be used for programmable and multiplexed imaging of DNA loci in living cells (e.g., by using multiple sgRNAs to target dCas9 and many copies of GFP to a DNA locus).

editors such as KRAB-dCas9, which has been shown to silence regulatory elements within a range of ~100 bp (Gilbert et al. 2014; Thakore et al. 2015; Fulco et al. 2016; Klann et al. 2017; Simeonov et al. 2017) or (2) engineering larger deletions by codelivering two sgRNAs per cell (Zhu et al. 2016; Diao et al. 2017; Gasperini et al. 2017). For example, Fulco and colleagues recently used CRISPRi (KRAB-dCas9) to tile 98,000 sgRNAs across 1.2 Mb and identified nine and two elements that quantitatively tuned *MYC* and *GATA1* expression, respectively (Fulco et al. 2016). Diao and colleagues used paired sgRNAs to tile ~2-kb deletions across a 2-Mb region and reported 45 distal elements that regulate *POU5F1* expression in human embryonic stem cells (Diao et al. 2017). Preliminary comparisons of these studies suggest that KRAB-dCas9 and 2-sgRNA deletions enable robust detection of *cis*-regulatory elements at a resolution of 100–1000 bp, whereas single-sgRNA deletion approaches enable finer resolution studies of regulatory sequences but are less robust in theory because of the small number of sgRNAs (often zero or one) interrogating each functional motif (Sanjana et al. 2016; Canver et al. 2017). Further work remains to understand the precise mechanisms of each perturbation method and to determine whether the screening approaches are adequately powered to detect repressive or activating elements with small effect sizes.

Technical considerations notwithstanding, CRISPR tiling studies are beginning to provide a sufficient number of noncoding perturbations to explore which elements act as enhancers, defined here as activating noncoding elements located >1 kb from any promoter region. (Notably, CRISPR tiling studies have also identified important roles for promoters in regulating other nearby genes, see Sec. 2.2.) Most of the activating elements identified in CRISPR-based screens, with a few exceptions, are marked by a chromatin signature previously thought to identify functional enhancers, including DNase accessibility and acetylation at histone 3 lysine 27 (H3K27ac) (Fulco et al. 2016; Rajagopal et al. 2016; Sanjana et al. 2016; Diao et al. 2017). However, only a small proportion of the DNase accessible elements detectably affect gene expression, suggesting that accessibility alone is not sufficient to identify a functional enhancer. Interestingly, the elements that do affect gene expression tend to have stronger quantitative accessibility and H3K27ac signals (Fulco et al. 2016), suggesting that the quantitative signals in these assays may be predictive of functional enhancers. Analysis of CRISPR tiling screens has also shown that enhancer elements frequently contact their target genes in 3D, as measured by chromosome conformation capture (Fulco et al. 2016; Sanjana et al. 2016), consistent with a model in which enhancers must physically contact promoters to activate transcription (Bulger and Groudine 2011; Spitz and Furlong 2012). Interestingly, in one study, a simple model incorporating a measure of enhancer activity (estimated from quantitative accessibility and H3K27ac signals) and a measure of contact frequency (estimated from Hi-C) was able to reasonably predict the quantitative effects of 93 noncoding elements on the expression of one gene (*MYC*) (Fulco et al. 2016). A later study integrated these measures in a combined H3K27ac ChIP-seq and Hi-C assay (HiChIP) and validated that this measurement accurately predicted the functions of several enhancers (Mumbach et al. 2017). Although further work will be required to test these conclusions across many genes, CRISPR perturbation-based studies have begun to reveal the regulatory architecture of enhancer-gene networks and determine how measurements of chromatin state and contact frequency predict the functional effects of noncoding elements on gene expression.

Additional CRISPR tools provide new ways to independently manipulate and test how chromatin state and 3D chromosome contacts influence enhancer function. For example, enhancer activity is thought to be determined by specific sequence motifs that recruit transcription factors and co-activators, including histone-modifying enzymes. Cas enzymes can be fused to these chromatin regulatory proteins to directly test how they affect gene expression in different genomic contexts (Fig. 5B). For example, Hilton and colleagues fused the catalytic domain of the acetyltransferase p300 to dCas9 and, by comparing the wild-type p300 domain to a mutated domain, showed that the acetyltransferase activity of p300 is required for gene activation at several loci (Hilton et al. 2015). In contrast, CRISPR-mediated recruitment of catalytically dead EZH2 (the enzymatic component of polycomb repressive complex 2) to the promoter of an active gene was unable to deposit H3K27me3 (a histone modification associated with gene silencing) but nonetheless was able to efficiently repress gene expression (O'Geen et al. 2017), indicating that dCas9-EZH2 fusions can modulate gene repression independent of the enzymatic activity of EZH2 and that H3K27me3 is not required for silencing at this locus. These CRISPR-based studies, by recruiting chromatin regulators to specific sites in the genome, circumvent the limitations of using knockdown or overexpression approaches to study the functions of chromatin regulators, which can lead to difficulties in separating direct and indirect effects on gene expression. Further studies, such as recruitment of individual or combinations of protein effectors at many locations in the genome, will be required to determine which complexes enable enhancers to activate gene transcription and how genomic context influences these regulatory functions.

CRISPR recruitment strategies can also be adapted to modulate and study 3D contact frequency between distal regulatory elements and gene promoters. The genomic determinants of contact frequency are thought to include properties of the chromatin fiber, loop-extrusion complexes, and protein–protein interactions (e.g., between distal CTCF-bound sites or interacting factors bound to enhancers and promoters [Dekker and Mirny 2016]). Recent studies have developed methods to directly manipulate these contacts (Fig. 5C), for example by using orthogonal Cas9 and Cas12 systems, each fused to a heterologous dimerization domain, to bring together two target DNA sequences (Morgan et al. 2017). In several examples, studies using CRISPR or other tools to tether two DNA sequences have shown that increasing the contact frequency between an enhancer and a target promoter leads to an increase in gene expression (Deng et al. 2012; Bartman et al. 2016; Morgan et al. 2017). Multiple CRISPR-based studies of individual CTCF sites have revealed that these sites do indeed control 3D contacts and gene expression, perhaps by tuning the contacts between genes and their enhancers (de Wit et al. 2015; Guo et al. 2015; Sanborn et al. 2015; Flavahan et al. 2016; Hnisz et al. 2016). Recent studies using CRISPR-GFP recruitment methods to image the locations and movement of DNA loci in living cells (Fig. 5D) have also provided new insights into mechanisms that control 3D chromatin contacts (Gu et al. 2018; Maass et al. 2018). Imaging of enhancers and promoters in the same 3D domain showed that enhancer activation correlates with an increased rate of chromatin diffusion in the same 3D domain in a transcription-dependent fashion, likely affecting the rate of enhancer–promoter contact formation (Gu et al. 2018). Imaging loci on different chromosomes revealed that gene loci on different chromosomes can stably associate and have stereotyped positions in the nucleus, perhaps influencing their regulation or function (Maass et al. 2018).

Together, the ability to target and manipulate specific DNA loci at scale is unlocking new insights into the functions of DNA regulatory elements, including how histone modifications, chromatin regulatory proteins, and 3D chromosome conformation influence gene expression. Because these properties of enhancers can vary dramatically across cell types, further exploration will be required to learn how these rules change in different cellular contexts. These studies to modify different aspects of transcriptional regulation also provide an important guide for understanding posttranscriptional regulation: similar CRISPR-based approaches using RNA-targeting tools could soon be applied to study the functions of RNA regulatory elements and the proteins, sequence and structural motifs, and RNA modification sites that mediate their functions.

3.2 Functions of Long Noncoding RNAs and Their Transcription

Long noncoding RNAs (lncRNAs) are a heterogeneous class of molecules, operationally defined as polyadenylated RNAs of >200 nt that do not encode peptides. Before the genomic era, biochemical and genetic studies identified a handful of lncRNAs with important molecular functions, including *Xist*, which orchestrates X-chromosome inactivation by spreading across the X chromosome in *cis* and recruiting multiple repressive chromatin regulatory complexes to silence gene transcription (Plath et al. 2002). RNA sequencing studies found that nearly the whole genome is transcribed to some extent in one cell type or another, and that these noncoding transcripts included thousands of polyadenylated and spliced lncRNAs (Kapranov et al. 2007; Djebali et al. 2012; ENCODE Project Consortium 2012). Although there are some notable exceptions, such as *Xist*, the function (if any) of most lncRNAs has remained elusive. CRISPR-based tools have now begun to address two key questions: Which genomic loci encoding lncRNAs have functional roles in the cell, and how do these noncoding loci accomplish these functions?

New tools for direct perturbation of lncRNA expression and sequence have provided an opportunity to systematically search for functional lncRNA loci via forward genetic screens. Like noncoding DNA elements, perturbations to lncRNAs involve different technical considerations than protein-coding genes. The relationship between RNA sequence and function is poorly understood, but is apparently flexible and can consist of functional sequences (such as sites that RNA-binding proteins recognize) interspersed with nonfunctional ones (Zappulla and Cech 2004; Guttman and Rinn 2012; Rinn and Chang 2012). High-throughput screening approaches to disrupt lncRNA function must either perturb broad regions of genomic sequence, for example to remove the promoter or some critical RNA domain of a lncRNA, or densely tile smaller mutations through each element to identify and perturb critical sequences.

Using these approaches, several groups have recently performed forward genetic screens by activating or inhibiting lncRNA promoters and assessing the effects on cellular functions (Zhu et al. 2016; Joung et al. 2017a; Liu et al. 2017). Liu and colleagues performed loss-of-function screens using CRISPRi (KRAB-dCas9) to target 16,401 lncRNA promoters and characterized their effects on cellular proliferation across nine cell lines (Liu et al. 2017). They found 499 loci where inhibition of the lncRNA promoter led to a significant effect on cellular proliferation in at least one cell line, 89% of which showed cell line–specific effects. Joung and colleagues designed gain-of-function screens using CRISPRa (SAM) to target 10,504 lncRNA promoters and identified 11 noncoding loci whose activation led to resistance to the melanoma drug vemurafinib (Joung et al. 2017a). Most of these effects appeared to be due to activation of nearby genes in *cis*. From these and other studies (Guttman et al. 2011; Rinn and Chang 2012; Luo et al. 2016), it is becoming clear that many noncoding genomic loci that produce lncRNAs play important roles in controlling gene expression and cellular functions, and that these roles are typically specific to certain cell types or contexts.

Importantly, however, perturbations to "lncRNA promoters" do not themselves establish that the "lncRNA transcripts" are functional. Rather, promoter perturbations might identify loci that affect a phenotype through a variety of mechanisms related to the lncRNA locus (Fig. 6), which can be broadly categorized as (1) nonlocal functions of a lncRNA, in which the RNA transcript acts at locations in the cell other than the genomic locus of the lncRNA; (2) local, sequence-specific functions of a lncRNA transcript to regulate the expression of nearby gene(s) on the same chromosome (in *cis*); (3) local functions to regulate nearby gene(s) via mechanisms of lncRNA transcription or splicing that do not require any specific sequences in the lncRNA transcript itself; and (4) local functions to regulate nearby gene(s) via the lncRNA promoter acting as an enhancer.

Although many lncRNAs have been hypothesized to act via the first two (RNA-mediated) mechanisms, CRISPR-based studies have surprisingly revealed that the third and fourth categories of functions—that is, effects on gene expression that are not mediated by the RNA transcript itself—occur in many loci that produce lncRNAs as well as in many loci that encode mRNAs. In a recent study, Engreitz and colleagues knocked out the promoters of 12 uncharacterized lncRNAs and six mRNAs and found, using allele-specific RNA sequencing, that promoter deletion led to an effect on the expression of a neighboring gene at five of these 12 lncRNA loci and four of the six mRNA loci (Engreitz et al. 2016). To distinguish between possible *cis*-regulatory mechanisms (Fig. 6), Engreitz and colleagues first inserted polyadenylation signals a few hundred bp downstream of the transcription start site, thereby leaving the DNA sequence of the promoter intact but eliminating downstream transcription and the mature lncRNA tran-

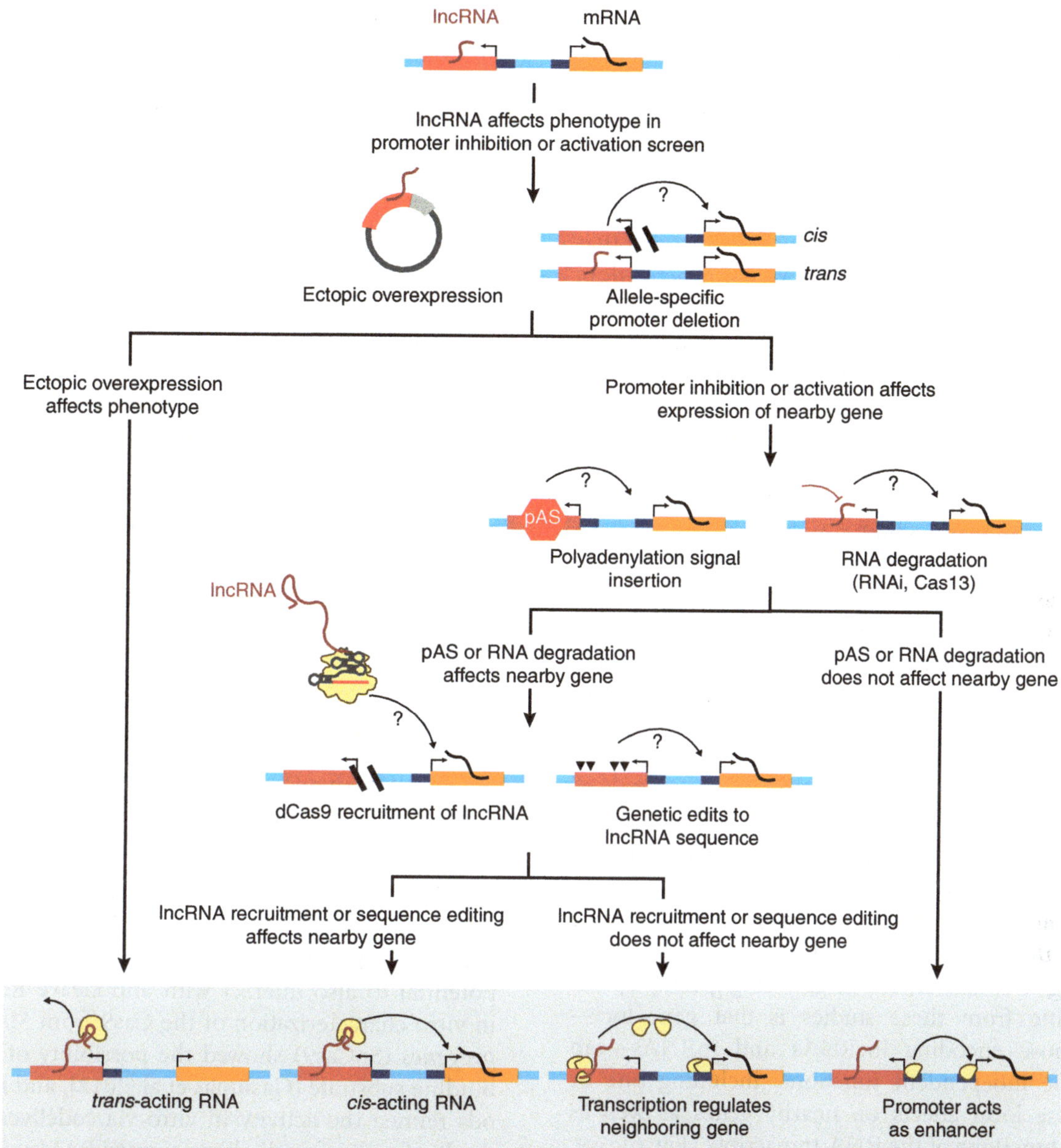

Figure 6. CRISPR tools to dissect the functions of genomic loci that produce long noncoding RNAs (lncRNAs). Flowchart (*top*) shows experimental approaches to determine how lncRNA loci regulate gene expression (*bottom*) via one or more of the following mechanisms: (1) producing a *trans*-acting lncRNA that acts nonlocally (away from its genomic locus); (2) producing a *cis*-acting lncRNA that acts locally at its genomic locus via a sequence-dependent function of the RNA molecule; (3) affecting a nearby gene via a sequence-independent function of lncRNA transcription or splicing; and (4) regulating a neighboring gene via proteins that bind its promoter or other overlapping *cis*-regulatory element. A series of targeted experiments can be used to distinguish between these functions. If ectopic overexpression of the lncRNA recapitulates or rescues the screen phenotype, this supports a *trans*-acting mechanism. Conversely, demonstrating that allele-specific deletion affects the expression of a nearby gene in an allele-specific manner is the best demonstration that a lncRNA locus acts to regulate gene expression in *cis*. To distinguish between multiple potential *cis*-acting mechanisms, a next step involves either inserting a polyadenylation signal (pAS) just downstream from the lncRNA promoter or using RNA interference (RNAi) or Cas13 to directly degrade the RNA. If the mature RNA and lncRNA transcription are successfully eliminated but do not affect the phenotype, then DNA-mediated effects of promoters are the most likely mechanism. If these perturbations do affect the phenotype, then dCas9 recruitment of the lncRNA in combination with genetic edits to the lncRNA sequence can be used to distinguish between sequence-specific effects of the lncRNA molecule or sequence-independent effects of lncRNA transcription.

script (Fig. 6). In five out of six cases, polyadenylation insertion near the promoter did not affect the expression of a gene neighbor, indicating that the regulatory effect was likely caused by the lncRNA promoter acting as an enhancer. For one lncRNA locus called *Blustr* (bivalent locus is upregulated by the splicing and transcription of an RNA), insertion of polyadenylation sites did in fact reduce the expression of a neighboring gene (*Sfmbt2*), indicating that transcription of *Blustr* was required for *Sfmbt2* activation. However, deleting individual exons or introns of the *Blustr* transcript with CRISPR–Cas9 did not affect *Sfmbt2* activation, suggesting that there were no specific sequences in the spliced lncRNA transcript that were individually necessary. Indeed, for none of the six loci studied did the RNA transcripts themselves appear to contribute to *cis* regulation of gene expression in a sequence-dependent fashion. Instead, activation depended on the lncRNA promoter acting as an enhancer or on the act of lncRNA transcription.

Together with an increasing number of other examples using similar genetic approaches—including *Airn* (Latos et al. 2012), *Hand2* (Anderson et al. 2016), and others (Shearwin et al. 2005; Yin et al. 2015; Paralkar et al. 2016; Joung et al. 2017a)—these systematic CRISPR-based dissections of lncRNA loci point to a pervasive role for promoters and the process of transcription in affecting gene expression in a surrounding region, re-orienting how we can view the vast number of noncoding transcripts produced by mammalian genomes. Indeed, the cell-type specificity seen in lncRNA promoter perturbation screens may result at least in part from the cell type–specific functions of these promoters acting as enhancers for neighboring genes. A principle emerging from these studies is that gene loci—including those encoding lncRNAs and mRNAs—can have multiple independent functions, including one or more of these local effects on nearby genes as well as *trans*-acting functions of the RNA transcripts (Fig. 6).

Given these observations, an important future direction for studies of the noncoding genome will be to expand the CRISPR forward genetics toolkit to perturb specific categories of regulatory mechanisms. For example, functional RNAs might be directly identified and distinguished from promoter-mediated mechanisms via knockdown with RNA-targeting CRISPR effectors, or by tailored screens to alter RNA sequence, RNA modifications, RNA localization, or RNA-binding proteins. *trans*-acting lncRNAs may be directly identified via lncRNA overexpression (via expression at an ectopic locus) in forward genetic screens, or first via CRISPRa/i screens followed by overexpression of specific candidates. *cis*-acting lncRNAs may be directly identified (and distinguished from sequence-independent functions of transcription) via CRISPR-mediated recruitment of RNAs to specific genomic sites and dense saturation mutagenesis of lncRNA sequences in situ. Technological advances in CRISPR systems, including RNA-targeting CRISPR systems (see next section), promise to enable such precise studies to dissect the functions of lncRNAs and noncoding transcription in mammalian genomes.

4 EMERGING CRISPR TOOLS TO DIRECTLY MANIPULATE RNA

The recent introduction of RNA-targeting CRISPR tools have opened a new realm of potential applications for studying RNA regulation via direct modulation or modification of RNA. In particular, Cas13 systems enable a new toolbox of easy-to-use technologies for perturbing RNA function, overcoming limitations of previous methods such as exogenous tagging of RNA with MS2 loops (Park et al. 2010) or engineering of Pumilio RNA-binding proteins (Abil and Zhao 2015). In this section, we discuss CRISPR-based RNA-targeting technologies and how they can be applied for studying RNA biology.

4.1 Methods to Target CRISPR–Cas Proteins to Specific RNA Sequences

Several CRISPR-based approaches have been developed to enable RNA-targeted degradation or programmable protein recruitment.

Cas9: Although Cas9 is typically regarded as a DNA-targeting enzyme, certain natural Cas9 variants or engineered modifications have been shown to have the potential to also interact with and cleave RNA. Initial in vitro characterization of the Cas9 from *Streptococcus pyogenes* (SpCas9) showed the possibility of RNA as a binding substrate (Gasiunas et al. 2012), and later methods refined the activity in vitro via codelivery of PAM duplex-forming small oligos, termed PAMmers (O'Connell et al. 2014). This engineered adaptation of SpCas9 for RNA binding was shown to function in vivo for RNA imaging or cleavage of repeat-containing RNA (Nelles et al. 2016; Batra et al. 2017). Certain Cas9 orthologs also possess endogenous RNase activity and can cleave RNA without a PAMmer in *Escherichia coli* (Strutt et al. 2018) or in vitro (Rousseau et al. 2018). RNA targeting with Cas9 in mammalian cells appears to require the introduction of the exogenous synthetic modified oligo (PAMmer) (Nelles et al. 2016), rendering Cas9 challenging to use for pooled screening approaches that require genetically encodable constructs.

Cas13: The single-effector RNA-targeting CRISPR system Type VI and the associated endoribonuclease Cas13 (Shmakov et al. 2015; Abudayyeh et al. 2016; Smargon

Cite as *Cold Spring Harb Perspect Biol* doi: 10.1101/cshperspect.a035386

et al. 2017) have provided an improved tool for RNA knockdown and RNA-targeted protein recruitment (Abudayyeh et al. 2017; Cox et al. 2017). The Cas13 family has diverse features, including different crRNA structures and different levels of activity (Shmakov et al. 2015, 2017; Abudayyeh et al. 2016; East-Seletsky et al. 2016; Smargon et al. 2017; Konermann et al. 2018; Yan et al. 2018). Cas13a, the first characterized member, has been shown to cleave RNA and reduce gene expression in bacterial, plant, and mammalian cells (Fig. 7A) (Abudayyeh et al. 2016, 2017). Catalytically dead Cas13—generated by mutating the endonuclease domains—provides a flexible platform for recruitment of proteins to target RNAs in mammalian cells, including

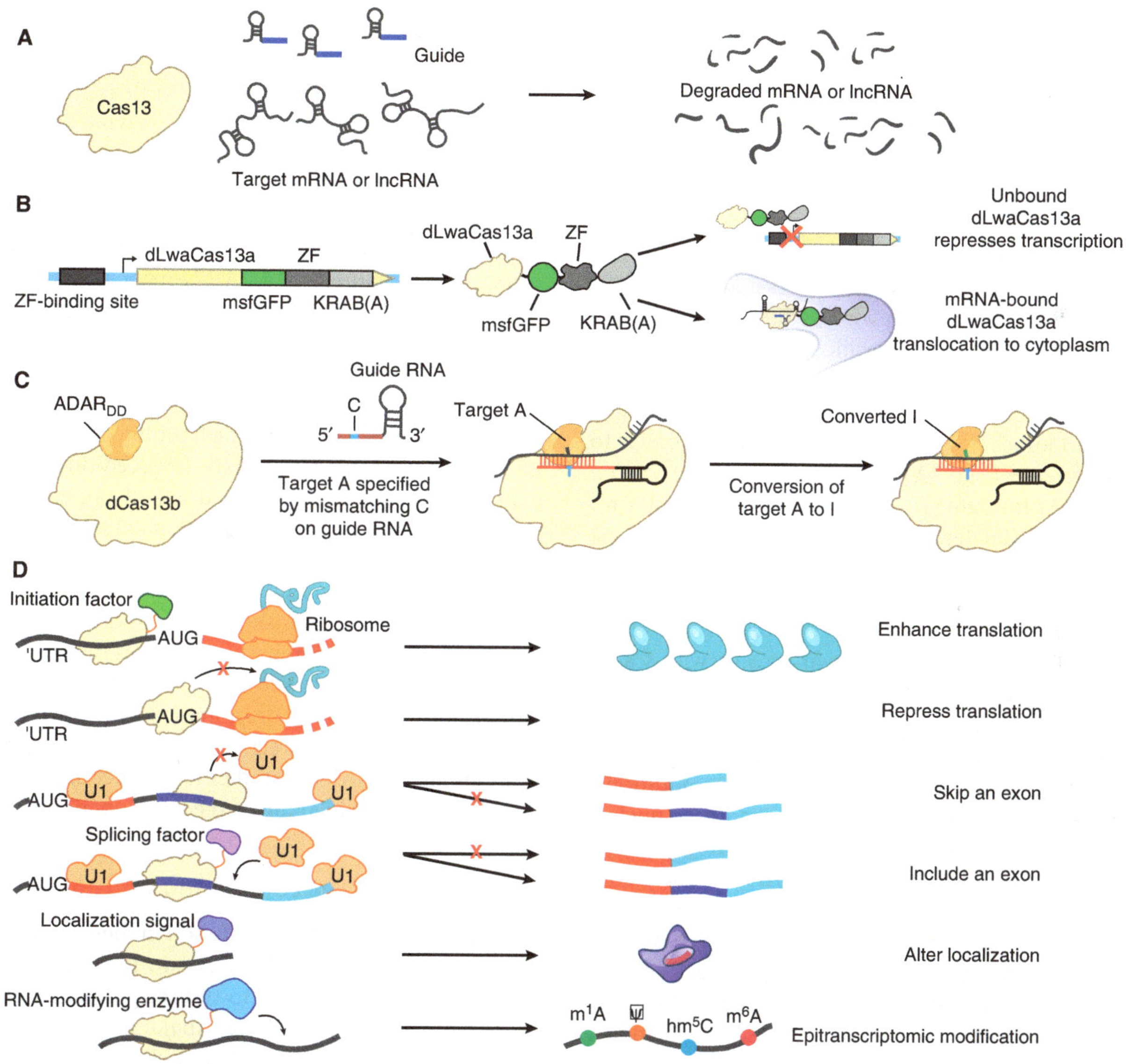

Figure 7. RNA targeting applications of Cas13. (*A*) RNA knockdown with Cas13. (*B*) Image tracking using dCas13-eGFP to fluorescently visualize transcripts in cells. Modifying the dCas13-EGFP fusion with a zinc finger–Krüppel associated box (KRAB) fusion can improve visualization and lower fluorescence background signal by repressing the expression of unbound dCas13-EGFP complexes via negative feedback. (*C*) RNA editing of adenines to inosines by a dCas13-ADAR2 fusion. Precise editing is accomplished by creating an A:C base pair bubble by encoding a mismatch in the guide RNA. (*D*) Further applications of Cas13 for studying RNA regulation. Translation could be activated by recruiting translation initiation factors or repressed by blocking start codons on transcripts. Splicing has been altered by blocking splice sites on transcripts or by recruiting splicing factors. Transcript localization could be modulated by fusing dCas13 to different localization sequences. Lastly, posttranscriptional RNA modifications could be introduced by recruiting RNA-modifying enzymes, such as pseudouridine synthase, to specific bases on RNA transcripts.

for imaging of RNA through fusion to GFP (Fig. 7B) (Abudayyeh et al. 2017), RNA editing through fusion to ADAR2 (Fig. 7C) (Cox et al. 2017), and alteration of isoforms by blocking splice sites (Konermann et al. 2018).

4.2 Opportunities for Systematic Studies of RNA Biology

These new RNA-targeted tools open new opportunities to study many aspects of RNA biology. Here, we highlight key studies that have developed proof-of-concept tools using Cas13.

4.2.1 RNA Knockdown

In light of the challenges in distinguishing the effects of DNA- and RNA-mediated mechanisms (see Sec. 2.2), technologies for RNA knockdown with CRISPR enzymes have the potential to enable more specific studies of RNA-mediated functions because they avoid manipulating DNA sequence. To this end, we recently showed that the Cas13 platform can knock down RNAs with efficiency similar to RNA interference (Fig. 7A) (Abudayyeh et al. 2017; Cox et al. 2017). At the same time, Cas13 knockdown of a reporter transcript (luciferase) had minimal effects on the rest of the transcriptome, whereas RNAi knockdown of the same transcript led to significant effects on ~900 other genes (Abudayyeh et al. 2017; Cox et al. 2017; Konermann et al. 2018). Thus, RNA knockdown screens with Cas13 have the potential to reduce the off-target effects that have confounded RNAi screens, and, in contrast to approaches that rely on endogenous RNAi proteins, the Cas13 enzymes can be further optimized and engineered for additional functionality or specificity. For example, adding a nuclear localization tag to Cas13a led to successful knockdown of nuclear-localized lncRNAs, while RNAi was ineffective (Abudayyeh et al. 2017). We also found that a second Cas13 enzyme, Cas13b from *Prevotella sp. P5-125*, was far more effective at RNA knockdown (>90% for each site targeted) than our original studies using LwaCas13a (Cox et al. 2017). Further optimizations of the system—including adapting other Cas13 proteins and guide RNAs for viral delivery, learning design rules for efficient guides, and exploring or optimizing Cas13 proteins for high stability and specificity—should soon enable RNA-targeted forward genetic screens to directly study RNA function.

4.2.2 Posttranscriptional RNA Modifications

Posttranscriptional modifications of RNA nucleotides—including N^6-methyladenosine (m^6A), pseudouridine, and inosine—appear to be a critical layer of posttranscriptional regulation (Schwartz 2016; Helm and Motorin 2017; Hsu et al. 2017; Lewis et al. 2017). RNA mapping technologies have revealed that these modifications can affect tens of thousands of sites in both coding and noncoding RNAs (Li et al. 2016; Schwartz 2016; Helm and Motorin 2017; Hsu et al. 2017; Lewis et al. 2017), but the functions of most of these modifications are poorly understood. We developed a Cas13 approach to perform adenine-to-inosine (A-to-I) editing, a natural modification that alters the sequence of RNA. In this approach, named RNA editing for programmable A to I replacement (REPAIR), we fused catalytically dead Cas13 to the ADAR2 enzyme to enable targetable A-to-I editing activity (Fig. 7C) (Cox et al. 2017). By extending the guide RNA and including a cytidine mismatch in the guide across from the targeted adenosine, a more optimal double-stranded RNA substrate with a bubble was created, maximizing ADAR2 enzymatic activity. This approach achieved up to 90% A-to-I editing of targeted sites on reporter transcripts and up to 40% on endogenous transcripts. Off-target editing observed with the initial system (>18,000 edited off-target sites) was reduced through a rational mutagenesis strategy, which produced a more specific version of Cas13 that retained on-target editing activity but reduced off-target activity (20 edited off-target sites) (Cox et al. 2017). Because of the high specificity, high efficiency, lack of sequence restriction, and precise modification site, REPAIR provides a useful tool for studying the effects of sequence variants and inosine modifications, and for developing therapeutics. Similar approaches with alternative RNA editing or modification enzymes, such as RNA methyltransferases, cytidine deaminases, and pseudouridine synthases, may enable precise introduction of RNA modifications and subsequent studies to connect these modifications with their cellular and molecular functions (Fig. 7D).

4.2.3 RNA Splicing, Localization, and More

RNA-targeted CRISPR tools are still in their infancy, and—just as an explosion of dCas9 fusion tools have begun to reveal biological insights into transcriptional regulation—dCas13 fusions will likely have a major impact in illuminating posttranscriptional regulation. Indeed, studies using previous RNA-targeting technologies have highlighted additional possibilities for manipulating RNA regulation with new CRISPR tools. For example, fusion of PUF proteins to the arginine- and serine-rich domain of SRSF1 or the glycine-rich domain of HNRNPA1 enabled the creation of programmable activators and repressors of splicing, respectively (Cheong and Hall 2006). RNA recruitment systems using MS2 and PP7 (or other systems that involve tagging an endogenous RNA with the recognition sequence for an ex-

Cite as *Cold Spring Harb Perspect Biol* doi: 10.1101/cshperspect.a035386

ogenous RNA-binding protein) have been used to test the sufficiency of specific RNA–protein interactions for RNA function (Chen and Varani 2013; Bos et al. 2016) and to track transcripts via fluorescence (Park et al. 2010). Early studies have shown the potential for Cas13 tools to modulate the expression of specific RNA isoforms by binding and blocking splice sites (Fig. 7D) (Konermann et al. 2018). The ability to combine each of these tools with transcriptome-wide screening will enable systematic perturbation studies of many aspects of RNA biology, including translation, splicing, RNA localization, and posttranscriptional modifications (Fig. 7D).

5 CONCLUDING REMARKS

CRISPR systems have remarkable diversity and flexibility, providing many opportunities for creative innovation to expand the CRISPR toolkit. Further exploration of natural CRISPR systems, as well as other natural enzymes and systems with novel functionalities, will help to uncover new tools for manipulating RNA and DNA. Directed evolution and rational engineering of CRISPR enzymes will expand their generalizability and increase specificity. Functionalizing CRISPR proteins with additional DNA- or RNA-modifying proteins will enable precisely perturbing a diverse array of regulatory processes in human cells.

Building on recent studies of DNA regulatory elements and lncRNAs, application of CRISPR tools has the potential to reveal new principles of molecular biology. In the past decade, many advances in our understanding of RNA biology were facilitated by genome-scale "observational" studies, including the application of new sequencing-based technologies to map the locations of RNA transcription, modifications, and RNA-binding proteins. In the next decade, we anticipate that an equally transformative contribution will come from genome-scale "perturbation" studies, including systematic interrogation of each of these layers of RNA regulation. These studies will establish general principles for regulatory molecules as well as illuminate how specific genomic or molecular contexts modify their functions. Integrating combinatorial perturbations and multiplexed molecular measurements will reveal molecular circuits that control cellular behavior. In tandem with the development of appropriate phenotypic assays and delivery methods, CRISPR systems will connect these circuits with cellular phenotypes relevant to human diseases.

CRISPR studies of RNA regulation will not only reveal the functions of key molecular processes but also will help us to directly manipulate them in humans. Preclinical development for CRISPR-based therapeutics is already underway, including approaches for genome editing ex vivo (where editing is performed outside the body and edited cells are transplanted into the patient) and in vivo (where CRISPR tools are delivered to edit cells directly in the tissue of interest). Many challenges remain, including optimizing the efficiency and specificity of CRISPR editing, developing new methods for efficient and precise delivery to target tissues and cell types, and minimizing unintended consequences of these manipulations on immune responses or target cell functions. Nonetheless, overcoming these challenges promises to enable a flexible class of therapeutics that leverage a programmable suite of CRISPR tools to manipulate DNA or RNA.

Although therapeutic development efforts have largely focused on modification of protein-coding genes, approaches for targeting DNA and RNA regulatory mechanisms may prove advantageous for certain indications. For example, therapeutic targeting of DNA enhancers, which can have cell type–specific activities, may enable more precise modulation of gene function in cell types relevant to disease (Canver et al. 2015). RNA-targeted therapeutic approaches may enable specific regulation of entities that are difficult to manipulate at the level of DNA, such as degradation of toxic repeat expansion RNAs (Batra et al. 2017) or selective activation of specific RNA isoforms (Palacino et al. 2015; Konermann et al. 2018).

Although much work remains to characterize and apply CRISPR tools, such studies may ultimately yield new insights into the regulatory machinery of the cell and new strategies for treating human disease.

ACKNOWLEDGMENTS

We thank Eric Lander and Rhiannon Macrae for comments on the manuscript and Sigrid Knemeyer for assistance with figures.

REFERENCES

Abil Z, Zhao H. 2015. Engineering reprogrammable RNA-binding proteins for study and manipulation of the transcriptome. *Mol Biosyst* **11:** 2658–2665.

Abudayyeh OO, Gootenberg JS, Konermann S, Joung J, Slaymaker IM, Cox DB, Shmakov S, Makarova KS, Semenova E, Minakhin L, et al. 2016. C2c2 is a single-component programmable RNA-guided RNA-targeting CRISPR effector. *Science* **353:** aaf5573.

Abudayyeh OO, Gootenberg JS, Essletzbichler P, Han S, Joung J, Belanto JJ, Verdine V, Cox DBT, Kellner MJ, Regev A, et al. 2017. RNA targeting with CRISPR–Cas13. *Nature* **550:** 280–284.

Adamson B, Norman TM, Jost M, Cho MY, Nunez JK, Chen Y, Villalta JE, Gilbert LA, Horlbeck MA, Hein MY, et al. 2016. A multiplexed single-cell CRISPR screening platform enables systematic dissection of the unfolded protein response. *Cell* **167:** 1867–1882.e1821.

Anantharaman V, Makarova KS, Burroughs AM, Koonin EV, Aravind L. 2013. Comprehensive analysis of the HEPN superfamily: Identification of novel roles in intra-genomic conflicts, defense, pathogenesis and RNA processing. *Biol Direct* **8:** 15.

Anderson KM, Anderson DM, McAnally JR, Shelton JM, Bassel-Duby R, Olson EN. 2016. Transcription of the non-coding RNA upperhand controls *Hand2* expression and heart development. *Nature* **539:** 433–436.

Bartman CR, Hsu SC, Hsiung CC-S, Raj A, Blobel GA. 2016. Enhancer regulation of transcriptional bursting parameters revealed by forced chromatin looping. *Mol Cell* **62:** 237–247.

Batra R, Nelles DA, Pirie E, Blue SM, Marina RJ, Wang H, Chaim IA, Thomas JD, Zhang N, Nguyen V, et al. 2017. Elimination of toxic microsatellite repeat expansion RNA by RNA-targeting Cas9. *Cell* **170:** 899–912.e810.

Bikard D, Jiang W, Samai P, Hochschild A, Zhang F, Marraffini LA. 2013. Programmable repression and activation of bacterial gene expression using an engineered CRISPR–Cas system. *Nucleic Acids Res* **41:** 7429–7437.

Bintu L, Yong J, Antebi YE, McCue K, Kazuki Y, Uno N, Oshimura M, Elowitz MB. 2016. Dynamics of epigenetic regulation at the single-cell level. *Science* **351:** 720–724.

Bos TJ, Nussbacher JK, Aigner S, Yeo GW. 2016. Tethered function assays as tools to elucidate the molecular roles of RNA-binding proteins. *Adv Exp Med Biol* **907:** 61–88.

Brummelkamp TR, Fabius AW, Mullenders J, Madiredjo M, Velds A, Kerkhoven RM, Bernards R, Beijersbergen RL. 2006. An shRNA barcode screen provides insight into cancer cell vulnerability to MDM2 inhibitors. *Nat Chem Biol* **2:** 202–206.

Bulger M, Groudine M. 2011. Functional and mechanistic diversity of distal transcription enhancers. *Cell* **144:** 327–339.

Caicedo JC, Singh S, Carpenter AE. 2016. Applications in image-based profiling of perturbations. *Curr Opin Biotechnol* **39:** 134–142.

Canver MC, Smith EC, Sher F, Pinello L, Sanjana NE, Shalem O, Chen DD, Schupp PG, Vinjamur DS, Garcia SP, et al. 2015. *BCL11A* enhancer dissection by Cas9-mediated in situ saturating mutagenesis. *Nature* **527:** 192–197.

Canver MC, Bauer DE, Orkin SH. 2017. Functional interrogation of non-coding DNA through CRISPR genome editing. *Methods* **121–122:** 118–129.

Chavez A, Scheiman J, Vora S, Pruitt BW, Tuttle M, Iyer EPR, Lin S, Kiani S, Guzman CD, Wiegand DJ, et al. 2015. Highly efficient Cas9-mediated transcriptional programming. *Nat Methods* **12:** 326–328.

Chen Y, Varani G. 2013. Engineering RNA-binding proteins for biology. *FEBS J* **280:** 3734–3754.

Chen B, Gilbert LA, Cimini BA, Schnitzbauer J, Zhang W, Li GW, Park J, Blackburn EH, Weissman JS, Qi LS, et al. 2013. Dynamic imaging of genomic loci in living human cells by an optimized CRISPR/Cas system. *Cell* **155:** 1479–1491.

Cheng AW, Wang H, Yang H, Shi L, Katz Y, Theunissen TW, Rangarajan S, Shivalila CS, Dadon DB, Jaenisch R. 2013. Multiplexed activation of endogenous genes by CRISPR-on, an RNA-guided transcriptional activator system. *Cell Res* **23:** 1163–1171.

Cheong CG, Hall TM. 2006. Engineering RNA sequence specificity of Pumilio repeats. *Proc Natl Acad Sci* **103:** 13635–13639.

Cong L, Ran FA, Cox D, Lin S, Barretto R, Habib N, Hsu PD, Wu X, Jiang W, Marraffini LA, et al. 2013. Multiplex genome engineering using CRISPR/Cas systems. *Science* **339:** 819–823.

Cox DBT, Gootenberg JS, Abudayyeh OO, Franklin B, Kellner MJ, Joung J, Zhang F. 2017. RNA editing with CRISPR–Cas13. *Science* **358:** 1019–1027.

Dahlman JE, Abudayyeh OO, Joung J, Gootenberg JS, Zhang F, Konermann S. 2015. Orthogonal gene knockout and activation with a catalytically active Cas9 nuclease. *Nat Biotechnol* **33:** 1159–1161.

Datlinger P, Rendeiro AF, Schmidl C, Krausgruber T, Traxler P, Klughammer J, Schuster LC, Kuchler A, Alpar D, Bock C. 2017. Pooled CRISPR screening with single-cell transcriptome readout. *Nat Methods* **14:** 297–301.

de Wit E, Vos ESM, Holwerda SJB, Valdes-Quezada C, Verstegen MJAM, Teunissen H, Splinter E, Wijchers PJ, Krijger PHL, de Laat W. 2015. CTCF binding polarity determines chromatin looping. *Mol Cell* **60:** 676–684.

Dekker J, Mirny L. 2016. The 3D genome as moderator of chromosomal communication. *Cell* **164:** 1110–1121.

Deltcheva E, Chylinski K, Sharma CM, Gonzales K, Chao Y, Pirzada ZA, Eckert MR, Vogel J, Charpentier E. 2011. CRISPR RNA maturation by *trans*-encoded small RNA and host factor RNase III. *Nature* **471:** 602–607.

Deng W, Lee J, Wang H, Miller J, Reik A, Gregory PD, Dean A, Blobel GA. 2012. Controlling long-range genomic interactions at a native locus by targeted tethering of a looping factor. *Cell* **149:** 1233–1244.

Diao Y, Li B, Meng Z, Jung I, Lee AY, Dixon J, Maliskova L, Guan K-L, Shen Y, Ren B. 2016. A new class of temporarily phenotypic enhancers identified by CRISPR/Cas9-mediated genetic screening. *Genome Res* **26:** 397–405.

Diao Y, Fang R, Li B, Meng Z, Yu J, Qiu Y, Lin KC, Huang H, Liu T, Marina RJ, et al. 2017. A tiling-deletion-based genetic screen for *cis*-regulatory element identification in mammalian cells. *Nat Methods* **14:** 629–635.

Dixit A, Parnas O, Li B, Chen J, Fulco CP, Jerby-Arnon L, Marjanovic ND, Dionne D, Burks T, Raychowdhury R, et al. 2016. Perturb-seq: Dissecting molecular circuits with scalable single-cell RNA profiling of pooled genetic screens. *Cell* **167:** 1853–1866.e1817.

Djebali S, Davis CA, Merkel A, Dobin A, Lassmann T, Mortazavi A, Tanzer A, Lagarde J, Lin W, Schlesinger F, et al. 2012. Landscape of transcription in human cells. *Nature* **489:** 101–108.

Doench JG. 2018. Am I ready for CRISPR? A user's guide to genetic screens. *Nat Rev Genet* **19:** 67–80.

Doench JG, Hartenian E, Graham DB, Tothova Z, Hegde M, Smith I, Sullender M, Ebert BL, Xavier RJ, Root DE. 2014. Rational design of highly active sgRNAs for CRISPR–Cas9-mediated gene inactivation. *Nat Biotechnol* **32:** 1262–1267.

Doudna JA, Charpentier E. 2014. Genome editing. The new frontier of genome engineering with CRISPR–Cas9. *Science* **346:** 1258096.

East-Seletsky A, O'Connell MR, Knight SC, Burstein D, Cate JH, Tjian R, Doudna JA. 2016. Two distinct RNase activities of CRISPR–C2c2 enable guide-RNA processing and RNA detection. *Nature* **538:** 270–273.

East-Seletsky A, O'Connell MR, Burstein D, Knott GJ, Doudna JA. 2017. RNA Targeting by functionally orthogonal type VI-A CRISPR–Cas enzymes. *Mol Cell* **66:** 373–383.e373.

ENCODE Project Consortium. 2012. An integrated encyclopedia of DNA elements in the human genome. *Nature* **489:** 57–74.

Engreitz JM, Haines JE, Perez EM, Munson G, Chen J, Kane M, McDonel PE, Guttman M, Lander ES. 2016. Local regulation of gene expression by lncRNA promoters, transcription and splicing. *Nature* **171:** 305–320.

Farzadfard F, Perli SD, Lu TK. 2013. Tunable and multifunctional eukaryotic transcription factors based on CRISPR/Cas. *ACS Synth Biol* **2:** 604–613.

Flavahan WA, Drier Y, Liau BB, Gillespie SM, Venteicher AS, Stemmer-Rachamimov AO, Suvà ML, Bernstein BE. 2016. Insulator dysfunction and oncogene activation in *IDH* mutant gliomas. *Nature* **529:** 110–114.

Fonfara I, Richter H, Bratovic M, Le Rhun A, Charpentier E. 2016. The CRISPR-associated DNA-cleaving enzyme Cpf1 also processes precursor CRISPR RNA. *Nature* **532:** 517–521.

Fulco CP, Munschauer M, Anyoha R, Munson G, Grossman SR, Perez EM, Kane M, Cleary B, Lander ES, Engreitz JM. 2016. Systematic mapping of functional enhancer-promoter connections with CRISPR interference. *Science* **354:** 769–773.

Gasiunas G, Barrangou R, Horvath P, Siksnys V. 2012. Cas9-crRNA ribonucleoprotein complex mediates specific DNA cleavage for adaptive immunity in bacteria. *Proc Natl Acad Sci* **109:** E2579–E2586.

Gasperini M, Findlay GM, McKenna A, Milbank JH, Lee C, Zhang MD, Cusanovich DA, Shendure J. 2017. CRISPR/Cas9-mediated scanning for regulatory elements required for HPRT1 expression via thousands of large, programmed genomic deletions. *Am J Hum Genet* **101:** 192–205.

Gaudelli NM, Komor AC, Rees HA, Packer MS, Badran AH, Bryson DI, Liu DR. 2017. Programmable base editing of A*T to G*C in genomic DNA without DNA cleavage. *Nature* **551:** 464–471.

Gilbert LA, Larson MH, Morsut L, Liu Z, Brar GA, Torres SE, Stern-Ginossar N, Brandman O, Whitehead EH, Doudna JA, et al. 2013. CRISPR-mediated modular RNA-guided regulation of transcription in eukaryotes. *Cell* **154:** 442–451.

Gilbert LA, Horlbeck MA, Adamson B, Villalta JE, Chen Y, Whitehead EH, Guimaraes C, Panning B, Ploegh HL, Bassik MC, et al. 2014. Genome-scale CRISPR-mediated control of gene repression and activation. *Cell* **159:** 647–661.

Gootenberg JS, Abudayyeh OO, Lee JW, Essletzbichler P, Dy AJ, Joung J, Verdine V, Donghia N, Daringer NM, Freije CA, et al. 2017. Nucleic acid detection with CRISPR–Cas13a/C2c2. *Science* **356:** 438–442.

Gootenberg JS, Abudayyeh OO, Kellner MJ, Joung J, Collins JJ, Zhang F. 2018. Multiplexed and portable nucleic acid detection platform with Cas13, Cas12a, and Csm6. *Science* **360:** 439–444.

Gu B, Swigut T, Spencley A, Bauer MR, Chung M, Meyer T, Wysocka J. 2018. Transcription-coupled changes in nuclear mobility of mammalian *cis*-regulatory elements. *Science* **7:** eaao3136.

Guo Y, Xu Q, Canzio D, Shou J, Li J, Gorkin DU, Jung I, Wu H, Zhai Y, Tang Y, et al. 2015. CRISPR inversion of CTCF sites alters genome topology and enhancer/promoter function. *Cell* **162:** 900–910.

Guttman M, Rinn JL. 2012. Modular regulatory principles of large noncoding RNAs. *Nature* **482:** 339–346.

Guttman M, Donaghey J, Carey BW, Garber M, Grenier JK, Munson G, Young G, Lucas AB, Ach R, Bruhn L, et al. 2011. lincRNAs act in the circuitry controlling pluripotency and differentiation. *Nature* **477:** 295–300.

Helm M, Motorin Y. 2017. Detecting RNA modifications in the epitranscriptome: Predict and validate. *Nat Rev Genet* **18:** 275–291.

Hentze MW, Castello A, Schwarzl T, Preiss T. 2018. A brave new world of RNA-binding proteins. *Nat Rev Mol Cell Biol* **19:** 327–341.

Hilton IB, D'Ippolito AM, Vockley CM, Thakore PI, Crawford GE, Reddy TE, Gersbach CA. 2015. Epigenome editing by a CRISPR–Cas9-based acetyltransferase activates genes from promoters and enhancers. *Nature Biotechnol* **33:** 510–517.

Hnisz D, Weintraub AS, Day DS, Valton A-L, Bak RO, Li CH, Goldmann J, Lajoie BR, Fan ZP, Sigova AA, et al. 2016. Activation of proto-oncogenes by disruption of chromosome neighborhoods. *Science* **351:** 1454–1458.

Hsu PD, Lander ES, Zhang F. 2014. Development and applications of CRISPR–Cas9 for genome engineering. *Cell* **157:** 1262–1278.

Hsu PJ, Shi H, He C. 2017. Epitranscriptomic influences on development and disease. *Genome Biol* **18:** 197.

Iorns E, Lord CJ, Turner N, Ashworth A. 2007. Utilizing RNA interference to enhance cancer drug discovery. *Nat Rev Drug Discov* **6:** 556–568.

Jaitin DA, Weiner A, Yofe I, Lara-Astiaso D, Keren-Shaul H, David E, Salame TM, Tanay A, van Oudenaarden A, Amit I. 2016. Dissecting immune circuits by linking CRISPR-pooled screens with single-cell RNA-seq. *Cell* **167:** 1883–1896.e1815.

Jinek M, Chylinski K, Fonfara I, Hauer M, Doudna JA, Charpentier E. 2012. A programmable dual-RNA-guided DNA endonuclease in adaptive bacterial immunity. *Science* **337:** 816–821.

Joung J, Engreitz JM, Konermann S, Abudayyeh OO, Verdine VK, Aguet F, Gootenberg JS, Sanjana NE, Wright JB, Fulco CP, et al. 2017a. Genome-scale activation screen identifies a lncRNA locus regulating a gene neighbourhood. *Nature* **548:** 343–346.

Joung J, Konermann S, Gootenberg JS, Abudayyeh OO, Platt RJ, Brigham MD, Sanjana NE, Zhang F. 2017b. Genome-scale CRISPR–Cas9 knockout and transcriptional activation screening. *Nat Protoc* **12:** 828–863.

Kapranov P, Cheng J, Dike S, Nix DA, Duttagupta R, Willingham AT, Stadler PF, Hertel J, Hackermüller J, Hofacker IL, et al. 2007. RNA maps reveal new RNA classes and a possible function for pervasive transcription. *Science* **316:** 1484–1488.

Kearns NA, Pham H, Tabak B, Genga RM, Silverstein NJ, Garber M, Maehr R. 2015. Functional annotation of native enhancers with a Cas9-histone demethylase fusion. *Nat Methods* **12:** 401–403.

Kiani S, Chavez A, Tuttle M, Hall RN, Chari R, Ter-Ovanesyan D, Qian J, Pruitt BW, Beal J, Vora S, et al. 2015. Cas9 gRNA engineering for genome editing, activation and repression. *Nat Methods* **12:** 1051–1054.

Kim H, Kim JS. 2014. A guide to genome engineering with programmable nucleases. *Nat Rev Genet* **15:** 321–334.

Klann TS, Black JB, Chellappan M, Safi A, Song L, Hilton IB, Crawford GE, Reddy TE, Gersbach CA. 2017. CRISPR–Cas9 epigenome editing enables high-throughput screening for functional regulatory elements in the human genome. *Nature Biotechnol* **7:** 46545.

Koike-Yusa H, Li Y, Tan EP, Velasco-Herrera Mdel C, Yusa K. 2014. Genome-wide recessive genetic screening in mammalian cells with a lentiviral CRISPR-guide RNA library. *Nat Biotechnol* **32:** 267–273.

Komor AC, Kim YB, Packer MS, Zuris JA, Liu DR. 2016. Programmable editing of a target base in genomic DNA without double-stranded DNA cleavage. *Nature* **533:** 420–424.

Komor AC, Badran AH, Liu DR. 2017. CRISPR-based technologies for the manipulation of eukaryotic genomes. *Cell* **169:** 559.

Konermann S, Brigham MD, Trevino A, Hsu PD, Heidenreich M, Cong L, Platt RJ, Scott DA, Church GM, Zhang F. 2013. Optical control of mammalian endogenous transcription and epigenetic states. *Nature* **500:** 472–476.

Konermann S, Brigham MD, Trevino AE, Joung J, Abudayyeh OO, Barcena C, Hsu PD, Habib N, Gootenberg JS, Nishimasu H, et al. 2015. Genome-scale transcriptional activation by an engineered CRISPR–Cas9 complex. *Nature* **517:** 583–588.

Konermann S, Lotfy P, Brideau NJ, Oki J, Shokhirev MN, Hsu PD. 2018. Transcriptome engineering with RNA-targeting type VI-D CRISPR effectors. *Cell* **173:** 665–676.

Korkmaz G, Lopes R, Ugalde AP, Nevedomskaya E, Han R, Myacheva K, Zwart W, Elkon R, Agami R. 2016. Functional genetic screens for enhancer elements in the human genome using CRISPR–Cas9. *Nat Biotechnol* **34:** 192–198.

Latos PA, Pauler FM, Koerner MV, Şenergin HB, Hudson QJ, Stocsits RR, Allhoff W, Stricker SH, Klement RM, Warczok KE, et al. 2012. *Airn* transcriptional overlap, but not its lncRNA products, induces imprinted *Igf2r* silencing. *Science* **338:** 1469–1472.

Lewis CJ, Pan T, Kalsotra A. 2017. RNA modifications and structures cooperate to guide RNA–protein interactions. *Nat Rev Mol Cell Biol* **18:** 202–210.

Li X, Xiong X, Yi C. 2016. Epitranscriptome sequencing technologies: Decoding RNA modifications. *Nat Methods* **14:** 23–31.

Liu XS, Wu H, Ji X, Stelzer Y, Wu X, Czauderna S, Shu J, Dadon D, Young RA, Jaenisch R. 2016. Editing DNA methylation in the mammalian genome. *Cell* **167:** 233–247.e217.

Liu SJ, Horlbeck MA, Cho SW, Birk HS, Malatesta M, He D, Attenello FJ, Villalta JE, Cho MY, Chen Y, et al. 2017. CRISPRi-based genome-scale identification of functional long noncoding RNA loci in human cells. *Science* **355:** aah7111.

Luo S, Lu JY, Liu L, Yin Y, Chen C, Han X, Wu B, Xu R, Liu W, Yan P, et al. 2016. Divergent lncRNAs regulate gene expression and lineage differentiation in pluripotent cells. *Cell Stem Cell* **18:** 637–652.

Ma H, Tu LC, Naseri A, Huisman M, Zhang S, Grunwald D, Pederson T. 2016. Multiplexed labeling of genomic loci with dCas9 and engineered sgRNAs using CRISPRainbow. *Nat Biotechnol* **34:** 528–530.

Maass PG, Barutcu AR, Weiner CL, Rinn JL. 2018. Inter-chromosomal contact properties in live-cell imaging and in Hi-C. *Mol Cell* **69:** 1039–1045.e1033.

Maeder ML, Linder SJ, Cascio VM, Fu Y, Ho QH, Joung JK. 2013. CRISPR RNA-guided activation of endogenous human genes. *Nat Methods* **10:** 977–979.

Mali P, Aach J, Stranges PB, Esvelt KM, Moosburner M, Kosuri S, Yang L, Church GM. 2013a. CAS9 transcriptional activators for target specificity screening and paired nickases for cooperative genome engineering. *Nat Biotechnol* **31:** 833–838.

Mali P, Yang L, Esvelt KM, Aach J, Guell M, DiCarlo JE, Norville JE, Church GM. 2013b. RNA-guided human genome engineering via Cas9. *Science* **339:** 823–826.

Marraffini LA. 2015. CRISPR–Cas immunity in prokaryotes. *Nature* **526:** 55–61.

Morgan SL, Mariano NC, Bermudez A, Arruda NL, Wu F, Luo Y, Shankar G, Jia L, Chen H, Hu JF, et al. 2017. Manipulation of nuclear architecture through CRISPR-mediated chromosomal looping. *Nat Commun* **8:** 15993.

Mumbach MR, Satpathy AT, Boyle EA, Dai C, Gowen BG, Cho SW, Nguyen ML, Rubin AJ, Granja JM, Kazane KR, et al. 2017. Enhancer connectome in primary human cells identifies target genes of disease-associated DNA elements. *Nat Genet* **49:** 1602–1612.

Nelles DA, Fang MY, O'Connell MR, Xu JL, Markmiller SJ, Doudna JA, Yeo GW. 2016. Programmable RNA tracking in live cells with CRISPR/Cas9. *Cell* **165:** 488–496.

Ngo VN, Davis RE, Lamy L, Yu X, Zhao H, Lenz G, Lam LT, Dave S, Yang L, Powell J, et al. 2006. A loss-of-function RNA interference screen for molecular targets in cancer. *Nature* **441:** 106–110.

Nishida K, Arazoe T, Yachie N, Banno S, Kakimoto M, Tabata M, Mochizuki M, Miyabe A, Araki M, Hara KY, et al. 2016. Targeted nucleotide editing using hybrid prokaryotic and vertebrate adaptive immune systems. *Science* **353:** aaf8729.

O'Connell MR, Oakes BL, Sternberg SH, East-Seletsky A, Kaplan M, Doudna JA. 2014. Programmable RNA recognition and cleavage by CRISPR/Cas9. *Nature* **516:** 263–266.

O'Geen H, Ren C, Nicolet CM, Perez AA, Halmai J, Le VM, Mackay JP, Farnham PJ, Segal DJ. 2017. dCas9-based epigenome editing suggests acquisition of histone methylation is not sufficient for target gene repression. *Nucleic Acids Res* **45:** 9901–9916.

Palacino J, Swalley SE, Song C, Cheung AK, Shu L, Zhang X, Van Hoosear M, Shin Y, Chin DN, Keller CG, et al. 2015. SMN2 splice modulators enhance U1-pre-mRNA association and rescue SMA mice. *Nat Chem Biol* **11:** 511–517.

Paralkar VR, Taborda CC, Huang P, Yao Y, Kossenkov AV, Prasad R, Luan J, Davies JOJ, Hughes JR, Hardison RC, et al. 2016. Unlinking an lncRNA from its associated *cis* element. *Mol Cell* **62:** 104–110.

Park HY, Buxbaum AR, Singer RH. 2010. Single mRNA tracking in live cells. *Methods Enzymol* **472:** 387–406.

Parnas O, Jovanovic M, Eisenhaure TM, Herbst RH, Dixit A, Ye CJ, Przybylski D, Platt RJ, Tirosh I, Sanjana NE, et al. 2015. A genome-wide CRISPR screen in primary immune cells to dissect regulatory networks. *Cell* **162:** 675–686.

Perez-Pinera P, Kocak DD, Vockley CM, Adler AF, Kabadi AM, Polstein LR, Thakore PI, Glass KA, Ousterout DG, Leong KW, et al. 2013. RNA-guided gene activation by CRISPR–Cas9-based transcription factors. *Nat Methods* **10:** 973–976.

Peterson VM, Zhang KX, Kumar N, Wong J, Li L, Wilson DC, Moore R, McClanahan TK, Sadekova S, Klappenbach JA. 2017. Multiplexed quantification of proteins and transcripts in single cells. *Nat Biotechnol* **35:** 936–939.

Plath K, Mlynarczyk-Evans S, Nusinow DA, Panning B. 2002. Xist RNA and the mechanism of X chromosome inactivation. *Annu Rev Genet* **36:** 233–278.

Qi LS, Larson MH, Gilbert LA, Doudna JA, Weissman JS, Arkin AP, Lim WA. 2013. Repurposing CRISPR as an RNA-guided platform for sequence-specific control of gene expression. *Cell* **152:** 1173–1183.

Rajagopal N, Srinivasan S, Kooshesh K, Guo Y, Edwards MD, Banerjee B, Syed T, Emons BJ, Gifford DK, Sherwood RI. 2016. High-throughput mapping of regulatory DNA. *Nat Biotechnol* **34:** 167–174.

Rinn JL, Chang HY. 2012. Genome regulation by long noncoding RNAs. *Annu Rev Biochem* **81:** 145–166.

Roadmap Epigenomics C, Kundaje A, Meuleman W, Ernst J, Bilenky M, Yen A, Heravi-Moussavi A, Kheradpour P, Zhang Z, Wang J, et al. 2015. Integrative analysis of 111 reference human epigenomes. *Nature* **518:** 317–330.

Rousseau BA, Hou Z, Gramelspacher MJ, Zhang Y. 2018. Programmable RNA cleavage and recognition by a natural CRISPR–Cas9 system from *Neisseria meningitidis*. *Mol Cell* **69:** 906–914.e904.

Sanborn AL, Rao SSP, Huang S-C, Durand NC, Huntley MH, Jewett AI, Bochkov ID, Chinnappan D, Cutkosky A, Li J, et al. 2015. Chromatin extrusion explains key features of loop and domain formation in wild-type and engineered genomes. *Proc Natl Acad Sci* **112:** E6456–E6465.

Sanjana NE, Wright J, Zheng K, Shalem O, Fontanillas P, Joung J, Cheng C, Regev A, Zhang F. 2016. High-resolution interrogation of functional elements in the noncoding genome. *Science* **353:** 1545–1549.

Schwartz S. 2016. Cracking the epitranscriptome. *RNA* **22:** 169–174.

Shalem O, Sanjana NE, Hartenian E, Shi X, Scott DA, Mikkelsen TS, Heckl D, Ebert BL, Root DE, Doench JG, et al. 2014. Genome-scale CRISPR–Cas9 knockout screening in human cells. *Science* **343:** 84–87.

Shearwin KE, Callen BP, Egan JB. 2005. Transcriptional interference—A crash course. *Trends in Genetics* **21:** 339–345.

Shechner DM, Hacisuleyman E, Younger ST, Rinn JL. 2015. Multiplexable, locus-specific targeting of long RNAs with CRISPR-Display. *Nat Methods* **12:** 664–670.

Shmakov S, Abudayyeh OO, Makarova KS, Wolf YI, Gootenberg JS, Semenova E, Minakhin L, Joung J, Konermann S, Severinov K, et al. 2015. Discovery and functional characterization of diverse Class 2 CRISPR–Cas systems. *Mol Cell* **60:** 385–397.

Shmakov S, Smargon A, Scott D, Cox D, Pyzocha N, Yan W, Abudayyeh OO, Gootenberg JS, Makarova KS, Wolf YI, et al. 2017. Diversity and evolution of class 2 CRISPR–Cas systems. *Nat Rev Microbiol* **15:** 169–182.

Simeonov DR, Gowen BG, Boontanrart M, Roth TL, Gagnon JD, Mumbach MR, Satpathy AT, Lee Y, Bray NL, Chan AY, et al. 2017. Discovery of stimulation-responsive immune enhancers with CRISPR activation. *Nature* **549:** 111–115.

Smargon AA, Cox DB, Pyzocha NK, Zheng K, Slaymaker IM, Gootenberg JS, Abudayyeh OA, Essletzbichler P, Shmakov S, Makarova KS, et al. 2017. Cas13b is a type VI-B CRISPR-associated RNA-guided RNase differentially regulated by accessory proteins Csx27 and Csx28. *Mol Cell* **65:** 618–630.e617.

Spitz F, Furlong EE. 2012. Transcription factors: From enhancer binding to developmental control. *Nat Rev Genet* **13:** 613–626.

Stewart SE, Menzies SA, Popa SJ, Savinykh N, Petrunkina Harrison A, Lehner PJ, Moreau K. 2017. A genome-wide CRISPR screen reconciles the role of N-linked glycosylation in galectin-3 transport to the cell surface. *J Cell Sci* **130:** 3234–3247.

Stoeckius M, Hafemeister C, Stephenson W, Houck-Loomis B, Chattopadhyay PK, Swerdlow H, Satija R, Smibert P. 2017. Simultaneous epitope and transcriptome measurement in single cells. *Nat Methods* **14:** 865–868.

Strutt SC, Torrez RM, Kaya E, Negrete OA, Doudna JA. 2018. RNA-dependent RNA targeting by CRISPR–Cas9. *eLife* doi: 10.7554/eLife.32724.

Swarts DC, van der Oost J, Jinek M. 2017. Structural basis for guide RNA processing and seed-dependent DNA targeting by CRISPR–Cas12a. *Cell* **66:** 221–233.e4.

Tanenbaum ME, Gilbert LA, Qi LS, Weissman JS, Vale RD. 2014. A protein-tagging system for signal amplification in gene expression and fluorescence imaging. *Cell* **159:** 635–646.

Thakore PI, D'Ippolito AM, Song L, Safi A, Shivakumar NK, Kabadi AM, Reddy TE, Crawford GE, Gersbach CA. 2015. Highly specific epigenome editing by CRISPR–Cas9 repressors for silencing of distal regulatory elements. *Nat Methods* **12:** 1143–1149.

Vojta A, Dobrinic P, Tadic V, Bockor L, Korac P, Julg B, Klasic M, Zoldos V. 2016. Repurposing the CRISPR–Cas9 system for targeted DNA methylation. *Nucleic Acids Res* **44:** 5615–5628.

Wang T, Wei JJ, Sabatini DM, Lander ES. 2014. Genetic screens in human cells using the CRISPR–Cas9 system. *Science* **343:** 80–84.

Wang T, Birsoy K, Hughes NW, Krupczak KM, Post Y, Wei JJ, Lander ES, Sabatini DM. 2015. Identification and characterization of essential genes in the human genome. *Science* **350:** 1096–1101.

Wu Y, Zhou L, Wang X, Lu J, Zhang R, Liang X, Wang L, Deng W, Zeng YX, Huang H, et al. 2016. A genome-scale CRISPR–Cas9 screening method for protein stability reveals novel regulators of Cdc25A. *Cell Discov* **2:** 16014.

Yamano T, Nishimasu H, Zetsche B, Hirano H, Slaymaker IM, Li Y, Fedorova I, Nakane T, Makarova KS, Koonin EV, et al. 2016. Crystal structure of Cpf1 in complex with guide RNA and target DNA. *Cell* **165:** 949–962.

Yan WX, Chong S, Zhang H, Makarova KS, Koonin EV, Cheng DR, Scott DA. 2018. Cas13d is a compact RNA-targeting type VI CRISPR effector positively modulated by a WYL-domain-containing accessory protein. *Mol Cell* **70:** 327–339.e5.

Yin Y, Yan P, Lu J, Song G, Zhu Y, Li Z, Zhao Y, Shen B, Huang X, Zhu H, et al. 2015. Opposing roles for the lncRNA *Haunt* and its genomic locus in regulating *HOXA* gene activation during embryonic stem cell differentiation. *Cell Stem Cell* **16:** 504–516.

Zalatan JG, Lee ME, Almeida R, Gilbert LA, Whitehead EH, La Russa M, Tsai JC, Weissman JS, Dueber JE, Qi LS, et al. 2015. Engineering complex synthetic transcriptional programs with CRISPR RNA scaffolds. *Cell* **160:** 339–350.

Zappulla DC, Cech TR. 2004. Yeast telomerase RNA: A flexible scaffold for protein subunits. *Proc Natl Acad Sci* **101:** 10024–10029.

Zetsche B, Gootenberg JS, Abudayyeh OO, Slaymaker IM, Makarova KS, Essletzbichler P, Volz SE, Joung J, van der Oost J, Regev A, et al. 2015. Cpf1 is a single RNA-guided endonuclease of a class 2 CRISPR–Cas system. *Cell* **163:** 759–771.

Zetsche B, Heidenreich M, Mohanraju P, Fedorova I, Kneppers J, DeGennaro EM, Winblad N, Choudhury SR, Abudayyeh OO, Gootenberg JS, et al. 2017. Multiplex gene editing by CRISPR–Cpf1 using a single crRNA array. *Nat Biotechnol* **35:** 31–34.

Zhou Y, Zhu S, Cai C, Yuan P, Li C, Huang Y, Wei W. 2014. High-throughput screening of a CRISPR/Cas9 library for functional genomics in human cells. *Nature* **509:** 487–491.

Zhu S, Li W, Liu J, Chen C-H, Liao Q, Xu P, Xu H, Xiao T, Cao Z, Peng J, et al. 2016. Genome-scale deletion screening of human long non-coding RNAs using a paired-guide RNA CRISPR–Cas9 library. *Nature Biotechnol* **34:** 1279–1286.

Drugging the RNA World

Matthew D. Disney, Brendan G. Dwyer, and Jessica L. Childs-Disney

Department of Chemistry, The Scripps Research Institute, Jupiter, Florida 33458

Correspondence: disney@scripps.edu

SUMMARY

Although we live in the remnants of an RNA world, the world of drug discovery and chemical probes is firmly protein-centric. Developing highly selective small molecules targeting RNA is often considered to be an insurmountable challenge. Our goal is to demystify the design of such compounds. In this review, we describe various approaches to design small molecules that target RNA from sequence and the application of these compounds in RNA biology, with a focus on inhibition of human RNA–protein complexes. We have developed a library-versus-library screening approach to define selective RNA–small-molecule binding partners and applied them to disease-causing RNAs, in particular noncoding oncogenic RNAs and expanded RNA repeats, to modulate their biology in cells and animals. We also describe the design of new types of small-molecule probes that could broadly decipher the mysteries of RNA in cells.

Outline

Cite as *Cold Spring Harb Perspect Biol* doi: 10.1101/cshperspect.a034769

1 INTRODUCTION

1.1 Sequence-Structure: The Tree of Life in the RNA World

Robert Holley and colleagues were the first to sequence a biomolecule, transfer RNA (tRNA)^{Ala}, after purifying modest amounts to study its role in protein synthesis. Phylogenetic sequence alignment and enzymatic probing suggested that tRNA adopted a cloverleaf structure (Holley 1965). Indeed, all tRNAs were found to adopt this general structure (Dudock et al. 1969). These insights were critical to the fundamental understanding of translation, an RNA-driven process. The interaction between the codon and anticodon, formed when the tRNA adopts its cloverleaf structure, decodes a messenger RNA (mRNA) into a polypeptide.

Although these investigations occurred more than 60 years ago, their reverberations are still felt today. The work of Holley and colleagues stimulated great efforts into developing improved methods for RNA sequencing (RNA-seq). These studies marched from (1) the plus and minus method of Sanger sequencing, to (2) Gilbert–Maxam sequencing, to (3) dideoxy sequencing, and, finally, to (4) the next-generation sequencing technology of today that can sequence an entire genome in a day for <$1000. Perhaps one of the most interesting aspects of sequencing has been the recognition that noncoding RNAs are numerous in cells, laying the foundation for future studies to explore their functions.

Holley's studies showed that the RNA sequence is only half of the equation; the structure adopted by tRNAs was essential to carry out its function in translation. Carl Woese, another giant in the history of science, was interested in deducing structure from sequence. Indeed, he obtained the sequence of ribosomal RNAs (rRNAs) from diverse organisms and used sequence alignment to deduce their secondary structures (Woese and Pace 1993). In particular, Woese was interested in identifying "molecular fossils" to rigorously trace the origins and associations of secondary structures. Associations of these molecular fossils to each other defined the tree of life, a controversial view at the time. Before Woese's work, organisms were simply classified by having or not having a nucleus. We now recognize that the phylogenetic tree of life can be classified into bacteria, archaea, and eukarya, not simply prokaryotes and eukaryotes. Holley's and Woese's studies as well as the discovery of catalytic RNAs (Stark et al. 1978; Zaug and Cech 1986) advanced the concept of an RNA world.

The studies described above inspired a keen interest in developing approaches to experimentally map RNA structure (Woese et al. 1980). The same Gilbert–Maxam reagents for DNA sequencing also react with RNA and can be tuned to identify unpaired bases (Peattie and Gilbert 1980). Studies by Cech (Inoue and Cech 1985) and Noller (Stern et al. 1986) used these reagents and showed that reverse transcriptase (RT) cannot proceed through modified bases, allowing one to use sequencing methods to read out unpaired bases. Computational approaches have also emerged to annotate RNA structure from sequence. These studies use the energetics of base pairing to predict the thermodynamically most stable RNA structure from its sequence and can also predict suboptimally folded (i.e., less energetically stable) structures. Indeed, computational algorithms have accurately predicted RNA structure as shown in the work of Turner (Jaeger et al. 1989) and Zuker (Zuker 1989). These advancements were followed by efforts to combine experimental and computational approaches to further refine these predictions, including those generated by cell-permeable chemical modification reagents that map RNA structure in cells (Wells et al. 2000; Mathews et al. 2004). The combination of computation and experiment has elucidated the structures of entire viral genomes (Deigan et al. 2009; Watts et al. 2009) and identified patterns in RNA structure within the human transcriptome (Bevilacqua et al. 2016; Lu and Chang 2016). Mapping the sequence and structure of RNAs depicted the tree of life in the RNA world and established the foundation for recent advancements in drugging RNA as we will discuss later in this review.

2 TARGETING RNA WITH SMALL MOLECULES

2.1 Ribosomal RNA and Riboswitches as Targets of Small Molecules

Streptomycin was the first-in-line treatment for tuberculosis infections and was isolated from bacteria, along with other natural products such as neomycin, in 1943 by Albert Schatz while he was a Ph.D. student in the laboratory of Selman Waksman (Jones et al. 1944). Both aminoglycoside antibiotics have served as invaluable drugs to treat otherwise fatal bacterial infections by inhibiting protein synthesis. Interestingly, streptomycin ushered in several historical advances in treatment and clinical trials. For example, it was used in the first double-blind trial ever completed to gain approval for use in humans. However, resistance of tuberculosis to streptomycin treatment was observed shortly after its introduction.

Noller and Moazed applied chemical probing methods to gain insight into the binding of antibiotics, including aminoglycosides, to the ribosome (Moazed and Noller 1987). Several features were identified among ligand binding sites, most notably (1) resistance correlated with mutations in the rRNA-binding site, and (2) these mutations were often located in functional sites, namely, the tRNA-binding site and the decoding site.

Cite as *Cold Spring Harb Perspect Biol* doi: 10.1101/cshperspect.a034769

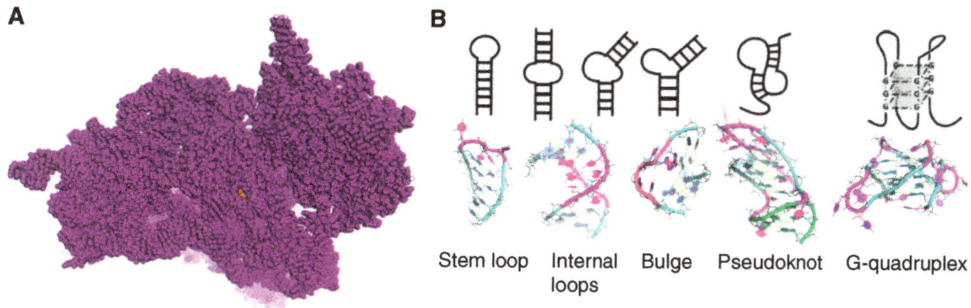

Figure 1. The ribosome is a well-known target of small molecules. (*A*) Various compounds bind to specific sites in the bacterial ribosome and protect them from chemical modification. High-resolution structures of the ribosomal subunits have been determined in the presence and absence of antibiotics. Shown here is the 30S subunit complexed with the aminoglycoside puromycin; the ribosome subunit is shown in purple and puromycin is shown in Corey, Pauling, and Kolton (CPK) colors. (*B*) Common secondary structural elements found in cellular RNAs.

Footprinting and biophysical analysis by Puglisi and colleagues further showed that the binding of antibiotics to the rRNA aminoacyl (A)-site affected decoding by stabilizing the binding of noncognate tRNAs (Yoshizawa et al. 1999). Subsequently, high-resolution structures of the rRNA particle in complex with antibiotics by the Ramakrishnan group confirmed Puglisi's studies completed with the A-site model (Fig. 1) (Carter et al. 2000). Interestingly, an A-site model was also used to show that antibiotic potency is correlated with altering A-site dynamics, not affinity (Kaul et al. 2006). Collectively, these studies established the sites of antibacterial action and showed that smaller RNA models are effective tools for studying small-molecule binding and can provide insight into therapeutic action.

Riboswitches have also emerged as tried-and-true targets of small molecules affecting RNA. Henkin and Breaker identified these small regulatory RNAs, primarily found in bacteria, that switch conformation in the presence of sufficient concentrations of small-molecule metabolites (Winkler et al. 2002). This conformational switch inhibits the synthesis of proteins responsible for production of the metabolite (Nahvi et al. 2002; Grundy and Henkin 2004). Structural mimicry approaches have been used to design riboswitch inhibitors that short-circuit metabolic synthesis pathways (Blount et al. 2007) and hence serve as antibacterials. Screening has also been used to identify inhibitors of riboswitches. For example, scientists at Merck identified ribocil as targeting bacterial riboflavin riboswitches by using a phenotypic screen (Howe et al. 2015). Studies show that the compound acts as a structurally distinct synthetic mimic of the natural ligand, flavin mononucleotide. Taken together, studies of the ribosome and riboswitches showed that RNA is a bona fide target of small molecules and that ligands can indeed bind discrete sites in RNA.

3 USING RNA STRUCTURE TO DESIGN SELECTIVE SMALL MOLECULES TARGETING RNA

3.1 The Transcriptome as a Source of Druggable RNAs: Potential and Perceived Risks

Although antibacterial agents targeting the ribosome have had an enormous impact on biomedical research and have provided invaluable chemical tools to understand biology, progress in targeting other types of RNAs has been modest to date. A large body of work has generated small molecules that target viral RNAs, particularly those from human immunodeficiency virus (HIV) and hepatitis C, among others (Hamy et al. 1997; Parsons et al. 2009), squarely cementing the applicability of small-molecule treatment to target RNAs involved in infectious disease.

There are few examples, however, of small molecules that target human RNAs associated with disease. Recently, Novartis and PTC/Roche identified compounds that modulate the splicing of a precursor mRNA (pre-mRNA) for the treatment of spinomuscular atrophy (Naryshkin et al. 2014; Palacino et al. 2015). Novartis' compound is thought to bind and stabilize an RNA structure at an exon–intron junction, thereby affecting exon inclusion. Other small molecules have been found to target pre-mRNAs and affect splice site selection (Luo and Disney 2014) or stimulate readthrough of premature termination codons in various diseases, particularly Duchenne muscular dystrophy (Welch et al. 2007), although the mechanism of action has been controversial (Auld et al. 2009).

Advancing small-molecule targeting of human RNAs is particularly timely as it is now appreciated that there are many more potential RNA targets than protein targets. Recent estimates suggest that 1%–2% of the human genome is translated into protein, yet >50% is transcribed into RNA (Clamp et al. 2007; Gerstein et al. 2007). Further, only ~15% of proteins are in families that are druggable

(Hopkins and Groom 2002). Although many RNAs may not be functional, these Encyclopedia of DNA Elements (ENCODE) studies have revealed that there is a large amount of druggable space in the transcriptome. The studies described herein are changing the perceived notion that RNA is undruggable.

Why has RNA been considered undruggable with small molecules in the past? The most highly explored and studied targets have complex three-dimensional structures with "protein-like" folds, which were thought to be required for selective small-molecule binding. Most RNAs, including microRNAs (miRNAs) and mRNAs, have defined secondary structure (see Fig. 1B for examples of secondary structures) but limited tertiary structure and were thus considered untargetable and not suitable for drug discovery efforts. It is an often-missed irony that many known drugs are not "drug-like" based on Lipinski guidelines (Lipinski 2004), which are often over- and/or misstated. Considering RNA as undruggable is of the same nature.

3.2 RNA Secondary Structure and Disease: RNA-Binding Proteins

RNA secondary structure regulates many cellular processes including (1) pre-mRNA alternative splicing (recognition of intron–exon junctions by splicing regulators), (2) translation (ribosomal loading controlled by secondary structure, including G-quadruplexes, and 5′ and 3′ untranslated regions [UTRs]), (3) cellular trafficking and delivery (by binding to RNA-binding proteins [RBPs]), (4) RNA editing, and (5) miRNA biogenesis. Notably, disease occurs when RNA secondary structure is altered by mutation or repeated too frequently, as observed in microsatellite (RNA repeat expansion) disorders. Thus, targeting RNA–protein interactions that lead to cellular dysfunction, in particular those that involve defined RNA secondary structure, could provide a therapeutic strategy. For a comprehensive review of how RNA secondary structure influences cellular processes and disease, please see Bernat and Disney (2015). Herein, we focus on small-molecule targeting of human RNAs that interact with proteins. Many excellent reviews are available that discuss targeting of other RNAs (Thomas and Hergenrother 2008; Guan and Disney 2012), including specialized reviews on targeting the bacterial ribosome (Yonath 2005; Blanchard et al. 2010), riboswitches (Blount and Breaker 2006), and viruses (Hermann 2016).

3.3 Identifying Selective RNA Motif–Small Molecules via Two-Dimensional Combinatorial Screening (2DCS)

Collectively, the studies that target mRNAs described above suggest that (1) targeting RNA secondary structure is relevant to a wide variety of diseases, particularly those in which a secondary structure binds a protein, and (2) human RNAs with extensive secondary structure but limited tertiary structure can indeed be targeted with small molecules. To do so by rational design, two types of data are required: the RNA's secondary structure, or composite of motifs, and a database of small molecules and the RNA motifs that are privileged for binding them. As described in the Introduction, accurate models of RNA secondary structure can be generated by a combination of free-energy minimization, experimental constraints, and phylogenetic alignment. Unfortunately, there has been a dearth of information about privileged chemical space for RNA and targetable RNA motifs. To provide data of this type, a library-versus-library screen approach dubbed two-dimensional combinatorial screening (2DCS) was developed (Fig. 2) (Disney et al. 2008). The 2DCS approach allows the assignment of the RNA motif–binding landscapes for small molecules (Velagapudi et al. 2010, 2017).

In a 2DCS experiment, a library of small molecules is immobilized onto an agarose-coated microarray. The agarose substrate was carefully chosen as it allows (1) spatial encoding of the small molecules, (2) increased loading of the compounds on the surface, and (3) excision of bound RNAs. The small-molecule microarrays are then incubated with an RNA motif library that displays a randomized region in the pattern of a discrete secondary structure, whether a bulge, internal loop, or hairpin (Fig. 1B). Importantly, incubations are completed under highly stringent conditions; that is, a large excess of oligonucleotides that mimic regions common to all members of the RNA library is included during hybridization. These competitor oligonucleotides are either single-stranded or duplex RNAs that cannot be amplified by primers selective for members of the RNA motif library. Bound RNAs are excised from the array surface, amplified, and sequenced.

Selected RNA motif–small-molecule interactions are assigned a fitness of binding via a scoring function, a method named structure–activity relationships through sequencing (StARTS) and its high-throughput variant using RNA-seq, HiT-StARTS (Velagapudi et al. 2010, 2017). A scoring function is defined by sequencing the starting library and comparing the frequency of each of the individual motifs with the frequency of the same motif in the selected RNA pool. This pooled population comparison affords a Z_{obs} score for each RNA motif–small-molecule pair. RNA–small-molecule motifs with larger Z-scores bind with higher affinity than ones with lower ones, defining a binding landscape that affords the scoring function for each selection. Further, relative Z_{obs} scores between

Cite as *Cold Spring Harb Perspect Biol* doi: 10.1101/cshperspect.a034769

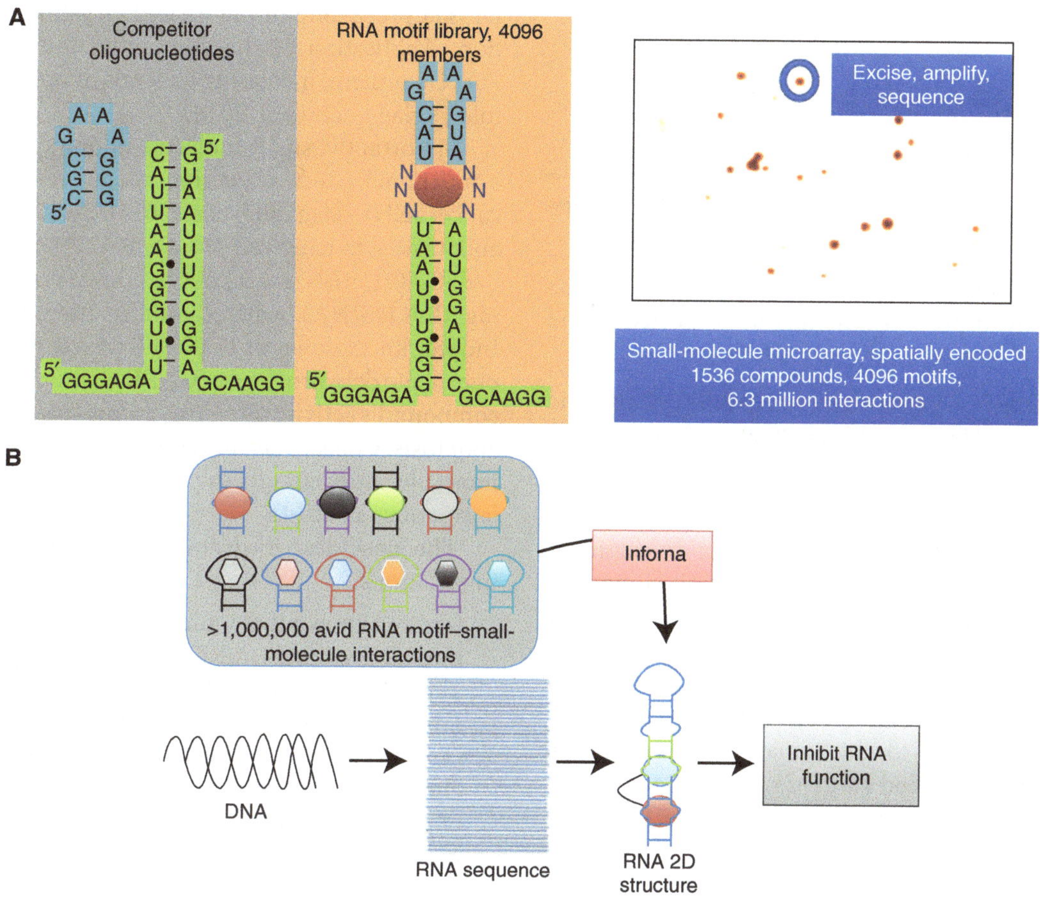

Figure 2. Sequence- and secondary structure-based approaches developed for small molecules targeting RNA. (*A*) Schematic of two-dimensional combinatorial screening (2DCS), a library-versus-library screen. A small-molecule library is immobilized onto a microarray and hybridized with an RNA motif library. The selected RNAs are sequenced and analyzed to form a chemical code for selectively targeting RNA. (*B*) Schematic of Inforna, a computational method to identify "druggable" RNA targets from sequence. The selected RNA motif–small-molecule partners from 2DCS are mined against the transcriptome to agnostically identify RNAs that can be targeted with these small molecules.

RNA motif–small-molecule partners can rapidly define ligand selectivity.

Collectively, 2DCS studies have shown that a myriad of ligands can bind RNA with affinities that range from low nanomolar to mid-micromolar (Disney et al. 2008; Velagapudi et al. 2010, 2017; Tran and Disney 2012). Furthermore, compound classes that bind RNAs range from aminoglycosides to heterocyclic and drug-like small molecules. Last, a broad swath of RNA motifs can selectively bind small molecules, including symmetric and asymmetric internal loops, bulges, and hairpins (Fig. 2). Thus, there are tremendous opportunities to exploit both a wide variety of chemistries and a diverse array of RNA structural motifs when drugging RNA, an irony given the aforementioned long history of skepticism concerning the "druggability" of RNA.

3.4 Merging the Results of 2DCS with Bioinformatics: Inforna, a Sequence-Based Approach to Design Small Molecules Targeting RNA

To design small molecules that specifically target RNA structures based on its sequence, we developed a computational approach dubbed Inforna (Velagapudi et al. 2014; Disney et al. 2016). Inforna integrates RNA motif–small-molecule interactions identified via 2DCS and structural data of target RNAs. That is, it generates lead compounds for an RNA of interest by comparing the motifs found in the RNA target's structure with the RNA motif–small-molecule interactions in our database. Because all RNAs have secondary structures, the approach is broadly applicable to identify rationally selective small-molecule chemical probes and lead medicines targeting RNA.

3.5 Targeting Protein-Binding Sites in Disease-Causing miRNAs

miRNAs are small, noncoding RNAs that regulate gene expression by binding to the 3′ UTRs of mRNAs with complementary sequences (Fire et al. 1998; Lagos-Quintana et al. 2001; Lee and Ambros 2001). Transcribed in a primary (pri-) form, miRNAs are processed in two steps:

1. Cleavage of 5′ and 3′ sequences by the nuclear nuclease Drosha to afford pre-miRNAs that are exported to the cytoplasm.
2. Cleavage of the pre-miRNA's hairpin by the cytoplasmic nuclease Dicer (Hammond et al. 2000).

Nuclease processing sites represent functional motifs, and inhibition of cleavage by Drosha or Dicer prevents biogenesis of the miRNA. Below, we describe the development of small molecules that have effectively inhibited maturation of disease-causing miRNAs by binding to Drosha or Dicer processing sites.

In the first broad application of Inforna (Velagapudi et al. 2014), all human pre-miRNAs were folded and compared with the RNA folds to which small-molecule-binding partners had been identified via 2DCS. Thus, the approach is target-agnostic rather than target-centric, the latter occurring when a single or a few target biomolecules are screened against a chemical library. Inforna identified compounds that target nearly two dozen disease-associated miRNAs that have broad therapeutic implications. Nearly half of these interactions were bioactive without lead optimization. Activities of the compounds ranged from triggering apoptosis in breast cancer cells (Velagapudi et al. 2014), short-circuiting hypoxic signaling in breast cancer cells (Costales et al. 2017), and inhibiting the invasive properties of hepatocellular carcinoma (Childs-Disney and Disney 2016).

The first target vetted by Inforna was a small molecule that targets the Drosha site of pri-miR-96 (Velagapudi et al. 2014). In triple-negative breast cancer, the miR-96-182-183 cluster is up-regulated. Two of the miRNAs, miR-96 and miR-182, repress the pro-apoptotic transcription factor forkhead box protein O1 (FOXO1). FOXO1 controls transcription of proapoptotic members of the B-cell lymphoma 2 (Bcl-2) family. In vitro, the lead compound bound the Drosha site with a 2 µM dissociation constant and inhibited Drosha processing of a model pri-miRNA transcript (Velagapudi et al. 2014). Treatment of Michigan Cancer Foundation-7 (MCF-7) breast cancer cells with 40 µM compound decreased levels of mature miR-96 by 90% but did not significantly affect any of the other miRNAs in the miR-96-182-183 cluster. Additionally, the compound increased the amount of the pri-miR-96, indicating that inhibition of Drosha processing of the desired target is the mode of action. Application of the compound (40 µM) derepressed FOXO1 protein, increasing its levels by approximately two-fold (Velagapudi et al. 2014).

Importantly, small-molecule inhibition of miR-96 had a functional effect, triggering apoptosis of MCF-7 breast cancer cells (Velagapudi et al. 2014). To investigate whether apoptosis was triggered via the FOXO1-miR-96 pathway, the FOXO1 mRNA was knocked down via a small interfering RNA (siRNA), which should ablate compound activity. Indeed, knockdown of FOXO1, but not glyceraldehyde 3-phosphate dehydrogenase (GAPDH), mRNA decreased the compound's activity by ~70%, as measured by induction of apoptosis. Collectively, these studies indicate that the small molecule short-circuits the FOXO1-miR-96 pathway (Velagapudi et al. 2014).

The data described above have important biological implications.

1. It is well known that multiple miRNAs can affect a single mRNA and single miRNAs can affect multiple mRNAs. Yet, in the case of miR-96 and cancer, it appears that inhibition of a single miRNA targeting FOXO1 is sufficient to reverse a disease-associated phenotype. Such effects would be difficult to determine using antisense oligonucleotides because the sequences of miR-96 and miR-183 are so similar that an oligonucleotide would target both.
2. We compared the selectivity of the small molecule with an antisense agent for reducing the level of mature miRNAs, and the small molecule is at least as selective, if not more selective at this level (Velagapudi et al. 2014).

Although our anti-miR-96 small molecule selectively inhibited biogenesis of miR-96, its activity was not sufficient to test in vivo. Thus, we used Inforna to identify compounds that bind to sites near the Drosha site of pri-miR-96 (Velagapudi et al. 2016). By linking the two compounds together, a higher affinity binder was designed named Targaprimir-96 (Fig. 3A). This compound had a >200-fold enhancement in avidity for the target sites in the pri-miR-96 with a K_d of 20 nM. Application of the Targaprimir-96 to MDA-MB-231 triple-negative breast cancer cells selectively inhibited production of miR-96 and boosted pri-miR-96 levels (Velagapudi et al. 2016). The molecule selectively induced apoptosis in breast cancer cells in which miR-96 is up-regulated but did not affect healthy breast cells in which miR-96 is expressed at normal levels. Importantly, this dimeric molecule slowed tumor growth in a xenograft mouse model, reduced mature miR-96 levels, and increased FOXO1 expression in tumors (Velagapudi et al. 2016). This is the first example of a small molecule that can inhibit tumor growth by targeting a noncoding RNA.

Cite as *Cold Spring Harb Perspect Biol* doi: 10.1101/cshperspect.a034769

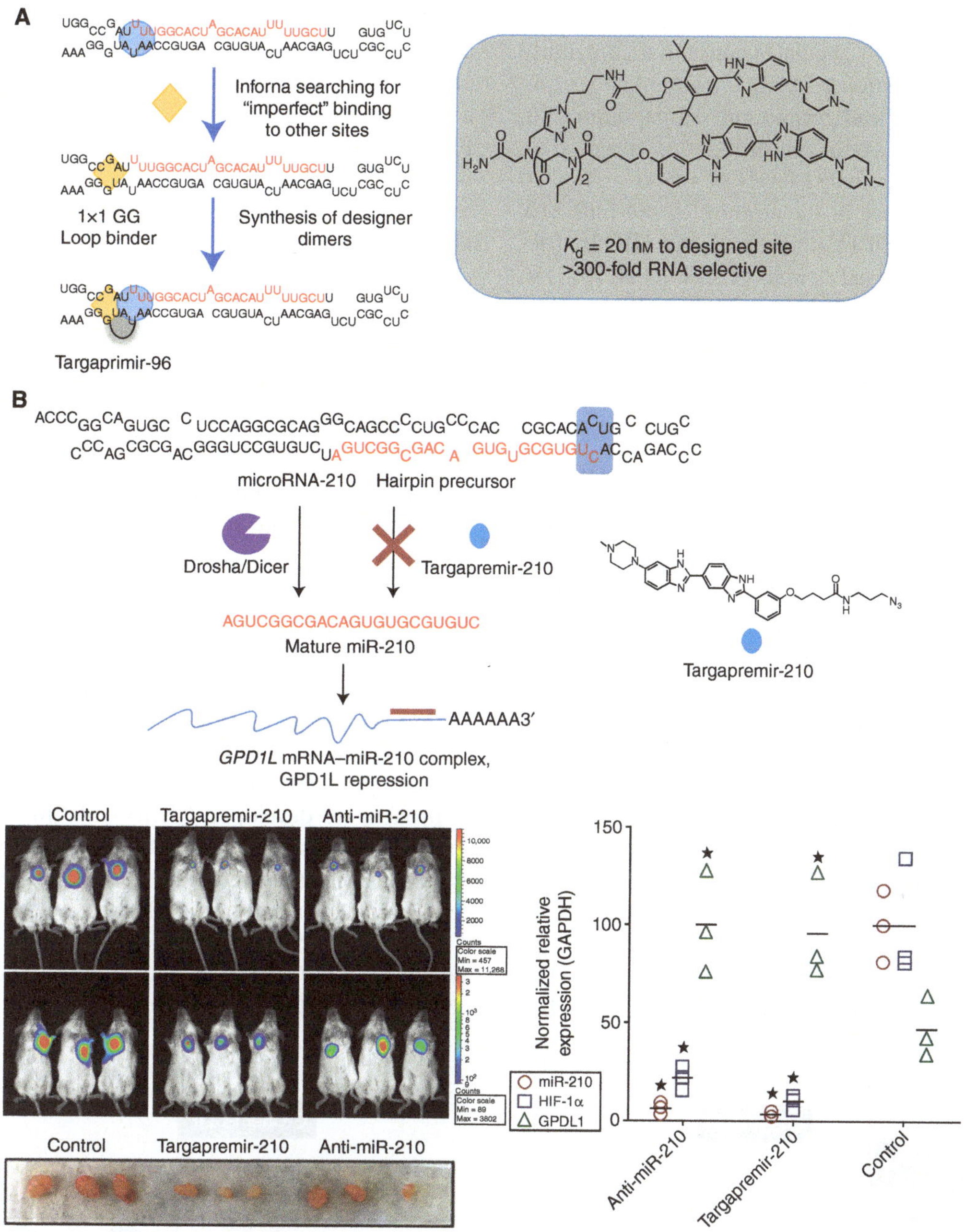

Figure 3. Examples of targeting RNA with small molecules that have been designed by using Inforna. (*A*) Targaprimir-96 targets the Drosha site in the microRNA-96 (miR-96) hairpin precursor. The compound inhibits miR-96 biogenesis and triggers apoptosis via a miR-96-forkhead box protein O1 (FOXO1) circuit. (*B*) Targapremir-210 silences miR-210, induces apoptosis of hypoxic, but not normoxic, cells, and reduces tumor burden in a xenograft mouse model.

miRNAs can also be up-regulated in response to changes in the cellular environment, for example when cancer cells are in a low oxygen (hypoxic) environment, which leads to resistance to chemotherapeutics. During the transition from normoxia to hypoxia, miR-210 is up-regulated and forms a hypoxic circuit with hypoxia-inducible factor 1-α (HIF-1α) (Semenza 2003). In particular, miR-210 stimulates a hypoxic response as it targets the 3′ UTR of glycerol-3-phosphate dehydrogenase 1-like enzyme (GPD1L); GPD1L stabilizes prolyl hydroxylase (PHD). Under normoxic conditions, PHD hydroxylates prolines on HIF-1α that causes it to be proteolyzed by the proteasome.

In the absence of stable PHD, HIF-1α dimerizes with HIF-1β, which transcriptionally regulates genes that contribute to cancer metastasis (Semenza 2003).

Inforna identified a lead compound, Targapremir-210, that bound to the Dicer site of pre-miR-210 (Costales et al. 2017). The compound reduced levels of mature miR-210 with concomitant increase in the amount of pre-miR-210. Expected changes in GDP1L (increased) and HIF-1α (decreased) levels were observed as well (Costales et al. 2017) to extents similar to those observed on anti-miR-210 antagomir treatment (an antagomir is an antisense oligonucleotide). Further, application of Targapremir-210 to hypoxic cells, but not to normoxic cells, caused the cells to undergo apoptosis (Fig. 3B). Profiling experiments showed that the small molecule was as selective as the oligonucleotide. Furthermore, forced overexpression of pre-miR-210 ablated the ability of Targapremir-210 to stimulate apoptosis (Costales et al. 2017). Thus, small molecules can target one portion of a biological circuit, in particular in a hypoxic circuit, and affect biological outcome, in this case stimulation of apoptosis.

One of the fastest routes to push a compound into clinical development is to repurpose a known drug with established tolerability and toxicological profiles. Thus, 2DCS was used to profile the RNA-binding abilities of known drug derivatives to infer new targets. Interestingly, 5′-azido neomycin B binds the Drosha processing site in pre-miR-525 (Velagapudi et al. 2014; Childs-Disney and Disney 2016). This miRNA is up-regulated in ~60% of liver cancers and confers invasive properties to hepatocellular carcinoma (HCC) cells by silencing zinc finger protein 395 (ZNF395). Application of 5′-azido neomycin B and its Food and Drug Administration (FDA)-approved derivative neomycin B to liver cancer cells silenced production of mature miR-525 and enhanced production of ZNF395 (Childs-Disney and Disney 2016). The compound also effectively modulated the invasive phenotype mediated by miR-525. Thus, FDA-approved drugs can inhibit miRNA-associated phenotypes in designed and programmable ways. Interestingly, neomycin B is a second-line agent for hepatic encephalopathy (HE) and bacterial infections due to cirrhosis.

3.6 Profiling RNA–Small-Molecule Interactions in Cells: Chemical Cross-Linking and Isolation by Pull-Down (Chem-CLIP)

One major void in the development of small molecules that target RNA is established cellular target recognition profiling methods. Work of the Darnell group has developed cross-linking and immunoprecipitation (CLIP), which cross-links proteins to RNA to identify binding partners (Ule et al. 2003, 2018). In CLIP, ultraviolet (UV) light is used to cross-link protein–RNA complexes, which are purified with antibody-functionalized beads and analyzed by sequencing. CLIP has been an invaluable tool used to study multiple pathways, including identification of protein-binding sites on pre-mRNAs that influence splicing outcomes.

As light can limit the applicability of a CLIP-based approach, we developed Chem-CLIP in which the cross-linking event is a proximity-based chemical reaction; that is, light is not required (Fig. 4) (Guan and Disney 2013). Small molecules that target RNA have been appended with a nucleic acid–reactive module (chlorambucil) and a biotin purification tag, or Chem-CLIP probes. The RNA-binding modules drive binding to their RNA targets in cells, effec-

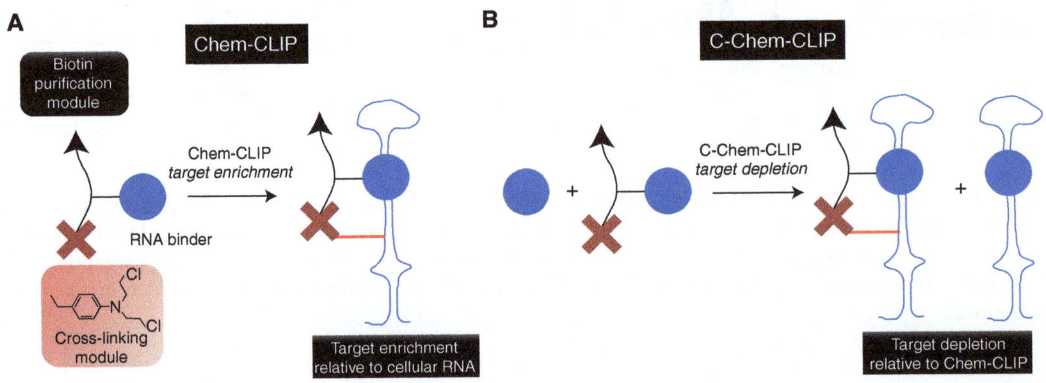

Figure 4. Schematic of chemical cross-linking isolated by pull-down (Chem-CLIP), an approach to identify the RNA targets of small molecules in cells. (*A*) Small molecules targeting RNA are appended with cross-linking and biotin modules. Application of Chem-CLIP probes to cells cross-links the small molecule to its cellular RNA targets. The biotin tag allows the targets to be purified from cells, which are identified by quantitative reverse transcription polymerase chain reaction (RT-qPCR) or RNA sequencing (RNA-seq). (*B*) In competitive Chem-CLIP (C-Chem-CLIP), cells are cotreated with the Chem-CLIP probe and the parent compound. Bona fide targets are depleted in the pulled-down fraction.

tively biotinylating them upon cross-linking. By using streptavidin, the small molecule–RNA complexes can be isolated and analyzed by using RNA-seq or quantitative reverse transcription polymerase chain reaction (RT-qPCR). Chem-CLIP not only identifies bound RNAs but also can identify ligand-binding sites within the transcripts in cells (Yang et al. 2015; Rzuczek et al. 2017). In fact, Chem-CLIP has shown that small molecules can be allele-selective when designed to target expanded RNA repeats (Rzuczek et al. 2017).

Chem-CLIP has been broadly applied to study the druggability of noncoding miRNA precursors. For example, the Targapremir-210 Chem-CLIP probe selectively reacted with pre-miR-210 in vitro, whereas a control compound lacking the RNA-binding module did not (Costales et al. 2017). In cells, the compound pulled down miR-210 from hypoxic triple-negative breast cancer cells, as detected by an enrichment in the pulled-down fraction as compared with unpurified lysate. To study the selectivity of Targapremir-210, we studied ligand occupancy via Chem-CLIP of various abundant noncoding RNAs, abundant mRNAs, hypoxia-associated miRNAs, and miRNAs that have potential ligand-binding sites for Targapremir-210. The latter we dub "RNA isoforms" or the RNA targets that have motifs predicted by Inforna to bind the lead small molecule, including even those that might have lower affinities. Importantly, these Chem-CLIP experiments showed that miR-210 is the most enriched RNA in the pulled-down fractions. For the hypoxia-associated miRNAs, miR-210 was the most pulled down, by at least threefold (Costales et al. 2017). Further, 14 other miRNA precursors contained potential binding sites for Targapremir-210, including miR-497, which has the same motif as miR-210 in its Dicer site. None of these miRNAs was enriched in the pull-down fraction (Costales et al. 2017).

To assess the selectivity of Targapremir-210 on a transcriptome-wide scale, we studied its effect on 92 highly abundant transcripts in the pulled-down fraction including rRNAs, small RNAs (sRNAs), tRNAs, and mRNAs. Only six RNAs were enriched in the pulled-down fractions (Costales et al. 2017). Competitive-Chem-CLIP (C-Chem-CLIP) experiments were completed to assess if any of these targets were pulled down owing to some lack of selectivity of the cross-linker. Bona fide targets will be depleted in the pull-down fraction when an excess of a competitive, unreactive ligand is added; targets that nonselectively reacted with the cross-linker will not be competed off by the parent compound. Indeed, a few of the enriched targets were not competed off, suggesting some nonspecific binding/cross-linking due to chlorambucil or biotin modules of the Chem-CLIP probe (Costales et al. 2017). Collectively, these results suggest that bioactivity of a small molecule is influenced by (1) its affinity for an RNA target (of note, the affinity of Targapremir-210 for miR-210's Dicer site is 165 nM, whereas its affinity for other motifs ranged from 2.5 to 11 μM); (2) the expression level of the target, as miR-497 is expressed at 10-fold lower levels than miR-210; and (3) whether the small-molecule binding site is in the Dicer or Drosha processing site.

3.7 Targeting Protein-Binding Sites in RNA Repeat Expansions

Microsatellite diseases are a class of neurological and neuromuscular disorders caused by repeat expansions (Orr and Zoghbi 2007). Repeat expansions are located in (1) open reading frames (ORFs; Huntington's disease [HD]), (2) UTRs (myotonic muscular dystrophy type 1 [DM1], fragile X–associated tremor ataxia syndrome [FXTAS], and fragile X syndrome [FXS]), and (3) introns (amyotrophic lateral sclerosis [ALS] and myotonic dystrophy type 2 [DM2]). For those diseases caused by sequences that lie in noncoding regions, a major driver of disease-associated toxicity can be the RNA transcribed from the repeat expansion, in particular its sequestration of RNA-binding proteins. In the cases of DM1, DM2, and FXTAS, the RNA repeat expansions bind and inactivate proteins that regulate alternative pre-mRNA splicing (Fardaei et al. 2002; Disney et al. 2012; Sellier et al. 2013). It is likely that the RNA repeat expansion that causes ALS also sequesters a splicing regulator; however, it is currently thought that toxicity is mainly driven by production of toxic dipeptide repeats via repeat-associated non-ATG (RAN) translation (Mori et al. 2013; Zu et al. 2013; Zhang et al. 2014) and defects in nuclear pore transport (Zhang et al. 2015). This disease pathomechanism, sequestration of RNA-binding proteins, provides a therapeutic strategy in which a small molecule binds the RNA repeat expansion and inhibits sequestration of RNA-binding proteins such that they can complete their normal cellular functions.

Various groups have tackled targeting RNA repeat expansions with small molecules, whether by design or screening. Indeed, compounds have been developed that improve disease-associated defects in cells and in *Drosophila* and mouse models. The most well-studied RNA repeat expansion is the trinucleotide repeat that causes DM1, $r(CUG)^{exp}$, which binds and sequesters the RBP splicing regulator muscleblind-like 1 (MBNL1), among others (Fardaei et al. 2002). Located in the 3′ UTR of the dystrophia myotonica protein kinase (DMPK) mRNA (Brook et al. 1992), $r(CUG)^{exp}$ folds into a hairpin structure with periodically repeated 1×1 UU internal loops, the binding sites for MBNL1. Compounds designed by various laboratories have focused on engineering selective binding to these loops. For example, using a structure of r(CUG) repeats, the Zimmerman laboratory designed a small molecule comprised of an acridine (well-known DNA intercalators) and a triaminotriazine that interacts with the RNA via Janus-wedge hydrogen bonding (Arambula et al. 2009). Optimization of this first-generation compound afforded three bioactive compounds that improved missplicing and formation of nuclear foci in a DM1 cellular model at concentrations between 50 and 100 µM; one second-generation compound improved defects in a *Drosophila* model (Jahromi et al. 2013).

Using Inforna, our laboratory has developed various compounds that improve DM1-associated defects in cellular (including DM1 patient-derived cells) and mouse models (Childs-Disney et al. 2012; Parkesh et al. 2012; Rzuczek et al. 2013, 2017). First-generation compounds were based on a *bis*-benzimidazole RNA-binding module (H) that were multivalently displayed on a peptoid scaffold; the optimal compounds display H modules at the precise distance that separated UU loops in r(CUG)^{exp}, or *n*H-4 (in which *n* indicates the number of H RNA-binding modules and 4 indicates that four propylamine spacing modules separate the Hs). Indeed, 2H-4 ameliorated a DM1-associated splicing defect with a half maximal inhibitory concentration (IC_{50}) of ~10 µM; note the H module itself was inactive (Childs-Disney et al. 2012). We also studied the ideal backbone for display of H RNA-binding modules and identified that the peptide tertiary amide 2H-K4NMe was superior (Rzuczek et al. 2013). The backbone was further improved to enhance metabolic stability to afford 2H-K4NMeS, a 100 nM inhibitor of DM1-associated toxicity in patient-derived cells (Rzuczek et al. 2017). Likewise, the H RNA-binding module was optimized via chemical similarity searching to afford H1, which improved DM1-associated defects in cellular and mouse models (Parkesh et al. 2012).

By far the most potent small molecule that targets r(CUG)^{exp} reacts to form multivalent compounds upon binding cellular r(CUG)^{exp} with an IC_{50} of ~10 nM in DM1 patient-derived cells (Rzuczek et al. 2017). That is, 2H-K4NMeS was modified to contain an alkyne and azide (Aak-2H-K4NMeS-N_3). When two or more Aak-2H-K4NMeS-N_3 bind adjacent sites in r(CUG)^{exp}, the azides and alkynes undergo a proximity-based click reaction; that is, the RNA serves as the catalyst, and no copper is required. This in situ click approach was first shown using the RNA that causes DM2.

Potency is likely driven by affinity of the small molecule for the RNA, the large amount of RNA's surface area occupied by the ligand, and the small molecule's k_{on} and k_{off} rates, particularly in comparison with the RBP. Importantly, Aak-2H-K4NMeS-N_3 is selective for r(CUG)^{exp} more than other RNAs containing short, nonpathogenic r(CUG) repeats including the wild-type (WT) DMPK allele (Rzuczek et al. 2017). Selectivity is likely mainly derived from Aak-2H-K4NMeS-N_3's recognition of the r(CUG)^{exp} structure; short r(CUG) repeats generally do not fold into an array of 1 × 1 UU internal loops. These factors that influence potency and selectivity are likely general for all RNA repeat expansions that sequester RNA-binding proteins.

3.8 Chemical Techniques to Image RNAs and to Affect Subcellular Localization and Function

Green fluorescent proteins (GFPs) have revolutionized the manner in which protein function and localization are studied (Stearns 1995). Various genetic approaches have been used to engineer RNA GFP mimics, the most successful of which appends an RNA of interest with a fluorescent aptamer (Paige et al. 2011). Chemical approaches to image cellular RNAs in their natural context, without genetic modification, are highly desirable. To this end, the on-site, click chemistry approach described above was modified to synthesize a turn-on fluorescence resonance energy transfer (FRET) sensor upon binding to r(CUG)^{exp} (Rzuczek et al. 2017). To allow for the synthesis of the fluorescent probes on-site, two derivatives of Aak-2H-K4NMeS-N_3 were synthesized: Aak-2H-K4NMeS-FAM (in which FAM is fluorescein) and TAMRA-2H-K4NMeS-N_3 (in which TAMRA is 5-carboxytetramethylrhodamine). Thus, on binding r(CUG)^{exp}, Aak-2H-K4NMeS-FAM and TAMRA-2H-K4NMeS-N_3 click together such that the two dyes are close together and FRET is observed (a readout of reaction). Robust reaction to form the FRET sensor was observed in DM1 patient-derived cells, whereas FRET was not observed in cells that did not express r(CUG)^{exp} as such (i.e., FRET was r(CUG)^{exp}-dependent). When cells were incubated with Aak-2H-K4NMeS-FAM and 2H-K4NMeS-N_3 (lacking FRET acceptor), no FRET was observed, showing that stochastic alignment of the compounds on the target did not result in FRET (Fig. 5) (Rzuczek et al. 2017). In DM1, sequestration of RBPs by r(CUG)^{exp} disallows the RNA's export from the nucleus and thus decreases translation of DMPK. Importantly, the FRET sensor shows that the small molecule alleviated this nucleocytoplasmic transport defect, trafficking the mutant DMPK RNA harboring r(CUG)^{exp} to the cytoplasm and increasing its translation (Rzuczek et al. 2017). Thus, small molecules can affect subcellular localization of mRNAs and impact biological function, such as translation.

3.9 Targeting Protein-Binding Sites in Human Telomerase RNA

First discovered by Greider and Blackburn (Greider and Blackburn 1985), telomerase is a ribonucleoprotein complex (RNP) responsible for addition of telomeric DNA to

Cite as *Cold Spring Harb Perspect Biol* doi: 10.1101/cshperspect.a034769

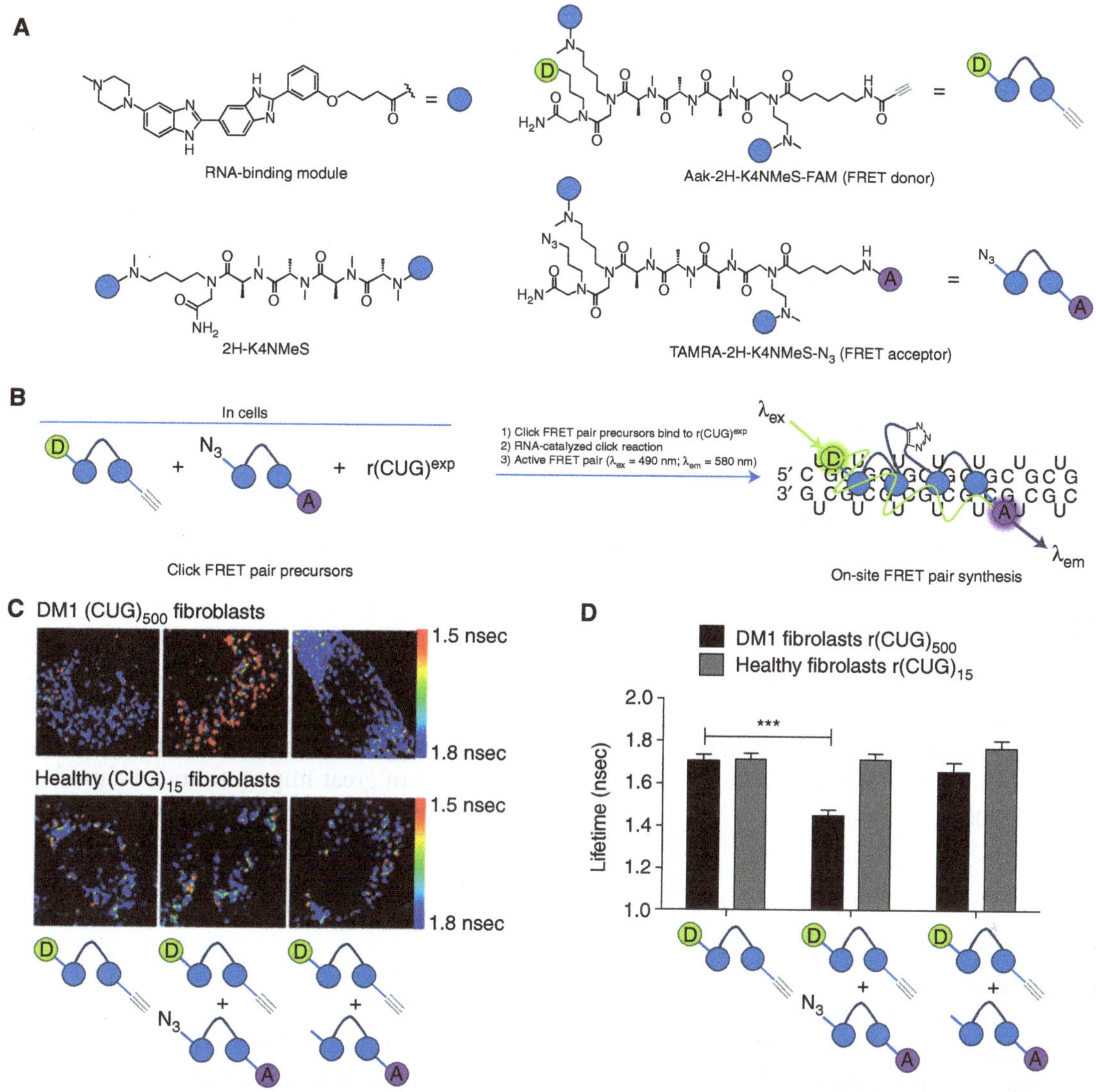

Figure 5. An example of on-site click chemistry to create a fluorescence resonance energy transfer (FRET) reporter to image RNA in cells without genetic modification. (*A*) Structures of an RNA-binding module (represented by a blue circle), 2H-K4NMeS, Aak-2H-K4NMeS-FAM (a green circle represents a fluorescein FRET donor), and TAMRA-2H-K4NMeS-N$_3$ (a purple circle represents TAMRA [5-arboxytetramethylrhodamine] FRET acceptor). (*B*) Scheme of on-site FRET pair synthesis that occurs after FRET pair precursors bind to r(CUG)^{exp} and undergo an RNA-catalyzed click reaction, allowing imaging of RNA in cells. (*C*) Images of myotonic muscular dystrophy type 1 (DM1) (CUG)$_{500}$ fibroblasts and healthy (CUG)$_{15}$ fibroblasts with Aak-2H-K4NMeS-FAM only (*left*), Aak-2H-K4NMeS-FAM with TAMRA-2H-K4NMeS-N$_3$ (*middle*), and Aak-2H-K4NMeS-FAM with TAMRA-2H-K4NMeS (lacks N^3 click partner) (*right*). (*D*) On-site click chemistry generates FRET reporters to show 2H-K4NMeS selectively binds to r(CUG)$_{500}$ in DM1 fibroblasts. $^{***}P < 0.001$, as determined by a two-tailed Student's *t*-test. (*A* and *B* are adapted, whereas *C* and *D* are directly reproduced, with permission, from Rzuczek et al. 2017.)

chromosome ends, providing genome stability. Telomerase RNA adopts a complex secondary structure, comprising the template for telomeric DNA synthesis and elements that bind telomerase proteins (Romero and Blackburn 1991). Interestingly, telomerase is highly expressed in stem cells and cancer cells but has very low expression and activity in somatic cells (Feng et al. 1995). Collectively, these data suggest that a small molecule that inactivates telomerase in cancer cells could serve as a chemotherapeutic and could do so selectively, that is, healthy cells would not be affected as the target is absent. Conversely, underexpression of telomerase leads to dyskeratosis congenita (DKC) and is often due to mutations in the RNA component (Theimer et al. 2003) that inhibit its binding to telomerase proteins.

Although DKC often manifests itself as changes in skin pigmentation and nail formation, it is often fatal as a result of bone marrow failure. Thus, an effective treatment for DKC could be achieved by stabilizing the functional fold of telomerase RNA with an oligonucleotide or small molecule.

Several laboratories have identified small molecules that bind regions within telomerase RNA to prevent binding to telomerase protein or to the telomeric DNA substrate. The James laboratory docked more than 3000 FDA-approved drugs into the structure of the telomerase P2b region (Theimer et al. 2003), affording 20 potential binders (Pinto et al. 2008). Thirteen compounds bind the RNA, as evidenced by nuclear magnetic resonance (NMR) spectroscopy. By using a combination of chemical similarity searching and docking, 24 additional small molecules were identified that selectively bound telomerase RNA. The selective small molecules fall into four main categories: chained aromatics, 4-aminoquinolines, phenothiazines, and 6-7-6 ring systems (Pinto et al. 2008), which could be further medicinally optimized. Other small molecules that bind telomerase RNA and inhibit protein or DNA binding include Hoechst 33258 (Dominick et al. 2004), various aminoglycosides (Dominick et al. 2004), Rivanol (DNA/RNA hybrid) (West et al. 2001), and acridines (DNA/RNA hybrid) (West et al. 2001).

3.10 Targeting Protein-Binding Sites in pre-mRNAs to Affect Splicing Outcomes

Many diseases are caused by aberrant alternative splicing of pre-mRNAs. For those caused by mutation of the RNA itself, a small molecule could offer a therapeutic option. For example, inherited frontotemporal dementia and parkinsonism linked to chromosome 17 (FTDP-17) is an incurable disease that is directly caused by increased inclusion of exon 10 in microtubule-associated protein τ (MAPT) mRNA (Jiang et al. 2000). Various studies have shown that mutations within a hairpin at the MAPT exon 10-intron junction destabilize the RNA's thermodynamic stability, increasing both binding to U1 small nuclear ribonucleoprotein complex (snRNP) and hence exon 10 inclusion (Jiang et al. 2000; Donahue et al. 2006). Therefore, small molecules were designed that bind and stabilize this mutant MAPT and improve deregulation of alternative splicing (Luo and Disney 2014), the first small molecules known to affect exon inclusion/exclusion in cells by binding directly to the RNA target. Furthermore, the compounds were selective, as introduction of a single nucleotide to the small-molecule-binding site eliminated binding and bioactivity (Luo and Disney 2014).

Spinal muscular atrophy (SMA) is caused by a loss of function of the survival motor neuron 1 protein (SMN1). Interestingly, the sequence of the survival motor neuron 2 (*SMN2*) gene differs from *SMN1* by a single nucleotide; this single nucleotide change increases the rate of exon 7 exclusion. The absence of exon 7 decreases SMN2 mRNA's half-life by twofold, as compared with SMN1. Thus, increasing exon 7 inclusion in SMN2 mRNA with a small molecule could be therapeutically beneficial. Indeed, this strategy is viable as first explored with antisense oligonucleotides by Adrian Krainer's laboratory (Cartegni and Krainer 2003). Since then, PTC Therapeutics/Roche and Novartis have separately developed small molecules that increase exon 7 inclusion in SMN2 mRNA and improve SMA-associated defects, including motor function, in a mouse model (Naryshkin et al. 2014; Palacino et al. 2015). Palacino et al. (Novartis) showed that the small molecule stabilized the interaction between SMN2 pre-mRNA and U1 snRNP in a sequence-selective manner, forming a ternary complex (Palacino et al. 2015). RNA-seq analysis showed that the compounds developed collaboratively by PTC/Roche affected the alternative splicing of <0.04% of genes (Naryshkin et al. 2014). Collectively, these studies suggest that small molecules can be developed to alter splicing outcomes, which could be a therapeutic strategy for many diseases. Thus, it is of great import to understand how to drive the molecular recognition of RNA motifs by small molecules from a fundamental perspective. In concert with structural data, small molecules could be designed to disrupt or enhance RNA–protein interactions.

4 CONCLUDING REMARKS

A myriad of roles for RNA in cellular biology have been uncovered. Notably RNA's interaction with RBPs is essential for many biological processes. Aberrant interactions with proteins can lead to disease, as observed in the RNA gain-of-function mechanism that causes microsatellite disorders and the enhanced binding of U1 snRNP to a MAPT mutation that causes FTDP-17. Importantly, to exact a therapeutic effect in such RNA gain-of-function diseases, a small-molecule drug would need to bind the mutant RNA to prevent protein binding. Binding to the protein itself would inhibit its normal cellular function and likely exacerbate disease. Conversely, in the case of loss-of-function diseases that decrease protein binding, a therapeutic small molecule needs to enhance protein binding, as shown for drugs that bind to SMN2 pre-mRNA and alter its alternative splicing.

These studies exemplify that RNA is indeed a viable target for drug-like small molecules, and they thus argue for a fundamental understanding of RNA–small-molecule interactions. The impact of such small-molecule probes could be transformative, as these tools can be used to define

Cite as *Cold Spring Harb Perspect Biol* doi: 10.1101/cshperspect.a034769

the pathways controlled by regulatory small RNAs (e.g., miRNAs) or to define the effects of long noncoding RNAs and toxic RAN peptides on transcription, gene silencing, cell fate, proliferation, survival, and transformation. Indeed, novel and transformative technologies have been developed that could enable the selective targeting of any RNA molecule by chemical probes. Proof-of-principle studies have established that this new generation of RNA-targeting chemical probes have tremendous potential to transform medicine for the treatment of a broad class of disorders and diseases.

ACKNOWLEDGMENTS

This work was funded by the National Institutes of Health (NIH; R01 GM097455; DP1 NS096898; P01 NS099114) and the Muscular Dystrophy Association (Grant ID 380467).

REFERENCES

*Reference is in this book.

Arambula JF, Ramisetty SR, Baranger AM, Zimmerman SC. 2009. A simple ligand that selectively targets CUG trinucleotide repeats and inhibits MBNL protein binding. *Proc Natl Acad Sci* **106:** 16068–16073.

Auld DS, Thorne N, Maguire WF, Inglese J. 2009. Mechanism of PTC124 activity in cell-based luciferase assays of nonsense codon suppression. *Proc Natl Acad Sci* **106:** 3585–3590.

Bernat V, Disney MD. 2015. RNA structures as mediators of neurological diseases and as drug targets. *Neuron* **87:** 28–46.

Bevilacqua PC, Ritchey LE, Su Z, Assmann SM. 2016. Genome-wide analysis of RNA secondary structure. *Ann Rev Genet* **50:** 235–266.

Blanchard SC, Cooperman BS, Wilson DN. 2010. Probing translation with small-molecule inhibitors. *Chem Biol* **17:** 633–645.

Blount KF, Breaker RR. 2006. Riboswitches as antibacterial drug targets. *Nat Biotechnol* **24:** 1558–1564.

Blount KF, Wang JX, Lim J, Sudarsan N, Breaker RR. 2007. Antibacterial lysine analogs that target lysine riboswitches. *Nat Chem Biol* **3:** 44–49.

Brook JD, McCurrach ME, Harley HG, Buckler AJ, Church D, Aburatani H, Hunter K, Stanton VP, Thirion JP, Hudson T, et al. 1992. Molecular basis of myotonic dystrophy: Expansion of a trinucleotide (CTG) repeat at the 3′ end of a transcript encoding a protein kinase family member. *Cell* **68:** 799–808.

Cartegni L, Krainer AR. 2003. Correction of disease-associated exon skipping by synthetic exon-specific activators. *Nat Struct Biol* **10:** 120–125.

Carter AP, Clemons WM, Brodersen DE, Morgan-Warren RJ, Wimberly BT, Ramakrishnan V. 2000. Functional insights from the structure of the 30S ribosomal subunit and its interactions with antibiotics. *Nature* **407:** 340–348.

Childs-Disney JL, Disney MD. 2016. Small molecule targeting of a microRNA associated with hepatocellular carcinoma. *ACS Chem Biol* **11:** 375–380.

Childs-Disney JL, Hoskins J, Rzuczek SG, Thornton CA, Disney MD. 2012. Rationally designed small molecules targeting the RNA that causes myotonic dystrophy type 1 are potently bioactive. *ACS Chem Biol* **7:** 856–862.

Clamp M, Fry B, Kamal M, Xie X, Cuff J, Lin MF, Kellis M, Lindblad-Toh K, Lander ES. 2007. Distinguishing protein-coding and noncoding genes in the human genome. *Proc Natl Acad Sci* **104:** 19428–19433.

Costales MG, Haga CL, Velagapudi SP, Childs-Disney JL, Phinney DG, Disney MD. 2017. Small molecule inhibition of microRNA-210 reprograms an oncogenic hypoxic circuit. *J Am Chem Soc* **139:** 3446–3455.

Deigan KE, Li TW, Mathews DH, Weeks KM. 2009. Accurate SHAPE-directed RNA structure determination. *Proc Natl Acad Sci* **106:** 97–102.

Disney MD, Labuda LP, Paul DJ, Poplawski SG, Pushechnikov A, Tran T, Velagapudi SP, Wu M, Childs-Disney JL. 2008. Two-dimensional combinatorial screening identifies specific aminoglycoside-RNA internal loop partners. *J Am Chem Soc* **130:** 11185–11194.

Disney MD, Liu B, Yang WY, Sellier C, Tran T, Charlet-Berguerand N, Childs-Disney JL. 2012. A small molecule that targets $r(CGG)^{exp}$ and improves defects in fragile X-associated tremor ataxia syndrome. *ACS Chem Biol* **7:** 1711–1718.

Disney MD, Winkelsas AM, Velagapudi SP, Southern M, Fallahi M, Childs-Disney JL. 2016. Inforna 2.0: A platform for the sequence-based design of small molecules targeting structured RNAs. *ACS Chem Biol* **11:** 1720–1728.

Dominick PK, Keppler BR, Legassie JD, Moon IK, Jarstfer MB. 2004. Nucleic acid-binding ligands identify new mechanisms to inhibit telomerase. *Bioorg Med Chem Lett* **14:** 3467–3471.

Donahue CP, Muratore C, Wu JY, Kosik KS, Wolfe MS. 2006. Stabilization of the τexon 10 stem loop alters pre-mRNA splicing. *J Biol Chem* **281:** 23302–23306.

Dudock BS, Katz G, Taylor EK, Holley RW. 1969. Primary structure of wheat germ phenylalanine transfer RNA. *Proc Natl Acad Sci* **62:** 941–945.

Fardaei M, Rogers MT, Thorpe HM, Larkin K, Hamshere MG, Harper PS, Brook JD. 2002. Three proteins, MBNL, MBLL and MBXL, co-localize in vivo with nuclear foci of expanded-repeat transcripts in DM1 and DM2 cells. *Hum Mol Genet* **11:** 805–814.

Feng J, Funk WD, Wang SS, Weinrich SL, Avilion AA, Chiu CP, Adams RR, Chang E, Allsopp RC, Yu J, et al. 1995. The RNA component of human telomerase. *Science* **269:** 1236–1241.

Fire A, Xu S, Montgomery MK, Kostas SA, Driver SE, Mello CC. 1998. Potent and specific genetic interference by double-stranded RNA in *Caenorhabditis elegans*. *Nature* **391:** 806–811.

Gerstein MB, Bruce C, Rozowsky JS, Zheng D, Du J, Korbel JO, Emanuelsson O, Zhang ZD, Weissman S, Snyder M. 2007. What is a gene, post-ENCODE? History and updated definition. *Genome Res* **17:** 669–681.

Greider CW, Blackburn EH. 1985. Identification of a specific telomere terminal transferase activity in *Tetrahymena* extracts. *Cell* **43:** 405–413.

Grundy FJ, Henkin TM. 2004. Regulation of gene expression by effectors that bind to RNA. *Curr Opin Microbiol* **7:** 126–131.

Guan L, Disney MD. 2012. Recent advances in developing small molecules targeting RNA. *ACS Chem Biol* **7:** 73–86.

Guan L, Disney MD. 2013. Covalent small-molecule–RNA complex formation enables cellular profiling of small-molecule–RNA interactions. *Angew Chem Int Ed Engl* **52:** 10010–10013.

Hammond SM, Bernstein E, Beach D, Hannon GJ. 2000. An RNA-directed nuclease mediates post-transcriptional gene silencing in *Drosophila* cells. *Nature* **404:** 293–296.

Hamy F, Felder ER, Heizmann G, Lazdins J, Aboul-ela F, Varani G, Karn J, Klimkait T. 1997. An inhibitor of the Tat/TAR RNA interaction that effectively suppresses HIV-1 replication. *Proc Natl Acad Sci* **94:** 3548–3553.

Hermann T. 2016. Small molecules targeting viral RNA. *Wiley Interdisc Rev RNA* **7:** 726–743.

Holley RW. 1965. Structure of an alanine transfer ribonucleic acid. *JAMA* **194:** 868–871.

Hopkins AL, Groom CR. 2002. The druggable genome. *Nat Rev Drug Discov* **1:** 727–730.

Howe JA, Wang H, Fischmann TO, Balibar CJ, Xiao L, Galgoci AM, Malinverni JC, Mayhood T, Villafania A, Nahvi A, et al. 2015. Selective small-molecule inhibition of an RNA structural element. *Nature* **526:** 672–677.

Inoue T, Cech TR. 1985. Secondary structure of the circular form of the *Tetrahymena* rRNA intervening sequence: A technique for RNA structure analysis using chemical probes and reverse transcriptase. *Proc Natl Acad Sci* **82:** 648–652.

Jaeger JA, Turner DH, Zuker M. 1989. Improved predictions of secondary structures for RNA. *Proc Natl Acad Sci* **86:** 7706–7710.

Jahromi AH, Fu Y, Miller KA, Nguyen L, Luu LM, Baranger AM, Zimmerman SC. 2013. Developing bivalent ligands to target CUG triplet repeats, the causative agent of myotonic dystrophy type 1. *J Med Chem* **56:** 9471–9481.

Jiang Z, Cote J, Kwon JM, Goate AM, Wu JY. 2000. Aberrant splicing of τ pre-mRNA caused by intronic mutations associated with the inherited dementia frontotemporal dementia with parkinsonism linked to chromosome 17. *Mol Cell Biol* **20:** 4036–4048.

Jones D, Metzger HJ, Schatz A, Waksman SA. 1944. Control of gram-negative bacteria in experimental animals by streptomycin. *Science* **100:** 103–105.

Kaul M, Barbieri CM, Pilch DS. 2006. Aminoglycoside-induced reduction in nucleotide mobility at the ribosomal RNA a-site as a potentially key determinant of antibacterial activity. *J Am Chem Soc* **128:** 1261–1271.

Lagos-Quintana M, Rauhut R, Lendeckel W, Tuschl T. 2001. Identification of novel genes coding for small expressed RNAs. *Science* **294:** 853–858.

Lee RC, Ambros V. 2001. An extensive class of small RNAs in *Caenorhabditis elegans*. *Science* **294:** 862–864.

Lipinski CA. 2004. Lead- and drug-like compounds: The rule-of-five revolution. *Drug Discov Today Technol* **1:** 337–341.

Lu Z, Chang HY. 2016. Decoding the RNA structurome. *Curr Opin Struct Biol* **36:** 142–148.

Luo Y, Disney MD. 2014. Bottom-up design of small molecules that stimulate exon 10 skipping in mutant MAPT pre-mRNA. *Chembiochem* **15:** 2041–2044.

Mathews DH, Disney MD, Childs JL, Schroeder SJ, Zuker M, Turner DH. 2004. Incorporating chemical modification constraints into a dynamic programming algorithm for prediction of RNA secondary structure. *Proc Natl Acad Sci* **101:** 7287–7292.

Moazed D, Noller HF. 1987. Interaction of antibiotics with functional sites in 16S ribosomal-RNA. *Nature* **327:** 389–394.

Mori K, Arzberger T, Grasser FA, Gijselinck I, May S, Rentzsch K, Weng SM, Schludi MH, van der Zee J, Cruts M, et al. 2013. Bidirectional transcripts of the expanded *C9orf72* hexanucleotide repeat are translated into aggregating dipeptide repeat proteins. *Acta Neuropathol* **126:** 881–893.

Nahvi A, Sudarsan N, Ebert MS, Zou X, Brown KL, Breaker RR. 2002. Genetic control by a metabolite binding mRNA. *Chem Biol* **9:** 1043–1049.

Naryshkin NA, Weetall M, Dakka A, Narasimhan J, Zhao X, Feng Z, Ling KK, Karp GM, Qi H, Woll MG, et al. 2014. Motor neuron disease. SMN2 splicing modifiers improve motor function and longevity in mice with spinal muscular atrophy. *Science* **345:** 688–693.

Orr HT, Zoghbi HY. 2007. Trinucleotide repeat disorders. *Annu Rev Neurosci* **30:** 575–621.

Paige JS, Wu KY, Jaffrey SR. 2011. RNA mimics of green fluorescent protein. *Science* **333:** 642–646.

Palacino J, Swalley SE, Song C, Cheung AK, Shu L, Zhang X, Van Hoosear M, Shin Y, Chin DN, Keller CG, et al. 2015. SMN2 splice modulators enhance U1-pre-mRNA association and rescue SMA mice. *Nat Chem Biol* **11:** 511–517.

Parkesh R, Childs-Disney JL, Nakamori M, Kumar A, Wang E, Wang T, Hoskins J, Tran T, Housman D, Thornton CA, et al. 2012. Design of a bioactive small molecule that targets the myotonic dystrophy type 1 RNA via an RNA motif-ligand database and chemical similarity searching. *J Am Chem Soc* **134:** 4731–4742.

Parsons J, Castaldi MP, Dutta S, Dibrov SM, Wyles DL, Hermann T. 2009. Conformational inhibition of the hepatitis C virus internal ribosome entry site RNA. *Nat Chem Biol* **5:** 823–825.

Peattie DA, Gilbert W. 1980. Chemical probes for higher-order structure in RNA. *Proc Natl Acad Sci* **77:** 4679–4682.

Pinto IG, Guilbert C, Ulyanov NB, Stearns J, James TL. 2008. Discovery of ligands for a novel target, the human telomerase RNA, based on flexible-target virtual screening and NMR. *J Med Chem* **51:** 7205–7215.

Romero DP, Blackburn EH. 1991. A conserved secondary structure for telomerase RNA. *Cell* **67:** 343–353.

Rzuczek SG, Gao Y, Tang ZZ, Thornton CA, Kodadek T, Disney MD. 2013. Features of modularly assembled compounds that impart bioactivity against an RNA target. *ACS Chem Biol* **8:** 2312–2321.

Rzuczek SG, Colgan LA, Nakai Y, Cameron MD, Furling D, Yasuda R, Disney MD. 2017. Precise small molecule recognition of a toxic CUG RNA repeat expansion. *Nat Chem Biol* **13:** 188–193.

Sellier C, Freyermuth F, Tabet R, Tran T, He F, Ruffenach F, Alunni V, Moine H, Thibault C, Page A, et al. 2013. Sequestration of DROSHA and DGCR8 by expanded CGG RNA repeats alters microRNA processing in fragile X-associated tremor/ataxia syndrome. *Cell Rep* **3:** 869–880.

Semenza GL. 2003. Targeting HIF-1 for cancer therapy. *Nat Rev Cancer* **3:** 721–732.

Stark BC, Kole R, Bowman EJ, Altman S. 1978. Ribonuclease P: An enzyme with an essential RNA component. *Proc Natl Acad Sci* **75:** 3717–3721.

Stearns T. 1995. Green fluorescent protein. The green revolution. *Curr Biol* **5:** 262–264.

Stern S, Wilson RC, Noller HF. 1986. Localization of the binding site for protein S4 on 16 S ribosomal RNA by chemical and enzymatic probing and primer extension. *J Mol Biol* **192:** 101–110.

Theimer CA, Finger LD, Trantirek L, Feigon J. 2003. Mutations linked to dyskeratosis congenita cause changes in the structural equilibrium in telomerase RNA. *Proc Natl Acad Sci* **100:** 449–454.

Thomas JR, Hergenrother PJ. 2008. Targeting RNA with small molecules. *Chem Rev* **108:** 1171–1224.

Tran T, Disney MD. 2012. Identifying the preferred RNA motifs and chemotypes that interact by probing millions of combinations. *Nat Commun* **3:** 1125.

Ule J, Jensen KB, Ruggiu M, Mele A, Ule A, Darnell RB. 2003. CLIP identifies Nova-regulated RNA networks in the brain. *Science* **302:** 1212–1215.

* Ule J, Hwang HW, Darnell RB. 2018. CLIP identifies Nova-regulated RNA networks in the brain. *Cold Spring Harb Persp Biol* **10:** a032243.

Velagapudi SP, Seedhouse SJ, Disney MD. 2010. Structure–activity relationships through sequencing (StARTS) defines optimal and suboptimal RNA motif targets for small molecules. *Angew Chem Int Ed Engl* **49:** 3816–3818.

Velagapudi SP, Gallo SM, Disney MD. 2014. Sequence-based design of bioactive small molecules that target precursor microRNAs. *Nat Chem Biol* **10:** 291–297.

Velagapudi SP, Cameron MD, Haga CL, Rosenberg LH, Lafitte M, Duckett DR, Phinney DG, Disney MD. 2016. Design of a small molecule against an oncogenic noncoding RNA. *Proc Natl Acad Sci* **113:** 5898–5903.

Velagapudi SP, Luo Y, Tran T, Haniff HS, Nakai Y, Fallahi M, Martinez GJ, Childs-Disney JL, Disney MD. 2017. Defining RNA–small molecule affinity landscapes enables design of a small molecule inhibitor of an oncogenic noncoding RNA. *ACS Cent Sci* **3:** 205–216.

Watts JM, Dang KK, Gorelick RJ, Leonard CW, Bess JW Jr, Swanstrom R, Burch CL, Weeks KM. 2009. Architecture and secondary structure of an entire HIV-1 RNA genome. *Nature* **460:** 711–716.

Welch EM, Barton ER, Zhuo J, Tomizawa Y, Friesen WJ, Trifillis P, Paushkin S, Patel M, Trotta CR, Hwang S, et al. 2007. PTC124 targets genetic disorders caused by nonsense mutations. *Nature* **447:** 87–91.

Wells SE, Hughes JM, Igel AH, Ares M Jr. 2000. Use of dimethyl sulfate to probe RNA structure in vivo. *Methods Enzymol* **318:** 479–493.

West C, Francis R, Friedman SH. 2001. Small molecule/nucleic acid affinity chromatography: Application for the identification of telomerase inhibitors which target its key RNA/DNA heteroduplex. *Bioorg Med Chem Lett* **11:** 2727–2730.

Winkler W, Nahvi A, Breaker RR. 2002. Thiamine derivatives bind messenger RNAs directly to regulate bacterial gene expression. *Nature* **419:** 952–956.

Woese C, Pace N. 1993. Probing RNA structure, function, and history by comparative analysis. In *The RNA world* (ed. Gesteland RF, Cech TR, Atkins JF), pp. 91–117. Cold Spring Harbor Laboratory Press, Cold Spring Harbor, NY.

Woese CR, Magrum LJ, Gupta R, Siegel RB, Stahl DA, Kop J, Crawford N, Brosius J, Gutell R, Hogan JJ, et al. 1980. Secondary structure model for bacterial 16S ribosomal RNA: Phylogenetic, enzymatic and chemical evidence. *Nucleic Acids Res* **8:** 2275–2293.

Yang WY, Wilson HD, Velagapudi SP, Disney MD. 2015. Inhibition of non-ATG translational events in cells via covalent small molecules targeting RNA. *J Am Chem Soc* **137:** 5336–5345.

Yonath A. 2005. Antibiotics targeting ribosomes: Resistance, selectivity, synergism and cellular regulation. *Ann Rev Biochem* **74:** 649–679.

Yoshizawa S, Fourmy D, Puglisi JD. 1999. Recognition of the codon-anticodon helix by ribosomal RNA. *Science* **285:** 1722–1725.

Zaug AJ, Cech TR. 1986. The intervening sequence RNA of *Tetrahymena* is an enzyme. *Science* **231:** 470–475.

Zhang Y-J, Jansen-West K, Xu Y, Gendron TF, Bieniek KF, Lin W-L, Sasaguri H, Caulfield T, Hubbard J, Daughrity L, et al. 2014. Aggregation-prone c9FTD/ALS poly(GA) RAN translated proteins cause neurotoxicity by inducing ER stress. *Acta Neuropathol* **128:** 505–524.

Zhang K, Donnelly CJ, Haeusler AR, Grima JC, Machamer JB, Steinwald P, Daley EL, Miller SJ, Cunningham KM, Vidensky S, et al. 2015. The *C9orf72* repeat expansion disrupts nucleocytoplasmic transport. *Nature* **525:** 56–61.

Zu T, Liu Y, Banez-Coronel M, Reid T, Pletnikova O, Lewis J, Miller TM, Harms MB, Falchook AE, Subramony SH, et al. 2013. RAN proteins and RNA foci from antisense transcripts in C9ORF72 ALS and frontotemporal dementia. *Proc Natl Acad Sci* **110:** E4968–E4977.

Zuker M. 1989. On finding all suboptimal foldings of an RNA molecule. *Science* **244:** 48–52.

RNA Switches for Synthetic Biology

Calvin M. Schmidt[1] and Christina D. Smolke[1,2]

[1]Department of Bioengineering, Stanford University, Stanford, California 94305

[2]Chan Zuckerberg Biohub, San Francisco, California 94158

Correspondence: csmolke@stanford.edu

SUMMARY

In addition to coding for protein sequences, RNA molecules encode a diverse set of gene-regulatory elements. RNA switches are one class of gene-regulatory elements that control protein expression in a manner that is dependent on the concentration of specific ligand molecules. These allosteric gene-regulatory elements have been shown as useful tools in engineering diverse cell types to display novel function. In particular, RNA switches have been used as genetically encoded biosensors and conditional controllers to direct cellular decisions based on the system's changing environment. A significant focus in the field has been the generation of novel RNA switches that are tailored for different biological systems. We review approaches that have been used to generate RNA switches, which leverage the unique physical properties of RNA and the myriad ways in which RNA can modulate gene expression.

Outline

Cite as *Cold Spring Harb Perspect Biol* doi: 10.1101/cshperspect.a032532

1 NATURALLY OCCURRING RIBOSWITCHES CAN SERVE AS A STARTING POINT FOR DESIGNING ENGINEERED RNA SWITCHES

Riboswitches were first discovered in natural systems. These naturally occurring RNA switches are usually targeted toward common metabolites, change genetic activity in response to the state of central metabolism, and are presumed to play an important role in maintaining homeostasis by providing positive or negative feedback regulation mechanisms (Winkler and Breaker 2003). In general, a riboswitch functions by coupling the binding of the target metabolite to a change in secondary structure conformation, which in turn alters the gene expression activity encoded within the switch. However, within this simple framework there are a variety of ways that natural switches couple a change in secondary structure conformation to a change in gene expression. Natural riboswitches have been discovered that alter the stability of secondary structures to change the rates of various gene expression processes, including transcription termination, translation initiation, and transcript degradation.

Naturally occurring RNA switches can be used as a starting point for generating switches that function in nonnatural contexts. In the extreme case, certain RNA switches have been taken from their natural context and used directly in different cellular contexts. For example, a c-di-GMP switch from *Bacillus cereus* was used to control the expression of target genes in *Bacillus subtilis* (Gao et al. 2014). In another example, a glucosamine 6-phosphate switch from *B. subtilis* was shown to be functional in *Saccharomyces cerevisiae* (Lee and Oh 2015). In both examples, no modification of the RNA switch was required; the switches were simply inserted into the 5′ and 3′ untranslated regions of the genes of interest, respectively, and shown to function in the new cellular contexts. These examples also show an advantage of using RNA-based switches over protein-based switches. RNA switches generally do not depend on other proteins for their activity and are often encoded on the same transcript as the gene target, which can reduce off-target effects. In this way, RNA switches can function more reliably than protein-based switches across different organisms (Ma et al. 2014).

In many instances, modifications to the switch sequence can be made to improve the functional activity of the switch in the new cellular context. These modifications can include modifying sequences to alter the relative stability of the different conformations adopted by the switch or the gene expression platform to better suit the new cellular environment. In one example, the preQ1 switch from *Thermoanaerobacter tengcongensis* was tested in *Mycobacterium smegmatis* and the switch was shown to display weak activity in the new cellular context (Vlack et al. 2017). By modifying the switch sequence to increase the thermodynamic stability of the gene expression off conformation, thereby decreasing the basal expression from the modified sequence, the activation ratio of the switch, or the ratio of gene expression activity in the presence to absence of ligand, was increased by fourfold in this new context. In other examples, the ligand-binding properties of naturally occurring switches have been used to build RNA switches in new cellular contexts. In one example, the flavin mononucleotide RNA switch from *Escherichia coli* was used as a ligand-binding domain, or aptamer, from which a new RNA switch was generated for *B. subtilis*. Specifically, the entire switch sequence was coupled to a ribozyme through a randomized linker region and a high-throughput screen was used to identify linker sequences that supported allosteric regulation of ribozyme cleavage and subsequent exposure of the ribosome binding site on binding flavin mononucleotide, which resulted in a 15-fold increase in gene expression (Meyer et al. 2015). The switch was subsequently used to identify riboflavin-producing *B. subtilis* strains with increased titers.

The numerous examples of naturally occurring RNA switches illustrate the many ways in which RNA can be used to regulate gene expression in response to a target compound's concentration. The mechanisms of action of naturally occurring RNA switches have been used to inform design choices for the rational design of engineered RNA switches.

2 RNA SWITCHES CAN BE RATIONALLY DESIGNED USING RULES-BASED APPROACHES

The functional activity of RNA switches is often largely dictated by their secondary structure. The likely secondary structure conformations and associated energetics of an RNA sequence can be predicted using a number of secondary structure prediction programs (Puton et al. 2013). Thus, these prediction programs can be used to design novel RNA switches de novo that adopt conformations likely to influence their gene regulatory activity. The ability to engineer novel switches from scratch and reengineer known switches to alter certain properties is another advantage that RNA switches hold over protein-based switches (Golynskiy et al. 2011).

Engineered RNA switches, like their naturally occurring counterparts, couple the binding of a target molecule to a change in RNA conformation, which ultimately affects the gene expression activity encoded within the RNA molecule. In general, designed RNA switches comprise at least two functional domains. The first domain is the sensor or the RNA sequence that binds to the target molecule, often referred to as an RNA aptamer. Aptamers can be generated de novo through in vitro selection methods, or SELEX (Ellington and Szostak 1990; Tuerk and Gold 1990), and subsequently used in the design of an RNA switch. The second domain is the actuator domain, which encodes the gene expression modulation activity of the switch. The two domains are physically coupled such that the RNA switch can adopt different conformations and binding of the target molecule to the aptamer will affect the conformation of the switch sequence.

A rational switch design process usually proceeds along several steps. First, an aptamer that binds to a target of interest is selected. In addition to the aptamer's sequence, other information regarding the aptamer's properties are important for incorporation of that element into the switch architecture such as the secondary structure of the aptamer, key nucleotides involved in ligand binding, binding affinity, kinetics, and specificities (Fig. 1A). Published RNA aptamers have been compiled into publicly available databases such as the Apta-Index offered by Aptagen (Aptamer Database with more than 500 entries; see aptagen.com/aptamer-index/aptamer-list.aspx), which contains more than 500 unique aptamers.

Next, the sequence for the actuator domain is selected. The choice of gene-regulatory element used in the actuator domain is often guided by properties of the biological system, such as the organism and/or target gene. A common property among the gene-regulatory elements used in switches is that they require a specific secondary structure to be active, such that altering the secondary structure of the element will affect its gene expression activity. One commonly used actuator domain is a ribosome binding site (RBS). The RBS promotes translation initiation in prokaryotic organisms when placed in the 5′ untranslated region of the gene of interest. The binding of the ribosome to the RBS and subsequent translation requires that the nucleotides of the RBS remain unbound by other nucleotides in the RNA's secondary structure. Thus, when the RBS is occluded in a secondary structure, translation initiation (and thus translation) of the gene of interest is inhibited (Fig. 1B).

Once the aptamer and the actuator domains are selected, the next step is to physically connect the two with a linker sequence in such a way that the secondary structure of the sensor domain affects the secondary structure of the actuator domain. Binding of the target molecule to the sensor domain would stabilize the ligand-bound conformation of the aptamer and thus the conformation of the gene-regulatory element associated with that aptamer conformation. For gene expression "on" switches, ligand binding stabilizes a conformation that results in increased gene expression, whereas for gene expression "off" switches, ligand

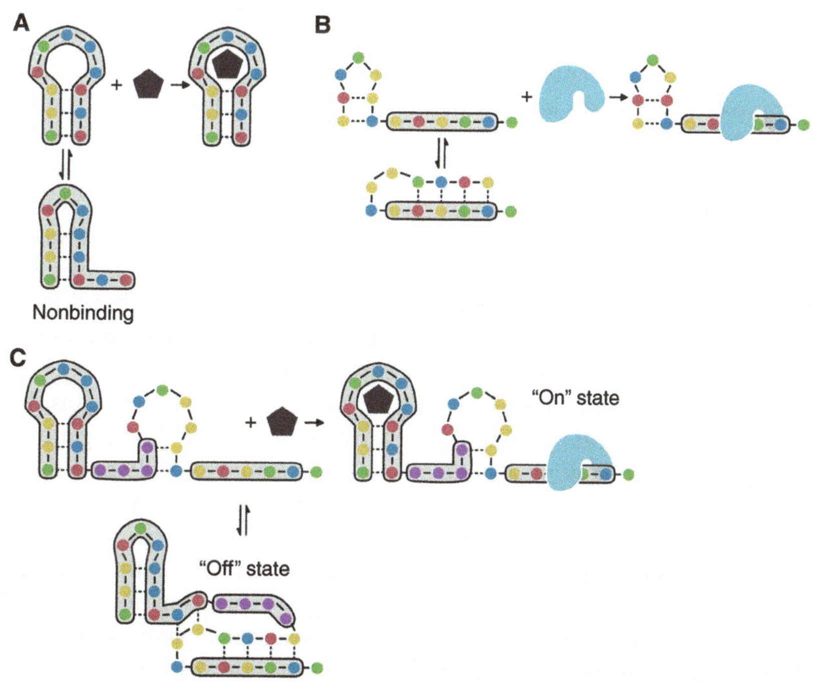

Figure 1. Diagrams showing the different components of a theoretical rationally designed conformational RNA switch. (*A*) An example aptamer (outlined) in both ligand-binding and non-ligand-binding conformations. (*B*) An example ribosome binding site (RBS) (outlined) in both ribosome-accessible and -inaccessible conformations. (*C*) The assembled RNA switch, in which a connecting sequence (outlined and with purple nucleotides) enables ligand binding to stabilize an RBS conformation that allows for ribosome binding.

binding stabilizes a conformation that results in decreased gene expression. In rational design approaches, optimal linker sequences are identified by iterating through different switch sequences in silico, computationally folding them one-by-one using structure prediction programs such as RNAstructure (Bellaousov et al. 2013) and Kinefold (Xayaphoummine et al. 2005). The predicted structures that have the lowest free energy are analyzed to identify sequences that form sets of structures with desired folding patterns.

As an example, we can take that design of a switch that turns on gene expression with the introduction of the target ligand using an RBS as the actuator domain. First, sequences are identified that can adopt two structures: one in which the aptamer domain is not in the correct conformation for ligand binding and the RBS is obstructed and a second structure with inverse properties—correct conformation for ligand binding and an unobstructed RBS (Fig. 1C). A sequence for which secondary structure prediction programs indicate that these two structures are among the most likely secondary structures that the sequence will adopt is a good candidate for a switch, as the binding of the ligand to the aptamer would stabilize it in the binding-competent structure and thus cause the RBS to be unbound and the gene to be translated.

In addition to these simple rules that govern whether a sequence is a switch candidate, other factors can be taken into consideration to further filter the number of possible sequences to test. For example, in general the two desired conformations of the switch should have a relatively small difference in free energy (approximately a few kcal/mol), such that the additional stabilization of one conformation caused by ligand binding would impact which conformation the molecule is more likely to adopt. In some designs, it is desired to have a faster transition between the two structures, such that relatively minor conformational rearrangements between the two structures are required. Examples of other factors that can be incorporated into the design are avoiding kinetic traps that might occur during transcription of the sequence that could prevent it from taking the desired final structure and avoiding sequence motifs within the switch that might cause it to interact with flanking genetic regions.

In one example, this design process was used to build theophylline-responsive RNA switches that regulate translation by controlling access to an internal ribosome entry site (IRES) (Ogawa 2011). The nucleotides upstream of an optimized IRES were modified by inserting sequences that were complementary to portions of the IRES, known as an anti-IRES (aIRES). Binding between the aIRES and IRES sequences would prevent ribosome binding and translation from the IRES. Upstream of the aIRES, an anti-aIRES (aaIRES), a theophylline aptamer, and a modulator sequence with partial complementarity to both the aaIRES and aptamer were inserted. The modulator sequence was designed such that it would bind the aaIRES, thereby freeing the aIRES to bind to the IRES and obstruct translation. However, when theophylline was bound to the aptamer, the modulator sequence would be unbound to the aaIRES, such that the aaIRES would bind to the aIRES, thereby freeing the IRES and allowing translation to occur. By altering the length of the modulator sequence, the researchers showed tuning of the activity of the RNA switch. Specifically, an optimal value for the energy of complementary binding of the modulator sequence to the aaIRES and the aptamer was determined, which was subsequently used to construct RNA switches that incorporated other aptamers, such as the tetracycline and sulforhodamine B aptamers. Activation ratios as high as 30 were achieved by optimizing the modulator sequence.

In addition to designs that are directed to controlling accessibility of RNA sequences, RNA switches have also been rationally designed that require the adoption of a specific structural conformation to cleave (Chi et al. 2008). As gene-regulatory elements in eukaryotes, ribozymes are commonly inserted into the 3′ untranslated region of a gene of interest, in which cleavage promotes transcript degradation. In one example, researchers used ligand binding to the theophylline aptamer to disrupt the ribozyme secondary structure required for cleavage (Win and Smolke 2007). The aptamer was physically coupled to the ribozyme through a small linker region that facilitated switching between two conformational states. For a gene expression "on" switch, the minimal free energy structure was designed such that the ribozyme was in a cleavage-capable conformation and the aptamer in a binding-incapable conformation, while in a structure that was less energetically favorable and therefore less likely to form in the absence of additional stabilization via ligand binding, the aptamer took a binding-capable conformation and the ribozyme a cleavage-incapable conformation.

After the initial switch was designed, which showed an activation ratio of 2, the gene expression activity of the switch was tuned by altering the free energy difference between the two conformational states of the switch. When the stability of the cleavage-incapable conformation was increased relative to the cleavage-capable conformation, the basal expression increased as the sequence adopted the cleavage-incapable state more often. In addition, the activation ratio decreased and eventually reached 0 as the cleavage-incapable state was strengthened beyond a certain free energy difference. Separately, increasing the stability of the cleavage-capable conformation relative to the cleavage-incapable conformation decreased basal expression. The decrease in basal expression caused the activation ratio to

Cite as *Cold Spring Harb Perspect Biol* doi: 10.1101/cshperspect.a032532

be increased to 3. However, as the cleavage-capable conformation was further stabilized beyond a certain free energy difference, the activation ratio decreased to 0 as even the increased stability of ligand binding to the cleavage-incapable conformation was not sufficient to overcome the differences in free energy between the two states. The dependence of the RNA switch on a specific energetic difference between binding-capable and binding-incapable states shows the importance of considering the energetics involved in these systems to create an RNA switch that is able to alternate between different conformations on ligand binding. In addition, the sensor domain of the switch could be replaced with other aptamers, including the tetracycline aptamer, highlighting the modular property of the design framework afforded by the rational design process.

In addition to the generation of RNA switches that can bind small molecule and protein targets, similar rational design strategies can be applied for switches that bind specific mRNA sequences. In these designs, the binding of complementary RNA sequences to the RNA switch are used in place of the binding of a molecular target to an aptamer. Both switch designs use binding to a molecular input, whether small molecule, protein, or RNA, to stabilize a switch conformation that changes the encoded gene expression activity. The design process for RNA-binding switches is flexible in that it does not rely on previously discovered aptamers, which can limit the potential molecular targets. In theory, the straightforward rules of RNA–RNA binding can be leveraged to design sensor domains to bind to any transcript. In one example, RNA switches were designed that used the binding of target transcripts to control RBS access (Green et al. 2014). This work highlighted the ability of the system to sense many different mRNA sequences simply by changing the sequence of the sensor domain based on complementary rules. To prevent cross talk, 417,216 pairwise combinations of 646 candidate switch sequences were analyzed to identify 144 sequences that would not interact with each other. In this example, the researchers tested 26 of their designed switches in vivo by expressing each switch along with its target RNA in *E. coli.* The switches ranged in activation ratios from 1 to 264, and 15% of the tested switch designs were activated by the target RNA sequences of other switches.

Despite the many examples of switches built through rules-based approaches, the rational design process does not always generate functional switches. The complex kinetics and thermodynamics involved in ligand binding, nucleic acid conformational switching, and gene-expression modulation often mean that the functional activity of rationally designed RNA switches can vary wildly. As a result, multiple rounds of iterative design and tuning are generally required before a switch with the desired level of activity is obtained.

3 COMPUTATIONAL TOOLS ENABLE A MORE EFFICIENT RNA SWITCH DESIGN PROCESS

The rational design of RNA switches guided by empirical rules can be a time-consuming process. Few of the possible sequences will result in a switch with desired functional activity, requiring iteration through many design cycles, comprising computational folding and ultimately experimentally validating each one individually. The incorporation of automation—using a computer program to simulate the folding of an RNA sequence, to analyze the predicted structures in silico, and to select sequences that fulfill all the design criteria—can speed up this process. For example, computational programs were developed for the design of ligand-sensitive terminators that use certain design rules: that the optimal sequences form a misfolded aptamer and a correctly folded terminator loop in one structure and a correctly folded aptamer and misfolded terminator loop in another structure (Domin et al. 2017). The programs searched through sequences that connected a theophylline aptamer to a transcription termination sequence with a randomized linker sequence, selecting those that followed the specified rules. The programs also incorporated other factors to further refine the selection of sequences for those that would provide the desired activities. For example, further filtering for sequences that had a difference in free energy between the two functional structures that was less than the binding energy of theophylline to the aptamer. As the number of filters increases to take into account the many different factors required for a functional switch, fewer designs pass all the filters. An automated process is able to iterate over many more sequences than a human researcher, and thus is more likely to identify sequences that satisfy all the criteria. Ultimately, these programs queried thousands of sequences and succeeded in developing theophylline and tetracycline switches that displayed activation ratios of ~4.

Programs that automate the rational design process make the designing of RNA switches accessible for researchers that do not have expertise in the field and reduce the time it takes to design a switch for researchers with expertise in functional RNA design. These tools can assist the rational design of RNA switches by performing the iteration and in silico evaluation of potential switch sequences to rapidly generate candidate switches for experimental validation. However, the designs produced by these tools are not guaranteed to show the desired functional activity. Although these tools can speed up and simplify the front-end design process, the generated candidate designs are generally similar to those that would be generated by an expert with the aid of RNA folding programs. In one example, only three of six tested theophylline RNA switch designs had an activation ratio of greater than 2 and the maximum activation ratio achieved was ~3 (Wachsmuth et al. 2013), which may limit the practical use of the generated switches depending on the desired application (Wang et al. 2013). Ultimately, these approaches are based on the same secondary structure prediction programs and empirical design rules, and thus have the same drawbacks as the rational design approaches in that they do not fully account for the complex thermodynamic and kinetic attributes involved in using dynamic RNA molecules to control gene expression activity.

To improve the accuracy of computational design programs based on RNA secondary structure prediction software, computational programs have been developed that incorporate biophysical models. The biophysical models are used to link the structural states of the RNA switch to levels of gene expression by taking into account the thermodynamics and kinetics of RNA folding, ligand binding, transcription, and translation to arrive at predictions concerning the functional activity of a specific RNA switch sequence. The RNA switches designed by these programs operate by the same principle as the rationally designed switches: that ligand binding to the sensor domain is transduced into a change in gene expression through an actuator domain. However, the program attempts to make the design process more reliable and robust by computationally predicting how the physical state of the RNA sequence will interplay with the different processes in the target biological system.

In one recent example, a computational program was developed to design RNA switches that control translation initiation by using ligand binding to control the obstruction of the RBS (Espah Borujeni et al. 2016). The program compares how strongly the ribosome will bind to the RBS (using a predicted free energy of binding) in the presence and absence of the target ligand, and uses these energies to calculate basal and activated expression levels (Fig. 2). This model was able to accurately predict ($R^2 = 0.68$) the maximum activation ratio under optimal conditions (in which there was excess ligand and no mRNA crowding). The model was further refined by taking in to account how mRNA and ligand concentrations affect the maximum activation ratio, which dramatically improved its performance predicting activities under limited ligand concentrations, macromolecular crowding, and unstable mRNA–ligand complexes from $R^2 = 0.04$ to $R^2 = 0.62$. RNA switches generated through this program showed activation ratios with-

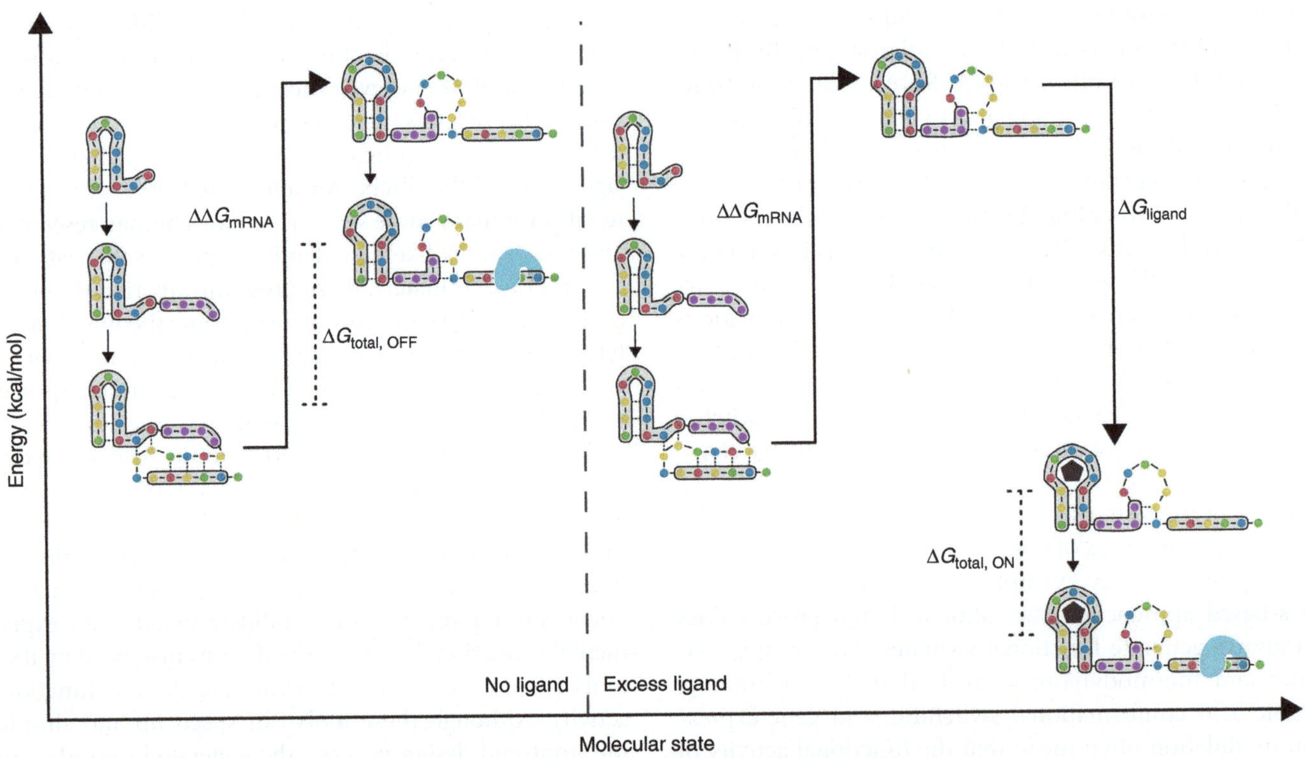

Figure 2. Reaction coordinate diagram showing the various RNA structure states that are modeled in Espah Borujeni et al. (2016), to produce a prediction of gene expression in the presence and absence of the ligand. ΔG_{total} reflects the difference in free energy between the minimal free energy conformation of the RNA molecule, bound to the ligand in the presence of excess ligand, and the free energy value of the ribosome-bound RNA molecule.

in twofold of predicted activation ratios in >60% of the cases. The design tool has been made accessible to the broader research community through a website that generates a set of RNA switch sequences when given a set of standard information concerning the surrounding sequence and the desired aptamer to be incorporated into the switch. The main limitation to this program is that the method of controlling gene expression on which the switch platform is based is compatible with only a certain number of prokaryotic organisms.

In another example, a physics-based model focused on the activity of ligand-dependent RNAi processing was used to advance RNA switch design. In this model, the distribution of RNA molecules is partitioned between two structures that have different RNAi processing efficiencies, with ligand binding stabilizing the structure with low processing efficiency. The model uses the partitioning constant between the two structures, the ligand's association constant, and the processing efficiency of the switch to predict how changing these parameters ultimately impacts the switch's gene-regulatory activity as a function of ligand concentration (Beisel et al. 2008). The model was able to quantitatively predict how modifications to the thermodynamic and kinetic parameters of the RNA switch impact the gene-regulatory activity of the switch.

Ultimately, the incorporation of biophysical models into computational design workflows can provide ways to more accurately predict how the free energy difference between RNA conformations will affect gene expression and increase the accuracy of forward design strategies of new RNA switches. Currently, such approaches are limited by a lack of knowledge about many of the biophysical parameters underlying these models, requiring assumptions that can detract from the accuracy of the models and their implementation for robust forward design pipelines.

4 HIGH-THROUGHPUT SCREENING OF RNA SWITCH LIBRARIES ALLOWS FOR GENERATION OF HIGHLY FUNCTIONAL SWITCH SEQUENCES

A third approach to generating RNA switches leverages increasing capacities to readily generate and screen many thousands of designs specific to a system to find the optimal design. RNA switch libraries can be generated through random mutagenesis or in a more rational manner through the ordering of libraries with specific nucleotides randomized. Typically, RNA switch libraries are designed with constant sensor and actuator sequences, and the regions linking these two domains are randomized to search sequence designs for those linker sequences that allow the binding of the molecular target to the sensor domain to affect the gene expression activity of the switch. Libraries designed in this manner are then screened using high-throughput experimental techniques to isolate the optimal sequences. The screening generally takes the form of two alternating stages: a "negative" screen in the absence of the target ligand and a "positive" screen in the presence of the target ligand, in which the sequences that show the desired activities under the particular ligand condition are selected to go on to the next stage of screening.

In vitro methods for screening have been used to generate RNA switches ultimately intended for in vivo applications. These methods offer advantages over in vivo screening approaches in that they are able to screen a much larger number of sequences. RNA libraries can be generated with very large numbers of unique sequences; however, bottlenecks associated with transformation efficiency and measurement techniques substantially limit the number of sequences that can be screened in vivo. A major drawback of in vitro screens is that sequences that function as switches in vitro may not retain that function when implemented in in vivo systems because of differences between the two environments (e.g., ion concentrations, transcription rates), which can impact how the RNA folds and ultimately the structures it can adopt. In vitro screens are established to identify RNA molecules that show changes in form or function in the presence and absence of ligand that may lead to changes in gene expression under an in vivo context.

In one example, a library of RNA molecules based on a ribozyme actuator was screened to identify switches that prevent self-cleavage in the absence of tetracycline (Wittmann and Suess 2011). The RNA library was designed by physically connecting a self-cleaving ribozyme sequence to a tetracycline aptamer through a small randomized linker region. The library was transcribed in vitro and the uncleaved members of the library were recovered via size separation and extraction through polyacrylamide gel electrophoresis (PAGE). This process was repeated multiple times to enrich for sequences that did not cleave in the absence of tetracycline. Then, the enriched library was transcribed in the presence of tetracycline, and cleaved members were recovered via a similar PAGE-based extraction approach. The final enriched library sequences that showed cleavage only in the presence of tetracycline were recovered through reverse transcription, cloned into a gene expression construct, and transformed into yeast for isolation of specific RNA switch sequences. The isolated sequences were characterized in vivo for their ability to control green fluorescent protein (GFP) expression in the presence and absence of tetracycline, measured with a fluorescence assay. Several of the enriched sequences showed the desired tetracycline-responsive gene-regulatory activity in yeast. However, when these sequences were subsequently tested in mammalian cells none showed the desired activity, highlighting the drawbacks to screening for RNA switches in contexts different from their intended cellular context.

In another example, an in vitro screening strategy was used to select for RNA sequences that left specific portions of an RBS sequence unbound in the presence of thiamine pyrophosphate (TPP) (Martini et al. 2016). An RNA library was created based on randomizing several regions of a natural RNA TPP switch. The library was transcribed, and then sequences that left the RBS sequence exposed to binding were removed via magnetic beads conjugated to a DNA strand encoded with a complementary sequence to the RBS. After several rounds of enrichment, TPP was added to the RNA library, and the sequences that bound to the beads in the presence of TPP were retained. The sequences were recovered through reverse transcription and tested by inserting the selected sequences into the 5′ region of a fluorescent protein coding sequence and using an in vitro transcription–translation system to produce the protein. All of the isolated switches showed at least a doubling of protein production on introduction of TPP into an in vitro transcription–translation reaction.

In vivo methods for screening RNA switches generally transform the RNA library into DNA form, such that the DNA-encoded switch library is cloned into an expression construct in which once transcribed it can impact expression of a reporter protein. The cells receive the expression construct and use their machinery to transcribe the encoded RNA, including the switch, and ultimately translate the encoded reporter protein. Each cell receives a single RNA switch and cells with functioning switches are recovered based on the expression of the reporter protein in the positive and negative conditions, and the sequences of the corresponding switches are ultimately recovered from those cells. Although in vivo screens are limited in the starting diversities of the RNA libraries relative to in vitro screens, they present the advantage that the switches are evaluated in an environment that is similar to that in which they will ultimately be used, and thus are more likely to translate to the intended application.

The throughput of in vivo screening techniques is generally dictated by the choice of reporter protein and the measurement techniques associated with that reporter protein. In one example, the β-galactosidase gene was used to screen for RBS-based switches that modulate translation initiation (Topp et al. 2010). The choice of β-galactosidase as the reporter protein enabled the use of an initial blue-white selection of colonies after the switch library was transformed into *E. coli*. The initial selection of white colonies identified clones in which reporter gene expression was turned off in the absence of the molecular target. These selected colonies were then grown up as clonal populations in 96-well plates and tested in media in the presence and absence of the ligand, to identify clones that showed increased β-galactosidase activity in the presence of ligand. Similar plate-based screens of clonal populations harboring RNA switch libraries have been shown with other reporter proteins encoding fluorescent proteins (Klauser et al. 2012; Davidson et al. 2013). However, the selection of an enzymatic reporter protein like β-galactosidase limits the throughput of the activity screen to clonal, plate-based assays, which show a lower throughput (several thousand constructs at most) than reporter proteins that are compatible with assays that enable the measurement and subsequent isolation of single cells in mixed populations, such as fluorescence-activated cell sorting (FACS). The integration of fluorescent reporter proteins and these types of single-cell assays have allowed for large RNA libraries to be rapidly screened and isolated in the same experiment.

In an early example, a FACS-based screen was used to identify RNA switches that controlled translation initiation in *E. coli* through the ligand-responsive regulation of RBS access (Lynch and Gallivan 2009). An RNA library was generated by coupling the theophylline aptamer to an RBS via a 12-nucleotide randomized linker and screened in *E. coli*. The library was placed upstream of the GFP gene in an expression construct and transformed cells were grown as a mixed population in the absence of ligand. FACS was used to iteratively isolate cells from the population that did not show fluorescence in the absence of ligand and showed high fluorescence levels in the presence of ligand. Using this FACS-based screening strategy, researchers evaluated 200,000 unique constructs, and switch sequences were identified that produced up to a 96-fold change in activity in the presence of ligand. FACS-based screening assays have subsequently been used by many groups to generate new RNA switches, including those that control transcription termination (Fowler et al. 2008) and translation activation (Xiu et al. 2017).

Although single-cell assays offer the advantage of higher-throughput screens, there can be significant cell-to-cell variation in gene expression, or noise, that can complicate the reliable quantification of gene expression activity from single cell measurements. This gene expression noise ultimately leads to challenges in reliably isolating switches with desired functional activities from FACS-based screens. However, researchers have shown that the cell-to-cell variation in expression can be measured by expressing a second fluorescent reporter gene under the control a constitutive promoter simultaneously with the fluorescent reporter gene under the control of the RNA switch (Liang et al. 2012). The ratio of fluorescence activity from these two reporter genes varies substantially less between cells of the same population than the expression of a single reporter gene, as both genes respond similarly to factors such as plasmid copy number and variation in transcription factor levels. This two-color assay was used in a FACS-based screen to select an improved ribozyme-based theophylline switch with 3.5-fold lower basal expression than a rationally designed parent theophylline ribozyme switch after undergoing only two rounds of sorting. The substantial increase in enrichment efficiency afforded by the two-color assay relative to the more traditional single-color assay was shown in a control sort. The latter assay did not lead to any observable enrichment in switches after two rounds of sorting, whereas the two-color assay enriched the population to near isolation.

As an alternative to screens, selections link the expression of a selection protein to a cell's growth rate. Thus, selections allow for cells harboring the desired switches to enrich themselves in a population through growth advantages conferred by the switches under different conditions. One important property for establishing a successful selection method is to identify a selection gene that can confer positive and negative growth advantages under different conditions, so that switches that increase gene expression in one state and decrease gene expression in another state can be enriched.

In one example, a switch-based selection in *E. coli* was developed based on regulation of the *tetA* gene, which causes cells to be resistant to the antibiotic tetracycline and susceptible to $NiCl_2$ (Sharma et al. 2008). Cells that did not express the selection gene in the absence of ligand were enriched by growing cells with $NiCl_2$, and cells that increased gene expression in the presence of ligand were then enriched by growing cells with tetracycline. Using this selection strategy, RBS-based switches that required both TPP and theophylline to turn on expression (i.e., AND gate) were generated. From an initial library of at least 100,000 variants the selection identified switches that showed activation ratios of >12 in the presence of thiamine and theophylline.

In another example, an RNA switch library was used to control the expression of a single transcription factor (Klauser et al. 2015). This transcription factor in turn controlled the expression of separate positive and negative selection genes, to enable a more generally applicable strategy for alternating between positive and negative selection conditions. An RNA switch library, comprising an aptamer connected to a ribozyme through a randomized linker, was cloned downstream from the transcription factor *GAL4*, which controlled the expression of three genes–*HIS3* (positive selection gene), *URA3* (negative selection gene), and *LacZ* (enzymatic reporter gene). Alternating rounds of positive and negative selection were used to search a sequence space of $\sim 10^7$ variants to identify novel theophylline and neomycin switches that inhibited gene expression up to fourfold on introduction of ligand. However, the best switches generated by the selection method came from a smaller library (~65,000 variants) that had an improved library design, which highlights how intelligent library design can improve the success of a screen or selection more so than simply increasing the number of sequences evaluated. From the theophylline library, 75 colonies were screened and two sequences were identified with an activation ratio of >2, whereas from the neomycin library 24 colonies were screened and seven sequences were identified with an activation ratio of >3. Although direct selection methods allow for enrichment of desired library members to occur through culturing the library population, they can be challenging to execute as the expression of the selective genes must be tuned so that the change in expression caused by the RNA switch results in an appreciable change in growth rate, which is further complicated in a library population with diverse activities represented.

One drawback of screens and selections is that they only provide activity information on the small number of switches that were enriched or isolated through the assays and subsequently taken through to further characterization and validation. High-throughput and data-rich strategies that provide activity information for all members of the library would provide valuable information that can be used to inform future designs and ultimately improve our understanding of RNA sequence–structure–function relationships. Recently, a FACS-Seq based approach was described that produces sequence–activity information for every sequence in RNA libraries as large as 100,000 unique sequences in a single experiment (Fig. 3A) (Townshend et al. 2015). The assay functions similarly to a two-color FACS-based screen, except that the initial library is split and grown under two separate conditions (i.e., presence and absence of ligand). The cell populations are then sorted on a FACS instrument into one of several activity bins, based on the cell's fluorescence ratio level (Fig. 3B). The sequences in each bin are recovered, barcoded, and the occurrence of each sequence in a given bin is counted using next-generation sequencing (NGS). The number of times a sequence occurs in the different activity bins is used to recreate a histogram of gene expression activity for that sequence under the different experimental conditions (Fig. 3C). This method was used to identify highly functional, ligand-responsive self-cleaving RNA switches in yeast. In addition, the rich datasets generated through these screens were used to identify consensus sequences required for active self-cleaving switches and provide greater insight into sequence–activity relationships of ribozyme switches.

In vivo screens and selections are based on coupling the RNA switch to the expression of a reporter or selection gene that can be used to directly isolate or enrich for sequences

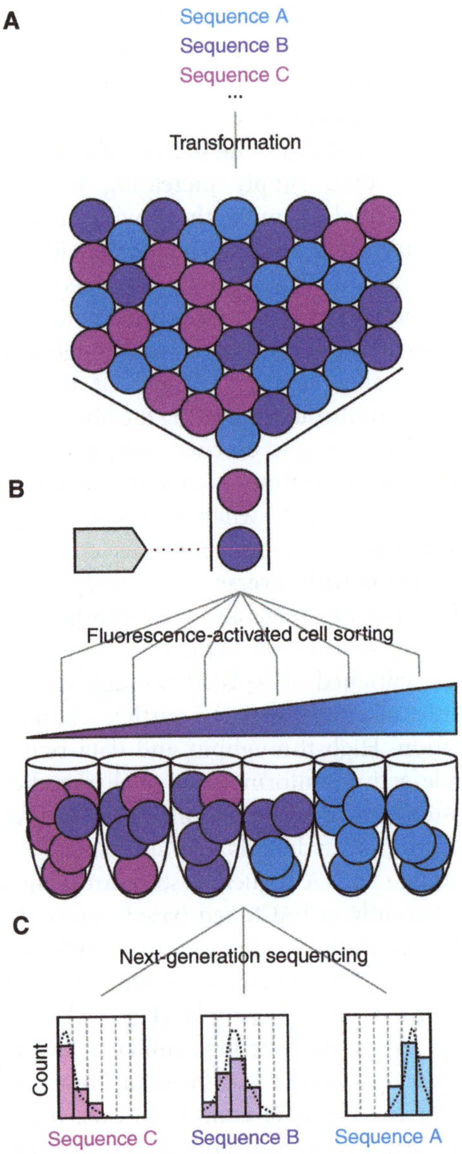

Figure 3. Schematic of the FACS-Seq workflow. (*A*) A library of genetic elements controlling a fluorescent reporter is transformed into cells. (*B*) The cells are grown under certain conditions and sorted into separate bins based on the expression levels of the fluorescent reporter. (*C*) The bins are sequenced to obtain the counts of each genetic element sequence in the library. The counts per bin for each sequence are used to recreate a distribution of expression levels for that sequence.

that cause the desired level of ligand-dependent expression. Strategies now enable high-throughput and data-rich approaches for RNA switch development by characterizing libraries of switches with targeted randomized regions. The major limitation to in vivo screening and selection systems is that the level of throughput is not large enough to sample the sequence space necessary to discover novel functional sensor activities, or aptamers, which requires the screening of on order 10^{14} unique sequences. Thus, the methods described in this section are limited to the generation of new switches based on known aptamer sequences.

5 CONCLUSIONS

RNA provides a flexible platform for the design of ligand-responsive gene-regulatory elements. Researchers have developed a broad suite of approaches that range from de novo computational design to high-throughput screening of very large sequence libraries for the generation of RNA switches tailored to the application of interest. Although there have been major advances in the field of engineering functional RNA, there are key challenges that remain in the design of RNA switches with desired quantitative activities and responsive to specified ligands. Advances in our understanding of RNA biology, including the prediction of RNA secondary and tertiary structure, cotranscriptional folding, and folding kinetics and energetics, are critical to improving the accuracy of forward design strategies. Further advances in the field will use computational analysis of the large sequence–activity data sets that are now being generated through high-throughput, massively parallel screens to ultimately generate new design rules. In addition, one can further streamline the process of developing RNA switches that can sense new ligands by combining de novo aptamer generation and switch generation into a single integrated selection process. As examples of RNA switches are increasingly implemented in diverse biological systems, expanded capacities for generating switches tailored for the intended application will be critical for their successful translation.

ACKNOWLEDGMENTS

We thank Matias Kaplan for valuable feedback in the preparation of the manuscript. This work was supported by the National Institutes of Health (grant to C.D.S., GM086663) and the National Institute of Standards and Technology (fellowship to C.M.S.). C.D.S. is a Chan Zuckerberg Biohub investigator.

REFERENCES

Beisel CL, Bayer TS, Hoff KG, Smolke CD. 2008. Model-guided design of ligand-regulated RNAi for programmable control of gene expression. *Mol Syst Biol* **4:** 224.

Bellaousov S, Reuter JS, Seetin MG, Mathews DH. 2013. RNAstructure: Web servers for RNA secondary structure prediction and analysis. *Nucleic Acids Res* **41:** W471–W474.

Chi Y-I, Martick M, Lares M, Kim R, Scott WG, Kim S-H. 2008. Capturing hammerhead ribozyme structures in action by modulating general base catalysis. *PLoS Biol* **6:** e234.

Davidson ME, Harbaugh SV, Chushak YG, Stone MO, Kelley-Loughnane N. 2013. Development of a 2,4-dinitrotoluene-responsive synthetic riboswitch in *E. coli* cells. *ACS Chem Biol* **8:** 234–241.

Domin G, Findeiß S, Wachsmuth M, Will S, Stadler PF, Mörl M. 2017. Applicability of a computational design approach for synthetic riboswitches. *Nucleic Acids Res* **45:** 4108–4119.

Ellington AD, Szostak JW. 1990. In vitro selection of RNA molecules that bind specific ligands. *Nature* **346:** 818–822.

Espah Borujeni A, Mishler DM, Wang J, Huso W, Salis HM. 2016. Automated physics-based design of synthetic riboswitches from diverse RNA aptamers. *Nucleic Acids Res* **44:** 1–13.

Fowler CC, Brown ED, Li Y. 2008. A FACS-based approach to engineering artificial riboswitches. *Chembiochem* **9:** 1906–1911.

Gao X, Dong X, Subramanian S, Matthews PM, Cooper CA, Kearns DB, Dann CE. 2014. Engineering of *Bacillus subtilis* strains to allow rapid characterization of heterologous diguanylate cyclases and phosphodiesterases. *Appl Environ Microbiol* **80:** 6167–6174.

Golynskiy MV, Koay MS, Vinkenborg JL, Merkx M. 2011. Engineering protein switches: Sensors, regulators, and spare parts for biology and biotechnology. *Chembiochem* **12:** 353–361.

Green AA, Silver PA, Collins JJ, Yin P. 2014. Toehold switches: De-novo-designed regulators of gene expression. *Cell* **159:** 925–939.

Klauser B, Saragliadis A, Ausländer S, Wieland M, Berthold MR, Hartig JS. 2012. Post-transcriptional Boolean computation by combining aptazymes controlling mRNA translation initiation and tRNA activation. *Mol Biosyst* **8:** 2242–2248.

Klauser B, Atanasov J, Siewert LK, Hartig JS. 2015. Ribozyme-based aminoglycoside switches of gene expression engineered by genetic selection in *S. cerevisiae*. *ACS Synth Biol* **4:** 516–525.

Lee S-W, Oh M-K. 2015. A synthetic suicide riboswitch for the high-throughput screening of metabolite production in *Saccharomyces cerevisiae*. *Metab Eng* **28:** 143–150.

Liang JC, Chang AL, Kennedy AB, Smolke CD. 2012. A high-throughput, quantitative cell-based screen for efficient tailoring of RNA device activity. *Nucleic Acids Res* **40:** e154.

Lynch SA, Gallivan JP. 2009. A flow cytometry-based screen for synthetic riboswitches. *Nucleic Acids Res* **37:** 184–192.

Ma AT, Schmidt CM, Golden JW. 2014. Regulation of gene expression in diverse cyanobacterial species by using theophylline-responsive riboswitches. *Appl Environ Microbiol* **80:** 6704–6713.

Martini L, Ellington AD, Mansy SS. 2016. An in vitro selection for small molecule induced switching RNA molecules. *Methods* **106:** 51–57.

Meyer A, Pellaux R, Potot S, Becker K, Hohmann H-P, Panke S, Held M. 2015. Optimization of a whole-cell biocatalyst by employing genetically encoded product sensors inside nanolitre reactors. *Nat Chem* **7:** 673–678.

Ogawa A. 2011. Rational design of artificial riboswitches based on ligand-dependent modulation of internal ribosome entry in wheat germ extract and their applications as label-free biosensors. *RNA* **17:** 478–488.

Puton T, Kozlowski LP, Rother KM, Bujnicki JM. 2013. CompaRNA: A server for continuous benchmarking of automated methods for RNA secondary structure prediction. *Nucleic Acids Res* **41:** 4307–4323.

Sharma V, Nomura Y, Yokobayashi Y. 2008. Engineering complex riboswitch regulation by dual genetic selection. *J Am Chem Soc* **130:** 16310–16315.

Topp S, Reynoso CMK, Seeliger JC, Goldlust IS, Desai SK, Murat D, Shen A, Puri AW, Komeili A, Bertozzi CR, et al. 2010. Synthetic riboswitches that induce gene expression in diverse bacterial species. *Appl Environ Microbiol* **76:** 7881–7884.

Townshend B, Kennedy AB, Xiang JS, Smolke CD. 2015. High-throughput cellular RNA device engineering. *Nat Methods* **12:** 989.

Tuerk C, Gold L. 1990. Systematic evolution of ligands by exponential enrichment: RNA ligands to bacteriophage T4 DNA polymerase. *Science* **249:** 505–510.

Vlack ERV, Topp S, Seeliger JC. 2017. Characterization of engineered PreQ1 riboswitches for inducible gene regulation in mycobacteria. *J Bacteriol* **199:** e00656–16.

Wachsmuth M, Findeiß S, Weissheimer N, Stadler PF, Mörl M. 2013. De novo design of a synthetic riboswitch that regulates transcription termination. *Nucleic Acids Res* **41:** 2541–2551.

Wang Y-H, Wei KY, Smolke CD. 2013. Synthetic biology: Advancing the design of diverse genetic systems. *Annu Rev Chem Biomol Eng* **4:** 69–102.

Win MN, Smolke CD. 2007. A modular and extensible RNA-based gene-regulatory platform for engineering cellular function. *Proc Natl Acad Sci* **104:** 14283–14288.

Winkler WC, Breaker RR. 2003. Genetic control by metabolite-binding riboswitches. *Chembiochem* **4:** 1024–1032.

Wittmann A, Suess B. 2011. Selection of tetracycline inducible self-cleaving ribozymes as synthetic devices for gene regulation in yeast. *Mol Biosyst* **7:** 2419–2427.

Xayaphoummine A, Bucher T, Isambert H. 2005. Kinefold web server for RNA/DNA folding path and structure prediction including pseudoknots and knots. *Nucleic Acids Res* **33:** W605–W610.

Xiu Y, Jang S, Jones JA, Zill NA, Linhardt RJ, Yuan Q, Jung GY, Koffas MAG. 2017. Naringenin-responsive riboswitch-based fluorescent biosensor module for *Escherichia coli* co-cultures. *Biotechnol Bioeng* **114:** 2235–2244.

Extensions, Extra Factors, and Extreme Complexity: Ribosomal Structures Provide Insights into Eukaryotic Translation

Melanie Weisser and Nenad Ban

Institute of Molecular Biology and Biophysics, Department of Biology, ETH Zürich, 8093 Zürich, Switzerland

Correspondence: ban@mol.biol.ethz.ch

SUMMARY

Although the basic aspects of protein synthesis are preserved in all kingdoms of life, there are many important structural and functional differences between bacterial and the more complex eukaryotic ribosomes. High-resolution cryo-electron microscopy (cryo-EM) and X-ray crystallography structures of eukaryotic ribosomes have revealed the complex architectures of eukaryotic ribosomes and species-specific variations in protein and ribosomal RNA (rRNA) extensions. They also enabled structural studies of a range of eukaryotic ribosomal complexes involved in translation initiation, elongation, and termination, revealing unique mechanistic features of the eukaryotic translation process, especially with respect to the identification and recognition of translation start and stop codons on messenger RNAs (mRNAs). Most recently, structural biology has provided insights into the eukaryotic ribosomal biogenesis pathway by visualizing several of its complex intermediates. This review highlights the past decade's structural work on eukaryotic ribosomes and its implications on our understanding of eukaryotic translation.

Outline

Cite as *Cold Spring Harb Perspect Biol* doi: 10.1101/cshperspect.a032367

1 RIBOSOMAL ARCHITECTURE

Ribosomes translate the information encoded in messenger RNA (mRNA) to synthesize proteins from amino acids. They are large macromolecular ribonucleoprotein complexes (RNPs) composed of several chains of ribosomal ribonucleic acids (rRNAs) and many ribosomal proteins (r-proteins), which are organized in two subunits, a small and a large one that are named according to their ultracentrifugation sedimentation coefficients expressed in Svedberg units (S) (Fig. 1). The large ribosomal subunit (LSU) harbors the active site of the ribosome, the "peptidyl transferase center" (PTC), that catalyzes peptide bond formation between amino acids during protein synthesis, whereas the small ribosomal subunit (SSU) is responsible for binding and decoding of the mRNA in the decoding center.

The process of protein synthesis is facilitated by a number of protein factors and occurs in several stages—initiation, elongation, termination, and recycling—during which LSUs and SSUs can reversibly dissociate and reassociate. They interact via intersubunit bridges at the predominantly rRNA-based subunit interface (Ben-Shem et al. 2011). Early low-resolution structures showed that both subunits contain three transfer RNA (tRNA)-binding sites at the subunit interface, termed "A-" (aminoacylation), "P-" (peptidyl), or "E-" (exit) site (Fig. 1). Protein synthesis starts with binding of an initiator tRNA to the P-site of the ribosome where it interacts with the start codon on the mRNA. During elongation, an aminoacylated tRNA is delivered to the A-site. After the peptide bond formation reaction, it moves to the P-site with the whole nascent peptide chain attached to it and after the next round of peptide bond formation, the tRNA leaves the ribosome in the deacylated state via the E-site. During the decoding process, the codon–anticodon matching is facilitated and monitored on the SSU, whereas peptide bond formation and nascent peptide exit take place on the LSU. Thus, tRNAs bind to the intersubunit space with their anticodon stems attached to the A-, P-, or E-site of the SSU, whereas their aminoacylated acceptor CCA 3′ ends point toward the LSU.

The most fundamental and functionally important features of LSUs are the PTC and the polypeptide exit tunnel that extends from the PTC to the universally conserved r-proteins uL23, uL29, and uL24 on the solvent-exposed side of the LSU (Ben-Shem et al. 2011; Klinge et al. 2011). Further characteristic features are the central protuberance (CP; including r-proteins uL18/uL5), the L1 stalk (including uL1) that coordinates the expulsion of E-site tRNAs from the ribosome, and the P1/P2 stalk and the sarcin-ricin loop (SRL) responsible for the recruitment and activation of ribosome-dependent guanosine triphosphatases (GTPases) to the ribosome (Fig. 1) (Diaconu et al. 2005; Voorhees et al. 2010; Ben-Shem et al. 2011).

The architecture of the SSU is marked by the groove of the mRNA channel, which divides the subunit into a "head" and a "body" domain (Wimberly et al. 2000; Rabl et al. 2011; Hussain et al. 2014). mRNAs enter the channel between the "beak" structure of the SSU head and the "shoulder" of the SSU body, in which r-proteins with intrinsic helicase activity or associated adenosine triphosphate (ATP)-driven RNA helicases unwind possible mRNA secondary structures (Takyar et al. 2005; Hashem et al. 2013a). After passing the so-called "mRNA entry latch" (Zhang et al. 2015), which can close and stabilize mRNAs inside the channel, mRNAs reach the decoding center of the SSU A-site where the base-pairing of an mRNA codon with an anticodon of incoming tRNAs takes place (Hussain et al. 2014; Shao et al. 2016). Finally, mRNAs exit the ribosome near the SSU E-site at the SSU "platform," formed by the central SSU rRNA domain and r-proteins (Fig. 1).

2 EUKARYOTIC RIBOSOMES—EXTENSIONS AND EXPANSIONS

Structural analysis of bacterial ribosomes by X-ray crystallography and electron microscopy has played an important role in revealing ribosomal architecture, thereby providing key insights into the mechanisms that underlie the translation process. High-resolution X-ray structures have shown the central role of rRNA in the selection of cognate tRNAs in the decoding center (Wimberly et al. 2000) and revealed that the PTC is exclusively composed of rRNA, which provided experimental proof that ribosomes are "ribozymes" and evolutionarily stem from an early RNA-based world (Nissen et al. 2000; Schmeing and Ramakrishnan 2009; Klinge et al. 2011).

Early structural research on ribosomes has been covered by many excellent reviews and book chapters (e.g., Schmeing and Ramakrishnan 2009; Voorhees and Ramakrishnan 2013), but predominantly focused on bacterial translation systems. Therefore, this review will focus on eukaryotic translation and the high-resolution structures of eukaryotic ribosomal complexes that became available in recent years.

Eukaryotic ribosomes have a molecular mass of ~4.3 MDa and a sedimentation coefficient of 80S and are therefore considerably larger than bacterial 70S ribosomes with a molecular mass of ~2.3 MDa (Melnikov et al. 2012). Eukaryotic SSU and LSU sediment at 40S and 60S (Klinge et al. 2011; Rabl et al. 2011), whereas bacterial ribosomal subunits sediment at 30S and 50S, respectively (Wimberly et al. 2000; Melnikov et al. 2012).

The first high-resolution crystal structures of eukaryotic ribosomes revealed how complex eukaryotic cytoplasmic ribosomes are in comparison with their bacterial and archaeal counterparts, with archaeal ribosomes representing

Cite as *Cold Spring Harb Perspect Biol* doi: 10.1101/cshperspect.a032367

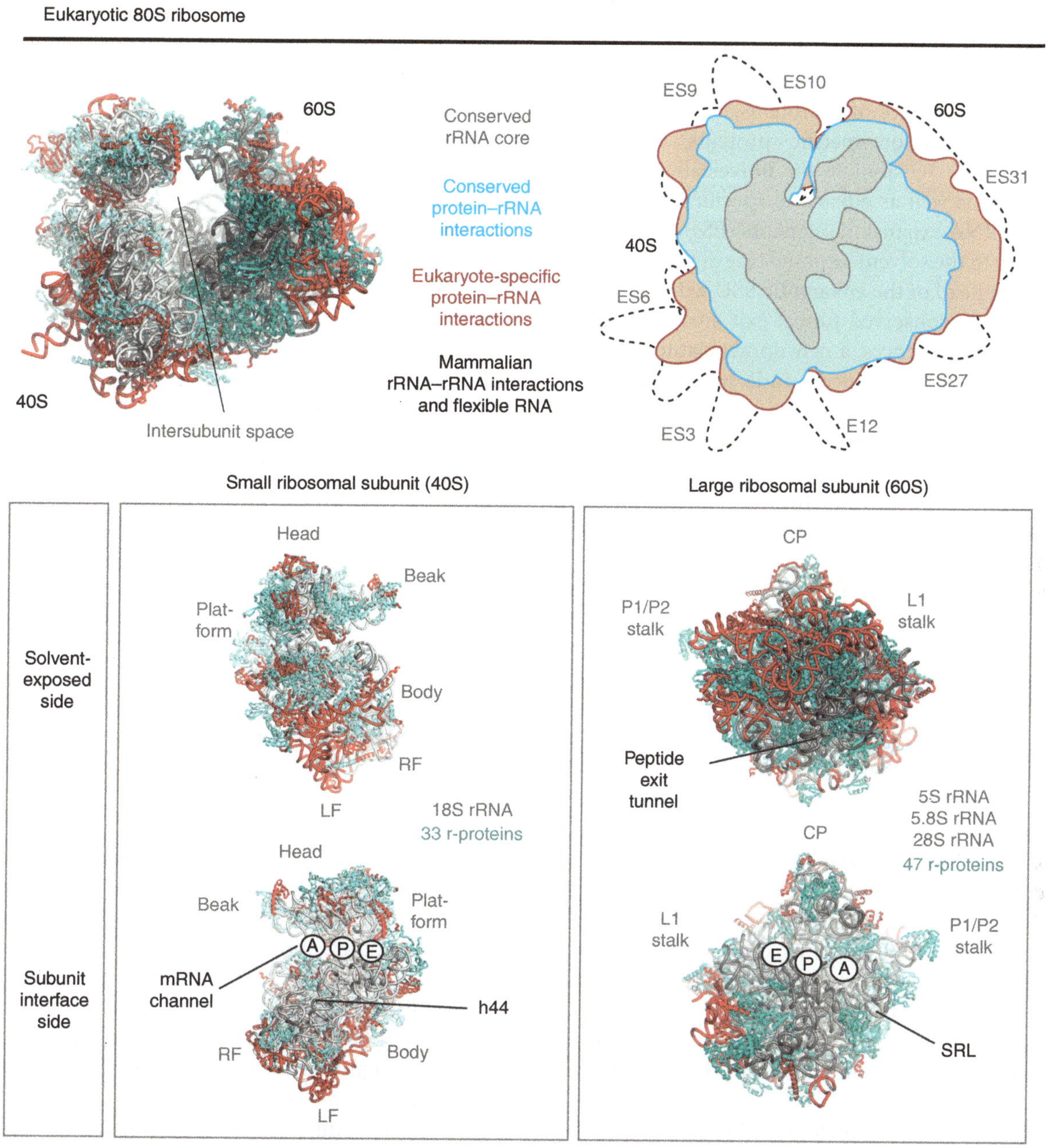

Figure 1. The eukaryotic 80 Svedberg unit (S) ribosome consists of a 40S and a 60S subunit. The 40S consists of an 18S rRNA (gray) and 33 proteins (cyan), whereas the 60S contains 5S, 5.8S, and 28S rRNA (gray) and 47 proteins (cyan). Eukaryotes have evolved eukaryote-specific proteins, protein extensions, and rRNA expansion segments (ESs) (red), with metazoans and particularly mammals displaying the greatest extent of complexity and ES length. As a result of this evolutionary process, eukaryotic ribosomes have a characteristic layered architecture (*top right* panel). The universally conserved ribosomal core is rRNA-based (gray) and wrapped in a conserved layer of ribosomal protein (r-protein) and rRNA interactions (cyan), which in eukaryotes is surrounded by an additional layer of eukaryote-specific r-protein–rRNA interactions (red). Because of their especially elongated ESs, mammals display an extra outer layer of rRNA–rRNA interactions between ESs and a periphery of flexible, tentacle-like ES extensions (black dotted lines). Both 40S and 60S are marked by characteristic conserved architectural landmarks (*lower* panel). Transfer RNAs (tRNAs) bind to ribosomal A- (aminoacylation), P- (peptidyl), and E- (exit) sites that are located on the subunit interface of both subunits. The small ribosomal subunit (SSU) can be divided into a head (with a beak) and a body with a platform, shoulder, left foot (LF), and right foot (RF). The large ribosomal subunit (LSU) has three characteristic lobes: the central protuberance (CP), the L1 stalk, and the P stalk, below which the guanosine triphosphatase (GTPase)-activating sarcin–ricin loop (SRL) is located.

an evolutionary intermediate (Ben-Shem et al. 2011; Klinge et al. 2011; Rabl et al. 2011). Despite the overall high degree of conservation of the rRNA core and the existence of a universally conserved set of r-proteins, eukaryotic ribosomes display eukaryote-specific proteins, eukaryote-specific tails, and extensions to the universally conserved r-protein cores, as well as eukaryote-specific rRNA insertions termed rRNA expansion segments (ESs), all of which localize mostly to the solvent-exposed face of the eukaryotic LSU and to the head of the eukaryotic SSU with the exception of the highly conserved tunnel exit area (Fig. 1). To rationalize these differences, a new nomenclature for ribosomal proteins was introduced, indicating whether an r-protein is universally conserved (uS/uL), specific for bacteria (bS/bL) or uniquely exist in eukaryotes (and archaea) (eS/eL) (Klinge et al. 2011; Ban et al. 2014).

Since these first breakthroughs, many high-resolution structures of ribosomes from different eukaryotic species have become available because of the technological and methodological advances in the field of cryo-electron microscopy (cryo-EM) (Bai et al. 2015). Structures of plant (Gogala et al. 2014), insect (Anger et al. 2013), pathogenic protozoa (Sun et al. 2015), kinetoplastid (Gao et al. 2005; Hashem et al. 2013; Shalev-Benami et al. 2016), yeast (Ben-Shem et al. 2011), ciliate (Klinge et al. 2011; Rabl et al. 2011), and mammalian ribosomes (Anger et al. 2013; Voorhees et al. 2014) have shown that even among eukaryotic species, which, except for yeast, all contain the same set of 80 r-proteins, ribosomes can differ significantly in their r-protein sequences and the length of rRNA ESs (Melnikov et al. 2012). This variability is further increased by the posttranslational modifications of r-proteins, differential expression levels of r-proteins, and the existence of r-protein paralogs, which indicate that specialized ribosomes could occur within different tissues, life cycles, and metabolic or developmental stages within the same organism (reviewed in Xue and Barna 2012).

With the exception of kinetoplastid ribosomes (Gao et al. 2005; Hashem et al. 2013; Shalev-Benami et al. 2016), metazoan and specifically mammalian ribosomes (Anger et al. 2013; Voorhees et al. 2014) are on average more complex and larger than protozoan ribosomes (Ben-Shem et al. 2011; Klinge et al. 2011; Rabl et al. 2011), which is in part attributable to an increase of mass on the protein level, but particularly to larger rRNA ESs.

Interestingly, these extra protein extensions can be found at functionally important ribosomal sites. On the LSU, extensions mainly affect regions in the vicinity of the L1 and P stalks and the peptide tunnel exit (eL22 and uL23 in *Drosophila*; Anger et al. 2013). On the SSU, changes are most pronounced at the mRNA channel entry and exit areas (eS26; Anger et al. 2013) where they are likely to have an effect on the binding of ribosome-associated factors and mRNAs.

Because the ESs are able to protrude like tentacles beyond the conserved layer of protein–rRNA interactions, they add an additional rRNA-based outer layer to metazoan ribosomes (Anger et al. 2013). The result is a characteristic layered architecture of cytosolic eukaryotic ribosomes (Fig. 1): The innermost core is rRNA-based and universally conserved. It is wrapped into a layer of universally conserved protein–rRNA interactions surrounded by an even more intricate network of eukaryote-specific protein–rRNA interactions followed by an outermost layer of metazoan-specific RNA–RNA inter-ES contacts and a periphery of flexibly protruding ES tips (Anger et al. 2013).

3 THE EUKARYOTIC TRANSLATION CYCLE

Because eukaryotic cytoplasmic ribosomes have long half-lives of several days (Nikolov et al. 1983) they are capable of translating many mRNAs during their life span. On each mRNA, they proceed through a four-stage translation cycle: The determination of the correct open reading frame (ORF) on an mRNA (initiation), repetitive cycles of peptide bond formation between amino acids (elongation), and the recognition of the end of the ORF (termination) when the ribosome is released from the mRNA (ribosomal recycling) allowing it to initiate another translation cycle (Fig. 2).

At each of the four stages, different regulatory translation initiation, elongation, or release factors associate with

Figure 2. The eukaryotic translation cycle. Ribosomal biogenesis (lavender) starts in the nucleolus and ends in the cytoplasm, where mature translation-competent cytoplasmic eukaryotic 40S and 60S ribosomal subunits are being released. Ribosomal protein translation proceeds in cycles of four stages: translation initiation (red), elongation (yellow), and termination (blue) followed by ribosomal recycling (green). During initiation, the SSU attaches to an mRNA, either via the canonical (eIF1–5) eukaryotic translation initiation factor-dependent pathway or via alternative eIF2D/malignant T-cell amplified protein 1 (MCT-1)/density-regulated protein (DENR) or internal ribosomal entry site (IRES)-mediated (re-) initiation, whereas the 60S ribosomal subunit is prevented from premature reassociation with the 40S by eIF6 until the correct start codon has been recognized by the initiator tRNA. Then, the two subunits associate and the 80S ribosome translates the open reading frame (ORF) of the mRNA in repetitive cycles of elongation/translocation until it either stalls prematurely in case of a truncated mRNA, which triggers the Hbs1–Pelota-dependent rescue pathway, or at a normal stop codon, leading to translation termination. (*Legend continues on following page.*)

Cite as *Cold Spring Harb Perspect Biol* doi: 10.1101/cshperspect.a032367

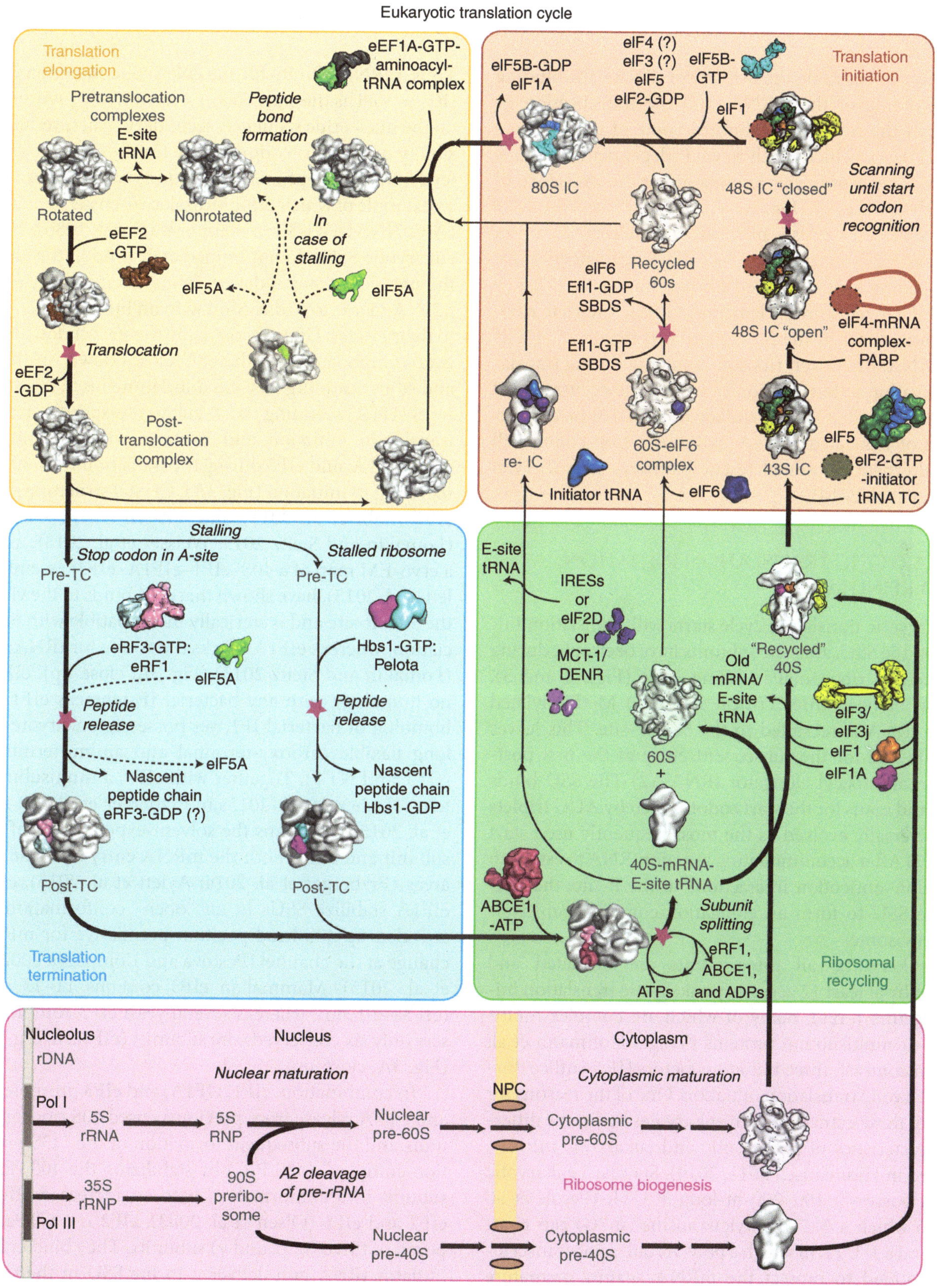

Figure 2. (*Legend continued from previous page.*) In both cases, the ribosomal subunits are being split by the ATPase ABCE1 and detached from mRNA and remaining deacylated tRNAs, such that they can engage in another round of translation initiation on either the same or a different mRNA. Some translation factors act as GTPases, which catalyze GTP-dependent reactions and therefore mark energy-consuming irreversible checkpoints along the trajectory of the translational cycle (pink stars). IC, initiation complex.

the ribosome catalyzing the delivery of tRNAs, the binding and unwinding of the mRNA, and the translocation or splitting of the ribosomal subunits. Some of these factors act as GTPases, which catalyze GTP-dependent reactions and mark irreversible checkpoints along the trajectory of the translational cycle (Fig. 2).

Translation can also proceed via alternative routes depending on the presence of short regulatory non-protein-coding upstream open reading frames (uORFs) or structured *cis*-regulatory elements (IRESs, internal ribosomal entry sites) in the 5′ untranslated region (5′ UTR) of an mRNA. Moreover, ribosomal stalling during translation elongation or termination caused by the occurrence of rare codons in the ORF (Melnikov et al. 2016) or the recognition of truncated mRNAs without a stop codon will trigger alternate ribosomal rescue pathways (Fig. 2) (van den Elzen et al. 2014).

4 EUKARYOTIC TRANSLATION INITIATION AND REINITIATION

The eukaryotic translation cycle starts with translation initiation on the SSU, after the subunits have been split during the preceding ribosomal recycling stage (Figs. 2 and 3). Initiation involves three critical steps: (1) Methionylated initiator tRNA is recruited to the SSU P-site. This leaves the A-site free for the subsequent entry of the first post-initiation aminoacyl elongator tRNA. (2) The SSU binds mRNA and scans for the start codon, whereby AUG triplets have universally evolved as the most frequently used start codon. (3) After accommodating initiator tRNA to establish start codon–anticodon interactions in the P-site, the LSU joins the SSU to form an elongation-competent mRNA-bound ribosome.

In eukaryotes, all three events are mediated and regulated by at least 12 accessory eukaryotic translation initiation factors (eIFs), many of which are complex multidomain or multisubunit proteins (Voigts-Hoffmann et al. 2012). In contrast, three initiation factors (IFs) suffice during prokaryotic translation initiation. One of the reasons for the use of these extra factors in eukaryotes lies in the different characteristics of prokaryotic and eukaryotic mRNAs (reviewed in Hinnebusch 2011, 2014). Standard eukaryotic mRNAs possess a 100–200-nt-long 5′ UTR (Pesole et al. 2001) to which a 5′ 7-methylguanosine (m7G) cap is attached and a 3′ UTR in which a poly(A) tail is posttranscriptionally added to protect the mRNA against premature nucleolytic degradation. Both features are recognized by eIFs to select intact mRNAs (reviewed in Hinnebusch 2011). Moreover, eukaryotic 5′ UTRs lack a Shine–Dalgarno sequence that is frequently used in prokaryotes to position the mRNA with the start codon in the P-site of the SSU. Thus, mRNA affinity for the eukaryotic initiation complex (IC) as well as the start codon selection rely predominantly on the nucleotide sequence, secondary structure, and accessibility of the start codon. These features are altogether referred to as the optimal start codon context (Kozak 1986), an example of which is the "Kozak consensus sequence" (5′-(A/G)NNAUGG-3′) in mammals (Kozak 1986). However, eukaryotic SSUs do not immediately bind in the vicinity of the start codon, instead, they "scan" for it moving along the mRNA in a 5′ to 3′ direction with an initiator tRNA bound to their P-sites. Upon recognizing the start codon, the tRNA is accommodated and the SSU closes down on the tRNA and stops scanning (reviewed in Hinnebusch 2011, 2014).

5′-UTR scanning is a eukaryote-specific feature of translation initiation and it starts with the binding of eIF1, eIF1A, and eIF3 during the transition from ribosomal recycling to initiation (Fig. 3A). Crystal structures of a 40S–eIF1 (Rabl et al. 2011) and 40S–eIF1–eIF1A complexes (Lomakin and Steitz 2013; Weisser et al. 2013), as well as a cryo-EM map of a 40S–eIF1–eIF1A–eIF3j assembly (Aylett et al. 2015), have shown that eIF1 binds to the vicinity of the SSU P-site and is sterically incompatible with 60S association, whereas eIF1A blocks the A-site for tRNA binding (Lomakin and Steitz 2013) (Fig. 3A, close-up). eIF1 bears no homology with any bacterial IF, whereas eIF1A is the homolog of bacterial IF1, yet possesses eukaryote-specific long flexible carboxy-terminal and amino-terminal tails (CTT and NTT). Together with eIF3, a multisubunit protein (Hashem et al. 2013a,b; Erzberger et al. 2014; Aylett et al. 2015) that spans the solvent-exposed side of the 40S subunit and binds both the mRNA entry and exit channel areas (Erzberger et al. 2014; Aylett et al. 2015), eIF1 and eIF1A stabilize SSUs in an "open" conformation with a swiveled-up 40S head position permissive for mRNA exchange at the channel (Pestova and Kolupaeva 2002; Llácer et al. 2015). Mammalian eIF3 contains 11–13 subunits (eIF3a–eIF3m), whereas several yeast eIF3 complexes possess only six conserved core subunits (eIF3a, b, c, g, I, and j) (Fig. 3A, close-up).

In combination, eIF1, eIF1A, and eIF3 promote mRNA and tRNA release from previously used 40S ribosomal subunits and the subsequent formation of the "43S pre-initiation complex" (43S PIC) by stabilizing the 40S ribosomal subunit in a conformation with an increased affinity for eIF2 and eIF5 (Olsen et al. 2003). eIF2 is a GTPase composed of three (α, β, and γ) subunits. They bind eukaryotic initiator tRNA and deliver it to the SSU in the form of a ternary eIF2–GTP–initiator tRNA complex (Schmitt et al. 2010). eIF5 is a two-lobed protein with a carboxy-terminal domain (CTD) and an amino-terminal domain (NTD) connected by a flexible linker (Bieniossek et al. 2006), which acts as the GTPase-activating protein (GAP) of eIF2, carry-

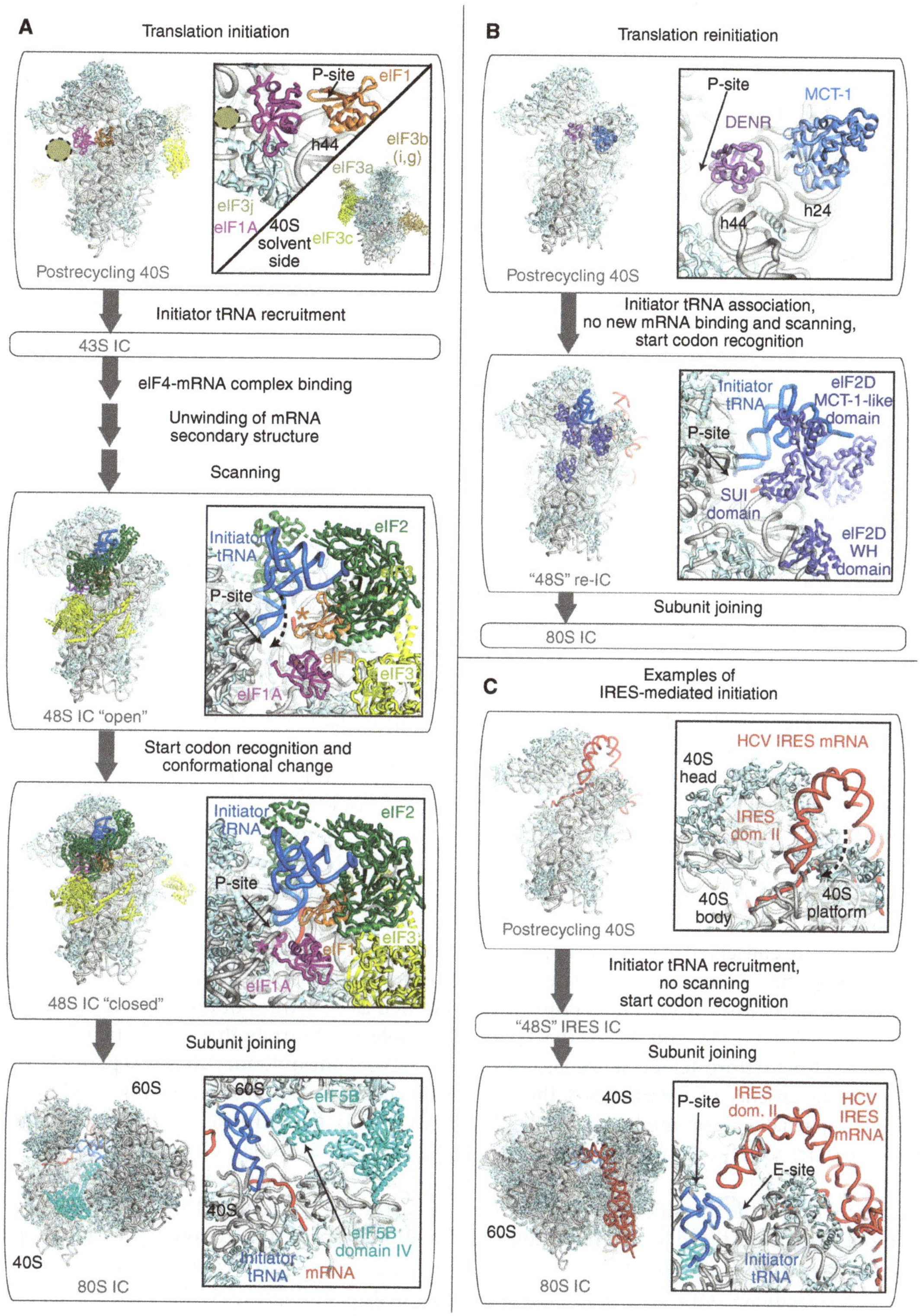

Figure 3. Eukaryotic translation initiation. (*A*) Classical eukaryotic translation initiation starts after subunit splitting with the assembly of eIF1, eIF1A, and eIF3 on the 40S ribosomal subunit and proceeds with the recruitment of the eIF2–GTP–methionine (Met)–initiator $tRNA_{MET}$ complex and eIF5, resulting in the formation of the 43S re-initiation complex (re-IC). After eIF4-mediated mRNA attachment, the open 48S IC is formed, which transitions to the closed 48S complex on start codon recognition and proper accommodation of the initiator tRNA in the ribosomal P-site. Subsequent initiation factor dissociation allows eIF5B binding, eIF5B–eIF1A-mediated subunit joining, and formation of the 80S IC. (*B*) Alternatively, translation reinitiation after the translation of a short upstream open reading frame (uORF) can be mediated either by the MCT-1–DENR heterodimer or eIF2D or (*C*) by IRESs like the hepatitis C virus IRES.

ing a GTPase-activating arginine finger in its NTD (Das and Maitra 2001).

eIF2γ is the core subunit of eIF2 that harbors the GTPase activity (Roll-Mecak et al. 2004; Schmitt et al. 2012) and displays a typical three-lobed GTPase structure (Roll-Mecak et al. 2004; Schmitt et al. 2012). Domain II of eIF2γ recognizes terminally methionylated initiator tRNA through a conserved binding groove, inducing a characteristically kinked conformation in the initiator tRNA acceptor end (Schmitt et al. 2012). In contrast to prokaryotes that discriminate between methionylated initiator and elongator tRNAs through initiator tRNA methionine formylation, eIF2γ specifically recognizes a conserved Watson–Crick base pair between A1 and U72 in eukaryotic initiator tRNAs, which does not exist in methionylated elongator tRNAs (Kapp et al. 2006). The eIF2α and eIF2β subunits interact with the eIF2γ core, but not with each other, and contribute to initiator tRNA binding (Schmitt et al. 2012; Llácer et al. 2015).

The ternary complex of eIF2–GTP–initiator tRNA and eIF5 assemble simultaneously or successively with the 40S–eIF1–eIF1A–eIF3 complex to form the 43S PIC. A high-resolution structure of the 43S PIC does not exist, but cryo-EM maps at lower resolutions have provided insights into its overall architectural features (Hashem et al. 2013). eIF2α binds to the E-site of the SSU, analogously to the binding of E-site tRNA, whereas the ribosomal contacts for eIF2γ and eIF2β remain debated (Hussain et al. 2014; Llácer et al. 2015). Together, the three subunits position the initiator tRNA into the 40S P-site, for which it has an increased affinity because of conserved interactions between three initiator tRNA-specific successive G:C base pairs (29:41, 30:40, and 31:39) in the tRNA anticodon stem loop with rRNA bases A790, G1338, and A1339 (bacterial numbering) of the SSU P-site (Korostelev et al. 2006; Kolitz and Lorsch 2010). In the absence of a start codon, the initiator tRNA adopts a partially accommodated metastable position in the P-site (termed "P_{OUT}") (reviewed in Hinnebusch 2011).

mRNA attachment causes the transition to an open 48S complex and is thought to be mediated by the multisubunit initiation factor complexes eIF3 and eIF4F (Jivotovskaya et al. 2006). eIF4F preassembles onto the mRNA causing it to adapt a "closed loop" conformation by simultaneously interacting with both mRNA ends: One eIF4F subunit called eIF4E binds the eukaryote-specific mRNA m7G cap, whereas another eIF4F subunit called eIF4G contacts the mRNA poly(A) tail via PABP (poly(A)-binding protein) (Wells et al. 1998). Multiple interactions between eIF4F and eIF3 subsequently recruit the mRNA to the SSU–PIC (Villa et al. 2013); however, they have not been structurally visualized in the context of an active complex at high resolution.

With the initiator tRNA bound to the P-site in a P_{OUT} conformation (Fig. 3A, middle), the open 48S complex starts to scan base per base along the mRNA to localize the correct start codon. Scanning is processive and occurs in a 5′ to 3′ direction at a speed of ~8 nt/sec (Vassilenko et al. 2011). It is supported by several copies of eIF4F subunit eIF4A, an ATP-driven RNA helicase, and eIF4B (or its homolog eIF4H), a stimulator of eIF4A activity, which drive the unfolding of mRNA secondary structure. Scanning usually also requires the assistance of accessory ATP-dependent DEAD-box RNA helicases, like DHX29 (Pisareva et al. 2008 and reviewed in Marintchev 2013).

eIF4A-driven mRNA unwinding is not processive and most likely happens "upstream" (at the mRNA entry site of the SSU). In its ATP-bound state, the helicase associates with double-stranded mRNA, whereas its ADP-bound state possesses a lower mRNA affinity and dissociates. Multiple rounds of attachment and detachment thereby allow the helicase domains to unwind the mRNA into single strands. Yet, it remains to be fully investigated how the mRNA is threaded into the mRNA channel of the SSU and whether the 48S PIC could scan along the mRNA while the mRNA 5′ cap remains stably associated with the IF complex (reviewed in Marintchev 2013).

The existing cryo-EM structure of an open 48S complex from yeast (Fig. 3A, middle) (Llácer et al. 2015) does not provide insights into this process because it was formed on a model mRNA oligonucleotide with a near-cognate start codon and does not contain eIF4F or accessory helicases. Still, it offers detailed insights into the three-dimensional arrangement of initiation factors eIF1, 1A, and 2 and, to some extent, also of eIF3 and eIF5. 48S complex structures, as well as structures of the 40S–eIF1 and 40S–eIF1–eIF1A complexes, illustrate that in an open, scanning-competent 40S, the P-site initiator tRNA is kept in a P_{OUT} position with full accommodation sterically hindered by eIF1 (Rabl et al. 2011) and the CTT of eIF1A (Pestova and Kolupaeva 2002; Yu et al. 2009). eIF1's basic β1 loop protrudes into the mRNA channel and contacts the mRNA in the P-site (Martin-Marcos et al. 2013), thereby restricting the conformational space for codon–anticodon duplex formation (Hussain et al. 2014). Thus, eIF1 and the eIF1A CTT prevent premature initiator tRNA binding to near-cognate start codons and ensure the fidelity of start codon recognition (Pestova and Kolupaeva 2002). This was previously suggested by genetic studies in yeast, which identified Sui^- (suppressor of initiation codon) mutations in eIF1 and in the eIF1A CTT that lead to an increase in leaky scanning (i.e., translation initiation on near-cognate start codons) (Cheung et al. 2007; Fekete et al. 2007; Lee et al. 2007).

Upon start codon recognition, the 48S complex transitions into a "closed" conformation with a less strongly rotated 40S head that is lowered over the mRNA channel, clamping onto the mRNA (Fig. 3A, middle) (Llácer et al. 2015). Premature closing can result in initiation at a noncognate start codon, whereas inhibiting the transition to the closed conformation results in decreased translation initiation efficiency at correct start codons (reviewed in Hinnebusch 2011).

The structure of the closed 48S complex (Fig. 3A, middle) (Hussain et al. 2014; Llácer et al. 2015) shows that the initiator tRNA has changed from its P_{OUT} to a stably bound P_{IN} position and becomes anchored to the start codon in the ribosomal P-site (reviewed in Hinnebusch 2011), thereby ejecting eIF1 and the CTT of eIF1A from their initial positions (Cheung et al. 2007). Interestingly, the new deep insertion of the codon–anticodon duplex into the P-site of the closed 48S complex is stabilized by interactions with the previously unstructured NTT of eIF1A (Hussain et al. 2014; Llácer et al. 2015). This supports results from yeast genetic studies, which identified mutations in the eIF1A NTT with Ssu^- (suppressor of Sui^-) phenotypes (Fekete et al. 2007; Saini et al. 2010).

eIF2β positions a helix–turn–helix motif and its zinc-binding domain between the tRNA, eIF1, eIF1A, and eIF2γ and contacts the 40S head across the mRNA channel. Its amino-terminal helix is stably anchored to the G domain of eIF2γ (Schmitt et al. 2012; Llácer et al. 2015). Nuclear magnetic resonance (NMR) and biochemical experiments revealed an intricate interaction network between eIF1, eIF2, and eIF5 (Reibarkh et al. 2008; Luna et al. 2012) and it was shown that overexpression of eIF5 promotes eIF1 dissociation from the IC (Nanda et al. 2009). Therefore, it was concluded that eIF5 and eIF1 might compete for binding to eIF2β (Luna et al. 2012). eIF1 dissociation allows eIF5 to adopt eIF1's original binding site next to eIF2β and positions the GTPase-activating NTD of eIF5 closer to the GTPase active site on eIF2, where it is stabilized through interactions with the repositioned CTT of eIF1A (Yu et al. 2009; Hussain et al. 2014). Although GTP hydrolysis on eIF2 may occur at earlier stages of initiation, the rearrangement of the eIF5 GAP domain possibly triggers phosphate release from eIF2 and thereby irreversibly commits the IC to translation (Algire et al. 2005; Cheung et al. 2007; Nanda et al. 2009). Conversely, failure to induce phosphate release after start codon "recognition" leads to a resumption of the scanning process (Terenin et al. 2016). By transitioning from the GTP- to the guanosine diphosphate (GDP)-bound state, eIF2 loses affinity for the initiator tRNA and the IC and is released from the IC, likely together with eIF5, eIF3, and eIF4.

Dissociation of eIF2/eIF5 from the complex allows the ribosome-dependent GTPase eIF5B, the ortholog of bacterial IF2/archaeal aIF2, to bind to the 40S subunit interface. It interacts with the core of eIF1A via a binding site that also mediates the contact between IF1 and IF2 in bacteria, but also via a conserved DIDDI motif in the eukaryote-specific eIF1A-CTT (Olsen et al. 2003; Acker et al. 2006; Fringer et al. 2007).

A structure of a mammalian eIF5B-initiator tRNA 80S complex without eIF1A shows that eIF5B binds to the 40S in a similar position as EF-Tu, IF2, and EF-G (Fig. 3A, bottom) (Fernandez et al. 2013). Although it is not completely clear how the 60S is recruited, it is known that eIF5B binding to the 40S subunit accelerates 60S association (Acker et al. 2006) and is dependent on both eIF1A (in particular the DIDDI motif) and eIF5B (Olsen et al. 2003; Fringer et al. 2007). In the 80S context, the interaction with eIF5B domain IV prevents the CCA end from contacting the 60S ribosomal P-site and the PTC, whereas the eIF5B G domain is positioned next to the ribosomal GTPase-activating center. GTP hydrolysis on eIF5B induces a conformational change within the IC that weakens eIF5B's affinity for the latter and ultimately leads to coupled release of eIF5B and eIF1A (Fringer et al. 2007). As a consequence, the initiator tRNA adapts a P/P state on the 80S ribosome, which can thus enter translation elongation.

5 ALTERNATIVE TRANSLATION INITIATION STRATEGIES

The fact that eukaryotic mRNAs possess 5′ UTRs opens up additional possibilities for regulating translation initiation via characteristic sequences and structured regions upstream of the coding sequence (reviewed in Hinnebusch et al. 2016). One well-investigated example of such *cis*-regulatory mRNA elements is IRESs that confer the ability to bind to the ribosomal subunit in the absence of a cap-binding eIF4F complex (Walsh and Mohr 2011). IRESs were first identified in viruses, but in recent years have also been detected in ~10% of all cellular mRNAs (Weingarten-Gabbay et al. 2016). Cap-independent translation initiation offers the possibility of maintaining normal translation efficiency even during cellular stress conditions, when eIF4F complex availability is down-regulated. Viruses make use of such alternative initiation strategies to escape the host cell stress response and bias the translation machinery toward preferential translation of viral transcripts.

Viral IRESs can be assigned into three classes based on their structure. Class I IRESs occur in viruses with a positive sense strand bicistronic RNA genome, where they precede the second ORF and are therefore also referred to as intergenic region IRESs (IGRs). IGRs mediate translation initiation in the absence of IFs, an AUG start codon, and cellular initiator tRNA, because they structurally mimic initiator tRNA binding (Costantino et al. 2008). Class II IRESs occur within RNA viruses with a positive sense strand genome like the hepatitis C virus (HCV) or the classical swine fever virus (CSFV). They bind directly to the 40S via a pseudoknot-like structure in a cap-independent and scanning-free manner because the AUG start codon is positioned into the 40S P-site on IRES–40S binding (Quade et al. 2015), yet they require initiation factor eIF2 for initiator tRNA recruitment and eIF3 for efficient translation initiation (Otto and Puglisi 2004). Class III IRESs usually occur in picornaviruses and have canonical IF requirements except for subunits eIF4E and the amino-terminal third of eIF4G that does not contain the eIF3-binding site (Redondo et al. 2013).

High-resolution structures of the class II HCV-IRES bound to the eukaryotic ribosome have shown that it is structured into four domains (Fig. 3C). Domain IV contains the start codon, whereas domain III mediates the interaction with eIF3 and the binding to 40S expansion segment ES7 and r-protein eS27 (Hashem et al. 2013; Quade et al. 2015), i.e., to a location that overlaps with eIF3's binding site on the 40S subunit. Domain II occupies a position typically occupied by the E-site tRNA or eIF2α (Fig. 3C, close-up) and appears to be critical for late steps of the initiation process, such as eIF5-induced hydrolysis of eIF2-bound GTP and joining of the LSU (Locker et al. 2007). Mimicking tRNA, the tip of domain II (IIb) appears to induce a conformational change in the 40S, which induces eIF2 release and promotes a transition from translation initiation to elongation (Locker et al. 2007).

Apart from IRES elements, eukaryotic cells make use of short uORFs in the 5′ UTR to fine-tune the translation initiation efficiency at the main ORF (reviewed in Hellens et al. 2016). uORFs usually consist of 100 or fewer codons and occur in all eukaryotes (Hellens et al. 2016). They are found in ~50% of all mammalian and ~13% of all yeast mRNAs (Calvo et al. 2009; Lawless et al. 2009). One possible mechanism for how uORFs can regulate translation is through "reinitiation" at the downstream main ORF with the help of canonical IFs (reviewed in Jackson et al. 2012; Skabkin et al. 2013). Reinitiation depends on the probability of reacquiring a new eIF2–GTP–initiator tRNA ternary complex after termination, which is influenced by the distance between the uORF stop codon and the main ORF start codon. Reinitiation efficiency is also affected by the length and secondary structure of mRNA in the uORF (Poyry et al. 2004). How ribosomes would assemble and reassemble between termination at the uORF stop codon and the new initiation start codon remains to be fully investigated. However, it is clear that for successful reinitiation the ribosome would have to stay bound to the mRNA, which would presumably be facilitated by some of the IFs staying bound to the ribosome; possibly eIF3 and eIF4G (de Breyne et al. 2009; Mohammad et al. 2017; and reviewed in Poyry et al. 2004; Jackson et al. 2012; Skabkin et al. 2013).

An alternative route of translation reinitiation was revealed with the identification of alternative translation initiation factors eIF2D (ligatin) and the homologous heterodimeric complex of oncogenic MCT-1 (malignant T-cell amplified protein or multiple copies of T-cell lymphoma) and DENR (density-regulated protein) (Fig. 3B) (Skabkin et al. 2010; Schleich et al. 2014). Like eIF1, eIF1A, and eIF3 (j), they can release mRNA and deacylated tRNA from the 40S during ribosomal recycling (Skabkin et al. 2013).

eIF2D and MCT-1–DENR are also able to recruit initiator tRNA to a start codon that is positioned in the 40S P-site and to promote subunit joining and translation initiation in the absence of eIF1, eIF1A, eIF2, eIF5, or eIF5B (Skabkin et al. 2010). They are thus capable of mediating translation reinitiation and also referred to as reinitiation factors (Schleich et al. 2014). Given that stress-induced phosphorylation of canonical initiation factor eIF2α reduces eIF2 ternary complex formation and sequesters eIF2 away from the ribosome, stress conditions could potentially favor reinitiation by alternative translation IFs by making ribosomal-binding sites and also initiator tRNAs available for the interaction with eIF2D, MCT-1–DENR or others.

Structures of the eIF2D- and MCT-1–DENR-containing ICs show that these re-initiation factors use several functional domains to position initiator tRNA in the P-site of the 40S subunit (Lomakin et al. 2017; Weisser et al. 2017). One of the domains binds to the 40S eIF1-binding site at the top of helix 44 (Fig. 3B) (Rabl et al. 2011) from where it monitors start codon recognition in the 40S P-site and prevents the association of the LSU. Another domain of eIF2D or MCT-1 binds and stabilizes the single-stranded CCA end of the initiator tRNA via its conserved RNA-binding groove but does not appear to establish conserved interactions with the aminoacyl moiety (Weisser et al. 2017), in agreement with the observation that eIF2D can also recruit deacylated tRNAs.

6 EUKARYOTIC TRANSLATION ELONGATION MECHANISMS

Once the correct ORF has been established on an mRNA during translation initiation and the 80S has been assembled with an initiator tRNA in the P-site, the elongation stage of the translation cycle starts. Polypeptide chain elongation by one amino acid requires three major steps that are highly conserved between prokaryotes and eukaryotes: (1) ternary complex recruitment to the ribosomal A-site and decoding, (2) peptide bond formation, and (3) subsequent translocation of the ribosome on the mRNA by one codon in 5′ to 3′ direction (reviewed in Voorhees and Ramakrishnan 2013). The eukaryotic ternary complex consists of the GTP-bound elongation factor eEF1A bound to an amino-

 Cite as *Cold Spring Harb Perspect Biol* doi: 10.1101/cshperspect.a032367

acylated elongator tRNA (Crepin et al. 2014). It is recruited to the ribosome via the flexible domains of the P1/P2 stalk and binds such that its GTPase domain interacts with the SRL on the 60S subunit. eEF1A positions the tRNA anticodon into the A-site of the 40S subunit for interactions with the mRNA codon. Upon formation of cognate codon–anticodon interactions, the rRNA in the decoding center changes conformation and causes a distortion of the tRNA into a strained "A/T" conformation (Fig. 4, top) (Shao et al. 2016).

Cognate tRNAs establish canonical complementary Watson–Crick base pairs with positions 1 and 2 of the mRNA codon, whereas position three is permissive for noncanonical wobble base pairs in accordance with the degenerate genetic code (Crick 1966). A1493 (bacterial numbering) from rRNA helix 44 of the 40S subunit establishes conserved hydrogen bond interactions with the minor groove of the codon–anticodon duplex at position 1 ("A-minor motif") and G530 from rRNA helix 18, A1492, and uS12 interact with the minor groove of position 2. The third, wobble position, is directly contacted by G530 and indirectly via a magnesium ion through uS12 (Ogle et al. 2001; Shao et al. 2016). These interactions energetically favor the binding of cognate tRNAs and a concerted conserved conformational change in the rRNA, referred to as "domain closure": On cognate codon recognition, the 40S head tilts toward the shoulder and the mRNA channel, whereas the shoulder rotates inward toward helix 44 (Ogle et al. 2002; Schmeing and Ramakrishnan 2009; Shao et al. 2016). This repositions eEF1A's G domain with respect to the SRL such that GTPase activation becomes possible and GTP hydrolysis is triggered. In its GDP-bound form, eEF1A loses its affinity for tRNAs and the ribosome, and dissociates from the complex (Voorhees and Ramakrishnan 2013; Shao et al. 2016). Noncognate tRNAs will be rejected because their accommodation will be energetically less favorable. They either do not trigger domain closure and GTP hydrolysis in the first place or they dissociate from the ribosome because their binding affinity for the ribosome and the mismatched codon is low (Shao et al. 2016).

After its release from eEF1A, the aminoacyl A-site tRNA relaxes from its constrained A/T conformation to allow the full accommodation of the A-site tRNA's aminoacylated end in the PTC of the 60S subunit (A/A state). There it interacts with the A-loop in an "induced fit" manner such that it is optimally positioned for nucleophilic attack of the P-site peptidyl-tRNA aminoacyl ester (Schmeing et al. 2005). eEF1A dissociation and tRNA accommodation are also accompanied by a characteristic eukaryote-specific rotation of the SSU around its longitudinal axis in the direction of the 60S L1 stalk. This movement is termed "subunit rolling" and is important for the subsequent tRNA repositioning (Budkevich et al. 2014). The resulting pretranslocation complexes predominantly adopt classical, nonrotated conformations with tRNAs being in A/A, P/P, and (in case of an existing E-site tRNA) E/E states (Fig. 4, upper middle) (Budkevich et al. 2011, 2014; Svidritskiy et al. 2014).

Once peptide bond formation has occurred, the nascent polypeptide chain is transferred to the A-site tRNA and the deacylated tRNA resides in the P-site. This allows repositioning of both tRNA acceptor arms with respect to the LSU, whereas their anticodon stems remain in position on the SSU. Because tRNA and intersubunit movements are coupled, the repositioning of the tRNAs is accompanied by a rotation of the SSU by ~10° with respect to the LSU, referred to as the "rotated state" (Agirrezabala et al. 2008). For both the unrotated and the rotated states, different conformations, "1 and 2," have been observed (Budkevich et al. 2011, 2014). Rotated states 1 and 2 differ in their tRNA positions, displaying tRNAs in A/A and P/E states or in A/P and P/E states (Fig. 4, upper middle) (Agirrezabala et al. 2008; Budkevich et al. 2011, 2014). Subunit rotation promotes the loss of an existing E-site tRNA because its affinity for the ribosome is lowered when the contact with the 60S E-site is lost. At the same time, the rotated state allows an inward movement of the L1 stalk, providing stabilizing interactions for the P/E–tRNA (Fig. 4, upper middle) (Fei et al. 2008).

To allow the elongation of the nascent polypeptide chain by another amino acid, the ribosome has to free the A-site for the delivery of the next aminoacyl-tRNA and translocate by one codon on the mRNA. The precise mechanism of mRNA movement has not been elucidated, but it appears to be coupled to tRNA movement in terms of step size because experiments with quadruplet tRNA anticodons promoted translocation by four instead of three nucleotides (reviewed in Voorhees and Ramakrishnan 2013).

In vivo, translocation is catalyzed by a multidomain GTPase called eEF2 (Spahn et al. 2004). Like the other ribosome-dependent GTPases, eEF2 binds the ribosome in its GTP-bound state and in an elongated conformation with its G domain next to the SRL of the 60S subunit (Fig. 4, lower middle) (Spahn et al. 2004; Anger et al. 2013; Voorhees et al. 2014). eEF2 binding causes a characteristic ribosomal movement: The SSU head "swivels" up to 20° and opens up the mRNA channel around the P- and E-sites of the SSU, which is a prerequisite for unrestricted tRNA/mRNA movement during translocation (Spahn et al. 2004; Ratje et al. 2010). Swiveling induces a repositioning of tRNAs because tRNAs establish interactions with the P- and E-site elements of the body of the SSU while still maintaining contacts with A- and P-site elements of the head (termed chimeric ap/P–pe/E states) (Ratje et al. 2010).

Upon GTP hydrolysis eEF2 relaxes into its GDP-bound conformation (Taylor et al. 2007), causing the ribosome to

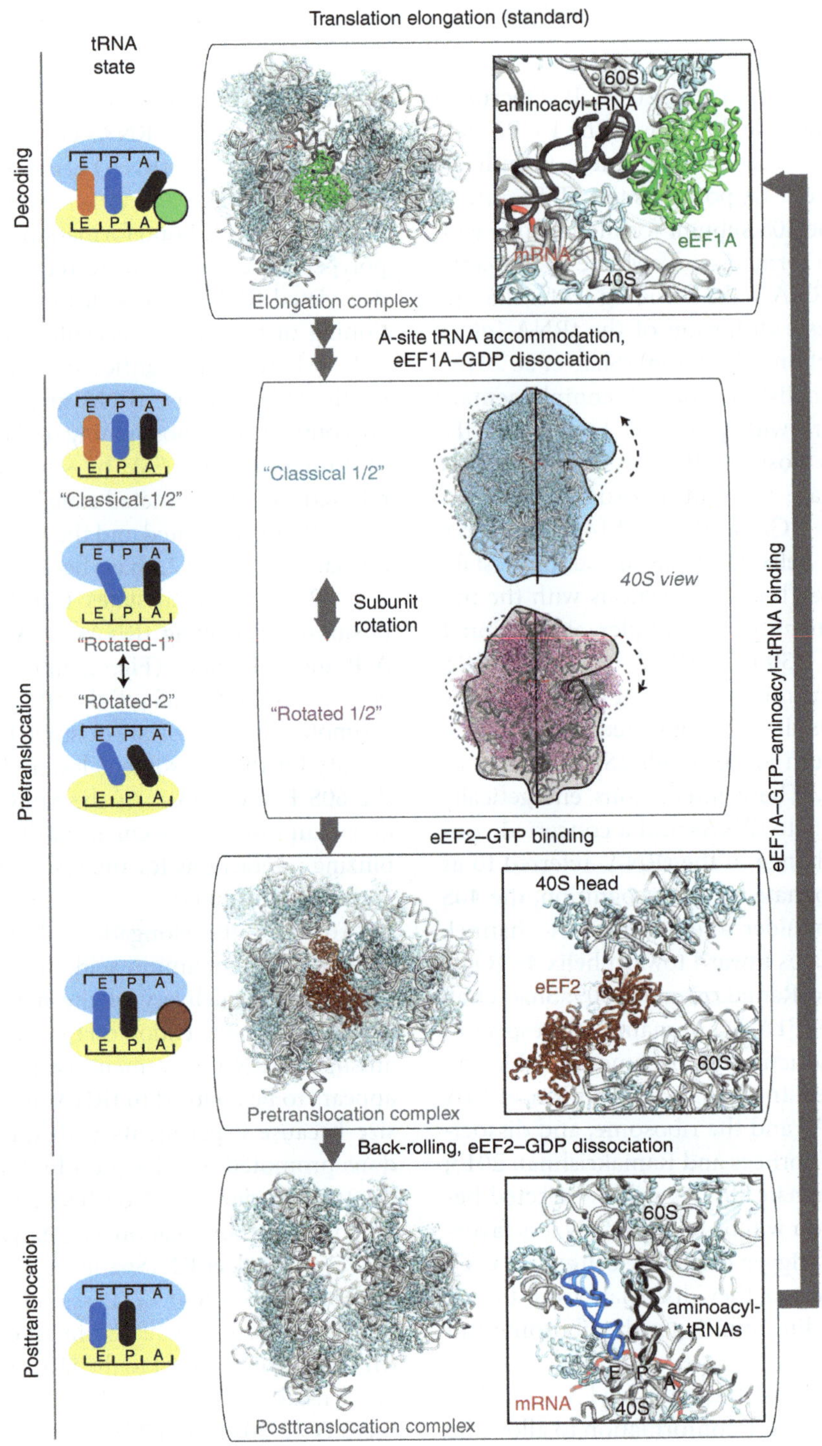

Figure 4. Eukaryotic translation elongation. For each mRNA codon within the open reading frame, an aminoacylated tRNA is delivered to the 80S ribosome A-site in a ternary complex with GTP-bound elongation factor eEF1A, initially adopting a strained A/T tRNA state (*top* panel). If cognate tRNA binding is verified through hydrogen bond interactions in the A-site decoding center of the 40S, the A-site tRNA fully accommodates (adopting an A/A tRNA state), whereas GTP hydrolysis by eEF1A promotes the factor's release from the complex. The pretranslocation complex can exist in rotated and nonrotated (classical) subunit conformations, from which the rotated conformation allows tRNAs to move more freely between the subunits and adopt hybrid tRNA states (A/P and P/E) (*upper middle* panel). Also, it is preferentially bound by the translocation factor GTPase eEF2. GTP hydrolysis by eEF2 couples a conformational relaxation in the factor with subunit back-rotation, promoting a tRNA-coupled mRNA movement by one base triplet on the ribosome (*lower middle* panel). After eEF2–GDP release, tRNAs adopt classical P/P and E/E states in the posttranslocation complex, freeing the A-site for another round of eEF1A-mediated aminoacyl tRNA recruitment (*bottom* panel).

revert back into a classical nonrotated posttranslocation state with peptidyl P/P–RNA and deacylated E/E–tRNA (Fig. 4, lower middle) (Budkevich et al. 2014; Svidritskiy et al. 2014 and reviewed in Voorhees and Ramakrishnan 2013). As soon as inorganic phosphate is released, eEF2–GDP dissociates from the ribosome, vacating the ribosomal A-site for binding of the next eEF1A–aminoacyl–tRNA ternary complex and the start of another round of elongation/translocation (Taylor et al. 2007).

7 EUKARYOTIC TRANSLATION TERMINATION AND RECYCLING

Once a stop codon reaches the A-site of a translating ribosome, it is recognized by eukaryotic release factor 1 (eRF1) instead of by the eEF1A–aminoacyl-tRNA complex. Eukaryotic termination relies on only one class I release factor, eRF1, a protein of three (N, M, and C) domains that shares no sequence homology with bacterial class I release factors RF1 and RF2, but functions similarly. In contrast to RF1/RF2, eRF1 recognizes all three types of stop codons (Brown et al. 2015). In solution, eRF1 adopts an extended conformation, but when bound to the ribosome-dependent GTPase eRF3, which is the equivalent of prokaryotic RF3, it adopts a compact fold suitable for efficient binding of the 40S subunit (Alkalaeva et al. 2006).

Association of the eRF1–eRF3–GTP ternary complex with the ribosome allows eRF3 to contact the ribosomal GTPase-binding site, whereas eRF1 binds analogously to an A-site tRNA (Fig. 5A, top) (Preis et al. 2014; Brown et al. 2015). The tip of the eRF1 N domain reaches into the A-site, where it induces a conformational change in the 18S rRNA of the 40S that compacts the mRNA and fixes the stop codon bases in defined positions, in which they can engage in a conserved hydrogen bonding network with eRF1. Structures of mammalian eRF1/eRF3 termination complexes bound to the three standard stop codons (UAG, UAA, and UGA) revealed how eukaryotes discriminate against nonstop codons (Brown et al. 2015). In contrast to the recognition of stop codons by prokaryotic RF1/RF2, eRF1 contacts stop codons via three previously identified conserved sequence motifs (GTS, NIKS, and YxCxxxF) and glutamate residue 55 (Glu55). Whereas purine residues and cytidines at position 1 are discriminated against by steric and hydrogen bonding requirements of the NIKS motif, respectively, bases at positions 2 and 3 can be recognized only at their Watson–Crick interface through hydrogen bonds with Glu55 if they are both purines, but not consecutive guanosines (Brown et al. 2015).

When eRF1 recognizes a stop codon, a conformational change occurs in the ribosome that stimulates eRF3's GTPase activity and ultimately leads to the dissociation of eRF3 (reviewed in Jackson et al. 2012). Upon eRF3 release, full accommodation of eRF1 becomes possible such that the eRF1 M domain can reach deeper into the PTC on the 60S subunit (Shao et al. 2016). As in RF1/RF2, this domain carries a conserved GGQ motif that mediates peptide release by promoting the hydrolysis of the ester that connects the polypeptide chain to the P-site tRNA (Frolova et al. 1999). eRF1's peptide release activity is stimulated by ABCE1, the eukaryotic recycling factor equivalent, which binds to SSU of the posttermination complex once that eRF3 has dissociated (Fig. 5A, middle) (Brown et al. 2015; Shao et al. 2016). Consequently, eRF1 acts as a binding platform for ABCE1 and enhances ABCE1's intrinsically low NTPase activity (reviewed in Jackson et al. 2012). Therefore, eukaryotic translation termination is coupled to ribosomal recycling and the directionality of both processes is guaranteed by the fact that ABCE1 and eRF3 are mutually exclusive and that ABCE1 does not bind vacant ribosomes (Pisarev et al. 2010). ATP hydrolysis by ABCE1 is coupled to a conformational change of the conserved rRNA-binding iron–sulfur (Fe–S) cluster domain of ABCE1 (Fig. 5A, bottom close-up) (Pisarev et al. 2010; Brown et al. 2015). This conformational change likely causes subunit splitting and subsequent release of ABCE1 and eRF1 (Fig. 5A, bottom) (Heuer et al. 2017b).

Subsequently, the translation initiation or reinitiation factor binding to the SSU facilitates the detachment of the translated mRNA, displacement of the remaining deacylated tRNA, and the transition to translation initiation on a new mRNA (Skabkin et al. 2013 and reviewed in Jackson et al. 2012). Premature reassociation of the SSU and the LSU is sterically prevented by eIF1 bound to the 40S and by eIF6 bound to the 60S (Klinge et al. 2011; Rabl et al. 2011).

Although eRF1/eRF3 catalyze termination on classical stop codons, the release factor paralogs Pelota (Dom34 in yeast)/Hbs1 evolved as mediators of premature translation termination on stalled mRNAs or mRNAs without a stop codon triggering "no-go decay" or "nonsense-mediated decay" pathways (Fig. 5B) (Kobayashi et al. 2010; Shao et al. 2016).

8 EUKARYOTIC RIBOSOMAL BIOGENESIS

Assembly and maturation of eukaryotic ribosomes are much more complex than in the case of prokaryotic ribosomes. The highly regulated process starts with the transcription of ribosomal pre-rRNA in the nucleolus and subsequently involves more than 200 assembly factors, including export factors, chaperones, and nucleases (reviewed in Gerhardy et al. 2014). The pre-rRNAs are transcribed by two polymerases: RNA polymerase I produces a 35S pre-rRNA transcript (yeast nomenclature), which comprises

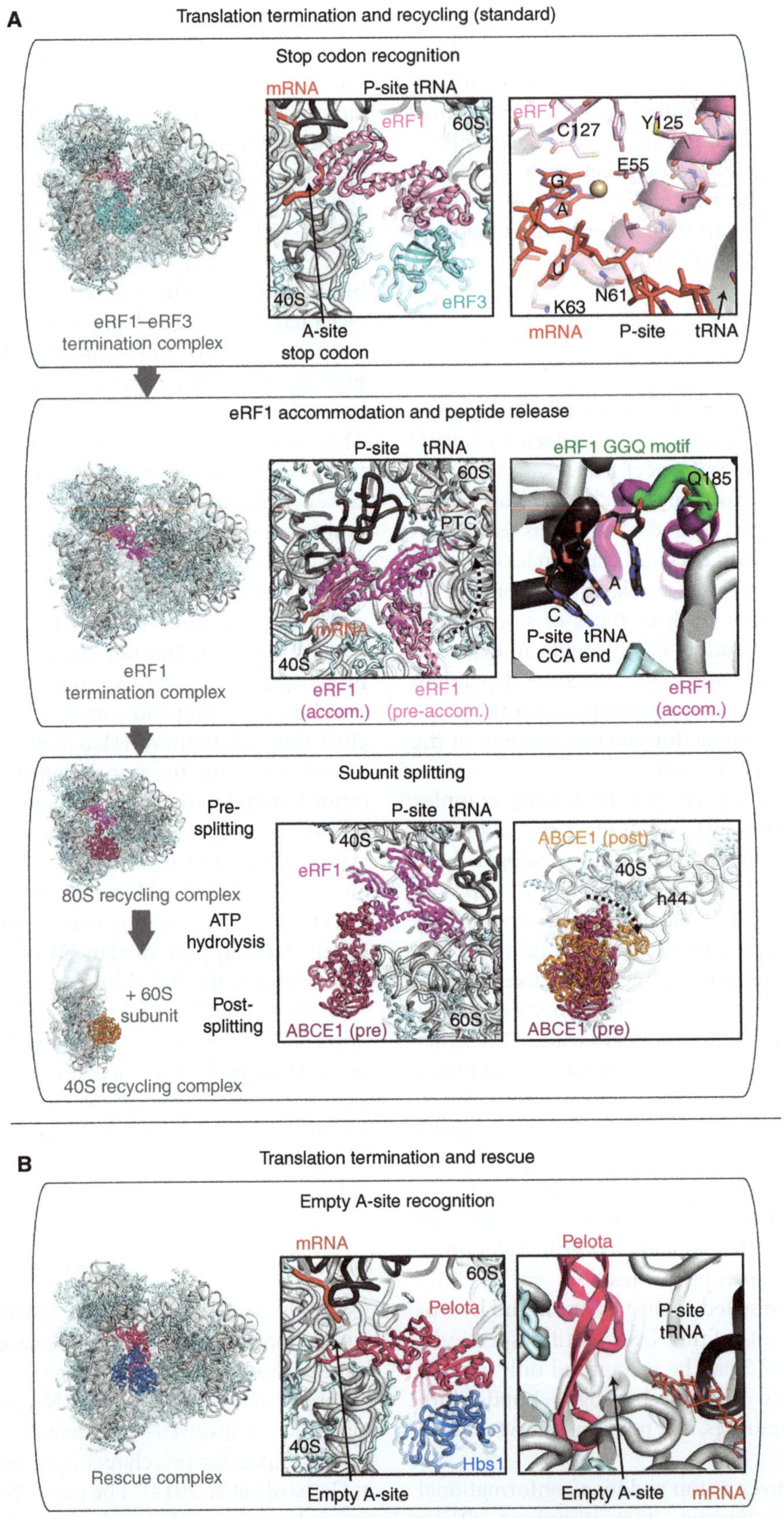

Figure 5. *(See legend on following page.)*

the 18S, 25S, and 5.8S sequences, separated and flanked by so called internal and external transcribed spacers (ITS and ETS), whereas RNA polymerase III transcribes 5S rRNA (reviewed in Nerurkar et al. 2015). The first r-proteins and assembly factors assemble cotranscriptionally onto the pre-35S rRNA and assist its folding process.

A central intermediate in ribosome biogenesis is the 5-MDa 90S preribosome or SSU processome, which encompasses 70 nonribosomal assembly factors and several small nucleolar RNAs (snoRNAs) that are organized into U3 snoRNP (small nucleolar ribonucleoprotein), UTP-A, UTP-B, UTP-C, Mpp10–Imp3–Imp4, and Bms1–Rcl1 subcomplexes (Kornprobst et al. 2016; Barandun et al. 2017). These factors modify (e.g., methylate and pseudouridinylate) and successively trim the nascent 35S pre-rRNA from the 5′end (A_0 and A_1 cleavage) to remove the 5′-ETS sequence (Fig. 6) (reviewed in Gerhardy et al. 2014; Nerurkar et al. 2015).

UTP-A and UTP-B are complexes of seven and six proteins, respectively, and among the earliest modules to assemble on the 5′ ETS of the nascent RNA transcript. UTP-A binds the rRNA first and then recruits UTP-B and U3 snoRNP, which also interacts with middle regions of the nascent 18S rRNA (Hunziker et al. 2016). They all act as chaperones and stabilize the initial fold of the nascent rRNA and guide subsequent cleavage steps (Hunziker et al. 2016). The third rRNA cleavage, thought to be mediated by the nuclease Utp24 (a pilus retraction protein amino terminal [PIN]-type nucleus) and takes place within ITS1 (A2 cleavage) (Wells et al. 2016). It separates the pre-20S rRNA that is part of the pre-40S particle from the pre-27S rRNA that is found in the pre-60S particle (Trapman et al. 1975). From here, pre-40S and pre-60S particles follow separate maturation pathways (Fig. 6). The pre-60S associates with the 5S RNP and undergoes at least three nucleolar processing steps including the 5′ trimming of the 27S pre-rRNA before transitioning to the nucleoplasm (reviewed in Nerurkar et al. 2015).

Two intermediates of the nucleoplasmic pre-60S maturation pathway have been structurally characterized in greater detail thus far: The "early Arx1 particle" (Wu et al. 2016) and the "late Rea1 particle" (Fig. 6) (Barrio-Garcia et al. 2016). In the early Arx1 particle, the endonucleolytic cleavage of ITS2, which separates the sequence of the 5.8S rRNA from the 25S rRNA by nuclease Las1, has not yet occurred. Therefore, the early Arx1 particle still displays the "foot structure" associated with ITS2 (comprising biogenesis factors Nop15, Cic1, Nop7, Rlp7, and Nop53) (Wu et al. 2016), which is lacking in the Rea1 complex where ITS2 removal has been completed (Fig. 6) (Barrio-Garcia et al. 2016).

The two particles also differ in the positioning of the 5S RNP at the CP that includes the A-site finger region (Leidig et al. 2014). In the early Arx1 particle (Wu et al. 2016), the 5S RNP is rotated by 180° compared with the Rea1 particle, in which it adopts its mature position (Fig. 6) (Leidig et al. 2014; Barrio-Garcia et al. 2016). The dynein-related ATPase Rea1 is one of the three AAA-type ATPases that guide pre-60S maturation and has been suggested to monitor the correct positioning of the 5S RNP and the associated CP remodeling (Bassler et al. 2010). Its ATPase activity mediates the release of GTPase Nog2 from the complex, which allows nuclear export factor Nmd3 to bind. Thus, Rea1-mediated release of biogenesis factors ensures directionality of pre-60S maturation and serves as a maturation checkpoint before nuclear export (Bassler et al. 2010).

In contrast to the pre-60S, the pre-40S is rather rapidly exported to the cytoplasm and does not acquire many additional factors before being exported: The kinase Hrr25, ATPase Rio2, nuclease Nob1, and factors Enp1, Ltv1, and Tsr1 are assembled already in the nuclear preribosome and remain associated with the particle throughout the nuclear export process (Schäfer et al. 2003).

Pre-60S and pre-40S use a set of shared export factors, such as Crm1 or Mex67/Mtr2 and Rrp12. However, there are also subunit-specific export factors that facilitate the transport of the particles across the nuclear pore complex. Nmd3, Bud20, Ecm1, and Arx1 act as export factors for the pre-60S, whereas Yrb2 (RanBP3) appears to be involved in

Figure 5. Eukaryotic translation termination. (*A*) All three commonly used stop codons are being recognized by eukaryotic release factor 1 (eRF1) that binds in complex with GTPase eRF3 such that its N domain, harboring the conserved sequence motif NIKS (residues 61 to 64), reaches into the A-site, where it stabilizes stop codon bases 2 and 3 in a stacked conformation such that they can establish hydrogen bonds with glutamate residue 55 (Glu55) (see *close-up*, *top* panel). After eRF3–GDP release from the complex, ABCE1 can bind eRF1 and the 40S stimulating eRF1's peptide release activity, which is mediated via the universally conserved GGQ motif in the PTC of the 60S (*close-up*, *middle* panel). ABCE1 is an ATPase that splits the 80S ribosome in 60S and 40S subunits through several cycles of ATP hydrolysis. In the released 40S complex, the iron–sulfur (Fe–S) cluster-containing domain of ABCE1 has moved and adopts a position on helix 44 that is incompatible with 60S association (*bottom* panel). (*B*) Vacant A-sites stemming from truncated mRNAs are recognized by the Hbs1–Pelota complex, which adopts a similar conformation on the ribosome as the eRF1/eRF3 complex. The insertion of Pelota into the A-site would be incompatible with the presence of a stop codon. In contrast to eRF1, Pelota possesses no peptide release activity, preventing polypeptide synthesis from truncated mRNAs.

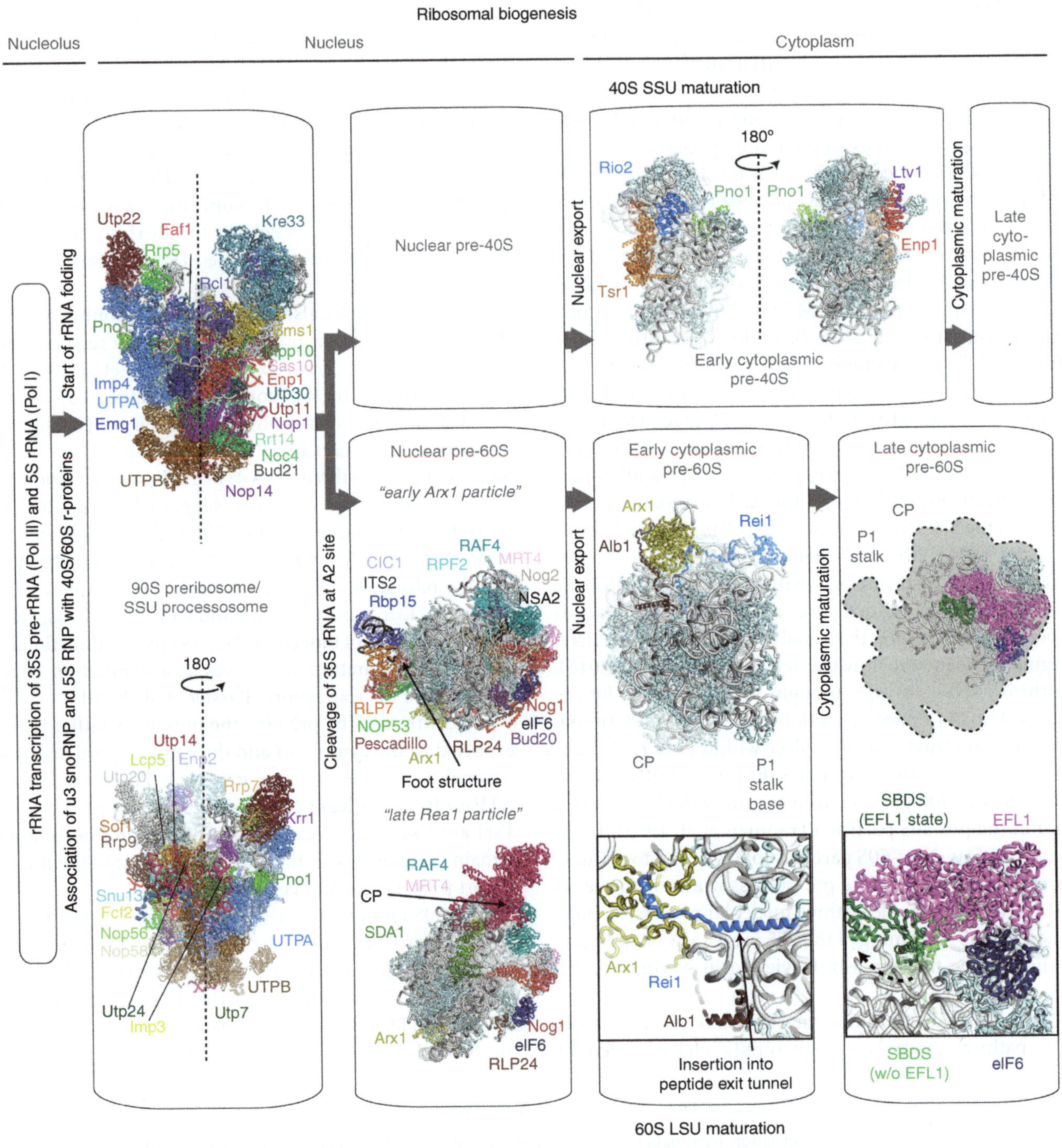

Figure 6. Eukaryotic ribosomal biogenesis. Eukaryotic ribosomal biogenesis of 40S and 60S subunits starts with the transcription of the 35S and 5S rRNA in the nucleolus and proceeds via a common precursor, the 90S preribosome (*left* panel), which facilitates the folding of the nascent rRNA and its cleavage into a 27S pre-rRNA as part of the pre-60S particle and a 20S pre-rRNA as part of a pre-40S subunit. Both subunits undergo nuclear maturation steps and acquire nuclear export factors before they are exported via the nuclear pore complex to the cytoplasm. Nuclear pre-60S maturation involves 5S RNP accommodation and central protuberance (CP) rearrangement between the nuclear "early Arx1" and the "late Rea1" particles. During cytoplasmic maturation, the functional centers of the pre-60S and pre-40S are proofread. The 60S peptide exit tunnel is probed by Rei1 and the sarcin–ricin loop (SRL) by SBDS/eIF6, whereas the proofreading of the 40S decoding center and the 40S beak involves Rio2/Tsr1/Dim1 and Enp1/Ltv1, respectively. Cytoplasmic pre-40S maturation also comprises the cleavage of 20S pre-rRNA into mature 18S rRNA and release of export and assembly factors.

 Cite as *Cold Spring Harb Perspect Biol* doi: 10.1101/cshperspect.a032367

pre-40S export (reviewed in Gerhardy et al. 2014; Nerurkar et al. 2015).

Cytoplasmic maturation steps involve the release of nuclear export factors as well as functional maturation checkpoints of the immature subunits. For example, the release of the Arx1 export factor from the pre-60S involves the cytoplasmic zinc-finger protein Rei1 (Meyer et al. 2010), the J-domain protein Jjj1 (Meyer et al. 2010), and the ATPase Ssa and stimulates proper formation of a functional peptide exit tunnel (Greber et al. 2012). The structure of the cytoplasmic Arx1–Alb1–Rei1 pre-60S complex revealed that the carboxy-terminal part of Rei1 enters the ribosomal tunnel to check its integrity; however, when this is not possible the maturation of the pre-60S subunit is arrested and Arx1 is not released (Fig. 6) (Greber et al. 2016).

The last cytoplasmic maturation step of the pre-60S subunit includes release of eIF6, which binds to the pre-60S already in the nucleus (Basu et al. 2001). eIF6 release is thought to be coupled to the proofreading of the SRL because the factor binds to the SRL on the subunit interface (Klinge et al. 2011). Dissociation of eIF6 is facilitated by EF-G/EF-2 homolog EFL1 and Shwachman–Bodian–Diamond syndrome protein (SBDS) (Fig. 6) (Weis et al. 2015). eIF6 release frees the 60S subunit interface and therefore allows the proofread 60S subunit to engage in translation.

Cytoplasmic maturation of the pre-40S subunit is governed by ribosomal assembly factors Tsr1, Ltv1, Enp1, Dim1, Dim2 (Pno1), Nob1, Rio2, and kinase Hrr25 and involves large-scale conformational rearrangements of the rRNA around the mRNA entry channel, the 40S beak, and helix 44. These rearrangements are coupled to proof-reading of the decoding center and ultimately to the cleavage of the 20S pre-rRNA to yield the mature 18S rRNA (D cleavage) (Scaiola et al. 2018).

Dim1 is a rRNA methylase that methylates the top of helix 44 on the pre-40S subunit. Upon Dim1 dissociation, helix 44 and the decoding center bases relocate closer to their positions in the mature subunit (Scaiola et al. 2018). Structures of the cytoplasmic pre-40S have revealed that Hrr25 kinase-triggered release of assembly factors Ltv1 and Enp1 allows the beak to adopt its mature conformation (Fig. 6) (Scaiola et al. 2018), whereas GTPase-like protein Tsr1 and ATPase Rio2 monitor A-site formation and prevent subunit association (Fig. 6) (Heuer et al. 2017a; Scaiola et al. 2018). Upon changes in the beak area and release of Tsr1 and Rio2 from the subunit interface area, helix 44 can adopt its mature conformation. This conformational change in the rRNA is propagated to the platform area where Pno1 is bound (Heuer et al. 2017a; Scaiola et al. 2018). Movement or dissociation of Pno1 is presumably triggered by conformational changes in the rRNA and aided by the cytoplasmic GTPase Fun12 (eIF5B) as well as the association with a mature 60S subunit into a so-called 80S-like particle (Turowski et al. 2014). It is a prerequisite for the final cleavage of the 20S pre-rRNA by the Nob1 endonuclease (Heuer et al. 2017a), yielding mature translation-competent ribosomal subunits.

9 FUTURE CHALLENGES

Although our understanding of eukaryotic translation has increased dramatically in recent years, there are still many exciting unanswered questions in the field, particularly with respect to translational control, translation initiation, ribosome assembly, and targeting of proteins to different cellular compartments. Despite the advances in cryo-EM methodology, some of the processes such as the 5′-UTR scanning, may involve continuous structural transitions and may therefore be better suited for structure-based experiments aimed at investigating dynamic aspects of the process. Furthermore, some of the ribosomal complexes may involve factors like eIF3 or eIF4 that exert their function through flexible attachments and connective tethers. To reveal the network of such interactions, hybrid structural biology approaches that involve cryo-EM, cross-linking, mass spectrometry, and X-ray crystallography combined with integrative computational modeling are likely to be used. Another challenge lies in studying ribosomal complexes with a high degree of compositional and conformational heterogeneity, such as ribosomal biogenesis intermediates, in which trapping and purification of native intermediates can be achieved by genetic manipulation. Considering that the repertoire of tools and methods at our disposal for studying molecular structure and function has never been greater, a new generation of scientists with interdisciplinary expertise will be in an excellent position to tackle the unsolved mysteries of the protein synthesis machinery.

REFERENCES

Acker MG, Shin BS, Dever TE, Lorsch JR. 2006. Interaction between eukaryotic initiation factors 1A and 5B is required for efficient ribosomal subunit joining. *J Biol Chem* **281:** 8469–8475.

Agirrezabala X, Lei J, Brunelle JL, Ortiz-Meoz RF, Green R, Frank J. 2008. Visualization of the hybrid state of tRNA binding promoted by spontaneous ratcheting of the ribosome. *Mol Cell* **32:** 190–197.

Algire MA, Maag D, Lorsch JR. 2005. Pi release from eIF2, not GTP hydrolysis, is the step controlled by start-site selection during eukaryotic translation initiation. *Mol Cell* **20:** 251–262.

Alkalaeva EZ, Pisarev AV, Frolova LY, Kisselev LL, Pestova TV. 2006. In vitro reconstitution of eukaryotic translation reveals cooperativity between release factors eRF1 and eRF3. *Cell* **125:** 1125–1136.

Anger AM, Armache JP, Berninghausen O, Habeck M, Subklewe M, Wilson DN, Beckmann R. 2013. Structures of the human and *Drosophila* 80S ribosome. *Nature* **497:** 80–85.

Aylett CH, Boehringer D, Erzberger JP, Schaefer T, Ban N. 2015. Structure of a yeast 40S-eIF1-eIF1A-eIF3-eIF3j initiation complex. *Nat Struct Mol Biol* **22:** 269–271.

Bai XC, McMullan G, Scheres SH. 2015. How cryo-EM is revolutionizing structural biology. *Trends Biochem Sci* **40:** 49–57.

Ban N, Beckmann R, Cate JH, Dinman JD, Dragon F, Ellis SR, Lafontaine DL, Lindahl L, Liljas A, Lipton JM, et al. 2014. A new system for naming ribosomal proteins. *Curr Opin Struct Biol* **24:** 165–169.

Barandun J, Chaker-Margot M, Hunziker M, Molloy KR, Chait BT, Klinge S. 2017. The complete structure of the small-subunit processome. *Nat Struct Mol Biol* **24:** 944–953.

Barrio-Garcia C, Thoms M, Flemming D, Kater L, Berninghausen O, Bassler J, Beckmann R, Hurt E. 2016. Architecture of the Rix1–Rea1 checkpoint machinery during pre-60S-ribosome remodeling. *Nat Struct Mol Biol* **23:** 37–44.

Bassler J, Kallas M, Pertschy B, Ulbrich C, Thoms M, Hurt E. 2010. The AAA-ATPase Rea1 drives removal of biogenesis factors during multiple stages of 60S ribosome assembly. *Mol Cell* **38:** 712–721.

Basu U, Si K, Warner JR, Maitra U. 2001. The *Saccharomyces cerevisiae TIF6* gene encoding translation initiation factor 6 is required for 60S ribosomal subunit biogenesis. *Mol Cell Biol* **21:** 1453–1462.

Ben-Shem A, Garreau de Loubresse N, Melnikov S, Jenner L, Yusupova G, Yusupov M. 2011. The structure of the eukaryotic ribosome at 3.0 Å resolution. *Science* **334:** 1524–1529.

Bieniossek C, Schutz P, Bumann M, Limacher A, Uson I, Baumann U. 2006. The crystal structure of the carboxy-terminal domain of human translation initiation factor eIF5. *J Mol Biol* **360:** 457–465.

Brown A, Shao S, Murray J, Hegde RS, Ramakrishnan V. 2015. Structural basis for stop codon recognition in eukaryotes. *Nature* **524:** 493–496.

Budkevich T, Giesebrecht J, Altman RB, Munro JB, Mielke T, Nierhaus KH, Blanchard SC, Spahn CM. 2011. Structure and dynamics of the mammalian ribosomal pretranslocation complex. *Mol Cell* **44:** 214–224.

Budkevich TV, Giesebrecht J, Behrmann E, Loerke J, Ramrath DJ, Mielke T, Ismer J, Hildebrand PW, Tung CS, Nierhaus KH, et al. 2014. Regulation of the mammalian elongation cycle by subunit rolling: A eukaryotic-specific ribosome rearrangement. *Cell* **158:** 121–131.

Calvo SE, Pagliarini DJ, Mootha VK. 2009. Upstream open reading frames cause widespread reduction of protein expression and are polymorphic among humans. *Proc Natl Acad Sci* **106:** 7507–7512.

Cheung YN, Maag D, Mitchell SF, Fekete CA, Algire MA, Takacs JE, Shirokikh N, Pestova T, Lorsch JR, Hinnebusch AG. 2007. Dissociation of eIF1 from the 40S ribosomal subunit is a key step in start codon selection in vivo. *Genes Dev* **21:** 1217–1230.

Costantino DA, Pfingsten JS, Rambo RP, Kieft JS. 2008. tRNA–mRNA mimicry drives translation initiation from a viral IRES. *Nat Struct Mol Biol* **15:** 57–64.

Crepin T, Shalak VF, Yaremchuk AD, Vlasenko DO, McCarthy A, Negrutskii BS, Tukalo MA, El'skaya AV. 2014. Mammalian translation elongation factor eEF1A2: X-ray structure and new features of GDP/GTP exchange mechanism in higher eukaryotes. *Nucleic Acids Res* **42:** 12939–12948.

Crick FH. 1966. Codon–anticodon pairing: The wobble hypothesis. *J Mol Biol* **19:** 548–555.

Das S, Maitra U. 2001. Functional significance and mechanism of eIF5-promoted GTP hydrolysis in eukaryotic translation initiation. *Prog Nucleic Acid Res Mol Biol* **70:** 207–231.

de Breyne S, Yu Y, Unbehaun A, Pestova TV, Hellen CU. 2009. Direct functional interaction of initiation factor eIF4G with type 1 internal ribosomal entry sites. *Proc Natl Acad Sci* **106:** 9197–9202.

Diaconu M, Kothe U, Schlunzen F, Fischer N, Harms JM, Tonevitsky AG, Stark H, Rodnina MV, Wahl MC. 2005. Structural basis for the function of the ribosomal L7/12 stalk in factor binding and GTPase activation. *Cell* **121:** 991–1004.

Erzberger JP, Stengel F, Pellarin R, Zhang S, Schaefer T, Aylett CH, Cimermancic P, Boehringer D, Sali A, Aebersold R, et al. 2014. Molecular architecture of the 40SeIF1eIF3 translation initiation complex. *Cell* **158:** 1123–1135.

Fei J, Kosuri P, MacDougall DD, Gonzalez RL Jr. 2008. Coupling of ribosomal L1 stalk and tRNA dynamics during translation elongation. *Mol Cell* **30:** 348–359.

Fekete CA, Mitchell SF, Cherkasova VA, Applefield D, Algire MA, Maag D, Saini AK, Lorsch JR, Hinnebusch AG. 2007. N- and C-terminal residues of eIF1A have opposing effects on the fidelity of start codon selection. *EMBO J* **26:** 1602–1614.

Fernandez IS, Bai XC, Hussain T, Kelley AC, Lorsch JR, Ramakrishnan V, Scheres SHW. 2013. Molecular architecture of a eukaryotic translational initiation complex. *Science* **342:** 1240585.

Fringer JM, Acker MG, Fekete CA, Lorsch JR, Dever TE. 2007. Coupled release of eukaryotic translation initiation factors 5B and 1A from 80S ribosomes following subunit joining. *Mol Cell Biol* **27:** 2384–2397.

Frolova LY, Tsivkovskii RY, Sivolobova GF, Oparina NY, Serpinsky OI, Blinov VM, Tatkov SI, Kisselev LL. 1999. Mutations in the highly conserved GGQ motif of class 1 polypeptide release factors abolish ability of human eRF1 to trigger peptidyl-tRNA hydrolysis. *RNA* **5:** 1014–1020.

Gao H, Ayub MJ, Levin MJ, Frank J. 2005. The structure of the 80S ribosome from *Trypanosoma cruzi* reveals unique rRNA components. *Proc Natl Acad Sci* **102:** 10206–10211.

Gerhardy S, Menet AM, Pena C, Petkowski JJ, Panse VG. 2014. Assembly and nuclear export of pre-ribosomal particles in budding yeast. *Chromosoma* **123:** 327–344.

Gogala M, Becker T, Beatrix B, Armache JP, Barrio-Garcia C, Berninghausen O, Beckmann R. 2014. Structures of the Sec61 complex engaged in nascent peptide translocation or membrane insertion. *Nature* **506:** 107–110.

Greber BJ, Boehringer D, Montellese C, Ban N. 2012. Cryo-EM structures of Arx1 and maturation factors Rei1 and Jjj1 bound to the 60S ribosomal subunit. *Nat Struct Mol Biol* **19:** 1228–1233.

Greber BJ, Gerhardy S, Leitner A, Leibundgut M, Salem M, Boehringer D, Leulliot N, Aebersold R, Panse VG, Ban N. 2016. Insertion of the biogenesis factor Rei1 probes the ribosomal tunnel during 60S maturation. *Cell* **164:** 91–102.

Hashem Y, des Georges A, Dhote V, Langlois R, Liao HY, Grassucci RA, Hellen CU, Pestova TV, Frank J. 2013a. Structure of the mammalian ribosomal 43S preinitiation complex bound to the scanning factor DHX29. *Cell* **153:** 1108–1119.

Hashem Y, des Georges A, Dhote V, Langlois R, Liao HY, Grassucci RA, Pestova TV, Hellen CU, Frank J. 2013b. Hepatitis-C-virus-like internal ribosome entry sites displace eIF3 to gain access to the 40S subunit. *Nature* **503:** 539–543.

Hashem Y, des Georges A, Fu J, Buss SN, Jossinet F, Jobe A, Zhang Q, Liao HY, Grassucci RA, Bajaj C, et al. 2013c. High-resolution cryo-electron microscopy structure of the *Trypanosoma brucei* ribosome. *Nature* **494:** 385–389.

Hellens RP, Brown CM, Chisnall MA, Waterhouse PM, Macknight RC. 2016. The emerging world of small ORFs. *Trends Plant Sci* **21:** 317–328.

Heuer A, Thomson E, Schmidt C, Berninghausen O, Becker T, Hurt E, Beckmann R. 2017a. Cryo-EM structure of a late pre-40S ribosomal subunit from *Saccharomyces cerevisiae*. *Elife* **6:** e30189.

Heuer A, Gerovac M, Schmidt C, Trowitzsch S, Preis A, Kotter P, Berninghausen O, Becker T, Beckmann R, Tampe R. 2017b. Structure of the 40S-ABCE1 post-splitting complex in ribosome recycling and translation initiation. *Nat Struct Mol Biol* **24:** 453–460.

Hinnebusch AG. 2011. Molecular mechanism of scanning and start codon selection in eukaryotes. *Microbiol Mol Biol Rev* **75:** 434–467, first page of Table of Contents.

Hinnebusch AG. 2014. The scanning mechanism of eukaryotic translation initiation. *Annu Rev Biochem* **83:** 779–812.

Hinnebusch AG, Ivanov IP, Sonenberg N. 2016. Translational control by 5′-untranslated regions of eukaryotic mRNAs. *Science* **352:** 1413–1416.

Hunziker M, Barandun J, Petfalski E, Tan D, Delan-Forino C, Molloy KR, Kim KH, Dunn-Davies H, Shi Y, Chaker-Margot M, et al. 2016. UtpA and UtpB chaperone nascent pre-ribosomal RNA and U3 snoRNA to initiate eukaryotic ribosome assembly. *Nat Commun* **7:** 12090.

Hussain T, Llácer JL, Fernandez IS, Munoz A, Martin-Marcos P, Savva CG, Lorsch JR, Hinnebusch AG, Ramakrishnan V. 2014. Structural changes enable start codon recognition by the eukaryotic translation initiation complex. *Cell* **159:** 597–607.

Jackson RJ, Hellen CU, Pestova TV. 2012. Termination and post-termination events in eukaryotic translation. *Adv Protein Chem Struct Biol* **86:** 45–93.

Jivotovskaya AV, Valasek L, Hinnebusch AG, Nielsen KH. 2006. Eukaryotic translation initiation factor 3 (eIF3) and eIF2 can promote mRNA binding to 40S subunits independently of eIF4G in yeast. *Mol Cell Biol* **26:** 1355–1372.

Kapp LD, Kolitz SE, Lorsch JR. 2006. Yeast initiator tRNA identity elements cooperate to influence multiple steps of translation initiation. *RNA* **12:** 751–764.

Klinge S, Voigts-Hoffmann F, Leibundgut M, Arpagaus S, Ban N. 2011. Crystal structure of the eukaryotic 60S ribosomal subunit in complex with initiation factor 6. *Science* **334:** 941–948.

Kobayashi K, Kikuno I, Kuroha K, Saito K, Ito K, Ishitani R, Inada T, Nureki O. 2010. Structural basis for mRNA surveillance by archaeal Pelota and GTP-bound EF1α complex. *Proc Natl Acad Sci* **107:** 17575–17579.

Kolitz SE, Lorsch JR. 2010. Eukaryotic initiator tRNA: Finely tuned and ready for action. *FEBS Lett* **584:** 396–404.

Kornprobst M, Turk M, Kellner N, Cheng J, Flemming D, Kos-Braun I, Kos M, Thoms M, Berninghausen O, Beckmann R, et al. 2016. Architecture of the 90S pre-ribosome: A structural view on the birth of the eukaryotic ribosome. *Cell* **166:** 380–393.

Korostelev A, Trakhanov S, Laurberg M, Noller HF. 2006. Crystal structure of a 70S ribosome-tRNA complex reveals functional interactions and rearrangements. *Cell* **126:** 1065–1077.

Kozak M. 1986. Point mutations define a sequence flanking the AUG initiator codon that modulates translation by eukaryotic ribosomes. *Cell* **44:** 283–292.

Lawless C, Pearson RD, Selley JN, Smirnova JB, Grant CM, Ashe MP, Pavitt GD, Hubbard SJ. 2009. Upstream sequence elements direct post-transcriptional regulation of gene expression under stress conditions in yeast. *BMC Genomics* **10:** 7.

Lee B, Udagawa T, Singh CR, Asano K. 2007. Yeast phenotypic assays on translational control. *Methods Enzymol* **429:** 105–137.

Leidig C, Thoms M, Holdermann I, Bradatsch B, Berninghausen O, Bange G, Sinning I, Hurt E, Beckmann R. 2014. 60S ribosome biogenesis requires rotation of the 5S ribonucleoprotein particle. *Nat Commun* **5:** 3491.

Llácer JL, Hussain T, Marler L, Aitken CE, Thakur A, Lorsch JR, Hinnebusch AG, Ramakrishnan V. 2015. Conformational differences between open and closed states of the eukaryotic translation initiation complex. *Mol Cell* **59:** 399–412.

Locker N, Easton LE, Lukavsky PJ. 2007. HCV and CSFV IRES domain II mediate eIF2 release during 80S ribosome assembly. *EMBO J* **26:** 795–805.

Lomakin IB, Steitz TA. 2013. The initiation of mammalian protein synthesis and mRNA scanning mechanism. *Nature* **500:** 307–311.

Lomakin IB, Stolboushkina EA, Vaidya AT, Zhao C, Garber MB, Dmitriev SE, Steitz TA. 2017. Crystal structure of the human ribosome in complex with DENR-MCT-1. *Cell Rep* **20:** 521–528.

Luna RE, Arthanari H, Hiraishi H, Nanda J, Martin-Marcos P, Markus MA, Akabayov B, Milbradt AG, Luna LE, Seo HC, et al. 2012. The C-terminal domain of eukaryotic initiation factor 5 promotes start codon recognition by its dynamic interplay with eIF1 and eIF2β. *Cell Rep* **1:** 689–702.

Marintchev A. 2013. Roles of helicases in translation initiation: A mechanistic view. *Biochim Biophys Acta* **1829:** 799–809.

Martin-Marcos P, Nanda J, Luna RE, Wagner G, Lorsch JR, Hinnebusch AG. 2013. β-hairpin loop of eukaryotic initiation factor 1 (eIF1) mediates 40 S ribosome binding to regulate initiator tRNA(Met) recruitment and accuracy of AUG selection in vivo. *J Biol Chem* **288:** 27546–27562.

Melnikov S, Ben-Shem A, Garreau de Loubresse N, Jenner L, Yusupova G, Yusupov M. 2012. One core, two shells: Bacterial and eukaryotic ribosomes. *Nat Struct Mol Biol* **19:** 560–567.

Melnikov S, Mailliot J, Shin BS, Rigger L, Yusupova G, Micura R, Dever TE, Yusupov M. 2016. Crystal structure of hypusine-containing translation factor eIF5A bound to a rotated eukaryotic ribosome. *J Mol Biol* **428:** 3570–3576.

Meyer AE, Hoover LA, Craig EA. 2010. The cytosolic J-protein, Jjj1 and Rei1 function in the removal of the pre-60 S subunit factor Arx1. *J Biol Chem* **285:** 961–968.

Mohammad MP, Munzarova Pondelickova V, Zeman J, Gunisova S, Valasek LS. 2017. In vivo evidence that eIF3 stays bound to ribosomes elongating and terminating on short upstream ORFs to promote reinitiation. *Nucleic Acids Res* **45:** 2658–2674.

Nanda JS, Cheung YN, Takacs JE, Martin-Marcos P, Saini AK, Hinnebusch AG, Lorsch JR. 2009. eIF1 controls multiple steps in start codon recognition during eukaryotic translation initiation. *J Mol Biol* **394:** 268–285.

Nerurkar P, Altvater M, Gerhardy S, Schutz S, Fischer U, Weirich C, Panse VG. 2015. Eukaryotic ribosome assembly and nuclear export. *Int Rev Cell Mol Biol* **319:** 107–140.

Nikolov EN, Dabeva MD, Nikolov TK. 1983. Turnover of ribosomes in regenerating rat liver. *Int J Biochem* **15:** 1255–1260.

Nissen P, Hansen J, Ban N, Moore PB, Steitz TA. 2000. The structural basis of ribosome activity in peptide bond synthesis. *Science* **289:** 920–930.

Ogle JM, Brodersen DE, Clemons WM Jr, Tarry MJ, Carter AP, Ramakrishnan V. 2001. Recognition of cognate transfer RNA by the 30S ribosomal subunit. *Science* **292:** 897–902.

Ogle JM, Murphy FV, Tarry MJ, Ramakrishnan V. 2002. Selection of tRNA by the ribosome requires a transition from an open to a closed form. *Cell* **111:** 721–732.

Olsen DS, Savner EM, Mathew A, Zhang F, Krishnamoorthy T, Phan L, Hinnebusch AG. 2003. Domains of eIF1A that mediate binding to eIF2, eIF3 and eIF5B and promote ternary complex recruitment in vivo. *EMBO J* **22:** 193–204.

Otto GA, Puglisi JD. 2004. The pathway of HCV IRES-mediated translation initiation. *Cell* **119:** 369–380.

Pesole G, Mignone F, Gissi C, Grillo G, Licciulli F, Liuni S. 2001. Structural and functional features of eukaryotic mRNA untranslated regions. *Gene* **276:** 73–81.

Pestova TV, Kolupaeva VG. 2002. The roles of individual eukaryotic translation initiation factors in ribosomal scanning and initiation codon selection. *Genes Dev* **16:** 2906–2922.

Pisarev AV, Skabkin MA, Pisareva VP, Skabkina OV, Rakotondrafara AM, Hentze MW, Hellen CU, Pestova TV. 2010. The role of ABCE1 in eukaryotic posttermination ribosomal recycling. *Mol Cell* **37:** 196–210.

Pisareva VP, Pisarev AV, Komar AA, Hellen CU, Pestova TV. 2008. Translation initiation on mammalian mRNAs with structured 5′UTRs requires DExH-box protein DHX29. *Cell* **135:** 1237–1250.

Poyry TA, Kaminski A, Jackson RJ. 2004. What determines whether mammalian ribosomes resume scanning after translation of a short upstream open reading frame? *Genes Dev* **18:** 62–75.

Preis A, Heuer A, Barrio-Garcia C, Hauser A, Eyler DE, Berninghausen O, Green R, Becker T, Beckmann R. 2014. Cryoelectron microscopic structures of eukaryotic translation termination complexes containing eRF1-eRF3 or eRF1-ABCE1. *Cell Rep* **8:** 59–65.

Quade N, Boehringer D, Leibundgut M, van den Heuvel J, Ban N. 2015. Cryo-EM structure of hepatitis C virus IRES bound to the human ribosome at 3.9-Å resolution. *Nat Commun* **6:** 7646.

Rabl J, Leibundgut M, Ataide SF, Haag A, Ban N. 2011. Crystal structure of the eukaryotic 40S ribosomal subunit in complex with initiation factor 1. *Science* **331:** 730–736.

Ratje AH, Loerke J, Mikolajka A, Brünner M, Hildebrand PW, Starosta AL, Dönhöfer A, Connell SR, Fucini P, Mielke T, et al. 2010. Head swivel on the ribosome facilitates translocation by means of intra-subunit tRNA hybrid sites. *Nature* **468:** 713–716.

Redondo N, Garcia-Moreno M, Sanz MA, Carrasco L. 2013. Translation of viral mRNAs that do not require eIF4E is blocked by the inhibitor 4EGI-1. *Virology* **444:** 171–180.

Reibarkh M, Yamamoto Y, Singh CR, del Rio F, Fahmy A, Lee B, Luna RE, Ii M, Wagner G, Asano K. 2008. Eukaryotic initiation factor (eIF) 1 carries two distinct eIF5-binding faces important for multifactor assembly and AUG selection. *J Biol Chem* **283:** 1094–1103.

Roll-Mecak A, Alone P, Cao C, Dever TE, Burley SK. 2004. X-ray structure of translation initiation factor eIF2γ: Implications for tRNA and eIF2α binding. *J Biol Chem* **279:** 10634–10642.

Saini AK, Nanda JS, Lorsch JR, Hinnebusch AG. 2010. Regulatory elements in eIF1A control the fidelity of start codon selection by modulating tRNA(i)(Met) binding to the ribosome. *Genes Dev* **24:** 97–110.

Scaiola A, Pena C, Weisser M, Bohringer D, Leibundgut M, Klingauf-Nerurkar P, Gerhardy S, Panse VG, Ban N. 2018. Structure of a eukaryotic cytoplasmic pre-40S ribosomal subunit. *EMBO J* **37:** e98499.

Schäfer T, Strauss D, Petfalski E, Tollervey D, Hurt E. 2003. The path from nucleolar 90S to cytoplasmic 40S pre-ribosomes. *EMBO J* **22:** 1370–1380.

Schleich S, Strassburger K, Janiesch PC, Koledachkina T, Miller KK, Haneke K, Cheng YS, Kuchler K, Stoecklin G, Duncan KE, et al. 2014. DENR-MCT-1 promotes translation re-initiation downstream of uORFs to control tissue growth. *Nature* **512:** 208–212.

Schmeing TM, Ramakrishnan V. 2009. What recent ribosome structures have revealed about the mechanism of translation. *Nature* **461:** 1234–1242.

Schmeing TM, Huang KS, Strobel SA, Steitz TA. 2005. An induced-fit mechanism to promote peptide bond formation and exclude hydrolysis of peptidyl-tRNA. *Nature* **438:** 520–524.

Schmitt E, Naveau M, Mechulam Y. 2010. Eukaryotic and archaeal translation initiation factor 2: A heterotrimeric tRNA carrier. *FEBS Lett* **584:** 405–412.

Schmitt E, Panvert M, Lazennec-Schurdevin C, Coureux PD, Perez J, Thompson A, Mechulam Y. 2012. Structure of the ternary initiation complex aIF2-GDPNP-methionylated initiator tRNA. *Nat Struct Mol Biol* **19:** 450–454.

Shalev-Benami M, Zhang Y, Matzov D, Halfon Y, Zackay A, Rozenberg H, Zimmerman E, Bashan A, Jaffe CL, Yonath A, et al. 2016. 2.8-Å cryo-EM structure of the large ribosomal subunit from the eukaryotic parasite *Leishmania*. *Cell Rep* **16:** 288–294.

Shao S, Murray J, Brown A, Taunton J, Ramakrishnan V, Hegde RS. 2016. Decoding mammalian ribosome-mRNA states by translational GTPase complexes. *Cell* **167:** 1229–1240.e15.

Skabkin MA, Skabkina OV, Dhote V, Komar AA, Hellen CU, Pestova TV. 2010. Activities of Ligatin and MCT-1/DENR in eukaryotic translation initiation and ribosomal recycling. *Genes Dev* **24:** 1787–1801.

Skabkin MA, Skabkina OV, Hellen CU, Pestova TV. 2013. Reinitiation and other unconventional posttermination events during eukaryotic translation. *Mol Cell* **51:** 249–264.

Spahn CMT, Gomez-Lorenzo MG, Grassucci RA, Jorgensen R, Andersen GR, Beckmann R, Penczek PA, Ballesta JPG, Frank J. 2004. Domain movements of elongation factor eEF2 and the eukaryotic 80S ribosome facilitate tRNA translocation. *EMBO J* **23:** 1008–1019.

Sun M, Li W, Blomqvist K, Das S, Hashem Y, Dvorin JD, Frank J. 2015. Dynamical features of the *Plasmodium falciparum* ribosome during translation. *Nucleic Acids Res* **43:** 10515–10524.

Svidritskiy E, Brilot AF, Koh CS, Grigorieff N, Korostelev AA. 2014. Structures of yeast 80S ribosome-tRNA complexes in the rotated and nonrotated conformations. *Structure* **22:** 1210–1218.

Takyar S, Hickerson RP, Noller HF. 2005. mRNA helicase activity of the ribosome. *Cell* **120:** 49–58.

Taylor DJ, Nilsson J, Merrill AR, Andersen GR, Nissen P, Frank J. 2007. Structures of modified eEF2 80S ribosome complexes reveal the role of GTP hydrolysis in translocation. *EMBO J* **26:** 2421–2431.

Terenin IM, Akulich KA, Andreev DE, Polyanskaya SA, Shatsky IN, Dmitriev SE. 2016. Sliding of a 43S ribosomal complex from the recognized AUG codon triggered by a delay in eIF2-bound GTP hydrolysis. *Nucleic Acids Res* **44:** 1882–1893.

Trapman J, Retel J, Planta RJ. 1975. Ribosomal precursor particles from yeast. *Exp Cell Res* **90:** 95–104.

Turowski TW, Lebaron S, Zhang E, Peil L, Dudnakova T, Petfalski E, Granneman S, Rappsilber J, Tollervey D. 2014. Rio1 mediates ATP-dependent final maturation of 40S ribosomal subunits. *Nucleic Acids Res* **42:** 12189–12199.

van den Elzen AM, Schuller A, Green R, Séraphin B. 2014. Dom34-Hbs1 mediated dissociation of inactive 80S ribosomes promotes restart of translation after stress. *EMBO J* **33:** 265–276.

Vassilenko KS, Alekhina OM, Dmitriev SE, Shatsky IN, Spirin AS. 2011. Unidirectional constant rate motion of the ribosomal scanning particle during eukaryotic translation initiation. *Nucleic Acids Res* **39:** 5555–5567.

Villa N, Do A, Hershey JWB, Fraser CS. 2013. Human eukaryotic initiation factor 4G (eIF4G) protein binds to eIF3c, -d, and -e to promote mRNA recruitment to the ribosome. *J Biol Chem* **288:** 32932–32940.

Voigts-Hoffmann F, Klinge S, Ban N. 2012. Structural insights into eukaryotic ribosomes and the initiation of translation. *Curr Opin Struct Biol* **22:** 768–777.

Voorhees RM, Ramakrishnan V. 2013. Structural basis of the translational elongation cycle. *Annu Rev Biochem* **82:** 203–236.

Voorhees RM, Schmeing TM, Kelley AC, Ramakrishnan V. 2010. The mechanism for activation of GTP hydrolysis on the ribosome. *Science* **330:** 835–838.

Voorhees RM, Fernandez IS, Scheres SHW, Hegde RS. 2014. Structure of the mammalian ribosome-Sec61 complex to 3.4 Å resolution. *Cell* **157:** 1632–1643.

Walsh D, Mohr I. 2011. Viral subversion of the host protein synthesis machinery. *Nat Rev Microbiol* **9:** 860–875.

Weingarten-Gabbay S, Elias-Kirma S, Nir R, Gritsenko AA, Stern-Ginossar N, Yakhini Z, Weinberger A, Segal E. 2016. Comparative genetics. Systematic discovery of cap-independent translation sequences in human and viral genomes. *Science* **351:** aad4939.

Weis F, Giudice E, Churcher M, Jin L, Hilcenko C, Wong CC, Traynor D, Kay RR, Warren AJ. 2015. Mechanism of eIF6 release from the nascent 60S ribosomal subunit. *Nat Struct Mol Biol* **22:** 914–919.

Weisser M, Voigts-Hoffmann F, Rabl J, Leibundgut M, Ban N. 2013. The crystal structure of the eukaryotic 40S ribosomal subunit in complex with eIF1 and eIF1A. *Nat Struct Mol Biol* **20:** 1015–1017.

Weisser M, Schäfer T, Leibundgut M, Bohringer D, Aylett CHS, Ban N. 2017. Structural and functional insights into human re-initiation complexes. *Mol Cell* **67:** 447–456.e7.

Wells SE, Hillner PE, Vale RD, Sachs AB. 1998. Circularization of mRNA by eukaryotic translation initiation factors. *Mol Cell* **2:** 135–140.

Wells GR, Weichmann F, Colvin D, Sloan KE, Kudla G, Tollervey D, Watkins NJ, Schneider C. 2016. The PIN domain endonuclease Utp24 cleaves pre-ribosomal RNA at two coupled sites in yeast and humans. *Nucleic Acids Res* **44:** 5399–5409.

Wimberly BT, Brodersen DE, Clemons WM Jr., Morgan-Warren RJ, Carter AP, Vonrhein C, Hartsch T, Ramakrishnan V. 2000. Structure of the 30S ribosomal subunit. *Nature* **407:** 327–339.

Wu S, Tutuncuoglu B, Yan K, Brown H, Zhang Y, Tan D, Gamalinda M, Yuan Y, Li Z, Jakovljevic J, et al. 2016. Diverse roles of assembly factors revealed by structures of late nuclear pre-60S ribosomes. *Nature* **534:** 133–137.

Xue S, Barna M. 2012. Specialized ribosomes: A new frontier in gene regulation and organismal biology. *Nat Rev Mol Cell Biol* **13:** 355–369.

Yu Y, Marintchev A, Kolupaeva VG, Unbehaun A, Veryasova T, Lai SC, Hong P, Wagner G, Hellen CU, Pestova TV. 2009. Position of eukaryotic translation initiation factor eIF1A on the 40S ribosomal subunit mapped by directed hydroxyl radical probing. *Nucleic Acids Res* **37:** 5167–5182.

Zhang F, Saini AK, Shin BS, Nanda J, Hinnebusch AG. 2015. Conformational changes in the P site and mRNA entry channel evoked by AUG recognition in yeast translation preinitiation complexes. *Nucleic Acids Res* **43:** 2293–2312.

Structural Biology of Telomerase

Yaqiang Wang, Lukas Sušac, and Juli Feigon

Department of Chemistry and Biochemistry, University of California Los Angeles (UCLA), Los Angeles, California 90095-1569

Correspondence: feigon@mbi.ucla.edu

SUMMARY

Telomerase is a DNA polymerase that extends the 3′ ends of chromosomes by processively synthesizing multiple telomeric repeats. It is a unique ribonucleoprotein (RNP) containing a specialized telomerase reverse transcriptase (TERT) and telomerase RNA (TER) with its own template and other elements required with TERT for activity (catalytic core), as well as species-specific TER-binding proteins important for biogenesis and assembly (core RNP); other proteins bind telomerase transiently or constitutively to allow association of telomerase and other proteins with telomere ends for regulation of DNA synthesis. Here we describe how nuclear magnetic resonance (NMR) spectroscopy and X-ray crystallography of TER and protein domains helped define the structure and function of the core RNP, laying the groundwork for interpreting negative-stain and cryo electron microscopy (cryo-EM) density maps of *Tetrahymena thermophila* and human telomerase holoenzymes. As the resolution has improved from ~30 Å to ~5 Å, these studies have provided increasingly detailed information on telomerase architecture and mechanism.

Outline

Cite as *Cold Spring Harb Perspect Biol* doi: 10.1101/cshperspect.a032383

1 INTRODUCTION

The linear chromosomes of higher organisms present a special problem for DNA replication. DNA replication normally starts with an RNA primer that pairs to the opposite strand and provides the free 3′-OH required for DNA polymerases to initiate synthesis. The RNA primers are subsequently removed and replaced by DNA. At the 3′ ends of a replicating chromosome, the removal of the RNA primer would result in loss of ~20 nucleotides (nt) for each round of DNA replication. Eventually, this loss of DNA leads to cellular senescence (Armanios and Blackburn 2012). This "end replication problem" led to the search for a specialized enzyme that might maintain the DNA at the ends of chromosomes and the discovery of telomerase (Greider and Blackburn 1989). The DNA at the ends of eukaryotic chromosomes has a repetitive sequence, the telomeric repeat, that is generally a 6–8-nt G-rich sequence (e.g., TTGGGG in ciliates, TTAGGG in vertebrates) on the 3′-end strand (G-strand) and ends with a single-stranded overhang of variable length (Blackburn and Collins 2011). Telomerase extends the 3′ end by synthesizing multiple copies of the telomere repeat using its specialized telomerase reverse transcriptase (TERT) and its telomerase RNA (TER), which contains the template that directs repeat synthesis (Blackburn et al. 2006; Blackburn and Collins 2011). The TER template is complementary to 1.5–1.8 telomere repeats and comprises an alignment region followed by the templating region (Greider and Blackburn 1987; Podlevsky and Chen 2016). TERT contains conserved reverse transcriptase (RT, palm and fingers) and carboxy-terminal element (CTE, thumb) domains found in other reverse transcriptases, as well as an RNA-binding domain (RBD), that together form a TERT ring structure (Gillis et al. 2008; Chan et al. 2017; Wang and Feigon 2017). It also contains a telomerase amino-terminal domain (TEN) that is connected to the RBD by a long linker. In contrast to the relatively conserved TERT, TER is highly divergent in length (varying from ~150 nt in ciliates to >2000 nt in some yeasts) and structural motifs (Podlevsky et al. 2008). This is explained by TER being a rapidly evolving noncoding (nc)RNA (Nelson and Shippen 2015; Podlevsky and Chen 2016). TER is an RNA polymerase III transcript in ciliates but is transcribed by RNA polymerase II in vertebrates, yeasts, and plants (Greider and Blackburn 1989; Egan and Collins 2012). This divergence has allowed multiple solutions for biogenesis and processing of TER, assembly of TER with TERT, cellular localization, and recruitment to telomeres (Fig. 1). However, TERs have two regions that are conserved to interact with TERT—the template/pseudoknot domain (t/PK), which forms a circle including the template and a pseudoknot, and the stem-terminus element (STE), which includes a hairpin (Figs. 1 and 2A,B) (Theimer and Feigon 2006; Zhang et al. 2011; Podlevsky and Chen 2016).

An understanding of telomerase structure and mechanism has become increasingly important as it has been recognized as a key determinant of human health, longevity, and tumorigenesis (Marrone et al. 2005; Fujii et al. 2009; Artandi and DePinho 2010; Armanios and Blackburn 2012; Bernardes de Jesus and Blasco 2013; Shay 2016). Even slight imbalances in telomerase levels can lead to debilitating diseases in humans. Telomerase insufficiencies are associated with premature aging and stem cell renewal disorders including dyskeratosis congenita, aplastic anemia, and pulmonary fibrosis (Townsley et al. 2014; Sarek et al. 2015; Shay 2016). In contrast, telomerase is up-regulated in >80% of human cancers, rendering tumor cells immortal and driving runaway cell proliferation (Blasco 2003; Stewart and Weinberg 2006; Artandi and DePinho 2010).

Despite its enormous medical interest, telomerase has been and remains a challenging target for structural studies, in part because of the low cellular abundance of telomerase in most cells (e.g., approximately 240 in human cells) (Xi and Cech 2014), the propensity for TERT to aggregate, and difficulties in assembling telomerase from its components in vitro. It has also been challenging to experimentally define the telomerase components and their stoichiometry from different organisms. The catalytic core of TERT and TER is sufficient in vitro for processive telomere repeat synthesis, but in vivo activity requires other species-specific proteins that bind alone or as complexes to TER, forming the core ribonucleoprotein (RNP) (Schmidt and Cech 2015; Wu et al. 2017b). Additional proteins associate constitutively or transiently during telomeric DNA synthesis for G-strand handling, recruitment of proteins for C-strand synthesis, and termination (Fig. 1) (Nandakumar and Cech 2013; Schmidt and Cech 2015; Chan et al. 2017; Wu et al. 2017b). Regions of TER outside the t/PK and STE vary greatly among organisms, as TER has evolved different solutions for telomerase biogenesis, assembly, and localization. In ciliates, a La-related protein group 7 (LARP7) (Bousquet-Antonelli and Deragon 2009) protein (p65 in *Tetrahymena thermophila*) (Witkin and Collins 2004) functions in 3′-end protection and assembly of TER with TERT. Human TER has a specialized H/ACA scaRNA domain (Mitchell et al. 1999a) that binds the H/ACA scaRNP proteins (Egan and Collins 2010). The much larger fission and budding yeast TERs bind different sets of proteins (Wu et al. 2017b), including various Sm and LSm proteins and, for budding yeast the Pop1–Pop6–Pop7 subcomplex from mitochondrial ribonuclease (RNase) P (Fig. 1C) (Lin et al. 2015; Lemieux et al. 2016), which will not be discussed further here. With the exception of a *Saccharomyces cerevisiae* Ku70/80–TER hairpin crystal structure (Chen et al.

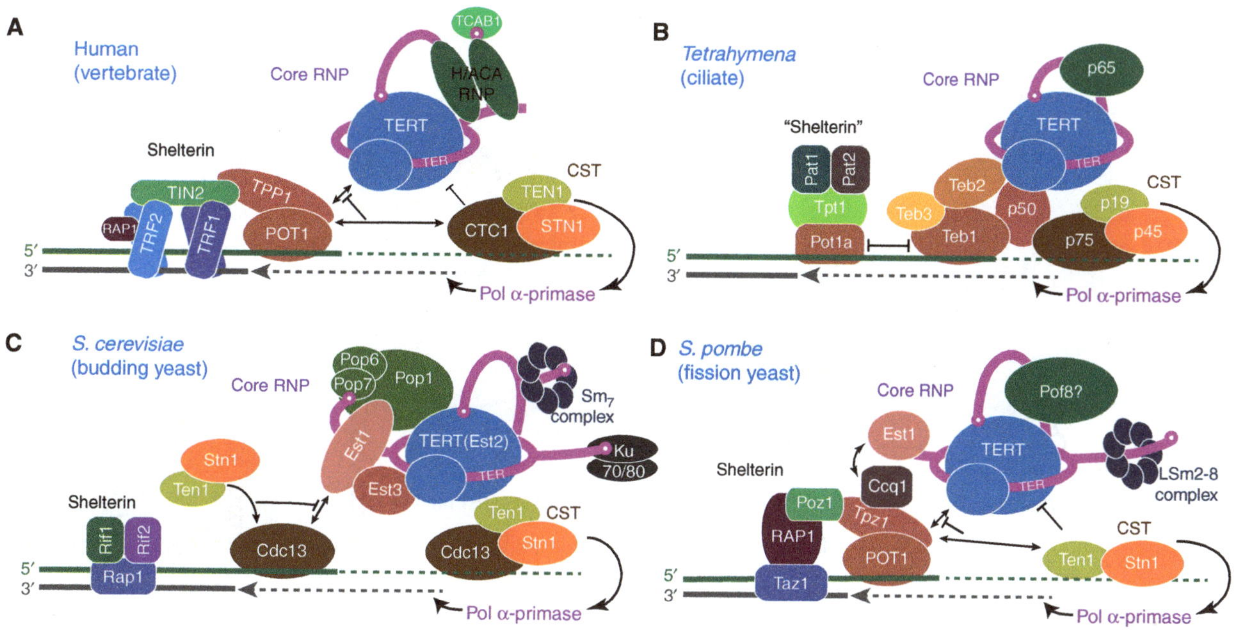

Figure 1. Telomerase ribonucleoprotein (RNP) complexes and their interactions at telomeres. Telomerase RNPs and telomere-associated proteins and interactions are illustrated schematically for (*A*) human, (*B*) *Tetrahymena thermophila*, (*C*) budding yeast (*Saccharomyces cerevisiae*), and (*D*) fission yeast (*Schizosaccharomyces pombe*). Protein identities are indicated and colors chosen to highlight structural or functional conservation between species. Three principal complexes are illustrated: The telomerase core RNP (TER–TERT catalytic core and TER-binding proteins), telomerase accessory proteins (CST complexes, p50/TPP1/Est3, TEB/POT1), and shelterin/shelterin-like complexes. Black lines with arrows represent recruitment and/or enhancement of activity, whereas blunted black lines denote inhibition or termination of activity.

2018), most have been studied structurally only in contexts other than telomerase, and no electron microscopy (EM) structures for yeast telomerase have been reported to date. For function at telomeres, human telomerase core RNP binds to telomere-associated proteins that enhance processivity, TPP1–POT1 (which are also components of telomere shelterin complex) (de Lange 2005; Palm and de Lange 2008), or inhibit telomerase and recruit DNA polymerase α-primase for synthesis of the C-strand $dCCCTAA_n$, CTC1–STN1–TEN1 (CST) (Fig. 1A) (Chen et al. 2012; Chen and Lingner 2013). As revealed by structural studies described below, homologs of these proteins associate constitutively with *Tetrahymena* telomerase (Fig. 1B) (Jiang et al. 2015; Chan et al. 2017).

The first studies of telomerase structure focused on defining TER domains, almost exclusively by nuclear magnetic resonance (NMR) spectroscopy, and TERT domains, by X-ray crystallography (Chan et al. 2017). As other protein components of telomerase were discovered and TER–TERT domain complexes defined, structural studies were extended to them, and recently many new protein domain structures have been reported (Chan et al. 2017; Chen et al. 2018). These studies have been paralleled by structural studies of telomere-interacting proteins (Price et al. 2010; Wan et al. 2015; Hoffman and Skordalakes 2016), which are beyond the scope of this review; some of these are also associated with telomerase at least during telomeric DNA synthesis. In 2013, the first negative-stain EM structures of telomerase, from *Tetrahymena* and human, at ~25–30 Å resolution, were reported (Jiang et al. 2013; Sauerwald et al. 2013). EM overcomes in part the problem of low cellular abundance of telomerase because it requires orders of magnitude less sample than NMR or crystallography. A cryo-electron microscopy (cryo-EM) structure of *Tetrahymena* telomerase at a resolution of 9 Å followed two years later in 2015 (Jiang et al. 2015). Cryo-EM has undergone a resolution revolution during the past few years, leading in favorable cases to atomic resolution structures of complexes such as the ribosome and spliceosome (Fica and Nagai 2017; Frank 2017). In 2018, both the first cryo-EM structure of human telomerase holoenzyme with DNA substrate, at a resolution of 8–10 Å (Nguyen et al. 2018), and a 4.8 Å resolution cryo-EM structure of *Tetrahymena* telomerase with telomeric DNA (Jiang et al. 2018), were published. In the following, we discuss how structural studies using NMR, crystallography, and EM of *Tetrahymena* and human telomerase have been applied to and evolved our understanding of telomerase RNP structure and function.

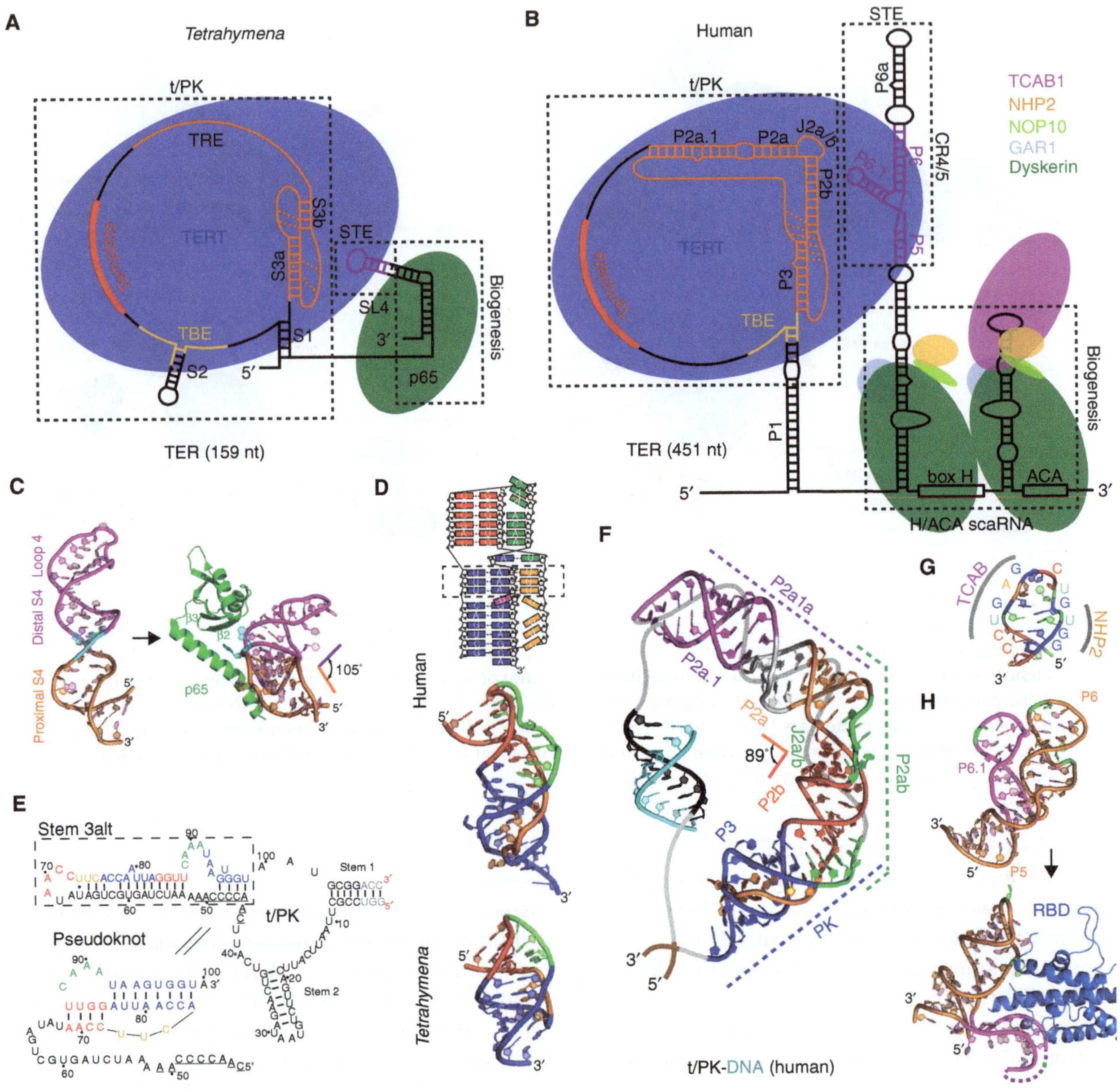

Figure 2. *Tetrahymena* and human telomerase core RNP and structures of TER domains. Schematics of telomerase core RNP of (*A*) *Tetrahymena* and (*B*) human telomerase. Proteins are shown at their approximate relative size and interactions, and the shape of TER is based on structural studies described in the text. (*C*) Nuclear magnetic resonance (NMR) structure of the free *Tetrahymena* stem-terminus element (STE; stem loop 4) (Protein Data Bank [PDB]: 2FEY) and model of the *Tetrahymena* p65 atypical RNA recognition motif (xRRM)–stem loop 4 complex based on the crystal structure of p65 xRRM–stem 4 (PDB: 4ERD) and NMR structure of loop 4 (PDB: 2M21). (*D*) Solution structures of minimal pseudoknots (PKs) from human (PDB: 2K95) and *Tetrahymena* (PDB: 5KMZ). Secondary structure schematic of the human P2b–P3 minimal PK is shown on the *top*. Dashed rectangle highlights the three U-A-U triples. (*E*) Comparison of secondary structures of free *Tetrahymena* TER and the folded PK, determined by NMR (Cash and Feigon 2017). The template residues are underlined. (*F*) Molecular model of human template/pseudoknot domain (t/PK). The solution NMR structure of the minimal PK (P2b/P3), P2b–J2a/b–P2a (P2ab), and residual dipolar coupling (RDC)–MC-Sym model of P2a.1–J2a.1–P2a (P2a1a) were computationally combined to model the full-length P2/P3 pseudoknot (Zhang et al. 2010). The single-stranded region (gray) containing the template (black) is shown bound to telomeric DNA (cyan). (*G*) Solution NMR structure of the 3′ apical stem loop (conserved region 7 [CR7]) containing the TCAB1 and NHP2 binding interfaces, determined by NMR and mutagenesis (Theimer et al. 2007). (*H*) NMR structure of free medaka (*Oryzias latipes*) STE (CR4/5) (PDB: 2MHI) compared with crystal structure of the CR4/5–RNA-binding domain (RBD) complex (PDB: 4O26).

2 SOLUTION NMR STUDIES OF TER STRUCTURE AND DYNAMICS

In parallel with the identification of a consensus structure for TER elements required for catalysis (Chen and Greider 2004), comprising the t/PK and STE domains that independently bind TERT (Fig. 2A,B), the first structures of domains of TER were reported (Fig. 2C–H) (Comolli et al. 2002; Leeper et al. 2003; Theimer et al. 2003; Leeper and Varani 2005; Theimer et al. 2005; Chen et al. 2006; Richards et al. 2006a; Richards et al. 2006b; Kim et al. 2008). NMR has proven particularly useful for structural studies of TER domains, although it was clear from the start that care needed to be taken to ensure that these structures, removed from the context of their protein partners and the full-length TER, were biologically relevant. In a typical study, the biological importance of structural features of the TER domain being studied were investigated by assaying the effect of nucleotide substitutions on structure, telomerase activity in the context of TERT and TER assembled in vitro and/or holoenzyme assembled in vivo, telomere length in vivo, and assembly with TERT using electrophoretic mobility shift assays.

Early studies identified a minimal pseudoknot (PK) in human TER, with most of the conserved nucleotides in the full-length PK, that contains the stem loop interactions (P2b/P3 in vertebrates) (Theimer et al. 2005). Ciliates have the smallest PK (~30 nt), whereas vertebrates and yeasts have PKs with an extended stem 1 (P2 in human) and loop 2 that are outside the minimal PK (Fig. 2A,B,D) (Theimer et al. 2005; Kim et al. 2008; Cash et al. 2013; Wang et al. 2016; Cash and Feigon 2017). In human TER, the full-length PK includes the P2b/P3 minimal PK, a central stem region with a 5–6-nt asymmetric internal loop (P2a–J2a/b–P2b), and an irregular helical extension (P2a.1–J2a.1) (Fig. 2B,F). The solution NMR structure of the human minimal PK established that its fold is stabilized by tertiary interactions between the loops and stems and also highlighted the structural importance of U-A-U base triples within the PK (Fig. 2D) (Theimer et al. 2005; Kim et al. 2008; Zhang et al. 2010). A correlation between PK stability and telomerase activity was shown by mutagenesis and thermodynamic analysis (Theimer et al. 2005). Structural and functional studies on ciliate, yeast, and other vertebrate (e.g., medaka fish [*Oryzias latipes*]) TER PKs have confirmed the importance of base triples (triple helix) for PK stability and telomerase function across various species (Shefer et al. 2007; Qiao and Cech 2008; Cash et al. 2013; Wang et al. 2016; Cash and Feigon 2017). The recent cryo-EM structures of *Tetrahymena* and human telomerase show that the solution NMR structures of the PKs fit well to the EM density maps of the holoenzymes (Jiang et al. 2015, 2018; Nguyen et al. 2018), confirming that the structures of the isolated domains obtained by NMR are biologically relevant. NMR structures of the other helical domains of *Tetrahymena* TER, stem loop 2 (Richards et al. 2006a), and stem loop 4 (Chen et al. 2006; Richards et al. 2006b) were also determined. The p65 carboxy-terminal domain binds and bends stem 4 at a conserved GA bulge (Fig. 2C) (Singh et al. 2012), as discussed in Section 3.

The structures and dynamics of the remaining subdomains of the TER full-length PK in vertebrates (human and medaka fish) have also been extensively investigated by solution NMR (Kim et al. 2010; Zhang et al. 2010; Kim et al. 2014; Wang et al. 2016). The human P2ab domain consists of P2a, P2b, and a 5-nt J2a/b pyrimidine-rich asymmetric bulge (Fig. 2B), conserved in location but not sequence. Its NMR structure showed that the J2a/b bulge loop has a characteristic S-shape and creates an ~90° bend with a surprisingly low twist (~10°) between the flanking helices (Zhang et al. 2010). The J2a/b bulge loop is intrinsically flexible but the interhelical motions across the loop are remarkably restricted. Nucleotide substitutions in J2a/b that affect the bend angle, direction, and interhelical dynamics are correlated with telomerase activity. The other PK subdomain, P2a.1, which was proposed to be specific to mammals, is connected to P2 by J2a.1. This region, which contains a flexible internal loop that forms an irregular helix, was modeled by combining the RNA modeling program MC-Sym (Parisien and Major 2008) with residual dipolar couplings (RDCs). A model of the complete human PK was then calculated by computationally combining the NMR structures encompassing the three subdomains (Fig. 2F). This model structure showed that the J2a/b bulge defines the overall arc-shaped topology of the full-length PK. The NMR analysis of medaka PK showed it has a similar topology, despite differences in sequence and size, including a cryptic P2a.1 (Wang et al. 2016).

Other vertebrate TER catalytic core domains that have been studied by NMR are the human and medaka conserved region 4/5 (CR4/5) (Kim et al. 2014), which rearranges when in complex with medaka TERT RBD (Fig. 2H) (Huang et al. 2014), and the P6.1 region of human CR4/5, which contains two pseudouracils (Kim et al. 2010). In the H/ACA domain, a structure–function study of the CR7 stem loop identified conserved elements for TCAB1 and NHP2 binding (Fig. 2G) (Theimer et al. 2007).

Recent NMR studies of *Tetrahymena* TER showed that the *Tetrahymena* PK does not form in the free TER. Instead, part of the PK sequence pairs with and sequesters the template (stem 3alt) (Fig. 2E) (Cash and Feigon 2017). This alternative structure may function in vivo to shield template and PK residues from unproductive interactions or cleavage by endonucleases while still leaving the high-affinity TER

RBD-binding sites available for assembly with TERT. Sequence analysis indicates that human t/PK may also form an alternative structure, and this is supported by single-molecule fluorescence resonance energy transfer (FRET) studies (Hengesbach et al. 2012).

3 USING NMR TO STUDY PROTEIN–RNA INTERACTIONS AND OPTIMIZE CONSTRUCTS FOR CRYSTALLOGRAPHY: p65–TER COMPLEX

NMR has also been used to study telomerase protein domains and interactions with TER and has proven useful in optimizing constructs for crystallography. Good examples are the studies of the *Tetrahymena* telomerase biogenesis and assembly protein p65 and its interaction with TER (Singh et al. 2012). The LARP7 protein p65 associates with the 3′ end of TER and is essential for assembly in vivo (Prathapam et al. 2005; O'Connor and Collins 2006; Singh et al. 2013). In addition to a La module (La motif and RNA recognition motif [RRM]) (Bousquet-Antonelli and Deragon 2009) that binds the TER 3′-UUUU tail, it has a carboxy-terminal domain that binds to and bends stem 4 (Stone et al. 2007; Berman et al. 2010; Akiyama et al. 2012; Singh et al. 2012). Many different protein constructs of the carboxy-terminal domain were made and NMR was used to evaluate whether they were stably folded, identify flexible regions, and define domain boundaries (Singh et al. 2012). The initial assay for this is a simple ^{1}H–^{15}N heteronuclear single-quantum correlation (HSQC) experiment, which provides one cross peak for each peptide NH. Once good spectra are obtained, backbone assignments can be determined from standard triple-resonance experiments. In the case of p65, a flexible carboxy-terminal tail and a large unstructured internal loop were deleted and truncated, respectively, ultimately yielding a construct that was suitable for NMR structure determination. Binding and assembly assays using isothermal titration calorimetry (ITC) and electrophoretic mobility shift showed that the carboxy-terminal tail, but not the internal loop, is important for TER binding and assembly of p65–TER with TERT. The solution structure identified the domain as a new type of atypical RRM, named xRRM, and the partially deleted internal loop at a highly flexible (but dispensable) region connecting the β2 and β3 strands (Fig. 2C) (Singh et al. 2013). NMR was also used to assay RNA constructs for structure and binding. This information guided the design of protein constructs to obtain a crystal structure of the p65 xRRM with TER. The carboxy-terminal tail, which is essential for RNA binding, was retained and various constructs with partial or full deletion of the β2–β3 loop were screened, ultimately yielding a crystal structure with TER stem 4, which revealed the structural basis of the conformational change in stem loop 4 induced by p65 that facilitates assembly of p65–TER with TERT (Fig. 2C).

Another example of the complementarity of NMR and crystallography are the studies of TERT TEN (Fig. 3E) (Jacobs et al. 2005; O'Connor et al. 2005; Jacobs et al. 2006). *Tetrahymena* TEN was identified as a soluble domain of TERT by high-throughput screening and characterized as an independently folded domain by NMR (Jacobs et al. 2005). It was subsequently crystallized (Jacobs et al. 2006). Recently the structure of another TEN domain from the thermotolerant yeast *Hansenula polymorpha* (hpTEN) was determined by both crystallography and NMR (Polshakov et al. 2016; Petrova et al. 2018). The protein core in the NMR and crystal structures has the same overall organization, with a backbone atom root-mean-square deviation (RMSD) of 0.8 Å, but there are flexible regions that differ between the two structures, including helices α5 and α6 present in only the crystal or NMR structure, respectively.

4 X-RAY CRYSTALLOGRAPHY OF TERT DOMAINS AND COMPLEXES WITH TER

TERT needs to be assembled with TER to be functional in synthesizing telomeric DNA repeats. In vitro, telomerase activity can be recovered from TERT and TER assembled in rabbit reticulocyte lysate (Min and Collins 2010); however, the vast majority of the protein is aggregated and not functional. Not surprisingly then, TERT has been notoriously difficult to purify and crystallize.

Almost all TERTs contain four major domains, two of which are unique to TERT (i.e., the TEN domain and RBD) and two domains found in other reverse transcriptases (i.e., RT [palm and fingers] and CTE [thumb]) (Fig. 3A) (Kelleher et al. 2002; Autexier and Lue 2006; Mason et al. 2011). A major breakthrough was the crystal structure of the putative TERT from the flour beetle *Tribolium castaneum* reported in 2008 (and with an RNA–DNA hairpin mimicking a template–DNA duplex in 2010), which revealed that the RBD–RT–CTE domains form a ring structure (Fig. 3G) (Gillis et al. 2008; Mitchell et al. 2010). *Tribolium* TERT is atypical in that it lacks a TEN domain (Fig. 3A) and it has no known TER. The genome sequence of *Tribolium* revealed that its telomeres are composed of a mix of retrotransposons and simple repeats that are mostly, but not exclusively, TCAGG, suggesting that telomerase, as a telomere maintenance mechanism, was lost in *Tribolium* (Tribolium Genome Sequencing Consortium et al. 2008); thus, *Tribolium* TERT is unlikely to be a true TERT. Nevertheless, the *Tribolium* TERT ring structure has proven useful for modeling into negative-stain and cryo-EM maps of *Tetrahymena* and human telomerase (Jiang et al. 2013, 2015; Nguyen et al. 2018). The *Tribolium* TERT structure has

Cite as *Cold Spring Harb Perspect Biol* doi: 10.1101/cshperspect.a032383

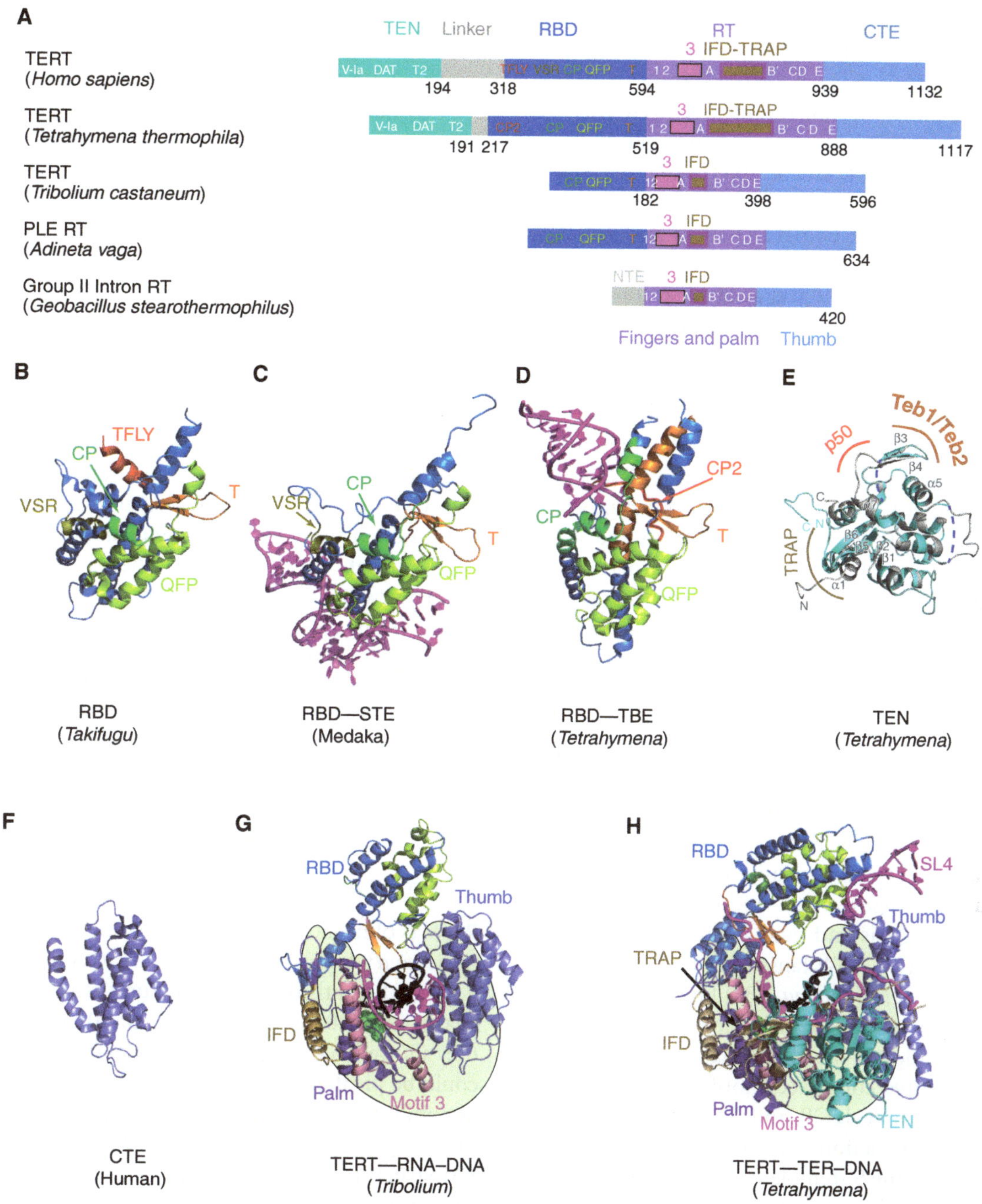

Figure 3. TERT domains and structures. (*A*) Schematics comparing domains of reverse transcriptases (RTs) from human telomerase, *Tetrahymena* telomerase, *Tribolium castaneum* telomerase, Penelope-like element (PLE; *Adineta vaga*), and Group II intron (*Geobacillus stearothermophilus*). Domains and conserved motifs are aligned using the RT domains. Note that *Tribolium* "TERT" is more similar to the *A. vaga* PLE RT than to the true TERTs from human and *Tetrahymena*. Crystal structures (*B–G*) of (*B*) *Takifugu rubripes* RBD (PDB: 4LMO), (*C*) medaka RBD–STE complex (PDB: 4O26), (*D*) *Tetrahymena* RBD–template boundary element (TBE) complex (PDB: 5C9H), and (*E*) free *Tetrahymena* telomerase amino-terminal domain (TEN) (PDB: 2B2A, cyan) superimposed with cryo-EM structure of TEN in *Tetrahymena* telomerase holoenzyme (PDB: 6D6V) (gray). Regions interacting with p50, TEB, and TRAP are indicated. Blue dashed lines represent the missing loops in the crystal structure. (*F*) Human carboxy-terminal element (CTE) (PDB: 5UGW) and (*G*) *Tribolium* TERT with an RNA–DNA hairpin mimicking a template–DNA duplex (PDB: 3KYL). (*H*) Cryo-EM structure of *Tetrahymena* TERT–TER with template–DNA duplex (PDB: 6D6V). TRAP is mostly covered by TEN, so is difficult to see in this view. The TRAP and TEN domains are unique to TERT, whereas motif 3 and IFD are found in closely related RTs from group II introns and PLEs. For *G* and *H*, the polymerase "hand" view is shown.

also provided structural and biochemical insights into the mechanism of nucleotide addition, revealing the location of many features found in other reverse transcriptases such as the primer grip, thumb loop, and thumb helix, as well as two previously unknown motifs, the insertion in fingers domain (IFD) (Lingner et al. 1997; Lue et al. 2003) and motif 3 (Xie et al. 2010) in the RT domain (Gillis et al. 2008; Mitchell et al. 2010). However, the two IFD helices and motif 3, originally thought to be TERT-specific, were subsequently also found in group II intron (Qu et al. 2016; Zhao and Pyle 2016; Stamos et al. 2017) and some Penelope-like element (PLE) RTs (Fig. 3A,G) (Gladyshev and Arkhipova 2007). Notably, structures of the TERT ring alone do not reveal how template boundaries are determined or how translocation occurs (e.g., what separates the template–DNA duplex and what keeps the telomeric DNA from dissociating during translocation). Comparison with the structure of the *Tetrahymena* TERT–TER catalytic core from the 4.8 Å cryo-EM map (discussed later) highlights the structural differences of a true TERT containing the telomerase unique TRAP and TEN domains, a larger RBD, and an integral TER (Fig. 3A,G,H).

The TERT RBD has distinct binding sites for the TER template boundary element (TBE) and STE domains (Lai et al. 2001), and these have been structurally characterized by EM (Jiang et al. 2015, 2018) and in greater atomic detail by crystallography (Fig. 3B–D) (Harkisheimer et al. 2013; Huang et al. 2014; Jansson et al. 2015). RNA-binding motifs were suggested from crystal structures of *Tetrahymena* and vertebrate fish (*Takifugu rubripes* and *O. latipes*) RBD, as well as sequence analysis and mutagenesis (Harkisheimer et al. 2013; Huang et al. 2014; Jansson et al. 2015). A 3.0 Å crystal structure of *Tetrahymena* RBD bound by the TBE shows that the single-stranded nucleotides on either side of stem 2 bind two sides of the RBD, making a network of polar contacts to three conserved motifs in the RBD (i.e., CP2 [TFLY in vertebrates], CP, and T) (Fig. 3D) (Jansson et al. 2015). These interactions anchor the TBE onto TERT, thereby determining the 5′-template boundary by physically preventing nucleotides beyond the last template nucleotide from moving into the active site.

The TER STE (distal stem loop 4 in *Tetrahymena* and P6.1 in CR4/5 in vertebrates) (Fig. 2A,B) binds to both the RBD and CTE (Bley et al. 2011; Huang et al. 2014; Jiang et al. 2015, 2018), which interface to close the TERT ring in the absence of TER in *Tribolium* TERT. A 3.0 Å crystal structure of medaka RBD bound to the CR4/5 domain provided details of the RBD part of the interaction (Fig. 3C) (Huang et al. 2014). The conserved motif QFP and the amino-terminal region of helix α2 interact extensively at a unique three-way junction in CR4/5. However, density for the P6.1 loop is missing and it appears that the missing loop could interact with the CTE (Nguyen et al. 2018).

The 2.2 Å crystal structure of the *Tetrahymena* TERT TEN domain revealed a mixed α/β-fold with a highly basic patch proposed to be a DNA-binding groove (Fig. 3E) (Jacobs et al. 2006; Zaug et al. 2008; Wu and Collins 2014). TEN has been proposed to function in binding TER (*Tetrahymena* loop 4 and the TRE) (O'Connor et al. 2005), as a telomeric DNA anchor site (Jacobs et al. 2006; Zaug et al. 2008; Wyatt et al. 2009; Jurczyluk et al. 2011; Eckert and Collins 2012), and more recently in stabilization of short template–DNA duplexes (Robart and Collins 2011; Wu and Collins 2014; Akiyama et al. 2015; Shastry et al. 2018). TERT TEN domain is connected to the RBD by a long at least partially unstructured linker, but its location relative to other TERT domains and TER was established only by the cryo-EM studies, which revealed unexpected interactions and functions (see Sec. 7) (Jiang et al. 2015, 2018).

Besides the RBD, the only other domain of vertebrate TERT for which a crystal structure has been reported is of a partial human CTE, missing the amino-terminal region containing the thumb loop, at 2.3 Å resolution (Fig. 3F) (Hoffman et al. 2017). The all-helical bundle structure is highly similar to the *Tribolium* CTE (thumb domain) and revealed the locations of some human mutations associated with the telomerase insufficiency disease dyskeratosis congenita.

5 NEGATIVE-STAIN EM OF TELOMERASE

Reports of negative-stain EM structures of human and *Tetrahymena* telomerase in 2013 at 25–30 Å resolution marked a major breakthrough in structural studies of telomerase (Jiang et al. 2013; Sauerwald et al. 2013), in part because they showed that sufficient quantities of functional enzymes could be purified at least for EM studies and suggested the possibility of higher resolution structures to follow (Akiyama and Stone 2013; Miracco et al. 2014). The negative-stain EM structure of *Tetrahymena* telomerase provided the architecture of a telomerase with locations of TERT, TER domains, p65, and other known accessory proteins (Fig. 4A) (Jiang et al. 2013). Although negative-stain EM structures have largely been supplanted by cryo-EM structures obtained with direct electron detectors, it is useful to discuss the method and what was learned from it, as it made possible the higher resolution cryo-EM structures subsequently reported. In addition, negative-stain EM is often useful to screen samples for homogeneity and concentration, especially because it may be difficult to find good cryogenic conditions for RNPs (e.g., sometimes RNPs do not survive the freezing process intact). Negative-stain EM is also useful for obtaining an initial unbiased model of a complex (i.e., no

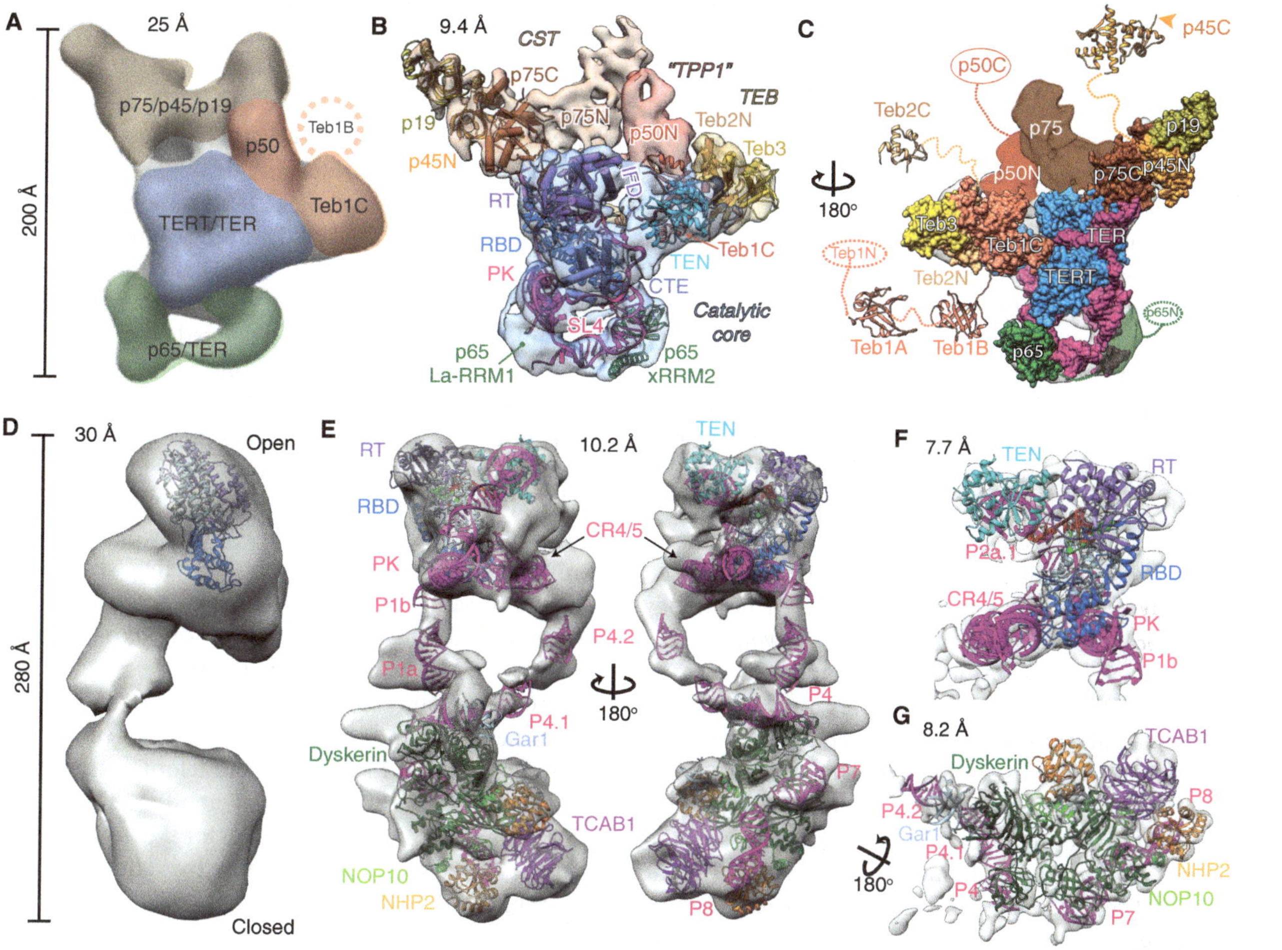

Figure 4. Comparison of negative-stain and cryo-EM density maps and models of *Tetrahymena* and human telomerase. (*A–C*) EM density maps and models of *Tetrahymena* telomerase. (*A*) A 25 Å negative-stain EM map with locations and estimated boundaries of subunits colored, based on affinity tagging of subunits, comparison of particles lacking all or part of a subunit, and modeling of the catalytic core (Jiang et al. 2013) (EMDB: 5804). (*B*) A 9.4 Å resolution cryo-EM map (core RNP, blue; CST, tan; TEB, straw; and p50, red) and pseudoatomic models of the core RNP and TEB and CST trimerization domains of three OB folds (Jiang et al. 2015). (*C*) An 180° rotated view of *B* with modeled domains shown as space-fill on the cryo-EM map. Model of the catalytic core and TEB is based on the 8.9 Å resolution map. Additional domains of Teb1, Teb2, p45, p65, and p50 are not visible in the cryo-EM map because of positional dynamics and are shown as crystal structures (Teb1A [PDB: 3U4V], Teb1B [PDB: 3U4Z], p45C [PDB: 5DFN]), homology models (Teb2C based on PDB: 1DPU), or ovals. (*D–G*) EM density maps and models of human telomerase. (*D*) A 30 Å negative-stain EM map originally proposed to be a dimer (EMDB: 2310) (Sauerwald et al. 2013). *Tribolium* TERT was automatically fit into the map using UCSF Chimera (Pettersen et al. 2004). (*E*) Two views of the 10.2 Å resolution cryo-EM map of human telomerase holoenzyme (EMDB: 7521) with a modeled catalytic core (*top*) and H/ACA scaRNP (*bottom*) (Nguyen et al. 2018). (*F*) A 7.7 Å resolution cryo-EM map (EMDB: 7518) from focused refinement and model of the catalytic core. (*G*) A 8.2 Å resolution map (EMDB: 7519) from focused refinement and model of the H/ACA scaRNP. TER is magenta in all models and TERT is blue. Proteins and TER domains are labeled.

assumed starting model is used) (Jiang et al. 2013). For negative-stain EM, aqueous samples are applied to an EM grid and then stained with a heavy metal solution such as uranyl formate.

Analysis of 3D reconstructions from negative-stain EM data on human and *Tetrahymena* telomerase provided the first views of the overall shape of telomerase (Fig. 4A,D). For *Tetrahymena* telomerase, detailed information on the locations of the seven known proteins and one RNA was obtained by collecting negative-stain EM data on telomerase holoenzymes with affinity tags on individual subunits (Fig. 4A). Strains with individual proteins tagged at the endogenous locus (in most cases) with tandem 3×FLAG (F) and two protein A (ZZ) tags separated by a TEV protease cleavage site allowed purification and biochemical characterization of an endogenously expressed enzyme (Min and Collins 2009). A method was developed to bind the antigen-binding fragment (Fab) of an anti-FLAG antibody during the purification, and one to three bound Fabs could then be visualized in negative-stain EM 2D class averages and 3D reconstructions, thereby revealing the location of the carboxyl (or amino, for ZZF tags) terminus of the protein where the Fab(s) bound. A different affinity tag, MS2 coat protein, identified the location of TER stem loop 2, where the loop was replaced by the MS2 coat protein-binding site. Using this method, locations of all the subunits except p45 were determined, whose tag was not visible in the class averages; this process was enhanced by comparison of 3D reconstructions of particles lacking all or part of a subunit. With the general subunit locations identified, known crystal and NMR structures of telomerase subunits or homology models (*Tribolium* TERT RT and CTE, *Tetrahymena* RBD and TEN, p65–stem loop 4, and Teb1C) could be fit into the negative-stain EM map. This analysis was notably successful; locations of all tagged subunits (except p45) were subsequently shown by cryo-EM to be correct.

Major discoveries in the negative-stain EM study were the identification of p50, previously thought to be substoichiometric, as a central hub binding TERT, Teb1C, and the p75–p45–p19 complex (Fig. 4A), and that p50 was essential to the processivity enhancement conferred by Teb1 (Hong et al. 2013; Jiang et al. 2013). Teb1 (Min and Collins 2009) is paralogous to the large subunit of replication protein A (RPA), a single-stranded DNA-binding trimeric complex that is involved in all aspects of DNA replication and repair (Prakash and Borgstahl 2012; Sugitani and Chazin 2015). The EM map also provided the location of p65 relative to TERT, defining the overall architecture of the core RNP. An interesting footnote to the affinity labeling of p45 is the subsequent discovery that the Fab was in fact bound but not visible in the EM 2D class averages attributable to domain dynamics (i.e., it was bound to a carboxy-terminal domain of p45 separated by a flexible linker to its amino-terminal domain) (Fig. 4C). In images of single particles, three Fabs could be seen in various locations relative to the rest of the particle (Jiang et al. 2015).

Human telomerase used for the negative-stain EM structure (Sauerwald et al. 2013) was purified from HEK293T cells transiently transfected with human TER and TERT (Cristofari et al. 2007). Analysis of mass spectrometry data identified TERT, dyskerin, and NOP10 as components of the enzyme. Based on molecular weight estimated from native polyacrylamide gel electrophoresis (PAGE) and activity assays on telomerases with two different affinity tags on TERT, the investigators concluded that human telomerase is an obligate dimer with two TERT and two TER molecules plus dyskerin and NOP10. The 3D reconstruction from negative-stain EM and tomography revealed a bilobal structure with an "open" and "closed" lobe at the two ends bridged by density ~25 Å in diameter (Fig. 4D). The location of TERT was identified from the position of nanogold-labeled bound telomeric DNA, which showed up as dots in none (13.5%), one (50%), or both (36.5%) of the lobes in the single particles. TERT was modeled into the computationally isolated open monomer, and it was concluded that there was one TERT in each lobe with the connecting density proposed to be a TER helix. However, studies from the Collins laboratory where each of the H/ACA scaRNP proteins had been individually tagged provided strong evidence that human telomerase was a monomer with one TERT and TER and two sets of H/ACA proteins bound to the scaRNA domain, which would be about the same molecular weight as the proposed dimer (Egan and Collins 2010). This was found to be unambiguously the case for the recently reported cryo-EM structure of human telomerase (Nguyen et al. 2018), discussed in detail below. Comparison of the negative-stain EM maps with the cryo-EM maps of human telomerase (Fig. 4) indicates that they are very similar when the difference in resolution is accounted for, although for the negative-stain EM sample it may be a mixture of active and inactive TERT assemblies (Nguyen et al. 2018).

6 SUBNANOMETER RESOLUTION CRYO-EM OF TELOMERASE

After the negative-stain EM structures of *Tetrahymena* and human telomerase reported in 2013, it took another two and five years, respectively, for cryo-EM structures, at 8–10 Å resolution, to be reported (Jiang et al. 2015; Nguyen et al. 2018).

6.1 *Tetrahymena* Telomerase

Taking advantage of the improvements in data collection for cryo-EM and in particular the emergence of direct detector cameras, cryo-EM structures of *Tetrahymena* telomerase holoenzyme were solved to 9.4 Å resolution for the entire particle (Fig. 4B,C) and to 8.9 Å resolution excluding the p75–p45–p19 complex, which negative-stain EM 3D reconstructions had shown to be flexibly positioned around p50 (Jiang et al. 2015). Processing of cryo-EM data using a focused refinement is useful for increasing the resolution of the part of the particle within a soft mask that excludes other regions, and was used for both *Tetrahymena* and human telomerase. It is also important to note that the reported EM resolutions use the "gold standard" approach with a Fourier shell correlation FSC = 0.143 cutoff (Rosenthal and Henderson 2003; Scheres and Chen 2012), but the resolution in different regions of the EM map varies depending on flexibility and other factors.

The identification of protein subunits and a pseudoatomic model of *Tetrahymena* telomerase catalytic core and most of the accessory proteins in the cryo-EM map was a tour de force that combined NMR spectroscopy, X-ray crystallography, and mass spectrometry with the cryo-EM data. The localization of subunits by affinity tagging and the unbiased model obtained from the negative-stain EM structure (Jiang et al. 2013) were crucial for obtaining and interpreting the 9 Å cryo-EM map. At this resolution, protein helices and β-barrels and RNA helices can be discerned in the density and crystal structures and the homology models fit with certainty. For the *Tetrahymena* telomerase core RNP, the known structures and models for TER were helical domains (including a newly determined NMR structure of the PK) (Richards et al. 2006a; Cash and Feigon 2017), the p65 xRRM–stem 4 complex (Singh et al. 2012) and loop 4 (Fig. 2) (Richards et al. 2006b), and for TERT the RT and CTE domains from the *Tribolium* TERT crystal structure (Mitchell et al. 2010) and crystal structures of *Tetrahymena* RBD (Rouda and Skordalakes 2007) and TEN domain (Fig. 3) (Jacobs et al. 2006). Rigid-body fitting of these TER, TERT, and p65 domains provided a pseudoatomic model of the RNP catalytic core, which showed that the t/PK encircles the TERT RBD–RT–CTE ring. The PK is positioned over the CTE and the TEN domain is stacked over the CTE on the other side of the TERT ring with the 3′ template–adjacent single-stranded TER running between the two (Fig. 4B). Stem 1 and stem loop 4 form a U-shape with only loop 4 contacting TERT, apparently inserting at the RBD–CTE interface. Loop 4 and the PK, both of which had been previously proposed to contribute directly to catalysis (O'Connor et al. 2005; Qiao and Cech 2008), are far from the active site. Based on their positions in the cryo-EM structure, the PK was proposed instead to act like a watchband ratchet clasp, with the PK fully folding only after the t/PK assembled onto TERT (Jiang et al. 2015), and loop 4 was proposed to stabilize closure of the TERT ring.

For the rest of the particle, only Teb1C had a previously known structure and the other proteins were thought to be specific to *Tetrahymena* (or ciliates). We note that although Teb1 has four domains, Teb1A, B, and N are connected by flexible linkers and are invisible in the cryo-EM map (Fig. 4C). Teb1C is a large OB fold with a Zn-ribbon motif that confers a characteristic C-shape (Zeng et al. 2011). Fitting of Teb1C into the cryo-EM map revealed a surprise—although its carboxyl terminus had been correctly located by anti-FLAG Fab tagging (Jiang et al. 2013), it fit only part of the assigned density in a "knob" and the remaining density could not be accounted for by any known *Tetrahymena* telomerase protein. Because Teb1 is an RPA1 paralog, the investigators surmised that the remaining density might arise from previously undiscovered RPA2 and RPA3 paralogs. A crystal structure of the RPA heterotrimer of three OB-folds (Fan and Pavletich 2012) fit the density, supporting this hypothesis, and the presence of these two previously unknown subunits (named Teb2 and Teb3) was confirmed by liquid chromatography–mass spectrometry (LC–MS/MS). A subsequent study showed that whereas Teb1 is unique to *Tetrahymena* telomerase, Teb2 and Teb3 are shared subunits with *Tetrahymena* RPA (Upton et al. 2017). Another surprise was the identification of the p75–p45–p19 complex as *Tetrahymena* homologs of human CST. CST is another RPA-related complex, whose STN1 and TEN1 subunits are similar to RPA2 and RPA3 proteins, respectively, whereas CTC1 (Cdc13 in yeasts) is more divergent from RPA1 (Chan et al. 2017). Crystal structures of p19 and of p45 carboxy-terminal domain revealed an OB-fold and winged helix (WH)–WH, respectively, that are structurally homologous to TEN1 and STN1 carboxy-terminal domain. An exhaustive 6D search algorithm for automated fitting revealed that the RPA trimeric core also fits into the EM density assigned to p75–p45–p19 in addition to the TEB density. The identification of p75–p45–p19 as CST was independently verified by structural and functional studies (Wan et al. 2015). Finally, p50 was identified as a β-barrel, and its location near TEN, Teb1, and p75 suggested that it is a homolog of TPP1, although it could not be modeled. In humans, TEN interacts directly with a region of TPP1 called the TEL patch, and this interaction is required for recruitment to telomeres (Zaug et al. 2010; Nandakumar et al. 2012; Sexton et al. 2012; Zhong et al. 2012; Schmidt et al. 2014). Thus, the integrative structural biology studies of *Tetrahymena* telomerase revealed that it is constitutively assembled with homologs of CST and TPP1–POT1 that in humans and yeasts only temporally associate with telomerase at telomeres. This makes *Tetrahymena* an ideal model organism for investigating how telomerase interacts with telomere-associated proteins (Feigon et al. 2016; Chan et al. 2017).

6.2 Human Telomerase

The cryo-EM structure of human telomerase with substrate DNA (T_{12}TTAGGG) at sub-nanometer resolution represents another landmark study, which defined the subunit composition and overall architecture of human telomerase and also provided the first model for a eukaryotic H/ACA RNP (Fig. 4E–G) (Nguyen et al. 2018). To overcome the problem of low cellular levels of telomerase, telomerase was reconstituted in vivo from human cells transiently transfected with tagged ZZ–TEV–Twin Strep–TERT and TER, similar to what was performed for the negative-stain EM structure, and purified. A key step was identifying the eluted telomerase fractions that were homogeneous by negative-stain EM, the so-called active telomerase. Another important step was identifying the locations of the catalytic core and the scaRNP protein TCAB1/WDR79 (tryptophan–aspartic acid repeat protein 79) in negative-stain 3D reconstructions by comparison with samples prepared with TER lacking the H/ACA domain and a TCAB1 mutant that does not bind the H/ACA RNP, respectively. The TER interface between the H/ACA RNP and the catalytic core of human telomerase is highly dynamic, limiting the overall resolution of a complete cryo-EM map to 10.2 Å (Fig. 4E). Focused refinement provided 3D reconstructions of the catalytic core at 7.7 Å (Fig. 4F) and the H/ACA scaRNP at 8.2 Å (Fig. 4G) resolution. The catalytic core was modeled by rigid-body fitting of crystal structures of RT and CTE from *Tribolium* TERT (Mitchell et al. 2010), RBD with TER CR4/5 (P5–P6.1–P6a) from medaka (Huang et al. 2014), and TEN from *Tetrahymena* (Fig. 3) (Jacobs et al. 2006), and of NMR structures from the human t/PK model described above (Fig. 2) (Zhang et al. 2010, 2011). A model of the human telomerase catalytic core, built by homology modeling human TERT and fitting the human t/PK NMR model of TERT based on the TER path in the cryo-EM structure of *Tetrahymena* telomerase (Wang et al. 2016; Wang and Feigon 2017), fits the human telomerase cryo-EM map remarkably well. Although the human t/PK has a much larger helical region, it encircles TERT in a similar way as does *Tetrahymena* t/PK, with the minimal PK on the CTE and the single-stranded template-adjacent TER between the TERT ring and TEN domain (Fig. 4E,F). The human STE element, CR4/5 domain, interacts with the RBD. In the cryo-EM maps of *Tetrahymena* telomerase, the STE element loop 4 inserts between the RBD and CTE (Jiang et al. 2015, 2018). Although the P6.1 stem loop of the CR4/5 domain is not modeled in the cryo-EM map of human telomerase, it is proposed to insert in a similar manner (Nguyen et al. 2018).

The H/ACA RNP, which is the human TER biogenesis domain, was first identified by Collins (Mitchell et al. 1999a), leading to the proposal that dyskerin mutations in dyskeratosis congenita are due to telomerase insufficiency (Mitchell et al. 1999b). Numerous other disease mutations associated with telomerase insufficiency have since been identified in most human telomerase components (Fogarty et al. 2003; Angrisani et al. 2014; Holohan et al. 2014; Wegman-Ostrosky and Savage 2017). Despite many structures of H/ACA proteins and various complexes (Hamma and Ferre-D'Amare 2004; Manival et al. 2006; Liang et al. 2007; Duan et al. 2009; Koo et al. 2011; Li et al. 2011a; Li et al. 2011b), as well as a complete single-hairpin archaeal H/ACA snoRNP used in the modeling (Li and Ye 2006), no structures of a complete eukaryotic H/ACA RNP have been previously reported. The structure of the human telomerase H/ACA RNP domain with its two sets of H/ACA proteins and TCAB1 (found in H/ACA scaRNPs) thus provides the first view of how the two hairpins are oriented with respect to each other and the intersubunit interactions (Fig. 4G). The 5′ and 3′ hairpins are approximately perpendicular to one another, and the two dyskerin molecules have an interacting surface that maps to several disease mutations, providing a structural basis for understanding their disease phenotype. Although dyskerin, NOP10, and NHP2 all contact the 3′ hairpin RNA, only dyskerin directly binds the 5′ hairpin, which lacks an NHP2-binding loop and is extended to form the STE. The human telomerase H/ACA RNP structure may also explain how Alu H/ACA RNPs with especially short or long 5′ hairpins assemble two sets of H/ACA proteins (Ketele et al. 2016).

7 APPROACHING ATOMIC RESOLUTION: TELOMERASE WITH TELOMERIC DNA

The cryo-EM structures of human and *Tetrahymena* telomerase discussed above show that the particle is remarkably dynamic; this has so far limited the resolution that can be obtained from cryo-EM maps. It took two and a half years after the ~9 Å structure of *Tetrahymena* telomerase was reported to achieve a resolution of 4.8 Å with telomeric DNA and 6.4 Å without DNA (Fig. 5A) (Jiang et al. 2018). In the 4.8 Å cryo-EM structure the telomeric DNA $(GTTGGG)_3G$ is bound midway through a cycle of telomere repeat synthesis, and the presence of DNA appears to stabilize the catalytic core somewhat. A soft mask was used to improve the resolution, and in the resulting density maps the dynamic p75–p45–p19 (*Tetrahymena* CST) complex was masked out. The cryo-EM maps were calculated starting with more than three million particles of which <2% were in the final data set for 3D reconstruction. At 4.8 Å resolution, it is just possible to do de novo modeling of secondary structure elements in the regions of the map where there is no starting crystal or NMR structure. Fortunately, related structures (albeit distantly for some) exist for most of the protein subunits or domains of *Tetrahymena* telomerase as described above, including the catalytic subunit TERT. Taking advantage of these evolutionary relationships in EM density-guided computational modeling using the Rosetta package (Song et al. 2013) allowed an ensemble of molecular models to be generated. Visual inspection of the highest scoring solutions, together with rebuilding select regions and subsequent real-space refine-

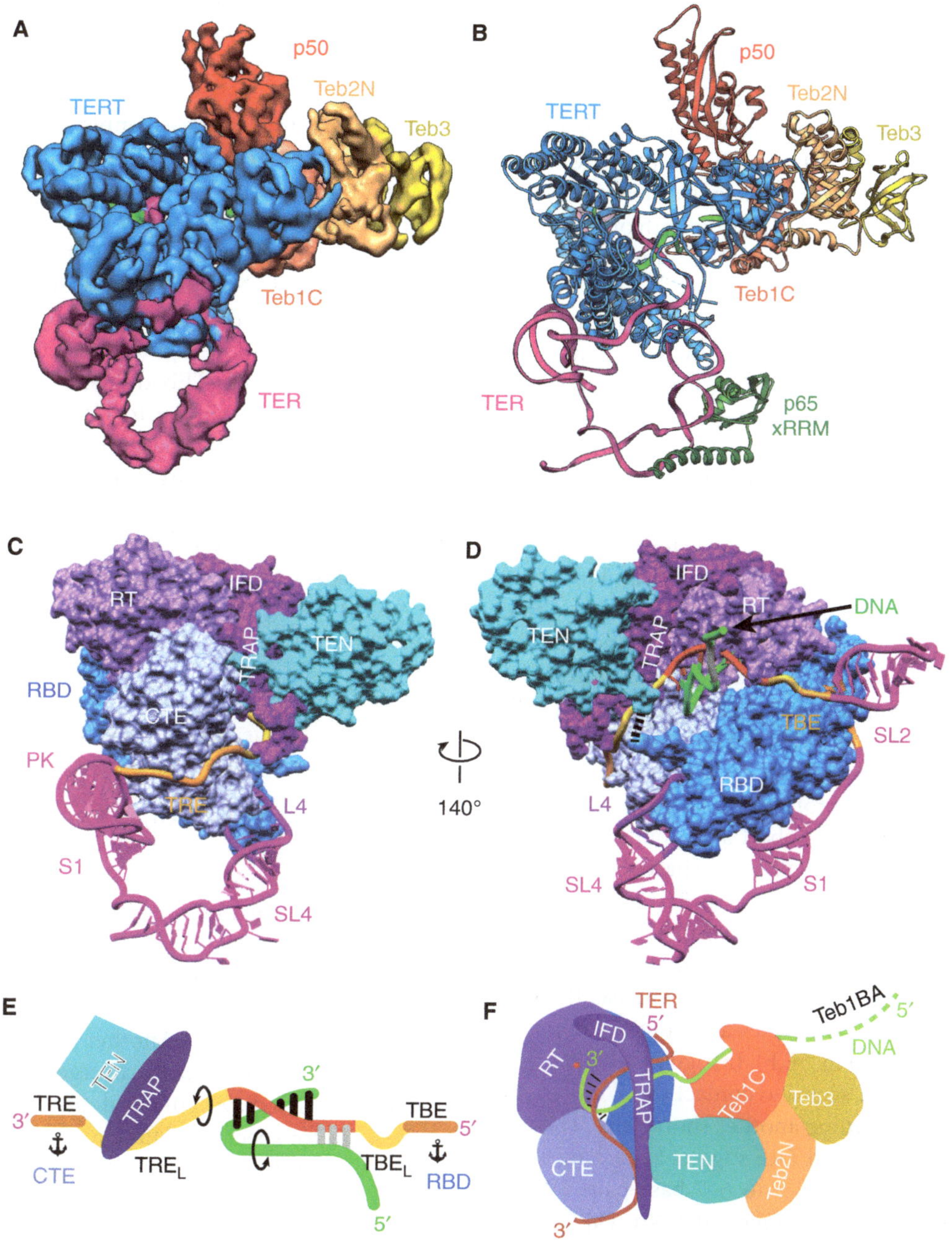

Figure 5. Cryo-EM of *Tetrahymena* telomerase at 4.8 Å resolution provides insights into the mechanism. (*A*) A 4.8 Å resolution cryo-EM density map of *Tetrahymena* telomerase with telomeric DNA. (*B*) Atomic model of the TERT–TER catalytic core, p65 xRRM, p50, and TEB. The model of p65 xRRM is based on the crystal structure (PDB: 4ERD). (*C-D*) Two views of the structure of the telomerase catalytic core with TERT as spacefill and TER as ribbon. TERT and TER domains are labeled. TEN and TRAP are connected to the RBD and RT domains, respectively, from opposite sides of the TERT ring, physically interlocking the t/PK and TERT. In *D*, black dashes between the TEN and RBD domains denote the linker between them. The nascent telomeric DNA is shown exiting the template. (*E*) Cartoon of template recognition element (TRE)–template–TBE interactions with TERT and telomeric DNA. Arrows indicate directions of rotation of TER and DNA during telomere repeat synthesis, and anchors correspond to TER anchor sites that define the template boundaries. TRE linker (TRE$_L$) and the TBE loop (TBE$_L$) are the single-stranded regions between the template and TRE and TBE, respectively. (*F*) Cartoon illustrating the paths of TRE–template–TBE on TERT and of telomeric DNA from the TERT active site to TEB; the TRAP–TEN interaction; the TRAP–TRE interaction. The protein p50 is omitted from this schematic for clarity. (Figure panels are modified from Jiang et al. 2018.)

ment of protein and modeled TER using PHENIX (Adams et al. 2010), ultimately allowed the determination of a complete nearly atomic model of the DNA-bound TERT-TER catalytic core, p50N, and Teb1C-Teb2N-Teb3 (Fig. 5B). This model provided the first complete structure of a true TERT, previously unknown details of the architecture of the catalytic core, structures and interactions of p50-TEB, the path of TER on TERT (Fig. 5C,D), and the path of telomeric DNA from the six base-pair template DNA duplex to its exit on Teb1C (Fig. 5E,F).

In addition to providing much more detail than the lower-resolution cryo-EM model of DNA-free telomerase, two previously unknown features of the catalytic core were identified. First, TERT and TER are intricately interlocked, with TER and TERT physically enclosing one another, explaining in part the difficulty of assembling telomerase in vitro (Fig. 5B–D). Second, a long linker between the two IFD helices, which among RTs is uniquely found in TERTs, was revealed to form a structured mostly β-strand interface with the TERT TEN domain (Fig. 5C–F). This region, named TRAP, explains how TEN, which is connected to the RBD by a long linker, can be stably positioned. The single-stranded 3′ template–adjacent TER (named TRE linker, TRE_L) runs between the CTE and TRAP, and the TRE binds a highly basic surface on CTE (Fig. 5C). The structure shows that roles proposed for TEN in regulating repeat addition processivity (Robart and Collins 2011; Wu and Collins 2014; Akiyama et al. 2015; Shastry et al. 2018) are mostly indirect through TRAP (Fig. 5E,F). The DNA–template duplex occupies a large cavity in the TERT ring. As the growing duplex exits toward TRAP, the looped out TRE_L appears to be captured by TRAP, whereas the single-stranded telomeric DNA makes a turn in the opposite direction toward Teb1C (Fig. 5E,F). Intriguingly, the single-stranded telomeric DNA seems to contact the 5′ end of the template, potentially forming another short helix. The opposing paths of the DNA and TRE suggest a mechanism to limit the template–DNA duplex to 6–7 base pairs (Wang et al. 1998; Forstemann and Lingner 2005; Wu et al. 2017a), because of destabilization of the helix end and accumulation of torsional stress (Fig. 5E). On the 5′ side of the template the TBE is anchored to the RBD (Jansson et al. 2015), and the TBE loop (TBE_L) could go from looped out to maximally stretched during synthesis of a telomere repeat. Thus, TRE–TRE_L and TBE_L–TBE are proposed to define the 3′ and 5′ template boundaries during each telomere repeat synthesis cycle by means of TRAP capture and release of TRE_L and TBE_L stretch and compaction, respectively, adjacent to anchor sites (TRE on the CTE and TBE on RBD) on each side. Other elements of TER apparently function to stabilize the closed structure of the TERT ring (loop 4) and in assembly of the t/PK on TERT (PK). The PK has a limited interaction surface with a basic patch on the CTE. This overall architecture appears to be conserved in human telomerase (Wang et al. 2016; Nguyen et al. 2018).

In addition to forming a complex with TRAP that physically encloses the t/PK on the TERT ring, the TEN domain binds p50 in the same way as predicted for human TEN–TPP1 recruitment of telomerase to telomeres. The modeled structure of p50 reveals that it contains an OB-fold oriented to interact with TEN in the same way as predicted for the TPP1 TEL-patch. The p50 interacting surface of TEN changes conformation significantly compared with the crystal structure of the TEN domain (Fig. 3E). The protein p50 also binds Teb1, similar to TPP1–POT1, although details of the interaction appear to be different (Chen et al. 2017; Wang and Feigon 2017). The TEB complex is further anchored to TERT through interactions with TEN (Fig. 3E), an interaction that likely does not exist for POT1. In summary, the 4.8 Å cryo-EM structure of *Tetrahymena* telomerase with telomeric DNA provided key new insights into telomerase mechanism and DNA handling.

8 CONCLUDING REMARKS: TELOMERASE IS A UNIQUE RNP

Unlike catalytic RNPs such as the ribosome and spliceosome in which the proteins are the scaffold and the RNA does the catalysis (Cech and Steitz 2014), it is now clear that the TERT protein does the actual catalysis. But in this unique RNP, the RT has evolved to require domains specific to TERT and elements of TER in addition to the template that work together for processive synthesis of telomere repeats (Jiang et al. 2018). Another unusual feature of telomerase is the species-specific divergence of TER biogenesis domains, which have co-opted protein complexes from other RNPs for specialized function in telomerase. The first substrate bound cryo-EM structures of *Tetrahymena* and human telomerase have provided a treasure trove of details of RNP architecture, mechanism, and interaction at telomeres. In the not too distant future, we expect atomic resolution structures of telomerase in which all side chains can be modeled to provide atomic details, elucidating mechanisms and enabling drug targeting and understanding of disease mutations.

ACKNOWLEDGMENTS

Telomerase studies in the Feigon laboratory are funded by National Institutes of Health (NIH) GM048123 and National Science Foundation (NSF) MCB1517625 grants. We acknowledge Department of Energy (DOE) grant DE-

FC0202ER63421 for support of the UCLA-DOE NMR, X-ray, and crystallization cores.

REFERENCES

Adams PD, Afonine PV, Bunkoczi G, Chen VB, Davis IW, Echols N, Headd JJ, Hung LW, Kapral GJ, Grosse-Kunstleve RW, et al. 2010. PHENIX: A comprehensive Python-based system for macromolecular structure solution. *Acta Crystallogr D Biol Crystallogr* **66:** 213–221.

Akiyama BM, Stone MD. 2013. Structural biology: A solution to the telomerase puzzle. *Nature* **496:** 177–178.

Akiyama BM, Loper J, Najarro K, Stone MD. 2012. The C-terminal domain of *Tetrahymena thermophila* telomerase holoenzyme protein p65 induces multiple structural changes in telomerase RNA. *RNA* **18:** 653–660.

Akiyama BM, Parks JW, Stone MD. 2015. The telomerase essential N-terminal domain promotes DNA synthesis by stabilizing short RNA-DNA hybrids. *Nucleic Acids Res* **43:** 5537–5549.

Angrisani A, Vicidomini R, Turano M, Furia M. 2014. Human dyskerin: Beyond telomeres. *Biol Chem* **395:** 593–610.

Armanios M, Blackburn EH. 2012. The telomere syndromes. *Nat Rev Genet* **13:** 693–704.

Artandi SE, DePinho RA. 2010. Telomeres and telomerase in cancer. *Carcinogenesis* **31:** 9–18.

Autexier C, Lue NF. 2006. The structure and function of telomerase reverse transcriptase. *Annu Rev Biochem* **75:** 493–517.

Berman AJ, Gooding AR, Cech TR. 2010. *Tetrahymena* telomerase protein p65 induces conformational changes throughout telomerase RNA (TER) and rescues telomerase reverse transcriptase and TER assembly mutants. *Mol Cell Biol* **30:** 4965–4976.

Bernardes de Jesus B, Blasco MA. 2013. Telomerase at the intersection of cancer and aging. *Trends Genet* **29:** 513–520.

Blackburn EH, Collins K. 2011. Telomerase: An RNP enzyme synthesizes DNA. *Cold Spring Harb Perspect Biol* **3:** a003558.

Blackburn EH, Greider CW, Szostak JW. 2006. Telomeres and telomerase: The path from maize, *Tetrahymena* and yeast to human cancer and aging. *Nat Med* **12:** 1133–1138.

Blasco MA. 2003. Telomeres and cancer: A tale with many endings. *Curr Opin Genet Dev* **13:** 70–76.

Bley CJ, Qi X, Rand DP, Borges CR, Nelson RW, Chen JJ. 2011. RNA-protein binding interface in the telomerase ribonucleoprotein. *Proc Natl Acad Sci* **108:** 20333–20338.

Bousquet-Antonelli C, Deragon JM. 2009. A comprehensive analysis of the La-motif protein superfamily. *RNA* **15:** 750–764.

Cash DD, Feigon J. 2017. Structure and folding of the *Tetrahymena* telomerase RNA pseudoknot. *Nucleic Acids Res* **45:** 482–495.

Cash DD, Cohen-Zontag O, Kim NK, Shefer K, Brown Y, Ulyanov NB, Tzfati Y, Feigon J. 2013. Pyrimidine motif triple helix in the *Kluyveromyces lactis* telomerase RNA pseudoknot is essential for function in vivo. *Proc Natl Acad Sci* **110:** 10970–10975.

Cech TR, Steitz JA. 2014. The noncoding RNA revolution—Trashing old rules to forge new ones. *Cell* **157:** 77–94.

Chan H, Wang Y, Feigon J. 2017. Progress in human and *Tetrahymena* telomerase structure determination. *Annu Rev Biophys* **46:** 199–225.

Chen JL, Greider CW. 2004. An emerging consensus for telomerase RNA structure. *Proc Natl Acad Sci* **101:** 14683–14684.

Chen LY, Lingner J. 2013. CST for the grand finale of telomere replication. *Nucleus* **4:** 277–282.

Chen Y, Fender J, Legassie JD, Jarstfer MB, Bryan TM, Varani G. 2006. Structure of stem-loop IV of *Tetrahymena* telomerase RNA. *EMBO J* **25:** 3156–3166.

Chen LY, Redon S, Lingner J. 2012. The human CST complex is a terminator of telomerase activity. *Nature* **488:** 540–544.

Chen C, Gu P, Wu J, Chen X, Niu S, Sun H, Wu L, Li N, Peng J, Shi S, et al. 2017. Structural insights into POT1-TPP1 interaction and POT1 C-terminal mutations in human cancer. *Nat Commun* **8:** 14929.

Chen H, Xue J, Churikov D, Hass EP, Shi S, Lemon LD, Luciano P, Bertuch AA, Zappulla DC, Geli V, et al. 2018. Structural insights into yeast telomerase recruitment to telomeres. *Cell* **172:** 331–343.e13.

Comolli LR, Smirnov I, Xu L, Blackburn EH, James TL. 2002. A molecular switch underlies a human telomerase disease. *Proc Natl Acad Sci* **99:** 16998–17003.

Cristofari G, Reichenbach P, Regamey PO, Banfi D, Chambon M, Turcatti G, Lingner J. 2007. Low- to high-throughput analysis of telomerase modulators with Telospot. *Nat Methods* **4:** 851–853.

de Lange T. 2005. Shelterin: The protein complex that shapes and safeguards human telomeres. *Genes Dev* **19:** 2100–2110.

Duan J, Li L, Lu J, Wang W, Ye K. 2009. Structural mechanism of substrate RNA recruitment in H/ACA RNA-guided pseudouridine synthase. *Mol Cell* **34:** 427–439.

Eckert B, Collins K. 2012. Roles of telomerase reverse transcriptase N-terminal domain in assembly and activity of *Tetrahymena* telomerase holoenzyme. *J Biol Chem* **287:** 12805–12814.

Egan ED, Collins K. 2010. Specificity and stoichiometry of subunit interactions in the human telomerase holoenzyme assembled in vivo. *Mol Cell Biol* **30:** 2775–2786.

Egan ED, Collins K. 2012. Biogenesis of telomerase ribonucleoproteins. *RNA* **18:** 1747–1759.

Fan J, Pavletich NP. 2012. Structure and conformational change of a replication protein A heterotrimer bound to ssDNA. *Genes Dev* **26:** 2337–2347.

Feigon J, Chan H, Jiang J. 2016. Integrative structural biology of *Tetrahymena* telomerase—Insights into catalytic mechanism and interaction at telomeres. *FEBS J* **283:** 2044–2050.

Fica SM, Nagai K. 2017. Cryo-electron microscopy snapshots of the spliceosome: Structural insights into a dynamic ribonucleoprotein machine. *Nat Struct Mol Biol* **24:** 791–799.

Fogarty PF, Yamaguchi H, Wiestner A, Baerlocher GM, Sloand E, Zeng WS, Read EJ, Lansdorp PM, Young NS. 2003. Late presentation of dyskeratosis congenita as apparently acquired aplastic anaemia due to mutations in telomerase RNA. *Lancet* **362:** 1628–1630.

Forstemann K, Lingner J. 2005. Telomerase limits the extent of base pairing between template RNA and telomeric DNA. *EMBO Rep* **6:** 361–366.

Frank J. 2017. Advances in the field of single-particle cryo-electron microscopy over the last decade. *Nat Protoc* **12:** 209–212.

Fujii H, Shao L, Colmegna I, Goronzy JJ, Weyand CM. 2009. Telomerase insufficiency in rheumatoid arthritis. *Proc Natl Acad Sci* **106:** 4360–4365.

Gillis AJ, Schuller AP, Skordalakes E. 2008. Structure of the *Tribolium castaneum* telomerase catalytic subunit TERT. *Nature* **455:** 633–637.

Gladyshev EA, Arkhipova IR. 2007. Telomere-associated endonuclease-deficient Penelope-like retroelements in diverse eukaryotes. *Proc Natl Acad Sci* **104:** 9352–9357.

Greider CW, Blackburn EH. 1987. The telomere terminal transferase of *Tetrahymena* is a ribonucleoprotein enzyme with two kinds of primer specificity. *Cell* **51:** 887–898.

Greider CW, Blackburn EH. 1989. A telomeric sequence in the RNA of *Tetrahymena* telomerase required for telomere repeat synthesis. *Nature* **337:** 331–337.

Hamma T, Ferre-D'Amare AR. 2004. Structure of protein L7Ae bound to a K-turn derived from an archaeal box H/ACA sRNA at 1.8 Å resolution. *Structure* **12:** 893–903.

Harkisheimer M, Mason M, Shuvaeva E, Skordalakes E. 2013. A motif in the vertebrate telomerase N-terminal linker of TERT contributes to RNA binding and telomerase activity and processivity. *Structure* **21:** 1870–1878.

Hengesbach M, Kim NK, Feigon J, Stone MD. 2012. Single-molecule FRET reveals the folding dynamics of the human telomerase RNA pseudoknot domain. *Angew Chem Int Ed Engl* **51:** 5876–5879.

Hoffman H, Skordalakes E. 2016. Crystallographic studies of telomerase. *Methods Enzymol* **573:** 403–419.

Hoffman H, Rice C, Skordalakes E. 2017. Structural analysis reveals the deleterious effects of telomerase mutations in bone marrow failure syndromes. *J Biol Chem* **292:** 4593–4601.

Holohan B, Wright WE, Shay JW. 2014. Cell biology of disease: Telomeropathies: An emerging spectrum disorder. *J Cell Biol* **205:** 289–299.

Hong K, Upton H, Miracco EJ, Jiang J, Zhou ZH, Feigon J, Collins K. 2013. Tetrahymena telomerase holoenzyme assembly, activation, and inhibition by domains of the p50 central hub. *Mol Cell Biol* **33:** 3962–3971.

Huang J, Brown AF, Wu J, Xue J, Bley CJ, Rand DP, Wu L, Zhang R, Chen JJ, Lei M. 2014. Structural basis for protein-RNA recognition in telomerase. *Nat Struct Mol Biol* **21:** 507–512.

Jacobs SA, Podell ER, Wuttke DS, Cech TR. 2005. Soluble domains of telomerase reverse transcriptase identified by high-throughput screening. *Protein Sci* **14:** 2051–2058.

Jacobs SA, Podell ER, Cech TR. 2006. Crystal structure of the essential N-terminal domain of telomerase reverse transcriptase. *Nat Struct Mol Biol* **13:** 218–225.

Jansson LI, Akiyama BM, Ooms A, Lu C, Rubin SM, Stone MD. 2015. Structural basis of template-boundary definition in *Tetrahymena* telomerase. *Nat Struct Mol Biol* **22:** 883–888.

Jiang J, Miracco EJ, Hong K, Eckert B, Chan H, Cash DD, Min B, Zhou ZH, Collins K, Feigon J. 2013. The architecture of *Tetrahymena* telomerase holoenzyme. *Nature* **496:** 187–192.

Jiang J, Chan H, Cash DD, Miracco EJ, Ogorzalek Loo RR, Upton HE, Cascio D, O'Brien Johnson R, Collins K, Loo JA, et al. 2015. Structure of *Tetrahymena* telomerase reveals previously unknown subunits, functions, and interactions. *Science* **350:** aab4070.

Jiang J, Wang Y, Sušac L, Chan H, Basu R, Zhou ZH, Feigon J. 2018. Structure of telomerase with telomeric DNA. *Cell* **173:** 1179–1190.e13.

Jurczyluk J, Nouwens AS, Holien JK, Adams TE, Lovrecz GO, Parker MW, Cohen SB, Bryan TM. 2011. Direct involvement of the TEN domain at the active site of human telomerase. *Nucleic Acids Res* **39:** 1774–1788.

Kelleher C, Teixeira MT, Forstemann K, Lingner J. 2002. Telomerase: Biochemical considerations for enzyme and substrate. *Trends Biochem Sci* **27:** 572–579.

Ketele A, Kiss T, Jady BE. 2016. Human intron-encoded AluACA RNAs and telomerase RNA share a common element promoting RNA accumulation. *RNA biology* **13:** 1274–1285.

Kim NK, Zhang Q, Zhou J, Theimer CA, Peterson RD, Feigon J. 2008. Solution structure and dynamics of the wild-type pseudoknot of human telomerase RNA. *J Mol Biol* **384:** 1249–1261.

Kim NK, Theimer CA, Mitchell JR, Collins K, Feigon J. 2010. Effect of pseudouridylation on the structure and activity of the catalytically essential P6.1 hairpin in human telomerase RNA. *Nucleic Acids Res* **38:** 6746–6756.

Kim NK, Zhang Q, Feigon J. 2014. Structure and sequence elements of the CR4/5 domain of medaka telomerase RNA important for telomerase function. *Nucleic Acids Res* **42:** 3395–3408.

Koo BK, Park CJ, Fernandez CF, Chim N, Ding Y, Chanfreau G, Feigon J. 2011. Structure of H/ACA RNP protein Nhp2p reveals *cis/trans* isomerization of a conserved proline at the RNA and Nop10 binding interface. *J Mol Biol* **411:** 927–942.

Lai CK, Mitchell JR, Collins K. 2001. RNA binding domain of telomerase reverse transcriptase. *Mol Cell Biol* **21:** 990–1000.

Leeper TC, Varani G. 2005. The structure of an enzyme-activating fragment of human telomerase RNA. *RNA* **11:** 394–403.

Leeper T, Leulliot N, Varani G. 2003. The solution structure of an essential stem-loop of human telomerase RNA. *Nucleic Acids Res* **31:** 2614–2621.

Lemieux B, Laterreur N, Perederina A, Noel JF, Dubois ML, Krasilnikov AS, Wellinger RJ. 2016. Active yeast telomerase shares subunits with ribonucleoproteins RNase P and RNase MRP. *Cell* **165:** 1171–1181.

Li L, Ye K. 2006. Crystal structure of an H/ACA box ribonucleoprotein particle. *Nature* **443:** 302–307.

Li S, Duan J, Li D, Ma S, Ye K. 2011a. Structure of the Shq1-Cbf5-Nop10-Gar1 complex and implications for H/ACA RNP biogenesis and dyskeratosis congenita. *EMBO J* **30:** 5010–5020.

Li S, Duan J, Li D, Yang B, Dong M, Ye K. 2011b. Reconstitution and structural analysis of the yeast box H/ACA RNA-guided pseudouridine synthase. *Genes Dev* **25:** 2409–2421.

Liang B, Xue S, Terns RM, Terns MP, Li H. 2007. Substrate RNA positioning in the archaeal H/ACA ribonucleoprotein complex. *Nat Struct Mol Biol* **14:** 1189–1195.

Lin KW, McDonald KR, Guise AJ, Chan A, Cristea IM, Zakian VA. 2015. Proteomics of yeast telomerase identified Cdc48-Npl4-Ufd1 and Ufd4 as regulators of Est1 and telomere length. *Nat Commun* **6:** 8290.

Lingner J, Hughes TR, Shevchenko A, Mann M, Lundblad V, Cech TR. 1997. Reverse transcriptase motifs in the catalytic subunit of telomerase. *Science* **276:** 561–567.

Lue NF, Lin YC, Mian IS. 2003. A conserved telomerase motif within the catalytic domain of telomerase reverse transcriptase is specifically required for repeat addition processivity. *Mol Cell Biol* **23:** 8440–8449.

Manival X, Charron C, Fourmann JB, Godard F, Charpentier B, Branlant C. 2006. Crystal structure determination and site-directed mutagenesis of the *Pyrococcus abyssi* aCBF5-aNOP10 complex reveal crucial roles of the C-terminal domains of both proteins in H/ACA sRNP activity. *Nucleic Acids Res* **34:** 826–839.

Marrone A, Walne A, Dokal I. 2005. Dyskeratosis congenita: Telomerase, telomeres and anticipation. *Curr Opin Genet Dev* **15:** 249–257.

Mason M, Schuller A, Skordalakes E. 2011. Telomerase structure function. *Curr Opin Struct Biol* **21:** 92–100.

Min B, Collins K. 2009. An RPA-related sequence-specific DNA-binding subunit of telomerase holoenzyme is required for elongation processivity and telomere maintenance. *Mol Cell* **36:** 609–619.

Min B, Collins K. 2010. Multiple mechanisms for elongation processivity within the reconstituted *Tetrahymena* telomerase holoenzyme. *J Biol Chem* **285:** 16434–16443.

Miracco EJ, Jiang J, Cash DD, Feigon J. 2014. Progress in structural studies of telomerase. *Curr Opin Struct Biol* **24:** 115–124.

Mitchell JR, Cheng J, Collins K. 1999a. A box H/ACA small nucleolar RNA-like domain at the human telomerase RNA 3′ end. *Mol Cell Biol* **19:** 567–576.

Mitchell JR, Wood E, Collins K. 1999b. A telomerase component is defective in the human disease dyskeratosis congenita. *Nature* **402:** 551–555.

Mitchell M, Gillis A, Futahashi M, Fujiwara H, Skordalakes E. 2010. Structural basis for telomerase catalytic subunit TERT binding to RNA template and telomeric DNA. *Nat Struct Mol Biol* **17:** 513–518.

Nandakumar J, Cech TR. 2013. Finding the end: Recruitment of telomerase to telomeres. *Nat Rev Mol Cell Biol* **14:** 69–82.

Nandakumar J, Bell CF, Weidenfeld I, Zaug AJ, Leinwand LA, Cech TR. 2012. The TEL patch of telomere protein TPP1 mediates telomerase recruitment and processivity. *Nature* **492:** 285–289.

Nelson AD, Shippen DE. 2015. Evolution of TERT-interacting lncRNAs: Expanding the regulatory landscape of telomerase. *Front Genet* **6:** 277.

Nguyen THD, Tam J, Wu RA, Greber BJ, Toso D, Nogales E, Collins K. 2018. Cryo-EM structure of substrate-bound human telomerase holoenzyme. *Nature* **557:** 190–195.

O'Connor CM, Collins K. 2006. A novel RNA binding domain in *Tetrahymena* telomerase p65 initiates hierarchical assembly of telomerase holoenzyme. *Mol Cell Biol* **26:** 2029–2036.

O'Connor CM, Lai CK, Collins K. 2005. Two purified domains of telomerase reverse transcriptase reconstitute sequence-specific interactions with RNA. *J Biol Chem* **280:** 17533–17539.

Palm W, de Lange T. 2008. How shelterin protects mammalian telomeres. *Annu Rev Genet* **42:** 301–334.

Parisien M, Major F. 2008. The MC-Fold and MC-Sym pipeline infers RNA structure from sequence data. *Nature* **452:** 51–55.

Petrova OA, Mantsyzov AB, Rodina EV, Efimov SV, Hackenberg C, Hakanpaa J, Klochkov VV, Lebedev AA, Chugunova AA, Malyavko AN, et al. 2018. Structure and function of the N-terminal domain of the yeast telomerase reverse transcriptase. *Nucleic Acids Res* **46:** 1525–1540.

Pettersen EF, Goddard TD, Huang CC, Couch GS, Greenblatt DM, Meng EC, Ferrin TE. 2004. UCSF Chimera—A visualization system for exploratory research and analysis. *J Comput Chem* **25:** 1605–1612.

Podlevsky JD, Chen JJ. 2016. Evolutionary perspectives of telomerase RNA structure and function. *RNA Biol* **13:** 720–732.

Podlevsky JD, Bley CJ, Omana RV, Qi X, Chen JJ. 2008. The telomerase database. *Nucleic Acids Res* **36:** D339–D343.

Polshakov VI, Petrova OA, Parfenova YY, Efimov SV, Klochkov VV, Zvereva MI, Dontsova OA. 2016. NMR assignments of the N-terminal domain of *Ogataea polymorpha* telomerase reverse transcriptase. *Biomol NMR Assign* **10:** 183–187.

Prakash A, Borgstahl GE. 2012. The structure and function of replication protein A in DNA replication. *Subcell Biochem* **62:** 171–196.

Prathapam R, Witkin KL, O'Connor CM, Collins K. 2005. A telomerase holoenzyme protein enhances telomerase RNA assembly with telomerase reverse transcriptase. *Nat Struct Mol Biol* **12:** 252–257.

Price CM, Boltz KA, Chaiken MF, Stewart JA, Beilstein MA, Shippen DE. 2010. Evolution of CST function in telomere maintenance. *Cell Cycle* **9:** 3157–3165.

Qiao F, Cech TR. 2008. Triple-helix structure in telomerase RNA contributes to catalysis. *Nat Struct Mol Biol* **15:** 634–640.

Qu G, Kaushal PS, Wang J, Shigematsu H, Piazza CL, Agrawal RK, Belfort M, Wang HW. 2016. Structure of a group II intron in complex with its reverse transcriptase. *Nat Struct Mol Biol* **23:** 549–557.

Richards RJ, Theimer CA, Finger LD, Feigon J. 2006a. Structure of the *Tetrahymena thermophila* telomerase RNA helix II template boundary element. *Nucleic Acids Res* **34:** 816–825.

Richards RJ, Wu H, Trantirek L, O'Connor CM, Collins K, Feigon J. 2006b. Structural study of elements of *Tetrahymena* telomerase RNA stem-loop IV domain important for function. *RNA* **12:** 1475–1485.

Robart AR, Collins K. 2011. Human telomerase domain interactions capture DNA for TEN domain-dependent processive elongation. *Mol Cell* **42:** 308–318.

Rosenthal PB, Henderson R. 2003. Optimal determination of particle orientation, absolute hand, and contrast loss in single-particle electron cryomicroscopy. *J Mol Biol* **333:** 721–745.

Rouda S, Skordalakes E. 2007. Structure of the RNA-binding domain of telomerase: Implications for RNA recognition and binding. *Structure* **15:** 1403–1412.

Sarek G, Marzec P, Margalef P, Boulton SJ. 2015. Molecular basis of telomere dysfunction in human genetic diseases. *Nat Struct Mol Biol* **22:** 867–874.

Sauerwald A, Sandin S, Cristofari G, Scheres SH, Lingner J, Rhodes D. 2013. Structure of active dimeric human telomerase. *Nat Struct Mol Biol* **20:** 454–460.

Scheres SH, Chen S. 2012. Prevention of overfitting in cryo-EM structure determination. *Nat Methods* **9:** 853–854.

Schmidt JC, Cech TR. 2015. Human telomerase: Biogenesis, trafficking, recruitment, and activation. *Genes Dev* **29:** 1095–1105.

Schmidt JC, Dalby AB, Cech TR. 2014. Identification of human TERT elements necessary for telomerase recruitment to telomeres. *eLife* **3**. doi: 10.7554/eLife.03563.

Sexton AN, Youmans DT, Collins K. 2012. Specificity requirements for human telomere protein interaction with telomerase holoenzyme. *J Biol Chem* **287:** 34455–34464.

Shastry S, Steinberg-Neifach O, Lue N, Stone MD. 2018. Direct observation of nucleic acid binding dynamics by the telomerase essential N-terminal domain. *Nucleic Acids Res* **46:** 3088–3102.

Shay JW. 2016. Role of telomeres and telomerase in aging and cancer. *Cancer Discov* **6:** 584–593.

Shefer K, Brown Y, Gorkovoy V, Nussbaum T, Ulyanov NB, Tzfati Y. 2007. A triple helix within a pseudoknot is a conserved and essential element of telomerase RNA. *Mol Cell Biol* **27:** 2130–2143.

Singh M, Wang Z, Koo BK, Patel A, Cascio D, Collins K, Feigon J. 2012. Structural basis for telomerase RNA recognition and RNP assembly by the holoenzyme La family protein p65. *Mol Cell* **47:** 16–26.

Singh M, Choi CP, Feigon J. 2013. xRRM: A new class of RRM found in the telomerase La family protein p65. *RNA Biol* **10:** 353–359.

Song Y, DiMaio F, Wang RY, Kim D, Miles C, Brunette T, Thompson J, Baker D. 2013. High-resolution comparative modeling with RosettaCM. *Structure* **21:** 1735–1742.

Stamos JL, Lentzsch AM, Lambowitz AM. 2017. Structure of a thermostable group II intron reverse transcriptase with template-primer and its functional and evolutionary implications. *Mol Cell* **68:** 926–939.e4.

Stewart SA, Weinberg RA. 2006. Telomeres: Cancer to human aging. *Annu Rev Cell Dev Biol* **22:** 531–557.

Stone MD, Mihalusova M, O'Connor CM, Prathapam R, Collins K, Zhuang X. 2007. Stepwise protein-mediated RNA folding directs assembly of telomerase ribonucleoprotein. *Nature* **446:** 458–461.

Sugitani N, Chazin WJ. 2015. Characteristics and concepts of dynamic hub proteins in DNA processing machinery from studies of RPA. *Prog Biophys Mol Biol* **117:** 206–211.

Theimer CA, Feigon J. 2006. Structure and function of telomerase RNA. *Curr Opin Struct Biol* **16:** 307–318.

Theimer CA, Finger LD, Trantirek L, Feigon J. 2003. Mutations linked to dyskeratosis congenita cause changes in the structural equilibrium in telomerase RNA. *Proc Natl Acad Sci* **100:** 449–454.

Theimer CA, Blois CA, Feigon J. 2005. Structure of the human telomerase RNA pseudoknot reveals conserved tertiary interactions essential for function. *Mol Cell* **17:** 671–682.

Theimer CA, Jady BE, Chim N, Richard P, Breece KE, Kiss T, Feigon J. 2007. Structural and functional characterization of human telomerase RNA processing and Cajal body localization signals. *Mol Cell* **27:** 869–881.

Townsley DM, Dumitriu B, Young NS. 2014. Bone marrow failure and the telomeropathies. *Blood* **124:** 2775–2783.

Tribolium Genome Sequencing Consortium, Richards S, Gibbs RA, Weinstock GM, Brown SJ, Denell R, Beeman RW, Gibbs R, Beeman RW, Brown SJ, Bucher G, et al. 2008. The genome of the model beetle and pest *Tribolium castaneum*. *Nature* **452:** 949–955.

Upton HE, Chan H, Feigon J, Collins K. 2017. Shared subunits of *Tetrahymena* telomerase holoenzyme and replication protein A have different functions in different cellular complexes. *J Biol Chem* **292:** 217–228.

Wan B, Tang T, Upton H, Shuai J, Zhou Y, Li S, Chen J, Brunzelle JS, Zeng Z, Collins K, et al. 2015. The *Tetrahymena* telomerase p75-p45-p19 subcomplex is a unique CST complex. *Nat Struct Mol Biol* **22:** 1023–1026.

Wang Y, Feigon J. 2017. Structural biology of telomerase and its interaction at telomeres. *Curr Opin Struct Biol* **47:** 77–87.

Wang H, Gilley D, Blackburn EH. 1998. A novel specificity for the primer-template pairing requirement in *Tetrahymena* telomerase. *EMBO J* **17:** 1152–1160.

Wang Y, Yesselman JD, Zhang Q, Kang M, Feigon J. 2016. Structural conservation in the template/pseudoknot domain of vertebrate telomerase RNA from teleost fish to human. *Proc Natl Acad Sci* **113:** E5125–E5134.

Wegman-Ostrosky T, Savage SA. 2017. The genomics of inherited bone marrow failure: From mechanism to the clinic. *Br J Haematol* **177:** 526–542.

Witkin KL, Collins K. 2004. Holoenzyme proteins required for the physiological assembly and activity of telomerase. *Genes Dev* **18:** 1107–1118.

Wu RA, Collins K. 2014. Human telomerase specialization for repeat synthesis by unique handling of primer-template duplex. *EMBO J* **33:** 921–935.

Wu RA, Tam J, Collins K. 2017a. DNA-binding determinants and cellular thresholds for human telomerase repeat addition processivity. *EMBO J* **36:** 1908–1927.

Wu RA, Upton HE, Vogan JM, Collins K. 2017b. Telomerase mechanism of telomere synthesis. *Annu Rev Biochem* **86:** 439–460.

Wyatt HD, Tsang AR, Lobb DA, Beattie TL. 2009. Human telomerase reverse transcriptase (hTERT) Q169 is essential for telomerase function in vitro and in vivo. *PLoS One* **4:** e7176.

Xi L, Cech TR. 2014. Inventory of telomerase components in human cells reveals multiple subpopulations of hTR and hTERT. *Nucleic Acids Res* **42:** 8565–8577.

Xie M, Podlevsky JD, Qi X, Bley CJ, Chen JJ. 2010. A novel motif in telomerase reverse transcriptase regulates telomere repeat addition rate and processivity. *Nucleic Acids Res* **38:** 1982–1996.

Zaug AJ, Podell ER, Cech TR. 2008. Mutation in TERT separates processivity from anchor-site function. *Nat Struct Mol Biol* **15:** 870–872.

Zaug AJ, Podell ER, Nandakumar J, Cech TR. 2010. Functional interaction between telomere protein TPP1 and telomerase. *Genes Dev* **24:** 613–622.

Zeng Z, Min B, Huang J, Hong K, Yang Y, Collins K, Lei M. 2011. Structural basis for *Tetrahymena* telomerase processivity factor Teb1 binding to single-stranded telomeric-repeat DNA. *Proc Natl Acad Sci* **108:** 20357–20361.

Zhang Q, Kim NK, Peterson RD, Wang Z, Feigon J. 2010. Structurally conserved five nucleotide bulge determines the overall topology of the core domain of human telomerase RNA. *Proc Natl Acad Sci* **107:** 18761–18768.

Zhang Q, Kim NK, Feigon J. 2011. Architecture of human telomerase RNA. *Proc Natl Acad Sci* **108:** 20325–20332.

Zhao C, Pyle AM. 2016. Crystal structures of a group II intron maturase reveal a missing link in spliceosome evolution. *Nat Struct Mol Biol* **23:** 558–565.

Zhong FL, Batista LF, Freund A, Pech MF, Venteicher AS, Artandi SE. 2012. TPP1 OB-fold domain controls telomere maintenance by recruiting telomerase to chromosome ends. *Cell* **150:** 481–494.

Cite as *Cold Spring Harb Perspect Biol* doi: 10.1101/cshperspect.a032383

Structural Basis of Nuclear pre-mRNA Splicing: Lessons from Yeast

Clemens Plaschka, Andrew J. Newman, and Kiyoshi Nagai

MRC Laboratory of Molecular Biology, Cambridge CB2 0QH, United Kingdom

Correspondence: kn@mrc-lmb.cam.ac.uk

SUMMARY

Noncoding introns are removed from nuclear precursor messenger RNA (pre-mRNA) in a two-step phosphoryl transfer reaction by the spliceosome, a dynamic multimegadalton enzyme. Cryo-electron microscopy (cryo-EM) structures of the *Saccharomyces cerevisiae* spliceosome were recently determined in eight key states. Combined with the wealth of available genetic and biochemical data, these structures have revealed new insights into the mechanisms of spliceosome assembly, activation, catalysis, and disassembly. The structures show how a single RNA catalytic center forms during activation and accomplishes both steps of the splicing reaction. The structures reveal how spliceosomal helicases remodel the spliceosome for active site formation, substrate docking, reaction product undocking, and spliceosome disassembly and how they facilitate splice site proofreading. Although human spliceosomes contain additional proteins, their cryo-EM structures suggest that the underlying mechanism is conserved across all eukaryotes. In this review, we summarize the current structural understanding of pre-mRNA splicing.

Outline

Cite as *Cold Spring Harb Perspect Biol* doi: 10.1101/cshperspect.a032391

1 INTRODUCTION

The discovery of "split genes" in adenovirus in 1977 (Berget et al. 1977; Chow et al. 1977) astonished the scientific community and galvanized four decades of intensive research on the underlying mechanism. It became clear early on that the majority of eukaryotic protein-coding genes contain noncoding sequences (introns), which must be removed from the precursor messenger RNA (pre-mRNA) so that the remaining coding sequences (exons) can be ligated to produce the functional mRNA—a process called pre-mRNA splicing. Introns are defined by three conserved sequences, the 5′ splice site (5′SS), the internal branch point (BP) sequence, and the 3′ splice site (3′SS). Early work established that introns are removed from pre-mRNA in two consecutive phosphoryl transfer reactions (Fig. 1A) (reviewed in Will and Lührmann 2011). In the first reaction (branching or step 1), the nucleophilic attack of the BP adenosine 2′-hydroxyl group at the phosphate of the 5′SS produces the free 5′ exon and the lariat intron-3′ exon intermediate. In the second phosphoryl transfer reaction, the 3′-terminal hydroxyl group of the free 5′ exon attacks the phosphate at the 3′SS, producing the ligated exons (mRNA) and a free intron lariat. The intron lariat is a unique structure essential for pre-mRNA splicing, in which the 2′-hydroxyl group of the BP adenosine covalently links to the 5′ phosphate of the first nucleotide of the intron.

This simple two-step reaction is catalyzed by a complex molecular machine known as the spliceosome (Brody and Abelson 1985), which comprises five snRNAs and approximately 70 proteins in the yeast *Saccharomyces cerevisiae* and more than 100 in humans (see Kastner et al. 2019). Each of the five small nuclear RNAs (U1, U2, U4, U5, and U6 snRNAs) organizes with a distinct set of proteins into small nuclear ribonucleoprotein particles (snRNPs) (Lerner and Steitz 1979), whereas other non-snRNP proteins, including the nineteen complex (NTC) and the nineteen-related (NTR) proteins, join during the splicing reaction (Fig. 2) (Will and Lührmann 2011). The RNA active site of the spliceosome is created from U6, U5, and U2 snRNAs after ordered assembly of the snRNPs on the pre-mRNA substrate and extensive conformational and compositional remodeling (Fig. 1B). These events include the dissociation of U1 and U4 snRNAs with their associated proteins, the binding of the NTC and NTR proteins (Chan and Cheng 2005) to form and stabilize the catalytic RNA core, and the binding of specific factors to enable the branching and exon ligation reactions. Eight conserved RNA helicases from the Ski2 (Brr2), DEAD-box (Sub2/UAP56, Prp5, Prp28), and DEAH-box (Prp2, Prp16, Prp22, Prp43) families promote the transitions between the different spliceosome states by remodeling the protein–RNA or RNA–RNA networks and by proofreading the selection of the correct intron sequences (Fig. 1C) (Burgess and Guthrie 1993; Semlow et al. 2016). At least 10 distinct spliceosome states can be defined during the splicing reaction, the E, A, pre-B, B, B^{act}, B*, C, C*, P, and intron lariat spliceosome (ILS) complexes, which differ in their RNA or protein composition and/or the state of the pre-mRNA substrate (Will and Lührmann 2011).

A detailed understanding of the molecular mechanism of splicing requires structural knowledge of the spliceosome in its various states. Until recently, structural information on the spliceosome was limited to crystal structures of proteins and RNA–protein complexes (Will and Lührmann 2011), as well as low-resolution electron microscopic images of the snRNPs in specific spliceosome states (reviewed in Stark and Lührmann 2006). Because of recent developments in cryo-electron microscopy (cryo-EM), the structures of biological macromolecules can now often be determined to near-atomic resolution (Fernandez-Leiro and Scheres 2016). This is largely due to the development of direct electron detectors, providing high-quality movies of the biological specimen, and to improved image processing software. In particular, the classification of complex mixtures into structurally homogeneous subsets in programs such as RELION has had a major impact on the determination of spliceosome structures (Wilkinson et al. 2018). The spliceosome is inherently flexible, and some of its domains undergo a continuous motion. Focused classification methods together with partial signal subtraction and focused refinement have enabled us to bring even highly mobile parts of the spliceosome into focus. These new structures are built on a foundation of decades of genetic and biochemical work, that established many of the methods to stall the spliceosome in its various states. The structures of the *S. cerevisiae* U4/U6.U5 tri-snRNP (Nguyen et al. 2015) and *Schizosaccharomyces pombe* ILS (Yan et al. 2015), both reported in 2015, presaged a flood of high-resolution cryo-EM structures of yeast and human spliceosomes. The resulting structures have substantially advanced our knowledge of splicing. They not only confirm and explain decades of functional observations, but they also dramatically enhance our understanding of the underlying molecular mechanism, suggesting a high conservation of the splicing mechanism across all eukaryotes. In this review, we will focus on the mechanistic insights gained from structural studies of the *S. cerevisiae* spliceosome and make only brief references to the human spliceosome, which is discussed in more depth in Kastner et al. (2019) and Yan et al. (2019). For details on cryo-EM and its application to the spliceosome, we refer the interested reader to other recent reviews (e.g., Wilkinson et al. 2018). PyMOL sessions of the yeast spliceosome structures discussed in this review

Cite as *Cold Spring Harb Perspect Biol* doi: 10.1101/cshperspect.a032391

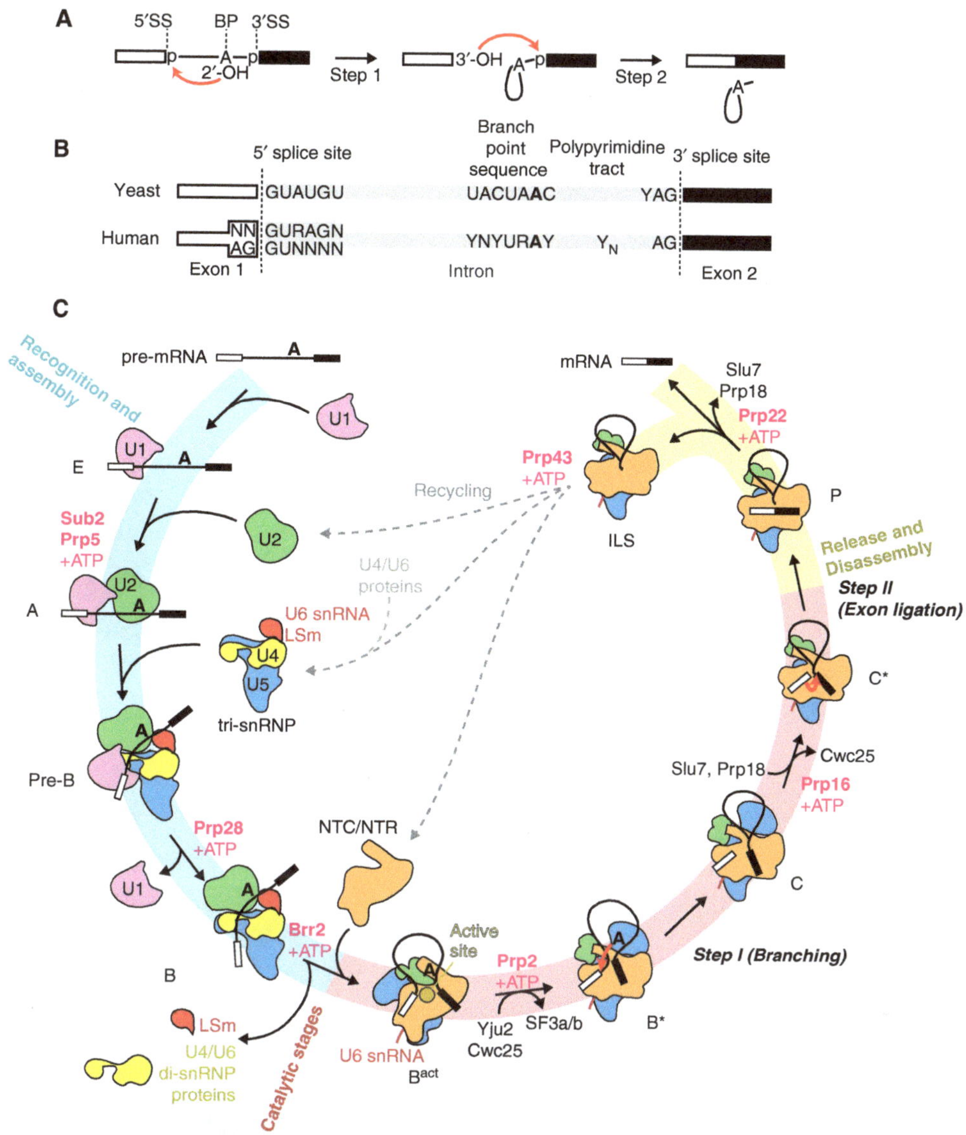

Figure 1. Schematic overview of the splicing reaction and the cycle. (*A*) Removal of introns from precursor messenger RNA (pre-mRNA) by the two-step phosphoryl transfer reaction: branching and exon-ligation reactions. (*B*) Yeast and human introns are identified by short conserved sequences at the 5′ and 3′ splice sites (5′SS and 3′SS) and the internal branch point (BP) sequence. (*C*) Schematic representation of the yeast splicing cycle. The spliceosome has been isolated in at least 10 distinct states, namely, E, A, pre-B, B, B^{act}, B*, C, C*, P, and intron lariat spliceosome (ILS) complexes. Transitions between these states are regulated by helicases or ATPases in the Ski2-like (Brr2), DEAD- (Sub2/UAP56, Prp5, Prp28) and DEAH-box (Prp2, Prp16, Prp22, Prp43) families. The 5′SS and BP sequence are recognized by the U1 and U2 small nuclear ribonucleoprotein particles (snRNPs) and the resulting A complex associates with the preassembled U4/U6.U5 tri-snRNP to form the fully assembled pre-B complex. The U1 snRNP dissociates from the pre-B complex by the activity of the DEAD-box helicase Prp28 to form the precatalytic B complex. The Brr2 helicase then unwinds the U4/U6 small nuclear RNA (snRNA) duplex allowing U4 snRNA to dissociate together with the U4/U6 di-snRNP proteins and the LSm proteins. U6 snRNA freed from U4 snRNA folds and associates with part of U2 snRNA to create the group II intron-like active site harboring the catalytic magnesium ions. The nineteen complex (NTC) and nineteen-related (NTR) proteins associate to stabilize the active site RNA structure in B^{act} complex. The U2 snRNP SF3a and SF3b complexes and the RES complex dissociate by the activity of the Prp2 helicase inducing the binding of the Prp16 helicase and the step 1 (branching) factors, Yju2, Isy1, and Cwc25, which dock the branch helix into the active site. In the B* complex, the BP adenosine 2′-OH group attacks the 5′SS producing the 5′exon and the lariat intron intermediate leading to the formation of the C complex. By its ATPase activity, the Prp16 helicase dissociates together with the step 1 factors, allowing the binding of the Prp22 helicase and step 2 factors, Prp18 and Slu7. In the resulting C* complex, the 3′SS can be docked into the active site enabling the 3′-OH group of the 5′ exon to attack the 3′SS, producing the ligated exons and the lariat intron. The Prp22 helicase activity releases the ligated exon (mRNA) from the postsplicing P complex and the remaining ILS is disassembled by the activity of the Prp43 helicase and the snRNPs are recycled for the next round of splicing.

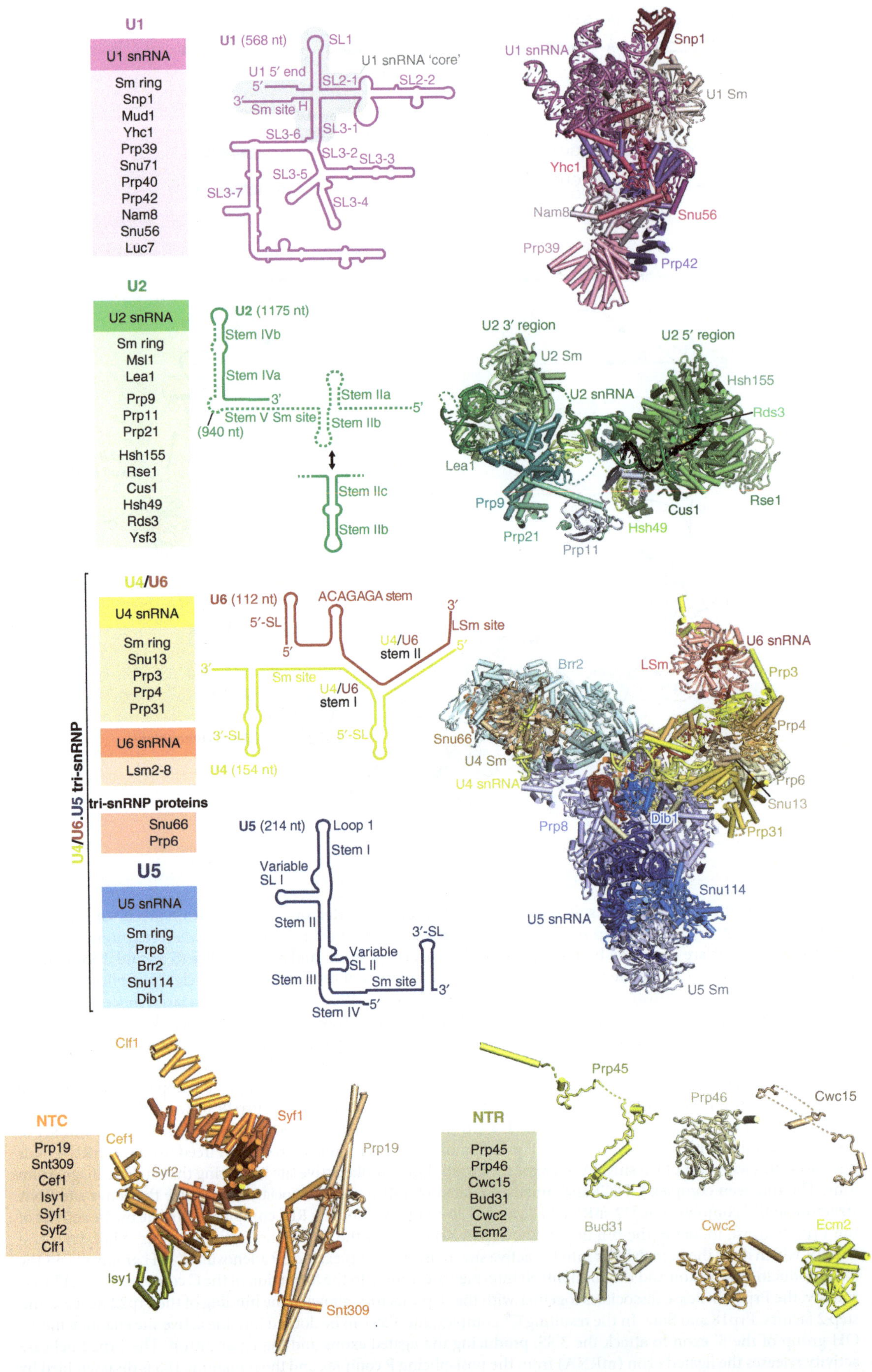

Figure 2. (*Legend on following page.*)

are available online (see www2.mrc-lmb.cam.ac.uk/groups/nagai/resources/).

2 SPLICE SITE RECOGNITION BY THE U1 AND U2 snRNPs AND PRESPLICEOSOME FORMATION

Much of our current understanding of the mechanism of pre-mRNA splicing is derived from genetic experiments using *S. cerevisiae* (hereafter termed "yeast," unless otherwise specified) and biochemical experiments using yeast whole cell extract or HeLa cell nuclear extract (Will and Lührmann 2011). *S. cerevisiae* contains approximately 5500 protein-coding genes of which approximately 300 contain introns. The human genome contains approximately 20,000 protein-coding genes, and the majority of these genes contain multiple introns, which are often alternatively spliced (Will and Lührmann 2011). In yeast, the 5′SS and BP sequences are stringently conserved to be /GUAUGU (/ is the splice junction) and UACUA**A**C (BP adenosine is bold) with few exceptions (Fig. 1B) (Parker et al. 1987; Spingola et al. 1999). In humans, most 5′SS sequences fall into two major classes: one with a G/GU(A or G)AG and the other AG/GUA (/ denotes the splice junction) consensus sequence. The metazoan BP sequence is YNCURAC (Will and Lührmann 2011) and precedes the 3′SS by 30–40 nucleotides. In both yeast and humans, the 3′SS is defined as YAG/ in all introns except those processed by the metazoan minor spliceosome (Tarn and Steitz 1996). In metazoans, the YAG is preceded by a poly-pyrimidine tract.

In the first step of spliceosome assembly, the U1 snRNP binds to the pre-mRNA 5′SS by base-pairing with the U1 snRNA 5′end ($m_3^{2,2,7}$GpppAUACΨΨACCUG- - -; Ψ, pseudo-uridine), which is conserved in most species including yeast and humans (see rfam.xfam.org/). The human U1 snRNP contains the seven-membered Sm ring, which also binds the U-rich tract of U2, U4, and U5 snRNAs, and three U1-specific proteins (U1-70k, U1A, and U1C) (Will and Lührmann 2011). In *S. cerevisiae*, U1 snRNA is expanded and the U1 snRNP contains seven additional yeast-specific proteins (Gottschalk et al. 1998), which, except for Snu56, have homologs in humans, where they function as alternative splicing factors (Fig. 3A) (Gottschalk et al. 1998; Li et al. 2017). The crystal structures of the recombinant U1 snRNP (Pomeranz Krummel et al. 2009) and the chymotrypsin-treated native U1 snRNP (Weber et al. 2010) together provided the complete image of the human U1 snRNP. The crystal structure of the functional core of U1 snRNP bound to a 5′SS RNA oligonucleotide showed that the Zn-finger domain of the U1C protein stabilizes 5′SS binding by making contacts with the sugar-phosphate backbone of the U1–5′SS duplex but not the bases (Fig. 3B) (Kondo et al. 2015). This suggests that 5′SSs are predominantly selected by the stability of the U1–5′SS base-pairing. A recent cryo-EM structure of the free yeast U1 snRNP (Li et al. 2017) shows that its functional core is highly similar to the human U1 snRNP crystal structure, whereas the yeast-specific RNA expansion is stabilized by yeast-specific proteins (Fig. 3A). The 5′ end of U1 snRNA is disordered in the free yeast U1 snRNP but it becomes ordered when it base-pairs with the 5′SS of pre-mRNA. The A complex structure (Plaschka et al. 2018) shows that 5′SS binding induces ordering of Luc7 and flexible parts of the U1C protein by making contacts with the duplex formed between the 5′SS and the 5′ end of U1 snRNA (Fig. 3C). Homologs of Luc7 in humans function as alternative splicing factors and thus the structure of the 5′SS bound yeast U1 snRNP suggests how human homologs of Luc7 stabilize binding of weak 5′SSs (Fig. 3D).

After 5′SS recognition, the BP sequence is recognized by the Msl5(BBP)–Mud2(U2AF65) heterodimer in yeast, whereby Msl5 binds to the branch site and interacts with U1 snRNP, bridging the interaction between the 5′SS and the branch site (Berglund et al. 1998). The DEAD-box helicases Sub2 and Prp5 then replace Msl5-Mud2 with the U2 snRNP and promote the interaction between U1 and U2 snRNPs to form the A complex (Shao et al. 2012). The region of U2 snRNA between stems I and IIa forms the branchpoint-interacting stem loop (BSL), in which the BP binding sequence is located in the loop (Perriman and Ares 2010). Prp5 associates with U2 snRNP by interacting with the BSL and is released upon the base-pairing of U2 with the branch site to allow the recruitment of the tri-snRNP (Liang

Figure 2. (*Continued.*) Composition of the spliceosome. Protein composition (*left*), small nuclear RNA (snRNA) secondary structure (*middle*), and three-dimensional structure (*right*) of yeast spliceosome components as they assemble on precursor messenger RNA (pre-mRNA) substrates. Each of the five small nuclear ribonucleoprotein particles (snRNPs) is shown with shades of distinct colors (U1, magenta; U2, green, U4, yellow; U5, blue; U6, red). For each snRNA, important secondary structure elements are labeled. The structure of the isolated U2 snRNP remains unknown, and in its stead, we show the yeast U2 snRNP structure bound to pre-mRNA (black) (yeast B complex, PDB 5NRL). For the nineteen complex (NTC) and nineteen-related (NTR), the three-dimensional structures found in the C complex are shown. The NTC forms a complex in isolation (Chan and Cheng 2005), whereas the NTR proteins are likely to join the spliceosome separately. The yeast U1 snRNP (PDB 5UZ5), the yeast U2 snRNP (PDB 5NRL), yeast tri-snRNP (PDB 5NRL), and the NTC and NTR in the B^{act} complex (PDB 5GM6) structures are shown. SL, stem loop.

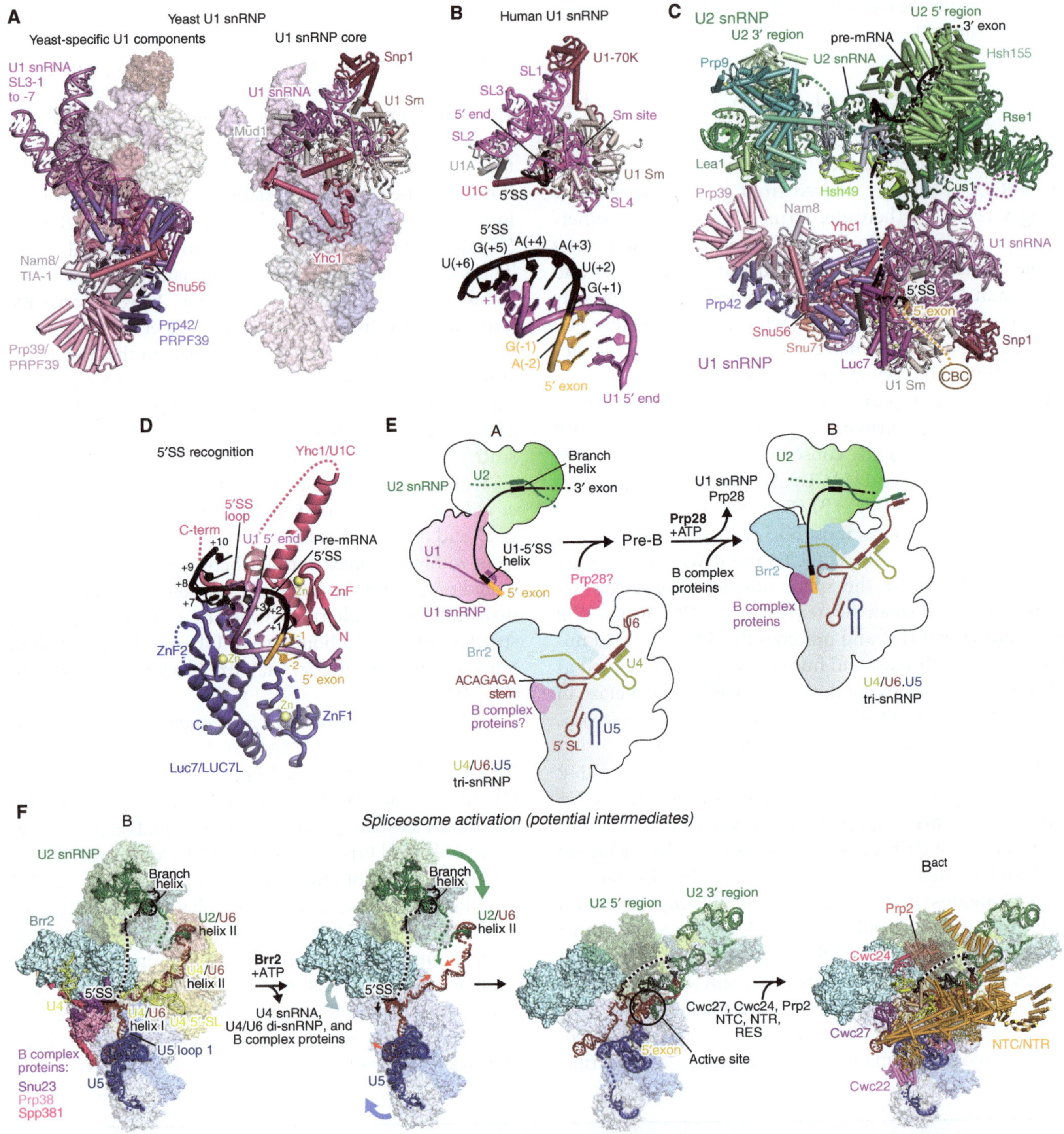

Figure 3. Spliceosome assembly and activation. (*A*) The yeast U1 small nuclear ribonucleoprotein particle (snRNP) structure (PDB 5UZ5) contains the conserved core of the U1 snRNP (*right*) and additionally contains yeast-specific RNA elements and eight proteins (*left*). (*B*) 5′ splice site (5′SS) recognition by the U1 small nuclear RNA (snRNA) 5′ end in the human minimal U1 snRNP crystal structure (PDB 4PJO). (*C*) Yeast A complex assembled on *UBC4* precursor messenger RNA (pre-mRNA) (PDB 6G90). (*D*) The RNA duplex formed between the 5′SS and the 5′ end of U1 snRNA interacts with Yhc1(U1C) and Luc7 in yeast A complex. The 5′ end of U1 snRNA and Luc7, completely disordered in the free U1 snRNP (Li et al. 2017), become ordered when pre-mRNA 5′SS binds to U1 snRNP to form the A complex (Plaschka et al. 2018). (*Legend continues on following page.*)

 Cite as *Cold Spring Harb Perspect Biol* doi: 10.1101/cshperspect.a032391

and Cheng 2015). The stability of the BSL is fine-tuned such that mutations in the stem affect the binding of suboptimal BP sequences. In *S. pombe*, Prp5 has been shown to interact directly with the SF3b HEAT repeat protein SF3B1 (Hsh155) and similar interactions may occur in other species. Mutations in SF3B1 are found in human myelodysplastic syndromes (MDSs) and other cancers. The equivalent mutations in Hsh155 in *S. cerevisiae* specifically change splicing of pre-mRNA substrates with suboptimal BP sequence (Tang et al. 2016). Mutations in Hsh155 and Prp5 change the direct physical interaction between Hsh155 and Prp5 and influence the loading (i.e., "fidelity") of the BS–U2 duplex into the SF3B complex during prespliceosome formation, thereby influencing branch site selection. The complete U2 snRNP structure with the bound BP sequence is now available as part of the A and B complexes (Plaschka et al. 2017, 2018). Consistent with negative stain EM analysis of the U2 snRNP, it has a bipartite structure consisting of the 5′ domain comprising SF3b (Rauhut et al. 2016; Yan et al. 2016; Plaschka et al. 2017) and the U2 snRNA 5′ end, as well as the 3′ region comprising the 3′ end of U2 snRNA, the SF3a subcomplex, the U2 Sm ring, and the U2B″(Msl1 in yeast)-U2A′(Lea1 in yeast) dimer (Plaschka et al. 2017). The BP sequence pairs with U2 snRNA to form the branch helix, which wedges into the ends of the horseshoe-shaped Hsh155 HEAT repeats, as discussed in more detail below.

3 U4/U6.U5 tri-snRNP

The U4/U6.U5 tri-snRNP is the largest preassembled spliceosomal complex and comprises the U5 snRNP, the U4/U6 di-snRNP, and tri-snRNP-specific proteins. In the U4/U6 di-snRNP, the U4 and U6 snRNAs are extensively base-paired to form a three-way junction, stabilized in a compact structure by di-snRNP proteins (Snu13, Prp31, Prp3, and Prp4) (Fig. 2) (Nguyen et al. 2015; 2016; Wan et al. 2016a). Only the U5 snRNP and U6 snRNA remain from the tri-snRNP after spliceosome activation and become part of the catalytic state of the spliceosome. The U5 snRNP comprises U5 snRNA, Prp8, Brr2, Snu114, and the U5 Sm ring.

Prp8 is the largest and most highly conserved protein in the spliceosome and contains four structurally distinct domains (amino-terminal, Large, RNaseH and Jab1/MPN domains), connected by flexible linkers (Fig. 4A) (Galej et al. 2013; Nguyen et al. 2015; 2016; Yan et al. 2015). These four domains extensively rearrange at different stages of splicing to organize the catalytic RNA core and regulate the docking and undocking of the substrate pre-mRNA (Fig. 5A,B) (reviewed in Fica and Nagai 2017; Shi 2017). The Prp8 Large domain consists of the helix bundle (HB), reverse transcriptase, Linker and endonuclease (Endo) domains, reminiscent of the group II intron encoded protein (maturase) and suggesting their common evolutionary origin (Galej et al. 2013; Zhao and Pyle 2016). Prp8 cross-links with each of the three splicing reactants in the pre-mRNA substrate (5′SS, BP, and 3′SS) and hence was thought to be at "the heart of the spliceosome" (Grainger and Beggs 2005). This has now been confirmed by the spliceosome structures. The active site cavity of the spliceosome was mapped to the Linker domain in the Prp8 crystal structure based on the Prp8–pre-mRNA crosslinks (Galej et al. 2013). The pre-mRNA branch point (BP+2) nucleotide was cross-linked to the disordered loop in the Linker domain, between residues 1575 and 1598, pinpointing the location of the active site (Galej et al. 2013). This region forms a helix-turn-helix in the yeast U4/U6.U5 tri-snRNP and was hence named the "α-finger" (Nguyen et al. 2016), although the α-finger region alters its structure in different states of the active site.

Brr2 contains two Ski2-type helicase domains. The crystal structure of human Brr2 showed how the enzymatically active amino-terminal helicase domain forms a tight complex with the enzymatically inactive carboxy-terminal domain (Fig. 4B) (Mozaffari-Jovin et al. 2013). Brr2 forms a stable complex with the Jab1 domain of Prp8 and is therefore a constituent subunit of the U5 snRNP. Brr2 remains a part of every spliceosome complex, unlike all other spliceosome helicases that bind only transiently. The crystal structures of human and yeast Brr2 in complex with the Prp8

Figure 3. (*Continued*) (*E*) Schematic model of early assembly events of the spliceosome. The A complex associates with the U4/U6.U5 tri-snRNP to form the pre-B complex, which is subsequently converted by Prp28 activity into the B complex spliceosome. For more details, see Plaschka et al. (2018). (*F*) Model for spliceosome activation, the B to B^{act} complex transition. Possible steps during activation are shown, starting with the cryo-electron microscopy (cryo-EM) structure of the yeast B complex (PDB 5NRL) and ending with the B^{act} complex (PDB 5GM6). During spliceosome activation, U4 snRNA and at least 22 proteins dissociate, 24 proteins join, and the RNA-based active site containing two catalytic magnesium ions forms. First, the Brr2 helicase unwinds the U4/U6 snRNA duplex to initiate the release of U4 snRNA and U4/U6 di-snRNP-specific, tri-snRNP-specific, and B complex proteins. After U4 snRNA dissociation, U6 snRNA folds to form the internal stem loop (ISL) and associates with part of U2 snRNA to form the catalytic RNA core. The nineteen complex (NTC) and nineteen-related (NTR) proteins and the B^{act}-specific proteins join to cradle the active site RNA core, which remains largely unchanged during the branching and exon-ligation reactions.

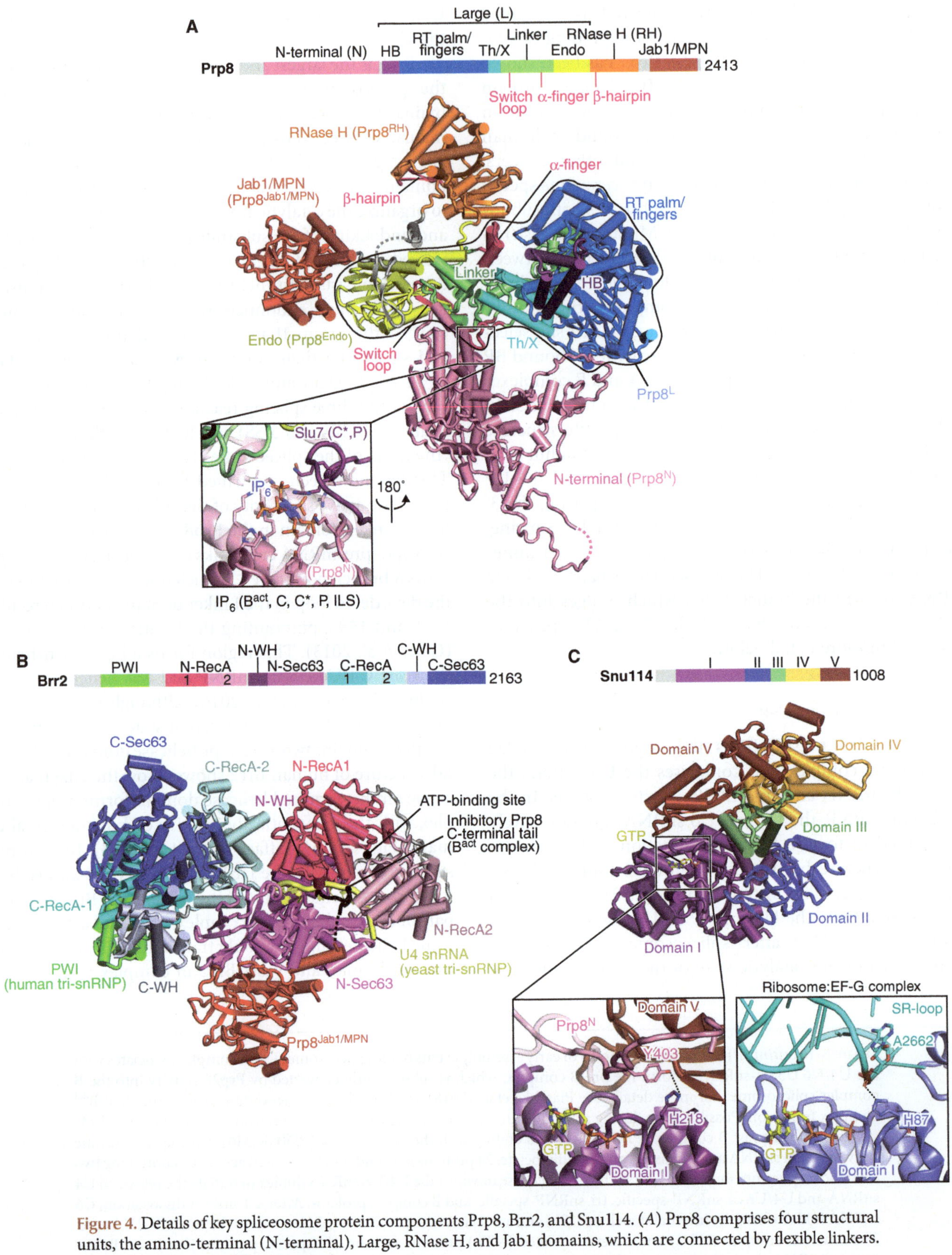

Figure 4. Details of key spliceosome protein components Prp8, Brr2, and Snu114. (*A*) Prp8 comprises four structural units, the amino-terminal (N-terminal), Large, RNase H, and Jab1 domains, which are connected by flexible linkers. (*Legend continues on following page.*)

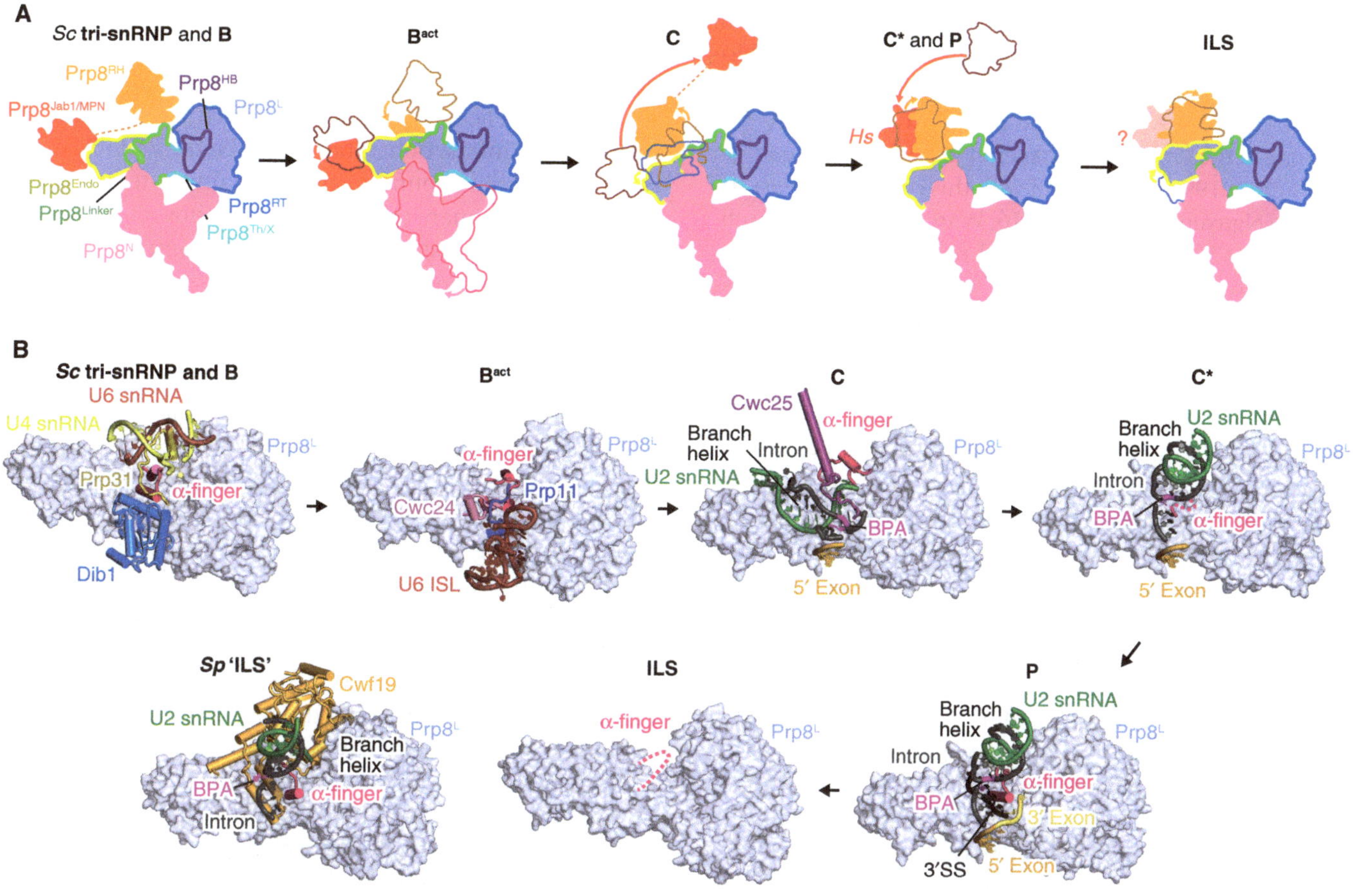

Figure 5. Prp8 rearrangements throughout the splicing cycle. (*A*) The four Prp8 domains (amino-terminal, Large, RNase H, and Jab) rearrange throughout the splicing cycle. All available spliceosome cryo-electron microscopy (cryo-EM) structures were aligned on the Prp8 Large domain and relative movements of the other domains are shown with a cartoon schematic. Although the amino-terminal domain moves only during the B to B^{act} transition, to clamp the 5′ exon, the RNase H and Jab1 domains move extensively between states. Thereby they contribute to the stabilization of exchanging spliceosome proteins (RNase H, Jab1) and movements of the branch helix (RNase H). See Figures 6 and 7. (*B*) Structural remodeling of the Prp8 α-finger throughout the splicing cycle. The Prp8 α-finger lies at the heart of the active site, where it changes its conformation to help organize the active site during multiple steps of splicing. See Figures 6 and 7.

Figure 4. (*Continued*) These domains change relative positions during the splicing cycle. The Large domain (encircled in black) contains the helical bundle (HB), reverse transcriptase (RT), thumb/X, linker, and endonuclease (En) domains and is reminiscent of the group II intron-encoded protein (IEP), which, however, lacks the amino-terminal HB domain. The catalytic RNA core of the spliceosome forms in the space between the Linker and HB domains. The *inset* shows the binding site for an IP_6 molecule, which is present in all catalytic stage spliceosome structures determined to date (yeast and human). In C* and P complexes (yeast), Slu7 is in part stabilized by binding the IP_6. (*B*) The Brr2 helicase comprises an amino-terminal region with a PWI domain and two Ski2-like helicase cassettes, where only the amino-terminal cassette is catalytically active. The Prp8 Jab1 domain wedges between the two helicase cassettes, anchoring Brr2 to the spliceosome. The Prp8 carboxyl terminus (red) forms an extended peptide, which can bind the amino-terminal helicase cassette and has been proposed to inhibit the binding of RNA substrates (Mozaffari-Jovin et al. 2013). (*C*) Snu114 is homologous to the translation translocases EF-G (bacterial) and eEF2 (eukaryotic). The *inset* shows the GTPase active sites of Snu114 and EF-G (PDB 4JUW), respectively. The catalytic His218 in Snu114 is hydrogen-bonded to Tyr403 in the amino-terminal domain of Prp8 and, unlike its EF-G counterpart, is not poised for GTP hydrolysis. Further, the Snu114 His218 is not essential for yeast viability (Nguyen et al. 2016) and its configuration is unchanged in all spliceosome structures determined, suggesting that if GTP hydrolysis occurs, it may instead play a role in spliceosome biogenesis. C-terminal, carboxy-terminal.

Jab1 domain show that the Jab1 domain is wedged between the amino- and carboxy-terminal helicase domains (Mozaffari-Jovin et al. 2013; Nguyen et al. 2013). Brr2 unwinds the U4/U6 snRNA duplex and plays a crucial role in spliceosome activation (Laggerbauer et al. 1998; Raghunathan and Guthrie 1998). Consistent with this role, in the yeast tri-snRNP (Nguyen et al. 2015; 2016; Wan et al. 2016a; Plaschka et al. 2017) the single-stranded region of U4 snRNA is loaded into the active site of the amino-terminal helicase cassette, ready to unwind the U4/U6 snRNA duplex to activate the spliceosome (Laggerbauer et al. 1998; Raghunathan and Guthrie 1998).

The U5 snRNP additionally contains Snu114, which is homologous to the ribosome translocases EF-G (prokaryotes) and EF-2 (eukaryotes) (Fig. 4C) (Fabrizio et al. 1997). Because of this homology, it was suggested that Snu114 induces conformational changes in the spliceosome upon GTP hydrolysis, as EF-G does during ribosome translocation. Biochemical experiments suggested that Snu114 may trigger conformational changes of the spliceosome on GTP binding (Small et al. 2006) or hydrolysis (Bartels et al. 2003) to activate Brr2. However, in all the available spliceosome cryo-EM structures GTP is bound to Snu114, except for the *S. pombe* intron lariat spliceosome (ILS) structure, in which GDP was modelled (Yan et al. 2015). However, close inspection of the density suggests GTP may be a better fit. In EF-G the catalytic His places a water molecule near the GTP γ-phosphate. The equivalent His-218 in Snu114 is hydrogen-bonded to Tyr-403 in the Prp8 amino-terminal domain. Neither mutation of His-218 nor Tyr-403 affect yeast viability or growth, showing that Snu114-bound GTP is not hydrolyzed during the splicing cycle (Nguyen et al. 2016). In contrast, mutations in the P-loop of Snu114 which abolish GTP binding are lethal (Bartels et al. 2003) showing GTP binding is important. The ternary complex between Prp8, Snu114, and Aar2, coexpressed in yeast, also contains GTP (L Strittmatter, W Galej, and K Nagai, unpubl.) and it remains possible that GTP binding or hydrolysis plays a role during U5 snRNP biogenesis.

In the yeast tri-snRNP (Nguyen et al. 2016), the U6 ACAGAGA sequence, which pairs with the 5′SS sequence during activation (Lesser and Guthrie 1993; Kandels-Lewis and Séraphin 1993), is occluded by forming a short stem loop (ACAGAGA stem) to prevent premature interactions with the 5′SS (Nguyen et al. 2016). The structure of the human U4/U6.U5 tri-snRNP (Agafonov et al. 2016) is substantially different from its yeast counterpart and the Brr2–Prp8 Jab1 domain complex is anchored near Snu114 by Sad1, in which the conserved Brr2 PWI domain contacts the Sad1 protein (Agafonov et al. 2016). This interaction keeps the Brr2 helicase away from its U4 snRNA substrate to prevent premature unwinding of the U4/U6 snRNA duplex. A similar tri-snRNP conformation may exist in yeast, in which Sad1 transiently interacts with the tri-snRNP and is essential for splicing (Huang et al. 2014). The Prp28 DEAD-box helicase facilitates the disruption of the U1 snRNP–5′SS interaction (Staley and Guthrie 1999; Chen et al. 2001) and is bound to the Prp8 amino-terminal domain in the human tri-snRNP (Agafonov et al. 2016), whereas in the yeast tri-snRNP the B complex proteins replace Prp28 at the same binding site.

4 SPLICEOSOME ASSEMBLY

The U4/U6.U5 tri-snRNP associates with the A complex to form the fully assembled spliceosome, the pre-B complex (Fig. 3C) (Boesler et al. 2016). The U2, U5, and U6 snRNAs, which form the catalytic RNA core during activation, as well as the branching substrates (5′SS and BP) are thus brought together into one assembly. The pre-B complex forms in an ATP-independent manner and is held together predominantly by an RNA–RNA (U2/U6 helix II) interaction (Fig. 3C) (Will and Lührmann 2011). In the subsequent step, the helicase Prp28 facilitates 5′SS transfer from the U1 snRNP to the U6 snRNA ACAGAGA sequence, forming the so-called ACAGAGA helix (Kandels-Lewis and Séraphin 1993; Lesser and Guthrie 1993). In the presence of an inactive Prp28 mutant, the human splicing reaction stalls after pre-B complex formation and just before 5′SS transfer. Negative-stain electron microscopy of the Prp28-stalled human pre-B complex revealed a morphology distinct from the B complex, which forms after 5′SS transfer (Boesler et al. 2016). Thus, the U1 and U2 snRNPs may be dynamically associated with tri-snRNP in the pre-B complex. On addition of a 5′SS RNA oligonucleotide, the pre-B complex is transformed into a B complex-like structure, suggesting that dissociation of the U1 snRNP or pairing of the U6 ACAGAGA region with the 5′SS leads to a large conformational rearrangement in the human pre-B complex (Boesler et al. 2016). In the human tri-snRNP, Brr2 bound to the Prp8 Jab1 domain is kept inactive through its location near Snu114. In the yeast tri-snRNP, the Brr2 helicase is similarly bound to the Prp8 Jab1 domain (Mozaffari-Jovin et al. 2013; Nguyen et al. 2013) but is located near the Prp8 Endo domain and the single-stranded region of U4 snRNA is already loaded in the active site of the amino-terminal Brr2 helicase domain (Nguyen et al. 2015; 2016; Wan et al. 2016a). Several different mechanisms have been proposed to prevent premature unwinding of the U4/U6 snRNA duplex in the yeast tri-snRNP (Mozaffari-Jovin et al. 2013). In both the yeast and human B-complex structures (Bertram et al. 2017a; Plaschka et al. 2017), Brr2 is found in a similar location on the U4 snRNA substrate as in the isolated yeast tri-snRNP (Nguyen et al. 2015; 2016; Wan et al. 2016a),

Cite as *Cold Spring Harb Perspect Biol* doi: 10.1101/cshperspect.a032391

suggesting that spliceosome activation proceeds through similar mechanisms. The timing and mechanism of Sad1 release from human pre-B complex are currently unknown.

The pre-B complex is transformed into the B complex, following Prp28 helicase activity and U1 snRNP release. In the human B complex structure, which was stalled with low Mg^{2+} ion concentration, the U6 ACAGAGA box is paired with the 5′SS to accommodate the 5′ exon in the "exon channel" (Bertram et al. 2017a). In contrast, in the yeast B complex structure, stalled at low ATP concentration, the 5′SS instead interacts with U6 snRNA upstream of the ACAGAGA box, in agreement with cross-linking results (Sawa and Abelson 1992; Chan and Cheng 2005). This suggests that the yeast B complex was stalled at a slightly earlier stage than its human counterpart. In both B complex structures and in the yeast tri-snRNP (Nguyen et al. 2016) the so-called B complex proteins (Snu23, Prp38, and Spp381/human MFAP1) bind at the Prp8 amino-terminal domain to stabilize the 5′SS–U6 snRNA interactions and to guide Brr2 onto its U4 snRNA substrate (Fig. 3D). In the yeast B complex structure (Plaschka et al. 2017), the Snu23 amino terminus reaches near the active amino-terminal Brr2 helicase cassette, and this may couple the productive 5′SS-U6 snRNA interaction with Brr2-mediated spliceosome activation.

5 SPLICEOSOME ACTIVATION

The cryo-EM structures of yeast B and B^{act} complexes provide important insights into the mechanism of spliceosome activation, which results in the formation of the RNA catalytic core of the spliceosome (Fig. 3D) (Rauhut et al. 2016; Yan et al. 2016; Plaschka et al. 2017). During activation, U4 snRNA and its associated U4/U6 di-snRNP-specific and tri-snRNP-specific proteins dissociate, and the active spliceosome forms around the remaining U5 snRNP components together with the newly associating NTC and NTR proteins. Spliceosome activation begins with translocation of Brr2 along U4 snRNA in a 3′ to 5′ direction. This disrupts the U4/U6 stem I, and U4/U6 stem II may also be unwound, although this remains unclear (Laggerbauer et al. 1998; Raghunathan and Guthrie 1998). During this process, at least 24 proteins are released (U4 snRNP proteins, the LSm ring, Dib1, Snu66, and Prp6) and 22 proteins associate (RES, NTC and NTR complexes, and the B^{act}-specific proteins Cwc24, Cwc27). U6 snRNA, freed from U4 snRNA, forms an internal stem loop (ISL), whereas the adjacent U6 snRNA sequence pairs with U2 snRNA to form U2/U6 helix Ia and Ib (Madhani and Guthrie 1992), producing the active center of the spliceosome (Figs. 3D, 6A–D) (Madhani and Guthrie 1992; Steitz and Steitz 1993; Fica et al. 2013; Rauhut et al. 2016; Yan et al. 2016). Three consecutive triple base pairs (catalytic triplex) in the RNA-based active site coordinate the two catalytic magnesium ions (Fig. 6D) (Steitz and Steitz 1993; Fica et al. 2013; Yan et al. 2015). This RNA structure is reminiscent of the group II intron active site, consistent with the idea that both splicing systems arose from a common ancestor (Toor et al. 2008).

In the B^{act} structure, the catalytic RNA core forms and is stabilized by the NTR/NTC proteins Cwc2, Bud31, Cef1, Ecm2, Clf1, and Syf2 (Rauhut et al. 2016; Yan et al. 2016). The structure of the catalytic RNA core "enzyme" remains unchanged throughout the catalytic phase of the splicing cycle (Fica and Nagai 2017; Shi 2017). The ACAGAGA helix, formed between the U6 snRNA ACAGAGA box and the GUAUGU sequence at the 5′ end of the intron, positions the exon–intron junction in a curved conformation to expose the scissile phosphate for nucleophilic attack by the BP adenosine (Fig. 6E). In the B^{act} structure (Rauhut et al. 2016; Yan et al. 2016), the BP adenosine is located 50 Å away from the 5′SS, and the SF3a subunit Prp11 and B^{act}-specific protein Cwc24 together protect the scissile bond at the 5′SS (Fig. 6A,B). The 5′ exon base-pairs with the conserved U5 snRNA loop 1 (Newman and Norman 1992; Sontheimer and Steitz 1993). The adjacent 5′ exon is loaded in the "exon channel" between the Prp8 amino-terminal and Large domains of Prp8, where it is stabilized by Cwc21 and Cwc22 and clamped by the Prp8 amino-terminal and Large domains.

The gap between the Prp8 amino-terminal and Large domains is wider in the B complex such that the 5′ exon can slide into this gap to replace the 5′ end of U6 snRNA (Plaschka et al. 2017). In the B^{act} complex, the gap between the two Prp8 domains closes, presumably as the result of NTC and NTR protein binding, and the switch loop of Prp8 (residues 1615–1639) together with Cwc21 fills the gap to clamp the 5′ exon between the two Prp8 domains (Rauhut et al. 2016; Yan et al. 2016) until the exon-ligation reaction is complete (Figs. 3D, 5A, 6B) (Bai et al. 2017; Liu et al. 2017; Wilkinson et al. 2017). In yeast, the GUAUGU sequence at the 5′ end of the intron is stringently conserved and the UGU within this sequence forms Watson–Crick base pairs with ACA of the ACAGAGA box (Kandels-Lewis and Séraphin 1993; Lesser and Guthrie 1993) and nucleates formation of the flanking non-Watson–Crick base pairs (Galej et al. 2016; Wan et al. 2016b). In humans, the 5′SS sequence is much more degenerate and only the G/GU sequence is stringently conserved. Therefore, in humans the pairing between the 5′SS and the ACAGAGA box (Will and Lührmann 2011) is unlikely to be driven by Watson–Crick pairs. This suggests that the U1 snRNP delivers the bound 5′SS directly to its site of interaction with U6, to load the 5′ exon into the gap between the amino-terminal and Large domains of Prp8.

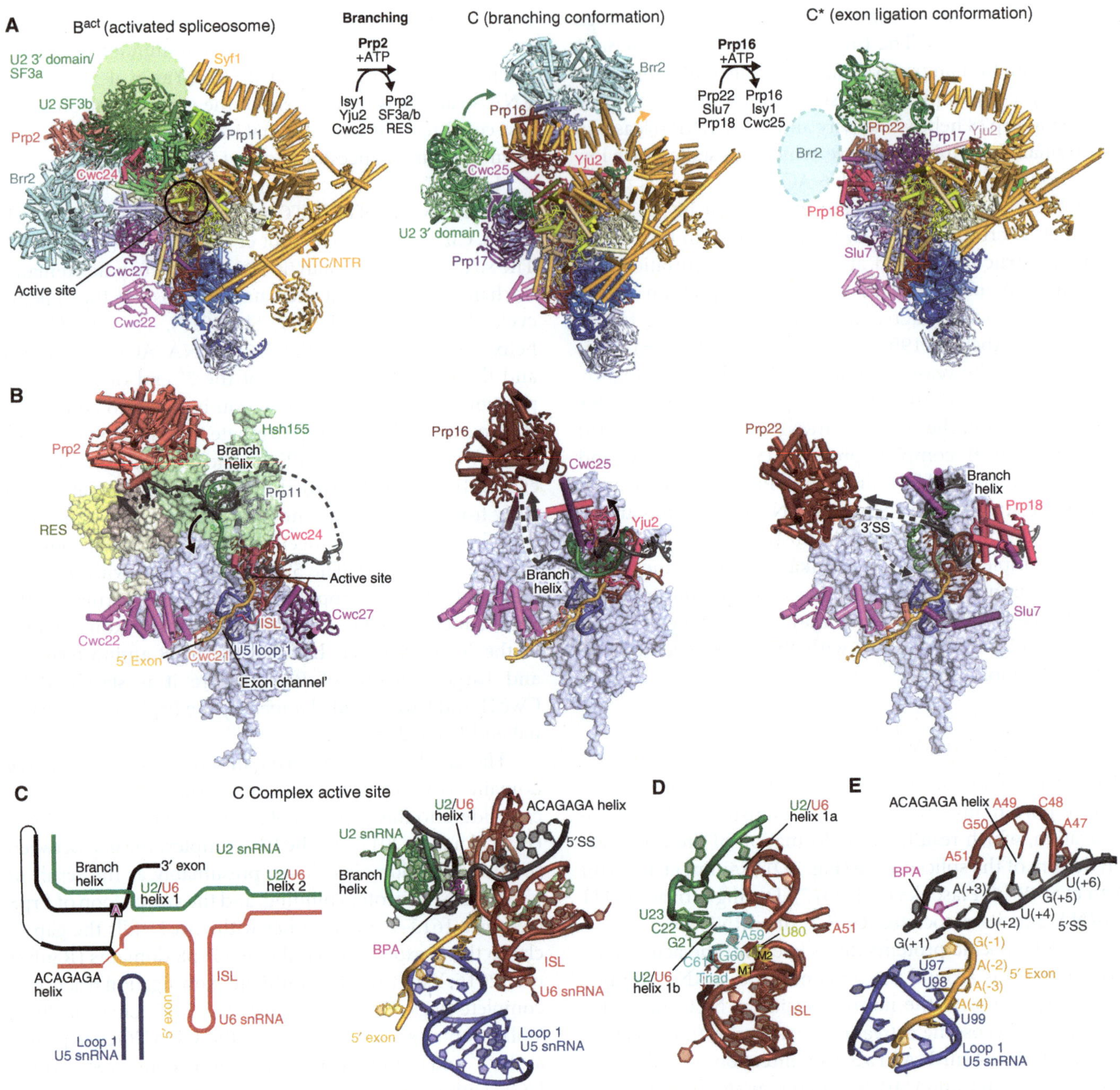

Figure 6. Branching and spliceosome remodeling. (*A*) Yeast B^{act}, C, and C* complex spliceosomes and step-specific factors are required to remodel the spliceosome between these states to move along the splicing pathway. The helicases Prp2 (salmon) and Prp16 (maroon) facilitate the remodeling from B^{act} to B* for branching and C to C* for exon ligation, respectively. Prp22 (light maroon) facilitates 3′SS proofreading in C* before exon ligation. The three structures, B^{act} (PDB 5GM6), C (PDB 5JL5), and C* (PDB 5MPS), are aligned using their Prp8 Large domains. (*B*) The location of the spliceosome active site, step-specific factors and the helicase-mediated remodeling in the B^{act}, C, and C* complex structures. Although the B^{act} and C complex structures show Prp2 (red) and Prp16 (dark red), respectively, poised for spliceosome remodeling, the C* structure, which lacks a "docked" 3′SS, reveals instead how Prp22 may contribute to 3′SS proofreading. Prp2 releases the branch helix from the U2 small nuclear ribonucleoprotein particle (snRNP) SF3b subcomplex, allowing it to bind in the active site for branching to occur. Prp16 subsequently undocks the branch helix together with step-specific factors Cwc25 and the Yju2 amino terminus, to allow docking of the 3′SS in C*. An arrow indicates the direction of RNA "translocation" (3′-5′ direction) of the helicases, which may remain bound to their respective locations and act at a distance (Semlow et al. 2016). (*Legend continues on following page.*)

The structure of the complete U2 snRNP was determined as part of the yeast B complex (Plaschka et al. 2017), comprising the SF3a/b subcomplexes and the U2 3′ domain, and this organization is unchanged in the B^{act} structure (Fig. 2). The U2 snRNP SF3b subcomplex consists of Rse1, containing three β-propeller domains, and the HEAT-repeat protein Hsh155, which together surround Ysf3 (Rcp10) and the Zn-containing Rds3 protein. Cus1 and Hsh49 associate on the periphery, to form the complete SF3b complex (Plaschka et al. 2017). The branch helix, which initially forms in the A complex between the pre-mRNA BP sequence and U2 snRNA, wedges into the split ends of the circular Hsh155 HEAT repeats within SF3b (Rauhut et al. 2016; Yan et al. 2016; Plaschka et al. 2017). The nucleotide base of the BP adenosine is flipped out from the branch helix and makes contact with Tyr35 of Rds3 and a pocket formed by Hsh155 HEAT repeats H15–H17.

Although the organization of the U2 snRNP SF3b complex in the B and B^{act} complexes is unchanged, the U2 snRNP undergoes a major repositioning during spliceosome activation. Comparisons between the B and B^{act} structures suggest that this large movement of the U2 snRNP is mediated by the energetically favorable folding of the U6 snRNA and remodeling of the U2 snRNA from the stem IIa/b to IIb/c conformation (Fig. 3D) (Hilliker et al. 2007; Perriman and Ares 2007). In the B complex, the U2 and U5 snRNPs are linked primarily through U6 snRNA, in particular through U2/U6 helix II. The U2 snRNA region that forms U2/U6 helix Ia and Ib as well as the Prp11 loop that folds over the 5′SS are disordered in the B complex, but become ordered in B^{act}. This suggests that as the RNA core forms during spliceosome activation, the U2 snRNP is pulled via the branch helix toward the active site, where it is held in place by association of the NTC and NTR complexes. Hence, although the helicase Brr2 is needed to overcome an initial barrier for spliceosome activation, the subsequent conformational remodeling is driven by U6 snRNA folding and new protein–RNA contacts within the spliceosome. The U2 3′ domain is held at one end of the Syf1/Clf1 arch and the other end is anchored by Syf2 near the U2/U6 stem 1. The solenoid structure of Syf1/Clf1 allows a large movement of the U2 3′ domain during different catalytic stages of the spliceosome while it is suspended above the active site (Fica and Nagai 2017).

After formation of the RNA-based active site in B^{act}, the BP adenosine is held 50 Å away from the active site. The branch helix must subsequently be inserted into the active site to enable the nucleophilic attack of the 2′ hydroxyl group of the BP adenosine at the scissile phosphate of the 5′SS. The Prp2 helicase is required to remodel B^{act} into this branching conformation, characteristic of the B* and C complexes. In the B^{act} structure, the intron sequence downstream from the BP winds through the SF3b and RES complexes to reach Prp2 (Warkocki et al. 2015; Rauhut et al. 2016; Yan et al. 2016). As Prp2 translocates along the intron in a 3′ to 5′ direction, the SF3a/b and RES complexes dissociate from the branch helix. This allows the Prp16 helicase and the branching factors Yju2, Cwc25, and Isy1 to bind (Fica and Nagai 2017).

6 STEP I (LARIAT FORMATION)

The structure of the yeast C complex was determined to high-resolution, revealing the spliceosome just after the branching reaction (Fig. 6A,B) (Galej et al. 2016; Wan et al. 2016b). Although the B* complex structure has not been determined so far, it is likely to closely resemble the C complex conformation. In B^{act}, the 3′ domain of U2 snRNP, comprising the Sm ring and the Msl1/Lea1 complex, binds the SF3a subcomplex, but in the C complex SF3a has dissociated and the U2 3′ domain bridges between the Prp8 RNaseH domain and the amino-terminal end of the Syf1 HAT domain. Thus, the U2 3′ domain may also be a sensor for spliceosome conformation. Upon release of the SF3b complex from B^{act}, U2 snRNA adopts the stem IIb/c configuration (Hilliker et al. 2007; Perriman and Ares 2007). Remarkably, the products of the branching reaction, the free 5′ exon and the lariat intron intermediate, remain close to the catalytic metal ions in the C complex structure (Galej et al. 2016; Wan et al. 2016b). This arrangement is consistent with the catalytic magnesium ion coordination inferred from metal rescue experiments (Fica et al. 2013). The 5′ phosphate of the first intron nucleotide (G+1) forms a 2′–5′ phosphodiester linkage with the 2′-hydroxyl group of the branch point adenosine. The branch helix is accommo-

Figure 6. (*Continued*) (*C*) Secondary structure diagram (*left*) and cryo-electron microscopy (cryo-EM) structure (*right*) of the RNA-based active site in C complex (PDB 5JL5). This active site conformation is characteristic of all catalytic stage complexes, except for the precise position of the branch helix and remodeling of the ACAGAGA helix. (*D*) Structure of the U6 small nuclear RNA (snRNA) internal stem loop (ISL) and U2/U6 helix I in C complex. The U2/U6 "catalytic" triad (cyan) and the U6 residue U80 (yellow) coordinate the two catalytic metal ions, which are essential for splicing (M1 and M2, not modeled). (*E*) Recognition of the precursor messenger RNA (pre-mRNA) 5′ splice site (5′SS) and 5′ exon in the C complex (PDB 5JL5). These interactions remain almost unchanged throughout the catalytic splicing stages.

dated in the active site cleft together with the step 1 factors, Cwc25, Yju2, and Isy1. Although the branch helix is in the canonical A-form in B and B^{act} complexes, in C complex the branch helix is highly distorted by the associated branching factors and thereby fits into the active site cleft (Galej et al. 2016). A long amino-terminal α-helix of Cwc25 interacts with the Prp8 RNaseH domain and the now unfolded Prp8 α-finger region and its amino terminus penetrates into the widened major groove of the branch helix. The Yju2 amino terminus wraps around the branch helix and makes intimate contacts with the 5′ exon. The Isy1 amino terminus also extends into the active site and interacts with the branch helix and U(+2) of the intron. In B and B^{act} complexes, the BP adenosine base is flipped out from the branch helix to interact with the surrounding SF3b proteins, whereas in the C complex the BP adenosine base is hydrogen bonded to intron nucleotide U68 (−2 from the BP), creating an unusual backbone conformation that allows the 2′-hydroxyl group of the BP adenosine to project toward the 5′SS for branching to occur.

The ATPase activity of Prp16 is essential for remodeling of the spliceosome for exon ligation (step 2). The structure by Wan et al. (2016b) lacks Prp16 whereas the position of Prp16 and Brr2 was revealed by focused classification in the C complex structure by Galej et al. (2016) providing structural insight into the Prp16-induced remodeling of the spliceosome. The sixth nucleotide downstream from the BP emerges from the spliceosome and the distance between the last ordered nucleotide and the Prp16 active site is consistent with Prp16–pre-mRNA crosslinks eighteen nucleotides downstream from the BP (Semlow et al. 2016). Prp16 is positioned near Prp8, Cwc25, and Yju2, in which the intron downstream from the BP emerges from the spliceosome. Prp16 translocates along the intron in a 3′ to 5′ direction on ATP hydrolysis, to undock the branch helix from the active site and to release the branching factors. The structure of human C complex was reported recently and reveals a similar arrangement of Prp16 showing that the remodeling mechanism is conserved in the human spliceosome (Zhang et al. 2018).

7 SPLICEOSOME REMODELING

Prp16 remodels the spliceosome from the branching (C complex) to the exon-ligation conformation (C* complex). The yeast C* structures (Fig. 6A,B) (Fica et al. 2017; Yan et al. 2017) confirm earlier observations that the spliceosome indeed harbors a single RNA-based active site (Fica et al. 2013). The 5′ exon participates in both branching and exon ligation and is held firmly in the exon channel for both catalytic steps. In contrast, the branch helix has to vacate the active site after branching, removing the BP, to allow the 3′SS to enter the active site for exon ligation. In the C* complex structure, Prp16 and the branching factors have dissociated, allowing the distorted branch helix to spring back into the canonical A form and to rotate 75° away from the active site. The step 2 (exon-ligation) factors, Prp18 and Slu7, bind to the Prp8 RNaseH domain and associate with the spliceosome together with the Prp22 helicase. The C to C* transition is accompanied by a rotation of the Prp8 RNaseH domain and a large movement of the Prp17 WD40 domain. The β-finger of the rotated RNaseH domain extends toward Cef1, and Prp17 bridges their interaction to lock the branch helix in its new orientation. This conformational change creates a binding site for the incoming 3′ exon. The conserved U5 snRNA loop 1 was shown genetically and biochemically (Newman and Norman 1992; Sontheimer and Steitz 1993) to align the 5′ and 3′ exons during exon ligation. As in the C complex the Prp8 RNaseH domain interacts with the 3′ domain of the U2 snRNP, which in turn interacts with one end of the HAT domain of Syf1. Hence, the rotation of the Prp8 RNaseH domain is accompanied by a large movement of the U2 3′ domain and the HAT domains of Syf1 and Clf1. The C* complex structures do not reveal how the 3′ exon docks into the active site before exon ligation. 3′ exon docking may be very transient and coincide with exon ligation, making its structural observation challenging. However, insights into 3′SS selection have been gained from the postsplicing P complex structure (see below). The binding of an inositol hexaphosphate molecule (IP_6) to Prp8 was first documented in the C* complex (PDB 5MPS) but putative IP_6 density is present in all yeast and human catalytic stage spliceosome structures (B^{act}, C, C*, and P and ILS complexes), but not in free Prp8, tri-snRNP, or B complex. In the C* and P complexes, Slu7 contacts the IP_6 molecule, suggesting a role in the stabilization of this exon-ligation factor (Fig. 4A).

8 STEP II (EXON LIGATION)

After 3′SS docking in C*, the exons are ligated and the product mRNA released. This depends on the ATPase activity of the DEAH-box helicase Prp22, which carries out 3′SS proofreading as well as mRNA release (Company et al. 1991; Schwer 2008). The cryo-EM structure of the post-splicing complex (P complex) was recently determined by three groups, yielding new insights into the exon-ligation reaction (Fig. 7A–C). All three groups used ATPase-deficient mutants of Prp22, which are unable to release the ligated exon (Bai et al. 2017; Liu et al. 2017; Wilkinson et al. 2017). In the structure, the ligated mRNA and the lariat intron, cleaved at the 3′SS, are bound in the active site (Fig. 7A). The 3′-hydroxyl group of the last intron nucleotide is close to the phosphorus atom at the exon

 Cite as *Cold Spring Harb Perspect Biol* doi: 10.1101/cshperspect.a032391

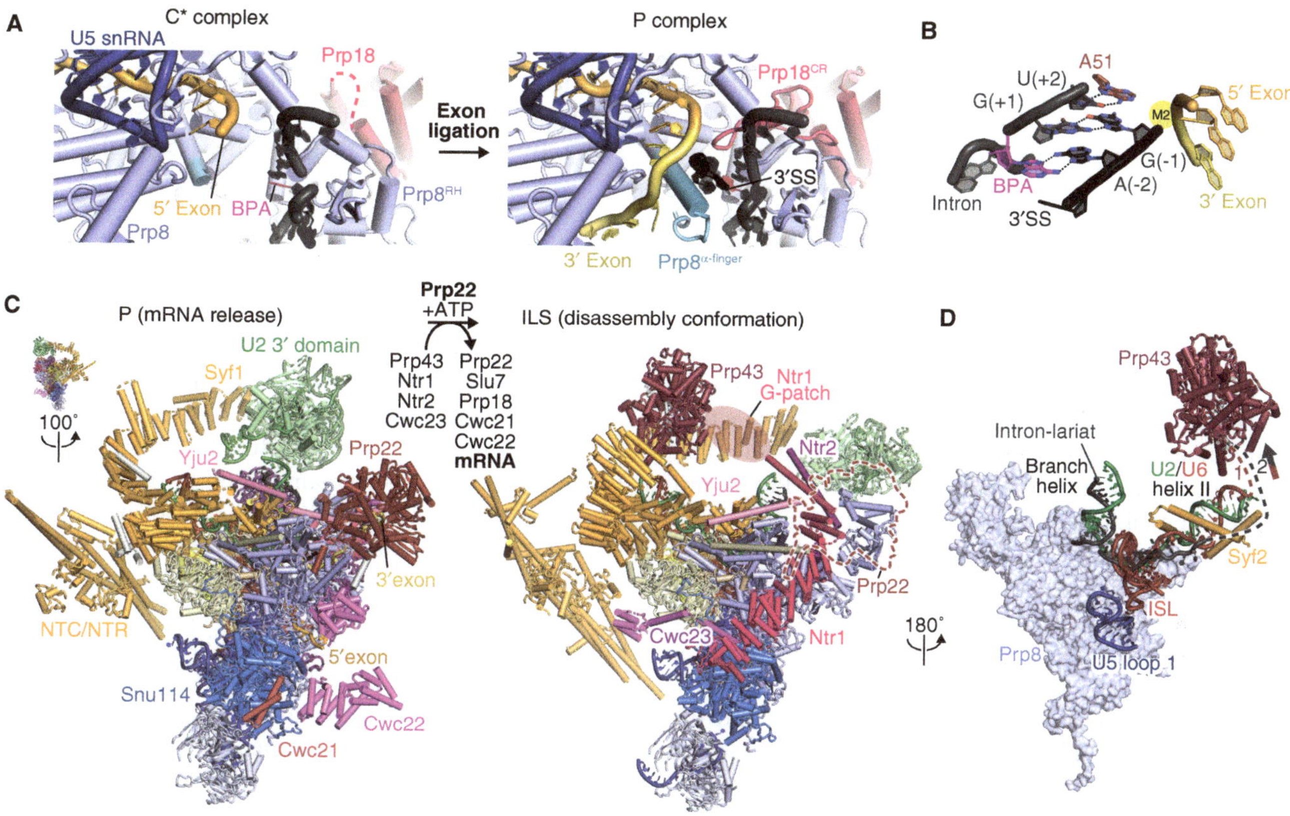

Figure 7. Exon ligation and spliceosome disassembly. (*A*) Comparison of the C* and P complex active sites. The extended Prp8 α-finger (light blue) and the Prp18 conserved region (CR) (pink) stabilize 3′SS docking in the active site. This positions the 3′-OH group of the 5′ exon to attack at the 3′SS for exon ligation (PDB 6EXN). (*B*) Recognition of the conserved YAG sequence at the 3′SS. The last intron nucleotide, G(−1), forms a non-Watson–Crick base pair with the first intron nucleotide G(+1) and the penultimate intron nucleotide A(−2) forms a symmetric Hoogsteen–Hoogsteen base pair with the branch point adenosine (PDB 6EXN). The third last pyrimidine (Y) is recognized by Gln-1594 of Prp8. (*C*) Transition from the P (PDB 6EXN) to ILS (intron lariat spliceosome) (PDB 5Y88). Prp22-mediated ligated-exon release facilitates the binding of Prp43/Ntr1/Ntr2/Cwc23 and the release of the step 2 factors (Prp18 and Slu7) and Cwc21 and Cwc22, which interact with 5′ exon. (*D*) Prp43 induces disassembly of ILS by pulling on either the intron or U6 small nuclear RNA (snRNA) (PDB 5Y88).

junction, showing that these atoms have not moved significantly after exon ligation. The invariant intron nucleotides (YAG where Y denotes pyrimidine) immediately preceding the 3′ splice junction dock into the active site and are specifically recognized (Fig. 7B). The Hoogsteen edge of the 3′SS G(−1) forms a base pair with the Watson–Crick edge of the first intron nucleotide G(+1) consistent with the analysis of Parker and Siliciano (1993). In two of the P complex structures (Liu et al. 2017; Wilkinson et al. 2017), the 3′SS A(−2) forms a symmetric Hoogsteen–Hoogsteen base pair with the BP adenosine, which is linked via its 2′-oxygen to the 5′SS G(+1). This is consistent with the stringent conservation of the penultimate A(−2) at the 3′SS. In contrast, in the P complex structure reported by Bai et al. (2017), A(−2) has a different geometry and does not appear to form a Hoogsteen–Hoogsteen base pair with the BP adenosine. In these structures, all invariant nucleotides at the 5′SS and 3′SS and BP come together and hydrogen bond with each other for exon ligation. The exon-ligation reaction could not occur unless the 5′SS and 3′SS substrates are precisely positioned by this network of hydrogen bonds, accounting for the stringent conservation of the GU and AG dinucleotides at the 5′SS and 3′SS and the identity of the branch nucleotide. In the P complex structures of Liu et al. (2017) and Wilkinson et al. (2017), the highly conserved Gln1594 forms a hydrogen bond with the O2 carbonyl group of U(−3). Cytosine at the −3 position could form an equivalent hydrogen bond accounting for the conservation of pyrimidines at −3 of the 3′SS. In the spliceosome, the intron folds back to bring the 3′SS and branch site into the active site. Interestingly, Liu et al. (2017) show that in a large number of pre-mRNAs, part of the intron between the BP sequence and the 3′SS may base-pair with itself, which may facilitate the insertion of the 3′ exon into the active site.

A network of protein–RNA interactions cradles the P complex active site, facilitating the exon-ligation reaction. Helix 8 in the α-finger region (residues 1570–1615) of the Prp8 Linker domain is extended in the P complex compared with the C* complex (Fica et al. 2017; Yan et al. 2017) and before exon ligation the intron-3′ exon nucleotides wrap around this extended α-helix to project the scissile phosphate toward the 3′ terminal hydroxyl group of the 5′ exon. The α-finger residue Arg1604 contacts the phosphate backbone of the 3′SS residues −3 and −4 and the mutation of this residue to Ala abolishes the exon-ligation reaction showing that Arg1604 is crucial in positioning the 3′SS residues in the active site. The step 2 factors, Prp18 and Slu7, are essential for exon ligation, but only the conserved region of Prp18 inserts into the active site and stabilizes the binding of the 3′SS (Liu et al. 2017; Wilkinson et al. 2017). Slu7 extends over a large surface of the spliceosome, and although its precise role in exon ligation is still unclear, it was proposed to assist with guiding the 3′SS into the active site when the distance between the branch point and the 3′SS exceeds nine nucleotides (Schwer 2008; Ohrt et al. 2013; Fica et al. 2017).

The study by Wilkinson et al. (2017) revealed an "undocked" state of the 3′SS in the P complex, in which the exons are ligated but the 3′SS is not docked in the active site. This provides insights into the mechanism of 3′SS discrimination and kinetic proofreading by the Prp22 helicase. Cognate and noncognate 3′SS sequences could potentially both enter the spliceosome active site but only the cognate 3′SS will induce the "docked" conformation, where the 3′SS, BP and 5′SS cooperatively pair with each other to align the 3′SS for nucleophilic attack by the 3′-hydroxyl group of the 5′ exon. This allows the exon ligation reaction to occur and Prp22 activity would subsequently release the ligated mRNA (Schwer 2008). However, if a noncognate 3′SS docks and does not induce the docked conformation to ligate within a "proofreading" window, then Prp22 would pull on the intron to remove the noncognate 3′SS from the active site and allow sampling of a new 3′SS (Mayas et al. 2010).

9 SPLICEOSOME DISASSEMBLY AND RECYCLING

After exon ligation, the ligated exon (mRNA) is released from the spliceosome and the remaining ILS is disassembled to recycle its snRNA and protein components for subsequent splicing rounds. The DEAH-box helicase Prp43 participates in this process (Arenas and Abelson 1997) and forms a stable complex with two NTC interacting proteins, Ntr1 and Ntr2 (Tsai et al. 2005). The purified Prp43-Ntr1-Ntr2 complex can catalyze the disassembly of ILS into U2, U5, U6, and NTC in an ATP-dependent manner. Prp43 interacts with the Ntr1 amino-terminal G-patch domain and this interaction is required to stimulate Prp43 ATPase activity (Tsai et al. 2005; Tanaka et al. 2007). It was suggested that Brr2 helicase is also involved in the disassembly of ILS (Small et al. 2006), yet recent in vitro experiments suggest that ILS disassembly does not depend on Brr2 helicase activity (Fourmann et al. 2016).

The cryo-EM structure of the *S. cerevisiae* ILS (Wan et al. 2017) has provided mechanistic insight into the disassembly process (Fig. 7C,D). In agreement with biochemical data (Fourmann et al. 2016), Cef1, Clf1, and Syf2 (NTC proteins) and Cwc2, Cwc15, Bud31, Ecm2, Prp45, and Prp46 (NTR proteins) remain bound to the U5 snRNP in the ILS, cradling the catalytic RNA core as in B^{act} (Rauhut et al. 2016; Yan et al. 2016), C (Galej et al. 2016; Wan et al. 2016b), C* (Fica et al. 2017; Yan et al. 2017), and P (Bai et al. 2017; Liu et al. 2017; Wilkinson et al. 2017) complexes. In contrast, Prp22, Cwc21, and Cwc22 have dissociated from the spliceosome. Prp22 likely dissociates together with the mRNA from the P complex, and this would release Cwc21 and Cwc22 from the exon channel. In metazoans, the exon-junction complex (EJC) assembles on Cwc22 to bind the mRNA, and also dissociates with the mRNA in P complex. This frees several spliceosome surfaces that allow the Prp43-Ntr1-Ntr2 complex to associate. In the cryo-EM structure of the *S. cerevisiae* ILS, Prp43 is bound to Syf1, near U2/U6 snRNA helix II. Ntr1 is anchored to Snu114 with its carboxy-terminal domain and the Ntr1 super-helical domain spans almost the entire length of the spliceosome, bridging between Prp43 and Snu114, which may serve as a marker for a disassembling spliceosome. The Ntr1 amino-terminal G patch domain, which binds Prp43 (Tsai et al. 2005), was not identified in the cryo-EM map. Although there is no RNA density bound to Prp43, it is located near the U6 snRNA 3′ end and the intron, suggesting that either of these RNAs may be the functional substrate of Prp43 (Fig. 7D). Similar to Prp16 in C complex, and Prp22 in C*, Prp43 is likely to pull on U6 snRNA or the intron to disassemble the ILS. No density was observed for Brr2 in the ILS structure, suggesting that it is mobile.

Unlike its *S. cerevisiae* counterpart, whole-cell extract from *S. pombe* does not support in vitro splicing, apparently because an endogenous U2/U6.U5 complex accumulates. This complex may correspond to another ILS-like intermediate. The cryo-EM structure of the *S. pombe* ILS (Yan et al. 2015) shows that its catalytic RNA core is very similar to its yeast counterpart, except that it lacks the Prp43-Ntr1-Ntr2 complex and Brr2, in agreement with earlier proteomic studies. Interestingly, the carboxy-terminal domain of Cwf19 (Cwf19L2 in human) is wedged between the Prp8

Cite as *Cold Spring Harb Perspect Biol* doi: 10.1101/cshperspect.a032391

Large and RNaseH domains in the *S. pombe* ILS. The carboxy-terminal domain of Cwf19 shows significant similarity to the carboxy-terminal domain of *S. cerevisiae* protein Drn1 whose amino-terminal domain is known to interact with the debranching enzyme Dbr1. It is possible that Drn1 binds to Prp8 in P complex and recruits Dbr1 to the lariat intron after the mRNA and step 2 factors dissociate from the complex. Alternatively, the *S. pombe* ILS may represent a novel and distinct dead-end complex, which explains why *S. pombe* splicing extract is inactive.

10 CONCLUSION

Since 2015, cryo-EM studies of the spliceosome have provided snapshots of splicing in its key steps. These have advanced our structural and mechanistic knowledge of the underlying reaction, and build on years of genetic, biochemical, and structural studies of the spliceosome, confirming the pioneering work by many laboratories, and leading to new molecular insights that can be tested in the future. The most important conclusions can be summarized in the following points.

1. The active site of the spliceosome is an RNA-based metalloenzyme cradled by many proteins.
2. A single active site performs the two *trans*-esterification reactions: branching and exon ligation.
3. The catalytic RNA core forms during the transition from the B to B^{act} complex and remains unchanged until the spliced mRNA is released.
4. RNA helicases play crucial roles not only in generating the active RNA core but also in docking and undocking the substrates (BP adenosine and the 3′SS) together with step-specific factors. By doing so, these helicases play a crucial role in proofreading the splice sites.
5. The spliceosome shows a remarkable structural and mechanistic conservation between yeast and human systems (Bertram et al. 2017b; Fica et al. 2017; Yan et al. 2017; Zhang et al. 2017).

ACKNOWLEDGMENTS

We thank the past and present members of the Nagai group for their contribution to our spliceosome project and critical reading of the manuscript. We are particularly grateful to Kelly Nguyen, Wojtek Galej, Sebastian Fica, Pei-Chun Lin, Max Wilkinson, Chris Oubridge, Lisa Strittmatter, Clément Charenton, and Chris Norman. We apologize for the omission of many important references because of the restriction on the number of references. Our work was supported by the Medical Research Council (MC_U105184330) and European Research Council Advanced Grant (AdG-693087-SPLICE3D). C.P. was supported by a European Molecular Biology Organization (EMBO) fellowship.

REFERENCES

* *Reference is in this book.*

Agafonov DE, Kastner B, Dybkov O, Hofele RV, Liu WT, Urlaub H, Lührmann R, Stark H. 2016. Molecular architecture of the human U4/U6.U5 tri-snRNP. *Science* **351:** 1416–1420.

Arenas JE, Abelson JN. 1997. Prp43: An RNA helicase-like factor involved in spliceosome disassembly. *Proc Natl Acad Sci* **94:** 11798–11802.

Bai R, Yan C, Wan R, Lei J, Shi Y. 2017. Structure of the post-catalytic spliceosome from *Saccharomyces cerevisiae*. *Cell* **171:** 1589–1598.

Bartels C, Urlaub H, Lührmann R, Fabrizio P. 2003. Mutagenesis suggests several roles of Snu114p in pre-mRNA splicing. *J Biol Chem* **278:** 28324–28334.

Berget SM, Moore C, Sharp PA. 1977. Spliced segments at the 5′ terminus of adenovirus 2 late mRNA. *Proc Natl Acad Sci* **74:** 3171–3175.

Berglund JA, Abovich N, Rosbash M. 1998. A cooperative interaction between U2AF65 and mBBP/SF1 facilitates branchpoint region recognition. *Genes Dev* **12:** 858–867.

Bertram K, Agafonov DE, Dybkov O, Haselbach D, Leelaram MN, Will CL, Urlaub H, Kastner B, Lührmann R, Stark H. 2017a. Cryo-EM structure of a pre-catalytic human spliceosome primed for activation. *Cell* **170:** 701–713.

Bertram K, Agafonov DE, Liu WT, Dybkov O, Will CL, Hartmuth K, Urlaub H, Kastner B, Stark H, Lührmann R. 2017b. Cryo-EM structure of a human spliceosome activated for step 2 of splicing. *Nature* **542:** 318–323.

Boesler C, Rigo N, Anokhina MM, Tauchert MJ, Agafonov DE, Kastner B, Urlaub H, Ficner R, Will CL, Lührmann R. 2016. A spliceosome intermediate with loosely associated tri-snRNP accumulates in the absence of Prp28 ATPase activity. *Nat Commun* **7:** 11997.

Brody E, Abelson J. 1985. The "spliceosome": Yeast pre-messenger RNA associates with a 40S complex in a splicing-dependent reaction. *Science* **228:** 963–967.

Burgess SM, Guthrie C. 1993. A mechanism to enhance mRNA splicing fidelity: The RNA dependent ATPase Prp16 governs usage of a discard pathway for aberrant lariat intermediates. *Cell* **73:** 1377–1391.

Chan SP, Cheng SC. 2005. The Prp19-associated complex is required for specifying interactions of U5 and U6 with pre-mRNA during spliceosome activation. *J Biol Chem* **280:** 31190–31199.

Chen JY, Stands L, Staley JP, Jackups RR Jr, Latus LJ, Chang TH. 2001. Specific alterations of U1-C protein or U1 small nuclear RNA can eliminate the requirement of Prp28p, an essential DEAD box splicing factor. *Mol Cell* **7:** 227–232.

Chow LT, Gelinas RE, Broker TR, Roberts RJ. 1977. An amazing sequence arrangement at the 5′ ends of adenovirus 2 messenger RNA. *Cell* **12:** 1–8.

Company M, Arenas J, Abelson J. 1991. Requirement of the RNA helicase-like protein PRP22 for release of messenger RNA from spliceosomes. *Nature* **349:** 487–493.

Fabrizio P, Laggerbauer B, Lauber J, Lane WS, Lührmann R. 1997. An evolutionarily conserved U5 snRNP-specific protein is a GTP-binding factor closely related to the ribosomal translocase EF-2. *EMBO J* **16:** 4092–4106.

Fernandez-Leiro R, Scheres SH. 2016. Unravelling biological macromolecules with cryo-electron microscopy. *Nature* **537:** 339–346.

Fica SM, Nagai K. 2017. Cryo-electron microscopy snapshots of the spliceosome: Structural insights into a dynamic ribonucleoprotein machine. *Nat Struct Mol Biol* **24:** 791–799.

Fica SM, Tuttle N, Novak T, Li NS, Lu J, Koodathingal P, Dai Q, Staley JP, Piccirilli JA. 2013. RNA catalyses nuclear pre-mRNA splicing. *Nature* **503:** 229–234.

Fica SM, Oubridge C, Galej WP, Wilkinson ME, Bai XC, Newman AJ, Nagai K. 2017. Structure of a spliceosome remodelled for exon ligation. *Nature* **542:** 377–380.

Fourmann JB, Dybkov O, Agafonov DE, Tauchert MJ, Urlaub H, Ficner R, Fabrizio P, Lührmann R. 2016. The target of the DEAH-box NTP triphosphatase Prp43 in *Saccharomyces cerevisiae* spliceosomes is the U2 snRNP-intron interaction. *eLife* **5:** e15564.

Galej WP, Oubridge C, Newman AJ, Nagai K. 2013. Crystal structure of Prp8 reveals active site cavity of the spliceosome. *Nature* **493:** 638–643.

Galej WP, Wilkinson ME, Fica SM, Oubridge C, Newman AJ, Nagai K. 2016. Cryo-EM structure of the spliceosome immediately after branching. *Nature* **537:** 197–201.

Gottschalk A, Tang J, Puig O, Salgado J, Neubauer G, Colot HV, Mann M, Séraphin B, Rosbash M, Lührmann R, et al. 1998. A comprehensive biochemical and genetic analysis of the yeast U1 snRNP reveals five novel proteins. *RNA* **4:** 374–393.

Grainger RJ, Beggs JD. 2005. Prp8 protein: At the heart of the spliceosome. *RNA* **11:** 533–557.

Hilliker AK, Mefford MA, Staley JP. 2007. U2 toggles iteratively between the stem IIa and stem IIc conformations to promote pre-mRNA splicing. *Genes Dev* **21:** 821–834.

Huang YH, Chung CS, Kao DI, Kao TC, Cheng SC. 2014. Sad1 counteracts Brr2-mediated dissociation of U4/U6.U5 in tri-snRNP homeostasis. *Mol Cell Biol* **34:** 210–220.

Kandels-Lewis S, Séraphin B. 1993. Involvement of U6 snRNA in 5′ splice site selection. *Science* **262:** 2035–2039.

* Kastner B, Will CL, Stark H, Lührmann R. 2019. Structural insights into nuclear pre-mRNA splicing in higher eukaryotes. *Cold Spring Harb Perspect Biol* doi: 10.1101/cshperspect.a032417.

Kondo Y, Oubridge C, van Roon AM, Nagai K. 2015. Crystal structure of human U1 snRNP, a small nuclear ribonucleoprotein particle, reveals the mechanism of 5′ splice site recognition. *eLife* **4** doi: 10.7554/eLife.04986.

Laggerbauer B, Achsel T, Lührmann R. 1998. The human U5-200kD DEXH-box protein unwinds U4/U6 RNA duplices in vitro. *Proc Natl Acad Sci* **95:** 4188–4192.

Lerner MR, Steitz JA. 1979. Antibodies to small nuclear RNAs complexed with proteins are produced by patients with systemic lupus erythematosus. *Proc Acad Natl Sci* **76:** 5495–5499.

Lesser CF, Guthrie C. 1993. Mutations in U6 snRNA that alter splice site specificity: Implications for the active site. *Science* **262:**1982–1988.

Li X, Liu S, Jiang J, Zhang L, Espinosa S, Hill RC, Hansen KC, Zhou ZH, Zhao R. 2017. CryoEM structure of *Saccharomyces cerevisiae* U1 snRNP offers insight into alternative splicing. *Nat Commun* **8:** 1035–1047.

Liang WW, Cheng SC. 2015. A novel mechanism for Prp5 function in prespliceosome formation and proofreading the branch site sequence. *Genes Dev* **29:** 81–93.

Liu S, Li X, Zhang L, Jiang J, Hill RC, Cui Y, Hansen KC, Zhou ZH, Zhao R. 2017. Structure of the yeast spliceosomal postcatalytic P complex. *Science* **358:** 1278–1283.

Madhani HD, Guthrie C. 1992. A novel base-pairing interaction between U2 and U6 snRNAs suggests a mechanism for the catalytic activation of the spliceosome. *Cell* **71:** 803–817.

Mayas RM, Maita H, Semlow DR, Staley JP. 2010. Spliceosome discards intermediates via the DEAH box ATPase Prp43p. *Proc Natl Acad Sci* **107:** 10020–10025.

Mozaffari-Jovin S, Wandersleben T, Santos KF, Will CL, Lührmann R, Wahl MC. 2013. Inhibition of RNA helicase Brr2 by the C-terminal tail of the spliceosomal protein Prp8. *Science* **341:** 80–84.

Newman AJ, Norman C. 1992. U5 snRNA interacts with exon sequences at 5′ and 3′ splice sites. *Cell* **68:** 743–754.

Nguyen TH, Li J, Galej WP, Oshikane H, Newman AJ, Nagai K. 2013. Structural basis of Brr2-Prp8 interactions and implications for U5 snRNP biogenesis and the spliceosome active site. *Structure* **21:** 910–919.

Nguyen TH, Galej WP, Bai XC, Savva CG, Newman AJ, Scheres SH, Nagai K. 2015. The architecture of the spliceosomal U4/U6.U5 tri-snRNP. *Nature* **523:** 47–52.

Nguyen THD, Galej WP, Bai XC, Oubridge C, Newman AJ, Scheres SHW, Nagai K. 2016. Cryo-EM structure of the yeast U4/U6.U5 tri-snRNP at 3.7 Å resolution. *Nature* **530:** 298–302.

Ohrt T, Odenwälder P, Dannenberg J, Prior M, Warkocki Z, Schmitzová J, Karaduman R, Gregor I, Enderlein J, Fabrizio P, et al. 2013. Molecular dissection of step 2 catalysis of yeast pre-mRNA splicing investigated in a purified system. *RNA* **19:** 902–915.

Parker R, Siliciano PG. 1993. Evidence for an essential non-Watson–Crick interaction between the first and last nucleotides of a nuclear pre-mRNA intron. *Nature* **361:** 660–662.

Parker R, Siliciano PG, Guthrie C. 1987. Recognition of the TACTAAC box during mRNA splicing in yeast involves base pairing to the U2-like snRNA. *Cell* **49:** 229–239.

Perriman RJ, Ares M Jr. 2007. Rearrangement of competing U2 RNA helices within the spliceosome promotes multiple steps in splicing. *Genes Dev* **21:** 811–820.

Perriman R, Ares M Jr. 2010. Invariant U2 snRNA nucleotides form a stem loop to recognize the intron early in splicing. *Mol Cell* **38:** 416–427.

Plaschka C, Lin PC, Nagai K. 2017. Structure of a pre-catalytic spliceosome. *Nature* **546:** 617–621.

Plaschka C, Lin PC, Charenton C, Nagai K. 2018. Prespliceosome structure provides insights into spliceosome assembly and regulation. *Nature* doi: 10.1038/nature.

Pomeranz Krummel DA, Oubridge C, Leung AK, Li J, Nagai K. 2009. Crystal structure of human spliceosomal U1 snRNP at 5.5 Å resolution. *Nature* **458:** 475–480.

Raghunathan PL, Guthrie C. 1998. RNA unwinding in U4/U6 snRNPs requires ATP hydrolysis and the DEIH-box splicing factor Brr2. *Curr Biol* **8:** 847–855.

Rauhut R, Fabrizio P, Dybkov O, Hartmuth K, Pena V, Chari A, Kumar V, Lee CT, Urlaub H, Kastner B, Lührmann R, et al. 2016. Molecular architecture of the *Saccharomyces cerevisiae* activated spliceosome. *Science* **353:** 1399–1405.

Sawa H, Abelson J. 1992. Evidence for a base-pairing interaction between U6 small nuclear RNA and 5′ splice site during the splicing reaction in yeast. *Proc Natl Acad Sci* **89:** 11269–11273.

Schwer B. 2008. A conformational rearrangement in the spliceosome sets the stage for Prp22-dependent mRNA release. *Mol Cell* **30:** 743–754.

Semlow DR, Blanco MR, Walter NG, Staley JP. 2016. Spliceosomal DEAH-Box ATPases remodel pre-mRNA to activate alternative splice sites. *Cell* **164:** 985–998.

Shao W, Kim HS, Cao Y, Xu YZ, Query CC. 2012. A U1-U2 snRNP interaction network during intron definition. *Mol Cell Biol* **32:** 470–478.

Shi Y. 2017. Mechanistic insights into precursor messenger RNA splicing by the spliceosome. *Nat Rev Mol Cell Biol* **18:** 655–670.

Small EC, Leggett SR, Winans AA, Staley JP. 2006. The EF-G-like GTPase Snu114p regulates spliceosome dynamics mediated by Brr2p, a DExD/H box ATPase. *Mol Cell* **23:** 389–399.

Sontheimer EJ, Steitz JA. 1993. The U5 and U6 small nuclear RNAs as active site components of the spliceosome. *Science* **262:**1989–1996.

Spingola M, Grate L, Haussler D, Ares M Jr. 1999. Genome-wide bioinformatic and molecular analysis of introns in *Saccharomyces cerevisiae*. *RNA* **5:** 221–234.

Staley JP, Guthrie C. 1999. An RNA switch at the 5′ splice site requires ATP and the DEAD box protein Prp28p. *Mol Cell* **3:** 55–64.

Stark H, Lührmann R. 2006. Cryo-electron microscopy of spliceosomal components. *Annu Rev Biophys Biomol Struct* **35:** 435–457.

Steitz TA, Steitz JA. 1993. A general two-metal-ion mechanism for catalytic RNA. *Proc Natl Acad Sci* **90:** 6498–6502.

Tanaka N, Aronova A, Schwer B. 2007. Ntr1 activates the Prp43 helicase to trigger release of lariat-intron from the spliceosome. *Genes Dev* **21:** 2312–2325.

Tang Q, Rodriguez-Santiago S, Wang J, Pu J, Yuste A, Gupta V, Moldón A, Xu YZ, Query CC. 2016. SF3B1/Hsh155 HEAT motif mutations affect interaction with the spliceosomal ATPase Prp5, resulting in altered branch site selectivity in pre-mRNA splicing. *Genes Dev* **30:** 2710–2723.

Tarn WY, Steitz JA. 1996. A novel spliceosome containing U11, U12, and U5 snRNPs excises a minor class (AT–AC) intron in vitro. *Cell* **84:** 801–811.

Toor N, Keating KS, Taylor SD, Pyle AM. 2008. Crystal structure of a self-spliced group II intron. *Science* **320:** 77–82.

Tsai RT, Fu RH, Yeh FL, Tseng CK, Lin YC, Huang YH, Cheng SC. 2005. Spliceosome disassembly catalyzed by Prp43 and its associated components Ntr1 and Ntr2. *Genes Dev* **19:** 2991–3003.

Wan R, Yan C, Bai R, Wang L, Huang M, Wong CC, Shi Y. 2016a. The 3.8 Å structure of the U4/U6.U5 tri-snRNP: Insights into spliceosome assembly and catalysis. *Science* **351:** 466–475.

Wan R, Yan C, Bai R, Huang G, Shi Y. 2016b. Structure of a yeast catalytic step I spliceosome at 3.4 Å resolution. *Science* **353:** 895–904.

Wan R, Yan C, Bai R, Lei J, Shi Y. 2017. Structure of an intron lariat spliceosome from *Saccharomyces cerevisiae*. *Cell* **171:** 120–132.

Warkocki Z, Schneider C, Mozaffari-Jovin S, Schmitzová J, Höbartner C, Fabrizio P, Lührmann R. 2015. The G-patch protein Spp2 couples the spliceosome-stimulated ATPase activity of the DEAH-box protein Prp2 to catalytic activation of the spliceosome. *Genes Dev* **29:** 94–107.

Weber G, Trowitzsch S, Kastner B, Lührmann R, Wahl MC. 2010. Functional organization of the Sm core in the crystal structure of human U1 snRNP. *EMBO J* **29:** 4172–4184.

Wilkinson ME, Fica SM, Galej WP, Norman CM, Newman AJ, Nagai K. 2017. Postcatalytic spliceosome structure reveals mechanism of 3′-splice site selection. *Science* **358:** 1283–1288.

Wilkinson ME, Lin PC, Plaschka C, Nagai K. 2018. Cryo-EM studies of pre-mRNA splicing: From sample preparation to model visualization. *Annu Rev Biophys* doi: 10.1146/annurev-biophys-070317-033410.

Will CL, Lührmann R. 2011. Spliceosome structure and function. *Cold Spring Harb Perspect Biol* **3:** a003707.

Yan C, Hang J, Wan R, Huang M, Wong CCL, Shi Y. 2015. Structure of a yeast spliceosome at 3.6-Angstrom resolution. *Science* **349:** 1182–1191.

Yan C, Wan R, Bai R, Huang G, Shi Y. 2016. Structure of a yeast activated spliceosome at 3.5 Å resolution. *Science* **353:** 904–911.

Yan C, Wan R, Bai R, Huang G, Shi Y. 2017. Structure of a yeast step II catalytically activated spliceosome. *Science* **355:** 149–155.

* Yan C, Wan R, Shi Y. 2019. Molecular mechanisms of pre-mRNA splicing through structural biology of the spliceosome. *Cold Spring Harb Perspect Biol* doi: 10.1101/cshperspect.032409.

Zhang X, Yan C, Hang J, Finci L, Lei J, Shi Y. 2017. An atomic structure of the human spliceosome. *Cell* **169:** 1–12.

Zhang X, Yan C, Zhan X, Li L, Lei J, Shi Y. 2018. Structure of the human activated spliceosome in three conformational states. *Cell Res* **28:** 307–322.

Zhao C, Pyle AM. 2016. Crystal structures of a group II intron maturase reveal a missing link in spliceosome evolution. *Nat Struct Mol Biol* **23:** 558–565.

Structural Insights into Nuclear pre-mRNA Splicing in Higher Eukaryotes

Berthold Kastner,[1] Cindy L. Will,[1] Holger Stark,[2] and Reinhard Lührmann[1]

[1]Department of Cellular Biochemistry, Max Planck Institute for Biophysical Chemistry, D-37077 Göttingen, Germany
[2]Department of Structural Dynamics, Max Planck Institute for Biophysical Chemistry, D-37077 Göttingen, Germany
Correspondence: reinhard.luehrmann@mpibpc.mpg.de

SUMMARY

The spliceosome is a highly complex, dynamic ribonucleoprotein molecular machine that undergoes numerous structural and compositional rearrangements that lead to the formation of its active site. Recent advances in cyroelectron microscopy (cryo-EM) have provided a plethora of near-atomic structural information about the inner workings of the spliceosome. Aided by previous biochemical, structural, and functional studies, cryo-EM has confirmed or provided a structural basis for most of the prevailing models of spliceosome function, but at the same time allowed novel insights into splicing catalysis and the intriguing dynamics of the spliceosome. The mechanism of pre-mRNA splicing is highly conserved between humans and yeast, but the compositional dynamics and ribonucleoprotein (RNP) remodeling of the human spliceosome are more complex. Here, we summarize recent advances in our understanding of the molecular architecture of the human spliceosome, highlighting differences between the human and yeast splicing machineries.

Outline

Cite as *Cold Spring Harb Perspect Biol* doi: 10.1101/cshperspect.a032417

1 INTRODUCTION

Pre-mRNA introns are defined by three short consensus sequences, the 5′ splice site (SS), the 3′SS, and the branch site (BS,) which is often followed by a pyrimidine-rich region (the PPT) (Fig. 1A) (Will and Lührmann 2011). Pre-mRNA splicing involves two sequential SN2-type transesterification reactions (Fig. 1B). First, the 2′-OH group of the branch site adenosine (BS-A) carries out a nucleophilic attack at the phosphodiester bond of the 5′SS, producing the cleaved 5′ exon and the intron lariat-3′ exon intermediates. During step 2, the 3′-OH group of the 5′ exon attacks the 3′SS, leading to exon ligation and intron excision. In higher eukaryotes, most pre-mRNAs undergo alternative splicing (Pan et al. 2008), during which different sets of exons from a pre-mRNA are spliced together, leading to different mRNA isoforms and vastly amplifying the palette of translated proteins. Furthermore, errors in splicing or its regulation are either the basis for or modulate the severity of many human diseases (Daguenet et al. 2015; Lee and Rio 2015).

Nuclear pre-mRNA introns contain relatively little conserved secondary or tertiary structure information and thus depend on the spliceosome to align their reactive sites (i.e., the 5′SS, BS, and 3′SS) for splicing catalysis. The major spliceosomal building blocks are the U1, U2, U4/U6, and U5 small nuclear ribonucleoproteins (snRNPs). Each consists of a small nuclear RNA (snRNA) (or two, in the case of U4/U6), seven Sm proteins that form a ring (or seven Lsm proteins in the case of U6), and a variable number of snRNP-specific proteins (Fig. 1F) (Will and Lührmann 2011). Under splicing conditions, the U4/U6 and U5 snRNPs form a 25S U4/U6.U5 tri-snRNP, in which the U4 and U6 snRNAs are base-paired, forming a three-way junction. The functionally important regions of the snRNAs are highly conserved, as is—for the most part—the protein composition of the snRNPs, although some proteins that are snRNP-associated in yeast are instead recruited to the human spliceosome as non-snRNP associated factors, and vice versa.

Spliceosome assembly proceeds stepwise and the generation of its active site requires extensive structural and compositional rearrangements (Fig. 1C) (Will and Lührmann 2011). These are driven mainly by eight conserved DEXD/H-box ATPases or RNA helicases that trigger RNA–RNA and RNA–protein rearrangements (Cordin and Beggs 2013). Initially, U1 snRNP interacts with the pre-mRNA, via base-pairing between the 5′ end of U1 snRNA and the 5′SS, and is stabilized by U1-associated proteins. Subsequently, U2 snRNP stably interacts with the BS, forming the spliceosomal A complex. In the latter, U2 snRNA base pairs with the BS, whereby the BS-A bulges out from the U2/BS helix. The U2–BS interaction is reinforced by the U2AF65/35 heterodimer, which binds the PPT, and by the U2-associated, heteromeric protein complexes SF3a and SF3b that contact the intron 6 to 26 nucleotides upstream of the BS-A (the so-called anchoring site) (Gozani et al. 1996). Next, the preformed U4/U6.U5 tri-snRNP joins the A complex, generating the pre-B complex in which the tri-snRNP is not yet stably bound (Boesler et al. 2016). Initial tri-snRNP docking is facilitated by the formation of a short duplex (U2/U6 helix II) between the 5′ and 3′ ends of the U2 and U6 snRNA, respectively (Fig. 1D). In humans, stable tri-snRNP integration requires the action of the helicase PRP28 (Boesler et al. 2016), which disrupts the U1/5′SS helix, allowing the latter to base-pair with the conserved U6 ACAGA box (Staley and Guthrie 1999). This destabilizes the U1 snRNP and yields the precatalytic B complex. In the latter, loop 1 of U5 snRNA interacts with

Figure 1. (*See figure on following page.*) Assembly pathway, RNA network, and compositional dynamics of the spliceosome. (*A*) Conserved sequences at the 5′ splice site (SS), branch site (BS), and 3′SS of U2-type (major) pre-mRNA introns in metazoans and the yeast *Saccharomyces cerevisiae*. (*B*) Schematic of the two-step splicing reaction. p, phosphates at the 5′ and 3′ splice sites, which are conserved in the splicing products. (*C*) Assembly, catalytic activation, and disassembly pathway of the spliceosome. For simplicity, the ordered interactions of the U1, U2, U4/U6, and U5 small nuclear ribonucleoproteins (snRNPs), but not non-snRNP spliceosomal proteins, are shown. Helicases required for splicing in both yeast and humans are indicated and include the Ski2-like helicase BBR2, the DEAD-box helicases UAP56, PRP5, and PRP28, and the DEAH-box helicases PRP2, PRP16, PRP22, and PRP43. (*D*) Dynamic network of RNA–RNA interactions in the spliceosomal B and B^{act} complexes. (*E*) 3D structure of the catalytic RNA network (in the human C* complex), showing the coordination of the catalytic magnesium ions M1 and M2. (*F*) Dynamic exchange of spliceosomal proteins during splicing. Proteins present in pre-B, B, B^{act}, C, or C* human spliceosomal complexes are indicated by a square, in which blue denotes highly abundant and gray moderately abundant proteins. Serine-arginine (SR) dipeptide-rich proteins and heterogeneous nuclear ribonucleoproteins (hnRNPs), as well as those present in very low amounts, are not shown. The table summarizes the proteomes of various, recently purified human spliceosomal complexes (Agafonov et al. 2011; De et al. 2015; Boesler et al. 2016; Bertram et al. 2017b), as well as our unpublished work. *Prp38 and Snu23 are associated with the tri-snRNP in *S. cerevisiae*. (*G*) Differential pre-organization of yeast nineteen complex (NTC) proteins in humans. (*H*) Alternative secondary structures of the U2 small nuclear RNA (snRNA).

Cite as *Cold Spring Harb Perspect Biol* doi: 10.1101/cshperspect.a032417

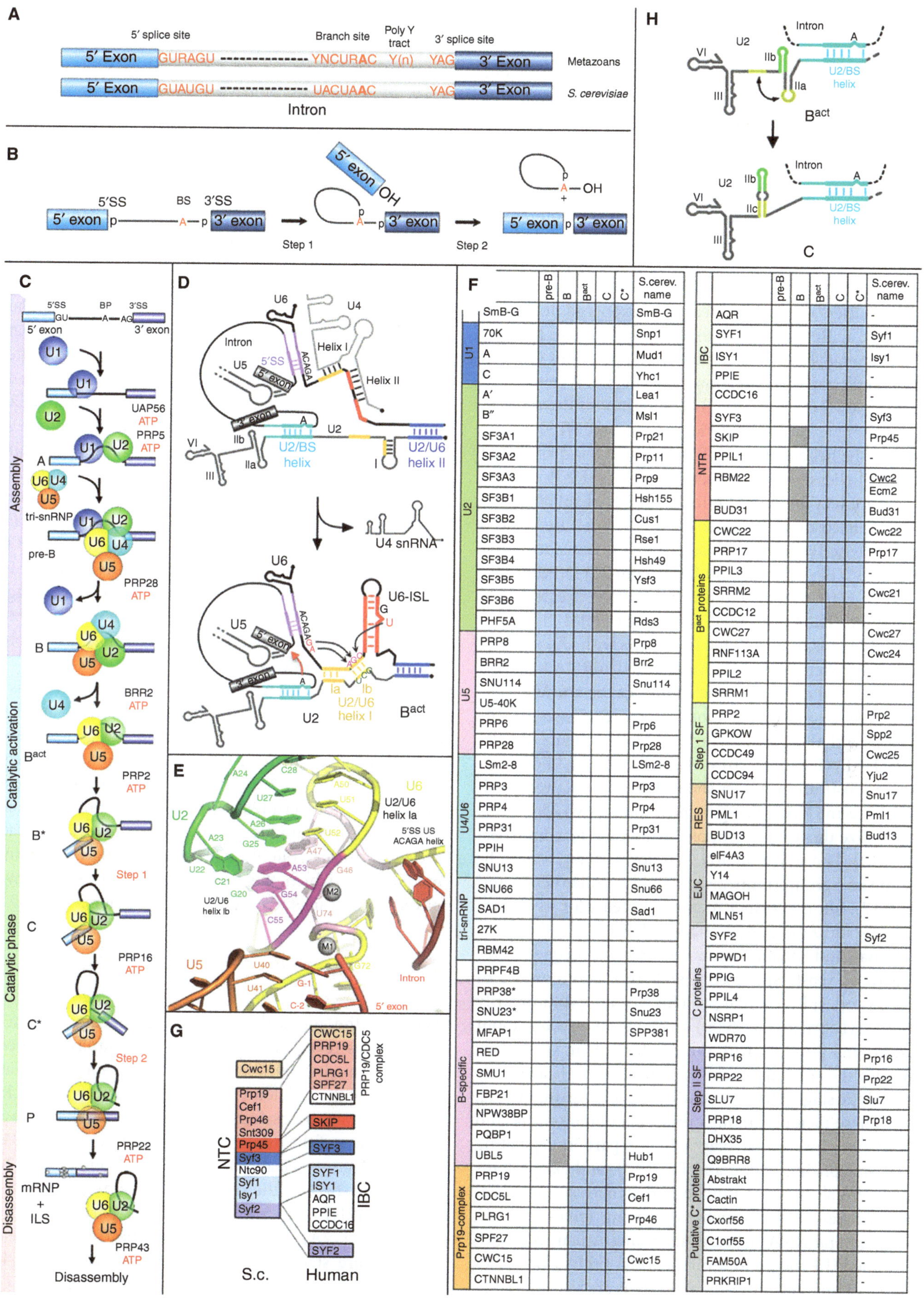

Figure 1. (*See legend on previous page.*)

the 3′-terminal nucleotides of the 5′ exon, and later aids in positioning both exons for step 2 of splicing (Newman and Norman 1992; Sontheimer and Steitz 1993).

Catalytic activation occurs stepwise, with RNA helicase BRR2 initially unwinding the U4/U6 snRNA duplex (Cordin and Beggs 2013). After displacement of U4 snRNP and integration of numerous proteins, the B^{act} complex is generated. Subsequent action of RNA helicase PRP2 transforms B^{act} into a catalytically competent B* complex, which catalyzes step 1 of splicing (branching), yielding the C complex. The latter contains the cleaved 5′ exon and the intron lariat-3′ exon. Remodeling of the C complex by the helicase PRP16 generates the C* complex, which performs the second catalytic step (exon ligation), yielding the P (postsplicing) complex (Horowitz 2012). The release of the mRNA, which generates the intron lariat spliceosome (ILS), is catalyzed by the helicase PRP22. Finally, the ILS is dismantled by PRP43 helicase, and the U2, U5, and U6 snRNPs that are released are recycled for additional rounds of splicing.

The spliceosomal RNA network is restructured into a catalytically active conformation during the transformation of B into B* (Fig. 1) (Will and Lührmann 2011). After disruption of U4/U6 stems I and II, U6 snRNA is extensively restructured. The U6 internal stem-loop (U6 ISL) is generated, and U6 forms additional base-pairs with U2, yielding the U2/U6 helices Ia and Ib. A catalytic U6 RNA triplex is also formed, whereby two nucleotides of the U6 ACAGAGAG sequence and U6-U74 form Hoogsteen base-pair interactions with the U6-catalytic triad AGC sequence of helix Ib (Fig. 1D,E) (Fica et al. 2014; Hang et al. 2015). Phosphorothioate substitution studies showed that U6 catalyzes both steps of pre-mRNA splicing by positioning two Mg^{2+} ions that stabilize the leaving groups during each reaction (Fica et al. 2013). The active site RNA conformations of the spliceosome and group II introns are highly similar (Zhao and Pyle 2017), but the former requires proteins for its formation and stabilization as described below.

Human and *Saccharomyces cerevisiae* (subsequently referred to as yeast) spliceosomes share a common set of about 85 core spliceosomal proteins that are evolutionarily conserved (Fabrizio et al. 2009). In higher eukaryotes, about 80 additional proteins, with no counterparts in yeast, copurify with the splicing machinery throughout its assembly and catalytic stages (Jurica and Moore 2003; Agafonov et al. 2011). These include, among others, (1) numerous SR and hnRNP proteins, which play a major part in regulating alternative splicing (Zhou and Fu 2013), and (2) proteins with potential enzymic functions that may act as molecular switches or conformational driving forces—that is, kinases, several RNA helicases including Aquarius (AQR, also termed IBP160), and seven peptidyl-prolyl isomerases (PPIases). Human spliceosomal proteins are also subject to substantially more posttranslational modification than yeast proteins. Many spliceosomal proteins, in addition to the snRNPs, are pre-organized into large functional ensembles, which simplifies the assembly landscape and enhances the kinetics of spliceosome assembly. These include the retention and splicing (RES) complex (Dziembowski et al. 2004), the splicing essential PRP19/CDC5L complex (Makarova et al. 2004), and the pentameric intron-binding complex (IBC) (De et al. 2015) (Fig. 1F). The yeast counterparts of some IBC and PRP19/CDC5L complex proteins are pre-organized in a different manner (Fig. 1G). That is, they are found in the large Prp19 complex (NTC, nineteen complex) together with additional spliceosomal proteins, some of which in human are recruited independently (i.e., SKIP, SYF2, SYF3) and/or at a different stage (SYF2) of splicing. Thus, not only is the assembly of the human spliceosome more complex, with likely additional assembly intermediates, but also more highly regulated.

The protein composition of the spliceosome changes from one spliceosomal complex to the next (Fig. 1F) (Will and Lührmann 2011). For example, during the pre-B to B complex transition, nine B-specific proteins are stably integrated. The greatest exchange of proteins occurs during the B to B^{act} transition, in which in humans about 20 proteins are released/destabilized, and more than 25 proteins are newly integrated. During this transition, U4/U6- and B-specific proteins are destabilized/released, whereas the PRP19/CDC5L and RES complexes, the IBC, and other proteins, including those operationally designated PRP19/CDC5L-related (nineteen-related, NTR) are stably integrated. Unlike PRP19/CDC5L and IBC proteins, PRP19-related proteins are not recruited as part of a large preformed complex. How this dramatic exchange of proteins takes place is presently unclear, but likely proceeds through several intermediates, in which the loss of certain proteins provides novel binding sites required to recruit others.

2 STRUCTURE OF HUMAN SPLICEOSOMAL snRNPs

Whereas the molecular architecture of the human U1 snRNP was revealed by X-ray crystallography (Pomeranz Krummel et al. 2009; Weber et al. 2010), a high-resolution structure of the isolated 17S U2 snRNP is currently lacking. Cryo-EM was successfully used to elucidate the molecular organization of the human (h) and yeast (y) U4/U6.U5 tri-snRNP (Nguyen et al. 2015, 2016; Agafonov et al. 2016; Wan et al. 2016a), the largest spliceosomal subunit. The U4/U6 snRNP is located in the upper part of the (h)tri-snRNP (Fig. 2A). Stems I and II of the U4/U6 three-way junction are stacked co-axially and stabilized by U4/U6 proteins that also extensively interact with each other. Con-

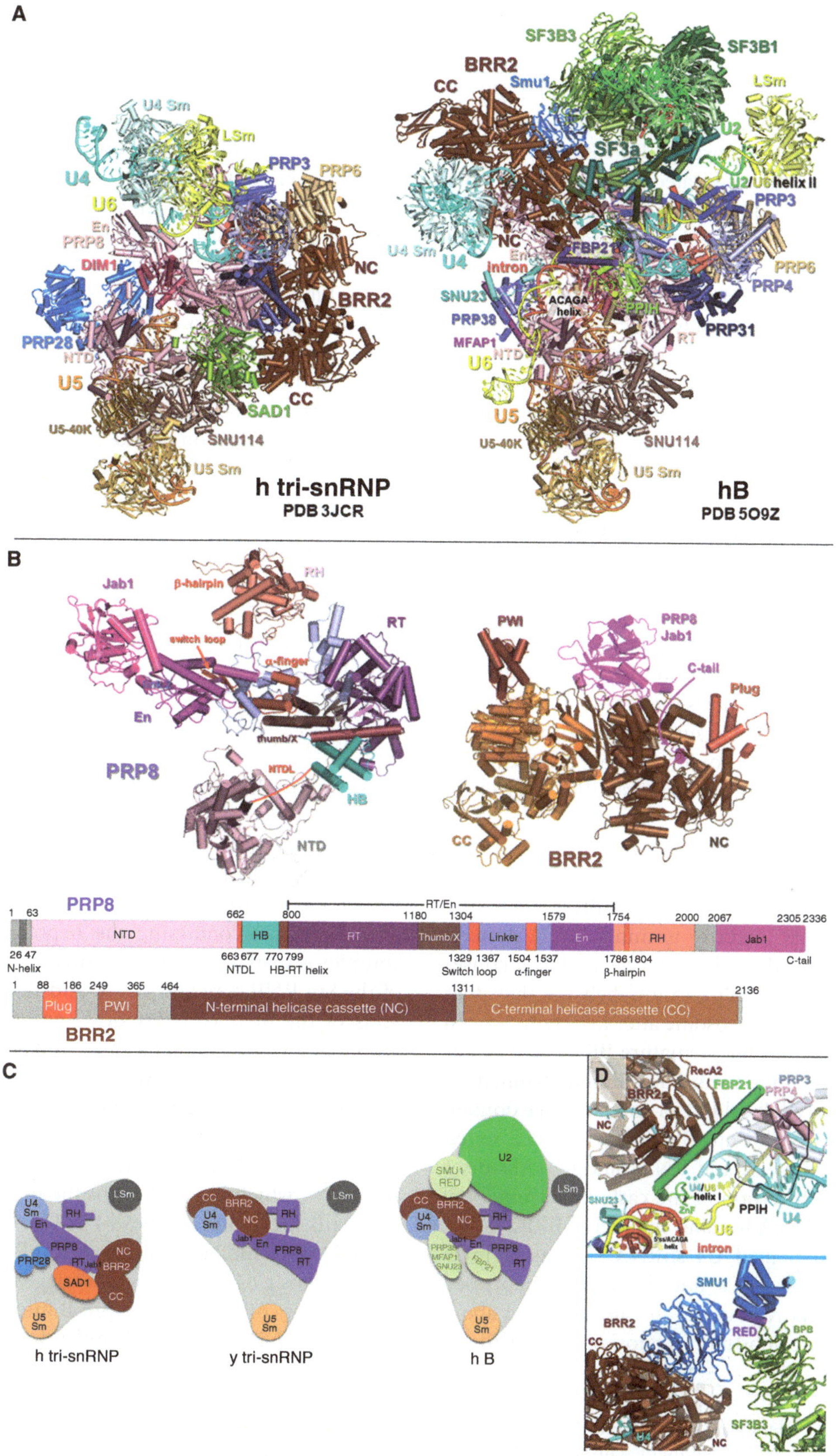

Figure 2. Structure of the human tri-snRNP and spliceosomal B complex. (*A*) Overview of the structure of the human U4/U6.U5 tri-snRNP and B complex. (*B*) Schematic of the domain organization of human PRP8 and BRR2. PRP8 contains an amino-terminal domain (NTD), helical bundle (HB), reverse transcriptase-like (RT) domain, and thumb/X domain, followed by a Linker region and an endonuclease-like (En) domain, plus carboxy (C)-terminal RNase-H-like (RH) and Jab1/MPN domains. (*Legend continues on following page.*)

sistent with previous studies (Schaffert et al. 2004), PRP6 plays a key bridging function in the tri-snRNP, contacting PRP4 and other U4/U6 proteins, and U5 PRP8.

U5 PRP8 and SNU114 are located in the central region of the tri-snRNP, whereas the U5 Sm core is located at the bottom (Fig. 2A). Prp8 is a large, highly conserved scaffolding protein that contains several functionally important domains (Fig. 2B). Cross-linking studies showed that Prp8 interacts with all of the chemically reactive groups of the pre-mRNA (5′SS, BS, and 3′SS) (Grainger and Beggs 2005), and these interactions were subsequently confirmed by cryo-EM of various spliceosomal complexes (Scheres and Nagai 2017; Shi 2017). In the isolated tri-snRNP (and after its integration into the spliceosome), the Prp8 amino-terminal domain (Prp8^{NTD}) tightly grasps U5 stem I, holding it in a rigid conformation and at a later stage positioning U5 loop 1 close to the active site. PRP8 also interacts extensively with the GTP-binding protein SNU114 mainly via its amino-terminal domain (NTD). SNU114 shares structural homology with the ribosome elongation factor EF-2/EF-G. In all currently available high-resolution cryo-EM structures of the spliceosome, Snu114 shows the same compact structure and appears to bind GTP, suggesting that it does not hydrolyse GTP to facilitate a conformational change in the spliceosome (for detailed discussion, see Plaschka et al. 2019).

The U5 helicase BRR2 contains two helicase cassettes (Fig. 2B), but only the amino-terminal cassette (NC) actively unwinds U4/U6 (Santos et al. 2012). BBR2^{NC} is tightly bound by the PRP8 carboxy-terminal Jab1 domain (PRP8^{Jab1}) throughout the splicing cycle, and as a consequence BRR2 is the only spliceosomal helicase that does not dissociate from the spliceosome after its action. Several mechanisms exist to prevent premature BRR2 activity including (1) auto-inhibition by its large amino-terminal region, which contains a helical PLUG and PWI-like domain (Fig. 2B) that folds back onto BRR2's helicase cassettes, preventing ATP hydrolysis and RNA binding (Absmeier et al. 2015) and (2) insertion of the carboxy-terminal tail of PRP8^{Jab1} into the RNA binding channel of BRR2^{NC}, inhibiting accommodation of U4 snRNA (Mozaffari-Jovin et al. 2013). In (h)tri-snRNPs, not only is BRR2 repressed by its amino-terminal region, but its NC is located about 10 nm from its U4/U6 substrate (Agafonov et al. 2016).

Surprisingly, although the region comprising the PRP8^{NTD}, SNU114, and U5 Sm core is structurally very similar, the remaining parts of yeast and human tri-snRNPs differ in their conformation or position to varying degrees (Fig. 2C). Most strikingly, in (y)tri-snRNPs, BRR2 is located ~20 nm away from its position in (h)tri-snRNPs (Fig. 2). Moreover, as yeast Brr2 is loaded onto U4 snRNA and is not auto-inhibited (Nguyen et al. 2016; Wan et al. 2016a), it appears to be ready to unwind the U4/U6 duplex. Indeed, unlike purified (h)tri-snRNPs, purified (y)tri-snRNPs dissociate in the presence of ATP in a BRR2-dependent manner (Nguyen et al. 2015; Agafonov et al. 2016). The different molecular organization of BRR2 could (1) reflect differences in the early stages of the spliceosome activation process in yeast versus humans and/or (2) be the result of the absence of Sad1 in (y)tri-snRNPs, which plays a key role in stabilizing the tri-snRNP in yeast extract (Huang et al. 2014). In (h)tri-snRNPs, SAD1 binds SNU114, BRR2's PWI domain and U4/U6-PRP31, thereby stabilizing the interaction between U5 and U4/U6, and also tethering BRR2 at a distance from U4/U6 (Agafonov et al. 2016).

3 STRUCTURE OF THE PRECATALYTIC B COMPLEX

Cryo-EM of human (Bertram et al. 2017a) and yeast (Plaschka et al. 2017) spliceosomal B complexes showed that their globular head domain is comprised of the U2 snRNP, and their triangular body contains mainly the U4/U6.U5 tri-snRNP (Fig. 2A). In human B (hB), the 5′ domain of U2 containing the heptameric SF3b protein complex is spatially separated from its 3′ domain comprised of the Sm RNP core and U2 A′ and B′ proteins, with proteins of the SF3a complex bridging both domains. SF3b sequesters the U2/BS helix in a similar manner to that in the B^{act} complex (see below). U2 snRNP is connected to the tri-snRNP by the U2/U6 helix II, which is stabilized by PRP3, and by bridges between BRR2 and SF3B3 involving SMU1 (Fig. 2A,D), and between the U2 SF3b domain and PRP8^{NTD}, which likely contains the intron and SF3A1.

In hB, intron nucleotides +4 to +6 base-pair with the U6 ACAGA-box and, accordingly, U1 snRNP and PRP28 are absent. This U6/5′SS interaction triggers a major structural change that converts human pre-B into a stable B complex (Boesler et al. 2016). U6 nucleotides directly upstream of the

Figure 2. (*Continued*) The RT, En, RH, and Jabl/MPN domains are deconstructed enzymes that do not show enzymatic activities. Within the Linker, several distinct structural elements play important roles within the spliceosome including the switch-loop and the α-finger. The 3D structure of PRP8 shown is from the hB complex, and that of BRR2 (except for its PWI domain) is from hB^{act}. (*C*) Cartoon of the organization of selected proteins in the human and yeast tri-snRNPs, and human B complex. (*D*) *Upper* panel: FBP21's zinc finger contacts the U6/intron helix, whereas its α-helix keeps U4/U6 stem I, via PPIH (shown as a black outline), at a distance from Brr2^{NC}. *Lower* panel: SMU1 and RED bridge BRR2 and SF3B3.

Cite as *Cold Spring Harb Perspect Biol* doi: 10.1101/cshperspect.a032417

ACAGA-box also form base pairs (including noncanonical ones) with additional intron nucleotides downstream from the 5′SS. This extended U6/intron helix, which is also found in human B^{act}, C, and C* complexes, appears to be a structural feature of spliceosomes solely in higher eukaryotes and may help to stabilize the short U6/5′SS helix. In hB, the extended U6/intron helix and, unlike in yeast also the 5′ region of U6, are accommodated by PRP8^{NTD} (Fig. 2A).

Most human tri-snRNP proteins undergo substantial structural rearrangements on tri-snRNP integration into the spliceosome (Fig. 2C) (Bertram et al. 2017a). The most dramatic is observed for BRR2 and its binding partner PRP8^{Jab1}, which are located in hB near the PRP8 En domain (Prp8^{En}) ~20 nm from their position in the isolated tri-snRNP. Moreover, BRR2^{NC} is loaded onto U4 snRNA, and is poised to unwind the U4/U6 duplex. SAD1, which tethers BRR2 in its pre-activation position, is largely displaced from the B complex, which may play a role in triggering tri-snRNP rearrangements. Finally, PRP8 adopts a partially closed conformation, in which several loops emanating from Prp8's Linker are juxtaposed with its NTD and the U5 DIM1 protein. Together, they form a pocket that binds the 5′SS GU nucleotides in an extended conformation, revealing an unexpected role for DIM1 in 5′SS recognition (Bertram et al. 2017a).

The structural organization of U5 and U4/U6 proteins, and PRP38 and SNU23, is highly similar in human and yeast B complexes, and, surprisingly, also in the purified yeast tri-snRNP (Fig. 2C) (Nguyen et al. 2016; Wan et al. 2016a). This suggests that, in contrast to the situation in humans, the yeast tri-snRNP is already rearranged before docking with the A complex. Although BRR2 binds its U4/U6 substrate and is no longer auto-inhibited in both B complexes, U4 and U6 snRNA dissociate solely from purified yeast B complexes in the presence of ATP (Bao et al. 2017), suggesting that BRR2 is still negatively regulated in the purified human, but not yeast B complex. This implies that tri-snRNP rearrangements leading to BRR2 substrate loading and the triggering of BRR2 activity are both more complex in higher eukaryotes.

The hB cyro-EM structure revealed the likely functions of several B-specific proteins, which include those lacking a yeast counterpart, such as SMU1, RED, and FBP21, and conserved proteins like PRP38, SNU23, and MFAP1. Because of their mutually exclusive binding sites, incorporation of the latter proteins in hB requires prior displacement of PRP28. PRP38 and SNU23 contact the U6/5′SS helix, and may facilitate its repositioning during activation, enabling the juxtapositioning of the 5′SS and U2/U6 catalytic center. The heterodimer SMU1/RED bridges U2 and the tri-snRNP (Fig. 2D), and via its interaction with BRR2^{NC}, it may potentially modulate BRR2 activity. Intriguingly, an α-helical element of FBP21, which was implicated in regulating BRR2 activity (Henning et al. 2017), is positioned between BRR2^{NC} and U4/U6 proteins that are bound to U4/U6 stem I (Fig. 2D). This architecture suggests that FBP21 may act to keep U4/U6 stem I at a distance from BRR2's NC, and that its release from the spliceosome, which occurs early during activation, may play a part in triggering BRR2 activity. Thus, the initial steps toward the formation of an activated spliceosome appear to differ between yeast and humans.

4 STRUCTURE OF THE ACTIVATED (B^{act}) SPLICEOSOME

The cryo-EM structure of the B^{act} complex (Fig. 3A) (Rauhut et al. 2016; Yan et al. 2016; Haselbach et al. 2018; Zhang et al. 2018) provides important insights into the mechanism of spliceosome activation, and elucidates the function of proteins integrated at this stage. As first indicated by biochemical studies, the catalytic U2/U6 RNA network is already in place in B^{act}—that is, the U6 ISL, U2/U6 helices Ia and Ib and catalytic RNA triple helix have formed. However, in contrast to catalytically active C or C* complexes, where catalytic metal ions M1 and M2 are positioned ~4.2 Å apart (Fig. 1E), in B^{act} density for M1 is either missing or displaced both in humans and yeast. Thus, B^{act} lacks a fully functional catalytic center.

The rearranged catalytic U2/U6 RNA core is stabilized to a large extent by interactions with PRP8 (Fig. 3B). During the B to B^{act} transition, PRP8 undergoes a further conformational change that leads to a closed conformation of its cavity that accommodates the U2/U6 RNA network. The U6 ILS fits snuggly into a basic pocket of PRP8^{NTD}, close to U5 loop 1, and is stabilized by PRP8^{NTDL}. The nearby PRP8 HB domain accommodates the U2 strand of U2/U6 helix Ib. The α-finger of the PRP8 Linker (also called the 1585-loop in yeast) further stabilizes the catalytic center by interacting with U2 nucleotides between U2/U6 helix I and the U2/BS helix. This leaves a gap between the U2/U6 catalytic center and the PRP8 Linker region in which part of RNF113A (Cwc24 in yeast) and SF3A2 are accommodated. During catalytic activation, RNF113A and SF3A2 are replaced in this gap by the U2/BS helix (Fig. 3B), thereby juxtapositioning the first-step reactants. PRP8^{NTD} plays an important part in orienting the scissile bond of the 5′SS toward the catalytic center. However, the bases of G+1 and U+2 of the 5′SS are contacted by the carboxy-terminal zinc finger of RNF113A and amino terminus of SF3A2, thus protecting the 5′SS from premature nucleophilic attack (Fig. 3C).

In B^{act}, the catalytic RNA network is also contacted by PRP19/CDC5L complex and PRP19-related proteins (i.e., CDC5L, CWC15, PLRG1, SKIP, and SYF3). These proteins

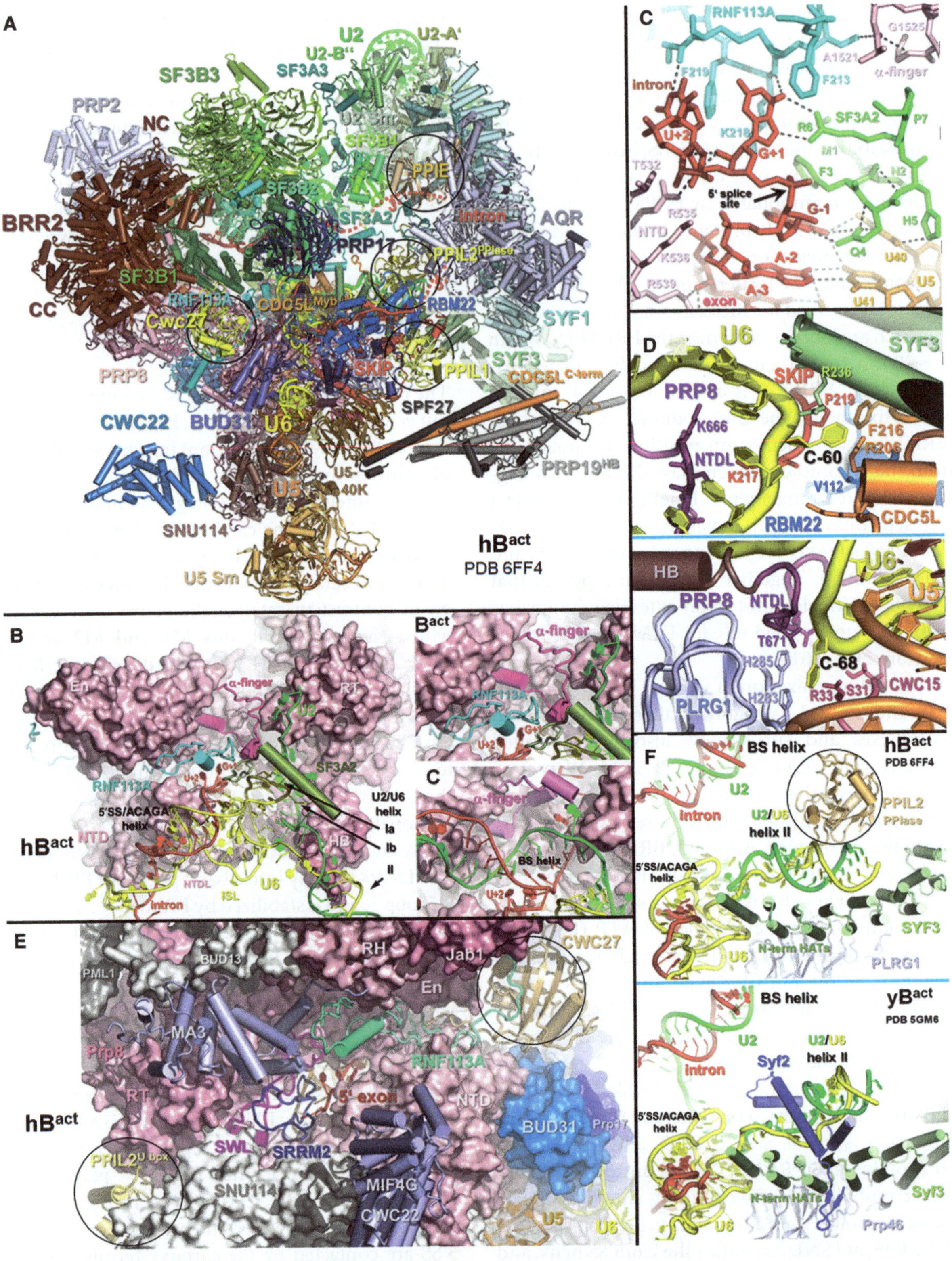

Figure 3. Molecular organization of RNA–protein interactions involving the catalytic RNA network, 5′SS, and 5′ exon in the human B^{act} complex. (*A*) Overview of the structural organization of the human B^{act} complex. The positions of PPIases are circled in black. (*B*) Accommodation of the catalytic RNA network that is first formed in B^{act} by the PRP8 amino-terminal domain (NTD) and helical bundle (HB) domains and α-finger of its Linker domain. (*Legend continues on following page.*)

make only limited direct contacts, interacting predominantly with flipped-out nucleotides of U6 snRNA such as U68 and C60 (Fig. 3D), consistent with studies showing that the yeast NTC plays a role in stabilizing the association of U6 with the spliceosome (Chan et al. 2003). The resulting protein–RNA network serves to both stabilize and fine-tune the conformation of the catalytic RNA network, and may transmit structural changes in the catalytic center to the periphery of the spliceosome and vice versa. As CDC5L, SKIP, CWC15, and PLRG1 interact with PRP8, they likely also indirectly help to maintain the RNA in a catalytically active conformation by stabilizing the closed conformation of PRP8. The extensive protein–RNA network that is established in the catalytic RNP core, which is conserved between yeast and humans (Rauhut et al. 2016; Yan et al. 2016), underscores the essential role of proteins in establishing a catalytically active RNA structure in the spliceosome. In fact, the spliceosome may be considered an RNP enzyme, in which the actual catalytic steps are facilitated by a catalytic metal–RNA center.

When PRP8 adopts its closed conformation in B^{act}, a narrow 5′ exon binding channel (EBC) is created between PRP8's RT/Linker and NTD (Fig. 3E). Exon nucleotides −5 to −7, upstream of where U5 loop 1 interacts with the 5′ exon, are contacted by loops of $PRP8^{Linker}$. The 5′ exon is then contacted by the amino-terminal region of SRRM2 (Cwc21 in yeast) and the Prp8 switch loop ($PRP8^{SWL}$), which has moved ~7 nm from the $PRP8^{En}$ in B, to $PRP8^{RT}$ in B^{act}. The EBC is straddled by CWC22 whose MIF4G and MA3 domains bind to SNU114 and $PRP8^{RT}$, respectively (Fig. 3E). In B^{act}, $PRP8^{En}$ is in an upright position relative to $PRP8^{RT}$, and does not interact with $PRP8^{NTD}$. Both domains are kept at a distance by the presence of amino-terminal regions of both RNF113A and CWC27 in a cleft between $PRP8^{En}$ and $PRP8^{NTD}$ (Fig. 3B,E). RNF113A tightly interacts with the PPIase domain of CWC27—the only human spliceosomal PPIase with a yeast counterpart. Release of these proteins during catalytic activation likely triggers the conformational change in the PRP8 RT/En domain relative to $PRP8^{NTD}$ that is observed during the B^{act} to C complex transition (see also Fig. 7B). This leads to tightening of PRP8's grasp on the 3′ end of the 5′ exon, thereby ensuring that it is stably bound after 5′SS cleavage.

In hB^{act}, the extended U6/intron helix binds to a positively charged edge of $PRP8^{NTD}$ and is further stabilized by contact with RBM22. Intron nucleotides downstream from this helix are enclosed in a positively charged channel formed by RBM22 and further downstream they are likely handed over to the RNA helicase AQR (Fig. 3A and see below). Thus, RBM22 plays an important part in chaperoning the 5′ region of the intron. In yB^{act}, similar tasks are performed by Ecm2 and Cwc2, which contain one or more domains homologous to those found in RBM22, indicating that the latter is a fused version of Ecm2 and Cwc2 (Rasche et al. 2012).

The conformation of the U2/U6 RNA and RNP network is largely similar in yeast and human B^{act}, and the structure of the spliceosome's core domain does not change substantially in the subsequently formed catalytically active spliceosomal complexes. However, U2/U6 helix II adopts a unique conformation in hB^{act} (Fig. 3F). In yeast B^{act} and in subsequently formed human and yeast complexes, U2/U6 helix II is bound by Syf2. In contrast, in hB^{act}, helix II is clamped between SYF3 and the PPIase domain of PPIL2, preventing SYF2 from binding. This explains, at least in part, why in humans SYF2 binds at a later stage and is not preorganized into a large complex like its yeast counterpart. SYF3 interacts with SYF1, which also shares a large interface with the RNA helicase AQR that is absent in yeast. Rearrangement of U2/U6 helix II and the incorporation of hSYF2 during the B^{act} to C transition in higher eukaryotes may thus involve remodeling by AQR, whose ATPase activity is required for efficient catalytic activation of human spliceosomes (De et al. 2015).

The bipartite U2 snRNP is located at the top of the hB^{act} structure (Fig. 3A). As in hB, the 5′ and 3′ domains of U2 are bridged by SF3a proteins that contact the U2 Sm ring and SF3b proteins in the 5′ domain. SF3B3, which contains

Figure 3. (*Continued*) Part of RNF113A (Cwc24) and SF3A2 (Prp11) are inserted into the gap between the RT/En domain and the catalytic RNA network and these proteins are replaced by the U2/BS helix in the C complex (see *insets*). (*C*) Close-up of interactions between the 5′SS nucleotides G+1 and U+2 and residues of RNF113A, and between SF3A2 and G+1 and G−1. (*D*) PRP19/CDC5L complex proteins and related proteins directly contact primarily flipped-out U6 nucleotides in the catalytic RNA core. For example, the base of U6-C60 in the lower stem of the ISL lies in a pocket formed by amino acids of SYF3, CDC5L, RBM22, and SKIP (*upper* panel) and flipped-out U6-U68 of the U6 ISL loop is bound cooperatively by amino acids of PLRG1 and CWC15 (*lower* panel). (*E*) Spatial organization of the 5′ exon binding channel in hB^{act}. Black circles indicate the position of PPIL2's U box domains and CWC27's PPIase domain. (*F*) U2/U6 helix II adopts a unique position in human (*upper* panel) compared with yeast (*lower* panel) B^{act}. In yB^{act}, helix II lies in a more upright position and interacts with Syf2, whereas in hB^{act} it lies more perpendicular (relative to the U6 ACAGA/5′SS helix) and contacts the PPIase domain of the PPIL2 protein, which is not present in the *S. cerevisiae* spliceosome. Black circle indicates the position of PPIL2's PPIase domain.

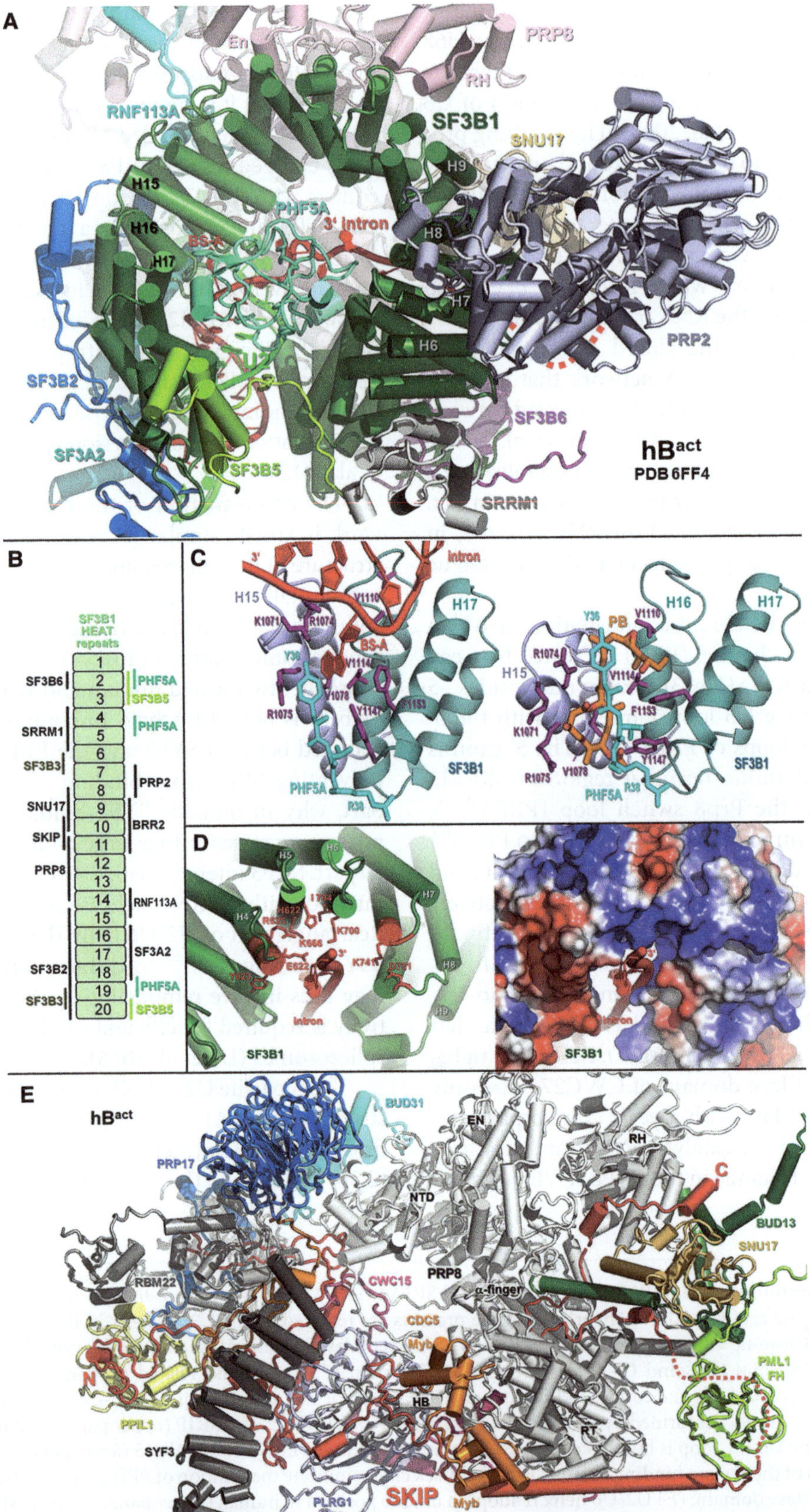

Figure 4. (*See legend on following page.*)

three β propeller domains (BPD) A-C, SF3B5, and PHF5A form an intricate protein network with the carboxy-terminal HEAT domain of SF3B1 ($SF3B1^{HEAT}$), similar to that observed in the crystal structure of the human SF3b core complex (Cretu et al. 2016). The $SF3B1^{HEAT}$ domain forms a superhelical structure that clamps the U2/BS helix between its terminal HEAT repeats (Fig. 4A), consistent with cross-linking studies showing SF3B1 contacts the pre-mRNA on both sides of the BS (Gozani et al. 1998). The base of the BS-A nucleotide resides in a pocket comprised of SF3B1 HEAT repeats H15–H17 and a Tyr side-chain of PHF5A (Fig. 4C). Thus, the first catalytic step nucleophile is sequestered by protein and also located ~5 nm from the 5′SS, preventing premature 5′SS cleavage.

X-ray crystallography (Cretu et al. 2018) and cryo-EM (Finci et al. 2018) of the recombinant SF3b core bound by the splicing inhibitor/modulator Pladienolide B (PB) or its derivative E7107, respectively, recently revealed that these antitumor compounds bind to SF3B1 and PHF5A residues that ultimately form the BS-A binding pocket, and that, when mutated, lead to compound resistance (Fig. 4C). Thus, inhibitor binding and accommodation of the BS-A appear to be mutually exclusive, suggesting that these SF3B inhibitors compete with the BS for SF3b binding. $SF3B1^{HEAT}$ shows a more open conformation in the apo SF3b core structure (Cretu et al. 2016), compared with its conformation in the B and B^{act} complex (Rauhut et al. 2016; Yan et al. 2016; Bertram et al. 2017a; Plaschka et al. 2017; Haselbach et al. 2018; Zhang et al. 2018). By binding to HEAT repeats H15–H17, PB appears to act as a wedge, "locking" $SF3B1^{HEAT}$ in the open form (Fig. 4C) (Cretu et al. 2018). In the absence of PB, interaction of the U2/BS helix with the open conformation likely induces the closed SF3B1 conformation, with H15–H17 acting as a conformational switch.

U2 and intron nucleotides upstream of the BS form an RNA helical structure comprised of many noncanonical base pairs. This extended U2/BS helix contains part of the U2 anchoring site and is contacted by SF3A2 and SF3A3, consistent with cross-linking studies (Gozani et al. 1996; Schneider et al. 2015). As SF3A2's amino terminus also contacts the 5′SS GU nucleotides, there is a direct link between the extended U2/BS helix and the 5′SS. The structure of SF3b and its interactions with the extended U2/BS helix are very similar in human and yeast B^{act}. However, in higher eukaryotes SF3b contains an additional subunit, the RRM protein SF3B6 (p14). In hB^{act}, SF3B6 is located close to HEAT repeats 1–2 on the convex (outer) side of $SF3B1^{HEAT}$ and is spatially separated by ~5 nm from the BS-A (Fig. 4A) (Haselbach et al. 2018; Zhang et al. 2018). As SF3B6 is cross-linked to the BS-A in early spliceosomal complexes (Will et al. 2001), it appears to be substantially repositioned before the formation of B^{act}.

$SF3B1^{HEAT}$ not only sequesters the U2/BS helix, but also acts as a scaffold, interacting not only with SF3a/b proteins, but also several other spliceosomal proteins (Fig. 4A,B). SNU17's RES complex binding partner BUD13 interacts with $PRP8^{RT}$ and thus the RES proteins bridge the major structural modules of the U2 (the SF3B complex) and U5 snRNP (PRP8) (Fig. 4E), consistent with recent studies indicating that they play an important structural role during catalytic activation (Bao et al. 2017).

5 STRUCTURAL BASIS OF PRP2-MEDIATED CATALYTIC ACTIVATION

PRP2 binds HEAT repeats H7–H8, ~6 nm from its actual target, the U2/BS helix (Fig. 4A). PRP2 requires ~25–30 intron nucleotides downstream from the BS to facilitate catalytic activation (Liu and Cheng 2012). Human B^{act} complexes analyzed by cryo-EM (Haselbach et al. 2018; Zhang et al. 2018) were formed on truncated pre-mRNAs with 10 or 19 nucleotides downstream from the BS. These downstream intron nucleotides contact PHF5A and exit the SF3B1 HEAT ring at repeats H4–H6, before they are bound

Figure 4. Sequestration of the U2/BS helix by SF3B1 and location of SF3B1 hotspot mutations and the Pladienolide B binding site. (*A*) The superhelical SF3B1 HEAT domain sequesters the U2/BS helix and acts as a scaffold that binds numerous proteins. Intron nucleotides downstream from the branch site (BS) exit the HEAT domain at repeats H4–H6 and are subsequently bound by the retention and splicing (RES) protein SNU17, followed by the helicase PRP2. SF3B3, which interacts with HEAT repeats H5 and H6 and H19 and H20, has been omitted. (*B*) Schematic of the SF3B1 HEAT repeats and their interaction partners. Proteins contacting both amino and carboxy-terminal HEAT repeats are colored as in *A*. (*C*) Overlapping binding site of Pladienolide B (PB) and the branch site adenosine (BS-A). (*Left*) Residues of SF3B1 HEAT repeats H15–H17 and PHF5A that form the BS-A binding pocket in the human B^{act} complex, in which the HEAT domain shows a closed conformation. (*Right*) The H15–H17 hinge region of SF3B1 bound by PB (orange sticks), which locks the SF3B1 HEAT domain in an open conformation. (*D*) SF3B1 residues frequently mutated in various human cancers (red lettering) cluster near the 3′ end of the intron where it exits the HEAT superhelix (*left*). (*Right*) Space filling model with surface potential (red, negatively charged amino acids; blue, positively charged) of the SF3B1 region shown in the *left* panel. Many hotspot mutations are nonconserved changes that alter the amino acid's charge. (*E*) Meandering path of the SKIP protein (red) in hB^{act} and its numerous interaction partners.

by SNU17, but do not extend to the adjacent PRP2 binding site (as observed in yB^{act}). Thus, PRP2's RecA domains cannot access the intron, providing a structural basis for why PRP2 cannot catalytically activate B^{act} complexes formed on these truncated pre-mRNAs. During B* formation, the U2/BS helix must be freed from $SF3B1^{HEAT}$, and also from other SF3b and SF3a proteins, to juxtapose the BS-A and the scissile bond of the 5′SS. At the same time, RNF113A and the amino terminus of SF3A2 must be repositioned to free the 5′SS and allow docking of the U2/BS helix into the catalytic center. The scaffolding function of $SF3B1^{HEAT}$ likely plays an important role in the coordinated release of multiple proteins during PRP2-mediated formation of B*. That is, PRP2 has been proposed to induce, in an intron-binding and ATP-dependent manner, a conformational change in the HEAT domain curvature (Rauhut et al. 2016) that would disrupt its interaction with the U2/BS helix, setting free the BS-A, and destabilize the interaction of proteins which bind SF3B1, including SF3A2 and RNF113A, as well as its binding partner CWC27. Indeed, previous biochemical studies showed that PRP2 action leads to the destabilization/release of these proteins (Warkocki et al. 2009; Lardelli et al. 2010; Ohrt et al. 2012). Thus, PRP2 facilitates remodeling of the U2/BS from a distance.

6 STRUCTURAL CONSEQUENCES OF SF3B1 CANCER-RELATED MUTATIONS

SF3B1 mutations that occur very frequently in patients with myelodysplastic syndromes and various types of cancer (i.e., "hot spot" mutations) are located predominantly within the intrarepeat loops of HEAT repeats H4–H7 (Fig. 4) (reviewed in Jenkins and Kielkopf 2017). In the B and B^{act} complexes, these cancer-related mutations are located close to where the 3′ region of the intron exits the HEAT superhelix (Fig. 4A,D). This region of SF3B1 forms a basic channel (Fig. 4D), and thus hot spot mutations such as K666E or K700E, will change the surface potential and thus potentially alter intron binding. In vivo, SF3B1 hot spot mutations cause alternative BS selection, leading to cryptic 3′SS usage and the formation of aberrantly spliced mRNAs (Darman et al. 2015; Alsafadi et al. 2016). The mechanism whereby these mutations affect AS is unclear, but could also be caused by changes (1) in the conformation of $SF3B1^{HEAT}$ that destabilize/alter its interaction with the U2/BS helix and/or the 3′ end of the intron (Jenkins and Kielkopf 2017) or (2) that alter its interaction with one or more protein, as recently shown for the RNA helicase Prp5 (Tang et al. 2016). As BS selection occurs predominantly during A complex formation, more detailed insights into the structural consequences of SF3B1 cancer-related mutations await the determination of the cryo-EM structure of the human A and earlier complexes.

7 INSIGHTS INTO THE FUNCTION OF PPIases AND IBC PROTEINS DURING SPLICING

Numerous human-specific spliceosomal proteins, including several PPIases, are located in peripheral domains of hB^{act} and their molecular organization provides insights into their functions (Fig. 3A). One function of the spliceosomal PPIases appears to be to bridge various protein modules. For example, PPIL1 helps to anchor the PRP19 helical bundle ($PRP19^{HB}$) to the B^{act} main body (Fig. 3A and below). The RRM domain of PPIE ($PPIE^{RRM}$) binds SYF1, and its PPIase domain interacts with $SF3B4^{RRM1}$ and the β-sandwich of SF3A2. $PPIL2^{PPIase}$ also interacts with SF3A2, while its amino-terminal U box domains bind to SNU114 and a nearby region of $PRP8^{RT}$ (Fig. 3A,E). The PPIase domains of PPIL2 and PPIE also interact with RNA, contacting U2/U6 helix II or the intron upstream of the extended U2/BS helix, respectively (Fig. 3A,F). Whether these PPIases act as molecular switches during splicing, isomerizing proline-peptide bonds, and thereby catalyzing protein/RNP rearrangements within the spliceosome, is not clear (Thapar 2015).

The helicase AQR shares a large interface with SYF1's amino-terminal HAT (half a TPR) repeats 2–8, which extend close to the U2 3′ domain (Figs. 3A, 5A). SYF1 and SYF3 form long, curved α-helical solenoids that cross one another close to U2/U6 helix II (Figs. 3A and 5B). Via SYF1 and its interaction partners, AQR is thus indirectly connected to the catalytic U2/U6 RNP core and U2/U6 helix II. Within spliceosomes, AQR cross-links to intron nucleotides 30–40 upstream of the BS-A (Hirose et al. 2006). Consistent with this, intron nucleotides upstream of the extended U2/BS helix are contacted by $SF3B4^{RRM1}$ and PPIE, before apparently being transferred to AQR in hB^{act} and hC (Figs. 3A and 5B) (Haselbach et al. 2018; Zhan et al. 2018; Zhang et al. 2018). AQR's β-barrel also directly contacts $RBM22^{RRM}$, which binds intron nucleotides downstream from the U6/intron helix. Thus, AQR connects and potentially fixes, directly or indirectly, intron nucleotides that lie upstream of the extended U2/BS helix and downstream from the extended U6/intron helix (Fig. 3A). This would allow the intron located between these regions, which potentially contains intronic small nucleolar RNAs (snoRNAs) that are processed during spliceosome assembly (Hirose et al. 2006), to loop out, and thereby prevent it from interfering with the spliceosome's core machinery.

Cryo-EM has revealed many examples of proteins that are intrinsically unfolded in the isolated state, but adopt ordered structures in the spliceosome, apparently in a coor-

Cite as *Cold Spring Harb Perspect Biol* doi: 10.1101/cshperspect.a032417

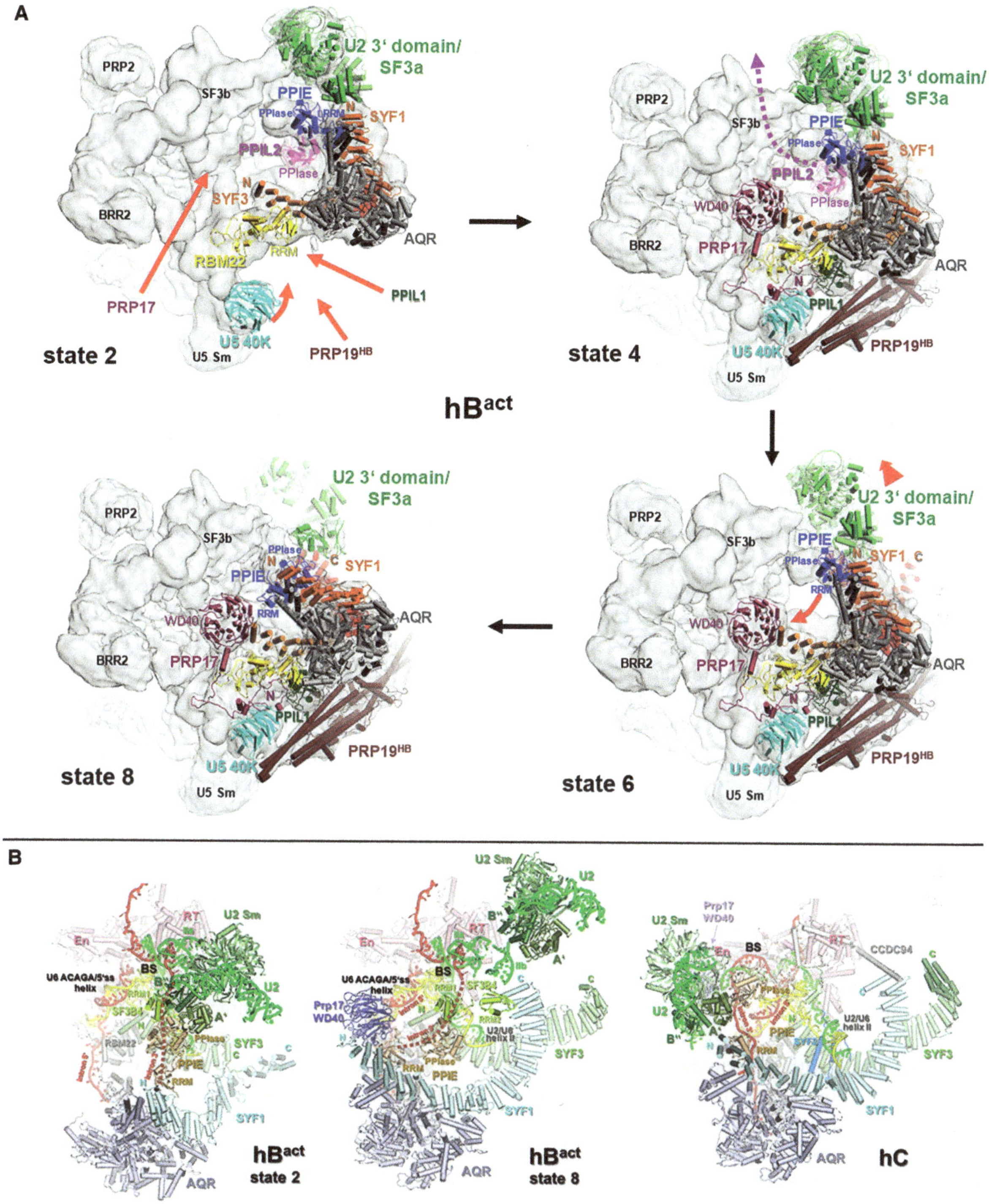

Figure 5. Conformational dynamics of the human B^{act} complex. (*A*) Eight major conformational states of the human B^{act} complex were obtained by computational sorting of images. Only states 2, 4, 6, and 8 are shown. The positions of the most mobile proteins/protein domains are indicated in the electron microscopic (EM) density of B^{act}. State 4 corresponds to the high-resolution hB^{act} structure from Haselbach et al. (2018) that is shown in Figures 3 and 4. Density for the PRP19^{HB}, PPIL1, and most of PRP17 cannot be recognized in states 1 and 2, even though there is evidence that these proteins, as well as nearly all other hB^{act} proteins, are already present in the B^{act} complex represented by stage 1. This means that they bind in a highly flexible manner and are simply not discernable by EM, as opposed to being first recruited to hB^{act} at a subsequent step. In state 4, U5-40K^{WD40} has shifted upward by 2 nm, generating a docking site for one end of the PRP19^{HB}, stabilizing its conformation and leading to its appearance in state 4. Similarly, density for PRP17 and PPIL1 is also first detected in state 4, in which PPIL1 now interacts with PRP19^{HB}. In state 6, PPIL2 is no longer detected, and in state 8, PPIE moves toward PRP17. The U2 3′ domain also moves by several nm from states 2 to 8. (*B*) Comparison of the spatial organization of the U2 3′ domain, SYF1, SYF3, AQR, PPIE, and the PRP17 WD40 domain in states 2 and 8 of hB^{act} and in the hC complex.

dinated manner with other spliceosomal proteins. Such proteins, which include RNF113A and SKIP, often form complex protein–protein networks. Indeed, SKIP meanders throughout B^{act}, contacting at least 10 other proteins, and is therefore thought to play an important part in the conformational maturation and stabilization of B^{act} (Fig. 4E). Proteins like SKIP are also well-suited for transferring information from the periphery of the spliceosome to the catalytic RNP core and thus, potentially, for influencing the kinetics of the splicing reaction. A particularly interesting case of long-range communication is observed with proteins of the PRP19/CDC5L complex. Whereas, the two amino-terminal Myb domains of CDC5L bind $PRP8^{RT}$ and help to stabilize the catalytic U2/U6 RNA structure, CDC5L's carboxy-terminal region forms, together with SPF27 and the coiled-coil regions of four copies of PRP19, an elongated helical bundle ($PRP19^{HB}$) (Figs. 3A and 5A). In B^{act}, the latter is docked to its main body, via PPIL1, SYF1 and the U5-40K WD40 domain (U5-$40K^{WD40}$), which communicate via PRP17 and SKIP with other regions of B^{act}, including its catalytic core. Via these multiple contacts, the $PRP19^{HB}$ potentially plays a role in transferring information from other molecular machines to the spliceosome's catalytic core.

8 ANALYSIS OF THE CONFORMATIONAL DYNAMICS OF hB^{act}

Whereas the spliceosome's core is generally stable and resolved at 3.0–4.5 Å in the various complexes analyzed by cryo-EM, peripheral regions are less well-resolved and the most highly mobile elements or protein domains are not visible in many of the EM density maps. Recently, the conformational dynamics of hB^{act} were investigated by a quantitative method (Haselbach et al. 2017) that combines iterative 3D classification with a newly developed approach based on 3D principal-component analysis (PCA) (Haselbach et al. 2018). Eight major conformational states of B^{act} present in the same cryo-EM sample were obtained (Haselbach et al. 2018). Transitions between these different conformational states can be achieved by the available thermal energy. Comparison of their cryo-EM structures (Fig. 5) allows insight into the conformational variability of the more mobile parts/proteins of B^{act}, which include the $PRP19^{HB}$, IBC proteins, the U2 snRNP 3′ domain, as well as SKIP, PRP17, PPIL1, and PPIE. During its procession from state 1 to 8, certain conformations of hB^{act} are first generated after the coordinated translocation or stable integration of other proteins (Fig. 5). Many of the newly created protein–protein and protein–RNA interactions, and protein conformations observed in state 8, are most similar to those in human C or C* complexes (Bertram et al. 2017b; Zhang et al. 2017; Zhan et al. 2018). For example, the position and conformation of the SYF1 and SYF3 HAT repeats change from state 2 to 8, but in state 8 they are most similar to those observed in C complexes (Fig. 5B). This supports the idea that the conformational maturation of B^{act} proceeds from state 1 to 8, with the latter likely representing the latest maturation state of hB^{act} before it is remodeled by PRP2 into a catalytically active complex. Thus, this novel method can potentially be used to both identify and temporally order the many distinct conformational states of the spliceosome during its assembly/work cycle.

9 STRUCTURE OF HUMAN C AND C* COMPLEXES

The cryo-EM structures of the human C and C* complexes (Bertram et al. 2017b; Zhang et al. 2017; Zhan et al. 2018) revealed the spliceosome's molecular architecture during the catalytic steps of splicing, and also RNP rearrangements likely mediated by the RNA helicase PRP16 during C* formation (Fig. 6). The catalytic RNA network is highly similar in both hC and yC complexes (Galej et al. 2016; Wan et al. 2016b; Zhan et al. 2018). Furthermore, in both complexes U2 snRNA has rearranged, now forming stem IIc instead of stem IIa (Fig. 1H), which in earlier complexes interacts with SF3b proteins. This U2 restructuring may thus be triggered by the release/destabilization of SF3b during the B^{act} to C transition. Stem IIc persists in the C* complex and, at least in yeast, in the postsplicing P complex (Bai et al. 2017; Liu et al. 2017; Wilkinson et al. 2017). Release of the U2/BS helix from SF3B1 by PRP2 allows it to dock into the catalytic center of the C complex, in which it is tethered by the step 1 splicing factors CCDC49 (Cwc25) and CCDC94 (Yju2), which also interact with each other. Residues in the amino terminus of CCDC94 contact ISY1 and $PRP8^{NTD}$, as well as the 5′ exon and U2/BS helix (Fig. 6B) (Zhan et al. 2018). CCDC49 is sandwiched between the U2/BS helix and U2/U6 helix Ia, and contacts the Prp8 α-finger. The close contact of CCDC49 and CCDC94 with the U2/BS helix is consistent with their important role in promoting step 1 of splicing (Liu et al. 2007; Chiu et al. 2009; Warkocki et al. 2009), where they likely help to position the BS-A for step 1 catalysis. Both proteins interact with BRR2 and PRP16, which are located at the periphery of hC, and thus link these helicases to the catalytic core.

Before step 2, the branched intron structure (BIS) generated during step 1 of splicing must be displaced from the catalytic center to allow juxtapositioning of the step 2 reactants, the 3′-OH of the 5′ exon and the 3′SS. PRP16 action leads to its own release and destabilization/reorganization of CCDC49 and CCDC94 (Horowitz 2012), and reorganization of ISY1, which remains bound to hC* (Bertram et al. 2017b). This in turn enables the repositioning of the BIS

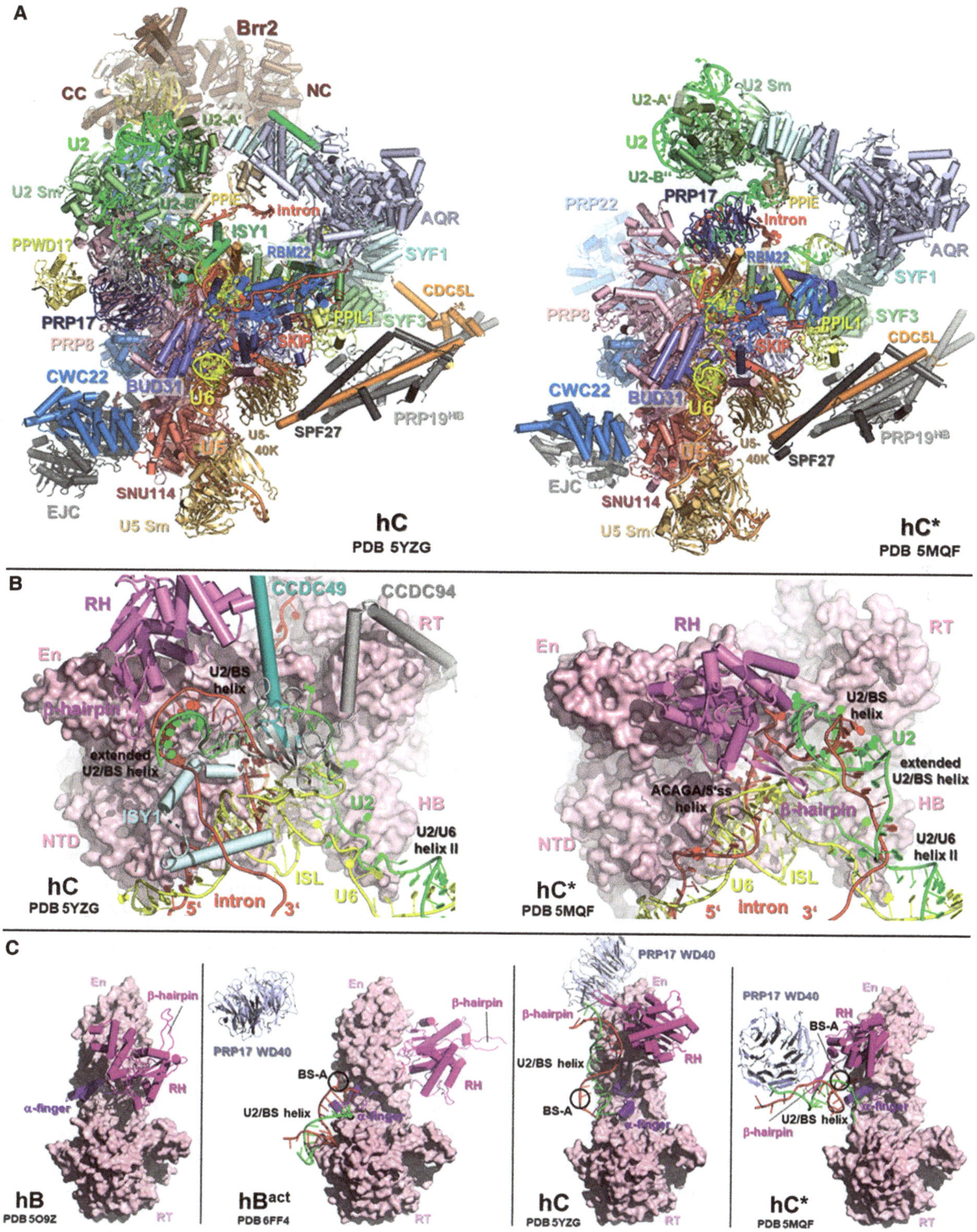

Figure 6. Molecular organization of the human C and C* complexes. (*A*) Overview of the structural organization of the human C and C* complexes. In higher eukaryotes, the exon junction complex (EJC) is deposited 20–24 nucleotides upstream of the exon–exon junction of spliced mRNAs (Le Hir et al. 2000). In hC*, eIF4AIII binds 6 nts of the 5′ exon at a position close to where the EJC should be deposited, and at the same time interacts with MAGOH and MLN51, and U5-SNU114. (*B*) Interaction of PRP8^{RH} with the U2/BS helix in the C (*left*) and C* complex (*right*), as well as interactions of CCDC49 (CWC25), CCDC94 (YJU2) and ISY1 within the RNP core of the C complex (*left*). (*C*) Conformational dynamics of the PRP8 RH domain and movements of the U2/BS helix. The position of the branched intron structure formed after step 1 of splicing is stabilized by contacts with PRP8^{RH} and Prp17. The branch site adenosine (BS-A) is indicated by a black circle.

~20 Å away from the catalytic center, creating space for the 3′ exon to dock near the 5′ exon (Fig. 6B) and the recruitment of the step 2 factors SLU7 and PRP18 and the RNA helicase PRP22. $PRP8^{RH}$ is a highly mobile domain whose position changes throughout the splicing cycle (Fig. 6C). In hC, it interacts with the extended region of the U2/BS helix, CCDC49 and the U2 3′ domain. During the C to C* transition, $PRP8^{RH}$ is rotated by 70 degrees and its β-hairpin is inserted between the U2/BS helix and the U6/intron helix; it thus appears to play an important part in stabilizing the conformation/position of the BIS in C* (Bertram et al. 2017b; Zhang et al. 2017). Whereas the position of the amino-terminal domain of PRP17—which has an extended conformation and contacts multiple proteins (Fig. 4E)—does not change, its WD40 domain ($PRP17^{WD40}$) moves ~70 Å, contacting the $PRP8^{RH}$ and the extended U2/BS helix and adjacent U2 nucleotides in hC* (Fig. 6). The step 2 factor SLU7 also interacts tightly with $PRP8^{RH}$, stabilizing its rotated conformation. Thus, both PRP17 and SLU7 likely aid in stabilizing the step 2 conformation of the spliceosome, consistent with them acting as step 2 factors (Horowitz 2012). Although the step 2 factor PRP18 was not detected in hC* complexes, in yC* it interacts with Slu7 and $PRP8^{RH}$, and thus also contributes to the formation of a step 2 competent spliceosome (Fica et al. 2017; Yan et al. 2017).

In both hC and hC*, the 5′ exon is tightly tethered to the spliceosome via base-pairing with U5 loop 1 nucleotides U40–U42 and via multiple contacts with the exon binding channel proteins (see above). Consistent with previous studies showing that CWC22 facilitates exon junction complex (EJC) formation (Barbosa et al. 2012), in C*, the amino-terminal MIF4G domain of CWC22 contacts the RNA helicase eIF4AIII of the EJC that additionally contains the proteins Y14, MAGOH, and MLN51 (Fig. 6A).

10 INSIGHTS INTO THE MECHANISM OF ACTION OF SPLICEOSOMAL HELICASES

CWC22 binds across the EBC and, together with its interaction partners SLU7 and PRP18, was shown in yeast to be released after step 2 by the action of PRP22, leading to release of the spliced mRNA (Fourmann et al. 2013). PRP22 is located at the periphery of the human C* complex and interacts via its carboxy-terminal OB domain with PRP8, and a long α-helix of SKIP (Fig. 6A). As PRP8 interacts with CWC22, PRP22 action may indirectly lead to release of CWC22. Cryo-EM revealed that PRP2, PRP16 and PRP22 bind sequentially to the periphery of the spliceosome, and all appear to be located far from their RNA targets (i.e., the U2/BS in the case of PRP2 or at/near the spliceosome's catalytic center in the case of PRP16/PRP22) (Fig. 7). This supports the idea that these helicases act indirectly, at a distance from their RNA targets (Rauhut et al. 2016; Semlow et al. 2016). The cryo-EM structures also revealed that each of these helicases interacts primarily with those proteins that are destabilized/released up on their action, and that their stable association with the spliceosome does not require RNA binding. Protein release or RNP remodeling by these helicases, which is dependent on their interaction with intronic RNA downstream from the BS, generates new conformational space that allows the subsequent integration of new spliceosomal proteins or the stabilization/rearrangement of protein domains or proteins already present in the spliceosome.

11 LARGE-SCALE MOVEMENTS OF SPLICEOSOMAL COMPONENTS

Cryo-EM has revealed that many spliceosomal proteins and large protein ensembles undergo large-scale movements during splicing. The major structural rearrangements that occur during the transitions from one spliceosomal complex to the next are summarized in Figure 7. These rearrangements require a dynamic network of molecular interactions among numerous spliceosomal components. The ~300-kDa U2 snRNP 3′ domain is a prime example, which unexpectedly undergoes complex translocations of 10–16 nm between complexes B, B^{act}, C, and C* (Fig. 7). SYF1, which contacts the U2 3′ domain via the U2-A′ protein in the latter three complexes, and SYF3 undergo movements in a concerted manner with the U2 3′ domain, suggesting they aid in its repositioning (Fig. 7A). The movement of the U2 3′ domain first toward the catalytic center during B^{act} formation, and then away from it after catalytic step 1, may be driven by the repositioning of the U2/BS helix, whose rotational movements closely mirror those of the U2 3′ domain (Figs. 6C and 7A). Four nucleotides of the human U2 snRNA (A29 to U32), located between the U2/BS helix and U2 nucleotides that form U2/U6 helix Ia, appear to act as a hinge and enable these complex U2 movements. Brr2, together with $PRP8^{Jab1}$ (which tethers BRR2 to the spliceosome) also undergoes large scale movements during most transitions from one spliceosomal complex to the next (Fig. 7A). As BRR2 helicase activity is negatively regulated in B^{act}, C, and C* complexes, its restructuring may reflect its potential role in stabilizing the position/conformation of its interaction partners at each of these stages.

Multiple domains of PRP8 undergo numerous rearrangements during spliceosome assembly and catalytic activation. As described above, its RT, Linker, and En domains undergo substantial changes relative to its NTD during the integration of the tri-snRNP and during the B to B^{act} and B^{act} to C transitions (Fig. 7B); these major PRP8 domains

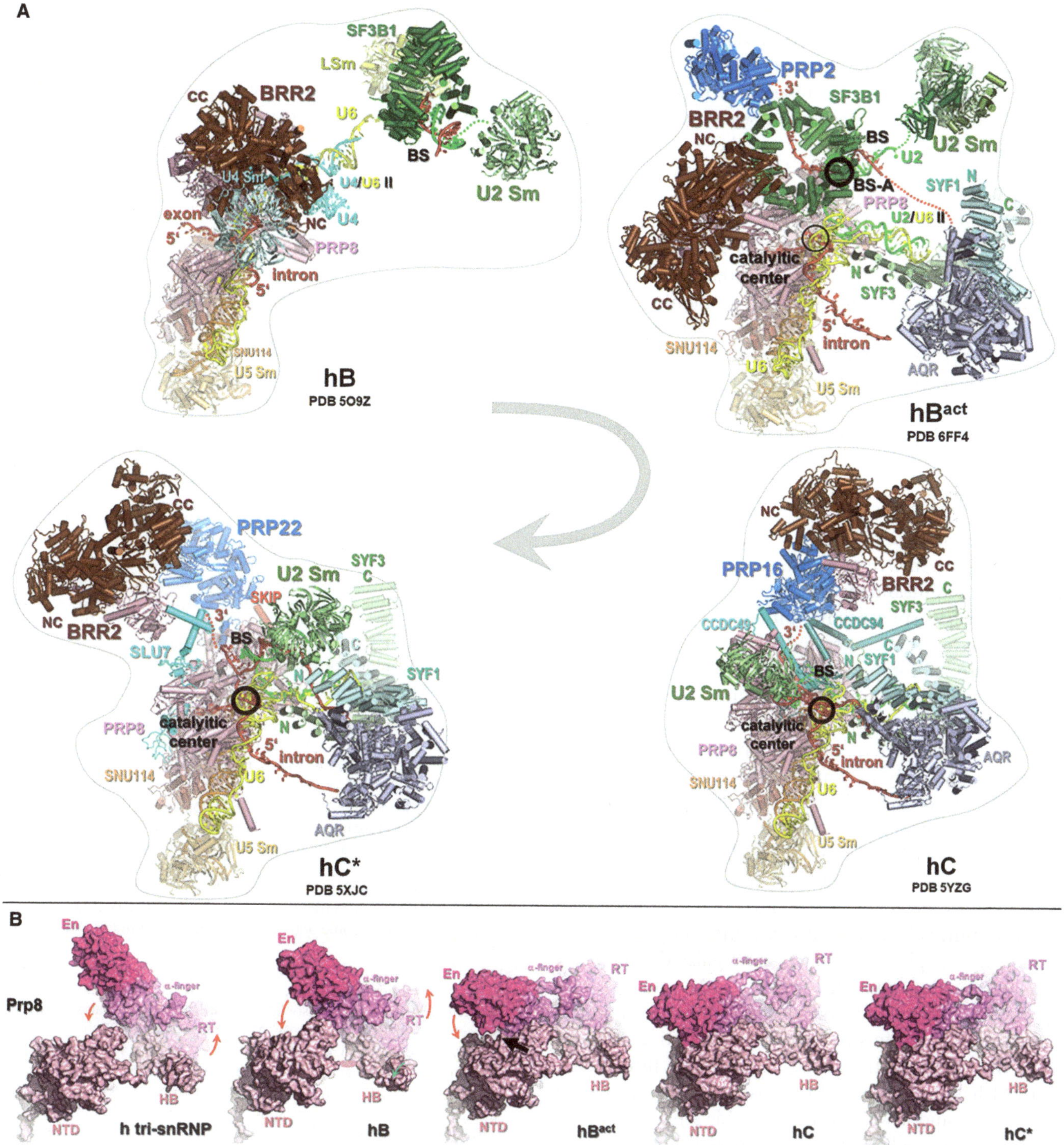

Figure 7. Structural dynamics of the spliceosome. (*A*) Cartoon depicting the major structural rearrangements that occur during the transitions from the human B to B^{act}, B^{act} to C, and C to C* complex. (*B*) Rearrangements of the Prp8 reverse transcriptase-like (RT) domain, Linker (α-finger), endonuclease-like (En) domain, helical bundle (HB), and amino-terminal (NTD) domain during B, B^{act}, and C complex formation (indicated by arrows in the preceding complex). Green arrow indicates rotation of HB relative to the RT domain. Black arrow indicates the cleft between the Prp8 NTD and En domains in the B^{act} complex that are occupied initially by regions of RNF113A and CWC27.

remain largely unchanged after formation of the C complex. $PRP8^{RH}$ is repositioned at each spliceosomal stage analyzed to date and both the PRP8 α-finger and switch loop (SWL) are repositioned and/or rearranged during the spliceosome's assembly and catalytic phases (Fig. 6C and see above). The SWL comprises part of the 5′ exon binding channel in hB^{act}, hC, and C* and toggles between two different conformations. In C* and the yeast post step 2 P complex, the α-finger contacts the 3′SS and nucleotides near the 5′ end of the 3′ exon (Bai et al. 2017; Liu et al. 2017; Wilkinson et al. 2017). Thus, together with U5 loop 1, the PRP8 α-finger aids in stabilizing/positioning the 3′SS and 3′ exon, both before and after exon ligation. Whether this and other functionally important interactions after exon ligation are conserved in humans and yeast is presently not clear, as high-resolution, cryo-EM structures of the human P complex and ILS remain to be determined.

12 FUTURE DIRECTIONS

Currently available cryo-EM structures of the human spliceosome and tri-snRNP provide important insights into the potential structural and mechanistic effects of disease-related mutations in spliceosomal proteins. These mutations may also have kinetic effects, altering the rate at which one spliceosomal complex is converted into the next, which may influence which proteins (including regulatory ones) ultimately bind. As many alternative splicing choices are made during the earliest stages of spliceosome assembly, future cryo-EM analyses of human spliceosomal complexes formed before complex B, will undoubtedly provide key insights into not only the structural basis for splice site recognition by early splicing factors, but also how mutations in these factors may alter splicing outcomes. Structural analyses of spliceosomes formed on alternatively spliced pre-mRNAs may also reveal the complex mechanisms whereby splice site choices are regulated. The structural information provided by cryo-EM can also be exploited to understand how splicing inhibitors/modulators act, and to design structure-based derivatives that enhance their therapeutic efficacy. In the past three years, cryo-EM has revolutionized the structural analysis of the spliceosome, and it will certainly provide a plethora of intriguing new insights into early spliceosome assembly events and the regulation of alternative splicing in higher eukaryotes in the future.

ACKNOWLEDGMENTS

We are grateful to Yigong Shi for providing the human C complex PDB. We thank Dmitry Agafonov and Reinhard Rauhut for help in preparing Figure 1F, and Constantin Cretu for help in preparing Figure 4C. We thank Leyla El Ayoubi and Dmitry Agafonov for providing unpublished data concerning the proteome of purified hC complexes stalled before Prp16 action. We apologize to our colleagues for omitting many references because of space limitations. Our work was supported by the Deutsche Forschungsgemeinschaft (Grants SFB 860 to H.S. and R.L. and LU294/15-1 to R.L.).

REFERENCES

Absmeier E, Wollenhaupt J, Mozaffari-Jovin S, Becke C, Lee CT, Preussner M, Heyd F, Urlaub H, Lührmann R, Santos KF, et al. 2015. The large N-terminal region of the Brr2 RNA helicase guides productive spliceosome activation. *Genes Dev* **29:** 2576–2587.

Agafonov DE, Deckert J, Wolf E, Odenwälder P, Bessonov S, Will CL, Urlaub H, Lührmann R. 2011. Semiquantitative proteomic analysis of the human spliceosome via a novel two-dimensional gel electrophoresis method. *Mol Cell Biol* **31:** 2667–2682.

Agafonov DE, Kastner B, Dybkov O, Hofele RV, Liu WT, Urlaub H, Lührmann R, Stark H. 2016. Molecular architecture of the human U4/U6.U5 tri-snRNP. *Science* **351:** 1416–1420.

Alsafadi S, Houy A, Battistella A, Popova T, Wassef M, Henry E, Tirode F, Constantinou A, Piperno-Neumann S, Roman-Roman S, et al. 2016. Cancer-associated *SF3B1* mutations affect alternative splicing by promoting alternative branchpoint usage. *Nat Commun* **7:** 10615.

Bai R, Yan C, Wan R, Lei J, Shi Y. 2017. Structure of the post-catalytic spliceosome from *Saccharomyces cerevisiae*. *Cell* **171:** 1589–1598.

Bao P, Will CL, Urlaub H, Boon KL, Lührmann R. 2017. The RES complex is required for efficient transformation of the precatalytic B spliceosome into an activated B^{act} complex. *Genes Dev* **31:** 2416–2429.

Barbosa I, Haque N, Fiorini F, Barrandon C, Tomasetto C, Blanchette M, Le Hir H. 2012. Human CWC22 escorts the helicase eIF4AIII to spliceosomes and promotes exon junction complex formation. *Nat Struct Mol Biol* **19:** 983–990.

Bertram K, Agafonov DE, Dybkov O, Haselbach D, Leelaram MN, Will CL, Urlaub H, Kastner B, Lührmann R, Stark H. 2017a. Cryo-EM structure of a pre-catalytic human spliceosome primed for activation. *Cell* **170:** 701–713.

Bertram K, Agafonov DE, Liu WT, Dybkov O, Will CL, Hartmuth K, Urlaub H, Kastner B, Stark H, Lührmann R. 2017b. Cryo-EM structure of a human spliceosome activated for step 2 of splicing. *Nature* **542:** 318–323.

Boesler C, Rigo N, Anokhina MM, Tauchert MJ, Agafonov DE, Kastner B, Urlaub H, Ficner R, Will CL, Lührmann R. 2016. A spliceosome intermediate with loosely associated tri-snRNP accumulates in the absence of Prp28 ATPase activity. *Nat Commun* **7:** 11997.

Chan SP, Kao DI, Tsai WY, Cheng SC. 2003. The Prp19-associated complex in spliceosome activation. *Science* **302:** 279–282.

Chiu YF, Liu YC, Chiang TW, Yeh TC, Tseng CK, Wu NY, Cheng SC. 2009. Cwc25 is a novel splicing factor required after Prp2 and Yju2 to facilitate the first catalytic reaction. *Mol Cell Biol* **29:** 5671–5678.

Cordin O, Beggs JD. 2013. RNA helicases in splicing. *RNA Biol* **10:** 83–95.

Cretu C, Schmitzová J, Ponce-Salvatierra A, Dybkov O, De Laurentiis EI, Sharma K, Will CL, Urlaub H, Lührmann R, Pena V. 2016. Molecular architecture of SF3b and structural consequences of its cancer-related mutations. *Mol Cell* **64:** 307–319.

Cretu C, Agrawal AA, Cook A, Will CL, Fekkes P, Smith PG, Lührmann R, Larsen N, Buonamici S, Pena V. 2018. Structural basis of splicing modulation by antitumor macrolide compounds. *Mol Cell* **70:** 265–273.

Daguenet E, Dujardin G, Valcárcel J. 2015. The pathogenicity of splicing defects: Mechanistic insights into pre-mRNA processing inform novel therapeutic approaches. *EMBO Rep* **16:** 1640–1655.

Darman RB, Seiler M, Agrawal AA, Lim KH, Peng S, Aird D, Bailey SL, Bhavsar EB, Chan B, Colla S, et al. 2015. Cancer-associated *SF3B1* hotspot mutations induce cryptic 3′ splice site selection through use of a different branch point. *Cell Rep* **13:** 1033–1045.

De I, Bessonov S, Hofele R, dos Santos K, Will CL, Urlaub H, Lührmann R, Pena V. 2015. The RNA helicase Aquarius exhibits structural adaptations mediating its recruitment to spliceosomes. *Nat Struct Mol Biol* **22:** 138–44.

Dziembowski A, Ventura AP, Rutz B, Caspary F, Faux C, Halgand F, Laprévote O, Séraphin B. 2004. Proteomic analysis identifies a new complex required for nuclear pre-mRNA retention and splicing. *EMBO J* **23:** 4847–4856.

Fabrizio P, Dannenberg J, Dube P, Kastner B, Stark H, Urlaub H, Lührmann R. 2009. The evolutionarily conserved core design of the catalytic activation step of the yeast spliceosome. *Mol Cell* **36:** 593–608.

Fica SM, Tuttle N, Novak T, Li NS, Lu J, Koodathingal P, Dai Q, Staley JP, Piccirilli JA. 2013. RNA catalyses nuclear pre-mRNA splicing. *Nature* **503:** 229–234.

Fica SM, Mefford MA, Piccirilli JA, Staley JP. 2014. Evidence for a group II intron-like catalytic triplex in the spliceosome. *Nat Struct Mol Biol* **21:** 464–471.

Fica SM, Oubridge C, Galej WP, Wilkinson ME, Bai XC, Newman AJ, Nagai K. 2017. Structure of a spliceosome remodelled for exon ligation. *Nature* **542:** 377–380.

Finci LI, Zhang X, Huang X, Zhou Q, Tsai J, Teng T, Agrawal A, Chan B, Irwin S, Karr C, et al. 2018. The cryo-EM structure of the SF3b spliceosome complex bound to a splicing modulator reveals a pre-mRNA substrate competitive mechanism of action. *Genes Dev* **32:** 309–320.

Fourmann JB, Schmitzová J, Christian H, Urlaub H, Ficner R, Boon KL, Fabrizio P, Lührmann R. 2013. Dissection of the factor requirements for spliceosome disassembly and the elucidation of its dissociation products using a purified splicing system. *Genes Dev* **27:** 413–428.

Galej WP, Wilkinson ME, Fica SM, Oubridge C, Newman AJ, Nagai K. 2016. Cryo-EM structure of the spliceosome immediately after branching. *Nature* **537:** 197–201.

Gozani O, Feld R, Reed R. 1996. Evidence that sequence-independent binding of highly conserved U2 snRNP proteins upstream of the branch site is required for assembly of spliceosomal complex A. *Genes Dev* **10:** 233–243.

Gozani O, Potashkin J, Reed R. 1998. A potential role for U2AF–SAP155 interactions in recruiting U2 snRNP to the branch site. *Mol Cell Biol* **18:** 4752–4760.

Grainger RJ, Beggs JD. 2005. Prp8 protein: At the heart of the spliceosome. *RNA* **11:** 533–557.

Hang J, Wan R, Yan C, Shi Y. 2015. Structural basis of pre-mRNA splicing. *Science* **349:** 1191–1198.

Haselbach D, Schrader J, Lambrecht F, Henneberg F, Chari A, Stark H. 2017. Long-range allosteric regulation of the human 26S proteasome by 20S proteasome-targeting cancer drugs. *Nat Commun* **8:** 15578.

Haselbach D, Komarov I, Agafonov DE, Hartmuth K, Graf B, Dybkov O, Urlaub H, Kastner B, Lührmann R, Stark H. 2018. Structure and conformational dynamics of the human spliceosomal B^{act} complex. *Cell* **172:** 454–464.

Henning LM, Santos KF, Sticht J, Jehle S, Lee CT, Wittwer M, Urlaub H, Stelzl U, Wahl MC, Freund C. 2017. A new role for FBP21 as regulator of Brr2 helicase activity. *Nucleic Acids Res* **45:** 7922–7937.

Hirose T, Ideue T, Nagai M, Hagiwara M, Shu MD, Steitz JA. 2006. A spliceosomal intron binding protein, IBP160, links position-dependent assembly of intron-encoded box C/D snoRNP to pre-mRNA splicing. *Mol Cell* **23:** 673–684.

Horowitz DS. 2012. The mechanism of the second step of pre-mRNA splicing. *Wiley Interdiscip Rev RNA* **3:** 331–350.

Huang YH, Chung CS, Kao DI, Kao TC, Cheng SC. 2014. Sad1 counteracts Brr2-mediated dissociation of U4/U6.U5 in tri-snRNP homeostasis. *Mol Cell Biol* **34:** 210–220.

Jenkins JL, Kielkopf CL. 2017. Splicing factor mutations in myelodysplasias: Insights from spliceosome structures. *Trends Genet* **33:** 336–348.

Jurica MS, Moore MJ. 2003. Pre-mRNA splicing: Awash in a sea of proteins. *Mol Cell* **12:** 5–14.

Lardelli RM, Thompson JX, Yates JR III, Stevens SW. 2010. Release of SF3 from the intron branchpoint activates the first step of pre-mRNA splicing. *RNA* **16:** 516–528.

Lee Y, Rio DC. 2015. Mechanisms and regulation of alternative splicing. *Annu Rev Biochem* **84:** 291–323.

Le Hir H, Izaurralde E, Maquat LE, Moore MJ. 2000. The spliceosome deposits multiple proteins 20–24 nucleotides upstream of mRNA exon–exon junctions. *EMBO J* **19:** 6860–6869.

Liu HL, Cheng SC. 2012. The interaction of Prp2 with a defined region of the intron is required for the first splicing reaction. *Mol Cell Biol* **32:** 5056–5066.

Liu YC, Chen HC, Wu NY, Cheng SC. 2007. A novel splicing factor, Yju2, is associated with NTC and acts after Prp2 in promoting the first catalytic reaction of pre-mRNA. *Mol Cell Biol* **27:** 5403–5413.

Liu S, Li X, Zhang L, Jiang J, Hill RC, Cui Y, Hansen KC, Zhou ZH, Zhao R. 2017. Structure of the yeast spliceosomal postcatalytic P complex. *Science* **358:** 1278–1283.

Makarova OV, Makarov EM, Urlaub H, Will CL, Gentzel M, Wilm M, Lührmann R. 2004. A subset of human 35S proteins, including Prp19, function prior to catalytic step 1 of splicing. *EMBO J* **23:** 2381–2391.

Mozaffari-Jovin S, Wandersleben T, Santos KF, Will CL, Lührmann R, Wahl MC. 2013. Inhibition of RNA helicase Brr2 by the C-terminal tail of the spliceosomal protein Prp8. *Science* **341:** 80–84.

Newman AJ, Norman C. 1992. U5 snRNA interacts with exon sequences at 5′ and 3′ splice sites. *Cell* **68:** 743–754.

Nguyen THD, Galej WP, Bai XC, Savva CG, Newman AJ, Scheres SH, Nagai K. 2015. The architecture of the spliceosomal U4/U6.U5 tri-snRNP. *Nature* **523:** 47–52.

Nguyen THD, Galej WP, Bai XC, Oubridge C, Newman AJ, Scheres SHW, Nagai K. 2016. Cryo-EM structure of the yeast U4/U6.U5 tri-snRNP at 3.7 Å resolution. *Nature* **530:** 298–302.

Ohrt T, Prior M, Dannenberg J, Odenwälder P, Dybkov O, Rasche N, Schmitzová J, Gregor I, Fabrizio P, Enderlein J, et al. 2012. Prp2-mediated protein rearrangements at the catalytic core of the spliceosome as revealed by dcFCCS. *RNA* **18:** 1244–1256.

Pan Q, Shai O, Lee LJ, Frey BJ, Blencowe BJ. 2008. Deep surveying of alternative splicing complexity in the human transcriptome by high-throughput sequencing. *Nat Genet* **40:** 1413–1415.

Plaschka C, Lin PC, Nagai K. 2017. Structure of a pre-catalytic spliceosome. *Nature* **546:** 617–621.

Plaschka C, Newman AJ, Nagai K. 2019. Structural basis of nuclear pre-mRNA splicing: Lessons from yeast. *Cold Spring Harb Perspect Biol* doi: 10.1101/cshperspect.a032391.

Pomeranz Krummel DA, Oubridge C, Leung AK, Li J, Nagai K. 2009. Crystal structure of human spliceosomal U1 snRNP at 5.5 Å resolution. *Nature* **458:** 475–480.

Rasche N, Dybkov O, Schmitzová J, Akyildiz B, Fabrizio P, Lührmann R. 2012. Cwc2 and its human homologue RBM22 promote an active conformation of the spliceosome catalytic centre. *EMBO J* **31:** 1591–1604.

Rauhut R, Fabrizio P, Dybkov O, Hartmuth K, Pena V, Chari A, Kumar V, Lee CT, Urlaub H, Kastner B, et al. 2016. Molecular architecture of the *Saccharomyces cerevisiae* activated spliceosome. *Science* **353:** 1399–1405.

Santos KF, Jovin SM, Weber G, Pena V, Lührmann R, Wahl MC. 2012. Structural basis for functional cooperation between tandem helicase cassettes in Brr2-mediated remodeling of the spliceosome. *Proc Natl Acad Sci* **109:** 17418–17423.

Schaffert N, Hossbach M, Heintzmann R, Achsel T, Lührmann R. 2004. RNAi knockdown of hPrp31 leads to an accumulation of U4/U6 di-snRNPs in Cajal bodies. *EMBO J* **23:** 3000–3009.

Scheres SH, Nagai K. 2017. CryoEM structures of spliceosomal complexes reveal the molecular mechanism of pre-mRNA splicing. *Curr Opin Struct Biol* **46:** 130–139.

Schneider C, Agafonov DE, Schmitzová J, Hartmuth K, Fabrizio P, Lührmann R. 2015. Dynamic contacts of U2, RES, Cwc25, Prp8 and Prp45 proteins with the pre-mRNA branch-site and 3′ splice site during catalytic activation and step1 catalysis in yeast spliceosomes. *PLoS Genet* **11:** e1005539.

Semlow DR, Blanco MR, Walter NG, Staley JP. 2016. Spliceosomal DEAH-box ATPases remodel pre-mRNA to activate alternative splice sites. *Cell* **164:** 985–998.

Shi Y. 2017. Mechanistic insights into precursor messenger RNA splicing by the spliceosome. *Nat Rev Mol Cell Biol* **18:** 655–670.

Sontheimer EJ, Steitz JA. 1993. The U5 and U6 small nuclear RNAs as active site components of the spliceosome. *Science* **262:** 1989–1996.

Staley JP, Guthrie C. 1999. An RNA switch at the 5′ splice site requires ATP and the DEAD box protein Prp28p. *Mol Cell* **3:** 55–64.

Tang Q, Rodriguez-Santiago S, Wang J, Pu J, Yuste A, Gupta V, Moldón A, Xu YZ, Query CC. 2016. SF3B1/Hsh155 HEAT motif mutations affect interaction with the spliceosomal ATPase Prp5, resulting in altered branch site selectivity in pre-mRNA splicing. *Genes Dev* **30:** 2710–2723.

Thapar R. 2015. Roles of prolyl isomerases in RNA-mediated gene expression. *Biomolecules* **5:** 974–999.

Wan R, Yan C, Bai R, Wang L, Huang M, Wong CC, Shi Y. 2016a. The 3.8 Å structure of the U4/U6.U5 tri-snRNP: Insights into spliceosome assembly and catalysis. *Science* **351:** 466–475.

Wan R, Yan C, Bai R, Huang G, Shi Y. 2016b. Structure of a yeast catalytic step I spliceosome at 3.4Å resolution. *Science* **353:** 895–904.

Warkocki Z, Odenwälder P, Schmitzová J, Platzmann F, Stark H, Urlaub H, Ficner R, Fabrizio P, Lührmann R. 2009. Reconstitution of both steps of *Saccharomyces cerevisiae* splicing with purified spliceosomal components. *Nat Struct Mol Biol* **16:** 1237–1243.

Weber G, Trowitzsch S, Kastner B, Lührmann R, Wahl MC. 2010. Functional organization of the Sm core in the crystal structure of human U1 snRNP. *EMBO J* **29:** 4172–4184.

Wilkinson ME, Fica SM, Galej WP, Norman CM, Newman AJ, Nagai K. 2017. Postcatalytic spliceosome structure reveals mechanism of 3′-splice site selection. *Science* **358:** 1283–1288.

Will CL, Lührmann R. 2011. Spliceosome structure and function. *Cold Spring Harb Perspect Biol* **3:** a003707.

Will CL, Schneider C, MacMillan AM, Katopodis NF, Neubauer G, Wilm M, Lührmann R, Query CC. 2001. A novel U2 and U11/U12 snRNP protein that associates with the pre-mRNA branch site. *EMBO J* **20:** 4536–4546.

Yan C, Wan R, Bai R, Huang G, Shi Y. 2016. Structure of a yeast activated spliceosome at 3.5 Å resolution. *Science* **353:** 904–911.

Yan C, Wan R, Bai R, Huang G, Shi Y. 2017. Structure of a yeast step II catalytically activated spliceosome. *Science* **355:** 149–155.

Zhao C, Pyle AM. 2017. Structural insights into the mechanism of group II intron splicing. *Trends Biochem Sci* **42:** 470–482.

Zhan X, Yan C, Zhang X, Lei J, Shi Y. 2018. Structure of a human catalytic step I spliceosome. *Science* **359:** 537–545.

Zhang X, Yan C, Hang J, Finci L, Lei J, Shi Y. 2017. An atomic structure of the human spliceosome. *Cell* **169:** 1–12.

Zhang X, Yan C, Zhan X, Li L, Lei J, Shi Y. 2018. Structure of the human activated spliceosome in three conformational states. *Cell Res* **28:** 307–322.

Zhou Z, Fu XD. 2013. Regulation of splicing by SR proteins and SR protein-specific kinases. *Chromosoma* **122:** 191–207.

Molecular Mechanisms of pre-mRNA Splicing through Structural Biology of the Spliceosome

Chuangye Yan,[1] Ruixue Wan,[1] and Yigong Shi[1,2]

[1]Beijing Advanced Innovation Center for Structural Biology, School of Life Sciences, Tsinghua University, Beijing 100084, China

[2]Institute of Biology, Westlake Institute for Advanced Study, Westlake University, Hangzhou 310064, Zhejiang Province, China

Correspondence: shi-lab@tsinghua.edu.cn; yancy05@mails.tsinghua.edu.cn; wrxruisnow@163.com

SUMMARY

Precursor messenger RNA (pre-mRNA) splicing is executed by the spliceosome. In the past 3 years, cryoelectron microscopy (cryo-EM) structures have been elucidated for a majority of the yeast spliceosomal complexes and for a few human spliceosomes. During the splicing reaction, the dynamic spliceosome has an immobile core of about 20 protein and RNA components, which are organized around a conserved splicing active site. The divalent metal ions, coordinated by U6 small nuclear RNA (snRNA), catalyze the branching reaction and exon ligation. The spliceosome also contains a mobile but compositionally stable group of about 13 proteins and a portion of U2 snRNA, which facilitate substrate delivery into the splicing active site. The spliceosomal transitions are driven by the RNA-dependent ATPase/helicases, resulting in the recruitment and dissociation of specific splicing factors that enable the reaction. In summary, the spliceosome is a protein-directed metalloribozyme.

Outline

Cite as *Cold Spring Harb Perspect Biol* doi: 10.1101/cshperspect.a032409

1 INTRODUCTION

Precursor messenger RNA (pre-mRNA) splicing, discovered 40 years ago (Berget et al. 1977; Chow et al. 1977), is executed by a multi-megaDalton, ribonucleoprotein (RNP) complex known as the spliceosome (Brody and Abelson 1985; Frendewey and Keller 1985; Grabowski et al. 1985). The components of the spliceosome constitute five distinct classes: the uridine-rich RNPs (U1, U2, U4, U5, and U6 snRNPs), the NineTeen complex (NTC) and the NTC-related complex (NTR), the splicing factors, the RNA-dependent ATPase/helicases, and other regulatory and/or auxiliary proteins such as kinases/phosphatases and peptidyl prolyl isomerases (PPIs). The involvement of snRNPs in pre-mRNA splicing was documented by the early 1980s (Lerner and Steitz 1979; Lerner et al. 1980; Rogers and Wall 1980; Yang et al. 1981) and subsequently confirmed (Mount et al. 1983; Padgett et al. 1983; Kramer et al. 1984; Black et al. 1985; DiMaria et al. 1985; Krainer and Maniatis 1985; Berget and Robberson 1986; Grabowski and Sharp 1986; Pikielny and Rosbash 1986). Only the precatalytic spliceosome (B complex) and the pre-B complex contain all five snRNPs but lack the NTC and NTR. Despite their relatively late discovery (Tarn et al. 1994; Chen et al. 2002; Ohi and Gould 2002; Chan et al. 2003; Chan and Cheng 2005), NTC and NTR are known to be indispensable in structurally shaping the splicing active site (Hang et al. 2015; Yan et al. 2015; Shi 2017b).

The assembled spliceosome sequentially assumes eight different compositional states during each cycle of the splicing reaction: pre-B, B, activated (B^{act}), catalytically activated (B*), step I spliceosome (C), step II activated (C*), postcatalytic (P), and intron lariat spliceosome (ILS) (Fig. 1A). The B complex is preceded by a recently characterized pre-B complex (Boesler et al. 2016), in which the 5′-splice site (5′SS) is still recognized by the U1 snRNP. Each cycle consists of two sequential steps: branching and exon ligation, which proceed in the B* and C* complexes, resulting in the C and P complexes, respectively. Therefore, the B*-to-C and C*-to-P transitions occur unassisted. In contrast, the B-to-B^{act}, B^{act}-to-B*, C-to-C*, and P-to-ILS transitions are driven by the RNA-dependent ATPase/helicases Brr2, Prp2, Prp16, and Prp22, respectively (Fig. 1A) (Cordin et al. 2012). The disassembly of the ILS complex is mediated by the ATPase/helicase Prp43 (Arenas and Abelson 1997; Tsai et al. 2005).

Both the branching and exon ligation reactions are of the bimolecular nucleophilic substitution (S_N2) type transesterification. In a pre-mRNA, two neighboring exons (5′-exon and 3′-exon) are separated by an intron, which contains three conserved sequence elements: the 5′SS, the branch point sequence (BPS), and the 3′-splice site (3′SS) (Fig. 1A). The branching reaction results in a free 5′-exon and an intron lariat-3′-exon intermediate, which has a 2′-5′ phosphodiester bond linking the invariant adenosine nucleotide of the BPS to the guanine nucleotide at the 5′-end of the 5′SS (Grabowski et al. 1984; Padgett et al. 1984; Ruskin et al. 1984). Exon ligation leads to the joining of the 5′-exon and the 3′-exon and the generation of a free intron lariat (Fig. 1A).

The splicing active site of the spliceosome was postulated to contain two divalent metal ions known as M1 and M2 (Steitz and Steitz 1993), which was supported by sulfur-substituted pre-mRNA splicing studies (Sontheimer et al. 1997). In vitro splicing of *Saccharomyces cerevisiae* using U6 small nuclear RNA (snRNA) with sulfur-substituted nonbridging phosphoryl oxygen unambiguously identified the nucleotide U80 as a coordinating ligand for the catalytic metals (Yean et al. 2000). In the presence of thiophilic ions such as Mn^{2+}, the spliceosome reconstituted with the sulfur-substituted U80(Sp), but not U80(Rp), U6 snRNA allowed the branching reaction to occur (Yean et al. 2000). U6 snRNA was subsequently shown to coordinate two divalent metals known as M1 and M2 (Fica et al. 2013), and structural analysis of the group II intron allowed prediction of conserved features in the active site of the spliceosome (Toor et al. 2008; Keating et al. 2010).

The entire U5 snRNP, the core of U2 snRNP, and U6 snRNA are present in all eight assembled spliceosomal complexes (Fig. 1A). Among all spliceosomal proteins, the central U5 snRNP component Prp8 is most highly conserved and plays a central role by anchoring the splicing active site. The step I splicing factors (Cwc25 and Yju2 in *S. cerevisiae*) only associate with the splicing active site surrounding the branching reaction, whereas the step II factors (Prp18 and Slu7 in *S. cerevisiae*) mainly act to promote exon ligation (Fig. 1A). In contrast, the general splicing factors Cwc21 and Cwc22 are present in the B^{act} through P complexes, whereas Cwc24 and Cwc27 are recruited into the B^{act} complex but dissociated in the B^{act}-to-B* transition (Ohrt et al. 2012).

2 STRUCTURAL STUDIES OF THE SPLICEOSOME

2.1 EM-Based Structural Studies before 2015

Mechanistic understanding of pre-mRNA splicing and its regulation requires detailed three-dimensional information on the intact spliceosome. X-ray crystallography allowed structural determination of a number of individual spliceosomal components and subcomplexes, including the U1 snRNP core (Pomeranz Krummel et al. 2009; Weber et al. 2010; Kondo et al. 2015), the U4 snRNP core (Leung et al. 2011), subcomplexes of the U2 and U6 snRNPs (Price et al. 1998; Sickmier et al. 2006; Lin and Xu 2012; Montemayor et al. 2014; Zhou et al. 2014), the exon junc-

 Cite as *Cold Spring Harb Perspect Biol* doi: 10.1101/cshperspect.a032409

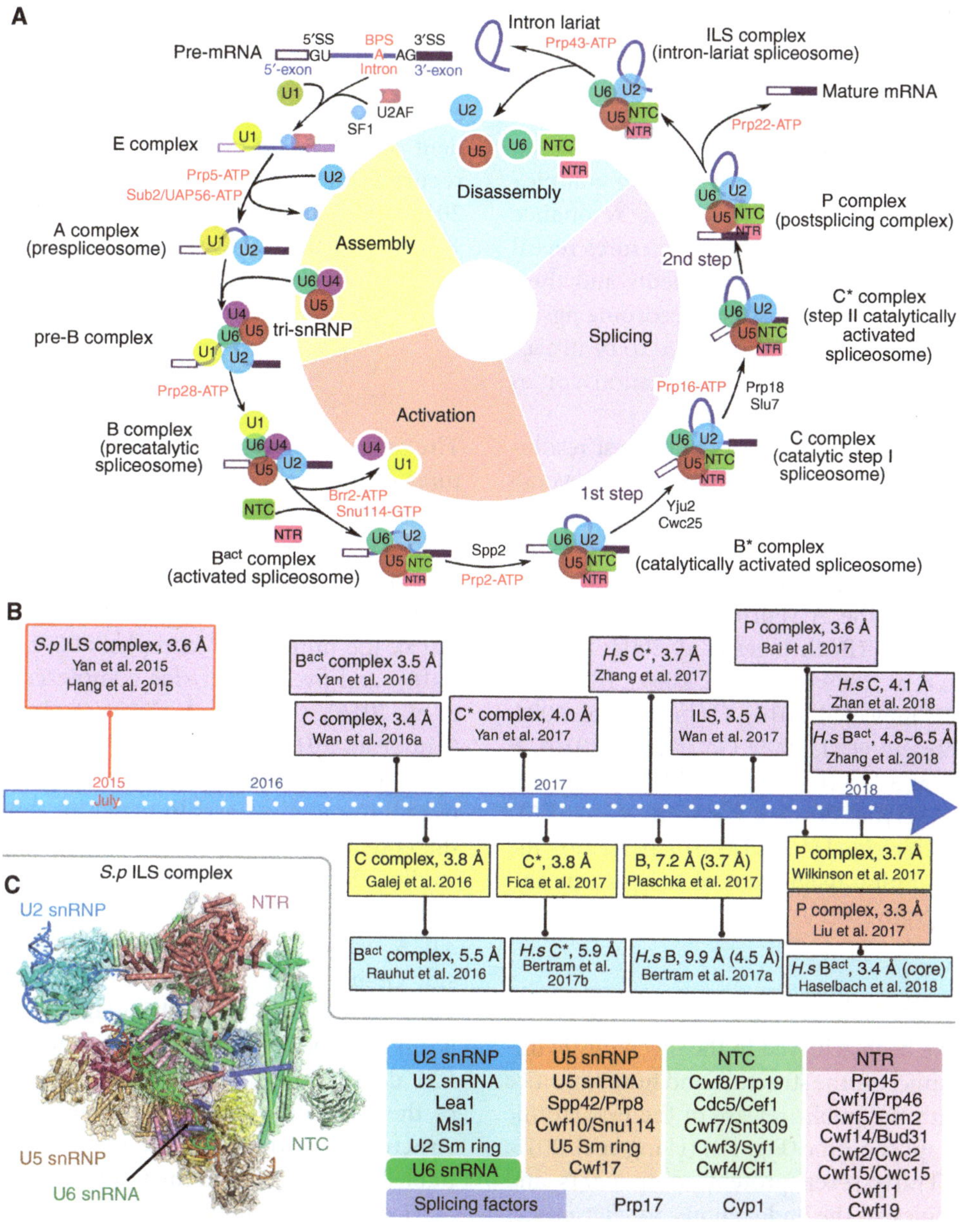

Figure 1. The precursor messenger RNA (pre-mRNA) splicing cycle and the major structures of the spliceosome. (*A*) The pre-mRNA splicing cycle. Each cycle includes three phases: assembly and activation of the spliceosome, execution of the splicing reaction, and disassembly of the spliceosome. The pre-B complex represents the first fully assembled spliceosome, in which all five snRNPs are present and the 5′SS is still recognized by U1 snRNP. In addition to the pre-B complex, the assembled spliceosome exists in seven compositionally distinct states: B, B^{act}, B*, C, C*, P, and intron lariat spliceosome (ILS). The unidirectional conversion of each of these neighboring spliceosomal complexes to the next is driven by the conserved ATPases/helicases Brr2 (B-to-B^{act}), Prp2 (B^{act}-to-B*), Prp16 (C-to-C*), and Prp22 (P-to-ILS). The splicing factors Cwc25 and Yju2 facilitate the branching reaction, whereas Prp18 and Slu7 assist exon ligation. Disassembly of the ILS complex is executed by the ATPase/helicase Prp43. (*B*) The major cryoelectron microscopy (cryo-EM) structures of the spliceosome since 2015. The time axis is drawn to scale, with each white dot representing a month. The first atomic model of an intact spliceosome was elucidated for the ILS complex from *Saccharomyces pombe* on the basis of an EM density map at an average resolution of 3.6 Å (Hang et al. 2015; Yan et al. 2015). The 18 cryo-EM structures of the spliceosome come from four independent laboratories: nine from the Shi group (shaded light purple), four from the Nagai group (shaded yellow), four from the Luhrmann/Stark group (shaded cyan), and one from the Zhao/Zhou group (shaded light brown). (*C*) The cryo-EM structure of the ILS complex from *S. pombe* (PDB code 3JB9) (Hang et al. 2015; Yan et al. 2015). The color-coded structure is shown on the *left*, and the individual components resolved in the structure are tabulated on the *right*.

tion complex bound to RNA (Andersen et al. 2006), a large fragment of Prp8 bound to Aar2 (Galej et al. 2013), Brr2 bound to a carboxy-terminal fragment of Prp8 (Mozaffari-Jovin et al. 2013; Nguyen et al. 2013), the SF3a complex (Lin and Xu 2012), and the core of the SF3b complex (Cretu et al. 2016). A structure of the retention and splicing complex (RES) was determined by nuclear magnetic resonance (NMR) (Wysoczanski et al. 2014). These structures reveal detailed features of the individual components and their mutual interactions within the colossal spliceosome jigsaw puzzle. Because the intact spliceosome proved to be intractable to crystallization, its structural information comes exclusively from electron microscopy (EM).

Until 2015, EM-based studies, with the highest resolution ~20–29 Å, had only allowed a glimpse of the overall appearance of the spliceosome at different splicing stages and tentative assignment of some snRNPs and protein components (Luhrmann and Stark 2009). These EM structures were reported for the A (Furman and Glitz 1995; Behzadnia et al. 2007), B (Boehringer et al. 2004; Deckert et al. 2006; Fabrizio et al. 2009; Wolf et al. 2009), B^{act} (Fabrizio et al. 2009; Bessonov et al. 2010), C (Jurica et al. 2004; Ohi et al. 2007; Fabrizio et al. 2009; Golas et al. 2010), P (Ilagan et al. 2013), and ILS (Chen et al. 2014) complexes. In 2015, the 5.9 Å cryoelectron microscopy (cryo-EM) structure of the *S. cerevisiae* U4/U6.U5 tri-snRNP, for the first time, allowed faithful assignment of the protein and RNA components (Nguyen et al. 2015).

2.2 EM-Based Structural Studies since 2015

Major advances have been achieved in the mechanistic understanding of the spliceosome structure and function since 2015, beginning with the public release of the first atomic model for an intact spliceosome (Fig. 1B) (Hang et al. 2015; Yan et al. 2015). Thirty-seven protein components and four snRNA molecules of the endogenous *Saccharomyces pombe* ILS complex were atomically modeled from the 3.6 Å cryo-EM reconstruction (Fig. 1C), in which the local resolution in the core regions of the spliceosome reached 2.9 Å (Yan et al. 2015). Most notably, structural insights derived from the *S. pombe* ILS complex are generally applicable to most other spliceosomes from both *S. cerevisiae* and human (Shi 2017a).

Since 2016, at least 17 independent cryo-EM structures of the assembled spliceosome have been reported, reflecting six distinct compositional states (Fig. 1B; Table 1). Among these structures, 11 were reported for the *S. cerevisiae* spliceosomal complexes: B (Plaschka et al. 2017), B^{act} (Rauhut et al. 2016; Yan et al. 2016), C (Galej et al. 2016; Wan et al. 2016a), C* (Fica et al. 2017; Yan et al. 2017), P (Bai et al. 2017; Liu et al. 2017; Wilkinson et al. 2017), and ILS (Wan et al. 2017); six were reported for the human spliceosomes: B (Bertram et al. 2017a), B^{act} (Haselbach et al. 2018; Zhang et al. 2018), C (Zhan et al. 2018), and C* (Fig. 1B) (Bertram et al. 2017b; Zhang et al. 2017). In addition, three independent structures of the U4/U6.U5 tri-snRNP have been reported for *S. cerevisiae* (Nguyen et al. 2016; Wan et al. 2016b) and for human (Agafonov et al. 2016), and the structure of the *S. cerevisiae* U1 snRNP was elucidated at 3.6 Å resolution (Table 1) (Li et al. 2017).

2.3 Quality Assessment of the Recent Cryo-EM Structures

The reported average resolution for the 18 cryo-EM structures of the entire spliceosome ranges between 3.3 Å and 9.9 Å resolution, with the highest limit reaching 2.8 Å (Table 1). Among these structures, only seven were determined at an average overall resolution of 3.6 Å or better, including the *S. pombe* ILS complex at 3.6 Å (Yan et al. 2015), the human B^{act} complex at 3.4 Å (Haselbach et al. 2018), and five *S. cerevisiae* complexes: B^{act} at 3.5 Å (Yan et al. 2016), C at 3.4 Å (Wan et al. 2016a), P at 3.3 Å (Liu et al. 2017) and 3.6 Å (Bai et al. 2017), and ILS at 3.4 Å (Fig. 1B; Table 1) (Wan et al. 2017). The average overall resolution refers to the entire spliceosome, not just the core or a specific region as evasively claimed in some studies. The local resolution for the core regions of the spliceosome in each of the six structures reaches at least 3.0 Å, which allows accurate assignment of amino acid side chains and placement of RNA nucleotides and metal ions. The ultimate test of the quality of a cryo-EM structure is not the reported resolution, but the quality of the original EM density map. Importantly, side-by-side comparison of the deposited EM density maps shows that the seven structures are indeed of the highest quality among all 18 published structures of the yeast or human spliceosome.

Two general methods were used for the isolation of the spliceosomes. One method involves direct purification of endogenous spliceosomal complexes from the nuclear extract using an affinity-tagged protein component (Ohi et al. 2007; Yan et al. 2015). This method was effectively applied to purification of the yeast, but not human, spliceosomes and allowed determination of seven of the 18 published spliceosome structures (Table 1). The pre-mRNA in these spliceosomes represents a mixture from endogenous genes; its atomic modeling was made possible by the strong sequence conservation of yeast intronic sequences (Spingola et al. 1999; Bon et al. 2003; Kupfer et al. 2004). In contrast, the other method relies on in vitro assembly of the spliceosome on MS2-tagged synthetic pre-mRNA (Jurica et al. 2002; Zhou et al. 2002). This method has been successfully applied to both human and yeast spliceosomes, although

Cite as *Cold Spring Harb Perspect Biol* doi: 10.1101/cshperspect.a032409

Table 1. Cryo-EM studies of the spliceosomes and the spliceosomal complexes

Species	State of the spliceosome	Resolution (Å) (highest limit)[a]	Method of assembly	Method of purification	Date of Epub	Reference(s)
Schizosaccharomyces pombe	ILS	3.6 (3.0)	Endogenous	Cdc5-TAPs	2015.08.20	Hang et al. 2015; Yan et al. 2015
Saccharomyces cerevisiae	B^{act}	3.5 (2.8)	Endogenous	Cef1-TAPs	2016.07.21	Yan et al. 2016
		5.8 (4.0)	In vitro assembling	Prp2-G551N; MS2-MBP; Grafix and PMPI cross-linking	2016.08.25	Rauhut et al. 2016
	C	3.4 (2.9)	Endogenous	Cef1-TAPs	2016.07.21	Wan et al. 2016a
		3.8 (3.5)	In vitro assembling	MS2-MBP	2016.07.26	Galej et al. 2016
	C*	4.0 (3.5)	Endogenous	Cef1-TAPs	2016.12.15	Yan et al. 2017
		3.8 (3.5)	In vitro assembling	Slu7-TAPs; BS3 cross-linking	2017.01.11	Fica et al. 2017
	B	7.2 (3.5)	In vitro assembling	Brr2-TAPs; MS2-MBP; BS3 cross-linking	2017.05.22	Plaschka et al. 2017
	ILS	3.5 (2.9)	Endogenous	Cef1 or Yju2 TAPs	2017.09.14	Wan et al. 2017
	P	3.6 (3.0)	Endogenous	Prp22-K512A; Cef1-TAPs	2017.11.16	Bai et al. 2017
		3.3 (3.0)	Endogenous	Prp22-H606A; Cef1-TAPs	2017.11.16	Liu et al. 2017
		3.8 (3.5)	In vitro assembling	Prp22-S635A; MS2-MBP	2017.11.16	Wilkinson et al. 2017
Homo sapiens	C*	5.9 (4.5)	In vitro assembling	MS2-MBP; Grafix	2017.01.11	Bertram et al. 2017b
		3.8 (3.0)	In vitro assembling	MS2-MBP; Grafix	2017.05.11	Zhang et al. 2017
	B	9.9 (3.5)	In vitro assembling	MS2-MBP; Grafix	2017.08.03	Bertram et al. 2017a
	C	4.1 (3.5)	In vitro assembling	MS2-MBP; Grafix	2018.01.04	Zhan et al. 2018
	B^{act}	3.4 (3.0)	In vitro assembling	MS2-MBP; Grafix	2018.01.17	Haselbach et al. 2018
		4.9 (4.0)	In vitro assembling	MS2-MBP; Grafix	2018.01.23	Zhang et al. 2018
Other spliceosomal complexes						
S. cerevisiae	Tri-snRNP	5.9 (4.5)	Endogenous	Brr2-TAPs	2015.06.24	Nguyen et al. 2015
		3.8 (3.0)	Endogenous	Prp6-TAPs	2016.01.07	Wan et al. 2016b
		3.7 (3.0)	Endogenous	Brr2-TAPs	2016.02.01	Nguyen et al. 2016
	U1 snRNP	3.6 (3.0)	Endogenous	U1A/Mud1-TAPs	2017.10.19	Li et al. 2017
H. sapiens	Tri-snRNP	7.0 (4.0)	Endogenous	m^3G-cap antibody; Grafix	2016.02.18	Agafonov et al. 2016

[a]The resolution shown refers to the average overall resolution of the entire spliceosome or the spliceosomal complex. The resolution in parentheses represents the highest local resolution in the spliceosome or the spliceosomal complex.

an additional step of chemical cross-linking is required for maintenance of the structural integrity for the human spliceosome (Table 1).

3 SHARED FEATURES OF THE SPLICEOSOMES

Except the pre-B and the B* complexes, six of the eight assembled spliceosomal complexes during each splicing cycle have been structurally characterized in recent cryo-EM studies. Among these, the pre-B and the B complex are the only spliceosomes that do not contain a well-formed splicing active site; they show a major difference compared with the other six spliceosomal complexes. The U4 snRNP remains associated and the NTC and NTR components are yet to be recruited. The other six spliceosomes share a set of strikingly similar structural features that were first observed in the cryo-EM structure of the *S. pombe* ILS complex (Hang et al. 2015; Yan et al. 2015).

The RNA elements in the splicing active site show a highly conserved conformation in the B^{act} through ILS complexes (Fig. 2A). The intramolecular stem loop (ISL) of U6 snRNA forms the center of the splicing active site and coordinates the two catalytic metal ions (M1 and M2). The ISL is stabilized by three conserved structural metal ions. Three consecutive nucleotides of the U2 snRNA from the helix Ib form a catalytic triplex with a duplex of U6 snRNA, which involves five nucleotides preceding the ISL and the key U6 nucleotide U80 in *S. cerevisiae* (Fig. 2B). The catalytic triplex facilitates the splicing reaction. Loop I of U5 snRNA, which recognizes the 5′-exon, is positioned in close proximity to the ISL. Beyond the active site, U5 and U6 snRNAs along their entire lengths and 30 consecutive

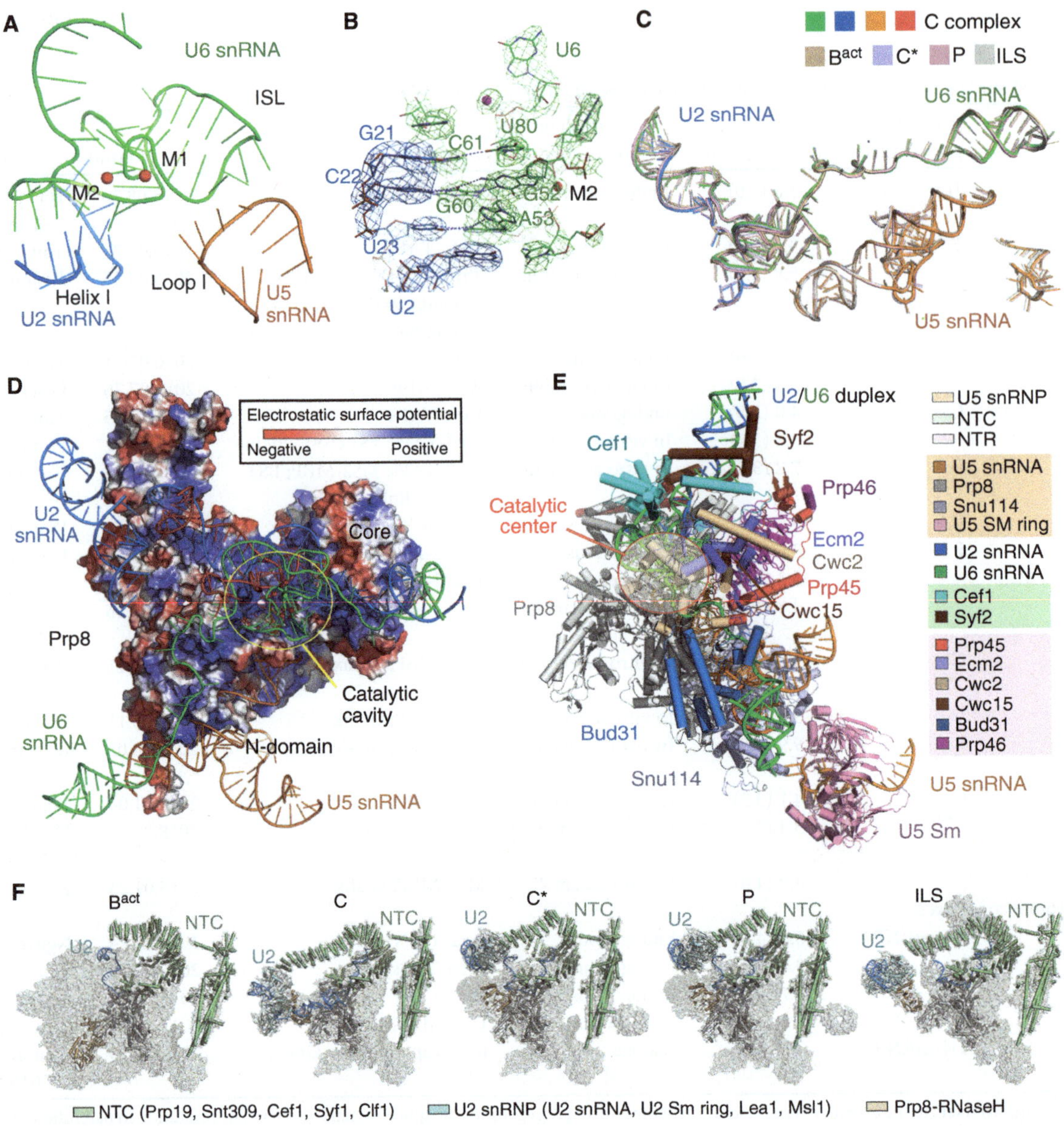

Figure 2. Shared structural features of the spliceosome during the splicing reaction. (*A*) The RNA elements at the conserved splicing active site. The active site comprises the intramolecular stem loop (ISL) of U6 small nuclear RNA (snRNA) and associated metal ions, the U2/U6 catalytic triplex, helix I of the U2/U6 duplex, and loop I of U5 snRNA. Although the active site shown here is derived from the *Saccharomyces cerevisiae* C complex (PDB code 5GMK) (Wan et al. 2016a), the overall conformation is identical in the B^{act} through intron lariat spliceosome (ILS) complexes. (*B*) Structure of the conserved catalytic triplex. The original electron microscopy (EM) density map, shown for the *S. cerevisiae* C complex (Wan et al. 2016a), allows unambiguous identification of the metal ions and the nucleobases. (*C*) U5 and U6 snRNAs along their entire lengths and the 5′ 30 nucleotides of U2 snRNA remain largely static in the B^{act} through ILS complexes. The RNA elements in the *S. cerevisiae* C complex are color coded; whereas those from the B^{act}, C*, P, and ILS complexes are each displayed in a single color. The PDB codes for the B^{act}, C, C*, P, and ILS complexes are 5GM6, 5GMK, 5WSG, 5YLZ, and 5Y88, respectively. (*D*) The conserved splicing active site RNA elements are anchored in the catalytic cavity of Prp8. The cavity is formed at the interface between the N-domain and the Prp8 core and is enriched in positively charged residues. (*E*) Sixteen additional protein components, for their partial or entire lengths, remain largely static in the B^{act} through ILS complexes. These proteins, together with Prp8, maintain the rigid conformation of the snRNA elements (U5 and U6 along their entire lengths and the 5′ 30 nucleotides of U2 snRNA). These proteins and the snRNAs from the *S. cerevisiae* C complex are color coded here. (*F*) A compositionally constant group of 13 proteins in the B^{act} through ILS complexes. These proteins constitute two groups: nine in the U2 snRNP (the heptameric Sm complex, Lea1, and Msl1) and four in the NTC (Prp19, Snt309, Cef1, Syf1, and Clf1). Each group rigidly maintains its collective structure but is translocated during the two steps of the splicing reaction. These proteins, together with the other 20 rigid components, give rise to the characteristic appearance of the spliceosome.

nucleotides at the 5′-end of U2 snRNA show the same structure in the B^{act} through ILS complexes (Fig. 2C). Thus, among the three snRNA molecules, the only mobile region is the sequence downstream from nucleotide 30 of U2 snRNA.

The relatively rigid conformation of the snRNAs in the splicing active site and beyond is maintained by at least 17 protein components, which in turn must share nearly identical structures in the B^{act} through ILS complexes (Shi 2017b). Among these proteins, Prp8 plays a pivotal role. A positively charged catalytic cavity, which is formed at the interface between its N-domain and the core (also known as the large domain; Galej et al. 2013), provides approximately half of the needed surface to host the conserved RNA elements at the splicing active site (Fig. 2D) (Hang et al. 2015; Yan et al. 2015). The other half is furnished by segments of the NTC components Cef1 (Cdc5 in *S. pombe*) and Syf2, the NTR components Cwc2 and Ecm2 (Cwf2 and Cwf5 in *S. pombe*), and the various splicing factors. In addition, Snu114, the heptameric Sm complex of the U5 snRNP, and six NTR components (Cwc2, Ecm2, Bud31, Cwc15, Prp45, and Prp46) display the same general conformation in the B^{act} through ILS complexes (Fig. 2E).

In addition to these 20 protein and RNA components that remain largely rigid in the B^{act} through ILS complexes, 13 additional protein components are included in each of these spliceosomes. These include nine components of the U2 snRNP (Lea1, Msl1, and the heptameric Sm complex), which interact with each other and bind the sequences around the Sm site of U2 snRNA, and four proteins of the NTC (Prp19, Snt309, Syf1, and Clf1), which together with a portion of Cef1 are intertwined to form a stable core of the NTC (Fig. 2F). The relative positioning of the components within the U2 snRNP core or the NTC core remains unchanged in the various spliceosomal complexes; but the core as a whole undergoes drastic translocation, particularly the U2 snRNP.

Together, the immobile 20 protein and RNA components, along with the compositionally stable but mobile 13 proteins, give rise to a characteristic appearance of the spliceosome that is conserved in the B^{act} through ILS complexes (Fig. 2F). In fact, for an experienced structural biologist, visual inspection of the two-dimensional averages of the particles not only allows identification of the spliceosomal complexes but also their specific stages in the splicing reaction. Such a structural hallmark also greatly facilitates atomic modeling of the various spliceosomal complexes. For example, the atomic coordinates of the *S. pombe* ILS complex (Yan et al. 2015) can be conveniently docked into the EM density map of the human C* complex (Zhang et al. 2017), followed by minor adjustment and residue replacement.

4 SPLICEOSOME AS A METALLORIBOZYME

The available structural evidence proves the notion that the spliceosome is a metalloribozyme. The roles of the two catalytic metal ions M1 and M2 are thought to reciprocate in the two transesterification reactions of each splicing cycle (Steitz and Steitz 1993; Fica et al. 2013). In the branching reaction that occurs in the B* complex, M2 activates the nucleophile (2′-OH of an adenine nucleotide in the BPS) and M1 stabilizes the leaving group (3′-OH of the 5′-exon). In the exon ligation that proceeds in the C* complex, M1 activates the nucleophile (3′-OH of the 5′-exon) and M2 stabilizes the leaving group (3′-OH of the intron lariat). In the *S. pombe* ILS complex (Hang et al. 2015; Yan et al. 2015), both M1 and M2 are observed and their locations are consistent with published biochemical characterization (Fica et al. 2013).

Table 2. Coordination of the M1 and M2 metals in different spliceosomal complexes in the yeast of *Saccharomyces cerevisiae*

		B^{act}		C		C*		P		ILS	
Ligand	Complex	M1	M2	M1	M2	M1	M2	M1	M2	M1	M2
Identity		–	+	+	+	+	+	+	+	+	+
U6 snRNA	A59: *pro-R*$_p$		√				√		√		
	A59: *pro-S*$_p$				√						
	G60: *pro-S*$_p$		√				√		√		
	G78: *pro-R*$_p$			√		√		√		√	
	U80: *pro-R*$_p$				√		√		√		√
	U80: *pro-S*$_p$			√		√		√		√	
5′-exon	G-1 at 3′ end: 3′ OH			√		√		√			
5′SS	G1 at 5′ end: *pro-R*$_p$			√							
3′SS	G at 3′ end: 2′/3′ OH								√√		
3′-exon	X at 5′ end: *pro-R*$_p$							√			
	Number of ligands		2	4	2	3	3	4	5	2	1

Presence of the M1 and M2 metals is indicated by red and blue check signs, respectively.

The M1 metal ion is uniformly coordinated by two phosphate oxygen atoms from G78 and U80 of U6 snRNA in the spliceosomal C, C*, P, and ILS complexes (Table 2). The coordination of the M1 metal, as illustrated in the splicing active site center of the *S. cerevisiae* C complex (Wan et al. 2016a), involves two more ligands: a phosphate oxygen atom from the 5′-end of the 5′SS and the 3′-OH at the 3′-end nucleotide of the 5′-exon (Table 2). These four ligands are arranged in a planar fashion (Fig. 3A). The putatively octahedral coordination suggests M1 to be a divalent ion, likely magnesium (Mg^{2+}). In addition to the two uniformly coordinating ligands from U6 snRNA, M1 is bound by the 3′-OH at the 3′-end nucleotide of the 5′-exon in both the C* (Yan et al. 2017) and P (Bai

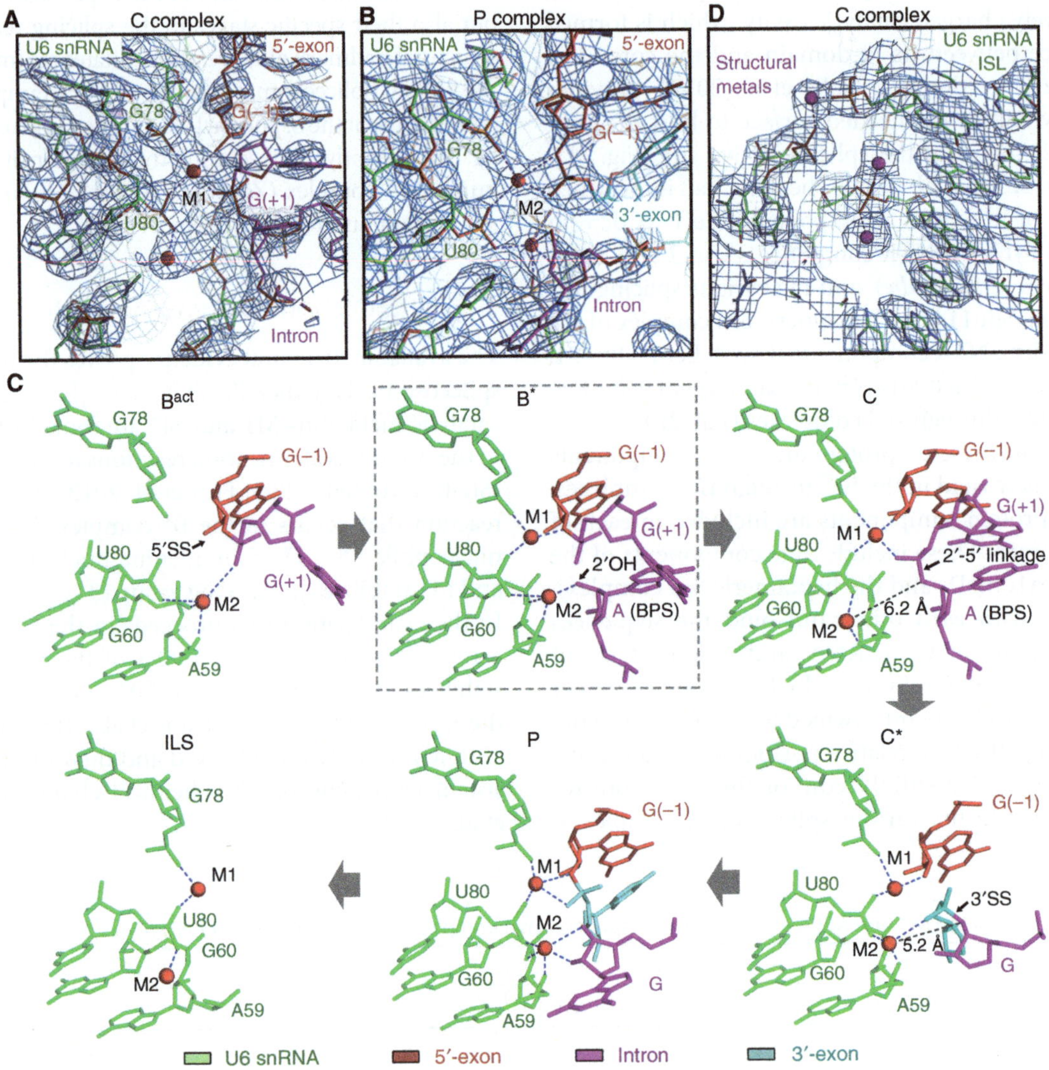

Figure 3. The spliceosome is a metalloribozyme. (*A*) Coordination of the catalytic metals M1 and M2. Shown here is metal coordination in the *Saccharomyces cerevisiae* C complex (Wan et al. 2016a). M1 is recognized by four ligands arranged in a planar fashion. The original electron microscopy (EM) density map is shown in panels *A*, *B*, and *D*. PDB code 5GMK and EMDB code EMD-9525 for the C complex; PDB code 5YLZ and EMDB code EMD-6839 for the P complex. (*B*) Coordination of M1 and M2 in the *S. cerevisiae* P complex (Bai et al. 2017). M2 is bound by five potential ligands. (*C*) Choreography of the catalytic metals during the splicing reaction. The coordination of M1 and M2 is shown for the spliceosomal B^{act}, B* (predicted), C, C*, P, and intron lariat spliceosome (ILS) complexes. Elements of the U6 small nuclear RNA (snRNA) are shown in the same orientation in the six complexes to allow meaningful comparison. Metal coordination in the B* complex is predicted based on the C complex and knowledge of the splicing reaction. (*D*) The structural metals are defined by an excellent EM density map (shown here for the *S. cerevisiae* C complex; Wan et al. 2016a) and presumably stabilize the intramolecular stem loop (ISL) fold by neutralizing the negative charges of the RNA phosphodiester backbone.

et al. 2017) complexes (Table 2). In the P complex (Bai et al. 2017), M1 is also recognized by a fourth ligand: a phosphate oxygen atom at the 5′-end of the 3′-exon. In the ILS complex (Wan et al. 2017), M1 loses all other ligands and is coordinated only by G78 and U80. Notably, M1 is yet to be loaded into the B^{act} complex (Yan et al. 2016).

M2 coordination appears to be more dynamic than M1. The M2 metal ion is stably coordinated by two phosphate oxygen atoms from A59 and U80 of U6 snRNA in the spliceosomal C, C*, and P complexes (Table 2). The coordination of the M2 metal is exemplified in the active site center of the P complex (Bai et al. 2017; Liu et al. 2017; Wilkinson et al. 2017), in which M2 is bound by five ligands: the 2′-OH and 3′-OH (the leaving group in exon ligation) at the 3′-end of the 3′SS and three phosphate oxygen atoms from the nucleotides A59, G60, and U80 of U6 snRNA (Fig. 3B; Table 2). Although M2 is recognized by only two ligands from A59 and U80 in the *S. cerevisiae* C complex (Wan et al. 2016a), it is coordinated by one more ligand—a phosphate oxygen atom from G60—in the *S. cerevisiae* C* complex (Table 2) (Fica et al. 2017; Yan et al. 2017). In the *S. cerevisiae* ILS complex (Wan et al. 2017), however, M2 is bound only by one ligand: a phosphate oxygen atom from U80. Although M2 is loaded in the B^{act} complex (Yan et al. 2016), it is recognized by the phosphate oxygen atoms of A59 and G60.

This analysis reveals a choreography of the catalytic metals during each cycle of splicing (Fig. 3C). In the B^{act} complex (Yan et al. 2016), M2 is loaded in a preactivated fashion and M1 is absent. Both catalytic metals are presumably loaded into the B* complex, which executes the branching reaction. In the resulting C complex (Wan et al. 2016a), both M1 and M2 are properly coordinated. In the C* complex, which executes the exon ligation (Fica et al. 2017; Yan et al. 2017), M1 loses one ligand (a phosphate oxygen atom at the 5′-end of the 5′SS) owing to the relocation of the lariat junction and M2 coordination also undergoes rearrangement. In the resulting P complex (Bai et al. 2017; Liu et al. 2017; Wilkinson et al. 2017), M1 gains one more ligand and M2 has two more ligands; the coordination of M1 and M2 appears to be most complete at this stage. Finally, after exon dissociation in the ILS complex (Wan et al. 2017) M1 loses two ligands from the exon, whereas M2 has only one ligand from U80 of U6 snRNA.

In contrast to the changing patterns of the catalytic metals, three structural metal ions are steadily present in the B^{act}, C, C*, P, and ILS complexes as judged by the local features of the EM density map (Fig. 3D). These metals appear to stabilize the delicate fold of the U6 ISL by neutralizing the negative charges of the RNA backbone phosphates. Two of the three metal ions are also present in the *S. pombe* ILS complex (Hang et al. 2015; Yan et al. 2015).

5 THE ROLE OF THE PROTEIN COMPONENTS

The protein components in each spliceosome can be divided into four distinct groups: the structural proteins, the splicing factors, the RNA-dependent ATPase/helicases, and other regulatory proteins. The structural proteins sustain the splicing active site conformation, support the overall characteristic appearance of the spliceosome, and provide the elasticity that is needed for the splicing reaction. The splicing factors facilitate assembly and activation of the spliceosome and assist the two steps of transesterification. The ATPase/helicases remodel the spliceosomal complexes to allow flux of the splicing factors and other proteins and RNA elements. The other regulatory proteins modulate the splicing of pre-mRNA and will be briefly discussed in a later section.

5.1 The Structural Proteins

Both the ribosome and the spliceosome are ribozymes. Unlike the ribosome in which the RNA components constitute the structural scaffold of the entire RNP complex (Ban et al. 2000; Nissen et al. 2000; Schluenzen et al. 2000; Wimberly et al. 2000), the snRNA elements in the spliceosome are flexible and are supported and directed by the surrounding protein components (Shi 2017b). The structural proteins include all protein components of the U2 snRNP core and the U5 snRNP, and virtually all proteins of the NTC and NTR. The structural proteins can be further classified into three distinct categories: the spring-like proteins, the rope-like proteins, and the rest.

The spring-like proteins are exemplified by the superhelical proteins Syf1 and Clf1 (Cwf3 and Cwf4 in *S. pombe*), which can bend and stretch, without being structurally disrupted, to allow large-degree movement of a subcomplex or a region of the spliceosome (Fig. 4A). During the C-to-C* transition, the U2 snRNP core is translocated by a distance of 60–100 Å, which is mediated by the elastic property of Syf1 (Fica et al. 2017; Yan et al. 2017). The rope-like proteins are represented by Prp45, the carboxy-terminal fragment of Cef1, and Cwc15; these proteins are flexible tethers that glue together multiple components of the spliceosome. For example, the NTR component Prp45 adopts an extended conformation, spans a distance of 150 Å, and simultaneously interacts with nine proteins and two snRNA elements in the spliceosome (Fig. 4B) (Yan et al. 2015). Other than the spring-like and rope-like proteins, each of the remaining protein components has a well-defined conformation and often contains well-folded globular domains.

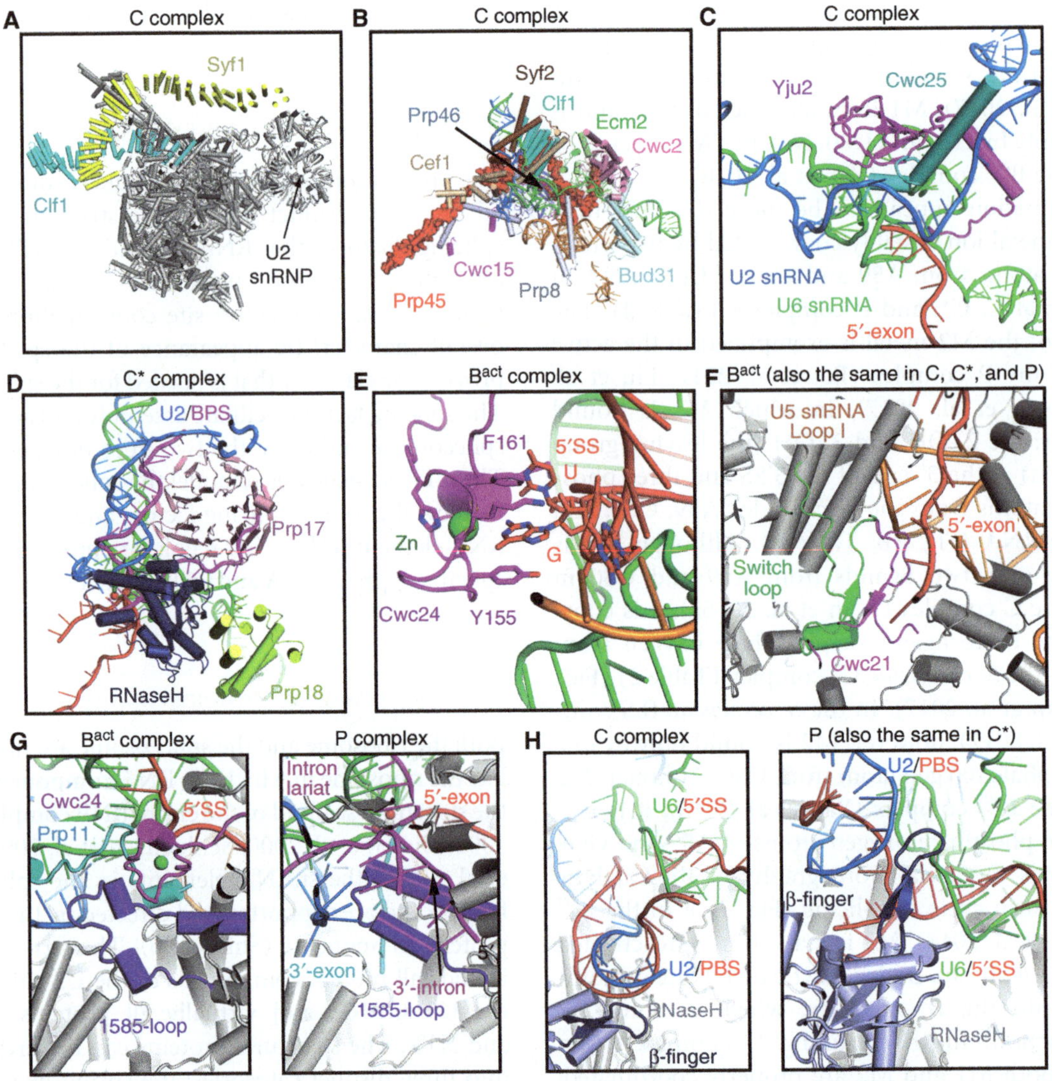

Figure 4. Protein components and catalytic motifs of the spliceosome. (*A*) The spring-like proteins, exemplified by Syf1 and Clf1 in *Saccharomyces cerevisiae*, allow large conformational rearrangements of the spliceosome while maintaining their structural integrity. The *S. cerevisiae* C complex (Wan et al. 2016a) is used for illustration in panels *A–C*. Syf1 and Clf1 are colored yellow and cyan, respectively. (*B*) The rope-like proteins, represented by Prp45 and Cwc15, stabilize spliceosomal organization through direct interactions with multiple proteins and RNA elements. Prp45, colored red and displayed in surface representation, simultaneously interacts with at least nine proteins and two small nuclear RNA (snRNA) elements (U2 and U6). (*C*) The step I factors Cwc25 and Yju2 constitute an integral part of the splicing active site during the branching reaction. The amino termini of both proteins reach into the center of the active site RNA elements and stabilize their conformations. (*D*) The step II factors Prp17 and Prp18 stabilize the active site conformation. Shown here is the *S. cerevisiae* C* complex (Yan et al. 2017). (*E*) The splicing factor Cwc24 protects the guanine nucleotide at the 5′-end of the 5′SS in the *S. cerevisiae* B^{act} complex (Yan et al. 2016). Two aromatic residues Tyr155 and Phe161 sandwich the guanine nucleobase. (*F*) The splicing factor Cwc21 stabilizes the binding of the 5′-exon to loop I of U5 snRNA. Cwc21 simultaneously interacts with nucleotides of the 5′-exon and the switch loop of Prp8 in the B^{act} through P complexes. The interactions, shown here for the *S. cerevisiae* B^{act} complex (Yan et al. 2016), are identically preserved in the C, C*, and P complexes. (*G*) The 1585-loop of Prp8 in the *S. cerevisiae* B^{act} (Yan et al. 2016) (*left* panel) and P (Bai et al. 2017) (*right* panel) complexes. (*H*) The β-finger (from the RNaseH-like domain of Prp8) in the *S. cerevisiae* C (Wan et al. 2016a) (*left* panel) and P (Bai et al. 2017) (*right* panel) complexes.

 Cite as *Cold Spring Harb Perspect Biol* doi: 10.1101/cshperspect.a032409

5.2 The Splicing Factors

The splicing factors constitute three classes: those that are specific for the branching reaction (step I factors) or the exon ligation (step II factors), and those that are generally required for splicing. The third class is further divided into two categories: those that are required for the early phase of active site formation (Cwc24 and Cwc27) and those that are present throughout both steps of the transesterification (Cwc21, Cwc22).

The step I factors Cwc25 and Yju2 play an indispensable role in stabilizing the active site RNA conformation (Chiu et al. 2009; Warkocki et al. 2009; Schneider et al. 2015), as observed in the cryo-EM structure of the *S. cerevisiae* C complex (Fig. 4C) (Galej et al. 2016; Wan et al. 2016a). Notably, for both Cwc25 and Yju2, the amino-terminal fragment is positioned close to the center of the active site, with conserved residues directly contacting the BPS/U2 duplex. Cwc25 also binds to the helix I of the U2/U6 duplex and the ISL. Yju2 interacts with the 5′-exon, Isy1, and Prp8. These elaborate interactions render Cwc25 and Yju2 integral parts of the splicing active site composite.

The step II factors Prp17, Prp18, and Slu7 play an important role in exon ligation (Jones et al. 1995; Aronova et al. 2007), although they appear to be less critically involved in the splicing active site compared with the step I factors. Indeed, the absence of Prp18 or Slu7 still allowed splicing of some pre-mRNA to proceed when the RNA sequence between the BPS and the 3′SS is shortened (Brys and Schwer 1996; Zhang and Schwer 1997; James et al. 2002). Nonetheless, these factors are located close to the splicing active site and may facilitate exon ligation through interactions with the protein and RNA elements. In the *S. cerevisiae* C* (Fica et al. 2017; Yan et al. 2017) and P (Bai et al. 2017; Liu et al. 2017; Wilkinson et al. 2017) complexes, the WD40 domain of Prp17 is located between the BPS/U2 duplex and the 5′SS/U6 duplex (Fig. 4D). Prp18 makes no physical interactions with the RNA elements but directly contacts the RNaseH-like domain of Prp8. The RNaseH-like domain, in turn, interacts with the BPS/U2 duplex in a nearly identical fashion in the spliceosomal C, C*, and P complexes (Galej et al. 2016; Wan et al. 2016a; Bai et al. 2017; Fica et al. 2017; Liu et al. 2017; Wilkinson et al. 2017; Yan et al. 2017). Both the RNaseH-like domain of Prp8 and the WD40 domain of Prp17 undergo marked translocation in the C-to-C* transition. Only a small portion of Slu7 was tentatively identified in the *S. cerevisiae* C* complex (Fica et al. 2017), although the bulk of this protein has been assigned in the human C* complex (Zhang et al. 2017).

The splicing factor Cwc24, which is recruited into the spliceosome in the B-to-B^{act} transition, plays an important role in the B^{act} complex by insulating the 5′-end guanine nucleobase of the 5′SS (Yan et al. 2016). The aromatic side chains of Tyr155 and Phe161 of Cwc24 sandwich the guanine nucleobase (Fig. 4E), whereas Lys160 mediate hydrogen bonds (H-bonds) to the nucleobase and the ribose. Through the action of Prp2, Cwc24 is dissociated in the B^{act}-to-B* transition. The splicing factors Cwc21 and Cwc22 are recruited into the B^{act} complex and dissociated in the P-to-ILS transition. The amino-terminal fragment of Cwc21 directly interacts with the 5′-exon to stabilize its binding to loop I of U5 snRNA (Fig. 4F). Cwc21, in turn, is stabilized by the switch loop of Prp8. The 5′-exon–Cwc21–switch loop interactions remain unchanged in the spliceosomal B^{act} through P complexes. Cwc22 is located close to Cwc21 and interacts with Prp8, whereas the splicing factor Cwc27 binds both Prp8 and Brr2 and is present only in the B^{act} complex.

5.3 The Catalytic Motifs of Prp8

Three prominent structural motifs, all from Prp8, appear to facilitate the splicing reaction through close interactions with the active site elements. These motifs include the switch loop and the 1585-loop (also known as α-finger; Nguyen et al. 2016) of the linker domain and the β-finger (Yang et al. 2008) of the RNaseH-like domain. The switch loop, which comprises an extended β-hairpin-containing structure, stabilizes 5′-exon binding. In all spliceosomes (B^{act}, C, C*, and P) in which the 5′-exon is loaded onto loop I of U5 snRNP, the switch loop is identically arranged (Fig. 4F). Compared with these spliceosomes, the switch loop is flipped by 180° in all structurally characterized complexes in which the interaction of the 5′-exon with U5 loop I is absent, including in the U4/U6.U5 tri-snRNP (Nguyen et al. 2016; Wan et al. 2016b) and in the B (Plaschka et al. 2017) and ILS (Hang et al. 2015; Yan et al. 2015; Wan et al. 2017) complexes.

In the B^{act} complex (Yan et al. 2016), the 1585-loop interacts with the U2/U6 duplex and binds the structural elements of Cwc24 and Prp11 that together protect the 5′-end guanine nucleotide of the 5′SS (Fig. 4G, left panel). In the C* (Fica et al. 2017; Yan et al. 2017) and P (Bai et al. 2017; Liu et al. 2017; Wilkinson et al. 2017) complexes, the tip of the 1585-loop is located close to the 5′-exon and the catalytic metals, contacting the ISL and the lariat junction (Fig. 4G, right panel). In addition, the 1585-loop directly interacts with the pre-mRNA nucleotides of the 3′SS and two nucleobases of the 3′-exon (Bai et al. 2017; Liu et al. 2017; Wilkinson et al. 2017). At the end of the splicing cycle, the 1585-loop is no longer needed and becomes disordered in the ILS complex (Hang et al. 2015; Yan et al. 2015; Wan et al. 2017).

In the B^{act} (Rauhut et al. 2016; Yan et al. 2016) and ILS (Hang et al. 2015; Yan et al. 2015; Bai et al. 2017) complexes, the β-finger remains unengaged, not interacting with any protein or RNA element. In the C (Galej et al. 2016; Wan et al. 2016a) and presumably B* complexes, the β-finger interacts with the BPS/U2 duplex (Fig. 4H, left panel). In the C* (Fica et al. 2017; Yan et al. 2017) and P (Bai et al. 2017; Liu et al. 2017; Wilkinson et al. 2017) complexes, the β-finger interacts with the lariat junction on the opposite side of the 1585-loop, located between the U2/BPS and U6/5′SS duplexes (Fig. 4H, right panel), and contacts the step II splicing factor Prp17.

5.4 The RNA-Dependent ATPase/Helicases

The cryo-EM structures of the various spliceosomes reveal mechanistic insights into the B-to-B^{act}, B^{act}-to-B*, C-to-C*, and P-to-ILS transitions. Except for the U5 snRNP, nearly all other protein and RNA components of the B complex undergo marked changes—either dissociation or translocation and conformational change—in the transition to the B^{act} complex. This transition is mediated by Brr2, which is thought to bind the single-stranded U4 snRNA sequences downstream from the U4/U6 duplex (Fig. 5A). Propelled by

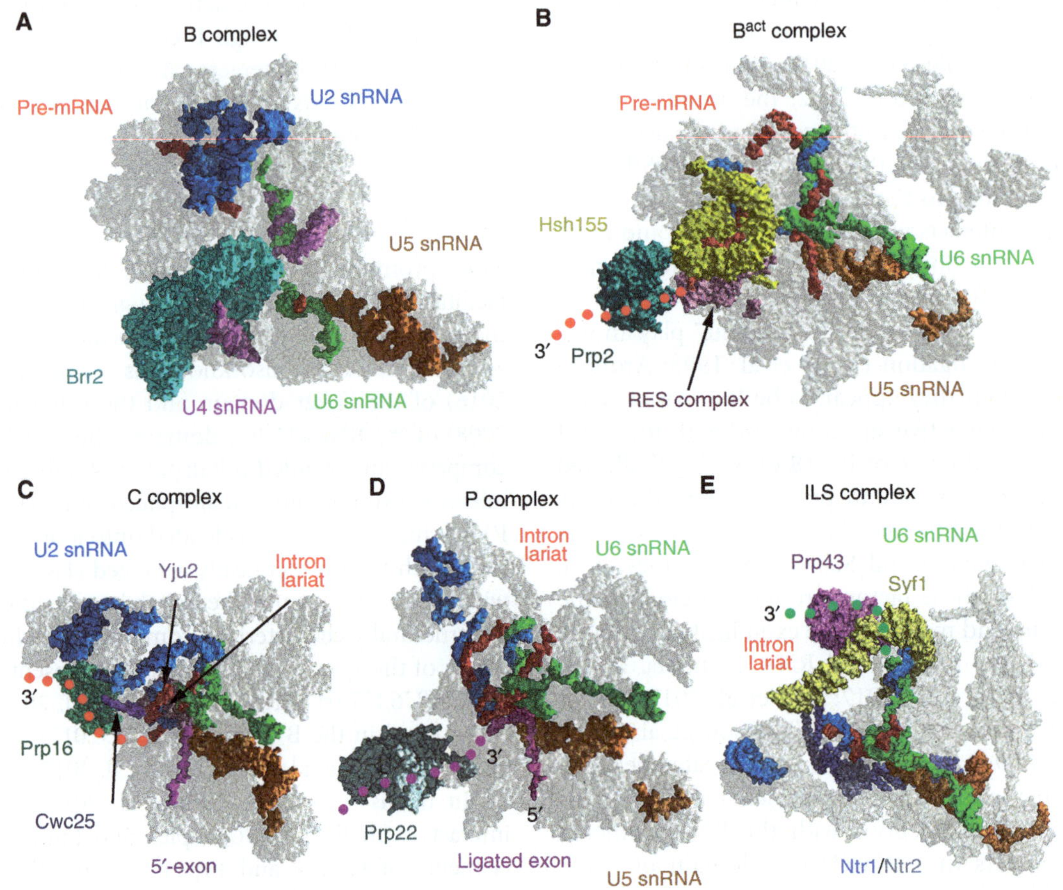

Figure 5. Spliceosomal transitions are driven by the ATPase/helicases. (*A*) The B-to-B^{act} transition is propelled by Brr2. Shown here is a surface view of the *Saccharomyces cerevisiae* B complex (Plaschka et al. 2017). Brr2 is thought to pull on the single-stranded U4 small nuclear RNA (snRNA) sequences, triggering the unwinding of the U4/U6 duplex and subsequent changes. (*B*) The B^{act}-to-B* transition is mediated by Prp2. Shown here is a surface view of the *S. cerevisiae* B^{act} complex (Yan et al. 2016). Prp2 likely pulls on the 3′-end sequences of the pre-mRNA, leading to dissociation of the retention and splicing complex (RES) complex, the SF3b and SF3a complexes, and the splicing factor Cwc24. (*C*) The C-to-C* transition is executed by Prp16. Shown here is a surface view of the *S. cerevisiae* C complex (Wan et al. 2016a). Prp16 presumably pulls on the 3′-end sequences of the intron lariat-3′-exon intermediate. (*D*) The P-to-intron lariat spliceosome (ILS) transition is driven by Prp22. Shown here is a surface view of the *S. cerevisiae* P complex (Bai et al. 2017). Dissociation of the ligated exon is mediated by Prp22 pulling on its 3′-end sequences. (*E*) The disassembly of the ILS complex is performed by Prp43. Shown here is a surface view of the *S. cerevisiae* ILS complex (Wan et al. 2017). Prp43 may function by binding to and pulling on either the 3′ end of U6 snRNA or the intron lariat. The PDB codes for the B, B^{act}, C, P, and ILS complexes are 5NRL, 5GM6, 5GMK/5LJ5, 5YLZ, and 5Y88, respectively.

ATP binding and hydrolysis, Brr2 presumably pulls on the single-stranded RNA sequence (Plaschka et al. 2017), leading to the dissociation of the associated protein factors and unwinding of the U4/U6 duplex. The B-to-B^{act} transition is speculated to occur in several distinct steps (Plaschka et al. 2017; Sidarovich et al. 2017; Zhang et al. 2018).

At the periphery of the B^{act} complex (Yan et al. 2016), Prp2 binds the SF3b protein Hsh155 and is located close to the RES complex (Fig. 5B). The RNA sequences downstream from the BPS are bound by the RES complex, with the disordered 3′-end sequences presumably recognized by Prp2 (Liu and Cheng 2012; Schneider et al. 2015). This structural arrangement is fully consistent with the notion that pulling on the 3′-end intron sequences may result in the dissociation of the RES complex and the SF3b complex during the B^{act}-to-B* transition. In the C complex (Galej et al. 2016), the RNA-binding site of Prp16 is about 70 Å away from the structurally identified nucleotide at the 3′ end of the intron (Fig. 5C). In the C-to-C* transition, Prp16 would presumably pull on the 3′-end sequences of the intron, leading to the dissociation of the step I factors Cwc25 and the amino-terminal fragment of Yju2 (Tseng et al. 2011) and subsequent recruitment of the step II factors Prp18 and Slu7 (Ansari and Schwer 1995; James et al. 2002). In the P complex (Fig. 5D), Prp22 directly interacts with the Linker and the RT Finger/Palm domains of Prp8; but the RNA-binding site of Prp22 is ~80 Å away from the structurally identified 3′-end nucleotide of the 3′-exon. This structural feature explains why the release of the ligated exon from the P complex is inhibited by shortening the 3′-exon sequence (Schwer 2008; Ilagan et al. 2013). In the P-to-ILS transition, Prp22 is thought to pull on the 3′-end RNA sequences to dissociate the ligated exon (Company et al. 1991; Schwer and Gross 1998; Semlow et al. 2016), triggering the dissociation of Cwc21, Cwc22, Prp18, Slu7, and the WD40 domain of Prp17 and subsequent recruitment of Cwc23 and the disassembly Ntr complex (Prp43, Ntr1, and Ntr2).

The disassembly of the spliceosome is mediated by the Ntr complex (Arenas and Abelson 1997; Tsai et al. 2005, 2007; Fourmann et al. 2016). The structure of the ILS complex suggests two potential mechanisms by which the spliceosome is disassembled by the ATPase/helicase Prp43 (Fig. 5E) (Wan et al. 2017). Prp43 binds the middle portion of the superhelical protein Syf1, whereas Ntr1 and Ntr2 are located away from Prp43. The 3′-end nucleotide of U6 snRNA that has been structurally identified is located only ~50 Å away from the RNA-binding site of Prp43. In contrast, the intron lariat is located 85–100 Å away. Therefore, it appears more likely that Prp43 grabs onto and pulls on the 3′-end of U6 snRNA as opposed to the intron lariat (Bohnsack et al. 2009).

These transitions are, by definition, dynamic and can be inferred only from comparative analysis of the static cryo-EM structures. The steps and mechanisms separating two neighboring spliceosomal complexes could be considerably more dynamic than the qualitative comparison. For example, the human B^{act} complex has at least three compositional states—early, mature, and late (Zhang et al. 2018)—and the mature state comprises at least eight distinct conformational states (Haselbach et al. 2018). Importantly, however, the essential features of the various transitions have been structurally observed. These features to a large extent confirm published biochemical and genetic data that have accumulated in the past three decades. Despite the tantalizing interpretation of the structural images, how the ATPase/helicases function mechanically remain to be elucidated, perhaps through single-molecule biophysical investigations. The need for in-depth biophysical studies is reinforced by the fact that these ATPase/helicases are located mostly in the periphery of the spliceosomes and are characterized by medium-to-low resolution, which does not allow accurate assignment of atomic features that are required for faithful mechanistic understanding.

6 RECOGNITION OF THE RNA ELEMENTS

The conserved sequence motifs of the pre-mRNA are mainly recognized by conserved RNA elements from both snRNA and pre-mRNA. Three nucleotides at the 3′-end of the 5′-exon are anchored on the loop I of U5 snRNA in the B^{act} through P complexes. The BPS and surrounding sequences of the intron form an elongated duplex with complementary sequences of the U2 snRNA in the A through ILS complexes. The dinucleotides GU at the 5′ end of the 5′SS are recognized by the Linker domain of Prp8 in the *S. cerevisiae* U4/U6.U5 tri-snRNP (Wan et al. 2016b) and the human B complex (Bertram et al. 2017a) and by Cwc24 in the B^{act} complex (Yan et al. 2016); the six consecutive nucleotides downstream from GU of the 5′SS are recognized by U6 snRNA in the B^{act} through ILS complexes. In the P (Bai et al. 2017; Liu et al. 2017; Wilkinson et al. 2017) and possibly C* complexes, the dinucleotides AG of the 3′SS are recognized by the lariat junction that is formed after the branching reaction (Fig. 6A). Notably, AG of the 3′SS pair up with the nucleophile-containing A of the BPS and the invariant G at the 5′-end of the 5′SS, respectively, through noncanonical Watson–Crick H-bonds.

The spliceosome is a protein-directed metalloribozyme. The RNA elements, both the immobile portion (U5, U6, the 5′-end portion of U2, and the 3′-end portion of the 5′-exon) and the mobile segments (U2 downstream from nucleotide 30, the bulk of the intron, and the 3′-exon), are specifically recognized by protein components. This subject has been

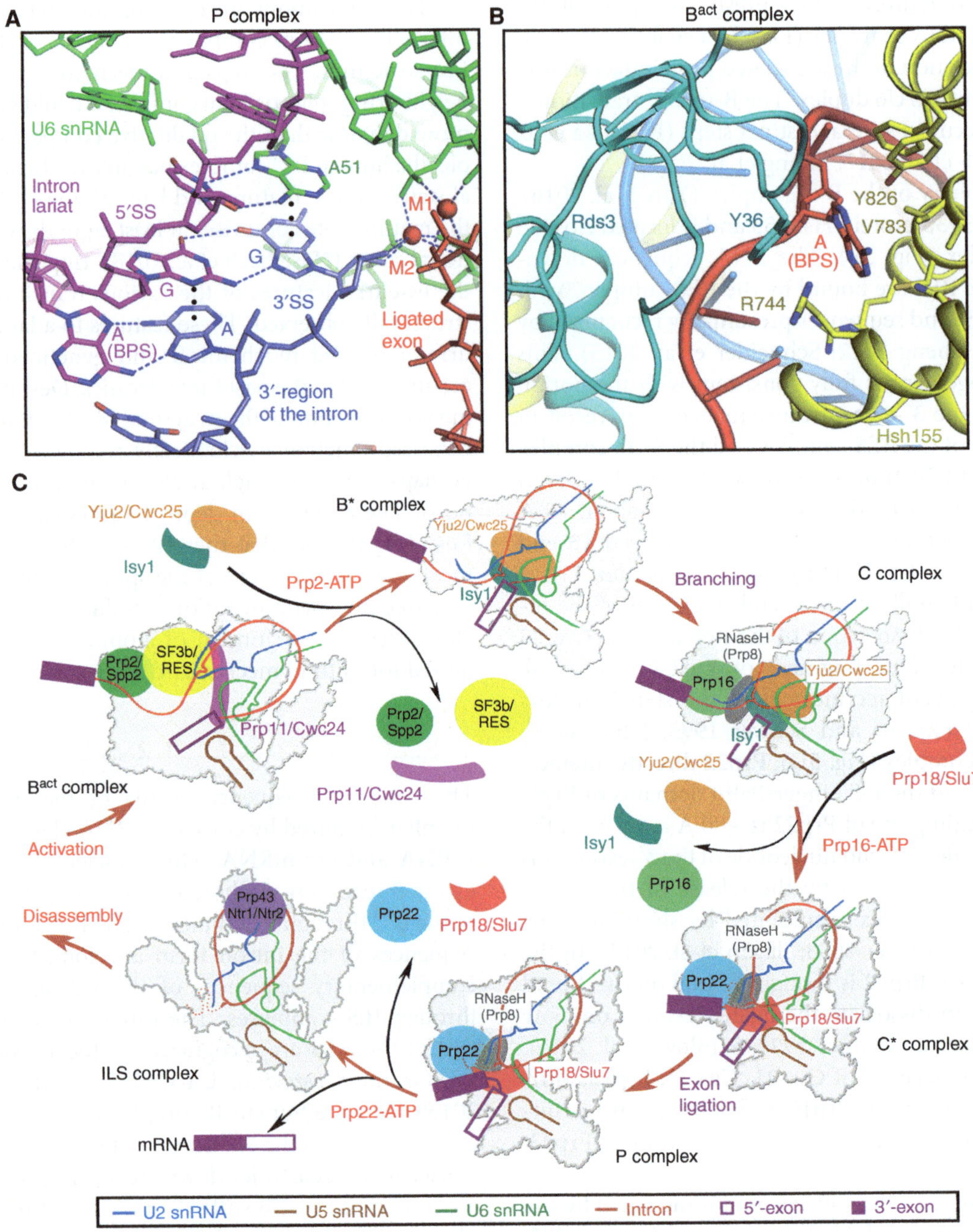

Figure 6. Choreography of the protein and RNA components during the splicing reaction. (*A*) Recognition of the AG dinucleotide at the 3′SS by the lariat junction in the P complex (Bai et al. 2017; Liu et al. 2017; Wilkinson et al. 2017). Notably, the AG nucleobases at the 3′SS form noncanonical Watson–Crick base-pairing interactions with two consecutive AG nucleobases of the lariat junction. (*B*) Recognition of the nucleophile-containing adenine nucleotide of the branch point sequence (BPS) by Hsh155. The adenine nucleobase is already flipped out of the registry of the intron/U2 duplex and is in close contact with a few residues of Hsh155 and Rds3. (*C*) A cartoon diagram depicting the activation of the B^{act} complex, the branching and exon ligation, and the disassembly of the intron lariat spliceosome (ILS) complex.

thoroughly reviewed recently (Shi 2017a) and is thus only briefly discussed here. The immobile portion is mainly recognized by the 17 static protein components. For example, the active site RNA elements are cradled in the positively charged catalytic cavity that is formed between the N-domain and the core of the Prp8 (Fig. 2D). The mobile RNA segments are recognized and directed by conserved protein elements. For example, in the B^{act} complex (Rauhut et al. 2016; Yan et al. 2016), the BPS/U2 duplex is surrounded by Hsh155 and the nucleophile-containing adenine nucleobase of the BPS is flipped out of the U2/BPS duplex registry and recognized by specific residues of Hsh155 (Fig. 6B). The 5′-end guanine nucleotide of the 5′SS is recognized by the splicing factor Cwc24 (Fig. 4E) and the SF3a protein Prp11.

7 CHOREOGRAPHY OF THE RNA AND PROTEIN COMPONENTS

Three snRNA molecules, U5, U6, and 30 nucleotides at the 5′-end of U2, remain largely static in the spliceosomal B^{act} through ILS complexes, with only minor shift (mostly <3 Å) for select local segments. The static segments of U2 and U6 form helix I and helix II. The 5′-exon of the pre-mRNA remains identically anchored to the loop I of U5 snRNA in the spliceosomal B^{act} through P complexes. The sequences downstream from nucleotide 30 of U2 snRNA, along with segments of the intron, undergo a series of dramatic movements that are designed to facilitate branching and exon ligation.

In the B^{act} complex (Rauhut et al. 2016; Yan et al. 2016), conserved U2 sequences form a duplex with the BPS, which is sequestered away from the splicing active site by components of the SF3b and SF3a complexes and the splicing factor Cwc24. In the B^{act}-to-B* transition, Prp2 presumably pulls on the 3′-end of the pre-mRNA, causing dissociation of the SF3a and SF3b complexes, the RES complex, and Cwc24 (Fig. 6C). The step I factors Cwc25 and Yju2 and the NTC component Isy1 are recruited into the splicing active site to stabilize the conformation required for the branching reaction. Segments of the U2 snRNA and the intron sequences are translocated by 50–100 Å, bringing the nucleophile—the invariant adenine nucleotide of the BPS—into the close proximity of the 5′SS as observed in the C complex (Fig. 6C) (Galej et al. 2016; Wan et al. 2016a).

In the C complex (Galej et al. 2016; Wan et al. 2016a), the lariat junction spatially blocks the entry of the 3′SS-3′-exon sequences into the splicing active site for exon ligation. In the C-to-C* transition, through the action of Prp16, the lariat junction is translocated by approximately 20 Å to vacate space for the 3′SS-3′-exon. Cwc25, Yju2, and Isy1 dissociate from the C complex, and the step II factors Prp18 and Slu7 are recruited to form the C* complex (Fig. 6C). Much of U2 snRNA undergoes drastic movement in the transition, and a large portion of the intron lariat is also flipped by up to 60 Å. In the P complex (Bai et al. 2017; Liu et al. 2017; Wilkinson et al. 2017), the dinucleotides AG of the 3′SS form base-pairing interactions with two consecutive nucleotides AG of the lariat junction, which represent the nucleophile-containing adenine nucleotide of the BPS and the 5′-end guanine nucleotide of the 5′SS.

In the P complex (Bai et al. 2017; Liu et al. 2017; Wilkinson et al. 2017), Prp22 binds the Linker domain of Prp8 at a corner of the spliceosome. In the P-to-ILS transition, Prp22 likely pulls on the 3′-end sequences of the ligated exon, leading to its release from the spliceosome and subsequent dissociation of the step II factors Prp18 and Slu7 (Fig. 6C). Finally, in the ILS complex (Wan et al. 2017), Prp43 may pull on the 3′-end nucleotide of U6 snRNA or the intron lariat, leading to the disassembly of the spliceosome (Fig. 6C).

8 THE HUMAN SPLICEOSOME

Compared with yeast, the human spliceosome is considerably more dynamic and less stable. Isolation of the intact human spliceosome toward structural studies requires application of chemical cross-linking reagents (Table 1). Cryo-EM investigation of the human spliceosome has yielded structures of the B (Bertram et al. 2017a), B^{act} (Haselbach et al. 2018; Zhang et al. 2018), C (Zhan et al. 2018), and C* (Bertram et al. 2017b; Zhang et al. 2017) complexes at resolutions that allow identification of structural details (Table 1). The overall features as well as the detailed interactions of the human spliceosomal complexes are very similar to those of the corresponding yeast complexes. These shared features include the overall organization, the splicing active site conformation, the location of catalytic and structural metal ions, the static protein and RNA components, and the vast majority of the protein–protein, protein–RNA, and RNA–RNA interactions. Here, we briefly describe only the unique features that have emerged from the cryo-EM structures of the human spliceosome. A detailed structural analysis of the human spliceosome and its comparison with the yeast spliceosome will be more meaningful after structural resolution of the human P and ILS complexes.

The human spliceosome generally contains more protein components than its yeast counterpart. Compared with the *S. cerevisiae* structure (Yan et al. 2016), the structure of the human mature B^{act} complex has 16 additional proteins: SF3a120, SF3a60, SF3b14a/p14, U5-40K, nine proteins (Lea1, Msl1, and the Sm complex) of the U2 snRNP core, and three proteins (Aquarius, RBM22, and PPIL1) of the NTR (Haselbach et al. 2018; Zhang et al. 2018). The structure of the human C complex contains 11 more proteins, which include four PPIs, four proteins of the exon junction complex, U5-40K, RBM22, and Aquarius (Zhan et al. 2018). Compared with the *S. cerevisiae* C* complex (Yan et al. 2017), the structure of the human complex has six additional proteins: Brr2, Slu7, the functionally mysterious component PRKRIP1, and three proteins (MAGOH, Y14, and MLN51) of the exon junction complex (EJC) (Zhang et al. 2017).

As exemplified by the six additional proteins that may assist exon ligation in the human C* complex (Zhang et al. 2017), these additional proteins apparently play an important role in the function of the corresponding human spliceosome. The human protein RBM22 appears to be a fusion between two yeast proteins Cwc2 and Ecm2 (Rasche et al. 2012). Notably, RBM22 is present in the mature, but not the early, B^{act} complex (Zhang et al. 2018). PRKRIP1 forms an elongated helix and links the U2 snRNP core to the splicing active site (Zhang et al. 2017). Compared with its yeast counterpart, the human spliceosome appears to have more conformational states within the same spliceosomal complex. For example, the human B^{act} complex assumes three compositionally different states: early, mature, and late B^{act} complexes (Zhang et al. 2018). The mature B^{act} complex has at least eight distinct conformational states (Haselbach et al. 2018).

9 CONCLUSION

During the three years since 2015, mechanistic understanding of spliceosome function through structural elucidation has taken a qualitative leap, beginning with the 3.6-Å cryo-EM structure of the *S. pombe* ILS complex (Hang et al. 2015; Yan et al. 2015). At present, with 18 independent structures of the spliceosome within the overall resolution range of 3.3–9.9 Å, we have developed a reasonably good understanding of the molecular mechanisms of spliceosome-mediated pre-mRNA splicing. With structures of the B, B^{act}, C, C*, P, and ILS complexes from *S. cerevisiae*, we have recapitulated the processes of spliceosome activation, the branching reaction and exon ligation, and disassembly of the spliceosome. We have the structure of the U4/U6.U5 tri-snRNP, but not the pre-spliceosome (A complex) or the early spliceosome (E complex). Importantly, the structure of the pre-B complex remains elusive. Consequently, we are yet to piece together the assembly process of the spliceosome.

The static structural images of the various spliceosomal complexes provide a physical basis for understanding the functions of the spliceosome. But these structures alone fail to answer some of the important mechanistic questions. For example, the RNA-dependent ATPase/helicases, which are responsible for RNP remodeling, are located at the periphery of the spliceosome and defined by relatively poor EM density maps at medium-to-low resolution, which disallows identification of potentially bound ATP molecules and/or RNA nucleotides. Consequently, how these ATPase/helicases function in the context of the spliceosome remains to be structurally characterized. Other biophysical approaches may provide advantages complementary to cryo-EM. For example, single-molecule fluorescence studies of the spliceosomal transitions by the ATPase/helicases have begun to reveal their underlying molecular mechanisms (Semlow et al. 2016). Adaptation of such investigative approaches to the mammalian PPIs may provide additional insights to regulation of the spliceosome.

The best local resolution achieved for an intact spliceosome is 2.8 Å for the stable core of the *S. cerevisiae* B^{act} complex (Yan et al. 2016), which displays clearly distinguishable features for amino acid side chains. At this resolution, water molecules and the metal ions are yet to be conclusively identified. Improved resolution may also allow conclusive identification of posttranslational protein modifications and small-molecule ligands that target the various spliceosomal proteins. This information, in turn, may guide rational design of splicing modulators with therapeutic potential.

For technical reasons, structural advances in 2015 and 2016 have been accomplished almost exclusively for the yeast spliceosome. Major progress was achieved in 2017 for the structures of the human B and C* complexes, followed by the structures of the B^{act} and C complexes in the beginning of 2018. It is anticipated that in the near future major advances for the human spliceosome may parallel those for the yeast spliceosome. We will be able to perform a comprehensive comparison of the yeast and human spliceosomes. Lessons gained from such a comparison will not only allow appreciation of the fine species distinctions but also engender discovery of potential modulators that selectively target the spliceosome from one, but not the other, species.

ACKNOWLEDGMENTS

We thank Rui Bai, Xiaofeng Zhang, and Xiechao Zhan for help in figure preparation and suggestions on text and references. This work was supported by funds from Beijing Advanced Innovation Center for Structural Biology at Tsinghua University, Beijing, China.

REFERENCES

Agafonov DE, Kastner B, Dybkov O, Hofele RV, Liu WT, Urlaub H, Luhrmann R, Stark H. 2016. Molecular architecture of the human U4/U6.U5 tri-snRNP. *Science* **351:** 1416–1420.

Andersen CB, Ballut L, Johansen JS, Chamieh H, Nielsen KH, Oliveira CL, Pedersen JS, Seraphin B, Le Hir H, Andersen GR. 2006. Structure of the exon junction core complex with a trapped DEAD-box ATPase bound to RNA. *Science* **313:** 1968–1972.

Ansari A, Schwer B. 1995. SLU7 and a novel activity, SSF1, act during the PRP16-dependent step of yeast pre-mRNA splicing. *EMBO J* **14:** 4001–4009.

Arenas JE, Abelson JN. 1997. Prp43: An RNA helicase-like factor involved in spliceosome disassembly. *Proc Natl Acad Sci* **94:** 11798–11802.

Aronova A, Bacikova D, Crotti LB, Horowitz DS, Schwer B. 2007. Functional interactions between Prp8, Prp18, Slu7, and U5 snRNA during the second step of pre-mRNA splicing. *RNA* **13:** 1437–1444.

Bai R, Yan C, Wan R, Lei J, Shi Y. 2017. Structure of the post-catalytic spliceosome from *Saccharomyces cerevisiae*. *Cell* **171:** 1589–1598.

Ban N, Nissen P, Hansen J, Moore PB, Steitz TA. 2000. The complete atomic structure of the large ribosomal subunit at 2.4 Å resolution. *Science* **289:** 905–920.

Behzadnia N, Golas MM, Hartmuth K, Sander B, Kastner B, Deckert J, Dube P, Will CL, Urlaub H, Stark H, et al. 2007. Composition and three-dimensional EM structure of double affinity-purified, human prespliceosomal A complexes. *EMBO J* **26:** 1737–1748.

Berget SM, Robberson BL. 1986. U1, U2, and U4/U6 small nuclear ribonucleoproteins are required for in vitro splicing but not polyadenylation. *Cell* **46:** 691–696.

Berget SM, Moore C, Sharp PA. 1977. Spliced segments at the 5′ terminus of adenovirus 2 late mRNA. *Proc Natl Acad Sci* **74:** 3171–3175.

Bertram K, Agafonov DE, Dybkov O, Haselbach D, Leelaram MN, Will CL, Urlaub H, Kastner B, Luhrmann R, Stark H. 2017a. Cryo-EM Structure of a pre-catalytic human spliceosome primed for activation. *Cell* **170:** 701–713.

Bertram K, Agafonov DE, Liu WT, Dybkov O, Will CL, Hartmuth K, Urlaub H, Kastner B, Stark H, Luhrmann R. 2017b. Cryo-EM structure of a human spliceosome activated for step 2 of splicing. *Nature* **542:** 318–323.

Bessonov S, Anokhina M, Krasauskas A, Golas MM, Sander B, Will CL, Urlaub H, Stark H, Luhrmann R. 2010. Characterization of purified human B^{act} spliceosomal complexes reveals compositional and morphological changes during spliceosome activation and first step catalysis. *RNA* **16:** 2384–2403.

Black DL, Chabot B, Steitz JA. 1985. U2 as well as U1 small nuclear ribonucleoproteins are involved in premessenger RNA splicing. *Cell* **42:** 737–750.

Boehringer D, Makarov EM, Sander B, Makarova OV, Kastner B, Luhrmann R, Stark H. 2004. Three-dimensional structure of a pre-catalytic human spliceosomal complex B. *Nat Struct Mol Biol* **11:** 463–468.

Boesler C, Rigo N, Anokhina MM, Tauchert MJ, Agafonov DE, Kastner B, Urlaub H, Ficner R, Will CL, Luhrmann R. 2016. A spliceosome intermediate with loosely associated tri-snRNP accumulates in the absence of Prp28 ATPase activity. *Nat Commun* **7:** 11997.

Bohnsack MT, Martin R, Granneman S, Ruprecht M, Schleiff E, Tollervey D. 2009. Prp43 bound at different sites on the pre-rRNA performs distinct functions in ribosome synthesis. *Mol Cell* **36:** 583–592.

Bon E, Casaregola S, Blandin G, Llorente B, Neuveglise C, Munsterkotter M, Guldener U, Mewes HW, Van Helden J, Dujon B, et al. 2003. Molecular evolution of eukaryotic genomes: Hemiascomycetous yeast spliceosomal introns. *Nucleic Acids Res* **31:** 1121–1135.

Brody E, Abelson J. 1985. The "spliceosome": Yeast pre-messenger RNA associates with a 40S complex in a splicing-dependent reaction. *Science* **228:** 963–967.

Brys A, Schwer B. 1996. Requirement for SLU7 in yeast pre-mRNA splicing is dictated by the distance between the branchpoint and the 3′ splice site. *RNA* **2:** 707–717.

Chan SP, Cheng SC. 2005. The Prp19-associated complex is required for specifying interactions of U5 and U6 with pre-mRNA during spliceosome activation. *J Biol Chem* **280:** 31190–31199.

Chan SP, Kao DI, Tsai WY, Cheng SC. 2003. The Prp19p-associated complex in spliceosome activation. *Science* **302:** 279–282.

Chen CH, Yu WC, Tsao TY, Wang LY, Chen HR, Lin JY, Tsai WY, Cheng SC. 2002. Functional and physical interactions between components of the Prp19p-associated complex. *Nucleic Acids Res* **30:** 1029–1037.

Chen W, Shulha HP, Ashar-Patel A, Yan J, Green KM, Query CC, Rhind N, Weng Z, Moore MJ. 2014. Endogenous U2.U5.U6 snRNA complexes in *S. pombe* are intron lariat spliceosomes. *RNA* **20:** 308–320.

Chiu YF, Liu YC, Chiang TW, Yeh TC, Tseng CK, Wu NY, Cheng SC. 2009. Cwc25 is a novel splicing factor required after Prp2 and Yju2 to facilitate the first catalytic reaction. *Mol Cell Biol* **29:** 5671–5678.

Chow LT, Gelinas RE, Broker TR, Roberts RJ. 1977. An amazing sequence arrangement at the 5′ ends of adenovirus 2 messenger RNA. *Cell* **12:** 1–8.

Company M, Arenas J, Abelson J. 1991. Requirement of the RNA helicase-like protein PRP22 for release of messenger RNA from spliceosomes. *Nature* **349:** 487–493.

Cordin O, Hahn D, Beggs JD. 2012. Structure, function and regulation of spliceosomal RNA helicases. *Curr Opin Cell Biol* **24:** 431–438.

Cretu C, Schmitzova J, Ponce-Salvatierra A, Dybkov O, De Laurentiis EI, Sharma K, Will CL, Urlaub H, Luhrmann R, Pena V. 2016. Molecular architecture of SF3b and structural consequences of its cancer-related cutations. *Mol Cell* **64:** 307–319.

Deckert J, Hartmuth M, Boehringer D, Behzadnia N, Will CL, Kastner B, Stark H, Urlaub H, Luhrmann R. 2006. Protein composition and electron microscopy structure of affinity-purified human spliceosomal B complexes isolated under physiological conditions. *Mol Cell Biol* **26:** 5528–5543.

DiMaria PR, Kaltwasser G, Goldenberg CJ. 1985. Partial purification and properties of a pre-mRNA splicing activity. *J Biol Chem* **260:** 1096–1102.

Fabrizio P, Dannenberg J, Dube P, Kastner B, Stark H, Urlaub H, Luhrmann R. 2009. The evolutionarily conserved core design of the catalytic activation step of the yeast spliceosome. *Mol Cell* **36:** 593–608.

Fica SM, Tuttle N, Novak T, Li NS, Lu J, Koodathingal P, Dai Q, Staley JP, Piccirilli JA. 2013. RNA catalyses nuclear pre-mRNA splicing. *Nature* **503:** 229–234.

Fica SM, Oubridge C, Galej WP, Wilkinson ME, Bai XC, Newman AJ, Nagai K. 2017. Structure of a spliceosome remodelled for exon ligation. *Nature* **542:** 377–380.

Fourmann JB, Dybkov O, Agafonov DE, Tauchert MJ, Urlaub H, Ficner R, Fabrizio P, Luhrmann R. 2016. The target of the DEAH-box NTP triphosphatase Prp43 in *Saccharomyces cerevisiae* spliceosomes is the U2 snRNP-intron interaction. *eLife* **5:** e15564.

Frendewey D, Keller W. 1985. Stepwise assembly of a pre-mRNA splicing complex requires U-snRNPs and specific intron sequences. *Cell* **42:** 355–367.

Furman E, Glitz DG. 1995. Purification of the spliceosome A-complex and its visualization by electron microscopy. *J Biol Chem* **270:** 15515–15522.

Galej WP, Oubridge C, Newman AJ, Nagai K. 2013. Crystal structure of Prp8 reveals active site cavity of the spliceosome. *Nature* **493:** 638–643.

Galej WP, Wilkinson ME, Fica SM, Oubridge C, Newman AJ, Nagai K. 2016. Cryo-EM structure of the spliceosome immediately after branching. *Nature* **537:** 197–201.

Golas MM, Sander B, Bessonov S, Grote M, Wolf E, Kastner B, Stark H, Luhrmann R. 2010. 3D cryo-EM structure of an active step I spliceosome and localization of its catalytic core. *Mol Cell* **40:** 927–938.

Grabowski PJ, Sharp PA. 1986. Affinity chromatography of splicing complexes: U2, U5, and U4 + U6 small nuclear ribonucleoprotein particles in the spliceosome. *Science* **233:** 1294–1299.

Grabowski PJ, Padgett RA, Sharp PA. 1984. Messenger RNA splicing in vitro: An excised intervening sequence and a potential intermediate. *Cell* **37:** 415–427.

Grabowski PJ, Seiler SR, Sharp PA. 1985. A multicomponent complex is involved in the splicing of messenger RNA precursors. *Cell* **42:** 345–353.

Hang J, Wan R, Yan C, Shi Y. 2015. Structural basis of pre-mRNA splicing. *Science* **349:** 1191–1198.

Haselbach D, Komarov I, Agafonov DE, Hartmuth K, Graf B, Dybkov O, Urlaub H, Kastner B, Luhrmann R, Stark H. 2018. Structure and conformational dynamics of the human spliceosomal B^{act} complex. *Cell* **172:** 454–464.e411.

Ilagan JO, Chalkley RJ, Burlingame AL, Jurica MS. 2013. Rearrangements within human spliceosomes captured after exon ligation. *RNA* **19:** 400–412.

James SA, Turner W, Schwer B. 2002. How Slu7 and Prp18 cooperate in the second step of yeast pre-mRNA splicing. *RNA* **8:** 1068–1077.

Jones MH, Frank DN, Guthrie C. 1995. Characterization and functional ordering of Slu7p and Prp17p during the second step of pre-mRNA splicing in yeast. *Proc Natl Acad Sci* **92:** 9687–9691.

Jurica MS, Licklider LJ, Gygi SR, Grigorieff N, Moore MJ. 2002. Purification and characterization of native spliceosomes suitable for three-dimensional structural analysis. *RNA* **8:** 426–439.

Jurica MS, Sousa D, Moore MJ, Grigorieff N. 2004. Three-dimensional structure of C complex spliceosomes by electron microscopy. *Nat Struct Mol Biol* **11:** 265–269.

Keating KS, Toor N, Perlman PS, Pyle AM. 2010. A structural analysis of the group II intron active site and implications for the spliceosome. *RNA* **16:** 1–9.

Kondo Y, Oubridge C, van Roon AM, Nagai K. 2015. Crystal structure of human U1 snRNP, a small nuclear ribonucleoprotein particle, reveals the mechanism of 5′ splice site recognition. *eLife* doi: 10.7554/eLife .04986.

Krainer AR, Maniatis T. 1985. Multiple factors including the small nuclear ribonucleoproteins U1 and U2 are necessary for pre-mRNA splicing in vitro. *Cell* **42:** 725–736.

Kramer A, Keller W, Appel B, Luhrmann R. 1984. The 5′ terminus of the RNA moiety of U1 small nuclear ribonucleoprotein particles is required for the splicing of messenger RNA precursors. *Cell* **38:** 299–307.

Kupfer DM, Drabenstot SD, Buchanan KL, Lai H, Zhu H, Dyer DW, Roe BA, Murphy JW. 2004. Introns and splicing elements of five diverse fungi. *Eukaryot Cell* **3:** 1088–1100.

Lerner MR, Steitz JA. 1979. Antibodies to small nuclear RNAs complexed with proteins are produced by patients with systemic lupus erythematosus. *Proc Natl Acad Sci* **76:** 5495–5499.

Lerner MR, Boyle JA, Mount SM, Wolin SL, Steitz JA. 1980. Are snRNPs involved in splicing? *Nature* **283:** 220–224.

Leung AK, Nagai K, Li J. 2011. Structure of the spliceosomal U4 snRNP core domain and its implication for snRNP biogenesis. *Nature* **473:** 536–539.

Li X, Liu S, Jiang J, Zhang L, Espinosa S, Hill RC, Hansen KC, Zhou ZH, Zhao R. 2017. CryoEM structure of *Saccharomyces cerevisiae* U1 snRNP offers insight into alternative splicing. *Nat Commun* **8:** 1035.

Lin PC, Xu RM. 2012. Structure and assembly of the SF3a splicing factor complex of U2 snRNP. *EMBO J* **31:** 1579–1590.

Liu HL, Cheng SC. 2012. The interaction of Prp2 with a defined region of the intron is required for the first splicing reaction. *Mol Cell Biol* **32:** 5056–5066.

Liu S, Li X, Zhang L, Jiang J, Hill RC, Cui Y, Hansen KC, Zhou ZH, Zhao R. 2017. Structure of the yeast spliceosomal postcatalytic P complex. *Science* **358:** 1278–1283.

Luhrmann R, Stark H. 2009. Structural mapping of spliceosomes by electron microscopy. *Curr Opin Struct Biol* **19:** 96–102.

Montemayor EJ, Curran EC, Liao HH, Andrews KL, Treba CN, Butcher SE, Brow DA. 2014. Core structure of the U6 small nuclear ribonucleoprotein at 1.7-Å resolution. *Nat Struct Mol Biol* **21:** 544–551.

Mount SM, Pettersson I, Hinterberger M, Karmas A, Steitz JA. 1983. The U1 small nuclear RNA-protein complex selectively binds a 5′ splice site in vitro. *Cell* **33:** 509–518.

Mozaffari-Jovin S, Wandersleben T, Santos KF, Will CL, Luhrmann R, Wahl MC. 2013. Inhibition of RNA helicase Brr2 by the C-terminal tail of the spliceosomal protein Prp8. *Science* **341:** 80–84.

Nguyen TH, Li J, Galej WP, Oshikane H, Newman AJ, Nagai K. 2013. Structural basis of Brr2-Prp8 interactions and implications for U5 snRNP biogenesis and the spliceosome active site. *Structure* **21:** 910–919.

Nguyen TH, Galej WP, Bai XC, Savva CG, Newman AJ, Scheres SH, Nagai K. 2015. The architecture of the spliceosomal U4/U6.U5 tri-snRNP. *Nature* **523:** 47–52.

Nguyen TH, Galej WP, Bai XC, Oubridge C, Newman AJ, Scheres SH, Nagai K. 2016. Cryo-EM structure of the yeast U4/U6.U5 tri-snRNP at 3.7 Å resolution. *Nature* **530:** 298–302.

Nissen P, Hansen J, Ban N, Moore PB, Steitz TA. 2000. The structural basis of ribosome activity in peptide bond synthesis. *Science* **289:** 920–930.

Ohi MD, Gould KL. 2002. Characterization of interactions among the Cef1p-Prp19p-associated splicing complex. *RNA* **8:** 798–815.

Ohi MD, Ren L, Wall JS, Gould KL, Walz T. 2007. Structural characterization of the fission yeast U5.U2/U6 spliceosome complex. *Proc Natl Acad Sci* **104:** 3195–3200.

Ohrt T, Prior M, Dannenberg J, Odenwalder P, Dybkov O, Rasche N, Schmitzova J, Gregor I, Fabrizio P, Enderlein J, et al. 2012. Prp2-mediated protein rearrangements at the catalytic core of the spliceosome as revealed by dcFCCS. *RNA* **18:** 1244–1256.

Padgett RA, Mount SM, Steitz JA, Sharp PA. 1983. Splicing of messenger RNA precursors is inhibited by antisera to small nuclear ribonucleoprotein. *Cell* **35:** 101–107.

Padgett RA, Konarska MM, Grabowski PJ, Hardy SF, Sharp PA. 1984. Lariat RNA's as intermediates and products in the splicing of messenger RNA precursors. *Science* **225:** 898–903.

Pikielny CW, Rosbash M. 1986. Specific small nuclear RNAs are associated with yeast spliceosomes. *Cell* **45:** 869–877.

Plaschka C, Lin PC, Nagai K. 2017. Structure of a pre-catalytic spliceosome. *Nature* **546:** 617–621.

Pomeranz Krummel DA, Oubridge C, Leung AK, Li J, Nagai K. 2009. Crystal structure of human spliceosomal U1 snRNP at 5.5 Å resolution. *Nature* **458:** 475–480.

Price SR, Evans PR, Nagai K. 1998. Crystal structure of the spliceosomal U2B″-U2A′ protein complex bound to a fragment of U2 small nuclear RNA. *Nature* **394:** 645–650.

Rasche N, Dybkov O, Schmitzova J, Akyildiz B, Fabrizio P, Luhrmann R. 2012. Cwc2 and its human homologue RBM22 promote an active conformation of the spliceosome catalytic centre. *EMBO J* **31:** 1591–1604.

Rauhut R, Fabrizio P, Dybkov O, Hartmuth K, Pena V, Chari A, Kumar V, Lee CT, Urlaub H, Kastner B, et al. 2016. Molecular architecture of the *Saccharomyces cerevisiae* activated spliceosome. *Science* **353:** 1399–1405.

Rogers J, Wall R. 1980. A mechanism for RNA splicing. *Proc Natl Acad Sci* **77:** 1877–1879.

Ruskin B, Krainer AR, Maniatis T, Green MR. 1984. Excision of an intact intron as a novel lariat structure during pre-mRNA splicing in vitro. *Cell* **38:** 317–331.

Schluenzen F, Tocilj A, Zarivach R, Harms J, Gluehmann M, Janell D, Bashan A, Bartels H, Agmon I, Franceschi F, et al. 2000. Structure of functionally activated small ribosomal subunit at 3.3 Å resolution. *Cell* **102:** 615–623.

Schneider C, Agafonov DE, Schmitzova J, Hartmuth K, Fabrizio P, Luhrmann R. 2015. Dynamic contacts of U2, RES, Cwc25, Prp8 and Prp45 proteins with the pre-mRNA branch-site and 3′ splice site during catalytic activation and step 1 catalysis in yeast spliceosomes. *PLoS Genet* **11:** e1005539.

Schwer B. 2008. A conformational rearrangement in the spliceosome sets the stage for Prp22-dependent mRNA release. *Mol Cell* **30:** 743–754.

Schwer B, Gross CH. 1998. Prp22, a DExH-box RNA helicase, plays two distinct roles in yeast pre-mRNA splicing. *EMBO J* **17:** 2086–2094.

Semlow DR, Blanco MR, Walter NG, Staley JP. 2016. Spliceosomal DEAH-box ATPases remodel pre-mRNA to activate alternative splice sites. *Cell* **164:** 985–998.

Shi Y. 2017a. Mechanistic insights into precursor messenger RNA splicing by the spliceosome. *Nat Rev Mol Cell Biol* **18:** 655–670.

Shi Y. 2017b. The spliceosome: A protein-directed metalloribozyme. *J Mol Biol* **429:** 2640–2653.

Sickmier EA, Frato KE, Shen H, Paranawithana SR, Green MR, Kielkopf CL. 2006. Structural basis for polypyrimidine tract recognition by the essential pre-mRNA splicing factor U2AF65. *Mol Cell* **23:** 49–59.

Sidarovich A, Will CL, Anokhina MM, Ceballos J, Sievers S, Agafonov DE, Samatov T, Bao P, Kastner B, Urlaub H, et al. 2017. Identification of a small molecule inhibitor that stalls splicing at an early step of spliceosome activation. *eLife* **6:** e23533.

Sontheimer EJ, Sun S, Piccirilli JA. 1997. Metal ion catalysis during splicing of premessenger RNA. *Nature* **388:** 801–805.

Spingola M, Grate L, Haussler D, Ares M Jr. 1999. Genome-wide bioinformatic and molecular analysis of introns in *Saccharomyces cerevisiae*. *RNA* **5:** 221–234.

Steitz TA, Steitz JA. 1993. A general two-metal-ion mechanism for catalytic RNA. *Proc Natl Acad Sci* **90:** 6498–6502.

Tarn WY, Hsu CH, Huang KT, Chen HR, Kao HY, Lee KR, Cheng SC. 1994. Functional association of essential splicing factor(s) with PRP19 in a protein complex. *EMBO J* **13:** 2421–2431.

Toor N, Keating KS, Taylor SD, Pyle AM. 2008. Crystal structure of a self-spliced group II intron. *Science* **320:** 77–82.

Tsai RT, Fu RH, Yeh FL, Tseng CK, Lin YC, Huang YH, Cheng SC. 2005. Spliceosome disassembly catalyzed by Prp43 and its associated components Ntr1 and Ntr2. *Genes Dev* **19:** 2991–3003.

Tsai RT, Tseng CK, Lee PJ, Chen HC, Fu RH, Chang KJ, Yeh FL, Cheng SC. 2007. Dynamic interactions of Ntr1-Ntr2 with Prp43 and with U5 govern the recruitment of Prp43 to mediate spliceosome disassembly. *Mol Cell Biol* **27:** 8027–8037.

Tseng CK, Liu HL, Cheng SC. 2011. DEAH-box ATPase Prp16 has dual roles in remodeling of the spliceosome in catalytic steps. *RNA* **17:** 145–154.

Wan R, Yan C, Bai R, Huang G, Shi Y. 2016a. Structure of a yeast catalytic step I spliceosome at 3.4 Å resolution. *Science* **353:** 895–904.

Wan R, Yan C, Bai R, Wang L, Huang M, Wong CC, Shi Y. 2016b. The 3.8 Å structure of the U4/U6.U5 tri-snRNP: Insights into spliceosome assembly and catalysis. *Science* **351:** 466–475.

Wan R, Yan C, Bai R, Lei J, Shi Y. 2017. Structure of an intron lariat spliceosome from *Saccharomyces cerevisiae*. *Cell* **171:** 120–132.

Warkocki Z, Odenwalder P, Schmitzova J, Platzmann F, Stark H, Urlaub H, Ficner R, Fabrizio P, Luhrmann R. 2009. Reconstitution of both steps of *Saccharomyces cerevisiae* splicing with purified spliceosomal components. *Nat Struct Mol Biol* **16:** 1237–1243.

Weber G, Trowitzsch S, Kastner B, Luhrmann R, Wahl MC. 2010. Functional organization of the Sm core in the crystal structure of human U1 snRNP. *EMBO J* **29:** 4172–4184.

Wilkinson ME, Fica SM, Galej WP, Norman CM, Newman AJ, Nagai K. 2017. Postcatalytic spliceosome structure reveals mechanism of 3′-splice site selection. *Science* **358:** 1283–1288.

Wimberly BT, Brodersen DE, Clemons WM Jr, Morgan-Warren RJ, Carter AP, Vonrhein C, Hartsch T, Ramakrishnan V. 2000. Structure of the 30S ribosomal subunit. *Nature* **407:** 327–339.

Wolf E, Kastner B, Deckert J, Merz C, Stark H, Luhrmann R. 2009. Exon, intron and splice site locations in the spliceosomal B complex. *EMBO J* **28:** 2283–2292.

Wysoczanski P, Schneider C, Xiang S, Munari F, Trowitzsch S, Wahl MC, Luhrmann R, Becker S, Zweckstetter M. 2014. Cooperative structure of the heterotrimeric pre-mRNA retention and splicing complex. *Nat Struct Mol Biol* **21:** 911–918.

Yan C, Hang J, Wan R, Huang M, Wong CC, Shi Y. 2015. Structure of a yeast spliceosome at 3.6-Å resolution. *Science* **349:** 1182–1191.

Yan C, Wan R, Bai R, Huang G, Shi Y. 2016. Structure of a yeast activated spliceosome at 3.5 Å resolution. *Science* **353:** 904–911.

Yan C, Wan R, Bai R, Huang G, Shi Y. 2017. Structure of a yeast step II catalytically activated spliceosome. *Science* **355:** 149–155.

Yang VW, Lerner MR, Steitz JA, Flint SJ. 1981. A small nuclear ribonucleoprotein is required for splicing of adenoviral early RNA sequences. *Proc Natl Acad Sci* **78:** 1371–1375.

Yang K, Zhang L, Xu T, Heroux A, Zhao R. 2008. Crystal structure of the β-finger domain of Prp8 reveals analogy to ribosomal proteins. *Proc Natl Acad Sci* **105:** 13817–13822.

Yean SL, Wuenschell G, Termini J, Lin RJ. 2000. Metal-ion coordination by U6 small nuclear RNA contributes to catalysis in the spliceosome. *Nature* **408:** 881–884.

Zhan X, Yan C, Zhang X, Lei J, Shi Y. 2018. Structure of a human catalytic step I spliceosome. *Science* **359:** 537–545.

Zhang X, Schwer B. 1997. Functional and physical interaction between the yeast splicing factors Slu7 and Prp18. *Nucleic Acids Res* **25:** 2146–2152.

Zhang X, Yan C, Hang J, Finci LI, Lei J, Shi Y. 2017. An atomic structure of the human spliceosome. *Cell* **169:** 918–929.e914.

Zhang X, Yan C, Zhan X, Lei J, Shi Y. 2018. Structure of the human activated spliceosome in three conformational states. *Cell Res* **28:** 307–322.

Zhou Z, Sim J, Griffith J, Reed R. 2002. Purification and electron microscopic visualization of functional human spliceosomes. *Proc Natl Acad Sci* **99:** 12203–12207.

Zhou L, Hang J, Zhou Y, Wan R, Lu G, Yin P, Yan C, Shi Y. 2014. Crystal structures of the Lsm complex bound to the 3′ end sequence of U6 small nuclear RNA. *Nature* **506:** 116–120.

Combining Mass Spectrometry (MS) and Nuclear Magnetic Resonance (NMR) Spectroscopy for Integrative Structural Biology of Protein–RNA Complexes

Alexander Leitner,[1] Georg Dorn,[2] and Frédéric H.-T. Allain[2]

[1]Institute of Molecular Systems Biology, Department of Biology, Eidgenössische Technische Hochschule (ETH) Zürich, 8093 Zürich, Switzerland

[2]Institute of Molecular Biology and Biophysics, Department of Biology, ETH Zürich, 8093 Zürich, Switzerland

Correspondence: allain@mol.biol.ethz.ch; leitner@imsb.biol.ethz.ch

SUMMARY

Deciphering complex RNA–protein interactions on a (near-)atomic level is a hurdle that hinders advancing our understanding of fundamental processes in RNA metabolism and RNA-based gene regulation. To overcome challenges associated with individual structure determination methods, structural information derived from complementary biophysical methods can be combined in integrative structural biology approaches. Here, we review recent advances in such hybrid structural approaches with a focus on combining mass spectrometric analysis of cross-linked protein–RNA complexes and nuclear magnetic resonance (NMR) spectroscopy.

Outline

Cite as *Cold Spring Harb Perspect Biol* doi: 10.1101/cshperspect.a032359

1 INTRODUCTION

1.1 Toward Integrative Structural Biology of Protein–RNA Complexes

With the increasingly recognized role of RNA in gene regulation, it has become clear that defining RNA–protein interactions at the atomic level is of utmost importance for understanding the underlying molecular mechanisms and finding therapies against diseases associated with gene regulation. Yet, solving structures of protein–RNA complexes remains a challenge as few structures of RNA-binding proteins (RBPs) bound to their cognate RNA have been solved in comparison to their abundance (5%–10% of the genome; Castello et al. 2012). High-throughput methodologies in RNA sequencing and mass spectrometry (MS) have recently allowed the description of the genome-wide positioning of individual RBPs (sometimes at the nucleotide resolution) and the identification of the RBPome (and the domains involved), respectively. Nevertheless, it remains largely elusive how RBPs assemble onto an RNA (pre-messenger RNA [mRNA], mRNA, noncoding RNA, etc.), shape its structure, and control its fate. The large gap between large-scale MS analysis and RBP structure is illustrated by the recent structures of the spliceosome in which one would not anticipate such structural complexity solely on the basis of the identification of components.

With the structures of the ribosome, the nuclear pore complex, and now of the spliceosome well characterized even for the most evolved species, the current challenge in the ribonucleoprotein (RNP) field is certainly to understand the regulatory mechanism controlling splicing, translation, decay, mRNA export, and transport. To this end, relying on a single methodology to investigate the structure of these regulatory RNPs is probably not a good avenue considering the expected complexity, dynamics, and versatility of these complexes. For several decades, structural biology relied primarily on two biophysical methods: X-ray crystallography and nuclear magnetic resonance (NMR) spectroscopy. The recent "resolution revolution" in cryo-electron microscopy (cryo-EM; Kühlbrandt 2014), the emergence of MS, electron paramagnetic resonance (EPR), Förster resonance energy transfer (FRET), small-angle X-ray scattering (SAXS), and small-angle neutron scattering (SANS) for structural applications and ever-increasing computational power have changed the landscape of biophysical methods that can be used in structural biology. Combining multiple biophysical methods in structural biology (so-called "integrative structural biology" or "hybrid approaches in structural biology"; Ward et al. 2013) is now clearly emerging and is likely to become the norm rather than the exception. A few examples of such hybrid approaches used for RNPs will be succinctly described to illustrate the need to combine multiple structural biology technologies.

There are multiple examples of combining X-ray crystallography and NMR spectroscopy in the literature, but a recent application that comes to mind is work from the Sprangers laboratory on decapping enzymes (Wurm et al. 2017), in which NMR spectroscopy measurements of the enzyme allowed clarification of which crystal structures were functionally relevant and which ones were not. When crystal packing makes the obvious interpretation of a crystal structure doubtful, it should become the norm to confirm the functionality of the structure using solution methods like NMR, SAXS, or SANS. Electron microscopy (EM) and crystallography have been long-term partners, in which—in earlier times—low-resolution 3D EM maps required crystal structures to interpret them—for example, to place the crystal structure of the Sm ring (Kambach et al. 1999) into the U1 small nuclear ribonucleoprotein complex (snRNP) low-resolution EM map (Stark et al. 2001). More recently, despite the EM revolution, several available crystal structures were used to interpret the 6 Å cryo-EM map of the tri-snRNP U4/U6.U5 (Nguyen et al. 2015), illustrating the need for high-resolution crystal structures of individual proteins or complexes to interpret cryo-EM maps at medium resolution. Cross-linking coupled to MS (XL-MS) has also been of great help for interpreting cryo-EM maps, in particular if the electron density is ambiguous despite high resolution and crystal structures of the fragments are not available. XL-MS data greatly improve the confidence of 3D model building. This is exemplified by structures of the mitochondrial ribosome in which MS data allowed the unambiguous placement of closely positioned ribosomal proteins into the ribosome 3D structure (Greber et al. 2014). Similarly, the combination of NMR and SAXS/SANS has been used in a number of RNP structural studies by the Sattler and Carlomagno groups. SAXS and SANS provide information about the 3D envelope, which can be used to restrain conformational space and help improve the precision of the structural ensemble by distinguishing conformations that fit NMR restraints equally well, as seen for the structure determination of U2AF bound to RNA (Huang et al. 2014). SANS and NMR were used to decipher a structural model of the box C/D RNA methylation enzyme bound to its RNA's substrate (Lapinaite et al. 2013). In that work, SANS enabled the protein shape to be differentiated from the RNA shape in the complex and helped in the modeling of a 3D structure. NMR probes of high-molecular weight (methyl groups in a deuterated environment) allowed probing conformational changes in the protein on complex formation and supported 3D model building. In addition, NMR spectroscopy allowed enzyme catalysis to be followed by observing ^{13}C 2′-*O*-methyl groups in the RNA.

Cite as *Cold Spring Harb Perspect Biol* doi: 10.1101/cshperspect.a032359

We also needed to combine NMR spectroscopy with other biophysical methods in some of our recent structural studies, especially when studying fairly large RNPs in the range of 50–80 kDa. Our first hybrid approach was to combine NMR and EPR spectroscopy when studying the 70-kDa complex of the protein ribosomal RNA small subunit methyltransferase E (RsmE) bound to the noncoding RNA RsmZ. RsmZ contains five stem loops (SLs) and can bind cooperatively up to five RsmE dimers (Duss et al. 2014a). Solving the solution structure of RsmZ bound to three RsmE dimers revealed two conformations for the complex (of almost equal population) that resulted from two different assembly pathways of the RNP. EPR data were essential to provide 21 long-range distances within the RNA, thereby revealing the existence of these two conformations. More recently, we also combined NMR, EPR, and single-molecule FRET to decipher the structure of the two double-stranded RNA-binding domains (dsRBDs) of human immunodeficiency virus transactivating response RNA-binding protein (TRBP) bound to a small interfering RNA (siRNA; Masliah et al. 2018). This work also revealed the presence of two different conformations for this complex, which could be unambiguously confirmed by EPR. With single-molecule FRET, the relative population of the two forms could be determined. We have also successfully combined NMR with long molecular dynamics simulations in a number of studies to understand hydration at protein–RNA interfaces (Krepl et al. 2017) and the role of aromatic side-chains in RNA recognition motif (RRM)–RNA recognition (Diarra Dit Konté et al. 2017). Last, but not least, we recently developed a novel approach that combines NMR spectroscopy and XL-MS and allows rapid generation of models for protein–RNA complexes at atomic resolution (Dorn et al. 2017). We believe that this new hybrid structural biology approach has very high potential in terms of speed and sensitivity. More details are given below.

2 MS IN STRUCTURAL BIOLOGY

MS is a method essential for the large-scale identification and quantification of diverse groups of biomolecules, including proteins and metabolites. For those classes of molecules that cannot be amplified before detection, it provides crucial benefits such as high sensitivity and sample throughput. This way, MS has emerged as the key enabling technology for proteomics and metabolomics research. In contrast, MS is not the method of choice for RNA profiling because techniques exist that are faster, more robust, and better suited for automation—although RNA can also be detected and sequenced by MS. In the last decade, MS has increasingly been applied to provide low-resolution structural information on proteins and protein complexes, including RNPs. Because MS may not be widely known as a structural biology tool, we focus in this section on the fundamentals of biological MS and introduce the concepts of structural MS methods.

2.1 Fundamentals of Biological MS

Ultimately, any mass spectrometer determines the mass-to-charge (m/z) ratio of ions in the gas phase and this information is used to infer the identity of compound(s) of interest. For this purpose, molecules such as peptides, proteins or RNA need to be transferred from a condensed phase (usually a liquid) into the gas phase and be ionized during the process. Nowadays, electrospray ionization (ESI) is the most commonly used ionization technique for large biomolecules (Wilm 2011). ESI generates ions through the formation of charged droplets in the presence of strong electric fields (several kV/cm), and free ions are formed from these droplets after droplet fission events and evaporation of the surrounding solvent. By virtue of their chemical nature, proteins and peptides are more easily converted into cations (protonation of basic residues), whereas oligonucleotides are best ionized as anions (deprotonation of phosphate groups).

The ionization process takes place at atmospheric pressure, and the generated ions need to be transferred into (ultra)high vacuum regions inside the mass spectrometer, where their m/z ratio can be determined. Detection is based on different physical principles, such as the frequency of the ions' movement in trapping devices (such as the Orbitrap analyzer; Eliuk and Makarov 2015) or the flight time of ions (time-of-flight analyzer). Different types of analyzers can be integrated in one instrument, and most mass spectrometers are also equipped with devices (e.g., collision cells) to fragment ions for further structure elucidation (tandem mass spectrometry or MS/MS). This is most commonly performed by collision with inert gas molecules. Bond cleavages between the monomeric units in proteins/peptide (Paizs and Suhai 2005) and in oligonucleotides (Schürch 2016) occur according to predictable rules, so that the sequence can be determined from MS/MS spectra (most often, software-assisted or completely automated).

2.2 Mass Spectrometric Workflows

Depending on the application and the complexity of the sample, different types of chromatographic separation may be performed off-line or in-line with MS detection. Alternatively, a (highly purified) sample is analyzed directly. For an overview of these workflows, see Figure 1.

The most commonly used workflow for high-throughput identification of proteins can be used to characterize a wide variety of sample types (e.g., cell lysates or crudely

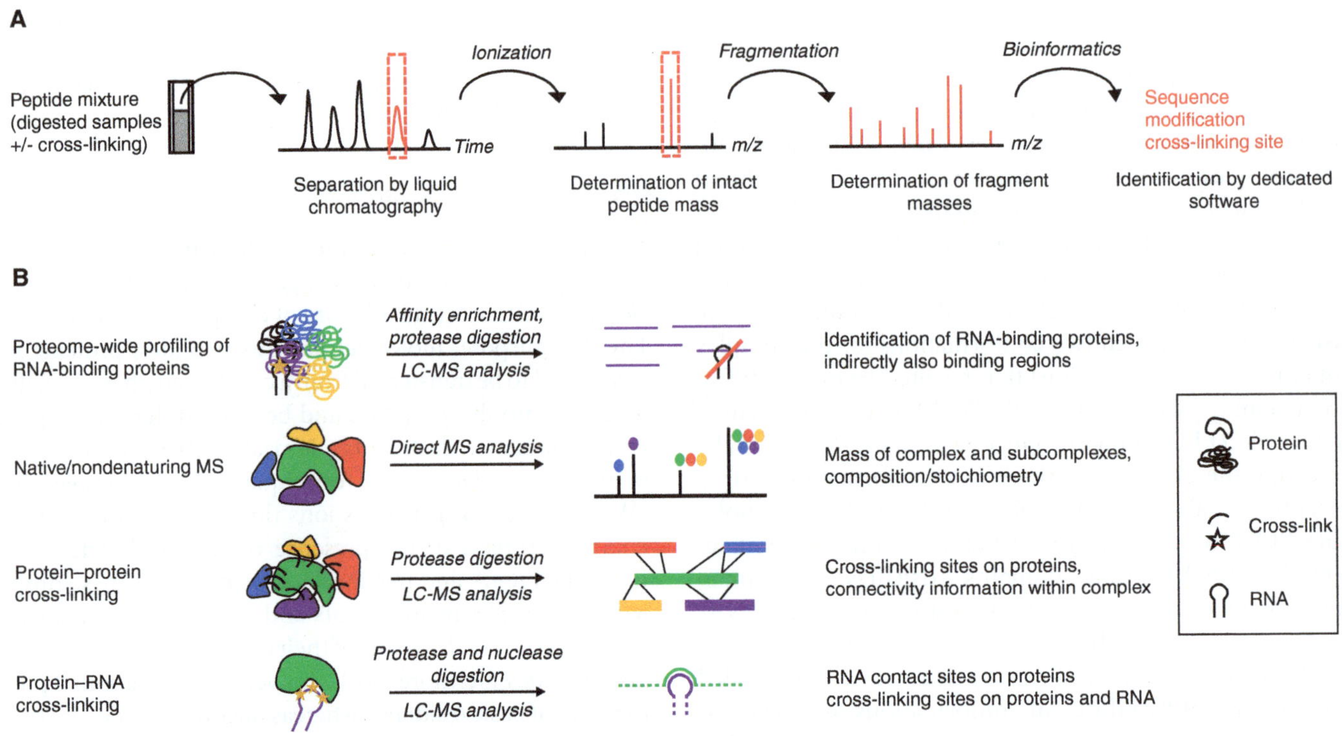

Figure 1. Characterization of protein–RNA interactions by mass spectrometry (MS). (*A*) Typical steps of a liquid chromatography (LC)-MS workflow. Proteins are digested into peptides, the resulting peptides are separated by LC and ionized and their mass-to-charge ratio (*m*/*z*) is determined. A subset of peptides is selected for fragmentation, allowing the identity of the peptides, possible modifications, and/or cross-linking (XL) sites to be derived through bioinformatic analysis. Samples can be of different complexity and can be prepared with or without XL steps. (*B*) Four different MS strategies for profiling protein–RNA interactions. From *top* to *bottom*: RNA-binding proteins (RBPs) may be identified from complex samples using XL with ultraviolet (UV) light (reflected by the orange star), and subsequent affinity purification. Upon enzymatic digestion, typically only peptides (visualized by horizontal bars on the *right*) that do not contain the RNA-binding site are identified. Samples of lower complexity such as purified complexes can be analyzed using structural MS methods, such as direct analysis by native/nondenaturing MS or protein–protein and protein–RNA XL. From such experiments, crucial information including complex mass, stoichiometry, spatial proximity of subunits, and even precise interaction sites can be obtained.

purified complexes isolated by [immuno]affinity enrichment). This way, RBPs may be profiled (see Sec. 3.2). Here, isolated proteins are first digested into peptides with a specific protease such as trypsin, which cleaves on the carboxy-terminal side of arginine (Arg) and lysine (Lys) residues and generates peptides of suitable length (typically 5–30 residues) for MS. The resulting peptide mixture is chromatographically separated before MS/MS analysis. Mass spectra are analyzed by dedicated software that assigns peptide sequences, and proteins are inferred from the identified peptide pool (Fig. 1A). This so-called "bottom-up" workflow—in which peptides are used as a proxy for the proteins from which they originate—is common to most proteomic methods, including those that provide structural information (see Sec. 2.3). Alternatively, proteins may be directly analyzed by MS without a digestion step ("top-down"), although this is less commonly performed because of challenges related to protein separation, fragmentation, and bioinformatic analysis.

2.3 Structural MS Methods

To connect MS analysis to structural biology, two different approaches are used. In the first approach, isolated biomolecules such as protein complexes, RNA, or RNPs are directly introduced into the mass spectrometer under near "native" or rather nondenaturing conditions (Leney and Heck 2017). Although the transition from liquid to gas phase and the energetic activation of molecules by the ionization step can lead to some artifacts, under carefully controlled conditions many ternary and quaternary structure elements are known to be preserved. This way, "native MS" may inform about the molecular mass of an intact assembly, the stoichiometry of its constituent subunits,

Cite as *Cold Spring Harb Perspect Biol* doi: 10.1101/cshperspect.a032359

and—by means of gas-phase dissociation—the composition of subcomplexes.

Most often, however, structural MS methods follow the bottom-up workflow outlined above (Fig. 1). In this case, structural information needs to be "encoded" in the sample before digestion (Leitner 2016a). This is usually performed by one of the following strategies: (1) exchange of labile hydrogens with deuterium; (2) covalent labeling with short-lived radicals ("footprinting"), such as hydroxyl radicals, or other labeling reagents; and (3) XL approaches. All of the above approaches result in a defined mass shift that can be detected in the mass spectrometer, but they provide orthogonal information about the sample. Hydrogen-deuterium (H/D) exchange mainly informs on solvent accessibility and hydrogen bonding, as well as changes in these properties if different samples are combined. Covalent labeling patterns also reflect solvent/surface exposure, whereas XL provides spatial proximity and (low-resolution) distance information. Such information makes XL the most practically useful method among the bottom-up structural MS methods, because it can be most easily integrated into hybrid structural biology workflows (Faini et al. 2016).

2.4 Protein–Protein and Protein–RNA XL

XL-MS can be applied in different ways to study the structure of protein and protein–RNA complexes (Leitner 2016b). Cross-links within or between proteins are most commonly formed using specific XL reagents such as succinimide esters that connect amino groups (Lys, amino terminus). Less frequently, XL approaches that connect carboxyl groups (aspartic acid [Asp], glutamic acid [Glu], carboxyl terminus) with each other or amines with carboxyl groups are used. Cross-linked samples are processed in a manner similar to that described above (proteolysis, chromatographic separation, MS/MS analysis). However, data analysis is more complex, because both of the peptides that become connected to each other by the cross-link need to be identified.

In contrast to the solution chemistry used for protein–protein XL, photochemistry is the preferred approach for forming cross-links between protein and RNA (Schmidt et al. 2012). UV irradiation at 254 nm leads to the formation of covalent links between nucleotides and amino acids, whereby uracil is most reactive on the RNA side, while most amino acids are found to be reactive. The yield of UV XL is fairly low but can be increased by incorporating nonnatural nucleotides such as 4-thio-uracil (4-sU) or 5-iodo-uracil, which react more efficiently at 365 nm. In all cases, both the protein and the RNA chain(s) need to be enzymatically cleaved before MS analysis. The size/length of the cleavage product on the protein side is again controlled by use of specific proteases such as trypsin, while a nuclease with comparable cleavage properties is not available. Therefore, a cocktail of different nucleases is commonly used to cut the oligonucleotide into remnants of one to four nucleotides. Upon MS/MS analysis, it is relatively straightforward to assign the corresponding peptide and thus protein, but the sequence context on the RNA side is typically lost because the short nucleotide sequence does not allow unequivocal identification of the cross-linked region. To overcome this limitation, we have recently introduced a new approach that is described in more detail in Section 5.

3 APPLICATIONS OF MS TO STUDY RNA AND PROTEIN–RNA INTERACTIONS

Using the experimental strategies outlined in Section 2 above and summarized in Figure 1B, MS has generated many insights into the structure and function of protein–RNA assemblies, as we illustrate below.

3.1 Direct Analysis of RNA by MS

The application of MS to the direct analysis of RNA has remained rather limited, although fundamental work has examined higher-order structures of RNA such as G-quadruplex formation (reviewed in Abi-Ghanem and Gabelica 2014; Schürch 2016). In addition, MS lends itself to the identification of natural and artificial/chemical RNA modifications (Gaston and Limbach 2014; Heiss and Kellner 2017; Limbach and Paulines 2017). Nonetheless, at the moment, these applications remain restricted to a few expert laboratories.

3.2 Identification of RBPs by MS-Based Proteomics

A more common approach related to RNA biology has been the application of proteomic methods to identify RBPs, either in targeted studies by using specific RNA motifs as baits for affinity purification workflows or in unbiased, proteome-wide screens (reviewed by Ascano et al. 2013; Jazurek et al. 2016). The methods for the latter application especially have dramatically improved in recent years. All use some form of photochemical XL to attach interacting proteins covalently to RNA in vivo; although the XL sites are not directly identified, the XL step serves to stabilize the interaction. In 2012, the first two landmark studies of large-scale profiling of RBPs were reported (Baltz et al. 2012; Castello et al. 2012). Both identified more than 800 RNA-interacting proteins in workflows that consisted of a UV XL step, followed by the lysis of cross-linked cells, the isolation of XL products by oligo-deoxythymine (oligo-d[T]) pull-down, and RNase/protease digestion before liquid chromatography (LC)-MS analysis. More recently, the Hentze laboratory reported an improved version of their RBDmap (mapping of RBDs) protocol that includes two steps of oligo-d(T) capture for enhanced coverage (Castello et al. 2016; Castello et al. 2017). This further refines the spatial localization of interacting sites, and 1,174 putative RNA-binding sites have been identified in HeLa cells using this approach. Recently, Nielsen and co-workers introduced pCLAP (peptide cross-linking and affinity profiling; Mullari et al. 2017), a method that compares favorably with RBDmap in terms of numbers of RBPs identified.

However, all the above approaches rely on the selection of poly(A) motifs in RNA, therefore excluding types of RNA that are not polyadenylated. To circumvent this limitation and taking advantage of the enhanced XL efficiency of 4sU-labeled cells, Bonasio and coworkers introduced their RBR-ID approach (identification of RNA-binding regions; He et al. 2016). RBR-ID relies on reduction of the abundance of peptides spanning RNA-interacting/binding regions after UV XL using quantitative proteomics workflows. Although the method is more generic than the abovementioned ones, the lack of any dedicated enrichment method makes it less sensitive in comparison. However, the reported data (interaction sites in >800 proteins from mouse embryonic stem cells) still suggest that it is competitive with and complementary to methods that use enrichment. Very recently, two methods that take advantage of bioorthogonal click chemistry approaches on metabolic incorporation of 5-ethynyluridine have been reported and enable poly(A)-independent enrichment steps (Bao et al. 2018; Huang et al. 2018).

Taking the spatial localization of RNA-binding sites on the protein level one step further, Urlaub and coworkers developed a series of methods and computational tools that directly identify peptides containing RNA remnants by MS after protease and nuclease digestion (reviewed in Schmidt et al. 2012). This is in contrast to using non-cross-linked peptides from RBPs for identification, which allows a protein to be categorized only as an RNA interactor. To enable the precise localization of interaction sites, peptides that are modified with short RNA segments (one to three nucleotides) are treated similarly to posttranslational protein modifications, such as phosphorylation, during data analysis. However, the detection of these peptide–RNA adducts is challenging for many reasons, including the low abundance of the products and the diversity of possible RNA attachments. Despite these challenges, the successful application of this strategy has been shown for purified protein complexes and even for whole cells. Notable recent examples include the application of this XL strategy to the spliceosome and clustered regularly interspaced palindromic repeat (CRISPR) systems (Cretu et al. 2016; Gleditzsch et al. 2016). Extending the strategy to the proteome scale has been made possible by the development of an integrated experimental and computational workflow (Kramer et al. 2014; Veit et al. 2016).

3.3 Structural Analysis of Protein–RNA Complexes by XL Coupled to MS

Protein–protein XL has been applied to various large macromolecular assemblies (reviewed by Leitner et al. 2016c) and has provided valuable insights into the architecture of these complexes. A typical example is the use of XL-MS to facilitate subunit positioning in combination with lower-resolution (cryo-)EM maps. Several protein–RNA complexes have been studied in this way. For example, XL was used to position subunits in the large subunit of the mammalian mitochondrial ribosome (Greber et al. 2014) and, later, in the complete 55S mitoribosome (Greber et al. 2015). XL-MS has also been used to investigate ribosomal assembly intermediates such as the yeast pre-60S particle containing numerous assembly factors (Wu et al. 2016) or the ribosome complexed with translation initiation factors (Erzberger et al. 2014; Eliseev et al. 2018). The spliceosome was also studied by a combination of XL-MS and EM in several studies focusing on the human U4/U6.U5 tri-snRNP subcomplex (Agafonov et al. 2016) or on activated states of the yeast (Rauhut et al. 2016) and human complex (Bertram et al. 2017).

The use of protein–protein XL in integrative structural biology has been facilitated by the relatively straightforward localization of the XL sites on the two interacting regions (both peptides). In contrast, site localization at the protein and RNA level for protein–RNA cross-links has been elusive for a long time. Recently, improvements have been made with the help of stable isotope-labeling strategies (Lelyveld et al. 2015; Dorn et al. 2017). In particular, CLIR–MS/MS (cross-linking of segmentally isotope-labeled RNA coupled to tandem MS analysis; Dorn et al. 2017) has been used to identify contacts at single-amino acid and single-nucleotide resolution in an integrated workflow that is described in more detail in Sections 5 and 6.

4 NMR SPECTROSCOPY OF LARGE PROTEIN–RNA COMPLEXES: THE "DIVIDE AND CONQUER" APPROACH AND ITS LIMITATIONS

NMR has clearly become a method of choice to investigate binding of RNA to medium size RBPs or RBDs (Dominguez et al. 2011; Daubner et al. 2013). Solving structures of protein–RNA complexes at atomic resolution is not an easy task as is evident from the low number of protein–RNA

Cite as *Cold Spring Harb Perspect Biol* doi: 10.1101/cshperspect.a032359

complex NMR structures deposited in the Protein Data Bank (PDB); only approximately 100 of such structures have been released since the first two were published more than 20 years ago (Allain et al. 1996; Battiste et al. 1996). Yet, NMR spectroscopy is a powerful method for detecting RNA binding and localizing the RNA-binding surface. Deuteration of proteins has also allowed the observation of large protein–RNA complexes as in the case of the 600-kDa box C/D methylation enzyme (Lapinaite et al. 2013).

Although solving structures of 10–15-kDa protein–RNA complexes is clearly feasible, this is still a lengthy process that requires two to three years of work by a dedicated investigator. Typically, the first year is needed to identify the correct RNA sequence and ideal buffer conditions for the complex, the second year to assign the resonances of the protein–RNA complex, and the third year to determine a precise structure. The time can double when the protein contains multiple RBDs. Yet, NMR spectroscopy allows structure determination of protein–RNA complexes with weak affinity (mM to μM K_d), which exist for many RNA regulatory processes such as splicing and translation.

To tackle the structure of larger protein–RNA complexes, additional methods like EPR, SAXS, SANS, and EM are required to complement NMR data if the mass of the complex exceeds 200 kDa. This so-called "divide and conquer" approach is exemplified by our study of RsmE bound to RsmZ (Duss et al. 2014a). NMR spectroscopy was used to solve the structure of several individual RNA SLs bound to RsmE ("divide") while EPR was crucial to obtain the long-range distances between the subcomplexes to calculate the whole 70-kDa complex ("conquer"; Fig. 2). Whereas solving the structures of RsmE bound to each individual SL was fairly straightforward but tedious (Duss et al. 2014b), considering the similarity with our previously published RsmE–SL complex (Schubert et al. 2007), inserting two spin labels in the different RsmE subcomplexes required preparation of the RNA in several parts, spin labeling them at 4-sU, and ligating the various pieces together (Duss et al. 2014c). For NMR spectroscopy of the 70-kDa complex, we also needed to isotope-label the RNA segmentally (Duss et al. 2010). Collecting 21 EPR distances between the three sub-complexes was then sufficient to calculate two

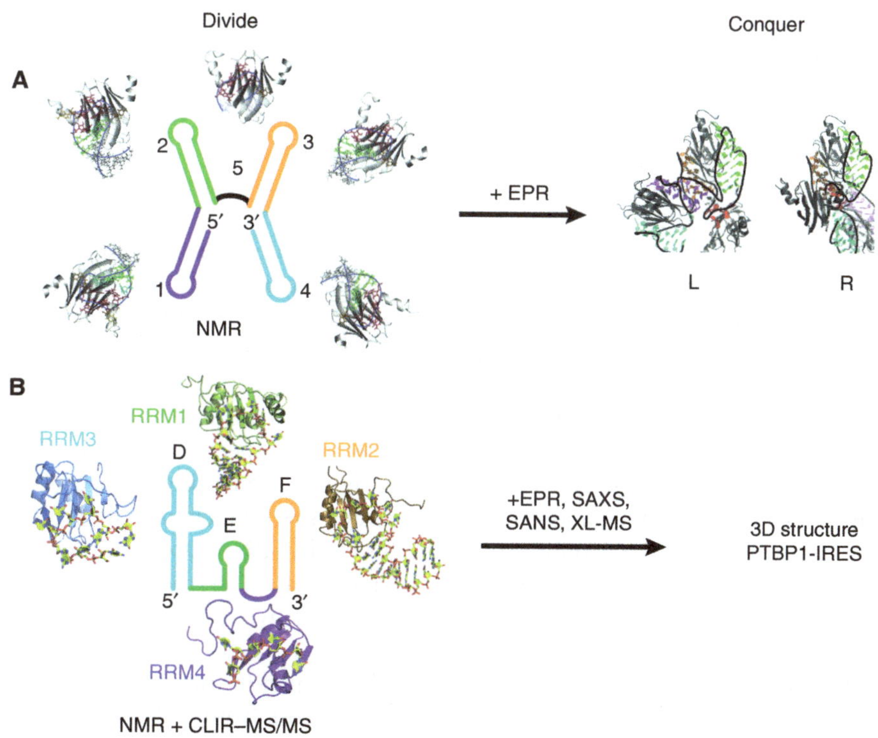

Figure 2. Examples of the "divide and conquer" approach used to solve the 3D structure of large RNPs in solution. (*A*) Combining nuclear magnetic resonance spectroscopy (NMR) and electron paramagnetic resonance (EPR) to solve the structure of the 70-kDa RsmE–RsmZ complex (Duss et al. 2014a,b). L and R represent two possible conformations, placing stem loop (SL) 1 (in violet) either to the left (L) or to the right (R) of SL 2 (in green). (*B*) Combining NMR, cross-linking of segmentally isotope-labeled RNA coupled to tandem MS analysis (CLIR–MS/MS), and other methods to solve the structure of the 85-kDa complex between polypyrimidine tract-binding protein 1 (PTBP1) and the encephalomyocarditis virus internal ribosomal entry site (EMCV IRES) (Dorn et al. 2017).

precise structural ensembles of this protein–RNA complex, because two conformations were observed by EPR.

Using the same approach, we initially considered it feasible to tackle the structure of the 58-kDa polypyrimidine tract-binding protein 1 (PTBP1) bound to an 85-nt RNA from the internal ribosomal entry site (IRES) of the encephalomyocarditis virus (EMCV), overall an 85-kDa complex (Fig. 2). We confirmed early on that RRM1 or RRM2 could bind SL E and F, respectively, but positioning RRM3 and RRM4 proved to be more difficult because of the large size of the complex and because RRM3 and RRM4 interact (Dorn 2017). We therefore asked whether MS could help us find precisely which RRM of PTBP1 binds where in the EMCV IRES. The idea of using segmentally labeled RNA to initially identify the peptides cross-linked to RNA came from our previous work with RsmZ (Duss et al. 2010). This approach proved very fruitful as described in detail below.

5 CLIR-MS/MS AS A TECHNIQUE FOR HIGH-RESOLUTION MAPPING OF INTERACTIONS IN PROTEIN–RNA COMPLEXES

UV-induced protein–RNA cross-linking is widely used for the transcriptome-wide identification of RNA targets of specific bait proteins (Ule et al. 2005; König et al. 2010; Sugimoto et al. 2015; Van Nostrand et al. 2016) and for the identification of new RBPs and their exact RNA interaction sites, respectively, by MS (Castello et al. 2012; Kramer et al. 2014), as discussed above. Although irradiation with UV light induces zero-length cross-links that therefore contain information about the local interfaces of both protein and RNA, single-residue resolution is achieved only for one of them. In particular, the unfavorable fragmentation properties of RNA (Kramer et al. 2014) prevent mass spectroscopic sequencing of cross-linked peptides when large nucleotide stretches are covalently attached; thus, cross-linked RNA moieties must be enzymatically digested. This, in turn, limits the accessible information about the cross-linked RNA to its overall mass, but only up to stretches of four nucleotides. Recently, we developed a new approach called CLIR-MS/MS that combines protein–RNA cross-linking and MS analysis with the use of segmental isotope labeling of RNA to pinpoint protein–RNA interactions (Dorn et al. 2017). The information obtained by the use of heavy isotopes is used to restrict the mapping possibilities for the cross-linked, short RNA stretches detected by MS to the isotopically labeled segment, thereby allowing the identification of the interface at single-amino acid and single-nucleotide resolution simultaneously. CLIR-MS/MS has no general limitation in terms of size, solubility, or crystallizability, but at present requires homogeneous protein–RNA complexes and depends critically on cross-linking efficiency (Dorn et al. 2017). We used CLIR-MS/MS to pinpoint contacts between PTBP1 and domains D–F of the EMCV IRES and combined these results with structural information derived from the PDB to generate local models for the four RRMs (hereafter, RRM1 for the amino-terminal RRM, etc.) of PTBP1 and their RNA-binding sites.

5.1 Segmental Labeling of RNA

Enzymatic RNA ligation has been used for decades to introduce site-specifically labeled or modified segments that enable insights into the structure and function of the cognate RNA segment in the context of a larger RNA (Moore and Sharp 1992; Xu et al. 1996). Reducing the complexity of NMR spectra by segmental isotope-labeling strategies has turned out to be essential for studying the structures of large RNAs by NMR spectroscopy (Kim et al. 2002; Lukavsky et al. 2003; Lu et al. 2011; Duss et al. 2014a), for instance allowing the determination of a solution structure for the noncoding RNA RsmZ in complex with RsmE protein dimers (Duss et al. 2014a).

RNA segments can be generated by chemical synthesis, which allows single-residue modifications (e.g., for spin-label attachment; Duss et al. 2014c), by in vitro transcription (Milligan et al. 1987; Gurevich et al. 1991; Pokrovskaya and Gurevich 1994; Tzakos et al. 2007; Nelissen et al. 2008; Lebars et al. 2014) or by enzymatic cleavage of a precursor RNA (Xu et al. 1996; Duss et al. 2010). Isotopically labeled nucleotides are commercially available and can be incorporated during transcription (Lu et al. 2010). Special care must be taken to avoid any inhomogeneity at the ligation site. For in vitro transcribed RNA, ribozymatic trimming of the RNA ends or enzymatic and sequence-specific cleavage of a large precursor RNA can yield homogeneous segments.

Sequence-specific cleavage of such precursor RNAs can be achieved using RNase H directed to the desired cleavage site by a chemically synthesized, complementary 2′-*O*-methyl-RNA/DNA chimera (Duss et al. 2015). Importantly, such chimeras contain about 10–12 residues of 2′-*O*-methyl-RNA to increase sequence specificity and exactly four deoxynucleotides that form the RNA–DNA hybrid, which in turn is recognized by RNase H. Cleavage occurs across from the 5′ end of these four deoxynucleotides (Fig. 3A). The concentration of RNase H and the optimal RNA:chimera ratios are typically optimized in a small-scale reaction to ensure complete cleavage and avoid nonspecific cleavage that can occur in unfavorable cases at certain RNA:chimera ratios (Duss et al. 2015). Recently, Kielpinski et al. characterized the substrate sequence preference of *E. coli* RNase H for efficient cleavage (Kiel-

Cite as *Cold Spring Harb Perspect Biol* doi: 10.1101/cshperspect.a032359

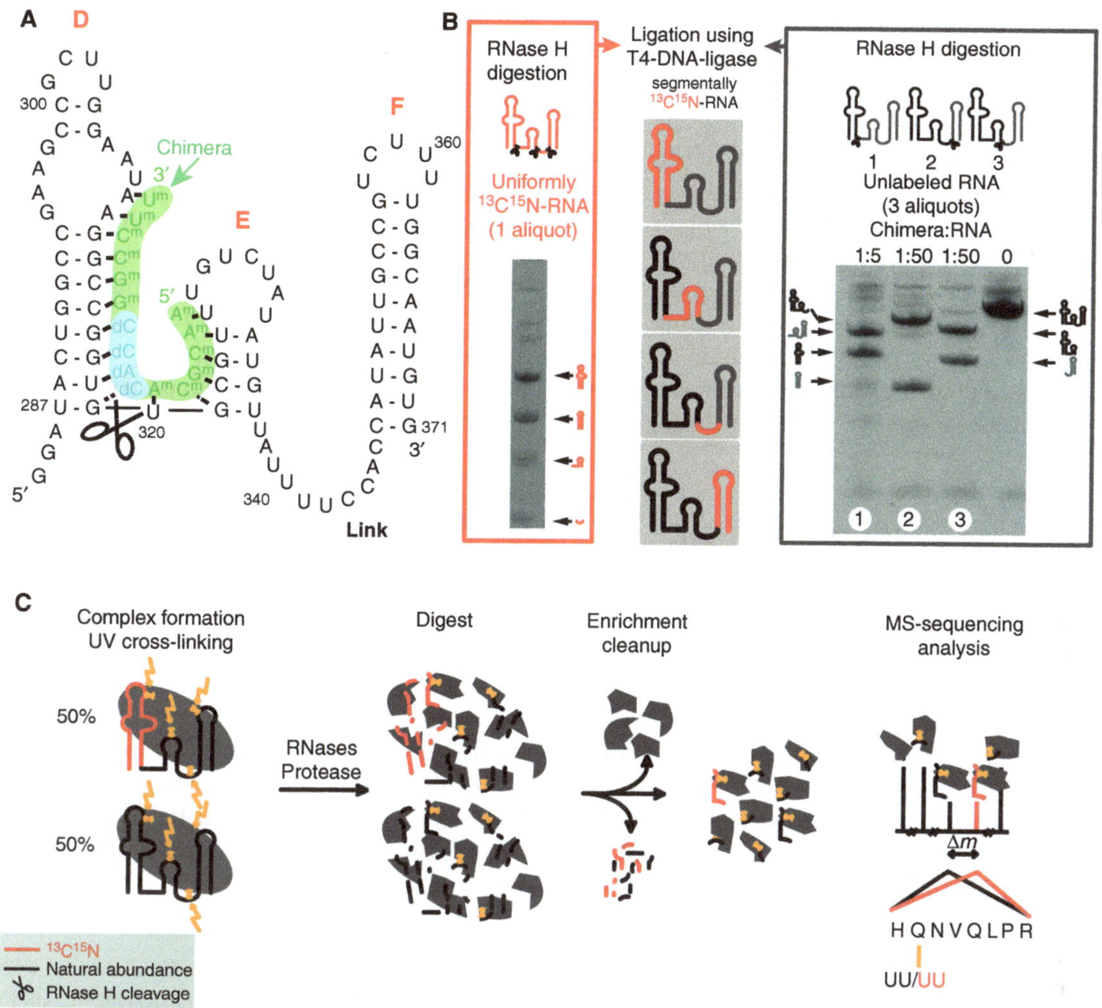

Figure 3. CLIR–MS/MS analysis of protein–RNA complexes. (*A*) Design of a 2′-*O*-methyl-RNA/DNA chimera used to induce RNase H cleavage between nucleotide 319 and 320 of the EMCV IRES RNA. (*B*) Segmental isotope-labeling strategy to label individual segments of the EMCV IRES RNA (domains D–F) using T4-DNA-ligase in a splinted ligation of RNase H-derived fragments. (*C*) CLIR–MS/MS analysis workflow to identify protein–RNA cross-links derived from the differentially labeled segment. (Modified, with permission, from Dorn et al. 2017.)

pinski et al. 2017). However, prediction of the optimal RNA:chimera ratio remains difficult as factors like RNA secondary structure might affect cleavage efficiency; in practice, a broad range of ratios should be screened. For segmental isotope labeling of the IRES domains D–F of EMCV and the U1 small nuclear RNA (snRNA), we used ratios ranging from 50:1 to 5:1 (Fig. 3B) or 2:1, respectively, leading to nearly complete RNase H cleavage (Dorn et al. 2017). To generate fragments for segmental isotope labeling of the EMCV IRES domains D, E, the sequence linking E and F (Link), and F, we in vitro transcribed three aliquots of unlabeled RNA (containing their natural abundance isotopes) and one aliquot of $^{13}C^{15}N$-labeled RNA. Each aliquot of unlabeled RNA was cleaved using one of the chimeras, while the isotopically labeled RNA was exposed to all three chimeras at the same time (Fig. 3B). Cleavage products were then separated by classical RNA purification methods (Edelmann et al. 2014), demanding that cleavage site selection, as well as the design of the precursor RNA, ideally produce optimal separation of fragments. Therefore, we designed cleavage sites such that RNase H digestion using all three chimeras at the same time results in fragments of 36, 17, 10, and 27 nt.

Several strategies are available for RNA ligation that differ specifically in the catalyst for RNA ligation and the use of a DNA splint for fragment association. RNA ligation using T4-DNA-ligase requires a 3′-hydroxyl and 5′-monophosphate group on substrates that are annealed to a DNA splint and has been widely used to incorporate functionalized or labeled RNA segments (Kershaw and O'Keefe

2012). Although the length of the splint influences the ligation efficiency (Kurschat et al. 2005), the splint can also be used to protect unstable RNA sequences (Duss et al. 2015). Ligations catalyzed by T4-RNA-ligase require the same functional groups on the 3′ and 5′ ends of substrates as for T4-DNA-ligase, but the use of a DNA splint is optional (Lu et al. 2010). Ligation of RNA segments using deoxyribozymes does not require a monophosphate group at the 5′ end of the substrate (Purtha et al. 2005). In our study, we used T4-DNA-ligase in a splinted two-piece or three-piece ligation to create samples for complex formation and subsequent UV cross-linking; ligation products were then purified by denaturing anion exchange chromatography (Dorn et al. 2017).

5.2 Strategy for CLIR–MS/MS-Derived High-Resolution Mapping of Protein–RNA Interactions

For CLIR–MS/MS, segmentally isotope-labeled RNA and unlabeled RNA are mixed at equimolar ratios before reconstitution of the protein–RNA complex of interest. UV cross-linking is conducted under conditions preoptimized to yield maximal amounts of cross-linked protein–RNA complex without inducing photodamage (Fig. 3C). Samples are digested with nonspecific RNases and specific proteases to generate peptides modified by the attachment of short RNA stretches (Dorn et al. 2017). As the majority of the resulting digest consists of unmodified peptides and RNA pieces, cross-linked peptides need to be enriched by solid-phase extraction and titanium dioxide affinity chromatography before MS/MS analysis (Leitner et al. 2010). Importantly, cross-links occurring between protein and nucleotides that reside in the labeled RNA segment will generate doublets in the precursor ion mass spectrum. The mass shift then corresponds to composition of the attached, differentially labeled nucleotide(s) and can be predicted for all possible modifications (Dorn et al. 2017). The predicted mass shift is used to identify modified peptides, the modified amino acid, and the modifications themselves using the software xQuest (Rinner et al. 2008; Walzthoeni et al. 2012). xQuest was originally developed to analyze chemically induced protein–protein cross-links and combines mass spectra from differentially isotope-labeled peptides (here due to the RNA adducts) to yield a consensus spectrum that is finally matched against a database of target proteins.

The identification of amino acids modified with differentially isotope-labeled nucleotides provides proof that the particular residue and consequently the corresponding protein domain is in close proximity to the isotopically labeled RNA segment. This information is already valuable for direct future experiments such as tests of deletion mutants, etc. Again, using PTBP1 as an example, we exclusively detected cross-linking for RRM2 and RRM4 to F and Link, respectively, while RRM1 and RRM3 were found to primarily cross-link to E and D, respectively. In our experience, RNase digestion is incomplete and detection of mono-, di-, and trinucleotide modifications can be expected. This infers the composition of a stretch of up to five nucleotides (if partial sequences overlap) and in turn restricts interaction mapping. For example, cross-links detected for PTBP1-RRM3 with D comprised only U and UU, thus restricting mapping to U303–U304 of D as this is the only UU sequence present. Assuming that the protein interacts with the RNA in one predominant conformation, different modifications detected at the same amino acid report the partial nucleotide sequence of the cross-linking site. His457 on PTBP1-RRM4 showed U, C, UC, and UU as modifications (Fig. 4). U and C as single nucleotide modifications indicate the positioning of His457 between these two nucleotides, allowing cross-linking to either of them on the Link sequence. This is consistent with the previously solved NMR structures of PTBP1-RRM34 complexed with single-stranded CU-6mer RNA (Oberstrass et al. 2005). Obtaining UC and UU as dinucleotide modifications indicates a sequence of (U)CUU or UUC(U), of which only UUC is present in the Link sequence. Thus, His457 resides between U342 and C343 (Figs. 3A and 4). This mapping is supported by cross-links of other amino acids at the RNA-binding surface of RRM4.

6 INTEGRATIVE MODELING USING CLIR–MS/MS-DERIVED DISTANCE RESTRAINTS AND STRUCTURAL INFORMATION DERIVED FROM NMR AND/OR AVAILABLE STRUCTURAL INFORMATION FROM THE PDB

As zero-length cross-links, CLIR–MS/MS results can be used as short-distance restraints for integrative modeling. We designed a modeling protocol based on the combined assignment and dynamics algorithm for NMR applications (CYANA) 3.0 (Guntert 2004), which uses both artificial restraints derived from previously published structures (Oberstrass et al. 2005) (RRM1, PDB: 2N3O) and the detected cross-links as ambiguous intermolecular distance restraints (Dorn et al. 2017). Although a detailed manual analysis of the detected modifications leads to a precise assignment of interacting amino acids and nucleotides, we anticipated that fewer data are available on other systems, resulting in ambiguity of the localization of cross-links throughout a coarsely mapped binding site. In contrast to rigid body docking programs, CYANA (Guntert 2004) uses a simulated annealing protocol to generate a bundle of structures that allows individual residues to alter their conformations to accommodate their binding partner.

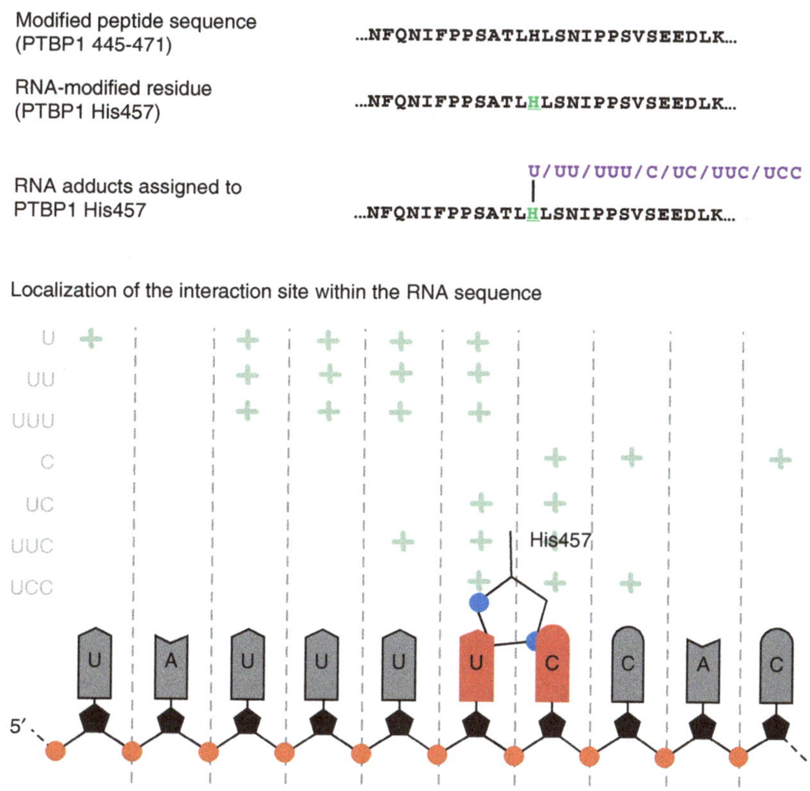

Figure 4. Single-residue resolution of protein–RNA interactions at the amino acid and nucleotide level by CLIR–MS/MS, exemplified by the identification of one PTBP1 peptide modified with multiple remnants of the EMCV IRES RNA. (Data from Dorn et al. 2017.)

Intra-RRM restraints extracted from the published structures comprise dihedral angle information and atom–atom distance restraints. Intra-RNA restraints were similarly derived from RNA SL structure predictions using MC-Fold and MC-Sym (Parisien and Major 2008). In an initial structure calculation run, we specified these intra-molecular restraints and ambiguous CLIR–MS/MS-derived restraints as the only intermolecular restraints. Because CLIR–MS/MS restraints are sparse, the resulting model was not precise but the ambiguity of CLIR–MS/MS restraints was resolved and binding registers could be extracted at this stage. In subsequent structure calculations, additional unambiguous intermolecular restraints were implemented that represent specific protein–RNA contacts extracted from the available complex structures (Dorn et al. 2017). CYANA 3.0 is a widely used program for NMR structure determination and, thus, the combination of CLIR–MS/MS-derived restraints with classical NMR-derived restraints is straightforward.

To check the quality and accuracy of the 3D model we obtained, we determined independently the structure of RRM2 bound to SL F using CLIR–MS/MS-derived distance constraints and more classically using NMR spectroscopy (Dorn et al. 2017). The results were quite unexpected because the precision of the structural ensemble was similar and the two structures showed very similar modes of RNA recognition; thus, CLIR–MS/MS appears to be a sufficiently powerful method to produce accurate structures. Note that with CLIR–MS/MS, the structure of RRM2 bound to SL F was determined in the context of the 85-kDa PTBP1–IRES complex, whereas with NMR the structure was determined for an isolated 15-kDa RRM2–SL F complex. Also, about six months of work were required to decipher the structure of RRM2 with CLIR–MS/MS (in parallel to the other three RRMs bound to RNA), whereas about two years were needed to obtain a similar 3D model by NMR. In terms of material needed for both experiments, 100–1000 times less material was needed for CLIR–MS/MS than for NMR. This illustrates the high potential of this new structural approach to decipher protein–RNA interactions. Yet, CLIR–MS/MS as a structural method is highly dependent on other biophysical methods, NMR spectroscopy, and X-ray crystallography. Indeed, while CLIR–MS/MS can provide the intermolecular constraints, the other methods are necessary to generate information on the 3D structure of the protein and the RNA. The other methods will also be needed to distinguish data resulting from less populated conformational states, which the CLIR–MS/MS method might detect.

In summary, CLIR–MS/MS offers a new avenue for integrative or hybrid structural biology approaches to protein–RNA complexes. Considering the large number of RBP structures determined in their apo state and the ever-increasing knowledge of the sequence specificity of RBPs, CLIR–MS/MS might become a method of choice to investigate rapidly the 3D structure of RBPs in vitro (and perhaps soon in cell extracts or cell culture). Although the method has been applied to only a few RBP complexes, we expect much wider applicability of the method as a very large fraction of the known RBPs can UV cross-link to RNA (Castello et al. 2012).

ACKNOWLEDGMENTS

This work was supported by the Swiss National Science Foundation (SNF) project 31003A_170130 and the SNF-National Center of Competence in Research (NCCR) RNA and Disease to F.H.-T.A.

REFERENCES

Abi-Ghanem J, Gabelica V. 2014. Nucleic acid ion structures in the gas phase. *Phys Chem Chem Phys* **16:** 21204–21218.

Agafonov DE, Kastner B, Dybkov O, Hofele RV, Liu W-T, Urlaub H, Lührmann R, Stark H. 2016. Molecular architecture of the human U4/U6.U5 tri-snRNP. *Science* **351:** 1416–1420.

Allain FH-T, Gubser CC, Howe PWA, Nagai H, Neuhaus D, Varani A. 1996. Specificity of ribonucleoprotein interaction determined by RNA folding during complex formation. *Nature* **380:** 646–650.

Ascano M, Gerstberger S, Tuschl T. 2013. Multi-disciplinary methods to define RNA-protein interactions and regulatory networks. *Curr Opin Gen Dev* **23:** 20–28.

Baltz AG, Munschauer M, Schwanhäusser B, Vasile A, Murakawa Y, Schueler M, Youngs N, Penfold-Brown D, Drew K, Milek M, et al. 2012. The mRNA-bound proteome and its global occupancy profile on protein-coding transcripts. *Mol Cell* **46:** 675–690.

Bao X, Guo X, Xin M, Tariq M, Lai Y, Kanwal S, Zhou J, Li N, Lv Y, Pulido-Quetglas C, et al. 2018. Capturing the interactome of newly transcribed RNA. *Nat Methods* **15:** 213–220.

Battiste JL, Mao HY, Rao NS, Tan RY, Muhandiram DR, Kay LE, Frankel AD, Williamson JR. 1996. α helix-RNA major groove recognition in an HIV-1 rev peptide RRE RNA complex. *Science* **273:** 1547–1551.

Bertram K, Agafonov DE, Liu W-T, Dybkov O, Will CL, Hartmuth K, Urlaub H, Kastner B, Stark H, Lührmann R. 2017. Cryo-EM structure of a human spliceosome activated for step 2 of splicing. *Nature* **542:** 318–323.

Castello A, Fischer B, Eichelbaum K, Horos R, Beckmann BM, Strein C, Davey NE, Humphreys DT, Preiss T, Steinmetz LM, et al. 2012. Insights into RNA biology from an atlas of mammalian mRNA-binding proteins. *Cell* **149:** 1393–1406.

Castello A, Fischer B, Frese CK, Horos R, Alleaume A-M, Foehr S, Curk T, Krijgsveld J, Hentze MW. 2016. Comprehensive identification of RNA-binding domains in human cells. *Mol Cell* **63:** 696–710.

Castello A, Frese CK, Fischer B, Järvelin AI, Horos R, Alleaume A-M, Foehr S, Curk T, Krijgsveld J, Hentze MW. 2017. Identification of RNA-binding domains of RNA-binding proteins in cultured cells on a system-wide scale with RBDmap. *Nat Protoc* **12:** 2447–2464.

Cretu C, Schmitzová J, Ponce-Salvatierra A, Dybkov O, De Laurentiis EI, Sharma K, Will CL, Urlaub H, Lührmann R, Pena V. 2016. Molecular architecture of SF3b and structural consequences of its cancer-related mutations. *Mol Cell* **64:** 307–319.

Daubner GM, Cléry A, Allain F H-T. 2013. RRM–RNA recognition: NMR or crystallography…and new findings. *Curr Opin Struct Biol* **23:** 100–108.

Diarra Dit Konté N, Krepl M, Damberger FF, Ripin N, Duss O, Sponer J, Allain FH-T. 2017. Aromatic side-chain conformational switch on the surface of the RNA recognition motif enables RNA discrimination. *Nat Commun* **8:** 654.

Dominguez C, Schubert M, Duss O, Ravindranathan S, Allain F H-T. 2011. Structure determination and dynamics of protein–RNA complexes by NMR spectroscopy. *Prog Nucl Magn Reson Spectrosc* **58:** 1–61.

Dorn G. 2017. "Structural investigation and hybrid structural modelling of polypyrimidine tract binding protein 1 in complex with the internal ribosomal entry site of encephalomyocarditis virus." PhD thesis, ETH Zurich. Dissertation no. 24054. doi: 10.3929/ethz-b-000165700.

Dorn G, Leitner A, Boudet J, Campagne S, Von Schroetter C, Moursy A, Aebersold R, Allain F H-T. 2017. Structural modeling of protein-RNA complexes using crosslinking of segmentally isotope-labeled RNA and MS/MS. *Nat Methods* **14:** 487–490.

Duss O, Maris C, Von Schroetter C, Allain F H-T. 2010. A fast, efficient and sequence-independent method for flexible multiple segmental isotope labeling of RNA using ribozyme and RNase H cleavage. *Nucl Acids Res* **38:** e188.

Duss O, Michel E, Yulikov M, Schubert M, Jeschke G, Allain F H-T. 2014a. Structural basis of the non-coding RNA RsmZ acting as a protein sponge. *Nature* **509:** 588–592.

Duss O, Michel E, Diarra Dit Konté N, Schubert M, Allain F H-T. 2014b. Molecular basis for the wide range of affinity found in Csr/Rsm protein–RNA recognition. *Nucl Acids Res* **42:** 5332–5346.

Duss O, Yulikov M, Jeschke G, Allain F H-T. 2014c. EPR-aided approach for solution structure determination of large RNAs or protein-RNA complexes. *Nat Commun* **5:** 3669.

Duss O, Diarra Dit Konté N, Allain F H-T. 2015. Cut and paste RNA for nuclear magnetic resonance, paramagnetic resonance enhancement, and electron paramagnetic resonance structural studies. *Methods Enzymol* **565:** 537–562.

Edelmann FT, Niedner A, Niessing D. 2014. Production of pure and functional RNA for in vitro reconstitution experiments. *Methods* **65:** 333–341.

Eliseev B, Yeramala L, Leitner A, Karuppasamy M, Raimondeau E, Huard K, Alkalaeva E, Aebersold R, Schaffitzel C. 2018. Structure of a human cap-dependent 48S translation pre-initiation complex. *Nucl Acids Res* **46:** 2678–2689.

Eliuk S, Makarov A. 2015. Evolution of Orbitrap mass spectrometry instrumentation. *Annu Rev Anal Chem* **8:** 61–80.

Erzberger JP, Stengel F, Pellarin R, Zhang S, Schaefer T, Aylett CHS, Cimermančič P, Boehringer D, Sali A, Aebersold R, et al. 2014. Molecular architecture of the 40S·eIF1·eIF3 translation initiation complex. *Cell* **158:** 1123–1135.

Faini M, Stengel F, Aebersold R. 2016. The evolving contribution of mass spectrometry to integrative structural biology. *J Am Soc Mass Spectrom* **27:** 966–974.

Gaston KW, Limbach PA. 2014. The identification and characterization of non-coding and coding RNAs and their modified nucleosides by mass spectrometry. *RNA Biol* **11:** 1568–1585.

Gleditzsch D, Müller-Esparza H, Pausch P, Sharma K, Dwarakanath S, Urlaub H, Bange G, Randau L. 2016. Modulating the Cascade architecture of a minimal Type I-F CRISPR–Cas system. *Nucl Acids Res* **44:** 5872–5882.

Greber BJ, Boehringer D, Leitner A, Bieri P, Voigts-Hoffmann F, Erzberger JP, Leibundgut M, Aebersold R, Ban N. 2014. Architecture of the large subunit of the mammalian mitochondrial ribosome. *Nature* **505:** 515–519.

Greber BJ, Bieri P, Leibundgut M, Leitner A, Aebersold R, Boehringer D, Ban N. 2015. The complete structure of the 55S mammalian mitochondrial ribosome. *Science* **348:** 303–308.

Guntert P. 2004. Automated NMR structure calculation with CYANA. *Methods Mol Biol* **278:** 353–378.

Gurevich VV, Pokrovskaya ID, Obukhova TA, Zozulya SA. 1991. Preparative in vitro mRNA synthesis using SP6 and T7 RNA polymerases. *Anal Biochem* **195:** 207–213.

He C, Sidoli S, Warneford-Thomson R, Tatomer DC, Wilusz JE, Garcia BA, Bonasio R. 2016. High-resolution mapping of RNA-binding regions in the nuclear proteome of embryonic stem cells. *Mol Cell* **64:** 416–430.

Heiss M, Kellner S. 2017. Detection of nucleic acid modifications by chemical reagents. *RNA Biol* **14:** 1166–1174.

Huang J-R, Warner LR, Sanchez C, Gabel F, Madl T, Mackereth CD, Sattler M, Blackledge M. 2014. Transient electrostatic interactions dominate the conformational equilibrium sampled by multidomain splicing factor U2AF65: A combined NMR and SAXS study. *J Am Chem Soc* **136:** 7068–7076.

Huang R, Han M, Meng L, Chen X. 2018. Transcriptome-wide discovery of coding and noncoding RNA-binding proteins. *Proc Natl Acad Sci* **115:** E3879–E3887.

Jazurek M, Ciesiolka A, Starega-Roslan J, Bilinska K, Krzyzosiak WJ. 2016. Identifying proteins that bind to specific RNAs—Focus on simple repeat expansion diseases. *Nucl Acids Res* **44:** 9050–9070.

Kambach C, Walke S, Young R, Avis JM, De La Fortelle E, Raker VA, Lührmann R, Li J, Nagai K. 1999. Crystal structures of two Sm protein complexes and their implications for the assembly of the spliceosomal snRNPs. *Cell* **96:** 375–387.

Kershaw CJ, O'Keefe RT. 2012. Splint ligation of RNA with T4 DNA ligase. *Methods Mol Biol* **941:** 257–269.

Kielpinski LJ, Hagedorn PH, Lindow M, Vinther J. 2017. RNase H sequence preferences influence antisense oligonucleotide efficiency. *Nucleic Acids Res* **45:** 12932–12944.

Kim I, Lukavsky PJ, Puglisi JD. 2002. NMR study of 100 kDa HCV IRES RNA using segmental isotope labeling. *J Am Chem Soc* **124:** 9338–9339.

König J, Zarnack K, Rot G, Curk T, Kayikci M, Zupan B, Turner DJ, Luscombe NM, Ule J. 2010. iCLIP reveals the function of hnRNP particles in splicing at individual nucleotide resolution. *Nat Struct Mol Biol* **17:** 909–915.

Kramer K, Sachsenberg T, Beckmann BM, Qamar S, Boon K-L, Hentze MW, Kohlbacher O, Urlaub H. 2014. Photo-cross-linking and high-resolution mass spectrometry for assignment of RNA-binding sites in RNA-binding proteins. *Nat Methods* **11:** 1064–1070.

Krepl M, Blatter M, Cléry A, Damberger FF, Allain F H-T, Sponer J. 2017. Structural study of the Fox-1 RRM protein hydration reveals a role for key water molecules in RRM-RNA recognition. *Nucl Acids Res* **45:** 8046–8063.

Kurschat WC, Muller J, Wombacher R, Helm M. 2005. Optimizing splinted ligation of highly structured small RNAs. *RNA* **11:** 1909–1914.

Kühlbrandt W. 2014. The resolution revolution. *Science* **343:** 1443–1444.

Lapinaite A, Simon B, Skjaerven L, Rakwalska-Bange M, Gabel F, Carlomagno T. 2013. The structure of the box C/D enzyme reveals regulation of RNA methylation. *Nature* **502:** 519–523.

Lebars IB, Vileno B, Bourbigot S, Turek P, Wolff P, Kieffer B. 2014. A fully enzymatic method for site-directed spin labeling of long RNA. *Nucleic Acids Res* **42:** e117.

Leitner A, Sturm M, Hudecz O, Mazanek M, Smått J-H, Lindén M, Lindner W, Mechtler K. 2010. Probing the phosphoproteome of HeLa cells using nanocast metal oxide microspheres for phosphopeptide enrichment. *Anal Chem* **82:** 2726–2733.

Leitner A. 2016a. Cross-linking and other structural proteomics techniques: How chemistry is enabling mass spectrometry applications in structural biology. *Chem Sci* **7:** 4792–4803.

Leitner A. 2016b. Application of chemical cross-linking/mass spectrometry to probe protein structures. In *Encyclopedia of analytical chemistry* (ed. Meyers RA), a9549. Wiley, New York.

Leitner A, Faini M, Stengel F, Aebersold R. 2016c. Crosslinking and mass spectrometry: An integrated technology to understand the structure and function of molecular machines. *Trends Biochem Sci* **41:** 20–32.

Lelyveld VS, Björkbom A, Ransey EM, Sliz P, Szostak JW. 2015. Pinpointing RNA-protein cross-links with site-specific stable isotope-labeled oligonucleotides. *J Am Chem Soc* **137:** 15378–15381.

Leney AC, Heck AJR. 2017. Native mass spectrometry: What is in the name? *J Am Soc Mass Spectrom* **28:** 5–13.

Limbach PA, Paulines MJ. 2017. Going global: The new era of mapping modifications in RNA. *WIREs RNA* **8:** e1367.

Lu K, Miyazaki Y, Summers MF. 2010. Isotope labeling strategies for NMR studies of RNA. *J Biomol NMR* **46:** 113–125.

Lu K, Heng X, Garyu L, Monti S, Garcia EL, Kharytonchyk S, Dorjsuren B, Kulandaivel G, Jones S, Hiremath A, et al. 2011. NMR detection of structures in the HIV-1 5′-leader RNA that regulate genome packaging. *Science* **334:** 242–245.

Lukavsky PJ, Kim I, Otto GA, Puglisi JD. 2003. Structure of HCV IRES domain II determined by NMR. *Nat Struct Biol* **10:** 1033–1038.

Masliah G, Maris C, König SLB, Yulikov M, Aeschimann F, Malinowska AL, Mabille J, Weiler J, Holla A, Hunziker J, et al. 2018. Structural basis of siRNA recognition by TRBP double-stranded RNA binding domains. *EMBO J* **37:** e97089.

Milligan JF, Groebe DR, Witherell GW, Uhlenbeck OC. 1987. Oligoribonucleotide synthesis using T7 RNA polymerase and synthetic DNA templates. *Nucleic Acids Res* **15:** 8783–8798.

Moore MJ, Sharp PA. 1992. Site-specific modification of pre-mRNA: The 2′-hydroxyl groups at the splice sites. *Science* **256:** 992–997.

Mullari M, Lyon D, Jensen LJ, Nielsen ML. 2017. Specifying RNA-binding regions in proteins by peptide cross-linking and affinity purification. *J Proteome Res* **16:** 2762–2772.

Nelissen FH, van Gammeren AJ, Tessari M, Girard FC, Heus HA, Wijmenga SS. 2008. Multiple segmental and selective isotope labeling of large RNA for NMR structural studies. *Nucleic Acids Res* **36:** e89.

Nguyen THD, Galej WP, Bai X-C, Savva CG, Newman AJ, Scheres SHW, Nagai K. 2015. The architecture of the spliceosomal U4/U6.U5 tri-snRNP. *Nature* **523:** 47–52.

Oberstrass FC, Auweter SD, Erat M, Hargous Y, Henning A, Wenter P, Reymond L, Amir-Ahmady B, Pitsch S, Black DL, et al. 2005. Structure of PTB bound to RNA: Specific binding and implications for splicing regulation. *Science* **309:** 2054–2057.

Paizs B, Suhai S. 2005. Fragmentation pathways of protonated peptides. *Mass Spectrom Rev* **24:** 508–548.

Parisien M, Major F. 2008. The MC-Fold and MC-Sym pipeline infers RNA structure from sequence data. *Nature* **452:** 51–55.

Pokrovskaya ID, Gurevich VV. 1994. In vitro transcription: Preparative RNA yields in analytical scale reactions. *Anal Biochem* **220:** 420–423.

Purtha WE, Coppins RL, Smalley MK, Silverman SK. 2005. General deoxyribozyme-catalyzed synthesis of native 3′–5′ RNA linkages. *J Am Chem Soc* **127:** 13124–13125.

Rauhut R, Fabrizio P, Dybkov O, Hartmuth K, Pena V, Chari A, Kumar V, Lee C-T, Urlaub H, Kastner B, et al. 2016. Molecular architecture of the *Saccharomyces cerevisiae* activated spliceosome. *Science* **353:** 1399–1405.

Rinner O, Seebacher J, Walzthoeni T, Mueller LN, Beck M, Schmidt A, Mueller M, Aebersold R. 2008. Identification of cross-linked peptides from large sequence databases. *Nat Methods* **5:** 315–318.

Schmidt C, Kramer K, Urlaub H. 2012. Investigation of protein–RNA interactions by mass spectrometry—Techniques and applications. *J Proteomics* **75:** 3478–3494.

Schubert M, Lapouge K, Duss O, Oberstrass FC, Jelesarov I, Hass D, Allain F H-T. 2007. Molecular basis of messenger RNA recognition by the specific bacterial repressing clamp RsmA/CsrA. *Nat Struct Mol Biol* **14:** 807–813.

Schürch S. 2016. Characterization of nucleic acids by tandem mass spectrometry—The second decade (2004–2013): From DNA to RNA and modified sequences. *Mass Spectrom Rev* **35:** 483–523.

Stark H, Dube P, Lührmann R, Kastner B. 2001. Arrangement of RNA and proteins in the spliceosomal U1 small nuclear ribonucleoprotein particle. *Nature* **409:** 539–542.

Sugimoto Y, Vigilante A, Darbo E, Zirra A, Militti C, D'Ambrogio A, Luscombe NM, Ule J. 2015. hiCLIP reveals the in vivo atlas of mRNA secondary structures recognized by Staufen 1. *Nature* **519:** 491–494.

Tzakos AG, Easton LE, Lukavsky PJ. 2007. Preparation of large RNA oligonucleotides with complementary isotope-labeled segments for NMR structural studies. *Nat Protoc* **2:** 2139–2147.

Ule J, Jensen K, Mele A, Darnell RB. 2005. CLIP: A method for identifying protein–RNA interaction sites in living cells. *Methods* **37:** 376–386.

Van Nostrand EL, Pratt GA, Shishkin AA, Gelboin-Burkhart C, Fang MY, Sundararaman B, Blue SM, Nguyen TB, Surka C, Elkins K, et al. 2016. Robust transcriptome-wide discovery of RNA-binding protein binding sites with enhanced CLIP (eCLIP). *Nat Methods* **13:** 508–514.

Veit J, Sachsenberg T, Chernev A, Aicheler F, Urlaub H, Kohlbacher O. 2016. LFQProfiler and RNPxl: Open-source tools for label-free quantification and protein–RNA cross-linking integrated into proteome discoverer. *J Proteome Res* **15:** 3441–3448.

Walzthoeni T, Claassen M, Leitner A, Herzog F, Bohn S, Förster F, Beck M, Aebersold R. 2012. False discovery rate estimation for cross-linked peptides identified by mass spectrometry. *Nat Methods* **9:** 901–903.

Ward AB, Sali A, Wilson IA. 2013. Integrative structural biology. *Science* **339:** 913–915.

Wilm M. 2011. Principles of electrospray ionization. *Mol Cell Proteomics* **10:** M111.009407.

Wu S, Tutuncuoglu B, Yan K, Brown H, Zhang Y, Tan D, Gamalinda M, Yuan Y, Li Z, Jakovljevic J, et al. 2016. Diverse roles of assembly factors revealed by structures of late nuclear pre-60S ribosomes. *Nature* **534:** 133–137.

Wurm JP, Holdermann I, Overbeck JH, Mayer PHO, Sprangers R. 2017. Changes in conformational equilibria regulate the activity of the Dcp2 decapping enzyme. *Proc Natl Acad Sci* **114:** 6034–6039.

Xu J, Lapham J, Crothers DM. 1996. Determining RNA solution structure by segmental isotopic labeling and NMR: Application to *Caenorhabditis elegans* spliced leader RNA 1. *Proc Natl Acad Sci* **93:** 44–48.

Group II Intron RNPs and Reverse Transcriptases: From Retroelements to Research Tools

Marlene Belfort[1] and Alan M. Lambowitz[2]

[1]Department of Biological Sciences and RNA Institute, University at Albany, State University of New York, Albany, New York 12222

[2]Institute for Cellular and Molecular Biology and Department of Molecular Biosciences, University of Texas at Austin, Austin, Texas 78712

Correspondence: mbelfort@albany.edu; lambowitz@austin.utexas.edu

SUMMARY

Group II introns, self-splicing retrotransposons, serve as both targets of investigation into their structure, splicing, and retromobility and a source of tools for genome editing and RNA analysis. Here, we describe the first cryo-electron microscopy (cryo-EM) structure determination, at 3.8–4.5 Å, of a group II intron ribozyme complexed with its encoded protein, containing a reverse transcriptase (RT), required for RNA splicing and retromobility. We also describe a method called RIG-seq using a retrotransposon indicator gene for high-throughput integration profiling of group II introns and other retrotransposons. Targetrons, RNA-guided gene targeting agents widely used for bacterial genome engineering, are described next. Finally, we detail thermostable group II intron RTs, which synthesize cDNAs with high accuracy and processivity, for use in various RNA-seq applications and relate their properties to a 3.0-Å crystal structure of the protein poised for reverse transcription. Biological insights from these group II intron revelations are discussed.

Outline

Cite as *Cold Spring Harb Perspect Biol* doi: 10.1101/cshperspect.a032375

1 INTRODUCTION

Group II introns are self-splicing RNAs and mobile retroelements that occupy a pivotal evolutionary niche as putative ancestors of spliceosomal introns and the spliceosome, retrotransposons, telomerase, and retroviruses (Lambowitz and Belfort 2015; Novikova and Belfort 2017). They also are important genome editing tools (Enyeart et al. 2014) and the source of reverse transcriptases (RTs) that have broad application in high-throughput characterization of RNAs (Mohr et al. 2013b). These group II intron ribozymes catalyze their self-excision from RNA precursors, and ligation of the flanking exons, with the aid of an intron-encoded protein (IEP), which is an ancient form of RT[3] (Fig. 1A) (Lambowitz and Zimmerly 2011). The ribozyme comprises a conserved, six-domain structure (DI–DVI) (Fig. 1B). Splicing occurs via two transesterification reactions, which are initiated by the 2′-OH of a bulged adenosine nucleophile in DVI and result in ligated exons and an excised intron lariat, in precise analogy to spliceosomal intron splicing (Fig. 1A). The IEP functions as a "maturase" to facilitate RNA splicing and reverse splicing of the excised intron RNA into a DNA site on the top strand. After endonuclease cleavage of the bottom strand, retrohoming proceeds when the RT reverse transcribes the reverse-spliced intron RNA into an intron cDNA that is incorporated into the host genome. Group II intron RTs have a DNA-binding domain (denoted D) and many also have a DNA endonuclease (EN) domain, both of which are involved in retromobility.

Structural studies have provided insight into the RNA component of the ribonucleoprotein (RNP) complex (Toor et al. 2008; Robart et al. 2014; Costa et al. 2016). DI, the largest intron RNA domain, folds first. DI docks the active site domain, DV, and also base-pairs its exon-binding sequences (EBSs) with the intron-binding sequences (IBSs) in the exons (Fig. 1B,C). These base-pairings determine the specificity of forward and reverse splicing reactions (Lambowitz and Zimmerly 2011; Marcia and Pyle 2012). DII and DIII contribute structurally to the catalytic complex, whereas DIV encompasses an unstructured open reading frame (ORF) that encodes the RT, which in turn binds tightly to a stem-loop preceding the ORF (DIVa for the Ll.LtrB intron shown in Fig. 1B). How the IEP interacts with the intron RNA remained elusive until a 3.8–4.5 Å cryo-electron microscopy (cryo-EM) structure of the RNP was solved (Qu et al. 2016). The RNP complex, purified from native host cells, revealed its major structural features including that of the RT and also underscored the relationship of the group II intron RNPs to the spliceosome (Fig. 2). High-resolution crystal structures of amino-terminal proteolytic fragments of group II intron RTs have also been reported, but with the RT active site region improperly folded (Zhao and Pyle 2016). A recent 3.0-Å crystal structure of a full-length group II intron RT in complex with an RNA-template–DNA primer substrate and incoming dNTP revealed the active structure of the enzyme for reverse transcription and a host of novel structural features that may contribute to the distinctive enzymatic properties of group II intron RTs (Stamos et al. 2017).

As progress was being made on discerning the biochemical reactions and different functionalities of group II introns, they began to be exploited as genome editing tools (Guo et al. 2000; Mohr et al. 2000; Karberg et al. 2001; Enyeart et al. 2014). The specificity of the IBS–EBS interactions was harnessed to target introns to preprogrammed positions in a wide variety of bacterial genomes. These "targetrons" were then used for a range of genome editing applications in bacteria. Subsequently, the robust properties of thermostable group II intron RTs (TGIRTs) were exploited for many different high-throughput RNA characterization purposes (Mohr et al. 2013b). Thus, these key elements in evolution also provide a toolbox for genetic manipulation and biochemical analysis.

2 TECHNIQUE 1. ISOLATION OF NATIVE RNPs FOR CRYO-EM STRUCTURE DETERMINATION

Cryo-EM structures were determined for the group II Ll.LtrB intron from the *ltrB* gene of *Lactococcus lactis*. To isolate RNPs from the native host, an elaborate expression vector was constructed, whereby the intron RNA and the IEP were expressed under separate nisin-inducible promoters (Fig. 2A) (Gupta et al. 2014). The IEP (denoted LtrA protein) was fused to an affinity tag, a chitin-binding domain (CBD), via an intein, to allow cleavage of the IEP from a chitin-affinity column. After induction by nisin, a polycyclic peptide, the *L. lactis* host cells containing the plasmid were lysed and the lysate applied to a chitin resin. To release the bound IEP and its associated intron RNA, the column was immersed in the reducing agent and nucleophile, dithiothreitol. For cryo-EM analysis, after purification by sucrose gradient sedimentation, residual genomic DNA was digested with DNase and the pure RNP was recovered by gel filtration (Qu et al. 2016).

For EM, the RNP complex was prepared for negative staining or frozen-hydrated for cryo-EM (Qu et al. 2016). In determining the structure of the RNP, negative staining was followed by random conical tilt reconstruction, in which two major class averages were established. Then micrographs of frozen-hydrated intron RNP particles were

[3]We use RT and IEP interchangeably, in keeping with nomenclature for other RTs that have multiple activities (e.g., RNase H).

Cite as *Cold Spring Harb Perspect Biol* doi: 10.1101/cshperspect.a032375

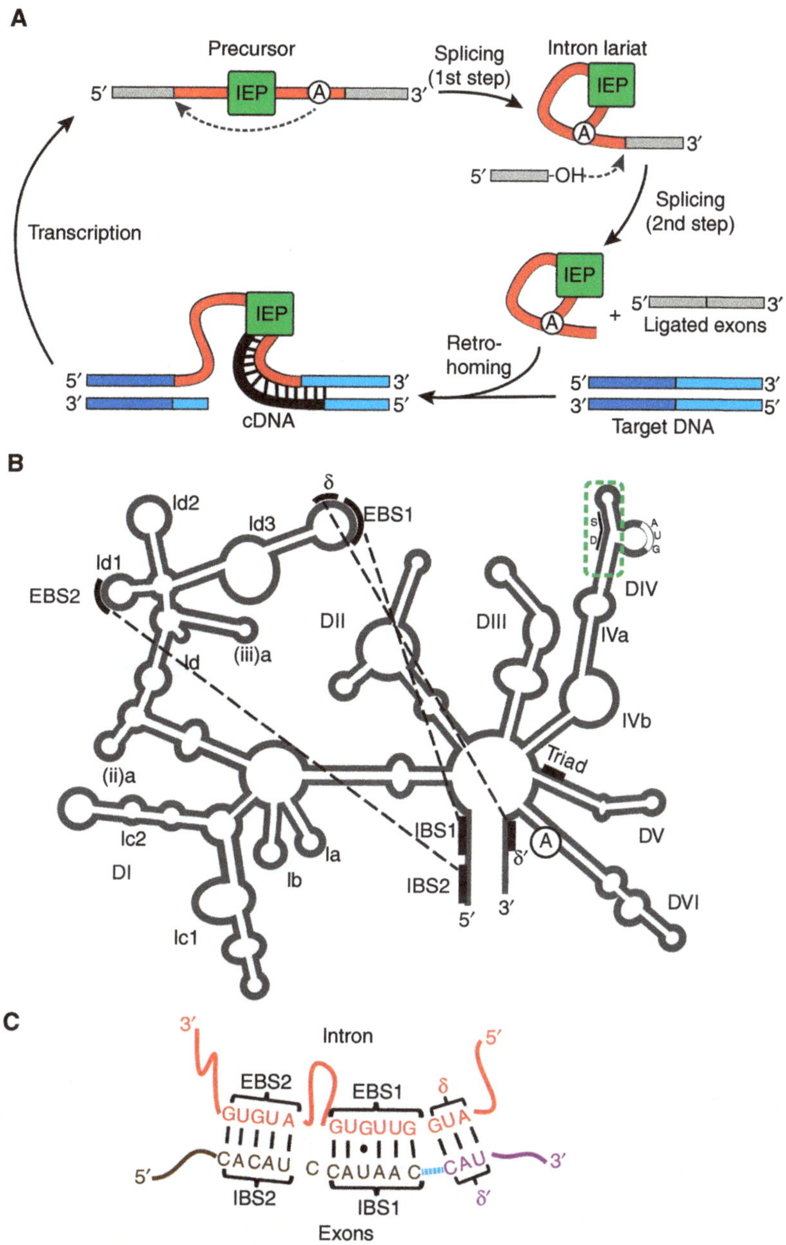

Figure 1. Group II intron structure and function. (*A*) Intron life cycle. Splicing is initiated by attack on the 5′ splice site by the 2′-OH of the bulged adenosine (circled A) in DVI to form a lariat intermediate (first step). Attack on the 3′ splice site by the 3′-OH of the upstream exon results in ligated exons and the intron lariat (second step). The intron lariat can reverse splice into target DNA (retrohoming). The endonuclease of the intron-encoded protein (IEP) then nicks the opposite DNA strand, and the reverse transcriptase (RT) uses the cleaved DNA strand as a primer to make a cDNA copy of the integrated intron. After degradation of the intron RNA, second-strand DNA synthesis and repair, leading to a fully integrated DNA copy of the intron in the genome (not shown), transcription can ensue to complete the cycle. (*B*) Secondary structure of a group II intron. The six domains (DI–DVI) radiating from a central hub are labeled along with various subdomains, the "exon-binding sequence"–"intron-binding sequence" (EBS–IBS) pairings, the IEP anchor site in DIV (dashed green box), the catalytic triad in DV, and the bulged adenosine in DVI. SD, Shine–Dalgarno sequence. (*C*) EBS–IBS pairings. EBS1–IBS1, EBS2–IBS2, and δ-δ′ are shown. Blue dashed line denotes ligation site.

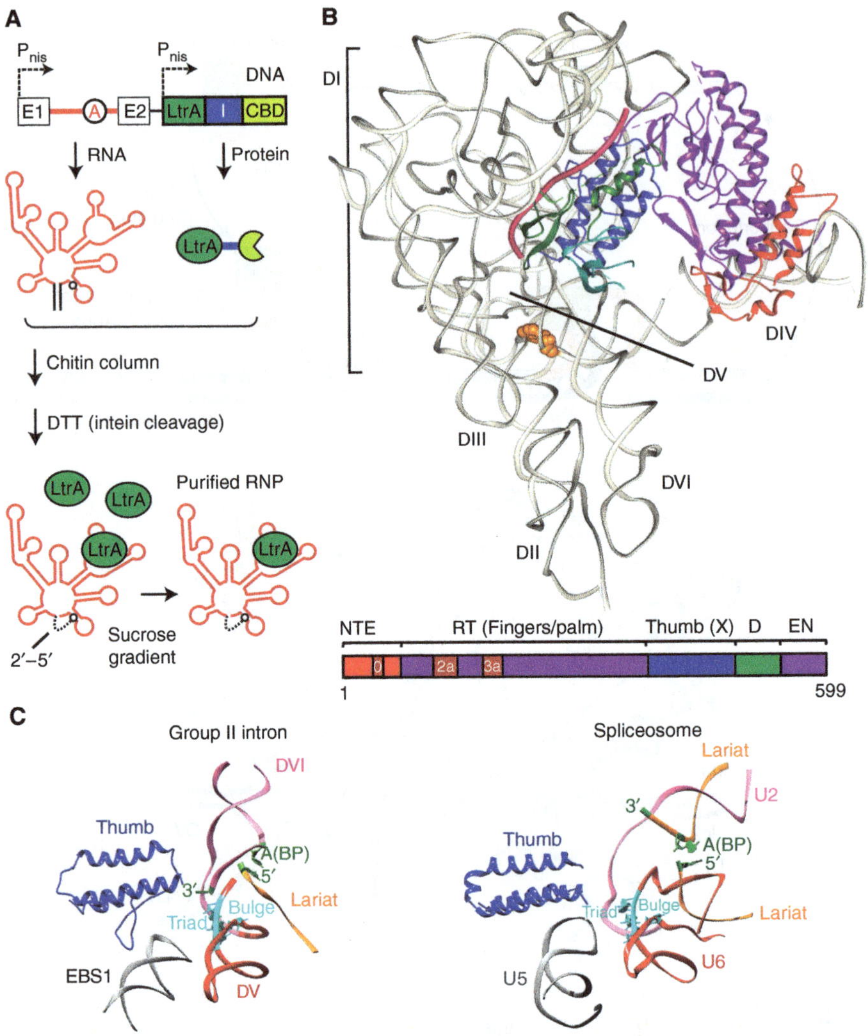

Figure 2. Cryo-electron microscopy (cryo-EM) structure of the Ll.LtrB group II intron. (*A*) Intein-based purification. The intron (red) and the intron-encoded protein (IEP) (termed LtrA, green) were expressed independently under nisin-inducible promoters (P_{nis}), with the IEP fused via an intein to a chitin-binding domain (CBD). The IEP promotes RNA splicing, and the resulting ribonucleoprotein (RNP) complex containing the IEP bound to excised intron lariat RNA was purified from lysed cells by binding to a chitin column. After dithiothreitol treatment, which induces intein cleavage, the released RNP was further purified on a sucrose gradient (Gupta et al. 2014; Qu et al. 2016). (*B*) Structure of the group II intron RNP determined by cryo-EM. The structure of the RNA (gray) and IEP (colored) was solved at a resolution of 3.8–4.5 Å. Cryo-EM was conducted by the laboratory of Dr. Hongwei Wang (Tsinghua University) and molecular interpretation of the cryo-EM map was a three-way collaboration among the Belfort and Wang laboratories and that of Dr. Raj Agrawal (Wadsworth Center) (Qu et al. 2016). RNA domains I–VI (DI–DVI) are shown (gray), along with the branch point adenosine (orange, space-filling) and trapped ligated exons (magenta). Below is a linear representation of the IEP, color-coded as in the cryo-EM structure, with amino-terminal extension (NTE), RT (fingers-palm), and thumb (X) domains, DNA-binding domain (D) and DNA endonuclease domain (EN). (*C*) Similarity of group II intron and spliceosome active centers. The group II intron (*left*) is shown alongside the *Schizosaccharomyces pombe* spliceosome (*right*) (Yan et al. 2015), illustrating the spatial similarity between the two. DV of the intron RNP and U6 of the U2/U6 spliceosome RNP (red) are aligned at their triad and bulge regions (cyan). The analogous structures of the group II intron and spliceosome are labeled: thumb domain in IEP and Prp8 (blue), the EBS1 loop of DI and U5 loop I (gray), DV and U6 of U2/U6 (red), DVI and U2 of U2/U6 (mauve), the lariat structures (orange), and their branch-point (BP) adenosines (green). (Adapted from Qu et al. 2016.)

generated from a 300-kV Titan Krios microscope. A 3D classification of the selected particles yielded three classes with strong LtrA protein occupancy (approximately 450,000 particles) and one class lacking density for the LtrA protein (approximately 100,000 particles). Image processing was then conducted to obtain a 3.8-Å resolution cryo-EM map of the group II intron RNP complex and a 4.5-Å resolution map of the IEP-depleted intron RNA. Soft masks around the most rigid portions of the RNP complex and IEP-depleted form were used for refinement to generate the final 3D reconstruction at the above average resolutions, followed by molecular interpretation of the cryo-EM map.

2.1 Biological Insights and Conclusions from Technique 1

The structures provide the first near-atomic model of the architecture of a group II intron RNP and the first structure of a group II intron RT and its interaction sites with the intron RNA. The six RNA domains form a Y-shaped structure with the IEP making contacts with the terminal stem-loop of DIVa and specific sites in DI (Fig. 1B). The Ll.LtrB group IIA intron RNA shows an overall similarity in structure to that seen in crystal structures of group IIB and IIC intron RNAs, which were derived from in vitro transcripts and lack the IEP (Fig. 2B, top) (Toor et al. 2008; Robart et al. 2014; Costa et al. 2016; Qu et al. 2016). The active-site center around DV, with its AGC catalytic triad, is the most highly conserved region across the different classes of group II introns and also the best-defined structure in the cryo-EM map. The 5′ and 3′ ends of the intron are proximal to the catalytic core, and the bulged A in DVI, which forms the intron lariat with the 5′ end, is adjacent. Remarkably, the structure of the spliced lariat contains the ligated-exon segment of the mRNA, highlighting its interactions with EBS1 and EBS2. The mRNA segment is absent from the IEP-depleted samples with concomitant conformational differences in EBS1 and EBS2, suggesting that the IEP contributes to trapping the mRNA in the RNP (Qu et al. 2016).

The structure gives us the first glimpse of an intact group II intron IEP RT, which has a typical polymerase “right-hand” organization. The IEP has four domains: an RT domain comprised of fingers and palm regions, a thumb domain (previously also referred to as domain X or the maturase domain), a DNA-binding domain (D), and a DNA endonuclease domain (EN), all of which are required for intron mobility (Fig. 2B, bottom). The RT domain contains the catalytic center for reverse transcription. It also has an amino-terminal extension (NTE), with a conserved sequence RT0, and two distinctive “insertion” regions RT2a and RT3a, which are present in group II intron RTs but not in retroviral RTs (Fig. 2B and see below). Protein density revealed some bulky amino acid side-chains but the D and EN domains appear to be dynamic and were less well resolved (Qu et al. 2016). The RT and thumb domains bind the intron RNA to promote both splicing and reverse splicing for retromobility in agreement with previous biochemical analyses (Matsuura et al. 2001; Cui et al. 2004; Gu et al. 2010). The NTE anchors the IEP to DIVa with contributions by RT3a, and the thumb and D domains bind to DI, stabilizing the EBS–IBS interactions for splicing and reverse splicing.

A conundrum raised by the structure is that biochemical and biophysical studies suggest that the IEP is a monomer in solution but binds precursor RNA and functions in splicing at a stoichiometry of ~2:1, raising the possibility that it functions in splicing as a dimer (Saldanha et al. 1999; Rambo and Doudna 2004). However, the IEP is a monomer in the cryo-EM reconstruction (Agrawal et al. 2016; Qu et al. 2016), and the GsI-IIC group II intron RT, described below, also binds an artificial template-primer substrate as a monomer (Stamos et al. 2017). Although biochemical and structural studies with the proteolytic fragments containing the fingers-palm domain of an IEP of *Roseburia intestinalis* (Ri) and *Eubacterium rectale* (Er) suggested dimerization, the dimer interface in those structures would be occluded by the thumb in the full-length protein and presents significant steric clashes when docked into the Ll.LtrB intron RNA (Agrawal et al. 2016; Zhao and Pyle 2016). Recent studies show temperature-dependent dimerization of Ll.LtrB precursor RNPs, although the spliced intron RNPs remained monomeric (Dong et al. 2018). Thus, the monomer–dimer discrepancy may relate to different stoichiometries in different states and will require further experimentation to resolve.

Based on similarities in RNA catalysis and structure, the hypothesis that the group II intron is the progenitor of the spliceosomal intron is widely accepted (reviewed in Lambowitz and Belfort 2015; Novikova and Belfort 2017). The cryo-EM structure provides supporting evidence by virtue of LtrA and Prp8, two related proteins, both directing 5′-splice-site recognition. The thumb domain of spliceosomal Prp8 interacts with the U5 snRNP in analogy to the thumb domain of the IEP interacting with the EBSs (Fig. 2C). Likewise, the branch-points are similarly disposed, where DVI presents the bulged A to the DV active site with its catalytic triad, much as U2 of the U2/U6 snRNP pair interacts with the intron to bring the bulged A to the catalytic triad (Agrawal et al. 2016; Piccirilli and Staley 2016; Qu et al. 2016).

The recent revolution in cryo-EM has made possible the close-to-atomic resolution structures of difficult-to-crystallize RNPs, such as the group II intron RNPs described here, and the spliceosome in various states (reviewed in Fica and Nagai 2017; Shi 2017; Galej et al. 2018). The cryo-EM structure of native telomerase holoenzyme has also been determined, but at somewhat lower resolution (Jiang et al. 2015; Nguyen et al. 2018). These cryo-EM studies have revealed the structural and functional relationships of these RNP assemblies.

3 TECHNIQUE 2. RETROMOBILITY INDICATOR GENES AND RIG-seq FOR HIGH-THROUGHPUT INTEGRATION PROFILING

Mechanisms of retrohoming and retrotransposition of group II introns have been well established. The frequency and diversity of retromobility events can be determined by using a retromobility indicator gene (RIG), patterned after that used to measure Ty1 retrotransposition (Fig. 3A) (Curcio and Garfinkel 1991). Refined analysis is facilitated by a high-throughput genomic retrotransposition detection system based on RIG, called RIG-seq. In *L. lactis*, the group II intron donor plasmid carries the intron (red) flanked by exons (gray) and under a nisin-inducible promoter (P_{nisA}) (Fig. 3A). The intron is interrupted by a RIG with the kanamycin resistance (kan^R) gene inserted in the antitranscriptional orientation to the intron and carrying its own promoter (P_{kan}). The kan^R gene is interrupted by a self-splicing group I intron (gpI) in the same orientation as Ll.LtrB (red arrow). Reconstitution of the kan^R gene with formation of a characteristic splice junction (SJ) sequence and kanamycin resistance is possible only through reverse transcription of an RNA intermediate that had lost the group I intron during retrotransposition (Fig. 3B).

High-throughput targeted sequencing of group II intron insertion loci is based on generation of the SJ sequence of the kan^R gene of RIG during retrotransposition (Fig. 3B, C). After ligation of Illumina-specific adapters (P5A and P7A), two tandem polymerase chain reactions (PCRs) are used to amplify 3′ flanks of retrotransposition events in

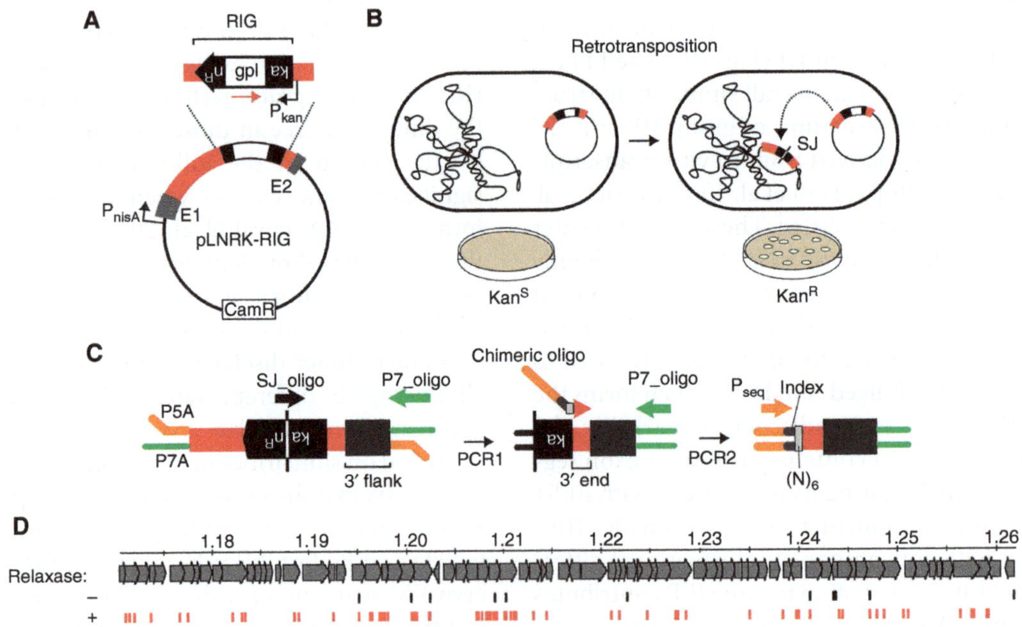

Figure 3. RIG-seq for high-throughput retrotransposon profiling. (*A*) RIG assay. The intron donor plasmid pLNRK-RIG carries the Ll.LtrB intron (red) flanked by exons (gray) under a nisin-inducible promoter (P_{nisA}). The intron is interrupted by a retromobility indicator gene (RIG) with the kan^R gene inserted in the anti-transcriptional orientation to the intron and carrying its own promoter (P_{kan}). The kan^R gene is interrupted by a self-splicing group I intron (gpI) in the same orientation as Ll.LtrB (red arrow). (*B*) Reconstitution of the kan^R gene with formation of a characteristic splice junction (SJ) sequence and kanamycin resistance is possible only through reverse transcription of an RNA intermediate that lost the group I intron during retrotransposition. Wavy line to the *left* is chromosomal DNA, whereas plasmid to the *right* is the RIG donor. (*C*) RIG-seq amplification scheme. High-throughput targeted sequencing of insertion loci was based on the generation of the SJ sequence of the kan^R gene during retrotransposition. After ligation of Illumina-specific adapters, P5A and P7A, two tandem polymerase chain reactions (PCRs) were used to amplify flanks of retrotransposition events in Illumina libraries: A specific SJ primer (SJ_oligo) with the P7_oligo is used in the first PCR; then a library-specific primer, the chimeric oligo, carrying library-specific indices, is used with the P7_oligo in the second PCR. Pseq is the sequencing primer. (*D*) Density of intron-insertions in part of the *Lactococcus lactis* chromosome is shown, in which numbers correspond to mbp from the origin of replication, in the presence (+, red) or absence (–, black) of relaxase. (Adapted from Novikova et al. 2014.)

 Cite as *Cold Spring Harb Perspect Biol* doi: 10.1101/cshperspect.a032375

Illumina libraries. The key to the method is the use of a SJ-specific primer (SJ_oligo) with a P7 primer (P7_oligo) in the first PCR. Then a chimeric library-specific primer (chimeric oligo) carrying library-specific indices and an $(N)_6$ sequence is used with the P7_oligo in the second PCR. The Pseq sequencing primer is then used to obtain Illumina single-end reads of group II intron integration junctions (Fig. 3C).

3.1 Biological Insights and Conclusions from Technique 2

Using the RIG construct, it was shown that retrotransposition of the Ll.LtrB intron to ectopic sites proceeds via reverse splicing into DNA targets (Ichiyanagi et al. 2002), that plasmid targets are favored (Ichiyanagi et al. 2003), and that pathway choice is dictated by cellular environment (Coros et al. 2005). A RIG was also useful for host mutant screens, which revealed both silencers and global stimulators of intron mobility (Coros et al. 2008, 2009). Thus, RNase E, an endoribonuclease that cleaves RNA in single-stranded A- and U-rich regions, was shown to depress retromobility. Conversely, amino acid and glucose starvation, acting through the alarmones ppGpp and cAMP, respectively, were shown to act as stimulators of intron mobility. Thereby, nutritional stress leads to bursts of intron integration.

To provide a comprehensive landscape of integration-site preferences, randomly sampled libraries of the RIG-seq reads were mapped along the *L. lactis* IL1403 chromosome (Novikova et al. 2014). Mapping of Illumina sequencing reads again showed preference for intron insertion into the intron donor plasmid (pLNRK-RIG). The relationship between retrotransposition of the group II intron and the conjugative plasmid, pRS01, on which it naturally resides, was then probed. This question was of particular interest because the intron is housed within the pRS01 *ltrB* gene, which encodes the relaxase required for specifically nicking pRS01 to facilitate transfer. Nicking occurs at the transfer origin (*oriT*), which allows conjugal transmission of pRS01 into the recipient cell. Using RIG-seq, we showed that relaxase introduces off-target nicks in the chromosomal DNA, stimulating both the frequency and density of retromobility events (Fig. 3D). Thus, relaxase, which is rendered functional through group II intron splicing, not only promotes horizontal conjugal transfer but also stimulates intron spread in donor and recipient cells (Belhocine et al. 2005).

The RIG-seq approach adds to the burgeoning high-throughput sequencing technologies for studying mobile-genetic element biology (Xing et al. 2013) and is uniquely and broadly applicable to performing retrotransposition profiling in diverse organisms.

4 TECHNIQUE 3. TARGETRONS FOR BACTERIAL GENE TARGETING APPLICATIONS

The finding that mobile group II introns recognize DNA target sites for retrohoming largely by base-pairing of the intron RNA led to their development into the first RNA-guided gene targeting tools, called targetrons (Guo et al. 2000; Mohr et al. 2000; Karberg et al. 2001; Perutka et al. 2004). Initial targetrons were based on the aforementioned Ll.LtrB group IIA intron, but additional introns, including group IIB introns EcI5 and RmIntI, and TeI3c from a bacterial thermophile, have also been used, as could any group II intron with suitably stringent DNA target-site specificity (Zhuang et al. 2009; Mohr et al. 2013a; Garcia-Rodriguez et al. 2014).

For gene targeting in bacteria, targetrons are typically transcribed from a plasmid expression vector that uses a strong and preferably inducible promoter (Fig. 4A). The targetron cassette expresses a precursor RNA containing a group II intron from which the RT ORF has been deleted (I-ΔORF) flanked by short 5′ and 3′ exons (E1 and E2, respectively), with the RT expressed in tandem downstream from the 3′ exon. The RT promotes splicing of the intron, forming lariat RNPs that recognize and integrate site-specifically into a DNA target site. Group IIA and IIB intron RNPs typically recognize DNA target sites via three base-pairing interactions (EBS1/IBS1, EBS2/IBS2, and δ-δ′ [IIA] or EBS3/IBS3 [IIB]), which provide most of the DNA target specificity (Figs. 1C and 4B). The group II intron RT contributes by recognizing a small number of specific bases in the distal 5′- and 3′-exon regions of the DNA target site to help promote local DNA melting for base-pairing of the intron RNA and bottom-strand cleavage (Singh and Lambowitz 2001; Perutka et al. 2004; Zhuang et al. 2009).

Targetrons are programmed to insert into desired sites with the aid of a commercially available computer algorithm that identifies optimal matches for the small number of nucleotide positions recognized by the IEP and then designs PCR primers for modifying the sequence elements in the intron RNA to base-pair to the IBS and δ sequences into the DNA target site (Perutka et al. 2004; Zhuang et al. 2009; Garcia-Rodriguez et al. 2014). Ll.LtrB and EcI5 targetrons designed by using such algorithms typically insert into DNA target sites with high efficiency (1%–100% without selection), with nonspecific integrations rarely detected. The insertion frequency of targetrons is usually high enough to detect desired integration events by PCR screening of colonies without selection, but if needed, a retrotransposition-activated genetic marker (RAM), equivalent to a RIG, can be incorporated into the intron RNA for genetic selections (Zhong et al. 2003). Polar effects on the expres-

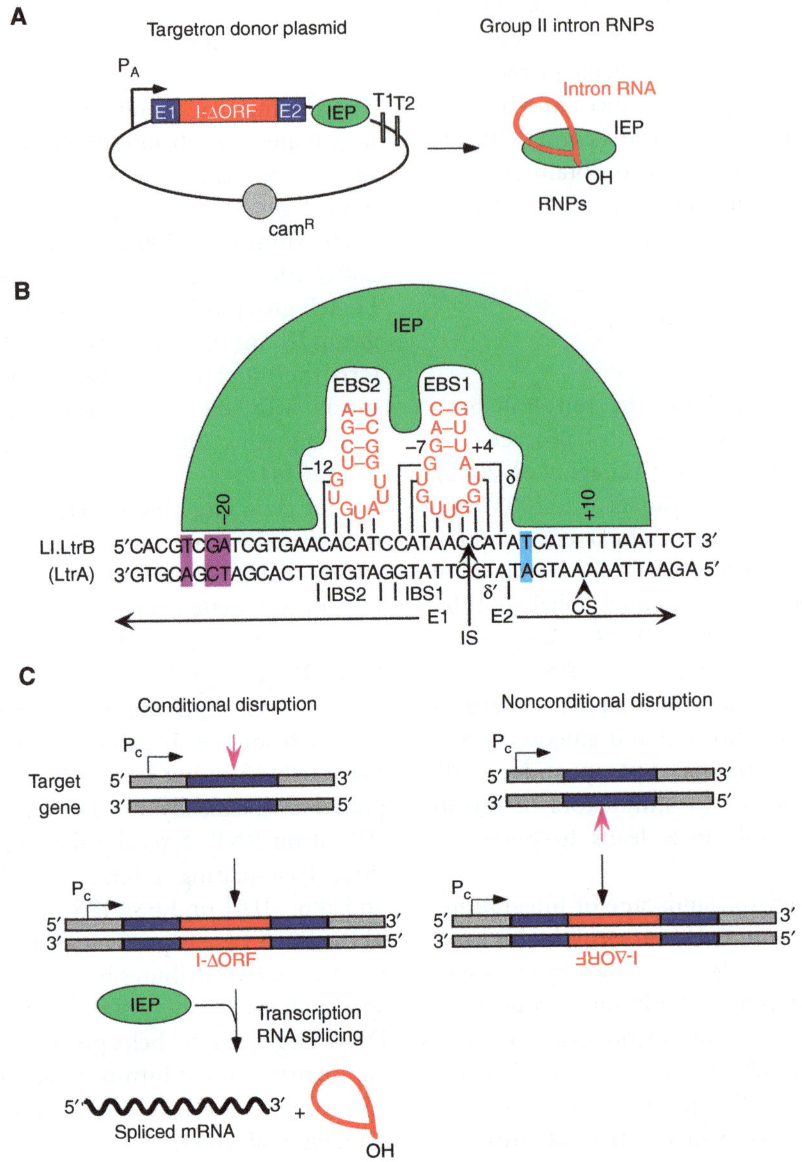

Figure 4. Group II introns as targetrons for bacterial gene targeting. (*A*) Targetron donor plasmid. The plasmid has a strong and preferably inducible active promoter (P_A) to express a targetron cassette consisting of a group II intron ribozyme with the reverse transcriptase (RT) open reading frame (ORF) deleted (I-ΔORF) flanked by short 5′ and 3′ exons (E1 and E2, respectively) and followed by an ORF encoding the intron-encoded protein (IEP). The IEP splices the group II intron RNA and remains bound to the excised intron lariat RNA in an RNP that can be preprogrammed to insert into desired target sites. (*B*) DNA target site interactions for the Ll.LtrB group IIA intron used for gene targeting involve base-pairing of EBS1, EBS2, and δ sequences in DI with IBS1, IBS2, and δ′ sequences flanking the intron insertion site (IS). The IEP recognizes specific nucleotides (purple) in the distal 5′-exon region upstream of IBS2 and promotes local DNA melting, enabling base-pairing of the intron RNA to the DNA target sequence. Additional interactions between the IEP and nucleotides in the 3′ exon (blue) are required for bottom-strand cleavage by the EN domain to generate the primer for reverse transcription of the intron RNA, leading to intron insertion via the retrohoming pathway. CS, bottom-strand cleavage site. (*C*) Use of targetrons for conditional or nonconditional disruptions. Targetrons can be made to insert in either orientation relative to transcription of the target gene (dark blue) by targeting a sequence (magenta arrow) in the top or bottom strand. A targetron that inserts in the antisense orientation yields an unconditional disruption, whereas a targetron that inserts in the sense orientation can yield a conditional disruption by linking splicing of the intron from the pre-mRNA to expression of the IEP from a separate construct. P_c, host chromosomal gene promoter. (Adapted from Enyeart et al. 2014.)

Cite as *Cold Spring Harb Perspect Biol* doi: 10.1101/cshperspect.a032375

sion of downstream genes can be mitigated by inserting a separate promoter near the 3′ end of the targetron.

Broad host-range vectors and promoters enable the use of the same targetron construct in a wide variety of bacteria (Yao and Lambowitz 2007). Targetrons can be made to insert in either the sense or antisense orientation relative to target gene transcription, depending on whether they are targeted to a sequence in the sense or antisense strands (Fig. 4C). A targetron that inserts in the antisense orientation yields an unconditional gene disruption, whereas a targetron that inserts in the sense orientation can be used to obtain a conditional disruption by linking splicing of the intron to expression of the IEP from a separate construct (Frazier et al. 2003; Yao et al. 2006). Targetrons can also be used to introduce a targeted double-strand break that can be repaired by homologous recombination with a cotransformed DNA fragment, enabling the seamless introduction of point mutations (Karberg et al. 2001; Mastroianni et al. 2008). A thermotargetron derived from a thermostable mobile group II intron from a bacterial thermophile was developed to enable gene targeting in thermophilic bacteria and archaea, which include biologically and commercially important species that had been recalcitrant to genetic manipulation (Mohr et al. 2013a; Hong et al. 2014). In recent applications, targetrons have been used to site-specifically position recombinase (lox) sites at two or more desired genomic regions, which can be used to obtain large deletions, insertions, and chromosome inversions after supplying Cre from a separate construct (Enyeart et al. 2013). Although targetrons afford the same benefits of largely RNA-guided design as clustered regularly interspersed short tandem repeat (CRISPR)–Cas gene-targeting systems, they work only inefficiently in eukaryotes because of the high Mg^{2+} requirements of the group II intron ribozyme (Mastroianni et al. 2008; Truong et al. 2015). Nevertheless, targetrons are useful for gene targeting in virtually any bacterium, and particularly those in which homology-based DNA repair outcomes are challenging (Enyeart et al. 2014).

4.1 Biological Insights and Conclusions from Technique 3

Targetrons have been sold commercially for 15 years and have been used in a wide variety of bacteria, including medically and commercially important species that were previously intractable to detailed genetic analysis or facile genetic engineering (reviewed in Enyeart et al. 2014). These include *Pseudomonas aeruginosa*, *Staphylococcus aureus*, *Francisella tularensis*, and a large variety of clostridial species. Recently, targetrons have also enabled the first directly targeted chromosomal mutations in the intracellular bacterial pathogen *Chlamydia trachomatis* (Johnson and Fisher 2013). Targetrons continue to be used in diverse bacteria for analysis of pathogenicity determinants (Buchan et al. 2009; Cole et al. 2016; Herrera et al. 2016; Kint et al. 2017), identification of new antibiotic targets (Zoraghi et al. 2011; Donnelly et al. 2017), construction of attenuated vaccine strains (Harper et al. 2016), and the engineering of industrially important bacteria for increased biofuel and chemical production (Hong et al. 2014; Dai et al. 2016; Xue et al. 2016).

5 TECHNIQUE 4. RTs WITH NOVEL ENZYMATIC PROPERTIES AS POTENT RESEARCH TOOLS

RTs are widely used for research and biotechnological applications that require cDNA synthesis, including RT-qPCR and high-throughput RNA sequencing (RNA-seq). Until recently, however, the only RTs available for such applications were retroviral RTs, and these have inherently low fidelity and processivity, presumably to help retroviruses evade host defenses by introducing and propagating mutational variations (Hu and Hughes 2012). In contrast, group II intron RTs function in retrohoming, which requires reverse transcription of a long, highly structured intron RNA with high fidelity and processivity (Cousineau et al. 1998; Conlan et al. 2005), properties that are desirable for biotechnological applications, particularly RNA-seq (Mohr et al. 2013b). Group II intron RTs also have a robust end-to-end template-switching activity, which enables RNA-seq adapter addition to target RNAs without RNA tailing or ligation, inefficient steps in many RNA-seq protocols (Mohr et al. 2013b). The development of new general methods for the bacterial expression of group II intron RTs with high yield and activity in a form that remains soluble when removed from the intron RNA have enabled their large-scale commercial production and use for biotechnological applications (Mohr et al. 2013b).

Thermostable group II intron RTs (TGIRTs) from bacterial thermophiles, which can carry out cDNA synthesis at high temperatures (≥60°C) that help melt out stable RNA secondary and tertiary structures, were the first to be adapted for biotechnological applications. Initial work focused on two TGIRTs, TeI4c RT from the thermophilic cyanobacterium *Thermosynechococcus elongatus* and GsI-IIC RT from the thermophilic soil bacterium *Geobacillus stearothermophilus* (Mohr et al. 2013b). These TGIRTs were found to have two- to fourfold higher fidelity in vitro than a retroviral RT assayed in parallel, as well as very high processivity, enabling them to continuously copy long and/or structured RNA templates without falling off (macroscopic processivity values defined as the average length of cDNA synthesized in the presence of a trap, 714 and 708 nt for TeI4c and GsI-IIC RT, respectively, fourfold higher than for SuperScript III, a widely used thermostable retrovirus-derived RT [Mohr et al. 2013b]). Nonthermostable group II intron RTs also have high-fidelity and processivity (macroscopic processivity value 616 nt [Zhao et al. 2018]), and it is likely that other group II intron RTs yet to be sampled will have similar properties.

A 3.0-Å X-ray crystal structure of a full-length TGIRT (GsI-IIC RT, a form of which is sold commercially as TGIRT-III) in complex with template-primer substrate and incoming dNTP has provided insight into structural features that may contribute to the beneficial enzymatic properties of group II intron RTs (Fig. 5) (Stamos et al. 2017). Thus, the group II intron RT-specific insertion regions NTE/RT0 and RT2a, which are lacking in retroviral RTs, contribute multiple additional interactions with the template-primer likely accounting for increased processivity. Additionally, the group II intron RT active site has more constrained binding pockets than in retroviral RTs for the templating base, the 3′ end of the primer, and incoming dNTP, potentially increasing fidelity. The RT3a insertion (also known as the insertion-in-fingers domain, IFD) contains a loop (also known as the α-loop), which occupied the RT active site in the misfolded RT fragment structures (Zhao and Pyle 2016; Zhao et al. 2018). Although also suggested to contribute to high processivity, this loop occupies a different position in the structure of the full-length GsI-IIC RT and does not contact the bound template-primer as would be expected for a direct role in processivity (Stamos et al. 2017). Notably, many of the distinctive structural features of group II intron RTs, including homologs of the NTE, RT2a, and RT3a and details of the RT active site, are also found in viral RNA-dependent RNA polymerases (RdRPs), which are potential evolutionary ancestors of RTs. In contrast, retroviral RTs, which are descended from these ancestral proteins, appear to have lost entire regions and active-site features that contribute to the high processivity and fidelity of group II intron RTs (Stamos et al. 2017). The crystal structure of a full-length group II intron RT in complex with template-primer substrate will enable detailed structure–function analysis and the engineering of group II intron RTs for improved performance in biotechnological applications.

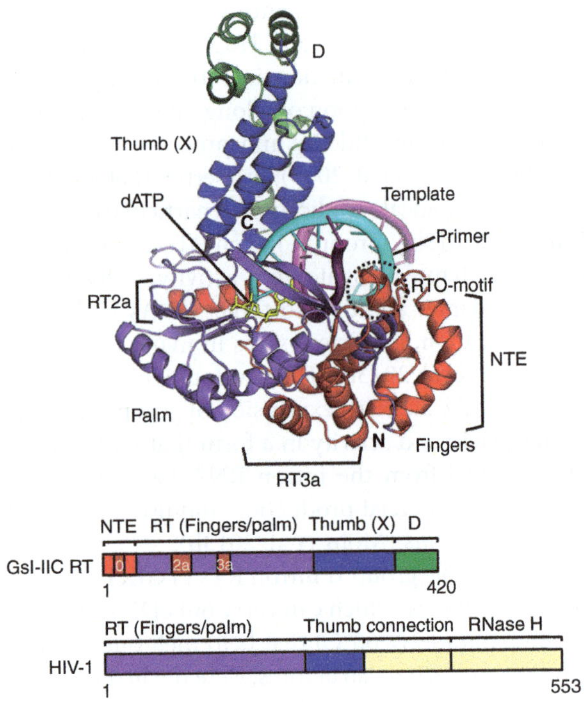

Figure 5. Crystal structure of the thermostable GsI-IIC reverse transcriptase (RT) bound to template-primer and incoming dNTP. The 3.0 Å structure shows full-length GsI-IIC RT bound to an 11-bp RNA template-DNA primer duplex with a single-stranded 5′ overhang (3 nt) on the RNA template strand and a dideoxynucleotide at the 3′ end of the DNA primer. The incoming dATP is bound at the RT active site poised for polymerization. The amino-terminal extension (NTE) and insert regions RT2a and RT3a not present in retroviral RTs are demarcated with brackets. N and C indicate the amino- and carboxyl-termini of the protein. The schematics below show the domain organization of GsI-IIC RT compared with that of HIV-1 RT, with regions color-coded as shown in the figure. The connection and RNase H domains of HIV-1 RT not present in group II intron RTs are shown in yellow. (Adapted from Stamos et al. 2017.)

The beneficial properties of TGIRT enzymes have been exploited in multiple ways for different applications, including three different methods for RNA-seq library construction (Fig. 6). In one method, cDNAs of human cell mRNAs were synthesized by initiating reverse transcription from an oligo(dT)$_{42}$ primer annealed to the 3′ poly(A) tail (Fig. 6A) (Mohr et al. 2013b). The cDNAs were then converted to dsDNAs by using a commercial kit, and the dsDNAs were fragmented by a transposon with simultaneous addition of RNA-seq adapters for Illumina sequencing. Because of the high processivity of TGIRTs, the resulting RNA-seq data sets showed dramatically more uniform 5′-to-3′ read coverage than those obtained in parallel with SuperScript III RT, which over-represented reads near the 3′ end of the mRNAs (Mohr et al. 2013b). Other applications in which TGIRTs have been used for reverse transcription from an annealed DNA primer include (1) synthesis of long ssDNAs from RNA templates for CRISPR genome editing (Li et al. 2017a); (2) reverse transcription and quantitation of GC-rich repeat expansions, which are clinically important in diseases such as myotonic dystrophy and some forms of amyotrophic lateral sclerosis (ALS) (Carrell et al. 2017); (3) RNA-structure mapping by DMS modification and SHAPE (Wu and Bartel 2017; Zubradt et al. 2017; Mohr et al. 2018); and (4) IR-CLIP for the identification of protein-bound RNAs, in which TGIRT gave four- to eightfold higher cDNA yields than SuperScript III (Zarnegar et al. 2016). TGIRT has also been used for in vitro

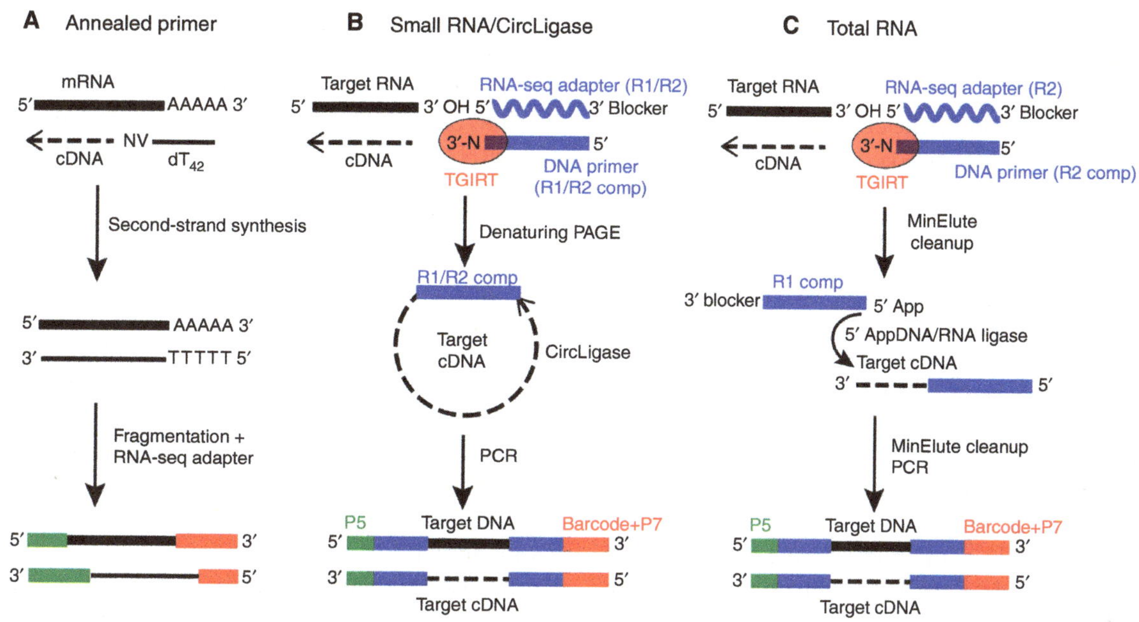

Figure 6. Three methods of using thermostable group II intron reverse transcriptase (TGIRT) enzymes for RNA-seq library construction (TGIRT-seq). (*A*) TGIRT-seq of eukaryotic mRNAs using an anchored oligo$(dT)_{42}$ primer for initiation of cDNA synthesis. The primer of sufficient length to stably anneal to mRNA poly(A) tails at high temperature is extended by TGIRTs at 60°C to synthesize cDNA copies of the mRNAs. After second-strand synthesis, dsDNAs are fragmented and RNA-seq adapters are added by conventional methods, such as the transposon-based Illumina Nextera method. (*B*) TGIRT-seq Small RNA/CircLigase method. The TGIRT initiates by template switching from an initial template-primer substrate comprised of an RNA oligonucleotide containing Illumina R1 and R2 adapter sequences annealed to a 5′-labeled DNA primer that leaves a single-nucleotide 3′ overhang (N, an equimolar mixture of A, C, G, and T) that can base-pair to the 3′ nucleotide of target RNAs. The single base-pair is sufficient to direct TGIRT template switching, even at 60°C (Mohr et al. 2013b; Qin et al. 2016). After reverse transcription, cDNAs are purified on a denaturing polyacrylamide gel to select specific size classes, circularized with CircLigase (Epicentre), and minimally polymerase chain reaction (PCR) amplified with primers that add capture sites and barcodes for Illumina sequencing. (*C*) TGIRT Total RNA-seq method for construction of comprehensive RNA-seq libraries without size selection. The TGIRT initiates from an initial template-primer substrate similar to that in panel *B* but containing only an Illumina R2 adapter sequence. After reverse transcription, cDNAs are cleaned up by using a MinElute column to remove unincorporated adapters and a second oligonucleotide containing the reverse complement of an Illumina R1 adapter is ligated to the 3′ end of the cDNA using thermostable 5′ AppDNA/RNA ligase (New England Biolabs). After an additional MinElute cleanup, the cDNAs with adapters on both ends are amplified by PCR with primers that add Illumina capture sites and barcodes. Before sequencing, the libraries are cleaned up using Ampure beads to remove adapter dimers (not shown).

evolution of the binding specificity of structured RNA aptamers, in which it was able to maintain the stable RNA scaffold structure through multiple rounds of selection, in contrast to retroviral RTs whose low fidelity and inability to reverse transcribe through structured RNAs resulted in the accumulation of mutations that destabilized the RNA scaffold structure, thereby compromising the selections (Porter et al. 2017).

The template switching activity of TGIRTs, which enables facile attachment of RNA-seq adapters to target RNA sequences (Mohr et al. 2013b), has been used to develop two new methods of RNA-seq library construction for Illumina sequencing, the Small RNA/CircLigase method and the Total RNA-seq method (Fig. 6B and C, respectively). In both methods, the TGIRT starts from a synthetic RNA template-DNA primer substrate that contains an RNA-seq adapter sequence and has a single nucleotide 3′ DNA overhang that directs template switching by base-pairing to the 3′ nucleotide of the target RNA (Mohr et al. 2013b). For RNA-seq library construction from a pool of RNAs in which minimal bias is required, the template-primer substrate is added in excess to the target RNA and the 3′ DNA overhang is an equimolar mix of A, C, G, and T (denoted N) that can base-pair to target RNAs in the pool having different 3′ termini. After template switching to the 3′ end of the target RNA, the TGIRT synthesizes a full-length cDNA in which the RNA-seq adapter is seamlessly linked to the target RNA sequence (Mohr et al. 2013b).

In the TGIRT Small RNA/CircLigase method, the initial RNA template/DNA primer substrate contains both Illumina Read 1 and Read 2 (R1 and R2) adapter sequences, and after template switching, the cDNAs with the attached adapter are size-selected on a gel and circularized with CircLigase (Epicentre) followed by PCR with primers that add Illumina barcodes and capture sites for sequencing (Fig. 6B). This variation of the method has been used for miRNA profiling (Mohr et al. 2013b), RIP-seq of protein-bound RNAs (Katibah et al. 2014), and quantitative tRNA sequencing (Katibah et al. 2014; Shen et al. 2015; Zheng et al. 2015). Recently, a further adaptation of the method was developed to measure tRNA aminoacylation levels by reverse transcription with TGIRT before and after removal of the attached amino acid, whose presence blocks template switching (Evans et al. 2017).

The second TGIRT-seq method, referred to as the Total RNA-seq method, enables the construction of comprehensive RNA-seq libraries from pools of RNAs without size selection (Fig. 6C). In this variation, the initial template/primer substrate used for template switching contains only an Illumina R2 sequence, and after cDNA synthesis a second adapter containing the reverse complement of an Illumina R1 sequence is ligated to the 3′ end of the cDNA by an efficient single-stranded DNA ligation (Nottingham et al. 2016; Qin et al. 2016). The resulting cDNAs, with different sizes, are then PCR amplified with primers that anneal to the RNA-seq adapters and add barcodes and capture sites for sequencing. Validation of the method using human reference RNA samples with External RNA Control Consortium (ERCC) spike-ins compared with the widely used strand-specific TruSeq v3 method for transcriptome profiling showed better quantitation of the relative abundance of cellular mRNAs and ERCC spike-ins, more uniform 5′-to-3′ coverage, even for fragmented RNAs, and larger numbers of detected SJs, particularly near the 5′ ends of mRNAs (Nottingham et al. 2016; Qin et al. 2016). Additionally, the TGIRT-seq method had higher strand specificity than TruSeq v3 and eliminated biases because of random hexamer primers, which are inherent in TruSeq. Finally, a major advantage of the Total RNA-seq method is that it enables simultaneous quantitation of mRNAs and structured small noncoding RNAs (ncRNAs) in the same RNA-seq instead of requiring separate preparations for long and short RNAs (Nottingham et al. 2016; Boivin et al. 2018).

In both the Small RNA/CircLigase and Total RNA-seq methods, the high processivity and strand displacement activities of TGIRTs combined with their ability to initiate directly at the 3′ nucleotide of RNA templates by template switching provide the ability to obtain full-length reads of tRNAs and other structured small ncRNAs (Katibah et al. 2014; Shen et al. 2015), which is not possible for retroviral RTs. This was first shown in Katibah et al. (2014) for RIP-seq of RNAs bound to the human interferon–induced protein IFIT5, which were found to include full-length mature and pre-tRNAs beginning at the PolIII initiation site. The method enabled detection of the 3′ CCA at the end of mature tRNAs, as well as poly(U) tails at the 3′ end of RNAs targeted for degradation. Further, the ability to reverse transcribe through tRNA structure is critical for resolving full-length tRNAs from tRNA fragments, which have biological and medical importance, but are difficult to distinguish from strong stops because of RNA secondary structures that impede conventional RTs (Qin et al. 2016). Additionally, in contrast to retroviral RTs, which tend to dissociate at RNA base modifications that affect base-pairing, TGIRTs pause at such modifications but eventually read through by misincorporation, with the spectrum of misincorporated nucleotides being distinctive for different modifications (Katibah et al. 2014). This ability of TGIRTs to read through posttranscriptional modifications that affect base-pairing by misincorporation has been used for high-throughput mapping of posttranscriptional modifications at single-nucleotide resolution in both tRNAs and mRNAs (Clark et al. 2016; Li et al. 2017b; Safra et al. 2017), as well as DMS-induced A and C modifications for RNA structure mapping in DMS-MaPseq (Wu and Bartel 2017; Zubradt et al. 2017). Because these methods map mutations rather than reverse transcription stops they can detect multiple modifications in the same RNA molecule.

The ability of the TGIRT template-switching activity to enable construction of comprehensive RNA-seq libraries from small amounts of starting material has been used for the analysis of extracellular RNAs present in human plasma or packaged in secreted membrane vesicles termed exosomes (Qin et al. 2016; Shurtleff et al. 2017). In both cases, TGIRTs revealed the presence of full-length tRNAs and other structured small ncRNAs, which could not be detected previously by retroviral RTs. Although highly purified exosomes contained only small amounts of mRNAs, the more complex plasma RNAs contain RNA fragments that map to large numbers of protein-coding and long ncRNAs (lncRNAs). These findings suggest that TGIRTs may provide advantages for liquid biopsy, including the ability to identify mRNA, lncRNA, and small ncRNA biomarkers in the same RNA-seq reaction and to capture RNA species that are refractory to retroviral RTs.

Finally, in another liquid biopsy application, a method has been developed for using TGIRT-template-switching activity for ssDNA-seq of human plasma DNAs, which are released into blood by apoptosis of blood lymphoid and myeloid cells or by necrosis of tumors and damaged tissues in cancer and other diseases (Wu and Lambowitz 2017). This TGIRT-ssDNA-seq method uses a workflow similar to the TGIRT-seq Total RNA method, but is optimized for copying of DNA templates, which under appropriate conditions TGIRTs can do robustly and with error rates after corrections approaching those for a conventional DNA-seq method (see Wu and Lambowitz 2017). In addition to a more streamlined workflow than other ssDNA-seq methods, by avoiding tailing steps, this method facilitates precise mapping of DNA ends for analysis of nucleosome positioning and bisulfite mapping of DNA methylation sites. The analysis of nucleosome positioning and mapping of transcription factor-binding and DNA methylation sites of human plasma DNAs can be used to identify their tissues of origin and to follow disease progression and response to treatment (Sun et al. 2015; Snyder et al. 2016; Wu and Lambowitz 2017).

5.1 Biological Insights and Conclusions from Technique 4

The novel properties of TGIRTs expand and improve RNA-seq in ways that are difficult or impossible for retroviral RTs and have provided a more complete picture of the RNA composition of human plasma, exosomes, and cells (Qin et al. 2016; Shurtleff et al. 2017; Boivin et al. 2018). The use of TGIRT for quantitative tRNA-seq of a yeast ribosome quality control complex identified $tRNA^{Ala}$ and $tRNA^{Thr}$ isoacceptor species that are recruited by a protein component to stalled ribosomes, in which they function in conjunction with free 60S subunits to promote non-mRNA-mediated addition of carboxy-terminal Ala and Thr extensions (CAT tails) helping to target nascent polypeptides for degradation (Shen et al. 2015). The method also contributed to establishing that codon optimality dictated by tRNA abundance regulates mRNA stability and translation during the maternal-to-zygote transition in zebrafish and likely across vertebrates (Bazzini et al. 2016). The ability of TGIRTs to read through RNA posttranscriptional modifications by misincorporation enabled high-throughput mapping and functional analysis of m1A base methylations in mammalian tRNAs and mRNAs, establishing roles for this modification in translational and developmental regulation (Clark et al. 2016; Li et al. 2017b; Safra et al. 2017). Likewise, the new methods for in vivo RNA structure mapping based on the ability of TGIRTs to read through and mutationally map DMS modifications enabled the first analysis of RNA structure within an animal tissue and revealed the widespread influence of 3′ end structures that juxtapose poly(A) signals and 3′ cleavage sites on mammalian mRNA processing and stability (Wu and Bartel 2017; Zubradt et al. 2017).

6 CONCLUDING REMARKS

Fundamental studies into group II intron structure and function have provided insight into the evolution of retroelements and formed the basis for applications in gene targeting and editing, as well as RT-based high-throughput RNA analysis. The group II intron RNP cryo-EM structure provided a number of "firsts," including an impression of group II RNP architecture at 3.8–4.5 Å resolution and a global view of similarities to the spliceosome. The RIG-seq methodology showed integration at spurious nicks in DNA, providing novel insights into intron spread. Targetrons, the first RNA-guided gene targeting system, have enabled genetic analysis and genome engineering in a wide variety of bacteria. Finally, group II intron RTs have enabled the development of new methods for RNA-seq, with unprecedented capabilities for comprehensive transcriptome profiling, RNA structure mapping, analysis of tRNAs and other small ncRNAs, mapping of posttranscriptional modifications, and potentially liquid biopsy.

COMPETING INTEREST STATEMENT

Targetrons and TGIRT enzymes and methods for their use are the subject of patents and patent applications that have been exclusively licensed by the Ohio State University, the University of Texas, and East Tennessee State University to InGex, LLC. A.M.L., M.B., the Ohio State University, and the University of Texas are minority equity holders in InGex, and A.M.L., the Ohio State University, the University of Texas, and some present and former Lambowitz laboratory members receive royalty payments from the sale of targetrons, TGIRT enzymes, and kits and the sublicensing of the intellectual property by InGex to other companies.

ACKNOWLEDGMENTS

We thank Matt Stanger, Olga Novikova, Eren (Xiaolong) Dong, Rebecca McCarthy, and Jennifer Stamos for help with the figures and manuscript and members of the Belfort and Lambowitz laboratories for their comments on the manuscript. Work in our laboratories is supported by National Institutes of Health (NIH) Grants GM39422 and GM44844 to M.B. and NIH Grant GM37949 and Welch Foundation Grant F-1607 to A.M.L.

REFERENCES

Agrawal RK, Wang HW, Belfort M. 2016. Forks in the tracks: Group II introns, spliceosomes, telomeres and beyond. *RNA Biol* **13:** 1218–1222.

Bazzini AA, Del Viso F, Moreno-Mateos MA, Johnstone TG, Vejnar CE, Qin Y, Yao J, Khokha MK, Giraldez AJ. 2016. Codon identity regulates mRNA stability and translation efficiency during the maternal-to-zygotic transition. *EMBO J* **35:** 2087–2103.

Belhocine K, Yam KK, Cousineau B. 2005. Conjugative transfer of the *Lactococcus lactis* chromosomal sex factor promotes dissemination of the Ll.LtrB group II intron. *J Bacteriol* **187:** 930–939.

Boivin V, Deschamps-Francoeur G, Couture S, Nottingham RM, Bouchard-Bourelle P, Lambowitz AM, Scott MS, Abou-Elela S. 2018. Simultaneous sequencing of coding and non-coding RNAs reveals a human transcriptome dominated by a small number of highly expressed non-coding genes. *RNA* doi: 10.1261/rna.064493.117.

Buchan BW, McCaffrey RL, Lindemann SR, Allen LA, Jones BD. 2009. Identification of migR, a regulatory element of the *Francisella tularensis* live vaccine strain iglABCD virulence operon required for normal replication and trafficking in macrophages. *Infect Immun* **77:** 2517–2529.

Carrell ST, Tang Z, Mohr S, Lambowitz AM, Thornton CA. 2017. Detection of expanded RNA repeats using thermostable group II intron reverse transcriptase. *Nucleic Acids Res* **46:** e1.

Clark WC, Evans ME, Dominissini D, Zheng G, Pan T. 2016. tRNA base methylation identification and quantification via high-throughput sequencing. *RNA* **22:** 1771–1784.

Cole AL, Muthukrishnan G, Chong C, Beavis A, Eade CR, Wood MP, Deichen MG, Cole AM. 2016. Host innate inflammatory factors and staphylococcal protein A influence the duration of human *Staphylococcus aureus* nasal carriage. *Mucosal Immunol* **9:** 1537–1548.

Conlan LH, Stanger MJ, Ichiyanagi K, Belfort M. 2005. Localization, mobility and fidelity of retrotransposed group II introns in rRNA genes. *Nucleic Acids Res* **33:** 5262–5270.

Coros CJ, Landthaler M, Piazza CL, Beauregard A, Esposito D, Perutka J, Lambowitz AM, Belfort M. 2005. Retrotransposition strategies of the *Lactococcus lactis* Ll.LtrB group II intron are dictated by host identity and cellular environment. *Mol Microbiol* **56:** 509–524.

Coros CJ, Piazza CL, Chalamcharla VR, Belfort M. 2008. A mutant screen reveals RNase E as a silencer of group II intron retromobility in *Escherichia coli*. *RNA* **14:** 2634–2644.

Coros CJ, Piazza CL, Chalamcharla VR, Smith D, Belfort M. 2009. Global regulators orchestrate group II intron retromobility. *Mol Cell* **34:** 250–256.

Costa M, Walbott H, Monachello D, Westhof E, Michel F. 2016. Crystal structures of a group II intron lariat primed for reverse splicing. *Science* **354.**

Cousineau B, Smith D, Lawrence-Cavanagh S, Mueller JE, Yang J, Mills D, Manias D, Dunny G, Lambowitz AM, Belfort M. 1998. Retrohoming of a bacterial group II intron: Mobility via complete reverse splicing, independent of homologous DNA recombination. *Cell* **94:** 451–462.

Cui X, Matsuura M, Wang Q, Ma H, Lambowitz AM. 2004. A group II intron-encoded maturase functions preferentially *in cis* and requires both the reverse transcriptase and X domains to promote RNA splicing. *J Mol Biol* **340:** 211–231.

Curcio MJ, Garfinkel DJ. 1991. Single-step selection for Ty1 element retrotransposition. *Proc Natl Acad Sci* **88:** 936–940.

Dai Z, Dong H, Zhang Y, Li Y. 2016. Elucidating the contributions of multiple aldehyde/alcohol dehydrogenases to butanol and ethanol production in *Clostridium acetobutylicum*. *Sci Rep* **6:** 28189.

Dong X, Ranganathan S, Qu G, Piazza CL, Belfort M. 2018. Structural accommodations accompanying splicing of a group II intron RNP. *Nucleic Acids Res* doi: 10.1093/nar/gky416.

Donnelly ML, Li W, Li YQ, Hinkel L, Setlow P, Shen A. 2017. A *Clostridium difficile*-specific, gel-forming protein required for optimal spore germination. *mBio* **8:** e02085.

Enyeart PJ, Chirieleison SM, Dao MN, Perutka J, Quandt EM, Yao J, Whitt JT, Keatinge-Clay AT, Lambowitz AM, Ellington AD. 2013. Generalized bacterial genome editing using mobile group II introns and Cre-*lox*. *Mol Syst Biol* **9:** 685.

Enyeart PJ, Mohr G, Ellington AD, Lambowitz AM. 2014. Biotechnological applications of mobile group II introns and their reverse transcriptases: Gene targeting, RNA-seq, and non-coding RNA analysis. *Mob DNA* **5:** 2.

Evans ME, Clark WC, Zheng G, Pan T. 2017. Determination of tRNA aminoacylation levels by high-throughput sequencing. *Nucleic Acids Res* **45:** e133.

Fica SM, Nagai K. 2017. Cryo-electron microscopy snapshots of the spliceosome: Structural insights into a dynamic ribonucleoprotein machine. *Nat Struct Mol Biol* **24:** 791–799.

Frazier CL, San Filippo J, Lambowitz AM, Mills DA. 2003. Genetic manipulation of *Lactococcus lactis* by using targeted group II introns: Generation of stable insertions without selection. *Appl Environ Microbiol* **69:** 1121–1128.

Galej WP, Toor N, Newman AJ, Nagai K. 2018. Molecular mechanism and evolution of nuclear pre-mRNA and group II intron splicing: Insights from cryo-electron microscopy structures. *Chem Rev* **118:** 4156–4176.

Garcia-Rodriguez FM, Hernandez-Gutierrez T, Diaz-Prado V, Toro N. 2014. Use of the computer-retargeted group II intron RmInt1 of *Sinorhizobium meliloti* for gene targeting. *RNA Biol* **11:** 391–401.

Gu SQ, Cui X, Mou S, Mohr S, Yao J, Lambowitz AM. 2010. Genetic identification of potential RNA-binding regions in a group II intron-encoded reverse transcriptase. *RNA* **16:** 732–747.

Guo H, Karberg M, Long M, Jones JP III, Sullenger B, Lambowitz AM. 2000. Group II introns designated to insert into therapeutically-relevant DNA target sites in human cells. *Science* **289:** 452–457.

Gupta K, Contreras LM, Smith D, Qu G, Huang T, Spruce LA, Seeholzer SH, Belfort M, Van Duyne GD. 2014. Quaternary arrangement of an active, native group II intron ribonucleoprotein complex revealed by small-angle X-ray scattering. *Nucleic Acids Res* **42:** 5347–5360.

Harper M, John M, Edmunds M, Wright A, Ford M, Turni C, Blackall PJ, Cox A, Adler B, Boyce JD. 2016. Protective efficacy afforded by live *Pasteurella multocida* vaccines in chickens is independent of lipopolysaccharide outer core structure. *Vaccine* **34:** 1696–1703.

Herrera A, Vu BG, Stach CS, Merriman JA, Horswill AR, Salgado-Pabon W, Schlievert PM. 2016. *Staphylococcus aureus* β-toxin mutants are defective in biofilm ligase and sphingomyelinase activity, and causation of infective endocarditis and sepsis. *Biochemistry* **55:** 2510–2517.

Hong W, Zhang J, Feng Y, Mohr G, Lambowitz AM, Cui GZ, Liu YJ, Cui Q. 2014. The contribution of cellulosomal scaffoldins to cellulose hydrolysis by *Clostridium thermocellum* analyzed by using thermotargetrons. *Biotechnol Biofuels* **7:** 80.

Hu WS, Hughes SH. 2012. HIV-1 reverse transcription. *Cold Spring Harbor Perspect Med* **2:** a006882.

Ichiyanagi K, Beauregard A, Lawrence S, Smith D, Cousineau B, Belfort M. 2002. Retrotransposition of the Ll.LtrB group II intron proceeds predominantly via reverse splicing into DNA targets. *Mol Microbiol* **46:** 1259–1272.

Ichiyanagi K, Beauregard A, Belfort M. 2003. A bacterial group II intron favors retrotransposition into plasmid targets. *Proc Natl Acad Sci* **100:** 15742–15747.

Jiang J, Chan H, Cash DD, Miracco EJ, Ogorzalek Loo RR, Upton HE, Cascio D, O'Brien Johnson R, Collins K, Loo JA, et al. 2015. Structure of *Tetrahymena* telomerase reveals previously unknown subunits, functions, and interactions. *Science* **350:** aab4070.

Johnson CM, Fisher DJ. 2013. Site-specific, insertional inactivation of incA in *Chlamydia trachomatis* using a group II intron. *PLoS ONE* **8:** e83989.

Karberg M, Guo H, Zhong J, Coon R, Perutka J, Lambowitz AM. 2001. Group II introns as controllable gene targeting vectors for the genetic manipulation of bacteria. *Nature Biotechnol* **19:** 1162–1167.

Katibah GE, Qin Y, Sidote DJ, Yao J, Lambowitz AM, Collins K. 2014. Broad and adaptable RNA structure recognition by the human interferon-induced tetratricopeptide repeat protein IFIT5. *Proc Natl Acad Sci* **111:** 12025–12030.

Kint N, Janoir C, Monot M, Hoys S, Soutourina O, Dupuy B, Martin-Verstraete I. 2017. The alternative sigma factor σ^B plays a crucial role in adaptive strategies of *Clostridium difficile* during gut infection. *Environ Microbiol* **19:** 1933–1958.

Lambowitz AM, Belfort M. 2015. Mobile bacterial group II introns at the crux of eukaryotic evolution. *Microbio Spectr* **3:** MDNA3-0050-2014.

Lambowitz AM, Zimmerly S. 2011. Group II introns: Mobile ribozymes that invade DNA. *Cold Spring Harbor Perspect Biol* **3:** a003616.

Li H, Beckman KA, Pessino V, Huange B, Weissman JS, Leonetti MD. 2017a. Design and specificity of long ssDNA donors for CRISPR-based knock-in. *BioRxiv* doi: 10.1101/178905.

Li X, Xiong X, Zhang M, Wang K, Chen Y, Zhou J, Mao Y, Lv J, Yi D, Chen XW, et al. 2017b. Base-resolution mapping reveals distinct m^1A methylome in nuclear- and mitochondrial-encoded transcripts. *Mol Cell* **68:** 993–1005.e9.

Marcia M, Pyle AM. 2012. Visualizing group II intron catalysis through the stages of splicing. *Cell* **151:** 497–507.

Mastroianni M, Watanabe K, White TB, Zhuang F, Vernon J, Matsuura M, Wallingford J, Lambowitz AM. 2008. Group II intron-based gene targeting reactions in eukaryotes. *PLoS ONE* **3:** e3121.

Matsuura M, Noah JW, Lambowitz AM. 2001. Mechanism of maturase-promoted group II intron splicing. *EMBO J* **20:** 7259–7270.

Mohr G, Smith D, Belfort M, Lambowitz AM. 2000. Rules for DNA target site recognition by a *Lactococcal* group II intron enable retargeting of the intron to specific DNA sequences. *Genes Dev* **14:** 559–573.

Mohr G, Hong W, Zhang J, Cui GZ, Yang Y, Cui Q, Liu YJ, Lambowitz AM. 2013a. A targetron system for gene targeting in thermophiles and its application in *Clostridium thermocellum*. *PLoS ONE* **8:** e69032.

Mohr S, Ghanem E, Smith W, Sheeter D, Qin Y, King O, Polioudakis D, Iyer VR, Hunicke-Smith S, Swamy S, et al. 2013b. Thermostable group II intron reverse transcriptase fusion proteins and their use in cDNA synthesis and next-generation RNA sequencing. *RNA* **19:** 958–970.

Mohr G, Kang SYS, Park SK, Qin Y, Grohman J, Yao J, Stamos JL, Lambowitz AM. 2018. A highly proliferative group IIC intron from *Geobacillus stearothermophilus* reveals new features of group II intron mobility and RNA splicing. *J Mol Biol* (in press).

Nguyen THD, Tam J, Wu RA, Greber BJ, Toso D, Nogales E, Collins K. 2018. Cryo-EM structure of substrate-bound human telomerase holoenzyme. *Nature* **557:** 190–195.

Nottingham RM, Wu DC, Qin Y, Yao J, Hunicke-Smith S, Lambowitz AM. 2016. RNA-seq of human reference RNA samples using a thermostable group II intron reverse transcriptase. *RNA* **22:** 597–613.

Novikova O, Belfort M. 2017. Mobile group II introns as ancestral eukaryotic elements. *Trends Genet* **33:** 773–783.

Novikova O, Smith D, Hahn I, Beauregard A, Belfort M. 2014. Interaction between conjugative and retrotransposable elements in horizontal gene transfer. *PLoS Genet* **10:** e1004853.

Perutka J, Wang W, Goerlitz D, Lambowitz AM. 2004. Use of computer-designed group II introns to disrupt *Escherichia coli* DExH/D-box protein and DNA helicase genes. *J Mol Biol* **336:** 421–439.

Piccirilli JA, Staley JP. 2016. Reverse transcriptases lend a hand in splicing catalysis. *Nat Struct Mol Biol* **23:** 507–509.

Porter EB, Polaski JT, Morck MM, Batey RT. 2017. Recurrent RNA motifs as scaffolds for genetically encodable small-molecule biosensors. *Nat Chem Biol* **13:** 295–301.

Qin Y, Yao J, Wu DC, Nottingham RM, Mohr S, Hunicke-Smith S, Lambowitz AM. 2016. High-throughput sequencing of human plasma RNA by using thermostable group II intron reverse transcriptases. *RNA* **22:** 111–128.

Qu G, Kaushal PS, Wang J, Shigematsu H, Piazza CL, Agrawal RK, Belfort M, Wang HW. 2016. Structure of a group II intron in complex with its reverse transcriptase. *Nat Struct Mol Biol* **23:** 549–559.

Rambo RP, Doudna JA. 2004. Assembly of an active group II intron–maturase complex by protein dimerization. *Biochemistry* **43:** 6486–6497.

Robart AR, Chan RT, Peters JK, Rajashankar KR, Toor N. 2014. Crystal structure of a eukaryotic group II intron lariat. *Nature* **514:** 193–197.

Safra M, Sas-Chen A, Nir R, Winkler R, Nachshon A, Bar-Yaacov D, Erlacher M, Rossmanith W, Stern-Ginossar N, Schwartz S. 2017. The m^1A landscape on cytosolic and mitochondrial mRNA at single-base resolution. *Nature* **551:** 251–255.

Saldanha R, Chen B, Wank H, Matsuura M, Edwards J, Lambowitz AM. 1999. RNA and protein catalysis in group II intron splicing and mobility reactions using purified components. *Biochemistry* **38:** 9069–9083.

Shen PS, Park J, Qin Y, Li X, Parsawar K, Larson MH, Cox J, Cheng Y, Lambowitz AM, Weissman JS, et al. 2015. Protein synthesis. Rqc2p and 60S ribosomal subunits mediate mRNA-independent elongation of nascent chains. *Science* **347:** 75–78.

Shi Y. 2017. The spliceosome: A protein-directed metalloribozyme. *J Mol Biol* **429:** 2640–2653.

Shurtleff MJ, Yao J, Qin Y, Nottingham RM, Temoche-Diaz MM, Schekman R, Lambowitz AM. 2017. Broad role for YBX1 in defining the small noncoding RNA composition of exosomes. *Proc Natl Acad Sci* **114:** E8987–E8995.

Singh NN, Lambowitz AM. 2001. Interaction of a group II intron ribonucleoprotein endonuclease with its DNA target site investigated by DNA footprinting and modification interference. *J Mol Biol* **309:** 361–386.

Snyder MW, Kircher M, Hill AJ, Daza RM, Shendure J. 2016. Cell-free DNA comprises an in vivo nucleosome footprint that informs its tissues-of-origin. *Cell* **164:** 57–68.

Stamos JL, Lentzsch AM, Lambowitz AM. 2017. Structure of a thermostable group II intron reverse transcriptase with template-primer and its functional and evolutionary implications. *Mol Cell* **68:** 926–939.

Sun K, Jiang P, Chan KC, Wong J, Cheng YK, Liang RH, Chan WK, Ma ES, Chan SL, Cheng SH, et al. 2015. Plasma DNA tissue mapping by genome-wide methylation sequencing for noninvasive prenatal, cancer, and transplantation assessments. *Proc Natl Acad Sci* **112:** E5503–E5512.

Toor N, Keating KS, Taylor SD, Pyle AM. 2008. Crystal structure of a self-spliced group II intron. *Science* **320:** 77–82.

Truong DM, Hewitt FC, Hanson JH, Cui X, Lambowitz AM. 2015. Retrohoming of a mobile group II intron in human cells suggests how eukaryotes limit group II intron proliferation. *PLoS Genet* **11:** e1005422.

Wu X, Bartel DP. 2017. Widespread influence of 3′-end structures on mammalian mRNA processing and stability. *Cell* **169:** 905–917. e11.

Wu DC, Lambowitz AM. 2017. Facile single-stranded DNA sequencing of human plasma DNA via thermostable group II intron reverse transcriptase template switching. *Sci Rep* **7:** 8421.

Xing J, Witherspoon DJ, Jorde LB. 2013. Mobile element biology: New possibilities with high-throughput sequencing. *Trends Genet* **29:** 280–289.

Xue Q, Yang Y, Chen J, Chen L, Yang S, Jiang W, Gu Y. 2016. Roles of three AbrBs in regulating two-phase *Clostridium acetobutylicum* fermentation. *Appl Microbiol Biotechnol* **100:** 9081–9089.

Yan C, Hang J, Wan R, Huang M, Wong CC, Shi Y. 2015. Structure of a yeast spliceosome at 3.6-angstrom resolution. *Science* **349:** 1182–1191.

Yao J, Lambowitz AM. 2007. Gene targeting in gram-negative bacteria by use of a mobile group II intron ("Targetron") expressed from a broad-host-range vector. *Appl Environ Microbiol* **73:** 2735–2743.

Yao J, Zhong J, Fang Y, Geisinger E, Novick RP, Lambowitz AM. 2006. Use of targetrons to disrupt essential and nonessential genes in *Staphylococcus aureus* reveals temperature sensitivity of Ll.LtrB group II intron splicing. *RNA* **12:** 1271–1281.

Zarnegar BJ, Flynn RA, Shen Y, Do BT, Chang HY, Khavari PA. 2016. irCLIP platform for efficient characterization of protein-RNA interactions. *Nat Methods* **13:** 489–492.

Zhao C, Pyle AM. 2016. Crystal structures of a group II intron maturase reveal a missing link in spliceosome evolution. *Nat Struct Mol Biol* **23:** 558–565.

Zhao C, Liu F, Pyle AM. 2018. An ultraprocessive, accurate reverse transcriptase encoded by a metazoan group II intron. *RNA* **24:** 183–195.

Zheng G, Qin Y, Clark WC, Dai Q, Yi C, He C, Lambowitz AM, Pan T. 2015. Efficient and quantitative high-throughput tRNA sequencing. *Nat Methods* **12:** 835–837.

Zhong J, Karberg M, Lambowitz AM. 2003. Targeted and random bacterial gene disruption using a group II intron (targetron) vector containing a retrotransposition-activated selectable marker. *Nucleic Acids Res* **31:** 1656–1664.

Zhuang F, Karberg M, Perutka J, Lambowitz AM. 2009. EcI5, a group IIB intron with high retrohoming frequency: DNA target site recognition and use in gene targeting. *RNA* **15:** 432–449.

Zoraghi R, Worrall L, See RH, Strangman W, Popplewell WL, Gong H, Samaai T, Swayze RD, Kaur S, Vuckovic M, et al. 2011. Methicillin-resistant *Staphylococcus aureus* (MRSA) pyruvate kinase as a target for bis-indole alkaloids with antibacterial activities. *J Biol Chem* **286:** 44716–44725.

Zubradt M, Gupta P, Persad S, Lambowitz AM, Weissman JS, Rouskin S. 2017. DMS-MaPseq for genome-wide or targeted RNA structure probing in vivo. *Nat Methods* **14:** 75–82.

Alternate RNA Structures

Marie Teng-Pei Wu and Victoria D'Souza

Department of Molecular and Cellular Biology, Harvard University, Cambridge, Massachusetts 02138

Correspondence: dsouza@mcb.harvard.edu

SUMMARY

RNA molecules fold into complex three-dimensional structures that sample alternate conformations ranging from minor differences in tertiary structure dynamics to major differences in secondary structure. This allows them to form entirely different substructures with each population potentially giving rise to a distinct biological outcome. The substructures can be partitioned along an existing energy landscape given a particular static cellular cue or can be shifted in response to dynamic cues such as ligand binding. We review a few key examples of RNA molecules that sample alternate conformations and how these are capitalized on for control of critical regulatory functions.

Outline

Cite as *Cold Spring Harb Perspect Biol* doi: 10.1101/cshperspect.a032425

1 INTRODUCTION

RNA plays an important role in a large number of cellular processes, and, in many cases, it is the RNA that drives function. RNA can fold into complex three-dimensional structures, enabling it to perform diverse biological functions beyond the simple transfer of genetic information. For example, in the case of mRNAs, these structural elements have been shown to regulate every aspect of the mRNA's voyage through the cell—from transcription, splicing, and nuclear export to trafficking, localization, translation, and stability. Even ribosomal RNA (rRNA), the most abundant form of RNA in cells, samples alternate structures to control translation (Zamir et al. 1969; McGinnis et al. 2015). Although we have for a long time appreciated the direct relationship between RNA structures and the functions they regulate, we are just starting to fathom the depth and extent of the regulatory information contained in RNA. Furthermore, we are just beginning to understand that, unlike proteins, an RNA with a particular sequence has the potential to form entirely different structures and can therefore be regulated to be partitioned into distinct subpopulations, with each population giving rise to a distinct biological outcome.

The potential for RNA to form energetically similar conformations comes from the chemical simplicity of RNA molecules, which are made up of combinations of nucleotides joined together by phosphodiester bonds. To begin with, there are only four nucleotides, making it possible to find alternate identical complementary sequences for a given stretch of RNA. Second, each nucleotide is capable of making hydrogen bonding pairs with each of the others (Leontis and Westhof 1998; Leontis and Westhof 2001), with some combinations rivaling the stability of Watson–Crick base pairs (Leontis et al. 2002; Sponer et al. 2005). Third, the canonical G-C and A-U Watson–Crick pairs are nearly isosteric (Westhof 2014), making base pairs in a stem interchangeable and allowing for multiple ways of forming RNA stems—the basic building blocks of RNA structures. Finally, a structure with a seemingly energetically suboptimal base-pairing pattern can still form because of stabilization by tertiary interactions, such as loops, bulges, helical junctions, and long-range interactions (Popenda et al. 2010; Petrov et al. 2013; Chojnowski et al. 2014; Parlea et al. 2016).

All of the above characteristics allow RNA molecules to sample alternate conformations that range from minor differences in tertiary structure dynamics to major differences in secondary structure. Accordingly, alternate RNA structures can be loosely grouped into categories based on the timescales at which the transitions between structures occur—for instance: structures that differ considerably in secondary structures occur at the millisecond to second timescales; structures that differ at the base pair or tertiary structure levels occur at the microsecond to second timescales; and structures that differ because of interhelical or loop dynamics occur at the picosecond to microsecond timescales (Mustoe et al. 2014). Nature has made use of all of these transitions to control regulatory networks, with new functions using alternate RNA structures being uncovered on a regular basis.

To coexist, alternate structures by necessity have to be nearly isoenergetic. Such structures can be a priori appropriately populated along an equilibrium given a particular static cellular environment, or the partitioning can be shifted in response to dynamic cues like ligand binding and temperature changes. Whereas the use of static cues allows for structural modulation within a given energetic landscape, the use of dynamic cues changes the fundamental landscape itself and allows access to previously inaccessible structures. Many examples have been found for regulation using dynamic cues, but to date only a few examples have been discovered for regulation using static cues. In this review, we will use a few key examples of interactions that occur primarily at the microsecond and higher timescales to highlight alternate RNA structures controlled by both static and dynamic cues, and how they can be used to fine tune regulatory functions.

2 ALTERNATE RNA STRUCTURES CONTROLLED BY STATIC CUES

Although it is intuitive that RNA elements can be partitioned into alternate structures in response to dynamic or changing cues, as will be discussed in more detail further on in this review, the partitioning of conformations based on a static environment is also possible and can be capitalized on for function. The simplest partitioning of alternate RNA structures is through equilibria set by fixed cellular parameters such as temperature, pH, thermodynamic stability, concentration of ions, etc. Furthermore, this partitioning can occur on every level of organization, from primary structures to tertiary structures to the binding of *trans* protein factors.

2.1 Primary Structure

A recent study by Kharytonchyk et al. on human immunodeficiency virus-1 (HIV-1) transcription shows the potential to use sequence heterogeneity during transcription initiation to leverage alternate RNA structures (Kharytonchyk et al. 2016). In this case, the subtle difference between starting the mRNA with one guanosine as opposed to two or three guanosines defines the biological fate of the mRNA.

Cite as *Cold Spring Harb Perspect Biol* doi: 10.1101/cshperspect.a032425

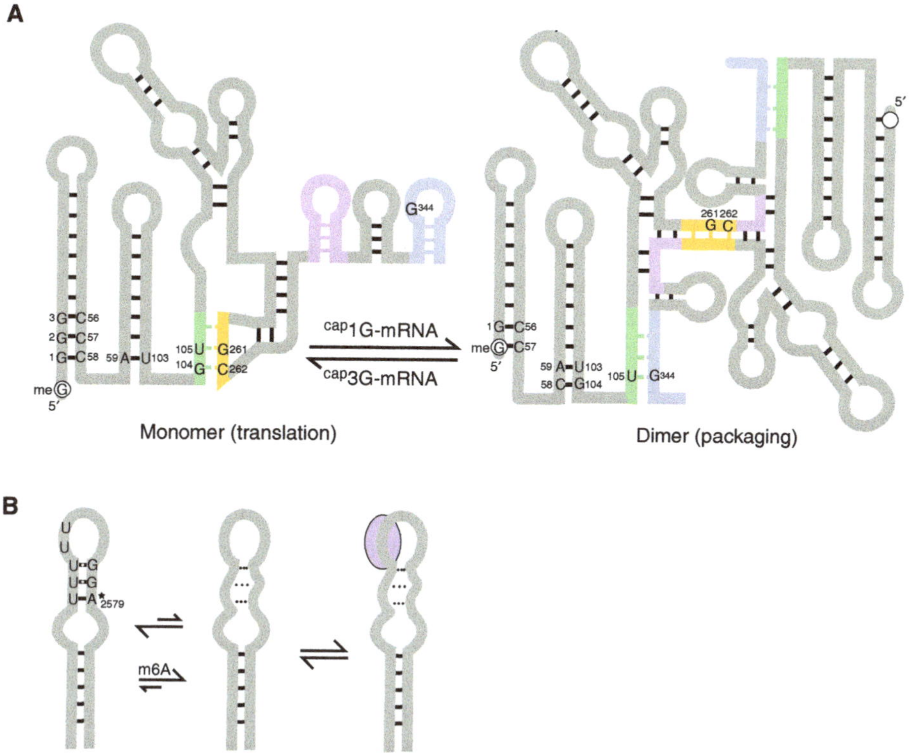

Figure 1. (*A*) The monomeric and dimeric states of the human immunodeficiency virus 1 (HIV-1) 5′ leader. Transcripts with ^{cap}2G or ^{cap}3G favor the monomeric form, whereas transcripts with ^{cap}1G favor the dimeric form, largely because of the dissolution or formation of the C58-G104 base pair, respectively. (*B*) Sampling of alternate conformations by the metastasis-associated lung adenocarcinoma transcript 1 long noncoding RNA (MALAT1 lncRNA). The RNA shows increased dynamics of the upper hairpin structure on N^6-methyladenosine (m^{6}A)-modification, as indicated by the dotted lines, allowing for increased sampling of the protein-binding conformation (lavender oval). The star indicates the location for the potential m^{6}A modification.

In brief, the full-length HIV mRNA plays a dual functional role: It is not only used as a transcript for the synthesis of essential Gag and Gag-Pol viral polyproteins, but also as the genetic material that gets packaged into virions during budding (D'Souza and Summers 2005). The full-length HIV mRNA also has two structural states: It not only exists in a monomeric state as per conventional mRNAs, but also in a dimeric state (Fig. 1A). It has been shown that the dual functions (translational vs. genomic) and the dual structures (monomer vs. dimer) of the viral RNA are correlated with each other. A region called the packaging signal (Ψ site) in the 5′ leader, containing nucleocapsid (NC, the genome packaging protein)-binding sites and a dimerization initiation site (DIS) (D'Souza and Summers 2005), is designed to dictate this correlation. Namely, the monomeric form has the DIS and NC-binding sites sequestered with the start codon exposed, making this mRNA conformation translation-competent. In contrast, the dimeric form exposes the NC-binding site and sequesters the start codon, making this mRNA conformation packaging-competent (Lu et al. 2011).

Kharytonchyk et al. found that the partitioning of the two conformations is defined by the variation in transcriptional start site usage (Kharytonchyk et al. 2016). Transcription of the HIV-1 genome can initiate at any of three guanosines at the transcription start site. The resulting transcripts will start with either ^{cap}1G, ^{cap}2G, or ^{cap}3G sequences immediately preceding the transactivation response (TAR) element in the 5′ leader. In the dimeric form, the terminal C58-G104 base pair of the poly(A) stem directly connects to both the TAR stem (via C58) and the DIS:U5 interaction stem (via G104) (Fig. 1A). On the other hand, in the monomeric form, G104 forms a base pair with C262. Kharytonchyk et al. showed that the equilibrium between the monomeric and dimeric forms can be manipulated simply by affecting the formation of the C58-G104 base pair. Transcripts with ^{cap}2G or ^{cap}3G at the start of the TAR stem can compete with the poly(A) stem for C58, thus freeing G104 to base-pair with C262. This extends the DIS:U5 interaction by one base pair and drives the equilibrium to the monomeric form. The extension of the DIS:U5 stem by this single additional base pair is estimated to stabilize the monomeric form by −3 to −10 kcal/mol compared with the dimeric form. Even at the lower end of this estimate, the structural equilibrium for transcripts that start with ^{cap}2G or ^{cap}3G will be shifted toward the monomeric form. Importantly, in the monomeric forms, the 5′ cap is exposed and would be available for cap recognition by the cap-binding protein eukaryotic translation initiation factor 4E (eIF4E). On the other hand, in the ^{cap}1G transcripts, the 5′ cap is sequestered as part of a base pair in the TAR stem and is not available for cap recognition. Instead, these transcripts are prone to dimerization and are packaged. It seems then that HIV-1 may have capitalized on the existence of sequence heterogeneity to partition its mRNAs into distinct pools (^{cap}3G ⇒ monomer ⇒ cap-exposed ⇒ translation, ^{cap}1G ⇒ dimer ⇒ cap-sequestered ⇒ packaging). Although the use of sequence heterogeneity as a regulatory mechanism has been shown only in HIV-1, it is possible that other retroviruses also use this mechanism, because the existence of alternate monomeric and dimeric conformations is a common theme in retroviral biology (D'Souza and Summers 2005).

The relative populations of structures could potentially be defined by the inherent initiation properties of the RNA polymerase, which viruses may have taken advantage of during evolution. It is also possible that the relative use of each transcription start site could be altered in response to modulation by other regulatory factors present in the cell that would change the initiation properties of the polymerase. In HIV-1, Kharytonchyk et al. suggest that such temporal changes could lead to the production of ^{cap}2G /^{cap}3G early in the viral life cycle, followed by an increase in the ^{cap}1G production for genome packaging (Kharytonchyk et al. 2016).

It is important to note that transcriptional start site heterogeneity is not limited to retroviruses. Many cellular genes also initiate transcription at heterogeneous start sites, the most pertinent example being the "twinned" TATA-driven promoters, which have start sites predominantly separated by 0–3 nucleotides (nt) (Carninci et al. 2006; Ponjavic et al. 2006). In addition, for dispersed promoters where the start sites are distributed over a larger region spanning between 50 and 100 nt in length (Juven-Gershon and Kadonaga 2010), the implications for structural heterogeneity at the 5′ end of transcripts, and the potential regulatory mechanisms at play, may be even more pronounced. The usage of specific transcriptional start sites has been shown to vary between different cell types and also between normal and disease states, suggesting that *trans* factors can also be involved (Ponjavic et al. 2006; Frith et al. 2008).

Although inhibition of translation by sequestration of the 5′ cap in HIV-1 may at first glance seem to be a unique system, a recent discovery of a translation inhibitory element (TIE) located in close proximity to the 5′ cap in the Hox mRNA suggests that the inhibition of cap-dependent translation may be more common than expected (Shi and Barna 2015). In this system, protein production occurs via specialized translation, with ribosomes accessing the mRNA through internal ribosome entry sites (IRES) concomitant with active inhibition of canonical translation by TIE. It has been hypothesized that differential regulation of the TIE and IRES motifs in the Hox mRNA provides the control required for ideal expression of the Hox transcripts (Shi and Barna 2015). Nevertheless, it will be interesting to see if this inhibition is also maintained by alternate RNA structures.

It is also worth noting that the concept of differences in primary sequence resulting in alternate RNA structures is not limited to transcription start sites; in fact, single point mutations can alter the entire energetic landscape. For example, riboSNitches are regulatory RNA structural elements in which single-nucleotide polymorphisms (SNPs) result in the formation of alternate RNA structures (Halvorsen et al. 2010; Ritz et al. 2012; Corley et al. 2015). RiboSNitches have been implicated in causing global RNA structural rearrangements that can be directly linked to disease states. In one example, Kutchko et al. used selective 2′-hydroxyl acylation with primer extension (SHAPE)-directed Boltzmann suboptimal sampling to show that the retinoblastoma 1 5′-untranslated region (RB1 5′ UTR) samples three major conformations (Kutchko et al. 2015). Single-nucleotide variants in RB1 that cause retinoblastoma were found to limit the amount of dynamic sampling and cause the stabilization of single conformational states, leading to an increase in protein expression (Kutchko et al. 2015). Other potential riboSNitches have been identified through combined analyses using human genome-wide association studies, RNA-folding algorithms, and genome-wide enzymatic structure probing (Halvorsen et al. 2010; Wan et al. 2014).

In the last few years, it has also become evident that posttranscriptional RNA modifications serve as another layer of primary structure-based regulation, with modifications implicated in splicing, localization, translation, and decay (Lewis et al. 2017; Roundtree et al. 2017). Some forms of RNA modification include N^6-methyladenosine (m^{6}A) (Lavi et al. 1977), N^1-methyladenosine (m^{1}A) (Dominissini et al. 2016), 5-methylcytosine (m^{5}C) (Desrosiers et al. 1974), 5-hydroxymethylcytosine (hm^{5}C) (Guerrero et al. 2014), pseudouridine (Ψ) (Cohn 1960), and 2′-*O*-methylated ribose (2′-OMe) (Schibler and Perry 1977), with m^{6}A as the most commonly found in mRNA. Such modifications change the folding landscape of RNA by altering their base-pairing and stacking capabilities, electrostatics, and sterics (Chawla et al. 2015). For example, in m^{6}A, Watson–Crick base-pairing leads to the methyl group being forced into an anticonformation, in which steric hindrance between the methyl group and the N7 leads to destabilization of the interaction (Roost et al. 2015). In conjunction, the m^6A modification also stabilizes single strands because of its enhanced base-stacking capabilities (Roost et al. 2015). Zhou et al. describe a mechanism in which the MALAT1 lncRNA (metastasis-associated lung adenocarcinoma transcript 1 long noncoding RNA) has a hairpin motif that samples alternate conformations (Fig. 1B) (Zhou et al. 2016). In the presence of m^6A modifications, the stem architecture is retained but the hairpin loop is more dynamic and samples a conformation similar to that of the protein heterogenous nuclear ribonucleoprotein C (hnRNPC)-bound form. This destabilized conformation allows for better access to the protein-binding site, thereby increasing protein affinity by eightfold. ($m^6A \Rightarrow \uparrow$loop dynamics $\Rightarrow$ $\uparrow$protein-binding site access $\Rightarrow \uparrow$protein affinity.) Mechanisms that leverage the sampling of alternate conformations within an altered folding landscape to drive regulation may be more general and need to be further investigated.

2.2 Secondary Structure

A few examples are beginning to emerge in which RNA domains sample conformations with significant differences in secondary structure. Additionally, structural complexity can involve more than just two equilibrating structures. One example of this involves mutually exclusive splicing, a form of alternative splicing in which only one exon in a set is spliced in at the exclusion of the others (Jin et al. 2011; Pohl et al. 2013). Splicing can dramatically increase the diversity of potential protein products that can be made from a single sequence. The most extensive example of mutually exclusive splicing is the insect *Dscam* gene, which contains four groups of mutually exclusive exons with 12, 48, 33, and two alternate exons, respectively, each group of which will select only one alternate exon (Schmucker et al. 2000). Mutually exclusive splicing can provide an advantage for helping proteins maintain similar overall folds while swapping out individual sequences. Examples of mutually exclusive splicing have been found throughout eukaryotes (Pohl et al. 2013), with a variety of mechanisms proposed including spliceosome incompatibility (Letunic et al. 2002), nonsense-mediated decay combined with splicing factors that enhance one site while suppressing another (Jones et al. 2001), steric hindrance on short introns (Smith and Nadal-Ginard 1989), and more recently, the use of alternative RNA structures (Graveley 2005; Jin et al. 2011; Yang et al. 2011).

Yang et al. show that the mutually exclusive splicing of exon 5a, 5b, or 5c in the *14-3-3ξ* mRNA in *Drosophila melanogaster* is regulated by three competing conformations (Fig. 2A) (Yang et al. 2011). Two of the conformations involve alternate base-pairing between intronic elements (IEs), in which IEa, located downstream from 5c, can pair with either IE1 or IE2, located downstream from 5a and 5b, respectively. The resulting alternate stem loops (SLs) vary in their relative locations of exons, with either 5b and 5c or only 5c in the loop. The exon located immediately proximal to the stem is activated, whereas exons sequestered in the loop are simultaneously suppressed. The interintronic base-pairings are strong enough that they can occur between partners as far as 1,000–12,000 nt away. The third conformation is formed when the base-pairings are suppressed altogether, which results in the activation of exon 5c. Other genes with mutually exclusive splicing, such as *Dscam* (Graveley 2005; Anastassiou et al. 2006; May et al. 2011; Yue et al. 2016), *Mhc* (Yang et al. 2011), and *MRP* (Yue et al. 2017), also use mechanisms involving the formation of structures that are mutually exclusive (Jin et al. 2018). As in the above examples, Yang et al. suggest that the particular usage of each splice site can be additionally modulated by enhancers and silencers that vary between cell types and developmental stages.

Although conformational equilibriums are being found in many systems, in some of these examples, the significance of the structural heterogeneity has yet to be defined. For example, in the case of the HIV-1 Rev Response Element (RRE), a unique motif responsible for exporting unspliced mRNA, two different structures with either four or five SLs form spontaneously in vitro and in vivo (Fig. 2B) (Sherpa et al. 2015). Both structures are present during infection in approximately equal amounts because of similar free energies of folding. However, the five SL structure has been shown to have higher infection rates, likely because of higher Rev/RRE-dependent mRNA export activity. Sherpa et al. suggest that the near neutral energetics provide a sensitive way to modulate the equilibrium, which may be used to dictate the course of infection in a particular cell by altering levels of cellular factors (Sherpa et al. 2015). Similarly, subtle differences in primary sequence during virus evolution can lead to differences in infection severity between patients (Sherpa et al. 2015).

Another example of a well characterized structural equilibrium is the HIV TAR element. In addition to the predominant "ground" state (GS) structure, TAR can also adopt two alternate "excited" states, ES1 and ES2 (Fig. 2C) (Lee et al. 2014). ES1 differs from the GS in the apical loop and is present at 13%, with a timescale of 45 µsec. ES2 differs from the GS by reshuffling its base pairs and is present at ~0.4% with a timescale of 2 msec. Although the functional significance of the complex dynamics shown by the TAR stem have yet to be determined, Lee et al. suggest

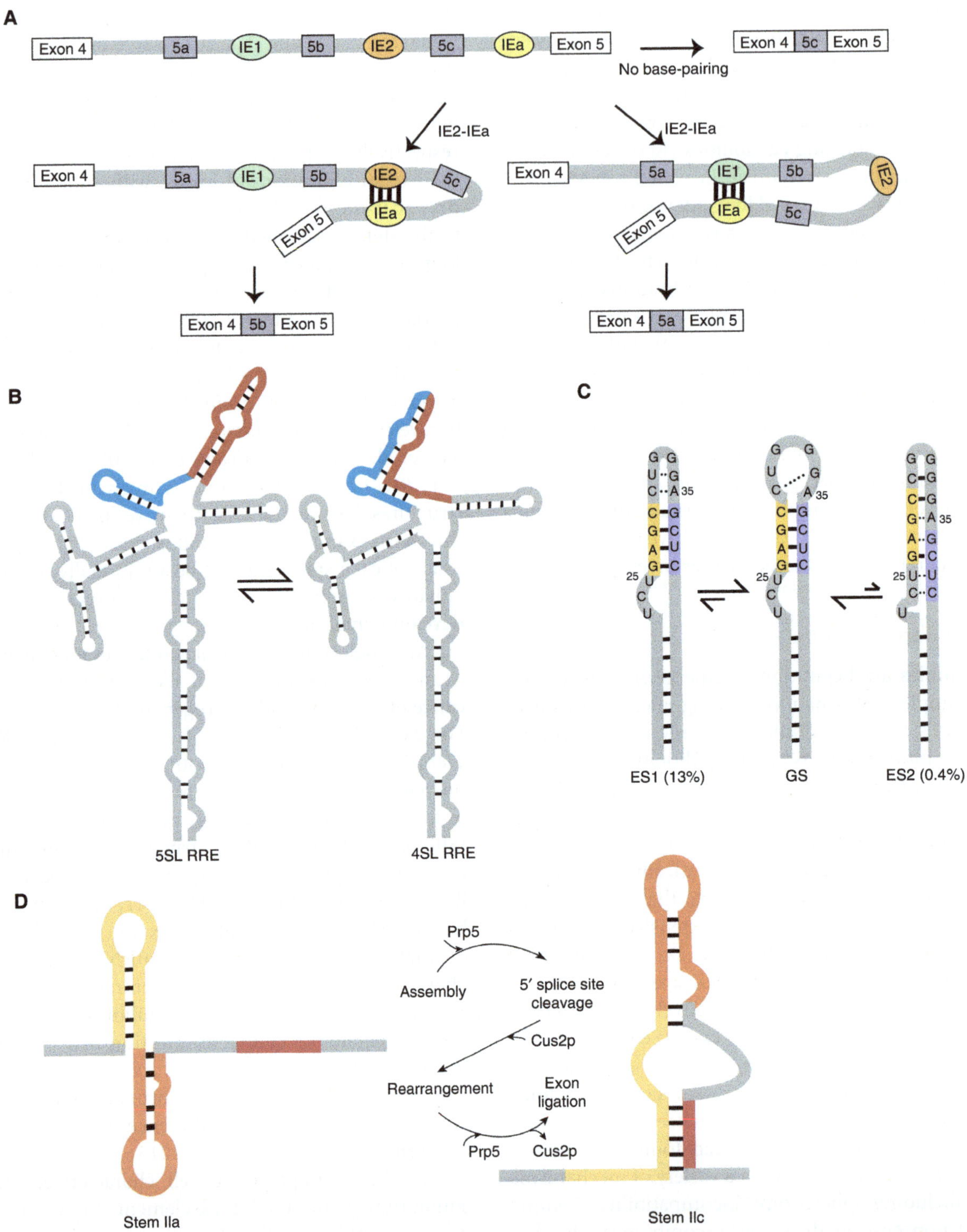

Figure 2. (*A*) Mutually exclusive splicing in the *14-3-3ξ* mRNA is controlled by three structures that differ in their base-pairing pattern: Intronic element a (IEa) forms base pairs with either IE1 or IE2 or neither. (*B*) The HIV-1 Rev response element (RRE) partitions into two equally populated structures with either four or five stem loops (SLs). (*C*) The HIV-1 transactivation response (TAR) element has two "excited" states that differ from the "ground" state (GS). Excited state 1 (ES1) maintains the same overall stem architecture but differs in tertiary interactions at the apical loop, whereas ES2 shows reshuffled base pairs, resulting in a large-scale repairing of the TAR secondary structure. (*D*) The stem II region on the spliceosome U2 small nuclear RNA (snRNA) can transition between two alternate conformations that each serve a different role in the process of splicing.

Cite as *Cold Spring Harb Perspect Biol* doi: 10.1101/cshperspect.a032425

that the excited structures may be involved in mediating the ability of the TAR element to perform its various critical functions, including transcriptional elongation and latency (Lee et al. 2014).

The above structural equilibriums have the potential to be modulated by protein effectors, and the resulting protein–RNA complexes can influence the relative conformational populations. In fact, examples exist in which such protein-derived stabilization of specific conformations along an energetic landscape is required for vital functions. For example, the stem II region of the spliceosome U2 small nuclear RNA (snRNA) spontaneously transitions between two mutually exclusive alternate conformations (stems IIa and IIc) that toggle back and forth throughout the process of splicing (Fig. 2D) (Rodgers et al. 2016). Although stem IIa is necessary for spliceosome assembly and formation of the active site, stem IIc is required for the two catalytic steps of splicing (Hilliker et al. 2007; Perriman and Ares 2007; Rodgers et al. 2016). Currently, the Cus2p (cold-sensitive U2 snRNA suppressor 2) and the DEAD-box adenosine triphosphatase (ATPase) Prp5p proteins are the two candidates that have been implicated in orchestrating these structural transitions (Rodgers et al. 2016). Rodgers et al. show that Cus2p stabilizes stem IIa and suggest that Prp5p may be involved in the removal of Cus2p, a model in which Cus2p and Prp5p may work in tandem to organize the interconversion of stems IIa and IIc during splicing. It is important to note that the equilibrium itself is preexisting and can be accessed without protein factors, even if protein factors present in the cell can help facilitate the transitions. Cus2p and Prp5p have yet to be shown to be dynamically changing.

2.3 Tertiary Structure

Many RNA viruses maintain appropriate viral protein ratios via translational recoding (Houck-Loomis et al. 2011; Walsh and Mohr 2011). For example, translational recoding in the murine leukemia virus (MLV) produces ~95% Gag (structural) polyprotein by canonical termination at the stop codon, and ~5% extended Gag-Pol (Pol, enzymatic) polyprotein by stop codon readthrough that occurs when ribosomes encounter a signal on the mRNA (Philipson et al. 1978; Houck-Loomis et al. 2011). Houck-Loomis et al. found that in MLV the frequency of stop codon readthrough is regulated by a dynamic structural equilibrium-based mechanism (Houck-Loomis et al. 2011), in which a pseudoknotted recoding signal on the mRNA is maintained in a preexisting, protonation-dependent equilibrium between two functional alternate conformations—an active conformation that allows recoding and an alternate inactive conformation that terminates translation. Importantly, the pK_a of the equilibrium is such that at physiological pH, the populations of the active and inactive forms correlate with the critical levels of recoding and termination, respectively—in other words, the RNA signal uses the physiological pH of the host cell to automatically partition its conformations into substates that govern explicit biological output.

In MLV, the protonation of the loop 1 adenine residue leads to the formation of hydrogen bond interactions with a base pair in stem 2, leading to the formation of a triple base interaction (Fig. 3A). This information is then transmitted through changes in the helical junction to loop 2 and stem 1, which together form a triple helix. Therefore, in this example the alternate structures have the same secondary structures, but dramatically different tertiary structures ($A_{17}H^+ \Rightarrow$ triple base $\Rightarrow$ tertiary structure $\Rightarrow$ stop codon read-through translation; $A_{17} \Rightarrow$ lack of tertiary structure $\Rightarrow$ translation termination at the stop codon). This type of proton-driven equilibrium may be a common mechanism for regulating recoding in diverse viruses. For example, this mechanism has been observed in the beet western yellows virus (BWYV) luteovirus (Houck-Loomis et al. 2011) and the severe acute respiratory syndrome (SARS) coronavirus (V D'Souza, unpubl.). Furthermore, both of these viruses genetically recode the mRNA message not by stop codon readthrough but by −1 programmed ribosomal frameshifting (Su et al. 1999; Thiel et al. 2003), suggesting that the control of translational recoding through structural partitioning may be a broader mechanism.

In general, adenines and cytosines are protonatable, but their pK_as (~3.7 and ~4.5, respectively) are far removed from physiological pH. However, the electrostatic environment surrounding a nucleotide in the folded structure of an RNA molecule can significantly shift its pK_a toward neutrality. The physiological pH of cells stays fairly constant, allowing the structural signal of the mRNA to be programmed with a pK_a value that is in line with the desired protein ratios. In the case of MLV and SARS, the pK_as of the protonation events are ~6.23 and ~6.65 (V D'Souza, unpubl.), respectively, which result in the desired ratios of 95:5 and 85:15 for structural and enzymatic proteins. The pK_a shift toward neutrality has been shown to be true for many biologically important RNAs (Cai and Tinoco 1996; Butcher et al. 2000; Huppler et al. 2002; Cornish et al. 2005; Gong et al. 2007; Houck-Loomis et al. 2011; Cash et al. 2013; Pechlaner et al. 2015). A recent example of a guanosine triphosphate (GTP)-sensing aptamer shows that the pK_a can be shifted as high as 8.7 (Wolter et al. 2017), which would in principle allow for a full range of modulation of alternate structures at physiological proton concentration.

In some cases, proteins assist in the alteration of the energy barriers between conformations along an existing

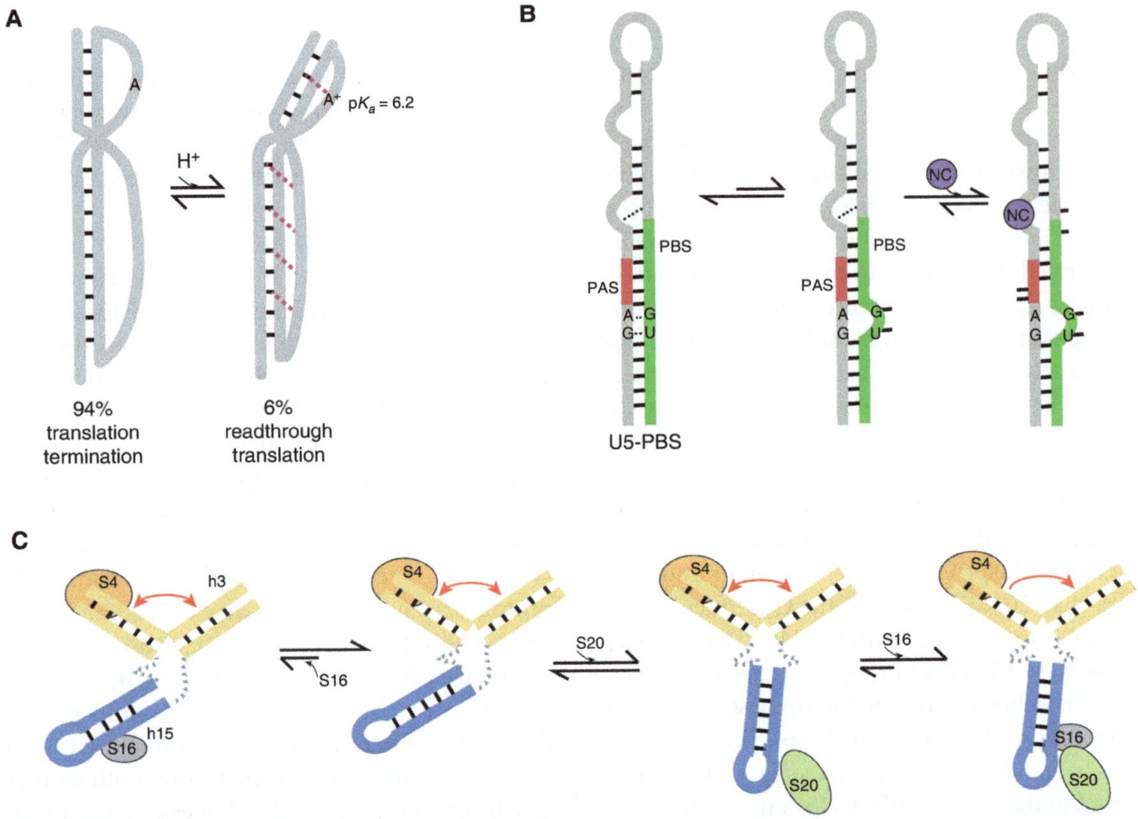

Figure 3. (*A*) The murine leukemia virus (MLV) readthrough RNA pseudoknot is partitioned into two alternate structures, depending on the protonation state. The inactive structure is unprotonated, with stem 2 stacked on top of stem 1. The active structure is protonated at the loop 1 adenine, forming a triple-base interaction that causes a bend at the junction, which allows for tertiary interactions to be formed between loop 2 and stem 1. (*B*) The U5–primer-binding site (PBS) of MLV exchanges between two conformations, one of which is bound by the nucleocapsid (NC). NC binding frees select nucleotides in the PBS and polyadenylation signal (PAS) regions for access by the appropriate complementary regions on the transfer RNA (tRNA; not pictured), to anneal and form the reverse transcription initiation complex. (*C*) The 16S ribosomal RNA (rRNA) samples many conformations including native and non-native forms. Protein binding simultaneously reduces conformational sampling and enriches the native forms.

landscape to enrich a particular conformation, as seen in the example of the MLV NC protein chaperone activity. In retroviruses, reverse transcription requires annealing of complementary sequences on a host transfer RNA (tRNA) primer and on the U5–primer-binding site (U5–PBS) region on the viral genome to form the reverse transcription initiation complex (Fig. 3B) (Harada et al. 1979; Wain-Hobson et al. 1985). The complementary sequences in both molecules are sequestered by intramolecular base pairs (Mougel et al. 1993; Paillart et al. 2004; Wilkinson et al. 2008) that must be broken before new intermolecular interactions can be formed, which is orchestrated by the binding of the NC protein (Levin et al. 2010). Miller et al. showed that the U5–PBS samples two conformations that differ at the tetraloop and at an internal loop. In the major conformation, both of the loops are structured, with the internal loop sequestered via cross-strand base-pairing, whereas in the minor conformation part of the tetraloop and part of the internal loop both extrude out of the structure (Miller et al. 2014). The NC tails are capable of pushing the structural equilibrium to the minor destabilized population, which allows the tRNA to access the appropriate complementary sequences on the U5–PBS for annealing. It is important to note that NC can shift only a preexisting equilibrium, without which it would not otherwise be able to access the U5–PBS for remodeling.

Another example is the U1A protein, which binds to the U1 snRNA hairpin II via conformational capture, whereby the free structure samples many conformations including a rigid state similar to the protein-bound state (Shajani et al. 2007). Similarly, the cyclic adenosine monophosphate response element-binding protein 2 (CBP2) protein binds the initially unstructured bI5 group I intronic RNA in a nonspecific manner and induces conformational

 Cite as *Cold Spring Harb Perspect Biol* doi: 10.1101/cshperspect.a032425

fluctuations, before binding specifically and stabilizing the final native structure (Bokinsky et al. 2006). Even in this example, the free RNA is able to transiently access the structured state in the absence of protein (Bokinsky et al. 2006).

Finally, some proteins can bind multiple conformations instead of just favoring one. For example, Abeysirigunawardena et al. show that ribosomal protein bS16 can bind both a native and non-native form of the 16S rRNA during ribosome assembly, although bS16 binding nevertheless drives the equilibrium to the required native structure (Fig. 3C) (Abeysirigunawardena et al. 2017). During ribosome assembly, the sequential binding of proteins stabilizes the rRNA into conformations that allow for binding of the next protein. Proteins uS4 and bS20 both bind the 16S 5′ domain, whereas protein bS16 binds the interface between the uS4 and bS20 domains. Upon uS4 binding, the complex shows a slow exchange equilibrium in which the h3 helix converts between the native and non-native "flipped" conformations. Binding of bS20 does not affect the equilibrium of the h3 helix, but switches the conformation of helix h15, which increases the probability of stable bS16 binding. Binding of bS16 stabilizes the native conformation of the h3 helix to allow for further assembly of 30S ribosome proteins. The binding of proteins does not lock in any particular RNA conformation, but rather limits the dynamics on the energy landscape to a narrower set of productive intermediate structures while selecting against unproductive conformations.

3 ALTERNATE RNA STRUCTURES CONTROLLED BY DYNAMIC CUES

Alternate RNA structures can also be triggered by dynamic cues, allowing for the accessing of states that would have otherwise been inaccessible; these types of alternate RNA structures are loosely grouped under the term riboswitches. Historically, riboswitches were defined as RNAs that bound metabolites such as purine (Batey 2012; Porter et al. 2014), lysine (Garst et al. 2008; Serganov et al. 2008; Smith-Peter et al. 2015), thiamine pyrophosphate (TPP) (Thore et al. 2006; Bocobza et al. 2007; Anthony et al. 2012; Li and Breaker 2013; Uhm et al. 2018), *S*-adenosyl methionine (SAM) (Price et al. 2014; Trausch et al. 2014; Wostenberg et al. 2015), preQ1 (Eichhorn et al. 2014), adenosyl cobalamin (Johnson et al. 2012; Peselis and Serganov 2012; Schaffer et al. 2014), glycine (Huang et al. 2010; Butler et al. 2011; Erion and Strobel 2011; Baird and Ferre-D'Amare 2013), fluoride (Baker et al. 2012; Ren et al. 2012; Li et al. 2013; Zhao et al. 2017), and tetrahydrofolate (Ames et al. 2010; Huang et al. 2011; Trausch et al. 2011). However, this definition has since been expanded to more broadly include RNAs that change structure in response to dynamic cues such as temperature (Kortmann and Narberhaus 2012; Krajewski and Narberhaus 2014), pH (Nechooshtan et al. 2009), tRNA binding (Zhang and Ferre-D'Amare 2013; Henkin 2014), and metal ion binding (Cromie et al. 2006; Dann et al. 2007; Furukawa et al. 2015; Price et al. 2015; Wedekind et al. 2017).

Riboswitches generally have two domains—a ligand-binding aptamer and a gene expression platform that can modify gene expression based on the structure of the ligand-binding aptamer (Schwalbe et al. 2007). In general, the most widely accepted mechanism is that riboswitches can exist as two conformational structures depending on whether they are ligand-bound or ligand-free. These two structures produce two different functions in response to changing metabolite levels in the cell, thereby enabling regulation. Usually the riboswitch functions as part of a feedback loop pathway for ligand concentration, in which the ligand sensed by the riboswitch is directly related to the function of the riboswitch-regulated gene, either to cause a decrease of expression of ligand-producing proteins or to cause an increase of expression of ligand-clearing proteins (Mandal and Breaker 2004; Mandal et al. 2004; Porter et al. 2014). Although most riboswitches loosely abide by this general model, recent studies have suggested that some riboswitches are not binary on/off switches, but function by finely tuning relative populations to act as molecular "dimmers" (Baird and Ferre-D'Amare 2010; Baird et al. 2010; Helmling et al. 2017).

Several different mechanisms for riboswitch function have been proposed, many of which were only able to be characterized using new techniques and technologies. In this review, we will not be covering the entire range of riboswitch structural and mechanistic diversity, as these have been extensively covered in other reviews (Haller et al. 2011b; Breaker 2012; Serganov and Nudler 2013; Peselis and Serganov 2014; Sherwood and Henkin 2016), but instead will focus on a few key examples in transcription, translation, and splicing to illustrate how these systems use alternate RNA structures to perform gene regulation and how new technologies have made some of these observations possible.

3.1 Transcription

Riboswitches can control transcription termination as a *cis*-acting switch using two mutually exclusive base-paired stems: One conformation (terminator stem) causes transcription termination, whereas the other (antiterminator stem) allows transcription to proceed (Fig. 4A) (Mandal et al. 2004; Porter et al. 2014). Many riboswitches use a

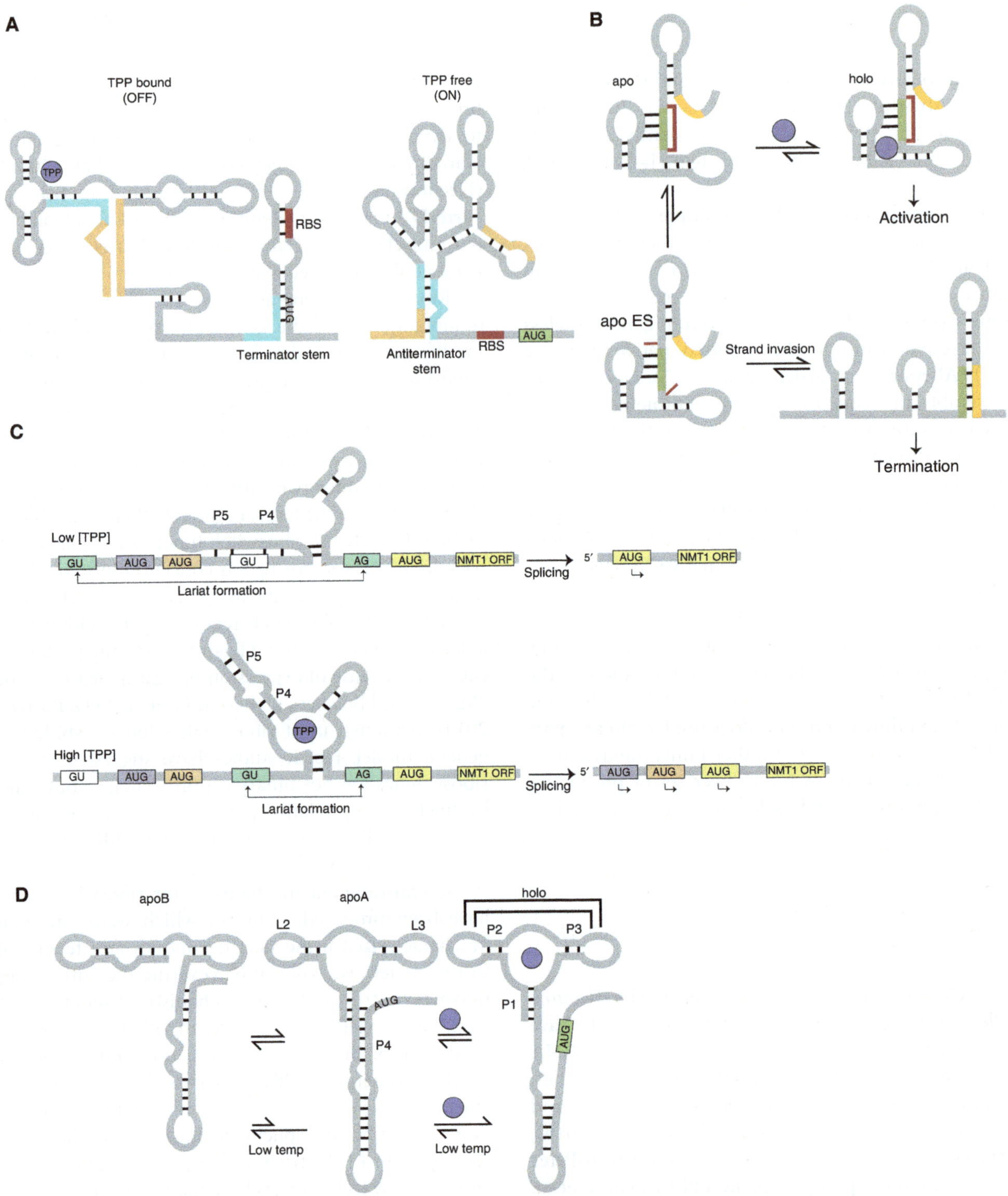

Figure 4. (*A*) The *Escherichia coli* thiamine pyrophosphate (TPP) *thiC* riboswitch has two conformations, ligand-bound (OFF) and ligand-free (ON). The ribosome-binding site (RBS) and the AUG start codon are sequestered in the ligand-bound state and exposed in the ligand-free state, respectively. (*B*) The *Bacillus cereus* fluoride riboswitch has apo and holo conformations with highly similar structures. The apo conformation can transiently access a short-lived excited state (apo ES) in which a highly conserved reverse Hoogsteen "linchpin" base pair is broken, enabling strand invasion and leading to formation of the terminator stem. (*C*) The *Neurospora crassa* TPP *NMT1* riboswitch regulates the expression of alternate upstream open reading frames (ORFs) by using alternate base-pairing involving residues in the P4 and P5 stems to change the formation of the lariat. (*D*) The *Vibrio vulnificus* adenine *add* riboswitch operates using a three-state mechanism rather than canonical two-state mechanism, allowing it to maintain its ligand concentration dependence at temperatures ranging from 37°C to 10°C. The apoA conformation is similar to the holo conformation, whereas apoB differs significantly.

version of the conformational capture mechanism. Once folded, the thermodynamic barrier between two competing structures is high enough that they cannot equilibrate within the timescale of transcription. Therefore, by necessity, whether transcription will terminate or proceed must be regulated by structural partitioning during co-transcriptional folding (Noeske et al. 2007; Haller et al. 2011a; Frieda and Block 2012; Lai et al. 2013; Peselis and Serganov 2014; Gong et al. 2015; Watters et al. 2016; Gong et al. 2017). In most instances, this decision-making occurs during transcriptional pausing, which can be used as a checkpoint for riboswitch-driven regulation (Wickiser et al. 2005a; Wickiser et al. 2005b; Greenleaf et al. 2008). We will first describe the general co-transcriptional pausing mechanism using the *Escherichia coli* TPP *thiC* riboswitch as an example, and then describe a variation on the mechanism using the *Bacillus cereus* fluoride riboswitch.

Chauvier et al. found that in the *thiC* riboswitch, the antiterminator stem conformation forms only in the presence of long pausing steps during transcription (Chauvier et al. 2017). In this riboswitch, ligand affinity is high prior to transcription reaching the transcription pause site (TPS), which is located near the translation start site. At the TPS, ligand affinity decreases by ~100-fold, ensuring that any ligand-free structures that reach the TPS will no longer preferentially bind ligand but can now form the alternate metastable antiterminator stem. This results in completion of transcription and subsequent translation. Even with a decreased ligand affinity, the pausing step is still necessary to provide enough time to fold into the metastable state (ligand-free at TPS ⇒ pausing ⇒ antiterminator stem ⇒ transcription). On the other hand, ligand-bound complexes that reach the TPS have folded into the terminator stem conformation and will primarily result in transcription termination. In addition, any ligand-bound complexes that do not terminate are nevertheless inhibited for translation initiation because of the sequestering of the translation start site in the terminator stem conformation. This provides a fail-safe mechanism to maximally ensure that the protein product is not made (ligand-bound at TPS ⇒ pausing ⇒ terminator stem ⇒ transcription termination/translation start site sequestered).

It is worth noting that transcriptional pausing itself can be modulated by ligand concentration, such as for the pH-responsive *alx* riboswitch involved in maintaining homeostasis in *E. coli* (Nechooshtan et al. 2009). In this example, ligand binding during cotranscriptional pausing regulates translation initiation rather than transcription termination. The RNA-folding dynamics during transcription are directly driven by pH: Transcriptional pausing is prolonged in alkaline conditions, allowing for better refolding of transcribed RNA into an active structure where the ribosome-binding site (RBS) is exposed rather than sequestered. Note that although the *alx* riboswitch is pH-dependent, it operates in an environment with a dynamically changing pH, as opposed to the previously discussed pH-dependent MLV readthrough system which used the unchanging physiological pH in the cell as a static cue.

In the *B. cereus* fluoride riboswitch, Zhao et al. discovered a switching mechanism counter to the current paradigm (Fig. 4B) (Zhao et al. 2017). Whereas most riboswitches involve the formation of alternate ligand-bound (holo) and ligand-free (apo) conformations with different secondary structures, in this example, Zhao et al. found that the two states have highly similar structures. The only difference between the two conformations is that the apo conformation has an additional structural equilibrium wherein it samples a comparatively short-lived (τ ~3 msec) ES that the holo conformation does not. Only a small population of the apo conformation (~1%) samples the ES, which unlocks a highly conserved reverse Hoogsteen "linchpin" base pair present in the apo and holo conformations. This unlocking event exposes the antiterminator stem to strand invasion and allows for the structural transition to the terminator state. In the absence of fluoride, the apo state (τ ~200 msec) cannot access the holo state; therefore, although only a small population samples the ES, because of the high favorability of strand invasion, the overall equilibrium proceeds toward termination (low fluoride ⇒ apo state ⇒ excited state ⇒ unlocked linchpin ⇒ terminator stem ⇒ transcription termination). On the other hand, the holo conformation is stabilized by ligand binding (τ ~2.4 sec) and therefore does not access the ES, but instead completes transcription (fluoride ⇒ holo state with antiterminator stem ⇒ transcription). Zhao et al. suggest that this mechanism could potentially be more widespread and that it would be prudent to reexamine previously characterized systems to determine if any are regulated by the presence of as yet unobserved ESs.

3.2 Splicing

Unlike bacterial TPP riboswitches that regulate transcription and translation (Miranda-Rios et al. 2001; Mironov et al. 2002; Winkler et al. 2002), eukaryotic TPP riboswitches are often located in the introns of pre-mRNAs and regulate splicing (Kubodera et al. 2003; Cheah et al. 2007; Li and Breaker 2013; Mukherjee et al. 2018). These are currently the only known class of eukaryotic riboswitches (Serganov and Nudler 2013). All of the mechanisms characterized to date involve alternate base-pairing interactions that selectively mask or expose splice sites (Cheah et al. 2007; Croft et al. 2007; Li and Breaker 2013). For instance, the TPP riboswitch located in an intron in the *NMT1* gene in the fungus *Neurospora crassa* regulates the expression of upstream open reading frames (ORFs) (Cheah et al. 2007). Two alternate conformations are formed with the second 5′ splice site either base-paired with residues in the P4 and P5 stems (ligand-free) or exposed (ligand-bound) (Fig. 4C). Sequestration of the second 5′ splice site allows the lariat to form between the first 5′ splice site and the 3′ splice site, giving rise to a short mRNA product and resulting in functional *NMT1* translation. On the other hand, exposure of the 5′ splice site under high TPP concentrations results in a longer spliced mRNA product which contains upstream ORFs that compete with the primary ORF to produce an overall lower *NMT1* expression. In another example, the TPP riboswitch found in the 3′ UTRs of the *THIC* gene in *Arabidopsis thaliana* and in a variety of other plant species, uses a similar mechanism but controls gene expression via alternate 3′ end processing (Wachter et al. 2007). This mechanism also uses the sequestration of splice sites, but operates by retaining or removing an intronic transcript-processing signal, thus leading to a longer or shorter 3′ UTR, respectively (Bocobza et al. 2007; Wachter et al. 2007).

3.3 Translation

Analogous to transcriptional riboswitches, many translational riboswitches use alternate structures that either sequester or expose the Shine–Dalgarno sequence in response to dynamic cues. However, unlike transcriptional riboswitches, translational riboswitches can recognize ligand recognition either during or after transcription and can equilibrate within the timescale of translation and the lifetime of the mRNA (Peselis and Serganov 2014). Most regulation occurs via a simple structural equilibrium with ligand-bound and ligand-free conformations. In this section, we will use the adenine *add* riboswitch discovered in the human pathogenic bacterium *Vibrio vulnificus* to illustrate an example of a noncanonical translation initiation riboswitch involving three alternate structures (Reining et al. 2013).

V. vulnificus occupies a range of habitats and can survive in temperatures ranging from ~10°C in marine habitats to 37°C in human hosts (Reining et al. 2013). It needs to maintain functional riboswitch activity through the entire range of temperatures. Reining et al. show that the *add* riboswitch uses a three-state mechanism, as opposed to the canonical two-state mechanism, to couple ligand and temperature sensing. In the three-state mechanism, there is one ligand-bound (holo) structure and two ligand-free (apoA and apoB) structures (Fig. 4D).

The holo conformation has three helices (P1, P2, P3) organized by a three-way junction and stabilized by long-range interactions between loops L2 and L3. Upon adenine binding, P1 is extended by three additional base pairs, which causes the dismantling of the P4 helix and exposes the Shine–Dalgarno sequence for ribosome binding. The apoA and apoB conformations differ significantly from each other although both are ligand-free structures. The apoA conformation is similar to the holo conformation and is binding-competent. However, it lacks L2–L3 interactions and contacts across the junction but has a partially formed P4 stem that sequesters the Shine–Dalgarno sequence. In contrast, the apoB conformation has key motifs rearranged to form a completely different secondary structure, which makes it insensitive to ligand binding.

Reining et al. measured and simulated how efficiently a riboswitch can switch between active and inactive conformations, and defined "switching efficiency" as the difference in the percentage of holo population at two different ligand concentrations. In a two-state mechanism, the riboswitch shows an increased ligand affinity at lower temperatures, leading to a corresponding increase in the holo population. In contrast, the three-state mechanism is able to use the equilibrium between apoB and apoA to counteract the increased ligand affinity of apoA as temperature decreases. Namely, at lower temperatures, the equilibrium favors apoB, thereby sequestering part of the population into a binding-incompetent state (↓temperature ⇒ ↑ligand affinity ⇒ ↑binding-incompetent apoB; ↑temperature ⇒ ↓ligand affinity ⇒ ↑binding-competent). In their simulation, the three-state mechanism is able to produce a 67% switching efficiency at 5°C compared with 83% at 30°C, whereas in the two-state mechanism, the riboswitch is only able to produce a 14% switching efficiency at 5°C. Therefore, the three-state mechanism is able to maintain a more consistent ligand concentration dependence over a wider range in temperatures.

4 NEW TECHNOLOGIES FOR INVESTIGATING ALTERNATE RNA STRUCTURES

In recent years, advances in new technology have enabled the characterization of previously unobserved alternate RNA conformations. Nuclear magnetic resonance (NMR) remains the most powerful technique for probing RNA structural dynamics and continues to be developed for visualizing alternate conformations even in large RNAs.

Typically, NMR experiments are challenging for oligonucleotides larger than approximately 50 residues, predominantly because of spectral overlap. Lu et al. developed a new NMR approach called long-range probing by adenosine interaction detection (lr-AID) and used it to characterize the monomeric (356 nt) and dimeric (712 nt) conformations of the HIV-1 5′ leader (described above) (Lu et al. 2011). Specifically, to confirm the DIS:AUG interaction in the dimer, they replaced three adjacent base pairs within the expected stem with AU base pairs (5′-AUU; 3′-AAU). This had two advantages: First, as expected, the adenine H2 from the central AU base pair experienced an upfield chemical shift away from the spectral overlap of the other nucleotides; second, this adenine H2 gave nuclear Overhauser effects (NOEs) to the H1s of the other two adenines, providing direct proof of stem formation. Use of these signature chemical shift patterns is widely applicable to the study of long-range interactions within large RNA structures; in theory, any number of base pairs within a stem not involved in tertiary interactions can be interchanged without altering the overall fold, because of the isosteric nature of Watson–Crick base pairs.

NMR is also a powerful technique for characterizing the timescales and populations of alternate RNA structures (µsec–msec). Whereas relaxation dispersion experiments are routinely used to study structures exchanging at fast timescales (µsec) (Hoogstraten et al. 2000; Blad et al. 2005), and their use in conjunction with a low spin lock field has been used to obtain information on medium timescales (µsec–msec) (Massi et al. 2004; Hansen et al. 2009; Dethoff et al. 2012), the challenge has been in observing transitions in slow timescales (msec). Slow timescales have recently become accessible through the use of chemical exchange saturation transfer (CEST) NMR combined with a low spin lock field (Zhao et al. 2014). In fact, as described above, Zhao et al. used CEST NMR to visualize a lowly populated ES in the *B. cereus* fluoride riboswitch and to describe the existence of a new kinetic switching mechanism (Zhao et al. 2017).

Beyond NMR, other techniques such as chemical modification-based experiments are being advanced to prove the existence of in vitro characterized equilibria in cells. One example is dimethyl sulfate mutational profiling with sequencing (DMS-MaPseq), a form of secondary structure determination distinct from previous structure probing techniques in that it encodes modifications as mutations instead of truncations (Zubradt et al. 2017). The recording of multiple mutation sites per RNA molecule via deep sequencing allows for the analysis of each individual RNA molecule rather than a population average. This enables clustering of distinct RNA substructures within a population, as was shown using the two human mitochondrial ribosomal protein small 21 (MRPS21) riboSNitch alleles (Zubradt et al. 2017).

Recently, an RNA duplex-mapping technique called PARIS (psoralen analysis of RNA interactions and structures) was used to identify new potential alternate base-pairing interactions in cells (Lu et al. 2016). PARIS uses chemical cross-linking combined with deep sequencing to determine which RNA sequences are paired to each other with near base pair resolution. PARIS can be used to obtain transcriptome-wide information, and was used to discover several alternate base-pairing partners in the 3′ UTR of tubulin β class I (TUBB) mRNA, and in the MALAT1 and X-inactive specific transcript (XIST) lncRNAs (Lu et al. 2016). Although there are no known functions associated with these alternate RNA structures as of yet, these observations provide new investigative directions.

The development of new computational techniques is also facilitating the identification of potential new alternate RNA structures, particularly within riboswitches. Many classes of riboswitches share common structural motifs (Breaker 2012) and can therefore be used for computational screening of transcriptomic data to identify new riboswitches (Drory Retwitzer et al. 2015). Also, evolutionary studies are being used to support the functional significance of alternate RNA structures by computational mapping of sequence co-variation across species (Ritz et al. 2013). In instances in which two conformations exist, covariation of base pairs for both conformations in homologous sequences are expected only if they are functionally important.

5 CONCLUDING REMARKS

Although we have long known about the richness of regulatory information contained in RNA, we have just begun to scrape the surface of the depth and extent of complexity that can be contained within individual RNA domains. The handful of examples described above gives a glimpse into the potential contained in alternate RNA structures to drive key biological functions. The development of new technologies on two major fronts has been key to opening up this field: Screening techniques have enabled us to realize how widespread the prevalence of alternate RNA structures is, whereas structural techniques have allowed for the identification and detailed investigation of such substructures. The combined power of these new technologies and their continued further development is promising for the future discovery of novel alternate RNA structures and the mechanisms by which they function.

ACKNOWLEDGMENTS

We would like to thank Aaron Fernandes for his assistance in the creation of the figures.

REFERENCES

Abeysirigunawardena SC, Kim H, Lai J, Ragunathan K, Rappe MC, Luthey-Schulten Z, Ha T, Woodson SA. 2017. Evolution of protein-coupled RNA dynamics during hierarchical assembly of ribosomal complexes. *Nat Commun* **8:** 492.

Ames TD, Rodionov DA, Weinberg Z, Breaker RR. 2010. A eubacterial riboswitch class that senses the coenzyme tetrahydrofolate. *Chem Biol* **17:** 681–685.

Anastassiou D, Liu H, Varadan V. 2006. Variable window binding for mutually exclusive alternative splicing. *Genome Biol* **7:** R2.

Anthony PC, Perez CF, Garcia-Garcia C, Block SM. 2012. Folding energy landscape of the thiamine pyrophosphate riboswitch aptamer. *Proc Natl Acad Sci* **109:** 1485–1489.

Baird NJ, Ferre-D'Amare AR. 2010. Idiosyncratically tuned switching behavior of riboswitch aptamer domains revealed by comparative small-angle X-ray scattering analysis. *RNA* **16:** 598–609.

Baird NJ, Ferre-D'Amare AR. 2013. Modulation of quaternary structure and enhancement of ligand binding by the K-turn of tandem glycine riboswitches. *RNA* **19:** 167–176.

Baird NJ, Kulshina N, Ferre-D'Amare AR. 2010. Riboswitch function: Flipping the switch or tuning the dimmer? *RNA Biol* **7:** 328–332.

Baker JL, Sudarsan N, Weinberg Z, Roth A, Stockbridge RB, Breaker RR. 2012. Widespread genetic switches and toxicity resistance proteins for fluoride. *Science* **335:** 233–235.

Batey RT. 2012. Structure and mechanism of purine-binding riboswitches. *Q Rev Biophys* **45:** 345–381.

Blad H, Reiter NJ, Abildgaard F, Markley JL, Butcher SE. 2005. Dynamics and metal ion binding in the U6 RNA intramolecular stem-loop as analyzed by NMR. *J Mol Biol* **353:** 540–555.

Bocobza S, Adato A, Mandel T, Shapira M, Nudler E, Aharoni A. 2007. Riboswitch-dependent gene regulation and its evolution in the plant kingdom. *Genes Dev* **21:** 2874–2879.

Bokinsky G, Nivon LG, Liu S, Chai G, Hong M, Weeks KM, Zhuang X. 2006. Two distinct binding modes of a protein cofactor with its target RNA. *J Mol Biol* **361:** 771–784.

Breaker RR. 2012. Riboswitches and the RNA world. *Cold Spring Harb Perspect Biol* **4:** a003566.

Butcher SE, Allain FH, Feigon J. 2000. Determination of metal ion binding sites within the hairpin ribozyme domains by NMR. *Biochemistry* **39:** 2174–2182.

Butler EB, Xiong Y, Wang J, Strobel SA. 2011. Structural basis of cooperative ligand binding by the glycine riboswitch. *Chem Biol* **18:** 293–298.

Cai Z, Tinoco I Jr. 1996. Solution structure of loop A from the hairpin ribozyme from tobacco ringspot virus satellite. *Biochemistry* **35:** 6026–6036.

Carninci P, Sandelin A, Lenhard B, Katayama S, Shimokawa K, Ponjavic J, Semple CA, Taylor MS, Engstrom PG, Frith MC, et al. 2006. Genome-wide analysis of mammalian promoter architecture and evolution. *Nat Genet* **38:** 626–635.

Cash DD, Cohen-Zontag O, Kim NK, Shefer K, Brown Y, Ulyanov NB, Tzfati Y, Feigon J. 2013. Pyrimidine motif triple helix in the *Kluyveromyces lactis* telomerase RNA pseudoknot is essential for function in vivo. *Proc Natl Acad Sci* **110:** 10970–10975.

Chauvier A, Picard-Jean F, Berger-Dancause JC, Bastet L, Naghdi MR, Dube A, Turcotte P, Perreault J, Lafontaine DA. 2017. Transcriptional pausing at the translation start site operates as a critical checkpoint for riboswitch regulation. *Nat Commun* **8:** 13892.

Chawla M, Oliva R, Bujnicki JM, Cavallo L. 2015. An atlas of RNA base pairs involving modified nucleobases with optimal geometries and accurate energies. *Nucleic Acids Res* **43:** 6714–6729.

Cheah MT, Wachter A, Sudarsan N, Breaker RR. 2007. Control of alternative RNA splicing and gene expression by eukaryotic riboswitches. *Nature* **447:** 497–500.

Chojnowski G, Walen T, Bujnicki JM. 2014. RNA Bricks—A database of RNA 3D motifs and their interactions. *Nucleic Acids Res* **42:** D123–D131.

Cohn WE. 1960. Pseudouridine, a carbon-carbon linked ribonucleoside in ribonucleic acids: Isolation, structure, and chemical characteristics. *J Biol Chem* **235:** 1488–1498.

Corley M, Solem A, Qu K, Chang HY, Laederach A. 2015. Detecting riboSNitches with RNA folding algorithms: A genome-wide benchmark. *Nucleic Acids Res* **43:** 1859–1868.

Cornish PV, Hennig M, Giedroc DP. 2005. A loop 2 cytidine-stem 1 minor groove interaction as a positive determinant for pseudoknot-stimulated-1 ribosomal frameshifting. *Proc Natl Acad Sci* **102:** 12694–12699.

Croft MT, Moulin M, Webb ME, Smith AG. 2007. Thiamine biosynthesis in algae is regulated by riboswitches. *Proc Natl Acad Sci* **104:** 20770–20775.

Cromie MJ, Shi Y, Latifi T, Groisman EA. 2006. An RNA sensor for intracellular Mg^{2+}. *Cell* **125:** 71–84.

D'Souza V, Summers MF. 2005. How retroviruses select their genomes. *Nat Rev Microbiol* **3:** 643–655.

Dann CE 3rd, Wakeman CA, Sieling CL, Baker SC, Irnov I, Winkler WC. 2007. Structure and mechanism of a metal-sensing regulatory RNA. *Cell* **130:** 878–892.

Desrosiers R, Friderici K, Rottman F. 1974. Identification of methylated nucleosides in messenger RNA from Novikoff hepatoma cells. *Proc Natl Acad Sci* **71:** 3971–3975.

Dethoff EA, Petzold K, Chugh J, Casiano-Negroni A, Al-Hashimi HM. 2012. Visualizing transient low-populated structures of RNA. *Nature* **491:** 724–728.

Dominissini D, Nachtergaele S, Moshitch-Moshkovitz S, Peer E, Kol N, Ben-Haim MS, Dai Q, Di Segni A, Salmon-Divon M, Clark WC, et al. 2016. The dynamic N_1-methyladenosine methylome in eukaryotic messenger RNA. *Nature* **530:** 441–446.

Drory Retwitzer M, Kifer I, Sengupta S, Yakhini Z, Barash D. 2015. An efficient minimum free energy structure-based search method for riboswitch identification based on inverse RNA folding. *PLoS One* **10:** e0134262.

Eichhorn CD, Kang M, Feigon J. 2014. Structure and function of preQ1 riboswitches. *Biochim Biophys Acta* **1839:** 939–950.

Erion TV, Strobel SA. 2011. Identification of a tertiary interaction important for cooperative ligand binding by the glycine riboswitch. *RNA* **17:** 74–84.

Frieda KL, Block SM. 2012. Direct observation of cotranscriptional folding in an adenine riboswitch. *Science* **338:** 397–400.

Frith MC, Valen E, Krogh A, Hayashizaki Y, Carninci P, Sandelin A. 2008. A code for transcription initiation in mammalian genomes. *Genome Res* **18:** 1–12.

Furukawa K, Ramesh A, Zhou Z, Weinberg Z, Vallery T, Winkler WC, Breaker RR. 2015. Bacterial riboswitches cooperatively bind Ni^{2+} or Co^{2+} ions and control expression of heavy metal transporters. *Mol Cell* **57:** 1088–1098.

Garst AD, Heroux A, Rambo RP, Batey RT. 2008. Crystal structure of the lysine riboswitch regulatory mRNA element. *J Biol Chem* **283:** 22347–22351.

Gong B, Chen JH, Chase E, Chadalavada DM, Yajima R, Golden BL, Bevilacqua PC, Carey PR. 2007. Direct measurement of a pK_a near neutrality for the catalytic cytosine in the genomic HDV ribozyme using Raman crystallography. *J Am Chem Soc* **129:** 13335–13342.

Gong S, Wang Y, Zhang W. 2015. Kinetic regulation mechanism of pbuE riboswitch. *J Chem Phys* **142:** 015103.

Gong S, Wang Y, Wang Z, Zhang W. 2017. Co-transcriptional folding and regulation mechanisms of riboswitches. *Molecules* **22:** 1169.

Graveley BR. 2005. Mutually exclusive splicing of the insect Dscam pre-mRNA directed by competing intronic RNA secondary structures. *Cell* **123:** 65–73.

Greenleaf WJ, Frieda KL, Foster DA, Woodside MT, Block SM. 2008. Direct observation of hierarchical folding in single riboswitch aptamers. *Science* **319:** 630–633.

Guerrero A, Dallas DC, Contreras S, Chee S, Parker EA, Sun X, Dimapasoc L, Barile D, German JB, Lebrilla CB. 2014. Mechanistic peptidomics: Factors that dictate specificity in the formation of endogenous peptides in human milk. *Mol Cell Proteomics* **13:** 3343–3351.

Haller A, Rieder U, Aigner M, Blanchard SC, Micura R. 2011a. Conformational capture of the SAM-II riboswitch. *Nat Chem Biol* **7:** 393–400.

Haller A, Souliere MF, Micura R. 2011b. The dynamic nature of RNA as key to understanding riboswitch mechanisms. *Acc Chem Res* **44:** 1339–1348.

Halvorsen M, Martin JS, Broadaway S, Laederach A. 2010. Disease-associated mutations that alter the RNA structural ensemble. *PLoS Genet* **6:** e1001074.

Hansen AL, Nikolova EN, Casiano-Negroni A, Al-Hashimi HM. 2009. Extending the range of microsecond-to-millisecond chemical exchange detected in labeled and unlabeled nucleic acids by selective carbon $R_{1\rho}$ NMR spectroscopy. *J Am Chem Soc* **131:** 3818–3819.

Harada F, Peters GG, Dahlberg JE. 1979. The primer tRNA for Moloney murine leukemia virus DNA synthesis. Nucleotide sequence and aminoacylation of tRNAPro. *J Biol Chem* **254:** 10979–10985.

Helmling C, Wacker A, Wolfinger MT, Hofacker IL, Hengesbach M, Furtig B, Schwalbe H. 2017. NMR structural profiling of transcriptional intermediates reveals riboswitch regulation by metastable RNA conformations. *J Am Chem Soc* **139:** 2647–2656.

Henkin TM. 2014. The T box riboswitch: A novel regulatory RNA that utilizes tRNA as its ligand. *Biochim Biophys Acta* **1839:** 959–963.

Hilliker AK, Mefford MA, Staley JP. 2007. U2 toggles iteratively between the stem IIa and stem IIc conformations to promote pre-mRNA splicing. *Genes Dev* **21:** 821–834.

Hoogstraten CG, Wank JR, Pardi A. 2000. Active site dynamics in the lead-dependent ribozyme. *Biochemistry* **39:** 9951–9958.

Houck-Loomis B, Durney MA, Salguero C, Shankar N, Nagle JM, Goff SP, D'Souza VM. 2011. An equilibrium-dependent retroviral mRNA switch regulates translational recoding. *Nature* **480:** 561–564.

Huang L, Serganov A, Patel DJ. 2010. Structural insights into ligand recognition by a sensing domain of the cooperative glycine riboswitch. *Mol Cell* **40:** 774–786.

Huang L, Ishibe-Murakami S, Patel DJ, Serganov A. 2011. Long-range pseudoknot interactions dictate the regulatory response in the tetrahydrofolate riboswitch. *Proc Natl Acad Sci* **108:** 14801–14806.

Huppler A, Nikstad LJ, Allmann AM, Brow DA, Butcher SE. 2002. Metal binding and base ionization in the U6 RNA intramolecular stem-loop structure. *Nat Struct Biol* **9:** 431–435.

Jin Y, Yang Y, Zhang P. 2011. New insights into RNA secondary structure in the alternative splicing of pre-mRNAs. *RNA Biol* **8:** 450–457.

Jin Y, Dong H, Shi Y, Bian L. 2018. Mutually exclusive alternative splicing of pre-mRNAs. *Wiley Interdiscip Rev RNA* **9:** e1468.

Johnson JE Jr., Reyes FE, Polaski JT, Batey RT. 2012. B12 cofactors directly stabilize an mRNA regulatory switch. *Nature* **492:** 133–137.

Jones RB, Wang F, Luo Y, Yu C, Jin C, Suzuki T, Kan M, McKeehan WL. 2001. The nonsense-mediated decay pathway and mutually exclusive expression of alternatively spliced FGFR2IIIb and -IIIc mRNAs. *J Biol Chem* **276:** 4158–4167.

Juven-Gershon T, Kadonaga JT. 2010. Regulation of gene expression via the core promoter and the basal transcriptional machinery. *Dev Biol* **339:** 225–229.

Kharytonchyk S, Monti S, Smaldino PJ, Van V, Bolden NC, Brown JD, Russo E, Swanson C, Shuey A, Telesnitsky A, et al. 2016. Transcriptional start site heterogeneity modulates the structure and function of the HIV-1 genome. *Proc Natl Acad Sci* **113:** 13378–13383.

Kortmann J, Narberhaus F. 2012. Bacterial RNA thermometers: Molecular zippers and switches. *Nat Rev Microbiol* **10:** 255–265.

Krajewski SS, Narberhaus F. 2014. Temperature-driven differential gene expression by RNA thermosensors. *Biochim Biophys Acta* **1839:** 978–988.

Kubodera T, Watanabe M, Yoshiuchi K, Yamashita N, Nishimura A, Nakai S, Gomi K, Hanamoto H. 2003. Thiamine-regulated gene expression of *Aspergillus oryzae thiA* requires splicing of the intron containing a riboswitch-like domain in the 5′-UTR. *FEBS Lett* **555:** 516–520.

Kutchko KM, Sanders W, Ziehr B, Phillips G, Solem A, Halvorsen M, Weeks KM, Moorman N, Laederach A. 2015. Multiple conformations are a conserved and regulatory feature of the RB1 5′ UTR. *RNA* **21:** 1274–1285.

Lai D, Proctor JR, Meyer IM. 2013. On the importance of cotranscriptional RNA structure formation. *RNA* **19:** 1461–1473.

Lavi U, Fernandez-Munoz R, Darnell JE Jr. 1977. Content of N-6 methyl adenylic acid in heterogeneous nuclear and messenger RNA of HeLa cells. *Nucleic Acids Res* **4:** 63–69.

Lee J, Dethoff EA, Al-Hashimi HM. 2014. Invisible RNA state dynamically couples distant motifs. *Proc Natl Acad Sci* **111:** 9485–9490.

Leontis NB, Westhof E. 1998. Conserved geometrical base-pairing patterns in RNA. *Q Rev Biophys* **31:** 399–455.

Leontis NB, Westhof E. 2001. Geometric nomenclature and classification of RNA base pairs. *RNA* **7:** 499–512.

Leontis NB, Stombaugh J, Westhof E. 2002. The non-Watson-Crick base pairs and their associated isostericity matrices. *Nucleic Acids Res* **30:** 3497–3531.

Letunic I, Copley RR, Bork P. 2002. Common exon duplication in animals and its role in alternative splicing. *Hum Mol Genet* **11:** 1561–1567.

Levin JG, Mitra M, Mascarenhas A, Musier-Forsyth K. 2010. Role of HIV-1 nucleocapsid protein in HIV-1 reverse transcription. *RNA Biol* **7:** 754–774.

Lewis CJ, Pan T, Kalsotra A. 2017. RNA modifications and structures cooperate to guide RNA-protein interactions. *Nat Rev Mol Cell Biol* **18:** 202–210.

Li S, Breaker RR. 2013. Eukaryotic TPP riboswitch regulation of alternative splicing involving long-distance base pairing. *Nucleic Acids Res* **41:** 3022–3031.

Li S, Smith KD, Davis JH, Gordon PB, Breaker RR, Strobel SA. 2013. Eukaryotic resistance to fluoride toxicity mediated by a widespread family of fluoride export proteins. *Proc Natl Acad Sci* **110:** 19018–19023.

Lu K, Heng X, Garyu L, Monti S, Garcia EL, Kharytonchyk S, Dorjsuren B, Kulandaivel G, Jones S, Hiremath A, et al. 2011. NMR detection of structures in the HIV-1 5′-leader RNA that regulate genome packaging. *Science* **334:** 242–245.

Lu Z, Zhang QC, Lee B, Flynn RA, Smith MA, Robinson JT, Davidovich C, Gooding AR, Goodrich KJ, Mattick JS, et al. 2016. RNA duplex map in living cells reveals higher-order transcriptome structure. *Cell* **165:** 1267–1279.

Mandal M, Breaker RR. 2004. Adenine riboswitches and gene activation by disruption of a transcription terminator. *Nat Struct Mol Biol* **11:** 29–35.

Mandal M, Lee M, Barrick JE, Weinberg Z, Emilsson GM, Ruzzo WL, Breaker RR. 2004. A glycine-dependent riboswitch that uses cooperative binding to control gene expression. *Science* **306:** 275–279.

Massi F, Johnson E, Wang C, Rance M, Palmer AG 3rd. 2004. NMR R1 ρ rotating-frame relaxation with weak radio frequency fields. *J Am Chem Soc* **126:** 2247–2256.

May GE, Olson S, McManus CJ, Graveley BR. 2011. Competing RNA secondary structures are required for mutually exclusive splicing of the Dscam exon 6 cluster. *RNA* **17:** 222–229.

McGinnis JL, Liu Q, Lavender CA, Devaraj A, McClory SP, Fredrick K, Weeks KM. 2015. In-cell SHAPE reveals that free 30S ribosome subunits are in the inactive state. *Proc Natl Acad Sci* **112:** 2425–2430.

Miller SB, Yildiz FZ, Lo JA, Wang B, D'Souza VM. 2014. A structure-based mechanism for tRNA and retroviral RNA remodelling during primer annealing. *Nature* **515:** 591–595.

Miranda-Rios J, Navarro M, Soberón M. 2001. A conserved RNA structure (*thi* box) is involved in regulation of thiamin biosynthetic gene expression in bacteria. *Proc Natl Acad Sci* **98:** 9736–9741.

Mironov AS, Gusarov I, Rafikov R, Lopez LE, Shatalin K, Kreneva RA, Perumov DA, Nudler E. 2002. Sensing small molecules by nascent RNA: A mechanism to control transcription in bacteria. *Cell* **111:** 747–756.

Mougel M, Tounekti N, Darlix JL, Paoletti J, Ehresmann B, Ehresmann C. 1993. Conformational analysis of the 5′ leader and the gag initiation site of Mo-MuLV RNA and allosteric transitions induced by dimerization. *Nucleic Acids Res* **21:** 4677–4684.

Mukherjee S, Retwitzer MD, Barash D, Sengupta S. 2018. Phylogenomic and comparative analysis of the distribution and regulatory patterns of TPP riboswitches in fungi. *Sci Rep* **8:** 5563.

Mustoe AM, Brooks CL, Al-Hashimi HM. 2014. Hierarchy of RNA functional dynamics. *Annu Rev Biochem* **83:** 441–466.

Nechooshtan G, Elgrably-Weiss M, Sheaffer A, Westhof E, Altuvia S. 2009. A pH-responsive riboregulator. *Genes Dev* **23:** 2650–2662.

Noeske J, Buck J, Furtig B, Nasiri HR, Schwalbe H, Wohnert J. 2007. Interplay of 'induced fit' and preorganization in the ligand induced folding of the aptamer domain of the guanine binding riboswitch. *Nucleic Acids Res* **35:** 572–583.

Paillart JC, Dettenhofer M, Yu XF, Ehresmann C, Ehresmann B, Marquet R. 2004. First snapshots of the HIV-1 RNA structure in infected cells and in virions. *J Biol Chem* **279:** 48397–48403.

Parlea LG, Sweeney BA, Hosseini-Asanjan M, Zirbel CL, Leontis NB. 2016. The RNA 3D motif atlas: Computational methods for extraction, organization and evaluation of RNA motifs. *Methods* **103:** 99–119.

Pechlaner M, Donghi D, Zelenay V, Sigel RK. 2015. Protonation-dependent base flipping at neutral pH in the catalytic triad of a self-splicing bacterial group II intron. *Angew Chem Int Ed Engl* **54:** 9687–9690.

Perriman RJ, Ares M Jr. 2007. Rearrangement of competing U2 RNA helices within the spliceosome promotes multiple steps in splicing. *Genes Dev* **21:** 811–820.

Peselis A, Serganov A. 2012. Structural insights into ligand binding and gene expression control by an adenosylcobalamin riboswitch. *Nat Struct Mol Biol* **19:** 1182–1184.

Peselis A, Serganov A. 2014. Themes and variations in riboswitch structure and function. *Biochim Biophys Acta* **1839:** 908–918.

Petrov AI, Zirbel CL, Leontis NB. 2013. Automated classification of RNA 3D motifs and the RNA 3D motif atlas. *RNA* **19:** 1327–1340.

Philipson L, Andersson P, Olshevsky U, Weinberg R, Baltimore D, Gesteland R. 1978. Translation of MuLV and MSV RNAs in nuclease-treated reticulocyte extracts: Enhancement of the gag-pol polypeptide with yeast suppressor tRNA. *Cell* **13:** 189–199.

Pohl M, Bortfeldt RH, Grutzmann K, Schuster S. 2013. Alternative splicing of mutually exclusive exons—A review. *Biosystems* **114:** 31–38.

Ponjavic J, Lenhard B, Kai C, Kawai J, Carninci P, Hayashizaki Y, Sandelin A. 2006. Transcriptional and structural impact of TATA-initiation site spacing in mammalian core promoters. *Genome Biol* **7:** R78.

Popenda M, Szachniuk M, Blazewicz M, Wasik S, Burke EK, Blazewicz J, Adamiak RW. 2010. RNA FRABASE 2.0: An advanced web-accessible database with the capacity to search the three-dimensional fragments within RNA structures. *BMC Bioinformatics* **11:** 231.

Porter EB, Marcano-Velazquez JG, Batey RT. 2014. The purine riboswitch as a model system for exploring RNA biology and chemistry. *Biochim Biophys Acta* **1839:** 919–930.

Price IR, Grigg JC, Ke A. 2014. Common themes and differences in SAM recognition among SAM riboswitches. *Biochim Biophys Acta* **1839:** 931–938.

Price IR, Gaballa A, Ding F, Helmann JD, Ke A. 2015. Mn^{2+}-sensing mechanisms of *yybP-ykoY* orphan riboswitches. *Mol Cell* **57:** 1110–1123.

Reining A, Nozinovic S, Schlepckow K, Buhr F, Furtig B, Schwalbe H. 2013. Three-state mechanism couples ligand and temperature sensing in riboswitches. *Nature* **499:** 355–359.

Ren A, Rajashankar KR, Patel DJ. 2012. Fluoride ion encapsulation by Mg^{2+} ions and phosphates in a fluoride riboswitch. *Nature* **486:** 85–89.

Ritz J, Martin JS, Laederach A. 2012. Evaluating our ability to predict the structural disruption of RNA by SNPs. *BMC Genomics* **13:** S6.

Ritz J, Martin JS, Laederach A. 2013. Evolutionary evidence for alternative structure in RNA sequence co-variation. *PLoS Comput Biol* **9:** e1003152.

Rodgers ML, Tretbar US, Dehaven A, Alwan AA, Luo G, Mast HM, Hoskins AA. 2016. Conformational dynamics of stem II of the U2 snRNA. *RNA* **22:** 225–236.

Roost C, Lynch SR, Batista PJ, Qu K, Chang HY, Kool ET. 2015. Structure and thermodynamics of N6-methyladenosine in RNA: A spring-loaded base modification. *J Am Chem Soc* **137:** 2107–2115.

Roundtree IA, Evans ME, Pan T, He C. 2017. Dynamic RNA modifications in gene expression regulation. *Cell* **169:** 1187–1200.

Schaffer MF, Choudhary PK, Sigel RK. 2014. The AdoCbl-riboswitch interaction investigated by in-line probing and surface plasmon resonance spectroscopy (SPR). *Methods Enzymol* **549:** 467–488.

Schibler U, Perry RP. 1977. The 5′-termini of heterogeneous nuclear RNA: A comparison among molecules of different sizes and ages. *Nucleic Acids Res* **4:** 4133–4149.

Schmucker D, Clemens JC, Shu H, Worby CA, Xiao J, Muda M, Dixon JE, Zipursky SL. 2000. *Drosophila* Dscam is an axon guidance receptor exhibiting extraordinary molecular diversity. *Cell* **101:** 671–684.

Schwalbe H, Buck J, Furtig B, Noeske J, Wohnert J. 2007. Structures of RNA switches: Insight into molecular recognition and tertiary structure. *Angew Chem Int Ed Engl* **46:** 1212–1219.

Serganov A, Nudler E. 2013. A decade of riboswitches. *Cell* **152:** 17–24.

Serganov A, Huang L, Patel DJ. 2008. Structural insights into amino acid binding and gene control by a lysine riboswitch. *Nature* **455:** 1263–1267.

Shajani Z, Drobny G, Varani G. 2007. Binding of U1A protein changes RNA dynamics as observed by 13C NMR relaxation studies. *Biochemistry* **46:** 5875–5883.

Sherpa C, Rausch JW, Le Grice SF, Hammarskjold ML, Rekosh D. 2015. The HIV-1 Rev response element (RRE) adopts alternative conformations that promote different rates of virus replication. *Nucleic Acids Res* **43:** 4676–4686.

Sherwood AV, Henkin TM. 2016. Riboswitch-mediated gene regulation: Novel RNA architectures dictate gene expression responses. *Annu Rev Microbiol* **70:** 361–374.

Shi Z, Barna M. 2015. Translating the genome in time and space: Specialized ribosomes, RNA regulons, and RNA-binding proteins. *Annu Rev Cell Dev Biol* **31:** 31–54.

Smith CW, Nadal-Ginard B. 1989. Mutually exclusive splicing of α-tropomyosin exons enforced by an unusual lariat branch point location: Implications for constitutive splicing. *Cell* **56:** 749–758.

Smith-Peter E, Lamontagne AM, Lafontaine DA. 2015. Role of lysine binding residues in the global folding of the lysC riboswitch. *RNA Biol* **12:** 1372–1382.

Sponer JE, Spackova N, Leszczynski J, Sponer J. 2005. Principles of RNA base pairing: Structures and energies of the *trans* Watson–Crick/sugar edge base pairs. *J Phys Chem B* **109:** 11399–11410.

Su L, Chen L, Egli M, Berger JM, Rich A. 1999. Minor groove RNA triplex in the crystal structure of a ribosomal frameshifting viral pseudoknot. *Nat Struct Biol* **6:** 285–292.

Thiel V, Ivanov KA, Putics A, Hertzig T, Schelle B, Bayer S, Weissbrich B, Snijder EJ, Rabenau H, Doerr HW, et al. 2003. Mechanisms and enzymes involved in SARS coronavirus genome expression. *J Gen Virol* **84:** 2305–2315.

Thore S, Leibundgut M, Ban N. 2006. Structure of the eukaryotic thiamine pyrophosphate riboswitch with its regulatory ligand. *Science* **312:** 1208–1211.

Trausch JJ, Ceres P, Reyes FE, Batey RT. 2011. The structure of a tetrahydrofolate-sensing riboswitch reveals two ligand binding sites in a single aptamer. *Structure* **19:** 1413–1423.

Trausch JJ, Xu Z, Edwards AL, Reyes FE, Ross PE, Knight R, Batey RT. 2014. Structural basis for diversity in the SAM clan of riboswitches. *Proc Natl Acad Sci* **111:** 6624–6629.

Uhm H, Kang W, Ha KS, Kang C, Hohng S. 2018. Single-molecule FRET studies on the cotranscriptional folding of a thiamine pyrophosphate riboswitch. *Proc Natl Acad Sci* **115:** 331–336.

Wachter A, Tunc-Ozdemir M, Grove BC, Green PJ, Shintani DK, Breaker RR. 2007. Riboswitch control of gene expression in plants by splicing and alternative 3′ end processing of mRNAs. *Plant Cell* **19:** 3437–3450.

Wain-Hobson S, Sonigo P, Danos O, Cole S, Alizon M. 1985. Nucleotide sequence of the AIDS virus, LAV. *Cell* **40:** 9–17.

Walsh D, Mohr I. 2011. Viral subversion of the host protein synthesis machinery. *Nat Rev Microbiol* **9:** 860–875.

Wan Y, Qu K, Zhang QC, Flynn RA, Manor O, Ouyang Z, Zhang J, Spitale RC, Snyder MP, Segal E, et al. 2014. Landscape and variation of RNA secondary structure across the human transcriptome. *Nature* **505:** 706–709.

Watters KE, Strobel EJ, Yu AM, Lis JT, Lucks JB. 2016. Cotranscriptional folding of a riboswitch at nucleotide resolution. *Nat Struct Mol Biol* **23:** 1124–1131.

Wedekind JE, Dutta D, Belashov IA, Jenkins JL. 2017. Metalloriboswitches: RNA-based inorganic ion sensors that regulate genes. *J Biol Chem* **292:** 9441–9450.

Westhof E. 2014. Isostericity and tautomerism of base pairs in nucleic acids. *FEBS Lett* **588:** 2464–2469.

Wickiser JK, Cheah MT, Breaker RR, Crothers DM. 2005a. The kinetics of ligand binding by an adenine-sensing riboswitch. *Biochemistry* **44:** 13404–13414.

Wickiser JK, Winkler WC, Breaker RR, Crothers DM. 2005b. The speed of RNA transcription and metabolite binding kinetics operate an FMN riboswitch. *Mol Cell* **18:** 49–60.

Wilkinson KA, Gorelick RJ, Vasa SM, Guex N, Rein A, Mathews DH, Giddings MC, Weeks KM. 2008. High-throughput SHAPE analysis reveals structures in HIV-1 genomic RNA strongly conserved across distinct biological states. *PLoS Biol* **6:** e96.

Winkler W, Nahvi A, Breaker RR. 2002. Thiamine derivatives bind messenger RNAs directly to regulate bacterial gene expression. *Nature* **419:** 952–956.

Wolter AC, Weickhmann AK, Nasiri AH, Hantke K, Ohlenschlager O, Wunderlich CH, Kreutz C, Duchardt-Ferner E, Wohnert J. 2017. A stably protonated adenine nucleotide with a highly shifted pK_a value stabilizes the tertiary structure of a GTP-binding RNA aptamer. *Angew Chem Int Ed Engl* **56:** 401–404.

Wostenberg C, Ceres P, Polaski JT, Batey RT. 2015. A highly coupled network of tertiary interactions in the SAM-I riboswitch and their role in regulatory tuning. *J Mol Biol* **427:** 3473–3490.

Yang Y, Zhan L, Zhang W, Sun F, Wang W, Tian N, Bi J, Wang H, Shi D, Jiang Y, et al. 2011. RNA secondary structure in mutually exclusive splicing. *Nat Struct Mol Biol* **18:** 159–168.

Yue Y, Yang Y, Dai L, Cao G, Chen R, Hong W, Liu B, Shi Y, Meng Y, Shi F, et al. 2016. Long-range RNA pairings contribute to mutually exclusive splicing. *RNA* **22:** 96–110.

Yue Y, Hou S, Wang X, Zhan L, Cao G, Li G, Shi Y, Zhang P, Hong W, Lin H, et al. 2017. Role and convergent evolution of competing RNA secondary structures in mutually exclusive splicing. *RNA Biol* **14:** 1399–1410.

Zamir A, Miskin R, Elson D. 1969. Interconversions between inactive and active forms of ribosomal subunits. *FEBS Lett* **3:** 85–88.

Zhang J, Ferre-D'Amare AR. 2013. Co-crystal structure of a T-box riboswitch stem I domain in complex with its cognate tRNA. *Nature* **500:** 363–366.

Zhao B, Hansen AL, Zhang Q. 2014. Characterizing slow chemical exchange in nucleic acids by carbon CEST and low spin-lock field $R_{1\rho}$ NMR spectroscopy. *J Am Chem Soc* **136:** 20–23.

Zhao B, Guffy SL, Williams B, Zhang Q. 2017. An excited state underlies gene regulation of a transcriptional riboswitch. *Nat Chem Biol* **13:** 968–974.

Zhou KI, Parisien M, Dai Q, Liu N, Diatchenko L, Sachleben JR, Pan T. 2016. N^6-methyladenosine modification in a long noncoding RNA hairpin predisposes is conformation to protein binding. *J Mol Biol* **428:** 822–833.

Zubradt M, Gupta P, Persad S, Lambowitz AM, Weissman JS, Rouskin S. 2017. DMS-MaPseq for genome-wide or targeted RNA structure probing in vivo. *Nat Methods* **14:** 75–82.

The Story of RNA Folding, as Told in Epochs

Daniel Herschlag,[1,2,3,4] Steve Bonilla,[1,2] and Namita Bisaria[5]

[1]Department of Biochemistry, Stanford University, Stanford, California 94305

[2]Department of Chemical Engineering, Stanford University, Stanford, California 94305

[3]Department of Chemistry, Stanford University, Stanford, California 94305

[4]Stanford ChEM-H (Chemistry, Engineering, and Medicine for Human Health), Stanford, California 94305

[5]Whitehead Institute for Biomedical Research, Cambridge, Massachusetts 02142

Correspondence: herschla@stanford.edu

SUMMARY

The past decades have witnessed tremendous developments in our understanding of RNA biology. At the core of these advances have been studies aimed at discerning RNA structure and at understanding the forces that influence the RNA folding process. It is easy to take the present state of understanding for granted, but there is much to be learned by considering the path to our current understanding, which has been tortuous, with the birth and death of models, the adaptation of experimental tools originally developed for characterization of protein structure and catalysis, and the development of novel tools for probing RNA. In this review we tour the stages of RNA folding studies, considering them as "epochs" that can be generalized across scientific disciplines. These epochs span from the discovery of catalytic RNA, through biophysical insights into the putative primordial RNA World, to characterization of structured RNAs, the building and testing of models, and, finally, to the development of models with the potential to yield generalizable predictive and quantitative models for RNA conformational, thermodynamic, and kinetic behavior. We hope that this accounting will aid others as they navigate the many fascinating questions about RNA and its roles in biology, in the past, present, and future.

Outline

Cite as *Cold Spring Harb Perspect Biol* doi: 10.1101/cshperspect.a032433

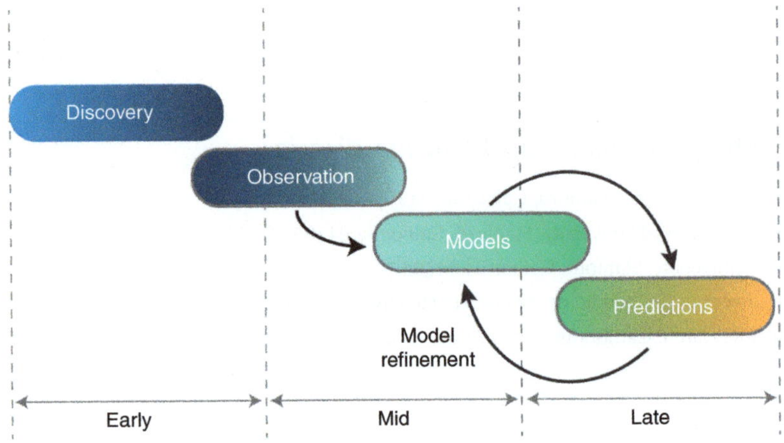

Figure 1. Schematic of progress in a scientific field in terms of "epochs."

1 INTRODUCTION

The reviews in this RNA World collection highlight new techniques and open questions in the field of RNA biology and emphasize how these technologies can generate new insights and further derive new questions. In this review, we break from this format as we cover a more mature scientific area, RNA folding, and we describe general insights drawn from RNA's molecular behavior and glean lessons from the study of the folding process that are broadly applicable to future studies across RNA biology. Just as geology and evolution can be categorized by epochs, so can the scientific process (Fig. 1). Enumerating the characteristics of each scientific epoch and designating such epochs within a particular scientific field can help inform individuals and communities when there is an opportunity to move toward the next stage.

This review takes the reader on a journey through the epochs in RNA folding research, with the goals of familiarizing the reader with the arc of this field's journey, conveying insights about RNA, technologies, and biology that have been gathered along the way, and laying out important lessons that will help us meet the current colossal challenge of understanding cellular RNAs, their structure and interactions, and their roles in the control of gene expression. As these scientific epochs (Fig. 1) are evidently general to the maturation of any scientific field, be it early astronomy (Crawford 1917) or enzymology (Copeland 2000), the case study provided here may be applicable to other aspects of biology.

Our journey through the epochs begins with observations of RNA's physical properties and the subsequent theories of an RNA World, progresses to a rapid discovery phase of uncovering and characterizing functional RNA structures, and closes with the field's recent entry into the phase in which these observations and principles can be codified into predictive models for designing functional RNAs for use in engineering and medicine.

2 THE RNA FOLDING EPOCHS: EARLY OBSERVATIONS AND THE RNA WORLD

Throughout biology, new traits and new species arise through evolutionary changes rooted in molecules and their interactions. Consequently, understanding these evolutionary solutions builds from an understanding of the properties and limitations of these molecules. The routes taken by evolution, by chance and guided by molecular properties, have led to—and are thus necessary to fully understand—modern-day biology. In this section, we focus on initial observations of the molecular properties of RNA that provided insights into why an RNA World may have been more probable than other options for starting life on Earth.

2.1 Why an RNA World?

The idea of an RNA World that predated life with proteins as the dominant catalysts was popularized by the discovery of catalytic RNA by Cech and by Altman and Pace (Kruger et al. 1982; Guerrier-Takada et al. 1983). But the idea arose more than a decade earlier, when Woese, Crick, and Orgel independently recognized that the "molecular properties" of RNA might endow it with the capabilities needed to support life (Woese 1967; Crick 1968; Orgel 1968). They noted that the ability of one RNA base to recognize another, A with U and G with C, could allow replication with some degree of specificity; the ability of RNA to form stable duplexes could allow templated synthesis of new RNA strands; and the more complex structure of transfer RNA (tRNA) resembled, to some extent, a folded protein and opened the possibility of RNAs as catalysts. Subsequently, the realization that the ribosome's catalytic core is made of RNA led many to consider this RNA core as a vestige from an RNA World that developed an ability to make proteins and was subsequently superseded by proteins for most functions (Cech 2000; Steitz and Moore 2003; Noller 2012).

A consideration of the molecular properties of RNA provides a foundation for understanding RNA folding and, moreover, supports the idea of an RNA World and the subsequent advancement of proteins. RNA secondary structure is highly stable such that a 6-base-pair helix can be stable for days, in contrast to an α helix of the same length, which unfolds on the microsecond timescale (Narlikar and Herschlag 1997). The ability to form a stable structure with short oligomeric strands would have facilitated development of function under primitive conditions. Consider a folding free-energy profile for formation of a simple three-way junction (Fig. 2A). The formation of each helix is energetically downhill, so that a milieu with short RNA strands would have a reasonable chance to fold intramolecularly or come together intermolecularly. In contrast, proteins fold more cooperatively and generally require longer sequences for stable folding. Of course, most structured molecules do not have function, but RNA can catalyze strand fusion simply via a templated duplex (Joyce et al. 1984; Adamala and Szostak 2013). At a more sophisticated level, catalytic RNAs that act via tertiary structures bring together residues involved in catalysis at junctions like the one in Figure 2A. As any RNA residue can make hydrogen bonding and stacking interactions with any other RNA residue (Tinoco 1993), there is a wealth of possibilities to create a functional site. Indeed, in vitro evolution experiments selected three-way junctions analogous to the one in Figure 2A that catalyze self-cleavage at a remarkably high probability (Salehi-Ashtiani and Szostak 2001). Selection experiments have also yielded a multitude of diverse RNA catalysts, and there has been impressive progress in developing an RNA-based RNA polymerase (Wilson and Szostak 1999; Horning and Joyce 2016).

RNA's stable secondary structure and promiscuous intermolecular interactions that are an asset for the development of life also present limitations, because RNA's stability creates an inherent tendency to also form alternative interactions and to be kinetically trapped in misfolded states (Fig. 2B, black line). RNA misfolding was first observed in the 1960s, by Sueoka and Fresco who independently obtained tRNAs that could not be charged by their respective aminoacyl synthetases until subject to denaturation and renaturation (Gartland and Sueoka 1966; Lindahl et al. 1966). Since then, those who work with RNA have made similar observations—observations that Uhlenbeck memorialized in his entertaining piece "Keeping RNA Happy" (Uhlenbeck 1995).

Because misfolding is an inherent property of RNA, modern biology too has to face this problem, and we reasoned that cells would require "RNA chaperones," proteins that facilitate unfolding and refolding of DNA either through simple binding or through ATP-driven conformational cycles (Fig. 2B, green line) (Herschlag 1995). Indeed, RNA chaperones are now known to be common cellular components, and so-called RNA helicases and dedicated RNA binding proteins serve to promote or mitigate RNA structure in every cellular process involving RNA, from precursor messenger RNA (pre-mRNA) splicing to mRNA

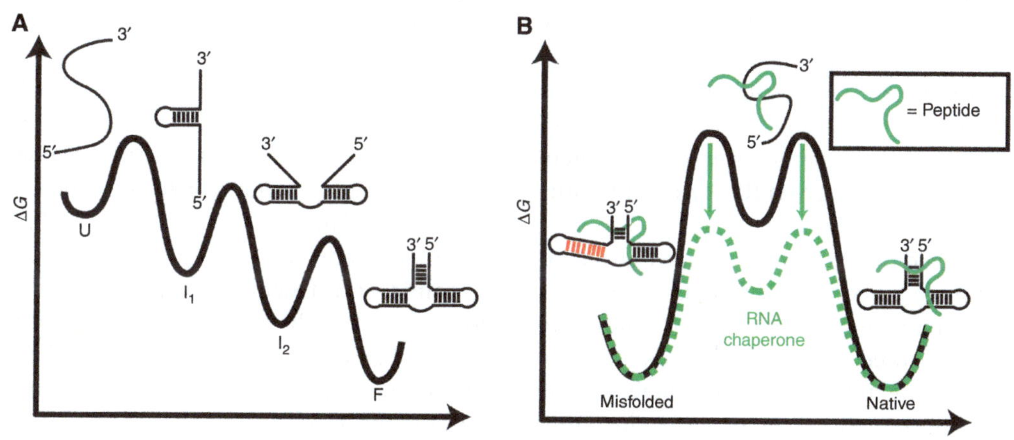

Figure 2. Propensity of RNA to form base-pair interactions and stable secondary structure. (*A*) Short RNAs can form stable structures as a result of high stability of base-pair interactions as shown for the formation of a three-way junction. (*B*) The high stability of secondary structure results in formation of stable misfolded states with alternative secondary structures. Short peptides (green) can act as RNA chaperones and may have aided RNA function in the RNA World and provided a path from the RNA World to the RNA–protein world (Herschlag 1995).

export, localization, translation, and decay (Russell 2008; Jarmoskaite and Russell 2014).

From the perspective of the RNA World, proteins acting as RNA chaperones may have endowed RNA with increased functionality, thereby providing a path out of the RNA World and first steps toward the protein world. The nucleocapsid protein from HIV was shown to increase the rate of conformational steps in hammerhead ribozyme catalysis, despite the fact that these molecules likely never coexist in the same biological context, and certain short peptides with positively charged side chains can also provide this function (Tsuchihashi et al. 1993; Herschlag et al. 1994). More recent work by Holliger and coworkers showed that small positively charged peptides enhance the function of in vitro selected RNA polymerase ribozymes (Tagami et al. 2017). As short peptides would be more readily synthesized by an early ribosome and as nonspecific interactions are more likely than the random occurrence of a peptide with specific recognition and function, it is reasonable that short, nonspecific peptides functioned as early RNA chaperones in an RNA World. The selective advantage from peptides would provide a driving force for evolution to improve the ribosome's accuracy and processivity in peptide synthesis, and with sufficient improvement and time, peptides and short proteins could have emerged that bound specifically to stabilize favorable RNA conformations and that provided one or more side chains that aided molecular recognition or catalysis (Fig. 2B) (Herschlag 1995).

But what would allow proteins to ultimately take over? Several ideas are popular and reasonable, including protein's favorable folding properties, relative to RNA, and the larger number of more catalytically effective side chains of proteins (Narlikar and Herschlag 1997). The diversity of protein side-chains facilitate folding, and the larger number of backbone degrees of freedom of RNA and its uniformly negatively charged backbone are impediments to folding and forming highly preorganized binding pockets and active sites. In addition, from the perspective of the "pathway" of evolution, it may have been easier to diversify the protein than RNA side-chain repertoire given RNA's role in information storage and transmission.

A recently recognized property that may have also contributed to the ultimate takeover by proteins is speed. A survey of association rate constants for ligands binding to RNA revealed that structured RNAs bind ligands orders of magnitude slower than the diffusional limit, whereas many proteins achieve association rates at or near this limit (Gleitsman et al. 2017). Given that slow association rates appear to be universal for RNA, the ability of proteins to more rapidly "grab" and use metabolites may have provided an additional driving force for the ascendancy of proteins. One of several physical mechanisms that may underlie RNA's slow molecular recognition is its propensity to make stable local interactions that may be difficult to escape from, kinetically and thermodynamically. Considering the roles that RNAs currently carry out in biology, RNA's strong local interactions that allow it to effectively bind and discriminate RNA (and DNA) sequences and its ability to form multiple quasi-stable states may account for its aptitude for and remaining functional roles in processes such as pre-mRNA splicing, telomere maintenance, and RNA interference.

Orgel and Crick emphasized the "principle of continuity" in their evolutionary discussions (Crick 1968; Orgel 1968), meaning that one must be able to draw a stepwise evolutionary path from one biological state to another. Here, we suggest that RNA's stability and local structural promiscuity may have allowed it to be the basis of early life, and later, with these strengths also manifesting as limitations, provided an opportunity for proteins to enter and outgrow RNA's use due to the chemical diversity, faster dynamics, and enhanced molecular recognition of proteins.

3 MIDDLE RNA FOLDING EPOCH: CHARACTERIZATION, MODEL BUILDING, AND REVISION USING CATALYTIC RNAs

Once catalytic RNAs were identified, it was clear that they must contain sophisticated tertiary structures. This realization led to intensive efforts to determine their structures and to dissect their catalytic activities. There existed a synergy between structural and catalytic studies, as catalysis indicated that an active structure had been achieved, guiding crystallization trials and empowering folding studies. In the face of challenges to understand RNA structure and folding, new approaches and concepts were developed and adapted from other fields.

We begin by presenting studies at the start of the middle RNA folding epoch (Fig. 1) that characterized the folding process and that led to initial folding models. These models were tested and further refined, as we describe in the second subsection. These advances, and their limits, clarified the need for an ability to obtain an integrated kinetic, thermodynamic, and structural framework for folding of a model RNA. We describe such a framework for a model RNA in Section 3.3, and how its development helped propel RNA folding research to the final epoch focused on predictive quantitative models.

3.1 Characterization of RNA Folding and Early Models

Early structure mapping of the *Tetrahymena* group I intron by hydroxyl radicals generated by Fe(II)•EDTA in solution identified protected and nonprotected groups and suggested that this RNA has an inside and outside, akin to a globular protein (Latham and Cech 1989). In addition, the lore that RNA is difficult to crystallize was challenged by several pioneering groups, ultimately leading to critical structural information from overall architecture down to the atomic level (Pley et al. 1994; Cate et al. 1996; Ferré-D'Amaré et al. 1998). Crystallographic studies of ribozymes emphasized active-site crevices and their apparent globular nature, akin to protein enzymes (Ferré-D'Amaré et al. 1998; Herschlag 1998; Kraut et al. 2003). The exception to this rule was the hammerhead ribozyme, but its crystal structure was shown to represent an inactive form that required a large-scale conformational change or an additional long-range tertiary interaction to form or stabilize the active conformation (Pley et al. 1994; Wang et al. 1999; Martick and Scott 2006; Nelson and Uhlenbeck 2006).

Returning to the *Tetrahymena* ribozyme, the development of time-resolved hydroxyl radical cleavage led to the identification of domains that folded at different times, suggesting the existence of multiple stable intermediates (Sclavi et al. 1998; Brenowitz et al. 2002). Evidence for folding intermediates also came from a time-resolved assay that used DNA oligos complementary to stretches of sequence throughout the ribozyme and RNase H to cleave regions that were exposed to form the DNA/RNA hybrids required for cleavage (Zarrinkar and Williamson 1994). Thus, this large ribozyme (~400 nt) did not show two-state folding like many small proteins. Other studies led to the discovery that the P4-P6 region of the ribozyme (Fig. 3A) is an independently folding, stable domain, and then to the P4-P6 X-ray structure (Fig. 3B) and its widespread use as a model in RNA folding studies (see Sec. 3.3).

As studies continued, evidence accumulated supporting models for kinetic traps in folding. In particular, the frequent observation of faster RNA folding on addition of urea and slower folding at higher concentrations of Mg^{2+} suggested the need to disrupt kinetically trapped species to rejoin a productive folding pathway (Pan and Sosnick 1997; Treiber and Williamson 1999; Woodson 2000). For the *Tetrahymena* ribozyme, a long-range helix—that is, one formed by strands distal in primary sequence—referred to as P3 ("paired" region 3; Fig. 3A) could be replaced by a local helix referred to as "altP3," and stabilization of altP3 enhanced formation of the kinetically trapped state or states (Pan and Woodson 1998; Russell et al. 2006). These observations were consistent with the high stability of RNA secondary structure leading to kinetically trapped states, as was emphasized in early tRNA folding studies.

Following the discovery of catalytic RNA, there was much discussion of the role of Mg^{2+} in RNA folding and function. One early model emerged from studies of the P4-P6 domain of the *Tetrahymena* ribozyme invoking nucleation of RNA structure around a core of site-specifically bound Mg^{2+} ions, analogous to a hydrophobic core in protein folding (Cate et al. 1997). This model arose from the structural observation that Mg^{2+} ions bound to a central component of P4-P6 that folds first and folds at lower Mg^{2+} than is needed to complete folding of P4-P6 (Celander and Cech 1991; Cate et al. 1996, 1997; Sclavi et al. 1998).

Emerging single-molecule fluorescence approaches could readily be applied to RNA, including early studies monitoring conformational changes of a four-way junction and monitoring the folding and catalysis by the *Tetrahymena* ribozyme (Zhuang et al. 2000; Tan et al. 2003). Subsequently, single-molecule fluorescence resonance energy transfer (smFRET) experiments were performed with numerous RNAs, including catalytic RNAs, later-discovered riboswitches (Breaker 2011), and RNA/protein complexes (Zhuang 2005). A general finding was that of "molecular individuality," meaning that different RNA molecules within the population displayed different kinetic, thermodynamic, or catalytic behavior (Zhao and Rueda 2013). Molecular individuality meant that RNA molecules did not equilibrate with one another—or swap behaviors—over timescales of minutes or hours. These observations, the preponderance of trapped folding intermediates, and the apparent ease of forming new trapped states led to a dominant model that RNA folding landscapes are "rugged" (Treiber and Williamson 1999; Ditzler et al. 2008).

3.2 Tests of RNA Folding Models in the Middle Epoch

As researchers gained control of model RNA systems and techniques matured there was an emphasis on testing emerging models. Below, we highlight some examples that helped further our understanding of the RNA folding process.

To test the model of nucleation of RNA structure around a Mg^{2+}-core, a mutant *Tetrahymena* ribozyme lacking the Mg^{2+}-binding domain (P5abc domain composed of helices P5a, P5b, and P5c; Fig. 3A) (Engelhardt et al. 2000) was generated and this truncated ribozyme was incubated in Mg^{2+}. Catalytic activity on addition of P5abc in *trans* arose faster than when folding was initiated by the simultaneous addition of Mg^{2+} and P5abc. Thus, the truncated ribozyme formed a folding intermediate that can bind P5abc and progress to its active form, faster than would occur if there were no intermediate or if unfolding of an off-pathway state were required (Russell and Herschlag 1999). Thus, Mg^{2+} core formation is not an obligate early folding step, and the ribozyme can fold along multiple pathways, minimally with P5abc folding early or late.

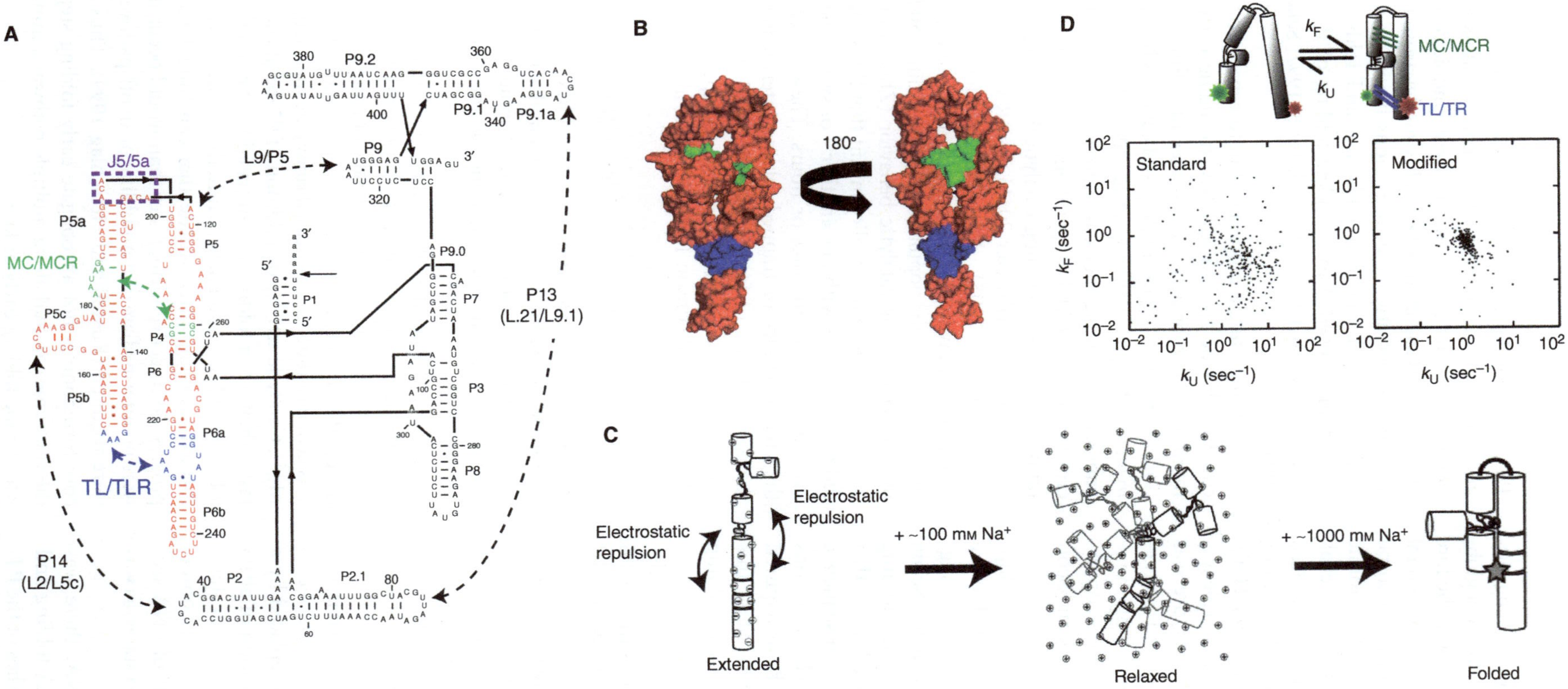

Figure 3. *Tetrahymena* group I intron ribozyme and its P4-P6 domain. (*A*) Secondary structure of the *Tetrahymena* group I intron ribozyme. Long–range contacts are marked with dashed arrows. The P4-P6 domain is highlighted in red, with its metal core/metal core receptor (MC/MCR) and tetraloop/tetraloop receptor (TL/TLR) tertiary contacts colored in green and blue, respectively. Junction J5/5a (purple box) bends and allows coaxial stacks to be roughly parallel to each other. (*B*) X-ray structure of the P4-P6 domain with tertiary contacts colored as in *A* (PDB 1GID; Cate et al. 1996). (*C*) Electrostatic relaxation provided by cations followed by stabilization of a compact state by the formation of tertiary contacts. (*D*) Heterogeneous kinetic behavior of P4-P6 RNA caused by covalent damage during heating and ultraviolet (UV) irradiation. P4-P6 is shown in cartoon form with a pair of donor–acceptor fluorescent dyes (green and red stars) attached for single-molecule fluorescence resonance energy transfer (smFRET) folding studies. Scatter plots show values of the folding and unfolding rate constants (k_F and k_U, respectively) for individual P4-P6 RNA molecules prepared by a standard protocol (*left*) that involves heat/cool renaturation and UV irradiation and by a modified protocol (*right*) in which heating and UV radiation were eliminated during construct preparation and purification. (*C*, from Takamoto et al. 2004; reprinted, with permission, from Elsevier. *D*, from Greenfeld et al. 2011, adapted, with permission, from The American Society for Biochemistry and Molecular Biology.)

The ability to carry out time-resolved small-angle X-ray scattering (SAXS) allowed monitoring of the overall shape of an RNA through early steps in the folding process (Russell et al. 2000). On addition of Mg^{2+}, the *Tetrahymena* ribozyme compacted rapidly (Russell et al. 2002). Compaction occurred well before time-resolved hydroxyl radical probing experiments showed evidence for formation of the first tertiary interactions, and this initial compaction phase was unaffected by ablation of the five long-range tertiary contacts of the *Tetrahymena* ribozyme. Thus, the initial compaction step was distinct from tertiary structure formation, whereas a second compaction step was eliminated for the tertiary-ablated construct (Das et al. 2003). These results revealed that Mg^{2+} is involved in more than one folding step.

To better understand the nature of these folding steps, conformational modeling of SAXS data, in conjunction with hydroxyl radical probing, was performed with simpler model systems (Takamoto et al. 2004). Under low-salt conditions, RNAs tend to adopt extended states, attributable to electrostatic repulsion between the negatively charged phosphates of the RNA backbone (Fig. 3C). Addition of any divalent cation causes compaction, not due to attractive forces but rather from lessening the repulsion that favors extended states and allowing exploration of the larger number of nonextended conformers. This process has been referred to as "electrostatic relaxation" (Das et al. 2003; Takamoto et al. 2004; Lipfert et al. 2014). Monovalent cations also promote electrostatic relaxation, but at far higher concentrations, consistent with expectations from polyelectrolyte theory (Draper et al. 2005; Lipfert et al. 2014).

Specific local metal-ion binding to RNA is of course also observed and is critical for many of RNA's functions. For the P4-P6 domain, its X-ray structure showed five "bound" or localized Mg^{2+} ions (Cate et al. 1996). To determine which of these were directly coupled to folding, Mg^{2+}-promoted folding studies of P4-P6 were performed in the presence of a high concentration of background monovalent cation (Das et al. 2005). The large number of monovalent cations screened nonspecific electrostatic interactions such that the smaller number of available Mg^{2+} would bind only to specific sites. The number of bound Mg^{2+} could be determined by "ion counting," a technique in which associated and free ions are distinguished and quantified by atomic emission spectroscopy or mass spectroscopy. The extent of folding as a function of Mg^{2+} concentration could be followed across the same conditions by hydroxyl radical footprinting. Ion counting revealed the binding of two Mg^{2+} upon folding, and footprinting gave a Hill slope of 2, showing a direct link between the physical association of Mg^{2+} and the thermodynamics of folding. Subsequent experiments using P4-P6 variants with phosphorothioate substitutions to site-specifically alter Mg^{2+} versus Mn^{2+} affinities mapped the identity of the Mg^{2+} sites and further developed thermodynamic models for Mg^{2+} binding (Frederiksen et al. 2012). This approach was borrowed from extensive ribozyme studies that used thio-substitutions to identify and dissect roles for metal ions in catalysis (Piccirilli et al. 1993; Weinstein et al. 1997).

The observation of molecular individuality from smFRET was the subject of much discussion and further tests. Experiments using the hairpin ribozyme showed that this behavior persisted whether molecules were tethered to the surface of a microscope slide or localized in micelles, suggesting that surface effects were not responsible for the observed heterogeneity (Okumus et al. 2004). Persistent molecular behavior over many minutes or even hours and upon denaturation and renaturation could result from deep energetic barriers between the differentially behaving conformations or could result from covalent differences between the molecules caused by synthesis errors or damage. Bulk approaches were used to assess whether covalent heterogeneity was present in populations of P4-P6 RNA molecules that gave persistent heterogeneous behavior in smFRET studies (Fig. 3D) (Greenfeld et al. 2011). Ultraviolet (UV) shadowing used in gel purification and heating used in RNA renaturation were shown to covalently modify the RNA and interfere with folding. A protocol to purify and prepare P4-P6 RNA for smFRET studies without these treatments greatly reduced heterogeneity observed in single-molecule studies close to noise levels (Fig. 3D; compare left and right scatter plots). Evidence for molecular individuality via conversion between states with distinct kinetics and thermodynamics has been shown only for the more complex *Tetrahymena* group I ribozyme, and even for this system interconversion was not complete, suggesting that a portion of the heterogeneous behavior of this molecule may also arise from covalent differences (Solomatin et al. 2010).

Early results, including those described above, identified the presence of and properties of multiple folding intermediates. The next challenge was to determine the order of these species. Russell and colleagues brought to bear multiple approaches to dissect the folding pathways of the *Tetrahymena* ribozyme: smFRET, to provide the time to the native state (N) and the fraction of molecules folding to N versus those folding to a long-lived misfolded state (M); SAXS, to provide information about overall compactness and shape; chemical probing with dimethyl sulfate (DMS), to provide local information about initial and final structural states; and mutations, to stabilize or destabilize known interactions. Depending on the starting monovalent salt conditions, ribozyme molecules preferentially folded along one of three pathways. The pathways are described as "channels" because once a molecule starts down one of the pathways it does not switch to a different pathway; in other words, the barriers between pathways are larger than those along the pathways (Russell et al. 2002). Such channels were also observed for RNase P folding (Pan et al. 1999).

With this information, conditions could be found to populate individual states and probe each to better understand the nature of the intermediates. The most mysterious state was M, as it was shown to have nearly the same overall shape, form the same set of long-range tertiary interactions, and have a nearly identical hydroxyl radical protection pattern as N (the active ribozyme form), yet interconversion between M and N took hours (Russell et al. 2006). Structural mapping and kinetic studies led to the model that M represented a topological isomer of N in which the strands in the core are crossed, relative to N. Subsequent experiments provided strong support for this model and evidence that the transition from M to N was slow because it required the peripheral, long-range tertiary interactions to be broken to allow the strands to disentangle and come together in the correct order (Russell et al. 2006; Wan et al. 2010; Mitchell et al. 2013).

3.3 The Value of an Integrated RNA Folding Framework

The results summarized above for the *Tetrahymena* ribozyme revealed complexities and details of its folding landscape and generalizable lessons. Yet, there is more missing even for this advanced model RNA than is known—for example, we do not know the order of formation of the five long-range tertiary contacts nor the quantitative barrier steps along the folding pathways, and structural information is limited to a few of the many states. In an effort to integrate structural, kinetic, and thermodynamic information, and to provide a more comprehensive and quantitative description of a folding process, we turned to the simple P4-P6 domain (Fig. 3).

P4-P6 RNA folds at a junction (J5/5a; purple box in Fig. 3A) that allows two sets of coaxially stacked helices to be placed side-by-side, as occurs in more complex RNAs, and this arrangement is stabilized by two tertiary contacts between these stacks (Fig. 3B). The discrete nature of the two P4-P6 tertiary contacts allowed each to be effectively ablated, without secondary effects. One of the tertiary interactions is a canonical tetraloop/tetraloop receptor (TL/TLR) motif, and the other is a metal core/metal core receptor (MC/MCR) (Fig. 3B, blue and green, respectively).

Its simplicity and the wealth of prior structural information rendered P4-P6 RNA an attractive target for the determination of an integrated folding framework. The availability of an X-ray structure of fully folded P4-P6 in combination with extensive chemical probing data allowed us to propose a set of structural intermediates on path to the folded state (Fig. 4). Additional intermediates with the P5abc subdomain in an alternative conformation was suggested by nuclear magnetic resonance (NMR), chemical probing, and mutagenesis of an isolated P5abc construct (Wu and Tinoco 1998; Silverman et al. 1999; Xue et al. 2016). Although the overall folding rate was known (Fig. 4, inset), it was not known which step or steps were responsible for the observed folding rate constants. To quantitatively dissect the kinetics and thermodynamics of the individual steps in P4-P6 folding, a set of P4-P6 mutants was generated and their folding properties characterized by smFRET under a variety of solution conditions.

Figure 4 summarizes the folding framework of the P4-P6 domain (Bisaria et al. 2016). Under the conditions studied, the native conformation of the P5abc domain dominates over its alternative conformation, the two critical Mg^{2+} ions bind rapidly and are retained in the thermodynamically favored unfolded state. Following Mg^{2+} binding, the MC/MCR forms first, followed by very rapid formation of the TL/TLR. Superficially, the formation of MC/MCR preceding formation of TL/TLR makes sense as it follows "contact order"—that is, the MC and MCR tertiary contact partners are closer to the J5/5a junction (Fig. 3A) compared with the TL and TLR. However, the scenario is not so simple. P4-P6 folds with ~90% of the RNA molecules forming MC/MCR first, and ~10% forming the more remote contact first, so the contact order preference is rather subtle. Further, a mutation in the MC tertiary partner that weakens the MC/MCR tertiary contact and does so in part by slowing the formation of the MC/MCR inverts the preferred folding pathway, such that the TL/TLR tertiary contact forms first ~90% of the time. Thus, the pathways and choice of pathways are influenced by collision frequencies and also by the properties of the individual tertiary elements. As we describe in the following section, a "reconstitution model" that builds from the properties of RNA structural components can account for, and ultimately predict, these preferences.

3.4 A Rugged Landscape Model for RNA Folding?

As we close this epoch in RNA folding studies, we are positioned to reconsider the model of a rugged folding landscape, which had provided a useful description of early observations. We discuss this model as it has become so prevalent that it has become rare to hear "RNA folding" without reference to its "rugged landscape" and because we can draw lessons from it for understanding the transitions between scientific epochs.

Analogy models, like the rugged landscape, can be beguiling, prompting their acceptance even in the absence of

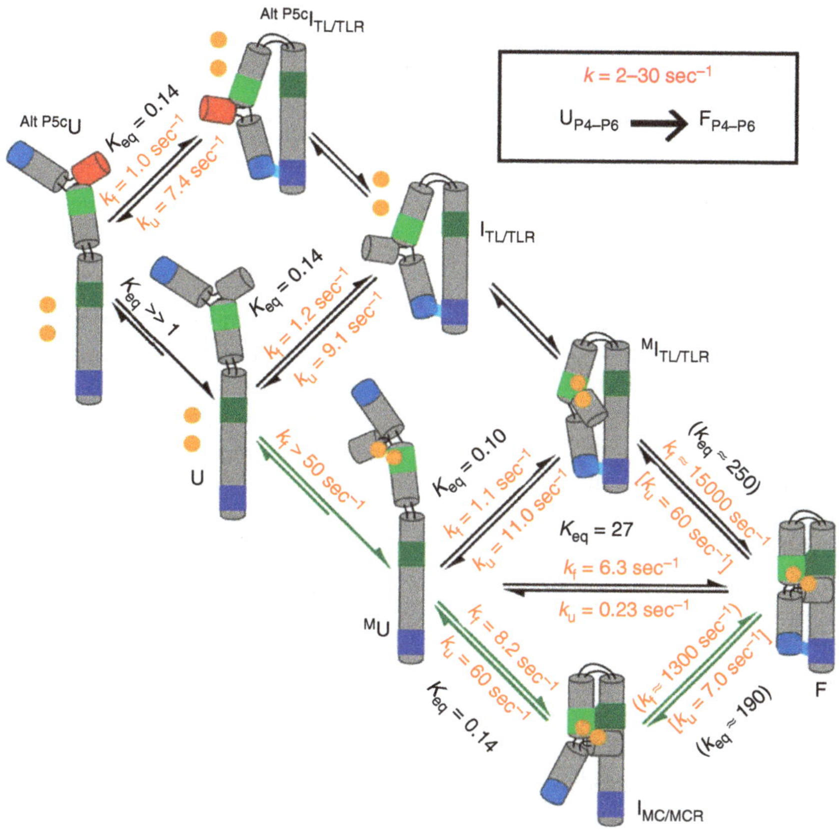

Figure 4. Kinetic and thermodynamic framework of the P4-P6 domain of the *Tetrahymena* group I intron (100 mM KCl, 5 mM $MgCl_2$, pH 8, 23°C). The *inset* shows the overall folding rate constants observed in several studies of overall folding (for references, see Bisaria et al. 2016). (From Bisaria et al. 2016, modified, with permission.)

defined predictions and tests. Indeed, it is not clear what the rugged landscape model predicts. A rugged landscape is consistent with the observation of kinetic traps. But how many, and how deep? We have a concept of rugged hills or mountains, landscape, but this ruggedness is irrelevant to a bird, and what looks smooth to us seems rugged to a smaller mammal or to an insect.

Interestingly, recent results suggest that RNA dynamics may act to smooth the RNA folding landscape. The P5abc subdomain of P4-P6 undergoes a cooperative transition involving a change in secondary structure coupled to tertiary structure formation (Wu and Tinoco 1998; Silverman et al. 1999; Koculi et al. 2012). Previously, it was thought that this transition is "fully" cooperative, such that one rearrangement could not happen without the other. In reality, there must be a limit to this cooperativity, as energetic barriers are finite. Indeed, recent NMR chemical shift exchange experiments have suggested that P5abc occupies a series of local and rare states that are favorable for tertiary structure formation, essentially breaking down a large-scale conformational process into smaller steps that can occur independently and with only modest kinetic and thermodynamic barriers (Xue et al. 2016). This model is supported by recent work that identified folding modules within P5abc that each contribute modestly to an overall cooperativity of ~4 kcal/mol (Gracia et al. 2018). Thus, structural modules within a complex RNA "experiment with" different subconformational states, which smooths the energy landscape and speeds folding transitions.

More broadly, analogies are useful, but only to a point. They help us organize information and think creatively, but analogies cannot substitute for molecular understanding and predictive models.

4 LATE RNA FOLDING EPOCH: TRANSITION TO PREDICTIVE STRUCTURAL AND ENERGETIC MODELS

In science teaching, there is an emphasis on experiments that provide distinct qualitative tests to distinguish between models, such as the molecular nature of genetic material or the semiconservative nature of DNA replication. Although these examples are truly elegant and inspiring, perhaps we underestimate the value of and need for quantitative models and tests. Quantitative predictive models are readily tested, are needed to describe how biological processes work, allow strongly discriminating tests of models, and can reveal missing features or factors in a process.

Although we would like a quantitative model that predicts RNA three-dimensional structure from sequence, this would not be enough. RNA molecules exist as dynamic ensembles, not as single static single conformations, and their dynamic properties dictate their function in complex and essential multistep processes. Thus, in addition to structural models we need quantitative models of RNA dynamics and energetics that can be used to predict relative stabilities of different states and the transition probabilities between them. It is only through a combination of structural, dynamic, and thermodynamic models that we will be able to predict the wealth of complex interactions and processes involving structured RNAs inside the cell.

4.1 RNA Structure Prediction and Design

We emphasized above lessons from the older and larger field of protein folding that were borrowed and used by the RNA community. For structure prediction, the central lesson was that the ability to make correct predictions cannot be determined without blind trials. Before such trials, solution to the protein folding problem seemed to be right around the corner as several approaches appeared to produce accurate "predictions." However, the limitations of these approaches became apparent through CASP (critical assessment of protein structure prediction) in which modelers were given time to make predictions before release of newly solved structures (Moult et al. 2014). The lesson from this history is that unknown sources of variation in complex problems, unconscious biases, and publication bias provide limitations such that it is insufficient to test models with "retrodictions"; instead, it is only through true predictions, as in CASP, that our understanding of a process can be tested.

Remarkably, structures for many proteins of modest size can be predicted in blind trials using ROSETTA, which uses energetic potentials based largely on empirical information from the large database of protein crystallographic structures (Das and Baker 2007; Kaufmann et al. 2010). Das and colleagues adapted ROSETTA for use in RNA structure prediction, and further integrated information from chemical structure mapping (Cheng et al. 2015). In the RNA-Puzzles blind trials, they have been able to predict overall RNA tertiary structure, in favorable cases to near-atomic resolution (Miao et al. 2017).

Before successful atomic-scale modeling, the presence of recurring RNA motifs was recognized and allowed modeling of RNA structures and engineering of new structures of modest complexity (Westhof et al. 1996; Jaeger and Leontis 2000). Now, similarly complex structured RNAs can be designed in an automated fashion from a diverse set of helix, junction, and tertiary motif elements (Yesselman et al. 2017).

4.2 Developing a Reconstitution Model for RNA Folding

An early goal in protein folding research was to obtain an additive accounting of energetics. Occasional additive contributions to stability were observed, general principles were established, and the conversion of these rules to approximate energy terms combined with empirical information from the protein data bank (PDB) has led to successes in protein structure prediction (Kaufmann et al. 2010; Huang et al. 2016). However, the larger goal of predicting protein folding energetics has proven elusive, apparently because of the many interacting groups within folded proteins that lead to a vast number of idiosyncratic energetic terms. In contrast, RNA's apparently simpler structural and energetic properties have allowed the development of a "reconstitution model" for RNA folding kinetics and thermodynamics from its component parts (Fig. 5A).

Three simplifying features of folded RNAs are foundational for the "reconstitution model."

1. RNA folding is hierarchical. RNA secondary structure is stable in the absence of tertiary contacts, in contrast to the case with proteins, allowing energy terms for tertiary interactions to be considered distinct from those for helix formation (Fig. 5B). Even in cases with alternative secondary structures (e.g., riboswitches), tertiary folding can be considered in the context of each of the limited number of secondary structure alternatives.

2. Tertiary contacts are sparsely distributed. Folded RNAs appear globular, like proteins; however, closer inspection reveals that most RNA residues interact only with their primary or secondary structure neighbors. Unlike proteins, very few residues are involved in tertiary interactions. This can be seen in the P4-P6 structure, where the side-by-side stacked helices come together only at two tertiary contact interfaces (Fig. 3B).

3. Structured RNAs are built from a limited set of structural motifs. Phylogenetic analysis and structural comparisons revealed the existence of recurring tertiary motifs in RNA (Costa and Michel 1995; Leontis et al. 2006) and Figure 5C shows one example. The TL/TLR in P4-P6 RNA described above is found in several other RNAs. These TL/TLRs, despite being embedded in distinct RNAs, closely superimpose. This structural modularity

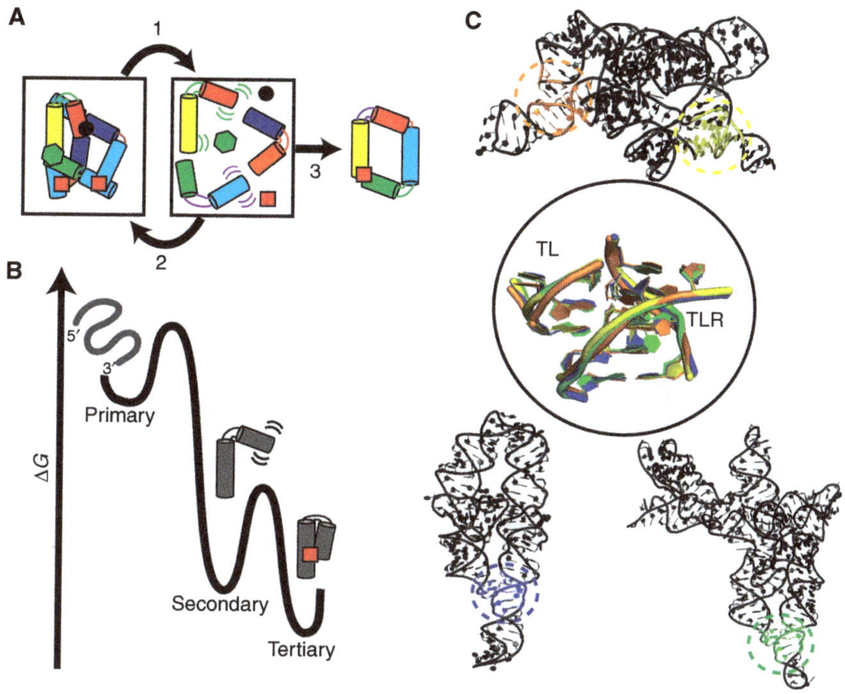

Figure 5. Hierarchical folding, sparsity of tertiary contacts, and structural modularity of RNA lead to a "reconstitution model" for RNA folding. (*A*) Schematic representation of a "reconstitution model" of RNA folding: (*1*) A complex RNA can be broken down into helix-junction-helix elements (HJH; cylinders and strings) and tertiary contact motifs (circles, squares, hexagons), and the energetic and conformational properties of these elements studied in isolation; (*2*) the behavior of the isolated elements can be reconstituted to understand their collective behavior in natural structured RNAs; and (*3*) novel RNAs can be constructed with structural motifs and can be engineered to have tunable energetic and conformational properties. (*B*) Free-energy diagram representing hierarchical folding of RNA. As an example, the folding of a simple RNA stabilized by a single tertiary contact is shown as a cartoon above the free-energy diagram. Helices are represented by cylinders, junctions as strings connecting the helices, and the single tertiary contact as a red square. (*C*) X-ray structures of the GAAA/11ntR tetraloop/tetraloop-receptor (TL/TLR) tertiary contact motif extracted from diverse structured RNAs and superimposed P4-P6 (PDB 1G1D; blue), RNase P (PDB 1NBS; green), and the *Azoarcus* group I intron (PDB 1ZZN; yellow and orange).

suggests an "energetic" modularity that could be foundational for a "reconstitution model."

4.3 The Reconstitution Model

Consider two tertiary contact partners separated by a secondary structure. Figure 6A shows a simple schematic example with two helices connected by a single junction, referred to as a helix-junction-helix (HJH) element, and with each helix containing a tertiary contact partner. The probability of forming the folded structure (F) is a function of two factors: (1) the probability that the two tertiary contact partners will align for formation of the tertiary contact (ΔG_{Align}) and (2) the strength of the tertiary interactions (ΔG_{Tert}; Fig. 6A).

The unfolded RNA (U) exists as an ensemble of conformations and as a narrower ensemble in the folded state (F), and the probability of folding is a function of these conformational spaces (Fig. 6B). For example, a more rigid junction (e.g., one that prefers stacked states) will be less likely to bend and thus less likely to form the folded state, whereas a junction that has internal interactions that favor a bent state will be more likely to fold, provided that the bent state overlaps with aligned states that allow the tertiary interaction to form. Experimentally, NMR residual dipolar coupling (RDC) and X-ray scattering interference can be used to determine RNA HJH conformational ensembles (Salmon et al. 2014; Shi et al. 2015).

Junctions play an obvious role in shaping conformational ensembles, but the conformational properties of helices also matter. For example, changing helix length rotates and translates the positions of the tertiary contact partners in relation to each other altering the probability of tertiary contact formation. Also, the formation of tertiary contacts between helices connected by floppier junctions will be less sensitive to changes in the conformational properties (e.g., length) of the helices as the flexibility of the junctions can "accommodate" misalignments more easily compared with

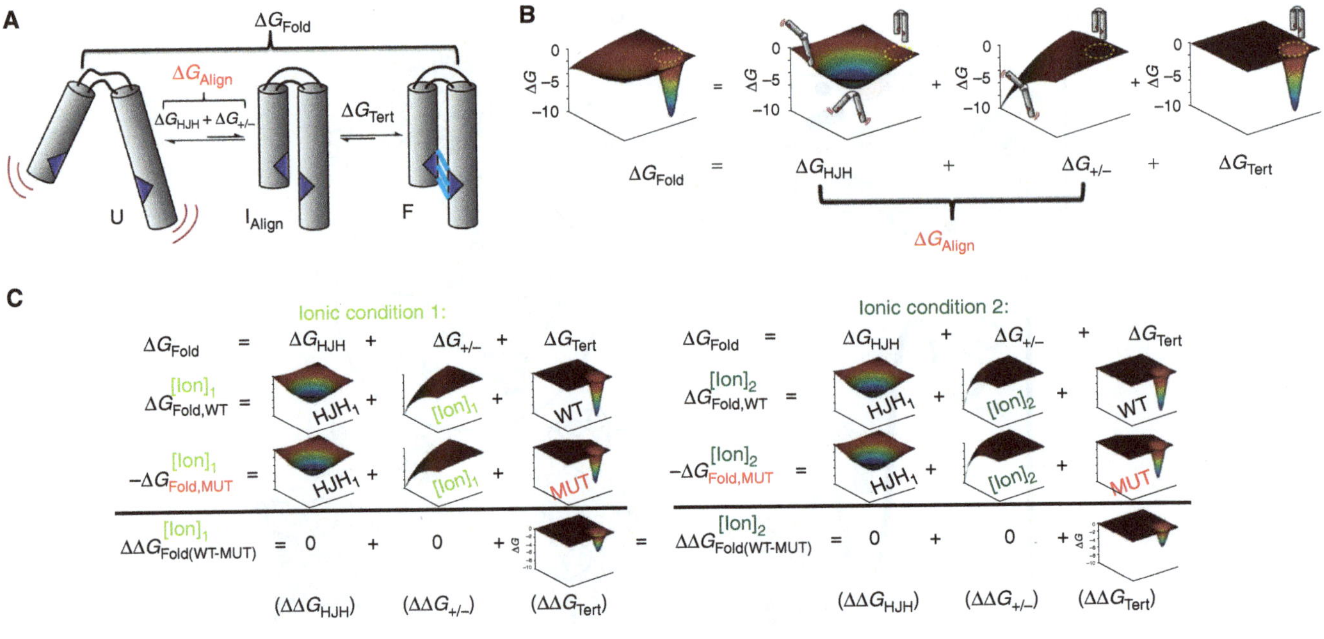

Figure 6. Schematic of a "reconstitution model" of RNA folding energetics. (*A*) Folding of a simple RNA involving the formation of a single tertiary contact (blue triangles) between two helices (cylinders) connected by a junction (strings). The free energy of folding (ΔG_{Fold}) is broken down into two terms: (1) the probability that the tertiary contact partners are found aligned for productive collision ($\Delta G_{\mathrm{Align}}$) and (2) the free energy associated with the formation of the tertiary contact (ΔG_{Tert}). (*B*) Each of the energy terms in *A* is represented in 3D free-energy landscapes. $\Delta G_{\mathrm{Align}}$ can be further broken down into an energy term intrinsic to the conformational preferences of the junction and the helices containing the tertiary contact partners (ΔG_{HJH}) and an energy term reflecting the electrostatic interactions between RNA and surrounding ions and between RNA helices ($\Delta G_{+/-}$). (*C*) Quantitative prediction of the "reconstitution model" of RNA. According to this model, the free energy of formation of a tertiary contact ΔG_{Tert} is separable from the free energies associated with the conformational search of the junction ΔG_{Tert} and the electrostatic interactions with ions $\Delta G_{+/-}$. Hence, this model predicts that the energetic effect of mutating a tertiary contact motif is constant across different ionic conditions. (From Bisaria et al. 2017, reprinted, with permission.)

rigid junctions. These effects have been illustrated using models with more mobile versus more constrained junctions (Chu et al. 2009).

Of course, the tertiary contact motif also matters for folding, most obviously because the folded state, F, is highly unlikely to be populated without stabilization from a tertiary interaction. In the "reconstitution model," the contribution of the tertiary motif to the stabilization of F is represented explicitly by the ΔG_{Tert} term. Thus, the more tertiary hydrogen bonds and stacking interactions between the tertiary contact partners, the more favorable this energy term will be; moreover, different tertiary motifs will also have distinct conformational preferences. Thus, an RNA containing HJH1 may be more stable with tertiary motif A than tertiary motif B, whereas an RNA containing a different HJH (HJH2) may instead be more stable with tertiary motif B than tertiary motif A, because of better matching of the conformational preferences of HJH2 and tertiary motif B.

Considering RNA secondary structure, we can predict duplex stability with reasonable accuracy from a simple "energetically additive" model that accounts for the identity of each base-pair and its nearest neighbors (Zuker et al. 1999). In contrast, the above considerations indicate that we do not expect this same simple energetic additivity for RNA tertiary structure—we cannot assign universal free-energy values to each structural element and add them to obtain the overall tertiary folding stability. Instead, we need to "add," or convolve, conformational ensembles of the individual elements—that is, the junctions, helices, and tertiary contact motifs present. The "reconstitution model" holds that there is "ensemble additivity" such that a convolution of ensembles determined for isolated RNA elements can describe the conformational ensemble of a complex RNA (after accounting for steric overlaps), and that this ensemble, along with a constant free-energy term indicative of the strength of each particular tertiary contact motif, can be used to "predict" the folding stability for an arbitrary RNA. Correspondingly, the "reconstitution model" readily handles complex scenarios with multiple HJHs and multiple tertiary contacts by combining multiple HJH ensembles with each tertiary interaction specified as formed or not to model each folding step.

4.4 Testing the Reconstitution Model

Free-energy landscapes are added in the "reconstitution model"; as free energy is directly convertible to probabilities, we can also think of this as combining (multiplying) probability landscapes (Fig. 6B). According to the "reconstitution model," a perturbation in one of the constituent energy landscapes should not alter the others. Thus, a mutation that alters the intrinsic stability of a tertiary contact motif (ΔG_{Tert}) is predicted to have no effect on the energetic consequences from changing the HJH element (ΔG_{HJH}) or from changing the ionic conditions ($\Delta G_{+/-}$). Note that the alignment term discussed above (i.e., $\Delta G_{\mathrm{Align}}$) is broken into two terms (Fig. 6), one representing the conformational preferences of the particular junction sequence (ΔG_{HJH}) and another related to the electrostatic interactions between RNA elements and surrounding ions ($\Delta G_{+/-}$).

These properties lead to predictions that were tested using smFRET (Bisaria et al. 2017). For example, the folding energetics for the P4-P6 domain with the WT TL/TLR tertiary contact motif and a mutant TL/TLR were measured under two different ionic conditions (Fig. 6C). The "reconstitution model" predicts that the overall effect of the mutation on folding ($\Delta\Delta G_{\mathrm{Fold}}$) is the same at each ionic condition, and this predicted result was observed across several ionic conditions and held for several different mutations (Bisaria et al. 2017). Analogously, these mutations to the tertiary motif gave the same energetic effect ($\Delta\Delta G_{\mathrm{Fold}}$) in different HJH contexts—that is, embedded in different RNAs (Bisaria et al. 2017). These results showed the separability of terms foundational to the "reconstitution model."

We have emphasized thermodynamic applications of the "reconstitution model," but it is equally applicable to folding kinetics, as folding kinetics can similarly be described in terms of free-energy landscapes. Indeed, the first evidence suggestive of this model was obtained from kinetic folding data (Bisaria et al. 2016). The folding framework for P4-P6 RNA (Fig. 4) intriguingly revealed the same unfolding rate constant, within error, for all steps in which the TL/TLR tertiary contact motif dissociated. Previously, Nesbitt and colleagues had extensively studied this same tertiary motif in isolation, by embedding the TL and TLR into a simple construct in which the TL and the TLR are connected by a A_7 tether (Hodak et al. 2005); indeed, this construct was used by Bisaria and colleagues in the thermodynamic tests of the "reconstitution model" described above. Remarkably, the unfolding of this simple construct had the same rate constant as those in P4-P6, and this was confirmed across a range of conditions (Bisaria et al. 2016, 2017). These observations suggested that the unfolding rate for a given tertiary motif is an intrinsic property of that motif. In contrast, folding rate constants vary considerably, as expected for the "reconstitution model" and described immediately below.

The kinetic version of the "reconstitution model" can be described as an RNA diffusion-collision model, a model borrowed from protein folding but likely more apt for RNA folding giving the cleaner separation between secondary and tertiary structure formation (Bisaria et al. 2016). As emphasized above, unfolding rate constants are local properties of the tertiary motif involved and thus are transferable across RNAs, for a given set of conditions. In contrast, the rate constants for each folding step are determined by the probability of the tertiary elements encountering one another, which is determined by the free-energy landscapes of the HJH elements and the ionic conditions, and by the probability that, on collision, the tertiary motif will form—that is, the probability of a productive collision.

We noted earlier in this review that RNA association rate constants are considerably slower than diffusion (Gleitsman et al. 2017). There is considerable evidence for conformational changes between free and bound RNA tertiary elements, including the canonical 11ntR TL/TLR highlighted herein (Fig. 7A) (Butcher et al. 1997; Gleitsman et al. 2017). Bonilla and colleagues used smFRET to dissect the folding behavior of this tertiary contact motif along with a set of TL/TLR variants under a wide range of solution conditions. These results in combination with structural characterization of the unbound canonical 11ntR (Butcher et al. 1997) and smFRET studies of the canonical GAAA/11ntR TL/TLR (Fiore et al. 2012) led to a model for the order of conformational changes along the folding pathway and the nature of the transition state for TL/TLR formation (Fig. 7B) (Bonilla et al. 2017). This study also identified an additional free-energy term to be incorporated into the "reconstitution model" ($\Delta G_{\mathrm{Tert},+/-}$), as the formation of some TL/TLR variants involve additional electrostatic and metal ion effects, an observation consistent with the considerable rearrangements observed on binding. Nevertheless, the electrostatic effects specific to a TL/TLR variant were conserved between different HJH contexts and is therefore intrinsic to the TL/TLR variant (Bonilla et al. 2017), consistent with the "reconstitution model."

In summary, there is strong evidence for the basic features of the "reconstitution model," a major step forward because this model is, in principle, predictive. Nevertheless, there are complexities, as noted by the specific ionic effects on certain tertiary contacts and as presented by the large number of RNA structural elements that require quantitative characterization to broadly apply the "reconstitution model." The following section briefly outlines a new

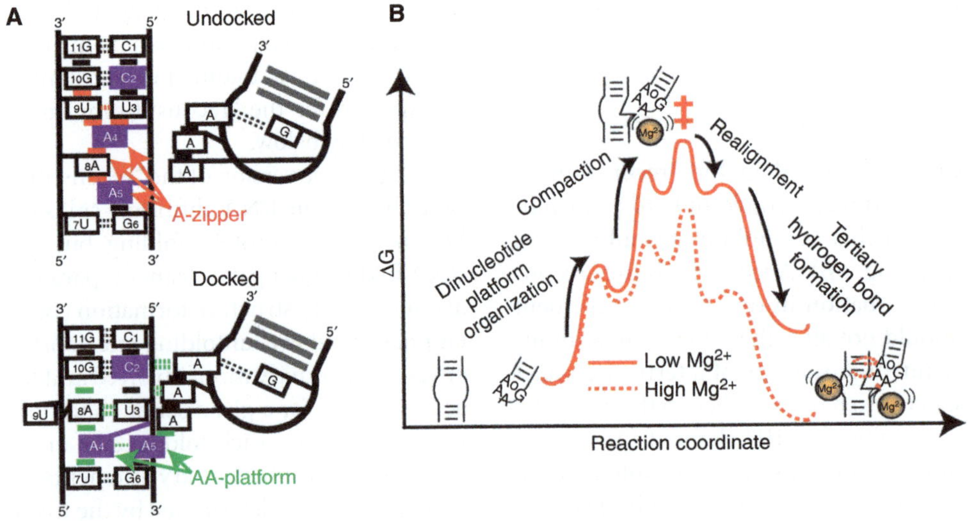

Figure 7. Conformational rearrangements on tetraloop/tetraloop receptor (TL/TLR) formation. (*A*) Comparison of undocked (*top*) and docked (*bottom*) structures of the canonical GAAA/11ntR TL/TLR motif indicate that the 11ntR tetraloop-receptor undergoes a series of conformational changes on binding to the GAAA tetraloop (Cate et al. 1996; Butcher et al. 1997; Davis et al. 2005). The structure of the GAAA tetraloop does not change significantly on docking (Heus and Pardi 1991; Fiore and Nesbitt 2013). (*B*) Single-molecule fluorescence resonance energy transfer (smFRET) dissection of the folding kinetics and thermodynamics of a set of TL/TLR sequence variants across a range of ionic conditions led to a model for the order of conformational changes along the folding pathway of the TL/TLR motif (Bonilla et al. 2017). (From Bonilla et al. 2017, reprinted, with permission, from the American Chemical Society.)

technology that is beginning to provide this needed information.

4.5 High-Throughput Characterization of RNA Elements and Their Ensemble Properties

The broad implementation of the "reconstitution model" rests on the availability of conformational ensembles for many structural elements. A simplifying feature of this model is that atomic-level data about junctions and tertiary contact elements are not required. Instead, what is needed are the preferred orientations of the helices emanating from junctions to determine the spatial probabilities of the tertiary contact elements connected to the helices. Additionally, for tertiary contacts, we need intrinsic stabilities and the orientation of the helices emanating from the formed tertiary contact, which depend on the conformational preferences of the tertiary contact itself. Current approaches to determine RNA conformational ensembles, NMR residual dipolar coupling, and X-ray scattering interference (Salmon et al. 2014; Shi et al. 2015) are powerful tools for obtaining the detailed conformational ensembles of structural elements of interest; however, they are low throughput and thus not capable of providing information for the wide array of RNA structural elements.

Recently, the high-throughput RNA-MaP platform developed by Greenleaf and colleagues (Denny and Greenleaf 2019) has been applied to this problem, yielding insights and models for each class of RNA structural element, helices, junctions, and tertiary contact motifs (Fig. 8) (Denny et al. 2018; Yesselman et al. 2018; S Bonilla, SK Denny, N Bisaria, et al., in prep.) The approach uses an engineered dimer, referred to as "TectoRNA," that assembles via formation of two distinct TL/TLR tertiary contacts (Fig. 8B) (Jaeger and Leontis 2000; Davis et al. 2005; Geary et al. 2008). The TectoRNA system was used as a scaffold for the insertion of a wealth of two-way junctions and components that could be systematically and broadly varied: helix length and content, junction type and sequence, and tertiary motif identity (Fig. 8C). Critically, the assay is direct, provides thermodynamic measurements, and can be performed rapidly, in parallel, and with high precision for tens of thousands of constructs.

4.6 Helices

Comparisons of helices in X-ray structures suggest sequence-specific differences in the conformational preferences of base pairs. Indeed, preliminary studies indicated that changes in the sequence of the TectoRNA helices alter dimerization by nearly 100-fold. Yesselman constructed a computational model that predicts the sequence dependence of the helix ensembles, based on the large numbers of conformations observed for each possible base-pair step in

Cite as *Cold Spring Harb Perspect Biol* doi: 10.1101/cshperspect.a032433

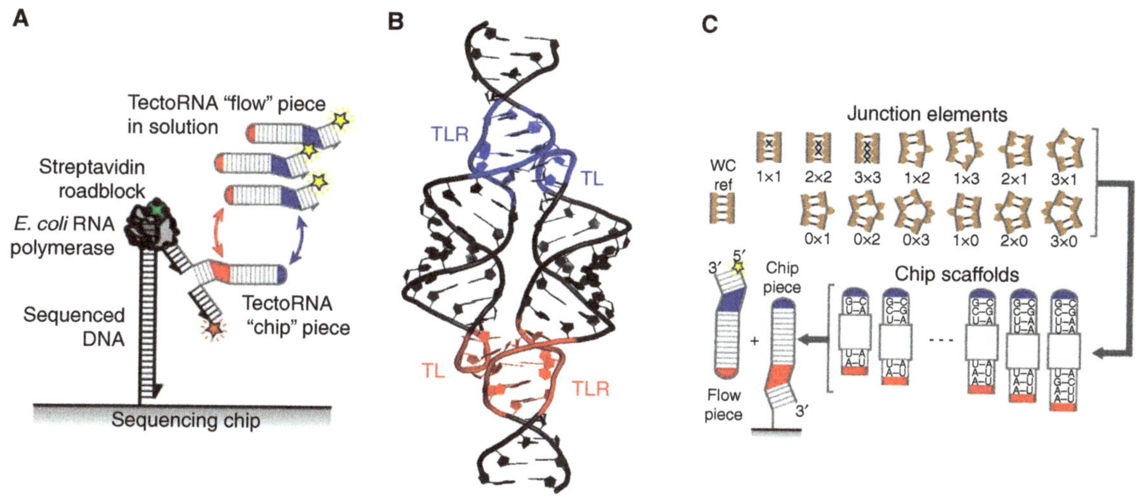

Figure 8. RNA-MaP method applied to the study of RNA tertiary elements. (*A*) The RNA-MaP platform with a bound expressed RNA ("chip" piece) and a "flow" piece RNA added in *trans* (Denny et al. 2018). These RNAs combine to form a tertiary "TectoRNA." (*B*) A model for the tertiary structure of the TectoRNA homodimer (PDB: 2ADT; Davis et al. 2005). A heterodimer in which one of the canonical GAAA/11ntR TL/TLRs is replaced by an orthogonal GGAA/R1 (Geary et al. 2008) was used as a "host" for different helix sequences and lengths, different junction elements, and different tertiary motif elements. (*C*) Examples of junction elements used in these studies. (*A*,*C*, Reproduced, with permission, from Denny et al. 2018.)

X-ray structures from the PDB. This model quantitatively predicted the dimerization affinity for thousands of TectoRNA sequences, with an R^2 of 0.71 and a root-mean-square deviation (RMSD) of 0.34 kcal/mol (Yesselman et al. 2018). Thus, helices can vary significantly in their conformational ensembles and substantially alter tertiary folding energetics, and an initial quantitative model of the ensemble of helix conformations is reasonably predictive.

4.7 Junctions

The TectoRNA was used as a scaffold to embed thousands of junctions. Each junction was embedded in many TectoRNA contexts (Fig. 8C), in which the junction conformational properties could stabilize or destabilize TectoRNA association. The collection of tectoRNA affinities for each junction provided a "thermodynamic fingerprint" for that junction that could be compared with the thermodynamic fingerprints of other junctions. Importantly, these thermodynamic fingerprints reflect the conformational preferences of the junctions; hence, two junctions with nearly identical thermodynamic fingerprints are suggested to have similar conformational ensembles (Denny et al. 2018). Junctions with the same number of unopposed residues tended to have similar thermodynamic fingerprints, and thus likely conformational behavior, as predicted by Al-Hashimi and colleagues (Bailor et al. 2010). Nevertheless, substantial sequence-dependent variation was observed. Further, although there is insufficient X-ray data for any individual junction to build ensembles akin to those developed for helices, Denny and colleagues could group junctions with similar thermodynamic fingerprints and known structures to obtain crude ensemble models for dozens of junctions. This model also had predictive power when combined with the helix model described above, whereas incorporating single X-ray junction structures gave poor predictive power (Denny et al. 2018; Yesselman et al. 2018).

4.8 Tertiary Contact Motifs

A range of tertiary contact motifs can be incorporated in TectoRNA constructs, allowing comparison of their thermodynamic properties. Investigation of more than 1000 TL/TLR variants has revealed different classes of conformational behaviors, most simply those that have conformational preferences similar as that of the canonical GAAA/11ntR, and those with distinct thermodynamic fingerprints (S Bonilla, SK Denny, N Bisaria, et al., in prep.). Whereas the available data suggest a very narrow conformational ensemble for the canonical GAAA/11ntR, the wide tolerance for alignment changes showed by several of the other TL/TLRs suggests broadened conformational landscapes.

The ability to carry out high-throughput and quantitative measurements provides extensive information about conformational ensembles and energetic properties of RNA helices, junctions, and tertiary contact motifs. Although this approach to broadly and deeply interrogate RNA structural elements is in its infancy, it is clear that we have transitioned to the predictive phase in RNA folding and dynamics. Future work will benefit from additional host systems, beyond the TectoRNA dimer, to more broadly explore conformational space and to profile junctions and tertiary contact motifs with very different conformational preferences. In addition, although the initial ensemble computational models have impressive predictive power, additional tests, including orthogonal methods such as NMR and X-ray scattering interference that give more direct structural and ensemble information, will allow these models to be improved.

5 CLOSING PERSPECTIVE

Scientific endeavors often feel like random walks punctuated by occasional discoveries. Discrete discoveries, such as the observation of new phenomena or the generation of structures of biological complexes are readily recognized as "quantized" steps forward; in contrast, results from characterization and dissection are harder to directly map to progress. Organization of the scientific process by epoch helps clarify the necessity of a breadth of studies, a necessity that is more apparent in retrospect (Fig. 1 and above discussion).

The field of RNA folding was marked by a continued characterization of additional structured RNAs at similar levels of description. The perspective of scientific epochs helped identify what next steps were needed to reach the ultimate goal of predictive models. Given this experience, we encourage those developing and using the powerful genomic-scale technologies and other cutting-edge approaches described in this collection and elsewhere to consider also how these tools may lead to predictive and testable models, to go beyond continued characterization of RNA biological phenomena.

Understanding RNA folding has, in essence by definition, required an atomic and molecular-based perspective. But often, biological questions are framed in terms of sequences, pathways, processes, and cellular features abstracted from their underlying molecular properties. Given this, one might ask how generally applicable are the findings and lessons from RNA folding to areas and questions seemingly more directly connected to biology and medicine. We offer the following perspectives.

We are heartened by the biological insights that have emerged from seemingly specialized and remote studies of RNA folding, including insights into the probability and potential properties of an RNA World, possible steps in the transition to protein-centric life on Earth, the occurrence and roles of RNA chaperones, and reasons for the use of RNA-based machines for several current-day processes involving nucleic acid recognition and complex controlled reaction cycles. More broadly, these molecules carry out biological functions through physical interactions and chemical reactions, so we should aim for understanding at that level; more practically, given the myriad complexity of biology and biological observations, the ability of molecular models to constrain and steer models, and the ability to alter molecules to incisively distinguish between models, argues for the adaptation of molecular perspectives, especially in this age of genomics and big data.

We see the need for and power of frameworks, here for RNA folding, that break a complex process into individual steps. Cellular processes are even more complex, suggesting that progress in understanding, and ultimately predicting cellular control and function, will benefit greatly from—and likely require—experiments that allow individual steps to be considered and probed. Finally, we reiterate the power of quantitative predictions, because of their ability to finely discriminate between models and the need for quantitation to model, understand, and predict processes and responses in biology.

ACKNOWLEDGMENTS

We thank Rick Russell, Joseph Yesselman, and members of the Herschlag laboratory for discussions and comments on the manuscript. S.B. and N.B were supported by the National Institutes of Health (NIH) Grant P01GM066275 to D.H.; S.B. was supported by the National Science Foundation (NSF) GRFP under grant DGE-114747.

REFERENCES

**Reference is in this book.*

Adamala K, Szostak JW. 2013. Nonenzymatic template-directed RNA synthesis inside model protocells. *Science* **342:** 1098–1100.

Bailor MH, Sun X, Al-Hashimi HM. 2010. Topology links RNA secondary structure with global conformation, dynamics, and adaptation. *Science* **327:** 202–206.

Bisaria N, Greenfeld M, Limouse C, Pavlichin DS, Mabuchi H, Herschlag D. 2016. Kinetic and thermodynamic framework for P4-P6 RNA reveals tertiary motif modularity and modulation of the folding preferred pathway. *Proc Natl Acad Sci* **113:** E4956–E4965.

Bisaria N, Greenfeld M, Limouse C, Mabuchi H, Herschlag D. 2017. Quantitative tests of a reconstitution model for RNA folding thermodynamics and kinetics. *Proc Natl Acad Sci* **114:** E7688–E7696.

Bonilla S, Limouse C, Bisaria N, Gebala M, Mabuchi H, Herschlag D. 2017. Single-molecule fluorescence reveals commonalities and distinctions among natural and in vitro-selected RNA tertiary motifs in a multistep folding pathway. *J Am Chem Soc* **139:** 18576–18589.

Breaker RR. 2011. Riboswitches and the RNA world. In *RNA worlds: From life's origins to diversity in gene regulation* (ed. Atkins JF, Gesteland RF, Cech TR), pp. 63–77. Cold Spring Harbor Laboratory Press, Cold Spring Harbor, NY.

Brenowitz M, Chance MR, Dhavan G, Takamoto K. 2002. Probing the structural dynamics of nucleic acids by quantitative time-resolved and equilibrium hydroxyl radical "footprinting." *Curr Opin Struct Biol* **12:** 648–653.

Butcher SE, Dieckmann T, Feigon J. 1997. Solution structure of a GAAA tetraloop receptor RNA. *EMBO J* **16:** 7490–7499.

Cate JH, Gooding AR, Podell E, Zhou K, Golden BL, Kundrot CE, Cech TR, Doudna JA. 1996. Crystal structure of a group I ribozyme domain: Principles of RNA packing. *Science* **273:** 1678–1685.

Cate JH, Hanna RL, Doudna JA. 1997. A magnesium ion core at the heart of a ribozyme domain. *Nat Struct Biol* **4:** 553–558.

Cech TR. 2000. The ribosome is a ribozyme. *Science* **289:** 878–879.

Celander DW, Cech TR. 1991. Visualizing the higher order folding of a catalytic RNA molecule. *Science* **251:** 401–407.

Cheng CY, Chou F-C, Das R. 2015. Modeling complex RNA tertiary folds with Rosetta. *Methods Enzymol* **553:** 35–64.

Chu VB, Lipfert J, Bai Y, Pande VS, Doniach S, Herschlag D. 2009. Do conformational biases of simple helical junctions influence RNA folding stability and specificity? *RNA* **15:** 2195–2205.

Copeland RA. 2000. A brief history of enzymology. In *Enzymes: A practical introduction to structure, mechanism, and data analysis*, pp. 1–10. John Wiley, New York.

Costa M, Michel F. 1995. Frequent use of the same tertiary motif by self-folding RNAs. *EMBO J* **14:** 1276–1285.

Crawford RT. 1917. The important epochs in the development of astronomy. *Publ Astro Soc Pac* **29:** 233–244.

Crick FH. 1968. The origin of the genetic code. *J. Mol Biol* **38:** 367–379.

Das R, Baker D. 2007. Automated de novo prediction of native-like RNA tertiary structures. *Proc Natl Acad Sci* **104:** 14664–14669.

Das R, Kwok LW, Millett IS, Bai Y, Mills TT, Jacob J, Maskel GS, Seifert S, Mochrie SGJ, Thiyagarajan P, et al. 2003. The fastest global events in RNA folding: Electrostatic relaxation and tertiary collapse of the *Tetrahymena* ribozyme. *J Mol Biol* **332:** 311–319.

Das R, Travers KJ, Bai Y, Herschlag D. 2005. Determining the Mg^{2+} stoichiometry for folding an RNA metal ion core. *J Am Chem Soc* **127:** 8272–8273.

Davis JH, Tonelli M, Scott LG, Jaeger L, Williamson JR, Butcher SE. 2005. RNA helical packing in solution: NMR structure of a 30 kDa GAAA tetraloop-receptor complex. *J Mol Biol* **351:** 371–382.

* Denny SK, Greenleaf WJ. 2019. Linking RNA sequence, structure, and function on massively parallel high-throughput sequencers. *Cold Spring Harb Perspect Biol* doi: 10.1101/cshperspect.a032300.

Denny SK, Bisaria N, Yesselman JD, Das R, Herschlag D, Greenleaf WJ. 2018. High-throughput investigation of diverse junction elements in RNA tertiary folding. *Cell* doi: 10.1016/j.cell.2018.05.038.

Ditzler MA, Rueda D, Mo J, Håkansson K, Walter NG. 2008. A rugged free energy landscape separates multiple functional RNA folds throughout denaturation. *Nucleic Acids Res* **36:** 7088–7099.

Draper DE, Grilley D, Soto AM. 2005. Ions and RNA folding. *Annu Rev Biophys Biomol Struct* **34:** 221–243.

Engelhardt MA, Doherty EA, Knitt DS, Doudna JA, Herschlag D. 2000. The P5abc peripheral element facilitates preorganization of the *Tetrahymena* group I ribozyme for catalysis. *Biochemistry* **39:** 2639–2651.

Ferré-D'Amaré AR, Zhou K, Doudna JA. 1998. Crystal structure of a hepatitis delta virus ribozyme. *Nature* **395:** 567–574.

Fiore JL, Nesbitt DJ. 2013. An RNA folding motif: GNRA tetraloop-receptor interactions. *Q Rev Biophys* **46:** 223–264.

Fiore JL, Holmstrom ED, Nesbitt DJ. 2012. Entropic origin of Mg^{2+}-facilitated RNA folding. *Proc Natl Acad Sci* **109:** 2902–2907.

Frederiksen JK, Li N-S, Das R, Herschlag D, Piccirilli JA. 2012. Metal-ion rescue revisited: Biochemical detection of site-bound metal ions important for RNA folding. *RNA* **18:** 1123–1141.

Gartland WJ, Sueoka N. 1966. Two interconvertible forms of tryptophanyl sRNA in *E. coli*. *Proc Natl Acad Sci* **55:** 948–956.

Geary C, Baudrey S, Jaeger L. 2008. Comprehensive features of natural and in vitro selected GNRA tetraloop-binding receptors. *Nucleic Acids Res* **36:** 1138–1152.

Gleitsman KR, Sengupta RN, Herschlag D. 2017. Slow molecular recognition by RNA. *RNA* **23:** 1745–1753.

Gracia B, Al-Hashimi HM, Bisaria N, Das R, Herschlag D, Russell R. 2018. Hidden structural modules in a cooperative RNA folding transition. *Cell Rep* **22:** 3240–3250.

Greenfeld M, Solomatin SV, Herschlag D. 2011. Removal of covalent heterogeneity reveals simple folding behavior for P4-P6 RNA. *J Biol Chem* **286:** 19872–19879.

Guerrier-Takada C, Gardiner K, Marsh T, Pace N, Altman S. 1983. The RNA moiety of ribonuclease P is the catalytic subunit of the enzyme. *Cell* **35:** 849–857.

Herschlag D. 1995. RNA chaperones and the RNA folding problem. *J Biol Chem* **270:** 20871–20874.

Herschlag D. 1998. RNA structure: Ribozyme crevices and catalysis. *Nature* **395:** 548.

Herschlag D, Khosla M, Tsuchihashi Z, Karpel RL. 1994. An RNA chaperone activity of non-specific RNA binding proteins in hammerhead ribozyme catalysis. *EMBO J* **13:** 2913–2924.

Heus HA, Pardi A. 1991. Structural features that give rise to the unusual stability of RNA hairpins containing GNRA loops. *Science* **253:** 191–194.

Hodak JH, Downey CD, Fiore JL, Pardi A, Nesbitt DJ. 2005. Docking kinetics and equilibrium of a GAAA tetraloop-receptor motif probed by single-molecule FRET. *Proc Natl Acad Sci* **102:** 10505–10510.

Horning DP, Joyce GF. 2016. Amplification of RNA by an RNA polymerase ribozyme. *Proc Natl Acad Sci* **113:** 9786–9791.

Huang P-S, Boyken SE, Baker D. 2016. The coming of age of de novo protein design. *Nature* **537:** 320–327.

Jaeger L, Leontis NB. 2000. Tecto-RNA: One-dimensional self-assembly through tertiary interactions. *Angew Chem Int Ed* **39:** 2521–2524.

Jarmoskaite I, Russell R. 2014. RNA helicase proteins as chaperones and remodelers. *Annu Rev Biochem* **83:** 697–725.

Joyce GF, Inoue T, Orgel LE. 1984. Non-enzymatic template-directed synthesis on RNA random copolymers. Poly(C, U) templates. *J Mol Biol* **176:** 279–306.

Kaufmann KW, Lemmon GH, Deluca SL, Sheehan JH, Meiler J. 2010. Practically useful: What the Rosetta protein modeling suite can do for you. *Biochemistry* **49:** 2987–2998.

Koculi E, Cho SS, Desai R, Thirumalai D, Woodson SA. 2012. Folding path of P5abc RNA involves direct coupling of secondary and tertiary structures. *Nucleic Acids Res* **40:** 8011–8020.

Kraut DA, Carroll KS, Herschlag D. 2003. Challenges in enzyme mechanism and energetics. *Annu Rev Biochem* **72:** 517–571.

Kruger K, Grabowski PJ, Zaug AJ, Sands J, Gottschling DE, Cech TR. 1982. Self-splicing RNA: Autoexcision and autocyclization of the ribosomal RNA intervening sequence of *Tetrahymena*. *Cell* **31:** 147–157.

Latham JA, Cech TR. 1989. Defining the inside and outside of a catalytic RNA molecule. *Science* **245:** 276–282.

Leontis NB, Lescoute A, Westhof E. 2006. The building blocks and motifs of RNA architecture. *Curr Opin Struct Biol* **16:** 279–287.

Lindahl T, Adams A, Fresco JR. 1966. Renaturation of transfer ribonucleic acids through site binding of magnesium. *Proc Natl Acad Sci* **55:** 941–948.

Lipfert J, Doniach S, Das R, Herschlag D. 2014. Understanding nucleic acid-ion interactions. *Annu Rev Biochem* **83:** 813–841.

Martick M, Scott WG. 2006. Tertiary contacts distant from the active site prime a ribozyme for catalysis. *Cell* **126:** 309–320.

Miao Z, Adamiak RW, Antczak M, Batey RT, Becka AJ, Biesiada M, et al. 2017. RNA-Puzzles Round III: 3D RNA structure prediction of five riboswitches and one ribozyme. *RNA* **23:** 655–672.

Mitchell D III, Jarmoskaite I, Seval N, Seifert S, Russell R. 2013. The long-range P3 helix of the *Tetrahymena* ribozyme is disrupted during folding between the native and misfolded conformations. *J Mol Biol* **425:** 2670–2686.

Moult J, Fidelis K, Kryshtafovych A, Schwede T, Tramontano A. 2014. Critical assessment of methods of protein structure prediction (CASP)—Round X. *Proteins* **82:** 1–6.

Narlikar GJ, Herschlag D. 1997. Mechanistic aspects of enzymatic catalysis: Lessons from comparison of RNA and protein enzymes. *Annu Rev Biochem* **66:** 19–59.

Nelson JA, Uhlenbeck OC. 2006. When to believe what you see. *Mol Cell* **23:** 447–450.

Noller HF. 2012. Evolution of protein synthesis from an RNA world. In *RNA worlds: From life's origins to diversity in gene regulation* (ed. Atkins JF, Gesteland RF, Cech TR), pp. 141–154. Cold Spring Harbor Laboratory Press, Cold Spring Harbor, New York.

Okumus B, Wilson TJ, Lilley DMJ, Ha T. 2004. Vesicle encapsulation studies reveal that single molecule ribozyme heterogeneities are intrinsic. *Biophys J* **87:** 2798–2806.

Orgel LE. 1968. Evolution of the genetic apparatus. *J Mol Biol* **38:** 381–393.

Pan T, Sosnick TR. 1997. Intermediates and kinetic traps in the folding of a large ribozyme revealed by circular dichroism and UV absorbance spectroscopies and catalytic activity. *Nat Struct Biol* **4:** 931–938.

Pan J, Woodson SA. 1998. Folding intermediates of a self-splicing RNA: Mispairing of the catalytic corel. *J Mol Biol* **280:** 597–609.

Pan T, Fang X, Sosnick T. 1999. Pathway modulation, circular permutation and rapid RNA folding under kinetic control. *J Mol Biol* **286:** 721–731.

Piccirilli JA, Vyle JS, Caruthers MH, Cech TR. 1993. Metal ion catalysis in the *Tetrahymena* ribozyme reaction. *Nature* **361:** 85–88.

Pley HW, Flaherty KM, McKay DB. 1994. Three-dimensional structure of a hammerhead ribozyme. *Nature* **372:** 68–74.

Russell R. 2008. RNA misfolding and the action of chaperones. *Front Biosci* **13:** 1–20.

Russell R, Herschlag D. 1999. New pathways in folding of the *Tetrahymena* group I RNA enzyme 1. *J Mol Biol* **291:** 1155–1167.

Russell R, Millett IS, Doniach S, Herschlag D. 2000. Small angle X-ray scattering reveals a compact intermediate in RNA folding. *Nat Struct Biol* **7:** 367–370.

Russell R, Zhuang X, Babcock HP, Millett IS, Doniach S, Chu S, et al. 2002. Exploring the folding landscape of a structured RNA. *Proc Natl Acad Sci* **99:** 155–160.

Russell R, Das R, Suh H, Travers KJ, Laederach A, Engelhardt MA, et al. 2006. The paradoxical behavior of a highly structured misfolded intermediate in RNA folding. *J Mol Biol* **363:** 531–544.

Salehi-Ashtiani K, Szostak JW. 2001. In vitro evolution suggests multiple origins for the hammerhead ribozyme. *Nature* **414:** 82–84.

Salmon L, Yang S, Al-Hashimi HM. 2014. Advances in the determination of nucleic acid conformational ensembles. *Annu Rev Phys Chem* **65:** 293–316.

Sclavi B, Sullivan M, Chance MR, Brenowitz M, Woodson SA. 1998. RNA folding at millisecond intervals by synchrotron hydroxyl radical footprinting. *Science* **279:** 1940–1943.

Shi X, Bonilla S, Herschlag D, Harbury P. 2015. Quantifying nucleic acid ensembles with X-ray scattering interferometry. *Methods Enzymol* **558:** 75–97.

Silverman SK, Zheng M, Wu M, Tinoco I Jr, Cech TR. 1999. Quantifying the energetic interplay of RNA tertiary and secondary structure interactions. *RNA* **5:** 1665–1674.

Solomatin SV, Greenfeld M, Chu S, Herschlag D. 2010. Multiple native states reveal persistent ruggedness of an RNA folding landscape. *Nature* **463:** 681–684.

Steitz TA, Moore PB. 2003. RNA, the first macromolecular catalyst: The ribosome is a ribozyme. *Trends Biochem* **28:** 411–418.

Tagami S, Attwater J, Holliger P. 2017. Simple peptides derived from the ribosomal core potentiate RNA polymerase ribozyme function. *Nat Chem* **9:** 325–332.

Takamoto K, Das R, He Q, Doniach S, Brenowitz M, Herschlag D, et al. 2004. Principles of RNA compaction: Insights from the equilibrium folding pathway of the P4-P6 RNA domain in monovalent cations. *J Mol Biol* **343:** 1195–1206.

Tan E, Wilson TJ, Nahas MK, Clegg RM, Lilley DMJ, Ha T. 2003. A four-way junction accelerates hairpin ribozyme folding via a discrete intermediate. *Proc Natl Acad Sci* **100:** 9308–9313.

Tinoco I Jr. 1993. Appendix 1: Structures of base pairs involving at least two hydrogen bonds. In *The RNA world* (ed. Gesteland RF, Atkins JF), pp. 603–607. Cold Spring Harbor Laboratory Press, Cold Spring Harbor, New York.

Treiber DK, Williamson JR. 1999. Exposing the kinetic traps in RNA folding. *Curr Opin Struct Biol* **9:** 339–945.

Tsuchihashi Z, Khosla M, Herschlag D. 1993. Protein enhancement of hammerhead ribozyme catalysis. *Science* **262:** 99–102.

Uhlenbeck OC. 1995. Keeping RNA happy. *RNA* **1:** 4–6.

Wan Y, Suh H, Russell R, Herschlag D. 2010. Multiple unfolding events during native folding of the *Tetrahymena* group I ribozyme. *J Mol Biol* **400:** 1067–1077.

Wang S, Karbstein K, Peracchi A, Beigelman L, Herschlag D. 1999. Identification of the hammerhead ribozyme metal ion binding site responsible for rescue of the deleterious effect of a cleavage site phosphorothioate. *Biochemistry* **38:** 14363–14378.

Weinstein LB, Jones BC, Cosstick R, Cech TR. 1997. A second catalytic metal ion in group I ribozyme. *Nature* **388:** 805–808.

Westhof E, Masquida B, Jaeger L. 1996. RNA tectonics: Towards RNA design. *Fold Des* **1:** R78–R88.

Wilson DS, Szostak JW. 1999. In vitro selection of functional nucleic acids. *Annu Rev Biochem* **68:** 611–647.

Woese CR. 1967. The present status of the genetic code. In *Progress in nucleic acid research and molecular biology* (ed. Davidson JN, Cohn WE), pp. 107–172. Academic, New York.

Woodson SA. 2000. Recent insights on RNA folding mechanisms from catalytic RNA. *Cell Mol Life Sci* **57:** 796–808.

Wu M, Tinoco I Jr. 1998. RNA folding causes secondary structure rearrangement. *Proc Natl Acad Sci* **95:** 11555–11560.

Xue Y, Gracia B, Herschlag D, Russell R, Al-Hashimi HM. 2016. Visualizing the formation of an RNA folding intermediate through a fast highly modular secondary structure switch. *Nat Commun* **7:** 11768.

Yesselman JD, Eiler D, Carlson ED, Ooms AN, Kladwang W, Shi X, Costantino DA, Herschlag D, Jewett MC, Kieft JS, Das R. 2017. Computational design of asymmetric three-dimensional RNA structures and machines. *bioRxiv* doi: 10/1101.223479.

Yesselman JD, Denny SK, Bisaria N, Herschlag D, Greenleaf WJ, Das R. 2018. RNA tertiary structure energetics predicted by an ensemble model of the RNA double helix. *bioRxiv* doi: 10.1101/341107.

Zarrinkar PP, Williamson JR. 1994. Kinetic intermediates in RNA folding. *Science* **265:** 918–924.

Zhao R, Rueda D. 2013. Memory effects in RNA folding dynamics revealed by single-molecule fluorescence. In *Biophysics of RNA folding* (ed. Russell R), pp. 117–133. Springer, New York.

Zhuang X. 2005. Single-molecule RNA science. *Annu Rev Biophys Biomol Struct* **34:** 399–414.

Zhuang X, Bartley LE, Babcock HP, Russell R, Ha T, Herschlag D, Chu S. 2000. A single-molecule study of RNA catalysis and folding. *Science* **288:** 2048–2051.

Zuker M, Mathews DH, Turner DH. 1999. Algorithms and thermodynamics for RNA secondary structure prediction: A practical guide. In *RNA biochemistry and biotechnology* (ed. Barciszewski J, Clark BFC), pp. 11–43. Springer, Dordrecht, The Netherlands.

Cite as *Cold Spring Harb Perspect Biol* doi: 10.1101/cshperspect.a032433

Coming Together: RNAs and Proteins Assemble under the Single-Molecule Fluorescence Microscope

Ameya P. Jalihal,[1,2] Paul E. Lund,[2] and Nils G. Walter[2,3]

[1]Cellular and Molecular Biology Graduate Program, University of Michigan, Ann Arbor, Michigan 48109

[2]Single Molecule Analysis Group, Department of Chemistry, University of Michigan, Ann Arbor, Michigan 48109

[3]Center for RNA Biomedicine, University of Michigan, Ann Arbor, Michigan 48109

Correspondence: nwalter@umich.edu

SUMMARY

RNAs, across their numerous classes, often work in concert with proteins in RNA–protein complexes (RNPs) to execute critical cellular functions. Ensemble-averaging methods have been instrumental in revealing many important aspects of these RNA–protein interactions, yet are insufficiently sensitive to much of the dynamics at the heart of RNP function. Single-molecule fluorescence microscopy (SMFM) offers complementary, versatile tools to probe RNP conformational and compositional changes in detail. In this review, we first outline the basic principles of SMFM as applied to RNPs, describing key considerations for labeling, imaging, and quantitative analysis. We then sample applications of in vitro and in vivo single-molecule visualization using the case studies of pre-messenger RNA (mRNA) splicing and RNA silencing, respectively. After discussing specific insights single-molecule fluorescence methods have yielded, we briefly review recent developments in the field and highlight areas of anticipated growth.

Outline

Cite as *Cold Spring Harb Perspect Biol* doi: 10.1101/cshperspect.a032441

1 INTRODUCTION

Contemplating the original "central dogma" of molecular biology (Crick 1970), one could easily come away with the impression that RNA serves only as a genetic messenger, conveying the information stored in DNA by helping convert it into functional protein actuators. Over the past few decades, RNA has, however, turned out to be much more than a mere genetic messenger, with the vast majority of RNA transcripts from eukaryotic genomes having no protein coding capacity. The many classes of such noncoding RNAs—including ribosomal RNA (rRNA), transfer RNA (tRNA), short noncoding RNA, such as small RNA (sRNA), microRNA (miRNA), and long noncoding RNA (lncRNA), to name a few—perform a multitude of catalytic, regulatory, and architectural functions. To help carry out their functions, each of these classes of RNA participates in a variety of interactions with one or more protein partners. As an example, the messenger RNAs (mRNAs) that inspired the central dogma are themselves dependent on interactions with a multitude of proteins throughout their life cycle, from biogenesis, editing, and regulation to decay. Given how intimately the biological activity of a given RNA is connected to interactions with these protein partners, discerning the nature of these interactions is vital to understanding the biology of RNA-dependent cellular pathways.

Many long-standing techniques for studying ribonucleoproteins (RNPs), such as electrophoretic gel shift assays or ultraviolet (UV) cross-linking, are most suitable for interactions that are tight and uniform in binding mode and composition, which is not the case for many complex biological processes. In fact, it is becoming increasingly clear

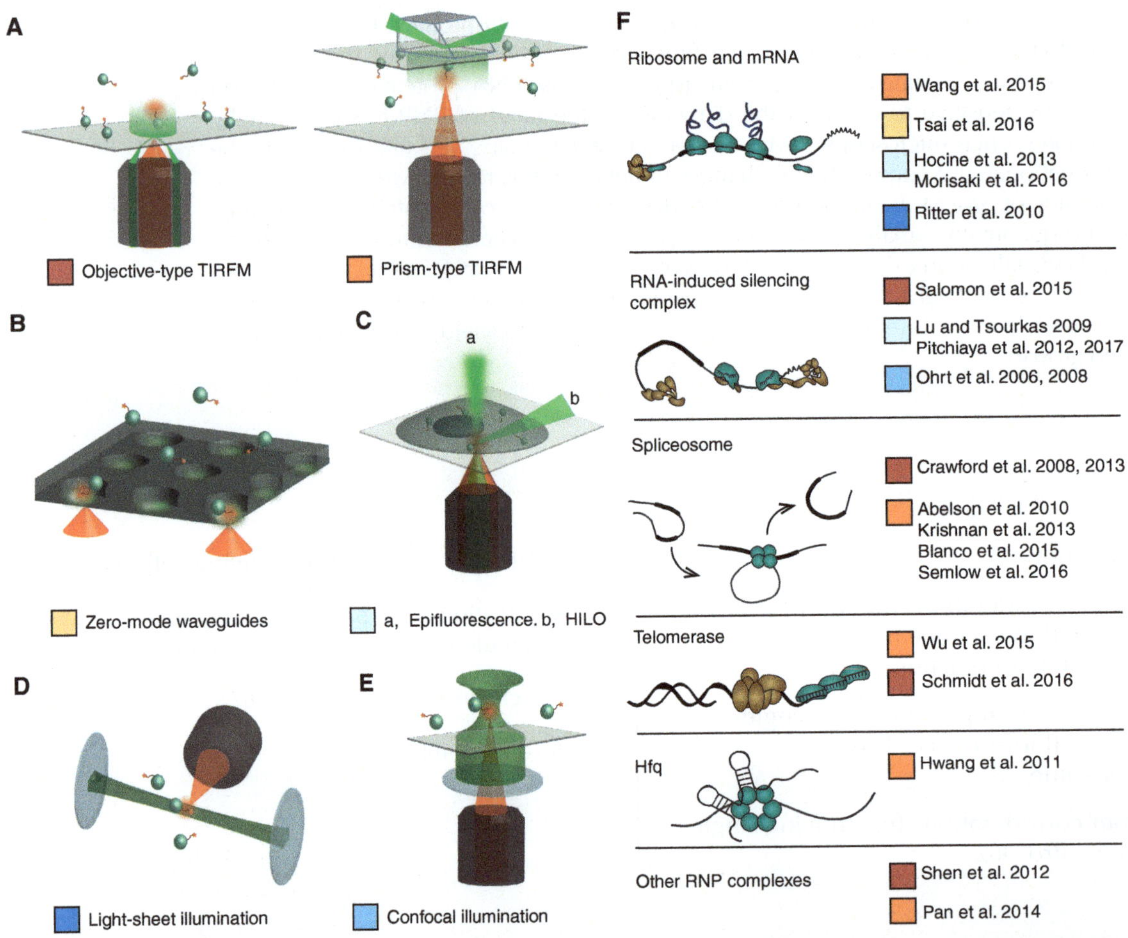

Figure 1. Overview of single-molecule microscopies applied to study ribonucleoproteins (RNPs). (*A*) Objective- and prism-type total internal reflection fluorescence microscopy (TIRFM) configurations. (*B*) Zero-mode waveguides. (*C*) a, Epifluorescence; b, highly inclined and laminated optical sheet (HILO) illumination. (*D*) Light-sheet illumination. (*E*) Confocal illumination. (*F*) Selected studies on RNPs that have used fluorescence-based, single-molecule methods. Colored boxes indicate the type of microscopy configuration used.

that the dynamic and transient nature of competing interactions between biomolecules is at the heart of cellular pathways, such as pre-mRNA splicing (Pan et al. 2014; Semlow et al. 2016), mRNA translation (Agirrezabala and Valle 2015; Tsai et al. 2016), RNA chaperoning (Shen et al. 2012), and RNA silencing and regulation (Hwang et al. 2011; Jens and Rajewsky 2015). Particularly as compositional and structural snapshots of large, multicomponent biomolecular machines such as the spliceosome come into focus thanks to new structural data, the need for a deeper understanding of the compositional and conformational dynamics of RNPs becomes increasingly apparent. Two decades of advances in single-molecule fluorescence microscopy (SMFM) have now given us the tools to address this need.

2 GENERAL CONSIDERATIONS FOR APPLYING SINGLE-MOLECULE FLUORESCENCE MICROSCOPY TO RNA–PROTEIN INTERACTIONS

SMFM techniques are well-suited to study molecular processes that occur in multiple steps, proceed via parallel reaction pathways, show transient excursions to distinct states, and/or contain varying components, all of which are frequently true of RNA–protein interactions (Wahl et al. 2009; Moffitt et al. 2010). The most important advantage is that these methods allow heterogeneous molecular "behaviors" to be uncovered in asynchronous populations of molecules proceeding through multiple reaction pathways at different rates (Lu et al. 1998; Larson et al. 2014). These single-molecule behaviors can be scrutinized at a spatiotemporal resolution inaccessible to bulk techniques, leading to insights that can be directly compared with bulk data by postacquisition time and ensemble averaging.

The general benefits and specifics of performing single-molecule studies are beyond the scope of this review, but have been reviewed elsewhere (Roy et al. 2008; Walter et al. 2008; Liu et al. 2015). We instead will focus on the salient features of SMFM and practical considerations for its application to the study of RNPs.

A variety of SMFM approaches exist today, the exact choice of which depends largely on the study design and experimental goals. SMFM has been used to investigate numerous RNPs, including those involved in transcription, splicing, RNA modification and editing, chaperone activity, and RNA interference (Fig. 1). Despite obvious differences in the underlying biology, key experimental elements common to all of these studies include illumination, a strategy for labeling the biomolecule(s) of interest, detection of fluorescence, and data analysis, as discussed below.

2.1 Illumination and Detection

By design, SMFM approaches aim to characterize the behavior of individual molecules. However, detecting the signal from an individual molecule in the background of other molecules, potential contaminants, and instrumental noise can be challenging. Fluorescence itself facilitates distinguishing genuine signal from background as light emitted from fluorescently tagged biomolecules is red-shifted relative to the wavelength of laser light used for excitation. This spectral separation (i.e., "Stokes shift") between the excitation and emission wavelength maxima makes genuine fluorescence signals easily distinguishable from background scattering of the sample medium.

In addition to spectral isolation of the relevant signal, limiting the volume of illumination improves signal-to-noise by decreasing the contribution of out-of-focus emission. Of the illumination modalities that achieve selective volume illumination, total internal reflection fluorescence microscopy (TIRFM) is perhaps the most widely implemented (Fig. 1A). In TIRFM, the excitation beam is totally internally reflected near the sample plane (Toomre 2012). An evanescent field of excitation light is generated at the surface that extends ~100 nM into the sample. This evanescent field illuminates molecules immobilized at the imaging surface, thus minimizing background fluorescence from molecules deeper in solution. The level of background fluorescence observed in TIRFM experiments is dependent on the concentration of excess fluorescently labeled biomolecules in solution and provides good signal-to-noise up to fluorophore concentrations of ~100 nM. For RNA and proteins that interact with high affinity, this effective concentration limit does not pose a problem. However, physiologically relevant interactions often involve micromolar affinities of binding partner(s). Zero-mode waveguide (ZMW) technology allows single-molecule resolution even in solutions with high concentrations of fluorescent components (Levene et al. 2003), and have been particularly useful in studies of the ribosome (Fig. 1B) (Tsai et al. 2016).

Another strategy to limit out-of-plane fluorescence by controlling the illumination area is highly inclined and laminated optical sheet (HILO) illumination (Fig. 1C) (Tokunaga et al. 2008). In HILO microscopy, the illuminated region is thicker than it is for TIRFM (e.g., ~7 µm for an illuminated area with a diameter of 20 µm), and contains a *z*-component, which allows for imaging of molecules that lie micrometers above the coverslip surface. These characteristics make HILO microscopy particularly suited for studying samples with three-dimensional spatial distributions, such as single molecules inside live cells. Even more sample penetration depth can be achieved by light-sheet fluorescence microscopy (LSFM), in which a focused sheet of light is used to illuminate only a thin section of a thick sample for single-molecule detection (Fig. 1D) (Ritter et al. 2010). Confocal microscopy uses pinholes to limit the volume in which single molecules are detected (Fig. 1E). In addition to two-dimensional (2D) scanning, confocal microscopy is typically used in the context of fluorescence correlation spectroscopy (FCS) and two-color fluorescence cross-correlation spectroscopy (FCCS), in which a high-sensitivity, photon-counting point detector generates time-resolved intensity measurements from the entire illuminated confocal volume (Liu et al. 2015; Gonzalez Bardeci et al. 2017).

2.2 Labeling Biomolecules of Interest

Another key requirement of SMFM is that one or more of the biological components in the system being studied must be fluorescently labeled.

The ever-increasing number of small-molecule fluorescent dyes available provide flexibility and choice for labeling. CyDye, Alexa Fluor, DyLight, and ATTO are perhaps the most popular in the single-molecule fluorescence field because of their brightness, photostability, and the commercial availability of derivatives for bioconjugation. Their pH tolerance also makes them suitable for fluorescent labeling in vivo (see Sec. 5). Additionally, multiple fluorophore pairs (e.g., Cy3 and Cy5; Dy547 and Cy5) have spectral properties that permit their use as single-molecule Förster resonance energy transfer (smFRET) reporters, allowing for sensitive monitoring of nanometer-scale changes in distance (Lakowicz 2006; McCann et al. 2010). Experiments using such small-molecule dyes typically use one of several oxygen-scavenging systems (OSSs) to improve dye photophysical performance and longevity (Rasnik et al. 2006; Aitken et al. 2008). However, the choice of an OSS must also keep the biological system in mind, including the potential effect of pH changes or sensitivity to nuclease contamination (Swoboda et al. 2012; Senavirathne et al. 2015). For example, a very common OSS based on glucose oxidase is inappropriate for experiments with whole-cell extracts (WCEs) because the activity of endogenous kinases quickly phosphorylates the glucose, depleting ATP from solution and thus inhibiting processes such as splicing (Crawford et al. 2008; Abelson et al. 2010).

It should be noted that in the context of RNP studies, although the most obvious choice may be to label the RNA and protein components that are directly interacting in multicomponent systems, it can be advantageous to label, for example, a strongly associated component whose presence or absence is then used to distinguish complexes of interest from other closely related species. In the following, we provide a brief overview of commonly used fluorescent-labeling strategies and provide particular considerations for the study of RNPs.

2.2.1 Labeling RNA Components

Labeling sites must be chosen so as to avoid influencing interactions between the RNA and its binding partner(s), or perturbing secondary or tertiary interactions within the RNA or its conformational dynamics (Al-Hashimi and Walter 2008; Mustoe et al. 2014; Darre et al. 2016). For example, 2′ modifications on the ribose sugar can alter the thermal stability of double-stranded regions (Hendrix et al. 1995; Fauster et al. 2012), whereas attachment of fluorophores at 5′ or 3′ termini can increase duplex stability through stacking of the fluorophores on the terminal base-pairs (Cisse et al. 2012; Liu and Lilley 2017). Such alterations to thermal stability can potentially have functional consequences. In the context of miRNA guide strand selection and loading into the RNA-induced silencing complex (RISC), for example, a 5′ terminal fluorophore disfavored use of the labeled strand as a guide (Custer and Walter 2017). Further, in contrast to DNA where canonical Watson–Crick base-pairing dominates, RNA secondary and tertiary structures show an astonishing variety of noncanonical base-pairing interactions (Leontis et al. 2002), including interactions with the sugar edge (Nissen et al. 2001). As such, it is useful to examine all available structural data when deciding which point of attachment (e.g., 2′ OH, phosphate, or pyrimidine C5; and position in the sequence) is likely to result in the least perturbation.

Chemical synthesis of oligonucleotides provides a direct route for preparing site-specifically labeled RNA. Fluorophore-labeled nucleotides can be directly incorporated during synthesis, or fluorophores can be conjugated after synthesis through the use of reactive groups present on modified nucleotides (Fig. 2A) (Paredes et al. 2011; Edwards and Sigurdsson 2014). Reactive groups such as primary amines and azides and common conjugation chemistries such as *N*-hydroxysuccinimide esters and "click" chemistry allow fluorophores to be attached after RNA synthesis (Fig. 2A,B) (Kolb et al. 2001; Rinaldi et al. 2015). Alternatively, a limited number of generalized labeling chemistries that use natural features of the RNA, such as the 5′ phosphate, naturally occurring 4-thiouridine in some tRNAs, and the 3′ vicinal diol, can be used for labeling (Fei et al. 2010; Hermanson 2013; Willkomm and Hartmann 2014). A significant limitation, however, is that chemical synthesis is only a viable option for relatively short RNAs (<80 nt). As a result, longer RNAs used for in vitro studies are frequently prepared by enzymatic, splint-mediated ligation, either from multiple chemically synthesized oligonucleotides

Cite as *Cold Spring Harb Perspect Biol* doi: 10.1101/cshperspect.a032441

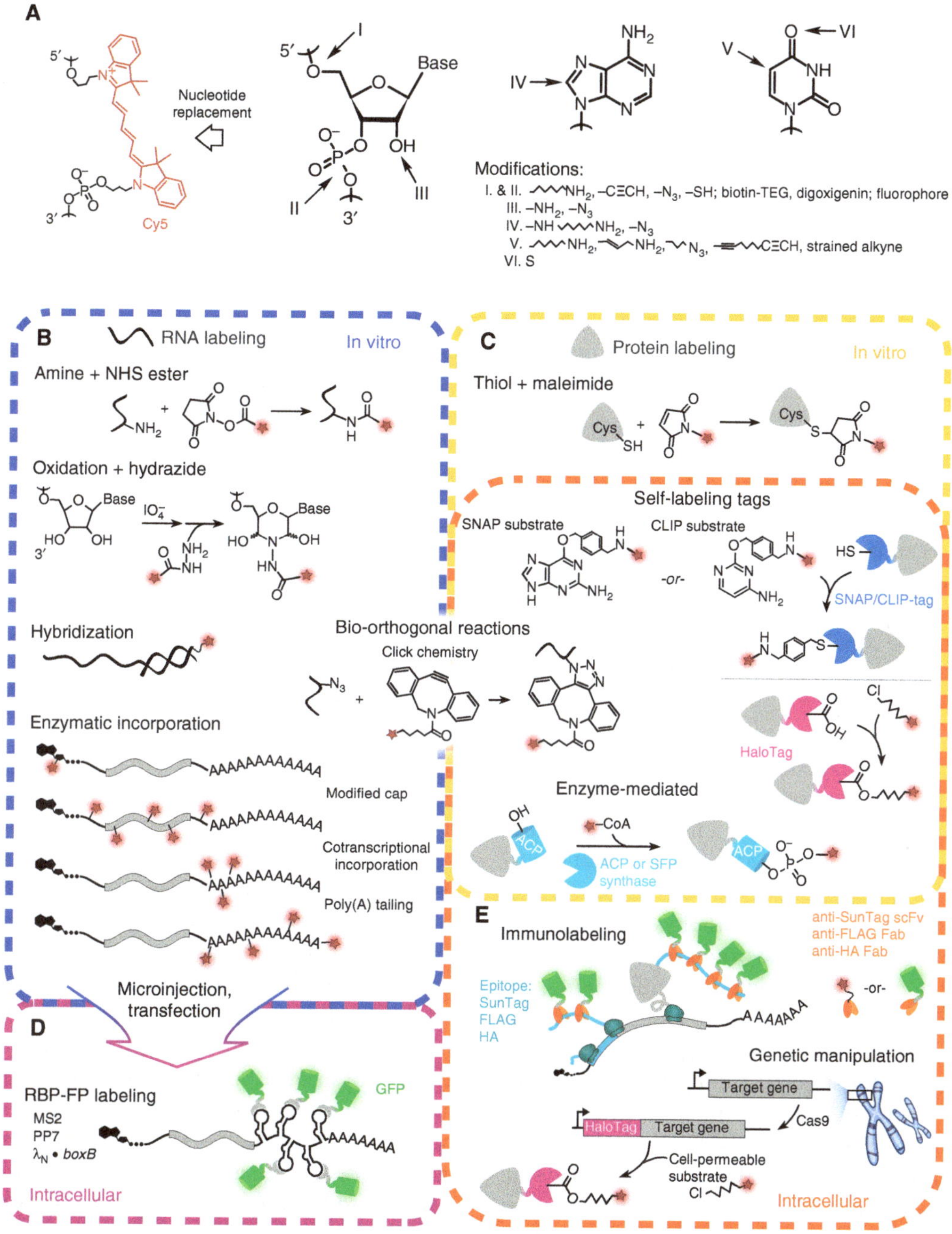

Figure 2. Strategies for labeling RNA and protein components for single-molecule fluorescence. (*A*) Modifications are possible at various positions on RNA molecules prepared through chemical synthesis, including the 5′ and 3′ termini (I and II, respectively), the 2′ position of the sugar (III), and various positions on the nucleobase (IV, V, and VI), among others. Modifications can include reactive chemical moieties that are used for subsequent conjugation reactions, small molecules like biotin and digoxigenin for immobilization, or direct attachment of fluorophore dyes or even replacement of entire nucleotides by fluorophores. (*B*) Example strategies for preparing fluorescently labeled RNA for in vitro studies. Fluorescent labels are shown as red stars. (*C*) Example strategies for preparing fluorescently labeled protein for in vitro studies. (*D*) For intracellular single-molecule studies, RNA can be prepared using many of the methods suitable for in vitro work (described in *B*) and subsequently introduced into cells via microinjection or transfection. Alternatively, RNA transcripts can be labeled by inserting stem-loop repeat sequences that are bound by their cognate RNA-binding proteins, which are expressed as fusions with intrinsically fluorescent proteins (RBP-FP). (*E*) Example strategies for preparing fluorescently labeled protein for intracellular studies. In contrast to RNA, there is greater overlap between strategies for labeling protein that are suitable for in vitro and intracellular work. A small number of strategies, such as click chemistry using strained alkynes, can be applied to both RNA and proteins.

and/or in vitro transcribed segments (Solomatin and Herschlag 2009; Turunen et al. 2014).

In addition to chemical synthesis and ligation, other labeling techniques that hold promise for RNA labeling include position-specific labeling of RNA (PLOR), fluorogenic RNA aptamers, 2′–5′ branched labeling, and engineered tRNA-modifying enzymes (Li et al. 2015; Carrocci et al. 2017; Trachman et al. 2017).

A simple alternative to covalent modification is to anneal the RNA of interest with one or more tightly binding complementary oligonucleotides, frequently DNA or locked nucleic acid (LNA) carrying a fluorescent label (Fig. 2B). The most prominent implementation of this strategy has been for studies involving large RNAs, especially the ribosome, in which a fluorescent oligonucleotide probe is annealed to short extensions engineered at various locations into the ribosomal RNA (Kim et al. 2014). In principle, such a probe can be designed for any RNA sequence, provided that the probe's hybridization is specific and only minimally perturbs the local and global RNA structure (Crawford et al. 2008; Rinaldi et al. 2016).

2.2.2 Labeling Protein Components

For proteins, fluorophore tags may be genetically appended, as is the case with fluorescent proteins (FPs) including green fluorescent protein (GFP), mCherry, and their many variants (Thorn 2017). However, FP tags are used almost exclusively in the context of intracellular studies (see Sec. 5.1) as their large size and unfavorable photophysical properties (e.g., propensity for blinking, low brightness) so far make them less desirable for in vitro single-molecule studies.

The shortcomings of FPs render proteins a more challenging target for site-specific labeling compared with RNA and the options for selective conjugation chemistries for amino acids are more limited compared with nucleotides. Fluorophore-NHS-ester derivatives readily react with primary amines and so can be used to label lysine residues as well as the amino terminus. For in vitro studies using recombinant protein, a more conventional strategy is to use fluorophore–maleimide derivatives, which have greater selectivity for the thiol side chain of rarer cysteine residues (Fig. 2C). This method typically requires that native cysteines be removed or significantly less reactive (and/or accessible) than the desired labeling position. Such was the case in a study of the role of initiation factor 2 (IF2) from *Escherichia coli*, in which removal of the three native cysteines in IF2 resulted in an unstable protein; however, labeling of an additional cysteine introduced through mutagenesis in the presence of the other native cysteines still resulted in site-specific labeling (Wang et al. 2015).

Among the other methods for site-selective labeling of proteins that have been developed (Adumeau et al. 2016), the use of unnatural amino acids to incorporate reactive handles, such as those for click chemistry, is one of the more promising avenues to achieve site specificity (Liu and Schultz 2010; Lee et al. 2016).

Purification of labeled protein from unincorporated fluorophore can be nontrivial because proteins, being more hydrophobic than nucleic acids, have a greater propensity to bind fluorescent dyes nonspecifically. The purification scheme therefore merits careful consideration. Although extensive dialysis or size-exclusion chromatography may be sufficient in some cases, for others unreacted fluorophore may be most robustly removed with ion-exchange chromatography under denaturing conditions (Hickerson et al. 2005).

Protein structure data are frequently unavailable and protein folding is difficult to predict. Furthermore, the complete details of a protein's RNA binding site and other interacting partners are frequently uncharacterized or ill-defined, thus precluding the rational choice of a labeling position. Despite these gaps in knowledge, there are many cases in which determining the presence or absence of a given protein is sufficient, requiring only that the protein be fluorescently labeled in a way that preserves function. However, the knowledge of a fluorophore's exact location on the three-dimensional structure of a protein is required if the protein serves as the donor or acceptor in an smFRET experiment (see Sec. 3.1).

A third alternative for protein labeling are self-labeling protein tags and enzyme-mediated labeling tags that are genetically encoded. These tags have been developed to allow covalent attachment of small-molecule fluorescent dyes, which can be used to label proteins in live cells, in cell extracts, or to label recombinant proteins (Fig. 2C). Of these, notable examples include the SNAP and CLIP tags, based on human O^6-alkylguanine-DNA alkyltransferase, and the engineered dehalogenase HaloTag, all of which catalyze covalent self-addition of their respective ligands (Gautier et al. 2008; Los et al. 2008; Sun et al. 2011). A conceptually similar, albeit noncovalent, self-labeling tag is *E. coli* dihydrofolate reductase (eDHFR) (Calloway et al. 2007).

2.2.3 Modification of Biomolecules for Surface Immobilization

For certain types of SMFM, namely, TIRFM and ZMW, biomolecules of interest must be immobilized at the imaging surface. This is commonly achieved by conjugating the biomolecule of interest with biotin, which is then captured noncovalently by surface-bound streptavidin. Other noncovalent interactions, such as digoxigenin–antidigoxigenin antibody, as well as covalent interactions (e.g., thiol-maleimide or click chemistry) are also used for immobilization.

For short nucleic acids, biotin and digoxigenin can be incorporated during chemical synthesis, whereas the most straightforward strategy for longer RNA is to use biotinylated complementary oligonucleotides that stably hybridize to the RNA of interest, often in added 5′ or 3′ extensions. For RNA prepared by in vitro transcription using T7 RNA polymerase, modified nucleotides can be incorporated at the 5′ end of the transcript (Huang et al. 2008) as well as through enzymatic ligation (Kinoshita et al. 1997; Turunen et al. 2014) and the conjugation chemistries described above.

For modifying proteins, random biotinylation of lysines and cysteines with biotin *N*-hydroxysuccinimide esters and maleimides, respectively, provides the most direct path but with the least control over the site(s) of labeling. For recombinant proteins, the biotin ligase BirA can be used to catalyze the site-specific biotinylation of the 15–amino acid Avitag (Tirat et al. 2006). Biotin substrates for the self-labeling tags (i.e., SNAP, CLIP, and HaloTag) (see Sec. 2.2.2) are also commercially available. Last, surface-bound antibodies raised against a protein of interest have also been used to immobilize single proteins or large RNPs such as the spliceosome, as discussed below.

3 APPLICATION EXAMPLE OF IN VITRO SINGLE-MOLECULE FLUORESCENCE: THE SPLICEOSOME

The majority of in vitro SMFM studies can be broadly categorized into smFRET and colocalization experiments. In the following sections, we discuss these two types of experiments in the context of the yeast spliceosome. Rather than focusing solely on the biological insights into splicing provided by SMFM (for which we refer the reader to DeHaven et al. 2016), this section will highlight select studies drawn from the work of our group as well as others that illustrate the insights that SMFM provides with added emphasis on elements enabling its application.

3.1 Single-Molecule FRET

The spliceosome, which in all eukaryotes acts on precursor mRNAs (pre-mRNAs) to remove introns and join exons, is a challenging subject of study because of the sheer number of components involved: innumerable pre-mRNA substrates, five small nuclear RNAs (snRNAs), each bound to protein components to yield small nuclear ribonucleoproteins (snRNPs), as well as numerous, transiently associated splicing factors, resulting in a highly compositionally dynamic macromolecular machine (Fig. 3A). The associated, coordinated conformational changes that occur in the spliceosome, specifically those occurring in the distance range of ~3–8 nm, are best studied by smFRET. Although smFRET can be used to measure absolute distances, doing so with high accuracy is nontrivial, particularly when there is uncertainty about the assumed isotropic relative orientation of the two fluorophores on the RNA (Milas et al. 2013) and/or the possibility of protein-induced fluorescence enhancement (Hwang and Myong 2014; Lerner et al. 2016), and thus smFRET is most frequently used as a relative measure (dos Remedios and Moens 1995; McCann et al. 2010).

For splicing to occur with high fidelity, specific sequences that are often separated by hundreds of nucleotides in the pre-mRNA, namely, the 5′ and 3′ splice sites (5′ splice site [SS] and 3′SS, respectively) as well as the branch site (BS), must be correctly identified by the splicing machinery and brought into close proximity, with the correct orientation and in the correct order, to permit the two isoenergetic transesterification reactions that the spliceosome catalyzes (Fig. 3A). To study pre-mRNA conformation, Abelson et al. (2010) labeled a model transcript based on the yeast Ubc4 pre-mRNA with a donor (Cy3, green star) and acceptor (Cy5, red star) fluorophore near the 5′ and 3′SSs, respectively (Fig. 3B), and monitored pre-mRNA conformational changes using FRET. Even in buffer alone, the pre-mRNA molecules showed fluctuations between various FRET states, indicating that pre-mRNAs adopted different conformations over time. Point mutations at the 3′SS had a significant impact on the conformations and dynamics of the RNA, underscoring the degree to which RNA sequence and structure are linked.

A follow-up study by Krishnan et al. (2013) used a similar smFRET strategy to examine the structural remodeling of the Ubc4 pre-mRNA in the context of a particular step of the splicing cycle, this time placing the donor and acceptor at the BS and 5′SS, respectively. One general feature of in vitro experiments is a high degree of control over reaction conditions and components. To preclude the possibility of pre-mRNA conformational changes brought about by RNA-binding proteins present in WCE but unrelated to splicing, Krishnan et al. (2013) used an immunoprecipitation-based immobilization strategy termed single-molecule pull-down (SiMPull) FRET (Kahlscheuer et al. 2015). This allowed them to selectively isolate splicing complexes and stall those complexes at the B^{act} stage (Fig. 3C). Leveraging this immobilization strategy, combined with the selective reintroduction of additional recombinant splicing factors and ATP, allowed for characterization of pre-mRNA conformational dynamics by smFRET at a specific step of splicing.

Analysis of the donor and acceptor fluorescence signals from individual immobilized spliceosomal complexes begins with calculation of the relative FRET efficiency

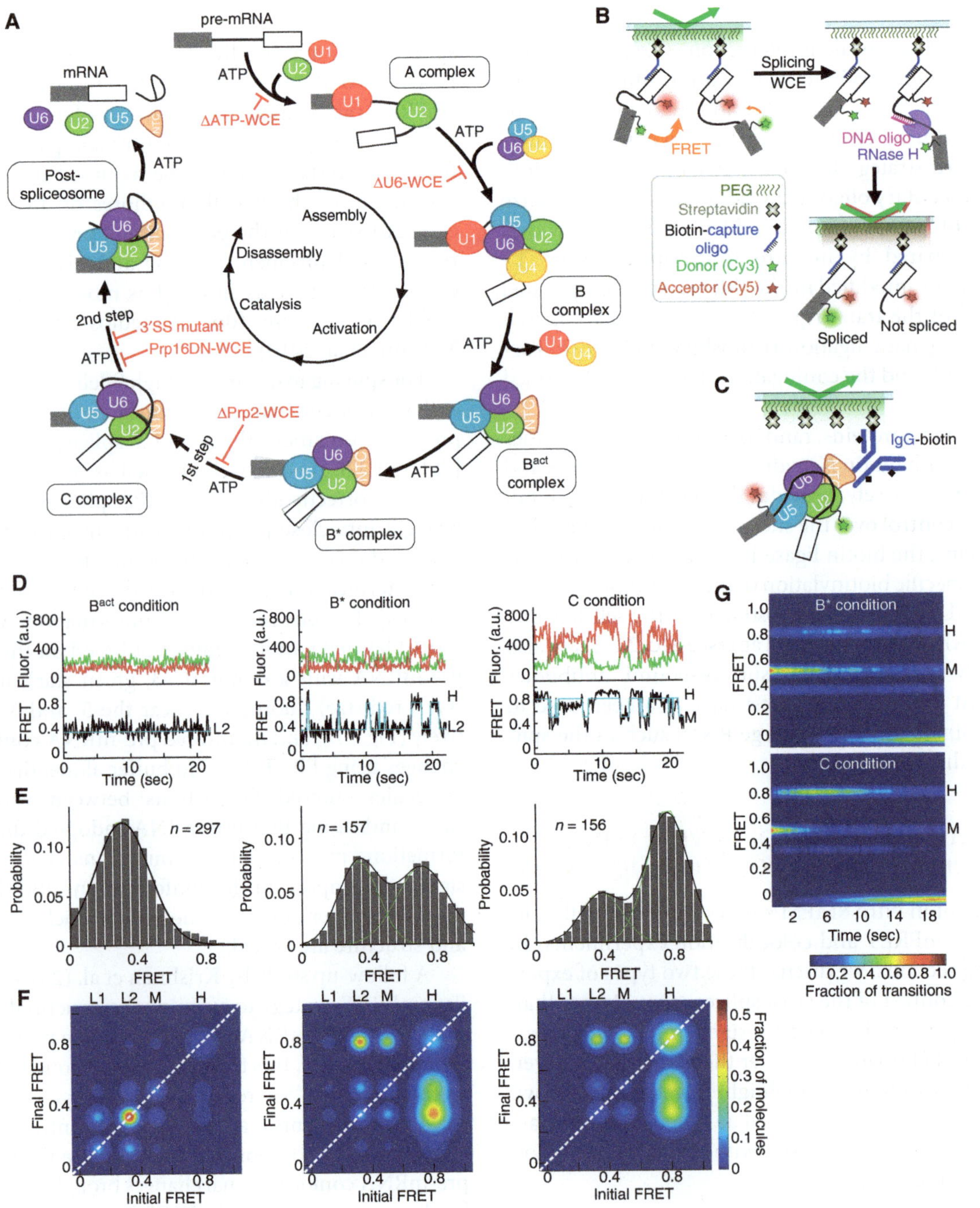

Figure 3. Insights into yeast pre-mRNA conformational dynamics from in vitro single-molecule Förster resonance energy transfer (smFRET) studies. (*A*) Overview of the splicing cycle, in which the 5′ (gray) and 3′ (white) exons of the pre-messenger RNA (mRNA) are ligated and the intron (black line) is removed. Colored circles represent the various small nuclear ribonucleoproteins (snRNPs) required to catalyze splicing. Red annotations represent experimental conditions that can be used to block the cycle at various points. WCE, whole-cell extract. (*B*) Experimental design used by Abelson et al. (2010) to monitor conformational changes in a doubly labeled pre-mRNA construct through changes in FRET. An in situ RNase H assay, using a DNA oligonucleotide that hybridizes to sequences in the intron, verified that modifications made to the RNA to adapt it to single-molecule observation did not inhibit splicing. RNA molecules observed to have both fluorophores under direct excitation had been successfully spliced, whereas those showing only the acceptor fluorophore had not. (*C*) Experimental smFRET design from Krishnan et al. (2013) in which single-molecule pull-down (SiMPull) was used to isolate the B^{act} complex assembled on a pre-mRNA labeled near the branch site (BS) and 5′ splice site (SS). (*D*) Example fluorescence–time traces from the donor (green) and acceptor (red) fluorophores attached to single pre-mRNA molecules under different experimental conditions designed to promote formation of the B^{act}, B*, or C complexes, respectively. (*Legend continues on following page.*)

(Fig. 3D) (Blanco and Walter 2010), which can then be used to create histograms of relative FRET efficiency that effectively represent the time-weighted conformational ensemble of molecules (Fig. 3E). Although the presence of multiple conformations is indicated by the presence of corresponding peaks, such histograms are largely unable to distinguish between molecules that show distinct kinetic behaviors, adopt varying numbers of states, or show overlapping, noise-broadened FRET distributions. Characterization of such nuance typically provides useful biological insight, particularly for highly reversible, multistep processes such as splicing, and requires appropriate analytical methods (Shi et al. 2008; Moffitt et al. 2010; Grima et al. 2014; Colomb and Sarkar 2015). smFRET time traces are often fit using hidden Markov models (HMMs) to idealize noisy and continuous FRET efficiency time-series data into a set of discrete states that facilitate the extraction of kinetic information (Blanco and Walter 2010; Liu et al. 2010). Fractions of molecules that undergo specific conformational changes, defined by a particular initial and final idealized FRET efficiency, can be identified using transition occupancy density plots (TODPs). This method was used to survey changes in the dynamic behavior of splicing complexes stalled at specific points in the splicing cycle (Fig. 3F) (Krishnan et al. 2013).

Although the catalytically active B* complex can undergo the first chemical reaction of splicing, this step is greatly enhanced by Cwc25, a protein known to be associated with the nineteen complex (NTC). Cwc25 is known to bias the spliceosome toward a configuration in which the 5′SS and BS are in close proximity by slowing the reverse reaction. This effect was directly visualized in this study using another common single-molecule analysis tool known as a postsynchronized histogram (Fig. 3G), where molecules starting in a medium FRET state fluctuate frequently between H and M states, but primarily transition to and stay in the H state in the presence of Cwc25. Cwc25 was thus found to act as a "pawl" to stabilize the otherwise thermally "ratcheting" pre-mRNA in its proximal, active conformation, promoting the transition to the C complex accompanied by the first chemical reactions of splicing.

The conclusions from Krishnan et al. (2013) about pre-mRNA conformational remodeling and dynamics, some of which are highly reversible, and the role of splicing factors in activating the spliceosome were in good agreement with complementary studies by Crawford et al. (2013), who combined observables from smFRET with colocalization single-molecule spectroscopy (CoSMoS) to correlate pre-mRNA conformational changes observed through smFRET with specific spliceosome compositions by monitoring colocalization of U1 and the NTC.

3.2 Single-Molecule Colocalization

In CoSMoS, multiple biological components are fluorescently labeled with spectrally distinct fluorophores and monitored simultaneously. The ability to observe multiple species provides a direct method of discriminating between parallel pathways that use alternative macromolecular assemblies of largely similar, but not identical, composition. Because of the binary nature of binding (i.e., either bound or unbound), rapid and large changes in fluorescence signal intensity are observed on binding and dissociation of a fluorescent molecule or complex, which can be used to yield kinetic information. Furthermore, coincidence of two fluorescence signals indicates the simultaneous presence of differentially labeled components.

In a series of multicolor CoSMoS experiments, Shcherbakova et al. (2013) examined the order of assembly of the U1 and U2 snRNPs on the 5′SS and BS for a number of pre-mRNAs that show different in vitro splicing efficiencies (Fig. 4A). Using genetically modified WCE from yeast that contained SNAP-, CLIP-, and eDHFR-tagged protein components, they prepared splicing-active extracts containing U1, U2, and U5 snRNPs where each was labeled with a unique fluorophore. Fluo-

Figure 3. (*Continued*) The FRET efficiency (black) is fit using hidden Markov modeling (HMM, cyan line) to identify unique FRET states. Fluor., fluorescence intensity; a.u., arbitrary units. (*E*) Histograms of FRET efficiency, generated here by combining the first 100 movie frames from each molecule in a given experimental condition, can be used to identify underlying FRET states. (*F*) Transition occupancy density plots (TODPs) showing the fraction of molecules in a given experimental condition that show transitions between a given initial and final FRET state. A molecule that remains in the same state during the entire observation period, such as the one shown in B^{act} condition in panel *D*, contributes to peaks on the diagonal. As splicing progresses from B^{act} to B* to C, there is an increase in the number of transitions to the highest FRET states (H), in which the 5′SS and BS are in close proximity. The fraction of molecules that stably reside in the high FRET state is greatest under C complex conditions, which is consistent with molecules that have undergone the first step of splicing. (*G*) Postsynchronized histograms of molecules collected under B* and C complex conditions were plotted such that they are synchronized to start in state M. Molecules reversibly transition between M and H states under B* conditions. In contrast, under C complex conditions (in which Cwc25 is present), molecules preferentially transition from M to H, where they largely reside until photobleaching (FRET values below zero). (*D–G* adapted, with permission, from Krishnan et al. 2013.)

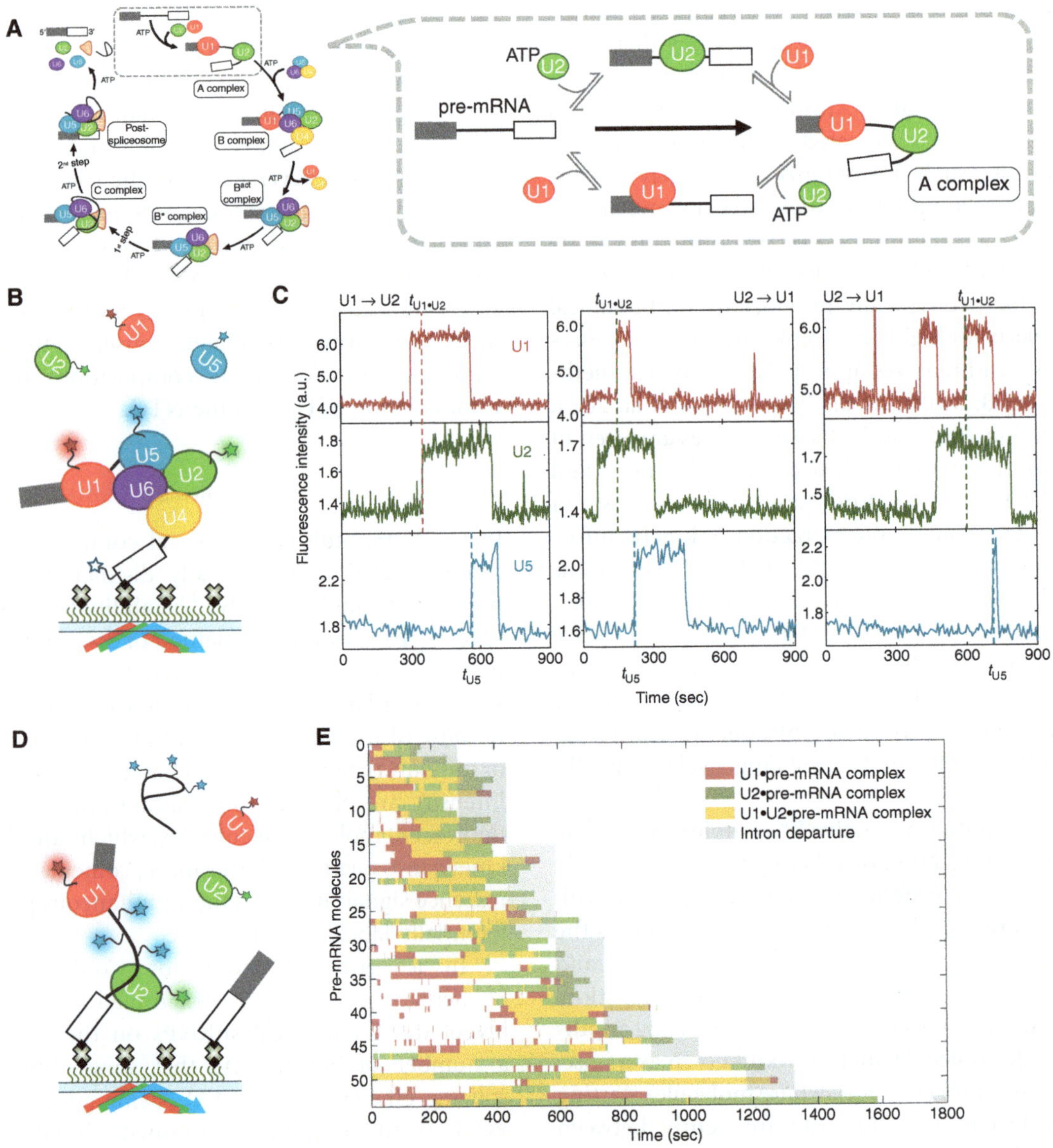

Figure 4. Insights into early spliceosome assembly in yeast from in vitro single-molecule fluorescence colocalization studies. (*A*) Detailed view of the early steps of yeast spliceosome assembly in context of the larger splicing cycle. In the earliest steps of splicing leading to formation of the A complex, the U1 and U2 small nuclear ribonucleoproteins (snRNPs) assemble on the pre-mRNA near the 5′ splice site (SS) and branch site (BS), respectively. (*B*) Experimental design used by Shcherbakova et al. (2013) to study the order of spliceosome assembly by simultaneously monitoring the arrival of fluorescently labeled U1, U2, and U5 snRNPs in whole-cell extracts (WCEs). The locations of the surface immobilized pre-mRNA were established at the beginning of the experiment using a fluorophore incorporated into the 3′ exon, which was subsequently photobleached (white star) before the introduction of WCE containing labeled snRNPs. (*C*) Example fluorescence–time traces for three pre-mRNA molecules, showing instances in which U1 arrival preceded arrival of U2 (*left*) or where U2 preceded U1 (*center, right*). The sudden increase in fluorescence in a given channel indicates the binding and stable association of the respective snRNP. Only pre-mRNA molecules that ultimately showed the arrival of U5, and thus progression to the B complex, were considered for analysis. (*D*) Experimental design used by Shcherbakova et al. (2013) to monitor the binding and dissociation of labeled U1 and U2 on a pre-mRNA containing multiple fluorophore in the intron. The high number of fluorophores (approximately eight) in the intron prolonged the viable observation window, as well as allowed dissociation of the intro lariat (sharp loss of all blue fluorescence) to be differentiated from photobleaching (stepwise loss of blue fluorescence). (*E*) Rastergram showing the residence time of U1 and U2, as well as the time until intron departure for 54 pre-mRNA molecules. (*C,E* reproduced, with permission, from Shcherbakova et al. 2013 and Elsevier.)

rescence of these snRNPs was then monitored at previously established locations of immobilized pre-mRNA molecules (Fig. 4B).

A key strength of CoSMoS is the fact that observation of one fluorophore-labeled component can indicate the presence of one or more secondary components that are constitutively present in the complex. By examining many multichannel fluorescence time traces, Shcherbakova et al. (2013) observed that the preferred order of binding of U1 and U2 varies for different pre-mRNAs (Fig. 4C). Further, the investigators found evidence that traces proceeding through the U2-first pathway did not represent dead-end complexes because they observed subsequent recruitment of U5 as a marker for the next splicing step. This ability to discern molecules traversing a particular reaction pathway within a highly heterogeneous reaction mixture, a hallmark of single-molecule methods, is particularly useful in a field such as splicing where the in vitro splicing efficiency of pre-mRNAs in crude cell extracts varies greatly, typically between 20% and 80%.

Shcherbakova et al. (2013) were also able to examine whether the order of U1 and U2 assembly affected the ability to advance through eventual intron removal by labeling U1, U2, and the intron of the pre-mRNA (Fig. 4D). They observed that molecules following U1-first or U2-first paths proceeded through the second step of splicing equally well. They also found that the interaction of U1 with the pre-mRNA was more dynamic (i.e., showed more frequent binding and dissociation) than that of U2, and neither snRNP bound after intron loss (Fig. 4E).

Given the relatively long timescales of splicing in vitro, photobleaching of fluorophores poses an obvious constraint in single-molecule fluorescence imaging. Intermittent laser illumination and a high number of fluorophores incorporated into the intron can prolong the observation window in CoSMoS and help distinguish true dissociation from photobleaching. This general strategy is frequently applied to RNP studies when loss of a component from a complex is of interest—for example, in Argonaute (Ago)-mediated RNA slicing (Salomon et al. 2015)—and is important for high-background specimens as encountered in intracellular studies (see Sec. 5.1).

In addition to identifying components of a complex and characterizing their binding and/or dissociation kinetics, the precise identity of a molecule or changes in its structural state can also be inferred from the binding/dissociation kinetics of a dynamic, state-specific reader probe (Johnson-Buck et al. 2015; Rinaldi et al. 2016). Taken together, these approaches exemplify the strengths of single-molecule microscopy for studying the interplay between multiple biological components.

3.3 Advanced Single-Molecule Analysis

Perhaps one of the most challenging aspects of studying complex RNPs using single-molecule experiments is determining how best to analyze the data to fully extract the details of interactions given that the entire scope of molecular behaviors is directly observed (within the experimental limit of accessible spatiotemporal resolution). Although strategies such as selective labeling and SiMPull-based enrichment help reduce the data complexity and focus the experiment on observables of interest, single-molecule fluorescence traces are frequently very feature-rich and heterogeneous in a system-specific manner, motivating continuous innovations in data analysis.

For example, Blanco et al. (2015) used the Ubc4 pre-mRNA labeled for smFRET at the 5′SS and BS and collected data from more than 10,000 molecules at three time points and under a variety of conditions designed to stall splicing at various steps (Fig. 3A, red annotations). To examine pre-mRNA dynamics during splicing in greater detail, facilitate finding commonalities among many idealized smFRET traces, and compare properties across a variety of experimental conditions, these investigators developed single-molecule cluster analysis (SiMCAn) (Blanco et al. 2015). In SiMCAn, each trace is represented by a FRET similarity matrix (FSM) that describes the occupancy of FRET states and the kinetics of transitions between these states. FSMs for all traces from one or more experimental conditions were then clustered based on their similarity (Fig. 5A), allowing individual molecules to be segregated on the basis of both their FRET states and kinetics. The result was a set of 35 clusters, 25 and 10 showing dynamic and static FRET, respectively, representing the conformational behaviors observed across all conditions in the large data set.

A second round of clustering was then performed to determine the prevalence of particular conformational behaviors. This grouping revealed behavioral trends along the splicing cycle. For example, clade VII was found to be comprised of a group of clusters typifying FRET states that correspond to a pre-mRNA conformation in which the 5′SS and BS are brought into close proximity (Fig. 5B). This single-molecule behavior was observed in either wild-type (WT) or mutated 3′SS pre-mRNA when using WCEs containing a dominant-negative form of the helicase Prp16 (Prp16DN), a mutant that is known to stall splicing just before the second chemical step, which is indeed expected to lead to the observed proximity between 5′SS and BS. That SiMCAn also can discover new states was shown by the observation of a time-dependent increase in the separation of the 5′SS and BS (reflected by increased occupancy in clade I, a static low-FRET state) for specifically the 3′SS mutant when splicing was allowed to progress in the presence of WT Prp16. This

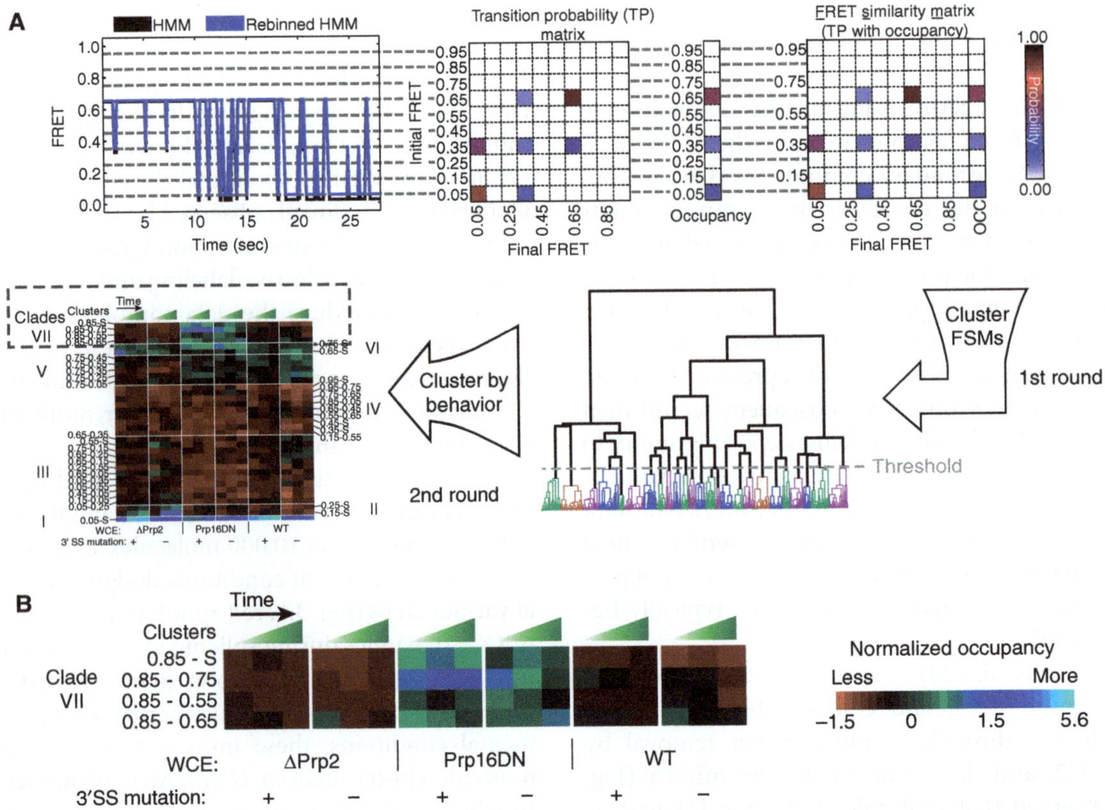

Figure 5. Advanced analysis of single-molecule Förster resonance energy transfer (smFRET) trajectory data—single-molecule cluster analysis (SiMCAn). (*A*) Workflow for SiMCAn. smFRET traces idealized with hidden Markov modeling (HMM) are rebinned to a discrete set of FRET values and then converted to a transition probability (TP) matrix, which describes the probability of observing transitions between various FRET states. The TP matrix for each molecule is combined with an additional occupancy column that describes the time spent in a given FRET state, yielding a FRET similarity matrix (FSM) for each molecule. The set of FSMs from all molecules are then clustered to identify common behaviors (clusters). The fractional representation of molecules from each experimental condition in a given cluster is then used in a second round of clustering to uncover particular behaviors (clusters) that are over- or underrepresented in a consistent way across a set of various experimental conditions. (*B*) Detailed view of clade VII. Experiments were performed using either wild-type (WT) yeast whole-cell extracts (WCEs), WCEs lacking Prp2, or WCEs containing a dominant-negative form of Prp16 (Prp16DN), on splicing complexes assembled on WT Ubc4 pre-messenger RNA (mRNA), or Ubc4 pre-mRNA containing a mutated 3′ splice site (SS). (Adapted, with permission, from Blanco et al. 2015.)

increased separation suggested a late-stage proofreading step involving Prp16 to allow for correct selection of the 3′SS, a finding subsequently characterized in greater detail by Semlow et al. (2016). SiMCAn thus facilitates comparison of data-rich single-molecule data sets to quickly identify unique behaviors that provide new biological insight.

4 DEEPER CONSIDERATIONS FOR STUDYING SINGLE RNPs IN VITRO

As the selected studies of the spliceosome described above show, RNPs can be highly dynamic and show heterogeneous behavior (Wahl et al. 2009). Although the biological process of interest can often be recapitulated in vitro, there is a risk of losing the "true" diversity, specificity, and speed of molecular behaviors that exist inside living cells. To further understand the scope and limitations of in vitro single-molecule studies of RNPs, in particular, some important design considerations for TIRFM-based imaging, the most widely used in vitro single-molecule imaging method, are discussed in the following.

The first consideration in designing any biophysical experiment to study RNPs is sample preparation. To minimize the variables influencing the process under study, in vitro single-molecule studies use minimal mixtures of purified RNAs and proteins. However, reactions involving RNPs often engage multiple, dynamically exchanging components whose stoichiometry is sometimes poorly defined.

Cite as *Cold Spring Harb Perspect Biol* doi: 10.1101/cshperspect.a032441

In such cases, it may be advantageous to instead isolate "native" holoenzyme complexes from whole cell lysates to recapitulate biological activity. This is exemplified by the strategy used by Krishnan et al. (2013), in which SiMPull was used to isolate the B^{act} complex. The drawback of this strategy for sample preparation is that the composition and stoichiometry of the isolated, individual complexes is undetermined. Importantly, molecular complexes obtained by native isolation and reconstitution may differ functionally. In some cases, such a discrepancy in activity can yield important biological insights, as single-molecule observations of the human telomerase complex illustrated (Wu et al. 2015). This RNP consists of the RNA-templated enzyme TERT, which adds repetitive nucleotide sequences to telomere ends in template-dependent manner. In vitro reconstituted minimal TERT RNPs were found to be monomers that were functionally active when assayed for repeat addition. In contrast, native holoenzymes isolated from cell lysates were found to exist as dimers and higher-order multimers, and showed lower repeat addition activity, pointing at intrinsic regulatory mechanisms (Wu et al. 2015). Additionally, native holoenzyme purification is often challenging and is not always conducive to facile fluorescent labeling.

The second consideration of SMFM experiments relates to the timescales of the reaction under study. To obtain useful statistics on reversible processes, such as those involving conformational changes or binding, it is important to be able to monitor a single molecule/complex over multiple reaction cycles, to adequately probe the residence times in each state (Moffitt et al. 2010). By design, TIRFM experiments extend observation windows for single molecules by anchoring them to a slide surface. Immobilization close to the surface, however, can occasionally introduce artifacts because of the effects of surface interactions on molecular conformation and folding kinetics (Yasuda et al. 2003). These effects can be overcome by using alternative immobilization strategies, such as encapsulating individual molecules or complexes in lipid vesicles that, in turn, are tethered to the surface (Okumus et al. 2004). Such alternate strategies, however, serve as practical solutions only when specific constraints or limitations are encountered on standard TIRFM-based imaging.

Given the compositional complexity of RNPs and their sensitivity to reaction conditions, it is arguably most informative to study them in their native context, in vivo. Observations of single-molecule interactions in living cells include the full context of biological pathways, arising from dynamic changes in RNP composition, effects of molecular chaperoning, molecular crowding and spatiotemporal organization. Using living cells themselves as reaction vessels to study RNPs trades precise control over the reaction conditions for new dimensions in single-molecule behavior. Studying the behavior of single RNPs in their cellular context represents a new paradigm for studying biology that harnesses the power and versatility of SMFM to unravel the roles of RNPs in shaping diverse cellular phenotypes.

5 SINGLE-RNP VISUALIZATION IN LIVING CELLS

The toolbox of complementary fluorescence-based approaches available today allows dynamic RNP interactions to be probed directly inside living cells. For example, fluorescence recovery after photobleaching (FRAP) and FCS both quantify diffusion of labeled molecules (McNally 2008; Bacia et al. 2014). The former measures the rate at which fluorophore-labeled molecules diffuse back into a small photobleached area (typically over minutes), whereas the latter measures fluorescence fluctuations in a small volume of the cell on subsecond timescales. These techniques have been used to study fluorescent molecules inside living cells for more than three decades, and although they do not necessarily provide single-molecule resolution, they have set the stage for intracellular SMFM.

More recently, the commercialization of photostable fluorophores, increasingly sensitive cameras, and super-resolution illumination strategies together have contributed to advances in a third technique, intracellular single-particle tracking (SPT) (Shen et al. 2017). SPT bridges the capabilities of ensemble approaches such as FRAP and FCS (diffusion information) with those of in vitro single-molecule analysis, such as colocalization (interaction) time and stepwise photobleaching analysis (for stoichiometry).

The principle behind SPT is that fluorescently labeled molecules, when present at sufficiently low densities, can be detected and tracked in time as bright, diffraction-limited "spots" using an illumination strategy such as HILO or TIRFM. The resulting trajectories represent the actual diffusive motion of the molecules and can reveal transient interactions with their local environments. Thus, intracellular SPT can be applied to study RNP reactions using cultured living cells themselves as reaction vessels, providing exquisite spatiotemporal resolution to make observations that enable both hypothesis- and discovery-driven approaches to studying the biophysical behavior of RNPs (Fig. 6).

The following sections briefly review methods for the fluorophore labeling and intracellular macromolecule delivery needed to perform intracellular SPT.

5.1 Fluorescent Labeling Strategies Compatible with Intracellular RNP Detection

One advantage of using living cells in fluorescence microscopy is the ability to genetically encode fluorophore tags in

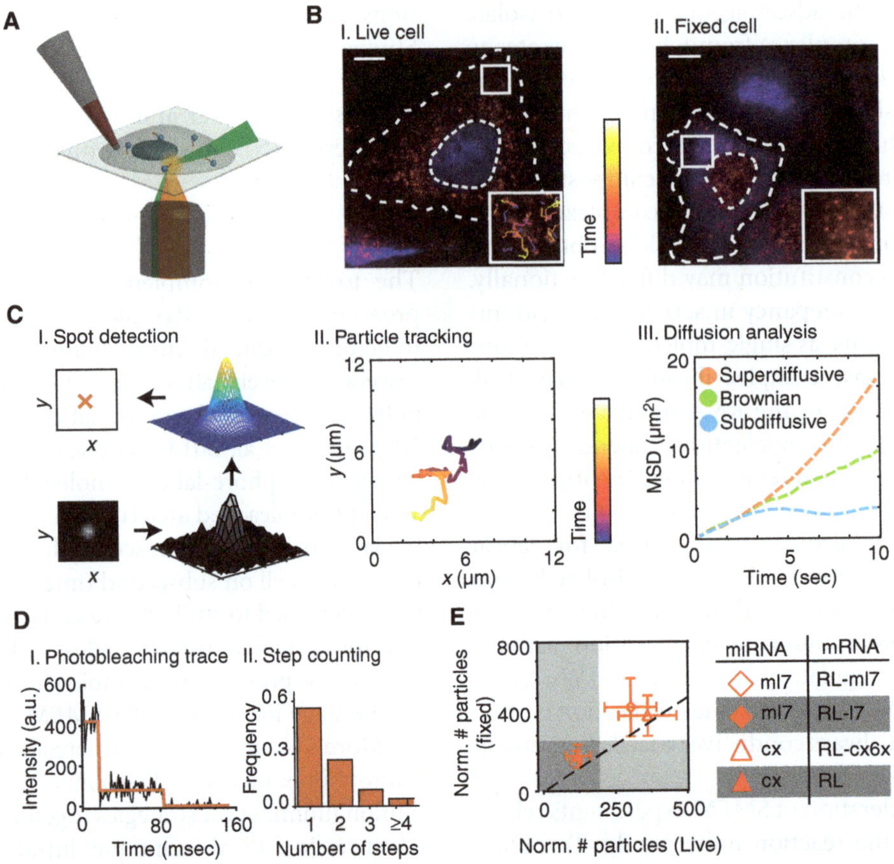

Figure 6. Intracellular single-molecule visualization. (*A*) Microinjection is an effective strategy to deliver labeled RNAs into living cells. The injected material can be imaged using various fluorescence microscopy techniques. Intracellular single-molecule high-resolution localization and counting (iSHiRLoC) uses highly inclined and laminated optical sheet (HILO) illumination to rapidly acquire images with single-molecule resolution for particle tracking. (*B*) Injected cells are distinguished from noninjected cells by the presence of an injection marker (a slowly diffusing, fluorescently labeled inert compound). Shown here are two cells injected into the nucleus (*I*) and cytosol (*II*) with Alexa488-labeled 100-kDa dextran. Labeled microRNAs (miRNAs) appear as diffusing, diffraction-limited spots on time-lapse imaging of live cells (*I, inset*) or as diffraction-limited spots in formaldehyde-fixed samples (*II, inset*). Scale bars, 10 µm. (*C*) (*I*) The intensity profiles of diffraction-limited spots, recorded as pixelated point-spread functions, can be analyzed to obtain the spot centers with subpixel accuracy, such as by fitting the intensity profile to two-dimensional (2D) Gaussians. (*II*) Particle-tracking of the spot centers results in 2D or 3D diffusion trajectories. (*III*) Displacements in successive frames yield mean-squared displacement (MSD) profiles, informing about diffusion types. (*D*) Photobleaching analysis. Cy5 fluorophores bleach rapidly and in a stepwise manner. The number of discrete step drops in intensity is a proxy for the number of molecules in a single diffraction-limited spot. (*E*) Correlating the number of fluorescent spots in fixed tissue with the number of diffusion-filtered trajectories yields an estimate of turnover and functionality of the labeled molecules. The open shapes represent counts of a mutant (mutant *let-7*, ml7) and artificial (*cxcr4*, cx) miRNA coinjected with messenger RNAs (mRNAs) bearing complementary binding sites, and the filled shapes represent conditions in which the coinjected mRNAs bore mismatched (*let-7*, l7) or no binding sites in the 3′ untranslated region (UTR) downstream from an open reading frame (ORF) encoding *Renilla* luciferase. (*D*, Adapted, with permission, from Pitchiaya et al. 2012; *E*, adapted, with permission, from the authors and John Wiley.)

the form of FPs. The most widely adopted intracellular RNA-labeling strategy has been the use of RNA-binding proteins (RBPs) fused with an FP (Tyagi 2009). This method, first demonstrated for single-RNA detection by Robert Singer's group, exploits the high specificity and affinity with which RBPs such as the viral coat proteins from the MS2 or PP7 phages bind their cognate RNA stem-loop structures. Since then, this technology has been expanded to other bacteriophages, such as the λ_N protein binding to the RNA stem loop of the *boxB* gene. Labeling is achieved by inserting these stem-loop sequences into untranslated regions (usually the 3′ UTRs) of mRNA transcripts, which are then bound by their respective RBP fused to an FP (Fig. 2D). Because the excess of unbound RBP-FP fusion protein creates a high fluorescence background, and because photobleaching is vastly accelerated in live cells as compared with in vitro, multiple copies of the stem-loop sequence (typically 8 to 96) are required for single-molecule sensitivity. Furthermore, multiple transcript species within the same cell can be simultaneously monitored by using combinations of RBPs tagged with distinct FPs (Hocine et al. 2013).

The biggest limitation of RBP-based detection is that their cognate stem loops are ~20 nucleotides long, which makes RBP-based methods unsuitable to study small noncoding RNAs such as miRNAs, small interfering RNAs (siRNAs), tRNAs, etc. whose function is sensitive to RNA length. Furthermore, the added molecular mass of bound RBP-FP fusion proteins can alter the diffusion properties, localization, and turnover of RNPs.

Another limitation of FPs is their relatively long maturation time. For studies specifically probing protein translation, using a classical fluorescent protein to study translation rate is of limited use because of the slow maturation of the FP (the fastest maturing GFP variants fold in ~10 min) relative to translation elongation (seconds to minutes for a given transcript) (Thorn 2017). However, moving the fluorescent tag to an antibody probe that can recognize a peptide epitope with specificity as it emerges from the ribosome's exit tunnel allows rapid processes such as translation elongation to be studied in real time at single-molecule resolution. Such fluorescent "immunolabeling" of proteins has been shown with the SunTag epitope and "spaghetti monster" fluorescent proteins (Fig. 2E). In these methods, single proteins are detected when multiple labeled antibody probes bind to a repeated epitope, in which detection is limited by diffusion of the probes and antibody–antigen affinity (Morisaki et al. 2016; Yan et al. 2016).

The relative simplicity of working with transiently or stably transfected cell lines has led genetically encoded FP methods for single-RNP visualization to become widely accepted. We refer the reader to reviews that cover the broad applications of RBP-based mRNA detection and live cell immunolabeling for intracellular single-molecule visualization (Buxbaum et al. 2015; Lyon and Stasevich 2017). An alternative approach to fluorescently label proteins uses self-modifying enzymatic tags (see Sec. 2.2.2 and Fig. 2E). Pulse labeling with a cell-permeable, fluorescent HaloTag substrate, for example, allows a sufficiently sparse set of the native proteins to be labeled. Schmidt et al. (2016) used this strategy to show that the telomerase RNP is dynamically recruited to the telomere ends of chromosomes.

An alternative RNP-labeling strategy that avoids genetic modification of RNAs/proteins for intracellular detection is to covalently attach synthetic fluorophore dyes to an exogenous RNA and subsequently deliver it into cells. For longer RNAs, click chemistry has been used to covalently attach fluorophores to modified nucleotides that are cotranscriptionally incorporated into in vitro transcribed RNA without the need for lengthy exogenous sequence motifs (compare Fig. 2B and D). For eukaryotic mRNAs, fluorophores can be incorporated into mRNAs in the cap structure (Mamot et al. 2017), randomly incorporated throughout the transcript (Schulz and Rentmeister 2014), in the poly(A) tail, or between the 3′ UTR and poly(A) tail (Custer and Walter 2017). Such covalent attachment of the fluorophore allows labeled molecules to be detected with greater sensitivity, without the background of free fluorophores that is a recurrent challenge of RBP-based RNA detection. As with RBP-mediated labeling, however, there can be functional consequences to these covalent labeling strategies; for example, random fluorophore incorporation throughout the transcript was shown to dramatically impair the coding function, and incorporation of fluorophores in the 3′ UTR was shown to stabilize mRNAs (Custer and Walter 2017).

The choice of delivery method determines the types of observations that can be performed after exogenous chemical labeling. The following section briefly surveys these methods.

5.2 Intracellular Delivery of Materials Labeled In Vitro

The wide array of available intracellular delivery methods presents several choices for RNA and/or protein delivery (Stewart et al. 2016). They fall on a spectrum defined by, on one hand, their ability to deliver precise amounts of material to individual cells and by, on the other hand, their throughput of cells. One subset of delivery strategies aims to temporarily disrupt the cell membrane using detergents, bacterial pore-forming toxins, or various physical disruption methods (bead loading, fluid shearing, electroporation, etc.), relying on diffusion of cargo along concentration gradients and the ability of cells to reseal breaks in their cell membrane. These strategies achieve efficient cargo delivery but are not suitable for applications requiring precise dosage control. Another delivery strategy in this class involves vesicle carriers and relies on fusion of lipid vesicles with the cell membrane or natural endocytic processes for cargo delivery. Lipid-based transfection is perhaps the most widely used strategy; however, it is unsuitable for RNA delivery for single-molecule applications as the cargo often remains trapped in endosomes and is only slowly released into the cell (Hirsch and Helm 2015), leading to the risk of artifacts and/or misinterpretation.

In contrast, methods that involve active forces, such as delivery by direct microinjection using microneedles or nanostraws, offer more precise control over the amount of material delivered to each cell, however, they can do so only with lower throughput. Microinjection, like other membrane disruption methods, relies on the ability of cells to rapidly reseal breaks in the cell membrane (Fig. 6A) and thus can encounter problems with cell viability if improperly executed (Pitchiaya et al. 2013). The microinjector needle is a fine glass capillary used to inject femtoliters of cargo solution into each cell, allowing for precise delivery into subcellular compartments such as the nucleus or the cytosol. Microinjection requires only small cargo quantities compared with any other bulk delivery strategy, making it a particularly valuable and preferred tool for delivering fluorescently labeled RNP components that may be technically challenging to label and purify.

5.3 Analysis and Interpretation of SPT Experiments

The first step for particle tracking is detection of the "spot" that represents the spatial location of a fluorescently labeled molecule of interest. The intensity levels of the brightest pixels of the spot together with the surrounding pixels approximate the point spread function (PSF) of the fluorophore(s) present on the molecule or complex. The challenge of spot detection lies in finding the center of the PSF, given the discrete pixel size and variable camera noise. A number of different algorithms exist today to perform this task (Fig. 6B) (Chenouard et al. 2014). Once detected, the spots can then be tracked (i.e., temporally linked) through successive frames to yield spatiotemporal trajectories, provided the spots are sufficiently visible in the focal plane in (almost) every frame, and the particles are present at sufficiently low density for the temporal linking to be unambiguous.

From these particle trajectories, the type of motion shown by individual molecules can be classified into one of various types of diffusion. The most common, simple, and useful tool to discriminate diffusive behaviors of single particles is the time-averaged mean-squared displacement (MSD) (Fig. 6C). This plot is used to characterize the trend of the displacement of the particle at increasing time intervals. For a given trajectory, the deviation of the MSD plot from linearity is used to classify the motion as diffusive (no deviation), subdiffusive (curved down), or superdiffusive (curved up). RNPs often show complex trajectories that can be modeled as mixtures of directed motion (such as during periods of active transport along axons) and Brownian (random) diffusion (Monnier et al. 2015). Together, such characterization can yield information regarding the molecular weight of the particle under study, and its interaction with its local environment, such as binding to and dissociation from unlabeled (and thus invisible) partners of distinct diffusivity (Knight et al. 2015).

Multicolor particle tracking can reveal even more about the intracellular milieu. By labeling molecular species using dyes with spectrally separated emission maxima, currently up to four different types of particles can be simultaneously tracked (Juette et al. 2014). The interaction between these differently labeled components can be read out as spatiotemporal colocalizations between trajectories of differently colored particles, similar to the in vitro techniques summarized in Section 3.2. Colocalizing trajectories can then be analyzed to characterize the kinetics and search dynamics of intracellular interactions, as has been the case for examining the intracellular search dynamics of RNP machines such as the Cas9 enzyme (Knight et al. 2015), telomerase (Schmidt et al. 2016), and miRNA–RISC (miRISC) complexes, as well as for studying the recruitment of small RNAs to sites of DNA double-strand breaks (Michelini et al. 2017).

The following section examines the example of intracellular single-molecule methods applied to the RNA silencing pathway.

6 APPLICATION EXAMPLE OF IN VIVO SINGLE-MOLECULE FLUORESCENCE: RNA SILENCING

The first reports of in situ single-miRNA visualization used complementary LNA hybridization probes to detect miRNAs in fixed cells and were applied to detect the subcellular localization and cell-to-cell variability in expression levels of miRNAs (Lu and Tsourkas 2009). Complementary ensemble-level observation of dynamic changes in subcellular localization of siRNA in living cells was reported by Ohrt et al. (2006). Examining the diffusion of chemically labeled siRNA strands using FCS, they found that the fluorophore labeling position on the double-stranded siRNA influenced intracellular diffusion and localization. Instead of using various commonly used lipid- and peptide-based transfection techniques, which can also influence siRNA localization (Chiu et al. 2004), the investigators used microinjection to deliver labeled, double-stranded siRNA variants directly into the cell. Microinjection also permitted them to start imaging as early as 10 min after introduction of the siRNAs, allowing them to study early steps of the siRNA pathway. Their observations corroborated earlier findings that the loading of the siRNA strand into RISC is sensitive to modifications at the 5′ end and overhangs of the guide strand. Observing the codiffusion of labeled siRNAs with GFP-tagged Argonaute protein using two-color FCCS, these investigators were able to compare the nuclear export pathways of "naked" and RISC-loaded siRNAs. FCS-based measurements of the molecular weights of nuclear and cytoplasmic RISC suggested that the two complexes are

Cite as *Cold Spring Harb Perspect Biol* doi: 10.1101/cshperspect.a032441

largely mRNA-bound and -unbound, respectively. By co-injecting labeled siRNA with purified mRNA target, they concluded that nuclear import of RISC is target-dependent (Ohrt et al. 2008).

Although FCS provides useful insight into diffusive processes and the concentration of fluorophores in a cell, it is limited by its confocal imaging geometry that only yields ensemble-averaged information on fluorophores passing through a relatively large (~$0.5 \times 0.5 \times 1\ \mu m^3$) open volume element. Alternatively, SPT can be used to detect the motion of small RNA molecules in a technique reported by Pitchiaya et al. (2012) termed intracellular single-molecule, high-resolution localization, and counting (iSHiRLoC). Based on their molecular weights, microinjected "naked" double-stranded miRNAs and those loaded into "minimal" RISC complexes are expected to diffuse rapidly (>10 $\mu m^2/s$ and ~5 $\mu m^2/s$, respectively [Ohrt et al. 2006]), blurring them out at the standard 100 ms exposure time required for sufficient fluorescent signal for SPT. Using microinjection and the HILO illumination strategy, Pitchiaya et al. (2012) reported that miRNAs chemically labeled on the 3′ end of the guide strand assembled into particles that could be tracked 1 h after microinjection into the cytosol. The resulting single-particle trajectories were quantified to yield the diffusion rates of miRNAs at a given time point after microinjection. The study revealed two slowly diffusing populations that were assigned as mRNA-bound RISC RNPs, either diffusing freely (at ~0.3 $\mu m^2/s$) or docking (at ~0.03 $\mu m^2/s$) onto processing bodies (Pitchiaya et al. 2012).

A second technique within the iSHiRLoC toolbox is particle counting in fixed cells. Formaldehyde fixation captures the instantaneous position of all molecules in the cell. The rapid and stochastic photobleaching of single Cy5 dye molecules can then be exploited to count the number of fluorophores in individual bright spots, and this count, in turn, can be used as a proxy for the number of labeled molecules (Fig. 6D). Counting singly fluorophore-labeled miRNAs in this manner revealed that miRNA-containing complexes assembled significantly into multimers only in the presence of mRNA targets. Using two-color colocalization analysis, a fraction of these multimers were found associated with FP-tagged P-bodies, supporting the notion that miRNAs are recruited to P-bodies.

RNA silencing involves miRNA trafficking, strand unwinding, RISC-loading, target binding, and eventual degradation (Ha and Kim 2014). To examine all of these steps by visualizing them in live cells, Pitchiaya et al. (2017) included a third metric in iSHiRLoC, correlating the number of trajectories of mature miRISC·mRNPs diffusing in living cells with the total number of miRNA in fixed cells. Correlating the counts of live- and fixed-cell miRISC showed that miRNAs that efficiently load into RISC and engage target are present in comparable numbers in live and fixed cells, whereas miRNAs that stay in minimal complexes or remain "naked" (not loaded into the RISC complex) are only visible in fixed cells and instead blur out by virtue of the 100 ms exposure time used in live cell imaging (Fig. 6E). Correlating the counts of miRNAs, in the presence and absence of mRNA targets, allowed the investigators to build a detailed kinetic model of miRNA activity and turnover, which revealed that miRNAs not engaging a target mRNA are rapidly degraded (Pitchiaya et al. 2017).

These intracellular single-molecule visualization methods are not without limitations. The fluorescent dyes currently available for labeling have finite photostability under the illumination conditions required for single-molecule sensitivity, limiting the observation window of imaging. SPT is restricted by temporal resolution of imaging and is unsuitable for the study of very rapid molecular motion. Finally, microinjection currently has a low throughput of cells that can be targeted in each experiment, limiting the amount of data obtained from each specimen.

7 CONCLUSIONS AND OUTLOOK

As outlined in this survey, in vitro single-molecule fluorescence imaging of RNPs enables detailed characterization of the kinetics of RNA–protein interactions and accompanying conformational rearrangements. In vitro assays allow for specific manipulations of the reaction conditions and composition, such as specific placement of fluorophores and addition of cofactors, facilitating quantitative biophysical measurements. In principle, any RNP that can be prepared using genetic, chemical, and biochemical approaches can thus be studied at the single-molecule level. As more pathways of RNA metabolism are enumerated, the unique insights offered by SMFM tools will undoubtedly motivate their widespread adoption. The future will also likely see the proliferation of observables as additional labeling colors and parameters such as force measurements become more widely adopted. In vitro SMFM tools are poised to be further strengthened with other in vitro tools, such as high-resolution electron microscopy, that will enable direct correlation and integration of results from different single-molecule observation modalities.

In parallel, complementary intracellular single-molecule tools such as iSHiRLoC have begun to reveal biological cornerstones of complex, multistep, and multicomponent RNA pathways such as RNA silencing. The key observables of intracellular tools such as the fluorescence intensities and diffusion trajectories of exogenous RNPs yield insights into cellular biology at defined probe concentrations, thereby overcoming the challenges of overexpression. The labeled components in these studies serve as molecular "baits" to probe the complexity of the cellular milieu, which potentially differs considerably from classical in vitro assay conditions. The intracellular single-molecule toolbox thus paves the way toward a holistic study of RNP molecular biology by integrating biochemical analysis with insights into cell biology.

iSHiRLoC and other intracellular single-molecule fluorescence strategies face several challenges that need to be addressed. It is only a matter of time, however, until the current limitations will be overcome by, among others, improvements in fluorophores, delivery strategies, automation, microscopes, and analysis methods. Such advances will then permit a broader repertoire of observables that are currently only available to in vitro studies, such as measurements of intraparticle distances by smFRET and binding and dissociation rate constants by coincidence analysis. The future is thus bright for studying RNA and protein assemblies under the single-molecule fluorescence microscope both in vitro and in vivo for a full understanding of vast, still-emerging RNA worlds.

ACKNOWLEDGMENTS

The authors thank Julia Widom for a critical reading of this manuscript and acknowledge funding by National Institutes of Health Grants R01 GM062357, R01 GM118524, R01 GM122803, and R21 CA204560, and University of Michigan Comprehensive Cancer Center/Biointerfaces Institute and M-TRAC pilot grants. A.P.J. is supported by Cellular and Molecular Biology Training Grant T32GM007315.

REFERENCES

Abelson J, Blanco M, Ditzler MA, Fuller F, Aravamudhan P, Wood M, Villa T, Ryan DE, Pleiss JA, Maeder C, et al. 2010. Conformational dynamics of single pre-mRNA molecules during in vitro splicing. *Nat Struct Mol Biol* **17:** 504–512.

Adumeau P, Sharma SK, Brent C, Zeglis BM. 2016. Site-specifically labeled immunoconjugates for molecular imaging—Part 2: Peptide tags and unnatural amino acids. *Mol Imaging Biol* **18:** 153–165.

Agirrezabala X, Valle M. 2015. Structural insights into tRNA dynamics on the ribosome. *Int J Mol Sci* **16:** 9866–9895.

Aitken CE, Marshall RA, Pulglisi JD. 2008. An oxygen scavenging system for improvement of dye stability in single-molecule fluorescence experiments. *Biophys J* **94:** 1826–1835.

Al-Hashimi HM, Walter NG. 2008. RNA dynamics: It is about time. *Curr Opin Struct Biol* **18:** 321–329.

Bacia K, Haustein E, Schwille P. 2014. Fluorescence correlation spectroscopy: Principles and applications. *Cold Spring Harb Protoc* **2014:** 709–725.

Blanco M, Walter NG. 2010. Analysis of complex single-molecule FRET time trajectories. *Methods Enzymol* **472:** 153–178.

Blanco MR, Martin JS, Kahlscheuer ML, Krishnan R, Abelson J, Laederach A, Walter NG. 2015. Single molecule cluster analysis dissects splicing pathway conformational dynamics. *Nat Meth* **12:** 1077–1084.

Buxbaum AR, Haimovich G, Singer RH. 2015. In the right place at the right time: Visualizing and understanding mRNA localization. *Nat Rev Mol Cell Biol* **16:** 95–109.

Calloway NT, Choob M, Sanz A, Sheetz MP, Miller LW, Cornish VW. 2007. Optimized fluorescent trimethoprim derivatives for in vivo protein labeling. *Chembiochem* **8:** 767–774.

Carrocci TJ, Lohe L, Ashton MJ, Hobartner C, Hoskins AA. 2017. Debranchase-resistant labeling of RNA using the 10DM24 deoxyribozyme and fluorescent modified nucleotides. *Chem Commun* **53:** 11992–11995.

Chenouard N, Smal I, de Chaumont F, Maska M, Sbalzarini IF, Gong Y, Cardinale J, Carthel C, Coraluppi S, Winter M, et al. 2014. Objective comparison of particle tracking methods. *Nat Meth* **11:** 281–289.

Chiu YL, Ali A, Chu CY, Cao H, Rana TM. 2004. Visualizing a correlation between siRNA localization, cellular uptake, and RNAi in living cells. *Chem Biol* **11:** 1165–1175.

Cisse II, Kim H, Ha T. 2012. A rule of seven in Watson–Crick base-pairing of mismatched sequences. *Nat Struct Mol Biol* **19:** 623–627.

Colomb W, Sarkar SK. 2015. Extracting physics of life at the molecular level: A review of single-molecule data analyses. *Phys Life Rev* **13:** 107–137.

Crawford DJ, Hoskins AA, Friedman LJ, Gelles J, Moore MJ. 2008. Visualizing the splicing of single pre-mRNA molecules in whole cell extract. *RNA* **14:** 170–179.

Crawford DJ, Hoskins AA, Friedman LJ, Gelles J, Moore MJ. 2013. Single-molecule colocalization FRET evidence that spliceosome activation precedes stable approach of 5′ splice site and branch site. *Proc Natl Acad Sci* **110:** 6783–6788.

Crick F. 1970. Central dogma of molecular biology. *Nature* **227:** 561–563.

Custer TC, Walter NG. 2017. In vitro labeling strategies for in cellulo fluorescence microscopy of single ribonucleoprotein machines. *Protein Sci* **26:** 1363–1379.

Darre L, Ivani I, Dans PD, Gomez H, Hospital A, Orozco M. 2016. Small details matter: The 2′-hydroxyl as a conformational switch in RNA. *J Am Chem Soc* **138:** 16355–16363.

DeHaven AC, Norden IS, Hoskins AA. 2016. Lights, camera, action! Capturing the spliceosome and pre-mRNA splicing with single-molecule fluorescence microscopy. *WIREs RNA* **7:** 683–701.

dos Remedios CG, Moens PD. 1995. Fluorescence resonance energy transfer spectroscopy is a reliable "ruler" for measuring structural changes in proteins: Dispelling the problem of the unknown orientation factor. *J Struct Biol* **115:** 175–185.

Edwards TE, Sigurdsson ST. 2014. Modified RNAs as tools in RNA biochemistry. In *Handbook of RNA biochemistry* (ed. Hartmann RK, Bindereif A, Schon A, Westhof E), pp. 36–52. Wiley, Weinheim, Germany.

Fauster K, Hartl M, Santner T, Aigner M, Kreutz C, Bister K, Ennifar E, Micura R. 2012. 2′-Azido RNA, a versatile tool for chemical biology: Synthesis, X-ray structure, siRNA applications, click labeling. *ACS Chem Biol* **7:** 581–589.

Fei J, Wang J, Sternberg SH, MacDougall DD, Elvekrog MM, Pulukkunat DK, Englander MT, Gonzalez RL Jr. 2010. A highly purified, fluorescently labeled in vitro translation system for single-molecule studies of protein synthesis. *Methods Enzymol* **472:** 221–259.

Gautier A, Juillerat A, Heinis C, Corrêa IR Jr, Kindermann M, Beaufils F, Johnsson K. 2008. An engineered protein tag for multiprotein labeling in living cells. *Chem Biol* **15:** 128–136.

Gonzalez Bardeci N, Angiolini JF, De Rossi MC, Bruno L, Levi V. 2017. Dynamics of intracellular processes in live-cell systems unveiled by fluorescence correlation microscopy. *IUBMB Life* **69:** 8–15.

Grima R, Walter NG, Schnell S. 2014. Single-molecule enzymology à la Michaelis–Menten. *FEBS J* **281:** 518–530.

Ha M, Kim VN. 2014. Regulation of microRNA biogenesis. *Nat Rev Mol Cell Biol* **15:** 509–524.

Hendrix C, Devreese B, Rozenski J, van Aerschot A, De Bruyn A, Van Beeumen J, Herdewijn P. 1995. Incorporation of 2′-amido-nucleosides in oligodeoxynucleotides and oligoribonucleotides as a model for 2′-linked conjugates. *Nucleic Acids Res* **23:** 51–57.

Hermanson GT. 2013. Nucleic acid and oligonucleotide modification and conjugation. In *Bioconjugate techniques*, 3rd ed. (ed. Hermanson GT), pp. 959–988. Academic, Boston.

Hickerson R, Majumdar ZK, Baucom A, Clegg RM, Noller HF. 2005. Measurement of internal movements within the 30 S ribosomal subunit using Förster resonance energy transfer. *J Mol Biol* **354:** 459–472.

Hirsch M, Helm M. 2015. Live cell imaging of duplex siRNA intracellular trafficking. *Nucleic Acids Res* **43:** 4650–4660.

Hocine S, Raymond P, Zenklusen D, Chao JA, Singer RH. 2013. Single-molecule analysis of gene expression using two-color RNA labeling in live yeast. *Nat Meth* **10:** 119–121.

Huang F, He J, Zhang Y, Guo Y. 2008. Synthesis of biotin-AMP conjugate for 5′ biotin labeling of RNA through one-step in vitro transcription. *Nat Protoc* **3:** 1848–1861.

Hwang H, Myong S. 2014. Protein induced fluorescence enhancement (PIFE) for probing protein–nucleic acid interactions. *Chem Soc Rev* **43:** 1221–1229.

Hwang W, Arluison V, Hohng S. 2011. Dynamic competition of DsrA and *rpoS* fragments for the proximal binding site of Hfq as a means for efficient annealing. *Nucleic Acids Res* **39:** 5131–5139.

Jens M, Rajewsky N. 2015. Competition between target sites of regulators shapes post-transcriptional gene regulation. *Nat Rev Genet* **16:** 113–126.

Johnson-Buck A, Su X, Giraldez MD, Zhao M, Tewari M, Walter NG. 2015. Kinetic fingerprinting to identify and count single nucleic acids. *Nat Biotechnol* **33:** 730–732.

Juette MF, Terry DS, Wasserman MR, Zhou Z, Altman RB, Zheng Q, Blanchard SC. 2014. The bright future of single-molecule fluorescence imaging. *Curr Opin Chem Biol* **20:** 103–111.

Kahlscheuer ML, Widom J, Walter NG. 2015. Single-molecule pull-down FRET to dissect the mechanisms of biomolecular machines. *Methods Enzymol* **558:** 539–570.

Kim H, Abeysirigunawarden SC, Chen K, Mayerle M, Ragunathan K, Luthey-Schulten Z, Ha T, Woodson SA. 2014. Protein-guided RNA dynamics during early ribosome assembly. *Nature* **506:** 334–338.

Kinoshita Y, Nishigaki K, Husimi Y. 1997. Fluorescence-, isotope- or biotin-labeling of the 5′-end of single-stranded DNA/RNA using T4 RNA ligase. *Nucleic Acids Res* **25:** 3747–3748.

Knight SC, Xie L, Deng W, Guglielmi B, Witkowsky LB, Bosanac L, Zhang ET, El Beheiry M, Masson JB, Dahan M, et al. 2015. Dynamics of CRISPR-Cas9 genome interrogation in living cells. *Science* **350:** 823–826.

Kolb HC, Finn MG, Sharpless KB. 2001. Click chemistry: Diverse chemical function from a few good reactions. *Angew Chem Int Ed Engl* **40:** 2004–2021.

Krishnan R, Blanco MR, Kahlscheuer ML, Abelson J, Guthrie C, Walter NG. 2013. Biased Brownian ratcheting leads to pre-mRNA remodeling and capture prior to first-step splicing. *Nat Struct Mol Biol* **20:** 1450–1457.

Lakowicz JR. 2006. *Principles of fluorescence spectroscopy*. Springer, New York.

Larson JD, Rodgers ML, Hoskins AA. 2014. Visualizing cellular machines with colocalization single molecule microscopy. *Chem Soc Rev* **43:** 1189–1200.

Lee TC, Kang M, Kim CH, Schultz PG, Chapman E, Deniz AA. 2016. Dual unnatural amino acid incorporation and click-chemistry labeling to enable single-molecule FRET studies of p97 folding. *Chembiochem* **17:** 981–984.

Leontis NB, Stombaugh J, Westhof E. 2002. The non-Watson–Crick base pairs and their associated isostericity matrices. *Nucleic Acids Res* **30:** 3497–3531.

Lerner E, Ploetz E, Hohlbein J, Cordes T, Weiss S. 2016. A quantitative theoretical framework for protein-induced fluorescence enhancement-Förster-type resonance energy transfer (PIFE-FRET). *J Phys Chem B* **120:** 6401–6410.

Levene MJ, Korlach J, Turner SW, Foquet M, Craighead HG, Webb WW. 2003. Zero-mode waveguides for single-molecule analysis at high concentrations. *Science* **299:** 682–686.

Li F, Dong J, Hu X, Gong W, Li J, Shen J, Tian H, Wang J. 2015. A covalent approach for site-specific RNA labeling in mammalian cells. *Angew Chem Int Ed Engl* **54:** 4597–4602.

Liu Y, Lilley DMJ. 2017. Crystal structures of cyanine fluorophores stacked onto the end of double-stranded RNA. *Biophys J* **113:** 2336–2343.

Liu CC, Schultz PG. 2010. Adding new chemistries to the genetic code. *Annu Rev Biochem* **79:** 413–444.

Liu Y, Park J, Dahmen KA, Chemla YR, Ha T. 2010. A comparative study of multivariate and univariate hidden Markov modelings in time-binned single-molecule FRET data analysis. *J Phys Chem B* **114:** 5386–5403.

Liu Z, Lavis LD, Betzig E. 2015. Imaging live-cell dynamics and structure at the single-molecule level. *Mol Cell* **58:** 644–659.

Los GV, Encell LP, McDougall MG, Hartzell DD, Karassina N, Zimprich C, Wood MG, Learish R, Ohana RF, Urh M, et al. 2008. HaloTag: A novel protein labeling technology for cell imaging and protein analysis. *ACS Chem Biol* **3:** 373–382.

Lu J, Tsourkas A. 2009. Imaging individual microRNAs in single mammalian cells in situ. *Nucleic Acids Res* **27:** e100.

Lu HP, Xun L, Xie XS. 1998. Single-molecule enzymatic dynamics. *Science* **282:** 1877–1882.

Lyon K, Stasevich TJ. 2017. Imaging translational and post-translational gene regulatory dynamics in living cells with antibody-based probes. *Trends Genet* **33:** 322–335.

Mamot A, Sikorski PJ, Warminski M, Kowalska J, Jemielity J. 2017. Azido-functionalized 5′ cap analogues for the preparation of translationally active mRNAs suitable for fluorescent labeling in living cells. *Angew Chem Int Ed Engl* **56:** 15628–15632.

McCann JJ, Choi UB, Zheng L, Weninger K, Bowen ME. 2010. Optimizing methods to recover absolute FRET efficiency from immobilized single molecules. *Biophys J* **99:** 961–970.

McNally JG. 2008. Quantitative FRAP in analysis of molecular binding dynamics in vivo. *Methods Cell Biol* **85:** 329–351.

Michelini F, Pitchiaya S, Vitelli V, Sharma S, Gioia U, Pessina F, Cabrini M, Wang Y, Capozzo I, Iannelli F, et al. 2017. Damage-induced lncRNAs control the DNA damage response through interaction with DDRNAs at individual double-strand breaks. *Nat Cell Biol* **19:** 1400–1411.

Milas P, Gamari BD, Parrot L, Krueger BP, Rahmanseresht S, Moore J, Goldner LS. 2013. Indocyanine dyes approach free rotation at the 3′ terminus of A-RNA: A comparison with the 5′ terminus and consequences for fluorescence resonance energy transfer. *J Phys Chem B* **117:** 8649–8658.

Moffitt JR, Chemla YR, Bustamante C. 2010. Methods in statistical kinetics. *Methods Enzymol* **475:** 221–257.

Monnier N, Barry Z, Park HY, Su KC, Katz Z, English BP, Dey A, Pan K, Cheeseman IM, Singer RH, et al. 2015. Inferring transient particle transport dynamics in live cells. *Nat Meth* **12:** 838–840.

Morisaki T, Lyon K, DeLuca KF, DeLuca JG, English BP, Zhang Z, Lavis LD, Grimm JB, Viswanathan S, Looger LL, et al. 2016. Real-time quantification of single RNA translation dynamics in living cells. *Science* **352:** 1425–1429.

Mustoe AM, Brooks CL, Al-Hashimi HM. 2014. Hierarchy of RNA functional dynamics. *Annu Rev Biochem* **83:** 441–466.

Nissen P, Ippolito JA, Ban N, Moore PB, Steitz TA. 2001. RNA tertiary interactions in the large ribosomal subunit: The A-minor motif. *Proc Natl Acad Sci* **98:** 4899–4903.

Ohrt T, Merkle D, Birkenfeld K, Echeverri CJ, Schwille P. 2006. In situ fluorescence analysis demonstrates active siRNA exclusion from the nucleus by exportin 5. *Nucleic Acids Res* **34:** 1369–1380.

Ohrt T, Mutze J, Staroske W, Weinmann L, Hock J, Crell K, Meister G, Schwille P. 2008. Fluorescence correlation spectroscopy and fluorescence cross-correlation spectroscopy reveal the cytoplasmic origination of loaded nuclear RISC in vivo in human cells. *Nucleic Acids Res* **36:** 6439–6449.

Okumus B, Wilson TJ, Lilley DM, Ha T. 2004. Vesicle encapsulation studies reveal that single molecule ribozyme heterogeneities are intrinsic. *Biophys J* **87:** 2798–2806.

Pan C, Potratz JP, Cannon B, Simpson ZB, Ziehr JL, Tijerina P, Russell R. 2014. DEAD-box helicase proteins disrupt RNA tertiary structure through helix capture. *PLoS Biol* **12:** e1001981.

Paredes E, Evans M, Das SR. 2011. RNA labeling, conjugation and ligation. *Methods* **54:** 251–259.

Pitchiaya S, Androsavich JR, Walter NG. 2012. Intracellular single molecule microscopy reveals two kinetically distinct pathways for microRNA assembly. *EMBO Rep* **13:** 709–715.

Pitchiaya S, Krishnan V, Custer TC, Walter NG. 2013. Dissecting noncoding RNA mechanisms in cellulo by single-molecule high-resolution localization and counting. *Methods* **63:** 188–199.

Pitchiaya S, Heinicke LA, Park JI, Cameron EL, Walter NG. 2017. Resolving subcellular miRNA trafficking and turnover at single-molecule resolution. *Cell Rep* **19:** 630–642.

Rasnik I, McKinney SA, Ha T. 2006. Nonblinking and long-lasting single-molecule fluorescence imaging. *Nat Meth* **3:** 891–893.

Rinaldi AJ, Suddala KC, Walter NG. 2015. Native purification and labeling of RNA for single molecule fluorescence studies. *Methods Mol Biol* **1240:** 63–95.

Rinaldi AJ, Lund PE, Blanco MR, Walter NG. 2016. The Shine–Dalgarno sequence of riboswitch-regulated single mRNAs shows ligand-dependent accessibility bursts. *Nat Commun* **7:** 8976.

Ritter JG, Veith R, Veenendaal A, Siebrasse JP, Kubitscheck U. 2010. Light sheet microscopy for single molecule tracking in living tissue. *PLoS ONE* **5:** e11639.

Roy R, Hohng S, Ha T. 2008. A practical guide to single-molecule FRET. *Nat Meth* **5:** 507–516.

Salomon WE, Jolly SM, Moore MJ, Zamore PD, Serebrov V. 2015. Single-molecule imaging reveals that Argonaute reshapes the binding properties of its nucleic acid guides. *Cell* **162:** 84–95.

Schmidt JC, Zaug AJ, Cech TR. 2016. Live cell imaging reveals the dynamics of telomerase recruitment to telomeres. *Cell* **166:** 1188–1197.

Schulz D, Rentmeister A. 2014. Current approaches for RNA labeling in vitro and in cells based on click reactions. *Chembiochem* **15:** 2342–2347.

Semlow DR, Blanco MR, Walter NG, Staley JP. 2016. Spliceosomal DEAH-box ATPases remodel pre-mRNA to activate alternative splice sites. *Cell* **164:** 985–998.

Senavirathne G, Liu J, Lopez MA Jr, Hanne J, Martin-Lopez J, Lee JB, Yoder KE, Fishel R. 2015. Widespread nuclease contamination in commonly used oxygen-scavenging systems. *Nat Meth* **12:** 901–902.

Shcherbakova I, Hoskins AA, Friedman LJ, Serebrov V, Corrêa IR Jr, Xu MQ, Gelles J, Moore MJ. 2013. Alternative spliceosome assembly pathways revealed by single-molecule fluorescence microscopy. *Cell Rep* **5:** 151–165.

Shen K, Arslan S, Akopian D, Ha T, Shan So. 2012. Activated GTPase movement on an RNA scaffold drives co-translational protein targeting. *Nature* **492:** 271–275.

Shen H, Tauzin LJ, Baiyasi R, Wang W, Moringo N, Shuang B, Landes CF. 2017. Single particle tracking: From theory to biophysical applications. *Chem Rev* **117:** 7331–7376.

Shi J, Dertouzos J, Gafni A, Steel D. 2008. Application of single-molecule spectroscopy in studying enzyme kinetics and mechanism. *Methods Enzymol* **450:** 129–157.

Solomatin S, Herschlag D. 2009. Methods of site-specific labeling of RNA with fluorescent dyes. *Methods Enzymol* **469:** 47–68.

Stewart MP, Sharei A, Ding X, Sahay G, Langer R, Jensen KF. 2016. In vitro and ex vivo strategies for intracellular delivery. *Nature* **538:** 183–192.

Sun X, Zhang A, Baker B, Sun L, Howard A, Buswell J, Maurel D, Masharina A, Johnsson K, Noren CJ, et al. 2011. Development of SNAP-tag fluorogenic probes for wash-free fluorescence imaging. *Chembiochem* **12:** 2217–2226.

Swoboda M, Henig J, Cheng HM, Brugger D, Haltrich D, Plumere N, Schlierf M. 2012. Enzymatic oxygen scavenging for photostability without pH drop in single-molecule experiments. *ACS Nano* **6:** 6364–6369.

Thorn K. 2017. Genetically encoded fluorescent tags. *Mol Biol Cell* **28:** 848–857.

Tirat A, Freuler F, Stettler T, Mayr LM, Leder L. 2006. Evaluation of two novel tag-based labelling technologies for site-specific modification of proteins. *Int J Biol Macromol* **39:** 66–76.

Tokunaga M, Imamoto N, Sakata-Sogawa K. 2008. Highly inclined thin illumination enables clear single-molecule imaging in cells. *Nat Meth* **5:** 159–161.

Toomre D. 2012. Cellular imaging using total internal reflection fluorescence microscopy: Theory and instrumentation. *Cold Spring Harb Protoc* **2012:** 414–424.

Trachman RJ 3rd, Truong L, Ferre-D'Amare AR. 2017. Structural principles of fluorescent RNA aptamers. *Trends Pharmacol Sci* **38:** 928–939.

Tsai A, Puglisi JD, Uemura S. 2016. Probing the translation dynamics of ribosomes using zero-mode waveguides. *Prog Mol Biol Transl Sci* **139:** 1–43.

Turunen JJ, Pavlova LV, Hengesbach M, Helm M, Müller S, Hartmann RK, Frilander MJ. 2014. RNA ligation. In *Handbook of RNA biochemistry* (ed. Hartmann RK, Bindereif A, Schön A, Westhof E), pp. 45–88. Wiley, Weinheim, Germany.

Tyagi S. 2009. Imaging intracellular RNA distribution and dynamics in living cells. *Nat Meth* **6:** 331–338.

Wahl MC, Will CL, Lührmann R. 2009. The spliceosome: Design principles of a dynamic RNP machine. *Cell* **136:** 701–718.

Walter NG, Huang CY, Manzo AJ, Sobhy MA. 2008. Do-it-yourself guide: How to use the modern single-molecule toolkit. *Nat Meth* **5:** 475–489.

Wang J, Caban K, Gonzalez RL Jr. 2015. Ribosomal initiation complex-driven changes in the stability and dynamics of initiation factor 2 regulate the fidelity of translation initiation. *J Mol Biol* **427:** 1819–1834.

Willkomm DK, Hartmann RK. 2014. 3′-Terminal attachment of fluorescent dyes and biotin. In *Handbook of RNA biochemistry* (ed. Hartmann RK, Bindereif A, Schön A, Westhof E), pp. 117–128. Wiley, Weinheim, Germany.

Wu RA, Dagdas YS, Yilmaz ST, Yildiz A, Collins K. 2015. Single-molecule imaging of telomerase reverse transcriptase in human telomerase holoenzyme and minimal RNP complexes. *eLife* **4:** e08363.

Yan X, Hoek TA, Vale RD, Tanenbaum ME. 2016. Dynamics of translation of single mRNA molecules in vivo. *Cell* **165:** 976–989.

Yasuda R, Masaike T, Adachi K, Noji H, Itoh H, Kinosita K Jr. 2003. The ATP-waiting conformation of rotating F1-ATPase revealed by single-pair fluorescence resonance energy transfer. *Proc Natl Acad Sci* **100:** 9314–9318.

Relating Structure and Dynamics in RNA Biology

Kevin P. Larsen,[1,2,4] Junhong Choi,[1,3,4] Arjun Prabhakar,[1,2] Elisabetta Viani Puglisi,[1] and Joseph D. Puglisi[1]

[1]Department of Structural Biology, Stanford University School of Medicine, Stanford, California 94305

[2]Biophysics Program, Stanford University, Stanford, California 94305

[3]Department of Applied Physics, Stanford University, Stanford, California 94305

Correspondence: puglisi@stanford.edu; epuglisi@stanford.edu

SUMMARY

Recent advances in structural biology methods have enabled a surge in the number of RNA and RNA–protein assembly structures available at atomic or near-atomic resolution. These complexes are often trapped in discrete conformational states that exist along a mechanistic pathway. Single-molecule fluorescence methods provide temporal resolution to elucidate the dynamic mechanisms of processes involving complex RNA and RNA–protein assemblies, but interpretation of such data often requires previous structural knowledge. Here we highlight how single-molecule tools can directly complement structural approaches for two processes—translation and reverse transcription—to provide a dynamic view of molecular function.

Outline

[4]These authors contributed equally to this work.

Cite as *Cold Spring Harb Perspect Biol* doi: 10.1101/cshperspect.a032474

1 INTRODUCTION

The past decades have seen an explosion in structures of RNA and RNA–protein assemblies, including the ribosome and spliceosome. These structures have revealed the basic molecular principles of RNA secondary and tertiary folding and recognition of RNAs by structured and unstructured protein domains. Yet these structures represent static snapshots of the dynamic RNA–protein machinery of the cell. What has been lacking are time-resolved methods to link detailed structural views into a mechanistic movie of biological systems. These systems evolve temporally in both macromolecular conformation and in chemical composition (ligands or factors bound) and may proceed along different mechanistic pathways. Single-molecule methods have provided a bridge between structure and dynamics, allowing real-time analysis of RNAs and their assemblies. Here we address the interplay of structure and dynamics, using examples from our recent work on translation by the ribosome and HIV reverse transcription.

The three major structural methods—X-ray crystallography, nuclear magnetic resonance (NMR) spectroscopy, and cryo-electron microscopy (cryo-EM)—provide different portals into dynamics (Fig. 1). X-ray crystallography

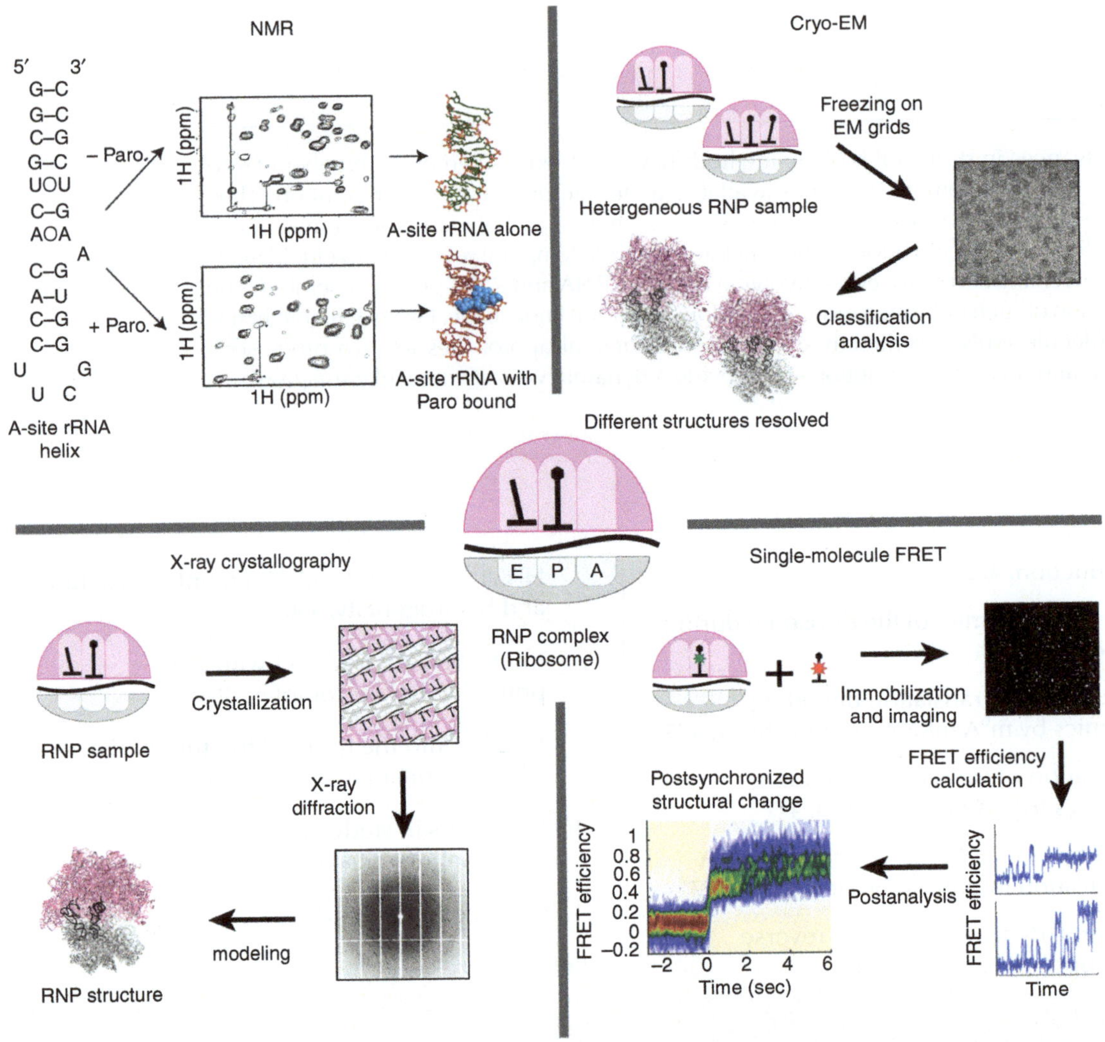

Figure 1. Major structural methods to study RNA–RNA–protein complexes. (*Top left*) Structures of ribosomal RNA (rRNA) helix with or without antibiotic paromomycin (Paro). (Adapted from Fourmy et al. 1998.) (*Bottom left*) High-resolution structure of ribosome using X-ray crystallography. (*Top right*) Cryo-electron microscopy (cryo-EM) structures of ribosome, with or without A-site transfer RNA (A-tRNA). (*Bottom right*) Single-molecule Förster resonance energy transfer (FRET) data showing binding of A-tRNA to the ribosome. (Data adapted from Fourmy et al. 1998 for nuclear magnetic resonance [NMR] spectroscopy [Protein Data Bank (PDB) IDs 1FYP, 1FYO], PDB ID 4V51 for X-ray crystallography, 5JTE for cryo-EM, and Choi et al. 2018 for single-molecule FRET.)

requires crystallization of macromolecular complexes into regular lattices, which in certain cases can skew conformations because of intermolecular packing interactions. Data in X-ray diffraction are derived from coherent scattering of the X-ray beam by all members of the crystalline lattice. Dynamic information is often inferred from so-called B-factors, or "temperature factors," that reflect uncertainty in the location of atoms—high B-factors can result from dynamic regions in a macromolecule or from errors during modeling. Time-resolved conformational changes have been observed but require triggering of all members of the lattice simultaneously. Synchronizing conformational changes within the crystalline lattice often requires an external cue, typically light, to initiate a reaction (Srajer et al. 1996; Jung et al. 2013). Recent advances in the field of coherent X-ray sources (X-ray free-electron lasers or XFELs) have enabled faster data acquisition on nanocrystalline materials (Chapman et al. 2011; Tenboer et al. 2014). The fast data acquisition at room temperature is coupled with small crystal size to allow rapid mixing with ligands that can enter the crystal lattice. This has led to X-ray studies on conformational changes induced by ligands that bind to RNA (Stagno et al. 2017). Although still in its infancy and often restricted to specific cases, X-ray crystallography using XFELs shows potential for studying complex biomolecular dynamics.

NMR spectroscopy is a powerful probe of chemical dynamics, over a broad range of timescales and processes. NMR relaxation measurements can probe residue-by-residue motions on timescales of 10^{-9} to 10^{-5} sec to reveal fast movements within biomolecules. Advances in using relaxation dispersion methods have allowed investigation of "excited," lowly populated states for RNAs and proteins at intermediate timescales; the work of Al-Hashimi and coworkers in particular has shown how RNA can sample alternate pairings on a micro to millisecond range (Al-Hashimi 2013). NMR can also monitor slower processes (10^{-3} sec and longer timescales) through distinctions in chemical shifts of resonances for different states (Zhou et al. 2016). However, NMR of RNAs, proteins, and their complexes suffer from size limitations on systems that can be investigated. As molecular weight increases, spectral complexity increases and resonances broaden because of slow molecular tumbling and efficient relaxation. These size limitations can be partially overcome by clever stable isotope-labeling strategies, novel NMR experiments that diminish relaxation mechanisms (like transverse relaxation optimized spectroscopy [TROSY]), or by solid state NMR.

Cryo-EM has undergone an impressive improvement allowing atomic resolution structures of many key RNA–protein complexes—ribosomes, spliceosomes, and RNA polymerase complexes among others (Bai et al. 2015; Cheng 2015; Yan et al. 2015; Robinson et al. 2016). Cryo-EM uses samples embedded in vitreous ice without additional staining, and imaging is performed using a low electron dose to avoid damage and preserve structural information. This results in images with low signal-to-noise ratio in which the contrast comes from the scattering of atoms in the biomolecules. RNA has an advantage compared with other biomolecules as high scattering is obtained from the phosphate backbone. Recent technological advances in both hardware and software have driven the improved resolution of cryo-EM structures into the realm of 3–5 Å resolution: Using a direct electron detection camera with increased sensitivity, cryo-EM data are taken as movies with many sequential frames at a low dose (Li et al. 2013). These frames are then computationally corrected for movements of the biomolecules in ice and analyzed to yield near-atomic resolution density maps. However, even at these resolutions, tracing an RNA chain can be difficult de novo. Further analysis of cryo-EM maps using computational modeling and fitting tools such as Rosetta will be powerful, as well as incorporating reliable base-pairing data from other methods such as NMR.

Unlike the other structural methods mentioned above, cryo-EM is inherently a single-molecule method, revealing insights into potential dynamics within biological systems. Structures are built by averaging more than 1,000–100,000 individual particles that are randomly oriented within the ice of an EM grid. Computational sorting into distinct classes can distinguish different conformational or compositional states of biomolecular assembly. This process can capture the biochemical and conformational heterogeneity of the input sample at the moment of freezing. As an extension of this capability in cryo-EM, Frank and coworkers have developed rapid-mixing and freezing approaches that allow time-resolved snapshots of the ensemble of structures in a biological reaction such as translation initiation ribosome recycling (Chen et al. 2015; Fu et al. 2016; Frank 2017a). In this process, temporal information provided by other methods is essential as dynamic observations by cryo-EM alone can be skewed by observations of only structured states.

Single-molecule fluorescence methods have revolutionized the investigation of biomolecular dynamics, providing structural dynamic information without ensemble averaging (Fig. 1). Traditionally, kinetic information has been extracted by bulk spectroscopy or by chemical kinetic methods, such as rapid-mixing, stopped-flow, or quenched-flow. These methods provide a time course of the chemical and compositional changes of the system as a function of time but require initial synchronization of a reaction and suffer from ensemble averaging that masks heterogeneous pathways. In contrast, single-molecule fluorescence methods can be used to monitor conformational and compositional changes of biological complexes, which later can be analyzed to reveal multistep kinetics or parsed to show different pathways.

In typical single-molecule fluorescence experiments, RNAs and proteins must first be fluorescently labeled with bright, extrinsic fluorophores. Multiple biomolecules can be individually labeled using fluorophores of different colors, which can be tracked using their respective emission channel. Nucleic acids are labeled using chemical modification of synthetic oligonucleotides or chemical reaction with specifically modified ends of RNAs (for example, through oxidation of the terminal *cis*-diol at the 3′ end of an RNA). Longer (>70-nt) labeled RNAs can be specifically labeled by hybridization of a dye-labeled DNA or RNA probe or created from shorter RNAs using efficient enzymatic ligation methods. For proteins, dyes can be attached using maleimide, *N*-hydroxysuccinimide (NHS), or click chemistry, where specific labeling sites are often engineered into proteins by mutagenesis. Alternatively, a short peptide sequence can be used to tag a fluorescently labeled full protein or peptide domain such as SNAP, HaloTag, and ybbR (Yin et al. 2005; Los et al. 2008; Mollwitz et al. 2012). For all these labeling approaches, the function of the modified protein must be tested, as modification of the protein (either the sequence, length, or dye labeling) may greatly perturb function. Such biochemical controls are absolutely essential for subsequent valid biophysical investigation. To prolong tracking of a single molecule over time, fluorescently labeled RNAs or their complexes are usually immobilized on an inert surface using biotin–streptavidin interactions for specificity and stability.

By observing the behavior of single molecules, complex dynamic behavior of biomolecular systems can be determined and sorted, free from ensemble averaging. Compositional dynamics can be directly monitored as a burst in fluorescence from labeled ligands that colocalize with the surface-immobilized biomolecule. To limit excitation to the immobilized molecules, total internal reflectance microscopy (TIRFM) is often used, which allows inclusion of the free fluorescent molecule in solution below ~20–50 nM in concentration before the signal is drowned by the background noise. Different instrumentation can be used to track fluorescently labeled ligands at higher concentration. Using nanophotonic devices called "zero-mode waveguides" (ZMWs) that reduce the excitation volume by orders of magnitude, free ligand concentrations up to μM (Levene et al. 2003) can be tolerated. The ZMWs are roughly cylindrical chambers of 40–180 nm in diameter, with functionalized quartz bottoms to immobilize biomolecules and conductive metal side walls. In both TIRFM and ZMW setups, propagating laser excitation is quenched as it moves away from the surface, leading to high excitation only near the immobilized biomolecules. Using these experimental setups, the kinetics of ligand binding and dissociation can be measured at the single-molecule level by analyzing the statistical distribution of binding event lifetimes (dissociation rate), or times between events (association rate). Thus, simple compositional dynamics can be measured for biomolecular systems from ligand occupancy as a function of time. These approaches have been applied extensively to investigate translation, splicing, and other RNA–ligand systems (Uemura et al. 2010; Hoskins et al. 2011; Friedman and Gelles 2015).

In addition to observing the compositional dynamics, conformational dynamics in RNAs can be observed directly using Förster resonance energy transfer (FRET) between fluorophores. FRET between a specific dye pair depends on the spectral overlap of the donor and acceptor dyes and is defined by an R_0 value (distance) in which energy transfer from donor to acceptor is 50%; for the standard cyanine 3 (Cy3)-Cy5 pair, R_0 is ~56 Å. FRET efficiency depends strongly on distance ($1/R^6$), such that transfer is efficient at distances well below R_0 (~20 Å) and near zero at distances greatly exceeding R_0 (~75–80 Å). Using FRET, RNA–RNA, RNA–protein, or protein–protein distance changes can be monitored in real time. As with many single-molecule approaches, timescales of conformational fluctuations are limited by camera speeds, which generally probe dynamics in the millisecond to second range. Combining FRET with fluorescently labeled ligand binding can reveal correlations between ligand occupancy and conformational changes (Wasserman et al. 2016; Prabhakar et al. 2017).

Computational approaches are central to our understanding of biomolecular dynamics. Molecular dynamics simulations have advanced greatly over the decades since their introduction by Lifson and Levitt in the 1960s. Molecular dynamics describe macromolecular motions over a range of timescales and can strongly complement experimental measurements of dynamic fluctuations in RNA. In addition, computation can fill in great detail for models of dynamics derived from a limited number of observables. Secondary structure prediction tools have improved, providing proposed secondary structures that may be in conformational equilibrium (Zuker 2003; Reuter and Mathews 2010). De novo RNA tertiary structure modeling methods have also improved, but their power will likely reside in their ability to integrate experimental data to generate accurate models of large assemblies (Cheng et al. 2015). Genomic methods using sequencing as a readout have provided dynamics of RNA processes in vivo, including structural changes, RNA–protein interactions, and ribosomal dynamics of translation.

A combined dynamic and structural view of RNA mechanism requires a confluence of experimental approaches. We discuss here two recent examples from our own work on RNA–protein complexes that highlight how structural and dynamic data provide mechanistic insights into HIV reverse transcription and translation.

2 STRUCTURAL DYNAMICS OF THE RIBOSOME DURING DECODING

Ribosomes represent the archetypical RNA–protein assembly for their complexity and functionality. From a structural point of view, prokaryotic and eukaryotic ribosomes are a massive machinery with scores of proteins decorating the core RNA molecules. The ribosome has deservedly gathered wide interest from both X-ray crystallographers and cryo-electron microscopists in the past couple of decades, revealing many functional states in different organisms (Bashan and Yonath 2008; Simonović and Steitz 2009; Voorhees and Ramakrishnan 2013; Frank 2017b; Yusupova and Yusupov 2017). The ribosome is not only a ribozyme with its catalytic activities driven by a conserved RNA core, but it also has a great specificity of selecting a right transfer RNA (tRNA) substrate out of a pool of structurally similar tRNAs via just three RNA base-pairing interactions. This remarkable specificity has led to the careful early enzymatic study of accuracy and efficiency of ribosomes (Gorini 1971), which inspired the concept of the kinetic proofreading that is applicable to many other biological systems involving a substrate selection mechanism (Hopfield 1974; Ninio 1975).

The combined kinetic and structural work has led to deep understanding of the ribosome and the decoding process of messenger RNA (mRNA). Yet initially, two different methodologies to study the ribosome have resulted in distinct views of ribosomal mechanics, in which two languages coexisted to describe the molecular process of translation. From the structural view, translation was described by distinct conformational changes and domain movements at different functional states. Description of these structural changes needed to be linked with the preexisting kinetic description of translation, in which multiple structurally distinct substates have been proposed to explain highly accurate and efficient translation. Single-molecule methods have been critical to link high-resolution structural and kinetics experiments. To show our case in point, we will discuss our recent works on the effect of mRNA modifications in translation, combining results from single-molecule, bulk kinetics, and X-ray crystallography methods (Fig. 2).

3 MODULATION OF TRANSLATION DECODING DYNAMICS BY m^6A mRNA MODIFICATIONS

Although no less than 160 distinct natural RNA modifications are known to exist in the transcriptome (Boccaletto et al. 2018), investigation into the roles of these modified

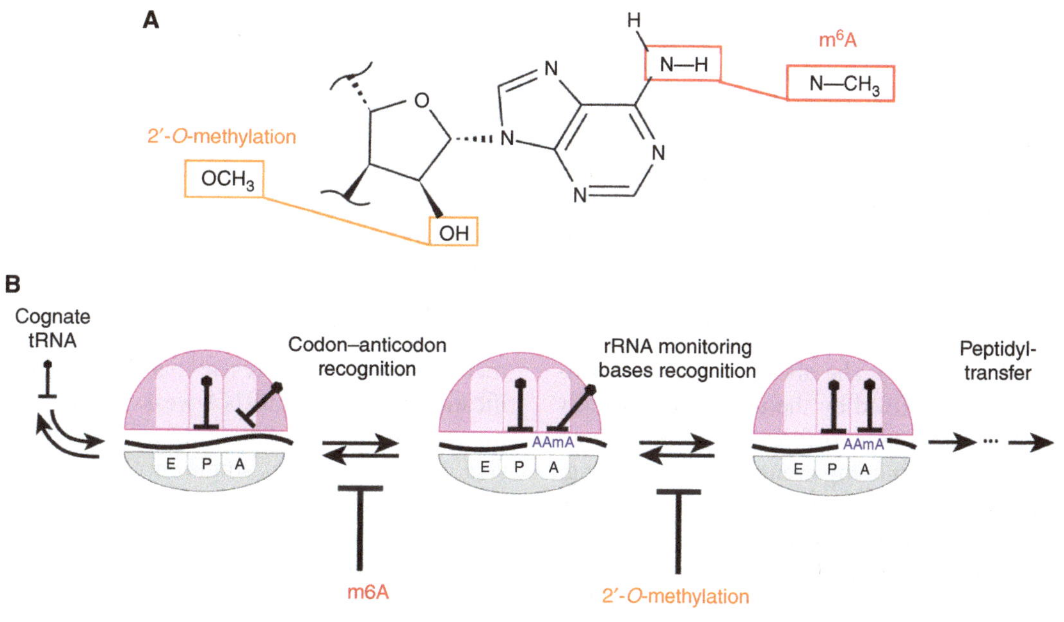

Figure 2. Effect of N^6-methyladenosine (m^6A) and 2′-*O*-methylation on translation elongation. (*A*) m^6A methylates one of two hydrogens on N^6, whereas 2′-*O*-methylation occurs on 2′-OH of RNA ribose. (*B*) m^6A inhibits initial codon–anticodon helix formation after binding of the cognate tRNA to the translating ribosome. 2′-*O*-methylation inhibits dynamics of hydrogen bonding between rRNA-monitoring bases and the mRNA–tRNA helix.

nucleotides in mRNA has recently exploded with the discovery of RNA modifications that were reversible. The first reversible modification in mRNA to be discovered was the modification of adenosine (A) to N^6-methyladenosine (m^6A) (Dominissini et al. 2012; Meyer et al. 2012), in which one of two hydrogens attached to the N^6 of A is methylated (Fig. 2A). The m^6A modification of mRNA was shown to be dynamically regulated by the modifier ("writer") and the demodifier ("eraser") proteins, which in turn, controls its recognition by a set of its associated proteins ("reader"). This possible dynamic regulation of mRNA modification has launched the field of epitranscriptomics (or RNA epigenetics) (He 2010; Saletore et al. 2012) and made m^6A a subject of intense studies. m^6A has been shown to participate in diverse translational regulation processes, in which its presence affects translation initiation, elongation, and the mRNA decay (Roundtree et al. 2017). Yet, in addition to recognition of m^6A by its reader proteins, the base-pairing between m^6A and uridine (U) is weaker than that between A and U, disrupting the formation of RNA secondary structures harboring m^6A within their RNA helix (Roost et al. 2015). Yet, such base-pairing interactions are essential for the codon recognition. Then, how would the translational machinery respond when there is m^6A present within the coding region?

To answer this question, compositional and conformational dynamics of the ribosome during in vitro translation were monitored using a single-molecule fluorescence assay (Choi et al. 2016). Previous structural methods have shown that when a codon is read by tRNA and the nascent peptide is elongated by one amino acid attached to the tRNA, the ribosome undergoes a conformational change from nonrotated state to rotated state (also referred to as the change from classical state to hybrid state) (Moazed and Noller 1989; Frank and Agrawal 2000). The resulting rotated state transitions back to the nonrotated state after translocation, which physically moves the ribosome to the next codon for decoding. By attaching a pair of FRET probes to the ribosome, the kinetics of its conformational change on a successful decoding event can be measured on each codon by measuring the lifetime of FRET states (Dorywalska et al. 2005; Aitken and Puglisi 2010). The ribosome conformational signal can be further fortified by the use of fluorescently labeled tRNA, which serves as an orthogonal probe for both successful and futile tRNA-binding events during decoding.

Slower decoding kinetics due to the increased rejection of tRNA was observed on a single-nucleotide m^6A-modified codon compared with that of the unmodified codon (Choi et al. 2016). The magnitude of change in kinetics varied by its context within the modified codon: The maximal effect of 15-fold slower decoding kinetics was observed on the m^6AAA codon, whereas modifications at the second or third position or the increase in G-C content of the codon decreased the effect. The sensitivity of the modification effect to the codon context was interesting, as it suggested the maximal accuracy of the particular codon–anticodon interaction is changing with the modification (Johansson et al. 2012). Using bulk kinetics methods, we showed a change in decoding kinetics caused by the presence of m^6A in a codon with higher precision and temporal resolution.

The next question was to understand how m^6A slows decoding kinetics structurally. Using X-ray crystallography, structures of the ribosomal decoding center were solved while programmed with the tRNA anticodon stem loop and with three different single-nucleotide m^6A-modified AAA codons at each nucleotide position (Choi et al. 2016). When compared with the canonical A-U base pair in the codon–anticodon helix, there was no observable difference between A and m^6A while base-paired with U: The methyl moiety of m^6A is simply turned away from the Watson–Crick edge between A and U. This result seemed quite incongruous with our kinetics results: How could m^6A affect kinetics of decoding while being structurally similar to A when base-pairing with the same U?

One possible explanation was that because only one N^6-hydrogen moiety is available for m^6A to participate in the hydrogen bond instead of two in the unmodified A, the m^6A-U base pairs are weaker than the A-U base pair in the codon–anticodon helix, similar to what was observed in the RNA helix by NMR (Roost et al. 2015). To test this hypothesis, the binding stability of tRNA to the ribosome was directly measured by fluorescently labeling two adjacent tRNAs with FRET probes while decoding the m^6A-modified codon. In addition, different small molecules were added to the experiment to probe various substeps of decoding. The most relevant small molecule was the antibiotic tetracycline, which prevents decoding after the initial codon recognition by the anticodon loop of tRNA (Blanchard et al. 2004). The m^6A-modified codon was observed to hasten the dissociation of tRNA in the presence of tetracycline, which suggested that the codon–anticodon interaction is weakened by the presence of m^6A-U base pair.

In this study, combinations of structural and kinetic methods were used to reveal that the presence of m^6A modification within the mRNA weakens the codon–anticodon interaction and slows down translation elongation kinetics (Fig. 2B). Particularly, single-molecule methods were used to bridge the contrasting observations made in the structural and bulk kinetics methods, by directly monitoring the kinetics of structural compositional changes during translation. Yet, it is still not clear how slower decoding kinetics are linked with the regulatory role of m^6A in the mRNA coding region. Perhaps, the modulation of translation elongation dynamics by mRNA modification coordinates the various cotranslational events such as protein folding,

Cite as *Cold Spring Harb Perspect Biol* doi: 10.1101/cshperspect.a032474

nascent peptide recognition, etc. In this scenario, m^{6}A may not be the only modification that is used to modulate the translation elongation kinetics, but other naturally occurring modifications within the coding region of mRNA may serve similar roles.

4 MODULATION OF TRANSLATION DECODING DYNAMICS BY 2′-*O*-METHYLATION

2′-*O*-methylation is a recent addition to the growing list of chemical modifications discovered to occur in mRNA. Unlike other known mRNA modifications, 2′-*O*-methylation alters the 2′-OH moiety of the ribose, rather than changing the nucleoside base and may arguably be the most interesting part of the molecule as it distinguishes RNA and DNA (Fig. 2A). Therefore, many distinct characteristics of RNA are linked with the 2′-OH group, such as preference for an A-form helical conformation, formation of the tertiary structure through hydrogen bonding with the 2′-OH group, and the enzymatic activity of RNA related to the hydrolysis via 2′-OH. The last point has been a particular disadvantage to work with RNA, which led to techniques such as substituting the entire 2′-OH moieties of RNA oligonucleotide with synthetically 2′-*O*-methylated nucleotides for resisting RNA hydrolysis (Kraynack 2006). Biophysical studies of 2′-*O*-methylated RNA has shown that it enhances the helix formation to a complementary RNA or 2′-*O*-methylated strand (Cummins et al. 1995). The resistance to chemical reaction by 2′-OH has been used to detect 2′-*O*-methylation within mRNA (Dai et al. 2017). The cellular mRNA is purified and subject to multiple cycles of a chemical treatment, which requires the 2′-OH moiety of mRNA to eliminate its presence before the high-throughput sequencing. This has resulted in the map of 2′-*O*-methylation occurring in mRNA, which is strongly enriched in the coding region, as well as codon preference and nucleotide preference within the enriched codon.

The enriched presence of 2′-*O*-methylation in the coding region of mRNA raises the question of its role during translation elongation. To understand how 2′-*O*-methylated codon affects translation, we took a similar approach as discussed above for m^{6}A: First, the decoding kinetics of a single-nucleotide 2′-*O*-methylated codon was measured using single-molecule and bulk kinetics assays and compared with that of an unmodified codon (Choi et al. 2018). Surprisingly, a substantially delayed tRNA accommodation on the 2′-*O*-methylated codon was observed, more than an order of magnitude slower than observed with the m^{6}A modification, reaching at most 1500-fold slower kinetics compared with that of the unmodified codon. The magnitude of delay by 2′-*O*-methylation was also dependent on its codon context, but its pattern was different from that of m^{6}A. The largest effect of 2′-*O*-methylation was observed when the second nucleotide position of a codon was modified, rather than at the first nucleotide position as in m^{6}A modification. The delayed decoding accompanied multiple tRNA sampling events in the order of 0.1 to 1 sec before a successful accommodation, unlike short tRNA sampling events with m^{6}A that were too transient to be directly observed with the temporal resolution of 0.1 sec. This suggested that 2′-*O*-methylation affects decoding after the initial codon–anticodon formation, in which the codon–anticodon interaction would lengthen the tRNA-bound time before its rejection.

How could this large perturbation to decoding kinetics be explained in terms of the ribosome structure? During tRNA decoding, the ribosome checks the geometry of the initial codon–anticodon helix formation via binding of ribosomal RNA (rRNA) to the 2′-OH of mRNA (Yoshizawa 1999; Ogle et al. 2001; Ogle et al. 2002). Thus, the substitution of 2′-OH with 2′-*O*-methyl moiety might alter the kinetics of ribosome recognition of a correctly formed codon–anticodon helix (Fig. 2B). This notion was supported by the structure of the ribosome decoding center harboring the 2′-*O*-methylated codon, obtained using X-ray crystallography (Choi et al. 2018). These structures showed a significant distortion between the interacting rRNA and the mRNA 2′-*O*-methyl group. The distortion was most pronounced when the second nucleotide position of the codon is 2′-*O*-methylated, matching the pattern of the pause duration as a function of position of the modification observed in the single-molecule assay.

Similar to the m^{6}A modification, the biological role of delaying translation elongation by 2′-*O*-methylation is not clear. The enrichment in the coding region, the codon preference as well as the preferred nucleotide position of 2′-*O*-methylation suggests its role during translation elongation (Dai et al. 2017). The presence of either m^{6}A or 2′-*O*-methylation modification may be used to fine-tune translation elongation kinetics and regulate the protein synthesis. The magnitude of the delay created by 2′-*O*-methylation, however, may exceed the tolerance of translation, and lead to other translation control processes that abort translation and subject the mRNA transcript to degradation (Zaher and Green 2009). Genes that are directly regulated by altering translational decoding kinetics have not yet been discovered, and the role of 2′-*O*-methylation within the coding region still remains elusive.

5 NEXT STEPS IN STUDYING THE ROLE OF mRNA MODIFICATION IN TRANSLATION DECODING

In the past decade, we have observed a rapid advance in methods to detect cellular RNA modifications (Linder and Jaffrey 2019). However, their biological roles in translation decoding remain unresolved. In addition to changing elongation dynamics, RNA modifications change the context of a modified codon and induce miscoding events (Karijolich and Yu 2011; Fernandez et al. 2013). It is plausible that different modes of regulation exist for controlling gene expression by posttranscriptional modifications. Different mRNA modifications may modulate structural dynamics of a translational complex. As such, we may learn more about mechanisms of translation by using mRNA modifications as a tunable perturbation to the canonical pathway. Combinations of structural and kinetic methods will be a powerful tool in this effort.

6 STRUCTURAL DYNAMICS OF HIV-1 REVERSE TRANSCRIPTION INITIATION

Reverse transcription, discovered in 1970, is the process by which RNA is copied into a double-stranded DNA (Baltimore 1970; Temin and Mizutani 1970). Although found in eukaryotic systems in the form of retrotransposons and telomere lengthening, the deepest study has been in retroviral replication, in particular for human immunodeficiency virus 1 (HIV-1) (Gilboa et al. 1979; Hu and Hughes 2012). In HIV-1, a virally encoded enzyme, reverse transcriptase (RT), binds an 18 base pair intermolecular primer-binding site (PBS) helix formed between a primer $tRNA^{Lys}_3$ and the positive-sense genomic viral RNA (vRNA) (Fig. 3A) (Ratner et al. 1985; Isel et al. 2010). Early studies on HIV-1 reverse transcription distinguished its initiation step from the later elongation phases owing to its unique kinetic properties (Isel et al. 1996; Lanchy et al. 1996). The polymerization rate of RT appears to be 100–500-fold slower during initiation than during elongation, with pausing events causing a buildup of discrete, partially extended products (Fig. 3B) (Lanchy et al. 1996; Suo and Johnson 1997a; Suo and Johnson 1997b; Liang et al. 1998; Goldschmidt et al. 2002). RT also has a higher dissociation rate from its initiation RNA–RNA substrate than later DNA–DNA and DNA–RNA elongation substrates (Kati et al. 1992; Lanchy et al. 1996; Suo and Johnson 1997b; Vaccaro et al. 1999; Goldschmidt et al. 2002; Marko et al. 2013; Coey et al. 2018).

The structural, biophysical, and biochemical basis for the RT polymerization cycle during elongation has been deeply established over the past decade (Sarafianos et al. 2009). Bulk steady state, pre–steady state, and single-turnover kinetics delineated the mechanism of deoxynucleotide triphosphate (dNTP) incorporation and fidelity (Kati et al. 1992; Isel et al. 1996; Lanchy et al. 1996; Suo and Johnson 1997a; Liang et al. 1998; Li et al. 2016). Structural studies defined the architecture of RT, a two-subunit (p51 and p66) polymerase, showed how double-stranded DNA substrates bound in a long cleft made by the p66 subunit, and how RT inhibitors bind and affect enzymatic function (Sarafianos et al. 2009). Despite this structural and biochemical context, how HIV-1 navigates the very initial phase of reverse transcription has remained unresolved.

The mechanisms governing the unique properties of the initiation phase of HIV-1 reverse transcription were initially explored using biochemical approaches. RNA mutagenesis, RNA chemical/enzymatic mapping studies, and kinetics assays were used to characterize the contributions of RNA sequence and structure in the vRNA–$tRNA^{Lys}_3$ binary complex (Isel et al. 1995; Isel et al. 1996; Lanchy et al. 1996; Liang et al. 1997; Lanchy et al. 1998; Liang et al. 1998; Isel et al. 1999; Vaccaro et al. 1999; Lanchy et al. 2000; Beerens et al. 2001; Beerens and Berkhout 2002b; Goldschmidt et al. 2002; Goldschmidt et al. 2003; Iwatani et al. 2003; Goldschmidt et al. 2004). HIV-1 vRNA has a region of 18 nucleotides (nt) that are complementary to the 3′ end of $tRNA^{Lys}_3$, which forms the PBS helix that serves as the initial binding site for RT. In addition to the PBS helix, these studies showed that HIV-1 vRNA is highly structured, and there was convincing evidence to suggest additional regions of sequence complementarity between the tRNA and vRNA that extended outside of the PBS sequence (Fig. 3C) (Isel et al. 1995; Isel et al. 1999; Lanchy et al. 2000; Goldschmidt et al. 2002; Goldschmidt et al. 2003; Goldschmidt et al. 2004; Paillart et al. 2004). Mutations within these sequences drastically altered initiation kinetics, decreased virus viability, and appeared to alter the vRNA–tRNA fold (Liang et al. 1997; Isel et al. 1998; Liang et al. 1998; Beerens and Berkhout 2002a; Beerens and Berkhout 2002b; Goldschmidt et al. 2002; Goldschmidt et al. 2003; Iwatani et al. 2003; Goldschmidt et al. 2004; Ooms et al. 2007). These data suggested that the vRNA and tRNA rearrange to form additional structures outside of the PBS helix that are vital in controlling initiation (Fig. 3C). The first and most widely studied of these structures was an A-rich loop sequence in the vRNA, which was proposed to pair with the anticodon of the $tRNA^{Lys}_3$ (Isel et al. 1993; Liang et al. 1997; Isel et al. 1998; Puglisi and Puglisi 1998; Bilbille et al. 2009). Additional sequences, like the primer activation signal (PAS) sequence within helix 1 (H1) of the vRNA, and a C-rich region in vRNA helix 2 (H2), were also identified and have complementary sequences respectively located within the tRNA T-loop and variable loop (Fig. 3C) (Beerens and Berkhout 2002b; Iwatani et al. 2003). However, attributable to the limited structural information of these bulk studies, an understanding of the timing of such rearrangements and the possibility of multiple coexisting RNA conformations remained speculative. More detailed structural information using modern methods were needed to parse RNA heterogeneity and the

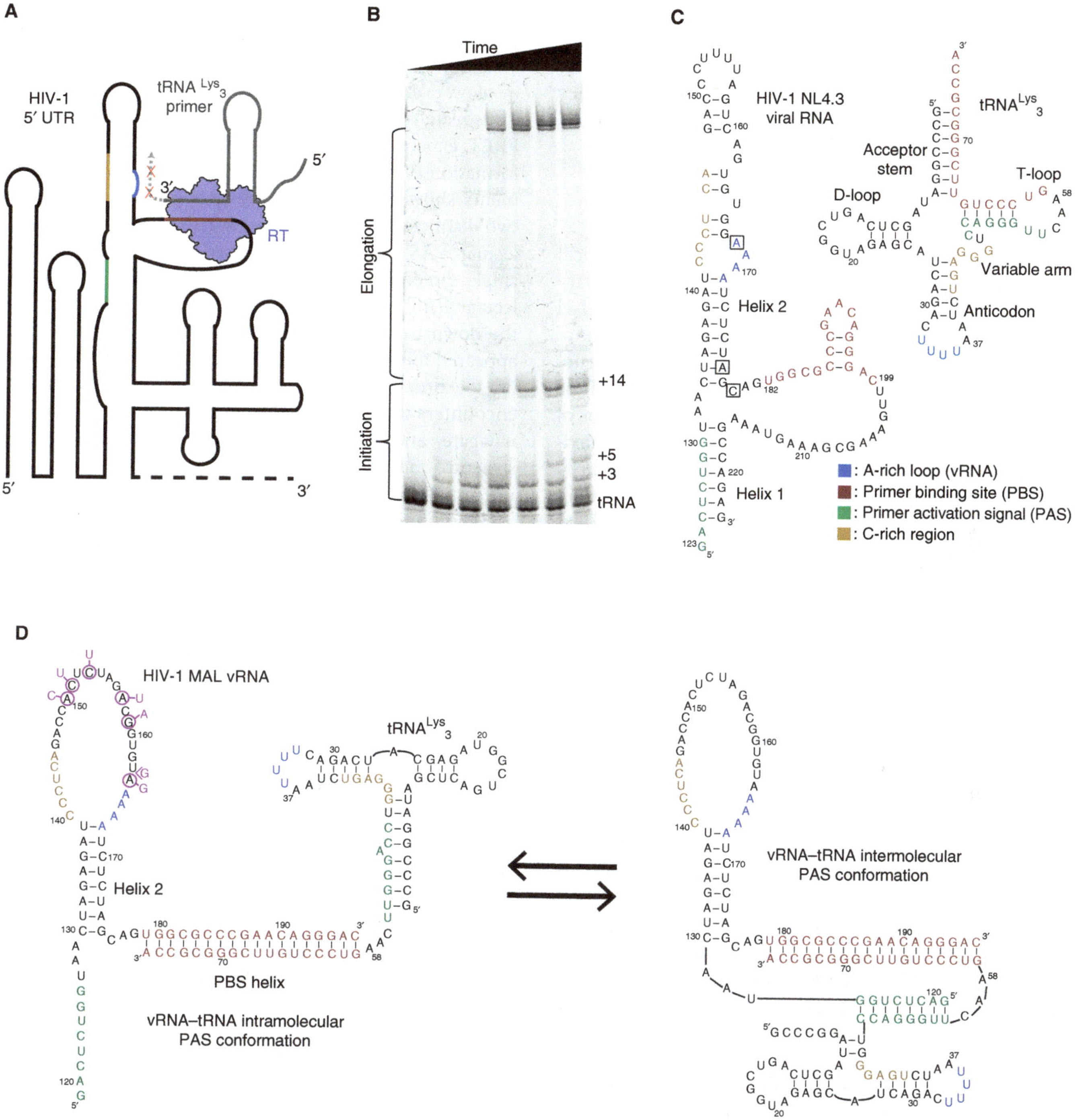

Figure 3. (*A*) Reverse transcriptase (RT) must bind an 18-base-pair helix formed between the tRNA primer and the 5′ untranslated region (UTR) of the viral RNA genome (vRNA). The primer-binding site (PBS) helix resides in a highly structured region of the vRNA genome. RT has a rapid dissociation rate from the PBS helix and shows discrete pausing events during initiation (red ×s). (*B*) A primer extension gel in which polymerase extension reactions of RT for the vRNA–tRNA complex have been quenched at different time points. Unextended tRNA runs at the *bottom* of the gel and fully extended primer runs near the *top*. The buildup of pausing products can be seen in the *lower* half of the gel. (*C*) Secondary structures of the $tRNA^{Lys}_3$ primer and the NL4.3 human immunodeficiency virus 1 (HIV-1) subtype genome PBS region vRNA. Regions of sequence complementarity have been colored. Nucleotide positions that correspond to the pausing points seen in *B* are boxed on the vRNA. (*D*) NMR-derived structures of the MAL HIV-1 subtype genome in complex with the tRNA primer. Complementary sequences are colored as in *C*. Nucleotide mutations in MAL compared with NL4.3 are indicated in purple.

timings of these structural rearrangements. We will discuss work from our laboratory and others that illustrates how traditional RNA structural methods can be coupled with single-molecule experimentation to elucidate the dynamics of a complex multicomponent system.

7 HIV-1 VIRAL RNA–tRNA$^{Lys}{}_3$ COMPLEX STRUCTURE AND HETEROGENEITY

The Marquet laboratory used a combination of chemical and enzymatic mapping methods to generate secondary structure models of the vRNA–tRNA$^{Lys}{}_3$ complex (Isel et al. 2010). These studies were often combined with functional assays in an attempt to determine the vRNA structures responsible for influencing initiation kinetics. These predicted models relied on comparison to the free forms of vRNA and tRNA$^{Lys}{}_3$ and were often interpreted to maximize the possible interactions between the two components. Initial efforts attempted to explore RNA dynamics and structural rearrangements by performing similar experiments on pre-extended or stalled complexes in the presence or absence of RT (Lanchy et al. 1998; Lanchy et al. 2000; Isel et al. 2010). From these groundbreaking experiments, the concept that the conformation of the initiation complex evolved structurally over time began to emerge. However, because of the bulk nature of these experiments, information on RNA structural heterogeneity was lost. A later study by Beerens et al., using single-molecule FRET, suggested that the vRNA–tRNA complex exists in several structures that can undergo conformational changes related to the PAS interaction (Beerens et al. 2013). The labeling scheme only monitored vRNA H1 formation and melting, thus this study did not directly reveal the formation of new stable vRNA–tRNA structures.

To probe the extent of structural heterogeneity and dynamics in the vRNA–tRNA complex, we have used a combination of NMR spectroscopy and single-molecule fluorescence methods (Puglisi and Puglisi 2011; Coey et al. 2016, 2018). Because of the size limitations of NMR spectroscopy, these studies focused on a minimal vRNA system containing the PAS and/or A-rich loop sequences (Fig. 3D). The first NMR study determined the secondary structure of the vRNA–tRNA complex, which showed extensive tRNA remodeling and the existence of a vRNA hairpin, H2, immediately downstream from the PBS (Puglisi and Puglisi 2011). However, the NMR experiments did not reveal additional vRNA–tRNA interactions outside of the PBS helix. This study was extended by increasing the size of the vRNA construct at its 5′ end to include the downstream PAS sequence (Coey et al. 2016). These NMR experiments identified two potential conformations of the vRNA–tRNA complex, with the first being similar to the results in Puglisi et al. However, an additional structure was identified that was consistent with formation of the intermolecular PAS helix between the vRNA and tRNA in slow exchange (slower than 100 msec timescale). To investigate the dynamics of this system, a single-molecule FRET experiment was designed to directly monitor the formation of the intermolecular PAS helix. These experiments showed that the vRNA–tRNA complex existed in two distinct conformations, with either intra- or intermolecular PAS formation (Fig. 3D). Additional NMR and single-molecule experiments confirmed that the intermolecular PAS helix coexisted with a formed PBS helix and the downstream vRNA H2. The two PAS conformations appeared to be extremely stable and in very slow exchange, on the order of minutes. These results suggest that RT encounters a variety of RNA conformations that it must navigate, and possibly remodel, during initiation. RNA structure could thus serve as a regulation mechanism during initiation.

The results discussed above showed the influence that RNA construct length can have in general on structural and biophysical studies. We subsequently lengthened the vRNA construct to replicate better the functional aspects of reverse transcription initiation (Coey et al. 2018). The vRNA sequence was extended both upstream of and downstream from the PBS region to include the full HIV-1 intramolecular PAS helix, dubbed H1 (Fig. 3C). To determine further the extent of the vRNA–tRNA complex dynamics, an additional FRET-labeling scheme was used to probe the spatial relationship between the tRNA 5′ end and vRNA H1. FRET probes were placed on the vRNA 5′ or 3′ ends (H1) and on the tRNA 5′ end. These experiments showed a diverse range of FRET values and states, indicative of conformational dynamics between the structural domains of the vRNA–tRNA complex (Fig. 4). The different conformations interconvert on short (2–4-sec) and long (>10-sec) timescales, supporting a model that the vRNA–tRNA complex is structurally dynamic in the absence of RT.

These experiments have begun to delineate the vRNA–tRNA structural landscape that exists in the absence of RT binding. NMR and single-molecule experiments directly revealed the ability of the vRNA–tRNA complex to form additional structures without perturbing the PBS helix and vRNA H2. Additional single-molecule experiments explored the structural dynamics of the vRNA–tRNA by monitoring the spatial relationship between vRNA H1 and the tRNA 5′ end (Fig. 4). This further revealed the extensive conformational heterogeneity and structural interconversion of the RNA elements. To delineate the interplay of structure and dynamics of the initiation complex, additional experiments introducing RT were required.

Cite as *Cold Spring Harb Perspect Biol* doi: 10.1101/cshperspect.a032474

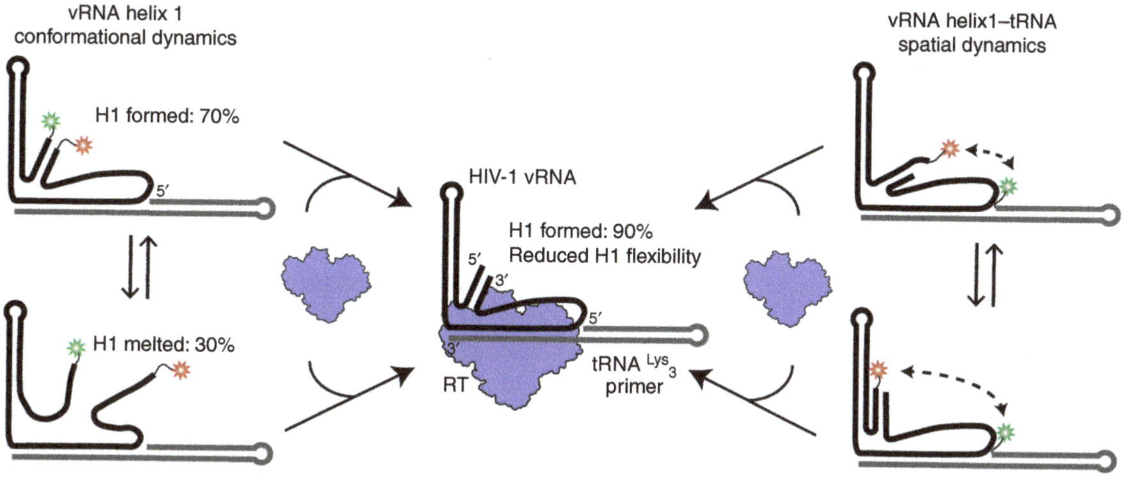

Figure 4. The conformational heterogeneity observed in the vRNA–tRNA complex is reduced on the introduction of reverse transcriptase (RT). (*Left*) Viral RNA helix 1 (vRNA H1) formation was monitored in the absence and presence of RT by labeling the vRNA on its 3′ and 5′ ends. Two main populations were observed. (*Right*) The spatial orientation of H1 and the tRNA was monitored by labeling vRNA H1 and the 5′ end of the tRNA. A variety of interconverting high and low FRET populations were observed, indicating conformational dynamics. (*Center*) Upon RT binding, we observed a reduction in the conformational and structural heterogeneity for vRNA H1. FRET populations shifted to indicate 90% H1 formation and the FRET populations for the vRNA–tRNA dynamics shifted to all low FRET states. (Data adapted from Coey et al. 2018.)

8 INTERACTIONS OF HIV-1 RT WITH THE PRIMER–TEMPLATE COMPLEX

HIV-1 RT is the enzyme responsible for not only reverse transcribing the entire viral RNA genome into DNA, but also degrading the RNA after copying has occurred. The prominent and key role RT plays in HIV-1 replication has caused it to be a widely studied drug target in AIDS therapies (Sarafianos et al. 2009). Thus, there are decades of structural and mechanistic data on RT and how it interacts with nucleic acid substrates. RT is a heterodimer with its smaller, structural p51 subunit being a truncation of its catalytic p66 subunit. The p66 subunit contains both of the enzyme's polymerase and RNase H active sites. The polymerase domain contains 4 subdomains (fingers, palm, thumb, and connection) that adopt a classic polymerase fold. Numerous crystal structures, with a wide range of inhibitors, DNA–DNA substrates, and DNA–RNA substrates, have revealed distinct functional states for RT (Huang et al. 1998; Sarafianos et al. 2001; Das et al. 2012; Das et al. 2014). This work has provided a detailed mechanistic understanding of how RT interacts with and extends elongation substrates. Although this repertoire of information is vast, the interaction of RT with the vRNA–tRNA complex during initiation has been poorly understood.

Extensive work has suggested that interactions of RT with the vRNA–tRNA complex, specifically those near the polymerase active site, govern the transition from initiation into elongation (Lanchy et al. 1998; Lanchy et al. 2000). Once the tRNA primer had been extended 6–9 nt, RT processivity increased because of inversely correlated changes in the polymerization rate (k_{pol}) and the dissociation rate (k_{off}) of RT for the extended substrates (Lanchy et al. 1998; Lanchy et al. 2000; Liu et al. 2010). However, the exact nature of the RT–nucleic acid interactions near the active site has remained a mystery. This is further convoluted because of the possible structural rearrangements in the vRNA–tRNA complex; the growing primer strand, extended by RT, perturbs the vRNA helixes downstream from the PBS. The nature of the helical structure with the RT-binding cleft also changes from an RNA–RNA duplex to RNA–DNA, which may also affect the accommodation of substrate in the active site. This suggests a highly dynamic process in which vRNA–tRNA conformation evolves temporally as RT alters primer composition through extension. An additional layer of complexity came in the form of a single-molecule FRET study by Liu et al. (2010), which showed that RT has two distinct binding modes on the vRNA–tRNA initiation complex. Their experimental setup monitored RT-binding orientations to the vRNA–tRNA complex in a variety of conditions. They concluded that there existed an "active" orientation with the polymerase active site located near the tRNA 3′ end or an "inactive" flipped orientation with the RNase H active located near the tRNA 3′ end. These binding preferences were correlated with polymerase activity, the enzyme preferring to reside in the flipped orientation on preextended complexes that showed a low k_{pol}. The transition between +5 and +6 showed an approximately fivefold increase on the equilibrium constant between the polymerase-competent and flipped orientations and an approximately fourfold increase in k_{pol}. The polymerase-competent orientation also appeared to be favored when vRNA H2 was disrupted and an additional FRET experiment showed that this disruption occurs after incorporation of the sixth nucleotide into the growing primer strand (Liu et al. 2010).

We have explored how RT can interact with the vRNA–tRNA complex to alter the dynamics of the HIV-1 initiation complex (Coey et al. 2018). Instead of focusing on RT-binding orientation, we determined the effect of RT binding on RNA structure. Experiments were performed using two FRET-labeling schemes that were used to study the free vRNA–tRNA complex above (Fig. 4). Using the vRNA 3′- and 5′-labeled scheme, we monitored the influence of RT binding on H1 formation within the vRNA–tRNA. In the absence of RT, helix formation as signified by high FRET, was found to occur in 70% of molecules. This population shifted to 90% H1 formation after the introduction of excess unlabeled RT. Such a shift suggested that RT binding was able to modulate the structural heterogeneity of the vRNA–tRNA complex. In addition, RT was added to FRET experiments that monitored the spatial orientation of vRNA H1 and the tRNA 5′ end (Fig. 4). In the absence of RT, these two RNA domains appeared to be highly dynamic with respect to each other with numerous interconverting FRET states. However, on the introduction of RT, the FRET population distribution shifted entirely to low FRET states. These sets of experiments highlighted the ability of RT to alter the conformational landscape of the vRNA–tRNA and reduce its structural heterogeneity in terms of secondary structure and spatial orientation (Fig. 4).

9 INSIGHTS INTO THE TERTIARY STRUCTURE OF THE HIV-1 RT INITIATION COMPLEX

As highlighted in the preceding sections, decades of work have attempted to understand the molecular mechanism of the HIV-1 reverse transcription initiation (Isel et al. 2010). The distinct aspects of initiation compared with elongation were underscored by kinetics measurements that showed its nonprocessive nature and discrete pauses. Detailed investigations of the sequence and secondary structure of the vRNA–tRNA complex yielded the discovery of additional interactions between regions on the vRNA and tRNA outside of the PBS helix. Single-molecule experiments underscored the structural variability and dynamics contained within the vRNA–tRNA and the vRNA–tRNA–RT complexes (Liu et al. 2010; Beerens et al. 2013; Coey et al. 2016; Coey et al. 2018). Despite these advances, there remained a lack of detailed structural information on the RT initiation complex (RTIC) outside of secondary structure models.

Recently, we determined the tertiary structure of a RTIC using cryo-EM and a cross-linking approach, similar to that used in Huang et al. (2000), to stabilize and trap a +1 extended state complex (Fig. 5A) (Larsen et al. 2018). This RTIC was formed using a 101-nt fragment of vRNA that encompasses all of the structural elements believed to interact with the tRNA primer during initiation (Figs. 3C and 5). The cryo-EM study resulted in two main electron density maps. The first, which best describes the global architecture of the HIV-1 RTIC, was limited to a global resolution of 8.0 Å because of the flexibility of the RNA elements. The second map, at a resolution of 4.5 Å, was derived by masking away the peripheral flexible RNA elements and focused on refining the core of RT and the RNA within its binding cleft. Together, these maps provided sufficient information to build a global model encompassing nearly all of the vRNA–tRNA elements as well as RT (Fig. 5A). The RNA domains were generated using Rosetta based on previously derived secondary structure models for the vRNA. Although this approach does not provide exact information on base pairs and detailed local conformations, it was able to shed light on the global fold of the RNA within the RTIC.

The global RTIC structure featured a long RNA helix emerging from RT cleft near the RNase H domain as an extension of the PBS helix. Based on this result, we proposed that additional base pairs form between the vRNA and tRNA, extending the PBS helix, and that the tRNA itself refolds to form a contiguous helix. This secondary structure model differs from the past models that identified a multi-stem structure for the tRNA. The RTIC structure also contained density for vRNA H1, H2, and the connecting loops. The two vRNA helices are located directly above the RT polymerase active site, highlighting their proposed function as a structural barrier to initiation. H1 and the connecting loop appear to position H2 while serving as a bridge linking RNA near the two RT active sites. Notably, the 3′ terminus of the tRNA is displaced from the RT polymerase active site residues, possibly because of the vRNA helices acting to prevent proper translocation and accommodation after nucleotide incorporation. Capturing of this pretranslocation +1 state may explain the low processivity observed during early stages of initiation.

The structural heterogeneity observed in the EM data, which could be modeled as RNA helical movements rather than alternative folds, appears to corroborate the results from Coey et al. (2018) that observed reduced RNA dynamics upon RT binding (Fig. 4). Although local flexibility hindered our ability to resolve the detailed structure of vRNA

Cite as *Cold Spring Harb Perspect Biol* doi: 10.1101/cshperspect.a032474

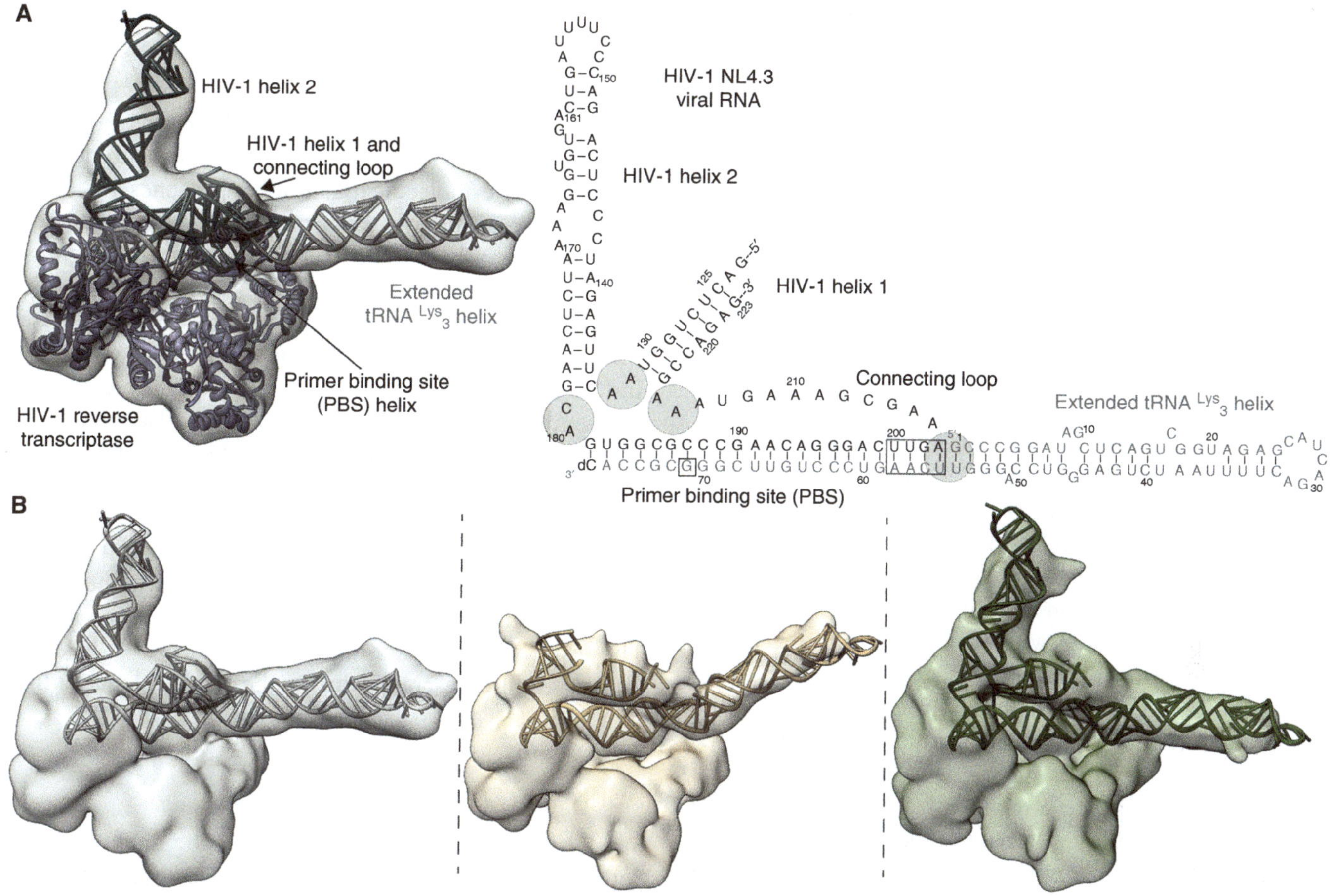

Figure 5. (*A*) Global tertiary and secondary structure of the RT initiation complex (RTIC) at the +1 extension state. The cryo-EM density encompasses all of vRNA H1, the PBS, and the tRNA. Density of H2 is weak near the apical end. (*B*) Three different global cryo-EM class averages of the RTIC. Alternative RNA models were generated by treating the H1, H2, and extended tRNA helices as rigid bodies. The flexible junctions are indicated in *A*. The center class showed very little density for the upper stem of H2. (Data adapted from Larsen et al. 2018.)

H1, it appears to occupy a defined region of space that varies slightly among all cryo-EM classes. This agrees with the single-molecule FRET data that showed reduced H1 global flexibility upon RT binding. The majority of the global heterogeneity for the +1 state of the RTIC involves the orientation of the extended tRNA and the apical portion of vRNA H2 with respect to the core (Fig. 5). These regions include the tRNA anti-PAS, anticodon, and the vRNA C-rich and A-rich loops. Although these domains may be flexible, they are linked together through the vRNA connecting loop and H1. Such a linkage between these RNA elements may limit their spatial orientations to allow the RNAs to refold and form intermolecular structures as reverse transcription progresses and disrupts the vRNA helices.

Although this structure of a +1 RTIC highlighted a previously unobserved vRNA–tRNA conformation and a possible explanation for the low processivity during early initiation, many questions regarding initiation remain open and higher-resolution structures are still needed. A single structure of an RNA assembly is just one snapshot of a molecular movie that links structures together through the dynamics of initiation. Single-molecule experiments, in particular smFRET, could be key in identifying the timing of the potential vRNA–tRNA structural rearrangements and revealing key intermediates that have only yet been hypothesized. The next section will focus on the next steps in studying HIV-1 reverse transcription initiation and how current methods can be harnessed to develop a more complete understanding of how the process evolves.

10 NEXT STEPS IN STUDYING HIV-1 REVERSE TRANSCRIPTION INITIATION

Investigation of the structure and dynamics of reverse transcription initiation has focused on discrete extension "states" (Lanchy et al. 1998; Lanchy et al. 2000; Liu et al.

2010; Coey et al. 2018; Larsen et al. 2018). Single-molecule experiments have directly monitored the interactions of RT with the vRNA–tRNA complex, but without simultaneously monitoring the polymerase activity of RT (Liu et al. 2010; Coey et al. 2018). RNA complexes are most frequently left unextended or preextended with a combination of deoxynucleotides and dideoxynucleotides to prevent further extension. Population changes can be monitored after adding an incoming nucleotide, which partially mimics extension via nucleotide binding. However, such experiments still rely on preextended primer–template complexes that may have already undergone conformational changes following incorporation. Real-time experiments, similar to those described above for translation, are needed.

All of the proposed vRNA–tRNA structural rearrangements have yet to be directly observed during the actual process of initiation. FRET-labeling schemes designed to monitor the formation these vRNA–tRNA structures revealed the PAS interaction in a simplified system, but these have yet to be combined with RT binding and extension. The ability to perform real-time RT extension assays using labeled nucleotides as seen in Vilfan et al. (2013), while simultaneously monitoring helix formation and destabilization, could be explored. Initiation has been described as a highly dissociative process, with weak affinity of RT for the vRNA–tRNA complex. Therefore, applying single-molecule methods in a manner that can correlate productive RT binding with RNA rearrangements will alleviate many mysteries surrounding the steps that govern initiation.

The structural heterogeneity observed in +1 RTIC data suggests that the RNA elements remain highly dynamic even in the absence of complete structural rearrangements (Larsen et al. 2018). Even with large data sets, the resolution of such conformers will likely be limited by multiple global and local RNA conformations. Designing single-molecule FRET experiments to probe the distances between RNA domains may serve as a vital tool to aid in more accurately applying molecular modeling tools like Rosetta (Das et al. 2010). The emergence of time-resolved and temperature-dependent (freezing after incubation at different temperatures) cryo-EM is appealing to the study of such a dynamic process (Zhang et al. 2013; Frank 2017a; Kaledhonkar et al. 2018). However, limitations in sample preparation for structural studies remain. The high dissociation rate of RT for the vRNA–tRNA complex limited our current structure approach to a cross-linking method, which traps extension states and would prevent any time-resolved studies (Huang et al. 1998; Huang et al. 2000; Larsen et al. 2018). Whereas such methodology will likely allow for solving the structures of additional RTIC states, single-molecule methods will aid in linking the transient structural snapshots.

11 CONCLUDING REMARKS

This is a wonderful moment for combined dynamic and structural studies of RNAs and their complexes. The explosion of structural methods, in particular cryo-EM with its acute ability to define conformational states, combined with powerful single-molecule dynamics approaches, has driven progress in creating molecular movies. The results presented here on translation and reverse transcription highlight the importance of detailed structural information by either cryo-EM or X-ray methods to form functional hypotheses. NMR and single-molecule methods allowed us to delineate and measure conformational dynamics in these systems, and computational approaches underpinned all our efforts. The results show the local and global plasticity of RNA, and how minor perturbations in structure or dynamics can lead to large functional consequences.

Moving forward, technological advances will improve structural resolution, power computational methods, and improve temporal resolution of dynamics experiments. The result will be a more holistic view of how RNAs and their assemblies rearrange during the delicate dance of their function.

ACKNOWLEDGMENTS

Single-molecule translation research in the Puglisi group is supported by the National Institutes of Health (NIH) grants GM51266, GM113078, AI047365, AI099506 to J.D.P. HIV-1 reverse transcription research is supported by NIH Grant GM082545 to E.V.P., K.P.L., and A.P. are funded by NIH Molecular Biophysics Training Grant (T32-GM008294). J.C. and A.P. are funded by the Stanford Interdisciplinary Graduate Fellowship (Bio-X). We thank H. Demirci for helping with the figure preparation.

REFERENCES

*Reference is in this book.

Aitken CE, Puglisi JD. 2010. Following the intersubunit conformation of the ribosome during translation in real time. *Nat Struct Mol Biol* **17:** 793–800.

Al-Hashimi HM. 2013. NMR studies of nucleic acid dynamics. *J Magn Reson* **237:** 191–204.

Bai XC, McMullan G, Scheres SH. 2015. How cryo-EM is revolutionizing structural biology. *Trends Biochem Sci* **40:** 49–57.

Baltimore D. 1970. RNA-dependent DNA polymerase in virions of RNA tumour viruses. *Nature* **226:** 1209–1211.

Bashan A, Yonath A. 2008. Correlating ribosome function with high-resolution structures. *Trends Microbiol* **16:** 326–335.

Beerens N, Berkhout B. 2002a. Switching the in vitro tRNA usage of HIV-1 by simultaneous adaptation of the PBS and PAS. *RNA* **8:** 357–369.

Beerens N, Berkhout B. 2002b. The tRNA primer activation signal in the human immunodeficiency virus type 1 genome is important for initiation and processive elongation of reverse transcription. *J Virol* **76:** 2329–2339.

Beerens N, Groot F, Berkhout B. 2001. Initiation of HIV-1 reverse transcription is regulated by a primer activation signal. *J Biol Chem* **276:** 31247–31256.

Beerens N, Jepsen MD, Nechyporuk-Zloy V, Kruger AC, Darlix JL, Kjems J, Birkedal V. 2013. Role of the primer activation signal in tRNA annealing onto the HIV-1 genome studied by single-molecule FRET microscopy. *RNA* **19:** 517–526.

Bilbille Y, Vendeix FA, Guenther R, Malkiewicz A, Ariza X, Vilarrasa J, Agris PF. 2009. The structure of the human $tRNA^{Lys}{}_3$ anticodon bound to the HIV genome is stabilized by modified nucleosides and adjacent mismatch base pairs. *Nucleic Acids Res* **37:** 3342–3353.

Blanchard SC, Gonzalez RL, Kim HD, Chu S, Puglisi JD. 2004. tRNA selection and kinetic proofreading in translation. *Nat Struct Mol Biol* **11:** 1008–1014.

Boccaletto P, Machnicka MA, Purta E, Piatkowski P, Baginski B, Wirecki TK, de Crecy-Lagard V, Ross R, Limbach PA, Kotter A, et al. 2018. MODOMICS: A database of RNA modification pathways. 2017 update. *Nucleic Acids Res* **46:** D303–D307.

Chapman HN, Fromme P, Barty A, White TA, Kirian RA, Aquila A, Hunter MS, Schulz J, DePonte DP, Weierstall U, et al. 2011. Femtosecond X-ray protein nanocrystallography. *Nature* **470:** 73–77.

Chen B, Kaledhonkar S, Sun M, Shen B, Lu Z, Barnard D, Lu TM, Gonzalez RL Jr., Frank J. 2015. Structural dynamics of ribosome subunit association studied by mixing-spraying time-resolved cryogenic electron microscopy. *Structure* **23:** 1097–1105.

Cheng Y. 2015. Single-particle cryo-EM at crystallographic resolution. *Cell* **161:** 450–457.

Cheng CY, Chou FC, Das R. 2015. Modeling complex RNA tertiary folds with Rosetta. *Methods Enzymol* **553:** 35–64.

Choi J, Ieong K-W, Demirci H, Chen J, Petrov A, Prabhakar A, O'Leary SE, Dominissini D, Rechavi G, Soltis SM, et al. 2016. N6-methyladenosine in mRNA disrupts tRNA selection and translation-elongation dynamics. *Nat Struct Mol Biol* **23:** 110–115.

Choi J, Indrisiunaite G, DeMirci H, Ieong K-W, Wang J, Petrov A, Prabhakar A, Rechavi G, Dominissini D, He C, et al. 2018. 2′-O-methylation in mRNA disrupts tRNA decoding during translation elongation. *Nat Struct Mol Biol* **25:** 208–216.

Coey A, Larsen K, Puglisi JD, Viani Puglisi E. 2016. Heterogeneous structures formed by conserved RNA sequences within the HIV reverse transcription initiation site. *RNA* **22:** 1689–1698.

Coey A, Larsen KP, Barrero D, Puglisi JD, Puglisi EV. 2018. Dynamic interplay of RNA and protein in the human immunodeficiency virus-1 reverse transcription initiation complex. *J Mol Biol* (in press).

Cummins LL, Owens SR, Risen LM, Lesnik EA, Freier SM, McGee D, Guinosso CJ, Cook PD. 1995. Characterization of fully 2′-modified oligoribonucleotide hetero- and homoduplex hybridization and nuclease sensitivity. *Nucleic Acids Res* **23:** 2019–2024.

Dai Q, Moshitch-Moshkovitz S, Han D, Kol N, Amariglio N, Rechavi G, Dominissini D, He C. 2017. Nm-seq maps 2′-*O*-methylation sites in human mRNA with base precision. *Nat Methods* **14:** 695–698.

Das R, Karanicolas J, Baker D. 2010. Atomic accuracy in predicting and designing noncanonical RNA structure. *Nat Methods* **7:** 291–294.

Das K, Martinez SE, Bauman JD, Arnold E. 2012. HIV-1 reverse transcriptase complex with DNA and nevirapine reveals non-nucleoside inhibition mechanism. *Nat Struct Mol Biol* **19:** 253–259.

Das K, Martinez SE, Bandwar RP, Arnold E. 2014. Structures of HIV-1 RT-RNA/DNA ternary complexes with dATP and nevirapine reveal conformational flexibility of RNA/DNA: Insights into requirements for RNase H cleavage. *Nucleic Acids Res* **42:** 8125–8137.

Dominissini D, Moshitch-Moshkovitz S, Schwartz S, Salmon-Divon M, Ungar L, Osenberg S, Cesarkas K, Jacob-Hirsch J, Amariglio N, Kupiec M, et al. 2012. Topology of the human and mouse m6A RNA methylomes revealed by m6A-seq. *Nature* **485:** 201–206.

Dorywalska M, Blanchard SC, Gonzalez RL, Kim HD, Chu S, Puglisi JD. 2005. Site-specific labeling of the ribosome for single-molecule spectroscopy. *Nucleic Acids Res* **33:** 182–189.

Fernandez IS, Ng CL, Kelley AC, Wu G, Yu YT, Ramakrishnan V. 2013. Unusual base pairing during the decoding of a stop codon by the ribosome. *Nature* **500:** 107–110.

Fourmy D, Yoshizawa S, Puglisi JD. 1998. Paromomycin binding induces a local conformational change in the A-site of 16 S rRNA. *J Mol Biol* **277:** 333–345.

Frank J. 2017a. Time-resolved cryo-electron microscopy: Recent progress. *J Struct Biol* **200:** 303–306.

Frank J. 2017b. The translation elongation cycle—Capturing multiple states by cryo-electron microscopy. *Philos Trans R Soc B Biol Sci* **372:** 20160180.

Frank J, Agrawal RK. 2000. A ratchet-like inter-subunit reorganization of the ribosome during translocation. *Nature* **406:** 318–322.

Friedman LJ, Gelles J. 2015. Multi-wavelength single-molecule fluorescence analysis of transcription mechanisms. *Methods* **86:** 27–36.

Fu Z, Kaledhonkar S, Borg A, Sun M, Chen B, Grassucci RA, Ehrenberg M, Frank J. 2016. Key intermediates in ribosome recycling visualized by time-resolved cryoelectron microscopy. *Structure* **24:** 2092–2101.

Gilboa E, Mitra SW, Goff S, Baltimore D. 1979. A detailed model of reverse transcription and tests of crucial aspects. *Cell* **18:** 93–100.

Goldschmidt V, Rigourd M, Ehresmann C, Le Grice SF, Ehresmann B, Marquet R. 2002. Direct and indirect contributions of RNA secondary structure elements to the initiation of HIV-1 reverse transcription. *J Biol Chem* **277:** 43233–43242.

Goldschmidt V, Ehresmann C, Ehresmann B, Marquet R. 2003. Does the HIV-1 primer activation signal interact with $tRNA^{Lys}{}_3$ during the initiation of reverse transcription? *Nucleic Acids Res* **31:** 850–859.

Goldschmidt V, Paillart JC, Rigourd M, Ehresmann B, Aubertin AM, Ehresmann C, Marquet R. 2004. Structural variability of the initiation complex of HIV-1 reverse transcription. *J Biol Chem* **279:** 35923–35931.

Gorini L. 1971. Ribosomal discrimination of tRNAs. *Nat New Biol* **234:** 261–264.

He C. 2010. Grand challenge commentary: RNA epigenetics? *Nat Chem Biol* **6:** 863–865.

Hopfield JJ. 1974. Kinetic proofreading: A new mechanism for reducing errors in biosynthetic processes requiring high specificity. *Proc Natl Acad Sci* **71:** 4135–4139.

Hoskins AA, Gelles J, Moore MJ. 2011. New insights into the spliceosome by single molecule fluorescence microscopy. *Curr Opin Chem Biol* **15:** 864–870.

Hu WS, Hughes SH. 2012. HIV-1 reverse transcription. *Cold Spring Harb Perspect Med* **2:** a006882.

Huang H, Chopra R, Verdine GL, Harrison SC. 1998. Structure of a covalently trapped catalytic complex of HIV-1 reverse transcriptase: Implications for drug resistance. *Science* **282:** 1669–1675.

Huang H, Harrison SC, Verdine GL. 2000. Trapping of a catalytic HIV reverse transcriptase*template:primer complex through a disulfide bond. *Chem Biol* **7:** 355–364.

Isel C, Marquet R, Keith G, Ehresmann C, Ehresmann B. 1993. Modified nucleotides of $tRNA^{Lys}{}_3$ modulate primer/template loop–loop interaction in the initiation complex of HIV-1 reverse transcription. *J Biol Chem* **268:** 25269–25272.

Isel C, Ehresmann C, Keith G, Ehresmann B, Marquet R. 1995. Initiation of reverse transcription of HIV-1: Secondary structure of the HIV-1 RNA/$tRNA^{Lys}{}_3$ (template/primer). *J Mol Biol* **247:** 236–250.

Isel C, Lanchy JM, Le Grice SF, Ehresmann C, Ehresmann B, Marquet R. 1996. Specific initiation and switch to elongation of human immunodeficiency virus type 1 reverse transcription require the post-transcriptional modifications of primer $tRNA^{Lys}{}_3$. *EMBO J* **15:** 917–924.

Isel C, Keith G, Ehresmann B, Ehresmann C, Marquet R. 1998. Mutational analysis of the $tRNA^{Lys}{}_3$/HIV-1 RNA (primer/template) complex. *Nucleic Acids Res* **26:** 1198–1204.

Isel C, Westhof E, Massire C, Le Grice SF, Ehresmann B, Ehresmann C, Marquet R. 1999. Structural basis for the specificity of the initiation of HIV-1 reverse transcription. *EMBO J* **18:** 1038–1048.

Isel C, Ehresmann C, Marquet R. 2010. Initiation of HIV reverse transcription. *Viruses* **2:** 213–243.

Iwatani Y, Rosen AE, Guo J, Musier-Forsyth K, Levin JG. 2003. Efficient initiation of HIV-1 reverse transcription in vitro. Requirement for RNA sequences downstream of the primer binding site abrogated by nucleocapsid protein-dependent primer–template interactions. *J Biol Chem* **278:** 14185–14195.

Johansson M, Zhang J, Ehrenberg M. 2012. Genetic code translation displays a linear trade-off between efficiency and accuracy of tRNA selection. *Proc Natl Acad Sci* **109:** 131–136.

Jung YO, Lee JH, Kim J, Schmidt M, Moffat K, Srajer V, Ihee H. 2013. Volume-conserving *trans-cis* isomerization pathways in photoactive yellow protein visualized by picosecond X-ray crystallography. *Nat Chem* **5:** 212–220.

Kaledhonkar S, Fu Z, White H, Frank J. 2018. Time-resolved cryo-electron microscopy using a microfluidic chip. *Methods Mol Biol* **1764:** 59–71.

Karijolich J, Yu YT. 2011. Converting nonsense codons into sense codons by targeted pseudouridylation. *Nature* **474:** 395–398.

Kati WM, Johnson KA, Jerva LF, Anderson KS. 1992. Mechanism and fidelity of HIV reverse transcriptase. *J Biol Chem* **267:** 25988–25997.

Kraynack BA. 2006. Small interfering RNAs containing full 2′-*O*-methylribonucleotide-modified sense strands display Argonaute2/eIF2C2-dependent activity. *RNA* **12:** 163–176.

Lanchy JM, Ehresmann C, Le Grice SF, Ehresmann B, Marquet R. 1996. Binding and kinetic properties of HIV-1 reverse transcriptase markedly differ during initiation and elongation of reverse transcription. *EMBO J* **15:** 7178–7187.

Lanchy JM, Keith G, Le Grice SF, Ehresmann B, Ehresmann C, Marquet R. 1998. Contacts between reverse transcriptase and the primer strand govern the transition from initiation to elongation of HIV-1 reverse transcription. *J Biol Chem* **273:** 24425–24432.

Lanchy JM, Isel C, Keith G, Le Grice SF, Ehresmann C, Ehresmann B, Marquet R. 2000. Dynamics of the HIV-1 reverse transcription complex during initiation of DNA synthesis. *J Biol Chem* **275:** 12306–12312.

Larsen KP, Mathiharan YK, Kappel K, Coey AT, Chen D-H, Barrero D, Madigan L, Puglisi JD, Skiniotis G, Puglisi EV. 2018. Architecture of an HIV-1 reverse transcriptase initiation complex. *Nature* **557:** 118–122.

Levene MJ, Korlach J, Turner SW, Foquet M, Craighead HG, Webb WW. 2003. Zero-mode waveguides for single-molecule analysis at high concentrations. *Science* **299:** 682–686.

Li X, Mooney P, Zheng S, Booth CR, Braunfeld MB, Gubbens S, Agard DA, Cheng Y. 2013. Electron counting and beam-induced motion correction enable near-atomic-resolution single-particle cryo-EM. *Nat Methods* **10:** 584–590.

Li A, Gong S, Johnson KA. 2016. Rate-limiting pyrophosphate release by HIV reverse transcriptase improves fidelity. *J Biol Chem* **291:** 26554–26565.

Liang C, Li X, Rong L, Inouye P, Quan Y, Kleiman L, Wainberg MA. 1997. The importance of the A-rich loop in human immunodeficiency virus type 1 reverse transcription and infectivity. *J Virol* **71:** 5750–5757.

Liang C, Rong L, Gotte M, Li X, Quan Y, Kleiman L, Wainberg MA. 1998. Mechanistic studies of early pausing events during initiation of HIV-1 reverse transcription. *J Biol Chem* **273:** 21309–21315.

* Linder B, Jaffrey SR. 2019. Discovering and mapping the modified nucleotides that comprise the epitranscriptome of mRNA. *Cold Spring Harb Perspect Biol* doi: 10.1101/cshperspect.a032201.

Liu S, Harada BT, Miller JT, Le Grice SF, Zhuang X. 2010. Initiation complex dynamics direct the transitions between distinct phases of early HIV reverse transcription. *Nat Struct Mol Biol* **17:** 1453–1460.

Los GV, Encell LP, McDougall MG, Hartzell DD, Karassina N, Zimprich C, Wood MG, Learish R, Ohana RF, Urh M, et al. 2008. HaloTag: A novel protein labeling technology for cell imaging and protein analysis. *ACS Chem Biol* **3:** 373–382.

Marko RA, Liu HW, Ablenas CJ, Ehteshami M, Gotte M, Cosa G. 2013. Binding kinetics and affinities of heterodimeric versus homodimeric HIV-1 reverse transcriptase on DNA–DNA substrates at the single-molecule level. *J Phys Chem B* **117:** 4560–4567.

Meyer KD, Saletore Y, Zumbo P, Elemento O, Mason Christopher E, Jaffrey Samie R. 2012. Comprehensive analysis of mRNA methylation reveals enrichment in 3′ UTRs and near stop codons. *Cell* **149:** 1635–1646.

Moazed D, Noller HF. 1989. Intermediate states in the movement of transfer RNA in the ribosome. *Nature* **342:** 142–148.

Mollwitz B, Brunk E, Schmitt S, Pojer F, Bannwarth M, Schiltz M, Rothlisberger U, Johnsson K. 2012. Directed evolution of the suicide protein O^6-alkylguanine-DNA alkyltransferase for increased reactivity results in an alkylated protein with exceptional stability. *Biochemistry* **51:** 986–994.

Ninio J. 1975. Kinetic amplification of enzyme discrimination. *Biochimie* **57:** 587–595.

Ogle JM, Brodersen DE, Clemons WM Jr., Tarry MJ, Carter AP, Ramakrishnan V. 2001. Recognition of cognate transfer RNA by the 30S ribosomal subunit. *Science* **292:** 897–902.

Ogle JM, Murphy FV, Tarry MJ, Ramakrishnan V. 2002. Selection of tRNA by the ribosome requires a transition from an open to a closed form. *Cell* **111:** 721–732.

Ooms M, Cupac D, Abbink TE, Huthoff H, Berkhout B. 2007. The availability of the primer activation signal (PAS) affects the efficiency of HIV-1 reverse transcription initiation. *Nucleic Acids Res* **35:** 1649–1659.

Paillart JC, Dettenhofer M, Yu XF, Ehresmann C, Ehresmann B, Marquet R. 2004. First snapshots of the HIV-1 RNA structure in infected cells and in virions. *J Biol Chem* **279:** 48397–48403.

Prabhakar A, Capece MC, Petrov A, Choi J, Puglisi JD. 2017. Post-termination ribosome intermediate acts as the gateway to ribosome recycling. *Cell Rep* **20:** 161–172.

Puglisi EV, Puglisi JD. 1998. HIV-1 A-rich RNA loop mimics the tRNA anticodon structure. *Nat Struct Biol* **5:** 1033–1036.

Puglisi EV, Puglisi JD. 2011. Secondary structure of the HIV reverse transcription initiation complex by NMR. *J Mol Biol* **410:** 863–874.

Ratner L, Haseltine W, Patarca R, Livak KJ, Starcich B, Josephs SF, Doran ER, Rafalski JA, Whitehorn EA, Baumeister K, et al. 1985. Complete nucleotide sequence of the AIDS virus, HTLV-III. *Nature* **313:** 277–284.

Reuter JS, Mathews DH. 2010. RNAstructure: Software for RNA secondary structure prediction and analysis. *BMC Bioinformatics* **11:** 129.

Robinson PJ, Trnka MJ, Bushnell DA, Davis RE, Mattei PJ, Burlingame AL, Kornberg RD. 2016. Structure of a complete mediator-RNA polymerase II pre-initiation complex. *Cell* **166:** 1411–1422.e16.

Roost C, Lynch SR, Batista PJ, Qu K, Chang HY, Kool ET. 2015. Structure and thermodynamics of N^6-methyladenosine in RNA: A spring-loaded base modification. *J Am Chem Soc* **137:** 2107–2115.

Roundtree IA, Evans ME, Pan T, He C. 2017. Dynamic RNA modifications in gene expression regulation. *Cell* **169:** 1187–1200.

Saletore Y, Meyer K, Korlach J, Vilfan ID, Jaffrey S, Mason CE. 2012. The birth of the Epitranscriptome: Deciphering the function of RNA modifications. *Genome Biol* **13:** 175.

Sarafianos SG, Das K, Tantillo C, Clark AD Jr., Ding J, Whitcomb JM, Boyer PL, Hughes SH, Arnold E. 2001. Crystal structure of HIV-1 reverse transcriptase in complex with a polypurine tract RNA:DNA. *EMBO J* **20:** 1449–1461.

Sarafianos SG, Marchand B, Das K, Himmel DM, Parniak MA, Hughes SH, Arnold E. 2009. Structure and function of HIV-1 reverse transcriptase: Molecular mechanisms of polymerization and inhibition. *J Mol Biol* **385:** 693–713.

Simonović M, Steitz TA. 2009. A structural view on the mechanism of the ribosome-catalyzed peptide bond formation. *Biochim Biophys Acta* **1789:** 612–623.

Srajer V, Teng T, Ursby T, Pradervand C, Ren Z, Adachi S, Schildkamp W, Bourgeois D, Wulff M, Moffat K. 1996. Photolysis of the carbon monoxide complex of myoglobin: Nanosecond time-resolved crystallography. *Science* **274:** 1726–1729.

Stagno JR, Bhandari YR, Conrad CE, Liu Y, Wang YX. 2017. Real-time crystallographic studies of the adenine riboswitch using an X-ray free-electron laser. *FEBS J* **284:** 3374–3380.

Suo Z, Johnson KA. 1997a. Effect of RNA secondary structure on the kinetics of DNA synthesis catalyzed by HIV-1 reverse transcriptase. *Biochemistry* **36:** 12459–12467.

Suo Z, Johnson KA. 1997b. RNA secondary structure switching during DNA synthesis catalyzed by HIV-1 reverse transcriptase. *Biochemistry* **36:** 14778–14785.

Temin HM, Mizutani S. 1970. RNA-dependent DNA polymerase in virions of Rous sarcoma virus. *Nature* **226:** 1211–1213.

Tenboer J, Basu S, Zatsepin N, Pande K, Milathianaki D, Frank M, Hunter M, Boutet S, Williams GJ, Koglin JE, et al. 2014. Time-resolved serial crystallography captures high-resolution intermediates of photoactive yellow protein. *Science* **346:** 1242–1246.

Uemura S, Aitken CE, Korlach J, Flusberg BA, Turner SW, Puglisi JD. 2010. Real-time tRNA transit on single translating ribosomes at codon resolution. *Nature* **464:** 1012–1017.

Vaccaro JA, Singh HA, Anderson KS. 1999. Initiation of minus-strand DNA synthesis by human immunodeficiency virus type 1 reverse transcriptase. *Biochemistry* **38:** 15978–15985.

Vilfan ID, Tsai YC, Clark TA, Wegener J, Dai Q, Yi C, Pan T, Turner SW, Korlach J. 2013. Analysis of RNA base modification and structural rearrangement by single-molecule real-time detection of reverse transcription. *J Nanobiotechnol* **11:** 8.

Voorhees RM, Ramakrishnan V. 2013. Structural basis of the translational elongation cycle. *Ann Rev Biochem* **82:** 203–236.

Wasserman MR, Alejo JL, Altman RB, Blanchard SC. 2016. Multiperspective smFRET reveals rate-determining late intermediates of ribosomal translocation. *Nat Struct Mol Biol* **23:** 333–341.

Yan C, Hang J, Wan R, Huang M, Wong CC, Shi Y. 2015. Structure of a yeast spliceosome at 3.6-angstrom resolution. *Science* **349:** 1182–1191.

Yin J, Straight PD, McLoughlin SM, Zhou Z, Lin AJ, Golan DE, Kelleher NL, Kolter R, Walsh CT. 2005. Genetically encoded short peptide tag for versatile protein labeling by Sfp phosphopantetheinyl transferase. *Proc Natl Acad Sci* **102:** 15815–15820.

Yoshizawa S. 1999. Recognition of the codon-anticodon helix by ribosomal RNA. *Science* **285:** 1722–1725.

Yusupova G, Yusupov M. 2017. Crystal structure of eukaryotic ribosome and its complexes with inhibitors. *Philos Trans R Soc B Biol Sci* **372:** 20160184.

Zaher HS, Green R. 2009. Quality control by the ribosome following peptide bond formation. *Nature* **457:** 161–166.

Zhang X, Sheng J, Plevka P, Kuhn RJ, Diamond MS, Rossmann MG. 2013. Dengue structure differs at the temperatures of its human and mosquito hosts. *Proc Natl Acad Sci* **110:** 6795–6799.

Zhou H, Kimsey IJ, Nikolova EN, Sathyamoorthy B, Grazioli G, McSally J, Bai T, Wunderlich CH, Kreutz C, Andricioaei I, et al. 2016. m(1)A and m(1)G disrupt A-RNA structure through the intrinsic instability of Hoogsteen base pairs. *Nat Struct Mol Biol* **23:** 803–810.

Zuker M. 2003. Mfold web server for nucleic acid folding and hybridization prediction. *Nucleic Acids Res* **31:** 3406–3415.

Single-Molecule Analysis of Reverse Transcriptase Enzymes

Linnea I. Jansson[1,3] and Michael D. Stone[2,3]

[1]Department of Molecular, Cell, and Developmental Biology, University of California, Santa Cruz, California 95064
[2]Department of Chemistry and Biochemistry, University of California, Santa Cruz, California 95064
[3]The Center for Molecular Biology of RNA, University of California, Santa Cruz, California 95064

Correspondence: mds@ucsc.edu

SUMMARY

The original discovery of enzymes that synthesize DNA using an RNA template appeared to contradict the central dogma of biology, in which information is transferred, in a unidirectional way, from DNA genes into RNA molecules. The paradigm-shifting discovery of RNA-dependent DNA polymerases, also called reverse transcriptases (RTs), reshaped existing views for how cells function; however, the scope of the impact RTs impose on biology had yet to be realized. In the decades of research since the early 1970s, the biomedical and biotechnological significance of retroviral RTs, as well as the evolutionarily related telomerase enzyme, has become exceedingly clear. One common theme that has emerged in the course of RT-related research is the central role of nucleic acid binding and dynamics during enzyme function. However, directly interrogating these dynamic properties is challenging because of the stochastic properties of biological macromolecules. In this review, we describe how the development of single-molecule biophysical techniques has opened new windows through which to observe the dynamic behavior of this remarkable class of enzymes. Specifically, we focus on how the powerful single-molecule Förster resonance energy transfer (FRET) method has been exploited to study the structure and function of the human immunodeficiency virus (HIV) RT and telomerase ribonucleoprotein (RNP) enzymes. These exciting studies have refined our understanding of RT catalysis, have revealed unforeseen structural rearrangements between RTs and their nucleic acid substrates, and have helped to characterize the mode of action of RT-inhibiting drugs. We conclude with a discussion of how the ongoing development of single-molecule technologies will continue to empower researchers to probe RT mechanisms in new and exciting ways.

Outline

Cite as *Cold Spring Harb Perspect Biol* doi: 10.1101/cshperspect.a032458

1 INTRODUCTION

Certain scientific discoveries have a way of broadly resonating with the basic research community. Such was the case with the initial discovery of viral reverse transcriptase (RT) enzymes in the 1970s by Baltimore and Temin (Baltimore 1970; Temin and Mizutani 1970). The capacity for an enzyme to synthesize DNA from an RNA template seemed to contradict the linearity of information flow put forth by Francis Crick in his famous lecture on the central dogma of biology (Crick 1970). Although Crick would personally contend that the existence of RT enzymes was in fact not at odds with his proposed model for information flow from nucleic acids to proteins, the discovery of these RNA-dependent DNA polymerases nevertheless altered the way in which scientists viewed the functional roles for RNA in biology and led to a Nobel Prize in Medicine just 5 years after the initial report was published on the discovery of RTs. A similar story can be told of the discovery of telomerase, the cellular RT that maintains telomeres in the majority of eukaryotic organisms. Telomeres are specialized chromatin structures that safeguard the genome from illicit recognition by DNA damage repair machinery (de Lange 2005). The mechanism of semiconservative DNA replication suggested that with each round of cell division there would be a gradual attrition of telomere length (Olovnikov 1971; Watson 1972; Allsopp et al. 1992). The formalization of this "end-replication problem" set into motion a search for an enzymatic activity that might maintain the ends of linear chromosomes. Greider and Blackburn discovered the telomerase enzyme in the mid-1980s (Greider and Blackburn 1987), for which they too were awarded the Nobel Prize in Medicine. Subsequent experiments showed that telomerase is a ribonucleoprotein (RNP) complex—part RNA, part protein. Interestingly, the catalytic telomerase reverse transcriptase (TERT) protein subunit shares many of the hallmark sequence motifs described previously for viral RT enzymes (Lingner et al. 1997; Nakamura and Cech 1998; Belfort et al. 2011). Once again, the discovery of an RT, this time one that carries its own RNA template as an integral enzyme subunit, motivated scientists to reevaluate existing views for functional roles of RNA in the modern world and, moreover, what implications these discoveries may have for the ancient RNA world and the origin of life.

2 IMPLICATIONS OF RT DISCOVERIES

The basic biology underlying the functions of each of these RTs is fascinating and continues to be the subject of intense research efforts. However, at the time of their initial discovery, few could have anticipated the far-reaching impacts of RTs on biotechnology and biomedical sciences. In the case of viral RTs, the AIDS epidemic, which broke out in the 1980s, made "retrovirus" a household word, and knowledge of the essential role of the RT enzyme in the viral life cycle contributed to the discovery of highly effective inhibitors of this class of enzymes, which to this day are a core component of human immunodeficiency virus (HIV) treatments (Kohlstaedt et al. 1992; Spence et al. 1995; El Safadi et al. 2007). In the case of telomerase, the discovery that ~90% of human cancer samples express abnormally high levels of telomerase placed the enzyme at the center of a massive research effort to better understand the role of telomerase in cancer progression, with the aim of producing novel diagnostic and therapeutic strategies (Kim et al. 1994). Although telomerase-based therapeutics have yet to realize their full potential, this research area remains a hotbed of biomedical inquiry.

With respect to biotechnology implications, the discovery of viral RTs has been particularly profound. Suddenly, through the use of robust RT enzymes, researchers were armed with a tool to convert all of the cellular information present in the more labile RNA form into a vastly more stable complementary DNA (cDNA) form. Once converted into cDNA, the entire RNA population of a cell, the transcriptome, could be sequenced and analyzed in ways that would be impossible without the RT enzyme. Thus, the case of the initial discovery of viral and cellular RTs serves as a quintessential example of how basic scientific inquiry can lead to paradigm shifts in biomedicine and biotechnology in ways that could not have been predicted at the outset.

3 ON THE EVOLUTION OF VIRAL RTs AND TELOMERASE

Phylogenetic analysis of RTs from retroviruses, telomerases, long-terminal repeat (LTR) retrotransposons, target-primed (TP) transposons, and group II introns permits one to group RT enzymes into evolutionarily related clusters (Curcio and Belfort 2007; Belfort et al. 2011; Gladyshev and Arkhipova 2011). However, the precise result of this type of analysis depends on how one roots the phylogenetic tree and therefore cannot provide conclusive information on the chronological progression of RT evolution. Nevertheless, two classes of models for the relationship between retroviral and telomerase RTs can be imagined: the RNA-first model and the protein-first model. In the RNA-first model, telomerase would have originated in an ancient RNA world from an ancestral catalytic RNA molecule. During the transition to DNA-based genomes, this ancient telomerase RNA (TR) molecule would have co-opted a protein RT enzyme (perhaps from a virus) to form a telomerase RNP complex similar to what we observe in modern cells. Support for this model comes from several observations. First, the discovery of catalytic RNAs makes possible an all RNA world, in which specialized RNAs performed many of the catalytic functions performed by proteins in the modern cell (Cech 1985). Second, although TRs from different organisms vary widely in both size and sequence, a conserved set of structural features has emerged (Theimer and Feigon 2006). Thus, RNA clearly plays functional roles beyond simply serving as a template for a protein RT, perhaps reflecting the possibility of an all RNA progenitor telomerase enzyme that later co-opted a protein RT subunit. The protein-first model provides an equally plausible theory for the evolution of RTs. In this model, an early protein RT would likely have evolved from an RNA-dependent RNA polymerase (RdRP) that may have emerged during the transition from the RNA world to the modern RNP world that now exists. RdRPs are often used to root phylogenetic trees of protein RTs based on the assumption that such an enzyme would have provided an advantage to organisms needing to efficiently copy RNA genomes (Xiong and Eickbush 1990). From this point, the polymerase active site would have gradually adapted to use deoxyribonucleoside triphosphates as selective pressures began to drive organisms toward DNA-based genomes. In the case of telomerase evolution, the preexisting RT would have coopted a cellular RNA as a stably associated enzyme component to facilitate the specialized telomerase catalytic cycle. Independent of which came first, protein RTs or an all RNA telomerase enzyme, it is nevertheless instructive to compare and contrast the properties of viral and telomerase RTs, both of which have recently been the subject of single-molecule analyses that have revealed remarkably dynamic nucleic acid-binding behaviors.

4 SINGLE-MOLECULE METHODS TO STUDY RT ENZYMES

The functional properties of biological systems emerge from highly dynamic and stochastic behaviors of individual molecules. Until relatively recently it was exceedingly difficult, if not impossible, to directly monitor the dynamic properties of a nucleic acid or protein molecule in real time. The challenge lies in the fact that macromolecular dynamics are typically obscured by the ensemble averaging intrinsic to most traditional biochemical and biophysical techniques which monitor the behavior of an asynchronous population of more than 10^{12} molecules in solution. Over the past several decades, the development of single-molecule tools has offered researchers an exciting new window into the complex behaviors of biological systems. These approaches have been used to study a wide range of systems, from protein synthesis by the ribosome to the folding dynamics of small catalytic RNAs (Zhuang et al. 2002; Tan et al. 2003; Wen et al. 2008; Perez and Gonzalez 2011). Among the most widely used single-molecule techniques are the force spectroscopy methods (i.e., optical trapping, magnetic tweezers, and atomic force microscopy) and the fluorescence-based approaches (i.e., fluorescence correlation spectroscopy and Förster resonance energy transfer, FRET). In the area of RT research, single-molecule FRET (smFRET) has been particularly powerful and will therefore be the focus of the present review. However, the interested reader is referred to any of a number of comprehensive technical reviews on single-molecule methods and their applications (Greenleaf et al. 2007; Deniz et al. 2008; Moffitt et al. 2008; Neuman and Nagy 2008; Roy et al. 2008; Tinoco and Gonzalez 2011).

smFRET techniques, which detect the efficiency of nonradiative energy transfer between a directly excited donor dye and a nearby acceptor dye (Stryer and Haugland 1967), emerged early on as a method of choice for researchers interested in studying dynamic biological systems. FRET is typically measured as the background-corrected ratio of the acceptor dye intensity divided by the sum of the donor plus the acceptor dye intensities. In this way, FRET ratios approaching a value of zero reflect the dyes being far from each other, whereas FRET ratios approaching a value of one indicate the two dyes are in close proximity. The popularity of smFRET is in part due to the relative simplicity of the optical setups required to make the measurements, as well as the ability to rapidly collect statistically meaningful results by measuring the behaviors of hundreds to thousands of single molecules in parallel (Deniz et al. 1999; Best et al. 2007; Merchant et al. 2007; Schuler and Eaton 2008). Although there exists a variety of microscopy tools with which to measure FRET, total internal reflection fluorescence (TIRF) microscopy is among the most widely used techniques (Zhuang et al. 2000; Ha 2001; Axelrod 2003). A standard TIRF microscope used for a smFRET experiment is equipped with specific detection optics that split the image into the donor and acceptor spectral channels, as well as a very sensitive electron-multiplying charge-coupled device (EMCCD) camera, which readily improves signal-to-noise ratios to levels that permit detection of photons emitted by individual dye molecules (Fig. 1). The capacity of a TIRF microscope to measure FRET at the single-molecule level was first demonstrated in pioneering studies that focused on DNA and RNA molecules, which could be readily conjugated to FRET dyes in a site-specific manner (Ha et al. 1996, Weiss 1999). Since this time, smFRET has been used to study a large number of diverse biological systems, attributable in part to the emergence of dye-coupling techniques with which to site-specifically label proteins and nucleic acids (Murphy et al. 2004; Rasnik et al. 2004; Jager et al. 2005; Jager et al. 2006). In the following sections, we will describe selected examples in which single-molecule

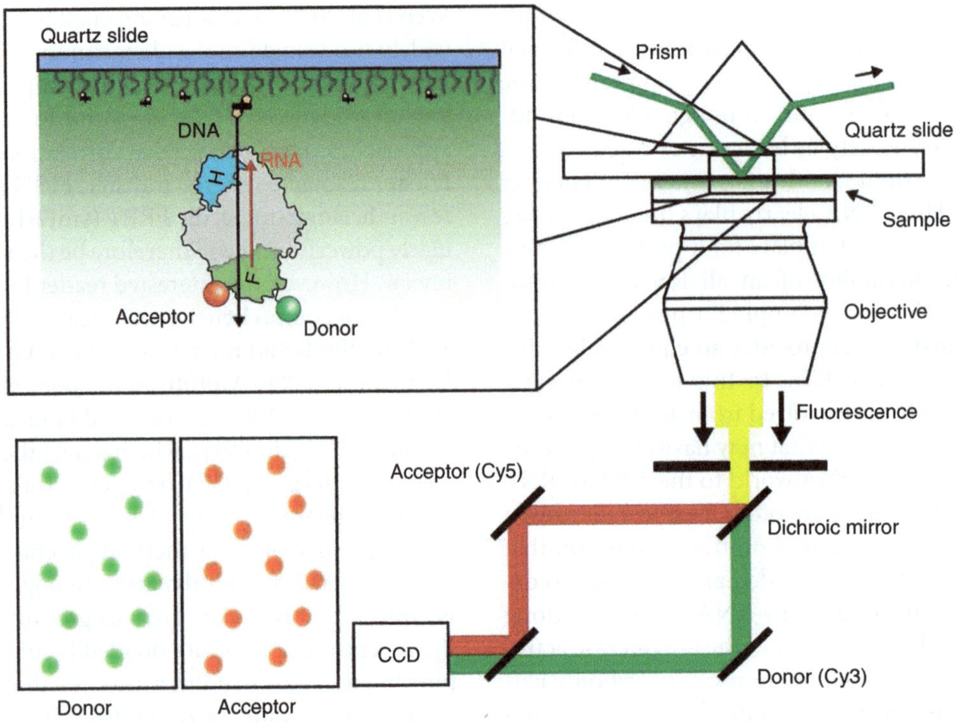

Figure 1. Single-molecule Förster resonance energy transfer (smFRET) microscope. smFRET measurements are often made using a total internal reflection fluorescence (TIRF) microscope. In the prism-type TIRF geometry, dye-labeled samples are immobilized onto a quartz microscope slide via biotin–streptavidin linkage. The appropriate excitation laser (532 nm shown here) is used to illuminate the sample at an angle, which causes total internal reflection of the incoming light because of the difference in the index of refraction of the quartz slide and the sample volume. In this way, the intensity of the evanescent field of light that propagates into the sample volume decays exponentially and dramatically enhances the single-to-noise ratio of the measurement. The emitted fluorescence of several hundred molecules in a single field of view is split using dichroic mirrors into separate acceptor and donor channels and captured by the electron-multiplying charge-coupled device (EMCCD) camera. Automated software routines are then used to analyze the raw movie data, and the FRET ratio can be calculated according to the simple expression $\mathrm{FRET} = I_A/(I_D + I_A)$, in which I_A and I_D are the background-corrected intensities of the acceptor and donor dyes, respectively. The schematic depicts a human immunodeficiency virus reverse transcriptase (HIV RT) enzyme immobilized onto a pegylated quartz slide using a biotinylated DNA substrate (black). HIV RT is shown in gray with the fingers and RNase H domains shown in green and blue, respectively. The RNA strand is shown in red. Biotinylated substrates are tethered to the slide surface using streptavidin (black rectangles). Biotin molecules are shown as yellow pentagons. An acceptor dye (red) is attached to the DNA and a donor dye (green) is attached to the enzyme.

analysis has provided unique insight into the mechanistic properties of viral and TERTs, with a particular emphasis on the dynamic interaction between these fascinating enzymes and their nucleic acid substrates.

5 HUMAN IMMUNODEFICIENCY VIRUS (HIV) RT

The type 1 HIV RT is a multifunction enzyme that catalyzes the conversion of the (+) strand viral RNA (vRNA) genome into a duplex DNA form that is competent for integration into the host genome (Fig. 2A) (Telesnitsky and Goff 1997). HIV RT is a heterodimeric protein complex that includes a small regulatory subunit (p51) and a large catalytic subunit (p66) that harbors two distinct enzyme active sites. The majority of the p66 polypeptide encodes a DNA polymerase domain comprised of the canonical fingers, palm, and thumb subdomains (Fig. 2B). In addition, p66 has a separate RNase H domain whose enzymatic activity is required to degrade the RNA within the RNA–DNA product intermediate before (+) strand DNA synthesis. During the viral life cycle, the single p66 subunit of HIV RT is responsible for both RNA-templated and DNA-templated DNA synthesis, as well as regulated RNA cleavage by the RNase H domain (Fig. 2A). Thus, during the various stages of proviral DNA synthesis, HIV RT interacts with a menagerie of nucleic acid substrates, including RNA, duplex DNA, and

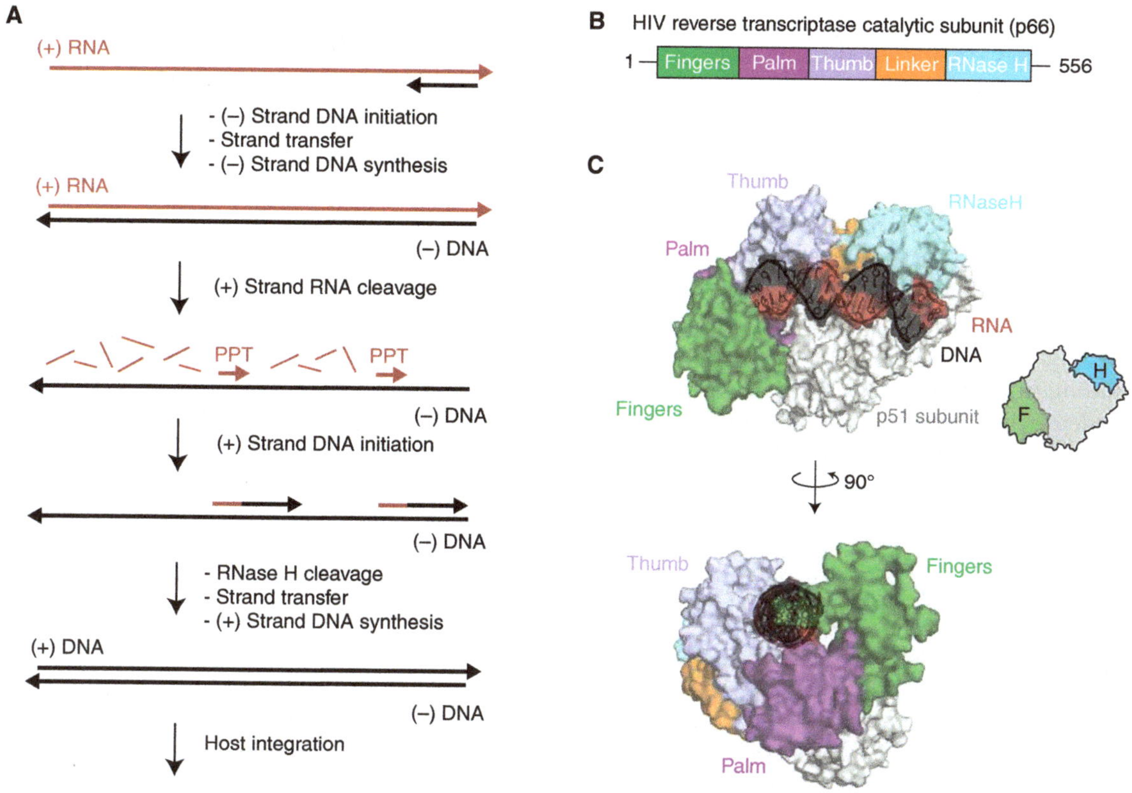

Figure 2. Human immunodeficiency virus reverse transcriptase (HIV RT) catalysis and structure. (*A*) HIV RT catalysis. Following initiation of (−) strand DNA synthesis by the binding of host-cell transfer RNA to the viral HIV RT RNA (not shown), a short DNA primer (black) anneals to the 3′ end of the (+) strand RNA template (red). On completion of (−) strand DNA synthesis by the RT DNA polymerase domain, cleavage of the (+) strand RNA is performed by the RNase H domain to generate a template for (+) strand DNA synthesis. Two short polypurine-rich RNA tracts (PPT) left behind after RNase H cleavage then serve as primers to initiate DNA synthesis by the RT domain. A second round of RNase H cleavage removes the RNA primers on completion of (+) strand DNA synthesis, ultimately generating an integration-competent DNA copy of the viral genome. (*B*) Conserved domain organization of HIV RT p66 subunit. The conserved fingers, palm, and thumb domain are shown in green, purple, and violet, respectively. The HIV-specific linker and RNase H domain are shown in orange and light blue, respectively. (*C*) Structure of HIV RT. The domains of the p66 subunit are colored as described in panel *B*, with the p51 subunit shown in light gray and an ideal RNA/DNA helix shown in red and black as indicated. The HIV RT cartoon has the fingers and RNase H domains highlighted in green and blue, respectively.

RNA–DNA hybrids. In addition, the specific structures engaged by the enzyme can also range from nicked duplexes, recessed ends, blunt ends, and 3′ or 5′ nucleic overhangs. Although biochemical and high-resolution structural studies continue to contribute to our understanding of the molecular recognition determinants between HIV RT and its nucleic acid substrates (Fig. 2C) (Sarafianos et al. 2009, Engelman and Cherepanov 2012), these static snapshots of the enzyme do not illuminate the mechanism for how HIV RT dynamically engages its substrates. To circumvent this limitation, researchers have recently used an elegant smFRET assay to directly monitor HIV RT-binding dynamics in real time (Abbondanzieri et al. 2008, Liu et al. 2008, Liu et al. 2010). In the following sections, we will describe several of these exciting findings, with a particular emphasis on how the use of single-molecule techniques revealed unforeseen properties of this fascinating and biomedically important enzyme.

6 CONFORMATIONAL DYNAMICS OF HIV RT ON NUCLEIC ACIDS

Given the diversity of nucleic acid-binding activities shown by HIV RT, and the a priori expectation for dynamic interconversion between distinct binding modes, the system was a natural candidate for single-molecule studies. Early work analyzed conformational heterogeneity of HIV RT in complex with a model RNA–DNA hybrid using a solution confocal microscope (Rothwell et al. 2003). Although informative, these initial experiments could not resolve time-dependent conformational changes within the RT–nucleic acid complexes. To address this challenge, Abbondanzieri et al. developed a powerful smFRET system with which to monitor real-time binding dynamics of HIV RT (Abbondanzieri et al. 2008). HIV RT enzymes were engineered to possess a single solvent-exposed cysteine residue located proximal to either the DNA polymerase active site or, conversely, the RNase H active site (Fig. 3A). By conjugating a FRET donor dye to one of these sites within HIV RT and placing an acceptor dye on the nucleic acid substrate, the investigators were able to unambiguously determine the binding orientation of the RT with respect to a variety of nucleic acid substrates (Fig. 3B). Initial experiments focused on understanding how the presence of an RNA, DNA, or chimeric primer bound to a DNA template-influenced RT binding (Fig. 3C). These experiments directly showed that HIV RT binds to RNA- and DNA-primed substrates with opposite binding orientations. If the primer is composed of DNA, the 3′ end of the primer is situated in the polymerase active site, poised to be extended by the RT. Conversely, if the primer was RNA, RT bound the RNA–DNA hybrid with the primer 3′ end proximal to the RNase H active site, in a DNA synthesis-incompetent state, and presumably poised for RNA cleavage. Taking the experiment a step further, the investigators next queried which part of the primer was being recognized by RT to direct the enzyme to bind in one orientation or the other. The finding was a surprise; the identity of the primer and, in turn, the binding orientation, was sensed on the primer 5′ end rather than the 3′ end. In other words, a chimeric primer that had nine DNA bases on the 5′ end followed by 10 RNA bases on the 3′ end bound to RT in a DNA synthesis-competent mode, a conclusion that was supported by accompanying biochemical experiments. These experiments provided the first demonstration of distinct substrate-dictated binding orientations of HIV RT, providing new insight into how RT efficiently mediates distinct forms of catalysis on different substrates.

After HIV RT has catalyzed the formation of the RNA–DNA hybrid and subsequently used its RNase H activity to digest away the (+) strand RNA template, a specific polypurine tract (PPT) of the original RNA template remains behind to serve as a primer for (+) strand DNA synthesis (Rausch and Le Grice 2004). In contrast to generic RNA–DNA hybrids, the hybrid formed by the specialized PPT sequence is both refractory to RNase H cleavage and competent to serve as an efficient primer for the DNA polymerase activity. Using the smFRET approach, the investigators next tested a model PPT hybrid for binding by HIV RT. Unlike the other RNA primer substrates, which predominantly bound RT in the DNA synthesis-incompetent orientation, the PPT primer exhibited a mixture of both binding orientations (Fig. 3D). This result explains how this unique RNA–DNA hybrid is specifically recognized to prime (+) strand DNA synthesis. Surprisingly, the investigators also reported that HIV RT can dynamically interconvert between each binding orientation, without dissociating from the template. This result was very unexpected given the intricate network of protein–nucleic acid contacts revealed by high-resolution crystal structures. To probe the mechanism of RT "flipping" on its substrates, the investigators tested the effects of adding cognate or noncognate deoxynucleotide triphosphates (dNTPs), as well as the clinically approved non-nucleoside RT inhibitor (NNRTI), nevirapine (Fig. 3D). Interestingly, the presence of the cognate dNTP altered the binding orientation equilibrium, biasing RT to bind in the DNA polymerization-competent conformation. On the other hand, the presence of the NNRTI, nevirapine, had the opposite effect and biased the binding of RT away from the DNA synthesis-competent orientation. The binding of a cognate dNTP or a NNRTI ligand is known to clamp down or open the polymerase fingers and thumb subdomains, respectively (Kohlstaedt et al. 1992; Spence et al. 1995; El Safadi et al. 2007). These structural impacts of ligand binding are consistent with the smFRET assay dynamic binding results. Thus, the use of smFRET to study the binding of HIV RT to diverse nucleic acid substrates not only revealed distinct binding orientations but also lead to the unexpected discovery of dynamic interconversion of RT between these states.

7 RT SLIDING, SUBSTRATE RECOGNITION, AND STRAND DISPLACEMENT SYNTHESIS

HIV RT must locate the 3′ end of the primer to promote DNA synthesis. This search process represents a considerable challenge during viral replication because RT is known to act distributively (Huber et al. 1989), synthesizing short segments of DNA while replicating the 9.2 kilobase viral genome. Thus, RT must repeatedly search and engage the 3′ end of the nascent DNA strand throughout the replication process. To investigate the mechanism of this search process, Liu et al. again used the smFRET technique to monitor RT-binding dynamics in real time (Liu et al. 2008). In this experiment, the investigators designed a set of model DNA primer-RNA template substrates of increasing length (Fig. 4A). When the DNA–RNA hybrid matched the RT footprint (~19 base pairs [bp]), the RT bound in the DNA synthesis-competent orientation with the 3′ end of the DNA primer proximal to the polymerase active site. However, as the length of the hybrid was increased, RT showed a bimodal binding distribution consistent with the ability to engage each of the ends of the hybrid in a mutually exclusive manner. Moreover, analysis of smFRET

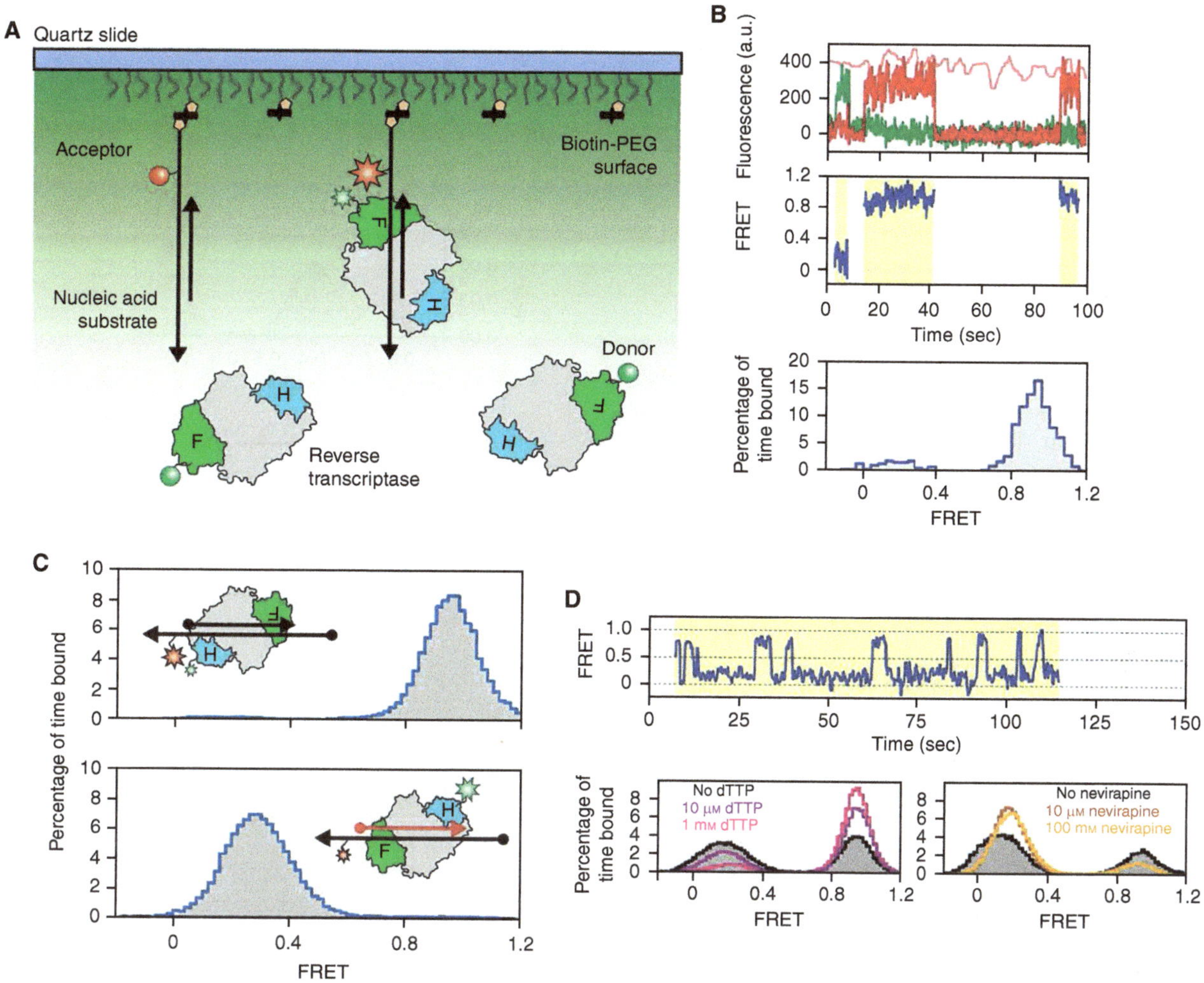

Figure 3. Conformational dynamics of human immunodeficiency virus reverse transcriptase (HIV RT). (*A*) Experimental setup for measuring dynamics of HIV RT binding to nucleic acid substrates. HIV RT is site-specifically labeled with a FRET donor dye (green circle). Biotinylated nucleic acid substrates (black) labeled with acceptor dye (red circle) are tethered to the slide surface, immersed in a solution containing FRET-labeled HIV RT, and imaged as described in Figure 1. (*B*) Individual HIV RT-binding events are detected by the onset of FRET between the donor and acceptor dyes (region of traces highlighted in yellow). Throughout the experiment, alternating excitation with a red laser that directly excites the acceptor dye is used to ensure the presence of an active acceptor dye (direct excitation of acceptor is shown as a light red trace). Data from multiple binding events taken in a particular experimental condition are compiled into a FRET histogram to reconstruct the distribution of FRET states sampled by HIV RT bound to the nucleic acid. (*C*) When the nucleic acid substrate is a DNA–DNA hybrid, HIV RT binds with the fingers domain proximal to the primer 3′ end, in the DNA polymerization-competent orientation, indicated as a high-FRET state when the donor dye is placed on the RNase H domain. In contrast, when the primer is comprised of RNA, the HIV RT binds in the opposite orientation, with the RNase H domain proximal to the RNA 3′ end, indicated by a low-FRET state. (*D*) HIV RT bound to a PPT RNA primer–DNA hybrid in the presence and absence of nucleotide or nonnucleoside RT inhibitors (NNRTI). The top panel shows a representative FRET trace with frequent interconversion between high- and low-FRET states indicating switching between the polymerase-competent and incompetent orientations when HIV RT is bound to the PPT RNA–DNA hybrid. The *bottom left* panel shows the FRET histogram for HIV RT bound to the PPT RNA–DNA hybrid in the presence of increasing concentrations of deoxythymidine triphosphate (dTTP) (purple and pink curves). Addition of the cognate dTTP biases the FRET distribution toward the higher FRET population, indicative of when the enzyme is in the polymerase-competent orientation. The bottom right panel shows the FRET histogram for HIV RT bound to the PPT RNA/DNA hybrid in the presence of increasing concentrations of the NNRTI, nevirapine (orange and yellow curves). Addition of nevirapine biases the FRET distribution toward the lower FRET population, indicative of when the enzyme is in the polymerase-incompetent orientation. (Adapted from Abbondanzieri et al. 2008, with permission, from Springer Nature ©2008.)

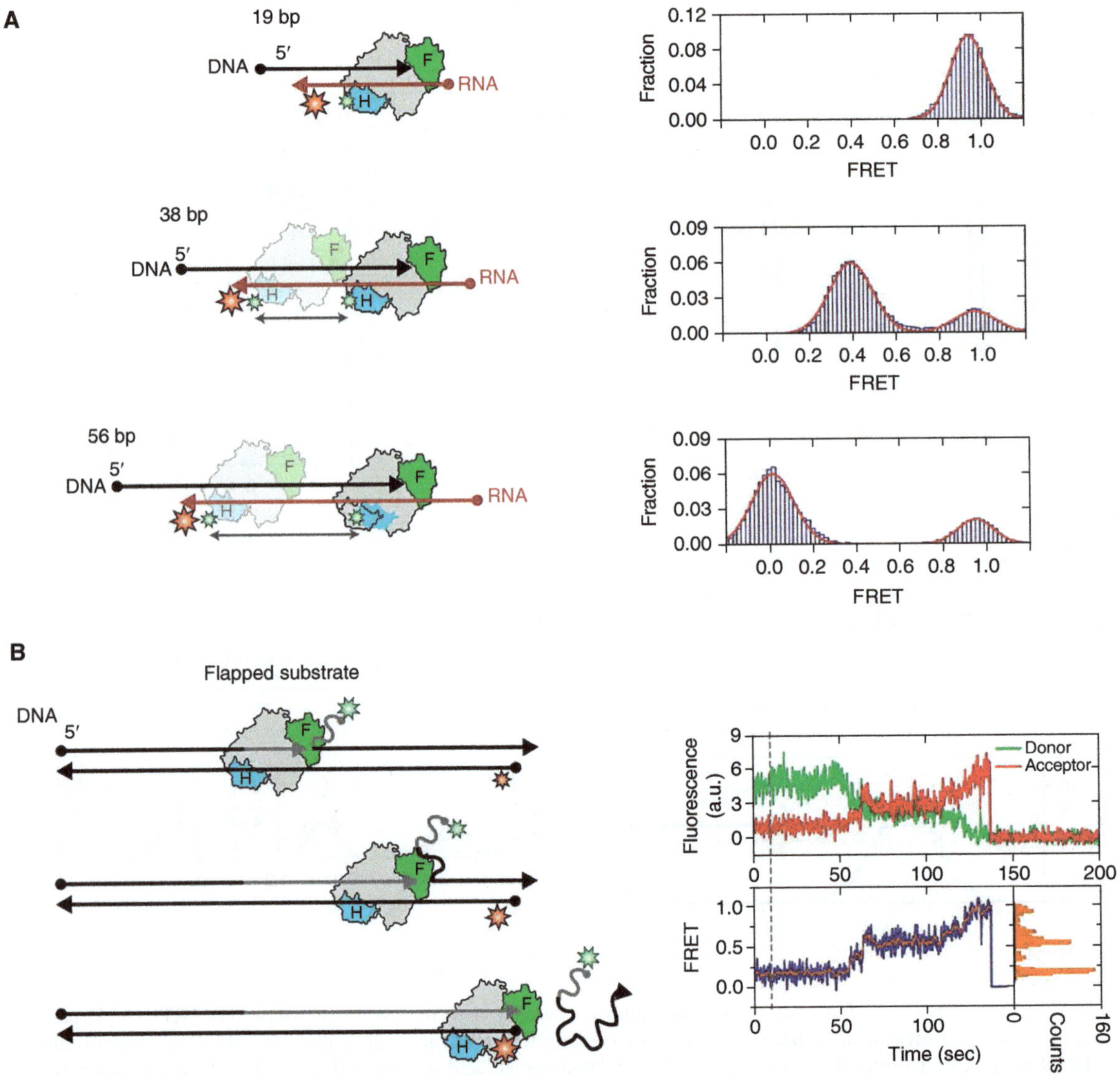

Figure 4. Sliding and strand displacement synthesis by HIV RT. (*A*) HIV RT sliding on DNA–RNA substrates. HIV RT is shown in gray with the fingers and RNase H domains shown in green and blue, respectively. DNA is shown in black and RNA in red. An acceptor dye (red) is attached to the DNA and a donor dye (green) is attached to the enzyme. The *top* cartoon and panel indicate the enzyme orientation when HIV RT is bound to a short 19-bp hybrid, which corresponds to the size of the HIV RT footprint. The high FRET peak indicates that the enzyme is primed at the 3′ end of the DNA substrate in the polymerase-competent orientation. When the hybrid length is changed to 38 bp (*middle* cartoon and panel) two FRET populations are observed, including a high-FRET and medium-FRET peak. This bimodal FRET population is indicative of HIV RT engaging either end of the hybrid, but with a preference for the DNA 3′ end where it is bound in the polymerase-competent orientation. The *bottom* cartoon and panel depict HIV RT bound to a 56-bp DNA–RNA hybrid. The FRET panel on the right reveals a bimodal FRET distribution, with the major FRET peak close to 0, again demonstrating that the enzyme can bind at either end of the DNA–RNA hybrid but preferentially binds in the polymerase-competent state at the 3′ end of the DNA. (*B*) HIV RT strand displacement of flapped DNA. The flapped DNA is labeled with a donor dye (green) on the 5′ end and the template DNA strand is labeled with an acceptor dye (red) on the 5′ end. HIV RT binds to the flapped DNA substrate and catalyzes strand displacement to complete DNA synthesis (gray arrow), resulting in a gradual increase in FRET. A representative FRET trace is shown on the *right* with raw acceptor and donor intensities shown in the *top* panel, calculated FRET ratio shown in the *bottom* panel, and the corresponding FRET histogram of the dynamically sampled FRET states (orange). Peaks in the FRET histogram suggest that HIV RT preferentially pauses at specific stages of the strand displacement synthesis reaction. (Adapted from Liu et al. 2008, with permission, from AAAS.)

time trajectories revealed dynamic "sliding" of the RT along the hybrid, with distinct pausing at preferred sites. To probe what type of structural changes were required to support RT sliding on DNA–RNA hybrids, the investigators again introduced either cognate dNTPs or the NNRTI, nevirapine. The presence of the cognate dNTP, which is known to clamp down the polymerase active site on the 3′ end of the DNA primer, stabilized RT in the DNA synthesis-competent binding mode. In contrast, the addition of nevirapine, which is known to loosen the grip of the polymerase active site for the hybrid, increased the frequency of back sliding along the DNA–RNA hybrid. In a separate experiment, a much longer duplex DNA substrate labeled at its terminus with a FRET acceptor dye was used. Because the DNA substrate was so large, the initial docking of a FRET donor-labeled HIV RT enzyme resulted in a zero FRET value. However, after a characteristic lag time the RT dynamically located the end of the DNA substrate. Strikingly, if the RT initially slid to the end of the DNA in a nonproductive-binding orientation for DNA synthesis, the enzyme dynamically flipped into the DNA synthesis-competent mode. These results provided a dramatic illustration of how a combination of DNA sliding and flipping can enhance the efficiency of the RT primer search process.

The single-stranded genomic vRNA transcript can fold into complex secondary structures. The propensity of RNA to fold into stable hairpins and other structures presents RT with yet another challenge during (−) strand DNA synthesis. As the RT progressively synthesizes the DNA product it must catalyze strand displacement of any structures that are encountered within the RNA template. This conundrum is magnified by the fact that RT frequently dissociates from its nucleic acid substrates, allowing the energetics of DNA and RNA nucleic acids to drive structural reorganization that can result in the production of "flapped" DNA substrates (Fig. 4B). Having characterized the ability of RT to slide on nucleic acid substrates, the investigators next asked whether this sliding activity may play a role in the productive binding and extension of flapped nucleic acids. In a series of elegant experiments, Liu et al. showed that the ability of RT to dynamically slide on its nucleic acid substrates facilitates rearrangement of the DNA–RNA hybrid structure to promote the DNA synthesis-competent state. Taken together, these studies further illuminated the complex and dynamic properties of RT binding to diverse nucleic acids, providing unique insights into how this multifaceted enzyme mediates various stages of the viral life cycle.

8 STRUCTURAL DYNAMICS OF THE HIV RT INITIATION COMPLEX

To initiate viral replication, HIV RT use a host cell–derived transfer RNA (tRNA) as a primer for the synthesis of (−) strand DNA (Marquet et al. 1995). This $tRNA_3^{Lys}$ engages a specific region of the vRNA genomic transcript at a region termed the primer binding site (PBS). Once formed, the complex of tRNA–vRNA is bound by HIV RT to begin DNA synthesis. Once again using their smFRET approach, Liu et al. directly monitored HIV RT-binding dynamics on the physiologically important initiation complex (Liu et al. 2010). Interestingly, the investigators found that HIV RT binds the tRNA–vRNA substrate in predominantly the flipped orientation, wherein the RNase H domain is proximal to the primer 3′ end (Fig. 5). This result is consistent

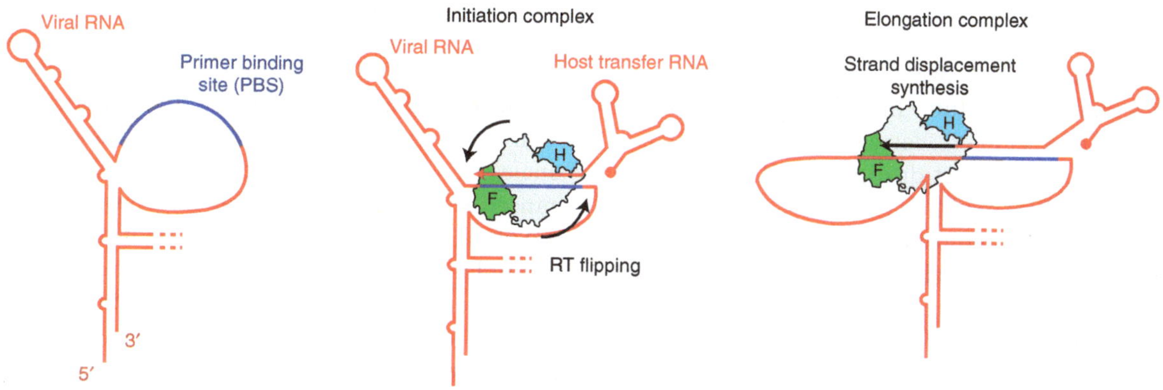

Figure 5. HIV RT initiation complex dynamics. HIV RT viral RNA is shown in red with the primer binding site (PBS) shown in blue (*left*). To form the initiation complex, a host transfer RNA anneals to the PBS and serves as a primer for HIV RT which binds to the hybrid and can dynamically interconvert between the polymerase-competent and incompetent orientations (*middle*). The presence of a stable RNA hairpin structure slows down the initiation process, in part by biasing the HIV RT-binding orientation toward the polymerase-incompetent state. After incorporation of six nascent deoxynucleotide triphosphates (dNTPs), the RNA secondary structure is destabilized, promoting rapid elongation of the product DNA by the HIV RT elongation complex (*right*).

with their previously reported experiments that showed RT binding to random RNA–DNA hybrids supports the same DNA synthesis-incompetent mode (Abbondanzieri et al. 2008). Once bound to the initiation complex, RT faces the energetic challenge of strand displacement synthesis through a stable RNA hairpin structure within the vRNA template. Indeed, this structural roadblock is thought to underlie the observation that HIV RT shows substantial pausing at early stages of the initiation process before transitioning to a higher processivity elongation complex (Isel et al. 1996; Lanchy et al. 1998). To dissect the role of vRNA template structure during the initiation process, the investigators altered the native vRNA template hairpin to either destabilize or stabilize the structure. Strikingly, destabilization of the vRNA template hairpin permitted RT to preferentially bind in the DNA synthesis-competent mode, whereas mutations that enhanced the stability of the vRNA template hairpin had the opposite effect of promoting the flipped binding orientation. Thus, the structural properties of the vRNA template directly modulate the binding orientation of HIV RT and, in turn, the initiation efficiency. To further probe the role of hairpin structure in governing the initiation process, the investigators systematically tested the binding orientation of RT on tRNA–DNA chimeras corresponding to each successive step in the early initiation process. These experiments revealed that once RT adds six DNA bases to the tRNA primer, it can then efficiently bind in the DNA synthesis-competent orientation. Moreover, the HIV RT-binding orientation equilibrium measured on each substrate correlated with the DNA synthesis rate measured in ensemble biochemical experiments. Taken together, these observations lead to the hypothesis that six bases of HIV RT-catalyzed DNA synthesis should be sufficient to disrupt the native vRNA template hairpin structure. To test this notion, the investigators engineered a vRNA template RNA with FRET dyes incorporated at the base of the RNA hairpin structure. In this way, they showed that a tRNA–DNA chimera corresponding to six bases of DNA synthesis did in fact disrupt the vRNA template hairpin structure. Why might HIV have evolved to possess this metastable vRNA template hairpin that apparently reduces the efficiency of viral replication initiation? One possibility put forth by Liu et al. is that the vRNA template hairpin serves the function of slowing down viral replication so that other aspects of viral production, such as translation of envelope proteins, can keep pace. This model is consistent with reports that mutations that destabilize the vRNA template hairpin and increase the rate of initiation can have negative impacts on virion production (Beerens et al. 2001). Once again, the use of smFRET to directly monitor binding dynamics of HIV RT, as well as structural remodeling of its nucleic acid substrates, provided new insights into the mechanism of viral replication initiation.

9 TELOMERASE

The ends of linear chromosomes are capped by repetitive G-rich stretches of DNA that are the foundation for telomere structure (de Lange 2005, 2010). Because of incomplete DNA replication of chromosome termini caused by the end-replication problem, telomere DNA gradually shortens with each cell division (Olovnikov 1971; Watson 1972; Allsopp et al. 1992). Without a mechanism to counteract this shortening, telomeres become critically short and serve as a signal for the cell to undergo senescence or programmed cell death. This senescent state and eventual cell death is a necessary mechanism to prevent unrestricted cell division in somatic cells (Hayflick 1985). In contrast, cells that need to maintain higher levels of proliferative potential, such as stem cells and germ cells, express the specialized telomerase enzyme to maintain telomere length (Blackburn et al. 1989). The same principle holds true for most human cancer cell types, making telomerase an important biomedical drug target (Kim et al. 1994).

Telomerase is a RNP complex composed of a protein subunit, TERT, and an integral long noncoding RNA component, TR (Feng et al. 1995; Nakamura et al. 1997). TERT and TR constitute the telomerase catalytic core, whereas a number of additional species-specific proteins can also associate with the telomerase complex (de Lange 2005). TERTs are generally well conserved and share significant sequence homology with viral RT enzymes (Peng et al. 2001; Belfort et al. 2011). Telomerase is a canonical RT in the sense that it catalyzes a reverse transcription reaction to synthesize telomere DNA from its RNA template. However, telomerase also has additional specialized features that support its ability to synthesize multiple telomere DNA repeats during a single binding event (Greider 1991; Zhao et al. 2011). During catalysis, telomerase docks on to the 3′ single-stranded end of the telomere DNA tail, and reverse transcribes the integral template region within TR onto the end of the chromosome (Fig. 6A). Following the addition of a complete telomeric DNA repeat, telomerase must realign the nascent DNA with downstream template residues to prime a subsequent round of telomere repeat synthesis. The unique property of telomerase to recycle its internal RNA template and synthesize multiple telomere DNA repeats without dissociating from the telomere is termed repeat addition processivity (RAP). The inherently dynamic rearrangements associated with RAP have motivated the use of smFRET to study the telomerase mechanism.

The TERT protein is composed of four conserved domains: the telomerase essential amino-terminal (TEN)

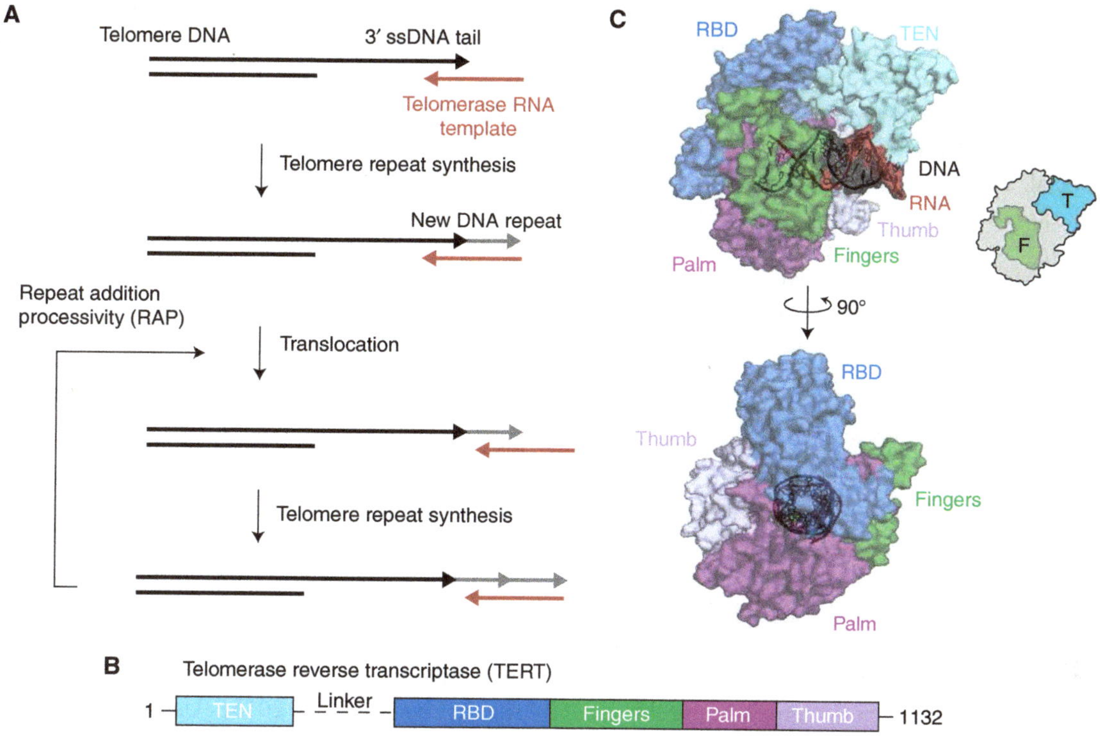

Figure 6. Telomerase catalysis and structural model. (*A*) The integral telomerase RNA (TR) template (red) anneals to the 3′ single-stranded (ss) tail of the telomere DNA (black). The RT domain of telomerase catalyzes a new telomere repeat (gray) followed by a translocation step to realign the nascent DNA repeat with the 3′ end of the RNA template. The reiterative process of telomere repeat synthesis followed by translocation is termed repeat addition processivity (RAP). (*B*) Conserved domain organization of telomerase reverse transcriptases (TERTs). The canonical fingers, palm, and thumb domain are shown in green, purple, and violet, respectively as represented for HIV RT in Figure 2. The telomerase-specific telomerase essential amino-terminal (TEN) domain and RNA-binding domain (RBD) are shown in light blue and dark blue, respectively. The telomerase linker region in between the TEN and RBD domains is indicated by a dotted line, which can be of varying length depending on the organism. (*C*) The structure shown is a homology model of human telomerase (Steczkiewicz et al. 2011) with the human TERT sequence threaded into the previously solved *Tribolium castaneum* structure of TERT bound to an ideal RNA–DNA hybrid (Mitchell et al. 2010) in complex with the *Tetrahymena thermophila* TEN domain (Jacobs et al. 2006). The domains of TERT are colored as described in *B*, with the ideal RNA–DNA helix shown in red and black as indicated. The telomerase cartoon has the fingers and TEN domains highlighted in green and blue, respectively.

domain, the RNA-binding domain (RBD), the RT domain, and the carboxy-terminal extension (CTE) (Fig. 6B). The TEN domain is critical for maintaining RAP but is not required during the addition of a single telomere repeat (Zaug et al. 2008; Robart and Collins 2011). The RBD is responsible for establishing RNA–protein interactions between TERT and TR and is absolutely essential for telomerase activity (Lai et al. 2001; Huang et al. 2014; Jansson et al. 2015). The RT and CTE domains are homologous to the palm and thumb domains of canonical polymerases, respectively (Fig. 6C) (Wyatt et al. 2007; Gillis et al. 2008). The RNA component of telomerase is also composed of several conserved structural domains, although the sequence and length of TR varies widely between species (Theimer and Feigon 2006). Two regions of TR that appear to be generally conserved are the pseudoknot (PK)/template domain and the stem terminal element (STE) (Theimer et al. 2005), both of which are required to support wild-type (WT) levels of telomerase function in vitro and in vivo (Theimer et al. 2005; Richards et al. 2006; Kim et al. 2008; Cash et al. 2013; Huang et al. 2014). Although the functional role of the RNA template is self-evident, the contributions of the TR PK and STE remain the subject of continued investigation.

Single-molecule techniques are particularly well suited to help tease apart the dynamics that occur during telomerase catalysis and have the added benefit of requiring a relatively small amount of sample material, a feature that is particularly relevant in telomerase research given the challenges associated with producing large amounts of homogeneous enzyme for biochemical and biophysical experiments. As in the case of HIV RT research, smFRET has so far emerged as the method of choice for studying the assembly and structural dynamics of telomerase at the single-molecule level (Wu et al. 2010; Berman et al. 2011; Hengesbach et al. 2012; Parks and Stone 2014; Akiyama et al. 2015; Parks et al. 2017; Shastry et al. 2018). In the following sections, we highlight selected examples in which smFRET has been used to study the structure and function of *Tetrahymena thermophila* and human telomerase.

10 STRUCTURAL DYNAMICS OF *Tetrahymena thermophila* TELOMERASE

Telomerase was initially discovered in the ciliated protozoan *Tetrahymena thermophila* and has since been extensively biochemically characterized. (Greider and Blackburn 1987). In the course of this work, several regions within *Tetrahymena* TR were shown to be critical for RAP (Greider and Blackburn 1989, Gilley and Blackburn 1999; Lai et al. 2002; Richards et al. 2006), including the template recognition element (TRE) downstream from the template and the template boundary element (TBE) just upstream of the template (Fig. 7A). In a recent study, Berman and Akiyama et al. designed an incisive experiment to show compression and extension of the TRE and TBE regions at different stages of telomerase activity (Berman et al. 2011). The investigators labeled either the TRE or the TBE with an acceptor dye and measured FRET to a donor dye incorporated into the DNA primer. In this way, individual binding events of telomerase from solution to surface-immobilized DNA primers could be detected by FRET in real time (Figs. 7B and C). DNA primers corresponding to different repeat extension lengths were used to measure the FRET distribution during different stages of catalysis. The experimental results showed that when the TBE is labeled with the acceptor, FRET dramatically decreases as a function of DNA-primer length, indicating that this region is being extended during catalysis (Fig. 7D). Conversely, when the acceptor dye is on the TRE, FRET was found to slightly increase as a function of DNA-primer length, indicating that the single-stranded region between the template and the TRE is being compressed during the addition of a repeat. Together, these data suggested that there are both compression and extension of the single-stranded RNA regions surrounding the template during the synthesis of each telomere repeat, leading the investigators to propose "the accordion model" for telomerase function (Fig. 7D). In subsequent experiments, the investigators showed that RNA backbone connectivity is required between the template and both upstream and downstream single-stranded regions of TER, indicating that the compaction of the TRE and the expansion of the single-stranded region between the TBE and the template must be coupled during catalysis. In more recent work, a high-resolution crystal structure of *Tetrahymena* TERT–RBD bound to the TBE RNA fragment was reported (Jansson et al. 2015). This structure was consistent with the notion that accumulating tension in the flexible RNA regions flanking the template serves to enforce proper template boundary definition in *Tetrahymena* telomerase. An interesting open question is whether a similar mechanism for template boundary definition exists in other telomerase systems.

Another focus of intense research during the past few years has been on understanding how nucleic acid dynamics are coupled with and regulated by TERT interactions. Specifically, the telomerase TEN domain has been implicated in making both RNA and DNA interactions during catalysis and is absolutely essential for RAP (O'Connor et al. 2005; Jacobs et al. 2006; Zaug et al. 2008; Robart and Collins 2011). An interesting study by Akiyama et al. used smFRET to probe the role of the *Tetrahymena* TEN domain in stabilizing RNA–DNA hybrids in the telomerase active site (Akiyama et al. 2015). Using a similar experimental strategy to that described earlier, the binding of dye-labeled telomerase from solution to a surface-immobilized DNA primer was monitored by smFRET. The aim of the study was to investigate how the DNA primer–binding dynamics varied as a function of template RNA–DNA primer hybrid length (Fig. 8). When using a DNA primer sequence that could only form a 5-bp hybrid with the RNA template, the investigators observed a predominantly stable high-FRET state indicative of the primer being docked in the telomerase active site (Fig. 8A). However, the investigators also noted that on occasion, a transient excursion to a much lower FRET state was observed, which they named an "alternative conformation," in which telomerase is stably bound to the DNA but the 3′ end of the primer is not in the "docked conformation." As expected, when a DNA- primer sequence with a greater degree of complementarity to the RNA template was used (9 bp), only a single FRET state was observed corresponding to the docked conformation (Fig. 8B). These experiments with WT telomerase on DNA primers of varying lengths showed the power of smFRET to reveal internal structural dynamics of nucleic acids bound to the telomerase enzyme.

Next, the investigators set out to investigate the impact of a previously reported separation of function mutation (L14A) in the TERT–TEN domain that was shown to significantly diminish RAP without impacting DNA binding (Zaug et al. 2008). When telomerase harboring the L14A TEN mutation bound to the DNA primer that formed a 5 bp hybrid, the investigators discovered a dramatic increase in the dynamics of the 3′ end of the DNA (Fig. 8A). In contrast, when the mutant TERT was bound to the primer

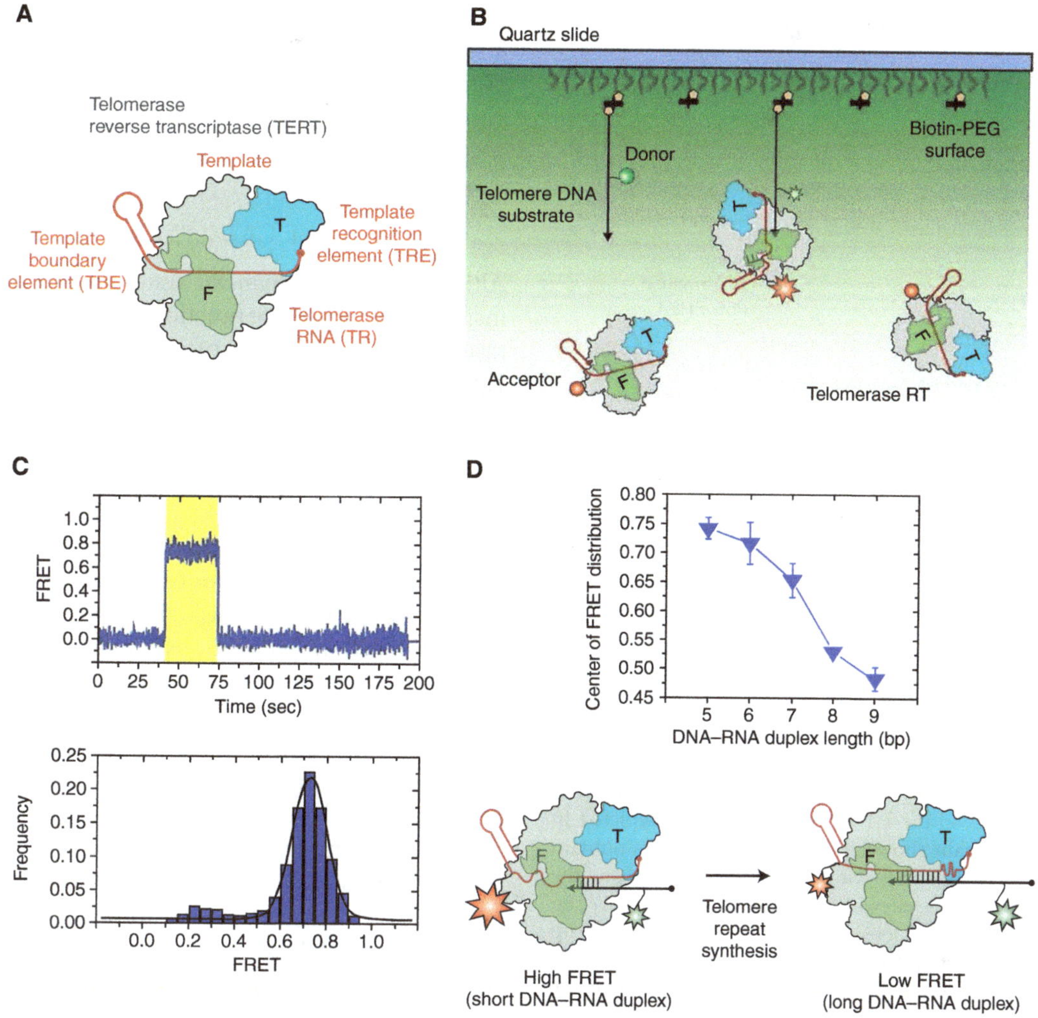

Figure 7. Structural dynamics within telomerase RNA during catalysis. (*A*) Telomerase ribonucleoprotein (RNP) composition. TERT is depicted in gray with the TEN domain shown in blue and the fingers domain shown in green. A highly oversimplified depiction of telomerase RNA is shown in red with the template, template recognition element (TRE), and template boundary element (TBE) indicated. (*B*) Experimental setup to detect binding of individual telomerase complexes to surface-immobilized biotinylated telomere DNA substrates. Telomere DNA primers are immobilized to a pegylated quartz surface via a biotin–streptavidin linkage. Streptavidin is shown as black rectangles and biotin is depicted as yellow pentagons. The DNA (black) is labeled with a donor dye (green circle) and the telomerase RNA subunit (red) is labeled with an acceptor dye (red circle). (*C*) Individual TERT-binding events are detected by the onset of FRET between the donor and acceptor dyes (region of trace highlighted in yellow). A representative smFRET trace showing the dynamic association of a telomerase enzyme from solution with a surface-immobilized telomere DNA primer (*top* panel). Many individual binding events are collected and compiled into a FRET distribution characterizing a particular binding interaction (*bottom* panel). (*D*) Observed FRET value decreases as the DNA–RNA hybrid is elongated, indicative of the RNA in the template and TBE being extended (*top* panel). In the *bottom* panel, the RNA accordion model is shown, which describes the reciprocal compression and extension of the single-stranded regions of telomerase RNA flanking the template observed during the telomere repeat synthesis reaction. (Adapted, with permission, from Berman et al. 2011.)

capable of forming the longer 9-bp hybrid, the FRET distribution matched the WT behavior. Taken together, these results suggested that one function of the TEN domain is to stabilize short DNA–RNA hybrids in the telomerase active site, which represents an obligatory intermediate during each successive round of telomere repeat addition. To test this prediction, the investigators turned to ensemble DNA-primer extension assays and showed that WT telomerase was able to extend short DNA–RNA hybrids, whereas the L14A mutant telomerase was not. At the moment, it remains unclear whether the TEN domain stabilizes short DNA–RNA hybrids in the telomerase active site via direct

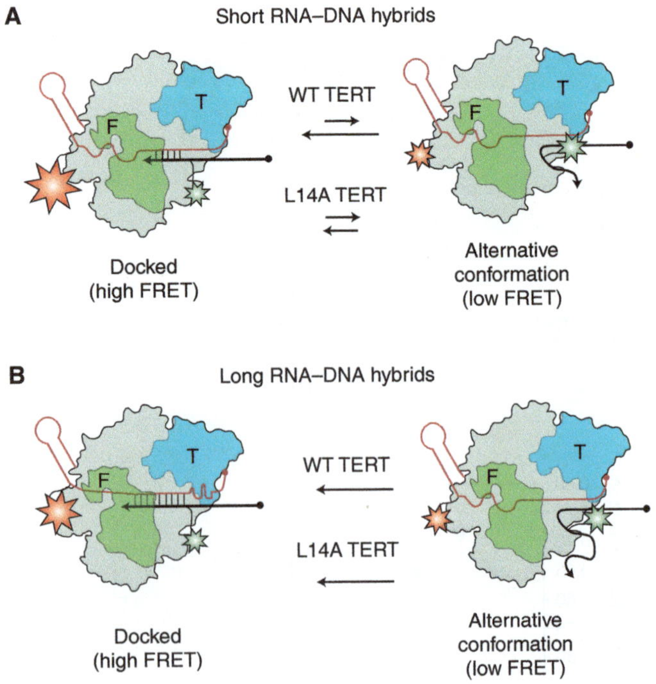

Figure 8. The telomerase essential amino-terminal (TEN) domain stabilizes short RNA–DNA hybrids. (*A*) Single-molecule experiments with wild-type (WT) *Tetrahymena* telomerase revealed DNA-binding dynamics in the RT active site. TERT is depicted in gray with the TEN domain shown in blue and the fingers domain shown in green. TR (red) is labeled with an acceptor dye (red) and the DNA primer (black) is labeled with a donor dye (green). When telomerase was bound to a DNA primer that can form a 5-bp RNA–DNA hybrid in the active site, telomerase predominantly bound with the 3′ end of the DNA positioned in the RT active site (high FRET). In addition, the DNA occasionally sampled an alternative conformation (low FRET) in which the enzyme remains bound but the DNA is not in the active site. When the same experiment was performed using telomerase harboring a mutation in the TEN domain (L14A) that disrupts RAP, the DNA was found to sample the alternative state with significantly higher frequency. (*B*) When a telomere DNA primer capable of making a 9-bp RNA–DNA hybrid was used, both the WT and mutant telomerases showed stable binding of the 3′ end of the DNA in the RT active site. Taken together, these results suggested that the TEN domain is required to stabilize the binding of a short RNA–DNA hybrid in the RT active site at the start of each round of telomere DNA repeat synthesis.

interactions or if the effect is mediated through additional protein contacts. Interestingly, smFRET was recently used to study the binding dynamics of a yeast TEN domain to different nucleic acid substrates (Shastry et al. 2018). In this system, it was found that the yeast TEN domain binds preferentially to DNA–RNA hybrids, consistent with the model that emerged from the earlier *Tetrahymena* experiments (Akiyama et al. 2015). In the case of the human TEN domain, biochemical data also suggest a similar role in mediating DNA–RNA hybrid-binding dynamics (Moriarty et al. 2004; Wyatt et al. 2007; Wu and Collins 2014), but further biophysical investigation is required to dissect the precise contributions of the TEN domain to telomerase function in these diverse systems.

11 NUCLEIC ACID DYNAMICS DURING HUMAN TELOMERASE CATALYSIS

Although the *Tetrahymena* system has provided valuable insight into telomerase function, studies of human telomerase are understandably of greater biomedical relevance. However, because of the larger size of human telomerase, in particular the RNA component (451 nucleotides compared with 159 in *Tetrahymena)*, human telomerase is more difficult to reconstitute into a functional enzyme in vitro. To circumvent this problem, a modified reconstitution protocol using only the two critical human telomerase RNA (hTR) domains required for catalysis, the PK/template and STE domains, was established (Chen and Greider 2003). This approach simplifies the assembly of human telomerase while still maintaining WT levels of telomerase activity, and also facilitates reconstitution of human telomerase with dye-labeled RNA, an advance that has led to several exciting studies focused on nucleic acid dynamics during human telomerase catalysis.

In one example, Parks and Stone used a series of strategically designed smFRET experiments to investigate DNA conformation and dynamics during the human telomerase catalytic cycle (Parks and Stone 2014). To study DNA dynamics, the investigators examined FRET distributions at different stages of telomerase catalysis using a stationary donor dye on the RNA and an acceptor dye on the human telomere primer $(TTAGGG)_3$ (Fig. 9A,B). This specific primer sequence was shown previously to support exceedingly long-lived telomerase-binding interactions (Wallweber et al. 2003), permitting the investigators to surface-immobilize prebound "stalled" telomerase–DNA complexes. To measure DNA dynamics at different stages of telomere repeat synthesis, a mixture of specific dNTPs and dideoxynucleotide triphosphates (ddNTPs) were added to surface-immobilized telomerase enzymes. The combination of various dNTPs and ddNTPs allowed the enzyme to extend to the desired extension product and then stall because of incorporation of a chain-terminating ddNTP. When telomerase was immobilized in the absence of dNTPs, a unimodal and stable FRET population was observed (Fig. 9B). Strikingly, after the addition of TTA to the primer an unexpected bimodal FRET population was observed (Fig. 9B). This distribution consisted of the previously observed high-FRET peak, corresponding to the DNA primer bound in the active site, as well as a new lower FRET population. Finally, incorporation of the sequence TTAG completed the synthe-

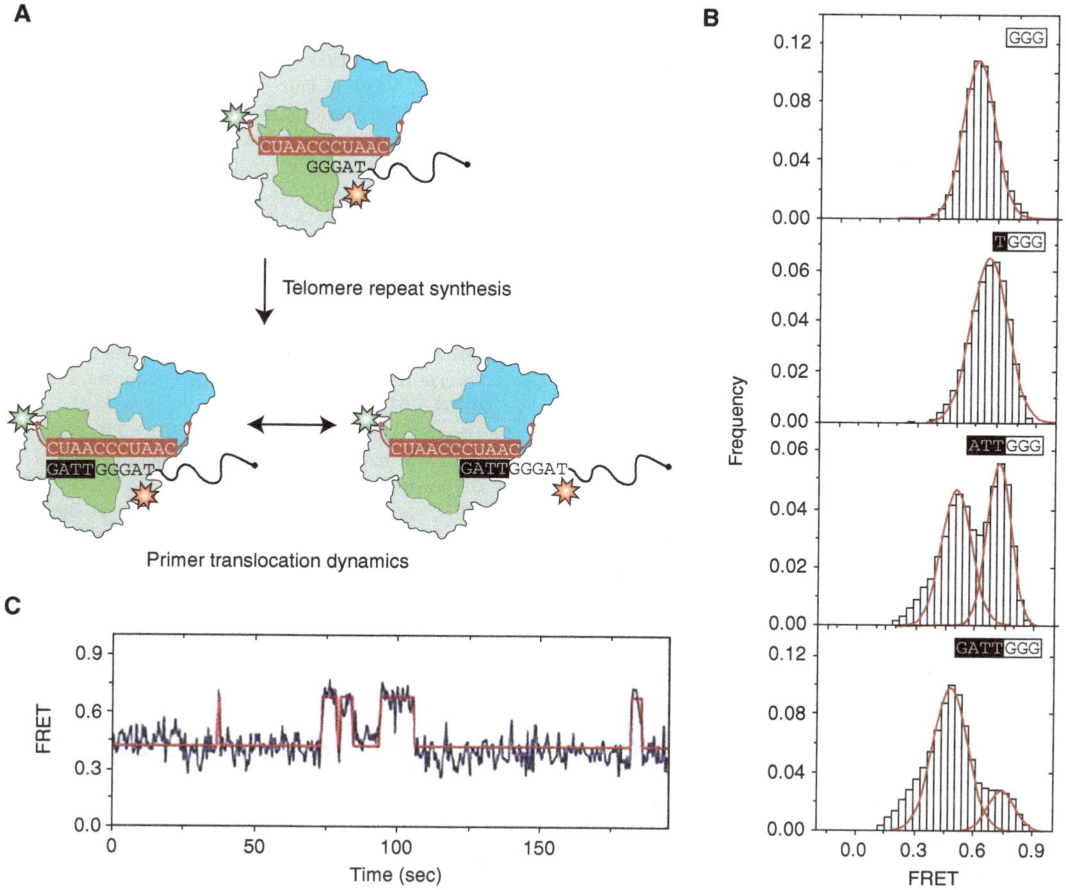

Figure 9. DNA dynamics during human telomerase catalysis. (*A*) Human telomerase was reconstituted and labeled with a donor dye upstream of the RNA template. Enzyme was prebound to a dye-labeled telomere DNA primer ending in the sequence -GGG, which binds with very high affinity. Enzyme–DNA complexes were then surface-immobilized and imaged in a TIRF microscope. TERT is shown in gray with the fingers and TEN domains shown in green and blue, respectively. The DNA primer (black) is labeled with an acceptor dye (red) and the RNA (red) is labeled with a donor dye (green). After addition of dNTPs and completion of a telomere repeat, the newly synthesized repeat adopts two distinct binding conformations, either annealed to the RNA template in the active site (*left*, high FRET) or realigned to the 5′ end of the template, poised for another round of telomere repeat addition (*right*, low FRET). (*B*) When complexes bound to the -GGG primer are imaged and data compiled into a FRET histogram, a single high-FRET population is observed (*top* panel). After addition of a single dideoxythymidine triphosphate (ddTTP) to the nascent telomere repeat, a single FRET peak is still observed (second panel from the *top*). In contrast, on addition of dTTP and dideoxyadenosine triphosphate (ddATP) to the growing DNA primer, a bimodal FRET distribution is observed, with a new lower FRET population emerging (second panel from the *bottom*). After addition of dTTP, deoxyadenosine triphosphate (dATP) and dideoxyguanosine triphosphate (ddGTP), a full repeat has been synthesized, and the major FRET population observed is the low-FRET peak corresponding to the telomere repositioned at the 3′ end of the template, primed for another round of telomere synthesis. (*C*) A representative FRET trace of a DNA primer that has been extended to complete a single repeat. The dynamic transitions observed in the trajectory represent the 3′ end of the DNA primer sampling different complementary registers of the TR template. The data were fit with a hidden Markov model (HMM) yielding idealized FRET trajectories that were used to analyze the kinetics of the DNA realignment process (red). (Adapted, with permission, from Parks and Stone 2014.)

sis of a telomere DNA repeat, which further biased the FRET distribution toward the new lower FRET state (Fig. 9B).

A series of carefully designed control experiments lead the investigators to interpret the lower FRET population that arises at the end of the repeat synthesis reaction to represent the dynamic repositioning of the primer 3′ end to the 3′ side of the RNA template, thus priming for another round of repeat addition (Fig. 9A). This realignment substep is often drawn as a necessary intermediate during telomerase RAP; however, the single-molecule studies provided the first direct observation of a telomere DNA primer sampling different binding registers of the RNA template

(Fig. 9C). Interestingly, kinetic analysis of individual FRET time traces revealed that the rate constant for realignment of the 3′ end of the nascent telomere DNA product was ∼100-fold faster than estimates for the rate constant governing the complete translocation process required for RAP (Latrick and Cech 2010). This result is interesting because many in the telomerase field anticipated that dissociation and realignment of the nascent DNA product would impose a large energetic barrier to RAP. However, the smFRET study suggested that some other step during translocation required for RAP, perhaps short DNA–RNA hybrid capture and or TERT protein conformational changes in the telomerase active site, are in fact rate limiting.

12 CONCLUSIONS AND PERSPECTIVES

In each of the cases described within this review, smFRET provided a new window into the complex binding properties between the RT and its nucleic acid substrates. Results from these studies show certain functional similarities between the viral and telomerase RTs, in that each must orchestrate distinct binding nucleic acid-binding modes depending on what stage of catalysis the enzyme is mediating. The data have been particularly illuminating in the case of HIV RT, in large part attributable to the extensive high-resolution structural data that exist. This point highlights the strong complementarity of single-molecule science with other structural and biochemical methods. As improved structural models for telomerase accumulate, we anticipate smFRET will continue to serve as a means to test specific models for how protein and RNA structural changes mediate telomerase function. Additionally, although both TERT and HIV RT have been well studied using smFRET, other related RTs such as the group II intron encoded protein (IEP) as well as the spliceosomal protein Prp8 are less well characterized in terms of their nucleic acid–handling properties. The recent discovery of the structures of several group II intron IEPs (Qu et al. 2016; Zhao and Pyle 2016) as well as key components of the spliceosome (Galej et al. 2013; Plaschka et al. 2017; Wilkinson et al. 2017) should provide the necessary framework required to design effective smFRET experiments to study these related RTs and their nucleic acid substrates. Indeed, as new and improved biophysical technologies emerge, researchers will become ever better prepared to probe deeper into the functional properties of RT enzymes, no doubt uncovering many unexpected discoveries along the way.

REFERENCES

Abbondanzieri EA, Bokinsky G, Rausch JW, Zhang JX, Le Grice SF, Zhuang X. 2008. Dynamic binding orientations direct activity of HIV reverse transcriptase. *Nature* **453:** 184–189.

Akiyama BM, Parks JW, Stone MD. 2015. The telomerase essential N-terminal domain promotes DNA synthesis by stabilizing short RNA-DNA hybrids. *Nucleic Acids Res* **43:** 5537–5549.

Allsopp RC, Vaziri H, Patterson C, Goldstein S, Younglai EV, Futcher AB, Greider CW, Harley CB. 1992. Telomere length predicts replicative capacity of human fibroblasts. *Proc Natl Acad Sci* **89:** 10114–10118.

Axelrod D. 2003. Total internal reflection fluorescence microscopy in cell biology. *Methods Enzymol* **361:** 1–33.

Baltimore D. 1970. RNA-dependent DNA polymerase in virions of RNA tumour viruses. *Nature* **226:** 1209–1211.

Beerens N, Groot F, Berkhout B. 2001. Initiation of HIV-1 reverse transcription is regulated by a primer activation signal. *J Biol Chem* **276:** 31247–31256.

Belfort M, Curcio MJ, Lue NF. 2011. Telomerase and retrotransposons: Reverse transcriptases that shaped genomes. *Proc Natl Acad Sci* **108:** 20304–20310.

Berman AJ, Akiyama BM, Stone MD, Cech TR. 2011. The RNA accordion model for template positioning by telomerase RNA during telomeric DNA synthesis. *Nat Struct Mol Biol* **18:** 1371–1375.

Best RB, Merchant KA, Gopich IV, Schuler B, Bax A, Eaton WA. 2007. Effect of flexibility and *cis* residues in single-molecule FRET studies of polyproline. *Proc Natl Acad Sci* **104:** 18964–18969.

Blackburn EH, Greider CW, Henderson E, Lee MS, Shampay J, Shippen-Lentz D. 1989. Recognition and elongation of telomeres by telomerase. *Genome* **31:** 553–560.

Cash DD, Cohen-Zontag O, Kim NK, Shefer K, Brown Y, Ulyanov NB, Tzfati Y, Feigon J. 2013. Pyrimidine motif triple helix in the *Kluyveromyces lactis* telomerase RNA pseudoknot is essential for function in vivo. *Proc Natl Acad Sci* **110:** 10970–10975.

Cech TR. 1985. Self-splicing RNA: Implications for evolution. *Int Rev Cytol* **93:** 3–22.

Chen JL, Greider CW. 2003. Template boundary definition in mammalian telomerase. *Genes Dev* **17:** 2747–2752.

Crick F. 1970. Central dogma of molecular biology. *Nature* **227:** 561–563.

Curcio MJ, Belfort M. 2007. The beginning of the end: Links between ancient retroelements and modern telomerases. *Proc Natl Acad Sci* **104:** 9107–9108.

de Lange T. 2005. Shelterin: The protein complex that shapes and safeguards human telomeres. *Genes Dev* **19:** 2100–2110.

de Lange T. 2010. How shelterin solves the telomere end-protection problem. *Cold Spring Harb Symp Quant Biol* **75:** 167–177.

Deniz AA, Dahan M, Grunwell JR, Ha T, Faulhaber AE, Chemla DS, Weiss S, Schultz PG. 1999. Single-pair fluorescence resonance energy transfer on freely diffusing molecules: Observation of Forster distance dependence and subpopulations. *Proc Natl Acad Sci* **96:** 3670–3675.

Deniz AA, Mukhopadhyay S, Lemke EA. 2008. Single-molecule biophysics: At the interface of biology, physics and chemistry. *J R Soc Interface* **5:** 15–45.

El Safadi Y, Vivet-Boudou V, Marquet R. 2007. HIV-1 reverse transcriptase inhibitors. *Appl Microbiol Biotechnol* **75:** 723–737.

Engelman A, Cherepanov P. 2012. The structural biology of HIV-1: Mechanistic and therapeutic insights. *Nat Rev Microbiol* **10:** 279–290.

Feng J, Funk WD, Wang SS, Weinrich SL, Avilion AA, Chiu CP, Adams RR, Chang E, Allsopp RC, Yu J, et al. 1995. The RNA component of human telomerase. *Science* **269:** 1236–1241.

Galej WP, Oubridge C, Newman AJ, Nagai K. 2013. Crystal structure of Prp8 reveals active site cavity of the spliceosome. *Nature* **493:** 638–643.

Gilley D, Blackburn EH. 1999. The telomerase RNA pseudoknot is critical for the stable assembly of a catalytically active ribonucleoprotein. *Proc Natl Acad Sci* **96:** 6621–6625.

Gillis AJ, Schuller AP, Skordalakes E. 2008. Structure of the *Tribolium castaneum* telomerase catalytic subunit TERT. *Nature* **455:** 633–637.

Gladyshev EA, Arkhipova IR. 2011. A widespread class of reverse transcriptase-related cellular genes. *Proc Natl Acad Sci* **108:** 20311–20316.

Greenleaf WJ, Woodside MT, Block SM. 2007. High-resolution, single-molecule measurements of biomolecular motion. *Annu Rev Biophys Biomol Struct* **36:** 171–190.

Greider CW. 1991. Telomerase is processive. *Mol Cell Biol* **11:** 4572–4580.

Greider CW, Blackburn EH. 1987. The telomere terminal transferase of *Tetrahymena* is a ribonucleoprotein enzyme with two kinds of primer specificity. *Cell* **51:** 887–898.

Greider CW, Blackburn EH. 1989. A telomeric sequence in the RNA of *Tetrahymena* telomerase required for telomere repeat synthesis. *Nature* **337:** 331–337.

Ha T. 2001. Single-molecule fluorescence resonance energy transfer. *Methods* **25:** 78–86.

Ha T, Enderle T, Ogletree DF, Chemla DS, Selvin PR, Weiss S. 1996. Probing the interaction between two single molecules: Fluorescence resonance energy transfer between a single donor and a single acceptor. *Proc Natl Acad Sci* **93:** 6264–6268.

Hayflick L. 1985. The cell biology of aging. *Clin Geriatr Med* **1:** 15–27.

Hengesbach M, Kim NK, Feigon J, Stone MD. 2012. Single-molecule FRET reveals the folding dynamics of the human telomerase RNA pseudoknot domain. *Angew Chem Int Ed Engl* **51:** 5876–5879.

Huang J, Brown AF, Wu J, Xue J, Bley CJ, Rand DP, Wu L, Zhang R, Chen JJ, Lei M. 2014. Structural basis for protein-RNA recognition in telomerase. *Nat Struct Mol Biol* **21:** 507–512.

Huber HE, McCoy JM, Seehra JS, Richardson CC. 1989. Human immunodeficiency virus 1 reverse transcriptase. Template binding, processivity, strand displacement synthesis, and template switching. *J Biol Chem* **264:** 4669–4678.

Isel C, Lanchy JM, Le Grice SF, Ehresmann C, Ehresmann B, Marquet R. 1996. Specific initiation and switch to elongation of human immunodeficiency virus type 1 reverse transcription require the post-transcriptional modifications of primer tRNA3Lys. *EMBO J* **15:** 917–924.

Jacobs SA, Podell ER, Cech TR. 2006. Crystal structure of the essential N-terminal domain of telomerase reverse transcriptase. *Nat Struct Mol Biol* **13:** 218–225.

Jager M, Michalet X, Weiss S. 2005. Protein–protein interactions as a tool for site-specific labeling of proteins. *Protein Sci* **14:** 2059–2068.

Jager M, Nir E, Weiss S. 2006. Site-specific labeling of proteins for single-molecule FRET by combining chemical and enzymatic modification. *Protein Sci* **15:** 640–646.

Jansson LI, Akiyama BM, Ooms A, Lu C, Rubin SM, Stone MD. 2015. Structural basis of template-boundary definition in *Tetrahymena* telomerase. *Nat Struct Mol Biol* **22:** 883–888.

Kim NK, Zhang Q, Zhou J, Theimer CA, Peterson RD, Feigon J. 2008. Solution structure and dynamics of the wild-type pseudoknot of human telomerase RNA. *J Mol Biol* **384:** 1249–1261.

Kim NW, Piatyszek MA, Prowse KR, Harley CB, West MD, Ho PL, Coviello GM, Wright WE, Weinrich SL, Shay JW. 1994. Specific association of human telomerase activity with immortal cells and cancer. *Science* **266:** 2011–2015.

Kohlstaedt LA, Wang J, Friedman JM, Rice PA, Steitz TA. 1992. Crystal structure at 3.5 Å resolution of HIV-1 reverse transcriptase complexed with an inhibitor. *Science* **256:** 1783–1790.

Lai CK, Mitchell JR, Collins K. 2001. RNA binding domain of telomerase reverse transcriptase. *Mol Cell Biol* **21:** 990–1000.

Lai CK, Miller MC, Collins K. 2002. Template boundary definition in *Tetrahymena* telomerase. *Genes Dev* **16:** 415–420.

Lanchy JM, Keith G, Le Grice SF, Ehresmann B, Ehresmann C, Marquet R. 1998. Contacts between reverse transcriptase and the primer strand govern the transition from initiation to elongation of HIV-1 reverse transcription. *J Biol Chem* **273:** 24425–24432.

Latrick CM, Cech TR. 2010. POT1-TPP1 enhances telomerase processivity by slowing primer dissociation and aiding translocation. *EMBO J* **29:** 924–933.

Lingner J, Hughes TR, Shevchenko A, Mann M, Lundblad V, Cech TR. 1997. Reverse transcriptase motifs in the catalytic subunit of telomerase. *Science* **276:** 561–567.

Liu S, Abbondanzieri EA, Rausch JW, Le Grice SF, Zhuang X. 2008. Slide into action: Dynamic shuttling of HIV reverse transcriptase on nucleic acid substrates. *Science* **322:** 1092–1097.

Liu S, Harada BT, Miller JT, Le Grice SF, Zhuang X. 2010. Initiation complex dynamics direct the transitions between distinct phases of early HIV reverse transcription. *Nat Struct Mol Biol* **17:** 1453–1460.

Marquet R, Isel C, Ehresmann C, Ehresmann B. 1995. tRNAs as primer of reverse transcriptases. *Biochimie* **77:** 113–124.

Merchant KA, Best RB, Louis JM, Gopich IV, Eaton WA. 2007. Characterizing the unfolded states of proteins using single-molecule FRET spectroscopy and molecular simulations. *Proc Natl Acad Sci* **104:** 1528–1533.

Mitchell M, Gillis A, Futahashi M, Fujiwara H, Skordalakes E. 2010. Structural basis for telomerase catalytic subunit TERT binding to RNA template and telomeric DNA. *Nat Struct Mol Biol* **17:** 513–518.

Moffitt JR, Chemla YR, Smith SB, Bustamante C. 2008. Recent advances in optical tweezers. *Annu Rev Biochem* **77:** 205–228.

Moriarty TJ, Marie-Egyptienne DT, Autexier C. 2004. Functional organization of repeat addition processivity and DNA synthesis determinants in the human telomerase multimer. *Mol Cell Biol* **24:** 3720–3733.

Murphy MC, Rasnik I, Cheng W, Lohman TM, Ha T. 2004. Probing single-stranded DNA conformational flexibility using fluorescence spectroscopy. *Biophys J* **86:** 2530–2537.

Nakamura TM, Cech TR. 1998. Reversing time: Origin of telomerase. *Cell* **92:** 587–590.

Nakamura TM, Morin GB, Chapman KB, Weinrich SL, Andrews WH, Lingner J, Harley CB, Cech TR. 1997. Telomerase catalytic subunit homologs from fission yeast and human. *Science* **277:** 955–959.

Neuman KC, Nagy A. 2008. Single-molecule force spectroscopy: Optical tweezers, magnetic tweezers and atomic force microscopy. *Nat Methods* **5:** 491–505.

O'Connor CM, Lai CK, Collins K. 2005. Two purified domains of telomerase reverse transcriptase reconstitute sequence-specific interactions with RNA. *J Biol Chem* **280:** 17533–17539.

Olovnikov AM. 1971. [Principle of marginotomy in template synthesis of polynucleotides]. *Dokl Akad Nauk SSSR* **201:** 1496–1499.

Parks JW, Stone MD. 2014. Coordinated DNA dynamics during the human telomerase catalytic cycle. *Nat Commun* **5:** 4146.

Parks JW, Kappel K, Das R, Stone MD. 2017. Single-molecule FRET-Rosetta reveals RNA structural rearrangements during human telomerase catalysis. *RNA* **23:** 175–188.

Peng Y, Mian IS, Lue NF. 2001. Analysis of telomerase processivity: Mechanistic similarity to HIV-1 reverse transcriptase and role in telomere maintenance. *Mol Cell* **7:** 1201–1211.

Perez CE, Gonzalez RL Jr, 2011. In vitro and in vivo single-molecule fluorescence imaging of ribosome-catalyzed protein synthesis. *Curr Opin Chem Biol* **15:** 853–863.

Plaschka C, Lin PC, Nagai K. 2017. Structure of a pre-catalytic spliceosome. *Nature* **546:** 617–621.

Qu G, Kaushal PS, Wang J, Shigematsu H, Piazza CL, Agrawal RK, Belfort M, Wang HW. 2016. Structure of a group II intron in complex with its reverse transcriptase. *Nat Struct Mole Biol* **23:** 549–557.

Rasnik I, Myong S, Cheng W, Lohman TM, Ha T. 2004. DNA-binding orientation and domain conformation of the *E. coli* rep helicase monomer bound to a partial duplex junction: Single-molecule studies of fluorescently labeled enzymes. *J Mol Biol* **336:** 395–408.

Rausch JW, Le Grice SF. 2004. 'Binding, bending and bonding': Polypurine tract-primed initiation of plus-strand DNA synthesis in human immunodeficiency virus. *Internatl J Biochem Cell Biol* **36:** 1752–1766.

Richards RJ, Wu H, Trantirek L, O'Connor CM, Collins K, Feigon J. 2006. Structural study of elements of *Tetrahymena* telomerase RNA stem-loop IV domain important for function. *RNA* **12:** 1475–1485.

Robart AR, Collins K. 2011. Human telomerase domain interactions capture DNA for TEN domain-dependent processive elongation. *Mol Cell* **42:** 308–318.

Rothwell PJ, Berger S, Kensch O, Felekyan S, Antonik M, Wohrl BM, Restle T, Goody RS, Seidel CA. 2003. Multiparameter single-molecule fluorescence spectroscopy reveals heterogeneity of HIV-1 reverse transcriptase:primer/template complexes. *Proc Natl Acad Sci* **100:** 1655–1660.

Roy R, Hohng S, Ha T. 2008. A practical guide to single-molecule FRET. *Nat Methods* **5:** 507–516.

Sarafianos SG, Marchand B, Das K, Himmel DM, Parniak MA, Hughes SH, Arnold E. 2009. Structure and function of HIV-1 reverse transcriptase: Molecular mechanisms of polymerization and inhibition. *J Mol Biol* **385:** 693–713.

Schuler B, Eaton WA. 2008. Protein folding studied by single-molecule FRET. *Curr Opin Struct Biol* **18:** 16–26.

Shastry S, Steinberg-Neifach O, Lue N, Stone MD. 2018. Direct observation of nucleic acid binding dynamics by the telomerase essential N-terminal domain. *Nucleic Acids Res* **46:** 3088–3102.

Spence RA, Kati WM, Anderson KS, Johnson KA. 1995. Mechanism of inhibition of HIV-1 reverse transcriptase by nonnucleoside inhibitors. *Science* **267:** 988–993.

Steczkiewicz K, Zimmermann MT, Kurcinski M, Lewis BA, Dobbs D, Kloczkowski A, Jernigan RL, Kolinski A, Ginalski K. 2011. Human telomerase model shows the role of the TEN domain in advancing the double helix for the next polymerization step. *Proc Natl Acad Sci* **108:** 9443–9448.

Stryer L, Haugland RP. 1967. Energy transfer: A spectroscopic ruler. *Proc Natl Acad Sci* **58:** 719–726.

Tan E, Wilson TJ, Nahas MK, Clegg RM, Lilley DM, Ha T. 2003. A four-way junction accelerates hairpin ribozyme folding via a discrete intermediate. *Proc Natl Acad Sci* **100:** 9308–9313.

Telesnitsky A, Goff SP. 1997. Reverse transcriptase and the generation of retroviral DNA. In *Retroviruses* (ed. Coffin JM, Hughes SH, Varmus HE). Cold Spring Harbor Laboratory Press, Cold Spring Harbor, NY.

Temin HM, Mizutani S. 1970. RNA-dependent DNA polymerase in virions of Rous sarcoma virus. *Nature* **226:** 1211–1213.

Theimer CA, Feigon J. 2006. Structure and function of telomerase RNA. *Curr Opin Struct Biol* **16:** 307–318.

Theimer CA, Blois CA, Feigon J. 2005. Structure of the human telomerase RNA pseudoknot reveals conserved tertiary interactions essential for function. *Mol Cell* **17:** 671–682.

Tinoco I Jr, Gonzalez RL Jr, 2011. Biological mechanisms, one molecule at a time. *Genes Dev* **25:** 1205–1231.

Wallweber G, Gryaznov S, Pongracz K, Pruzan R. 2003. Interaction of human telomerase with its primer substrate. *Biochemistry* **42:** 589–600.

Watson JD. 1972. Origin of concatemeric T7 DNA. *Nat New Biol* **239:** 197–201.

Weiss S. 1999. Fluorescence spectroscopy of single biomolecules. *Science* **283:** 1676–1683.

Wen JD, Lancaster L, Hodges C, Zeri AC, Yoshimura SH, Noller HF, Bustamante C, Tinoco I. 2008. Following translation by single ribosomes one codon at a time. *Nature* **452:** 598–603.

Wilkinson ME, Fica SM, Galej WP, Norman CM, Newman AJ, Nagai K. 2017. Postcatalytic spliceosome structure reveals mechanism of 3′-splice site selection. *Science* **358:** 1283–1288.

Wu RA, Collins K. 2014. Human telomerase specialization for repeat synthesis by unique handling of primer-template duplex. *EMBO J* **33:** 921–935.

Wu JY, Stone MD, Zhuang X. 2010. A single-molecule assay for telomerase structure–function analysis. *Nucleic Acids Res* **38:** e16.

Wyatt HD, Lobb DA, Beattie TL. 2007. Characterization of physical and functional anchor site interactions in human telomerase. *Mol Cell Biol* **27:** 3226–3240.

Xiong Y, Eickbush TH. 1990. Origin and evolution of retroelements based upon their reverse transcriptase sequences. *EMBO J* **9:** 3353–3362.

Zaug AJ, Podell ER, Cech TR. 2008. Mutation in TERT separates processivity from anchor-site function. *Nat Struct Mol Biol* **15:** 870–872.

Zhao C, Pyle AM. 2016. Crystal structures of a group II intron maturase reveal a missing link in spliceosome evolution. *Nat Struct Mol Biol* **23:** 558–565.

Zhao Y, Abreu E, Kim J, Stadler G, Eskiocak U, Terns MP, Terns RM, Shay JW, Wright WE. 2011. Processive and distributive extension of human telomeres by telomerase under homeostatic and nonequilibrium conditions. *Mol Cell* **42:** 297–307.

Zhuang X, Bartley LE, Babcock HP, Russell R, Ha T, Herschlag D, Chu S. 2000. A single-molecule study of RNA catalysis and folding. *Science* **288:** 2048–2051.

Zhuang X, Kim H, Pereira MJ, Babcock HP, Walter NG, Chu S. 2002. Correlating structural dynamics and function in single ribozyme molecules. *Science* **296:** 1473–1476.

Protocells and RNA Self-Replication

Gerald F. Joyce[1] and Jack W. Szostak[2]

[1]The Salk Institute for Biological Studies, La Jolla, California 92037

[2]Howard Hughes Medical Institute and Department of Molecular Biology, Massachusetts General Hospital, Boston, Massachusetts 02114

Correspondence: gjoyce@salk.edu; szostak@molbio.mgh.harvard.edu

SUMMARY

The general notion of an "RNA world" is that, in the early development of life on the Earth, genetic continuity was assured by the replication of RNA, and RNA molecules were the chief agents of catalytic function. Assuming that all of the components of RNA were available in some prebiotic locale, these components could have assembled into activated nucleotides that condensed to form RNA polymers, setting the stage for the chemical replication of polynucleotides through RNA-templated RNA polymerization. If a sufficient diversity of RNAs could be copied with reasonable rate and fidelity, then Darwinian evolution would begin with RNAs that facilitated their own reproduction enjoying a selective advantage. The concept of a "protocell" refers to a compartment where replication of the primitive genetic material took place and where primitive catalysts gave rise to products that accumulated locally for the benefit of the replicating cellular entity. Replication of both the protocell and its encapsulated genetic material would have enabled natural selection to operate based on the differential fitness of competing cellular entities, ultimately giving rise to modern cellular life.

Outline

Cite as *Cold Spring Harb Perspect Biol* doi: 10.1101/cshperspect.a034801

1 INTRODUCTION

The first cells must have been simple in comparison with modern highly evolved cells and therefore are often referred to as "protocells." Protocells must have been simple enough to self-assemble spontaneously in a chemically rich environment under appropriate physical conditions but sufficiently complex that they were poised to evolve to greater complexity, ultimately giving rise to all of modern biology (Szostak et al. 2001). What kind of cell structure and composition would have been consistent with the dual requirements of original simplicity and future complexity?

Given that all modern cells are bounded by a semipermeable cell membrane, it is reasonable to postulate that protocells were also membrane-bound structures, albeit with very different membranes than modern cells (Deamer 1997). Similarly, all modern cells contain diverse RNA molecules that encode information and fulfill many biochemical functions. The central role of RNA in modern cellular biochemistry, and especially the role of the large ribosomal RNA as the catalyst of protein synthesis (Steitz and Moore 2003), provides a powerful argument in support of RNA as the dominant genetic and functional biopolymer at an early stage in the evolution of life. Whether RNA, a close relative of RNA, or some much simpler genetic material formed the original basis of protocell heredity remains controversial. However, recent advances in prebiotic chemistry suggest that RNA itself may have emerged directly from the chemistry of the young Earth and may therefore have been the genetic polymer of the first protocells (Sutherland 2016).

Many aspects of modern biology are universal, including the translation apparatus and instructed protein synthesis, enzyme-catalyzed metabolism, and protein-mediated membrane transport. However, all of these complex processes are the result of extensive evolution and therefore could not have been present in protocells. In contrast, short noncoded peptides and a wide range of other products of prebiotic chemistry likely were present and may have played important roles in the complex environment required to nurture the assembly, growth, and division of protocells.

Protocells may be viewed as the products of a series of self-assembly processes that took place in specific early Earth environments that provided the necessary chemical building blocks and sources of energy (Fig. 1). The classical bilayer membrane structure results from the spontaneous self-assembly of separate amphiphilic molecules into a lower energy aggregate state. The self-assembly of RNA is more complex, but given suitably activated ribonucleotides, RNA polymers can form spontaneously on mineral surfaces

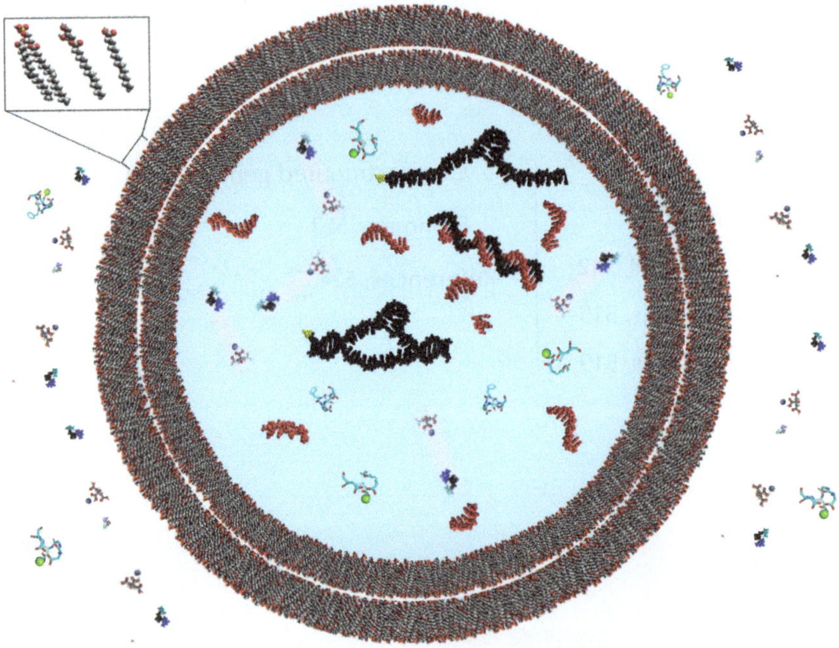

Figure 1. Schematic diagram of a protocell. Protocells bounded by multiple bilayer membranes composed of simple amphiphilic molecules would have been permeable to nucleotides and metal ions complexed with citrate or other ligands. Larger genomic and functional RNA molecules would have been trapped within the protocell interior. Protocell replication would involve growth and division of the membrane boundary as well as replication of the genomic RNA.

Cite as *Cold Spring Harb Perspect Biol* doi: 10.1101/cshperspect.a034801

(Ferris et al. 1996) or simply upon freezing because of the concentration of ribonucleotides in the eutectic phase between ice crystals (Kanavarioti et al. 2001).

As membrane sheets form, the high energy of the membrane edge drives closure into vesicles, which will trap molecules such as RNA that are present in the surrounding solution. Seen in this way, the assembly of protocells appears as a straightforward and perhaps inevitable process, given the right environment. The real problem is not protocell formation, but protocell replication, including growth and division of the protocell membrane and replication of the encapsulated genetic material. Furthermore, these processes must be mutually compatible, and the sources of energy required to drive protocell growth and division must also be compatible with the integrity of the protocell components. The nature of these replication processes and the way that protocell replication led to the evolution of functional RNAs and thus to the complexity of the RNA world is the subject of this review.

2 COMPARTMENTALIZED SYSTEMS

What evolutionary advantages does compartmentalization provide compared with the replication of informational molecules in solution? Simple processes of natural selection can occur in solution. Even without replication, chemical and physical factors will control and alter the sequences that are generated and persist (Chen and Nowak 2012). In an environment that allows for nonenzymatic replication, sequences that are more efficiently replicated will have a selective advantage. Repeated cycles of replication would therefore be expected to shape the structure of a population by influencing base composition, template length, secondary structure, and other properties. But what about the evolution of functional properties such as molecular recognition and catalysis? Numerous in vitro selection experiments have shown that, even without compartmentalization, RNAs can be obtained that bind a ligand or catalyze a self-modification reaction.

In nature, selective pressures might favor binding to mineral surfaces, for example, to prevent RNAs from being washed out of a favorable local environment. However, for more complex functions, such as RNA-catalyzed replicative or metabolic reactions, some form of compartmentalization becomes essential. Consider an RNA that catalyzes formation of a metabolite that contributes to the replication of that RNA, for example, an activated nucleotide. In solution, the products of that reaction will diffuse away from the ribozyme that generated them, thus failing to benefit the RNA itself and perhaps benefiting other neighboring RNAs. A similar argument can be made with respect to an RNA replicase, which in free solution might replicate other surrounding RNAs, thus failing to benefit itself. Compartmentalization enables the positive feedback required for Darwinian evolution by keeping useful products close to the catalysts that generated them and by keeping molecules related by descent physically closer, on average, compared with more distantly related molecules.

Compartmentalization is also essential to prevent system crashes caused by the evolution of parasites. The activity of a ribozyme RNA replicase, or indeed any ribozyme, relies on its folded structure, but the formation of a stable folded structure is at odds with the properties of an ideal template, which is unfolded and readily available to be copied. In free solution, parasitic RNAs that are better templates will increase at the expense of functional replicase RNAs, leading to a population crash. In a population of small compartments, however, the emergence of a parasitic sequence in one compartment will not affect the population as a whole, even if it dooms the descendants of that compartment. Experimentally, even transient compartmentalization is sufficient to prevent extinction caused by parasites (Bansho et al. 2016; Matsumura et al. 2016). These considerations imply that compartmentalization is an essential step on the path to biology. But of the many ways that compartmentalization can be achieved, which is most relevant to the origin of life?

2.1 Pros and Cons of Membrane Systems

The formation of compartments through the assembly of bilayer membranes has a major conceptual advantage over other types of compartments, namely, its continuity with known biology. If the first protocells were membrane-bound compartments that could grow and divide, together with the replication of their encapsulated genetic material, then there could be a continuous line of descent from the first protocells to all subsequent cellular biology. Other forms of compartmentalization would have had to undergo a transition to membrane-based compartments. Such a transition would have become increasingly difficult as the system of replicating and functional RNAs became more highly adapted to any prior form of compartmentalization. For example, compartments that are more open to the environment might enable the evolution of ribozymes that use large, polar, or charged substrates that would become inaccessible during a transition to membrane encapsulation.

There are, however, several conceptual challenges with a membrane-based protocell structure. Chief among these is that bilayer membranes are a barrier to free access to the substrates needed for genome replication, most notably charged nucleotides. Modern cell membranes, composed of phospholipids, sterols, and other complex lipids, are nearly complete barriers to the permeation of nucleotides and even divalent cations such as Mg^{2+} (Deamer 1997). Because primitive cells would have lacked the sophisticated transport machinery that controls molecular trafficking in modern cells, protocells must have been bounded by simpler, more permeable membranes. Membranes formed from single-chain amphiphiles, such as fatty acids and their glycerol esters as well as fatty alcohols and related compounds, tend to be much more permeable than membranes composed of two-chain lipids. Shorter alkyl chains also contribute to increased permeability but require a higher critical concentration for bilayer membrane stability.

The prebiotic synthesis of membrane-forming amphiphilic molecules is poorly studied. Highly unsaturated carbon chains are abundant in interstellar molecular clouds (Duley and Hu 2009), and these molecules may be among the source materials for the heterogeneous collections of amphiphiles present in organic-rich meteorites. Lipids have been extracted from carbonaceous chondrites, such as Murchison, and have been shown to assemble into vesicle-like compartments (Deamer 1985). Whether linear fatty acids could be synthesized by robust chemical pathways on the early Earth is less clear. The best studied pathways involve Fischer–Tropsch-type chemistry, in which carbon monoxide and hydrogen react under high temperature and pressure on the surface of transition metal minerals to generate terminally oxygenated, saturated linear chains (Rushdi and Simoneit 2001). However, the relevance of this chemistry to early Earth conditions is debated. Such reactions tend to give chain length distributions that fall off exponentially with increasing length, suggesting that the resulting membranes would be dominated by short-chain fatty acids and alcohols, but stabilized by a small fraction of long-chain species (Budin et al. 2014).

Protocell membranes must be compatible with the physical and chemical conditions necessary for RNA chemistry. Because fatty acid–based membranes, nonenzymatic RNA copying, and ribozyme activity all tend to be optimal at moderate salt concentrations and neutral to mildly alkaline pH, this requirement might not seem to be a major concern. However, nonenzymatic copying of RNA templates requires very high concentrations of divalent cations, typically Mg^{2+}, and most ribozymes require high concentrations of Mg^{2+} for optimal activity. In contrast, fatty acid–based membranes become unstable in ≥4 mM Mg^{2+}, indicating an apparent incompatibility between the two systems.

One potential resolution of this conflict involves chelation of Mg^{2+} by citrate, which protects membranes from the disruptive effects of free Mg^{2+} while allowing nonenzymatic RNA copying chemistry to proceed with only a moderate reduction of activity (Adamala and Szostak 2013a). As an added benefit, the Mg^{2+}-catalyzed degradation of RNA is prevented by citrate. However, it is unlikely that sufficient citrate was present in the prebiotic environment, and thus alternatives must be sought. A short acidic peptide in the active site of cellular RNA polymerase binds Mg^{2+} and positions it for catalysis (Zhang et al. 1999; Cramer et al. 2001). Could acidic peptides synthesized prebiotically have played a similar role in early RNA copying? The condensation of acidic amino acids is catalyzed by positively charged mineral surfaces (Ferris et al. 1996), suggesting that such peptides might have been available on the early Earth. However, catalysis of templated RNA polymerization by acidic peptides has not been shown. Alternatively, RNA copying strategies that do not require a high concentration of divalent cations would circumvent the problem.

2.2 Alternatives to Membrane Systems

A fundamental requirement for the evolution of complex adaptive traits is to restrict the free diffusion of genomic and functional polymers and their metabolic products so that mutations that lead to improved activity have the opportunity to benefit the mutant more than nearby wild-type individuals. In addition, compartmentalization is essential to prevent a fatal increase in parasitic sequences. Although these requirements could be met by a system of growing and dividing vesicles, other mechanisms of spatial isolation could, in principle, provide a solution. Such alternatives include porous rocks, particulate mineral surfaces, noncovalent assemblies of oligonucleotides, emulsion droplets, and phase-separated droplets, including coacervates.

The porous rocks seen in hydrothermal vents have long been proposed as a possible site for the origin of life (Lane and Martin 2012). As the heated vent fluids cool, dissolved minerals form precipitates and build up a network of interconnected microscale channels. Transverse temperature gradients across such channels have been proposed to support a combination of thermophoresis and convection, resulting in very strong concentration gradients (Baaske et al. 2007).

Laboratory experiments using glass capillaries as analogs of porous rock channels have shown the simultaneous concentration of dilute fatty acids and DNA, leading to the assembly of vesicles with encapsulated DNA near the bottom of the capillary tube (Budin et al. 2009). Subsequent experiments have shown that temperature gradients across narrow channels can concentrate oligonucleotides, facilitating the assembly of longer DNA strands (Mast et al. 2013). Concentration as a result of thermophoresis can also preferentially enrich longer products relative to shorter ones, providing a means to overcome the replication bias that favors shorter templates (Kreysing et al. 2015). Because thermophoresis is strongly inhibited by high salt concen-

 Cite as *Cold Spring Harb Perspect Biol* doi: 10.1101/cshperspect.a034801

trations, this concentration mechanism would only be operative in freshwater hydrothermal vents, such as those currently found in Lake Yellowstone (Inskeep et al. 2015). How nucleic acids that evolved within rock compartments would transition to membrane compartments is not clear. Nevertheless, the remarkable concentration effects achieved by thermophoresis deserve further study.

Another scenario for spatial isolation involves the assembly and replication of RNA on mineral surfaces. The ability of montmorillonite clay to catalyze the polymerization of activated mononucleotides is well known (Ferris et al. 1996; Ferris 2002). The resulting RNAs are tightly bound to the mineral surface, suggesting that mineral particles might function like compartments, with each particle bearing a spatially isolated population of RNA molecules. If RNAs could replicate while remaining at least partially bound to the mineral surface, the system could in principle support the Darwinian evolution of functional RNAs. Although some early work supported the idea that RNA templates can be copied while adsorbed onto a mineral surface (Gibbs et al. 1980; Schwartz and Orgel 1985), further work is needed to confirm the template-directed synthesis of mineral-immobilized RNA.

Compartments based on liquid–liquid phase separation, including coacervate droplets, are a third alternative to membrane vesicles. The strong electrostatic interactions between polyanionic nucleic acids and various polycations leads to the formation of complex coacervates, which, depending on their composition and the ambient conditions, may remain as liquid droplets. The contents of these droplets are highly concentrated relative to the bulk solution, and there is little barrier to the uptake of nutrients from the environment into the coacervate droplets. However, phase-separated droplets are subject to Ostwald ripening, leading to droplet growth, and to fusion events, leading to coalescence. Fusion events can be blocked by a surfactant coating, but this would eliminate the presumed advantage of enhanced access to Mg^{2+} and other charged species (Tang et al. 2014; Aumiller et al. 2016). Furthermore, in some coacervates, the RNA components exchange rapidly between droplets, so there may not be the degree of isolation required for Darwinian behavior (Jia et al. 2014). Further study is needed to determine whether phase-separated droplets can act as isolated compartments and whether RNA replication can be performed in aqueous two-phase systems.

Emulsion droplets provide a fourth alternative to membrane vesicles. The local accumulation of hydrocarbon oils on the early Earth is not implausible. Given sufficient turbulent mixing and the presence of stabilizing surfactants, water-in-oil emulsion droplets would be likely to form. Such droplets could grow through fusion or Ostwald ripening, followed by shear-induced division, leading to a cycle of growth and division. Although attractive as a simple model of protocell growth and division, the limited accessibility of aqueous emulsion droplets to sources of nutrients would be a major problem. However, even if emulsion droplets are not directly relevant to origin of life scenarios, they provide an interesting model system for the laboratory study of evolutionary phenomena (Ichihashi et al. 2013).

Finally, nucleic acids might provide the basis for their own compartmentalization. The growing field of DNA (and RNA) origami provides many examples of compartments assembled from nucleic acids (for reviews, see Pinheiro et al. 2011; Hong et al. 2017). An early example was the assembly of an octahedral structure from a defined single-stranded DNA (Shih et al. 2004). In principle, genomic strands that replicate within such a compartment could encode the compartment itself. There are examples of DNA origami that undergo self-replication based on growth and division processes analogous to those of membrane systems (Schulman et al. 2012). Thus far, however, the properties of encapsulation and replication have not been combined within a single system.

2.3 Pathways for Vesicle Division

Early work by Luisi and colleagues explored processes that could potentially lead to the growth and division of micelles and vesicles. By encapsulating ferritin particles in preformed 1-palmitoyl-2-oleoyl-*sn*-glycerol-3-phosphatidylcholine (POPC) vesicles, electron cryomicroscopy evidence was obtained suggesting that the addition of oleate to POPC vesicles led to vesicle growth and subsequent division (Berclaz et al. 2001). Experiments using fluorescent markers of membrane area and vesicle content confirmed that oleate addition to preformed vesicles led to growth, but division was not observed unless the vesicles were forced through small pores (Hanczyc et al. 2003). However, the addition of excess alkaline oleate micelles to preformed oleate vesicles, buffered at a lower pH, led to an increase in vesicle number. Strikingly, the size of the newly formed vesicles was close to that of the original preformed vesicles (Rasi et al. 2004).

More recently, vesicle growth has been observed directly by video microscopy, following the addition of oleate to larger multilamellar vesicles ~4 µm in diameter (Zhu and Szostak 2009). Surprisingly, vesicle growth proceeded through the extrusion of a thin filament from the outermost bilayer. Over time, this filament grew in length, and equilibration with inner bilayers resulted in a filamentous multilamellar vesicle under conditions in which surface area growth is much faster than volume growth owing to osmotic constraints. Filamentous vesicles are subject to a pearling instability that likely contributes to the ease with which they are fragmented under mild shear stress. Gentle agitation is sufficient to bring about vesicle division without loss of the vesicle contents.

Alternatively, in an environment rich in organic thiols, photochemical oxidative processes that lead to disulfide formation can induce extreme pearling in such vesicles, resulting in spontaneous division into a large number of small daughter vesicles (Zhu et al. 2012). In contrast to the behavior of multilamellar vesicles, large unilamellar vesicles are fragile and tend to rupture with extensive loss of contents under shear stress. These features combine to make multilamellar vesicles and filamentous growth attractive as a means of simple, environmentally driven cycles of growth and division, requiring only episodic delivery of additional amphiphiles and a moderately turbulent environment.

Competition between vesicles for limiting membrane components could provide a simple scenario for protocell growth that obviates the need for periodic input of new amphiphilic material. If some vesicles in a population are able to grow by obtaining membrane components from surrounding vesicles, many generations of growth and division could occur within a given locale. Furthermore, the resulting competition for resources could drive natural selection and thus Darwinian evolution.

The first experimental demonstration of competition between protocells arose from studies of osmotically driven growth (Chen et al. 2004). Vesicles with a high concentration of encapsulated RNA are osmotically swollen and the membrane is under tension. This high-energy state can be relaxed by the incorporation of additional fatty acid molecules into the membrane. Because single-chain amphiphiles exchange rapidly between vesicles, fatty acid vesicles containing RNA grow in the presence of empty (relaxed) fatty acid vesicles. This simple physical mechanism could lead to a coupling between RNA replication and membrane growth and to competition between protocells. However, it is energetically expensive to cause osmotically swollen vesicles to divide, making it difficult to couple competitive growth to a faster overall cell cycle.

Subsequent studies revealed that a small fraction of phospholipid in a largely fatty acid membrane can also lead to competitive growth (Budin and Szostak 2011). In this case, competition results from the slower dissociation of fatty acid molecules from membranes that contain phospholipids, which results in growth of phospholipid-containing vesicles at the expense of vesicles that do not contain (or contain less) phospholipid. Thus, any vesicle that also contained a heritable catalyst of phospholipid synthesis would experience a strong selective advantage. Other molecules that have a similar effect in slowing fatty acid dissociation, such as hydrophobic peptides, also confer a competitive growth advantage, implying that a heritable RNA that catalyzes the synthesis of hydrophobic peptides could also provide a selective advantage through competition at the cellular level (Adamala and Szostak 2013b).

An alternative to protocell growth through molecular accretion is stepwise growth through vesicle fusion. Cycles of fusion and division could allow replicating RNAs to spread throughout a population of vesicles, with strong selection for replication efficiency. Under this scenario, selection for increased efficiency of RNA-catalyzed RNA replication would come first, followed by the evolution of additional ribozymes because of the competition between protocells. Wet/dry cycles transform vesicles into lamellar stacks that upon rehydration again form vesicles. Whether wet/dry cycles could lead to the spread of replicating RNAs through a vesicle population remains unclear. Further studies of the fate of encapsulated RNAs following wet/dry cycles may shed light on the potential role of such processes in early RNA evolution.

3 NATURE OF THE FIRST GENETIC MATERIAL

3.1 RNA as the First Genetic Material

The prebiotic synthesis of nucleotides could have occurred in several ways. The simplest, conceptually, would be to synthesize a nucleoside base, couple it to ribose, and finally to phosphorylate the resulting nucleoside. Recent studies, however, suggest that other routes may be more feasible, most notably those that involve the coassembly of the base and sugar phosphate.

Many sugars, including the four pentoses, react readily with cyanamide to form stable bicyclic amino-oxazolines (Fig. 2A) (Sanchez and Orgel 1970). The free sugars are unstable (Larralde et al. 1995) and cannot be synthesized directly in the presence of cyanamide. However, the sugars could be synthesized in one locale through the "formose" reaction (Mizuno and Weiss 1974; Müller et al. 1990; Ricardo et al. 2004; Kim et al. 2011), then combined with cyanamide in a different locale to give the sugar amino-oxazolines. When this reaction is performed with a complex mixture of sugars, the ribose derivative strikingly crystallizes from aqueous solution, segregating it from the other sugars (Springsteen and Joyce 2004).

Sutherland and colleagues (Ingar et al. 2003; Anastasi et al. 2007; Powner et al. 2009) avoided the complication of the separate synthesis of sugars by starting simply with glycolaldehyde and cyanamide, which in the presence of 1 M phosphate at neutral pH gives 2-amino-oxazole an excellent yield (Fig. 2B). The phosphate both buffers and catalyzes the reaction, directing glycolaldehyde toward 2-amino-oxazole, rather than a complex mixture of prod-

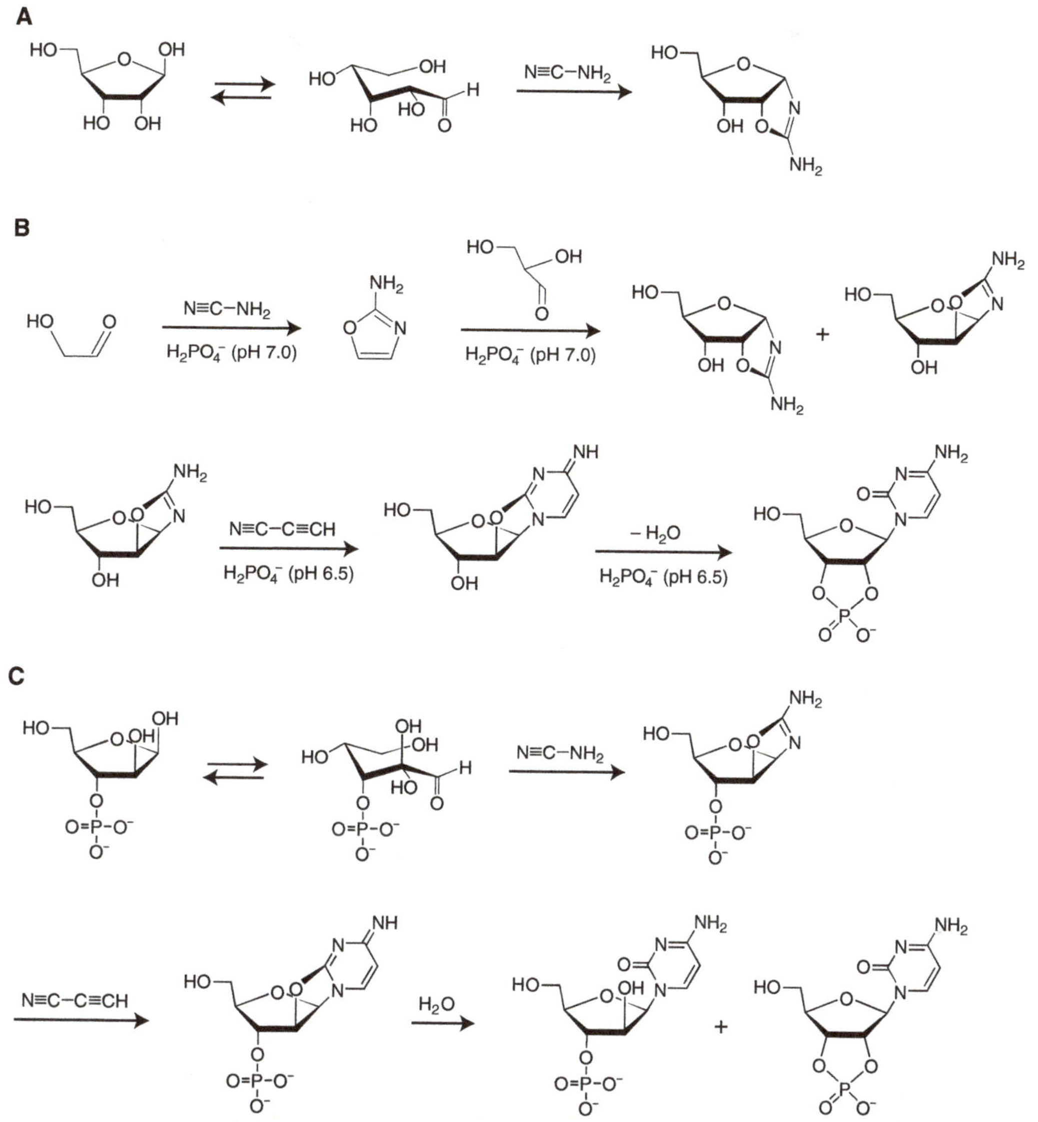

Figure 2. Potential prebiotic synthesis of pyrimidine nucleosides. (*A*) Reaction of ribose with cyanamide to form a bicyclic product, with cyanamide joined at both the anomeric carbon and 2-hydroxyl. (*B*) Reaction of glycolaldehyde with cyanamide in neutral phosphate buffer, followed by addition of glyceraldehyde, to form ribose and arabinose amino-oxazoline (and lesser amounts of the xylose and lyxose compounds). Arabinose amino-oxazoline then reacts with cyanoacetylene to give cytosine 2′,3′-cyclic phosphate as the major product. (*C*) Analogous reaction of arabinose-3-phosphate to form a bicyclic product, which then reacts with cyanoacetylene to form a tricyclic intermediate that hydrolyzes to give a mixture of cytosine arabinoside-3′-phosphate and cytosine 2′,3′-cyclic phosphate.

ucts. Glyceraldehyde is then added, resulting in formation of the various pentose amino-oxazolines, including the arabinose compound. Arabinose amino-oxazoline in turn can react with cyanoacetylene, also in phosphate buffer, to form cytosine 2′,3′-cyclic phosphate as the major product. This approach still requires sequential reactions. However, glycolaldehyde and glyceraldehyde can react with 2-aminothiazole to give the corresponding aminals, which selectively precipitate to give crystalline reservoirs of these key intermediates (Islam et al. 2017).

A potential route from arabinose amino-oxazoline to pyrimidine nucleotides first was explored nearly 40 years ago (Tapiero and Nagyvary 1971) and has been reexamined in light of alternative routes to the pentose amino-oxazolines (Ingar et al. 2003). Like arabinose, arabinose 3-phosphate reacts with cyanamide to give the corresponding amino-oxazoline (Fig. 2C). This in turn reacts with cyanoacetylene to form a tricyclic intermediate that hydrolyzes to produce a mixture of cytosine arabinoside-3′-phosphate and cytosine 2′,3′-cyclic phosphate. Ribose amino-oxazoline

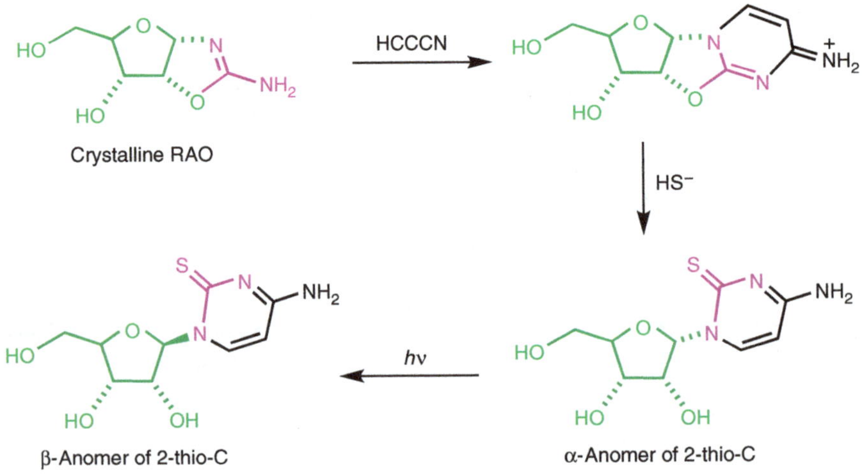

Figure 3. Potential prebiotic synthesis of 2-thio-C starting from ribose amino-oxazoline (RAO). RAO, which is generated by the reaction of glyceraldehyde with 2-amino-oxazole, crystallizes readily, leading to its accumulation as a reservoir of purified material. Subsequent reaction with cyanoacetylene generates an anhydronucleoside intermediate, which upon attack by hydrosulfide gives the α-anomer of 2-thio-C. Exposure to ultraviolet light results in photoanomerization to give the β-anomer of 2-thio-C, which then desulfurizes to form cytosine.

(RAO) also reacts with cyanoacetylene, generating the 2,2′-anhydro-ribonucleoside, but in this case as the α-anomer. However, ring opening of this compound with sulfide generates α-2-thio-C, which readily photoanomerizes to give the β-anomer, which in turn desulfurizes to give C or deaminates to give 2-thio-U and U (Fig. 3) (Xu et al. 2017).

A similar approach has been applied to the chemoselective assembly of purine nucleotide precursors (Powner et al. 2010). It has long been known for the purines, but not the pyrimidines, that simply heating the nucleobase with ribose or ribose-phosphate results in the synthesis of the corresponding ribonucleoside or ribonucleotide, but in low yield (Fuller et al. 1972). Two alternative approaches that have been explored recently involve either ribosylation of formamidopyrimidine, which is selective for the N9 position, but gives a mixture of α/β-furanosides and pyranosides (Becker et al. 2016), and a route from 2-thiooxazole to the pentose amino-oxazolidinone thiones and ultimately to the corresponding 8-oxopurine nucleotides (Stairs et al. 2017).

The story remains incomplete, however, because these syntheses still require temporally separated reactions using high concentrations of just the right reactants and would be disrupted by the presence of other closely related compounds. The reactions channel material toward the desired products, but other fractionation processes are required to provide the correct starting materials at the requisite time and place. In addition to the selective crystallization processes described above, it has been proposed that precipitation of ferrocyanide salts could generate a concentrated reservoir of starting materials that can be liberated by geothermal activity (Patel et al. 2015). Other potential fractionation processes, involving either selective synthesis or selective degradation, are actively being pursued.

Even chemical fractionation could not achieve on a macroscopic scale one desirable separation, the resolution of *D*-ribonucleotides from their *L*-enantiomers. This is a serious problem because experiments on the nonenzymatic template-directed polymerization of imidazole-activated mononucleotides suggest that the polymerization of the *D*-enantiomer is strongly inhibited by the *L*-enantiomer (Joyce et al. 1984). This difficulty may not be insuperable; perhaps with a different mode of phosphate activation, the inhibition would be less severe. However, enantiomeric cross-inhibition is certainly a serious problem if life arose in a racemic environment.

It is possible that the locale for life's origins was not racemic, although the global chemical environment contained nearly equal amounts of each pair of stereoisomers. There likely were biases in the inventory of compounds delivered to the Earth by comets and meteorites (Cronin and Pizzarello 1997; Engel and Macko 1997; Pizzarello et al. 2003; Glavin and Dworkin 2009). These in turn could have biased terrestrial syntheses, although the level of enantiomeric enrichment generally declines with successive chemical reactions. A special exception involves a remarkable set of reactions and fractionation processes that amplify a slight chiral imbalance, even to the level of local homochirality (Kondepudi et al. 1990; Soai et al. 1995; Viedma 2005; Klussmann et al. 2006; Noorduin et al. 2008; Viedma et al. 2008). As discussed elsewhere in this collection (Blackmond 2019), these systems have in common both a catalytic process for

 Cite as *Cold Spring Harb Perspect Biol* doi: 10.1101/cshperspect.a034801

amplification of same-handed molecules and an inhibition process for suppression of opposite-handed molecules.

3.2 Potential Alternatives to RNA

The problems that arise when one tries to understand how an RNA world could have arisen de novo on the primitive Earth are sufficiently challenging that one must consider other possibilities. What kind of alternative genetic systems might have preceded the RNA world? How could they have "invented" the RNA world? These topics have generated a good deal of speculative interest and some relevant experimental data.

Eschenmoser and colleagues undertook a systematic study of the properties of analogs of nucleic acids in which ribose is replaced by some other sugar, or in which the furanose form of ribose is replaced by the pyranose form (Fig. 4) (Eschenmoser 1999). Strikingly, polynucleotides based on the pyranosyl analog of ribose (p-RNA) form Watson–Crick-paired double helices that are more stable than RNA, and p-RNAs are less likely than the corresponding RNAs to form multiple-strand competing structures (Pitsch et al. 1993, 1995, 2003). Furthermore, the helices twist much more gradually than those of standard nucleic acids, which should make it easier to separate strands of p-RNA during replication. p-RNA appears to be an excellent choice as a genetic system; in some ways it seems an improvement compared with the standard nucleic acids. However, p-RNA does not interact with normal RNA to form base-paired double helices.

Most double-helical structures reported in the literature are characterized by a backbone with a six-atom repeat. Eschenmoser and colleagues made the surprising discovery that an RNA-like structure based on threose nucleotide analogs (TNA; Fig. 4C), although it involves a five-atom repeat, can still form a stable double-helical structure with standard RNA (Schöning et al. 2000). This provides an example of a pairing system based on a sugar that could be formed more readily than ribose. Tetroses are the unique products of the dimerization of glycolaldehyde, whereas pentoses are formed along with tetroses and hexoses from glycolaldehyde and glyceraldehyde. A structural simplification of Eschenmoser's TNA has been achieved by Meggers and colleagues (Zhang et al. 2005). They replaced threose by its open chain analog, glycol, in the backbone of TNA, resulting in glycol nucleic acid (GNA; Fig. 4D). Complementary oligomers of GNA form antiparallel, double helices with surprisingly high duplex stabilities. The activated monomer of GNA is susceptible to intramolecular cyclization, but this might be overcome by using dimeric or short oligomeric building blocks.

These and similar studies suggest that there are many ways of linking together the nucleotide bases into chains that are capable of forming base-paired double helices. It is not clear that it is much easier to synthesize the monomers of p-RNA, TNA, or GNA than to synthesize the standard nucleotides. However, it is possible that a base-paired structure of this kind will be discovered that can be synthesized readily under prebiotic conditions. It also may be fruitful to explore a broader range of potential precursors to RNA, changing the recognition elements as well as the backbone. A strong candidate for the first genetic material would be any informational macromolecule that is replicable in a sequence-general manner and derives from compounds that would have been abundant on the primitive Earth, and preferably has the ability to cross-pair with RNA.

4 CHEMICAL RNA REPLICATION SYSTEMS

4.1 Templated Polymerization of Mononucleotides

The nonenzymatic template-directed polymerization of activated nucleotides has been studied for decades as a model

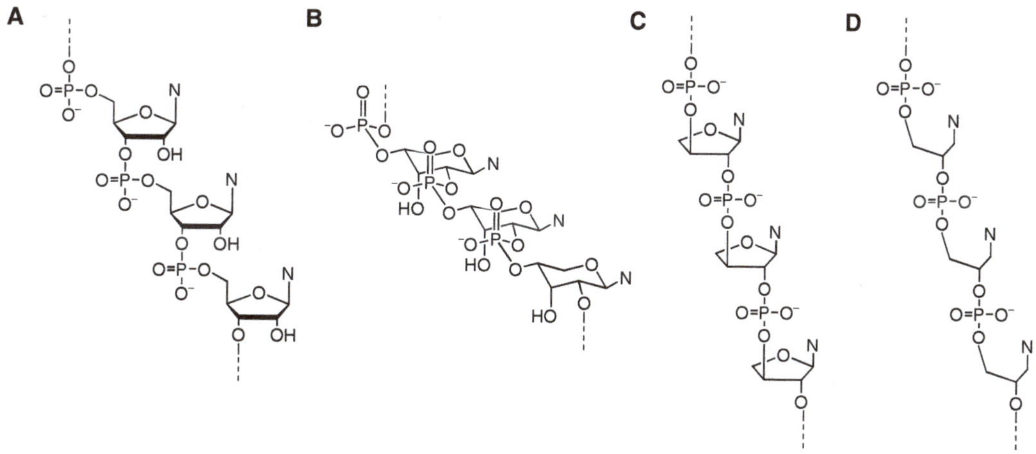

Figure 4. The structures of (*A*) RNA; (*B*) p-RNA; (*C*) TNA; and (*D*) GNA.

for the emergence of replicating informational systems during the origin of life. Rather than recount the detailed history of this field, which has been extensively reviewed (Joyce 2002; Orgel 2004), this review focuses on recent developments that have led to significant improvements in the extent and generality of RNA-templated synthesis of RNA, concluding with a discussion of the remaining barriers to the development of prebiotically plausible replication chemistry.

Template-directed primer extension with imidazole-activated monomers is attractive as a prebiotic means of copying RNA templates, in part for its similarity to the universal biological processes of enzymatic RNA and DNA replication. Since the discovery in 1981 that 2-methylimidazole (2MeI) is an especially good nucleotide activating agent (Inoue and Orgel 1981, 1983), nucleoside 5′-phosphoro-2-methylimidazolides have been widely used for primer extension experiments. The mechanism of the reaction between a growing primer chain and an incoming imidazole-activated monomer was assumed to parallel that of enzymatic primer extension, which proceeds via in-line S_N2 nucleophilic substitution. However, the very slow rate at which the last nucleotide of a template is copied relative to internal sites (Wu and Orgel 1992a), suggested that other factors influenced the reaction. Indeed, the same investigators showed that the reaction of the primer with an activated nucleotide is catalyzed by a downstream, activated nucleotide that is base-paired to the template at the primer +2 position. Furthermore, the identity of the leaving group on the 5′-phosphate of the downstream nucleotide has a large effect on the rate of the reaction between the primer and the adjacent monomer (Wu and Orgel 1992b). Only recently has the mechanism behind these effects become clear.

In the course of comparing the rates of oligonucleotide ligation and primer extension, Prywes and Szostak rediscovered the catalytic effect of a downstream, activated nucleotide and showed that this effect is much stronger for an oligonucleotide than a mononucleotide at the downstream position (Prywes et al. 2016). Initially, catalysis was thought to result from noncovalent interactions between the leaving groups of adjacent activated nucleotides, but subsequent experiments suggested that the relevant interaction was covalent and involved the off-template formation of a reactive intermediate. Nuclear magnetic resonance experiments identified this intermediate as a 5′,5′-imidazolium-bridged dinucleotide, generated by an attack of the nucleophilic nitrogen of an unprotonated phosphoroimidazolide on the phosphate of a protonated imidazolide (Fig. 5). The template specificity of primer extension suggested that the intermediate could bind to the template through two adjacent Watson–Crick base pairs (Walton and Szostak 2016). The Richert laboratory also noted the formation of an imidazolium-bridged dinucleotide and its high reactivity in primer extension reactions (Kervio et al. 2016). More recent kinetic studies of primer extension with 2-aminoimidazole-(2AI) activated monomers showed that all detectable primer extension occurs via the 2-aminoimidazolium intermediate and not through the reaction of a primer with an activated monomer (Walton and Szostak 2017).

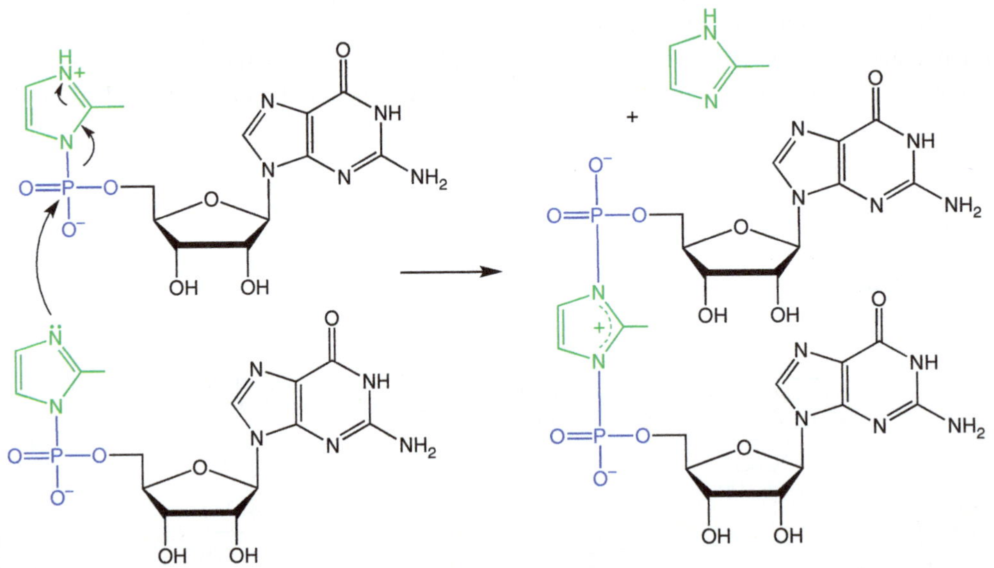

Figure 5. Formation of a 5′,5′-linked imidazolium-bridged dinucleotide, which is the reactive intermediate in template-directed polymerization of imidazole-activated nucleotides.

4.2 Alternative Means of Monomer Activation

The Richert laboratory has explored a wide range of phosphate-activating groups in an effort to improve the rate and generality of primer extension chemistry. One of the problems with nonenzymatic primer extension is that the 3′-hydroxyl of the primer must be deprotonated during nucleophilic attack on the phosphate of the incoming monomer. Conversely, an imidazole-based activating group must be protonated to function as a leaving group, and this typically requires a lower pH because the pK_a of this group is in the range of 6–8. Therefore, a leaving group that departs as an anion (with low pK_a) could provide a much faster reaction rate at elevated pH. 1-Hydroxy-7-azabenzotriazole (HOAt) functions in this manner and results in rapid primer extension attributable to direct reaction of the primer with the HOAt-activated monomer (Kervio et al. 2016).

A related approach is to use an *N*-alkyl-imidazole as an activating agent because it can leave as a neutral species that does not require protonation. Following this approach, the Richert laboratory used *N*-ethyl-imidazole (NEI) as an organocatalyst to facilitate in situ nucleotide activation with 1-ethyl-3-(3-dimethylaminopropyl)carbodiimide (EDC). The nucleoside 5′-phosphate reacts first with EDC to form a reactive isourea intermediate, then displacement by NEI yields NEI-activated monomers that can then react with the primer (Jauker et al. 2015). Although the highly reactive activated nucleotides are labile to hydrolysis at elevated pH, the continuous addition of EDC could in principle lead to continuing primer extension. An efficient and prebiotically plausible means of performing nucleotide activation and primer extension in the same reaction mixture would greatly simplify the conditions required for protocell replication.

In view of the catalytic effect on primer extension by an activated downstream nucleotide, a series of substituted imidazoles were tested for their influence on the rate of reaction, in all cases with 2MeI-activated guanylate as the incoming monomer. The magnitude of the catalytic effect correlated positively with the pK_a of the imidazole and with a smaller size of the imidazole substituent. 2AI has both of these properties and proved to be a superior activating group (Li et al. 2017). This effect is now understood to reflect, at least in part, the greater stability of the 2-amino-imidazolium-bridged dinucleotide intermediate compared with the relatively unstable 2-methylimidazolium intermediate. This allows the 2AI intermediate to build up to higher levels, significantly enhancing the rate and extent of primer extension (Walton and Szostak 2017). The potential prebiotic synthesis of 2AI is intriguing because it can be generated in the same reaction mixture as the related heterocycle 2-aminooxazole, a precursor of the pyrimidine nucleotides (Fahrenbach et al. 2017).

4.3 Alternative Modes of Strand Elongation

Primer extension with activated monomers is a biologically inspired mode of template copying. In principle, however, template copying could also be accomplished by the ligation of activated oligonucleotides. Unfortunately, nonenzymatic RNA ligation reactions tend to have low yield and poor fidelity. An interesting exception comes from a study of template-directed ligation of DNA oligonucleotides mediated by cyanogen bromide (James and Ellington 1997). Although still low yielding, when random hexamers were annealed to a defined template, most of the ligation products had perfect complementarity to the template. This study suggests that oligonucleotide ligation could play a significant role in nonenzymatic RNA-templated synthesis, if more efficient ligation chemistry could be defined.

The ligation of oligonucleotides activated at the 2′(3′) end has traditionally been hampered by the fact that activation of a 2′- or 3′-phosphate, for example, by reaction with cyanoimidazole or EDC, leads to rapid formation of a 2′,3′-cyclic phosphate, which is only weakly activated and yields predominately 2′,5′-linked ligation products. This outcome can be partially avoided if the 3′-phosphate is first acetylated by reaction with *N*-acetyl-imidazole or other acetylating reagents, in which case the acetyl group is preferentially transferred to the 2′-hydroxyl (Bowler et al. 2013). Subsequent activation of the 3′-phosphate allows for ligation and formation of a 3′,5′-phosphodiester. In a final step, the 2′-hydroxyl is regenerated by slow hydrolytic loss of the acetyl group. This process can be iterated to replace labile 2′,5′ linkages, which lead to strand cleavage by transesterification, with a progressively higher proportion of the more stable 3′,5′ linkages (Mariani and Sutherland 2017).

Surprisingly, the rate of ligation of a primer to a 5′-imidazole-activated oligonucleotide is very slow compared with the rate of primer extension with similarly activated monomers (Prywes et al. 2016). This is now explained by the fact that primer extension with monomers proceeds through the highly reactive imidazolium-bridged dinucleotide intermediate discussed above (Walton and Szostak 2017). Because the ligation reaction is slow, the final yield is determined by competition with hydrolysis of the activated phosphate. This suggests that chemistry that would reactivate hydrolyzed substrates, such as the combination of EDC and an *N*-alkyl imidazole explored by the Richert laboratory (Jauker et al. 2015), could lead to a high ligation yield.

Because oligonucleotides can have a long residence time on the template, a slow rate of ligation is not necessarily a problem, so long as the reaction can go to completion. An intriguing possibility is that a hybrid process, in which short oligonucleotides act as primers at multiple sites on a template, followed by gap filling by primer extension with monomers, and a final stage of oligonucleotide ligation, could result in effective copying of long RNA templates (Szostak 2011). Such a hybrid process would avoid some of the difficulties inherent to template copying solely by ligation of oligonucleotide substrates, such as the formation of gaps or overhangs.

4.4 Fidelity and the Error Threshold

The concept of an error threshold, that is, an upper limit to the frequency of copying errors that can be tolerated by a replicating macromolecule, was first introduced by Eigen (1971). Eigen's model envisaged a population of replicating polynucleotides that draw on a limited supply of activated mononucleotides to produce additional copies of themselves. In this model, the rate of synthesis of new copies of a particular replicating RNA is proportional to its concentration, resulting in autocatalytic growth. The net rate of production is the difference between the rate of formation of error-free copies and the rate of decomposition of existing copies of the RNA. For an advantageous RNA to outgrow its competitors, its net rate of production must exceed the mean rate of production of all other RNAs in the population. Only the error-free copies of the advantageous RNA contribute to its net rate of production, but all the copies of the other RNAs contribute to their collective production. Thus, the relative advantage enjoyed by the advantageous individual compared with the rest of the population (often referred to as the "superiority" of the advantageous individual) must exceed the probability of producing an error copy of that advantageous individual.

The proportion of copies of an RNA that are error free is determined by the fidelity of the component condensation reactions that are required to produce a complete copy. For simplicity, consider a self-replicating RNA that is formed by n condensation reactions, each having mean fidelity q. The probability of obtaining a completely error-free copy is given by q^n, which is the product of the fidelity of the component condensation reactions. If an advantageous individual is to outgrow its competitors, q^n must exceed the superiority, s, of that individual. Expressed in terms of the number of reactions required to produce the advantageous individual,

$$n < |\ln s| \,/\, |\ln q|.$$

For $s > 1$ and $q > 0.9$, this equation simplifies to

$$n < \ln s \,/\, (1 - q).$$

This is the "error threshold," which describes the inverse relationship between the fidelity of replication, q, and the maximum allowable number of component condensation reactions, n. The maximum number of component reactions is highly sensitive to the fidelity of replication but depends only weakly on the superiority of the advantageous individual. For a self-replicating RNA that is formed by the template-directed condensation of activated mononucleotides, a total of $2n - 2$ condensation reactions are required to produce a complete copy. This takes into account the synthesis of both complementary strands.

When chemical replication was first established, fidelity was likely to be poor and there would have been strong selection pressure favoring improvement of fidelity. As fidelity improves, a larger genome can be maintained. This allows exploration of a larger number of possible sequences, some of which may lead to further improvement in fidelity, which in turn allows a still larger genome size, and so on. If one assumes that the first genomes encoded a small but relatively efficient ribozyme, for which the identity of roughly 25 nucleotides must be specified, then a fidelity of ~0.98 would be required to enable inheritance of the corresponding information.

An analysis of error rates in the RNA-templated polymerization of imidazole-activated nucleotides indicated that most errors result from G•U wobble pairing (Rajamani et al. 2010). Calculated overall error rates from that study were 5%–15%, which would severely limit the size of a primitive genome. However, primer extension following a mismatch was generally very slow, depending on the specific mismatch. This stalling effect slows template copying, but as a consequence, the first copies to be completed are more likely to be accurate compared with those completed at later times. If full-length copies can serve as templates for the next round of copying as soon as they are completed, then there will be enrichment for accurate copies during the course of exponential amplification.

Template copying fidelity can be improved by the use of nucleotide analogs. The weak base-pairing of U with A allows for the misincorporation of C opposite A, and the similar strength of the U•A and U•G base pairs allows for the misincorporation of U opposite G. Remarkably, replacing U with 2-thio-U reduces both of these sources of errors because the 2-thio-U•A base pair is almost as strong as a C•G base pair. Template copying experiments with 2MeI-activated monomers suggest that this substitution reduces wobble-pairing errors from 4%–5% to ~2% at template G positions, suggesting that an overall error rate of 1%–2% may be possible (Heuberger et al. 2015).

Replacement of U by 2-thio-U also results in faster and more efficient copying of templates that contain A residues. The combined effect of faster and more accurate copying makes the 2-thio substitution attractive and raises the question of the prebiotic plausibility of 2-thio-U. Recent

Cite as *Cold Spring Harb Perspect Biol* doi: 10.1101/cshperspect.a034801

advances in the prebiotic chemistry of pyrimidine nucleotides suggest that a mixture of U and 2-thio-U might be obtained through the synthesis of 2-thio-C, followed by spontaneous deamination (Fig. 5) (Xu et al. 2017). The 2-thio-U then would be preferentially incorporated into RNA. Although speculative at this point, the hypothesis that primordial RNA contained 2-thio-U rather than U is worth considering.

4.5 Complete Cycles of Replication

The inheritance of RNA-based genetic information requires cycles of both plus- and minus-strand synthesis. Because the product of template copying is a double-stranded RNA, there must be some means of either strand separation or strand displacement synthesis. Transient temperature fluctuations could lead to thermal strand separation, but long RNA duplexes (≥30 base pairs) are difficult to denature thermally. Many environmental factors affect the melting temperature of RNA duplexes and could facilitate thermal strand separation, such as low ionic strength or the presence of high concentrations of formamide or urea.

A more difficult problem is that strand reannealing is very fast relative to the time scale of nonenzymatic template copying. Strand reannealing can be slowed by dilution but achieving a reannealing time of hours would require strand concentrations in the picomolar range, which likely would be too low to maintain RNA replication within a protocell. Thus, environmental conditions that might drastically slow reannealing kinetics are of great interest. Recent work from Hud and colleagues has shown that the combination of template secondary structure and high solvent viscosity can significantly slow the reannealing of long templates (300–500 nucleotides), while still allowing shorter oligonucleotides (~30 nucleotides) to diffuse rapidly and anneal to complementary sites on the template where ligation can occur (He et al. 2017). Whether this approach can be extended to shorter templates and copying with monomers or short oligonucleotides remains to be seen.

Recently, studies of liquid–liquid phase separation suggest that this phenomenon might be relevant to protocell structure and function (Koga et al. 2011; Frankel et al. 2016). The high viscosity of phase-separated droplets could slow reannealing, whereas the colocalization of RNA templates and substrates could facilitate template copying (Aumiller et al. 2016). The recent observation of phase-separated droplets containing only RNA is intriguing in this regard (Jain and Vale 2017). These droplets form when the RNAs contain certain triplet repeats, which enable the formation of a highly branched interconnected network. The droplets seem to be liquid-like upon initial phase separation, but rapidly mature to a gel-like state. It is possible that template strands trapped within such a matrix are unable to diffuse and therefore reanneal but could still be copied by short oligonucleotide or monomer substrates. Any physical means of immobilizing RNA strands after strand separation could prevent reannealing, so binding of RNA to surfaces such as membranes, mineral particles, or self-assembling peptide filaments could potentially facilitate replication.

The alternative solution of strand displacement synthesis is appealing because this is the strategy that is universally used in biology. However, biological strand displacement relies on helicases and/or polymerases that can transduce some of the chemical energy used to drive primer extension into the mechanical energy needed for strand separation. It is unclear how this could be achieved in a prebiotic setting devoid of enzymes.

4.6 Current Status of Laboratory Protocell Replication Systems

The experimental realization of a complete protocell model will require replication of mixed-sequence templates long enough to encode functional ribozymes within replicating vesicles. This goal has been gradually approached over the past decade. The first demonstration of nonenzymatic template copying within fatty acid vesicles was achieved by copying an RNA template with 5′-activated 2′-amino-nucleotides. This model had the advantage of rapid primer extension in the absence of Mg^{2+} (Mansy et al. 2008). Five years later, chelation of Mg^{2+} with citrate was found to protect fatty acid membranes, while still allowing templated RNA polymerization to proceed. Addition of citrate enabled the copying of an RNA homopolymer template inside fatty acid vesicles. Subsequently, the use of the 2AI leaving group combined with activated downstream trinucleotides enabled the efficient copying of mixed-sequence templates containing up to seven nucleotides. This approach was recently adapted to template copying within vesicles by increasing the permeability of fatty acid membranes using Mg^{2+}-citrate. As a result, it has been possible to copy short RNA templates inside fatty acid vesicles following the addition of activated monomers and trimers to the outside of the vesicles (O'Flaherty et al. 2018).

5 RNA-CATALYZED REPLICATION SYSTEMS

5.1 The Replicase Model

The notion of the RNA world places emphasis on an RNA molecule that catalyzes its own replication. Such a molecule must function as an RNA-dependent RNA polymerase, acting on itself (or copies of itself) to produce complementary RNAs, and acting on the complementary RNAs to produce additional copies of itself. The efficiency and fidelity of this process must be sufficient to produce viable "progeny" RNA molecules at a rate that exceeds the rate of decomposition of the "parents." Beyond these requirements, the details of the replication process are not highly constrained.

The RNA-first view of the origin of life assumes that a supply of activated nucleotides was available through some set of abiotic processes. Furthermore, it assumes that a means existed to convert the activated nucleotides to an ensemble of random sequence polynucleotides, a subset of which had the ability to undergo nonenzymatic replication. From this population of replicating RNAs, particular RNAs would have gained in relative abundance if they enjoyed a selective advantage because of a higher net rate of production. There would have been selection for the more favorable template sequences and for RNAs that promote the replication process. The most rudimentary form of replicase function would involve assistance of chemical replication to improve that rate, accuracy, or sequence generality of RNA synthesis. With ongoing selective pressure, this chemical replication process would become supplanted by a biochemical replication process. The latter likely used a less-reactive, but energetically favorable, activating group on the mononucleotides, such as a polyphosphate, so that the RNA-catalyzed reaction dominated over spontaneous polymerization.

Although it is unclear how the first RNA replicase ribozyme arose, it is not difficult to imagine how such a molecule, once developed, would function. The polymerization of activated nucleotides proceeds via nucleophilic attack by the 3′-hydroxyl of a template-bound oligonucleotide at the α-phosphorus of an adjacent template-bound nucleotide derivative. This reaction could be assisted by favorable orientation of the reactive groups, deprotonation of the nucleophilic 3′-hydroxyl, stabilization of the trigonal-bipyramidal transition state, or charge neutralization of the leaving group. All of these tasks might be performed by RNA (Narlikar and Herschlag 1997; Emilsson et al. 2003), acting either alone (Strobel and Ortoleva-Donnelly 1999) or with the help of a suitably positioned metal cation or other cofactor (Shan et al. 1999, 2001).

5.2 From Ligase to Polymerase

The possibility that an RNA replicase ribozyme could have existed has been made abundantly clear by work involving ribozymes that have been developed through in vitro evolution (Bartel and Szostak 1993; Ekland et al. 1995; Ekland and Bartel 1996; Jaeger et al. 1999; Robertson and Ellington 1999; Johnston et al. 2001; Rogers and Joyce 2001; McGinness and Joyce 2002; Ikawa et al. 2004; Wochner et al. 2011; Sczepanski and Joyce 2014; Horning and Joyce 2016). Bartel and Szostak (1993), for example, began with a large population of random sequence RNAs and evolved the "class I" RNA ligase ribozyme, an optimized version of which is about 100 nucleotides in length and catalyzes the joining of two template-bound oligonucleotides. Condensation occurs between the 3′-hydroxyl of one oligonucleotide and the 5′-triphosphate of another, forming a 3′,5′-phosphodiester linkage and releasing inorganic pyrophosphate. This reaction is classified as ligation owing to the nature of the oligonucleotide substrates but involves the same chemical transformation as is catalyzed by modern RNA polymerase enzymes.

X-ray crystal structures of two RNA ligase ribozymes, the L1 and class I ligases, provide a glimpse into mechanistic strategies a ribozyme could use to catalyze RNA polymerization (Fig. 6) (Robertson and Scott 2007; Shechner et al. 2009). Both ligases are dependent on Mg^{2+} ions for their activity. The L1 structure (Fig. 6A) shows a bound metal ion in the active site, coordinated by three nonbridging phosphate oxygens, one of which belongs to the newly formed phosphodiester linkage. This Mg^{2+} ion is favorably positioned to help neutralize the increased negative charge of the transition state and, potentially, to activate the 3′-hydroxyl nucleophile and to help orient the α-phosphate for a more optimal in-line alignment. In the class I structure (Fig. 6B), no catalytic metal ion is seen in the vicinity of the active site, although there is what appears to be an empty metal-binding site formed by two nonbridging phosphate oxygens, positioned directly opposite the ligation junction in a manner similar to that observed for the L1 ligase. These structures point to a universal catalytic strategy, very similar to that used by modern protein-based RNA polymerases.

Subsequent to its isolation as a ligase, the class I ribozyme was shown to catalyze a polymerization reaction in which the 5′-triphosphate-bearing oligonucleotide is replaced by one or more nucleoside 5′-triphosphates (NTPs) (Ekland and Bartel 1996). This reaction proceeds with high fidelity ($q = 0.92$), although the reaction rate drops sharply with successive nucleotide additions.

Bartel and colleagues performed further in vitro evolution experiments to convert the class I ligase to a bona fide RNA polymerase that operates on a separate RNA template (Johnston et al. 2001). To the 3′ end of the class I ligase, they added 76 random sequence nucleotides that were evolved to form an accessory domain that assists in the polymerization of template-bound NTPs. The polymerization reaction is applicable to a variety of template sequences, and for well-behaved sequences proceeds with an average fidelity of 0.97. This would be sufficient to support a genome length of 20–30 nucleotides, although the ribozyme itself contains

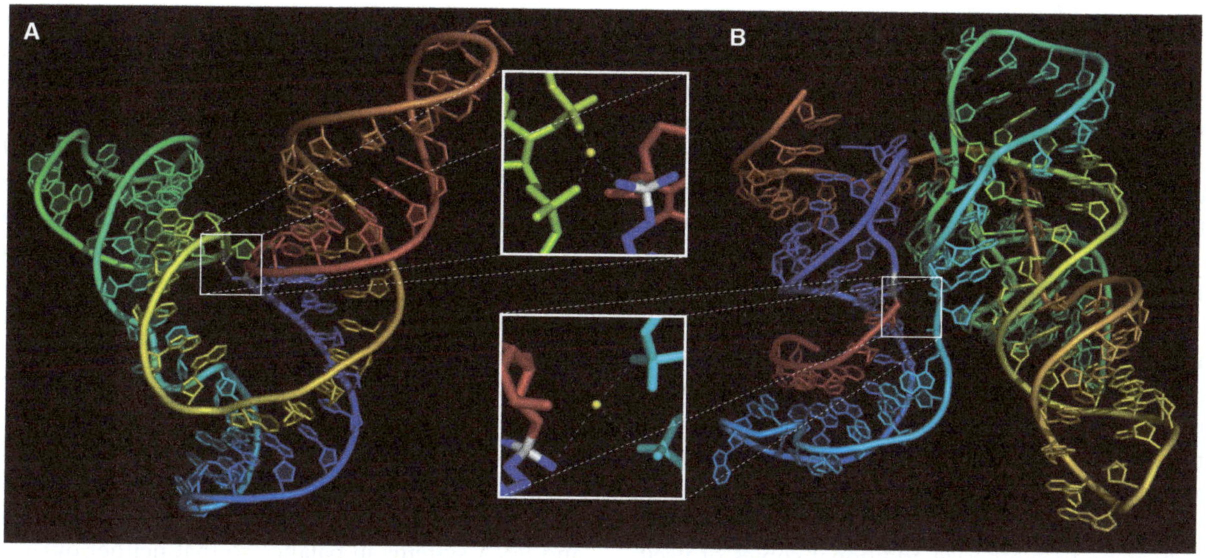

Figure 6. X-ray crystal structure of the (*A*) L1 ligase, and (*B*) class I ligase ribozymes. *Insets* show the putative magnesium ion-binding sites at the respective ligation junctions. The structures are rendered in rainbow continuum, with the 5′-triphosphate-bearing end of the ribozyme colored violet and the 3′-hydroxyl-bearing end of the substrate colored red. The phosphate at the ligation junction is shown in white, and the proximate magnesium ion (modeled for the class I ligase) is shown as a yellow sphere, with dashed lines indicating coordination contacts.

about 190 nucleotides. The ribozyme has a catalytic rate for NTP addition of at least 1.5 min^{-1}, but its K_m is so high that, even in the presence of micromolar concentrations of oligonucleotides and millimolar concentrations of NTPs, it requires about 2 h to complete each NTP addition (Lawrence and Bartel 2003). The ribozyme operates best under conditions of high Mg^{2+} concentration but becomes degraded under those conditions after 24 h, by which time it has added no more than 14 NTPs (Johnston et al. 2001).

Further evolutionary optimization of the class I polymerase ribozyme has led to substantial improvement of its biochemical properties. By directly selecting for extension of an external primer on a separate template, Zaher and Unrau (2007) were able to improve the maximum length of template-dependent polymerization to >20 nucleotides, at a rate approximately threefold faster than that of the parent for the first nine monomer additions and up to 75-fold faster for additions beyond 10 nucleotides. In addition, this ribozyme variant displays significantly improved fidelity, particularly with respect to discrimination against G•U wobble pairs. This improved fidelity appears to be the underlying source for the improvements in the maximum length of extension and the rate of polymerization.

5.3 From Polymerase to Replicase

Using a sophisticated emulsion-based in vitro selection technique, the activity of the class I polymerase was greatly improved so that it can synthesize long RNA products, in some cases exceeding 100 nucleotides in length (Wochner et al. 2011). The improved ribozyme contains a 5′-terminal "processivity tag" that binds to a complementary region of the template, thus overcoming the problem of a high K_m through pseudointramolecularity. It continues to show a good average fidelity of 0.97. When the reaction is performed in the eutectic phase of water ice, even longer polymers can be obtained because of the increased concentration of reactants and reduced rate of RNA degradation under these near-frozen conditions (Attwater et al. 2013). However, the optimized polymerase strongly prefers cytidine-rich templates that lack any secondary structure, which precludes the synthesis of most functional RNAs.

The class I polymerase was further optimized by selecting for its ability to react quickly and to copy "difficult" templates, including those that are cytidine-poor or contain regions of stable secondary structure (Horning and Joyce 2016). This resulted in a polymerase ribozyme that requires less than a minute to complete each NTP addition and can synthesize a variety of complex structured RNAs. It still struggles with highly structured templates or runs of consecutive template A residues. It also has a reduced fidelity of 0.92, primarily because of an increased frequency of G•U wobble mutations. Nonetheless, because of its much greater sequence generality, this optimized polymerase can catalyze the reciprocal synthesis of both an RNA and its complement, enabling the exponential amplification of short RNA templates in an all-RNA form of the polymerase chain reaction.

This most highly optimized class I polymerase also has the ability to synthesize DNA polymers on an RNA template (Samanta and Joyce 2017). The ribozyme can incorporate all four deoxynucleoside 5′-triphosphates at a rate of about one addition per minute and can generate products containing up to 32 deoxynucleotides. Reverse transcriptase activity likely was necessary for the transition from RNA to DNA genomes during the early history of life on Earth. The fact that an RNA polymerase ribozyme also can function as a reverse transcriptase suggests that a bridge may have existed between RNA and DNA as a selective adaptation of the RNA world. DNA is much more stable compared with RNA and thus provides a larger and more secure repository for genetic information.

Generalized RNA-catalyzed RNA replication has not yet been achieved and it is not possible for the class I polymerase ribozyme to synthesize additional copies of itself. The most advanced form of the polymerase cannot replicate long RNA sequences because attempts to do so are thwarted by the emergence of shorter amplicons that are copied more efficiently. There must be a selective advantage in maintaining the full-length amplicon and that advantage must exceed the probability of producing an error copy. It is reasonable to expect that a general RNA replicase will eventually be developed in the laboratory, thus providing a working model of RNA-based life. However, like the search for life elsewhere in the universe, such expectations must be met through direct observation rather than speculation.

Despite falling short of the goal of generalized RNA-catalyzed RNA replication, there is one example of an RNA enzyme that catalyzes its own amplification and can undergo limited Darwinian evolution. This replication system uses a pair of ligase ribozymes that each catalyze the formation of the other, using a mixture of four different substrate oligonucleotides (Lincoln and Joyce 2009). In reaction mixtures containing only these RNA substrates, $MgCl_2$, and buffer, a small starting number of ribozymes gives rise to many additional ribozymes through RNA-catalyzed exponential amplification. The replication process can be sustained indefinitely by replenishing the supply of substrates, in one case achieving an overall amplification factor of 10^{100}-fold in 37.5 h (Robertson and Joyce 2014).

6 COMPARTMENTED GENETICS

6.1 Integration of Genetic and Compartment Systems

Assuming that the chemistry of RNA replication can be fully realized in a manner compatible with protocell replication, what other challenges must be addressed for the system to be capable of undergoing Darwinian evolution? Because functional RNAs are likely to provide significant selective advantage, it is helpful to understand the conditions necessary to obtain robust ribozyme function within model protocells. Such studies have generally used simple ribozymes, such as the self-cleaving hammerhead ribozyme, for which activity within fatty acid–based vesicles has been shown (Chen et al. 2005). Because most ribozymes require high concentrations of Mg^{2+} for optimal activity, it is necessary to use a lipid composition that maintains membrane stability in the presence of Mg^{2+}. Alternatively, it is possible that the first ribozymes did not require high concentrations of Mg^{2+}, instead relying on monovalent cations or other cofactors.

New questions arise when considering the integration of replicating nucleic acids with replicating vesicles. For example, what mechanisms might serve to keep the membrane and RNA systems in balance so that neither out-replicates the other? One potential answer to this problem of homeostasis is the presence of short oligonucleotides that are complementary to a replicase ribozyme, generated as either replication intermediates or degradation fragments, that inactivate the replicase when present in high concentrations. Upon dilution, as would occur upon vesicle growth (Engelhart et al. 2016), this inhibition would be relieved and RNA replication could continue.

6.2 Membrane Permeability, Trafficking, and the Origins of Metabolism

Assuming that the first protocells were heterotrophic, the protocell membranes must have been permeable to nutrients in the external environment. Fatty acid–based membranes show high permeability to ions, including divalent cations, as well as to larger polar and charged molecules, such as nucleotides. Environmental factors, including elevated temperature and the presence of divalent cations, can further enhance membrane permeability. High membrane permeability is likely to be necessary for protocell growth but is not a demanding physical requirement. However, high permeability can also limit the ability of a protocell to evolve metabolic functions because any metabolites synthesized within the protocell could leak out and be lost to the environment or captured by competing protocells. One possibility is that protocells did not evolve a metabolism until membrane permeability had decreased, perhaps driven by increasing phospholipid content resulting from the selective growth advantage provided by those phospholipids (Budin and Szostak 2011). Many compounds of biological metabolism are charged carboxylates or phosphate esters, which may reflect the evolution of metabolism at a time when cellular membranes were more permeable than in modern cells.

Cite as *Cold Spring Harb Perspect Biol* doi: 10.1101/cshperspect.a034801

6.3 Remaining Challenges

The experiments described above bring the field tantalizingly close to a replicating and evolving protocell, but several additional problems must be solved to reach that goal. Divalent metal cations appear to be essential for efficient RNA copying, but the poor affinity of the catalytic metal for the reaction center means that very high concentrations of these ions are required, which causes problems for both the RNA (degradation, hydrolysis of activated monomers) and for the fatty acid–based membranes. RNA polymerase enzymes solve these problems by binding and precisely positioning the metal ion for catalysis (Zhang et al. 1999; Cramer et al. 2001; Robertson and Scott 2007; Shechner et al. 2009). A prebiotically plausible means of achieving effective metal ion catalysis at low ambient concentration would greatly simplify the development of model protocells. Another key problem is the mobilization of chemical energy to activate (and reactivate) nucleotides for polymerization in a specific manner and under mild conditions. It also is not clear how to achieve multiple generations of nucleic acid replication within the context of a protocell. If and only if all of these challenges can be met will it be possible to synthesize an evolving protocell.

6.4 Implications for Synthetic Biology and Biotechnology

The main purpose of research into both nonenzymatic and RNA-catalyzed RNA replication is to discern plausible pathways leading from chemistry to biology, and thus to provide potential explanations for the origin of life on Earth. However, the development of replicating protocells also will provide new approaches for the compartmentalized evolution of novel functional RNAs. One of the longstanding limitations of conventional in vitro evolution methods is that selection depends on self-modification of the catalytic RNA and thus does not enrich for high catalytic turnover. Replicating protocells may provide a new way to obtain efficient ribozymes, so long as their function can be linked to the selective advantage of the protocell.

Many research groups are involved in the construction of artificial cells, with the goal of reconstituting some (or all) of the complex subsystems of microbial life. Most such systems therefore include the protein synthesis machinery, as either a crude extract or a reconstituted translation system. Reconstituting a fully functional and replicating cell from purified biological components is a major challenge and has inspired efforts to simplify aspects of cellular metabolism that seem gratuitous. For example, many chemical modifications of ribosomal RNA (rRNA) are essential for ribosome assembly, but recently Ichihashi and colleagues were able to evolve a variant small subunit rRNA that lacks modifications and can still be efficiently assembled into ribosomes (Murase et al. 2018). If this approach can be extended to the large subunit rRNA, ribosome assembly will be greatly simplified. Similar approaches are aimed at evolving a simpler ribosome, comparable with evolutionary intermediates in the path from the first peptidyl transferase ribozyme to the modern ribosome. Such approaches may contribute to the development of a comprehensive picture of possible paths from simple protocells to complex RNA world cells, and ultimately to the evolution of modern cells that contain DNA genomes, instructed protein synthesis, and a complex metabolism.

7 SUMMARY

The constraints that must have been met to originate a self-sustained evolving system are reasonably well understood. One can sketch out a logical order of events, beginning with prebiotic chemistry and ending with protocellular life. However, it must be said that many details of this process remain obscure, and even for reactions that have been convincingly shown in the laboratory, it is difficult to know whether those same reactions were operable in the historical origins of life.

The presumed RNA world should be viewed as a milestone, a plateau in the early history of life on Earth. So, too, the concept of an RNA world has been a milestone in the scientific study of life's origins. Although this concept does not fully explain how life originated, it has helped to guide scientific thinking and has served to focus experimental efforts. Further progress will depend primarily on new experimental results as chemists, biochemists, and molecular biologists work together to address problems concerning molecular replication, ribozyme enzymology, and RNA-based cellular processes.

ACKNOWLEDGMENTS

J.W.S. is an Investigator of the Howard Hughes Medical Institute. This work was supported by research grants 287624 (G.F.J.) and 290363 (J.W.S.) from the Simons Foundation, research grants 80NSSC17K0462 (G.F.J.) and NNX15AL18G (J.W.S.) from the National Aeronautics and Space Administration (NASA), and research grant CHE-1607034 (J.W.S.) from the National Science Foundation (NSF). Some portions of the text were published in prior editions of *The RNA World*. The contributions of Leslie Orgel to those prior versions and to the scientific literature of the origins of life are gratefully acknowledged.

REFERENCES

*Reference is in this book.

Adamala K, Szostak JW. 2013a. Nonenzymatic template-directed RNA synthesis inside model protocells. *Science* **342:** 1098–1100.

Adamala K, Szostak JW. 2013b. Competition between model protocells driven by an encapsulated catalyst. *Nat Chem* **5:** 495–501.

Anastasi C, Crowe MA, Sutherland JD. 2007. Two-step potentially prebiotic synthesis of α-D-cytidine-5′-phosphate from D-glyceraldehyde-3-phosphate. *J Am Chem Soc* **129:** 24–25.

Attwater J, Wochner A, Holliger P. 2013. In-ice evolution of RNA polymerase ribozyme activity. *Nat Chem* **5:** 1011–1018.

Aumiller WM Jr, Pir Cakmak F, Davis BW, Keating CD. 2016. RNA-based coacervates as a model for membraneless organelles: Formation, properties, and interfacial liposome assembly. *Langmuir* **32:** 10042–10053.

Baaske P, Weinert FM, Duhr S, Lemke KH, Russell MJ, Braun D. 2007. Extreme accumulation of nucleotides in simulated hydrothermal pore systems. *Proc Natl Acad Sci* **104:** 9346–9351.

Bansho Y, Furubayashi T, Ichihashi N, Yomo T. 2016. Host-parasite oscillation dynamics and evolution in a compartmentalized RNA replication system. *Proc Natl Acad Sci* **113:** 4045–4050.

Bartel DP, Szostak JW. 1993. Isolation of new ribozymes from a large pool of random sequences. *Science* **261:** 1411–1418.

Becker S, Thoma I, Amrei Deutsch A, Gehrke T, Mayer P, Zipse H, Carell T. 2016. A high-yielding, strictly regioselective prebiotic purine nucleoside formation pathway. *Science* **352:** 833–836.

Berclaz N, Muller M, Walde P, Luisi PL. 2001. Growth and transformation of vesicles studied by ferritin labeling and cryotransmission electron microscopy. *J Phys Chem B* **105:** 1056–1064.

* Blackmond DG. 2019. The origin of biological homochirality. *Cold Spring Harb Perspect Biol* doi: 10.1101/cshperspect.a032540.

Bowler FR, Chan CK, Duffy CD, Gerland B, Islam S, Powner MW, Sutherland JD, Xu J. 2013. Prebiotically plausible oligoribonucleotide ligation facilitated by chemoselective acetylation. *Nat Chem* **5:** 383–389.

Budin I, Szostak JW. 2011. Physical effects underlying the transition from primitive to modern cell membranes. *Proc Natl Acad Sci* **108:** 5249–5254.

Budin I, Bruckner RJ, Szostak JW. 2009. Formation of protocell-like vesicles in a thermal diffusion column. *J Am Chem Soc* **131:** 9628–9629.

Budin I, Prwyes N, Zhang N, Szostak JW. 2014. Chain-length heterogeneity allows for the assembly of fatty acid vesicles in dilute solutions. *Biophys J* **107:** 1582–1590.

Chen IA, Nowak MA. 2012. From prelife to life: How chemical kinetics become evolutionary dynamics. *Acc Chem Res* **45:** 2088–2096.

Chen IA, Roberts RW, Szostak JW. 2004. The emergence of competition between model protocells. *Science* **305:** 1474–1476.

Chen IA, Salehi-Ashtiani K, Szostak JW. 2005. RNA catalysis in model protocell vesicles. *J Am Chem Soc* **127:** 13213–13219.

Cramer P, Bushnell DA, Kornberg RD. 2001. Structural basis of transcription: RNA polymerase II at 2.8 Å resolution. *Science* **292:** 1863–1876.

Cronin JR, Pizzarello S. 1997. Enantiomeric excesses in meteoritic amino acids. *Science* **275:** 951–955.

Deamer DW. 1985. Boundary structures are formed by organic components of the Murchison carbonaceous chondrite. *Nature* **317:** 792–794.

Deamer DW. 1997. The first living systems: A bioenergetic perspective. *Microbiol Mol Biol Rev* **61:** 239–261.

Duley WW, Hu A. 2009. Polyynes and interstellar carbon nanoparticles. *Astrophys J* **698:** 808–811.

Eigen M. 1971. Selforganization of matter and the evolution of biological macromolecules. *Naturwiss* **58:** 465–523.

Ekland EH, Bartel DP. 1996. RNA-catalysed RNA polymerization using nucleoside triphosphates. *Nature* **382:** 373–376.

Ekland EH, Szostak JW, Bartel DP. 1995. Structurally complex and highly active RNA ligases derived from random RNA sequences. *Science* **269:** 364–370.

Emilsson GM, Nakamura S, Roth A, Breaker RR. 2003. Ribozyme speed limits. *RNA* **9:** 907–918.

Engel M, Macko S. 1997. Isotopic evidence for extraterrestrial non-racemic amino acids in the Murchison meteorite. *Nature* **389:** 265–268.

Engelhart AE, Adamala KP, Szostak JW. 2016. A simple physical mechanism enables homeostasis in primitive cells. *Nat Chem* **8:** 448–453.

Eschenmoser A. 1999. Chemical etiology of nucleic acid structure. *Science* **284:** 2118–2124.

Fahrenbach C, Giurgiu C, Tam CP, Li L, Hongo Y, Aono M, Szostak JW. 2017. Common and potentially prebiotic origin for precursors of nucleotide synthesis and activation. *J Am Chem Soc* **139:** 8780–8783.

Ferris JP. 2002. Montmorillonite catalysis of 30–50 mer oligonucleotides: Laboratory demonstration of potential steps in the origin of the RNA world. *Orig Life Evol Biosph* **32:** 311–332.

Ferris JP, Hill AR Jr, Liu R, Orgel LE. 1996. Synthesis of long prebiotic oligomers on mineral surfaces. *Nature* **381:** 59–61.

Frankel EA, Bevilacqua PC, Keating CD. 2016. Polyamine/nucleotide coacervates provide strong compartmentalization of Mg^{2+}, nucleotides, and RNA. *Langmuir* **32:** 2041–2049.

Fuller WD, Sanchez RA, Orgel LE. 1972. Studies in prebiotic synthesis. VI. Synthesis of purine nucleosides. *J Mol Biol* **67:** 25–33.

Gibbs D, Lohrmann R, Orgel LE. 1980. Template-directed synthesis and selective adsorption of oligoadenylates on hydroxyapatite. *J Mol Evol* **15:** 347–354.

Glavin DP, Dworkin JP. 2009. Enrichment of the amino acid L-isovaline by aqueous alteration on CI and CM meteorite parent bodies. *Proc Natl Acad Sci* **106:** 5487–5492.

Hanczyc MM, Fujikawa SM, Szostak JW. 2003. Experimental models of primitive cellular compartments: Encapsulation, growth, and division. *Science* **302:** 618–622.

He C, Gállego I, Laughlin B, Grover MA, Hud NV. 2017. A viscous solvent enables information transfer from gene-length nucleic acids in a model prebiotic replication cycle. *Nat Chem* **9:** 318–324.

Heuberger BD, Pal A, Del Frate F, Topkar VV, Szostak JW. 2015. Replacing uridine with 2-thiouridine enhances the rate and fidelity of nonenzymatic RNA primer extension. *J Am Chem Soc* **137:** 2769–2775.

Hong F, Zhang F, Liu Y, Yan H. 2017. DNA origami: Scaffolds for creating higher order structures. *Chem Rev* **117:** 12584–12640.

Horning D, Joyce GF. 2016. Amplification of RNA by an RNA polymerase ribozyme. *Proc Natl Acad Sci* **113:** 9786–9791.

Ichihashi N, Usui K, Kazuta Y, Sunami T, Matsuura T, Yomo T. 2013. Darwinian evolution in a translation-coupled RNA replication system within a cell-like compartment. *Nat Commun* **4:** 2494.

Ikawa Y, Tsuda K, Matsumura S, Inoue T. 2004. *De novo* synthesis and development of an RNA enzyme. *Proc Natl Acad Sci* **101:** 13750–13755.

Ingar A-A, Luke RWA, Hayter BR, Sutherland JD. 2003. Synthesis of a cytidine ribonucleotide by stepwise assembly of the heterocycle on a sugar phosphate. *Chembiochem* **4:** 504–507.

Inoue T, Orgel LE. 1981. Substituent control of the poly(C)-directed oligomerization of guanosine 5′-phosphoroimidazolide. *J Am Chem Soc* **103:** 7666–7667.

Inoue T, Orgel LE. 1983. A nonenzymatic RNA polymerase model. *Science* **219:** 859–862.

Inskeep WP, Jay ZJ, Macur RE, Clingenpeel S, Tenney A, Lovalvo D, Beam JP, Kozubal MA, Shanks WC, Morgan LA, et al. 2015. Geomicrobiology of sublacustrine thermal vents in Yellowstone Lake: Geochemical controls on microbial community structure and function. *Front Microbiol* **6:** 1044.

Islam S, Dejan-Krešimir B, Powner MW. 2017. Prebiotic selection and assembly of proteinogenic amino acids and natural nucleotides from complex mixtures. *Nat Chem* **9:** 584–589.

Jaeger L, Wright MC, Joyce GF. 1999. A complex ligase ribozyme evolved in vitro from a group I ribozyme domain. *Proc Natl Acad Sci* **96:** 14712–14717.

Jain A, Vale RD. 2017. RNA phase transitions in repeat expansion disorders. *Nature* **546:** 243–247.

James KD, Ellington AD. 1997. Surprising fidelity of template-directed chemical ligation of oligonucleotides. *Chem Biol* **4:** 595–605.

Jauker M, Griesser H, Richert C. 2015. Copying of RNA sequences without pre-activation. *Angew Chemie* **54:** 14559–14563.

Jia TZ, Hentrich C, Szostak JW. 2014. Rapid RNA exchange in aqueous two-phase system and coacervate droplets. *Orig Life Evol Biosph* **44:** 1–12.

Johnston WK, Unrau PJ, Lawrence MS, Glasner ME, Bartel DP. 2001. RNA-catalyzed RNA polymerization: Accurate and general RNA-templated primer extension. *Science* **292:** 1319–1325.

Joyce GF. 2002. The antiquity of RNA-based evolution. *Nature* **418:** 214–221.

Joyce GF, Visser GM, van Boeckel CAA, van Boom JH, Orgel LE, van Westrenen J. 1984. Chiral selection in poly(C)-directed synthesis of oligo(G). *Nature* **310:** 602–604.

Kanavarioti A, Monnard PA, Deamer DW. 2001. Eutectic phases in ice facilitate nonenzymatic nucleic acid synthesis. *Astrobiology* **1:** 271–281.

Kervio E, Sosson M, Richert C. 2016. The effect of leaving groups on binding and reactivity in enzyme-free copying of DNA and RNA. *Nucleic Acids Res* **44:** 5504–5514.

Kim HJ, Ricardo A, Illangkoon H, Kim MJ, Carrigan MA, Frye F, Benner SA. 2011. Synthesis of carbohydrates in mineral-guided prebiotic cycles. *J Am Chem Soc* **133:** 9457–9468.

Klussmann M, Iwamura H, Mathew SP, Wells DH Jr, Pandya U, Armstrong A, Blackmond DG. 2006. Thermodynamic control of asymmetric amplification in amino acid catalysis. *Nature* **441:** 621–623.

Koga S, Williams DS, Perriman AW, Mann S. 2011. Peptide-nucleotide microdroplets as a step towards a membrane-free protocell model. *Nat Chem* **3:** 720–724.

Kondepudi DK, Kaufman RJ, Singh N. 1990. Chiral symmetry breaking in sodium chlorate crystallization. *Science* **250:** 975–976.

Kreysing M, Keil L, Lanzmich S, Braun D. 2015. Heat flux across an open pore enables the continuous replication and selection of oligonucleotides towards increasing length. *Nat Chem* **7:** 203–208.

Lane N, Martin WF. 2012. The origin of membrane bioenergetics. *Cell* **151:** 1406–1416.

Larralde R, Robertson MP, Miller SL. 1995. Rates of decomposition of ribose and other sugars: Implications for chemical evolution. *Proc Natl Acad Sci* **92:** 8158–8160.

Lawrence MS, Bartel DP. 2003. Processivity of ribozyme-catalyzed RNA polymerization. *Biochemistry* **42:** 8748–8755.

Li L, Prywes N, Tam CP, O'Flaherty DK, Lelyveld VS, Izgu EC, Pal A, Szostak JW. 2017. Enhanced nonenzymatic RNA copying with 2-aminoimidazole activated nucleotides. *J Am Chem Soc* **139:** 1810–1813.

Lincoln TA, Joyce GF. 2009. Self-sustained replication of an RNA enzyme. *Science* **323:** 1229–1232.

Mansy SS, Schrum JP, Krishnamurthy M, Tobé S, Treco DA, Szostak JW. 2008. Template-directed synthesis of a genetic polymer in a model protocell. *Nature* **454:** 122–125.

Mariani A, Sutherland JD. 2017. Non-enzymatic RNA backbone proofreading through energy-dissipative recycling. *Angew Chemie* **56:** 6563–6566.

Mast CB, Schink S, Gerland U, Braun D. 2013. Escalation of polymerization in a thermal gradient. *Proc Natl Acad Sci* **110:** 8030–8035.

Matsumura S, Kun Á, Ryckelynck M, Coldren F, Szilágyi A, Jossinet F, Rick C, Nghe P, Szathmáry E, Griffiths AD. 2016. Transient compartmentalization of RNA replicators prevents extinction due to parasites. *Science* **354:** 1293–1296.

McGinness KE, Joyce GF. 2002. RNA-catalyzed RNA ligation on an external RNA template. *Chem Biol* **9:** 297–307.

Mizuno T, Weiss AH. 1974. Synthesis and utilization of formose sugars. *Adv Carbohyd Chem Biochem* **29:** 173–227.

Müller D, Pitsch S, Kittaka A, Wagner E, Wintner CE, Eschenmoser A. 1990. Chemie von α-aminonitrilen. Aldomerisierung von Glykolaldehydphosphat zu *racemischen* hexose-2,4,6-triphosphaten und (in gegenwart von formaldehyd). *Racemischen* pentose-2,4-diphosphaten: Rac-allose-2,4,6-triphosphat und rac-ribose-2,4-diphosphat sind die reaktionshauptprodukte. *Helv Chim Acta* **73:** 1410–1468.

Murase Y, Nakanishi H, Tsuji G, Sunami T, Ichihashi N. 2018. In vitro evolution of unmodified 16S rRNA for simple ribosome reconstitution. *ACS Synth Biol* **7:** 576–583.

Narlikar GJ, Herschlag D. 1997. Mechanistic aspects of enzymatic catalysis: Lessons from comparison of RNA and protein enzymes. *Annu Rev Biochem* **66:** 19–59.

Noorduin WL, Izumi T, Millemaggi A, Leeman M, Meekes H, Van Enckevort WJP, Kellogg RM, Kaptein B, Vlieg E, Blackmond DG. 2008. Emergence of a single solid chiral state from a nearly racemic amino acid derivative. *J Am Chem Soc* **130:** 1158–1159.

O'Flaherty DK, Kamat NP, Mirza FN, Li L, Prywes N, Szostak JW. 2018. Copying of mixed-sequence RNA templates inside model protocells. *J Am Chem Soc* **140:** 5171–5178.

Orgel LE. 2004. Prebiotic chemistry and the origin of the RNA world. *Crit Rev Biochem Mol Biol* **39:** 99–123.

Patel BH, Percivalle C, Ritson DJ, Duffy CD, Sutherland JD. 2015. Common origins of RNA, protein and lipid precursors in a cyanosulfidic protometabolism. *Nat Chem* **7:** 301–307.

Pinheiro AV, Han D, Shih WM, Yan H. 2011. Challenges and opportunities for structural DNA nanotechnology. *Nat Nanotechnol* **6:** 763–772.

Pitsch S, Wendeborn S, Jaun B, Eschenmoser A. 1993. Why pentose- and not hexose-nucleic acids? Pyranosyl-RNA ("p-RNA"). *Helv Chim Acta* **76:** 2161–2183.

Pitsch S, Krishnamurthy R, Bolli M, Wendeborn S, Holzner A, Minton M, Lesueur C, Schlönvogt I, Jaun B, Eschenmoser A. 1995. Pyranosyl-RNA ("p-RNA"): Base-pairing selectivity and potential to replicate. *Helv Chim Acta* **78:** 1621–1635.

Pitsch S, Wendeborn S, Krishnamurthy R, Holzner A, Minton M, Bolli M, Miculka C, Windhab N, Micura R, Stanek M, et al. 2003. Pentopyranosyl oligonucleotide systems. 9th communication. The β-D-ribopyranosyl-(4′→2′)-oligonucleotide system ("pyranosyl-RNA"): Synthesis and resumé of base-pairing properties. *Helv Chim Acta* **86:** 4270–4363.

Pizzarello S, Zolensky M, Turk KA. 2003. Nonracemic isovaline in the Murchison meteorite: Chiral distribution and mineral association. *Geochim Cosmochim Acta* **67:** 1589–1595.

Powner MW, Gerland B, Sutherland JD. 2009. Synthesis of activated pyrimidine ribonucleotides in prebiotically plausible conditions. *Nature* **459:** 239–242.

Powner MW, Sutherland JD, Szostak JW. 2010. Chemoselective multicomponent one-pot assembly or purine precursors in water. *J Am Chem Soc* **132:** 16677–16688.

Prywes N, Blain JC, Del Frate F, Szostak JW. 2016. Nonenzymatic copying of RNA templates containing all four letters is catalyzed by activated oligonucleotides. *eLife* **5:** e17756.

Rajamani S, Ichida JK, Antal T, Treco DA, Leu K, Nowak MA, Szostak JW, Chen IA. 2010. Effect of stalling after mismatches on the error catastrophe in nonenzymatic nucleic acid replication. *J Am Chem Soc* **132:** 5880–5885.

Rasi S, Mavelli F, Luisi PL. 2004. Matrix effect in oleate micelles–vesicles transformation. *Orig Life Evol Biosph* **34:** 215–224.

Ricardo A, Carrigan MA, Olcott AN, Benner SA. 2004. Borate minerals stabilize ribose. *Science* **303:** 196.

Rushdi AI, Simoneit BR. 2001. Lipid formation by aqueous Fischer–Tropsch-type synthesis over a temperature range of 100 to 400 degrees C. *Orig Life Evol Biosph* **31:** 103–118.

Robertson MP, Ellington AD. 1999. In vitro selection of an allosteric ribozyme that transduces analytes into amplicons. *Nat Biotechnol* **17:** 62–66.

Robertson MP, Joyce GF. 2014. Highly efficient self-replicating RNA enzymes. *Chem Biol* **21:** 238–245.

Robertson MP, Scott WG. 2007. The structural basis of ribozyme-catalyzed RNA assembly. *Science* **315:** 1549–1553.

Rogers J, Joyce GF. 2001. The effect of cytidine on the structure and function of an RNA ligase ribozyme. *RNA* **7:** 395–404.

Samanta B, Joyce GF. 2017. A reverse transcriptase ribozyme. *eLife* **6:** e31153.

Sanchez RA, Orgel LE. 1970. Studies in prebiotic synthesis. V. Synthesis and photoanomerization of pyrimidine nucleosides. *J Mol Biol* **47:** 531–543.

Shechner DM, Grant RA, Bagby SC, Koldobskaya Y, Piccirilli JA, Bartel DP. 2009. Crystal structure of the catalytic core of an RNA polymerase ribozyme. *Science* **326:** 1271–1275.

Schöning K, Scholz P, Guntha S, Wu X, Krishnamurthy R, Eschenmoser A. 2000. Chemical etiology of nucleic acid structure: The α-threofuranosyl-(3′→2′) oligonucleotide system. *Science* **290:** 1347–1351.

Schulman R, Yurke B, Winfree E. 2012. Robust self-replication of combinatorial information via crystal growth and scission. *Proc Natl Acad Sci* **109:** 6405–6410.

Schwartz AW, Orgel LE. 1985. Template-directed polynucleotide synthesis on mineral surfaces. *J Mol Evol* **21:** 299–300.

Sczepanski JT, Joyce GF. 2014. A cross-chiral RNA polymerase ribozyme. *Nature* **515:** 440–442.

Shan S, Yoshida A, Sun S, Piccirilli JA, Herschlag D. 1999. Three metal ions at the active site of the *Tetrahymena* group I ribozyme. *Proc Natl Acad Sci* **96:** 12299–12304.

Shan S, Kravchuk AV, Piccirilli JA, Herschlag D. 2001. Defining the catalytic metal ion interactions in the *Tetrahymena* ribozyme reaction. *Biochemistry* **40:** 5161–5171.

Shih WM, Quispe JD, Joyce GF. 2004. A 1.7-kilobase single-stranded DNA that folds into a nanoscale octahedron. *Nature* **427:** 618–621.

Soai K, Shibata T, Morioka H, Choji K. 1995. Asymmetric autocatalysis and amplification of enantiomeric excess of a chiral molecule. *Nature* **378:** 767–768.

Springsteen G, Joyce GF. 2004. Selective derivatization and sequestration of ribose from a prebiotic mix. *J Am Chem Soc* **126:** 9578–9583.

Stairs S, Nikmal A, Bučar DK, Zheng SL, Szostak JW, Powner MW. 2017. Divergent prebiotic synthesis of pyrimidine and 8-oxo-purine ribonucleotides. *Nat Commun* **8:** 15270.

Steitz TA, Moore PB. 2003. RNA, the first macromolecular catalyst: The ribosome is a ribozyme. *Trends Biochem Sci* **28:** 411–418.

Strobel SA, Ortoleva-Donnelly L. 1999. A hydrogen-bonding triad stabilizes the chemical transition state of a group I ribozyme. *Chem Biol* **6:** 153–165.

Sutherland JD. 2016. The origin of life—Out of the blue. *Angew Chemie* **55:** 104–121.

Szostak JW. 2011. An optimal degree of physical and chemical heterogeneity for the origin of life? *Philos Trans R Soc Lond B Biol Sci* **366:** 2894–2901.

Szostak JW, Bartel DP, Luisi PL. 2001. Synthesizing life. *Nature* **409:** 387–390.

Tang TYD, Hak CRC, Thompson AJ, Kuimova MK, Williams DS, Perriman AW, Mann S. 2014. Fatty acid membrane assembly on coacervate microdroplets as a step towards a hybrid protocell model. *Nat Chem* **6:** 527–533.

Tapiero CM, Nagyvary J. 1971. Prebiotic formation of cytidine nucleotides. *Nature* **231:** 42–43.

Viedma C. 2005. Chiral symmetry breaking during crystallization: Complete chiral purity induced by nonlinear autocatalysis and recycling. *Phys Rev Lett* **94:** 065504.

Viedma C, Ortiz JE, de Torres T, Izumi T, Blackmond DG. 2008. Evolution of solid phase homochirality for a proteinogenic amino acid. *J Am Chem Soc* **130:** 15274–15275.

Walton T, Szostak JW. 2016. A highly reactive imidazolium-bridged dinucleotide intermediate in nonenzymatic RNA primer extension. *J Am Chem Soc* **138:** 11996–12002.

Walton T, Szostak JW. 2017. A kinetic model of nonenzymatic RNA polymerization by cytidine-5′-phosphoro-2-aminoimidazolide. *Biochemistry* **56:** 5739–5747.

Wochner A, Attwater J, Coulson A, Holliger P. 2011. Ribozyme-catalyzed transcription of an active ribozyme. *Science* **332:** 209–212.

Wu T, Orgel LE. 1992a. Nonenzymatic template-directed synthesis on oligodeoxycytidylate sequences in hairpin oligonucleotides. *J Am Chem Soc* **114:** 317–322.

Wu T, Orgel LE. 1992b. Nonenzymatic template-directed synthesis on hairpin oligonucleotides. II. Templates containing cytidine and guanosine residues. *J Am Chem Soc* **114:** 5496–5501.

Xu J, Tsanakopoulou M, Magnani CJ, Szabla R, Šponer JE, Šponer J, Góra RW, Sutherland JD. 2017. A prebiotically plausible synthesis of pyrimidine β-ribonucleosides and their phosphate derivatives involving photoanomerization. *Nat Chem* **9:** 303–309.

Zaher HS, Unrau PJ. 2007. Selection of an improved RNA polymerase ribozyme with superior extension and fidelity. *RNA* **13:** 1017–1026.

Zhang G, Campbell EA, Minakhin L, Richter C, Severinov K, Darst SA. 1999. Crystal structure of *Thermus aquaticus* core RNA polymerase at 3.3 Å resolution. *Cell* **98:** 811–824.

Zhang L, Peritz A, Meggers E. 2005. A simple glycol nucleic acid. *J Am Chem Soc* **127:** 4174–4175.

Zhu TF, Szostak JW. 2009. Coupled growth and division of model protocell membranes. *J Am Chem Soc* **131:** 5705–5713.

Zhu TF, Adamala K, Zhang N, Szostak JW. 2012. Photochemically driven redox chemistry induces protocell membrane pearling and division. *Proc Natl Acad Sci* **109:** 9828–9832.

Cite as *Cold Spring Harb Perspect Biol* doi: 10.1101/cshperspect.a034801

The Origin of Biological Homochirality

Donna G. Blackmond

Department of Chemistry, The Scripps Research Institute, La Jolla, California 92037

Correspondence: blackmond@scripps.edu

SUMMARY

The fact that sugars, amino acids, and the biological polymers they construct exist exclusively in one of two possible mirror-image forms has fascinated scientists and laymen alike for more than a century. Yet, it was only in the late 20th century that experimental studies began to probe how biological homochirality, a signature of life, arose from a prebiotic world that presumably contained equal amounts of both mirror-image forms of these molecules. This review discusses experimental studies aimed at understanding how chemical reactions, physical processes, or a combination of both may provide prebiotically relevant mechanisms for the enrichment of one form of a chiral molecule over the other to allow for the emergence of biological homochirality.

Outline

Cite as *Cold Spring Harb Perspect Biol* doi: 10.1101/cshperspect.a032540

1 INTRODUCTION

The biopolymers that characterize life on Earth, and the molecular building blocks from which they are constructed, are both chiral and single-handed: that is, they have the potential to exist as either one of two nonsuperimposable mirror-image forms (Fig. 1), and they are selectively biosynthesized in only one of the two forms. Abiotic synthesis of chiral molecules in the absence of some type of chiral directing force produces an equal (racemic) mixture of both mirror-image molecules (enantiomers). The molecular "chirality" (a term originally coined by Kelvin 1904) of biological molecules is thus considered to be a fundamental signature of life (Blackmond 2010). Understanding how modern biology emerged from a presumably racemic prebiotic world represents an important step in understanding the origin of life.

In the history of chirality at both the molecular and macroscopic levels, investigations of chemical reactions that produce chiral molecules have been interwoven with studies of their physical properties. Although it is commonly agreed that the concept of molecular chirality was first introduced in Pasteur's famous 1848 physical separation of the mirror-image crystals of a tartrate salt (Pasteur 1848), the molecular basis of chirality was not in fact established for a further quarter century, when Le Bel and van't Hoff independently established the tetrahedral carbon atom (Le Bel 1874; van't Hoff 1874). This finding clarified that such an atom with four different substituents would be a chiral center with two mirror-image forms. Similarly, the physical phase behavior of amino acids was well-studied (Roozeboom 1899) a half century before the macroscopic helicity of biopolymers was characterized (Pauling et al. 1951; Dunitz 2001), which in turn preceded our certain knowledge of which of the two hands of the monomers (called enantiomers) is favored by biology (Cahn et al. 1956).

RNA, DNA, and polypeptides all form supramolecular right-handed α-helix structures, even while the amino acid molecules that make up peptides are designated as left-handed (L), and the molecular building blocks that comprise RNA and DNA—the sugars ribose and deoxyribose—are called right-handed (D) (Fig. 1). Although the focus of RNA world chemistry tends to fall on these macroscopic structures without questioning how single-handedness arose at the molecular level, an important consideration is how the relative amounts of the two hands of the monomers, given as percent enantiomeric excess (ee) (% ee = 100% ∗ (D − L)/(D + L)), influences the production of the biopolymers. Polymerization of RNA is effectively inhibited if a monomeric nucleotide of the wrong chirality is ligated to a growing chain (Joyce et al. 1984). Proteins, in contrast, are rather more promiscuous; incorporation of amino acid residues of opposite chirality in nonribosomal peptides can occur and has even been shown in some cases to add structural and functional diversification as well as to impart chemical robustness. However, this opposite chirality is achieved not by incorporating unnatural D-amino acid molecular building blocks but through posttranslational conversion of the natural hand of an amino acid molecule in a peptide chain to its enantiomer (Kreil 1997). Thus, the establishment of a highly enantioenriched, if not enantiopure (ee = 100%) pool of D-sugar and L-amino acid molecular building blocks would

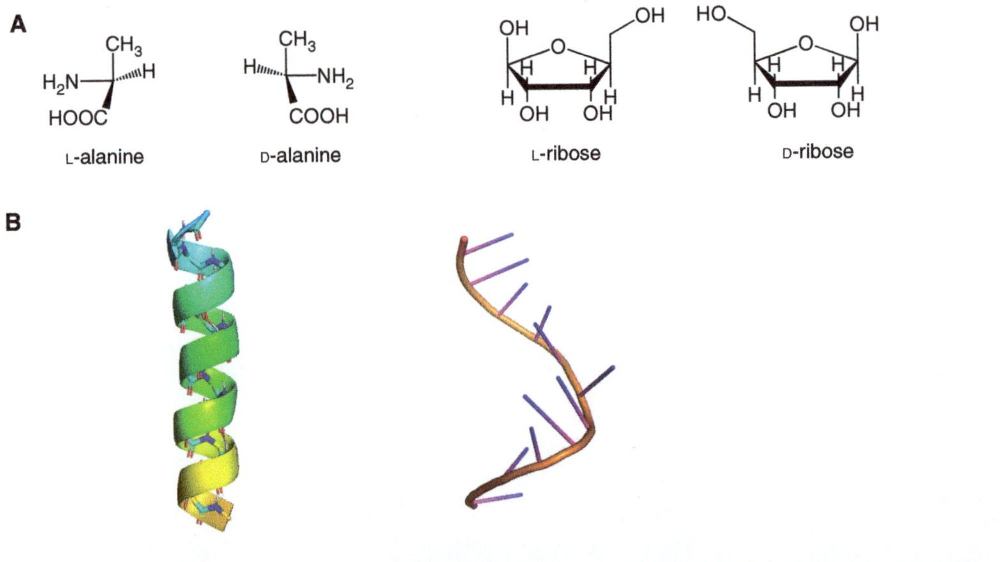

Figure 1. (*A*) Mirror-image molecules of alanine, a proteinogenic amino acid, and ribose, a pentose sugar. (*B*) Right-handed α-helix structures of a peptide from amino acids (*left*) and RNA from ribonucleotides (*right*).

appear to be a prerequisite for formation of the informational polymers implicated in the origin of life.

2 MODELS FOR THE ORIGIN OF HOMOCHIRALITY

Models for how homochirality might have emerged at the molecular level have been proposed since the early 1950s (Frank 1953). Interestingly, theoretical treatments preceded direct experimental investigations by nearly half a century. The question has two parts. The first concerns "symmetry breaking": What could have served as the original template for biasing production of one enantiomer over the other in the prebiotic world? The second concerns "amplification": How could a small imbalance of enantiomers be sustained and propagated to give us the biological world of single chirality that surrounds us?

2.1 Symmetry Breaking

Symmetry breaking is a term that has different meanings in different branches of science. To a physicist, it is a phenomenon in which infinitesimally small fluctuations acting on a system decide the system's fate by choosing which branch of a bifurcation is taken when crossing a critical point. To an outside observer unaware of the fluctuations (or "noise"), the choice will appear arbitrary. To a biologist, symmetry breaking is the process by which uniformity or invariance in a system is broken, linked to functional diversification from molecular assemblies to subcellular structures to cell types. To a chemist, symmetry breaking is the spontaneous generation of an excess of one enantiomer of a chiral molecule over the other in the absence of an external chiral source.

The chemist's definition of chirality meets that of the physicist's in the concept of a parity inversion, an operation that transforms a physical phenomenon (or a molecule) into its mirror image. Parity is conserved in electromagnetism, gravity, and strong interactions, but in the mid-20th century, it was discovered that parity is not conserved ("parity violation") in weak interactions. One of the consequences of parity violation is that a very small energy difference exists between enantiomers that previously had been considered identical. This finding led many to consider possible implications for the origin of biological homochirality. Work aimed at estimating the magnitude of this (exceedingly small) energy difference is ongoing, and although the question is not yet settled (Quack 2002), a relationship between biological homochirality and parity violation energy difference of enantiomers is not yet supported by either theoretical or experimental findings.

2.2 Chiral Amplification

Evidence of small enantiomeric excesses toward L-amino acids found in chondritic meteorite deposits (Elsila et al. 2016) allows the hypothesis that the initial imbalance between enantiomers on earth may have been seeded from an extraterrestrial source. A small imbalance is, however, a necessary but not sufficient condition to explain the emergence of homochirality, which requires some type of amplification or propagation mechanism following symmetry breaking. Thus, whether symmetry breaking to provide an imbalance in enantiomers was preordained or came about by chance, arising on earth or elsewhere, an amplification mechanism remains the key to increasing enantiomeric excess and ultimately to approaching the homochiral state. Experimental studies aimed at uncovering such processes, which began in earnest in the last decade of the 20th century, may be divided into either chemical or physical processes. The remainder of this review will highlight some of these experimental findings.

3 CHEMICAL MODELS

Studies exploring how enantioenrichment may result from chemical transformations fall into a number of categories, including autocatalytic "far-from-equilibrium" chiral amplification (Fig. 2) and reactions mediated by chiral catalysts or chiral auxiliaries (Fig. 3). The latter group of reactions includes the formose carbohydrate synthesis, the Strecker

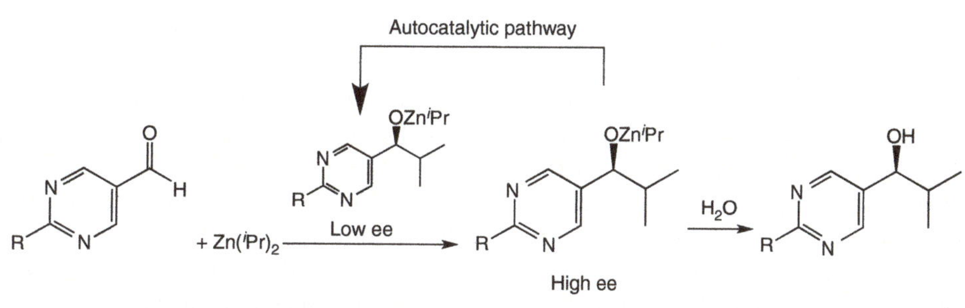

Figure 2. Autocatalytic amplification of enantiomeric excess in the Soai alkylation of pyrimidyl aldehydes.

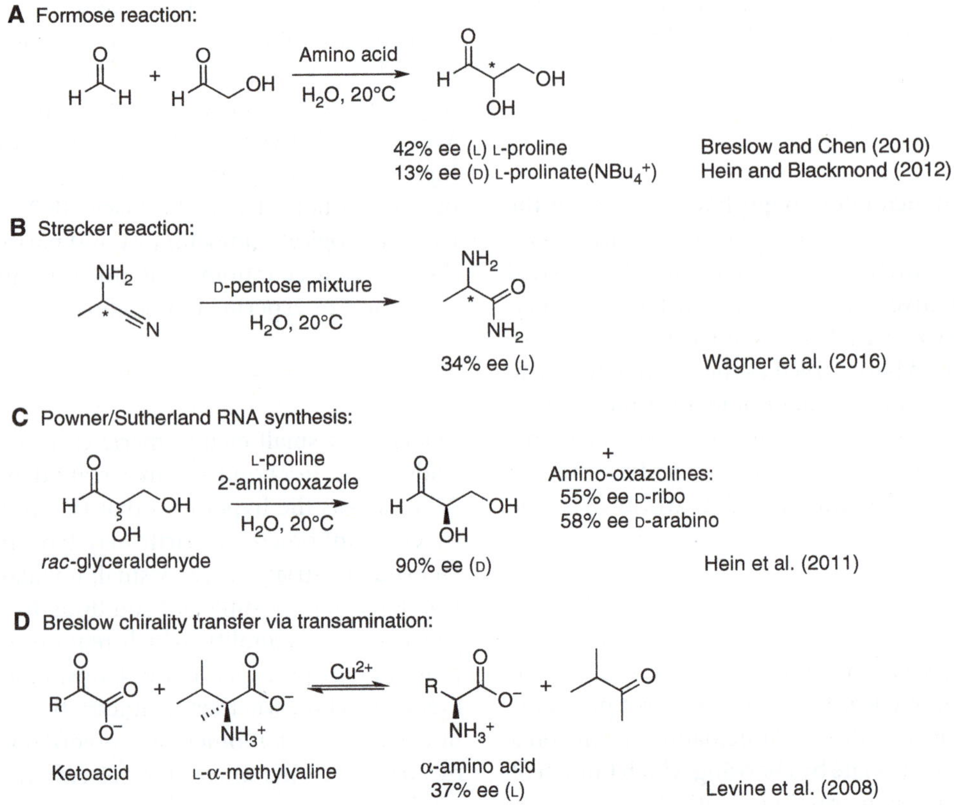

Figure 3. Prebiotically plausible reactions showing enantioenrichment of either sugars or amino acids.

amino acid synthesis, the Powner/Sutherland synthesis of RNA precursors, and chirality transfer reactions such as transaminations, and all involve chemistry that might occur under prebiotically plausible conditions. In contrast, the Soai alkylation of pyrimidyl aldehydes (Soai et al. 1995), which is the sole documented example of chiral amplification via autocatalysis, requires conditions far removed from those thought to exemplify the prebiotic world. However, because this reaction provided the first experimental proof-of-concept of earlier theoretical proposals, its study serves as a model for understanding how homochirality might have emerged and may help us search for prebiotically plausible reactions with similar characteristics.

3.1 Homochirality via Asymmetric Autocatalysis

In 1953 (a good year for research reports related to the origin of life), a mathematical model for the evolution of homochirality was proposed based on a chiral molecule that catalyzes its own self-production and at the same time acts to suppress synthesis of its enantiomer (Frank 1953). The last sentence of this theoretical work, which captivated several generations of chemists, contained an exhortation to experimental chemists: "a laboratory demonstration may not be impossible." More than 40 years passed before Soai and coworkers met this challenge with a report of the autocatalytic alkylation of pyrimidyl aldehydes with diisopropylzinc (Soai et al. 1995). The rate of the reaction is not only accelerated by addition of catalytic amounts of its own product, but this autocatalytic product may ultimately be formed in very "high" enantiomeric excess starting from a very "low" enantiomeric excess in the original catalyst (Fig. 2). Since this initial discovery, Soai's group has gone on to present remarkable further observations of asymmetric amplification in the reaction that now bears his name. Enantiomeric excesses as high as 85% were reported for a reaction initiated with a 0.1% ee mixture of chiral molecules produced from exposure to circularly polarized light (Shibata et al. 1998). Asymmetric amplification has also been observed for the reaction initiated by inorganic chiral materials such as quartz (Soai et al. 1999). Most recently, Soai has shown that the reaction may be selectively triggered solely by the minute mirror-image difference provided by isotope chirality of an initiator molecule (Kawasaki et al. 2009), showing that the reaction needs only an extremely small nudge to direct it consistently to the left or to the right.

The first mechanistic rationalization of this extraordinary reaction was reported in 2001 (Blackmond et al. 2001).

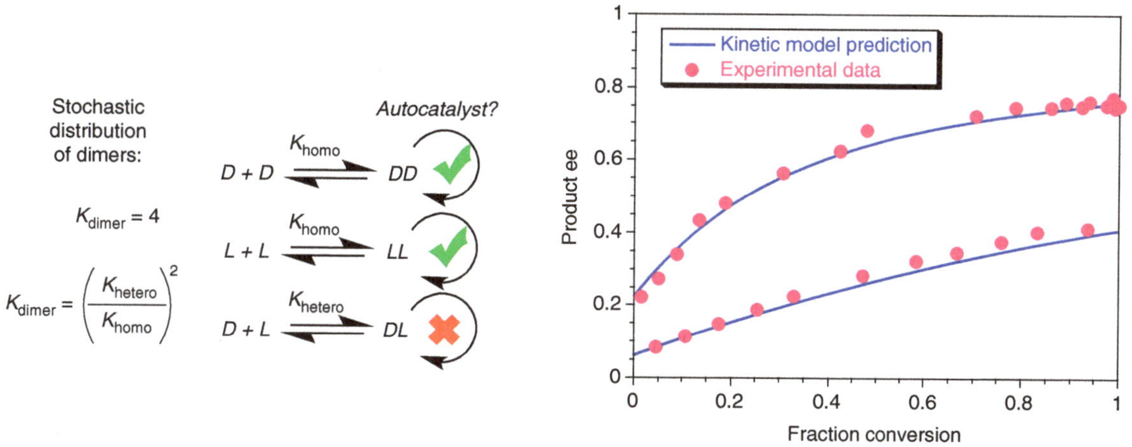

Figure 4. Blackmond/Brown model (*left*) and product enantiomeric excess as a function of fraction conversion of substrate (*right*) predicted from reaction rate profiles fit to the kinetic model (blue lines) and corroborated by analysis of experimental enantiomeric excess (ee) data.

A kinetic model developed based on highly accurate in situ measurements of the reaction's progress independently predicted both the temporal degree of asymmetric amplification, confirmed by compositional analysis, and the relative concentrations of the catalyst species, confirmed by nuclear magnetic resonance (NMR) spectroscopy. The reaction product dimerizes stochastically to form the active autocatalyst in the case of homochiral dimers (DD and LL), with the heterochiral dimer (DL) as an inactive reservoir. Figure 4 shows how the kinetic model compares with experimental data for asymmetric amplification in the Soai reaction performed out with two different initial catalyst ee values. This work showed that the Soai reaction couples autocatalysis with a form of "mutual antagonism" between enantiomers, thus evoking the main features of the Frank model. Further mechanistic studies have refined this reaction as a model for symmetry breaking and autocatalytic amplification of ee. The structure and possible role of tetrameric catalysts first proposed by Brown and coworkers (Gridnev and Brown 2004) was supported by kinetic and spectroscopic measurements by Blackmond, Brown, and coworkers (Buono and Blackmond 2003, Quaranta et al. 2010) before Soai and coworkers (Matsumoto et al. 2015) ultimately obtained a crystal structure that confirmed the tetrameric species first proposed by Brown. Most recently, a mechanistic rationalization for the role of isotopically chiral molecules in the Soai reaction has shed light on both the processes of symmetry breaking and asymmetric amplification via autocatalysis (Hawbaker and Blackmond 2018).

Because small-molecule autocatalysis may be thought of as the simplest form of self-replication, it remains an appealing scenario for the RNA community accustomed to believing about macromolecular self-replication. The discovery of a prebiotically relevant reaction, or a network of reactions, that obey the precepts of Frank model autocatalysis remains something of a "Holy Grail" for researchers probing the origin of biological homochirality (Blackmond 2004).

3.2 Reactions Mediated by Chiral Entities

A number of different transformations have been documented that produce enantioenriched sugar or amino acid building blocks under conditions with greater prebiotic relevance than the Soai autocatalytic system discussed above. Two venerable reactions known for more than a century and a half provide routes to the two classes of biological molecules for which chirality is key: the formose sugar synthesis, discovered in 1861 (Butlerow 1861) and unveiled mechanistically by Breslow in 1959 (Breslow 1959), and the Strecker amino acid synthesis discovered in 1850 (Strecker 1850). Although most reports do not address chirality, recent studies have begun to show how chiral catalysts or auxiliaries could provide partial product enantioenrichment. Interestingly, these reactions invoke a kind of synergy between the sugars and amino acids, as one of these two classes of molecule helps to direct enantioselectivity in the synthesis of the other: amino acids act as catalysts to enantioenrich sugars in the formose reaction (Fig. 3A), and sugars enantioselectively resolve amino acid precursors in the Strecker hydration of aminonitriles (Fig. 3B).

The Strecker reaction between aldehydes, cyanide, and ammonia represents a prebiotically plausible route to amino acids, forming first a racemic mixture of aminonitriles. Directing the hydration of these molecules in a dynamic kinetic resolution has been proposed as a pathway for enantioenrichment, and C4-C6 aldehydic carbohydrates can act as catalysts in this transformation. Derivatives of these sugars have been found in meteorites with significant excesses toward the natural D-sugars. Strikingly, mixtures of the natural D-pentoses produce enantioenriched L-aminoamides, which are then further hydrolyzed to the natural L-amino acids (Fig. 3B). This work reiterates the "systems chemistry" aspect of emergence of enantioenrichment from complex mixtures, and it hints at a prebiotic role for other pentose sugars that would have been present in quantities similar to ribose and deoxyribose before these sugars were ultimately chosen for their role in biological polymers (Wagner et al. 2016).

The Powner/Sutherland synthesis of RNA (Powner et al. 2009) represented a leap forward for the RNA world hypothesis because it showed for the first time a synthesis of activated RNA nucleotide monomers from simple prebiotically plausible feedstock molecules. Starting from D-glyceraldehyde, the synthesis bypasses the problematic step of combining ribose with a nucleobase. The work also highlighted "systems chemistry" aspects of emergent reactivity in complex mixtures. Enantioenrichment in this reaction network was established later when it was shown (Hein et al. 2011) that natural L-amino acids preferentially sequester unnatural L-glyceraldehyde, leaving D-glyceraldehyde free for RNA synthesis (Fig. 3C). Even partial enantioenrichment suffices to produce homochiral RNA building blocks; enantiopure crystals of the ribo-amino-oxazoline naturally precipitate from reaction mixtures above 20% ee.

Transfer of chirality from nonracemic meteoritic amino acids to proteinogenic amino acids was shown by Breslow and coworkers (Levine et al. 2008) in transamination reactions catalyzed by Cu ions (Fig. 3D). This work suggests how chirality first seeded extraterrestrially could provide some level of enantioenrichment in the building blocks of life. However, because the enantiomeric excess of the reaction products is lower than that of the molecules from which they are transferred, further amplification mechanisms would be required to produce significant enantioenrichment.

4 PHYSICAL MODELS

Understanding how the physical properties of chiral molecules might be exploited for enantioenrichment requires an understanding of their physical phase behavior—their solubilities and the manner in which they crystallize (Jacques et al. 1994). These concepts, which have been studied extensively for more than a century, are illustrated in Figure 5. When both enantiomers of a molecule are present in solution, crystallization proceeds either with both enantiomers mixed in a single crystal or with enantiomers forming separate crystals. The formation of separate D and L crystals, as in those Pasteur separated with his tweezers, are known as "conglomerates" (Fig. 5, right panel). When the mixed enantiomers form a regular 1:1 D:L matrix, the molecule is called a "racemic compound" (Fig. 5, left panel). The difference in the two types of crystals comes about as a result of characteristics of the molecules that make their solid phase stability greater with heterochiral interactions or homochiral interactions. The former makes up ~90% of all known chiral compounds, and indeed 17 of the 19 proteinogenic amino acids are racemic compounds.

When a system containing an unequal number of D and L molecules in a solvent reaches equilibrium between solid and solution phase molecules, the Gibbs phase rule for a ternary system (two enantiomers and one solvent) with three phases (two solid phases and one solution phase) dictates that, at fixed temperature and pressure, the solution composition is fixed at a composition characteristic of that system known as its eutectic composition. An important implication of the phase rule is that a partitioning of enantiomers occurs between the solution and solid in both cases. For conglomerates, because the solubilities of the two enantiomers are identical, the saturated solution will contain an equal number of D and L molecules regardless of how great the overall imbalance between D and L, suggesting the potential for enantioenrichment in the solid phase. For racemic compounds with an unequal number of D and L molecules, the solid phase will form the heterochiral

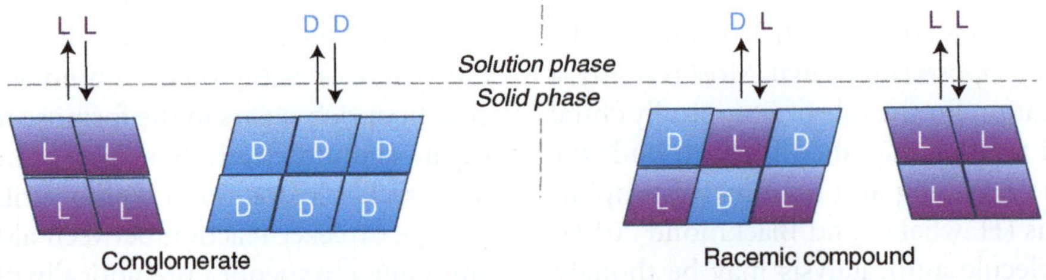

Figure 5. Equilibrium between chiral crystals and their saturated solution phases for systems comprised of an unequal number of D and L molecules. (*Right*) Racemic compound, in which heterochiral interactions are stronger than homochiral interactions; (*left*) conglomerate, in which homochiral interactions dominate.

1:1 D:L solid with all of the minor enantiomer present in the solid phase, while the remaining solid phase molecules of the major enantiomer will be left to form homochiral crystals. The relative solubility of the D:L solid compared with that of the homochiral solid dictates the magnitude of the solution phase imbalance between D and L. In this case, enantioenrichment in the solution phase is common. Below, we discuss models developed for the emergence of homochirality for each type of crystal in turn.

4.1 "Eve Crystal" Model for Conglomerates

It has been known for more than a hundred years that certain molecules such as $NaClO_3$ are achiral in solution but crystallize as conglomerates, typically forming equal amounts of left and right-handed crystals (Kipping and Pope 1898). Under some conditions, the system breaks symmetry during the crystallization and produces more of one hand than the other, even in the absence of a detectable chiral source. The most striking example of this phenomenon was reported by Kondepudi (Kondepudi et al. 1990) who showed that when a homogeneous solution of achiral $NaClO_3$ was cooled to initiate crystallization, the process could be directed if it was accompanied by rapid stirring. In this case, all of the crystals formed were of single chirality, randomly left-handed or right-handed in repeated experiments. This was rationalized by considering the dynamics of crystallization: Formation of the very first crystal (primary nucleation) from a homogeneous supersaturated solution is energetically unfavorable, although crystal growth by adding molecules to crystals already present (secondary nucleation) occurs much more readily. Under rapid stirring, if the first crystal formed, or "Eve" crystal, is broken by into thousands of smaller crystals by impact with the stirring bar, all of these "daughter" crystals are necessarily of the same chiral form, and they grow rapidly by secondary nucleation from solution molecules. If secondary nucleation occurs fast enough before nucleation of any new primary crystals, a single chiral solid state may result (McBride and Carter 1991). This phenomenon, shown in Figure 6 bears resemblance to "far-from-equilibrium" autocatalytic reaction processes such as the Soai reaction discussed earlier.

4.2 Attrition-Enhanced Deracemization of Conglomerates

The same achiral salt $NaClO_3$ became the subject of another intriguing process resulting in solid phase homochirality about fifteen years after Kondepudi's Eve crystal model was introduced. A report by Viedma (Viedma 2005) described a remarkable experiment that starts at what would be the end of a conventional crystallization with an equal number of right- and left-hand crystals of $NaClO_3$ in saturated aqueous solution. Because the system is not supersaturated, no new crystals nucleate; the only processes occurring in the flask are the continual dissolution and reaccretion of $NaClO_3$ molecules to and from existing crystals, and these rates are balanced at equilibrium. Thermodynamics tells us that no net change should occur, and that is exactly what has been found in such experiments for more than a hundred years. Viedma changed all this when he observed that the system evolved over time to "one" of the two chiral solids in a random fashion, provided that he added glass beads to the stirred vial. Attrition by glass beads produces a greater number of smaller crystals, which show higher solubility according to the Gibbs-Thomson rule, causing a slight supersaturation of $NaClO_3$ in solution. Not sufficiently supersaturated to support primary nucle-

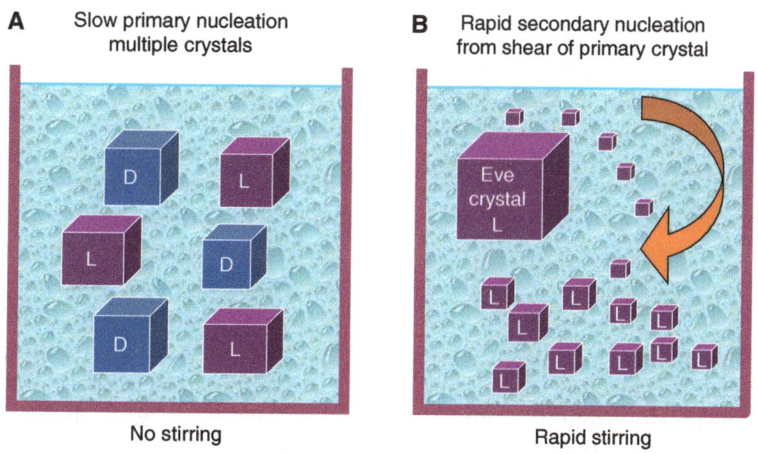

Figure 6. Effect of stirring on rates of primary (*A*) versus secondary (*B*) nucleation in the "Eve crystal" model of conglomerate-forming $NaClO_3$.

Figure 7. Evolution of solid phase homochirality based on dissolution and reaccretion of molecules from small crystals (shown in blue) of one enantiomer onto larger crystals (shown in purple) of the other enantiomer, through the conduit of solution phase interconversion between enantiomers.

ation, the system strives to redress the balance between solid and solution by increasing the rate of reaccretion of solution phase $NaClO_3$ onto existing crystals. A key point is that once a molecule of $NaClO_3$ dissolves from a crystal, it no longer possesses chirality and it retains no memory of the chiral form it previously showed as part of a crystal. Solution-phase $NaClO_3$ thus has no preference for reaccreting to a left- or right-handed crystal. What solution phase molecules do have, however, is a preference for adding to larger crystals over smaller ones, a phenomenon known as Ostwald ripening. If, by chance, the system contains a predominance of large crystals of one hand, solution phase $NaClO_3$ will preferentially add to these crystals, and the quantity of this enantiomorphic solid will increase relative to its mirror-image form (Fig. 7). Evolution of solid phase homochirality in this case requires only a chance imbalance in the crystal size of left versus right-hand crystals, with the solution phase serving as the conduit through which the molecules forget their solid-state chiral history and are free to choose a new solid-state chiral destiny.

These results led many to consider a possible extension of this process from the achiral $NaClO_3$ to systems more relevant to the origin of life, intrinsically chiral molecules that form conglomerate solids. In this case, the solution conduit between the two solids arises by the process through which chiral molecules rapidly interconvert between their left and right forms, known as racemization. This was successfully shown first for an amino acid derivative (Noorduin et al. 2008) and then for the proteinogenic amino acid aspartic acid (Viedma et al. 2008).

4.3 Crystal Engineering of Racemic Compounds

Morowitz was the first to recognize that solution-phase enantioenrichment is possible for racemic compounds if the heterochiral crystal is relatively insoluble compared with the homochiral crystal (Morowitz, 1969). This work was elaborated by measuring the eutectic composition for a variety of amino acids and nucleosides (Klussmann et al. 2006a; Breslow and Levine 2006; Breslow and Cheng 2009). The degree of chiral amplification is dictated by the thermodynamic properties of the particular compound. The proteinogenic amino acid serine shows a eutectic of 99% ee, meaning that a nearly homochiral solution is achievable even when the overall ee of the solid and solution combined is close to racemic. Valine, on the other hand, shows a eutectic of 47% ee, which means that a much lower level of chiral amplification is possible for this amino acid. Further work discovered that the cocrystallization of a variety of small molecules with amino acids can alter the relative heterochiral/homochiral solubility and thus "engineer" the eutectic composition (Klussmann et al. 2006b; 2007). This finding helped to generalize the eutectic model, revealing a number of prebiotically plausible cocrystallizations that raised the eutectic composition of amino acids. For example, when valine cocrystallizes with fumaric acid, its eutectic increases to 99% ee, affording a path to solution–phase homochirality. Similarly, proline's eutectic composition increases from 50% ee to 99% ee when chloroform cocrystallizes with the racemic compound (Fig. 8). Similar concepts have been invoked to rationalize enantioenrichment during sublimation of amino acids (Fletcher et al. 2007).

5 COMBINING PHYSICAL AND CHEMICAL APPROACHES

A number of examples illustrate how prebiotic chemistry resulting in amplification of enantiomeric excesses of amino acids and sugars has been aided by physical processes. The selective thermodynamic partitioning of amino acid enantiomers between solution and solid phases has been invoked as a possible prebiotic mechanism for solution phase amplification of enantiomeric excess.

A combination of these physical and chemical amplification processes was shown for the proline-mediated reaction of glyceraldehyde and concomitant production of enantioenriched ribo-amino-oxazoline (Fig. 4C). Enantioenrichment of 1% ee L-proline was achieved by forming

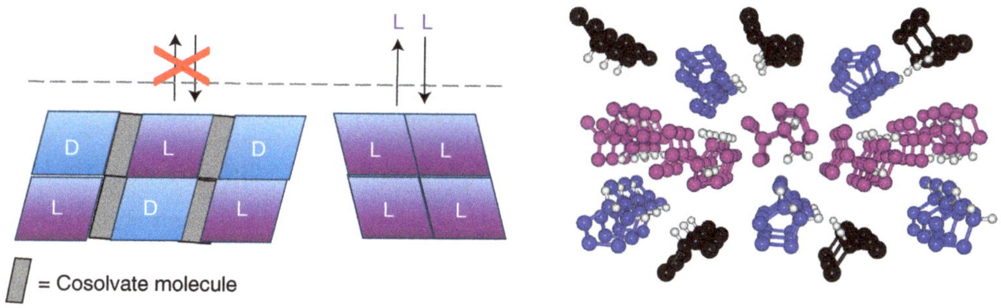

Figure 8. Cosolvate enriches the solution phase enantiomeric excess by suppressing the solutility of the racemic compound. (*Right*) Crystal structure of LD proline with CH_3Cl.

the $CHCl_3$-DL-proline cosolvate (Klussmann et al. 2006b). Nearly enantiopure solution phase proline was recovered and mixed with racemic glyceraldehyde and 2-amino-oxazole in aqueous solution to produce enantioenriched ribo- and arabino-amino-oxazolines in solution, from which enantiopure crystals of the ribo-amino-oxazoline precipitated (Hein et al. 2011). Thus, chemical amplification of ee is augmented by two physical amplification processes: enantioenrichment of the amino acid before the chemical reaction, and the selective crystallization of enantiopure ribo-amino-oxazoline following the chemical reaction.

A likely prebiotic scenario suggests that a complex pool of organic molecules would likely be present in any potential reaction medium. To test the robustness of these chirality amplification processes in multicomponent mixtures, the reaction of racemic glyceraldehyde with 2-amino-oxazole was performed in mixtures of a variety of different amino acids. Enantiopure crystals of ribo-amino-oxazoline were successfully obtained directly from crude reaction mixtures containing as many as fourteen of the nineteen chiral, proteinogenic amino acids (Hein et al. 2011). The emergence of molecules of single chirality from complex, multicomponent mixtures is a proof of "systems chemistry" (von Kiedrowski 2005) and supports the robustness of the processes under potential prebiotic conditions.

6 CONCLUDING REMARKS

The implications of the single chirality of biological molecules may be viewed in the context of increasing complexity at both the molecular and macroscopic levels. In the molecular sciences research, this has been codified as the field of "systems chemistry" (von Kiedrowski, 2005), which seeks to understand the chemical roots of biological organization by studying the emergence of system properties that may be different from those showed individually by the components in isolation. The abiotic construction of homochiral genetic polymers from molecular building blocks implies at least some minimal degree of enantioenrichment at the molecular level, if not molecular homochirality, to ensure efficiency as RNA or peptide chains grow in length. This review has described a variety of work invoking either chemical or physical processes, or both, that might allow the emergence of biological homochirality.

REFERENCES

Blackmond DG. 2004. Asymmetric autocatalysis and its implications for the origin of homochirality. *Proc Natl Acad Sci* **101:** 5732–5736.

Blackmond DG. 2010. The origin of biological homochirality. *Cold Spring Harb Perspect Biol* **2:** a002147.

Blackmond DG, McMillan CR, Ramdeehul S, Schorm A, Brown JM. 2001. Origins of asymmetric amplification in autocatalytic alkylzinc additions. *J Am Chem Soc* **123:** 10103–10104.

Breslow R. 1959. On the mechanism of the formose reaction. *Tetrahedron Lett* 22–36.

Breslow R, Cheng Z-L. 2009. On the origin of terrestrial homochirality for nucleosides and amino acids. *Proc Natl Acad Sci* **106:** 9144–9146.

Breslow R, Cheng Z-L. 2010. L-Amino acids catalyze the formation of an excess of D-glyceraldehyde, and thus of other D sugars, under credible prebiotic conditions. *Proc Natl Acad Sci* **107:** 5723–5725.

Breslow R, Levine M. 2006. Amplification of enantiomeric concentrations under credible prebiotic conditions. *Proc Natl Acad Sci* **103:** 12979–12980.

Buono FG, Blackmond DG. 2003. Kinetic evidence for a tetrameric transition state in the asymmetric alkylation of pyrimidyl aldehydes. *J Am Chem Soc* **125:** 8978–8979.

Butlerow A. 1861. Bildung einer zuckerartigen substanz durch synthese. *Justus Liebigs Annalen der Chemie* **120:** 295–298.

Cahn RS, Ingold CK, Prelog V. 1956. The specification of asymmetric configuration in organic chemistry. *Experientia* **12:** 81–94.

Dunitz JD. 2001. Pauling's left-handed α-helix. *Angew Chemie Int Ed.* **40:** 4167–4173.

Elsila JE, Aponte JC, Blackmond DG, Burton AS, Dworkin JP, Glavin DP. 2016. Meteoritic amino acids: Diversity in compositions reflects parent body histories. *ACS Central Sci* **2:** 370–379.

Fletcher SP, Jagt RBC, Feringa BL. 2007. An astrophysically-relevant mechanism for amino acid enantiomer enrichment. *Chem Commun* 2578–2580.

Frank FC. 1953. On spontaneous asymmetric synthesis. *Biochimica et Biophysica Acta* **11:** 459–463.

Gridnev ID, Brown JM. 2004. Solution structure and reagent binding of the zinc alkoxide catalyst in the Soai asymmetric autocatalytic reaction. *Proc Natl Acad Sci* **101:** 5727–5731.

Hawbaker N, Blackmond DG. 2018. Rationalization of asymmetric amplification via autocatalysis triggered by isotopically chiral molecules. *ACS Cent Sci* doi: 10.1021/acscentsci.8b00297.

Hein JE, Blackmond DG. 2012. On the origin of single chirality of amino acids and sugars in biogenesis. *Acc Chem Res* **45:** 2045–2054.

Hein JE, Tse E, Blackmond DG. 2011. A route to enantiopure RNA precursors from nearly racemic starting materials. *Nat Chem* **3:** 704–706.

Jacques J, Collet A, Wilen SH. 1994. *Enantiomers, racemates and resolution*. Krieger, Malabar, FL.

Joyce GF, Visser GM, van Boekel CAA, van Boom JH, Orgel LE. 1984. Chiral selection in poly(C)-directed synthesis of oligo(G). *Nature* **310:** 602–604.

Kawasaki T, Matsumura Y, Tsutsumi T, Suzuki K, Ito M, Soai K. 2009. Asymmetric autocatalysis triggered by carbon isotope (13C/12C) chirality. *Science* **324:** 492–495.

Kelvin WT. 1904. *Baltimore lectures on molecular dynamics and the wave theory of light*. C.J. Clay, London.

Kipping WS, Pope WJ. 1898. Enantiomorphism. *J Chem Soc Trans* **73:** 606–617.

Klussmann M, Iwamura H, Mathew SP, Wells DH Jr, Pandya U, Armstrong A, Blackmond DG. 2006a. Thermodynamic control of asymmetric amplification in amino acid catalysis. *Nature* **441:** 621–623.

Klussmann M, White AJP, Armstrong A, Blackmond DG. 2006b. Rationalization and prediction of solution enantiomeric excess in ternary phase systems. *Angew Chem Int Ed* **47:** 7985–7989.

Klussmann M, Izumi T, White AJP, Armstrong A, Blackmond DG. 2007. Emergence of solution-phase homochirality via crystal engineering of amino acids. *J Am Chem Soc* **123:** 7657–7660.

Kondepudi DK, Kaufman RJ, Singh N. 1990. Chiral symmetry breaking in sodium chlorate crystallization. *Science* **250:** 975–976.

Kreil G. 1997. D-amino acids in animal peptides. *Annu Rev Biochem* **66:** 337–345.

Le Bel JA. 1874. Sur des relations qui existent entre les formules atomiques des corps organiques et la pouvoir rotatoire de leur dissolutions. *Bull Soc Chim Fr* **22:** 337–347.

Levine M, Kenesky CS, Mazori D, Breslow R. 2008. Enantioselective synthesis and enantiomeric amplification of amino acids under prebiotic conditions. *Org Lett* **10:** 2433–2436.

Matsumoto A, Abe T, Hara A, Tobita T, Sasagawa T, Kawasaki T, Soai K. 2015. Crystal structure of the isopropylzinc alkoxide of pyrimidyl alkanol: Mechanistic insights for asymmetric autocatalysis with amplification of enantiomeric excess. *Angew Chemie Int Ed* **127:** 15433–15436.

McBride JM, Carter RL. 1991. Spontaneous resolution by stirred crystallization. *Angew Chem Intl Ed* **30:** 293–295.

Morowitz M. 1969. A mechanism for the amplification of fluctuations in racemic Mixtures. *J Theor Biol* **25:** 491–494.

Noorduin WL, Izumi T, Millemaggi A, Leeman M, Meekes H, Van Enckevort WJP, Kellogg RM, Kaptein B, Vlieg E, Blackmond DG. 2008. Emergence of a single solid chiral state from a nearly racemic amino acid derivative. *J Am Chem Soc* **130:** 1158–1159.

Pasteur L. 1848. Relation qui peut exister entre la forme crystalline et la compostion chimique, et sur las cause de la polarization rotatoire. *Ann Chim Phys* **26:** 535–538.

Pauling L, Corey RB, Branson HR. 1951. The structure of proteins: Two hydrogen-bonded helical configurations of the polypeptide chain. *Proc Natl Acad Sci* **37:** 205–211.

Powner MW, Gerland B, Sutherland JD. 2009. Synthesis of activated pyrimidine ribonucleotides in prebiotically plausible conditions. *Nature* **459:** 239–242.

Quack M. 2002. How important is parity violation for molecular and biomolecular chirality? *Angew Chem Int Ed* **41:** 4618–4630.

Quaranta M, Gehring T, Odell B, Brown JM, Blackmond DG. 2010. Unusual inverse temperature dependence on reaction rate in the asymmetric autocatalytic alkylation of pyrimidyl aldehydes. *J Am Chem Soc* **132:** 15104–15107.

Roozeboom HWB. 1899. Löslichkeit und schmelzpunkt als kriterion für racemische verbindungen, speuoracemische mischkrystalle und inactive conglomerate. *Z Phys Chem* **28:** 494–517.

Shibata J, Yamamoto T, Matsumoto N, Yonekubo S, Osanai S, Soai K. 1998. Amplification of a slight enantiomeric imbalance in molecules based on asymmetric autocatalysis. The first correlation between high enantiomeric enrichment in a chiral molecule and chircularly polarized light. *J Am Chem Soc* **120:** 12157–12158.

Soai K, Shibata T, Morioka H, Choji K. 1995. Asymmetric autocatalysis and amplification of enatiomeric excess of a chiral molecule. *Nature* **378:** 767–768.

Soai K, Osanai S, Kadowaki K, Yonekubo S, Shibata T, Sato I. 1999. D- and L-Quartz-promoted highly enantioselective synthesis of a chiral compound. *J Am Chem Soc* **121:** 11235–11236.

Strecker A. 1850. Ueber die künstliche Bildung der Milchsäure und einen neuen, dem Glycocoll homologen Körper. *Ann Chem Pharm* **75:** 27–45.

van't Hoff JH. 1874. *Voorstel tot uitbredning der tegenwoordig*. J. Greven, Utrecht.

Viedma C. 2005. Chiral symmetry breaking during crystallization: Complete chiral purity induced by nonlinear autocatalysis and recycling. *Phys Rev Lett* **94:** 065504.

Viedma C, Ortiz JE, de Torres T, Izumi T, Blackmond DG. 2008. Evolution of solid phase homochirality for a proteinogenic amino acid. *J Am Chem Soc* **130:** 15274–15275.

von Kiedrowski G. 2005. Systems Chemistry, Workshop at the European Center for Living Technology, Venice International University, Venice, Oct 3–4, 2005.

Wagner AJ, Zubarev D Yu, Asparu-Guzik A, Blackmond DG. 2016. Chiral sugars drive enantioenrichment in prebiotic amino acid synthesis. *ACS Central Sci* **3:** 322–328.

Cite as *Cold Spring Harb Perspect Biol* doi: 10.1101/cshperspect.a032540

Beyond DNA and RNA: The Expanding Toolbox of Synthetic Genetics

Alexander I. Taylor, Gillian Houlihan, and Philipp Holliger

Medical Research Council Laboratory of Molecular Biology, Cambridge CB2 0QH, United Kingdom

Correspondence: ph1@mrc-lmb.cam.ac.uk

SUMMARY

The remarkable physicochemical properties of the natural nucleic acids, DNA and RNA, define modern biology at the molecular level and are widely believed to have been central to life's origins. However, their ability to form repositories of information as well as functional structures such as ligands (aptamers) and catalysts (ribozymes/DNAzymes) is not unique. A range of nonnatural alternatives, collectively termed xeno nucleic acids (XNAs), are also capable of supporting genetic information storage and propagation as well as evolution. This gives rise to a new field of "synthetic genetics," which seeks to expand the nucleic acid chemical toolbox for applications in both biotechnology and molecular medicine. In this review, we outline XNA polymerase and reverse transcriptase engineering as a key enabling technology and summarize the application of "synthetic genetics" to the development of aptamers, enzymes, and nanostructures.

Outline

Cite as *Cold Spring Harb Perspect Biol* doi: 10.1101/cshperspect.a032490

1 INTRODUCTION

The capacity of DNA and RNA for accessible high-density information storage and propagation separates nucleic acids from other sequence-defined biopolymers (including proteins and peptides). This provides both the medium and the mechanism for Darwinian evolution and is as yet unmatched by any other polymer or other molecular system. Indeed, DNA can store as much as 200 petabytes of information per gram with high chemical stability. This function is underpinned by a unique chemistry including the polyanionic phosphodiester backbone, which dominates the physicochemical behavior and thereby decouples base sequence (i.e., information content) from molecular properties and Watson–Crick base-pairing (a combination of hydrogen-bonding and stacking interactions), which allows encoding and decoding of information in a redundant fashion.

1.1 Natural Chemical Variation

Given the stringency of molecular requirements for function, one might expect the chemical makeup of natural nucleic acids to be completely uniform. However, considerable chemical variation from the cognate DNA and RNA chemistry is found in nature. Such variation is both diverse and widespread, including a range of epigenetic and other markers in both prokaryotic and eukaryotic DNA. Although most occur sporadically, some modifications can completely replace the canonical bases in some bacteriophage genomes (reviewed in Weigele and Raleigh 2016). Arguably, such bacteriophages represent genetics based on alternative, genetic polymers, suggesting that integration of expanded chemistry sets into living systems (xenobiology) may be possible (Herdewijn and Marlière 2009).

In RNA, in addition to a diverse array of modifications found in transfer RNAs (tRNAs) (reviewed in Schimmel 2018), an expanding range of epigenetic messenger RNA (mRNA) markers modulating translation initiation, RNA half-life, and even decoding by the ribosome have been discovered (reviewed in Meyer and Jaffrey 2017).

1.2 Artificial Chemical Variation and Synthetic Genetics

Beyond the natural variations, organic chemistry has explored a much wider range of alternative backbones, sugar congeners, and base chemistries with the aim to better define the key molecular parameters required for nucleic acid function (Eschenmoser 1999). This has recently been extended to their potential for genetic information storage, propagation, and evolution. Termed "synthetic genetics," this approach promises both new insights into the chemical boundary conditions of genetics as well as new tools to study and modify biological processes. A prime example is the modification of nucleobases, utilizing alternative hydrogen bonding patterns, hydrophobic and/or geometric compatibility, or metal ion chelation, toward an expansion of the genetic alphabet in vitro and in vivo, as well as yielding secondary structure motifs beyond those of DNA and RNA (reviewed in Lee et al. 2018). Furthermore, in vitro replication and evolution of nucleic acid variants comprising backbone chemistries not found in nature (here referred to collectively as xeno nucleic acids [XNAs]) yields ligands (XNA aptamers) and enzymes (XNAzymes), as well as simple nanostructures with novel properties including increased biostability (Pinheiro et al. 2012; Yu et al. 2012; Taylor et al. 2015; 2016). Similarly, expansion of DNA molecular diversity (primarily through pyrimidine C5 substituent chemistry) has accelerated aptamer discovery (Gold et al. 2010) and improved nucleic acid catalysts (Zhou et al. 2016; Wang et al. 2018). Here, we provide an overview of the synthetic genetics toolbox, with a special focus on the impact of replacing the canonical ribofuranose sugar of DNA and RNA with synthetic congeners, informing broader chemical possibilities for biotechnology and genetic function (Fig. 1).

2 POLYMERASE ENGINEERING FOR SYNTHETIC GENETICS

High-fidelity polymerase enzymes catalyze synthesis and decoding of DNA sequence information with error rates as low as 1×10^{-7} coupled with an exquisite substrate specificity enforced by both minor groove readout and active site geometrical constraints. This provides for both accurate substrate selection and exclusion of damaged and/or noncognate nucleotides. Indeed, DNA polymerases must stringently select cognate deoxyribonucleotide triphosphates (dNTPs) from a pool that may contain between 10- and 100-fold higher concentration of ribonucleotide triphosphates (NTPs) as incorporation of NTPs into the genome leads to genomic instability. The molecular basis of NTP discrimination lies in bulky amino acid side chains (e.g., tyrosine in the case of the archaeal polB family polymerases like *Thermococcus gorgonarius* [Tgo]: Y409) (Cozens et al. 2012), which excludes nucleotides with 2′-OH substituents from the polymerase active site. This "steric gate" residue has analogs in all DNA polymerase families and is paradigmatic of the mechanisms polymerases use to ensure substrate specificity, understanding of which will be key for progress in synthetic genetics (Fig. 1).

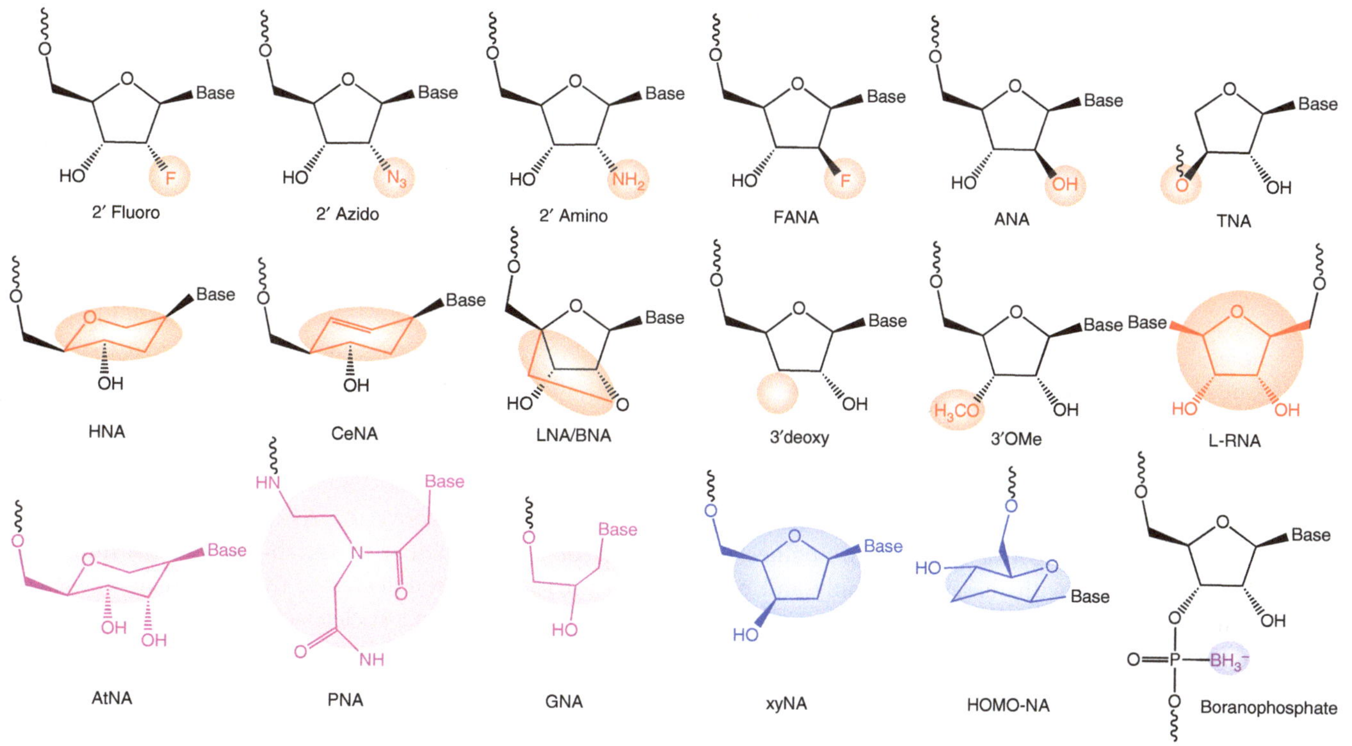

Figure 1. Chemical structures of unnatural nucleic acids investigated for synthetic genetics. Sugar modifications incorporated by polymerases are shown in red, whereas nucleic acids that currently cannot be synthesized by polymerases are shown in pink. Unnatural chemistries that do not hybridize to natural nucleic acids are shown in blue, and modifications to the phosphodiester backbone are shown in purple.

2.1 Engineering Primer-Dependent RNA Polymerases

The comparable simplicity of such molecular checkpoints has inspired efforts to engineer DNA polymerases into primer-dependent RNA polymerases (Houlihan et al. 2017a). Replacement of the steric gate with less bulky amino acids can improve NTP incorporation kinetics, up to levels comparable to dNTP incorporation kinetics by DNA polymerases, but still precludes RNA synthesis beyond short oligomers. Indeed, a second checkpoint in the thumb domain was recently discovered that enabled primer-dependent synthesis of longer RNAs (Cozens et al. 2012). Primer-dependent RNA synthesis has a number of applications, as it obviates the need to initiate transcription with a 5′-pppG and allows a wide range of chemical groups to be added during solid-phase synthesis of primers (e.g., divergent 5′-cap chemistries and cap-proximal epigenetic modification such as N^6,2′-*O*-dimethyladenosine [m^6A_m] and 7-methylguanosine [m^7G] [reviewed in Meyer and Jaffrey 2017]). Further engineering enabled incorporation of regioisomeric 2′-5′ linkages into both DNA and RNA by replacement of purine dNTPs with 3′-deoxy or 3′-*O*-methyl (3′OMe) analogs (Cozens et al. 2015).

2.2 Engineering Modified DNA and RNA Polymerases

To incorporate 2′ modified nucleotide substrates, much work has gone into engineering T7 RNA polymerase (T7RP) (e.g., for incorporation of dNTPs as well as 2′F- and 2′N_3-pyrimidine triphosphates [YTPs]). One such T7RP is available commercially as DuraScribe (Lucigen Corp.); it enables incorporation of 2′OMe-, 2′F-, and 2′N_3-YTPs or all four 2′OMe-NTPs provided GTP is also present to enable the initiation step (Burmeister et al. 2005).

DNA polymerases from the polA family have also been extensively engineered for 2′-modified substituents. Phage display was adapted to evolve the Stoffel fragment of *Taq* polymerase for 2′OMe-CTP and -ATP incorporation (Chen et al. 2016a). Limited synthesis was possible with a polymerase (SFM19) with a mutation at residue 615 (E615G), which had previously been identified as a "steric gate" mutation enabling NTP incorporation (Ong et al. 2006). Further evolution resulted in polymerase SFM4-3 (SFM19: V518A, N583S, D655N, E681K, E742Q, M747R), which was able to transcribe fully modified 2′OMe 60-mer templates (Chen et al. 2016a).

The success of this approach suggests that mutations in the finger domain can provide efficiency gains possibly by stabilizing its interaction with the thumb subdomain, thus favoring the catalytically competent closed complex.

2.3 Engineering XNA Polymerases and Reverse Transcriptases

Going beyond 2′ substitution on the RNA ribofuranose ring, wholesale replacement with alternative ring structures (congeners) has also been explored. Such nucleic acids are not found in nature and have therefore been termed xeno nucleic acids (XNAs). Such modifications can engender profound changes in the nucleic acid helix structure, stability, hybridization kinetics, and biological and chemical behavior (Anosova et al. 2016). Engineering polymerases compatible with these structurally and conformationally divergent chemistries from the canonical nucleic acids required the development of novel polymerase selection and screening methods (reviewed in Houlihan et al. 2017a,b). These enabled the discovery of polymerases for a range of XNAs, including 1,5-anhydrohexitol nucleic acid (HNA), cyclohexene nucleic acid (CeNA), 2′-*O*,4′-C-methylene-β-D-ribonucleic acids (locked or bridged nucleic acid, LNA/BNA), and arabino- (ANA) and 2′-fluoro-arabinonucleic acid (FANA) (Pinheiro et al. 2012). Another XNA chemistry, α-L-threofuranosyl nucleic acid (TNA), based on a threose scaffold (a possible "simpler" precursor of RNA) has also been explored, with mutants of the archaeal DNA polymerases Tgo and 9°N both able to synthesize TNA polymers (Pinheiro et al. 2012; Larsen et al. 2016).

Genetic information encoded in RNA can be retrieved using reverse transcriptases (RTs) derived from retroviruses. However, natural polymerases and RTs generally cannot reverse transcribe XNA strands into DNA. Notable exceptions include Bst (the large fragment of the *Bacillus stearothermophilus* DNA polymerase), which was recently shown to faithfully reverse transcribe TNA polymers containing TNA-A, -G, and -T and a fluorescent TNA-C nucleotide (Mei et al. 2017), and RTs from avian myeloblastosis virus (AMV) and Moloney murine monkey virus (M-MLV), which are able to reverse transcribe templates containing 2′OMe (Pan et al. 1995; Burmeister et al. 2005).

Although not as developed as polymerase engineering for forward synthesis, XNA-RT activity has also been discovered by screening or in vitro evolution strategies. Statistical coupling analysis (SCA) has been used to map polymerase positions involved in function; screening SCA-informed mutants of the Tgo polymerase led to the discovery of the critical active site- and C-motif-proximal 521 position. A subtle but crucial mutational change (I521L) enabled reverse transcription of a range of XNAs including HNA, ANA, FANA, and TNA (Pinheiro et al. 2012); introduction of E664K into the Tgo pol (shown to increase primer-template affinity [Cozens et al. 2012]) further extended XNA reverse transcription to LNA and CeNA. Recently, an emulsion-based selection method was adapted to evolve *Thermococcus kodakaraenis* (KOD) DNA polymerase, yielding a mutant (RTX) capable of high-fidelity RNA reverse transcription (Ellefson et al. 2016), with a threefold reduction in error rate compared with the M-MLV RT. RTX (with 3′-5′ exonuclease activity inactivated) could also reverse transcribe a fully modified 2′OMe oligomer.

Although the above XNAs hybridize specifically to DNA or RNA (albeit with variable affinity), a number of chemistries have been described that hybridize very weakly or not at all to natural nucleic acids (Anosova et al. 2016). These include glycol nucleic acid (GNA): R-GNA and S-GNA enantiomers pair with themselves, but not each other, and neither is able to cross-pair with DNA. Similarly, other orthogonal self-pairing systems that do not pair with DNA include xylo nucleic acid (xyNA), β-Homo-DNA, enantio-DNA, and the mirror image L-forms (L-DNA, L-RNA) of the natural D-form nucleic acids. Such chemistries cannot currently be replicated via a DNA intermediate and have proved challenging for polymerase engineering. However, S-GNA synthesis by Bst polymerase has been reported, suggesting that the polymerase active site can enforce and stabilize weak primer–template interactions (Tsai et al. 2007).

In summary, polymerase engineering has progressed rapidly, and both the synthesis and reverse transcription of a range of unnatural or modified nucleic acids have become routine. With multiple selection formats available (Houlihan et al. 2017a,b), the prospects for adapting polymerases to ever more diverse chemistries are good. Conversely, medicinal chemistry approaches (adapting nucleotide substrate to the polymerase) have proven remarkably successful, most notably in the discovery of unnatural base pairs (UBPs) (reviewed in Feldman and Romesberg 2017). However, a number of challenges remain. Catalytic efficiency and fidelity of polymerases with unnatural triphosphates is generally lower than with cognate dNTPs or NTPs. A recent crystal structure of the ternary complex of KOD-RI with a DNA primer–template duplex and a TNA triphosphate (tATP) bound revealed severe distortion of the nascent base pair when compared with the binary structure (Chim et al. 2017), suggesting an important role for structural investigations to guide future efforts in synthetic genetics.

3 EVOLVING FUNCTION USING SYNTHETIC GENETIC POLYMERS

3.1 Functional Nucleic Acids

Beyond the classical double-helix conformation, nucleic acids (and in particular RNA) are able to form a wide range of three-dimensional folds and structures (including inter alia hairpin loops, internal bulges, pseudoknots, and G-quartets), underpinned by canonical Watson–Crick as well as noncanonical intramolecular bonds (e.g., Hoogsteen pairs, base triples, GU wobble pairs). Such nucleic acid structures can, in some cases, match the ability of proteins to form sophisticated interaction surfaces (paratopes), comprising shape–charge complementarity, salt bridges, hydrogen bond networks, van der Waals forces, and metal ion binding pockets (Gelinas et al. 2016). The diverse ranges of structures and functions in modern biology have been formulated in the concept of the cell as an "RNA machine."

3.2 In Vitro Evolution of Functional Nucleic Acids

Nucleic acids are unique in their capacity to encompass both genotype (i.e., sequence information) and phenotype (i.e., positioning functional chemical groups in a defined 3D fold), thus enabling Darwinian evolution at the molecular level, as first shown in vitro by Spiegelman (reviewed in Joyce 2007). In vitro evolution experiments to identify synthetic functional nucleic acids by selection from diverse repertoires are typically termed "systematic evolution of ligands by exponential enrichment" (SELEX) methods (reviewed in Darmostuk et al. 2015). Fundamentally, SELEX relies on the generation of very large ($>10^{14}$) repertoires of single-stranded, random-sequence RNA or DNA repertoires that are panned for binding to a target of choice, allowing functional library members to be enriched from the bulk population through iterative cycles of selection and amplification.

Using this approach, a wide variety of synthetic functional nucleic acids have been developed, including RNA and DNA ligands (aptamers) (Dunn et al. 2017) or catalysts—ribozymes/DNAzymes (Wachowius et al. 2017).

3.3 Aptamers

Nucleic acid aptamers can be selected to bind a wide range of molecular targets including small molecules, proteins, or RNA motifs immobilized on solid phase, or extracellular structures on bacteria, viruses, or eukaryotic cells. Selections for cell-penetrating aptamers (Tawiah et al. 2017) and even selections directly in live animals have also been performed (Mi et al. 2010). In recent years, the field has grown exponentially and is increasingly recognized as a powerful biotechnology for biosensing, proteomics, diagnostics, imaging, and targeted drug delivery (Zhou and Rossi 2017; Hori et al. 2018).

In comparison with antibodies or other protein-based affinity reagents, aptamers offer a number of advantages, including automation of discovery and solid-phase synthesis. Unlike protein biologics, aptamers tolerate repeated cycles of heating, drying, and freezing without irreversible aggregation and have a very long shelf life without dependence on a temperature-controlled cold chain. Furthermore, aptamers show low immunogenicity (with the exception of CpG DNA motifs) and their small size (typically 5–20 kDa) offers improved tissue penetration compared with whole antibodies.

Aptamers can be incorporated into multimeric complexes (such as multispecific constructs, concatemers, dendrimers, and nanoparticles) in a relatively straightforward manner, allowing directed cross-linking of target molecules, or design of systems for hybridization-mediated signal amplification, as well as integration with nanotechnology designs for assembly of biosensors, enzyme arrays, and nanorobots (Jo and Ban 2016). However, a number of significant challenges to their widespread use in biotechnology have been identified. For one, the functional scope of aptamers is constrained by systemic shortcomings inherent in nucleic acid chemistry, including a comparatively uniform chemical makeup with the steric, electronic, and physicochemical properties of the four canonical bases spanning a rather limited range. This has spurred efforts to expand the functionality of aptamers (and functional nucleic acids in general) by expanding their chemical diversity through the addition of extraneous substituent chemistries.

3.4 Modifying Aptamers by Postselection Medicinal Chemistry

Despite setbacks in the clinical progression of aptamers as therapeutic agents, key hurdles for in vivo applications have begun to be addressed; many of the lessons learned from the antisense/short-interfering RNA (siRNA) community (reviewed in Wittrup and Lieberman 2015) have improved the prospects for aptamers—in particular, including diversification of backbone and substituent chemistry to enhance and fine-tune functionality as well as chemical conjugation to improve organ-specific or intracellular delivery and reduce renal clearance.

One of the most significant roadblocks to in vivo applications for all types of functional nucleic acid remains their poor biostability. Exo- and endonucleases that rapidly degrade naked DNA and RNA polymers are common components of all extracellular as well as intracellular (both endosomal and cytoplasmic) compartments. Typically, the half-life of single-stranded DNA or RNA aptamers in serum is minutes to hours, although a combination of adding small terminal DNA hairpins and engineering increased G-C content of stem structures can (at least in some cases) offer improved stability. Traditionally, medicinal chemistry cycles have been used to test each nucleotide position for substitution with chemical analogs that enhance biostability (Röthlisberger and Hollenstein 2018). Although such modifications can stabilize secondary structures and increase binding affinity in some contexts, extensive substitution can disrupt aptamer 3D structure or folding and consequently reduce target binding affinity.

3.5 Partially Modified DNA and RNA Aptamer Selections (modSELEX)

To expand beyond the limitations of the natural chemistries and bypass costly and time-consuming medicinal chemistry cycles, much effort has been expended on incorporating modified nucleotides into the SELEX process. Selections based on direct partitioning can allow discovery from modified libraries prepared by solid-phase synthesis, and for close steric and electronic analogs of the natural nucleotides, natural polymerases may be sufficient for both synthesis and reverse transcription. However, more divergent modifications, such as 2′-modified sugars or ring congeners have generally required the engineering of polymerases and RTs tolerant to alternative NTP analogs (see Sec. 2).

Although the value of expanded chemistries has been shown in antisense and siRNA therapeutics (improving target binding, increasing resistance to nucleases, and reducing immunogenicity) (Khvorova and Watts 2017), comparatively few of these chemistries have so far been incorporated into successful aptamer selections. Furthermore, the majority of reports to date have involved selections with partially modified libraries (modSELEX), with some DNA or RNA components retained during the selection step. Despite this, modSELEX has proven to be a powerful approach for selecting aptamers with improved properties for in vivo applications (Dellafiore et al. 2016; Lipi et al. 2016); although phosphorothioate (αS)- and boranophosphate (αBH_3)-linked RNA backbones have been explored, modified sugars have become a preferred chemistry, such as 2′-amino- and 2′-fluoro-pyrimidine analogs (2′F-dYTPs) in particular, because of the ease of chemical synthesis of both triphosphates and oligonucleotides. DNA aptamer selection using the analogous purines (2′F-dRTPs) has recently been shown (Thirunavukarasu et al. 2017), and polymerases capable of using all four 2′F analogs (2′F-dNTPs) have been developed (Cozens et al. 2012), thus aptamer selections at full-2′F substitution should now be feasible. Polymerase engineering (see Sec. 2) has also allowed another standard chemistry, 2′-*O*-methyl (2′OMe) RNA, to be used in selections. 2′OMe is an attractive substitution as it is biostable as well as nontoxic, occurring naturally as a modification of ribosomal RNA and tRNAs. A significant milestone involved a selection for anti-VEGF aptamers using a T7RP mutant to incorporate a mixture of 2′OMe-A, -C, -G, -U, spiked with GTP to enable polymerase initiation (Burmeister et al. 2005).

Although the first crop of 2′OMe-modified therapeutic aptamers proved insufficiently effective in clinical trials (Zhou and Rossi 2017), recent successes of 2′-modifications in the context of designed antisense reagents has generated renewed interest in the therapeutic potential of nucleic acids. More than 30 antisense therapeutics are in clinical trials, close to approval or about to enter the clinic, most notably those containing 2′-methoxyethyl- (2′MOE) RNA such as Spinraza/Nusinersen and IONIS-HTTRx (Ionis Pharmaceuticals, Inc., in partnership with Biogen/Hoffman-La Roche AG, respectively), as well as uncharged phosphorodiamidate morpholino (PMO) oligonucleotides, such as ExonDys 51/Eteplirsen (Sarepta Therapeutics, Inc.).

Other modifications that have been explored in modSELEX include 4′-thioribose pyrimidines (4′S-YTPs), which conferred a 50-fold increase in resistance to RNase A degradation. Locked or bridged nucleic acids (LNA/2′4′BNA), in which the 2′O and 4′C of the deoxyribose ring are linked by a (oxy)methylene bridge, are known to be even more resistant to nuclease degradation, and confer exceptional duplex stability (Dellafiore et al. 2016; Lipi et al. 2016). So far, only one instance has been reported in which an LNA TTP replaced dTTP in the randomized region of a DNA library for aptamer selection (Kasahara et al. 2013). Engineered polymerases capable of synthesis and reverse transcription of fully substituted LNA polymers have been described (Pinheiro et al. 2012), suggesting that in principle all-LNA aptamer selection should also be possible.

3.6 Aptamer Selections Using Modified Bases and Novel Base Pairs

Although modified sugars can improve target binding as well as biostability, diversification of the nucleobase chemistry (Fig. 2) has proven to be a more effective strategy for expansion of the chemical diversity of target binding surfaces (the paratope). Various side chains were explored early on; however, the majority of studies have focused on the attachment of hydrophobic groups to the C5 position of uridine or deoxyuridine. A key advance in this area are slow off-rate modified aptamers (SOMAmers) comprising an amide bond linkage between the pyrimidine C5 position and the substituent side chain, which enhanced aptamer

Cite as *Cold Spring Harb Perspect Biol* doi: 10.1101/cshperspect.a032490

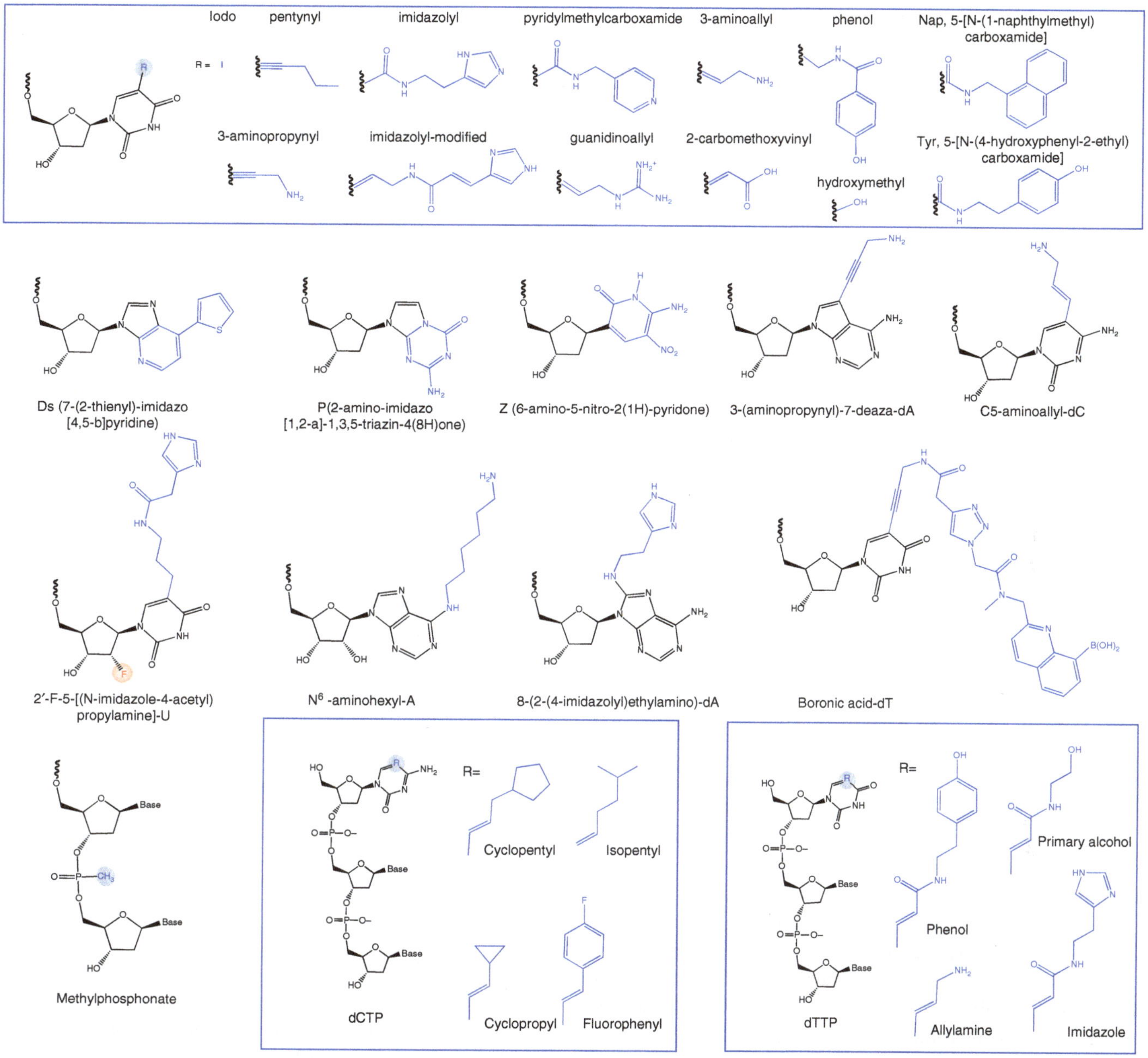

Figure 2. Chemical structures of a range of base-modified nucleic acids used in aptamer and nucleic acid catalyst selections and in nanostructures. Base modifications are indicated in blue, whereas sugar modifications are shown in red.

discovery success rate, especially against previously challenging targets (Gold et al. 2010). Unlike the freely rotating alkyne linkages used previously, the amide is planar, thereby restricting free rotation and limiting entropic costs upon binding to target. SOMAmers look especially promising for diagnostic proteomics, in which the hydrophobic substituents (e.g., benzyl derivatives) provide extensive nonpolar interactions with target molecules, expanding the range of targetable epitopes beyond that of unmodified DNA and RNA (Gelinas et al. 2016). This is an interesting approach as tyrosine side chains are overrepresented in the complementarity-determining regions (CDRs) of antibodies, and high-affinity antibodies have been isolated from repertoires comprising a simple binary Tyr/Ser binding code (Koide and Sidhu 2009).

Although incorporation of a range of C5-modified pyrimidine triphosphates has been achieved using natural (or minimally engineered) polymerases, high-density modification remains challenging. An alternative approach involves postsynthetic modification, whereby only a single modified nucleotide (for each backbone chemistry) comprising the requisite chemical handle need be synthesized. In the recently developed click-SELEX, 5-ethynyl-substituted dUTP is incorporated and postsynthetically reacted with azido group–containing substituents using click chemistry (Pfeiffer et al. 2018). Again, as described above, this approach may also benefit from the restricted rotation of the resulting 1,2,3-triazole ring.

In a different approach, "synthetic translation" has the potential to offer an even wider range of substituent chemistries: DNA triplets bearing one of eight different side chains can be aligned on DNA templates and ligated (by T4 DNA ligase) to prepare heavily modified DNA, which was recently used to select aptamers against IL-6 and PCSK9 (Chen et al. 2018).

Distinct from the modification of the natural bases, the development of novel orthogonal base-pairing systems can be exploited for aptamer technology. Indeed, an expanded alphabet should allow site-specific incorporation of defined functional groups or an expansion of the side chain repertoire in aptamers (Benner et al. 2016). Five- and six-letter systems (A,T,C,G plus one or two orthogonal base analogs) have been independently developed, termed "genetic alphabet expansion SELEX" (ExSELEX) and "artificially expanded genetic information systems" (AEGIS), respectively. Both systems have been used to select expanded-alphabet DNA aptamers specific for targets that include whole cells and toxins (Lee et al. 2018). The AEGIS pair (Z and P) exploits alternative patterns of hydrogen bond donor and acceptor groups and can be sequenced by specific differential chemical conversion (akin to bisulfite sequencing). Even in the absence of nonnatural functional groups (apart from the nitro group on Z; 6-amino-5-nitro-3(1′-β-D-2′deoxyribofuranosyl)-2(1H)-pyridone), the six-letter system appears to increase the number of functional aptamers within a repertoire, presumably because of an increase in the diversity of possible aptamer structures, as well as a reduction of alternative folds of each sequence (i.e., limiting the conformational ambiguity). The ExSELEX system uses a different pair (Ds and Px), which achieves orthogonality through steric complementarity, and only one of the letters (Ds) is incorporated into DNA libraries; these positions were initially fixed, but recent aptamer selections have allowed them to vary (Matsunaga et al. 2017).

3.7 Fully Modified Xeno Nucleic Acid (XNA) Selections (X-SELEX)

The modSELEX approaches described above still involve DNA or RNA components and thus retain vulnerability to nucleases, but complete replacement of all four natural nucleotides during aptamer selections is possible (Fig. 3).

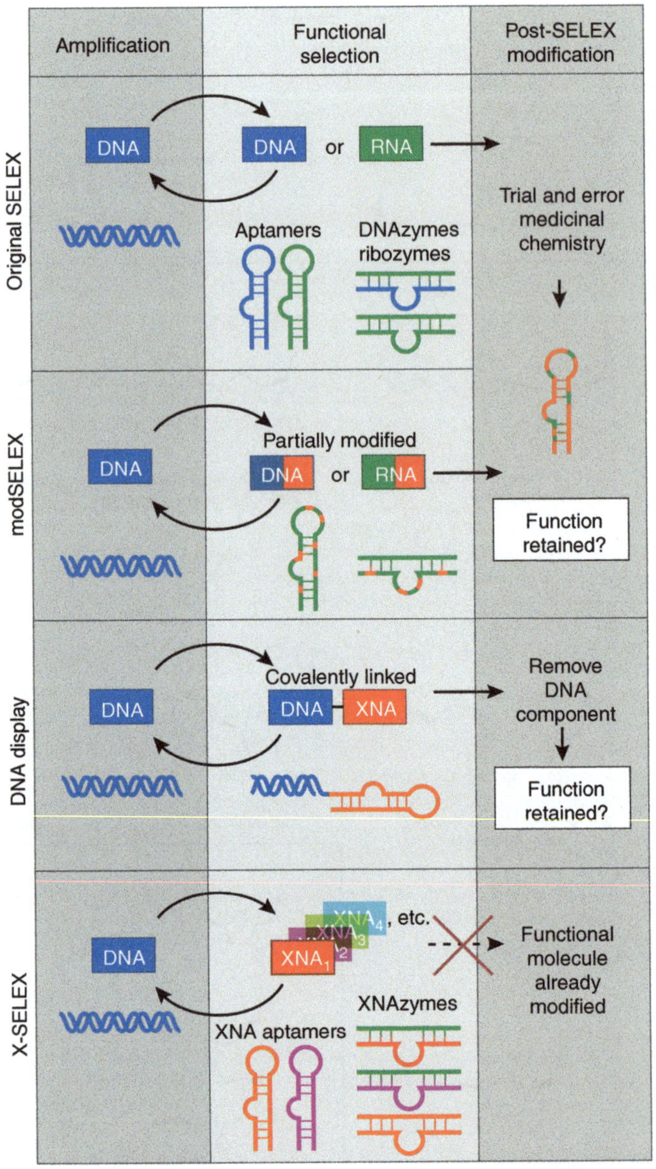

Figure 3. Overview of methods for directed evolution of functional nucleic acids (aptamers and enzymes) composed of modified DNA or RNA, or xeno nucleic acids (XNAs), using systematic evolution of ligands by exponential enrichment (SELEX) strategies.

Polymerase engineering by in vitro evolution and screening (see Sec. 2) have recently enabled DNA-templated synthesis and reverse transcription of a range of nonnatural nucleotide analogs composed of modified or alternative backbones, sugars, or ring congeners not found in nature (xeno nucleic acids [XNAs]). Systems that allow XNA replication (via a DNA intermediate [DNA→XNA→DNA]) enable selections for functional molecules directly from diverse repertoires of fully modified nucleic acid analogs. Transcribing between DNA and XNA in a selection cycle (cross-chemistry SELEX [X-SELEX]) (Pinheiro et al. 2012)

Cite as *Cold Spring Harb Perspect Biol* doi: 10.1101/cshperspect.a032490

allows selection steps to be performed on XNA libraries, whereas cDNA is used for amplification, cloning, and sequencing as XNAs cannot yet be directly sequenced, although the advent of nanopore sequencing may allow this in the near future.

To date, XNA aptamer selections have been reported using a range of chemistries bearing 2′-modified ribose or alternative ring structures. Before the development of an appropriate RT for TNA, DNA display was used to select antithrombin TNA aptamers from a three-letter TNA alphabet (synthesized from tA-, tC-, and tTTP) (Yu et al. 2012). Using engineered polymerases for synthesis and reverse transcription, HNA aptamers against hen egg lysozyme (HEL) or a truncated version of the TAR RNA motif from HIV were selected directly from diverse repertoires (up to 10^{14} variants) (Pinheiro et al. 2012). The anti-TAR HNA aptamers were shown to be specific for different RNA structural motifs (and not binding TAR RNA by antisense complementarity), whereas the anti-HEL HNA aptamers could stain mouse plasmacytoma cells expressing HEL as a transmembrane protein. Crucially, HNA aptamers could be exposed to acid, base, and nuclease treatment with no measurable effect on activity, showing that such XNAs can yield aptamers with both extreme biostability and chemical inertness.

2′OMe-RNA, although technically not an XNA (as it is found as a posttranslational modification of cellular RNA), has also continued to be explored. Engineering of T7RP allowed selections to be performed on a mixture of all four 2′OMe nucleotides plus GTP (Burmeister et al. 2005). Subsequent efforts have shown that a mixture of 2′OMe-dA, -dC-, -dUTPs, and 2′F-GTP can be used for selection of fully substituted aptamers with extended half-lives in 10% serum (Friedman et al. 2015). Recently, mutants of the Stoffel fragment of *Taq* DNA polymerase were described that could synthesize and reverse transcribe fully 2′OMe-RNA in the context of a selection for binders to human neutrophil elastase (Liu et al. 2017).

An alternative approach, adapted from the mirror-image phage selection strategy originally described by Kim (Schumacher et al. 1996) is the selection of RNA or DNA aptamers against enantiomeric target molecules, followed by chemical synthesis of the nonnatural L-RNA or L-DNA isomers of the aptamers, termed "Spiegelmers." Although Spiegelmers are composed of unmodified DNA or RNA, the L-forms are not recognized by natural nucleases, and so show greatly extended half-lives in serum. Various Spiegelmers have been developed, including (remarkably) L-RNA aptamers able to recognize natural D-RNA motifs such as microRNAs (Sczepanski and Joyce 2015), and various candidate therapeutics are in clinical trials (Vater and Klußmann 2015). One drawback of this approach is that mirror-image targets must first be prepared (by chemical synthesis of D-(poly)peptides or L-nucleic acid targets), precluding the use of whole cells or large proteins in Spiegelmer selections. However, the preparation of L-DNA polymerases composed of D-amino acids (Wang et al. 2016) suggests selection of Spiegelmers directly from L-form aptamer libraries will be possible.

Another strategy in development seeks to leverage DNA template–directed nonenzymatic assembly of chemical entities to assemble sequence-defined polymers, such as peptide nucleic acid (PNA), an uncharged nonpolar backbone chemistry with very high resistance to both nucleases and proteases. Approaches such as this have the potential to extend in vitro evolution into chemistries increasingly distant from the natural nucleic acids (Lutz et al. 2013).

3.8 Modified Ribozymes and DNAzymes

As described above, the diverse folded structures of nucleic acids not only engender specific ligand binding but can form catalytic motifs analogous to protein enzymes, which can be evolved de novo from random sequence repertoires. Many such ribozymes and DNA catalysts (DNAzymes) have been isolated (Silverman 2016; Wachowius et al. 2017), including ribozymes with RNA polymerase activity. Initially limited to short templates, years of directed evolution and RNA engineering have enhanced RNA polymerase ribozyme activity toward synthesis of ever longer RNAs, although mostly on favorable RNA templates. Nevertheless, the most advanced polymerase ribozymes are now able to synthesize RNA aptamers (Horning and Joyce 2016), synthesize simple ribozymes (Wochner et al. 2011), and even support exponential amplification (ribo-PCR) (Horning and Joyce 2016). Although functional nucleic acids are able to catalyze a wide spectrum of chemical transformations (Silverman 2016), they appear to be most adept at phosphoester chemistry, with the most powerful catalysts being RNA endonucleases and ligase ribozymes, including the class I ligase ribozyme from which the RNA polymerase ribozymes derive (Wachowius et al. 2017).

Although both ribozymes and DNAzymes have begun to find applications in synthetic biology as biosensors, molecular computational systems or synthetic regulators of gene expression (e.g., synthetic riboswitches), the majority of research involving synthetic nucleic acid catalysts has taken two avenues: (1) the expansion of chemical functionality to broaden the scope of their enzymatic activity (Silverman 2016); and (2) the improvement of biostability and activity under physiological conditions for potential usage in precision molecular medicine, such as sequence-specific targeting of viral or disease-associated RNAs (Fokina et al. 2015).

From these perspectives, several groups have explored an expansion of the chemical functionality beyond the standard nucleotides. Typically, as with aptamers, postselection medicinal chemistry has been a major focus. A considerable body of work has also been performed on the classic "10–23" and "8–17" RNA-endonuclease DNAzymes (Santoro and Joyce 1997). A range of variants with modified bases, sugars, and backbones have been prepared by solid-phase synthesis, with a view to augment in vivo stability, substrate binding, or catalytic power.

A fascinating approach has also been to explore simpler evolutionary systems with reduced chemical (and informational) diversity. Ligase ribozymes were selected from genetic alphabets consisting of just three (by deaminating Cs to Us using bisulphite), and subsequently two (replacing A with 2,6-diaminopurine), chemical letters (Reader and Joyce 2002). The observed reductions in catalytic rate suggest that catalytic power (at least for these simple ribozymes) does scale with informational and chemical complexity, boding well for synthetic genetic approaches to expanding the chemical diversity of nucleic acid backbones, base substituent chemistries as well as informational diversity through expansion of the genetic alphabet itself (Benner et al. 2016).

Substitutions with modified nucleotides are necessary for in vivo applications, but generally it has not been possible to replace all RNA or DNA residues in pre-existing ribozymes or DNAzymes (particularly their catalytic cores) without compromising activity. The challenge of replacing all residues with robust analogs while retaining catalytic activity has posed a major obstacle to the development of effective endogenously delivered ribozyme and DNAzyme therapeutics. Although several ribozymes have progressed to clinical trials, such as "angiozyme" (an RNA endonuclease targeting mRNA encoding the vascular endothelial growth factor [VEGF] receptor), and appeared to be well-tolerated in vivo (Morrow et al. 2012), in general efficacy was limited, in part due to reduced biostability resulting from unmodified ribonucleotide positions. To date, no nucleic acid catalyst has achieved clinical approval.

The majority of research to advance DNAzymes for biological applications has centered on the 10–23 RNA endonuclease DNAzyme, due to its modularity and exceptional catalytic efficiency. Variants targeting disease-associated RNAs have been reported to show activity in vivo, and clinical applications for knockdown of pro-inflammatory RNAs in the lung and skin are under investigation (Garn and Renz 2017). However, inactive or photoactivatable versions of 10–23 show similar effects suggesting that knockdown activity may be due to antisense-mediated mechanisms rather than catalytic activity of the DNAzyme itself (Young et al. 2010).

The use of alternative nucleotide chemistries in ribozyme or DNAzyme selections has the potential to overcome some of these difficulties, offering the prospect of more biostable backbones, formation of more stable secondary structures and improved activities under physiological conditions. Base modifications have also been explored (Fig. 2) including C5-imidazoyl-dU, which yielded a DNAzyme with an impressive catalytic rate in the presence of Zn^{2+} (Santoro et al. 2000). Extending this approach, RNA endonuclease DNAzymes with up to three nucleotide analogs in which the bases display amino acid-like side chains have been isolated, including the recent report of an RNase A mimic that functions independently of Mg^{2+} (Wang et al. 2018). A DNAzyme capable of amide bond hydrolysis was recently realized, again using protein-like side chains, a chemical transformation that could not be performed by an unmodified nucleic acid catalyst (Zhou et al. 2016).

As with Spiegelmer aptamers, modification of catalysts with L-RNA or L-DNA offers an alternative route to biostability. However, this requires (for now) selection of D-RNA/D-DNA libraries on mirror-image substrates unless these are achiral. Such cross-chiral catalysis, specifically a cross-chiral RNA polymerase ribozyme, has also been explored to overcome chiral inhibition of prebiotic RNA-catalyzed RNA polymerization (Sczepanski and Joyce 2014).

3.9 Catalysts from Synthetic Genetic Polymers (XNAzymes)

The advent of synthetic genetics presents new opportunities to extend nucleic acid catalyst selections to fully modified libraries composed of XNA chemistries (i.e., X-SELEX for catalysts [Taylor and Holliger 2015]). Using broadly similar selection strategies used for ribo- and DNAzyme discovery, XNA catalysts (XNAzymes) were isolated from four different backbone chemistries: ANA, CeNA, FANA, and HNA (Taylor et al. 2015). XNAzymes were identified with RNA endonuclease activity, as well as RNA and XNA ligase activity (in the FANA chemistry). Prediction of XNA secondary structures using standard software are fraught with difficulty as the requisite force fields and structural and conformational parameters are largely unknown (Anosova et al. 2016). However, secondary structures of several of the FANAzymes in complex with their RNA substrates could be determined using a combination of SHAPE (on non-cleavable variants of their RNA substrates in the case of an RNA endonuclease FANAzyme) and mutational profiling (RING-MaP) by reacting FANA (as well as substrate RNA) nucleotides with DMS and reverse transcribing to DNA for deep sequencing using an engineered polymerase

(Taylor et al. 2015). Using this information, catalytic core residues and substrate-binding guide sequences could be unambiguously identified and subsequently reengineered. That catalysts can be discovered by in vitro evolution processes in the sequence and chemical spaces beyond those used in biology has broad implications for our understanding of the boundary conditions necessary for life or its hallmark processes.

4 NANOSCALE ENGINEERING USING SYNTHETIC GENETIC POLYMERS

4.1 DNA and RNA Nanotechnology

Watson–Crick hybridization rules allow hierarchical, programmable formation of duplexes into various 2D and 3D nanoscale objects and devices. Many studies have shown the potential usage of DNA (and RNA) nanotechnology as, for example, in vivo logic devices, cargo delivery vehicles, or topological guides for synthesis of polymers (Seeman and Sleiman 2017). However, DNA and RNA impose constraints on in vivo applications such as poor biostability and immunogenicity. Although several studies have focused on the impact of alternative design strategies using unmodified DNA and RNA, a range of chemical modification to 2D and (in some cases) 3D designs have also been explored (such as GNA, PNA, methylphosphonate backbones, and alternative bases capable of photo-cross-linking or metal coordination), either by solid-phase synthesis of modified oligonucleotides or by appending functional groups following nanostructure assembly (Gothelf 2017).

One recent advance of note is the application of mirror-image L-DNA. L-DNA versions of a tetrahedral structure were more resistant to nucleases and could be used to deliver a toxic doxorubicin payload to tumor cells in a xenograft mouse model (Kim et al. 2016). In an impressive expansion of nanostructure function using alternative chemistries, hydrophobic moieties and phosphorothioate-linked oligonucleotides (reacted postassembly with ethyliodide) could be used to add nonpolar motifs to DNA barrel-like architectures, allowing them to insert into lipid bilayers, a strategy that has provided a range of engineered DNA nanopores (Howorka 2017).

4.2 Incorporating Synthetic Genetic Polymers into DNA Nanotechnology

Recently, an alternative approach has sought the wholesale replacement of natural nucleic acids as the building material with XNAs. Encouragingly, partially 2′F-substituted Φ29 bacteriophage ∼80-nt packaging RNA (pRNA) nanomotors were found to retain their ability to assemble into a functional complex, while acquiring increased resistance to RNAse degradation (Liu et al. 2011).

Engineered polymerases (see Sec. 2) can enable access to XNA strands that cannot be obtained by solid-phase synthesis (e.g., a 1.7-kb FANA strand [Taylor et al. 2016]). This has enabled assembly of four-strand tetrahedra composed of 100% 2′-F-DNA, FANA, HNA, and CeNA, as well as hybrid tetrahedral structures comprising DNA/FANA, DNA/2′F-DNA, and 2′F-DNA/HNA components. In the case of HNA, tetrahedra were shown to be completely resistant to degradation in serum-containing media at 37°C for more than a week. However, it was notable that tetrahedral designs presenting two-nucleotide single-stranded vertices were more amenable to XNA assembly, presumably allowing a relaxation of conformational strain. In a similar vein, a fully modified 1.7-kb FANA strand, together with five 40-nt FANA staple strands, was shown to assemble into an octahedral design that could be resolved by transmission electron microscopy (TEM) single-particle reconstruction (Taylor et al. 2016). Deviations from the DNA design were apparent, presumably caused by the divergent FANA helical geometries, although the overall topology of the structure was maintained. A deeper understanding of the structural and energetic parameters of the various XNAs may be required to access more complex XNA nanostructure designs.

5 CONCLUDING REMARKS: WORLDS BEYOND RNA?

There is considerable merit to the argument that natural nucleic acids appear to be near-optimal solutions for information storing molecules and may hint at general principles likely to be common among functional genetic systems (Benner et al. 2004). However, the ability of a range of unnatural nucleic acid polymers (XNAs), in addition to natural DNA and RNA variants, to form functional molecules (Fig. 4), including aptamers, enzymes, and nanostructures as described, suggests that for nucleic acids (and biopolymers in general) to evolve function in biology, they may fundamentally only require (1) a molecular framework capable of precisely positioning chemically reactive groups, and (2) a link between this phenotypic arrangement and heritable information. Consequently, a whole range of alternative sequence-defined polymers not found in nature might in principle be capable of storing information and forming structures that underpin biological processes. Indeed, as we have discussed, efforts are ongoing to explore the sequence space of ever more divergent sequence defined polymers for the discovery of new materials and drugs. One might therefore ask if "nature's choice" of RNA (and DNA) is based on clear functional advantages over simple alter-

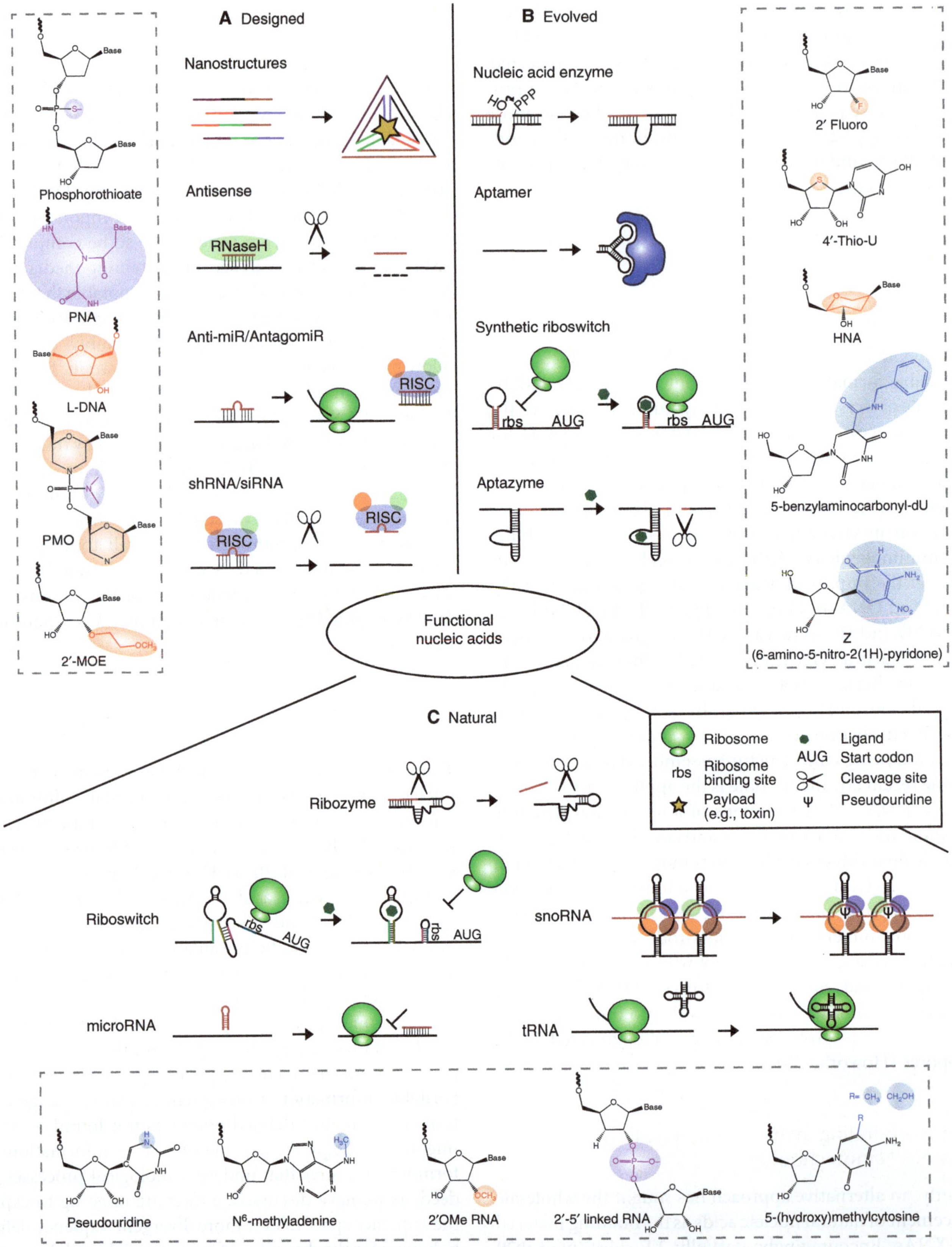

Figure 4. A summary schematic of (*A*) designed, (*B*) evolved, and (*C*) natural functional nucleic acids with expanded chemistry and their respective functions as discussed in this review.

Cite as *Cold Spring Harb Perspect Biol* doi: 10.1101/cshperspect.a032490

natives or rather on historical contingencies such as the constraints of terran prebiotic chemistry. The question of the degree to which nucleic acids present unique solutions to the problem of genetic information storage has been an ongoing subject of investigation in organic and medicinal chemistry, which has synthesized a wide range of backbone, sugar ring, and base variants. Synthetic genetics is now beginning to probe their functional and genetic potential, and their interaction with the natural genetic system (Chaput et al. 2012; Benner et al. 2016; Chen et al. 2016b; Houlihan et al. 2017a,b).

Expansion of the genetic alphabet probes fundamental aspects of molecular information encoding and the approaches taken and the difficulties encountered are illustrative. Efforts to expand the genetic alphabet by one (or several) novel, orthogonal base pairs were initiated by Benner and colleagues by shuffling of H-bond donor and acceptor pairs on the natural purine and pyrimidine rings to potentially yield four novel pairs (Benner et al. 2016). Alternative approaches have involved exploiting hydrophobic interactions (i.e., base stacking) and shape complementarity. Some of these novel pairs have already allowed the in vitro evolution of higher affinity aptamers to protein or cell-surface targets (see Sec. 3.6). Even more striking, some of the hydrophobic pairs can be both replicated and transcribed in a living organism, enabling new triplet codons for in vivo genetic code expansion (Ledbetter et al. 2018). Although the current generation of artificial base pairs works best when integrated sporadically into DNA (or RNA), future iterations may be able to function at higher multiplicity. However, the manifold challenges that had to be overcome to arrive at functional, orthogonal pairs may hint at possible causes for the "sparse" natural four-letter alphabet. For example, the original base-pair designs with modified H-bond donor–acceptor patterns suffered from unfavorable tautomerization or epimerization equilibria that required further chemical modification (Benner et al. 2016). Likewise, the design of successful hydrophobic pairs lacking H-bonds required systematic screening of a large number of candidate compounds. Thus, a four-letter alphabet may not just reflect constraints of prebiotic chemistry or mutual orthogonality (Szathmáry 2003), but also the rarity of suitable heterocycles that support stable pairing patterns.

Furthermore, stable pairing appears to be an insufficient criterion by itself as some simple heterocycles capable of doing so are not used (or used only sparingly) in biology. For example, 2,6-diaminopurine would appear to have some advantages over adenine, notably providing a more stable pairing with uridine or thymidine, which can aid RNA-catalyzed RNA synthesis by RNA polymerase ribozymes (Attwater et al. 2013) and might have simplified some genomic transactions by equalizing base-pairing energetics among different base pairs. Likewise, many A- and U-rich anticodons bear chemical modifications to stabilize interactions and enable more efficient readout of the genetic code (Grosjean and Westhof 2016). Nevertheless, 2,6-diaminopurine appears to be only found in the genome of a cyanophage (Weigele and Raleigh 2016). This and other modifications may reflect vestiges of a range of prebiotic base-pairing systems, which did not become stably incorporated into biology.

As discussed above, XNAs bearing a range of variations to the sugar ring chemistry (e.g., 2′F, 2′OMe) including nonglycosidic ring congeners (e.g., HNA, FANA) as well as acyclic analogs (GNA) have been found to be compatible with at least rudimentary genetic function. Broadly, this finding could be interpreted as an indication that life's genetic system based on phosphodiester and ribofuranose backbones may reflect constraints of prebiotic chemistry rather than stringent functional requirements. However, at least some XNAs, including TNA and ANA, appear compatible with some prebiotic synthesis schemes suggesting that a hypothetical "pre-RNA World" could have had multiple chemical inhabitants from which only the most functional survive in biology. Furthermore, it cannot be excluded that perhaps XNAs could support genetic systems (XNA Worlds) elsewhere in the cosmos.

However, the ribofuranose chemistry of RNA may be special in at least one respect, the vicinal *cis*-diol configuration. This vicinal diol configuration appears to aid import of ribose sugar into protocellular vesicles (Sacerdote and Szostak 2005) and, in the context of the RNA backbone, activates the proximal phosphodiester linkage for transesterification (either cleavage or recombination). This structure underlies the instability of RNA oligomers at high (alkaline) pH and elevated temperatures especially in the presence of bivalent metal ions. Although such hydrolytic instability may be considered undesirable for a genetic material, it accelerates recombination and splicing reactions among RNA strands, and this may have aided early RNA evolution. Thus, RNA's "Achilles' heel" that renders it susceptible to hydrolysis might be an important aspect of its functional potential as a genetic polymer.

REFERENCES

Anosova I, Kowal EA, Dunn MR, Chaput JC, Van Horn WD, Egli M. 2016. The structural diversity of artificial genetic polymers. *Nucleic Acids Res* **44:** 1007–1021.

Attwater J, Tagami S, Kimoto M, Butler K, Kool ET, Wengel J, Herdewijn P, Hirao I, Holliger P. 2013. Chemical fidelity of an RNA polymerase ribozyme. *Chem Sci* **4:** 2804–2814.

Benner SA, Ricardo A, Carrigan MA. 2004. Is there a common chemical model for life in the universe? *Curr Opin Chem Biol* **8:** 672–689.

Benner SA, Karalkar NB, Hoshika S, Laos R, Shaw RW, Matsuura M, Fajardo D, Moussatche P. 2016. Alternative Watson–Crick synthetic genetic systems. *Cold Spring Harb Perspect Biol* **8:** a023770.

Burmeister PE, Lewis SD, Silva RF, Preiss JR, Horwitz LR, Pendergrast PS, McCauley TG, Kurz JC, Epstein DM, Wilson C, et al. 2005. Direct in vitro selection of a 2′-*O*-methyl aptamer to VEGF. *Chem Biol* **12:** 25–33.

Chaput JC, Yu H, Zhang S. 2012. The emerging world of synthetic genetics. *Chem Biol* **19:** 1360–1371.

Chen T, Hongdilokkul N, Liu Z, Adhikary R, Tsuen SS, Romesberg FE. 2016a. Evolution of thermophilic DNA polymerases for the recognition and amplification of C2′-modified DNA. *Nat Chem* **8:** 556–562.

Chen T, Hongdilokkul N, Liu Z, Thirunavukarasu D, Romesberg FE. 2016b. The expanding world of DNA and RNA. *Curr Opin Chem Biol* **34:** 80–87.

Chen Z, Lichtor PA, Berliner AP, Chen JC, Liu DR. 2018. Evolution of sequence-defined highly functionalized nucleic acid polymers. *Nat Chem* **10:** 420–427.

Chim N, Shi C, Sau SP, Nikoomanzar A, Chaput JC. 2017. Structural basis for TNA synthesis by an engineered TNA polymerase. *Nat Commun* **8:** 1810.

Cozens C, Pinheiro VB, Vaisman A, Woodgate R, Holliger P. 2012. A short adaptive path from DNA to RNA polymerases. *Proc Natl Acad Sci* **109:** 8067–8072.

Cozens C, Mutschler H, Nelson GM, Houlihan G, Taylor AI, Holliger P. 2015. Enzymatic synthesis of nucleic acids with defined regioisomeric 2′–5′ linkages. *Angew Chem Int Ed* **54:** 15570–15573.

Darmostuk M, Rimpelova S, Gbelcova H, Ruml T. 2015. Current approaches in SELEX: An update to aptamer selection technology. *Biotechnol Adv* **33:** 1141–1161.

Dellafiore MAA, Montserrat JM, Iribarren AM. 2016. Modified nucleoside triphosphates for in-vitro selection techniques. *Front Chem* **4:** 18.

Dunn MR, Jimenez RM, Chaput JC. 2017. Analysis of aptamer discovery and technology. *Nat Rev Chem* **1:** 0076.

Ellefson JW, Gollihar J, Shroff R, Shivram H, Iyer VR, Ellington AD. 2016. Synthetic evolutionary origin of a proofreading reverse transcriptase. *Science* **352:** 1590–1593.

Eschenmoser A. 1999. Chemical etiology of nucleic acid structure. *Science* **284:** 2118–2124.

Feldman AW, Romesberg FE. 2017. Expansion of the genetic alphabet: A chemist's approach to synthetic biology. *Acc Chem Res* **51:** 394–403.

Fokina AA, Stetsenko DA, François J-C. 2015. DNA enzymes as potential therapeutics: Towards clinical application of 10–23 DNAzymes. *Expert Opin Biol Ther* **15:** 689–711.

Friedman AD, Kim D, Liu R. 2015. Highly stable aptamers selected from a 2′-fully modified fGmH RNA library for targeting biomaterials. *Biomaterials* **36:** 110–123.

Garn H, Renz H. 2017. GATA-3-specific DNAzyme—A novel approach for stratified asthma therapy. *Eur J Immunol* **47:** 22–30.

Gelinas AD, Davies DR, Janjic N. 2016. Embracing proteins: Structural themes in aptamer–protein complexes. *Curr Opin Struct Biol* **36:** 122–132.

Gold L, Ayers D, Bertino J, Bock C, Bock A, Brody EN, Carter J, Dalby AB, Eaton BE, Fitzwater T, et al. 2010. Aptamer-based multiplexed proteomic technology for biomarker discovery. *PLoS ONE* **5:** e15004.

Gothelf KV. 2017. Chemical modifications and reactions in DNA nanostructures. *MRS Bulletin* **42:** 897–903.

Grosjean H, Westhof E. 2016. An integrated, structure- and energy-based view of the genetic code. *Nucleic Acids Res* **44:** 8020–8040.

Herdewijn P, Marlière P. 2009. Toward safe genetically modified organisms through the chemical diversification of nucleic acids. *Chem Biodivers* **6:** 791–808.

Hori S-I, Herrera A, Rossi JJ, Zhou J. 2018. Current advances in aptamers for cancer diagnosis and therapy. *Cancers* **10:** 9.

Horning DP, Joyce GF. 2016. Amplification of RNA by an RNA polymerase ribozyme. *Proc Natl Acad Sci* **113:** 9786–9791.

Houlihan G, Arangundy-Franklin S, Holliger P. 2017a. Engineering and application of polymerases for synthetic genetics. *Curr Opin Biotechnol* **48:** 168–179.

Houlihan G, Arangundy-Franklin S, Holliger P. 2017b. Exploring the chemistry of genetic information storage and propagation through polymerase engineering. *Acc Chem Res* **50:** 1079–1087.

Howorka S. 2017. Building membrane nanopores. *Nat Nanotech* **12:** 619–630.

Jo H, Ban C. 2016. Aptamer–nanoparticle complexes as powerful diagnostic and therapeutic tools. *Exp Mol Med* **48:** e230.

Joyce GF. 2007. Forty years of in vitro evolution. *Angew Chem Int Ed Engl* **46:** 6420–6436.

Kasahara Y, Irisawa Y, Fujita H, Yahara A, Ozaki H, Obika S, Kuwahara M. 2013. Capillary electrophoresis-systematic evolution of ligands by exponential enrichment selection of base-and sugar-modified DNA aptamers: Target binding dominated by 2′-*O*,4′-C-methylene-bridged/locked nucleic acid primer. *Anal Chem* **85:** 4961–4967.

Khvorova A, Watts JK. 2017. The chemical evolution of oligonucleotide therapies of clinical utility. *Nat Biotechnol* **35:** 238–248.

Kim K-R, Kim HY, Lee Y-D, Ha JS, Kang JH, Jeong H, Bang D, Ko YT, Kim S, Lee H, et al. 2016. Self-assembled mirror DNA nanostructures for tumor-specific delivery of anticancer drugs. *J Control Release* **243:** 121–131.

Koide S, Sidhu SS. 2009. The importance of being tyrosine: Lessons in molecular recognition from minimalist synthetic binding proteins. *ACS Chem Biol* **4:** 325–334.

Larsen AC, Dunn MR, Hatch A, Sau SP, Youngbull C, Chaput JC. 2016. A general strategy for expanding polymerase function by droplet microfluidics. *Nat Commun* **7:** 11235.

Ledbetter MP, Karadeema RJ, Romesberg FE. 2018. Reprograming the replisome of a semisynthetic organism for the expansion of the genetic alphabet. *J Am Chem Soc* **140:** 758–765.

Lee KH, Hamashima K, Kimoto M, Hirao I. 2018. Genetic alphabet expansion biotechnology by creating unnatural base pairs. *Curr Opin Biotechnol* **51:** 8–15.

Lipi F, Chen S, Chakravarthy M, Rakesh S, Veedu RN. 2016. In vitro evolution of chemically-modified nucleic acid aptamers: Pros and cons, and comprehensive selection strategies. *RNA Biol* **13:** 1232–1245.

Liu J, Guo S, Cinier M, Shlyakhtenko LS, Shu Y, Chen C, Shen G, Guo P. 2011. Fabrication of stable and RNase-resistant RNA nanoparticles active in gearing the nanomotors for viral DNA packaging. *ACS Nano* **5:** 237–246.

Liu Z, Chen T, Romesberg FE. 2017. Evolved polymerases facilitate selection of fully 2′-OMe-modified aptamers. *Chem Sci* **8:** 8179–8182.

Lutz J-F, Ouchi M, Liu DR, Sawamoto M. 2013. Sequence-controlled polymers. *Science* **341:** 1238149.

Matsunaga K-I, Kimoto M, Hirao I. 2017. High-affinity DNA aptamer generation targeting von Willebrand factor A1-domain by genetic alphabet expansion for systematic evolution of ligands by exponential enrichment using two types of libraries composed of five different bases. *J Am Chem Soc* **139:** 324–334.

Mei H, Shi C, Jimenez RM, Wang Y, Kardouh M, Chaput JC. 2017. Synthesis and polymerase activity of a fluorescent cytidine TNA triphosphate analogue. *Nucleic Acids Res* **45:** 5629–5638.

Meyer KD, Jaffrey SR. 2017. Rethinking m^6A readers, writers, and erasers. *Annu Rev Cell Dev Biol* **33:** 319–342.

Mi J, Liu Y, Rabbani ZN, Yang Z, Urban JH, Sullenger BA, Clary BM. 2010. In vivo selection of tumor-targeting RNA motifs. *Nat Chem Biol* **6:** 22.

Morrow PK, Murthy RK, Ensor JD, Gordon GS, Margolin KA, Elias AD, Urba WJ, Weng DE, Rugo HS, Hortobagyi GN. 2012. An open-label, phase 2 trial of RPI.4610 (angiozyme) in the treatment of metastatic breast cancer. *Cancer* **118:** 4098–4104.

Ong JL, Loakes D, Jaroslawski S, Too K, Holliger P. 2006. Directed evolution of DNA polymerase, RNA polymerase and reverse transcriptase activity in a single polypeptide. *J Mol Biol* **361:** 537–550.

Pan W, Craven RC, Qiu Q, Wilson CB, Wills JW, Golovine S, Wang JF. 1995. Isolation of virus-neutralizing RNAs from a large pool of random sequences. *Proc Natl Acad Sci* **92:** 11509–11513.

Pfeiffer F, Tolle F, Rosenthal M, Brändle GM, Ewers J, Mayer G. 2018. Identification and characterization of nucleobase-modified aptamers by click-SELEX. *Nat Protoc* **13:** 1153–1180.

Pinheiro VB, Taylor AI, Cozens C, Abramov M, Renders M, Zhang S, Chaput JC, Wengel J, Peak-Chew S-Y, McLaughlin SH, et al. 2012. Synthetic genetic polymers capable of heredity and evolution. *Science* **336:** 341–344.

Reader JS, Joyce GF. 2002. A ribozyme composed of only two different nucleotides. *Nature* **420:** 841–844.

Röthlisberger P, Hollenstein M. 2018. Aptamer chemistry. *Adv Drug Deliv Rev* doi: 10.1016/j.addr.2018.04.007.

Sacerdote MG, Szostak JW. 2005. Semipermeable lipid bilayers exhibit diastereoselectivity favoring ribose. *Proc Natl Acad Sci* **102:** 6004–6008.

Santoro SW, Joyce GF. 1997. A general purpose RNA-cleaving DNA enzyme. *Proc Natl Acad Sci* **94:** 4262–4266.

Santoro SW, Joyce GF, Sakthivel K, Gramatikova S, Barbas CF. 2000. RNA cleavage by a DNA enzyme with extended chemical functionality. *J Am Chem Soc* **122:** 2433–2439.

Schimmel P. 2018. The emerging complexity of the tRNA world: Mammalian tRNAs beyond protein synthesis. *Nat Rev Mol Cell Biol* **19:** 45–58.

Schumacher T, Mayr LM, Minor DL. 1996. Identification of D-peptide ligands through mirror-image phage display. *Science* **271:** 1854–1857.

Sczepanski JT, Joyce GF. 2014. A cross-chiral RNA polymerase ribozyme. *Nature* **515:** 440–442.

Sczepanski JT, Joyce GF. 2015. Specific inhibition of microRNA processing using l-RNA aptamers. *J Am Chem Soc* **137:** 16032–16037.

Seeman NC, Sleiman HF. 2017. DNA nanotechnology. *Nat Rev Mater* **3:** 17068.

Silverman SK. 2016. Catalytic DNA: Scope, applications, and biochemistry of deoxyribozymes. *Trends Biochem Sci* **41:** 595–609.

Szathmáry E. 2003. Why are there four letters in the genetic alphabet? *Nat Rev Genet* **4:** 995–1001.

Tawiah KD, Porciani D, Burke DH. 2017. Toward the selection of cell targeting aptamers with extended biological functionalities to facilitate endosomal escape of cargoes. *Biomedicines* **5:** 51.

Taylor AI, Holliger P. 2015. Directed evolution of artificial enzymes (XNAzymes) from diverse repertoires of synthetic genetic polymers. *Nat Protoc* **10:** 1625–1642.

Taylor AI, Pinheiro VB, Smola MJ, Morgunov AS, Peak-Chew S, Cozens C, Weeks KM, Herdewijn P, Holliger P. 2015. Catalysts from synthetic genetic polymers. *Nature* **518:** 427–430.

Taylor AI, Beuron F, Peak-Chew S-Y, Morris EP, Herdewijn P, Holliger P. 2016. Nanostructures from synthetic genetic polymers. *ChemBioChem* **17:** 1107–1110.

Thirunavukarasu D, Chen T, Liu Z, Hongdilokkul N, Romesberg FE. 2017. Selection of 2′-fluoro-modified aptamers with optimized properties. *J Am Chem Soc* **139:** 2892–2895.

Tsai C-H, Chen J, Szostak JW. 2007. Enzymatic synthesis of DNA on glycerol nucleic acid templates without stable duplex formation between product and template. *Proc Natl Acad Sci* **104:** 14598–14603.

Vater A, Klußmann S. 2015. Turning mirror-image oligonucleotides into drugs: The evolution of Spiegelmer® therapeutics. *Drug Discov Today* **20:** 147–155.

Wachowius F, Attwater J, Holliger P. 2017. Nucleic acids: Function and potential for abiogenesis. *Q Rev Biophys* **50:** e4.

Wang Z, Xu W, Liu L, Zhu TF. 2016. A synthetic molecular system capable of mirror-image genetic replication and transcription. *Nat Chem* **8:** 698–704.

Wang Y, Liu E, Lam CH, Perrin DM. 2018. A densely modified M^{2+}-independent DNAzyme that cleaves RNA efficiently with multiple catalytic turnover. *Chem Sci* **9:** 1813–1821.

Weigele P, Raleigh EA. 2016. Biosynthesis and function of modified bases in bacteria and their viruses. *Chem Rev* **116:** 12655–12687.

Wittrup A, Lieberman J. 2015. Knocking down disease: A progress report on siRNA therapeutics. *Nat Rev Genet* **16:** 543–552.

Wochner A, Attwater J, Coulson A, Holliger P. 2011. Ribozyme-catalyzed transcription of an active ribozyme. *Science* **332:** 209–212.

Young DD, Lively MO, Deiters A. 2010. Activation and deactivation of DNAzyme and antisense function with light for the photochemical regulation of gene expression in mammalian cells. *J Am Chem Soc* **132:** 6183–6193.

Yu H, Zhang S, Chaput JC. 2012. Darwinian evolution of an alternative genetic system provides support for TNA as an RNA progenitor. *Nat Chem* **4:** 183–187.

Zhou J, Rossi J. 2017. Aptamers as targeted therapeutics: Current potential and challenges. *Nat Rev Drug Discov* **16:** 181–202.

Zhou C, Avins JL, Klauser PC, Brandsen BM, Lee Y, Silverman SK. 2016. DNA-Catalyzed amide hydrolysis. *J Am Chem Soc* **138:** 2106–2109.

Index

T

U